MICROWAVE FILTERS, IMPEDANCE-MATCHING NETWORKS, AND COUPLING STRUCTURES

For a complete listing of the *Artech House Microwave Library*,
turn to the back of this book.

MICROWAVE FILTERS, IMPEDANCE-MATCHING NETWORKS, AND COUPLING STRUCTURES

GEORGE L. MATTHAEI

Department of Electrical
& Computer Engineering
University of California
Santa Barbara, CA

LEO YOUNG

Naval Research Laboratory
Washington, DC

E. M. T. JONES

Technology for Communications
International
Mountain View, CA

ARTECH HOUSE

ARTECH HOUSE, INC.
685 Canton Street
Norwood, MA 02062
Reprint of the edition published by McGraw-Hill Book Company, Inc., 1964

Library of Congress Catalog Card Number: 80-68976
Standard Book Number 978-0-89006-099-5

20 19 18 17

PREFACE

The organization of this book has three general objectives. The first objective is to present fundamental concepts, techniques, and data that are of general use in the design of the wide range of microwave structures discussed in this book. The second objective is to present specialized data in more or less handbook form so that a designer can work out practical designs for structures having certain specific configurations, without having to recreate the design theory or the derivation of the equations. (However, the operation of most of the devices discussed herein is sufficiently complex that knowledge of some of the basic concepts and techniques is usually important.) The third objective is to present the theory upon which the various design procedures are based, so that the designer can adapt the various design techniques to new and unusual situations, and so that researchers in the field of microwave devices may use some of this information as a basis for deriving additional techniques. The presentation of the material so that it can be adapted to new and unusual situations is important because many of the microwave filter techniques described in this book are potentially useful for the design of microwave devices not ordinarily thought of as having anything to do with filters. Some examples are tubes, parametric devices, and antennas, where filter structures can serve as efficient impedance-matching networks for achieving broad-band operation. Filter structures are also useful as slow-wave structures or time-delay structures. In addition, microwave filter techniques can be applied to other devices not operating in the microwave band of frequencies, as for instance to infrared and optical filters.

The three objectives above are not listed in any order of importance, nor is this book entirely separated into parts according to these objectives. However, in certain chapters where the material lends itself to such organization, the first section or the first few sections discuss general principles which a designer should understand in order to make best use of the design data in the chapter, then come sections giving design data for specific types of structures, and the end of the chapter discusses the derivations of the various design equations. Also, at numerous places cross references are made to other portions of the book where information useful for the design of the particular structure under consideration can be found. For example, Chapter 11 describes procedures for measuring the unloaded Q and external Q of resonators, and for measuring the coupling coefficients between resonators. Such procedures have wide application in the practical development of many types of band-pass filters and impedance-matching networks.

Chapter 1 of this book describes the broad range of applications for which microwave filter structures are potentially useful. Chapters 2 through 6 contain reference data and background information for the rest of the book. Chapter 2 summarizes various concepts and equations that are particularly useful in the analysis of filter structures. Although the image point of view for filter design is made use of only at certain points in this book, some knowledge of image design methods is desirable. Chapter 3 gives a brief summary of the image design concepts which are particularly useful for the purposes of this book. Chapters 1 to 3 should be especially helpful to readers whose background for the material of this book may have some gaps.

Most of the filter and impedance-matching network design techniques described later in the book make use of a low-pass prototype filter as a basis for design. Chapter 4 discusses various types of lumped-element, low-pass, prototype filters, presents tables of element values for such filters, discusses their time-delay properties, their impedance-matching properties, and the effects of dissipation loss upon their responses. In later chapters these low-pass prototype filters and their various properties are employed in the design of low-pass, high-pass, band-pass, and band-stop microwave filters, and also in the design of microwave impedance-matching networks, and time-delay networks.

Various equations, graphs, and tables relevant to the design of coaxial line, strip-line, waveguide, and a variety of resonators, coupling structures, and discontinuities, are summarized for easy reference in Chapter 5. Chapter 6 discusses the design of step transformers and presents tables of designs for certain cases. The step transformers in Chapter 6 are not only for use in conventional impedance-transformer applications, but also for use as prototypes for certain types of band-pass or pseudo high-pass filters discussed in Chapter 9.

Design of low-pass filters and high-pass filters from the semi-lumped-element point of view are treated in Chapter 7. Chapters 8, 9, and 10 discuss band-pass or pseudo-high-pass filter design using three different design approaches. Which approach is best depends on the type of filter structure to be used and the bandwidth required. A tabulation of the various filter structures discussed in all three chapters, a summary of the properties of the various filter structures, and the section number where design data for the various structures can be found, are presented at the beginning of Chapter 8.

Chapter 11 describes various additional techniques which are useful to the practical development of microwave band-pass filters, impedance-matching networks, and time-delay networks. These techniques are quite general in their application and can be used in conjunction with the filter structures and techniques discussed in Chapters 8, 9, and 10, and elsewhere in the book.

Chapter 12 discusses band-stop filters, while Chapter 13 treats certain types of directional couplers. The TEM-mode, coupled-transmission-line, directional couplers discussed in Chapter 13 are related to certain types of directional filters discussed in Chapter 14, while the branch-guide directional couplers can be designed using the step-transformer prototypes in Chapter 6. Both waveguide and strip-line directional filters are discussed in Chapter 14, while high-power filters are treated in Chapter 15. Chapter 16 treats multiplexers and diplexers, and Chapter 17 deals with filters that can be tuned either mechanically or by varying a biasing magnetic field.

It is hoped that this book will fill a need (which has become increasingly apparent in the last few years) for a reference book on design data, practical development techniques, and design theory, in a field of engineering which has been evolving rapidly.

ACKNOWLEDGMENTS

The preparation of this book was largely supported by the Signal Corps, under Contract DA 36-039 SC-87398; its preparation was also partially supported by Stanford Research Institute, and by off-work time contributed by the authors. Many of the design techniques described in this book are the result of research programs sponsored by the Signal Corps under Contracts DA 36-039 SC-63232, DA 36-039 SC-64625, DA 36-039 SC-74862, and DA 36-039 SC-87398.

Mr. Nathan Lipetz of the U.S. Army Electronics Research Laboratory, Ft. Monmouth, N. J., because of his belief in the importance of research work in the area of microwave filters and coupling structures, and in the potential value of a book such as this, did much to make this book possible. Mr. John P. Agrios and Mr. William P. Dattilo, both of the U.S. Army Electronics Research Laboratory also were of great help. Mr. Agrios maintained a close interest in this project throughout the preparation of this book, as did Mr. Dattilo, who reviewed all of the chapters as they were prepared. He made numerous suggestions which have resulted in valuable improvement in the clarity and usefulness of this book.

Dr. S. B. Cohn, formerly of Stanford Research Institute and presently of the Rantec Corporation, led the microwave filter and coupling structure research at Stanford Research Institute during the earlier Signal Corps filter programs at SRI. In many places this book presents research results, or reflects points of view, which are due to him.

The authors' colleagues at SRI, and numerous researchers elsewhere have made many valuable contributions to the subject area of this book, and many results of their work have been incorporated into this book.

The authors thank the various journals, book publishers, and electronics firms who have given permission for the use of numerous figures and tables.

And finally, the authors thank the many people at SRI who took a special interest in the huge job of preparing this book. Mrs. Edith Chambers spent countless painstaking hours supervising the preparation of the staggering number of illustrations in this book, and helped greatly in insuring illustrations of high quality and clarity. Mrs. Mary F. Armstrong supervised the Varityping of the text. The authors' thanks also go to the editors, secretaries, and report production staff at SRI who all were very cooperative in the production of this book.

CONTENTS

CONTENTS

MICROWAVE FILTERS, IMPEDANCE-MATCHING NETWORKS, AND COUPLING STRUCTURES

CHAPTER 1

SOME GENERAL APPLICATIONS OF FILTER STRUCTURES IN MICROWAVE ENGINEERING

SEC. 1.01, INTRODUCTION

Most readers will be familiar with the use of filters as discussed in Sec. 1.02 below. However, the potential applications of the material in this book goes much beyond these classical filter applications to cover many other microwave engineering problems which involve filter-type structures but are not always thought of as being filter problems.

Thus, the purpose of this chapter is to make clear to the reader that this book is not addressed only to filter design specialists, but also to antenna engineers who may need a broadband antenna feed, to microwave tube engineers who may need to obtain broadband impedance matches in and out of microwave tubes, to system engineers who may need a microwave time-delay network, and to numerous others having other special microwave circuit design problems.

SEC. 1.02, USE OF FILTERS FOR THE SEPARATION OR SUMMING OF SIGNALS

The most obvious application of filter structures, of course, is for the rejection of unwanted signal frequencies while permitting good transmission of wanted frequencies. The most common filters of this sort are designed for either low-pass, high-pass, band-pass or band-stop attenuation characteristics such as those shown in Fig. 1.02-1. Of course, in the case of practical filters for the microwave or any other frequency range, these characteristics are only achieved approximately, since there is a high-frequency limit for any given practical filter structure above which its characteristics will deteriorate due to junction effects, resonances within the elements, etc.

Filters are also commonly used for separating frequencies in diplexers or multiplexers. Figure 1.02-2 shows a multiplexer which segregates signals within the 2.0 to 4.0 Gc band into three separate

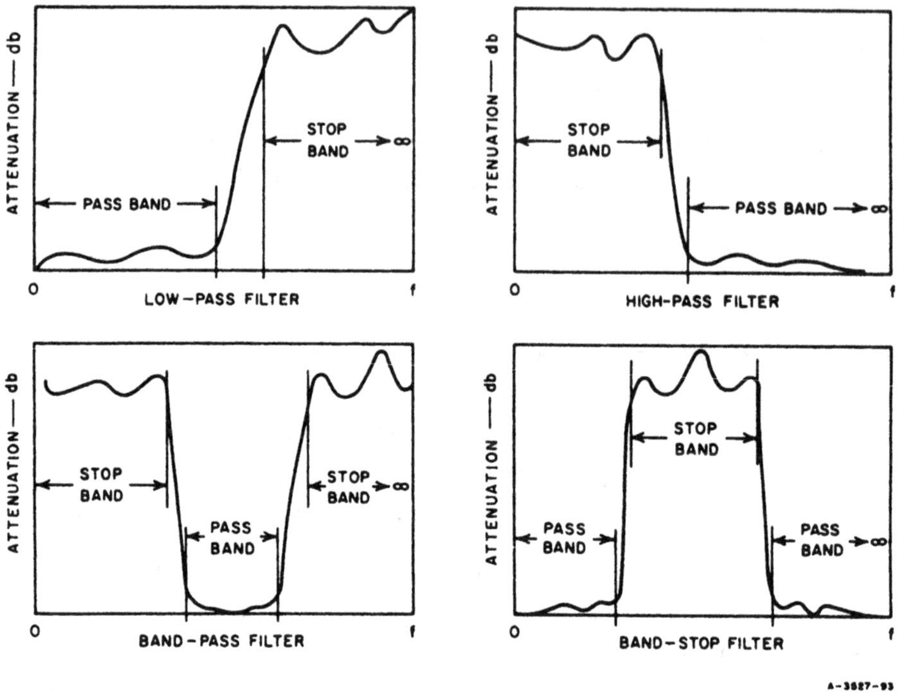

FIG. 1.02-1 FOUR COMMON TYPES OF FILTER CHARACTERISTICS

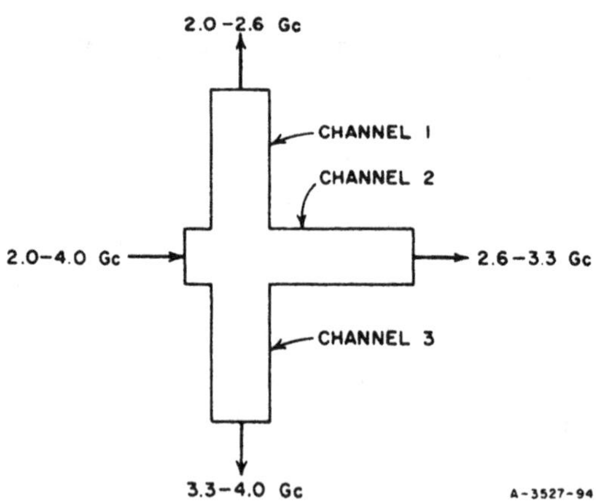

FIG. 1.02-2 A THREE-CHANNEL MULTIPLEXING FILTER GROUP

channels according to their frequencies. A well designed multiplexer of this sort would have very low VSWR at the input port across the 2.0 to 4.0 Gc input band. To achieve this result the individual filters must be designed specially for this purpose along with a special junction-matching network. These matters are treated in Chapter 16.

Another way that diplexers or multiplexers are often used is in the summing of signals having different frequencies. Supposing that the signal-flow arrowheads in Fig. 1.02-2 are reversed; in this event, signals entering at the various channels can all be joined together with negligible reflection or leakage of energy so that all of the signals will be superimposed on a single output line. If signals in these various channel frequency ranges were summed by a simple junction of transmission lines (*i.e.*, without a multiplexer), the loss in energy at the single output line would, of course, be considerable, as a result of reflections and of leakage out of lines other than the intended output line.

SEC. 1.03, IMPEDANCE-MATCHING NETWORKS

Bode[1] first showed what the physical limitations were on the broadband impedance matching of loads consisting of a reactive element and a resistor in series or in parallel. Later, Fano[2] presented the general limitations on the impedance matching of any load. Fano's work shows that efficiency of transmission and bandwidth are exchangeable quantities in the impedance matching of any load having a reactive component.

To illustrate the theoretical limitations which exist on broadband impedance matching, consider the example shown in Fig. 1.03-1 where the load to be matched consists of a capacitor C_1 and a resistor R_0 in parallel. A lossless impedance-matching network is to be inserted between the generator and the load, and the reflection coefficient between the generator and the impedance-matching network is

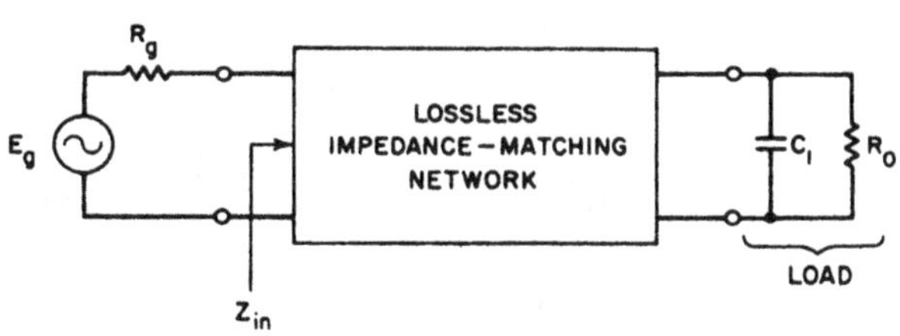

FIG. 1.03-1 EXAMPLE OF AN IMPEDANCE-MATCHING PROBLEM

$$\Gamma = \frac{Z_{in} - R_g}{Z_{in} + R_g} \tag{1.03-1}$$

The work of Bode[1] and that of Fano[2] shows that there is a physical limitation on what Γ can be as a function of frequency. The *best possible* results are limited as indicated by the relation*

$$\int_0^{\infty} ln \left|\frac{1}{\Gamma}\right| d\omega = \frac{\pi}{R_0 C_1} \tag{1.03-2}$$

Recall that for a passive circuit $0 \leqq |\Gamma| \leqq 1$, for total reflection $|\Gamma| = 1$, and that for perfect transmission $|\Gamma| = 0$. Thus, the larger $ln\ |1/\Gamma|$ is the better the transmission will be. But Eq. (1.03-2) says that the area under the curve of $ln\ |1/\Gamma|$ *vs* ω can be no greater than $\pi/(R_0 C_1)$.

If a good impedance match is desired from frequency ω_a to ω_b, best results can be obtained if $|\Gamma| = 1$ at all frequencies except in the band from ω_a to ω_b. Then $ln\ |1/\Gamma| = 0$ at all frequencies except in the ω_a to ω_b band, and the available area under the $ln\ |1/\Gamma|$ curve can all be concentrated in the region where it does the most good. With this specification, Eq. (1.03-2) becomes

$$\int_{\omega_a}^{\omega_b} ln \left|\frac{1}{\Gamma}\right| d\omega = \frac{\pi}{R_0 C_1} \tag{1.03.3}$$

and if $|\Gamma|$ is assumed to be constant across the band of impedance match, $|\Gamma|$ as a function of frequency becomes

$$\begin{aligned} |\Gamma| &= e^{\frac{-\pi}{(\omega_b - \omega_a) R_0 C_1}} \quad \text{for} \quad \omega_a \leqq \omega \leqq \omega_b \\ |\Gamma| &= 1 \quad \text{for} \quad 0 \leqq \omega \leqq \omega_a \quad \text{and} \quad \omega_b \leqq \omega \leqq \infty \end{aligned} \tag{1.03-4}$$

---

* This relation holds if the impedance matching network is designed so that the reflection coefficient between R_0 and the circuit to the left of R_0 in Fig. 1.03-1 has all of its zeros in the left half plane.[1,2]

Equation (1.03-4) says that an ideal impedance-matching network for the load in Fig. 1.03-1 would be a band-pass filter structure which would cut off sharply at the edges of the band of impedance match. The curves in Fig. 1.03-2 show how the $|\Gamma|$ *vs* ω curve for practical band-pass impedance-matching filters might look. The curve marked Case 1 is for the impedance matching of a given load over the relatively narrow band from ω_a to ω_b, while the curve marked Case 2 is for the impedance matching of the same load over the wider band from ω_c to ω_d using the same number of elements in the impedance-matching network. The rectangular $|\Gamma|$ characteristic indicated by Eq. (1.03-4) is that which would be achieved by an optimum band-pass matching filter with an infinite number of elements.*

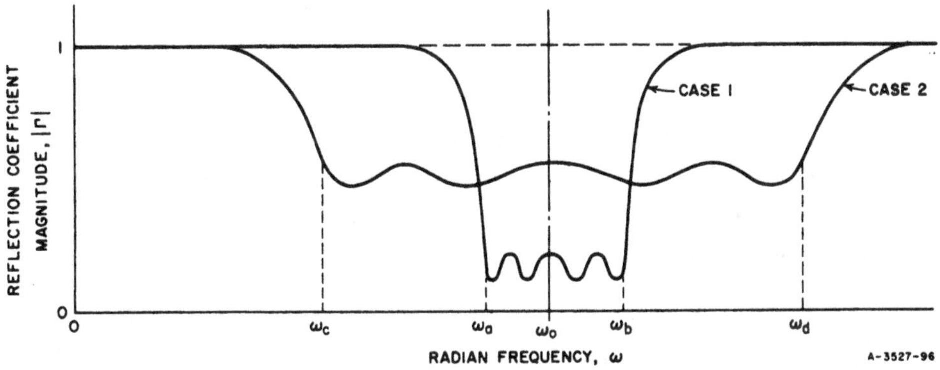

FIG. 1.03-2 CURVES ILLUSTRATING RELATION BETWEEN BANDWIDTH AND DEGREE OF IMPEDANCE MATCH POSSIBLE FOR A GIVEN LOAD HAVING A REACTIVE COMPONENT

The work of Fano[2] shows that similar conditions apply no matter what the nature of the load (as long as the load is not a pure resistance). *Thus, for this very fundamental reason, efficient broadband impedance-matching structures are necessarily filter structures.* See Secs. 4.09 through 4.11, 7.08, and 11.08 through 11.10 for discussions of how the filter structures treated in this book may be designed for use as impedance-matching networks.

---

* Simple matching networks can give very great improvements in impedance match, and as the number of matching elements is increased the improvement per additional element rapidly becomes smaller and smaller. For this reason fairly simple matching networks can give performance which comes close to the theoretically optimum performance for an infinite number of impedance-matching elements.

SEC. 1.04, COUPLING NETWORKS FOR TUBES AND NEGATIVE-RESISTANCE AMPLIFIERS

A pentode vacuum tube can often be simulated at its output as an infinite-impedance current generator with a capacitor shunted across the terminals. Broadband output circuits for such tubes can be designed as a filter to be driven by an infinite-impedance current generator at one end with only one resistor termination (located at the other end of the filter). Then the output capacitance of the tube is utilized as one of the elements required for the filter, and in this way the deleterious effects of the shunt capacitance are controlled.[3] Data presented later in this book will provide convenient means for designing microwave broadband coupling circuits for possible microwave situations of a similar character where the driving source may be regarded as a current or voltage generator plus a reactive element.

In some cases the input or output impedances of an oscillator or an amplifying device may be represented as a resistance along with one or two reactive elements. In such cases impedance-matching filters as discussed in the preceding section are necessary if optimum broadband performance is to be approached.

Negative-resistance amplifiers are yet another class of devices which require filter structures for optimum broadband operation. Consider the circuit in Fig. 1.04-1, where we shall define the reflection coefficient at the left as

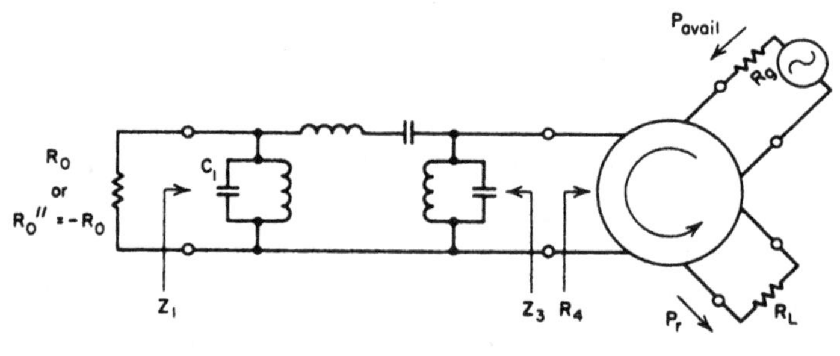

FIG. 1.04-1 CIRCUIT ILLUSTRATING THE USE OF FILTER STRUCTURES IN THE DESIGN OF NEGATIVE-RESISTANCE AMPLIFIERS

$$\Gamma_1 = \frac{Z_1 - R_0}{Z_1 + R_0} \tag{1.04-1}$$

and that at the right as

$$\Gamma_3 = \frac{Z_3 - R_4}{Z_3 + R_4} \tag{1.04-2}$$

Since the intervening band-pass filter circuit is dissipationless,

$$|\Gamma_1| = |\Gamma_3| \tag{1.04-3}$$

though the phases of Γ_1 and Γ_3 are not necessarily the same. The available power entering the circulator on the right is directed into the filter network, and part of it is reflected back to the circulator where it is finally absorbed in the termination R_L. The transducer gain from the generator to R_L is

$$\frac{P_r}{P_{avail}} = |\Gamma_3|^2 \tag{1.04-4}$$

where P_{avail} is the available power of the generator and P_r is the power reflected back from the filter network.

If the resistor R_0 on the left in Fig. 1.04-1 is positive, the transducer gain characteristic might be as indicated by the Case 1 curve in Fig. 1.04-2. In this case the gain is low in the pass band of the filter since $|\Gamma_1| = |\Gamma_3|$ is small then. However, if R_0 is replaced by a negative resistance $R_0'' = -R_0$, then the reflection coefficient at the left becomes

$$\Gamma_1'' = \frac{Z_1 - R_0''}{Z_1 + R_0''} = \frac{Z_1 + R_0}{Z_1 - R_0} \quad . \tag{1.04-5}$$

As a result we then have

$$|\Gamma_3''| = |\Gamma_1''| = \frac{1}{|\Gamma_1|} \quad . \tag{1.04-6}$$

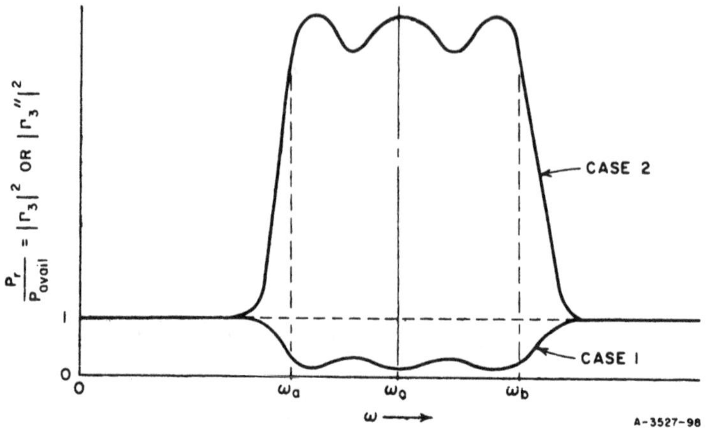

FIG. 1.04-2 TRANSDUCER GAIN BETWEEN GENERATOR IN FIG. 1.04-1 AND THE CIRCULATOR OUTPUT
Case 1 is for R_0 Positive while Case 2 is for R_0 Replaced by $R_0'' = -R_0$

Thus, replacing R_0 by its negative corresponds to $|\Gamma_3|$ being replaced by $|\Gamma_3''| = 1/|\Gamma_3|$, and the transducer gain is as indicated by the curve marked Case 2 in Fig. 1.04-2. Under these circumstances the output power greatly exceeds the available power of the generator for frequencies within the pass band of the filter.

With the aid of Eqs. (1.04-1) and (1.04-6) coupling networks for negative-resistance amplifiers are easily designed using impedance-matching filter design techniques. Practical negative-resistance elements such as tunnel diodes are not simple negative resistances, since they also have reactive elements in their equivalent circuit. In the case of tunnel diodes the dominant reactive element is a relatively large capacitance in parallel with the negative resistance. With this large capacitance present satisfactory operation is impossible at microwave frequencies unless some special coupling network is used to compensate for its effects. In Fig. 1.04-1, C_1 and R_0'' on the left can be defined as the tunnel-diode capacitance and negative resistance, and the remainder of the band-pass filter circuit serves as a broadband coupling network.

Similar principles also apply in the design of broadband coupling networks for masers and parametric amplifiers. In the case of parametric amplifiers, however, the design of the coupling filters is complicated somewhat by the relatively complex impedance transforming effects of the time-varying element.[4,7]

The coupling network shown in Fig. 1.04-1 is in a lumped-element form which is not very practical to construct at microwave frequencies. However, techniques which are suitable for designing practical microwave filter structures for such applications are discussed in Sec. 11.10, using results developed in Secs. 11.08 and 11.09. Low-pass prototypes for negative-resistance-amplifier coupling networks are tabulated in Sec. 4.11.

SEC. 1.05, TIME-DELAY NETWORKS AND SLOW-WAVE STRUCTURES

Consider the low-pass filter network in Fig. 1.05-1(a) which has a voltage transfer function E_0/E_6. The transmission phase is defined as

$$\phi = \arg \frac{E_0}{E_6} \qquad \text{radians} \quad . \tag{1.05-1}$$

The *phase delay* of this network at any given frequency ω is

$$t_p = \frac{\phi}{\omega} \qquad \text{seconds} \tag{1.05-2}$$

while its *group delay* is

$$t_d = \frac{d\phi}{d\omega} \qquad \text{seconds} \tag{1.05-3}$$

where ϕ is in radians and ω is in radians per second. Under different circumstances either phase or group delay may be important, but it is

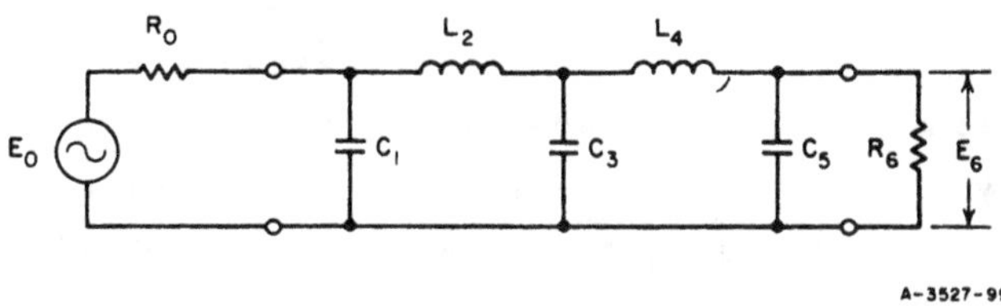

FIG. 1.05-1(a) LOW-PASS FILTER DISCUSSED IN SEC. 1.05

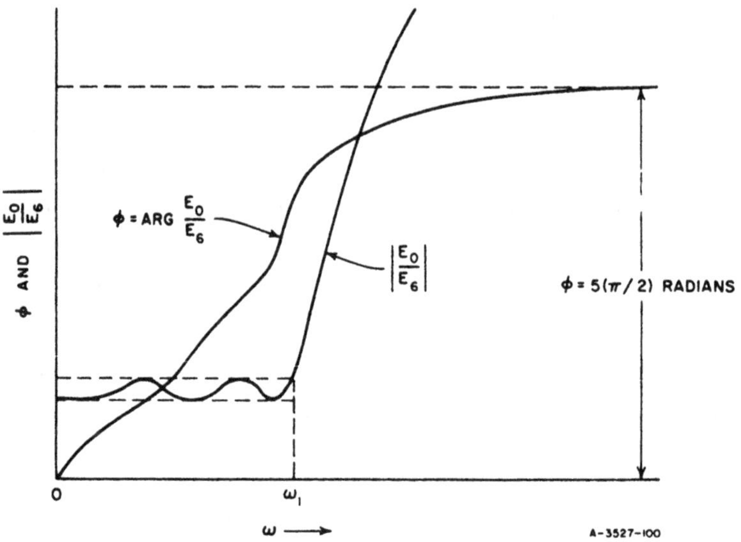

FIG. 1.05-1(b) A POSSIBLE $|E_0/E_6|$ CHARACTERISTIC FOR THE FILTER IN FIG. 1.05-1(a), AND AN APPROXIMATE CORRESPONDING PHASE CHARACTERISTIC

group delay which determines the time required for a signal to pass through a circuit.[5,6]*

Low-pass ladder networks of the form in Fig. 1.05-1(a) have zero transmission phase for $\omega = 0$, and as ω becomes large

$$\varphi\Big|_{\omega\to\infty} \to \frac{n\pi}{2} \qquad \text{radians} \qquad (1.05\text{-}4)$$

where n is the number of reactive elements in the circuit. Figure 1.05-1(b) shows a possible $|E_0/E_6|$ characteristic for the filter in Fig. 1.05-1(a) along with the approximate corresponding phase characteristic. Note that most of the phase shift takes place within the pass band $\omega = 0$ to $\omega = \omega_1$. This is normally the case, hence a rough estimate of the group time delay in the pass band of filters of the form in Fig. 1.05-1(a) can be obtained from

* That is, if there is no amplitude distortion and $d\phi/d\omega$ is constant across the frequency band of the signal, then the output signal will be an exact replica of the input signal but displaced in time by t_d seconds.

$$t_d \approx \frac{n\pi}{2\omega_1} \quad \text{seconds} \qquad (1.05\text{-}5)$$

where n is again the number of reactive elements in the filter. Of course, in some cases t_d may vary appreciably within the pass band, and Eq. (1.05-5) is very approximate.*

Figure 1.05-2(a) shows a five-resonator band-pass filter while Fig. 1.05-2(b) shows a possible phase characteristic for this filter. In this case the total phase shift from $\omega = 0$ to $\omega = \infty$ is $n\pi$ radians,

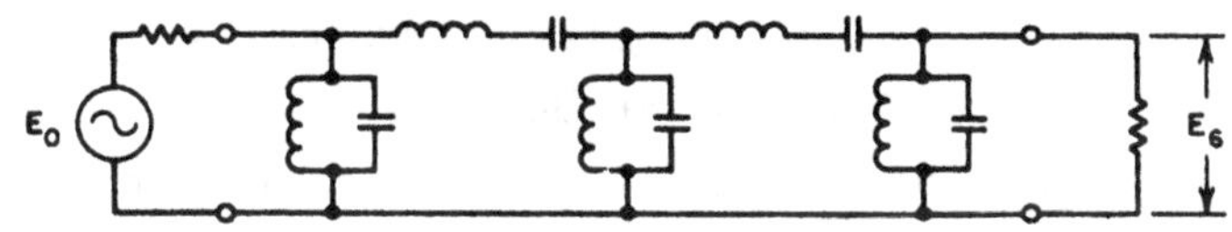

FIG. 1.05-2(a) A BAND-PASS FILTER CORRESPONDING TO THE LOW-PASS FILTER IN FIG. 1.05-1(a)

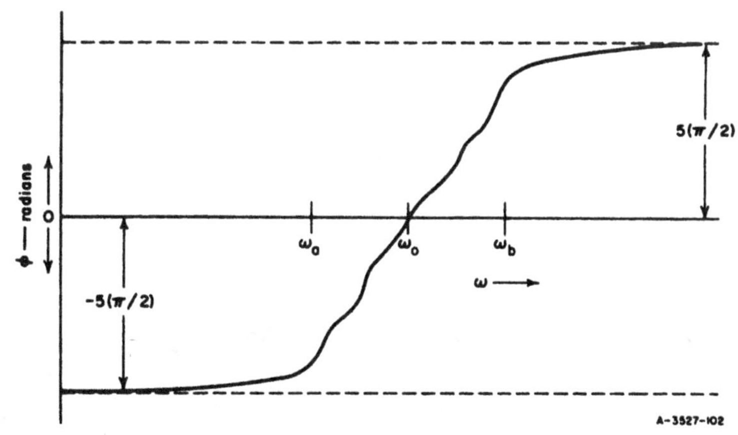

FIG. 1.05-2(b) A POSSIBLE PHASE CHARACTERISTIC FOR THE FILTER IN FIG. 1.05-2(a)

* Since not all of the phase shift occurs within the pass band, Eq. (1.05-5) will usually overestimate the time delay, often by as much as a factor of two.

where n is the number of resonators, and a rough estimate of the pass-band group time delay is*

$$t_d \approx \frac{n\pi}{\omega_b - \omega_a} \quad \text{seconds} \tag{1.05-6}$$

where ω_a and ω_b are the radian frequencies of the pass-band edges.

In later chapters more precise information on the time delay characteristics of filters will be presented. Equations (1.05-5) and (1.05-6) are introduced here simply because they are helpful for giving a feel for the general time delay properties of filters. Suppose that for some system application it is desired to delay pulses of S-band energy 0.05 microseconds, and that an operating bandwidth of 50 Mc is desired to accommodate the signal spectrum and to permit some variation of carrier frequency. If this delay were to be achieved with an air-filled coaxial line, 49 feet of line would be required. Equation (1.05-6) indicates that this delay could be achieved with a five-resonator filter having 50 Mc bandwidth. An S-band filter designed for this purpose would typically be less than a foot in length and could be made to be quite light.

In slow-wave structures usually *phase velocity*

$$v_p = \frac{l}{t_p} \tag{1.05-7}$$

or *group velocity*

$$v_g = \frac{l}{t_d} \tag{1.05-8}$$

is of interest, where l is the length of the structure and t_p and t_d are as defined in Eqs. (1.05-2) and (1.05-3). Not all structures used as slow-wave structures are filters, but very many of them are. Some examples of slow-wave structures which are basically filter structures are waveguides periodically loaded with capacitive or inductive irises, interdigital lines, and comb lines. The methods of this book should be quite helpful in the design of such slow-wave structures which are basically filters. The use of filters as time delay networks is discussed further in Secs. 4.07, 4.08, 6.15, 7.09, and 11.11.

*As for Eqs. (1.05-5), this equation will usually give an over estimate of the time delay.

SEC. 1.06, GENERAL USE OF FILTER PRINCIPLES IN THE DESIGN OF MICROWAVE COMPONENTS

As can be readily seen by extrapolating from the discussions in preceding sections, microwave filter design techniques when used in their most general way are fundamental to the efficient design of a wide variety of microwave components. In general, these techniques are basic to precision design when selecting, rejecting, or channeling of energy of different frequencies is important; when achieving energy transfer with low reflection over a wide band is important; or when achieving a controlled time delay is important. The possible specific practical situations where such considerations arise are too numerous and varied to permit any attempt to treat them individually herein. However, a reader who is familiar with the principles to be treated in this book will usually have little trouble in adapting them for use in the many special design situations he will encounter.

REFERENCES

1. H. W. Bode, *Network Analysis and Feedback Amplifier Design*, pp. 360-371 (D. Van Nostrand Co. New York, N.Y., 1945).

2. R. M. Fano, "Theoretical Limitations on the Broadband Matching of Arbitrary Impedances," *Journal of the Franklin Institute*, Vol. 249, pp. 57-84 and 139-154 (January-February 1950).

3. G. E. Valley and H. Wallman, *Vacuum Tube Amplifiers*, Chapters 2 and 4 (McGraw-Hill Book Co., New York, N.Y., 1948).

4. G. L. Matthaei, "A Study of the Optimum Design of Wide-Band Parametric Amplifiers and Up-Converters," *IRE Trans. PGMTT*-9, pp. 23-38 (January 1961).

5. E. A. Guillemin, *Communication Networks*, Vol. 2, pp. 99-106 and 490-498 (John Wiley and Sons, Inc., New York, N.Y., 1935).

6. M. J. Di Toro, "Phase and Amplitude Distortion in Linear Networks," *Proc. IRE*, Vol. 36, pp. 24-36 (January 1948).

7. E. S. Kuh and M. Fukada, "Optimum Synthesis of Wide-Band Parametric Amplifiers and Converters," *IRE Trans. PGCT-8*, pp. 410-415 (December 1961).

CHAPTER 2

SOME USEFUL CIRCUIT CONCEPTS AND EQUATIONS

SEC. 2.01, INTRODUCTION

The purpose of this chapter is to summarize various circuit theory concepts and equations which are useful for the analysis of filters. Though much of this material will be familiar to many readers, it appears desirable to gather it together for easy reference. In addition, there will undoubtedly be topics with which some readers will be unfamiliar. In such cases the discussion given here should provide a brief introduction which should be adequate for the purposes of this book.

SEC. 2.02, COMPLEX FREQUENCY AND POLES AND ZEROS

A "sinusoidal" voltage

$$e(t) = |E_m| \cos (\omega t + \phi) \qquad (2.02\text{-}1)$$

may also be defined in the form

$$e(t) = \mathrm{Re}\,[E_m e^{j\omega t}] \qquad (2.02\text{-}2)$$

where t is the time in seconds, ω is frequency in radians per second, and $E_m = |E_m| e^{j\phi}$ is the complex amplitude of the voltage. The quantity E_m, of course, is related to the root-mean-square voltage E by the relation $E = E_m/\sqrt{2}$.

Sinusoidal waveforms are a special case of the more general waveform

$$e(t) = |E_m| e^{\sigma t} \cos (\omega t + \phi) \qquad (2.02\text{-}3)$$

$$= R_e\,[E_m e^{pt}] \qquad (2.02\text{-}4)$$

where $E_m = |E_m| e^{j\phi}$ is again the complex amplitude. In this case

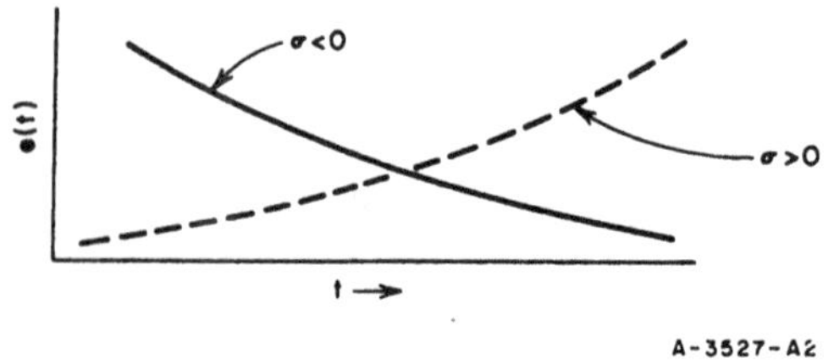

FIG. 2.02-1(a) SHAPE OF COMPLEX-FREQUENCY WAVEFORM WHEN $p = \sigma + j0$

$$p = \sigma + j\omega \tag{2.02-5}$$

is the *complex frequency*. In this general case the waveform may be a pure exponential function as illustrated in Fig. 2.02-1(a), it may be an exponentially-varying sinusoid as illustrated in Fig. 2.02-1(b), or it may be a pure sinusoid if $p = j\omega$.

In linear, time-invariant circuits such as are discussed in this book complex-frequency waveforms have fundamental significance not shared by other types of waveforms. Their basic importance is exemplified by the following properties of linear, time-invariant circuits:

(1) If a "steady-state" driving voltage or current of complex frequency p is applied to a circuit the steady-state response seen at any point in the circuit* will also have a complex-frequency waveform with the same frequency p. The amplitude and phase angle will, in general, be different at different points throughout the circuit. But at any given point in the circuit the response amplitude and the phase angle are both linear functions of the driving-signal amplitude and phase.

(2) The various possible natural modes of vibration of the circuit will have complex-frequency waveforms. (The natural modes are current and voltage vibrations which can exist after all driving signals are removed.)

The concepts of impedance and transfer functions result from the first property listed above, since these two functions represent ratios between the complex amplitudes of the driving signal and the response. As a result of Property (2), the transient response of a network will contain a superposition of the complex-frequency waveforms of the various natural modes of vibration of the circuit.

The impedance of a circuit as a function of complex frequency p will take the form

---

* Unless stated otherwise, a linear, time-invariant circuit will be understood.

$$Z(p) = \frac{E_m}{I_m} = \frac{E}{I} = \frac{a_n p^n + a_{n-1} p^{n-1} + \ldots + a_1 p + a_0}{b_n p^n + b_{n-1} p^{n-1} + \ldots + b_1 p + b_0} \quad . \tag{2.02-6}$$

By factoring the numerator and denominator polynomials this may be written as

$$Z(p) = \left(\frac{a_n}{b_n}\right) \frac{(p - p_1)(p - p_3)(p - p_5) \ldots}{(p - p_2)(p - p_4)(p - p_6) \ldots} \quad . \tag{2.02-7}$$

At the frequencies $p = p_1, p_3, p_5, \ldots$ etc., where the numerator polynomial goes to zero the impedance function will be zero; these frequencies are thus known as the *zeros* of the function. At the frequencies $p = p_2, p_4, p_6, \ldots$, etc., where the denominator polynomial is zero the impedance function will be infinite; these frequencies are known as the *poles* of the function. The poles and zeros of a transfer function are defined in a similar fashion.

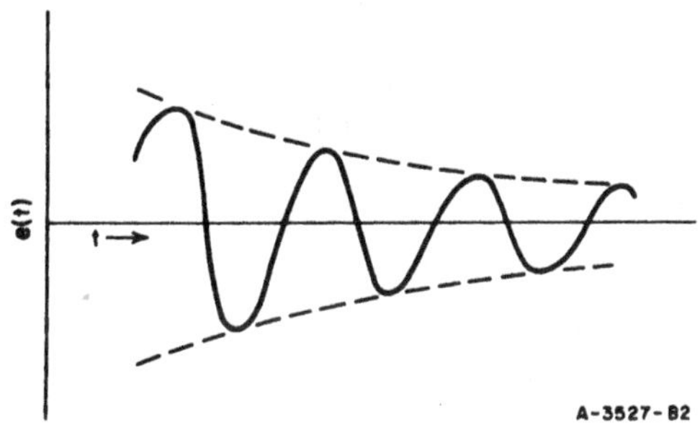

FIG. 2.02-1(b) SHAPE OF COMPLEX-FREQUENCY WAVEFORM WHEN $p = \sigma + j\omega$ AND $\sigma < 0$

A circuit with a finite number of lumped, reactive elements will have a finite number of poles and zeros. However, a circuit involving *distributed elements* (which may be represented as an infinite number of infinitesimal lumped-elements) will have an infinite number of poles and zeros. Thus, circuits involving transmission lines will have impedance functions that are transcendental, *i.e.*, when expressed in the form in Eq. (2.02-7) they will be infinite-product expansions. For example, the input impedance to a lossless, short-circuited transmission line which is one-quarter wavelength long at frequency ω_0 may be expressed as

$$Z(p) = Z_0 \tanh\left(\frac{\pi p}{2\omega_0}\right)$$

$$= Z_0 \frac{\pi p}{2\omega_0} \prod_{k=1}^{\infty} \left(\frac{2k - 1}{2k}\right)^2 \frac{[p + j2k\omega_0][p - j2k\omega_0]}{[p + j(2k - 1)\omega_0][p - j(2k - 1)\omega_0]} \tag{2.02-8}$$

where Z_0 is the characteristic impedance of the line. This circuit is seen to have poles at $p = \pm j(2k - 1)\omega_0$ and zeros at $p = 0$ and $\pm j2k\omega_0$, where $k = 1, 2, 3, \ldots, \infty$.

Regarding frequency as the more general $p = \sigma + j\omega$ variable instead of the unnecessarily restrictive $j\omega$ variable permits a much broader point of view in circuit analysis and design. Impedance and transfer functions become functions of a complex variable (*i.e.*, they become functions of the variable $p = \sigma + j\omega$) and all of the powerful tools in the mathematical theory of functions of a complex variable become available. It becomes very helpful to define the properties of an impedance or transfer function in terms of the locations of their poles and zeros, and these poles and zeros are often plotted in the *complex-frequency plane* or *p-plane*. The poles are indicated by crosses and the zeros by circles. Figure 2.02-2(a) shows such a plot for the lossless transmission line input impedance in Eq. (2.02-8) while Fig. 2.02-2(b) shows a sketch of the shape of the magnitude of this function for $p = j\omega$. The figure also shows what happens to the poles and zeros if the line has loss: the poles and zeros are all moved to the left of the $j\omega$ axis, and the $|Z(j\omega)|$ characteristic becomes rounded off.

The concepts of complex frequency and poles and zeros are very helpful in network analysis and design. Discussions from this point of view will be found in numerous books on network analysis and synthesis, including those listed in References 1 to 5. Poles and zeros also have an electrostatic analogy which relates the magnitude and phase of an impedance or transfer function to the potential and flux, respectively, of an analogous electrostatic problem. This analogy is useful both as a tool for mathematical reasoning and as a means for determining magnitude and phase by measurements on an analog setup. Some of these matters are discussed in References 2, 3, 6, and 7. Further use of the concepts of complex frequency and poles and zeros will be discussed in the next two sections.

SEC. 2.03, NATURAL MODES OF VIBRATION AND THEIR RELATION TO INPUT-IMPEDANCE POLES AND ZEROS

The natural modes of vibration of a circuit are complex frequencies at which the voltages and currents in the circuit can "vibrate" if the circuit is disturbed. These vibrations can continue even after all driving signals have been set to zero. It should be noted that here the

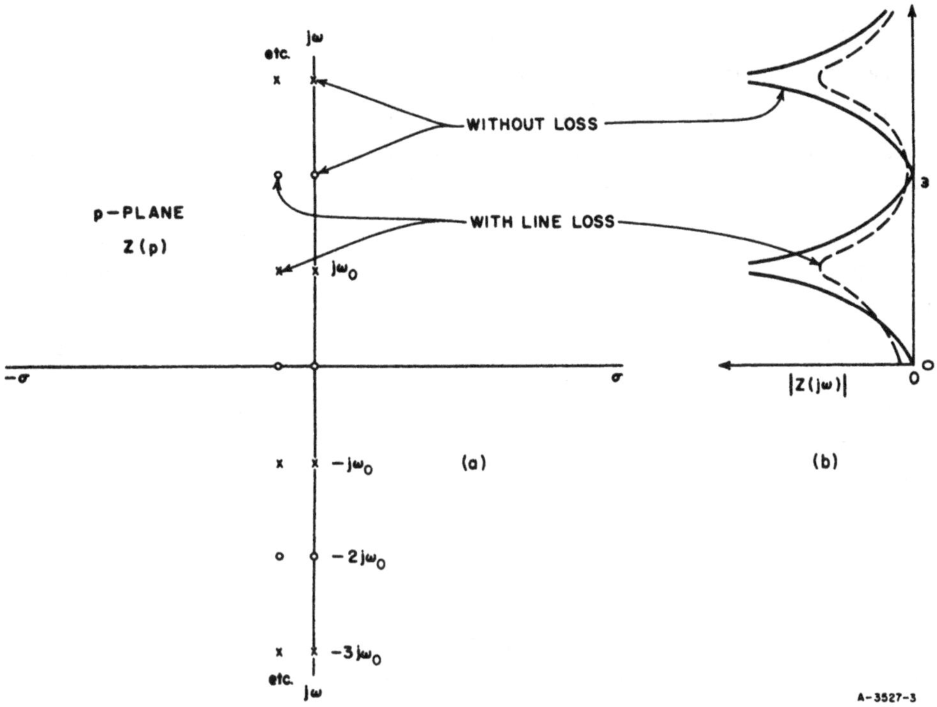

FIG. 2.02-2 THE LOCATIONS OF THE POLES AND ZEROS OF A SHORT-CIRCUITED TRANSMISSION LINE WHICH IS A QUARTER-WAVELENGTH LONG WHEN $p = j\omega_0$
The Magnitude of the Input Impedance for Frequencies $p = j\omega$ is also Sketched

word *vibration* is used to include natural modes having exponential waveforms of frequency $p = \sigma$ as well as oscillatory waveforms of frequency $p = \sigma + j\omega$.

Suppose that the input impedance of a circuit is given by the function

$$Z(p) = \frac{E}{I} = \frac{(p - p_1)(p - p_3)(p - p_5) \ldots}{(p - p_2)(p - p_4)(p - p_6) \ldots} \quad . \qquad (2.03\text{-}1)$$

Rearranging Eq. (2.03-1),

$$E = I\,Z(p) \qquad (2.03\text{-}2)$$

If the input terminals of this circuit are open-circuited and the circuit is vibrating at one of its natural frequencies, there will be a complex-frequency voltage across $Z(p)$ even though $I = 0$. By Eq. (2.03-2) it is seen that the only way in which the voltage E can be non-zero while $I = 0$ is for $Z(p)$ to be infinite. Thus, if $Z(p)$ is open-circuited, natural vibration can be observed only at the frequencies p_2, p_4, p_6, etc., which are the poles of the input impedance function $Z(p)$. Also, by analogous reasoning it is seen that if $Z(p)$ is short-circuited, the natural frequencies of vibration will be the frequencies of the *zeros* of $Z(p)$.

Except for special cases where one or more natural modes may be stifled at certain points in a circuit, if any natural modes are excited in any part of the circuit, they will be observed in the voltages and currents throughout the entire circuit. The frequency $p_n = \sigma_n + j\omega_n$ of each natural mode must lie in the left half of the complex-frequency plane, or on the $j\omega$ axis. If this were not so the vibrations would be of exponentially increasing magnitude and energy, a condition which is impossible in a passive circuit. Since under open-circuit or short-circuit conditions the poles or the zeros, respectively, of an impedance function are natural frequencies of vibration, any impedance of a linear, passive circuit must have all of its poles and zeros in the left half plane or on the $j\omega$ axis.

SEC. 2.04, FUNDAMENTAL PROPERTIES OF TRANSFER FUNCTIONS

Let us define the *voltage attenuation function* E_g/E_L for the network in Fig. 2.04-1 as

$$T(p) = \frac{E_g}{E_L} = \frac{c(p - p_1)(p - p_3)(p - p_5)\ \cdots}{(p - p_2)(p - p_4)(p - p_6)\ \cdots} \tag{2.04-1}$$

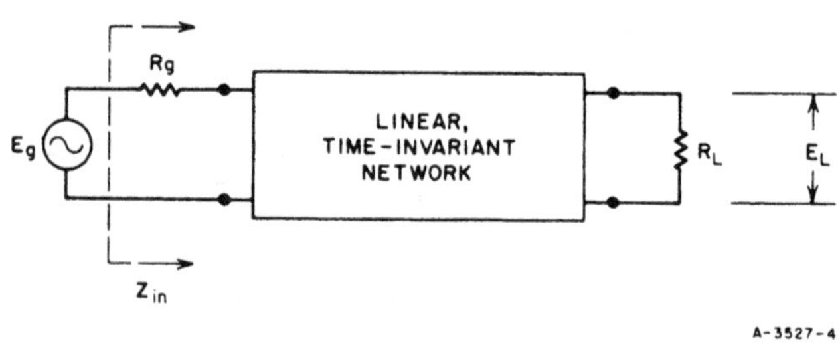

FIG. 2.04-1 NETWORK DISCUSSED IN SECTION 2.04

where c is a real constant and p is the complex-frequency variable. We shall now briefly summarize some important general properties of linear, passive circuits in terms of this transfer function and Fig. 2.04-1.

(1) The zeros of $T(p)$, *i.e.*, p_1, p_3, p_5, ... are all frequencies of natural modes of vibration for the circuit. They are influenced by all of the elements in the circuit so that, for example, if the value for R_g or R_L were changed, generally the frequencies of all the natural modes will change also.

(2) The poles of $T(p)$, *i.e.*, p_2, p_4, p_6, ..., along with any poles of $T(p)$ at $p = 0$ and $p = \infty$ are all *frequencies of infinite attenuation*, or "poles of attenuation." They are properties of the network alone and will not be changed if R_g or R_L is changed. Except for certain degenerate cases, if two networks are connected in cascade, the resultant over-all response will have the poles of attenuation of both component networks.

(3) In a ladder network, a pole of attenuation is created when a *series* branch has *infinite* impedance, or when a *shunt* branch has *zero* impedance. If at a given frequency, infinite impedance occurs in series branches simultaneously with zero impedance in shunt branches, a higher-order pole of attenuation will result.

(4) In circuits where there are two or more transmission channels in parallel, poles of attenuation are created at frequencies where the outputs from the parallel channels have the proper magnitude and phase to cancel each other out. This can happen, for example, in bridged-T, lattice, and parallel-ladder structures.

(5) The natural modes [zeros of $T(p)$] must lie in the left half of the p-plane (or on the $j\omega$ axis if there are no loss elements).

(6) The poles of attenuation can occur anywhere in the p-plane.

(7) If E_g is a zero impedance voltage generator, the zeros of Z_{in} in Fig. 2.04-1 will be the natural frequencies of vibration of the circuit. These zeros must therefore correspond to the zeros of the attenuation function $T(p)$. (Occasionally this fact is obscured because in some special cases cancellations can be carried out between coincident poles and zeros of $T(p)$ or of Z_{in}. Assuming that no such cancellations have been carried out even when they are possible, the above statement always holds.)

(8) If the zero impedance voltage generator E_g were replaced by an infinite impedance current generator I_g, then the natural frequencies of vibration would correspond to the *poles* of Z_{in}. Redefining $T(p)$ as $T'(p) = I_g/E_L$, the zeros of $T'(p)$ would in this case still be the natural frequencies of vibration but they would in this case be the same as the *poles* of Z_{in}.

Let us now consider some examples of how some of the concepts in the statements itemized above may be applied. Suppose that the box in Fig. 2.04-1 contains a lossless transmission line which is one-quarter wavelength long at the frequency ω_0. Let us suppose further that $R_g = R_L \neq Z_0$, where Z_0 is the characteristic impedance of the line. Under these conditions the voltage attenuation function $T(p)$ would have a p-plane plot as indicated in Fig. 2.04-2(a). Since the transmission line is a distributed circuit there are an infinite number of natural

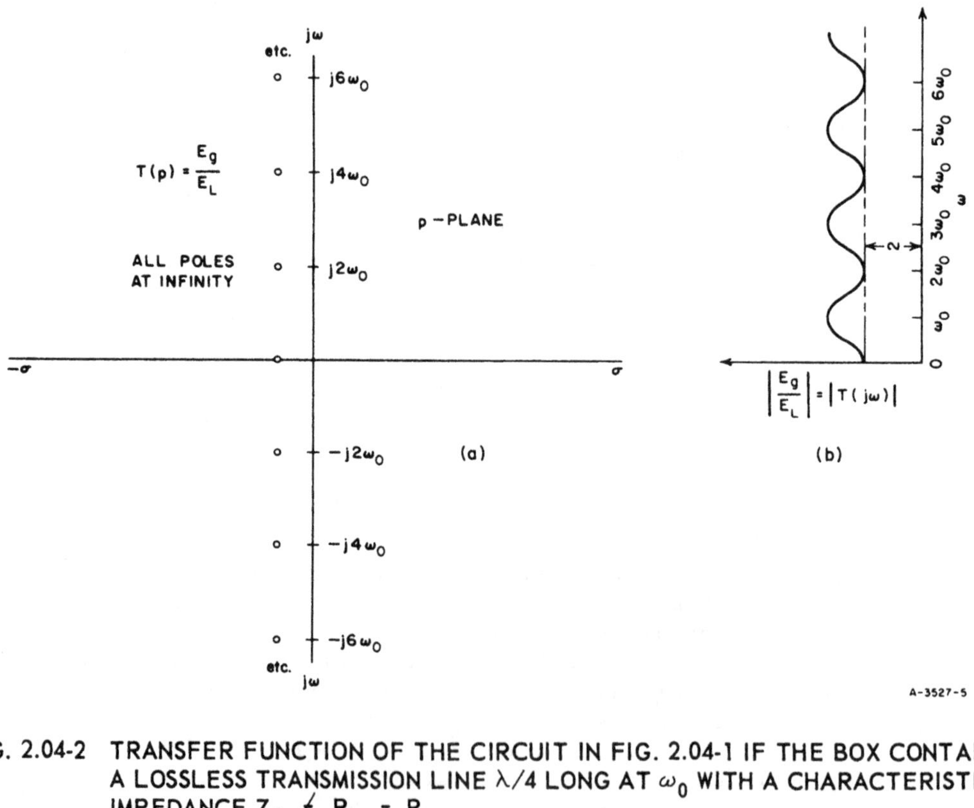

FIG. 2.04-2 TRANSFER FUNCTION OF THE CIRCUIT IN FIG. 2.04-1 IF THE BOX CONTAINS A LOSSLESS TRANSMISSION LINE λ/4 LONG AT ω_0 WITH A CHARACTERISTIC IMPEDANCE $Z_0 \neq R_g = R_L$

modes of vibration, and, hence there are an infinite array of zeros to $T(p)$. In all impedance and transfer functions the number of poles and zeros must be equal if the point at $p = \infty$ is included. In this case there are no poles of attenuation on the finite plane; they are all clustered at infinity. As a result of the periodic array of zeros, $|T(j\omega)|$ has an oscillatory behavior *vs* ω as indicated in Fig. 2.04-2(b). As the value of $R_g = R_L$ is made to approach that of Z_0, the zeros of $T(p)$ will move to the left, the poles will stay fixed at infinity, and the variations in $|T(j\omega)|$ will become smaller in amplitude. When $R_g = R_L = Z_0$, the zeros will have moved toward the left to minus infinity, and the transfer function becomes simply

$$\frac{E_g}{E_L} = T(p) = 2e^{\pi p/2\omega_0} \tag{2.04-2}$$

which has $|T(j\omega)|$ equal to two for all $p = j\omega$.

From the preceding example it is seen that the transcendental function e^p has an infinite number of poles and zeros which are all clustered at infinity. The poles are clustered closest to the $p = +\sigma$ axis so that if we approach infinity in that direction e^p becomes infinite. If we approach infinity via the $p = -\sigma$ axis e^p goes to zero. On the other hand, if we approach infinity along the $p = j\omega$ axis, e^p will always have unit magnitude but its phase will vary. This unit magnitude results from the fact that the amplitude effects of the poles and zeros counter balance each other along the $p = j\omega$ axis. The infinite cluster of poles and zeros at infinity forms what is called an *essential singularity*.

Figure 2.04-3 shows a band-pass filter using three transmission-line resonators which are a quarter-wavelength long at the frequency ω_0, and Fig. 2.04-4(a) shows a typical transfer function for this filter. In the example in Fig. 2.04-4, the response is periodic and has an infinite number of poles and zeros. The natural modes of vibration [*i.e.*, zeros of $T(p)$] are clustered near the $j\omega$ axis near the frequencies $j\omega_0$, $j3\omega_0$, $j5\omega_0$, etc., for which the lines are an odd number of quarter wavelengths long. At $p = 0$, and the frequencies $p = j2\omega_0$, $j4\omega_0$, $j6\omega_0$, etc., for which the lines are an even number of quarter-wavelengths long, the circuit functions like a short-circuit, followed by an open-circuit, and then another short-circuit. In accordance with Property (3) above, this creates *third-order* poles of attenuation as indicated in Fig. 2.04-4(a).

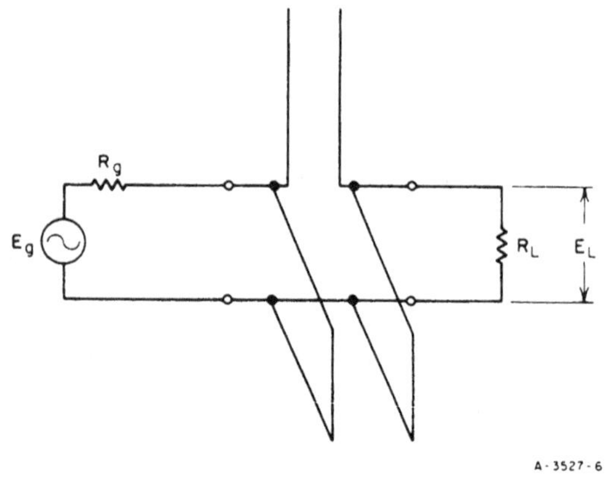

FIG. 2.04-3 A THREE-RESONATOR, BAND-PASS FILTER USING RESONATORS CONSISTING OF AN OPEN-CIRCUITED STUB IN SERIES AND TWO SHORT-CIRCUITED STUBS IN SHUNT

The approximate shape of $|T(j\omega)|$ is indicated in Fig. 2.04-4(b). If the termination values R_g and R_L were changed, the positions of the natural modes [zeros of $T(p)$] would shift and the shape of the pass bands would be altered. However, the positions of the poles of attenuation would be unaffected [see Property (2)].

The circuit in Fig. 2.04-3 is not very practical because the open-circuited series stub in the middle is difficult to construct in a shielded structure. The filter structure shown in Fig. 2.04-5 is much more common and easy to build. It uses short-circuited shunt stubs with connecting lines, the stubs and lines all being one-quarter wavelength long at frequency ω_0. This circuit has the same number of natural modes as does the circuit in Fig. 2.04-3, and can give similar pass-band responses for frequencies in the vicinity of $p = j\omega_0$, $j3\omega_0$, etc. However, at $p = 0$, $j2\omega_0$, $j4\omega_0$, etc., the circuit operates like three short circuits in parallel (which are equivalent to one short-circuit), and as a result the poles of attenuation at these frequencies are *first-order poles* only. It can thus be seen that this filter will not have as fast a rate of cut-off as will the filter in Fig. 2.04-3 whose poles on the $j\omega$ axis are *third*-order poles. The connecting lines between shunt stubs introduce poles of attenuation also, but as for the case in Fig. 2.04-2, the poles they introduce are all at infinity where they do little good as far as creating a fast rate of cutoff is concerned since there are an equal number of zeros (*i.e.*, natural modes) which are much closer, hence more influential.

These examples give brief illustrations of how the natural modes and frequencies of infinite attenuation occur in filters which involve transmission-line elements. Reasoning from the viewpoints discussed above can often be very helpful in deducing what the behavior of a given filter structure will be.

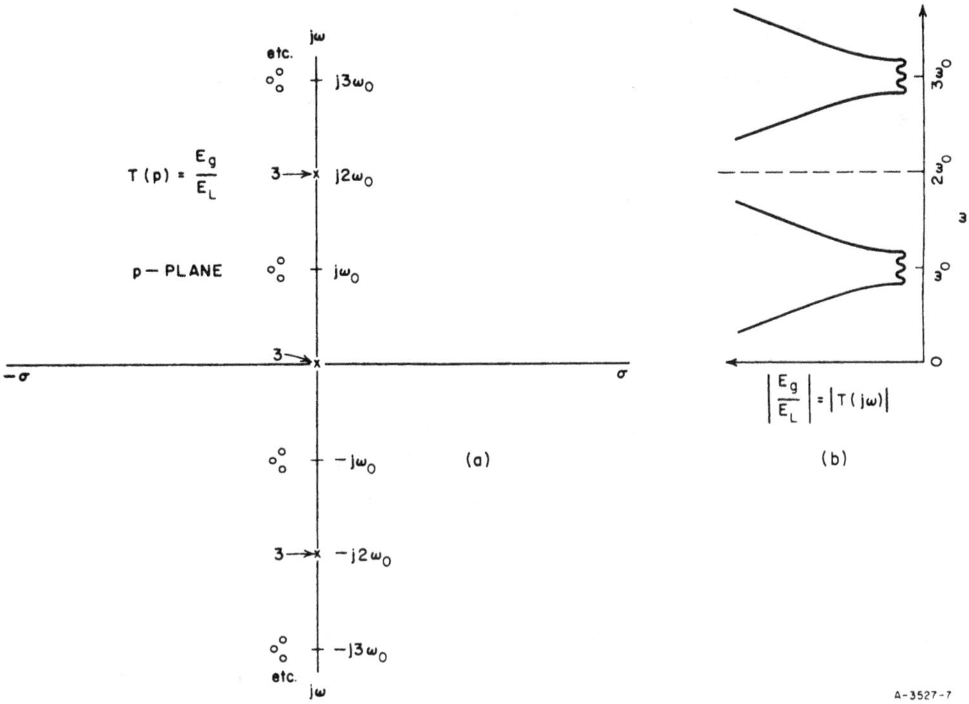

FIG. 2.04-4 VOLTAGE ATTENUATION FUNCTION PROPERTIES FOR THE FILTER IN FIG. 2.04-3
The Stubs are One-Quarter Wavelength Long at Frequency ω_0

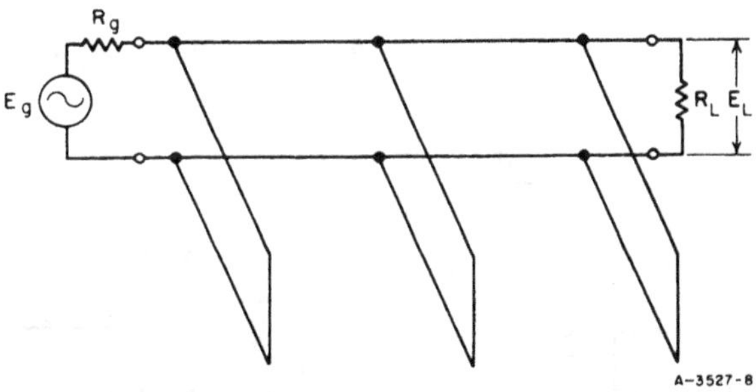

FIG. 2.04-5 A BAND-PASS FILTER CIRCUIT USING SHORT-CIRCUITED STUBS WITH CONNECTING LINES ALL OF WHICH ARE A QUARTER-WAVELENGTH LONG AT THE MIDBAND FREQUENCY ω_0

In terms of Fig. 2.05-1, the *general circuit parameters* are defined by the equations

$$\begin{aligned} E_1 &= AE_2 + B(-I_2) \\ I_1 &= CE_2 + D(-I_2) \end{aligned} \qquad (2.05\text{-}1)$$

or in matrix notation

$$\begin{bmatrix} E_1 \\ I_1 \end{bmatrix} = \begin{bmatrix} A & B \\ C & D \end{bmatrix} \begin{bmatrix} E_2 \\ -I_2 \end{bmatrix} \qquad (2.05\text{-}2)$$

These parameters are particularly useful in relating the performance of cascaded networks to their performance when operated individually. The general circuit parameters for the two cascaded networks in Fig. 2.05-2 are given by

$$\begin{aligned} \begin{bmatrix} A & B \\ C & D \end{bmatrix} &= \begin{bmatrix} A_a & B_a \\ C_a & D_a \end{bmatrix} \begin{bmatrix} A_b & B_b \\ C_b & D_b \end{bmatrix} \\ &= \begin{bmatrix} (A_aA_b + B_aC_b) & (A_aB_b + B_aD_b) \\ (C_aA_b + D_aC_b) & (C_aB_b + D_aD_b) \end{bmatrix} \end{aligned} \qquad (2.05\text{-}3)$$

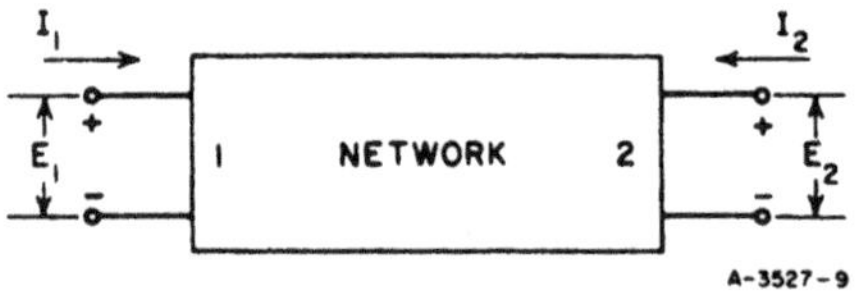

FIG. 2.05-1 DEFINITION OF CURRENTS AND VOLTAGES FOR TWO-PORT NETWORKS

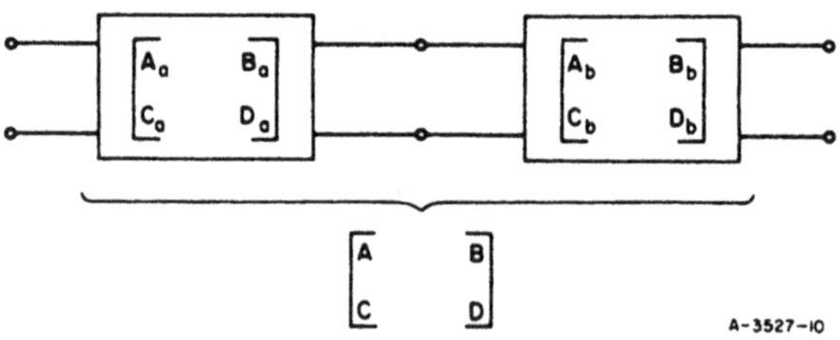

FIG. 2.05-2 CASCADED TWO-PORT NETWORKS

By repeated application of this operation the general circuit parameters can be computed for any number of two-port networks in cascade. Figure 2.05-3 gives the general circuit parameters for a number of common structures.

Under certain conditions the general circuit parameters are inter-related in the following special ways: If the network is reciprocal

$$AD - BC = 1 \quad . \tag{2.05-4}$$

If the network is symmetrical

$$A = D \quad . \tag{2.05-5}$$

If the network is lossless (*i.e.*, without dissipative elements), then for frequencies $p = j\omega$, A and D will be purely real while B and C will be purely imaginary.

If the network in Fig. 2.05-1 is turned around, then the square matrix in Eq. (2.05-2) is

$$\begin{bmatrix} A_t & B_t \\ C_t & D_t \end{bmatrix} = \begin{bmatrix} D & B \\ C & A \end{bmatrix} \tag{2.05-6}$$

where the parameters with t subscripts are for the network when turned around, and the parameters without subscripts are for the network with its original orientation. In both cases, E_1 and I_1 are at the terminals at the left and E_2 and I_2 are at the terminals at the right.

By use of Eqs. (2.05-6), (2.05-3), and (2.05-4), if the parameters A', B', C', D' are for the left half of a reciprocal symmetrical network, the general circuit parameters for the entire network are

$$\begin{bmatrix} A & B \\ C & D \end{bmatrix} = \begin{bmatrix} (1 + 2B'C') & (2A'B') \\ (2C'D') & (1 + 2B'C') \end{bmatrix} \quad . \tag{2.05-7}$$

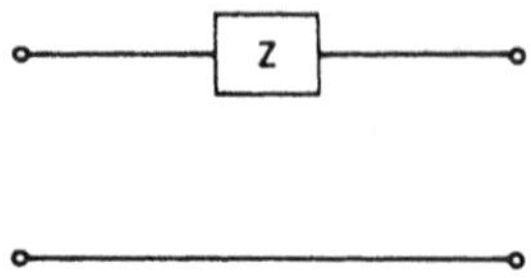

$A = 1\,, \quad B = Z$

$C = 0\,, \quad D = 1$

(a)

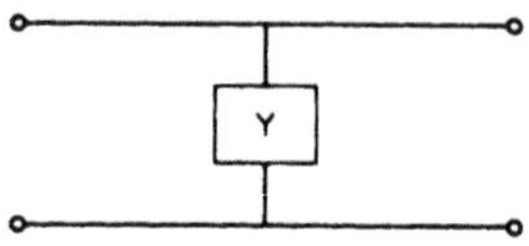

$A = 1\,, \quad B = 0$

$C = Y\,, \quad D = 1$

(b)

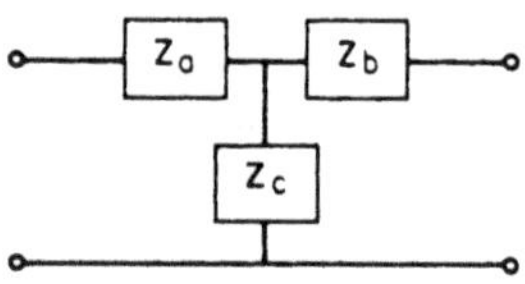

$A = 1 + \dfrac{Z_a}{Z_c}\,, \quad B = Z_a + Z_b + \dfrac{Z_a Z_b}{Z_c}$

$C = \dfrac{1}{Z_c}\,, \quad D = 1 + \dfrac{Z_b}{Z_c}$

(c)

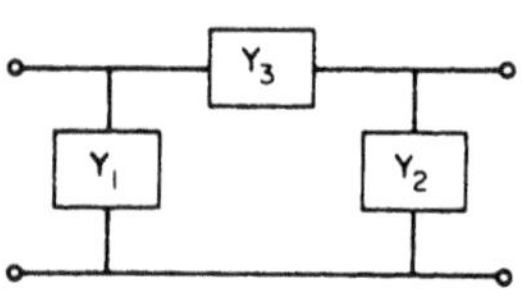

$A = 1 + \dfrac{Y_2}{Y_3}\,, \quad B = \dfrac{1}{Y_3}$

$C = Y_1 + Y_2 + \dfrac{Y_1 Y_2}{Y_3}\,, \quad D = 1 + \dfrac{Y_1}{Y_3}$

(d)

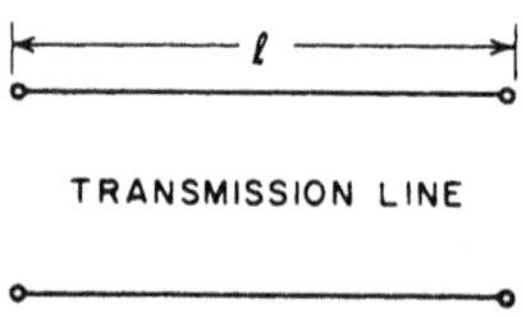

$A = \cosh \gamma_t \ell\,, \quad B = Z_0 \sinh \gamma_t \ell$

$C = \dfrac{\sinh \gamma_t \ell}{Z_0}\,, \quad D = \cosh \gamma_t \ell$

$\gamma_t = \alpha_t + j\beta_t$ = PROPAGATION CONSTANT, PER UNIT LENGTH

Z_0 = CHARACTERISTIC IMPEDANCE

(e)

B-3527-11

FIG. 2.05-3 GENERAL CIRCUIT PARAMETERS OF SOME COMMON STRUCTURES

SEC. 2.06, OPEN-CIRCUIT IMPEDANCES AND SHORT-CIRCUIT ADMITTANCES

In terms of Fig. 2.05-1, the *open-circuit impedances* of a two-port network may be defined by the equations

$$\begin{aligned} z_{11}I_1 + z_{12}I_2 &= E_1 \\ z_{21}I_1 + z_{22}I_2 &= E_2 \end{aligned} \quad . \qquad (2.06\text{-}1)$$

Physically, z_{11} is the input impedance at End 1 when End 2 is open-circuited. The quantity z_{12} could be measured as the ratio of the voltage E_1/I_2 when End 1 is open-circuited and current I_2 is flowing in End 2. The parameters z_{21} and z_{22} may be interpreted analogously.

In a similar fashion, the *short-circuit admittances* may be defined in terms of Fig. 2.05-1 and the equations

$$\begin{aligned} y_{11}E_1 + y_{12}E_2 &= I_1 \\ y_{21}E_1 + y_{22}E_2 &= I_2 \end{aligned} \quad . \qquad (2.06\text{-}2)$$

In this case y_{11} is the admittance at End 1 when End 2 is short-circuited. The parameter y_{12} could be computed as the ratio I_1/E_2 when End 1 is short-circuited and a voltage E_2 is applied at End 2.

Figure 2.06-1 shows the open-circuit impedances and short-circuit admittances for a number of common structures. For reciprocal networks $z_{12} = z_{21}$ and $y_{12} = y_{21}$. For a lossless network (*i.e.*, one composed of reactances), the open-circuit impedances and the short-circuit admittances are all purely imaginary for all $p = j\omega$.

SEC. 2.07, RELATIONS BETWEEN GENERAL CIRCUIT PARAMETERS AND OPEN- AND SHORT-CIRCUIT PARAMETERS

The relationships between the general circuit parameters, the open-circuit impedances, and the short-circuit admittances defined in Secs. 2.05 and 2.06 are as follows:

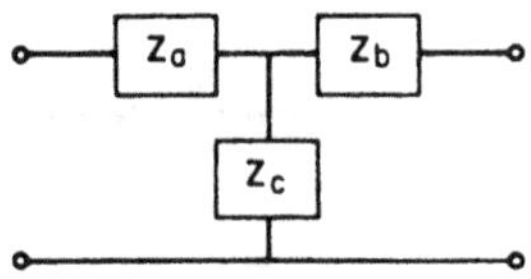

$$z_{11} = (Z_a + Z_c)\ , \quad z_{12} = Z_c$$

$$z_{21} = Z_c\ , \quad z_{22} = (Z_c + Z_b)$$

$$\Delta_z = Z_a Z_b + Z_b Z_c + Z_c Z_a$$

(a)

$$y_{11} = \frac{Z_b + Z_c}{\Delta_z}\ , \quad y_{12} = \frac{-Z_c}{\Delta_z}$$

$$y_{21} = \frac{-Z_c}{\Delta_z}\ , \quad y_{22} = \frac{Z_a + Z_c}{\Delta_z}$$

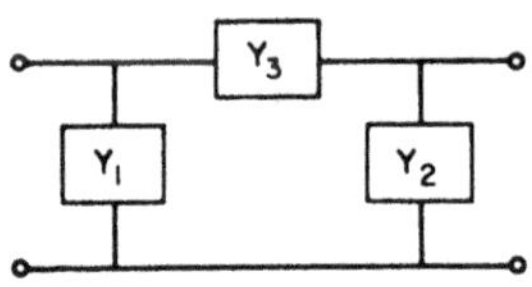

$$y_{11} = Y_1 + Y_3\ , \quad y_{12} = -Y_3$$

$$y_{21} = -Y_3\ , \quad y_{22} = Y_2 + Y_3$$

$$\Delta_y = Y_1 Y_2 + Y_2 Y_3 + Y_3 Y_1$$

(b)

$$z_{11} = \frac{Y_2 + Y_3}{\Delta_y}\ , \quad z_{12} = \frac{Y_3}{\Delta_y}$$

$$z_{21} = \frac{Y_3}{\Delta_y}\ , \quad z_{22} = \frac{Y_1 + Y_3}{\Delta_y}$$

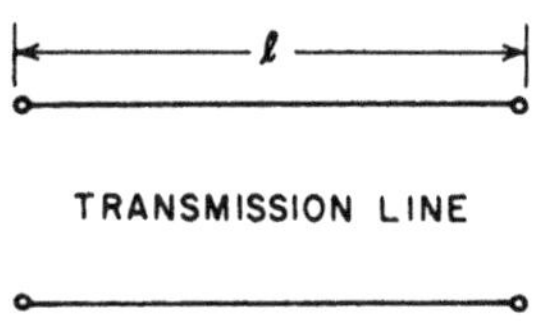

$$z_{11} = Z_0 \coth \gamma_t \ell\ , \quad z_{12} = \frac{Z_0}{\sinh \gamma_t \ell}$$

$$z_{21} = \frac{Z_0}{\sinh \gamma_t \ell}\ , \quad z_{22} = Z_0 \coth \gamma_t \ell$$

(c)

$$y_{11} = \frac{\coth \gamma_t \ell}{Z_0}\ , \quad y_{12} = \frac{-1}{Z_0 \sinh \gamma_t \ell}$$

$$y_{21} = \frac{-1}{Z_0 \sinh \gamma_t \ell}\ , \quad y_{22} = \frac{\coth \gamma_t \ell}{Z_0}$$

$\gamma_t = \alpha_t + j\beta_t$ = PROPAGATION CONSTANT, PER UNIT LENGTH

Z_0 = CHARACTERISTIC IMPEDANCE

B-3527-12

FIG. 2.06-1 OPEN-CIRCUIT IMPEDANCES AND SHORT-CIRCUIT ADMITTANCES OF SOME COMMON STRUCTURES

$$A = \frac{z_{11}}{z_{21}} = \frac{-y_{22}}{y_{21}} = \frac{n_{so}}{m_{21}}$$

$$B = \frac{\Delta_z}{z_{21}} = \frac{-1}{y_{21}} = \frac{n_{ss}}{m_{21}}$$

(2.07-1)

$$C = \frac{1}{z_{21}} = \frac{-\Delta_y}{y_{21}} = \frac{n_{oo}}{m_{21}}$$

$$D = \frac{z_{22}}{z_{21}} = \frac{-y_{11}}{y_{21}} = \frac{n_{os}}{m_{21}}$$

$$z_{11} = \frac{y_{22}}{\Delta_y} = \frac{A}{C} = \frac{n_{so}}{n_{oo}}$$

$$z_{12} = \frac{-y_{12}}{\Delta_y} = \frac{\Delta}{C} = \frac{m_{12}}{n_{oo}}$$

(2.07-2)

$$z_{21} = \frac{-y_{21}}{\Delta_y} = \frac{1}{C} = \frac{m_{21}}{n_{oo}}$$

$$z_{22} = \frac{y_{11}}{\Delta_y} = \frac{D}{C} = \frac{n_{os}}{n_{oo}}$$

$$y_{11} = \frac{z_{22}}{\Delta_z} = \frac{D}{B} = \frac{n_{os}}{n_{ss}}$$

$$y_{12} = \frac{-z_{12}}{\Delta_z} = \frac{-\Delta}{B} = \frac{-m_{12}}{n_{ss}}$$

(2.07-3)

$$y_{21} = \frac{-z_{21}}{\Delta_z} = \frac{-1}{B} = \frac{-m_{21}}{n_{ss}}$$

$$y_{22} = \frac{z_{11}}{\Delta_z} = \frac{A}{B} = \frac{n_{so}}{n_{ss}}$$

where

$$\left.\begin{aligned}\Delta &= AD - BC = \frac{m_{21}m_{12}}{(m_{21})^2} = \frac{m_{12}}{m_{21}} \\ &= 1 \text{ (for reciprocal networks)}\end{aligned}\right\}, \qquad (2.07\text{-}4)$$

$$\Delta_z = z_{11}z_{22} - z_{12}z_{21} = \frac{n_{ss}n_{oo}}{(n_{oo})^2} = \frac{n_{ss}}{n_{oo}} \qquad (2.07\text{-}5)$$

$$\Delta_y = y_{11}y_{22} - y_{12}y_{21} = \frac{n_{oo}n_{ss}}{(n_{ss})^2} = \frac{n_{oo}}{n_{ss}} \quad . \qquad (2.07\text{-}6)$$

If any of these various circuit parameters are expressed as a function of complex frequency p, they will consist of the ratio of two polynomials, each of which may be put in the form

$$\text{polynomial} = c(p - p_1)(p - p_2)(p - p_3) \ldots \qquad (2.07\text{-}7)$$

where c is a real constant and the p_k are the roots of the polynomial. As should be expected from the discussions in Secs. 2.02 to 2.04, the locations of the roots of these polynomials have physical significance. The quantities on the right in Eqs. (2.07-1) to (2.07-6) have been introduced to clarify this physical significance.

The symbols n_{ss}, n_{oo}, n_{os}, and n_{so} in the expressions above represent polynomials of the form in Eq. (2.07-7) whose roots are natural frequencies of vibration of the circuit under conditions indicated by the subscripts. Thus, the roots of n_{ss} are the natural frequencies of the circuit in Fig. 2.05-1 when both ports are short-circuited, while the roots of n_{oo} are natural frequencies when both ports are open-circuited. The roots of n_{os} are natural frequencies when the left port is open-circuited while the right port is short-circuited, and the inverse obtains for n_{so}. The symbols m_{12} and m_{21} represent polynomials whose roots are poles of attenuation (see Sec. 2.04) of the circuit, except for those poles of attenuation at $p = \infty$. The polynomial m_{12} has roots corresponding to the poles of attenuation for transmission *to* End 1 *from* End 2 in Fig. 2.05-1, while the polynomial m_{21} has roots which are poles of attenuation for transmission to End 2 from End 1. If the network is reciprocal, $m_{12} = m_{21}$. These polynomials for a given circuit are interrelated by the expression[8]

$$n_{oo}n_{ss} = n_{so}n_{os} - m_{12}m_{21} \tag{2.07-8}$$

and, as is discussed in Ref. 8, they can yield certain labor-saving advantages when they themselves are used as basic parameters to describe the performance of a circuit.

As is indicated in Eqs. (2.07-4) to (2.07-6), when the determinants Δ, Δ_z and Δ_y are formed as a function of p, the resulting rational function will necessarily contain cancelling polynomials. This fact can be verified by use of Eqs. (2.07-1) to (2.07-3) along with (2.07-8). Removal of the cancelling polynomials will usually cut the degree of the polynomials in these functions roughly in half. Analogous properties exist when the network contains distributed elements, although the polynomials then become of infinite degree (see Sec. 2.02) and are most conveniently represented by transcendental functions such as sinh p and cosh p. For example, for a lossless transmission line

$$\begin{aligned}
n_{so} &= n_{os} = Z_0 \cosh \frac{\pi p}{2\omega_0} = Z_0 \prod_{k=1}^{\infty} \left[\frac{1}{(2k-1)\omega_0}\right]^2 \{[p + j(2k-1)\omega_0][p - j(2k-1)\omega_0]\} \\
n_{oo} &= \sinh \frac{\pi p}{2\omega_0} = \frac{\pi p}{2\omega_0} \prod_{k=1}^{\infty} \left(\frac{1}{2k\omega_0}\right)^2 [(p + j2k\omega_0)(p - j2k\omega_0)] \\
n_{ss} &= Z_0^2 \sinh \frac{\pi p}{2\omega_0} \\
m_{12} &= m_{21} = Z_0
\end{aligned} \tag{2.07-9}$$

where Z_0 is the characteristic impedance of the line, and ω_0 is the radian frequency for which the line is one-quarter wavelength long. In this case, $m_{12} = m_{21}$ is a constant since all of the poles of attenuation are at infinity (see Sec. 2.04 and Ref. 8). The choice of constant multipliers for these "polynomials" is arbitrary to a certain extent in that any one multiplier may be chosen arbitrarily, but this then fixes what the other constant multipliers must be.[8]

SEC. 2.08, INCIDENT AND REFLECTED WAVES, REFLECTION COEFFICIENTS, AND ONE KIND OF TRANSMISSION COEFFICIENT

Let us suppose that it is desired to analyze the transmission across the terminals 2-2′ in Fig. 2.08-1 from the wave point of view. By definition

$$E_i + E_r = E \tag{2.08-1}$$

where E_i is the amplitude of the incident voltage wave emerging from the network, E_r is the reflected voltage wave amplitude, and E is the transmitted voltage wave amplitude (which is also the voltage that would be measured across the terminals 2-2′). If $Z_s = Z_L$, there will be no reflection* so that $E_r = 0$ and $E_i = E$. Replacing the network and generator at (a) in Fig. 2.08-1 by a Thevenin equivalent generator as shown at (b), it is readily seen that since for $Z_s = Z_L$, $E_i = E$, then

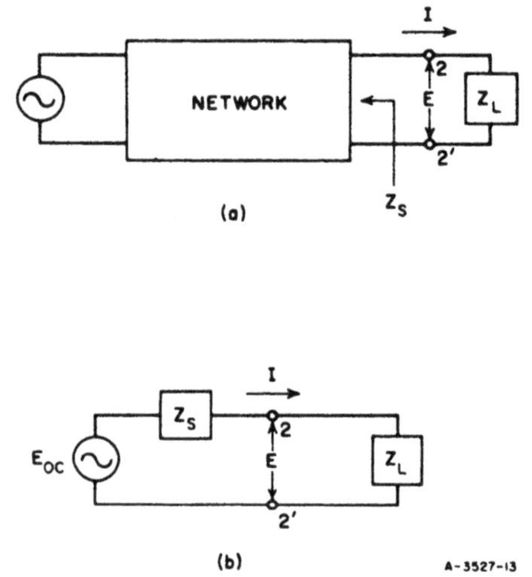

FIG. 2.08-1 CIRCUITS DISCUSSED IN SEC. 2.08 FROM THE WAVE-ANALYSIS VIEWPOINT

$$E_i = \frac{E_{oc}}{2} \tag{2.08-2}$$

where E_{oc} is the voltage which would be measured at terminals 2-2′ if they were open-circuited. Using Eqs. (2.08-1) and (2.08-2) the voltage *reflection coefficient* is defined as

$$\Gamma = \frac{E_r}{E_i} = \frac{Z_L - Z_s}{Z_L + Z_s} \tag{2.08-3}$$

An analogous treatment for current waves proceeds as follows:

$$I_i + I_r = I \tag{2.08-4}$$

---

* Note that no reflection of the voltage wave does not necessarily imply maximum power transfer. For no reflection of the voltage wave $Z_s = Z_L$, while for maximum power transfer $Z_s = Z_L^*$ where Z_L^* is the complex conjugate of Z_L.

where I_i is the incident current amplitude, I_r is the reflected current amplitude, and I is the transmitted current amplitude which is also the actual current passing through the terminals 2-2'. The incident current is

$$I_i = \frac{I_{sc}}{2} \tag{2.08-5}$$

where I_{sc} is the current which would pass through the terminals 2-2' if they were short-circuited together. The *current reflection coefficient* is then defined as

$$\Gamma_c = \frac{I_r}{I_i} = \frac{Y_L - Y_s}{Y_L + Y_s} = -\Gamma \tag{2.08-6}$$

where $Y_s = 1/Z_s$ and $Y_L = 1/Z_L$.

In addition, sometimes the *voltage transmission coefficient*

$$\tau = \frac{E}{E_i} = \frac{2Z_L}{Z_L + Z_s} = 1 + \Gamma \tag{2.08-7}$$

is used. The corresponding *current transmission coefficient* is

$$\tau_c = \frac{I}{I_i} = \frac{2Y_L}{Y_L + Y_s} = 1 + \Gamma_c \quad . \tag{2.08-8}$$

It will be noted that these transmission coefficients τ and τ_c are not the same as the transmission coefficient t discussed in Sec. 2.10.

SEC. 2.09, CALCULATION OF THE INPUT IMPEDANCE OF A TERMINATED, TWO-PORT NETWORK

The input impedance $(Z_{in})_1$ defined in Fig. 2.09-1 can be computed from Z_2 and any of the circuit parameters used to describe the performance of a two-port network. In terms of the general circuit parameters, the open-circuit impedances, and the short-circuit admittances,

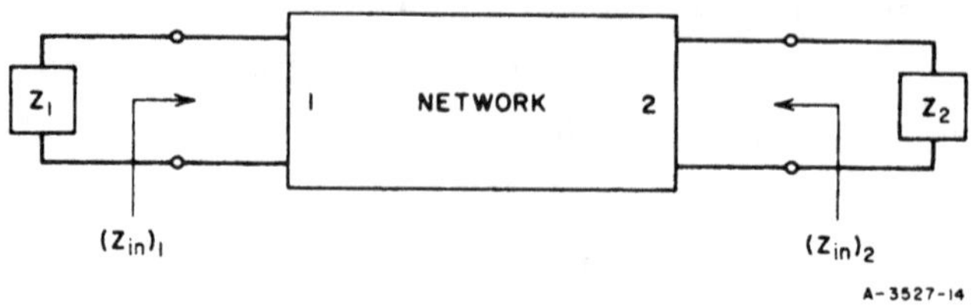

FIG. 2.09-1 DEFINITION OF INPUT IMPEDANCES COMPUTED IN SEC. 2.09

$$(Z_{in})_1 = \frac{AZ_2 + B}{CZ_2 + D} \tag{2.09-1}$$

$$= z_{11} - \frac{z_{12}z_{21}}{z_{22} + Z_2} \tag{2.09-2}$$

$$= \frac{y_{22} + Y_2}{y_{11}(y_{22} + Y_2) - y_{12}y_{21}} \tag{2.09-3}$$

respectively, where $Y_2 = 1/Z_2$. Similarly, for the impedance $(Z_{in})_2$ in Fig. 2.09-1

$$(Z_{in})_2 = \frac{DZ_1 + B}{CZ_1 + A} \tag{2.09-4}$$

$$= z_{22} - \frac{z_{12}z_{21}}{z_{11} + Z_1} \tag{2.09-5}$$

$$= \frac{y_{11} + Y_1}{y_{22}(y_{11} + Y_1) - y_{12}y_{21}} \tag{2.09-6}$$

where $Y_1 = 1/Z_1$.

SEC. 2.10, CALCULATION OF VOLTAGE TRANSFER FUNCTIONS

The transfer function E_g/E_2 for the circuit in Fig. 2.10-1 can be computed if any of the sets of circuit parameters discussed in Secs. 2.05 to 2.07 are known for the network in the box. The appropriate equations are

$$\frac{E_g}{E_2} = \frac{AR_2 + B + CR_1R_2 + DR_1}{R_2} , \qquad (2.10\text{-}1)$$

$$= \frac{(z_{11} + R_1)(z_{22} + R_2) - z_{12}z_{21}}{z_{21}R_2} , \qquad (2.10\text{-}2)$$

and

$$= \frac{(y_{11} + G_1)(y_{22} + G_2) - y_{12}y_{21}}{-y_{21}G_1} . \qquad (2.10\text{-}3)$$

Transfer functions such as the E_g/E_2 function presented above are commonly used but have a certain disadvantage. This disadvantage is that, depending on what the relative size of R_1 and R_2 are, complete energy transfer may correspond to any of a wide range of $|E_g/E_2|$ values. Such confusion is eliminated if the transfer function

$$\frac{(E_2)_{avail}}{E_2} = \frac{1}{2}\sqrt{\frac{R_2}{R_1}}\left(\frac{E_g}{E_2}\right) \qquad (2.10\text{-}4)$$

is used instead. The quantity

$$(E_2)_{avail} = \frac{1}{2}\sqrt{\frac{R_2}{R_1}}\,E_g \qquad (2.10\text{-}5)$$

will be referred to herein as the *available voltage*, which is the voltage across R_2 when the entire available power of the generator is absorbed by R_2. Thus, for complete energy transfer $|(E_2)_{avail}/E_2| = 1$ regardless

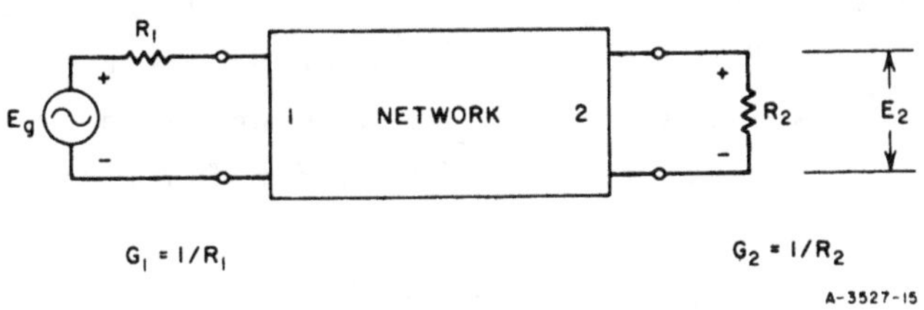

FIG. 2.10-1 A CIRCUIT DISCUSSED IN SEC. 2.10

of the relative sizes of R_2 and R_1. Note that $(E_2)_{avail}$ has the same phase as E_g.

In the literature a *transmission coefficient* t is commonly used where

$$t = \frac{1}{[(E_2)_{avail}/E_2]} = 2\sqrt{\frac{R_1}{R_2}}\left(\frac{E_2}{E_g}\right) \quad . \qquad (2.10\text{-}6)$$

Note that this is not the same as the transmission coefficients τ or τ_c discussed in Sec. 2.08. The transmission coefficient t is the same, however, as the scattering coefficients $S_{12} = S_{21}$, discussed in Sec. 2.12. Also note that t is an output/input ratio of a "voltage gain" ratio, while the function in Eq. (2.10-4) is an input/output ratio or a "voltage attenuation" ratio.

SEC. 2.11, CALCULATION OF POWER TRANSFER FUNCTIONS AND "ATTENUATION"

One commonly used type of power transfer function is the *insertion loss* function

$$\frac{P_{20}}{P_2} = \left(\frac{R_2}{R_1 + R_2}\right)^2 \left|\frac{E_g}{E_2}\right|^2 \qquad (2.11\text{-}1)$$

where R_1, R_2, E_g, and E_2 are defined in Fig. 2.10-1, and $|E_g/E_2|$ can be computed by use of Eqs. (2.10-1) to (2.10-3). The quantity P_2 is the power absorbed by R_2 when the network in Fig. 2.10-1 is in place, while P_{20} is the power in R_2 when the network is removed and R_2 is connected directly to R_1 and E_g.

Insertion loss functions have the same disadvantage as the E_g/E_2 function discussed in Sec. 2.10, *i.e.*, complete power transfer may correspond to almost any value of P_{20}/P_2 depending on the relative sizes of R_1 and R_2. For this reason the power transfer function

$$\frac{P_{avail}}{P_2} = \frac{1}{4}\left(\frac{R_2}{R_1}\right)\left|\frac{E_g}{E_2}\right|^2 = \frac{1}{|t|^2} \qquad (2.11\text{-}2)$$

will be used in this book instead of insertion loss. The function P_{avail}/P_2 is known as a *transducer loss ratio*, where P_2 is again the power delivered to R_2 in Fig. 2.10-1 while

$$P_{avail} = \frac{|E_g|^2}{4R_1} \qquad (2.11\text{-}3)$$

is the available power of the generator composed of E_g and the internal resistance R_1. Thus, for complete power transfer $P_{avail}/P_2 = 1$ regardless of the relative size of R_1 and R_2. Note that t in Eq. (2.11-2) is the transmission coefficient defined by Eq. (2.10-6).

It will often be desirable to express P_{avail}/P_2 in db so that

$$L_A = 10 \log_{10} (P_{avail}/P_2) \quad \text{db} \quad . \qquad (2.11\text{-}4)$$

Herein, when *attenuation* is referred to, the transducer loss (*i.e.*, transducer attenuation) in db as defined in Eq. (2.11-4) will be understood, unless otherwise specified.

If we define $L_i = 10 \log_{10} P_{20}/P_2$ as *insertion loss in db*, then the attenuation in db is

$$L_A = L_i + 10 \log_{10} \frac{(R_1 + R_2)^2}{4R_1R_2} \quad \text{db} \quad . \qquad (2.11\text{-}5)$$

Note that if $R_1 = R_2$, then insertion loss and transducer attenuation are the same.

If the network in Fig. 2.11-1 contains dissipative elements which cannot be neglected, then L_A may be computed by use of Eqs. (2.11-4), (2.11-2), and any of Eqs. (2.10-1) to (2.10-3). However, if the network in the box may be regarded as lossless (*i.e.*, without any

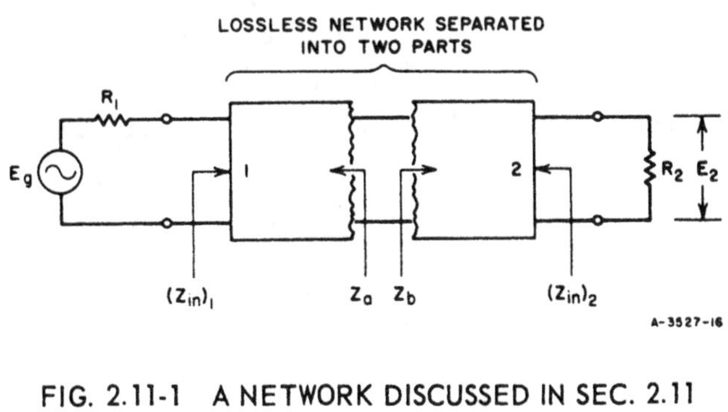

FIG. 2.11-1 A NETWORK DISCUSSED IN SEC. 2.11

dissipative elements), then some simplifications can be taken advantage of. For example, as discussed in Sec. 2.05, for a dissipationless network A and D will be purely real while B and C will be purely imaginary for frequencies $j\omega$. Because of this, the form

$$\frac{P_{avail}}{P_2} = \frac{1}{4R_1R_2}\left[(AR_2 + DR_1)^2 + \left(\frac{B + CR_1R_2}{j}\right)^2\right] \qquad (2.11\text{-}6)$$

becomes convenient for computation. This expression applies to dissipationless reciprocal networks and also to non-reciprocal dissipationless networks for the case of transmission from left to right. If we further specify that $R_1 = R_2 = R$, that the network is reciprocal (*i.e.*, $AD - BC = 1$), and that the network is symmetrical (*i.e.*, $A = D$), then Eq. (2.11-6) becomes

$$\frac{P_{avail}}{P_2} = 1 + \frac{1}{4}\left(\frac{B}{jR} - \frac{CR}{j}\right)^2 \quad . \qquad (2.11\text{-}7)$$

Furthermore, it is convenient in such cases to compute the general circuit parameters A', B', C', D' for the left half of the network only. Then by Eqs. (2.05-7) and (2.11-7), the transducer loss ratio for the over-all network is

$$\frac{P_{avail}}{P_2} = 1 + \left(\frac{A'B'}{jR} - \frac{C'D'R}{j}\right)^2 \quad . \qquad (2.11\text{-}8)$$

In the case of dissipationless networks such as that shown in Fig. 2.11-1, the power transmission is easily computed from the generator parameters and the input impedance of the dissipationless network terminated in R_2. This is because any power absorbed by $(Z_{in})_1$ must surely end up in R_2. The computations may be conveniently made in terms of the voltage reflection coefficient Γ discussed in Sec. 2.08. In these terms

$$\frac{P_{avail}}{P_2} = \frac{1}{|t|^2} = \frac{1}{1 - |\Gamma_1|^2} \qquad (2.11\text{-}9)$$

where

$$\Gamma_1 = \frac{(Z_{in})_1 - R_1}{(Z_{in})_1 + R_1} \tag{2.11-10}$$

and R_1 and $(Z_{in})_1$ are as defined in Fig. 2.11-1. The reflection coefficient at the other end is

$$\Gamma_2 = \frac{(Z_{in})_2 - R_2}{(Z_{in})_2 + R_2} \tag{2.11-11}$$

and for a dissipationless network

$$|\Gamma_1| = |\Gamma_2| \tag{2.11-12}$$

so that the *magnitude* of either reflection coefficient could be used in Eq. (2.11-9). [It should be understood in passing that the phase of Γ_1 is not necessarily the same as that of Γ_2 even though Eq. (2.11-12) holds.]

The reflection coefficient

$$\Gamma_{ab} = \frac{Z_b - Z_a}{Z_b + Z_a} \tag{2.11-13}$$

between Z_a and Z_b in Fig. 2.11-1 *cannot* be used in Eq. (2.11-9) if both Z_a and Z_b are complex. However, it can be shown that

$$|\Gamma_1| = |\Gamma_2| = \left| \frac{Z_b^* - Z_a}{Z_b + Z_a} \right| \tag{2.11-14}$$

where Z_b^* is the complex conjugate of Z_b. Thus, if $Z_a = R_a + jX_a$ and $Z_b = R_b + jX_b$, by use of Eqs. (2.11-14) and (2.11-9) we obtain

$$\frac{P_{avail}}{P_2} = \frac{(R_a + R_b)^2 + (X_a + X_b)^2}{4R_aR_b} \quad . \tag{2.11-15}$$

For cases where $Z_a = Z_b$ such as occurs at the middle of a *symmetrical* network, Eq. (2.11-15) reduces to

$$\frac{P_{\text{avail}}}{P_2} = 1 + \left(\frac{X_a}{R_a}\right)^2 \qquad (2.11\text{-}16)$$

Another situation which commonly occurs in filter circuits is for the structure to be *antimetrical** about its middle. In such cases, if Z_a and Z_b in Fig. 2.11-1 are at the middle of a antimetrical network, then for all frequencies

$$Z_b = \frac{R_h^2}{Z_a} \qquad (2.11\text{-}17)$$

where R_h is a real, positive constant. Defining Z_a again as $R_a + jX_a$, by Eqs. (2.11-17), (2.11-14), and (2.11-9),

$$\frac{P_{\text{avail}}}{P_2} = \frac{1}{4}\left\{\left(\frac{R_a}{R_h} + \frac{R_h}{R_a}\right)^2 + 2\left[\left(\frac{X_a}{R_h}\right)^2 - \left(\frac{X_a}{R_a}\right)^2\right] + \left(\frac{X_a}{R_a}\right)^2\left(\frac{X_a}{R_h}\right)^2\right\} . \qquad (2.11\text{-}18)$$

The quantity R_h is obtained most easily by evaluating

$$R_h = \sqrt{Z_a Z_b} = \text{real, positive} \qquad (2.11\text{-}19)$$

at a frequency where Z_a and Z_b are both known to be real. The maximally flat and Tchebyscheff low-pass prototype filter structure whose element values are listed in Tables 4.05-1(a), (b) and 4.05-2(a), (b) are symmetrical for an odd number n of reactive elements, and they are antimetrical for an even number n of reactive elements. The step transformers discussed in Chapter 6 are additional examples of antimetrical circuits.

SEC. 2.12, SCATTERING COEFFICIENTS

In this book there will be some occasion to make use of scattering coefficients. Scattering coefficients are usually defined entirely from a wave point of view. However, for the purposes of this book it will be sufficient to simply extrapolate from previously developed concepts.

---

* This term was coined by Guillemin. See pp. 371 and 467 of Ref. 2.

The performance of any linear two-port network with terminations can be described in terms of four scattering coefficients: S_{11}, S_{12}, S_{21}, and S_{22}. With reference to the two-port network in Fig. 2.11-1, $S_{11} = \Gamma_1$ and $S_{22} = \Gamma_2$ are simply the reflection coefficients at Ends 1 and 2, respectively, as defined in Eqs. (2.11-10) and (2.11-11). The scattering coefficient S_{21} is simply the transmission coefficient, t, for transmission to End 2 from End 1 as defined in Eqs. (2.10-5) and (2.10-6). The scattering coefficient, S_{12}, is likewise the same as the transmission coefficient, t, for transmission to End 1 from End 2. Of course, if the network is reciprocal $S_{21} = S_{12}$. The relations in Sec. 2.11 involving t, Γ_1, and Γ_2, of course, apply equally well to $t = S_{12} = S_{21}$, $\Gamma_1 = S_{11}$, and $\Gamma_2 = S_{22}$, respectively.

Thus, it is seen that as far as two-port networks are concerned, the scattering coefficients are simply the reflection coefficients or transmission coefficients discussed in Secs. 2.10 and 2.11. However, scattering coefficients may be applied to networks with an arbitrary number of ports. For example, for a three-port network there are nine scattering coefficients, which may be displayed as the matrix

$$[S] = \begin{bmatrix} S_{11} & S_{12} & S_{13} \\ S_{21} & S_{22} & S_{23} \\ S_{31} & S_{32} & S_{33} \end{bmatrix} \quad . \tag{2.12-1}$$

For an n-port network there are n^2 coefficients. In general, for any network with resistive terminations,

$$S_{jj} = \Gamma_j = \frac{(Z_{in})_j - R_j}{(Z_{in})_j + R_j} \tag{2.12-2}$$

is the reflection coefficient between the input impedance $(Z_{in})_j$ at Port j and the termination R_j at that port. For the other coefficients, analogously to Eqs. (2.10-5) and (2.10-6),

$$S_{jk}\Big|_{j \neq k} = \frac{E_j}{(E_j)_{avail}} \tag{2.12-3}$$

where

$$(E_j)_{avail} = \frac{1}{2}\sqrt{\frac{R_j}{R_k}}\,(E_g)_k \qquad (2.12\text{-}4)$$

The voltage E_j is the response across termination R_j at Port j due to a generator of voltage $(E_g)_k$ and internal impedance R_k at Port k. In computing the coefficients defined by Eqs. (2.12-2) to (2.12-4), all ports are assumed to always be terminated in their specified terminations R_j.

If an n-port network is reciprocal,

$$S_{jk} = S_{kj} \quad . \qquad (2.12\text{-}5)$$

By Eqs. (2.11-9)and (2.11-12) for a dissipationless reciprocal two-port network

$$1 = |S_{11}|^2 + |S_{12}|^2 \quad , \qquad (2.12\text{-}6)$$

$$|S_{11}| = |S_{22}| \quad , \qquad (2.12\text{-}7)$$

and

$$S_{12} = S_{21} \quad . \qquad (2.12\text{-}8)$$

The analogous relation for the general n-port, dissipationless, reciprocal network is

$$[I] = [S]^* [S] \qquad (2.12\text{-}9)$$

where $[S]$ is the *scattering matrix* of scattering coefficients [as illustrated in Eq. (2.12-1) for the case of $n = 3$], $[S]^*$ is the same matrix with all of its complex numbers changed to their conjugates, and $[I]$ is an nth-order unit matrix. Since the network is specified to be reciprocal, Eq. (2.12-5)applies and $[S]$ is symmetrical about its principal diagonal.

For any network with resistive terminations,

$$|S_{j,k}|^2_{j \neq k} = \frac{P_j}{(P_{avail})_k} \qquad (2.12\text{-}10)$$

where P_j is the power delivered to the termination R_j at Port j, and $(P_{avail})_k$ is the available power of a generator at Port k. In accord with Eq. (2.11-4) the db attenuation for transmission from Port k to Port j (with all specified terminations connected) is

$$L_A = 20 \log_{10}\left|\frac{1}{S_{jk}}\right| \quad \text{db} \tag{2.12-11}$$

Further discussion of scattering coefficients will be found in Ref. 9.

SEC. 2.13, ANALYSIS OF LADDER CIRCUITS

Ladder circuits often occur in filter work, some examples being the low-pass prototype filters discussed in Chapter 4. The routine outlined below is particularly convenient for computing the response of such networks.

The first step in this routine is to characterize each *series* branch by its *impedance* and the *current* flowing through the branch, and each *shunt* branch by its *admittance* and the *voltage* across the branch. This characterization is illustrated in Fig. 2.13-1. Then, in general terms we define

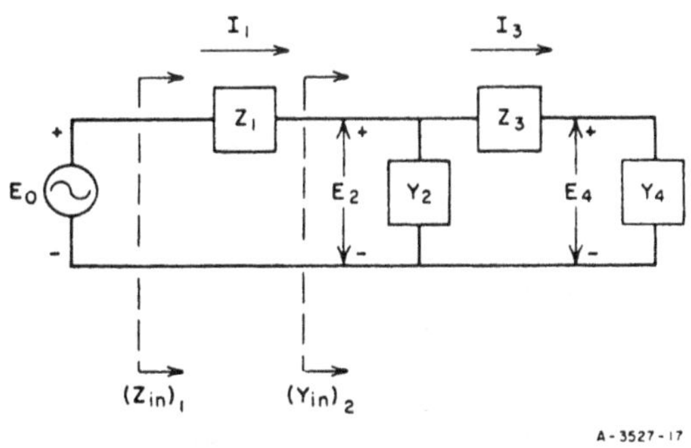

FIG. 2.13-1 A LADDER NETWORK EXAMPLE DISCUSSED IN SEC. 2.13

F_k = series impedance or shunt admittance of Branch k (2.13-1)

U_k = series-branch current or shunt-branch voltage of Branch k (2.13-2)

U_m = series-branch current or shunt-branch voltage for the last branch on the right (2.13-3)

U_0 = current or voltage associated with the driving generator on the left. (2.13-4)

In general, if Branch 1 is in shunt, U_0 should be the current of an infinite-impedance current generator; if Branch 1 is in series, U_0 should be the voltage of a zero-impedance voltage generator. Then, for all cases,

$$A_{m+1} = 1 = \frac{U_m}{U_m}$$

$$A_m = F_m A_{m+1} = \frac{U_{m-1}}{U_m}$$

$$A_{m-1} = F_{m-1} A_m + A_{m+1} = \frac{U_{m-2}}{U_m}$$

$$\cdots \tag{2.13-5}$$

$$A_k = F_k A_{k+1} + A_{k+2} = \frac{U_{k-1}}{U_m}$$

$$\cdots$$

$$A_1 = F_1 A_2 + A_3 = \frac{U_0}{U_m} \quad .$$

Thus A_1 is the transfer function from the generator on the left to Branch m on the right. If we define

$(F_{in})_k$ = *impedance* looking right through Branch k if Branch k is in series, or *admittance* looking right through Branch k if Branch k is in shunt, (2.13-6)

then

$$(F_{in})_k = \frac{A_k}{A_{k+1}} \quad . \tag{2.13-7}$$

To illustrate this procedure consider the case in Fig. 2.13-1. There m = 4 and

$$A_5 = 1 = \frac{E_4}{E_4}$$

$$A_4 = Y_4 A_5 = \frac{I_3}{E_4}$$

$$A_3 = Z_3A_4 + A_5 = \frac{E_2}{E_4}$$

$$A_2 = Y_2A_3 + A_4 = \frac{I_1}{E_4}$$

$$A_1 = Z_1A_2 + A_3 = \frac{E_0}{E_4}$$

Thus, A_1 is the transfer function between E_0 and E_4. The impedance $(Z_{in})_1$ and admittance $(Y_{in})_2$ defined in the figure are

$$(Z_{in})_1 = \frac{A_1}{A_2}$$

$$(Y_{in})_2 = \frac{A_2}{A_3}$$

REFERENCES

1. E. A. Guillemin, *Introductory Circuit Theory* (John Wiley and Sons, New York City, 1953).
2. E. A. Guillemin, *Synthesis of Passive Networks* (John Wiley and Sons, New York City, 1957).
3. D. F. Tuttle, Jr., *Network Synthesis*, Vol. 1 (John Wiley and Sons, New York City, 1958).
4. M. E. Van Valkenburg, *Network Analysis* (Prentice-Hall, Inc., New York City, 1955).
5. E. S. Kuh and D. O. Pederson, *Principles of Circuit Synthesis* (McGraw-Hill Book Co. Inc., New York City, 1959).
6. W. W. Hansen and O. C. Lundstrom, "Experimental Determination of Impedance Functions by the Use of an Electrolytic Tank," *Proc. IRE*, **33**, pp. 528-534 (August 1945).
7. R. E. Scott, "Network Synthesis by the Use of Potential Analogs," *Proc. IRE*, **40**, pp. 970-973 (August 1952).
8. G. L. Matthaei, "Some Simplifications for Analysis of Linear Circuits," *IRE Trans. on Circuit Theory*, **CT-4**, pp. 120-124 (September 1957).
9. H. J. Carlin, "The Scattering Matrix in Network Theory," *IRE Trans. on Circuit Theory*, **CT-3**, pp. 88-97 (June 1956).

CHAPTER 3

PRINCIPLES OF THE IMAGE METHOD FOR FILTER DESIGN

SEC. 3.01, INTRODUCTION

Although the image method for filter design will not be discussed in detail in this book, it will be necessary for readers to understand the image method in order to understand some of the design techniques used in later chapters. The objective of this chapter is to supply the necessary background by discussing the physical concepts associated with the image method and by summarizing the most useful equations associated with this method. Derivations will be given for only a few equations; more complete discussions will be found in the references listed at the end of the chapter.

SEC. 3.02, PHYSICAL AND MATHEMATICAL DEFINITION OF IMAGE IMPEDANCE AND IMAGE PROPAGATION FUNCTION

The image viewpoint for the analysis of circuits is a wave viewpoint much the same as the wave viewpoint commonly used for analysis of transmission lines. In fact, for the case of a uniform transmission line the *characteristic impedance* of the line is also its *image impedance*, and if γ_t is the propagation constant per unit length then $\gamma_t l$ is the *image propagation* function for a line of length l. However, the terms *image impedance* and *image propagation function* have much more general meaning than their definition with regard to a uniform transmission line alone would suggest.

Consider the case of a two-port network which can be symmetrical, but which, for the sake of generality, will be assumed to be unsymmetrical with different impedance characteristics at End 1 than at End 2. Figure 3.02-1 shows the case of an infinite number of identical networks of this sort all connected so that at each junction either End 1s are connected together or End 2s are connected together. Since the chain of networks extends to infinity in each direction, the same impedance Z_{I1} is

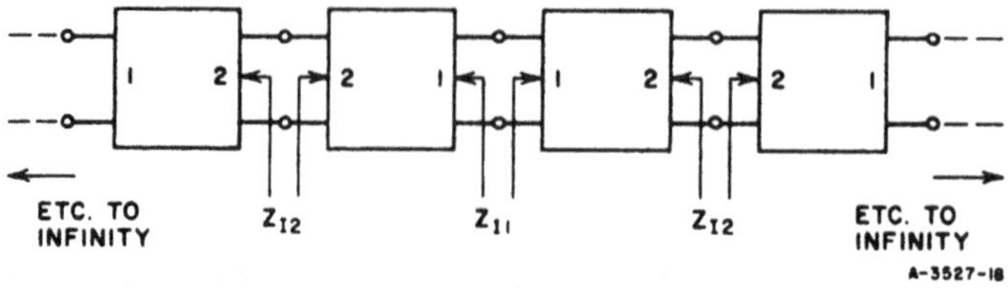

FIG. 3.02-1 INFINITE CHAIN OF IDENTICAL NETWORKS USED FOR DEFINING IMAGE IMPEDANCES AND THE IMAGE PROPAGATION FUNCTION

seen looking both left and right at a junction of the two End 1s, while at a junction of two End 2s another impedance Z_{I2} will be seen when looking either left or right. The impedances Z_{I1} and Z_{I2}, defined as indicated in Fig. 3.02-1, are the image impedances for End 1 and End 2, respectively, of the network. For an unsymmetrical network they are generally unequal.

Note that because of the way the infinite chain of networks in Fig. 3.02-1 are connected, the impedances seen looking left and right at each junction are always equal, hence there is never any reflection of a wave passing through a junction. Thus, from the wave point of view, the networks in Fig. 3.02-1 are all perfectly matched. If a wave is set to propagating towards the right, through the chain of networks, it will be attenuated as determined by the propagation function of each network, but will pass on from network to network without reflection. Note that the image impedances Z_{I1} and Z_{I2} are actually impedance of *infinite* networks, and as such they should be expected to have a mathematical form different from that of the rational impedance functions that are obtained for finite, lumped-element networks. In the cases of lumped-element filter structures, the image impedances are usually irrational functions; in the cases of microwave filter structures which involve transmission line elements, the image impedances are usually both irrational and transcendental.

An equation for the image impedance is easily derived in terms of the circuit in Fig. 3.02-2. If Z_L is made to be equal to Z_{I1} then the impedance Z_{in} seen looking in from the left of the circuit will also be equal to Z_{I1}. Now, if A, B, C, and D are the general circuit parameters for the box on the left in Fig. 3.02-2, assuming that the network is reciprocal, the general circuit parameters A_s, B_s, C_s, and D_s for the two boxes connected as shown can be computed by use of Eq. (2.05-7). Then by Eq. (2.09-1)

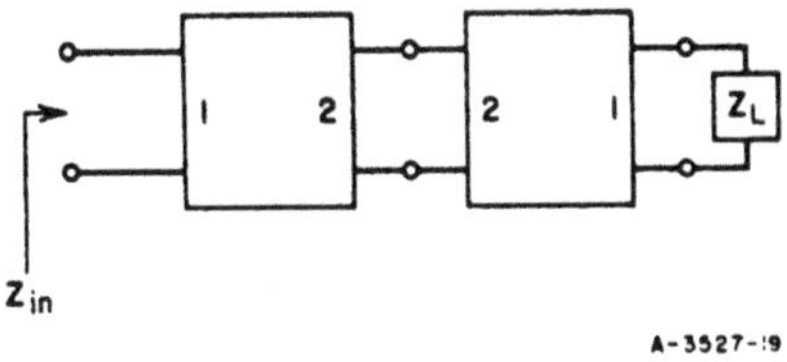

FIG. 3.02-2 CIRCUIT DISCUSSED IN SEC. 3.02

$$Z_{in} = \frac{A_s Z_L + B_s}{C_s Z_L + D_s} \quad . \qquad (3.02\text{-}1)$$

Setting $Z_{in} = Z_L = Z_{I1}$ and solving for Z_{I1} in terms of A, B, C, and D gives

$$Z_{I1} = \sqrt{\frac{AB}{CD}} \quad . \qquad (3.02\text{-}2)$$

The same procedure carried out with respect to End 2 gives

$$Z_{I2} = \sqrt{\frac{DB}{CA}} \quad . \qquad (3.02\text{-}3)$$

Figure 3.02-3 shows a network with a generator whose internal impedance is the same as the image impedance at End 1 and with a load impedance on the right equal to the image impedance at End 2. With the terminations matched to the image impedances in this manner it can be shown that

$$\frac{E_1}{E_2} = \sqrt{\frac{Z_{I1}}{Z_{I2}}}\, e^{\gamma} \qquad (3.02\text{-}4)$$

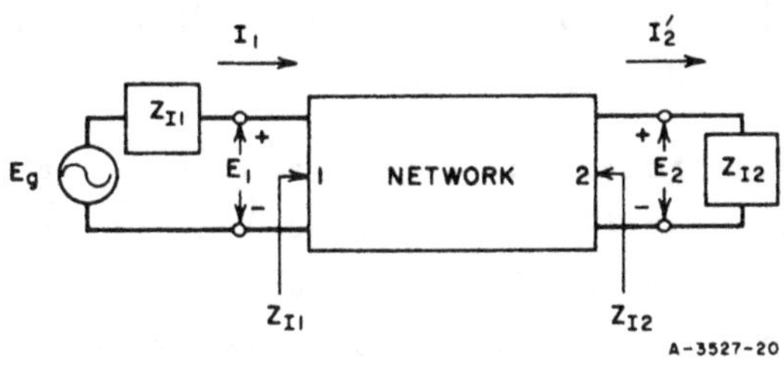

FIG. 3.02-3 NETWORK HAVING TERMINATIONS WHICH ARE MATCHED ON THE IMAGE BASIS

or

$$\frac{E_1}{E_2}\sqrt{\frac{Z_{I2}}{Z_{I1}}} = e^{\gamma} \tag{3.02-5}$$

where

$$\gamma = \alpha + j\beta = ln\ [\sqrt{AD} + \sqrt{BC}] \tag{3.02-6}$$

is the *image propagation function*, α is the *image attenuation* in nepers,* and β is the *image phase* in radians. Note that the $\sqrt{Z_{I2}/Z_{I1}}$ factor in Eq. (3.02-5) has the effect of making γ independent of the relative impedance levels at Ends 1 and 2, much as does the $\sqrt{R_2/R_1}$ factor in Eq. (2.10-4). An alternative form of Eq. (3.02-5) is

$$\gamma = \alpha + j\beta = ln\sqrt{\frac{E_1 I_1}{E_2 I_2'}} \tag{3.02-7}$$

where $I_1 = E_1/Z_{I1}$ and $I_2' = E_2/Z_{I2}$ are as defined in Fig. 3.02-3.

It should be emphasized that the image propagation function defines the transmission through the circuit as indicated by Eq. (3.02-4), (3.02-5), or (3.02-7) *only if the terminations match the image impedances as in Fig. 3.02-3*. The effects of mismatch will be discussed in Sec. 3.07. For a reciprocal network the image propagation function is the same for propagation in either direction even though the network may not be symmetrical.

SEC. 3.03, RELATION BETWEEN THE IMAGE PARAMETERS AND GENERAL CIRCUIT PARAMETERS, OPEN-CIRCUIT IMPEDANCES, AND SHORT-CIRCUIT ADMITTANCES

The transmission properties of a linear two-port network can be defined in terms of its image parameters as well as in terms of the various parameters discussed in Secs. 2.05 to 2.07. Any of these other parameters can be computed from the image parameters and vice versa. These various relationships are summarized in Tables 3.03-1 and 3.03-2. For simplicity, only equations for reciprocal networks are included.

* To change nepers to decibels multiply nepers by 8.686.

Table 3.03-1

IMAGE PARAMETERS IN TERMS OF GENERAL CIRCUIT PARAMETERS, OPEN-CIRCUIT IMPEDANCES, OR SHORT-CIRCUIT ADMITTANCES

IMAGE PARAMETER	IN TERMS OF A,B,C,D	IN TERMS OF $z_{11}, z_{12} = z_{21}, z_{22}$	IN TERMS OF $y_{11}, y_{12} = y_{21}, y_{22}$	IN CONVENIENT MIXED FORM
Z_{I1}	$\sqrt{\frac{AB}{CD}}$	$\sqrt{\frac{z_{11}\,\Delta_z}{z_{22}}}$	$\sqrt{\frac{y_{22}}{y_{11}\,\Delta_y}}$	$\sqrt{\frac{z_{11}}{y_{11}}}$
Z_{I2}	$\sqrt{\frac{DB}{CA}}$	$\sqrt{\frac{z_{22}\,\Delta_z}{z_{11}}}$	$\sqrt{\frac{y_{11}}{y_{22}\,\Delta_y}}$	$\sqrt{\frac{z_{22}}{y_{22}}}$
$\gamma = \alpha + j\beta$	$\coth^{-1}\sqrt{\frac{AD}{BC}}$	$\coth^{-1}\sqrt{\frac{z_{11}z_{22}}{\Delta_z}}$	$\coth^{-1}\sqrt{\frac{y_{11}y_{22}}{\Delta_y}}$	$\coth^{-1}\sqrt{z_{11}y_{11}}$ $= \coth^{-1}\sqrt{z_{22}y_{22}}$
$\gamma = \alpha + j\beta$	$\cosh^{-1}\sqrt{AD}$	$\cosh^{-1}\left(\frac{\sqrt{z_{11}z_{22}}}{z_{21}}\right)$	$\cosh^{-1}\left(\frac{\sqrt{y_{11}y_{22}}}{y_{21}}\right)$	
$\gamma = \alpha + j\beta$	$\sinh^{-1}\sqrt{BC}$	$\sinh^{-1}\left(\frac{\sqrt{\Delta_z}}{z_{21}}\right)$	$\sinh^{-1}\left(\frac{\sqrt{\Delta_y}}{y_{21}}\right)$	

where $\Delta_z = z_{11}z_{22} - z_{12}^2$,

$\Delta_y = y_{11}y_{22} - y_{12}^2$,

Table 3.03-2

GENERAL CIRCUIT PARAMETERS, OPEN-CIRCUIT IMPEDANCES, AND SHORT-CIRCUIT ADMITTANCES IN TERMS OF IMAGE PARAMETERS

$$A = \sqrt{\frac{Z_{I1}}{Z_{I2}}}\cosh\gamma \ , \qquad B = \sqrt{Z_{I1}Z_{I2}}\sinh\gamma$$

$$C = \frac{\sinh\gamma}{\sqrt{Z_{I1}Z_{I2}}} \ , \qquad D = \sqrt{\frac{Z_{I2}}{Z_{I1}}}\cosh\gamma$$

$$z_{11} = Z_{I1}\coth\gamma \ , \qquad z_{12} = \frac{\sqrt{Z_{I1}Z_{I2}}}{\sinh\gamma}$$

$$z_{21} = z_{12} \ , \qquad z_{22} = Z_{I2}\coth\gamma$$

$$y_{11} = Y_{I1}\coth\gamma \ , \qquad y_{12} = \frac{-\sqrt{Y_{I1}Y_{I2}}}{\sinh\gamma}$$

$$y_{21} = y_{12} \ , \qquad y_{22} = Y_{I2}\coth\gamma$$

where $Y_{I1} = \frac{1}{Z_{I1}}$ and $Y_{I2} = \frac{1}{Z_{I2}}$.

SEC. 3.04, IMAGE PARAMETERS FOR SOME COMMON STRUCTURES

The image parameters of the L-section network in Fig. 3.04-1 are given by

$$Z_{I1} = \sqrt{Z_a(Z_a + Z_c)} \qquad (3.04\text{-}1)$$

$$= \sqrt{Z_a Z_c}\sqrt{1 + \frac{Z_a}{Z_c}} \qquad (3.04\text{-}2)$$

$$Z_{I2} = \frac{Z_a Z_c}{\sqrt{Z_a(Z_a + Z_c)}} \qquad (3.04\text{-}3)$$

$$= \frac{\sqrt{Z_a Z_c}}{\sqrt{1 + \frac{Z_a}{Z_c}}} \qquad (3.04\text{-}4)$$

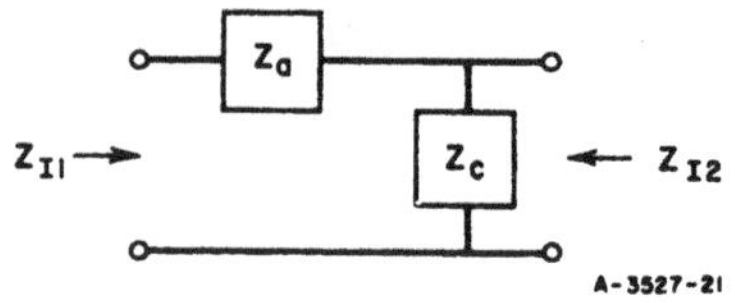

FIG. 3.04-1 AN L-SECTION NETWORK

$$\gamma = \coth^{-1}\sqrt{1 + \frac{Z_c}{Z_a}} \qquad (3.04\text{-}5)$$

$$= \cosh^{-1}\sqrt{1 + \frac{Z_a}{Z_c}} \qquad (3.04\text{-}6)$$

$$= \sinh^{-1}\sqrt{\frac{Z_a}{Z_c}} \qquad (3.04\text{-}7)$$

Note that by Eqs. (3.04-2) and (3.04-4)

$$Z_{I1} = \frac{Z_a Z_c}{Z_{I2}} \qquad (3.04\text{-}8)$$

For the symmetrical T-section in Fig. 3.04-2

$$Z_{I1} = Z_{I2} = \sqrt{Z_a(Z_a + 2Z_b)} \qquad (3.04\text{-}9)$$

$$\gamma = 2\coth^{-1}\sqrt{1+\frac{2Z_b}{Z_a}} \qquad (3.04\text{-}10)$$

$$= 2\cosh^{-1}\left(1+\frac{Z_a}{2Z_b}\right) \qquad (3.04\text{-}11)$$

$$= 2\sinh^{-1}\sqrt{\frac{Z_a}{2Z_b}} \qquad (3.04\text{-}12)$$

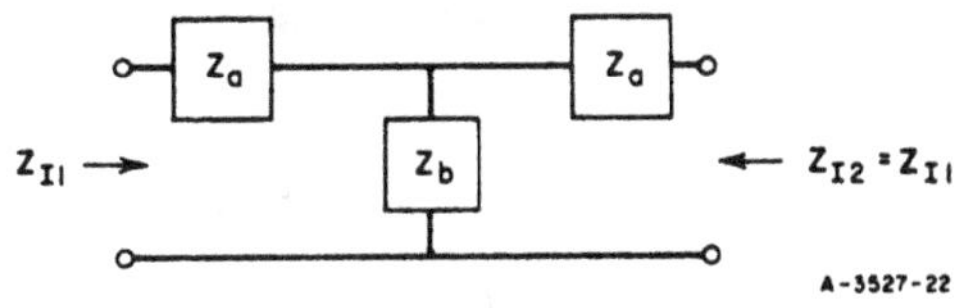

FIG. 3.04-2 A SYMMETRICAL T-SECTION NETWORK

Note that the circuit in Fig. 3.04-2 can be formed by two L-sections as in Fig. 3.04-1 put back to back so that Z_b in Fig. 3.04-2 is one-half of Z_c in Fig. 3.04-1. Then Z_{I1} will be the same for both networks and γ for the T-section is twice that for the L-section.

For the π-section in Fig. 3.04-3 the image admittances are

$$Y_{I1} = Y_{I2} = \sqrt{Y_1(Y_1 + 2Y_3)} \qquad (3.04\text{-}13)$$

and

$$\gamma = 2\coth^{-1}\sqrt{1+\frac{2Y_3}{Y_1}} \qquad (3.04\text{-}14)$$

$$= 2\cosh^{-1}\left(1+\frac{Y_1}{2Y_3}\right) \qquad (3.04\text{-}15)$$

$$= 2\sinh^{-1}\sqrt{\frac{Y_1}{2Y_3}} \qquad (3.04\text{-}16)$$

A π-section can also be constructed from two half sections back to back, so that $Y_1 = 1/Z_c$ and $Y_3 = 1/(2Z_a)$. For Fig. 3.04-1, $Y_{I2} = 1/Z_{I2}$ will then be the same as $Y_{I2} = Y_{I1}$ in Fig. 3.04-3, while γ for Fig. 3.04-3 will again be twice that for Fig. 3.04-1.

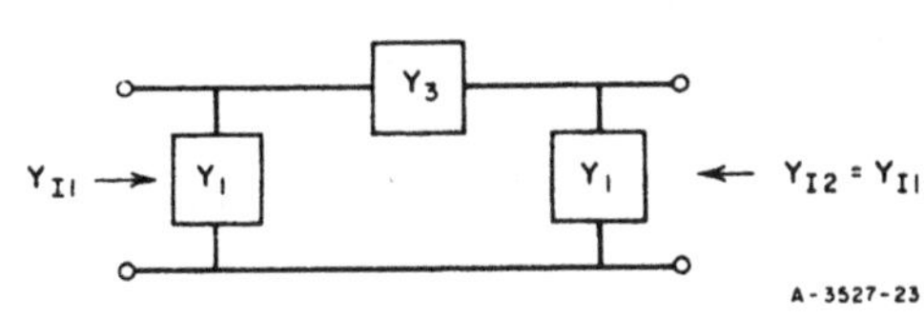

FIG. 3.04-3 A SYMMETRICAL π-SECTION NETWORK

For a uniform transmission line of length l, characteristic impedance Z_0, and propagation constant $\gamma_t = \alpha_t + j\beta_t$ per unit length,

$$Z_{I1} = Z_{I2} = Z_0 \qquad (3.04\text{-}17)$$

$$\gamma = \gamma_t l = \alpha_t l + j\beta_t l \qquad (3.04\text{-}18)$$

SEC. 3.05, THE SPECIAL IMAGE PROPERTIES OF DISSIPATIONLESS NETWORKS

By Table 3.03-1

$$Z_{I1} = \sqrt{\frac{z_{11}}{y_{11}}} \qquad (3.05\text{-}1)$$

while

$$\gamma = \alpha + j\beta = \coth^{-1} \sqrt{z_{11}y_{11}} \quad . \qquad (3.05\text{-}2)$$

For a dissipationless network, we may write for frequencies $p = j\omega$

$$z_{11} = j(X_{oc})_1 \qquad (3.05\text{-}3)$$

and

$$y_{11} = \frac{1}{j(X_{sc})_1} \qquad (3.05\text{-}4)$$

where $j(X_{oc})_1$ is the impedance at End 1 of the network with End 2 open-circuited, and $j(X_{sc})_1$ is the impedance at End 1 with End 2 short-circuited. Then by Eqs. (3.05-1) to (3.05-4), for dissipationless networks

$$Z_{I1} = \sqrt{-(X_{oc})_1(X_{sc})_1} \qquad (3.05\text{-}5)$$

and

$$\gamma = \alpha + j\beta = \coth^{-1}\sqrt{\frac{(X_{oc})_1}{(X_{sc})_1}} \quad . \qquad (3.05\text{-}6)$$

The inverse, hyperbolic cotangent function in Eq. (3.05-6) is a multivalued function, whose various possible values all differ by multiples of $j\pi$. For this reason, it is convenient to write Eq. (3.05-6) in the form

$$\gamma = \alpha + j\beta = \coth^{-1}\sqrt{\frac{(X_{oc})_1}{(X_{sc})_1}} + jn\pi \tag{3.05-7}$$

where the inverse hyperbolic function is to be evaluated to give an imaginary part having minimum magnitude, and where the appropriate value for the integer n must be determined by examination of the circuit under consideration. Equation (3.05-7) also has the equivalent form

$$\gamma = \alpha + j\beta = \tanh^{-1}\sqrt{\frac{(X_{oc})_1}{(X_{sc})_1}} + j(2n - 1)\frac{\pi}{2} \quad . \tag{3.05-8}$$

Two distinct cases occur in the evaluation of Eq. (3.05-5) and Eq. (3.05-7) or (3.05-8) depending on whether $(X_{oc})_1$ and $(X_{sc})_1$ have the same sign or opposite signs. These two cases will be summarized separately.

Case A, Condition for a Pass Band—In this case $(X_{oc})_1$ and $(X_{sc})_1$ have opposite signs and

$$Z_{I1} = \sqrt{-(X_{oc})_1 (X_{sc})_1} = \text{real and positive.} \tag{3.05-9}$$

It can be shown that, at the same time, the condition

$$Z_{I2} = \sqrt{-(X_{oc})_2 (X_{sc})_2} = \text{real and positive.} \tag{3.05-10}$$

must also exist, where $(X_{oc})_2$ and $(X_{sc})_2$ are the open- and short-circuit impedances measured from End 2. Under these conditions, Eqs. (3.05-7) and (3.05-8) yield for α and β,

$$\alpha = 0 \tag{3.05-11}$$

$$\beta = -\cot^{-1}\sqrt{\left|\frac{(X_{oc})_1}{(X_{sc})_1}\right|} \quad \text{radians}$$

$$= \left[\tan^{-1}\sqrt{\left|\frac{(X_{oc})_1}{(X_{sc})_1}\right|}\right] - \frac{\pi}{2} \quad \text{radians} \quad . \tag{3.05-12}$$

Note that for this pass-band case, the attenuation is zero while the phase is generally non-zero and varying with frequency. In Eqs. (3.05-11) and (3.05-12) the $n\pi$ term has been omitted since the multivalued nature of these inverse trigonometric functions will be familiar to the reader (though perhaps the multivalued nature of inverse hyperbolic functions may not).

Case B, Conditions for a Stop Band—In this case $(X_{oc})_1$ and $(X_{sc})_1$ [and also $(X_{oc})_2$ and $(X_{sc})_2$] have the same sign. Then

$$Z_{I1} = \sqrt{-(X_{oc})_1 (X_{sc})_1} = jX_{I1} \tag{3.05-13}$$

and

$$Z_{I2} = \sqrt{-(X_{oc})_2 (X_{sc})_2} = jX_{I2} \tag{3.05-14}$$

are both purely imaginary. Both X_{I1} and X_{I2} must have positive slopes *vs.* frequency, in accord with Foster's reactance theorem. If $(X_{oc})_1 > (X_{sc})_1$ Eq. (3.05-7) should be used to obtain α and β:

$$\alpha = \coth^{-1}\sqrt{\frac{(X_{oc})_1}{(X_{sc})_1}} \quad \text{nepers} \tag{3.05-15}$$

and

$$\beta = n\pi \quad \text{radians} \tag{3.05-16}$$

If $(X_{oc})_1 < (X_{sc})_1$, Eq. (3.05-8) should be used, and it gives

$$\alpha = \tanh^{-1}\sqrt{\frac{(X_{oc})_1}{(X_{sc})_1}} \qquad \text{nepers} \qquad (3.05\text{-}17)$$

and

$$\beta = (2n - 1)\frac{\pi}{2} \qquad \text{radians} \quad . \qquad (3.05\text{-}18)$$

Note that for this stop-band case the image attenuation is non-zero and will vary with frequency. Meanwhile, the image phase is constant *vs.* frequency at some multiple of π, or odd multiple of $\pi/2$. However, it will be found that the image phase can make discrete jumps at points in the stop band where there are poles of attenuation for frequencies $j\omega$.

A similar analysis for dissipationless networks can be carried out using the various other expressions for the image parameters in Secs. 3.03 and 3.04. The various equations given for the image propagation constant will involve inverse hyperbolic functions of a purely real or purely imaginary argument. Due to the multivalued nature of these inverse hyperbolic functions care must be taken in evaluating them. Table 3.05-1 should prove helpful for this purpose. Note that in some cases a different equation must be used depending on whether $|u|$ or $|v|$ is greater or less than one. This is because, for example, $\cosh^{-1}w$ when taken to be a function of a real variable cannot be evaluated for $w = |u| < 1$; if however, w is a function of a complex variable the above example has a value, namely, $j(\cos^{-1}u)$. The proper value of the integer n to be used with the various equations in Table 3.05-1 must be determined by examination of the circuit at some frequency where the transmission phase is easily established. As was done in the case of Eqs. (3.05-11) and (3.05-12), the $n\pi$ terms have been omitted for forms involving inverse trigonometric functions since their multivalued nature is much more widely familiar than is that of inverse hyperbolic functions.

Table 3.05-1

EVALUATION OF SOME INVERSE HYPERBOLIC FUNCTIONS FOR PURELY REAL OR PURELY IMAGINARY ARGUMENTS

In general, $W = U + jV$, $w = u + jv$, and n is an integer (positive, negative or zero)

Function	Case of $w = u$	Case of $w = jv$
$W = \coth^{-1} w$	If $\lvert u \rvert > 1$, $W = \coth^{-1} u + jn\pi$	$W = 0 + j(-\cot^{-1} v)$
	If $\lvert u \rvert < 1$, $W = \tanh^{-1} u + j(2n - 1)\frac{\pi}{2}$	$= 0 + j\left[(\tan^{-1} v) - \frac{\pi}{2}\right]$
$W = \sinh^{-1} w$	$W = (-1)^{\lvert n \rvert} \sinh^{-1} u + jn\pi$	If $\lvert v \rvert \geq 1$, $W = \cosh^{-1} v + j(2n - 1)\frac{\pi}{2}$ $\;$ n = odd if $v > 1$, n = even if $v < -1$
		If $\lvert v \rvert \leq 1$, $W = 0 + j \sin^{-1} v$
$W = \cosh^{-1} w$	If $\lvert u \rvert > 1$, $W = \cosh^{-1} u + jn\pi$ $\;$ n = even if $u > 1$, n = odd if $u < -1$.	$W = \sinh^{-1} v + j(2n - 1)\frac{\pi}{2}$ $\;$ n = odd if $v > 0$. n = even if $v < 0$.
	If $\lvert u \rvert < 1$ $W = 0 + j \cos^{-1} u$	

SEC. 3.06, CONSTANT-k AND m-DERIVED FILTER SECTIONS

Constant-k and m-derived filters are classic examples of filters which are designed from the image point of view. Their properties will be briefly summarized in order to illustrate some of the image properties of dissipationless networks discussed in the preceding section, and to provide reference data. The filter sections shown are all normalized so that their image impedance is $R_0' = 1$ ohm at $\omega' = 0$ and their cutoff frequency occurs at $\omega_1' = 1$ radian/sec. However, these normalized circuits can easily be changed to other impedance and frequency scales. Each resistance, inductance, or capacitance is scaled using

$$R = \left(\frac{R_0}{R_0'}\right)R' \tag{3.06-1}$$

$$L = \left(\frac{R_0}{R_0'}\right)\left(\frac{\omega_1'}{\omega_1}\right)L' \tag{3.06-2}$$

or

$$C = \left(\frac{R_0'}{R_0}\right)\left(\frac{\omega_1'}{\omega_1}\right)C' \tag{3.06-3}$$

where R', L', and C' are for the normalized circuit and R, L, and C are corresponding elements for the scaled circuit. The ratio R_0/R_0' defines the change in impedance level while ω_1/ω_1' defines the change in frequency scale.

Figure 3.06-1(a) shows a normalized constant-k filter half section. Its image impedances are

$$Z_{IT} = \sqrt{1 - (\omega')^2} \tag{3.06-4}$$

and

$$Z_{I\pi} = \frac{1}{\sqrt{1 - (\omega')^2}} = \frac{1}{Z_{IT}} \tag{3.06-5}$$

Its propagation function is

$$\gamma = \alpha + j\beta = 0 + j \sin^{-1}\omega' \tag{3.06-6}$$

for the $0 \leqq \omega' \leqq 1$ pass band, and

$$\gamma = \alpha + j\beta = \cosh^{-1}\omega' + j\,\frac{\pi}{2} \tag{3.06-7}$$

for the $1 \leqq \omega' \leqq \infty$ stop band, where α is in nepers and β is in radians.

Figures 3.06-1(b), (c) show sketches of the image impedance and attenuation characteristics of this structure. Note that, as discussed

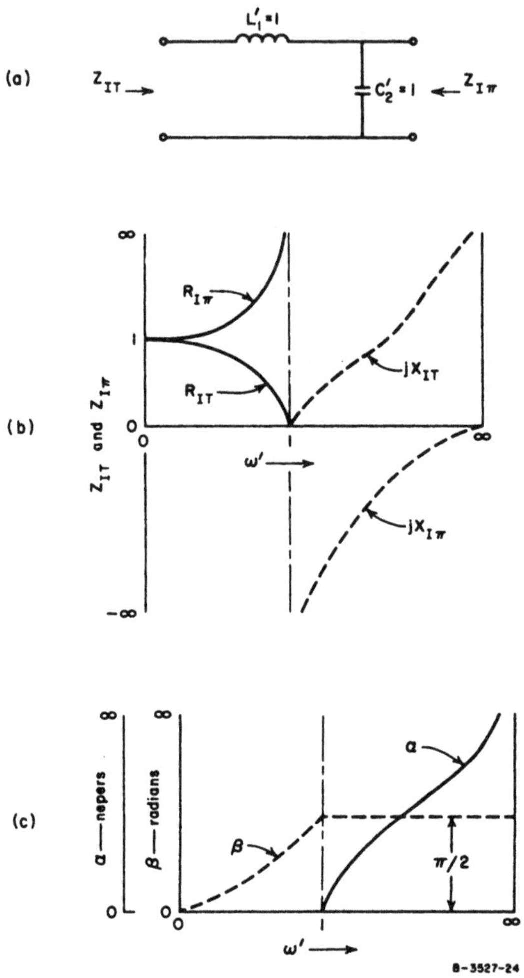

FIG. 3.06-1 THE IMAGE PROPERTIES OF NORMALIZED, CONSTANT-k HALF SECTIONS

in Sec. 3.05, Z_{IT} and $Z_{I\pi}$ are purely real in the pass band and purely imaginary in the stop band. Also note that $\alpha = 0$ in the pass band while β is constant in the stop band.

Figure 3.06-2(a) shows a "series, *m*-derived" half section. Its image impedances are

$$Z_{IT} = \sqrt{1 - (\omega')^2} \qquad (3.06\text{-}8)$$

$$Z_{I\pi m} = \frac{1 - \left(\frac{\omega'}{\omega'_\infty}\right)^2}{\sqrt{1 - \omega'^2}} \qquad (3.06\text{-}9)$$

where

$$\omega'_\infty = \frac{1}{\sqrt{1 - m^2}} \quad . \qquad (3.06\text{-}10)$$

Note that Eq. (3.06-8) is identical to Eq. (3.06-4), but Eq. (3.06-9) differs from Eq. (3.06-5). The propagation function is

$$\gamma = \alpha + j\beta = 0 + j\,\frac{1}{2}\cos^{-1}\left[1 - \frac{2m^2}{\left(\frac{1}{\omega'}\right)^2 - (1 - m^2)}\right] \qquad (3.06\text{-}11)$$

in the $0 \leqq \omega' \leqq 1$ pass band,

$$\gamma = \frac{1}{2}\cosh^{-1}\left[\frac{2m^2}{\left(\frac{1}{\omega'}\right)^2 - (1 - m^2)} - 1\right] + j\,\frac{\pi}{2} \qquad (3.06\text{-}12)$$

in the $1 = \omega' \leqq \omega'_\infty$ stop band, and

$$\gamma = \frac{1}{2}\cosh^{-1}\left[1 - \frac{2m^2}{\left(\frac{1}{\omega'}\right)^2 - (1 - m^2)}\right] + j0 \qquad (3.06\text{-}13)$$

in the $\omega'_\infty \leqq \omega' \leqq \infty$ stop band.

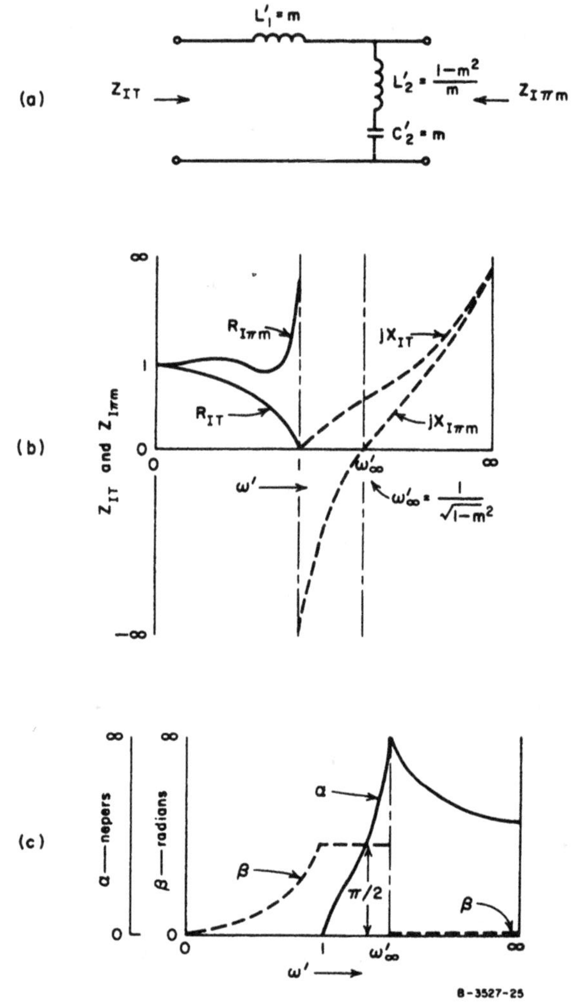

FIG. 3.06-2 NORMALIZED, SERIES, m–DERIVED HALF-SECTION CHARACTERISTICS

Figures 3.06-2(b) and (c) show sketches of the image impedance and propagation characteristics of this structure. Note that introducing a series resonance in the shunt branch in Fig. 3.06-2(a) has produced a pole of attenuation at the frequency ω'_∞ where the shunt branch short-circuits transmission. (See discussion in Sec. 2.04.) Note that $Z_{I\pi m} = R_{I\pi m}$ in the pass band in Fig. 3.06-2(b) is more nearly constant than is $R_{I\pi}$ in Fig. 3.06-1(b). This property of m-derived image impedances makes them helpful for improving the impedance match to resistor terminations.

The "shunt m-derived" half section in Fig. 3.06-3(a) is the dual of that in Fig. 3.06-2(a). The image impedances are

$$Z_{ITm} = \frac{\sqrt{1 - (\omega')^2}}{1 - \left(\dfrac{\omega'}{\omega'_\infty}\right)^2} \tag{3.06-14}$$

$$Z_{I\pi} = \frac{1}{\sqrt{1 - (\omega')^2}} \tag{3.06-15}$$

where again

$$\omega'_\infty = \frac{1}{\sqrt{1 - m^2}} \quad , \tag{3.06-16}$$

In this case Z_{ITm} in Eq. (3.06-14) differs from Z_{IT} in Eq. (3.06-4), but Eqs. (3.06-15) and (3.06-5) are identical. The image propagation function for this section is the same as that in Eqs. (3.06-11) to (3.06-13). Figures 3.06-3(b) and (c) show sketches of the image characteristics of this filter section. In this case, a pole of attenuation is produced at the frequency ω'_∞ where the series branch has a pole of impedance which blocks all transmission. The image impedance Z_{ITm} is seen to be more nearly constant in the pass band than was Z_{IT} in Fig. 3.06-1(b). Thus, m-derived half sections of this type are also useful for improving the impedance match to resistor terminations.

Figure 3.06-4(a) and (b) show how constant-k and m-derived half sections may be pieced together to form a sizeable filter. In this case, three constant-k half sections are used along with two, series, m-derived,

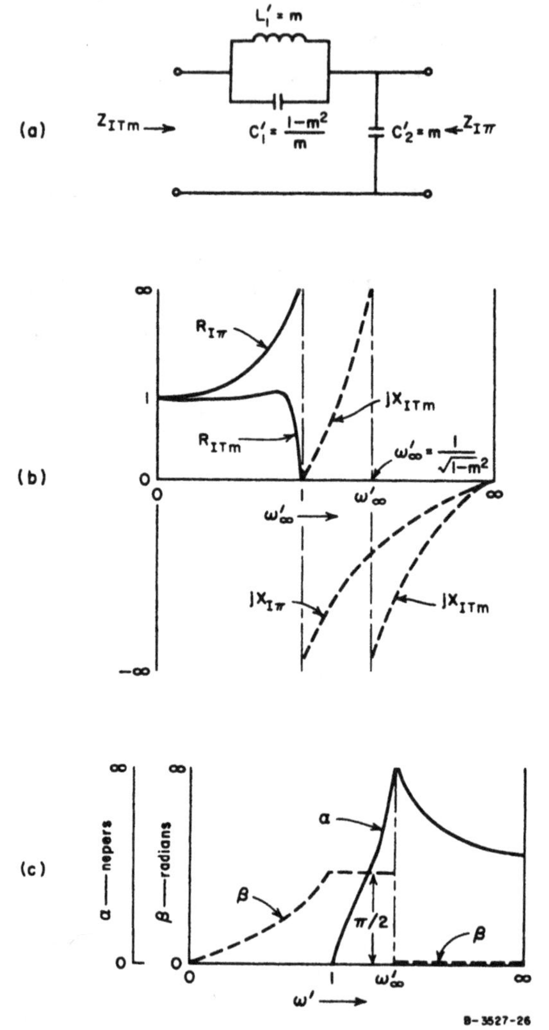

FIG. 3.06-3 NORMALIZED, SHUNT, m–DERIVED HALF–SECTION CHARACTERISTICS

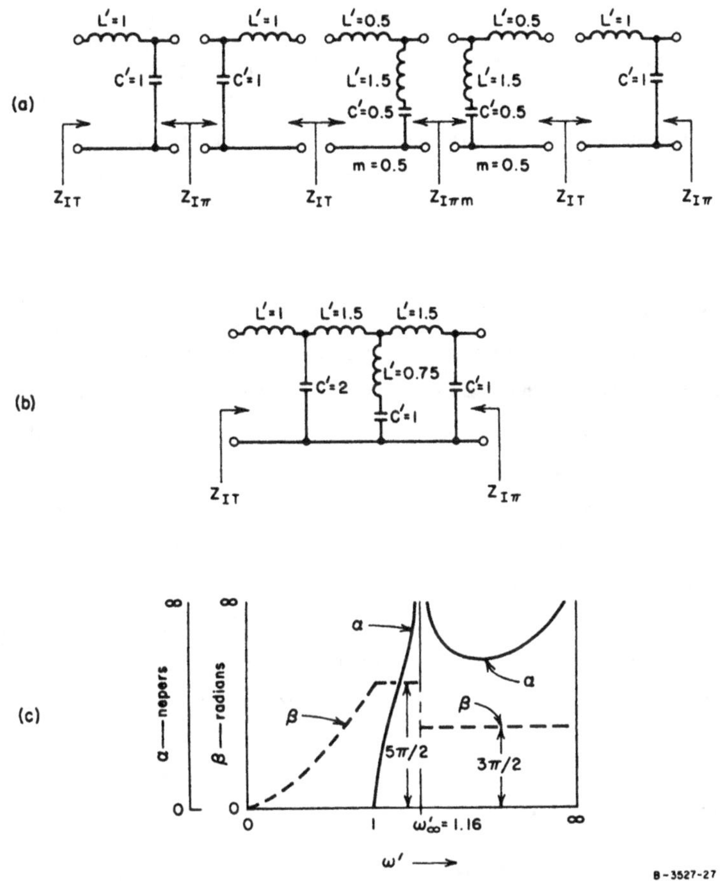

FIG. 3.06-4 A FILTER PIECED TOGETHER FROM THREE CONSTANT–k AND TWO m–DERIVED HALF SECTIONS
The resulting image propagation function is sketched at (c)

half sections. The two m-derived sections have $m = 0.5$, which introduces a pole of attenuation at $\omega'_\infty = 1.16$ and greatly increases the rate of cutoff of the filter. As indicated in Fig. 3.06-4(a) the sections are all chosen so that the image impedances match at each junction. Under these conditions when the sections are all joined together, the image attenuation and the image phase for the entire structure are simply the sum of the image attenuation and phase values for the individual sections. Likewise, with all of the sections matched to each other, the image impedances seen at the ends are the same as the image impedances of the end sections before they were connected to the interior sections.

The circuit in Fig. 3.06-4(b) would have the transmission characteristics indicated in Fig. 3.06-4(c) if it were terminated in its image impedances at both ends. However, since in practice resistor terminations are generally required, this transmission characteristic will be considerably altered (mainly in the pass band) due to the reflections at both ends of the filter. In order to reduce the magnitude of these reflections effects, it is customary with filters of this type to introduce m-derived half-sections at each end of the filter with the impedance Z_{ITm} or $Z_{I\pi m}$ next to the termination resistor. With $m = 0.6$, these image impedances are relatively constant in the pass band and it becomes possible to greatly reduce the reflection effects over much of the pass band. These matters will be discussed further in Secs. 3.07 and 3.08.

SEC. 3.07, THE EFFECTS OF TERMINATIONS WHICH MISMATCH THE IMAGE IMPEDANCES

The resistance terminations used on dissipationless filter structures cannot match the image impedance of the structure except at discrete frequencies in the pass band. As a result of the multiple reflections that occur, the performance of the filter may be considerably altered from that predicted by the image propagation function. This alteration is most severe in the pass band and in the stop band near cutoff. Formulas which account for the effects of such terminal reflections are summarized below.

Consider the circuit in Fig. 3.07-1 whose image impedances, Z_{I1} and Z_{I2}, may differ considerably from R_1 and R_2. The voltage attenuation ratio, E_g/E_2, may be calculated from the image parameters and the terminations using the equation

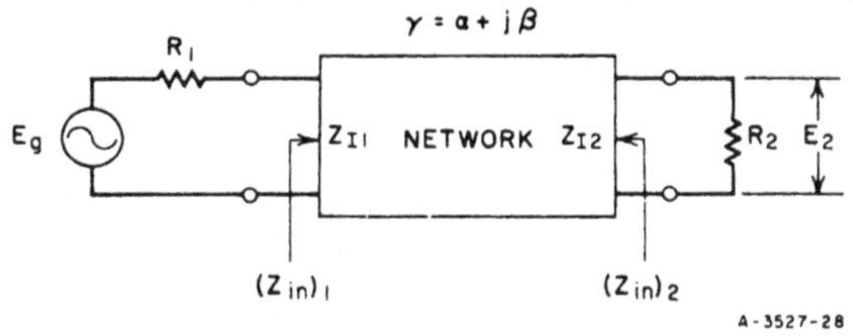

FIG. 3.07-1 NETWORK DISCUSSED IN SEC. 3.07

$$\frac{E_g}{E_2} = 2\sqrt{\frac{Z_{I1}}{Z_{I2}}}\, e^{\gamma} \left[\frac{1 - e^{-2\gamma}\Gamma_{I1}\Gamma_{I2}}{\tau_{I1}\tau_{I2}}\right] \tag{3.07-1}$$

where

$$\Gamma_{I1} = \frac{R_1 - Z_{I1}}{R_1 + Z_{I1}} \tag{3.07-2}$$

and

$$\Gamma_{I2} = \frac{R_2 - Z_{I2}}{R_2 + R_{I2}} \tag{3.07-3}$$

are the reflection coefficients at Ends 1 and 2 while

$$\tau_{I1} = \frac{2Z_{I1}}{R_1 + Z_{I1}} \tag{3.07-4}$$

and

$$\tau_{I2} = \frac{2R_2}{R_2 + Z_{I2}} \tag{3.07-5}$$

are the transmission coefficients (see Sec. 2.08). Note that these reflection and transmission coefficients are defined with respect to the image impedances rather than with respect to the actual input impedances $(Z_{in})_1$ and $(Z_{in})_2$.

The actual input impedance seen looking in End 1 with End 2 terminated in R_2 is

$$(Z_{in})_1 = Z_{I1}\left[\frac{1 + \Gamma_{I2}e^{-2\gamma}}{1 - \Gamma_{I2}e^{-2\gamma}}\right] . \tag{3.07-6}$$

By analogy, $(Z_{in})_2$ in Fig. 3.07-1 is

$$(Z_{in})_2 = Z_{I2}\left[\frac{1 + \Gamma_{I1}e^{-2\gamma}}{1 - \Gamma_{I1}e^{-2\gamma}}\right] . \tag{3.07-7}$$

Equations (3.07-1) to (3.07-7) apply whether the circuit has dissipation or not.

For a dissipationless network at pass band frequencies where $\gamma = 0 \pm jn\pi|_{n=1,2,3,\ldots}$ Eq. (3.07-6) shows that

$$(Z_{in})_1 = \frac{Z_{I1}}{Z_{I2}} R_2 \tag{3.07-8}$$

while at frequencies where $\gamma = 0 \pm j(2n - 1)(\pi/2)|_{n=1,2,3,\ldots}$,

$$(Z_{in})_1 = \frac{Z_{I1}Z_{I2}}{R_2} \tag{3.07-9}$$

where Z_{I1} and Z_{I2} will be purely real. Analogous expressions also exist for $(Z_{in})_2$.

Equation (3.07-1) is quite general, and it can be used with Eqs. (2.11-2) and (2.11-4) for computing the attenuation of a network. However, simpler expressions (about to be presented) can be used if the network is dissipationless. Such expressions become especially simple if the dissipationless network is symmetrical (*i.e.*, $Z_{I1} = Z_{I2}$) and has symmetrical terminations (*i.e.*, $R_1 = R_2$). Another case of relative simplicity is that of a dissipationless antimetrical network (see Sec. 2.11) with antimetrical terminations. Such a filter will satisfy the conditions

$$Z_{I1} = \frac{R_0^2}{Z_{I2}} \qquad (3.07\text{-}10)$$

$$R_1 = \frac{R_0^2}{R_2} \qquad (3.07\text{-}11)$$

at all frequencies, where R_0 is a positive, real constant. The constant-k half section in Fig. 3.06-1 is an example of an antimetrical network. The filter in Fig. 3.06-4 also satisfies the antimetry condition given by Eq. (3.07-10).

For dissipationless symmetrical networks with symmetrical terminations, $Z_{I1} = R_{I1}$ in the pass band and the attenuation is

$$L_A = 10 \log_{10} \left[1 + \frac{1}{4} \left(\frac{R_{I1}}{R_1} - \frac{R_1}{R_{I1}} \right)^2 \sin^2 \beta \right] \quad \text{db} \qquad (3.07\text{-}12)$$

while in the stop band $Z_{I1} = jX_{I1}$ and

$$L_A = 10 \log_{10} \left[1 + \frac{1}{4} \left(\frac{X_{I1}}{R_1} + \frac{R_1}{X_{I1}} \right)^2 \sinh^2 \alpha \right] \quad \text{db} \quad . \qquad (3.07\text{-}13)$$

Similarly for dissipationless antimetrical networks with antimetrical terminations, in the pass band

$$L_A = 10 \log_{10} \left[1 + \frac{1}{4} \left(\frac{R_{I1}}{R_1} - \frac{R_1}{R_{I1}} \right)^2 \cos^2 \beta \right] \quad \text{db} \qquad (3.07\text{-}14)$$

while in the stop band Eq. (3.07-13) applies just as for the symmetrical case. For the symmetrical case

$$\Gamma_{I1} = \Gamma_{I2} \qquad (3.07\text{-}15)$$

while for the antimetrical case

$$\Gamma_{I1} = -\Gamma_{I2} \quad . \qquad (3.07\text{-}16)$$

For the dissipationless symmetrical case the stop-band image phase is a multiple of π radians, while in the dissipationless antimetric case it is an *odd* multiple of $\pi/2$ radians.

The actual pass-band attenuation which will result from mismatched image impedances is seen by Eqs. (3.07-12) and (3.07-14) to depend strongly on the image phase, β. For given Z_{I1} and R_1 it is easily shown that the maximum possible pass-band attenuation in a dissipationless symmetrical or antimetrical network with symmetrical or antimetrical terminations, respectively, is

$$L_A = 20 \log_{10} \left(\frac{a^2 + 1}{2a} \right) \quad \text{db} \qquad (3.07\text{-}17)$$

where

$$a = \frac{Z_{I1}}{R_1} \text{ or } \frac{R_1}{Z_{I1}}$$

with either definition giving the same answer. For symmetrical networks, the value given by Eq. (3.07-17) applies when $\beta = (2n - 1)\pi/2$ radians while $L_A = 0$ when $\beta = n\pi$ radians (where n is an integer). For antimetrical networks Eq. (3.07-17) applies when $\beta = n\pi$ radians while $L_A = 0$ when $\beta = (2n - 1)\pi/2$ radians. Figure 3.07-2 shows a plot of maximum L_A *vs.* a, and also shows the corresponding input VSWR.

SEC. 3.08, DESIGN OF MATCHING END SECTIONS TO IMPROVE THE RESPONSE OF FILTERS DESIGNED ON THE IMAGE BASIS

As mentioned in Sec. 3.06, one way in which the pass-band response of constant-*k* filters can be improved is to use *m*-derived half sections at the ends. Experience shows that a half section with *m* about 0.6 will cause Z_{ITm} or $Z_{I\pi m}$ to give the best approximation of a constant resistance in the pass band, and hence will cause the ends of the filter to give the best match to resistor terminations. As an example, Fig. 3.08-1 shows the normalized filter structure in Fig. 3.06-4(b) with matching sections added to improve the pass-band match to the one-ohm terminations shown. The matching sections also introduce poles of attenuation at $\omega'_\infty = 1.25$, which will further sharpen the cutoff characteristics of the filter.

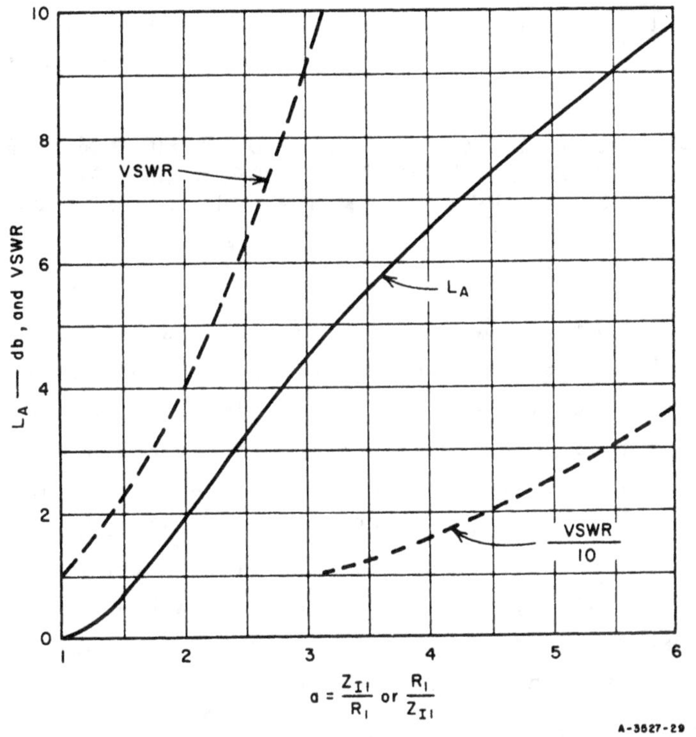

FIG. 3.07-2 MAXIMUM POSSIBLE PASS-BAND ATTENUATION AND VSWR FOR DISSIPATIONLESS SYMMETRICAL NETWORKS WITH SYMMETRICAL TERMINATIONS, OR DISSIPATIONLESS ANTIMETRICAL NETWORKS WITH ANTIMETRICAL TERMINATIONS

These values will apply if $\beta = (2n - 1)(\pi/2)\big|_{n=1,2,3,\ldots}$ for the symmetrical case or $\beta = n\pi\big|_{n=1,2,3,\ldots}$ for the antimetrical case

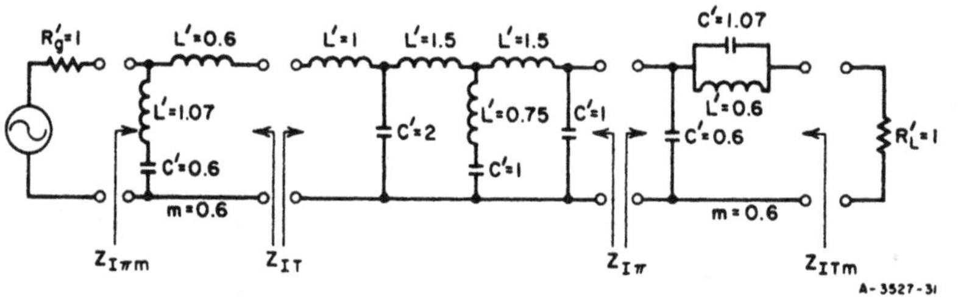

FIG. 3.08-1 THE NORMALIZED FILTER CIRCUIT IN FIG. 3.06-4(b) WITH m-DERIVED HALF SECTIONS ADDED TO IMPROVE THE PASS-BAND IMPEDANCE MATCH TO RESISTOR TERMINATIONS

In the design of microwave filter structures on the image basis, it is often desirable that the matching end sections be of the same general form as the main part of the filter. Consider the case of a wide band, band-pass filter to be constructed using filter sections as shown in Fig. 3.08-2(a). The filter sections have image characteristics as shown in Fig. 3.08-2(b), (c). Figure 3.08-3 shows the left half of a symmetrical filter formed from such sections. In this filter the interior sections of the filter are all alike, but two sections at each end are different in order to improve the pass-band match to the terminations. The design of such end sections will now be considered.

As is seen from Fig. 3.08-2(c), each section of the filter has a midband image phase shift of $\beta = \pi/2$. The total midband image phase shift for the end matching network in Fig. 3.08-3 at f_0 is thus $\beta = \pi$. At midband, then, the end matching network will operate similarly to a half-wavelength transmission line, and in Fig. 3.08-3

$$Z_{in}\Big|_{f=f_0} = R_g \quad . \tag{3.08-1}$$

Thus, if Z_I is the image impedance of the interior sections of the filter, and Z_{Ie} is the image impedance of the sections in the end matching network, then if

$$Z_I\Big|_{f=f_0} = R_g \tag{3.08-2}$$

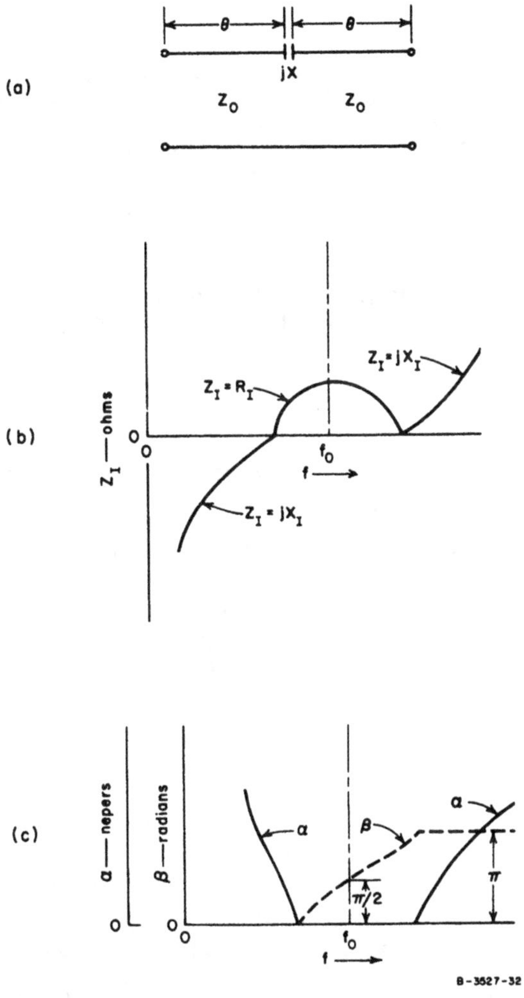

FIG. 3.08-2 A BAND-PASS FILTER SECTION USING TRANSMISSION LINES, AND ITS IMAGE CHARACTERISTICS

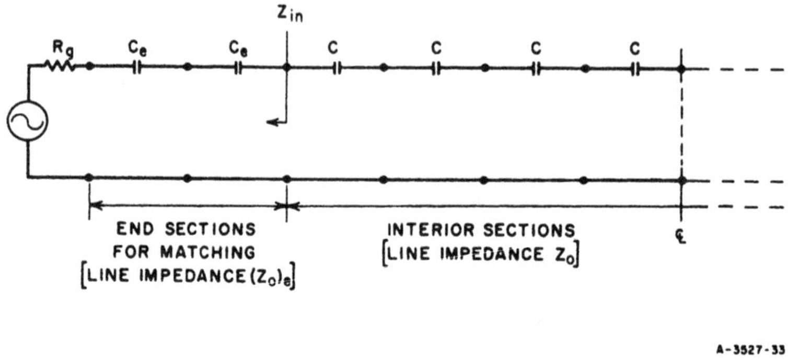

FIG. 3.08-3 ONE-HALF OF A SYMMETRICAL FILTER COMPOSED OF SECTIONS OF THE TYPE IN FIG. 3.08-2

a perfect match is assured at f_0, regardless of the size of Z_{Ie} at that frequency. At pass-band frequencies $f_{\pi/2}$ and $f_{3\pi/2}$, where the image phase shift of the end matching network is $\beta = \pi/2$ and $3\pi/2$, respectively,

$$Z_{in} = \frac{(Z_{Ie})^2}{R_g} \qquad (3.08\text{-}3)$$

similarly to Eq. (3.07-9). Thus, setting $Z_{in} = Z_I$ and solving for Z_{Ie} gives

$$Z_{Ie} = \sqrt{Z_I R_g} \qquad (3.08\text{-}4)$$

as the condition for a perfect impedance match when $\beta = \pi/2$ or $3\pi/2$ for the end matching network. By such procedures a perfect impedance match can be assured when the end matching network has $\pi/2$, π, or $3\pi/2$ radians image phase.

Figure 3.08-4 shows how the image impedance of the end matching network might compare with the image impedance of the interior sections for a practical design. In this case R_g is made a little less than Z_I for the interior sections at f_0, but Z_I and Z_{Ie} are both made to be equal to R_g at f_a and f_b, a little to each side of f_0, so that a perfect match will be achieved at those two frequencies. This procedure will result in a small mismatch in the vicinity of f_0, but should improve the over-all re- ·s. The end matching network is made to be more broadband than the

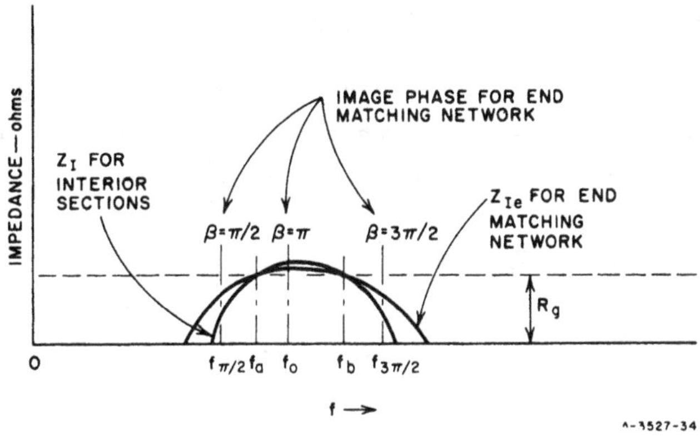

FIG. 3.08-4 RELATIVE IMAGE IMPEDANCE CHARACTERISTICS FOR THE END MATCHING NETWORK AND INTERIOR SECTIONS OF A PROPOSED FILTER OF THE FORM IN FIG. 3.08-3

interior sections of the filter so that the $\beta = \pi/2$ and $3\pi/2$ phase shift points will occur near the cutoff frequencies of the interior sections. The end matching network is designed so that Eq. (3.08-4) will be satisfied, at least approximately, at these two frequencies in order to give a good impedance match close to the cutoff frequencies of the filter. In this particular example there are only three degrees of freedom in the design of the end matching network, namely the size of C_e, the size of $(Z_0)_e$, and the length of the transmission lines in the sections of the end matching network. One degree of freedom is used in fixing the center frequency of the response, another may be used for setting $Z_{Ie} = R_g$ at frequency f_a in Fig. 3.08-4, and another may be used for satisfying Eq. (3.08-4) at $f_{\pi/2}$. Although matching conditions are not specifically forced at frequencies f_b and $f_{3\pi/2}$ in Fig. 3.08-4, they will be approximately satisfied because of the nearly symmetrical nature of the response about f_0.

The design procedure described above provides a perfect impedance match at certain frequencies and assures that the maximum mismatch throughout the pass band will not be large. In addition it should be recalled that perfect transmission will result at pass-band frequencies where the image phase of the over-all filter structure is a multiple of π radians, as well as at points where the image impedances are perfectly matched. These same principles also apply for the design of matching sections for other types of filters.

Occasionally it will be desirable to measure the image parameters of a circuit. A general method is to measure the input impedance at one end for open- and short-circuit terminations at the other end. Then

$$Z_{I1} = \sqrt{(Z_{oc})_1 (Z_{sc})_1} \tag{3.09-1}$$

$$Z_{I2} = \sqrt{(Z_{oc})_2 (Z_{sc})_2} \tag{3.09-2}$$

and for a reciprocal network

$$\gamma = \coth^{-1}\sqrt{\frac{(Z_{oc})_1}{(Z_{sc})_1}} \tag{3.09-3}$$

In these equations $(Z_{oc})_1$ and $(Z_{sc})_1$ are impedances measured at End 1 with End 2 open-circuited and short-circuited, respectively. Impedances $(Z_{oc})_2$ and $(Z_{sc})_2$ are corresponding impedances measured from End 2 with End 1 open-circuited or short-circuited.

If the network has negligible dissipation and is symmetrical, a convenient method due to Dawirs[5] can be used. Using this method the network is terminated at one port in a known resistive load R_L and its input impedance $Z_{in} = R_{in} + jX_{in}$ is measured at the other port. Then the image impedance Z_I can be computed from Z_{in} and R_L by the equation[5]

$$Z_I = R_L \sqrt{\frac{\left(\frac{R_{in}}{R_L}\right)^2 + \left(\frac{X_{in}}{R_L}\right)^2 - \left(\frac{R_{in}}{R_L}\right)}{\left(\frac{R_{in}}{R_L}\right) - 1}} \tag{3.09-4}$$

which applies for both the pass and stop bands.

Dawirs[5] has expressed this method in terms of a very useful chart which is reproduced in Fig. 3.09-1. This chart should be thought of as being superimposed on top of a Smith chart[6,7] with the zero "wavelengths toward generator" point coinciding with that of the Smith chart. Then

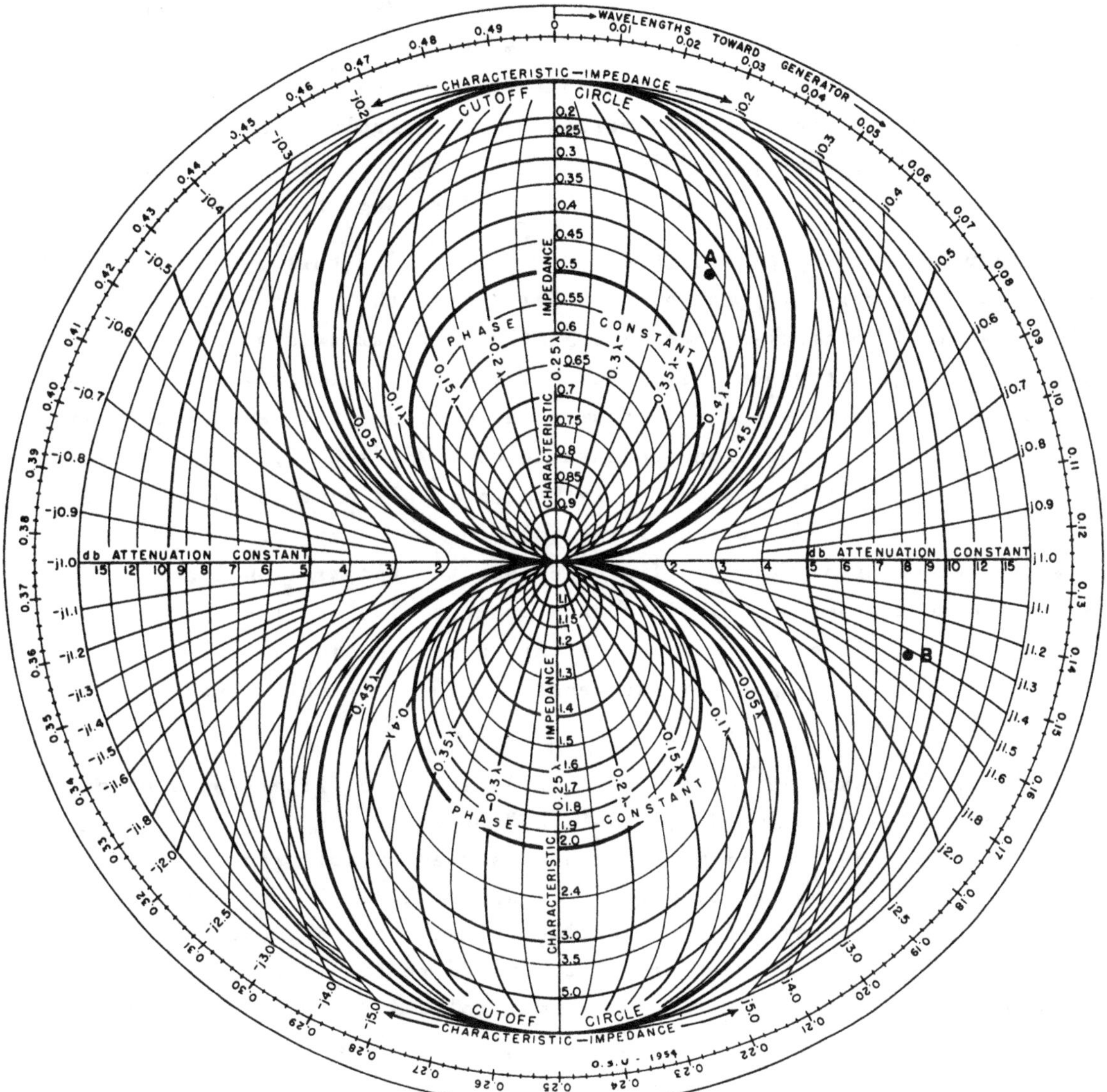

RA-3527-200

SOURCE: By courtesy of H. N. Dawirs and *Proc. IRE*.[5]

FIG. 3.09-1 DAWIRS' CHART FOR DETERMINING THE IMAGE PARAMETERS OF SYMMETRICAL, DISSIPATIONLESS NETWORKS

to obtain the image parameters, Z_{in} measured as discussed above, is normalized with respect to R_L. Next, the point Z_{in}/R_L is first plotted on a Smith chart, and then scaled to the same point on this chart by use of a scale and cursor. In the pass band the Z_{in}/R_L points will fall *within* either of the two heavy circles marked "cutoff circle," while in the stop band the Z_{in}/R_L points will fall *outside* of these circles. Further details of the use of the chart are perhaps best illustrated by examples.

Suppose that $Z_{in}/R_L = 0.20 + j\ 0.25$. Plotting this point on a Smith chart and then rescaling it to this chart gives the point shown at *A* in Fig. 3.09-1. The circles intersecting the vertical axis at right angles give the image impedance while the nearly vertical lines give the phase constant. Following the circle from point *A* around to the vertical axis gives a normalized image impedance value of $R_I/R_L = 0.35$, while the phase constant is seen to be approximately 0.37 λ. This chart uses the term "characteristic impedance" for image impedance and expresses the image phase in wavelengths for specific reference to transmission lines. However, the more general image impedance concept also applies and the corresponding image phase in radians (within some unknown multiple of π) is simply 2π times the number of wavelengths. Thus, in this case $\beta = 0.37(2\pi) + n\pi$ radians.

If Z_{in}/R_L gave the point *B* in Fig. 3.09-1, the filter would be cut off, hence, the image impedance would be imaginary and α would be non-zero. In this case the image impedance is read by following the line to the outer edge of the chart to read $jX_I/R_L = j\ 1.4$, while the image attenuation in db is read from the horizontal axis of the chart as being about 8.5 db. Since the network is specified to be symmetrical, the stop-band image phase will be zero or some multiple of π radians (see Sec. 3.07).

REFERENCES

1. T. E. Shea, *Transmission Networks and Wave Filters* (D. Van Nostrand Co., New York City, 1929).

2. E. A. Guillemin, *Communication Networks*, Vol. 2, Chapters 4, and 7 to 10 (John Wiley and Sons, New York City, 1935).

3. Harvard Radio Research Laboratory Staff, *Very High-Frequency Techniques*, Vol. 2, Chapters 26 and 27 by S. B. Cohn (McGraw-Hill Book Co., Inc., New York City, 1947).

4. M. E. Van Valkenburg, *Network Analysis*, Chapter 13 (Prentice-Hall, Inc., Englewood Cliffs, N.J., 1955).

5. H. N. Dawirs, "A Chart for Analyzing Transmission-Line Filters from Input Impedance Characteristics," *Proc. IRE* **43**, pp. 436-443 (April 1955).

6. P. H. Smith, "A Transmission Line Calculator," *Electronics* **12**, pp. 29-31 (January 1939).

7. E. L. Ginzton, *Microwave Measurements*, pp. 228-234 (McGraw-Hill Book Co., Inc., New York City, 1957).

CHAPTER 4

LOW-PASS PROTOTYPE FILTERS OBTAINED BY NETWORK SYNTHESIS METHODS

SEC. 4.01, INTRODUCTION

Many of the filter design methods to be discussed in later chapters of this book will make use of the lumped-element, low-pass prototype filters discussed in this chapter. Most of the low-pass, high-pass, band-pass, or band-stop microwave filters to be discussed will derive their important transmission characteristics from those of a low-pass prototype filter used in their design. Element values for such low-pass prototype filters were originally obtained by network synthesis methods of Darlington and others.[1-3] However, more recently concise equations[4-11] which are convenient for computer programming have been found for the element values of the types of prototype filters of interest in this book, and numerous filter designs have been tabulated. Some of the tables in this book were obtained from the work of Weinberg,[8,9] while others were computed at Stanford Research Institute for the purposes of this book. No discussion of formal network synthesis methods will be included in this book since these matters are discussed extensively elsewhere (see Refs. 1 to 3, for example), and since the availability of tabulated designs makes such discussion unnecessary. The main objectives of this chapter are to make clear the properties of the tabulated prototype filters, delay networks, and impedance-matching networks so that they may be used intelligently in the solution of a wide variety of microwave circuit design problems of the sorts discussed in Chapter 1.

It should be noted that the step transformers in Chapter 6 can also be used as prototypes for the design of certain types of microwave filters as is discussed in Chapter 9.

SEC. 4.02, COMPARISON OF IMAGE AND NETWORK SYNTHESIS METHODS FOR FILTER DESIGN

As was discussed in Chapter 3, the image impedance and attenuation function of a filter section are defined in terms of an infinite chain of identical filter sections connected together. Using a finite, dissipationless filter network with resistor terminations will permit the image impedances to be matched only at discrete frequencies, and the reflection effects can cause sizeable attenuation in the pass band, as well as distortion of the stop-band edges.

In Sec. 3.08 principles were discussed for the design of end sections which reduce these reflection effects. Although such methods will definitely reduce the size of reflections in filters designed by the image method, they give no assurance as to how large the peak reflection loss values may be in the pass band. Thus, though the image method is conceptually simple, it requires a good deal of "cut and try" or "know how" if a precision design with low pass-band reflection loss and very accurately defined band edges is required.

Network synthesis methods[1,2,3] for filter design generally start out by specifying a transfer function [such as the transmission coefficient t, defined by Eq. (2.10-6)] as a function of complex frequency p. From the transfer function the input impedance to the circuit is found as a function of p. Then, by various continued-fraction or partial-fraction expansion procedures, the input impedance is expanded to give the element values of the circuit. The circuit obtained by these procedures has the same transfer function that was specified at the outset, and all guess work and "cut and try" is eliminated. Image concepts never enter such procedures, and the effects of the terminations are included in the initial specifications of the transfer function.

In general, a low-pass filter designed by the image method and an analogous filter designed for the same application by network synthesis methods will be quite similar. However, the filter designed by network synthesis methods will have somewhat different element values, to give it the specified response.

The Tchebyscheff and maximally flat transfer functions discussed in the next section are often specified for filter applications. The filters whose element values are tabulated in Sec. 4.05 will produce responses discussed in Sec. 4.03 exactly. However, in designing *microwave filters* from low-pass, lumped-element prototypes approximations will be involved. Nevertheless, the approximations will generally be very good over sizeable frequency ranges, and the use of such prototypes in determining the parameters of the microwave filter will eliminate the guess work inherent in the classical image method.

SEC. 4.03, MAXIMALLY FLAT AND TCHEBYSCHEFF FILTER ATTENUATION CHARACTERISTICS

Figure 4.03-1 shows a typical *maximally flat*,* low-pass filter attenuation characteristic. The frequency ω_1', where the attenuation is L_{Ar}, is defined as the pass-band edge. This characteristic is expressed mathematically as

$$L_A(\omega') = 10 \log_{10}\left[1 + \epsilon\left(\frac{\omega'}{\omega_1'}\right)^{2n}\right] \quad \text{db} \tag{4.03-1}$$

where

$$\epsilon = \left[\text{antilog}_{10} \frac{L_{Ar}}{10}\right] - 1 \quad . \tag{4.03-2}$$

The response in Fig. 4.03-1 can be achieved by low-pass filter circuits such as those discussed in Secs. 4.04 and 4.05, and the parameter n in Eq. (4.03-1) corresponds to the number of reactive elements required in the circuit. This attenuation characteristic acquires its name *maximally flat* from the fact that the quantity within the square brackets in Eq. (4.03-1) has $(2n - 1)$ zero derivatives at $\omega' = 0$.

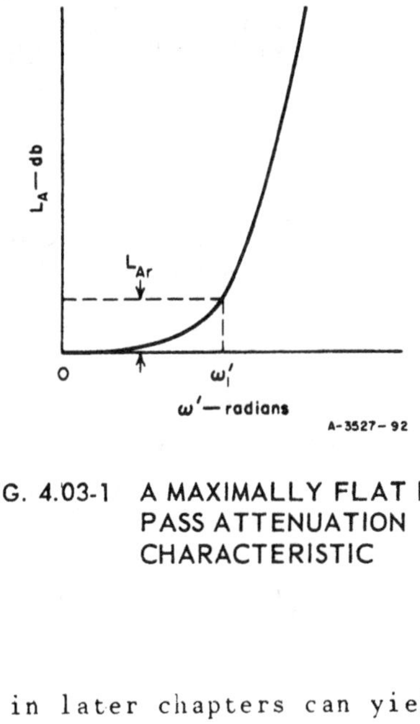

FIG. 4.03-1 A MAXIMALLY FLAT LOW-PASS ATTENUATION CHARACTERISTIC

In most cases ω_1' for maximally flat filters is defined as the 3-db band-edge point. Figure 4.03-2 shows plots of the stop-band attenuation characteristics of maximally flat filters where $L_{Ar} = 3$ db, for $n = 1$ to 15. Note that for convenience in plotting the data $|\omega'/\omega_1'| - 1$ was used for the abscissa. The magnitude sign is used on ω'/ω_1' because the low-pass to band-pass or band-stop mappings to be discussed in later chapters can yield negative values of ω'/ω_1' for which the attenuation is interpreted to be the same as for positive values of ω'/ω_1'.

Another commonly used attenuation characteristic is the *Tchebyscheff* or "equal-ripple" characteristic shown in Fig. 4.03-3. In this case L_{Ar}

---

* This characteristic is also known as a *Butterworth* filter characteristic.

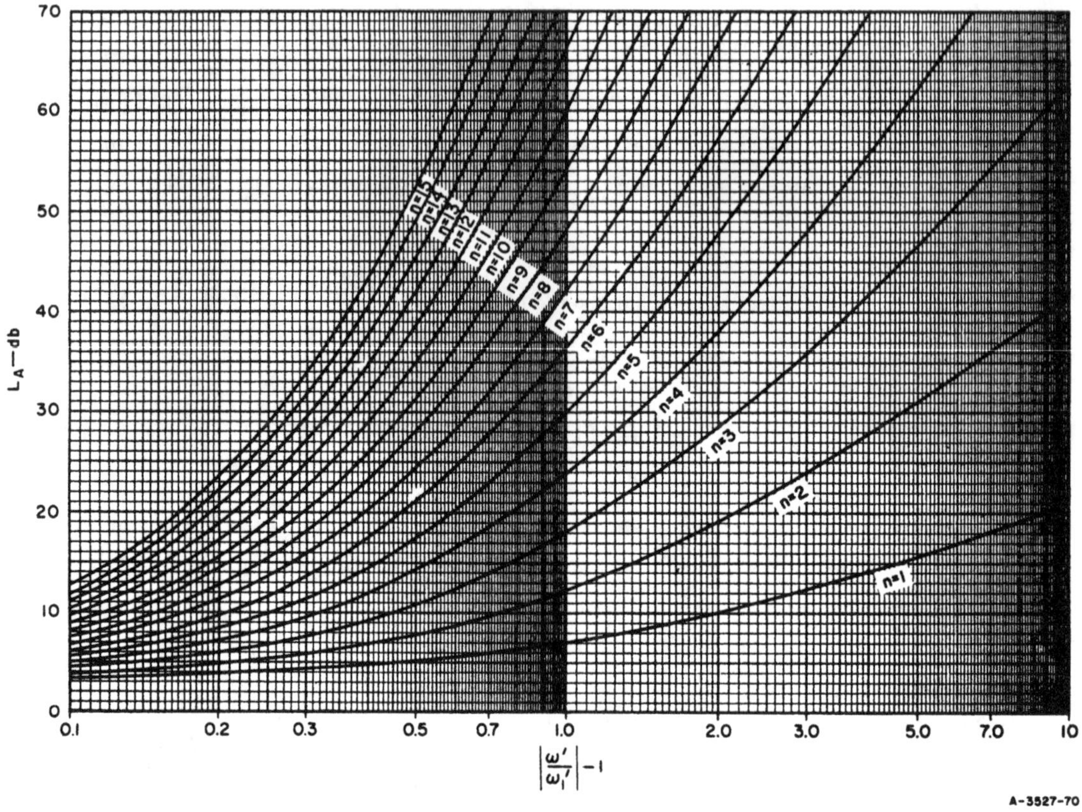

FIG. 4.03-2 ATTENUATION CHARACTERISTICS OF MAXIMALLY FLAT FILTERS
The Frequency ω_1' is the 3-db Band-Edge Point

is again the maximum db attenuation in the pass band, while ω_1' is the equal-ripple band edge. Attenuation characteristics of the form in Fig. 4.03-3 may be specified mathematically as

$$L_A(\omega') = 10 \log_{10} \left\{1 + \epsilon \cos^2\left[n \cos^{-1}\left(\frac{\omega'}{\omega_1'}\right)\right]\right\}_{\omega' \leqq \omega_1'} \quad (4.03\text{-}3)$$

and

$$L_A(\omega') = 10 \log_{10} \left\{1 + \epsilon \cosh^2\left[n \cosh^{-1}\left(\frac{\omega'}{\omega_1'}\right)\right]\right\}_{\omega' \geqq \omega_1'} \quad (4.03\text{-}4)$$

where

$$\epsilon = \left[\text{antilog}_{10}\left(\frac{L_{Ar}}{10}\right)\right] - 1 \quad . \tag{4.03-5}$$

This type of characteristic can also be achieved by the filter structures described in Secs. 4.04 and 4.05, and the parameter n in Eqs. (4.03-3) and (4.03-4) is again the number of reactive elements in the circuit. If n is even there will be $n/2$ frequencies where $L_A = 0$ for a low-pass Tchebyscheff response, while if n is odd there will be $(n + 1)/2$ such frequencies. Figures 4.03-4 to 4.03-10 show the stop-band attenuation characteristics of Tchebyscheff filters having L_{Ar} = 0.01, 0.10, 0.20, 0.50, 1.00, 2.00, and 3.00 db pass-band ripple. Again, $|\omega'/\omega'_1| - 1$ is used as the abscissa.

It is interesting to compare the maximally flat attenuation characteristics in Fig. 4.03-2 with the Tchebyscheff characteristics in Figs. 4.03-4 to 4.03-10. It will be seen that for a given pass-band attenuation tolerance, L_{Ar}, and number of reactive elements, n, that a Tchebyscheff filter will give a much sharper rate of cutoff. For example, the maximally flat characteristics in Fig. 4.03-2 and the Tchebyscheff characteristics in Fig. 4.03-10 both have L_{Ar} = 3 db. For the n = 15 maximally flat case, 70 db attenuation is reached at ω' = 1.7 ω'_1; for the n = 15 Tchebyscheff case, 70 db attenuation is reached at ω' = 1.18 ω'_1. Because of their sharp cutoff, Tchebyscheff characteristics are often preferred over other possible characteristics; however, if the reactive elements of a filter have appreciable dissipation loss the shape of the pass-band response of any type of filter will be altered as compared with the lossless case, and the effects will be particularly large in a Tchebyscheff filter. These matters will be discussed in Sec. 4.13. Maximally flat filters have often been reputed to have less delay distortion than Tchebyscheff

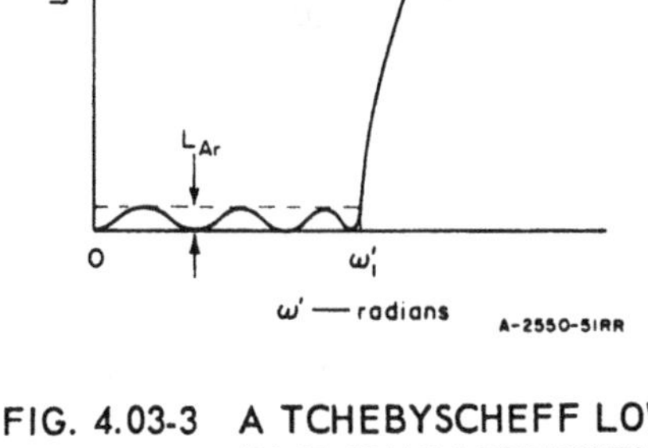

FIG. 4.03-3 A TCHEBYSCHEFF LOW-PASS CHARACTERISTIC

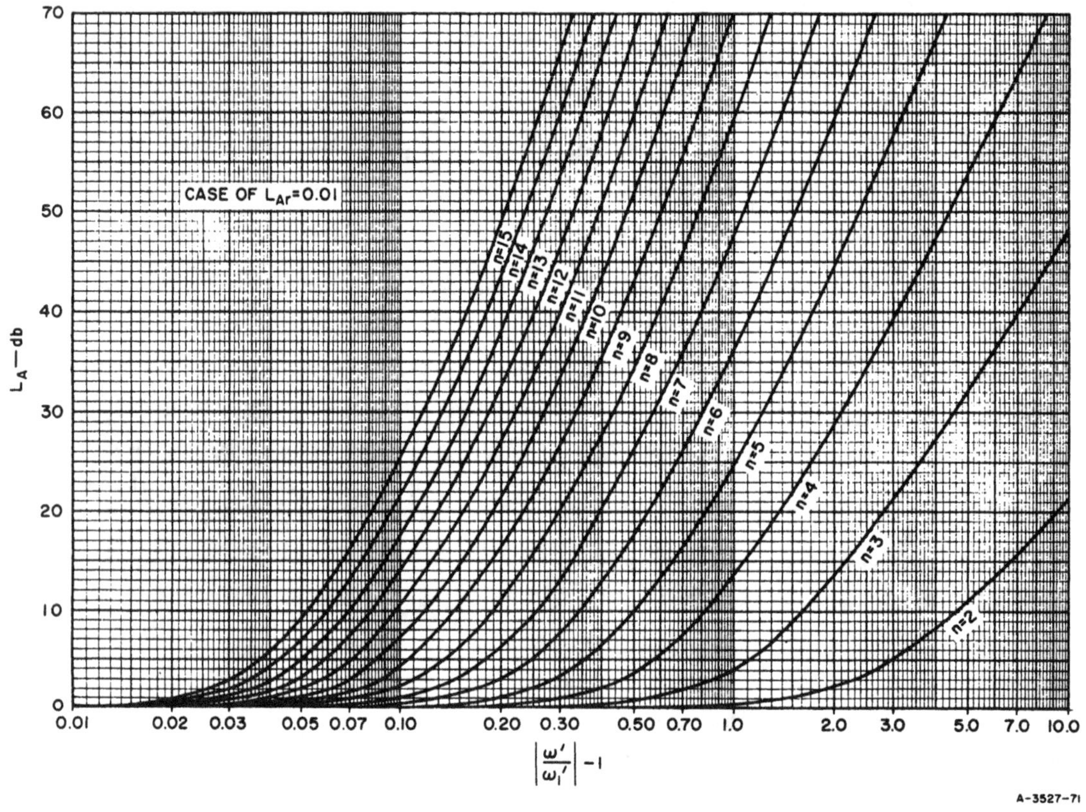

FIG. 4.03-4 0.01-db-RIPPLE TCHEBYSCHEFF FILTER CHARACTERISTICS

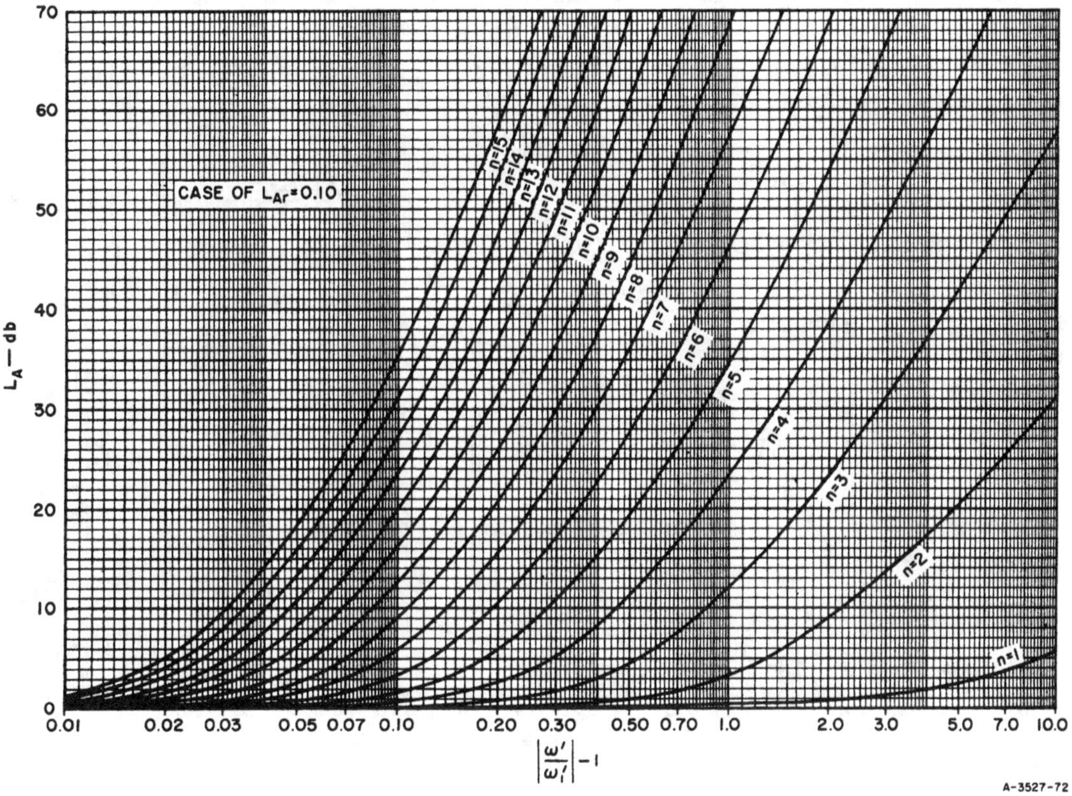

FIG. 4.03-5 0.10-db-RIPPLE TCHEBYSCHEFF FILTER CHARACTERISTICS

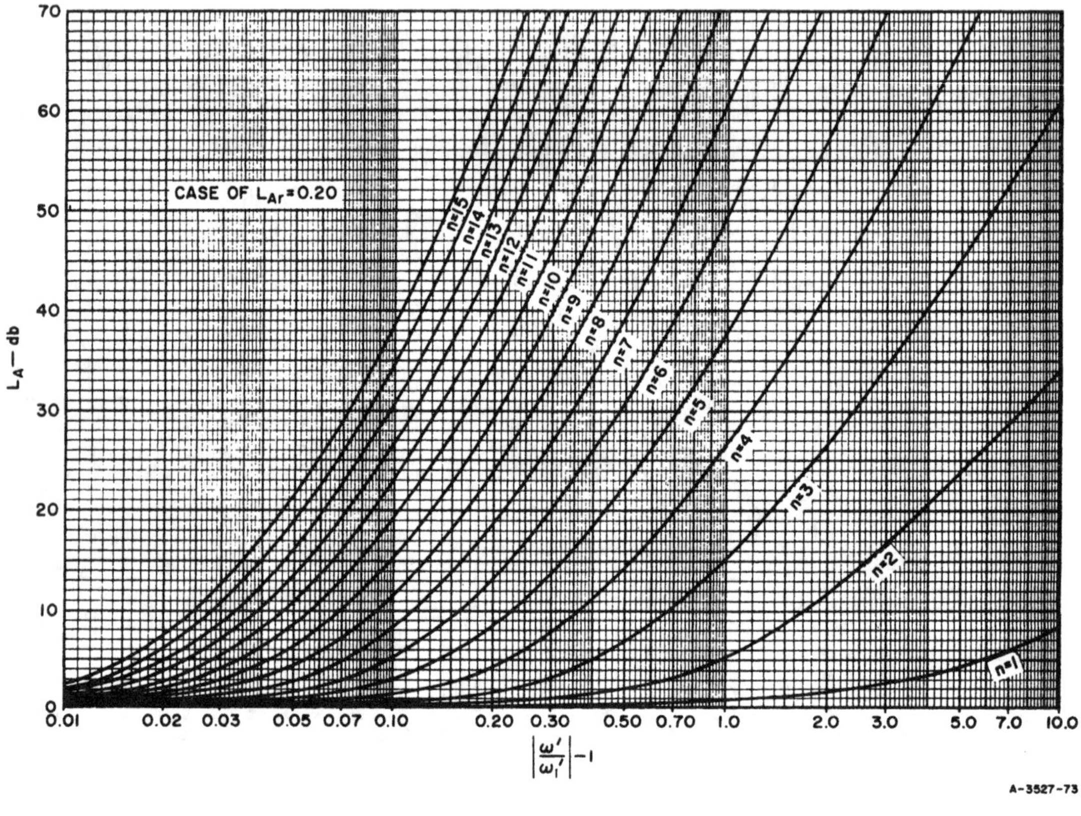

FIG. 4.03-6 0.20-db-RIPPLE TCHEBYSCHEFF FILTER CHARACTERISTICS

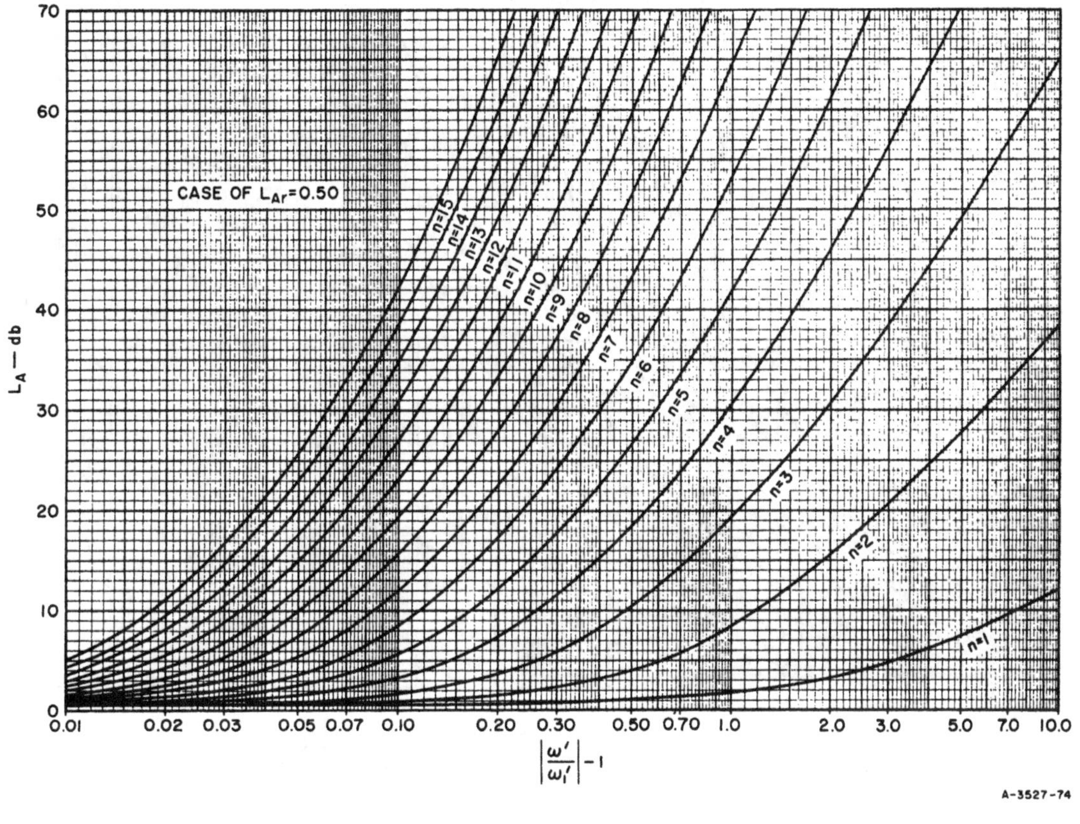

FIG. 4.03-7 0.50-db-RIPPLE TCHEBYSCHEFF FILTER CHARACTERISTICS

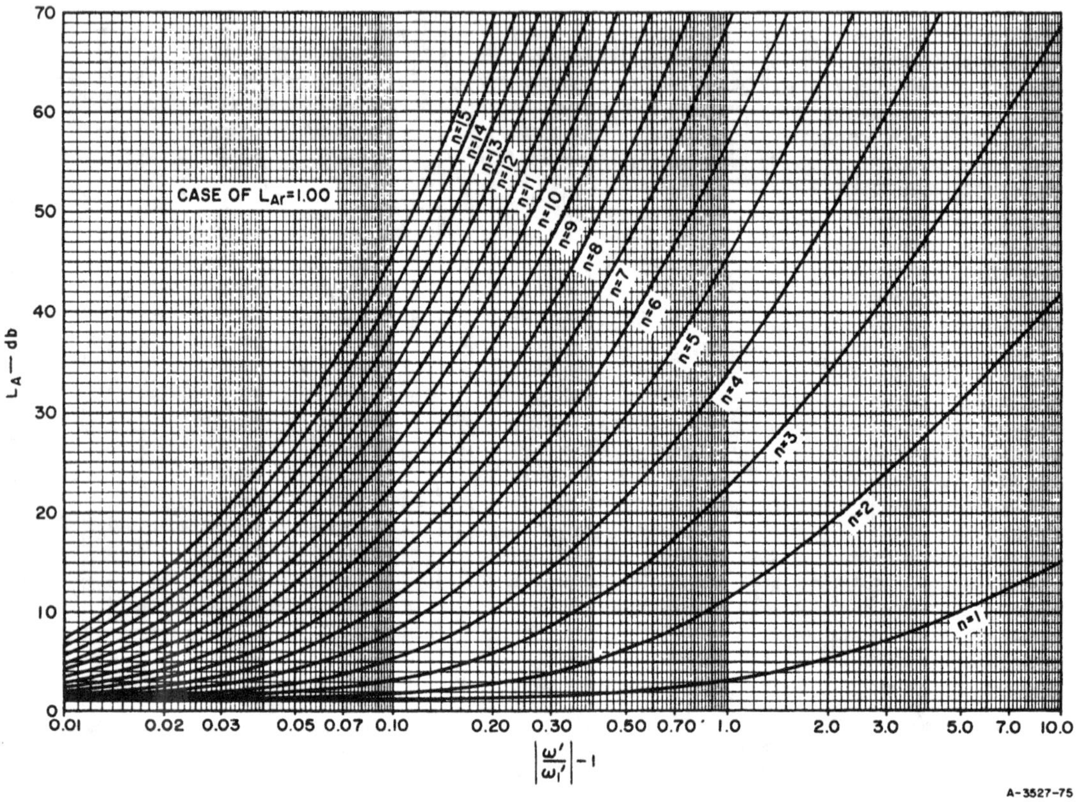

FIG. 4.03-8 1.00-db-RIPPLE TCHEBYSCHEFF FILTER CHARACTERISTICS

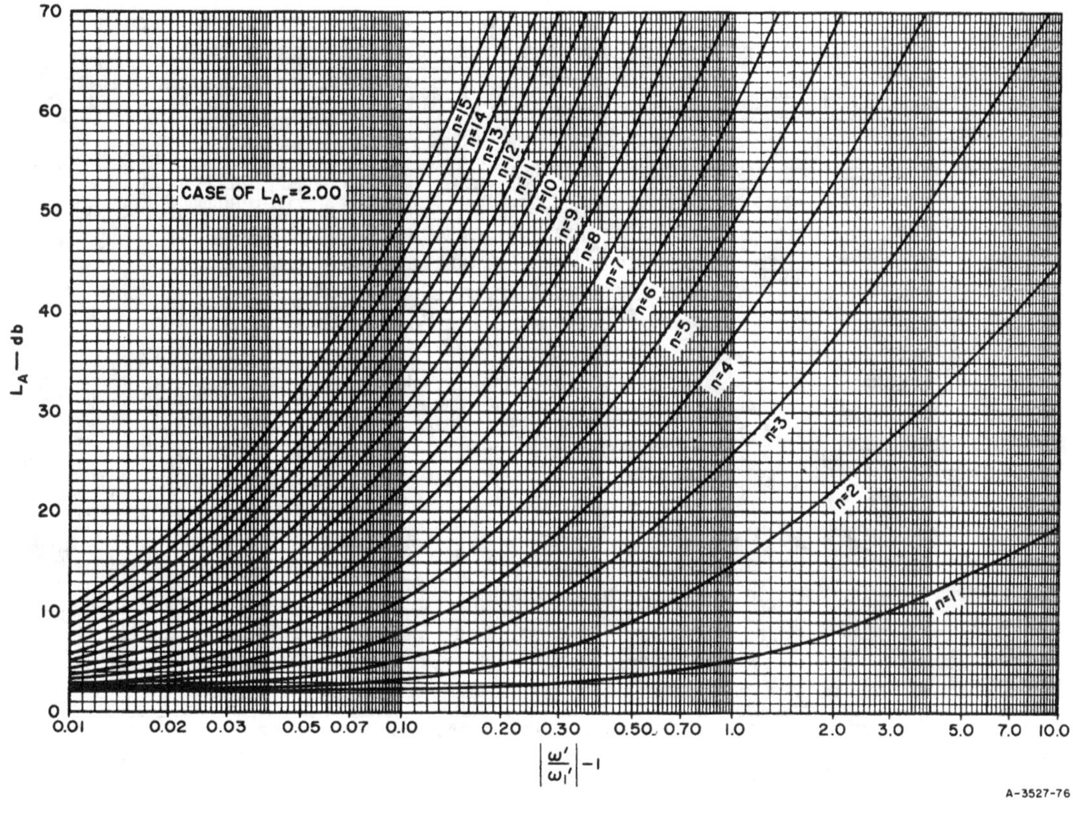

FIG. 4.03-9 2.00-db-RIPPLE TCHEBYSCHEFF FILTER CHARACTERISTICS

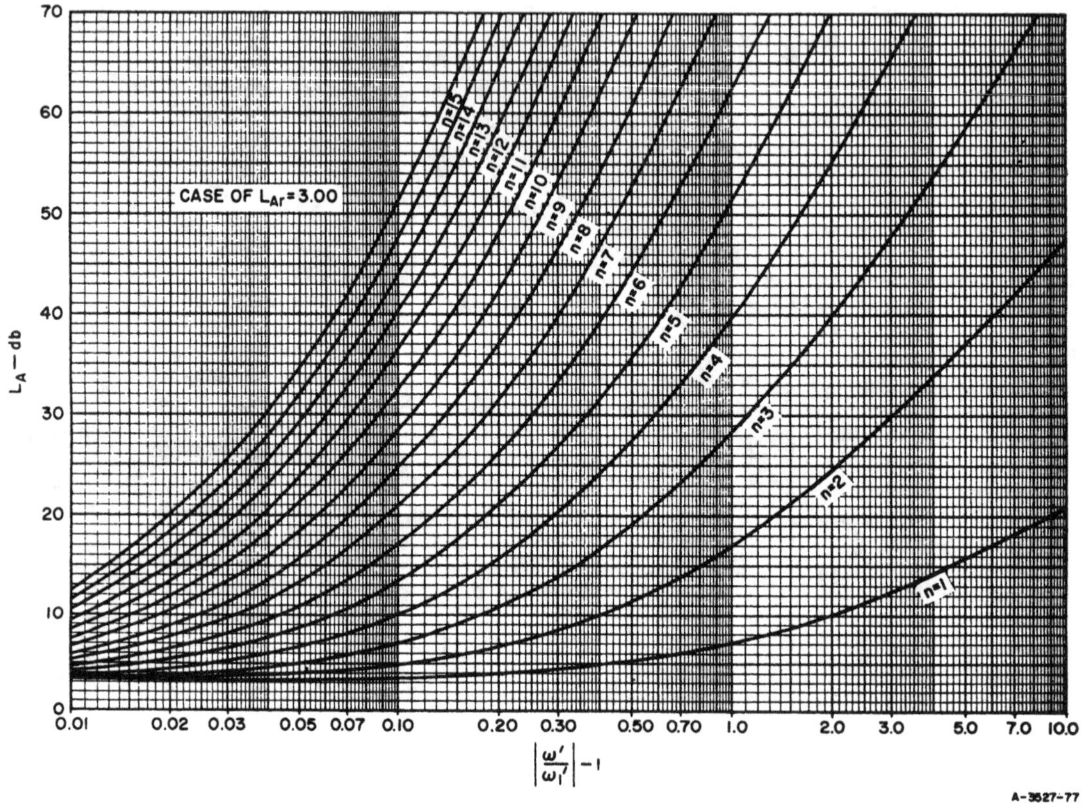

FIG. 4.03-10 3.00-db-RIPPLE TCHEBYSCHEFF FILTER CHARACTERISTICS

filters; however, as discussed in Sec. 4.08, this may not be true, depending on the size of L_{Ar}.

The maximally flat and Tchebyscheff characteristics in Figs. 4.03-1 and 4.03-3 are not the only possible characteristics of this type. For example, the Tchebyscheff characteristics of the impedance-matching-network prototypes to be discussed in Secs. 4.09 and 4.10 will be similar in shape, but L_A will not touch zero at the bottom of the ripples. Sometimes Tchebyscheff filters are designed to have both an equal-ripple characteristic in the pass band, and an "equal-ripple" approximation of a specified attenuation level in the stop band. Although such filters are used at low frequencies, they are very difficult to design precisely for use at microwave frequencies. One possible exception is the type of microwave filter discussed in Sec. 7.03.

SEC. 4.04, DEFINITION OF CIRCUIT PARAMETERS FOR LOW-PASS PROTOTYPE FILTERS

The element values $g_0, g_1, g_2, \ldots, g_n, g_{n+1}$ of the low-pass prototype filters discussed in this chapter are defined as shown in Fig. 4.04-1.

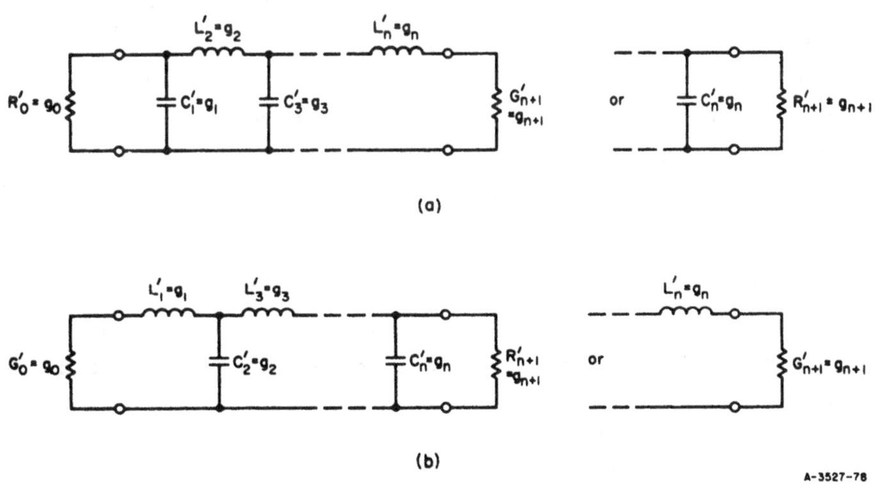

FIG. 4.04-1 DEFINITION OF PROTOTYPE FILTER PARAMETERS $g_0, g_1, g_2, \ldots, g_n, g_{n+1}$
A prototype circuit is shown at (a) and its dual is shown at (b). Either form will give the same response.

One possible form of a prototype filter is shown at (a) while its dual is shown at (b). Either form may be used, since both give identical responses. Since the networks are reciprocal, either the resistor on the left or the one on the right may be defined as the generator internal impedance. It should be noted that in Fig. 4.04-1 the following conventions are observed:

$$g_k\big|_{k=1 \text{ to } n} = \begin{cases} \text{the inductance of a series coil,} \\ \text{or the capacitance of a shunt capacitor} \end{cases}$$

$$g_0 = \begin{cases} \text{the generator resistance } R_0' \text{ if } g_1 = C_1' \text{, but is} \\ \text{defined as the generator conductance } G_0' \text{ if } g_1 = L_1' \end{cases} \qquad (4.04\text{-}1)$$

$$g_{n+1} = \begin{cases} \text{the load resistance } R_{n+1}' \text{ if } g_n = C_n'\text{, but is} \\ \text{defined as the load conductance } G_{n+1}' \text{ if } g_n = L_n' \end{cases}$$

The reason for using these conventions is that they lead to equations of identical form whether a given circuit or its dual is used. Besides the circuit element values, g_k, an additional prototype parameter, ω_1', will also be used. The parameter ω_1' is the radian frequency of the pass-band edge, which is defined in Figs. 4.03-1 and 4.03-3 for maximally flat and Tchebyscheff filters of the sort discussed here. Its definition in the case of maximally flat time-delay filters is discussed in Sec. 4.07.

The element values of the prototype filters discussed in this chapter are all normalized to make $g_0 = 1$ and $\omega_1' = 1$. These prototypes are easily changed to other impedance levels and frequency scales by the following transformations applied to the circuit elements. For resistances or conductances,

$$R = \left(\frac{R_0}{R_0'}\right)R' \quad \text{or} \quad G = \left(\frac{G_0}{G_0'}\right)G \qquad (4.04\text{-}2)$$

For inductances,

$$L = \left(\frac{R_0}{R_0'}\right)\left(\frac{\omega_1'}{\omega_1}\right) L' = \left(\frac{G_0'}{G_0}\right)\left(\frac{\omega_1'}{\omega_1}\right)L' \quad . \qquad (4.04\text{-}3)$$

And, for capacitances,

$$C = \left(\frac{R_0'}{R_0}\right)\left(\frac{\omega_1'}{\omega_1}\right) C' = \left(\frac{G_0}{G_0'}\right)\left(\frac{\omega_1'}{\omega_1}\right) C' \quad . \tag{4.04-4}$$

In these equations the primed quantities are for the normalized prototype and the unprimed quantities are for the corresponding scaled circuit. As indicated from the preceding discussion, for the prototypes in this chapter, $g_0 = R_0' = 1$ or $g_0 = G_0' = 1$.

As an example of how this scaling is accomplished, suppose that we have a low-pass prototype with $R_0' = 1.000$ ohm, $C_1' = 0.8430$ farad, $L_2' = 0.6220$ henry, and $G_3' = 1.3554$ mho. These element values are for a Tchebyscheff filter with 0.10-db ripple and an equal-ripple band edge of $\omega_1' = 1$ radian. [See the case of 0.10-db ripple and $n = 2$ in Table 4.05-2(a).] Assuming that it is desired to scale this prototype so that $R_0 = 50$ ohms and so that the equal-ripple band edge occurs at $f_1 = 1000$ Mc, then $(R_0/R_0') = 50$, and $(\omega_1'/\omega_1) = 1/(2\pi 10^9) = 0.159 \times 10^{-9}$. Next, by Eqs. (4.04-2) to (4.04-4), $R_0 = 50$ ohm, $C_1 = (1/50)\,(0.159 \times 10^{-9})\,(0.8430) = 2.68 \times 10^{-12}$ farad, $L_2 = 50\,(0.159 \times 10^{-9})\,(0.6220) = 4.94 \times 10^{-10}$ henry, and $G_3 = (1/50)\,(1.3554) = 0.0271$ mho.

SEC. 4.05, DOUBLY TERMINATED, MAXIMALLY FLAT AND TCHEBYSCHEFF PROTOTYPE FILTERS

For maximally flat filters having resistor terminations at both ends a response of the form of that in Fig. 4.03-1 with $L_{Ar} = 3$ db, $g_0 = 1$, and $\omega_1' = 1$, the element values may be computed as follows:[5]

$$\begin{aligned} g_0 &= 1 \\ g_k &= 2 \sin\left[\frac{(2k-1)\pi}{2n}\right] , \quad k = 1, 2, \ldots, n \\ g_{n+1} &= 1 \quad . \end{aligned} \tag{4.05-1}$$

Table 4.05-1(a) gives element values for such filters having $n = 1$ to 10 reactive elements, while Table 4.05-1(b) presents corresponding filters with $n = 11$ to 15 reactive elements.

Table 4.05-1(a)

ELEMENT VALUES FOR FILTERS WITH MAXIMALLY FLAT ATTENUATION HAVING $g_0 = 1$, $\omega_1' = 1$, and $n = 1$ to 10
The responses are of the form in Fig. 4.03-1 with $L_{Ar} = 3$ db

VALUE OF n	g_1	g_2	g_3	g_4	g_5	g_6	g_7	g_8	g_9	g_{10}	g_{11}
1	2.000	1.000									
2	1.414	1.414	1.000								
3	1.000	2.000	1.000	1.000							
4	0.7654	1.848	1.848	0.7654	1.000						
5	0.6180	1.618	2.000	1.618	0.6180	1.000					
6	0.5176	1.414	1.932	1.932	1.414	0.5176	1.000				
7	0.4450	1.247	1.802	2.000	1.802	1.247	0.4450	1.000			
8	0.3902	1.111	1.663	1.962	1.962	1.663	1.111	0.3902	1.000		
9	0.3473	1.000	1.532	1.879	2.000	1.879	1.532	1.000	0.3473	1.000	
10	0.3129	0.9080	1.414	1.782	1.975	1.975	1.782	1.414	0.9080	0.3129	1.000

Table 4.05-1(b)

ELEMENT VALUES FOR FILTERS WITH MAXIMALLY FLAT ATTENUATION HAVING $g_0 = 1$, $\omega_1' = 1$, and $n = 11$ to 15
The responses are of the form in Fig. 4.03-1 with $L_{Ar} = 3$ db

VALUE OF n	g_1	g_2	g_3	g_4	g_5	g_6	g_7	g_8
11	0.2846	0.8308	1.3097	1.6825	1.9189	2.0000	1.9189	1.6825
12	0.2610	0.7653	1.2175	1.5867	1.8477	1.9828	1.9828	1.8477
13	0.2410	0.7092	1.1361	1.4970	1.7709	1.9418	2.0000	1.9418
14	0.2239	0.6605	1.0640	1.4142	1.6934	1.8877	1.9874	1.9874
15	0.2090	0.6180	1.0000	1.3382	1.6180	1.8270	1.9563	2.0000

	g_9	g_{10}	g_{11}	g_{12}	g_{13}	g_{14}	g_{15}	g_{16}
11	1.3097	0.8308	0.2846	1.0000				
12	1.5867	1.2175	0.7653	0.2610	1.0000			
13	1.7709	1.4970	1.1361	0.7092	0.2410	1.0000		
14	1.8877	1.6934	1.4142	1.0640	0.6605	0.2239	1.0000	
15	1.9563	1.8270	1.6180	1.3382	1.0000	0.6180	0.2090	1.0000

For Tchebyscheff filters having resistor terminations at both ends, with responses of the form shown in Fig. 4.03-3 having L_{Ar} db pass-band ripple, $g_0 = 1$, and $\omega_1' = 1$, the element values may be computed as follows:[4,5] first compute

$$\beta = ln\left(\coth \frac{L_{Ar}}{17.37}\right)$$

$$\gamma = \sinh\left(\frac{\beta}{2n}\right)$$

$$a_k = \sin\left[\frac{(2k-1)\pi}{2n}\right] \quad , \quad k = 1, 2, \ldots, n$$

$$b_k = \gamma^2 + \sin^2\left(\frac{k\pi}{n}\right) \quad , \quad k = 1, 2, \ldots, n \qquad (4.05\text{-}2)$$

then compute

$$g_1 = \frac{2a_1}{\gamma}$$

$$g_k = \frac{4a_{k-1}a_k}{b_{k-1}g_{k-1}} \quad , \qquad k = 2, 3, \ldots, n$$

$$g_{n+1} = 1 \text{ for } n \text{ odd}$$

$$= \coth^2\left(\frac{\beta}{4}\right) \text{ for } n \text{ even} \quad .$$

Table 4.05-2(a) gives element values for such filters for various L_{Ar} and $n = 1$ to 10 reactive elements. Table 4.05-2(b) gives corresponding data for filters having $n = 11$ to 15 reactive elements.

It will be noted that all of the filter prototypes discussed in this section are symmetrical if n is odd. If n is even, they have the property of antimetry mentioned in Secs. 2.11 and 3.07. Under this condition one half of the network is the reciprocal of the other half of the network with respect to a positive real constant R_h, where R_h may be defined as

$$R_h = \sqrt{R_0' R_{n+1}'} \qquad (4.05\text{-}3)$$

Table 4.05-2(a)

ELEMENT VALUES FOR TCHEBYSCHEFF FILTERS HAVING $g_0 = 1$, $\omega_1' = 1$, AND RESPONSES OF THE FORM IN FIG. 4.03-3 WITH VARIOUS db RIPPLE

Cases of n = 1 to 10

VALUE OF n	g_1	g_2	g_3	g_4	g_5	g_6	g_7	g_8	g_9	g_{10}	g_{11}
0.01 db ripple											
1	0.0960	1.0000									
2	0.4488	0.4077	1.1007								
3	0.6291	0.9702	0.6291	1.0000							
4	0.7128	1.2003	1.3212	0.6476	1.1007						
5	0.7563	1.3049	1.5773	1.3049	0.7563	1.0000					
6	0.7813	1.3600	1.6896	1.5350	1.4970	0.7098	1.1007				
7	0.7969	1.3924	1.7481	1.6331	1.7481	1.3924	0.7969	1.0000			
8	0.8072	1.4130	1.7824	1.6833	1.8529	1.6193	1.5554	0.7333	1.1007		
9	0.8144	1.4270	1.8043	1.7125	1.9057	1.7125	1.8043	1.4270	0.8144	1.0000	
10	0.8196	1.4369	1.8192	1.7311	1.9362	1.7590	1.9055	1.6527	1.5817	0.7446	1.1007
0.1 db ripple											
1	0.3052	1.0000									
2	0.8430	0.6220	1.3554								
3	1.0315	1.1474	1.0315	1.0000							
4	1.1088	1.3061	1.7703	0.8180	1.3554						
5	1.1468	1.3712	1.9750	1.3712	1.1468	1.0000					
6	1.1681	1.4039	2.0562	1.5170	1.9029	0.8618	1.3554				
7	1.1811	1.4228	2.0966	1.5733	2.0966	1.4228	1.1811	1.0000			
8	1.1897	1.4346	2.1199	1.6010	2.1699	1.5640	1.9444	0.8778	1.3554		
9	1.1956	1.4425	2.1345	1.6167	2.2053	1.6167	2.1345	1.4425	1.1956	1.0000	
10	1.1999	1.4481	2.1444	1.6265	2.2253	1.6418	2.2046	1.5821	1.9628	0.8853	1.3554
0.2 db ripple											
1	0.4342	1.0000									
2	1.0378	0.6745	1.5386								
3	1.2275	1.1525	1.2275	1.0000							
4	1.3028	1.2844	1.9761	0.8468	1.5386						
5	1.3394	1.3370	2.1660	1.3370	1.3394	1.0000					
6	1.3598	1.3632	2.2394	1.4555	2.0974	0.8838	1.5386				
7	1.3722	1.3781	2.2756	1.5001	2.2756	1.3781	1.3722	1.0000			
8	1.3804	1.3875	2.2963	1.5217	2.3413	1.4925	2.1349	0.8972	1.5386		
9	1.3860	1.3938	2.3093	1.5340	2.3728	1.5340	2.3093	1.3938	1.3860	1.0000	
10	1.3901	1.3983	2.3181	1.5417	2.3904	1.5536	2.3720	1.5066	2.1514	0.9034	1.5386
0.5 db ripple											
1	0.6986	1.0000									
2	1.4029	0.7071	1.9841								
3	1.5963	1.0967	1.5963	1.0000							
4	1.6703	1.1926	2.3661	0.8419	1.9841						
5	1.7058	1.2296	2.5408	1.2296	1.7058	1.0000					
6	1.7254	1.2479	2.6064	1.3137	2.4758	0.8696	1.9841				
7	1.7372	1.2583	2.6381	1.3444	2.6381	1.2583	1.7372	1.0000			
8	1.7451	1.2647	2.6564	1.3590	2.6964	1.3389	2.5093	0.8796	1.9841		
9	1.7504	1.2690	2.6678	1.3673	2.7239	1.3673	2.6678	1.2690	1.7504	1.0000	
10	1.7543	1.2721	2.6754	1.3725	2.7392	1.3806	2.7231	1.3485	2.5239	0.8842	1.9841

Table 4.05-2(a) *Concluded*

VALUE OF n	g_1	g_2	g_3	g_4	g_5	g_6	g_7	g_8	g_9	g_{10}	g_{11}
1.0 db ripple											
1	1.0177	1.0000									
2	1.8219	0.6850	2.6599								
3	2.0236	0.9941	2.0236	1.0000							
4	2.0991	1.0644	2.8311	0.7892	2.6599						
5	2.1349	1.0911	3.0009	1.0911	2.1349	1.0000					
6	2.1546	1.1041	3.0634	1.1518	2.9367	0.8101	2.6599				
7	2.1664	1.1116	3.0934	1.1736	3.0934	1.1116	2.1664	1.0000			
8	2.1744	1.1161	3.1107	1.1839	3.1488	1.1696	2.9685	0.8175	2.6599		
9	2.1797	1.1192	3.1215	1.1897	3.1747	1.1897	3.1215	1.1192	2.1797	1.0000	
10	2.1836	1.1213	3.1286	1.1933	3.1890	1.1990	3.1738	1.1763	2.9824	0.8210	2.6599
2.0 db ripple											
1	1.5296	1.0000									
2	2.4881	0.6075	4.0957								
3	2.7107	0.8327	2.7107	1.0000							
4	2.7925	0.8806	3.6063	0.6819	4.0957						
5	2.8310	0.8985	3.7827	0.8985	2.8310	1.0000					
6	2.8521	0.9071	3.8467	0.9393	3.7151	0.6964	4.0957				
7	2.8655	0.9119	3.8780	0.9535	3.8780	0.9119	2.8655	1.0000			
8	2.8733	0.9151	3.8948	0.9605	3.9335	0.9510	3.7477	0.7016	4.0957		
9	2.8790	0.9171	3.9056	0.9643	3.9598	0.9643	3.9056	0.9171	2.8790	1.0000	
10	2.8831	0.9186	3.9128	0.9667	3.9743	0.9704	3.9589	0.9554	3.7619	0.7040	4.0957
3.0 db ripple											
1	1.9953	1.0000									
2	3.1013	0.5339	5.8095								
3	3.3487	0.7117	3.3487	1.0000							
4	3.4389	0.7483	4.3471	0.5920	5.8095						
5	3.4817	0.7618	4.5381	0.7618	3.4817	1.0000					
6	3.5045	0.7685	4.6061	0.7929	4.4641	0.6033	5.8095				
7	3.5182	0.7723	4.6386	0.8039	4.6386	0.7723	3.5182	1.0000			
8	3.5277	0.7745	4.6575	0.8089	4.6990	0.8018	4.4990	0.6073	5.8095		
9	3.5340	0.7760	4.6692	0.8118	4.7272	0.8118	4.6692	0.7760	3.5340	1.0000	
10	3.5384	0.7771	4.6768	0.8136	4.7425	0.8164	4.7260	0.8051	4.5142	0.6091	5.8095

Table 4.05-2(b)

ELEMENT VALUES FOR TCHEBYSCHEFF FILTERS HAVING $g_0 = 1$, $\omega_1' = 1$, AND RESPONSES OF THE FORM IN FIG. 4.03-3 WITH VARIOUS db RIPPLE.

Cases of n = 11 to 15

VALUE OF n	g_1	g_2	g_3	g_4	g_5	g_6	g_7	g_8	g_9	g_{10}	g_{11}	g_{12}	g_{13}	g_{14}	g_{15}	g_{16}
0.01 db ripple																
11	0.8234	1.4442	1.8298	1.7437	1.9554	1.7856	1.9554	1.7437	1.8298	1.4442	0.8234	1.0000				
12	0.8264	1.4497	1.8377	1.7527	1.9684	1.8022	1.9837	1.7883	1.9293	1.6695	1.5957	0.7508	1.1007			
13	0.8287	1.4540	1.8437	1.7594	1.9777	1.8134	2.0014	1.8134	1.9777	1.7594	1.8437	1.4540	0.8287	1.0000		
14	0.8305	1.4573	1.8483	1.7644	1.9845	1.8214	2.0132	1.8290	2.0048	1.8029	1.9422	1.6792	1.6041	0.7545	1.1007	
15	0.8320	1.4600	1.8520	1.7684	1.9897	1.8272	2.0216	1.8394	2.0216	1.8272	1.9897	1.7684	1.8520	1.4600	0.8320	1.0000
0.10 db ripple																
11	1.2031	1.4523	2.1515	1.6332	2.2378	1.6559	2.2378	1.6332	2.1515	1.4523	1.2031	1.0000				
12	1.2055	1.4554	2.1566	1.6379	2.2462	1.6646	2.2562	1.6572	2.2200	1.5912	1.9726	0.8894	1.3554			
13	1.2074	1.4578	2.1605	1.6414	2.2521	1.6704	2.2675	1.6704	2.2521	1.6414	2.1605	1.4578	1.2074	1.0000		
14	1.2089	1.4596	2.1636	1.6441	2.2564	1.6745	2.2751	1.6786	2.2696	1.6648	2.2283	1.5963	1.9784	0.8919	1.3554	
15	1.2101	1.4612	2.1660	1.6461	2.2598	1.6776	2.2804	1.6839	2.2804	1.6776	2.2598	1.6461	2.1660	1.4612	1.2101	1.0000
0.20 db ripple																
11	1.3931	1.4015	2.3243	1.5469	2.4014	1.5646	2.4014	1.5469	2.3243	1.4015	1.3931	1.0000				
12	1.3954	1.4040	2.3289	1.5505	2.4088	1.5713	2.4176	1.5656	2.3856	1.5136	2.1601	0.9069	1.5386			
13	1.3972	1.4059	2.3323	1.5532	2.4140	1.5758	2.4276	1.5758	2.4140	1.5532	2.3323	1.4059	1.3972	1.0000		
14	1.3986	1.4073	2.3350	1.5553	2.4178	1.5790	2.4342	1.5821	2.4294	1.5714	2.3929	1.5176	2.1653	0.9089	1.5386	
15	1.3997	1.4085	2.3371	1.5569	2.4207	1.5813	2.4388	1.5862	2.4388	1.5813	2.4207	1.5569	2.3371	1.4085	1.3997	1.0000
0.50 db ripple																
11	1.7572	1.2743	2.6809	1.3759	2.7488	1.3879	2.7488	1.3759	2.6809	1.2743	1.7572	1.0000				
12	1.7594	1.2760	2.6848	1.3784	2.7551	1.3925	2.7628	1.3886	2.7349	1.3532	2.5317	0.8867	1.9841			
13	1.7610	1.2772	2.6878	1.3802	2.7596	1.3955	2.7714	1.3955	2.7596	1.3802	2.6878	1.2772	1.7610	1.0000		
14	1.7624	1.2783	2.6902	1.3816	2.7629	1.3976	2.7771	1.3997	2.7730	1.3925	2.7412	1.3558	2.5362	0.8882	1.9841	
15	1.7635	1.2791	2.6920	1.3826	2.7654	1.3991	2.7811	1.4024	2.7811	1.3991	2.7654	1.3826	2.6920	1.2791	1.7635	1.0000
1.00 db ripple																
11	2.1865	1.1229	3.1338	1.1957	3.1980	1.2041	3.1980	1.1957	3.1338	1.1229	2.1865	1.0000				
12	2.1887	1.1241	3.1375	1.1974	3.2039	1.2073	3.2112	1.2045	3.1849	1.1796	2.9898	0.8228	2.6599			
13	2.1904	1.1250	3.1403	1.1987	3.2081	1.2094	3.2192	1.2094	3.2081	1.1987	3.1403	1.1250	2.1904	1.0000		
14	2.1917	1.1257	3.1425	1.1996	3.2112	1.2108	3.2245	1.2123	3.2207	1.2073	3.1908	1.1815	2.9944	0.8239	2.6599	
15	2.1928	1.1263	3.1442	1.2004	3.2135	1.2119	3.2282	1.2142	3.2282	1.2119	3.2135	1.2004	3.1442	1.1263	2.1928	1.0000
2.00 db ripple																
11	2.8863	0.9195	3.9181	0.9682	3.9834	0.9737	3.9834	0.9682	3.9181	0.9195	2.8863	1.0000				
12	2.8886	0.9203	3.9219	0.9693	3.9894	0.9758	3.9967	0.9740	3.9701	0.9575	3.7695	0.7052	4.0957			
13	2.8904	0.9209	3.9247	0.9701	3.9936	0.9771	4.0048	0.9771	3.9936	0.9701	3.9247	0.9209	2.8904	1.0000		
14	2.8919	0.9214	3.9269	0.9707	3.9967	0.9781	4.0101	0.9791	4.0062	0.9758	3.9761	0.9587	3.7739	0.7060	4.0957	
15	2.8930	0.9218	3.9287	0.9712	3.9990	0.9788	4.0139	0.9803	4.0139	0.9788	3.9990	0.9712	3.9287	0.9218	2.8930	1.0000
3.00 db ripple																
11	3.5420	0.7778	4.6825	0.8147	4.7523	0.8189	4.7523	0.8147	4.6825	0.7778	3.5420	1.0000				
12	3.5445	0.7784	4.6865	0.8155	4.7587	0.8204	4.7664	0.8191	4.7381	0.8067	4.5224	0.6101	5.8095			
13	3.5465	0.7789	4.6896	0.8162	4.7631	0.8214	4.7751	0.8214	4.7631	0.8162	4.6896	0.7789	3.5465	1.0000		
14	3.5480	0.7792	4.6919	0.8166	4.7664	0.8222	4.7808	0.8229	4.7766	0.8204	4.7444	0.8076	4.5272	0.6107	5.8095	
15	3.5493	0.7795	4.6938	0.8170	4.7689	0.8227	4.7847	0.8238	4.7847	0.8227	4.7689	0.8170	4.6938	0.7795	3.5493	1.0000

and where R'_0 and R'_{n+1} are the resistances of the terminations at the ends of the filter. If Z'_k is the impedance of one branch of the filter ladder network, then

$$Z'_{n+1-k} = \frac{R_h^2}{Z'_k} \qquad (4.05\text{-}4)$$

where Z'_{n+1-k} is the dual branch at the other end of the filter. By Eq. (4.05-4) it will be seen that the inductive reactances at one end of the filter are related to the capacitive susceptances at the other end by

$$\omega C'_{n+1-k} = \frac{\omega L'_k}{R_h^2} \quad . \qquad (4.05\text{-}5)$$

Also,

$$\omega L'_{n+1-k} = R_h^2 \omega C'_k \qquad (4.05\text{-}6)$$

so that it is possible to obtain the element values of the second half of the filter from those of the first half if the filter is antimetrical, (as well as when the filter is symmetrical).

It will be found that the symmetry and antimetry properties discussed above will occur in maximally flat and Tchebyscheff filters of the form in Fig. 4.04-1 having terminations at both ends, provided that the filter is designed so that $L_A = 0$ at one or more frequencies in the pass band as shown in Figs. 4.03-1 and 4.03-3. The maximally flat and Tchebyscheff filters discussed in Secs. 4.06, 4.09, and 4.10 do not have this property. The maximally flat time-delay filters in Sec. 4.07 are not symmetrical or antimetrical, even though $L_A = 0$ at $\omega' = 0$.

In some rare cases designs with n greater than 15 may be desired. In such cases good approximate designs can be obtained by augmenting an $n = 14$ or $n = 15$ design by repeating the two middle elements of the filter. Thus, suppose that an $n = 18$ design is desired. An $n = 14$ design can be augmented to $n = 18$ by breaking the circuit immediately following the g_7 element, repeating elements g_6 and g_7 twice, and then continuing on with element g_8 and the rest of the elements. Thus, letting primed g's indicate element values for the $n = 18$ filter, and unprimed g's indicate element values from the $n = 14$ design, the $n = 18$ design would have the element values

$$g'_0 = g_0 \;,\quad g'_1 = g_1 \;,\quad g'_2 = g_2, \ldots, g'_6 = g_6 \;,$$

$$g'_7 = g_7 \;,\quad g'_8 = g_6 \;,\quad g'_9 = g_7 \;,\quad g'_{10} = g_6 \;,\quad g'_{11} = g_7$$

$$g'_{12} = g_8 \;,\quad g'_{13} = g_9, \ldots, \quad g'_{18} = g_{14} \;,\quad g'_{19} = g_{15} \;.$$

This is, of course, an approximate procedure, but it is based on the fact that for a given Tchebyscheff ripple the element values in a design change very little as n is varied, once n is around 10 or more. This is readily seen by comparing the element values for different values of n, down the columns at the left in Table 4.05-2(b).

SEC. 4.06, SINGLY TERMINATED MAXIMALLY FLAT AND TCHEBYSCHEFF FILTERS

All of the prototype filters discussed in Sec. 4.05 have resistor terminations at both ends. However, in some cases it is desirable to use filters with a resistor termination at one end only. Figure 4.06-1 shows an example of such a filter with a resistor termination on the left and a zero internal impedance voltage generator on the right to drive the circuit. In this case the attenuation L_A defined by Eq. (2.11-4) does not apply, since a zero internal impedance voltage generator has infinite available power. The power absorbed by the circuit is

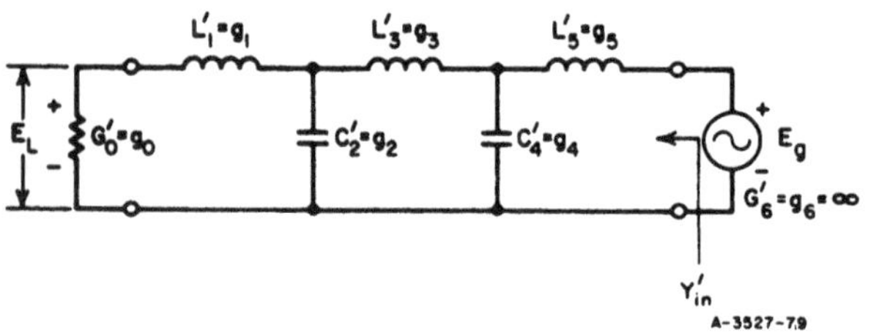

FIG. 4.06-1 AN n = 5 REACTIVE ELEMENT SINGLY TERMINATED FILTER DRIVEN BY A ZERO-IMPEDANCE VOLTAGE GENERATOR

$$P = |E_g|^2 \text{ Re } Y'_{in} \tag{4.06-1}$$

where Y'_{in} and E_g are defined in the Fig. 4.06-1. Since all of the power must be absorbed in G'_0,

$$|E_g|^2 \text{ Re } Y'_{in} = |E_L|^2 G'_0 \tag{4.06-2}$$

and

$$\left|\frac{E_g}{E_L}\right| = \sqrt{\frac{G'_0}{\text{Re } Y'_{in}}} \quad . \tag{4.06-3}$$

Thus in this case it is convenient to use the *voltage attenuation* function

$$L_E = 20 \log_{10} \left| \frac{E_g}{E_L} \right| = 10 \log_{10} \frac{G'_0}{\text{Re } Y'_{in}} \qquad \text{db} \quad . \qquad (4.06\text{-}4)$$

Figure 4.06-2 shows the dual case to that in Fig. 4.06-1. In this latter case the circuit is driven by an *infinite-impedance current generator* and it is convenient to use the *current attenuation* function defined as

$$L_I = 20 \log_{10} \left| \frac{I_g}{I_L} \right| = 10 \log_{10} \frac{R'_0}{\text{Re } Z'_{in}} \qquad \text{db} \qquad (4.06\text{-}5)$$

where I_g, I_L, R'_0, and Z'_{in} are as defined in Fig. 4.06-2. If L_A and L_{Ar} in Sec. 4.03 are replaced by analogous quantities L_E and L_{Er}, or L_I and L_{Ir}, all of the equations and charts in Sec. 4.03 apply to the singly terminated maximally flat or Tchebyscheff filters of this section as well as to the doubly terminated filters in Sec. 4.05.

Equation (4.06-1) shows that for a given generator voltage, E_g, the power transmission through the filter is controlled entirely by Re Y'_{in}. Thus, if the filter in Fig. 4.06-1 is to have a maximally flat or Tchebyscheff transmission characteristic, Re Y'_{in} must also have such a characteristic. Figure 4.06-3 shows the approximate shape of Re Y'_{in} and Im Y'_{in} for the circuit in Fig. 4.06-1 if designed to give a Tchebyscheff transmission characteristic. The curves in Fig. 4.06-3 also apply to the circuit in Fig. 4.06-2 if Y'_{in} is replaced by Z'_{in}. As will be discussed in Chapter 16, this property of Re Y'_{in} or Re Z'_{in} for singly loaded filters makes them quite useful in the design of diplexers and multiplexers. Prototypes of this sort will also be useful for the design of filters to be driven by energy sources that look approximately like a zero-impedance voltage generator or an infinite-impedance current generator. A typical example is a pentode tube which, from its plate circuit may resemble a current generator with a capacitor in parallel. In such cases a broadband response can be obtained if the shunt capacitance is used as the first element of a singly terminated filter.

Orchard[5] gives formulas for singly terminated maximally flat filters normalized so that $g_0 = 1$, and $\omega'_1 = 1$ at the band-edge point where $L_I = L_{Ir}$ or $L_E = L_{Er}$ is 3 db. They may be written as follows:

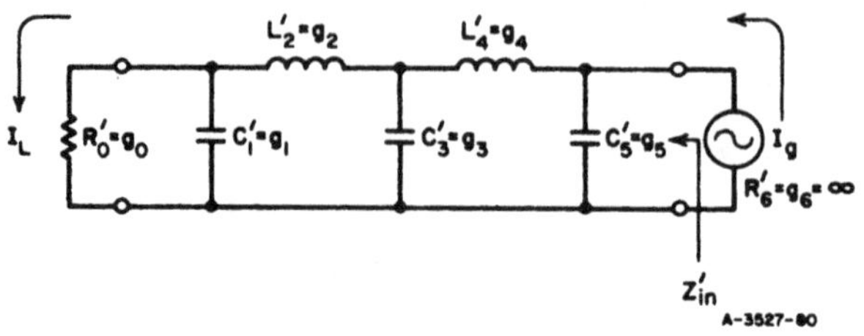

FIG. 4.06-2 THE DUAL CIRCUIT TO THAT IN FIG. 4.06-1
In this case the generator is an infinite-impedance current generator.

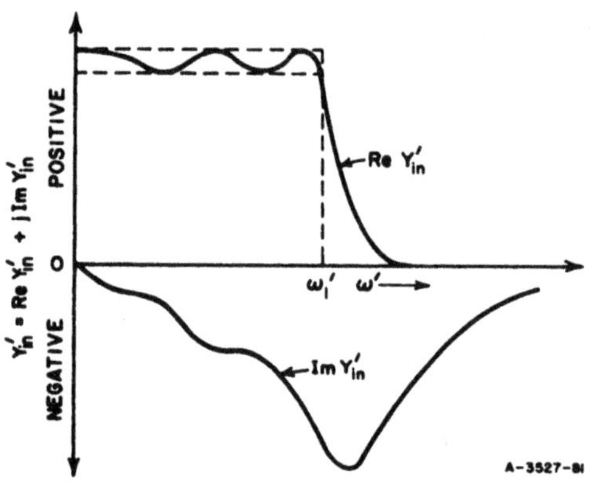

FIG. 4.06-3 THE APPROXIMATE FORM OF THE INPUT ADMITTANCE Y_{in}' IN FIG. 4.06-1 FOR AN $n = 5$ REACTIVE-ELEMENT, SINGLY TERMINATED TCHEBYSCHEFF FILTER

$$a_k = \sin \frac{\pi}{2} \frac{(2k - 1)}{n} , \quad k = 1, 2, \ldots, n$$

$$c_k = \cos^2 \left(\frac{\pi k}{2n} \right) , \quad k = 1, 2, \ldots, n$$

with the element values

$$g_1 = a_1 \tag{4.06-6}$$

$$g_k = \frac{a_k a_{k-1}}{c_{k-1} g_{k-1}} , \quad k = 2, 3, \ldots, n$$

$$g_{n+1} = \infty$$

$$[\text{check: } g_n = n g_1]$$

where the g_k defined above are to be interpreted as in Fig. 4.04-1(a) and (b). Table 4.06-1 gives element values for such filters for the cases of $n = 1$ to $n = 10$.

Table 4.06-1

ELEMENT VALUES FOR SINGLY TERMINATED MAXIMALLY FLAT FILTERS HAVING $g_0 = 1$, $g_{n+1} = \infty$, AND $\omega_1' = 1$

VALUE OF n	g_1	g_2	g_3	g_4	g_5	g_6	g_7	g_8	g_9	g_{10}	g_{11}
1	1.0000	∞									
2	0.7071	1.4142	∞								
3	0.5000	1.3333	1.5000	∞							
4	0.3827	1.0824	1.5772	1.5307	∞						
5	0.3090	0.8944	1.3820	1.6944	1.5451	∞					
6	0.2588	0.7579	1.2016	1.5529	1.7593	1.5529	∞				
7	0.2225	0.6560	1.0550	1.3972	1.6588	1.7988	1.5576	∞			
8	0.1951	0.5776	0.9370	1.2588	1.5283	1.7287	1.8246	1.5607	∞		
9	0.1736	0.5155	0.8414	1.1408	1.4037	1.6202	1.7772	1.8424	1.5628	∞	
10	0.1564	0.4654	0.7626	1.0406	1.2921	1.5100	1.6869	1.8121	1.8552	1.5643	∞

Note: Data by courtesy of L. Weinberg and the Journal of the Franklin Institute[9]

For singly loaded Tchebyscheff filters having $g_0 = 1$, $\omega_1' = 1$, and L_{Ar} or L_{Er} db pass-band ripple, Orchard's equations[5] give

$$\beta = ln\left[\coth \frac{(L_{Ar} \text{ or } L_{Er})}{17.37}\right]$$

$$\gamma = \sinh\left(\frac{\beta}{2n}\right)$$

$$a_k = \sin\frac{\pi(2k-1)}{2n}, \quad k = 1, 2, \ldots, n$$

$$d_k = \left(\gamma^2 + \sin^2\frac{\pi k}{2n}\right)\cos^2\frac{\pi k}{2n}, \quad k = 1, 2, \ldots, n-1$$

(4.06-7)

with element values

$$g_1 = \frac{a_1}{\gamma}$$

$$g_k = \frac{a_k a_{k-1}}{d_{k-1} g_{k-1}}, \quad k = 1, 2, \ldots, n$$

$$g_{n+1} = \infty \quad .$$

Table 4.06-2 presents element values for singly terminated filters for various amounts of Tchebyscheff ripple.

SEC. 4.07, MAXIMALLY FLAT TIME-DELAY PROTOTYPE FILTERS

The voltage attenuation ratio $(E_2)_{avail}/E_2$ (see Sec. 2.10) for a normalized, maximally flat, time-delay filter may be defined as[10,9]

$$\frac{(E_2)_{avail}}{E_2} = cp' y_n(1/p') \tag{4.07-1}$$

where $p' = \sigma' + j\omega'$ is the normalized complex-frequency variable, c is a real, positive constant, and

$$y_n(1/p') = \sum_{k=0}^{n} \frac{(n+k)!}{(n-k)!\,k!\,(2p')^k} \tag{4.07-2}$$

Table 4.06-2

ELEMENT VALUES FOR SINGLY TERMINATED TCHEBYSCHEFF FILTERS HAVING
$g_0 = 1$, $g_{n+1} = \infty$, AND $\omega_1' = 1$

VALUE OF n	g_1	g_2	g_3	g_4	g_5	g_6	g_7	g_8	g_9	g_{10}	g_{11}
0.10 db ripple											
1	0.1526	∞									
2	0.4215	0.7159	∞								
3	0.5158	1.0864	1.0895	∞							
4	0.5544	1.1994	1.4576	1.2453	∞						
5	0.5734	1.2490	1.5562	1.5924	1.3759	∞					
6	0.5841	1.2752	1.5999	1.6749	1.7236	1.4035	∞				
7	0.5906	1.2908	1.6236	1.7107	1.7987	1.7395	1.4745	∞			
8	0.5949	1.3008	1.6380	1.7302	1.8302	1.8070	1.8163	1.4660	∞		
9	0.5978	1.3076	1.6476	1.7423	1.8473	1.8343	1.8814	1.7991	1.5182	∞	
10	0.6000	1.3124	1.6542	1.7503	1.8579	1.8489	1.9068	1.8600	1.8585	1.4964	∞
0.20 db ripple											
1	0.2176	∞									
2	0.5189	0.8176	∞								
3	0.6137	1.1888	1.1900	∞							
4	0.6514	1.2935	1.5615	1.2898	∞						
5	0.6697	1.3382	1.6541	1.6320	1.4356	∞					
6	0.6799	1.3615	1.6937	1.7083	1.7870	1.4182	∞				
7	0.6861	1.3752	1.7149	1.7401	1.8590	1.7505	1.5161	∞			
8	0.6902	1.3840	1.7276	1.7571	1.8880	1.8144	1.8623	1.4676	∞		
9	0.6930	1.3899	1.7360	1.7675	1.9034	1.8393	1.9257	1.7974	1.5512	∞	
10	0.6950	1.3941	1.7418	1.7744	1.9127	1.8523	1.9500	1.8560	1.8962	1.4914	∞
0.50 db ripple											
1	0.3493	∞									
2	0.7014	0.9403	∞								
3	0.7981	1.3001	1.3465	∞							
4	0.8352	1.3916	1.7279	1.3138	∞						
5	0.8529	1.4291	1.8142	1.6426	1.5388	∞					
6	0.8627	1.4483	1.8494	1.7101	1.9018	1.4042	∞				
7	0.8686	1.4596	1.8675	1.7371	1.9712	1.7254	1.5982	∞			
8	0.8725	1.4666	1.8750	1.7508	1.9980	1.7838	1.9571	1.4379	∞		
9	0.8752	1.4714	1.8856	1.7591	2.0116	1.8055	2.0203	1.7571	1.6238	∞	
10	0.8771	1.4748	1.8905	1.7645	2.0197	1.8165	2.0432	1.8119	1.9816	1.4539	∞
1.00 db ripple											
1	0.5088	∞									
2	0.9110	0.9957	∞								
3	1.0118	1.3332	1.5088	∞							
4	1.0495	1.4126	1.9093	1.2817	∞						
5	1.0674	1.4441	1.9938	1.5908	1.6652	∞					
6	1.0773	1.4601	2.0270	1.6507	2.0491	1.3457	∞				
7	1.0832	1.4694	2.0437	1.6736	2.1192	1.6489	1.7118	∞			
8	1.0872	1.4751	2.0537	1.6850	2.1453	1.7021	2.0922	1.3691	∞		
9	1.0899	1.4790	2.0601	1.6918	2.1583	1.7213	2.1574	1.6707	1.7317	∞	
10	1.0918	1.4817	2.0645	1.6961	2.1658	1.7306	2.1803	1.7215	2.1111	1.3801	∞
2.00 db ripple											
1	0.7648	∞									
2	1.2441	0.9766	∞								
3	1.3553	1.2740	1.7717	∞							
4	1.3962	1.3389	2.2169	1.1727	∞						
5	1.4155	1.3640	2.3049	1.4468	1.9004	∞					
6	1.4261	1.3765	2.3383	1.4974	2.3304	1.2137	∞				
7	1.4328	1.3836	2.3551	1.5159	2.4063	1.4836	1.9379	∞			
8	1.4366	1.3881	2.3645	1.5251	2.4332	1.5298	2.3646	1.2284	∞		
9	1.4395	1.3911	2.3707	1.5304	2.4463	1.5495	2.4386	1.4959	1.9553	∞	
10	1.4416	1.3932	2.3748	1.5337	2.4538	1.5536	2.4607	1.5419	2.3794	1.2353	∞
3.00 db ripple											
1	0 9976	∞									
2	1.5506	0.9109	∞								
3	1.6744	1.1739	2.0302	∞							
4	1.7195	1.2292	2.5272	1.0578	∞						
5	1.7409	1.2501	2.6227	1.3015	2.1491	∞					
6	1.7522	1.2606	2.6578	1.3455	2.6309	1.0876	∞				
7	1.7591	1.2666	2.6750	1.3614	2.7141	1.3282	2.1827	∞			
8	1.7638	1.2701	2.6852	1.3690	2.7436	1.3687	2.6618	1.0982	∞		
9	1.7670	1.2726	2.6916	1.3733	2.7577	1.3827	2.7414	1.3380	2.1970	∞	
10	1.7692	1.2744	2.6958	1.3761	2.7655	1.3893	2.7683	1.3774	2.6753	1.1032	∞

NOTE: Most of the data in this table were obtained by courtesy of L. Weinberg and the Journal of the Franklin Institute.[9]

is a Bessel polynomial function of $1/p'$. Equations (4.07-1) and (4.07-2) reduce to a simple polynomial of the form

$$\frac{(E_2)_{\mathrm{avail}}}{E_2} = P_n(p') = (p')^n a_n + (p')^{n-1} a_{n-1} + \ldots + p' a_1 + a_0 \quad . \tag{4.07-3}$$

Let

$$\phi' = \arg \left. \frac{(E_2)_{\mathrm{avail}}}{E_2} \right|_{p'=j\omega'} = \tan^{-1} \frac{\mathrm{Im}\ P_n(j\omega)}{\mathrm{Re}\ P_n(j\omega)} \quad \text{radians} \quad . \tag{4.07-4}$$

Then, as was discussed in Sec. 1.05, the time delay (*i.e.*, group delay) is

$$t'_d = \frac{d\phi'}{d\omega'} \quad \text{secs} \tag{4.07-5}$$

where ω' is in radians per second. The transfer function, defined by Eqs. (4.07-1) and (4.07-2) has the property that its group delay, t'_d, has the maximum possible number of zero derivatives with respect to ω' at $\omega' = 0$, which is why it is said to have maximally flat time delay. The time delay, t'_d, may be expressed as[10,9]

$$t'_d = t'_{d0}\left(1 - \frac{1}{\left(\dfrac{\omega'}{\omega'_1}\right)^2 \left\{\dfrac{\pi\omega'_1}{2\omega'}\left[J^2_{-n-\frac{1}{2}}\left(\dfrac{\omega'}{\omega'_1}\right) + J^2_{n+\frac{1}{2}}\left(\dfrac{\omega'}{\omega'_1}\right)\right]\right\}}\right) \tag{4.07-6}$$

where $J_{-n-\frac{1}{2}}(\omega'/\omega'_1)$ and $J_{n+\frac{1}{2}}(\omega'/\omega'_1)$ are Bessel functions of ω'/ω'_1, and

$$t'_{d0} = \frac{1}{\omega'_1} \tag{4.07-7}$$

is the group delay as $\omega' \to 0$. The magnitude of $(E_2)_{\mathrm{avail}}/E_2$ is

$$\left|\frac{(E_2)_{avail}}{E_2}\right| = c\left(\frac{\omega'}{\omega_1'}\right)^{n+1}\left\{\frac{\pi\omega_1'}{2\omega'}\left[J^2_{-n-1/2}\left(\frac{\omega'}{\omega_1'}\right) + J^2_{n+1/2}\left(\frac{\omega'}{\omega_1'}\right)\right]\right\}^{1/2} \tag{4.07-8}$$

and for increasing n the attenuation approaches the Gaussian form[11,10,9]

$$L_A = \frac{10\left(\frac{\omega'}{\omega_1'}\right)^2}{(2n - 1)\ ln\ 10} \quad \text{db} \quad . \tag{4.07-9}$$

For $n \geqq 3$ the 3 db bandwidth is nearly

$$\left(\frac{\omega'}{\omega_1'}\right)_{3\,db} \approx \sqrt{(2n - 1)\ ln\ 2} \quad . \tag{4.07-10}$$

Weinberg[9] has prepared tables of element values for normalized maximally flat time-delay filters, and the element values in Table 4.07-1 are from his work. These element values are normalized so that $t'_{d0} = 1/\omega_1' =$ 1 second, and $g_0 = 1$. In order to obtain a different time delay, t_{d0}, the frequency scale must be changed by the factor

$$\frac{\omega_1}{\omega_1'} = \frac{t'_{d0}}{t_{d0}} \tag{4.07-11}$$

using the scaling procedure discussed in Sec. 4.04. Weinberg also presents some computed data showing time delay and attenuation in the

Table 4.07-1

ELEMENT VALUES FOR MAXIMALLY FLAT TIME DELAY FILTERS HAVING $g_0 = 1$ and $\omega_1' = 1/t'_{d0} = 1$

VALUE OF n	g_1	g_2	g_3	g_4	g_5	g_6	g_7	g_8	g_9	g_{10}	g_{11}	g_{12}
1	2.0000	1.0000										
2	1.5774	0.4226	1.0000									
3	1.2550	0.5528	0.1922	1.0000								
4	1.0598	0.5116	0.3181	0.1104	1.0000							
5	0.9303	0.4577	0.3312	0.2090	0.0718	1.0000						
6	0.8377	0.4116	0.3158	0.2364	0.1480	0.0505	1.0000					
7	0.7677	0.3744	0.2944	0.2378	0.1778	0.1104	0.0375	1.0000				
8	0.7125	0.3446	0.2735	0.2297	0.1867	0.1387	0.0855	0.0289	1.0000			
9	0.6678	0.3203	0.2547	0.2184	0.1859	0.1506	0.1111	0.0682	0.0230	1.0000		
10	0.6305	0.3002	0.2384	0.2066	0.1808	0.1539	0.1240	0.0911	0.0557	0.0187	1.0000	
11	0.5989	0.2834	0.2243	0.1954	0.1739	0.1528	0.1296	0.1039	0.0761	0.0465	0.0154	1.0000

Note: Data by courtesy of L. Weinberg and the Journal of the Franklin Institute.[9]

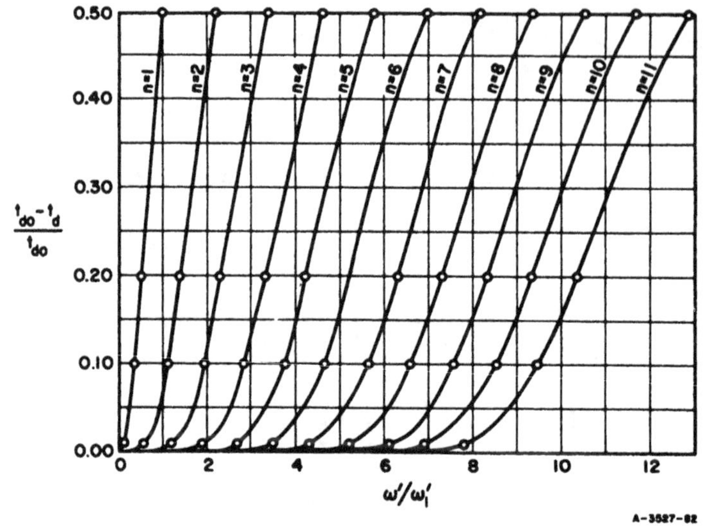

NOTE: Plotted from data prepared by L. Weinberg and published in the *Journal of the Franklin Institute.*[9]

FIG. 4.07-1 TIME-DELAY CHARACTERISTICS OF MAXIMALLY FLAT TIME-DELAY FILTERS

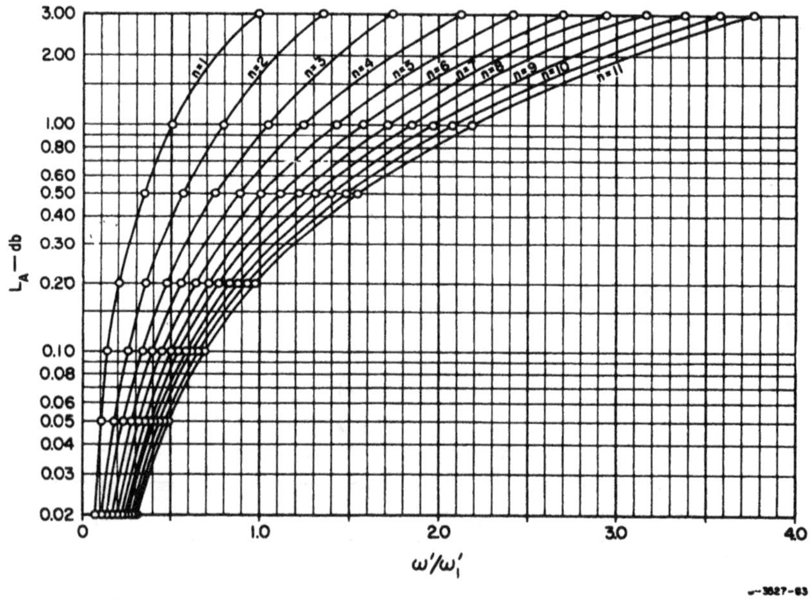

NOTE: Plotted from data prepared by L. Weinberg and published in the *Journal of the Franklin Institute.*[9]

FIG. 4.07-2 PASS-BAND ATTENUATION CHARACTERISTICS OF MAXIMALLY FLAT TIME-DELAY FILTERS

vicinity of the pass band for filters with n = 1 to 11. His data have been plotted in Figs. 4.07-1 and 4.07-2, and curves have been drawn in to aid in interpolating between data points. Although the time-delay characteristics are very constant in the pass-band region, these filters will be seen to have low-pass filter attenuation characteristics which are generally inferior to those of ordinary maximally flat attenuation or Tchebyscheff filters having the same number of reactive elements.

SEC. 4.08, COMPARISON OF THE TIME-DELAY CHARACTERISTICS OF VARIOUS PROTOTYPE FILTERS*

If the terminations of a prototype filter are equal or are not too greatly different, the group time delay as $\omega' \to 0$ can be computed from the relation†

$$t'_{d0} = \left.\frac{d\phi}{d\omega'}\right|_{\omega' \to 0} = \frac{1}{2}\sum_{k=1}^{n} g_k \qquad \text{seconds} \qquad (4.08\text{-}1)$$

where $g_1, g_2, \ldots, g_n$ are the prototype element values as defined in Fig. 4.04-1. Also in Table 4.13-1 and Fig. 4.13-2 a coefficient C_n is tabulated for maximally flat and Tchebyscheff prototype filters where

$$t'_{d0} = C_n \qquad \text{seconds} \qquad (4.08\text{-}2)$$

which is exact.

If the frequency scale of a low-pass prototype is altered so that ω'_1 becomes ω_1, then the time scale is altered so that as $\omega \to 0$ the delay is

$$t_{d0} = t'_{d0}\,\frac{\omega'_1}{\omega_1} \qquad \text{seconds} \quad . \qquad (4.08\text{-}3)$$

If a band-pass filter is designed from a low-pass prototype, then the *midband* time delay is (at least for narrow-band cases)§

* The time delay characteristics of some additional structures will be found in Sec. 6.15.

† This equation is due to S. B. Cohn and can be derived by use of Eqs. (4.13-9) and (4.13-11) to follow.

§ This is the approximate delay for a lumped-element band-pass filter consisting of a ladder of series and shunt resonators. If transmission line circuits are used there may be additional time delay due to the physical length of the filter.

$$t_{d0} = \frac{2\omega_1'}{\omega_2 - \omega_1} t_{d0}' \tag{4.08-4}$$

where ω_1 and ω_2 are the pass-band edges of the band-pass response corresponding to ω_1' for the low-pass response.

In order to determine the time delay at other frequencies it is necessary to work from the transfer functions. For all of the prototype filters discussed in this chapter the voltage attenuation ratio $(E_2)_{avail}/E$ defined in Sec. 2.10 can be represented by a polynomial $P_n(p')$ so that

$$\frac{(E_2)_{avail}}{E_2} = P_n(p')$$

where $p' = \sigma' + j\omega'$ is the complex frequency variable. In the case of prototype filters with maximally flat attenuation, n reactive elements, $\omega_1' = 1$, and $L_{Ar} = 3$ db (see Fig. 4.03-1), $P_n(p')$ is for n even

$$P_n(p') = c \prod_{m=1}^{n/2} \left\{ (p')^2 + \left[2 \cos \frac{\pi(2m-1)}{2n} \right] p' + 1 \right\} \tag{4.08-5}$$

and for n odd

$$P_n(p') = c(p+1) \prod_{m=1}^{(n-1)/2} \left[(p')^2 + \left(2 \cos \frac{\pi m}{n} \right) p' + 1 \right] \tag{4.08-6}$$

where c is a real constant.

For Tchebyscheff prototype filters having n reactive elements, $\omega_1' = 1$, and L_{Ar} db ripple (see Fig. 4.03-3), $P_n(p')$ is for n even

$$P_n(p', x) = c \prod_{m=1}^{n/2} \left\{ (p')^2 + \left[2x \cos \frac{\pi(2m-1)}{2n} \right] p' + x^2 + \sin^2 \frac{\pi(2m-1)}{2n} \right\} \tag{4.08-7}$$

and for n odd with $n \geqq 3$

$$P_n(p',x) = c(p' + x) \prod_{m=1}^{(n-1)/2} \left[(p')^2 + \left(2x \cos \frac{\pi m}{n}\right) p' + x^2 + \sin^2 \frac{\pi m}{n}\right] \tag{4.08-8}$$

where

$$x = \sinh\left[\frac{1}{n} \sinh^{-1} \sqrt{\frac{1}{\left(\text{antilog}_{10} \dfrac{L_{Ar}}{10}\right) - 1}}\right] \tag{4.08-9}$$

and c is again a real constant. The constants c in Eqs. (4.08-5) to (4.08-9) are to be evaluated so as to fix the minimum attenuation of the response. For example, for the Tchebyscheff response in Fig. 4.03-3, c would be evaluated so as to make, $L_A = 20 \log_{10} (E_2)_{avail}/E_2 = 0$ at the bottom of the pass-band ripples. However, for the Tchebyscheff response in the impedance-matching filter response to be presented in Fig. 4.09-2 a different value of c would be required since L_A never goes to zero in this latter case. Both cases would, however, have identical phase shift and time delay characteristics.

The phase shift and group time delay for filters with maximally flat or Tchebyscheff attenuation characteristics can be computed by use of Eqs. (4.08-5) to (4.08-9) above and Eqs. (4.07-4) and (4.07-5). Cohn[12] has computed the phase and time delay characteristics for various prototype filters with $n = 5$ reactive elements in order to compare their relative merit in situations where time-delay characteristics are important. His results are shown in Fig. 4.08-1 to 4.08-3.

Figure 4.08-1 shows the phase characteristics of Tchebyscheff filters having 0.01-db and 0.5-db ripple with $\omega_1' = 1$, and a maximally flat attenuation filter with its 3-db point at $\omega_1' = 1$. The 3-db points of the Tchebyscheff filters are also indicated. Note that the 0.5-db ripple filter has considerably more curvature in its phase characteristic than either the 0.01-db ripple or maximally flat attenuation filters. It will be found that in general the larger the ripple of a Tchebyscheff filter the larger the curvature of the phase characteristic will be in the vicinity of ω_1'. As a result, the larger the ripple, the more the delay distortion will be near cutoff.

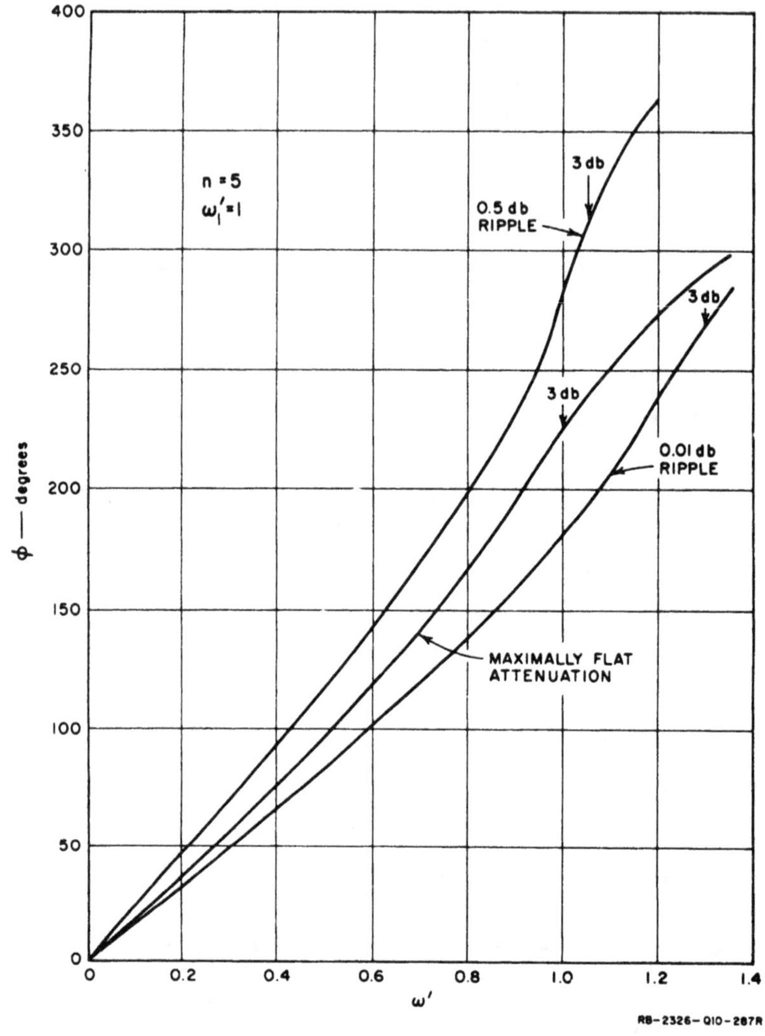

SOURCE: Final Report, Contract DA 36-039 SC-74862, Stanford Research Institute, reprinted in *The Microwave Journal* (see Ref. 13 by S. B. Cohn).

FIG. 4.08-1 PHASE-SHIFT CHARACTERISTICS OF FILTERS WITH MAXIMALLY FLAT OR TCHEBYSCHEFF ATTENUATION RESPONSES AND n = 5

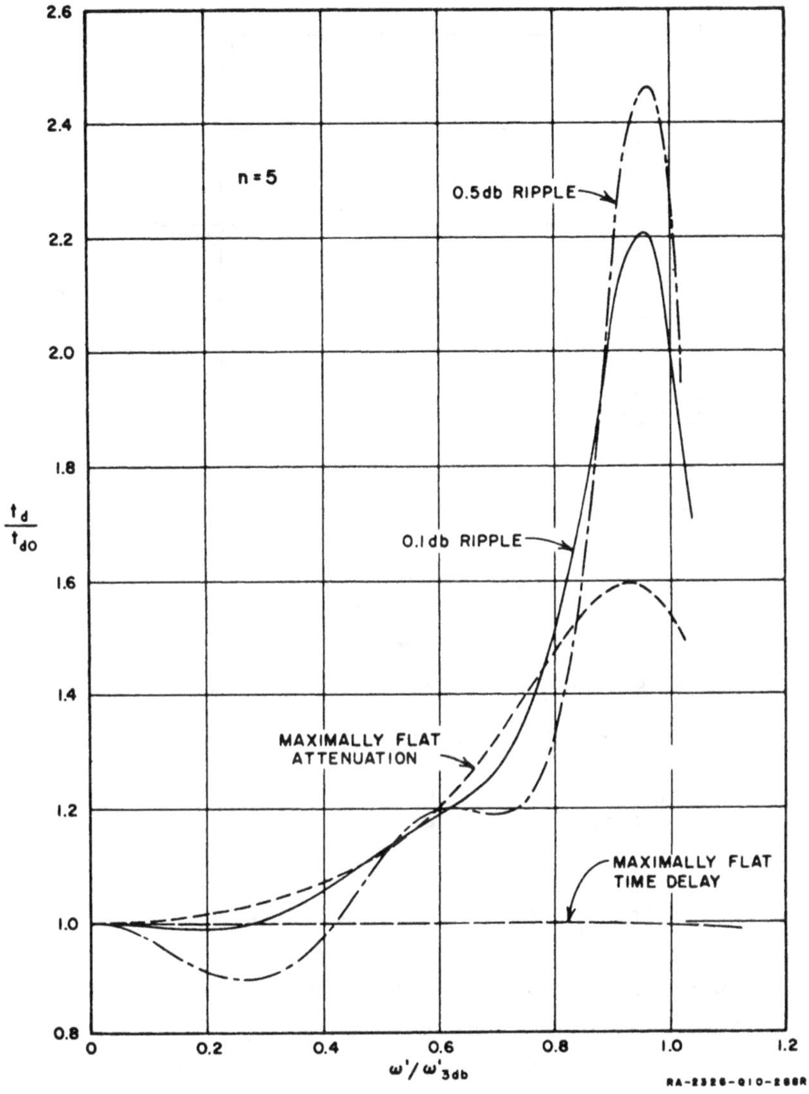

SOURCE: Final Report, Contract DA 36-039 SC-74862, Stanford Research Institute, reprinted in *The Microwave Journal* (see Ref. 13 by S. B. Cohn).

FIG. 4.08-2 NORMALIZED TIME DELAY vs. ω'/ω'_{3db} FOR VARIOUS PROTOTYPE FILTERS

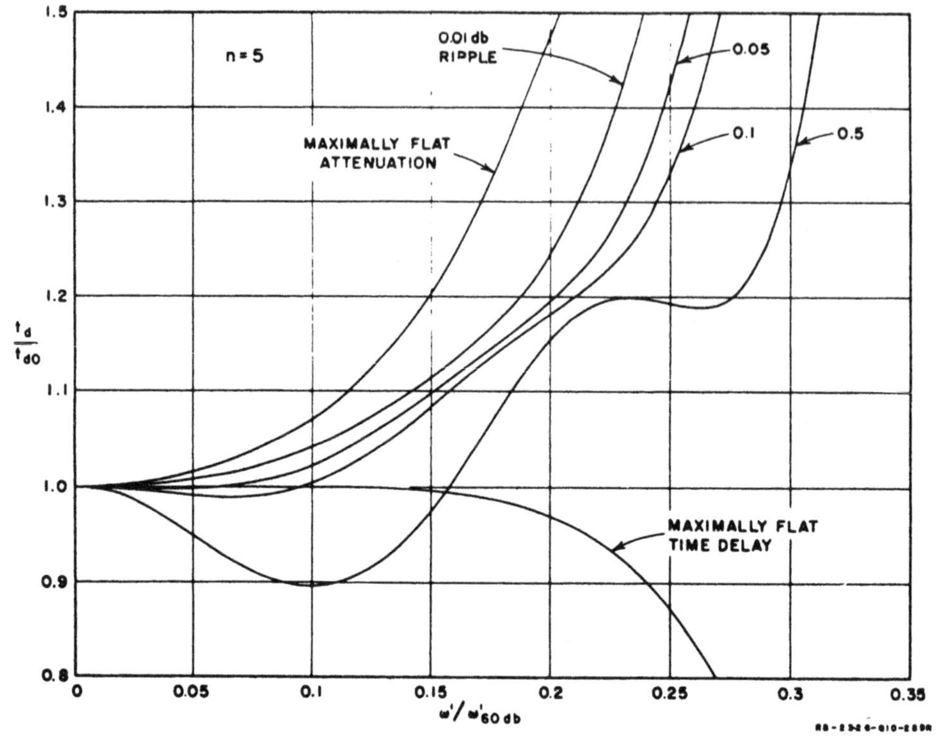

SOURCE: Final Report, Contract DA 36-039 SC-74862, Stanford Research Institute, reprinted in *The Microwave Journal* (see Ref. 13 by S. B. Cohn).

FIG. 4.08-3 NORMALIZED TIME DELAY vs. ω'/ω'_{60db} FOR VARIOUS PROTOTYPE FILTERS

Figure 4.08-2 shows the time delay characteristics of 0.1- and 0.5-db ripple Tchebyscheff filters, of a maximally flat attenuation filter, along with that of a maximally flat time delay filter. The scale of t_d is normalized to the time delay t_{d0}, obtained as $\omega' \to 0$, and the frequency scale is normalized to the frequency $\omega'_{3\ db}$ where $L_A = 3$ db for each case. Note that the time-delay characteristic of the 0.5-db ripple filter is quite erratic, but that delay characteristics for the 0.1-db ripple filter are superior to those of the maximally flat attenuation filter. The 0.1-db-ripple curve is constant within ±1 percent for $\omega'/\omega'_{3\ db} \leqq 0.31$ while the maximally flat filter is within this tolerance only for $\omega'/\omega'_{3\ db} \leqq 0.16$. The maximally flat time-delay filter is seen to have by far the most constant time delay of all. However, the equal-ripple band for the 0.1-db-ripple filter extends to

0.88 $\omega'/\omega'_{3\,db}$ while the maximally flat time delay filter has about 2.2 db attenuation at that frequency. (See Fig. 4.07-2.) Thus, it is seen that maximally flat time-delay filters achieve a more constant time delay at the cost of a less constant attenuation characteristic.

In some cases a band of low loss and low distortion is desired up to a certain frequency and then a specified high attenuation is desired at an adjacent higher frequency. Figure 4.08-3 shows the time-delay characteristics of various prototype filters with the frequency scale normalized to the 60-db attenuation frequency $\omega'_{60\,db}$ for each filter. For a ±1 percent tolerance on t_d, a 0.1-db-ripple filter is found to be usable to 0.106 $\omega'_{60\,db}$ while a maximally flat attenuation filter is within this tolerance only to 0.040 $\omega'_{60\,db}$. For a ±10 percent tolerance on t_d a 0.5-db-ripple filter is usable to 0.184 $\omega'_{60\,db}$ while the maximally flat attenuation filter is usable only to 0.116 $\omega'_{60\,db}$. The maximally flat time-delay filter again has by far the broadest usable band for a given time-delay tolerance; however, its reflection loss will again be an important consideration. For example, for $\omega' = 0.1\ \omega'_{60\,db}$ its attenuation is 1.25 db and its attenuation is 3 db for $\omega' = 0.15\ \omega'_{60\,db}$. In contrast the 0.1-db-ripple prototype filter has 0.1 db attenuation or less out to $\omega' = 0.294\ \omega'_{60\,db}$.

The choice between these various types of filters will depend on the application under consideration. In most cases where time delay is of interest in microwave filters, the filters used will probably be band-pass filters of narrow or moderate bandwidth. Such filters can be designed from prototype filters or step transformers by methods discussed in Chapters 8, 9, and 10.* For cases where the spectrum of a signal being transmitted is appreciable as compared with the bandwidth of the filter, variations in either time delay or pass-band attenuation within the signal spectrum will cause signal distortion.[13] However, for example, a maximally flat time-delay filter which has very little delay distortion and a monotonically increasing attenuation will tend to round a pulse out without overshoot or ringing, while a filter with a sharp cutoff (such as a Tchebyscheff filter) will tend to cause ringing.[13] The transient response requirements for the given application will be dominant considerations when choosing a filter type for such cases where the signal spectrum and filter pass band are of similar bandwidth.

---

* As is discussed in Sec. 1.05, most microwave filters will have extra time delay over that of their prototypes because of the electrical length of their physical structures.

In other situations the signal spectrum may be narrow compared with the bandwidth of the filter so that the spectral components of a given signal see essentially constant attenuation and delay for any common filter response, and distortion of the signal shape may thus be negligible. In such cases a choice of filter response types may depend on considerations of allowable time delay tolerance over the range of possible frequencies, allowable variation of attenuation in the carrier operating band, and required rate of cutoff. For example, if time-delay constancy was of major importance and it didn't matter whether signals with different carrier frequencies suffered different amounts of attenuation, a maximally flat time-delay filter would be the best choice.

SEC. 4.09, PROTOTYPE, TCHEBYSCHEFF IMPEDANCE-MATCHING NETWORKS GIVING MINIMUM REFLECTION

In this section the low-pass impedance matching of loads representable as a resistance and inductance in series, and of loads representable as a resistor and capacitance in parallel will be discussed. A load of the former type with a matching network of the sort to be treated is shown in Fig. 4.09-1. In general, the elements g_0, and g_1 in the circuits in Fig. 4.04-1(a), (b) may be regarded as loads, and the remainder of the reactive elements regarded as impedance-matching networks. For convenience it will be assumed that the impedance level of the load to be matched has been normalized so that the resistor or conductance is equal to one, and that the frequency scale has been normalized so that the edge of the desired band of good impedance match is $\omega_1' = 1$.

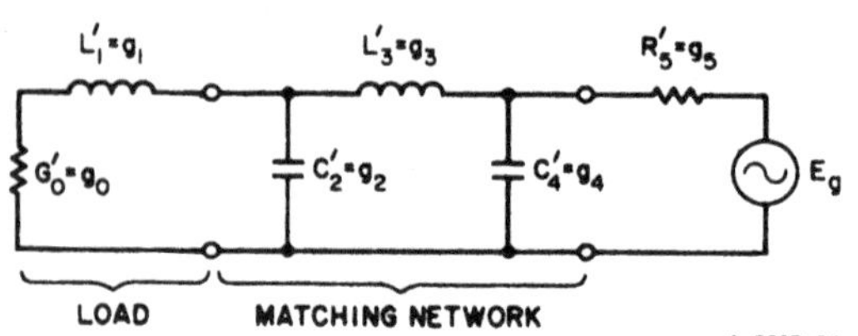

FIG. 4.09-1 A LOAD WITH A LOW-PASS IMPEDANCE-MATCHING NETWORK (Case of n = 4)

As was discussed in Sec. 1.03, if an impedance having a reactive part is to be matched over a band of frequencies, an optimum impedance-matching network must necessarily have a filter-like characteristic. Any degree of impedance match in frequency regions other than that for which a good match is required will detract from the performance possible in the band where good match is required. Thus, the sharper the cutoff of a properly designed matching network, the better its performance can be.

Another important property of impedance-matching networks is that if the load has a reactive part, perfect power transmission to the load is possible only at discrete frequencies, and not over a band of frequencies. Furthermore, it will usually be found that the over-all transmission can be improved if at least a small amount of power is reflected at all frequencies. This is illustrated in Fig. 4.09-2, where it will be assumed that the designer's objective is to keep $(L_A)_{max}$ as small as possible from $\omega' = 0$ to $\omega' = \omega_1'$, where the db attenuation L_A refers to the attenuation of the power received by the load with respect to the available power of the generator (see Sec. 2.11). If $(L_A)_{min}$ is made very small so as to give very efficient transmission at the bottoms of the pass-band ripples, the excessively good transmission at these points must be compensated for by excessively poor transmission at the crests of the ripples, and as a result, $(L_A)_{max}$ will increase. On the

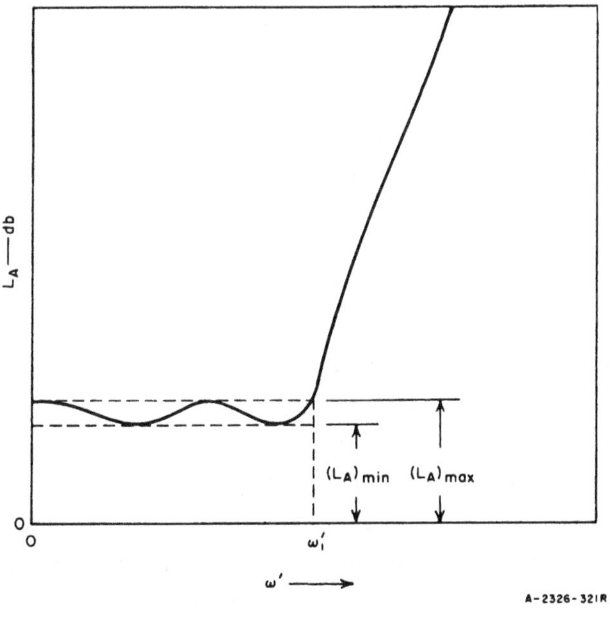

FIG. 4.09-2 DEFINITION OF $(L_A)_{max}$ AND $(L_A)_{min}$ FOR TCHEBYSCHEFF IMPEDANCE MATCHING NETWORKS DISCUSSED HEREIN

other hand if $(L_A)_{min}$ is specified to be nearly equal to $(L_A)_{max}$, the small pass-band ripple will result in a reduced rate of cutoff for the filter; as indicated above, this reduced rate of cutoff will degrade the performance and also cause $(L_A)_{max}$ to increase. Thus, it is seen that for a given load, a given number of impedance-matching elements, and a given impedance-matching bandwidth, there is some definite value of Tchebyscheff pass-band ripple $(L_A)_{max} - (L_A)_{min}$ that goes with a minimum value of $(L_A)_{max}$. The prototype impedance-matching networks discussed in this section are optimum in this sense, *i.e.*, they do minimize $(L_A)_{max}$ for a load and impedance-matching network of the form in Fig. 4.09-1 or its generalization in terms of Figs. 4.04-1(a), (b).

It is convenient to characterize the loads under consideration by their *decrement*, which is defined as

$$\delta = \frac{1}{g_0 g_1 \omega_1'} \tag{4.09-1}$$

$$= \frac{1}{G_0' L_1' \omega_1'} \quad \text{or} \quad \frac{1}{R_0' C_1' \omega_1'} \quad ,$$

where the various quantities in this equation are as indicated in Figs. 4.09-1, 4.09-2, and 4.04-1(a), (b). Note that δ is the reciprocal of the Q of the load evaluated at the edge of the impedance-matching band and *that δ evaluated for the un-normalized load is the same as that for the normalized load.* Figure 4.09-3 shows the minimum value of $(L_A)_{max}$ *vs* δ for circuits having $n = 1$ to $n = 4$ reactive elements (also for case of $n = \infty$). Since one of the reactive elements in each case is part of the load, the $n = 1$ case involves no L or C impedance-matching elements, the optimum result being determined only by optimum choice of driving-generator internal impedance. Note that for a given value of δ, $(L_A)_{max}$ is decreased by using more complex matching networks (*i.e.*, larger values of n). However, a point of diminishing returns is rapidly reached so that it is usually not worthwhile to go beyond $n = 3$ or 4. Note that $n = \infty$ is not greatly better than $n = 4$.

Figure 4.09-4 shows the db Tchebyscheff ripple *vs* δ for minimum $(L_A)_{max}$. Once again, going to larger values of n will give better results, since when n is increased, the size of the ripple is reduced for a given δ. For $n = \infty$ the ripple goes to zero.

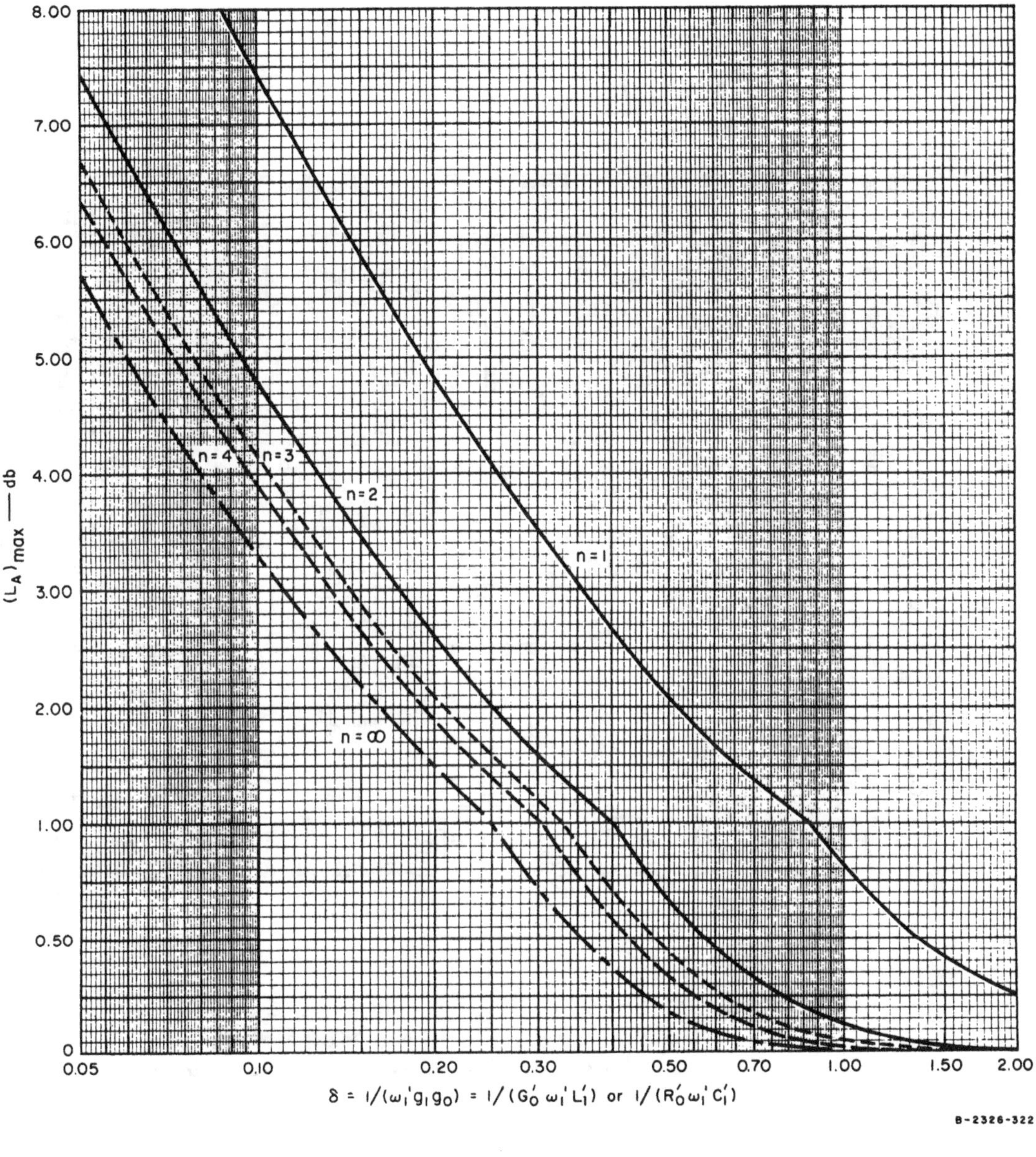

FIG. 4.09-3 $(L_A)_{max}$ vs. δ FOR THE IMPEDANCE-MATCHING NETWORKS WHOSE ELEMENT VALUES ARE GIVEN IN FIGS. 4.09-5 TO -8

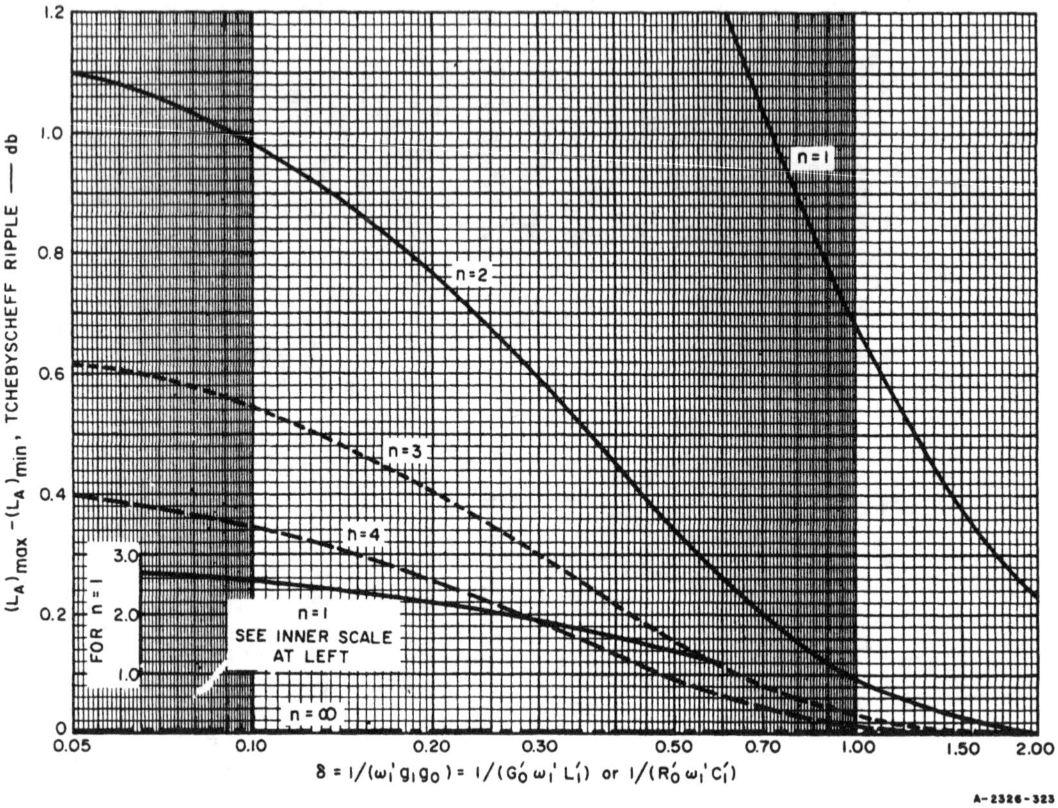

FIG. 4.09-4 TCHEBYSCHEFF RIPPLE IN db vs. δ FOR THE IMPEDANCE-MATCHING NETWORKS WHOSE ELEMENT VALUES ARE GIVEN IN FIGS. 4.09-5 TO 4.09-8

Figures 4.09-5 to 4.09-8 show charts of element values *vs* δ for optimum Tchebyscheff matching networks. Their use is probably best illustrated by an example. Suppose that an impedance match is desired to a load which can be represented approximately by a 50-ohm resistor (G_0 = 0.020 mho) in series with an inductance $L_1 = 3.98 \times 10^{-8}$ henry, and that good impedance match is to extend up to f_1 = 1 Gc so that $\omega_1 = 2\pi f_1 = 6.28 \times 10^9$. Then the decrement is $\delta = 1/(G_0\omega_1 L_1) = 1/(0.020 \times 6.28 \times 10^9 \times 3.98 \times 10^{-8}) = 0.20$. After consulting Figs. 4.09-3 and 4.09-4 for $\delta = 0.20$ let us suppose that $n = 4$ is chosen which calls for $(L_A)_{max}$ = 1.9 db and a ripple of about 0.25 db. Then by Fig. 4.09-8 (which is for $n = 4$) we obtain for $g_0 = 1$, $\omega_1' = 1$, and $\delta = 0.20$: $g_1/10 = 0.50$, $g_2 = 0.445$, $g_3/10 = 0.54$, $g_4 = 0.205$, and $g_5/10 = 0.39$. This corresponds to the circuit in Fig. 4.09-1 with $g_0 = G_0' = 1$, $g_1 = 5.00 = L_1'$, $g_2 = 0.445 = C_2'$, $g_3 = 5.40 = L_3'$, $g_4 = 0.205 = C_4'$, and $g_5 = 3.90 = R_5'$. Un-normalizing this by use of Eqs. (4.04-2) to (4.04-4) with $(G_0/G_0') = 0.020/1$ and $\omega_1'/\omega_1 = 1/(6.28 \times 10^9) = 1.59 \times 10^{-10}$ gives: G_0 = 0.020 mho, $L_1 = 3.98 \times 10^{-8}$ henry, $C_2 = 1.415 \times 10^{-12}$ farad, $L_3 = 4.29 \times 10^{-8}$ henry, $C_4 = 6.52 \times 10^{-13}$ farad, and R_5 = 195 ohms. Note that G_0 and L_1 are the original elements given for the load. The physical realization of microwave structures for such an application can be accomplished using techniques discussed in Chapter 7.

It is interesting to note how much the impedance-matching network design discussed above actually improves the power transfer to the load. If the *R-L* load treated above were driven directly by a generator with a 50-ohm internal impedance, the loss would approach 0 db as $f \to 0$, but it would be 8.6 db at f_1 = 1 Gc. By Figs. 4.09-3 to 4.09-5, the optimum $n = 1$ design for this case would call for the generator internal impedance to be about 256 ohms, which would give about 2.6 db loss as $f \to 0$ and 5.9 db loss at 1 Gc (a reduction of 2.7 db from the preceding case). Thus, the $n = 4$ design with only 1.9 db maximum loss and about 0.25 db variation across the operating band is seen to represent a major improvement in performance. Going to larger values of n would give still greater improvement, but even with $n = \infty$, $(L_A)_{max}$ would still be about 1.46 db.

In most microwave cases band-pass rather than low-pass impedance matching networks are desired. The design of such networks is discussed in Secs. 11.08 and 11.09 working from the data in this section. A special feature of band-pass impedance-matching networks is that they are easily designed to permit any desired value of generator internal resistance,

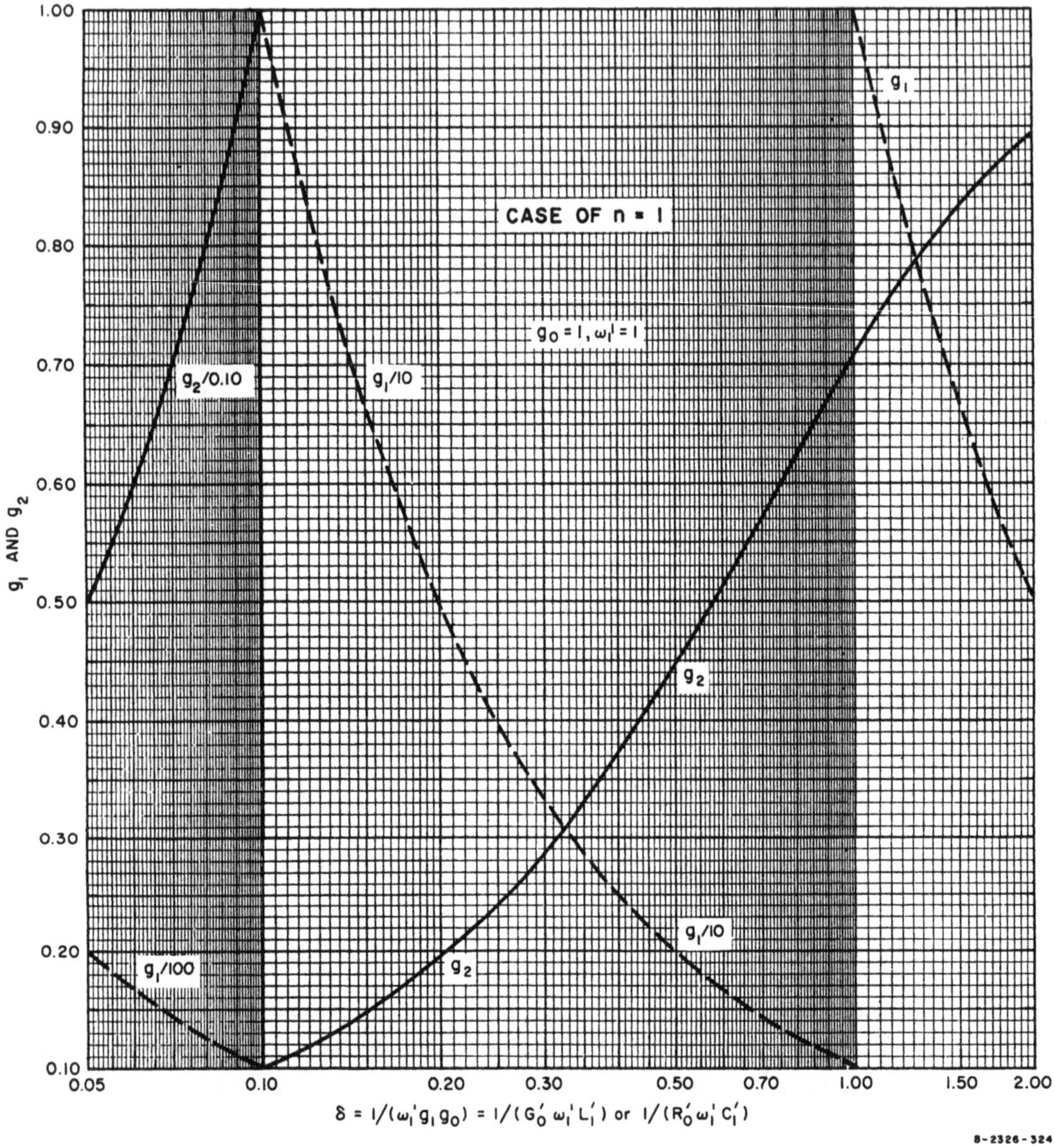

FIG. 4.09-5 ELEMENT VALUES vs. δ FOR TCHEBYSCHEFF IMPEDANCE-MATCHING NETWORKS THAT MINIMIZE $(L_A)_{max}$

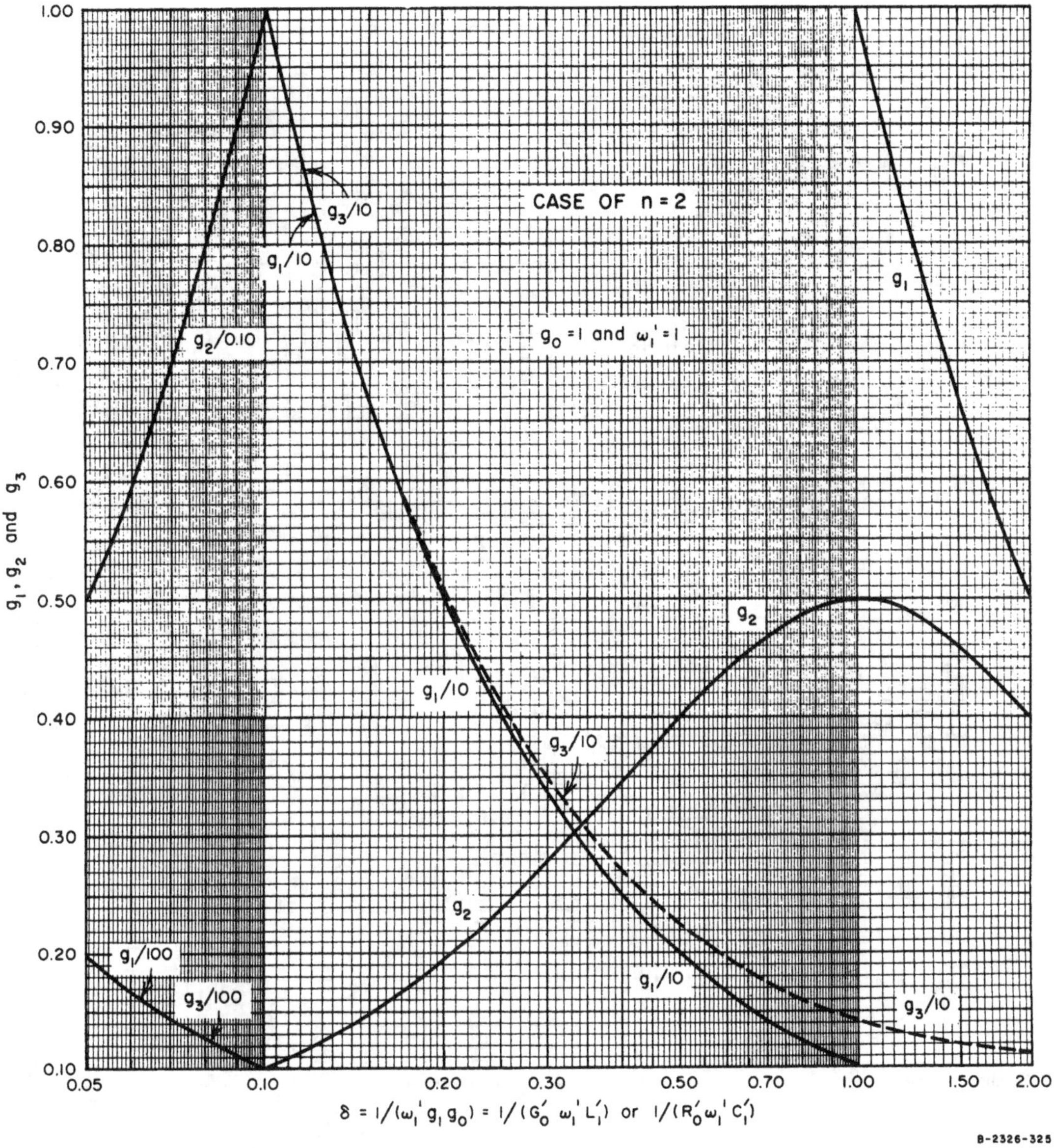

FIG. 4.09-6 ELEMENT VALUES vs. δ FOR TCHEBYSCHEFF IMPEDANCE-MATCHING NETWORKS THAT MINIMIZE $(L_A)_{max}$

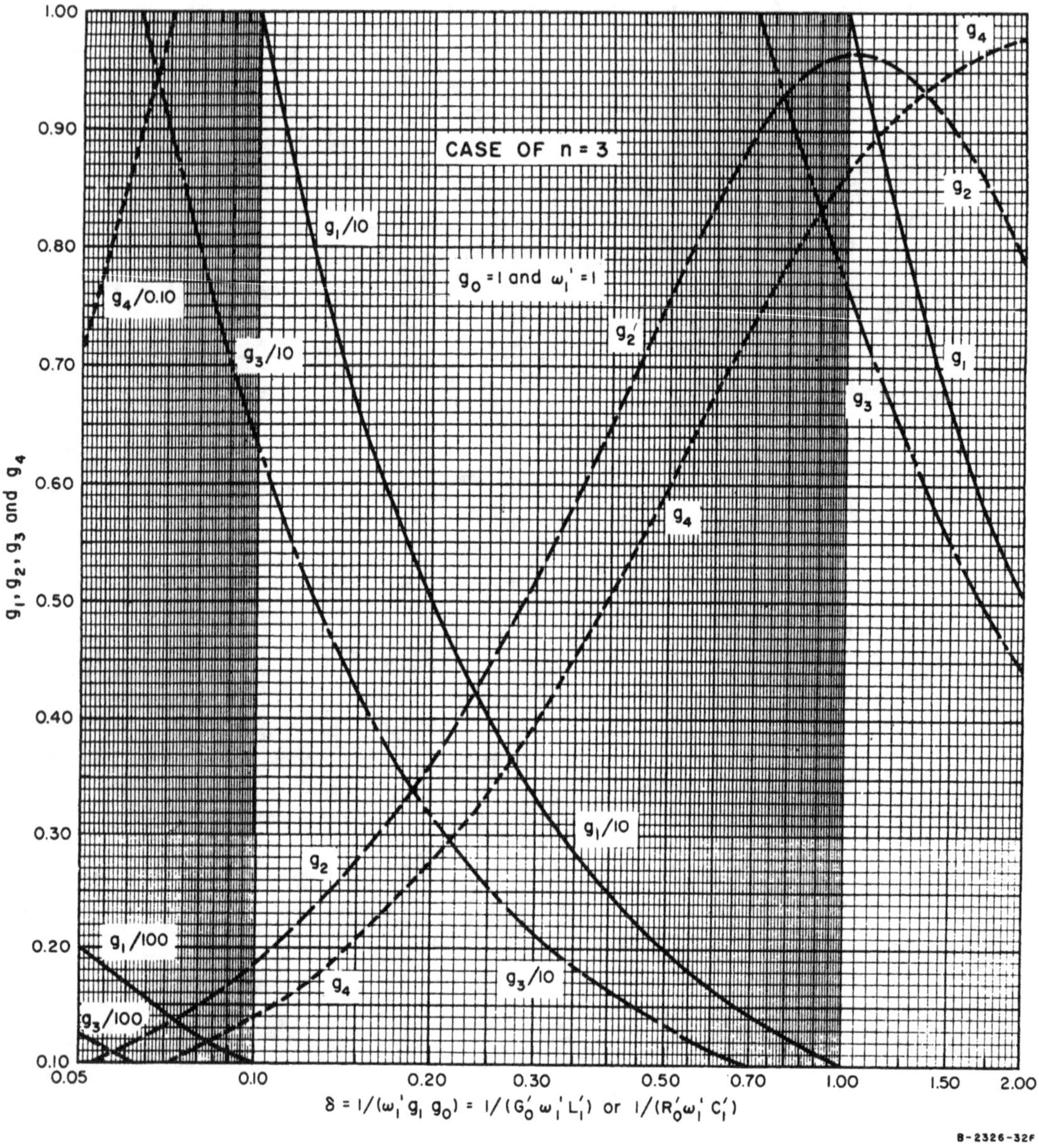

FIG. 4.09-7 ELEMENT VALUES vs. δ FOR TCHEBYSCHEFF IMPEDANCE-MATCHING NETWORKS THAT MINIMIZE $(L_A)_{max}$

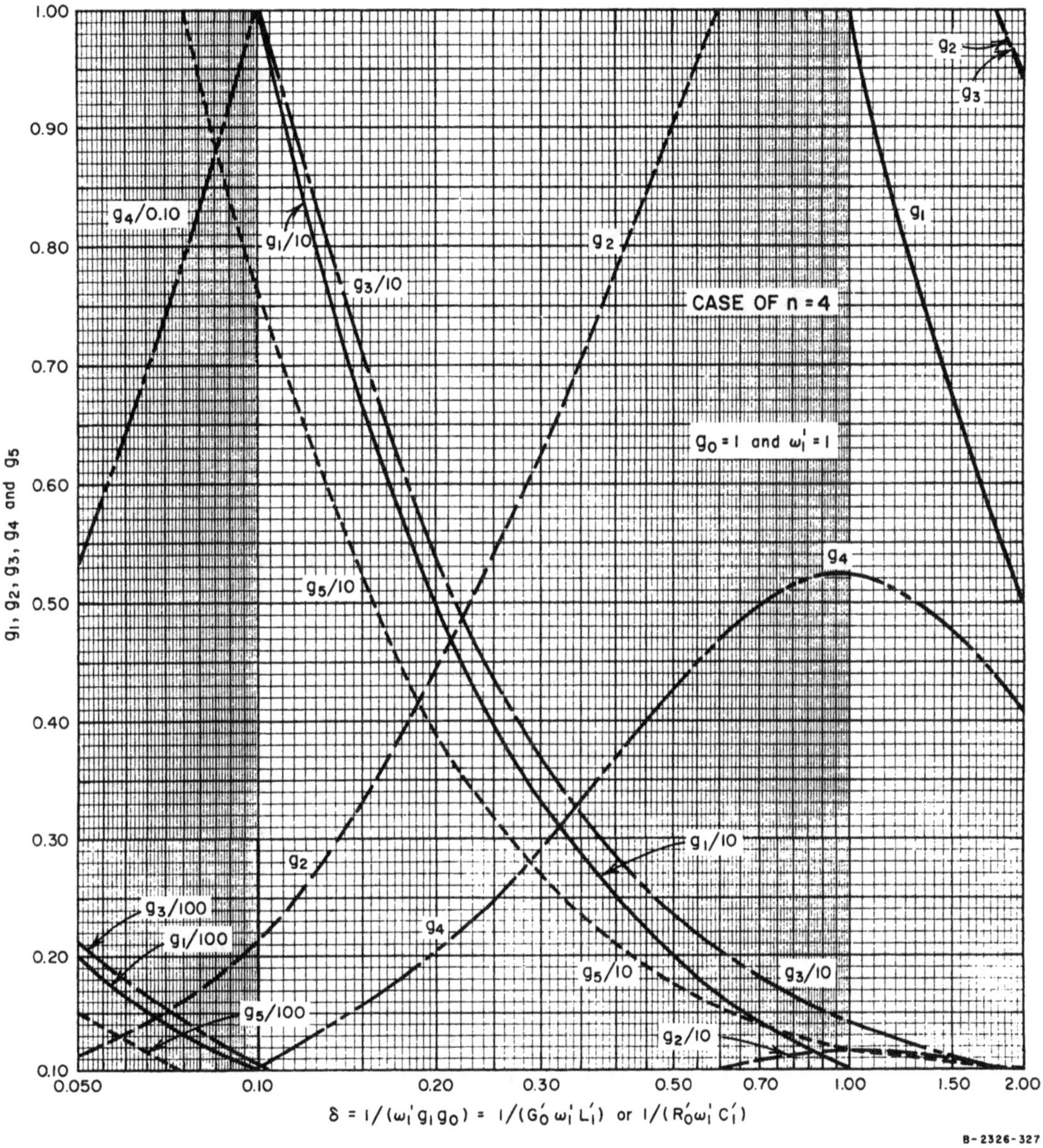

FIG. 4.09-8 ELEMENT VALUES vs. δ FOR TCHEBYSCHEFF IMPEDANCE-MATCHING NETWORKS THAT MINIMIZE $(L_A)_{max}$

whereas low-pass matching networks must have a specified generator internal resistance for optimum design.

The attenuation characteristics of the impedance-matching networks discussed in this section and in Sec. 4.10 may be computed by

$$L'_A = L_A + (L_A)_{min} \quad \text{db} \qquad (4.09\text{-}2)$$

where L'_A is the attenuation of the impedance-matching network and L_A is obtained by Eqs. (4.03-3) to (4.03-5), or by Figs. 4.03-4 to 4.03-10 for the appropriate db Tchebyscheff ripple $L_{Ar} = (L_A)_{max} - (L_A)_{min}$.

In the next section the calculation of prototype impedance-matching networks so as to give a specified Tchebyscheff ripple [at the cost of a larger $(L_A)_{max}$] will be discussed. The method by which Figs. 4.09-3 to 4.09-8 were prepared will also be outlined.

SEC. 4.10, COMPUTATION OF PROTOTYPE IMPEDANCE-MATCHING NETWORKS FOR SPECIFIED RIPPLE OR MINIMUM REFLECTION

The networks discussed in the preceding section were specified so that $(L_A)_{max}$ was to be as small as possible. Under that condition, it was necessary to accept whatever pass-band Tchebyscheff ripple the charts might call for in the case of any given design. Alternatively, we may specify the pass-band Tchebyscheff ripple and accept whatever value of $(L_A)_{max}$ may result. Since in some cases keeping the pass-band attenuation constant may be the major consideration, computation of prototype matching-network element values for a specified Tchebyscheff ripple will be briefly outlined.

Prototype circuits for specified decrement $\delta = 1/(g_0 g_1 \omega'_1)$ and db ripple may be obtained as follows. First compute[14]

$$H = \operatorname{antilog}_{10} \frac{(\text{db Tchebyscheff ripple})}{10} \qquad (4.10\text{-}1)$$

and

$$d = \sinh\left[\frac{\sinh^{-1}\sqrt{\dfrac{1}{H-1}}}{n}\right] \qquad (4.10\text{-}2)$$

where n is the number of reactive elements in the prototype. Next compute

$$e = d - 2\delta \sin\left(\frac{\pi}{2n}\right) \tag{4.10-3}$$

and the maximum, pass-band reflection coefficient value

$$|\Gamma|_{max} = \frac{\cosh\ (n \sinh^{-1} e)}{\cosh\ (n \sinh^{-1} d)} \quad . \tag{4.10-4}$$

Then the $(L_A)_{max}$ value which must be accepted is

$$(L_A)_{max} = 10 \log_{10} \frac{1}{1 - |\Gamma|^2_{max}} \quad . \tag{4.10-5}$$

Figure 4.10-1 shows a plot of $(L_A)_{max}$ *vs* δ for various values of n and various amounts of Tchebyscheff ripple amplitude $[(L_A)_{max} - (L_A)_{min}]$. Suppose that $\delta = 0.10$ and 0.10-db ripple is desired with $n = 2$. This chart shows that $(L_A)_{max}$ will then be 5.9 db. By Figs. 4.09-3 and 4.09-4 it is seen that for the same δ, when $(L_A)_{max}$ is minimized, $(L_A)_{max}$ = 4.8 db while the ripple is 0.98 db. Thus, the price for reducing the ripple from 0.98 db to 0.10 db is an increase in $(L_A)_{max}$ of about 1.1 db.

Green's work[6,7] appears to provide the easiest means for determining the element values. Using his equations altered to the notation of this chapter, we obtain

$$D = \frac{d}{\delta \sin\left(\frac{\pi}{2n}\right)} - 1 = \frac{g_0 g_1}{g_{n+1} g_n} \tag{4.10-6}$$

where the g_j's are as defined in Fig. 4.04-1. The element values are then computed by use of the equations

$$g_1 = \frac{1}{\delta \omega_1' g_0} \tag{4.10-7}$$

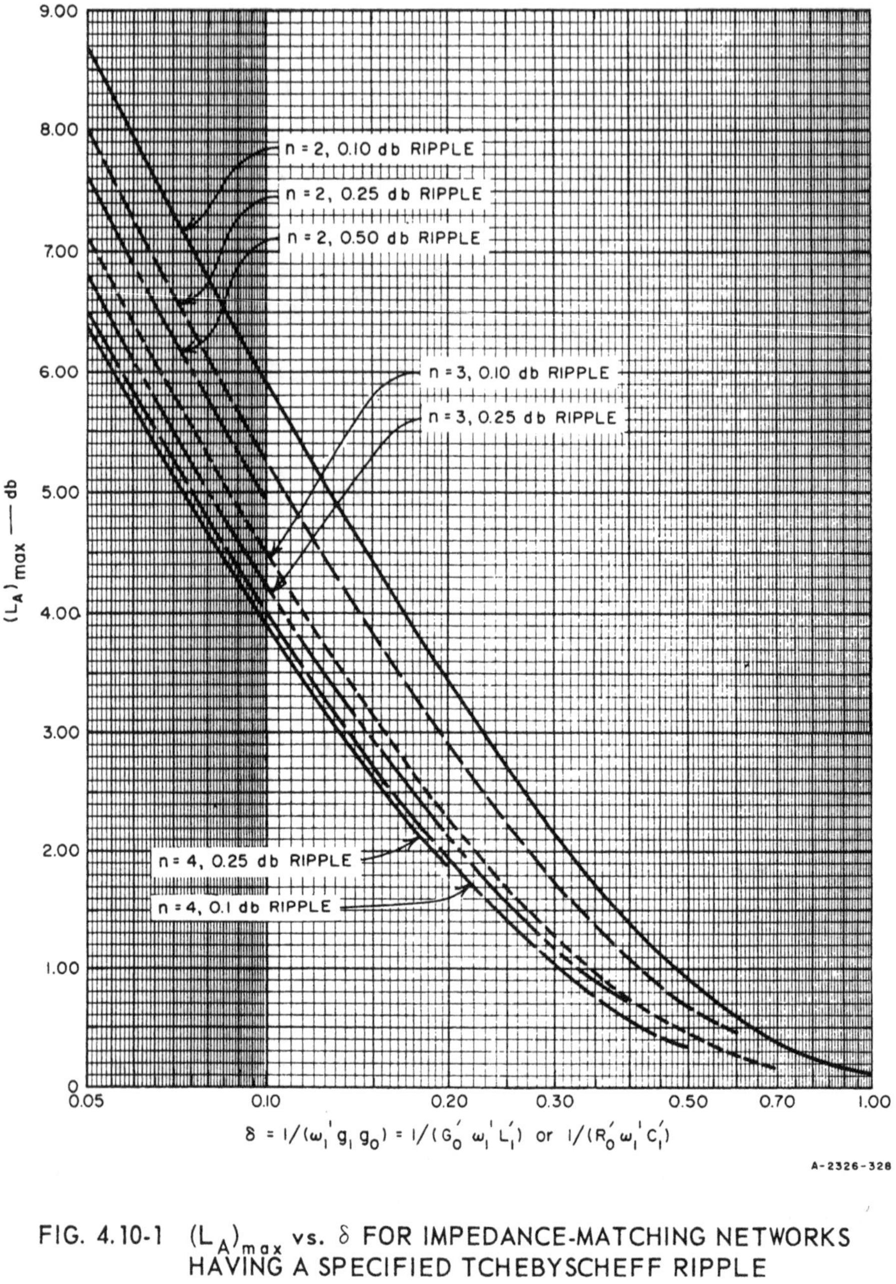

FIG. 4.10-1 $(L_A)_{max}$ vs. δ FOR IMPEDANCE-MATCHING NETWORKS HAVING A SPECIFIED TCHEBYSCHEFF RIPPLE

$$g_j \Big|_{j=2 \text{ to } n} = \frac{1}{g_{j-1}(k_{j-1,j})^2(\omega_1')^2} \tag{4.10-8}$$

$$g_{n+1} = \frac{1}{D\delta g_n \omega_1'} \tag{4.10-9}$$

where the $k_{j-1,j}$ are coupling coefficients to be evaluated as shown below.

Green's equations for the $k_{j-1,j}$ are[6,7]

$n = 2$

$$k_{12} = \sqrt{\frac{1 + (1 + D^2)\delta^2}{2}} \tag{4.10-10}$$

$n = 3$

$$k_{12} = \sqrt{\frac{3}{8}\left[1 + \left(1 + \frac{D^2}{3}\right)\delta^2\right]} \tag{4.10-11}$$

$$k_{23} = \sqrt{\frac{3}{8}\left[1 + \left(\frac{1}{3} + D^2\right)\delta^2\right]} \tag{4.10-12}$$

$n = 4$

$$k_{12} = \sqrt{\frac{1}{2\sqrt{2}}\left[1 + \left(1 + \frac{8D^2}{\alpha^4}\right)\delta^2\right]} \tag{4.10-13}$$

$$k_{23} = \sqrt{\frac{2}{\alpha^2}\left[1 + \frac{2}{\alpha^2}(1 + D^2)\delta^2\right]} \tag{4.10-14}$$

$$k_{34} = \sqrt{\frac{1}{2\sqrt{2}}\left[1 + \left(\frac{8}{\alpha^4} + D^2\right)\delta^2\right]} \tag{4.10-15}$$

where

$$\alpha^2 = 2(2 + \sqrt{2}) \approx 6.83$$

Also, *for n arbitrary,*

$$k_{r,r+1} = \sqrt{\frac{\sin^2 r\theta \cos^2 r\theta + (\cos^2 r\theta + D^2 \sin^2 r\theta)(\sin^2 \theta)\delta^2}{\sin (2r - 1)\theta \sin (2r + 1)\theta}} \qquad (4.10\text{-}16)$$

where

$$\theta = 2\pi/n \quad .$$

It is usually convenient to normalize the prototype design so that $g_0 = 1$ and $\omega_1' = 1$, as has been done with the tabulated designs in this chapter.

The element values for the prototype matching networks discussed in Sec. 4.09 and plotted in Figs. 4.09-5 to 4.09-8 could have been obtained using Green's charts[7] of coupling coefficients and D values along with Eqs. (4.10-7) to (4.10-9).* However, in order to ensure high accuracy, to add the $n = 1$ case, and to cover a somewhat wider range of decrements than was treated by Green, the computations for the charts in Sec. 4.09 were carried out from the beginning. The procedure used was that described below.

Fano[14] has shown that, for low-pass networks of the type under consideration, $(L_A)_{max}$ will be as small as possible if

$$\frac{\tanh na}{\cosh a} = \frac{\tanh nb}{\cosh b} \qquad (4.10\text{-}17)$$

where

$$a = \sinh^{-1} d \qquad (4.10\text{-}18)$$

$$a = \sinh^{-1} e \qquad (4.10\text{-}19)$$

and d and e are as indicated in Eqs. (4.10-2) and (4.10-3). By Eqs. (4.10-18), (4.10-19), and (4.10-3),

$$b = \sinh^{-1} \left[\sinh a - 2\delta \sin \frac{\pi}{2n}\right] \quad . \qquad (4.10\text{-}20)$$

---

* Barton (see Ref. 15) has independently also computed charts equivalent to the coupling-coefficient charts of Green. Barton, however, includes the maximally flat case in addition.

A computer program was set up to find values of a and b that satisfy Eq. (4.10-17) under the constraint given by Eq. (4.10-20). From these a and b values for various δ, values for d and e were obtained by $d = \sinh a$ and $e = \sinh b$. When values of d and e had been obtained for various δ, the element values for the networks were computed using Eqs. (4.10-6) to (4.10-15).

The data for the charts in Fig. 4.09-3 were obtained by using the values of a and b vs δ obtained above, and then computing $(L_A)_{max}$ by use of Eqs. (4.10-18), (4.10-19), (4.10-4), and (4.10-5). The data in Fig. 4.09-4 were obtained by solving Eqs. (4.10-18), (4.10-19), (4.10-1) and (4.10-2) for the db ripple as a function of a and b.

Lossless impedance matching networks for some more general forms of loads are discussed in Refs. 14, 16, 17, and 18. However, much work remains to be done on the practical, microwave realization of the more complicated forms of matching networks called for in such cases. At the present time the prototype networks in Sec. 4.09 and this section appear to have the widest range of usefulness in the design of low-pass, high-pass, and band-pass microwave impedance matching networks in the forms discussed in Chapters 7, and 11.

SEC. 4.11, PROTOTYPES FOR NEGATIVE-RESISTANCE AMPLIFIERS

As was discussed in Sec. 1.04, if a dissipationless filter with resistor terminations has one termination replaced by a negative resistance of the same magnitude, the circuit can become a negative-resistance amplifier. It was noted that, if $\Gamma_1(p)$ is the reflection coefficient between a positive resistance R_0 and the filter, when R_0 is replaced by $R_0'' = -R_0$, the reflection coefficient at that end of the filter becomes

$$\Gamma_1''(p) = \frac{1}{\Gamma_1(p)} \tag{4.11-1}$$

where $p = \sigma + j\omega$ is the complex frequency variable. Then, referring to Figs. 1.04-1 and 1.04-2, the gain of the amplifier as measured at a circulator will be

$$\frac{P_r}{P_{avail}} = |\Gamma_1''(p)|^2_{p=j\omega} = |\Gamma_3''(p)|^2_{p=j\omega} \tag{4.11-2}$$

where P_r is the power reflected into the circulator by the negative-resistance amplifier. If L_A is the attenuation in db (*i.e.*, transducer loss, as defined in Sec. 2.11) between R_g and R_0 in Fig. 1.04-1 for the dissipationless filter with *positive* terminations, then the squared magnitude of the reflection coefficient for positive terminations is given by

$$|\Gamma_1|^2 = 1 - \frac{1}{\text{antilog}_{10} \frac{L_A}{10}} \quad . \tag{4.11-3}$$

When R_0 on the left in Fig. 1.04-1 is replaced by its negative, $R_0'' = -R_0$, then Eq. (4.11-3) can be replaced by

$$G_t = 10 \log_{10} \frac{1}{|\Gamma_1|^2} = 10 \log_{10} \frac{P_r}{P_{avail}}$$

$$= \frac{\text{antilog}_{10} \frac{L_A}{10}}{\text{antilog}_{10} \frac{L_A}{10} - 1} \quad \text{db} \tag{4.11-4}$$

where G_t is the transducer gain at the circulator, in db. Figure 4.11-1 shows the attenuation for a filter with positive-resistance terminations *vs.* the db transducer gain of the corresponding negative-resistance amplifier with a circulator, as determined using the above relations.

The prototype impedance-matching filters discussed in Secs. 4.09 and 4.10 can also be used as prototypes for negative-resistance amplifiers. With regard to their use, some consideration must be given to the matter of stability. Let us define $\Gamma_1(p)$ as the reflection coefficient between any of the filters in Fig. 4.04-1 and the termination $g_0 = R_0$ or G_0 at the left and $\Gamma_n(p)$ as the reflection coefficient at the other end. It can be shown that the poles of a reflection coefficient function are the frequencies of natural vibration of the circuit (see Secs. 2.02 to 2.04); hence, they must lie in the left half of the complex-frequency plane if the circuit is passive. However, the zeros of $\Gamma_1(p)$, or of $\Gamma_n(p)$, can lie in either the left or right half of the p-plane. Since $\Gamma_1''(p) = 1/\Gamma_1(p)$, the *zeros* of $\Gamma_1(p)$ for the passive filter become the *poles* of

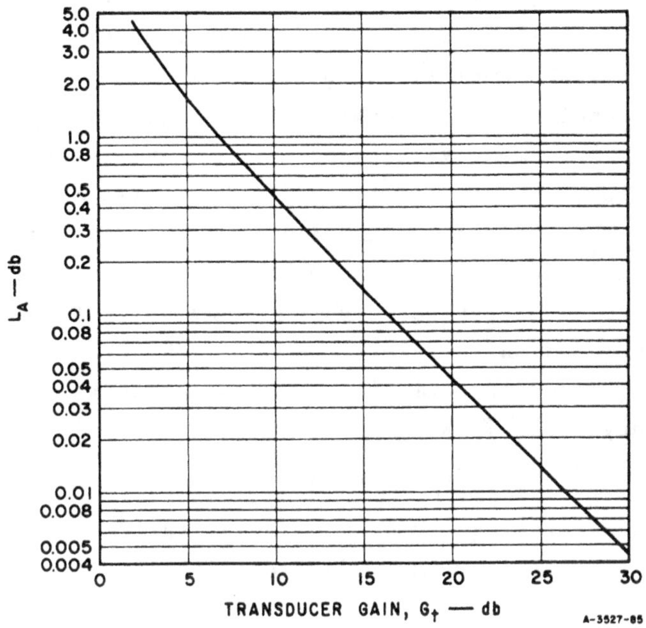

FIG. 4.11-1 ATTENUATION OF A FILTER WITH POSITIVE-RESISTANCE TERMINATIONS vs. TRANSDUCER GAIN OF THE CORRESPONDING NEGATIVE-RESISTANCE AMPLIFIER USING A CIRCULATOR

$\Gamma_1''(p)$ for the negative-resistance amplifier. Thus, in choosing a filter as a prototype for a negative-resistance amplifier, it is important that $\Gamma_1(p)$ have its zeros in the left half plane since if they are not, when these zeros become poles of $\Gamma_1''(p)$ for the negative-resistance amplifier they will cause exponentially increasing oscillations (*i.e.*, until some non-linearity in the circuit limits the amplitude).

The mathematical data given in Secs. 4.09 and 4.10 for filter prototypes of the various forms in Fig. 4.04-1 are such that the reflection coefficient $\Gamma_1(p)$ involving the termination g_0 on the left will have all of its zeros in the left half of the p-plane, while the reflection coefficient $\Gamma_n(p)$ involving the termination g_{n+1} on the right will have all of its zeros in the right half plane.* For this reason it is seen that the termination g_0 at the left must be the one which is replaced by its negative, *never* the termination g_{n+1} at the right.

* An exception to this occurs when $\epsilon = 0$ in Eq. (4.10-3) which leads to $(L_A)_{min} = 0$ in Fig. 4.09-2. Then the zeros of $\Gamma_1(p)$ and $\Gamma_n(p)$ are all on the $p = j\omega$ axis of the p-plane.

By use of Eqs. (4.11-1) to (4.11-4) along with the equations in Sec. 4.10 it is possible to compute prototype impedance-matching networks for negative-resistance amplifiers so as to give a specified gain and pass-band Tchebyscheff ripple. However, Getsinger[21] has prepared extensive tables of prototype element values for negative-resistance amplifiers, and his tables are reproduced in Tables 4.11-1 to 4.11-3 herein. Figure 4.11-2 shows how the element values in Tables 4.11-1 to 4.11-3 should be interpreted. The conductance or resistance on the left is always negative, and it is scaled so as to always be equal to minus one. Note that an idealized circulator is included on the right side of each circuit. Figure 4.11-3 shows a typical gain characteristic for these circuits. It should be noted that the frequency scale is normalized so that the equal-ripple band-edge frequency ω_1' is equal to one.

With the normalizations indicated above, the decrements for the normalized prototype circuits in Fig. 4.11-2 and Tables 4.11-1 to 4.11-3 are given by

$$\delta = \frac{1}{g_0 g_1 \omega_1'} = \frac{-1}{g_1} \quad . \tag{4.11-5}$$

This is the same definition of decrement as was used in Sec. 4.09, but, of course, since the resistance is negative the decrement is negative.

Practical negative-resistance devices have a parasitic reactive impedance component as well as negative resistance. If a negative-resistance device (such as a tunnel diode) may be represented by a negative-resistance R_0 in parallel with a shunt capacitance C_1, or by a negative conductance G_0 in series with an inductance L_1, then in either case if an operating range from $\omega = 0$ to ω_1 is desired, the decrement of the device is

$$\delta = \frac{1}{R_0 C_1 \omega_1} \quad \text{or} \quad \frac{1}{G_0 L_1 \omega_1} \quad . \tag{4.11-6}$$

This decrement is unchanged if the impedance level or frequency scale of the circuit is changed, so the desired equivalent element values for the device can be obtained by selecting a design from Tables 4.11-1 to 4.11-3 which has the required decrement value, and then altering the impedance level and frequency scale of the design as described in Sec. 4.04 so as to make the prototype circuit correspond to the desired design. By use

Table 4.11-1

LOW-PASS PROTOTYPE, NEGATIVE-RESISTANCE AMPLIFIER DESIGN PARAMETERS AND ELEMENT VALUES FOR $n = 2$

For these designs $g_0 = -1$ and $\omega_1' = 1$

$(G_t)_{min}$ (db)	G_{tr} (db)	g_1	g_2	g_3	$(G_t)_{min}$ (db)	G_{tr} (db)	g_1	g_2	g_3
5.000000	0.050000	2.047029	0.090762	3.569771	15.000000	0.050000	0.469574	0.139802	1.432581
	0.100000	2.373910	0.110079			0.100000	0.551911	0.167702	
	0.200000	2.717309	0.134542			0.200000	0.644755	0.201783	
	0.500000	3.142281	0.178296			0.500000	0.779488	0.259173	
	1.000000	3.366818	0.224112			1.000000	0.881797	0.314573	
	1.500000	3.423646	0.257965			1.500000	0.935940	0.352678	
	2.000000	3.419884	0.285850			2.000000	0.969092	0.382409	
	2.500000	3.389207	0.309936			2.500000	0.990582	0.406960	
	3.000000	3.345843	0.331297			3.000000	1.004796	0.427896	
8.000000	0.050000	1.123240	0.118139	2.322851	20.000000	0.050000	0.298027	0.131194	1.222222
	0.100000	1.311097	0.142574			0.100000	0.351245	0.156965	
	0.200000	1.515839	0.173040			0.200000	0.411985	0.188153	
	0.500000	1.792064	0.226176			0.500000	0.502288	0.239812	
	1.000000	1.972788	0.280055			1.000000	0.573930	0.288488	
	1.500000	2.049030	0.318820			1.500000	0.614000	0.321199	
	2.000000	2.083000	0.350145			2.000000	0.640025	0.346245	
	2.500000	2.095366	0.376789			2.500000	0.658075	0.366593	
	3.000000	2.095468	0.400110			3.000000	0.671031	0.383694	
10.000000	0.050000	0.834764	0.129751	1.924951	25.000000	0.050000	0.200869	0.113609	1.119170
	0.100000	0.976940	0.156244			0.100000	0.237197	0.135667	
	0.200000	1.134010	0.189038			0.200000	0.279000	0.162179	
	0.500000	1.352354	0.245540			0.500000	0.342139	0.205553	
	1.000000	1.504684	0.301886			1.000000	0.393617	0.245674	
	1.500000	1.576134	0.341831			1.500000	0.423354	0.272154	
	2.000000	1.613766	0.373748			2.000000	0.443296	0.292131	
	2.500000	1.633451	0.400644			2.500000	0.457604	0.308152	
	3.000000	1.642492	0.423994			3.000000	0.468262	0.321458	
12.000000	0.050000	0.649071	0.136702	1.670900	30.000000	0.050000	0.140150	0.093649	1.065311
	0.100000	0.761136	0.164326			0.100000	0.165732	0.111676	
	0.200000	0.886156	0.198319			0.200000	0.195341	0.133228	
	0.500000	1.063612	0.256292			0.500000	0.240559	0.168160	
	1.000000	1.192752	0.313285			1.000000	0.278115	0.200021	
	1.500000	1.257234	0.353164			1.500000	0.300269	0.220757	
	2.000000	1.294108	0.384704			2.000000	0.315426	0.236220	
	2.500000	1.315985	0.411054			2.500000	0.326523	0.248494	
	3.000000	1.328730	0.433757			3.000000	0.334966	0.258593	

SOURCE: *IEEE Trans. PTGMTT* (see Ref. 21 by W. J. Getsinger).

Table 4.11-2

LOW-PASS PROTOTYPE, NEGATIVE-RESISTANCE AMPLIFIER DESIGN PARAMETERS AND ELEMENT VALUES FOR $n = 3$

For these designs $g_0 = -1$ and $\omega_1' = 1$

$(G_t)_{min}$ (db)	G_{tr} (db)	g_1	g_2	g_3	g_4	$(G_t)_{min}$ (db)	G_{tr} (db)	g_1	g_2	g_3	g_4
5.000000	0.050000	3.331969	0.267590	1.480225	0.282907	15.000000	0.050000	0.899471	0.492406	0.424202	0.699589
	0.100000	3.609226	0.298983	1.712289	0.285538		0.100000	0.988304	0.551044	0.485895	0.701047
	0.200000	3.863582	0.333932	1.985458	0.290802		0.200000	1.078864	0.615791	0.558720	0.703954
	0.500000	4.109295	0.387519	2.401165	0.306513		0.500000	1.193476	0.711024	0.675120	0.712550
	1.000000	4.161764	0.437078	2.718835	0.332345		1.000000	1.265789	0.790323	0.778783	0.726424
	1.500000	4.111606	0.471951	2.875535	0.357693		1.500000	1.296596	0.839485	0.844033	0.739733
	2.000000	4.032773	0.500368	2.959788	0.382527		2.000000	1.310987	0.875496	0.891002	0.752488
	2.500000	3.945864	0.524956	3.004295	0.406821		2.500000	1.316946	0.903975	0.926958	0.764703
	3.000000	3.858565	0.546898	3.024945	0.430554		3.000000	1.318019	0.927511	0.955484	0.776393
8.000000	0.050000	1.937111	0.371582	0.853981	0.432964	20.000000	0.050000	0.612714	0.498175	0.312927	0.819179
	0.100000	2.111857	0.415407	0.984394	0.435286		0.100000	0.675675	0.557797	0.357161	0.820118
	0.200000	2.281328	0.463897	1.138997	0.439927		0.200000	0.741056	0.623680	0.409027	0.821989
	0.500000	2.472640	0.536446	1.382740	0.453732		0.500000	0.826882	0.720318	0.491559	0.827505
	1.000000	2.561531	0.599948	1.587860	0.476278		1.000000	0.885209	0.799773	0.565434	0.836360
	1.500000	2.576011	0.642064	1.705653	0.498219		1.500000	0.913192	0.848009	0.612641	0.844802
	2.000000	2.564324	0.674885	1.782205	0.519542		2.000000	0.928808	0.882565	0.647258	0.852843
	2.500000	2.541110	0.702323	1.834596	0.540238		2.500000	0.937854	0.909293	0.674295	0.860500
	3.000000	2.512514	0.726152	1.871329	0.560301		3.000000	0.942946	0.930906	0.696198	0.867787
10.000000	0.050000	1.486554	0.422623	0.662305	0.521697	25.000000	0.050000	0.443772	0.465624	0.247119	0.894128
	0.100000	1.625214	0.472625	0.761901	0.523777		0.100000	0.490786	0.521566	0.281204	0.894700
	0.200000	1.762385	0.527874	0.879988	0.527930		0.200000	0.540262	0.583450	0.320888	0.895840
	0.500000	1.924898	0.609885	1.067795	0.540258		0.500000	0.606828	0.674200	0.383550	0.899194
	1.000000	2.012176	0.680178	1.230320	0.560313		1.000000	0.654147	0.748423	0.439410	0.904561
	1.500000	2.038022	0.725586	1.327783	0.579736		1.500000	0.678291	0.793017	0.475181	0.909657
	2.000000	2.041053	0.760176	1.394269	0.598523		2.000000	0.692808	0.824582	0.501548	0.914494
	2.500000	2.033243	0.788539	1.442316	0.616675		2.500000	0.702113	0.848687	0.522276	0.919083
	3.000000	2.019759	0.812767	1.478165	0.634194		3.000000	0.708212	0.867927	0.539190	0.923436
12.000000	0.050000	1.190940	0.459836	0.540560	0.600417	30.000000	0.050000	0.333414	0.415205	0.201964	0.939052
	0.100000	1.304959	0.514395	0.620709	0.602244		0.100000	0.369610	0.465235	0.229242	0.939389
	0.200000	1.419374	0.574643	0.715620	0.605890		0.200000	0.408094	0.520639	0.260785	0.940060
	0.500000	1.559391	0.663636	0.867159	0.616693		0.500000	0.460818	0.601944	0.310167	0.942034
	1.000000	1.641159	0.738814	1.000439	0.634205		1.000000	0.499471	0.668329	0.353841	0.945186
	1.500000	1.671082	0.786424	1.082476	0.651093		1.500000	0.519967	0.708021	0.381704	0.948172
	2.000000	1.681114	0.822034	1.140072	0.667362		2.000000	0.532819	0.735944	0.402225	0.951000
	2.500000	1.681295	0.850757	1.183006	0.683019		2.500000	0.541475	0.757121	0.418363	0.953677
	3.000000	1.676024	0.874936	1.216132	0.698073		3.000000	0.547515	0.773901	0.431546	0.956211

SOURCE: *IEEE Trans. PTGMTT* (see Ref. 21 by W. J. Getsinger).

Table 4.11-3

LOW-PASS PROTOTYPE, NEGATIVE-RESISTANCE AMPLIFIER DESIGN PARAMETERS AND ELEMENT VALUES FOR $n = 4$

For these designs $g_0 = -1$ and $\omega_1' = 1$

$(G_t)_{min}$ (db)	G_{tr} (db)	g_1	g_2	g_3	g_4	g_5
5.000000	0.050000	4.067117	0.371272	3.823667	0.128714	3.569771
	0.100000	4.265940	0.397333	4.163568	0.148113	
	0.200000	4.430001	0.424924	4.499190	0.172666	
	0.500000	4.545024	0.466360	4.872849	0.216973	
	1.000000	4.498242	0.506094	5.019656	0.264118	
	1.500000	4.393585	0.535595	5.013455	0.299444	
	2.000000	4.278915	0.560596	4.957323	0.328835	
	2.500000	4.166251	0.582841	4.882189	0.354422	
	3.000000	4.059411	0.603106	4.800493	0.377260	
8.000000	0.050000	2.430347	0.534019	2.304947	0.177435	2.322851
	0.100000	2.563192	0.572286	2.518041	0.202431	
	0.200000	2.682265	0.612109	2.737978	0.233399	
	0.500000	2.796735	0.668730	3.014286	0.287590	
	1.000000	2.823660	0.717907	3.175971	0.343263	
	1.500000	2.802370	0.751468	3.232275	0.383893	
	2.000000	2.766292	0.778450	3.248102	0.417098	
	2.500000	2.725038	0.801621	3.244283	0.445613	
	3.000000	2.682429	0.822198	3.229971	0.470780	
10.000000	0.050000	1.894074	0.619039	1.828252	0.202109	1.924951
	0.100000	2.002472	0.663986	1.999482	0.229662	
	0.200000	2.102309	0.710558	2.178931	0.263443	
	0.500000	2.206311	0.775551	2.413300	0.321650	
	1.000000	2.245017	0.829690	2.564675	0.380348	
	1.500000	2.242173	0.865059	2.629503	0.422548	
	2.000000	2.225317	0.892597	2.659667	0.456673	
	2.500000	2.202606	0.915679	2.672077	0.485731	
	3.000000	2.177419	0.935793	2.674328	0.511195	
12.000000	0.050000	1.539498	0.685375	1.522322	0.220959	1.670900
	0.100000	1.630827	0.735758	1.666094	0.250258	
	0.200000	1.716537	0.787861	1.818352	0.285860	
	0.500000	1.810401	0.859716	2.022419	0.346399	
	1.000000	1.852870	0.917789	2.162431	0.406462	
	1.500000	1.859130	0.954403	2.228930	0.449070	
	2.000000	1.852538	0.982108	2.265234	0.483190	
	2.500000	1.840129	1.004799	2.285589	0.512017	
	3.000000	1.824882	1.024200	2.296465	0.537112	

$(G_t)_{min}$ (db)	G_{tr} (db)	g_1	g_2	g_3	g_4	g_5
15.000000	0.050000	1.187226	0.752175	1.227945	0.239468	1.432581
	0.100000	1.260775	0.808430	1.344798	0.270104	
	0.200000	1.331198	0.866578	1.469873	0.306909	
	0.500000	1.412175	0.945950	1.641896	0.368431	
	1.000000	1.454822	1.008145	1.766806	0.428181	
	1.500000	1.467311	1.045794	1.831425	0.469817	
	2.000000	1.468609	1.073275	1.870707	0.502725	
	2.500000	1.464518	1.095092	1.896230	0.530232	
	3.000000	1.457540	1.113247	1.913305	0.553962	
20.000000	0.050000	0.836990	0.791503	0.945424	0.249706	1.222222
	0.100000	0.891859	0.852249	1.035868	0.280095	
	0.200000	0.945608	0.915190	1.133698	0.316034	
	0.500000	1.010563	1.000585	1.271672	0.374707	
	1.000000	1.049369	1.065684	1.377209	0.430021	
	1.500000	1.064820	1.103448	1.435786	0.467610	
	2.000000	1.071251	1.129874	1.474223	0.496770	
	2.500000	1.073112	1.150033	1.501452	0.520778	
	3.000000	1.072355	1.166192	1.521607	0.541223	
25.000000	0.050000	0.627756	0.769122	0.778152	0.242959	1.119170
	0.100000	0.670809	0.829501	0.852613	0.271317	
	0.200000	0.713673	0.892317	0.933607	0.304422	
	0.500000	0.767156	0.977611	1.049401	0.357433	
	1.000000	0.801375	1.041965	1.140409	0.406203	
	1.500000	0.816766	1.078513	1.192674	0.438665	
	2.000000	0.824708	1.103482	1.228146	0.463465	
	2.500000	0.828774	1.122060	1.254155	0.483627	
	3.000000	0.830548	1.136584	1.274107	0.500612	
30.000000	0.050000	0.488870	0.713195	0.662674	0.227225	1.065311
	0.100000	0.523683	0.770292	0.725929	0.252817	
	0.200000	0.558781	0.829963	0.794933	0.282369	
	0.500000	0.603598	0.911298	0.894321	0.328930	
	1.000000	0.633575	0.972573	0.973601	0.370901	
	1.500000	0.647992	1.007074	1.019958	0.398356	
	2.000000	0.656150	1.030372	1.051959	0.419062	

SOURCE: *IEEE Trans. PTGMTT* (see Ref. 21 by W. J. Getsinger).

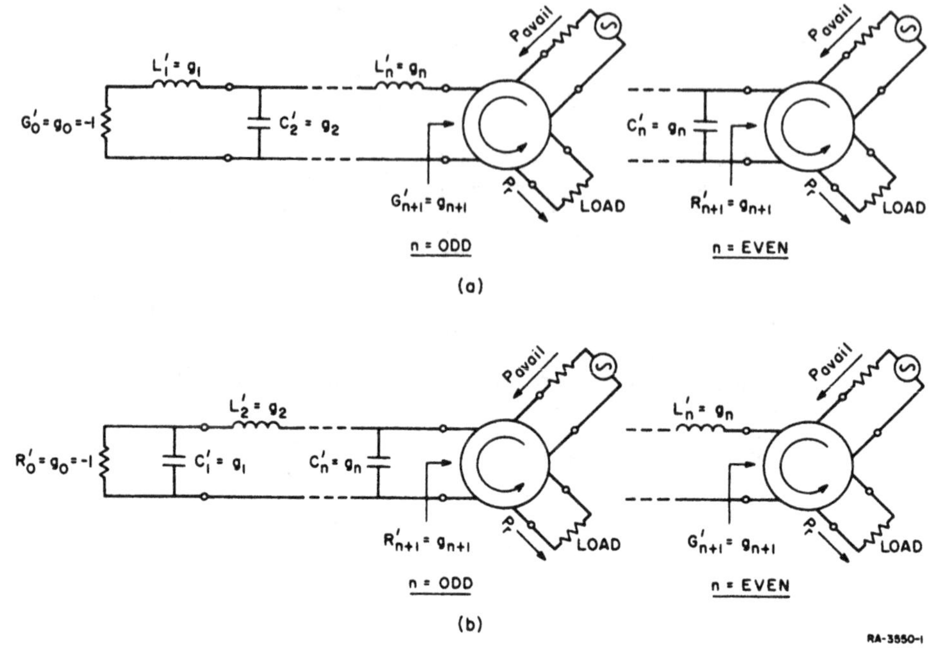

FIG. 4.11-2 DUAL INTERPRETATIONS OF THE LOW-PASS, NEGATIVE-RESISTANCE AMPLIFIER DESIGNS IN TABLES 4.11-1 TO 4.11-3

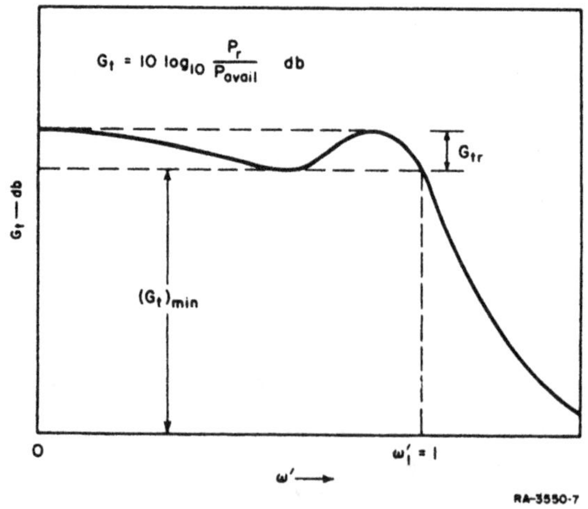

FIG. 4.11-3 A TYPICAL TCHEBYSCHEFF TRANSDUCER GAIN CHARACTERISTIC FOR THE LOW-PASS PROTOTYPE NEGATIVE-RESISTANCE AMPLIFIERS IN TABLES 4.11-1 TO 4.11-3
The number of pass-band ripples shown corresponds to the case of n = 3

of Eq. (4.11-5), if the required negative decrement δ has been determined, the designs in Tables 4.11-1 through 4.11-3 which can be used are those that have a value of g_1 equal to or larger than

$$g_1 = \frac{-1}{\delta} \quad . \tag{4.11-7}$$

A value of g_1 greater than the value given by Eq. (4.11-7) is acceptable because to achieve this value it is merely necessary to add more shunt capacitance (or more series inductance) to the negative-resistance device, and this is always possible to do. However, since δ was assumed to have been computed from the device in its fundamental form, from which no capacitance or inductance can be removed, it is not possible to make use of designs calling for values of g_1 that are smaller than that indicated by Eq. (4.11-7). To obtain the maximum gain corresponding to a given bandwidth, choose a design from the table (or by interpolation from the table) so that g_1 is exactly the value given by Eq. (4.11-7).

Since the design procedure herein assumes the use of a circulator, and since practical circulators have limited bandwidth, the data in this section is of most interest as a source of low-pass prototype designs to be used in the design of band-pass amplifiers as described in Sec. 11.10. However, in order to further clarify the selection of prototype designs, let us consider a hypothetical low-pass example for use with an idealized circulator of unlimited bandwidth.

Let us suppose that a negative-resistance device can be represented by a negative resistance of −50 ohms in parallel with a capacitance of 5 $\mu\mu$f, and that it is desired to use this device in an amplifier which is to have minimum pass-band gain $(G_t)_{min}$ of 15 db, and a Tchebyscheff ripple of G_{tr} = 1 db. Let us suppose that very high performance is needed, so that an n = 4 circuit from Table 4.11-3 is chosen in preference to the simpler and lower performance circuits in Tables 4.11-1 and 4.11-2. From Table 4.11-3 the design with the desired gain parameters has $g_1 = 1.4548$, $g_2 = 1.0081$, $g_3 = 1.7668$, $g_4 = 0.4282$, and $g_5 = 1.4326$.

By Eqs. (4.11-5) and (4.11-6)

$$\omega_1 = \frac{1}{R_0 C_1 \delta} = \frac{-g_1}{R_0 C_1} = \frac{-1.4548}{-50(5 \times 10^{-12})} = 5.82 \times 10^9 \quad . \tag{4.11-8}$$

Thus, this design will give the desired 15-db minimum gain with 1-db Tchebyscheff ripple up to $f_1 = (\omega_1/2\pi) = 926$ Mc. The prototype circuit has the form in Fig. 4.11-2(b) (for n even) with $R_0' = -1$, $C_1' = 1.4548$, and $\omega_1' = 1$. Using the methods of Sec. 4.04, the impedance of this circuit is next scaled up by a factor $R_0/R_0' = (-50)/(-1) = 50$, while the frequency scale is altered by the factor $\omega_1/\omega_1' = (5.82)(10^9)/1$. These scale changes make $R_0' = -1$ become $R_0 = -50$, and $C_1' = 1.4548$ become $C_1 = 5 \times 10^{-12}$, and make the other elements take corresponding values. Then in the scaled circuit, R_0 and C_1 on the left [see Fig. 4.11-2(b)] are replaced by the negative-resistance device which is to be used, while L_2, C_3, and L_4 comprise the elements of the circuit required to give the desired broad-band response shape up to 926 Mc.

The physical structure of the low-pass filter structure for the above example could be designed using methods described in Chapter 5. However, as mentioned above, this low-pass amplifier example is rather hypothetical because of the bandwidth limitations of circulators, and the results of this section are of most practical interest for use in the design of band-pass amplifiers as described in Sec. 11.10.

SEC. 4.12, CONVERSION OF FILTER PROTOTYPES TO USE IMPEDANCE- OR ADMITTANCE-INVERTERS AND ONLY ONE KIND OF REACTIVE ELEMENT

In deriving design equations for certain types of band-pass and band-stop filters it is desirable to convert the prototypes in Fig. 4.04-1 which use both inductances and capacitances to equivalent forms which use only inductances or only capacitances. This can be done with the aid of the idealized inverters which are symbolized in Fig. 4.12-1.

An idealized *impedance inverter* operates like a quarter-wavelength line of characteristic impedance K at all frequencies. Therefore, if it is terminated in an impedance Z_b on one end, the impedance Z_a seen looking in at the other end is

$$Z_a = \frac{K^2}{Z_b} \quad . \tag{4.12-1}$$

An idealized *admittance inverter* as defined herein is the admittance representation of the same thing, *i.e.*, it operates like a quarter-wavelength line of characteristic admittance J at all frequencies. Thus, if an admittance Y_b is attached at one end, the admittance Y_a seen looking in the other end is

$$Y_a = \frac{J^2}{Y_b} \quad . \tag{4.12-2}$$

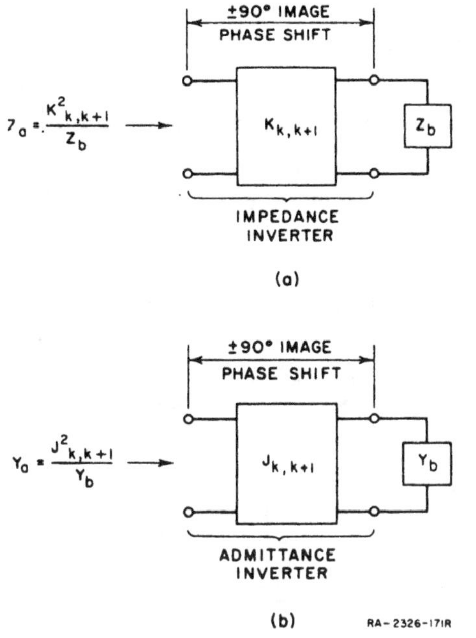

SOURCE: Final Report, Contract DA 36-039 SC-74862, Stanford Research Institute, reprinted in *IRE Trans., PGMTT* (see Ref. 1 of Chapter 10, by G. L. Matthaei).

FIG. 4.12-1 DEFINITION OF IMPEDANCE INVERTERS AND ADMITTANCE INVERTERS

As indicated in Fig. 4.12-1, an inverter may have an image phase shift of either ±90 degrees or an odd multiple thereof.

Because of the inverting action indicated by Eqs. (4.12-1) and (4.12-2) a series inductance with an inverter on each side looks like a shunt capacitance from its exterior terminals. Likewise, a shunt capacitance with an inverter on both sides looks like a series inductance from its external terminals. Making use of this property, the prototype circuits in Fig. 4.04-1 can be converted to either of the equivalent forms in Fig. 4.12-2 which have identical transmission characteristics to those prototypes in Fig. 4.04-1. As can be seen from Eqs. (4.12-1) and (4.12-2), inverters have the ability to shift impedance or admittance levels depending on the choice of the K or J parameters. For this reason in Fig. 4.12-2(a) the sizes of R_A, R_B, and the inductances L_{ak} may be chosen arbitrarily and the response will be identical to that of the original prototype as in Fig. 4.04-1 provided that the inverter parameters $K_{k,k+1}$ are specified as indicated by the equations in Fig. 4.12-2(a). The same holds for the circuit in Fig. 4.12-2(b) only on the dual basis. Note that the g_k values referred to in the equations in Fig. 4.12-2 are the prototype element values as defined in Fig. 4.04-1.

A way that the equations for the $K_{k,k+1}$ and $J_{k,k+1}$ can be derived will now be briefly considered. A fundamental way of looking at the relation between the prototype circuits in Figs. 4.04-1(a), (b) and the corresponding circuit in, say, Fig. 4.12-2(a) makes use of the concept of duality. A given circuit as seen through an impedance inverter looks like the dual of that given circuit. Thus, the impedances seen from inductor L_{a1} in Fig. 4.12-2(a) are the same as those seen from inductance L_1' in Fig. 4.04-1(b), except for an impedance scale factor. The impedances seen from inductor L_{a2} in Fig. 4.12-2(a) are identical to those seen from inductance L_2' in Fig. 4.04-1(a), except for a possible impedance scale change. In this manner the impedances in any point of the circuit in Fig. 4.12-2(a) may be quantitatively related to the corresponding impedances in the circuits in Fig. 4.04-1(a), (b).

Figure 4.12-3(a) shows a portion of a low-pass prototype circuit that has been open-circuited just beyond the capacitor C_{k+1}. The dual circuit is shown at (b), where it should be noted that the open circuit

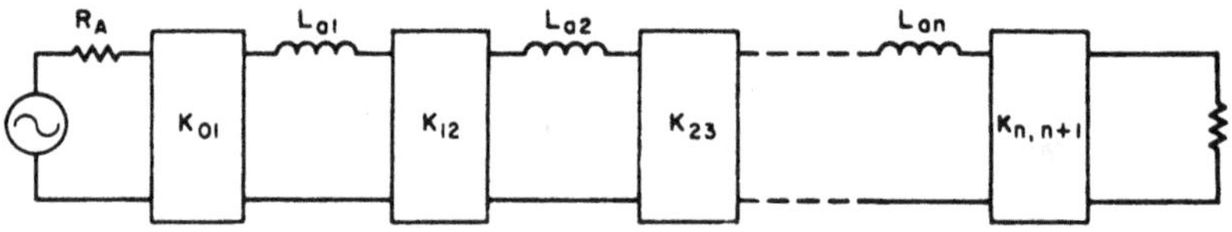

$$K_{01} = \sqrt{\frac{R_A L_{a1}}{g_0 g_1}} \qquad K_{k,k+1}\Big|_{k=1 \text{ to } n-1} = \sqrt{\frac{L_{ak} L_{a(k+1)}}{g_k\, g_{k+1}}} , \qquad K_{n,n+1} = \sqrt{\frac{L_{an} R_B}{g_n\, g_{n+1}}}$$

(a) MODIFIED PROTOTYPE USING IMPEDANCE INVERTERS

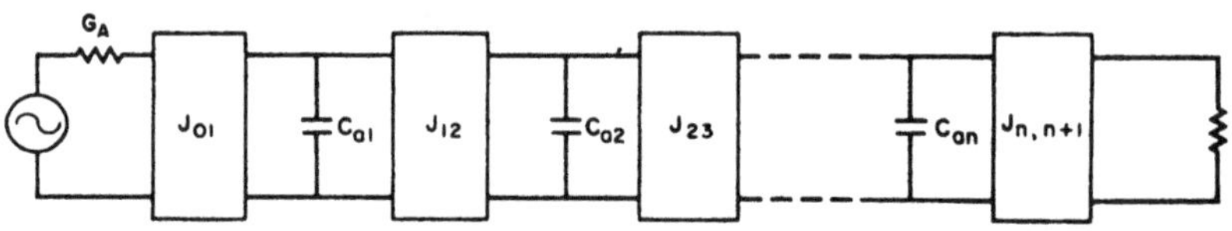

$$J_{01} = \sqrt{\frac{G_A C_{a1}}{g_0 g_1}} , \qquad J_{k,k+1}\Big|_{k=1 \text{ to } n-1} = \sqrt{\frac{C_{ak} C_{a(k+1)}}{g_k\, g_{k+1}}} , \qquad J_{n,n+1} = \sqrt{\frac{C_{an} G_B}{g_n\, g_{n+1}}}$$

(b) MODIFIED PROTOTYPE USING ADMITTANCE INVERTERS

RB-2326-T7-172R

SOURCE: Final Report, Contract DA 36-039 SC-74862, Stanford Research Institute, reprinted in *IRE Trans., PGMTT* (see Ref. 1 of Chapter 10, by G. L. Matthaei).

FIG. 4.12-2 LOW-PASS PROTOTYPES MODIFIED TO INCLUDE IMPEDANCE INVERTERS OR ADMITTANCE INVERTERS
The $g_0, g_1, \ldots, g_{n+1}$ are obtained from the original prototype as in Fig. 4.04-1, while the $R_A, L_{a1}, \ldots, L_{an}$, and R_B or the $G_A, C_{a1}, \ldots, C_{an}$ and G_B may be chosen as desired.

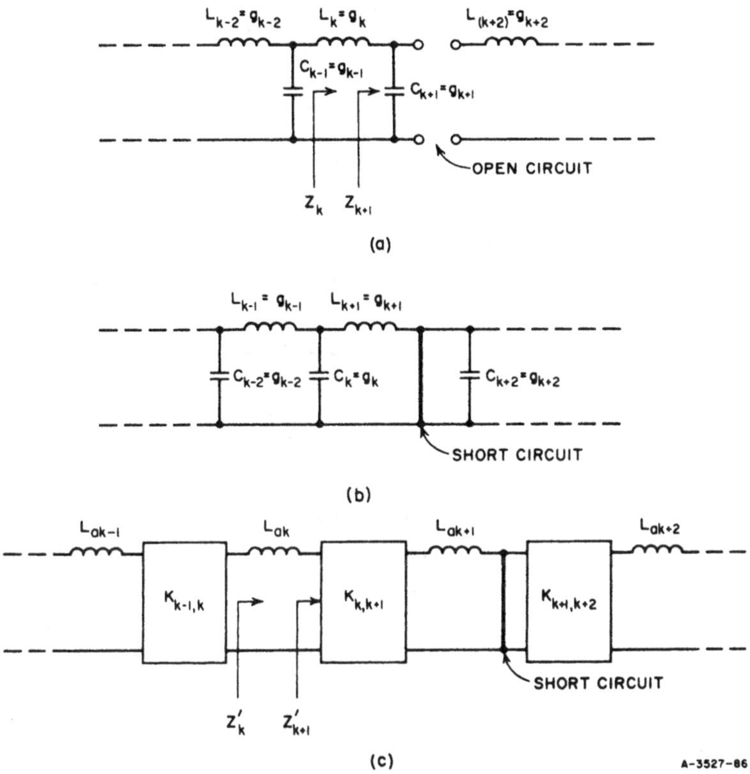

FIG. 4.12-3 SOME CIRCUITS DISCUSSED IN SEC. 4.12
A ladder circuit is shown at (a), and its dual is shown at (b). The analogous K-inverter form of these two circuits is shown at (c).

shown at (a) becomes a short circuit in the dual case. The corresponding circuit using all series inductors and K inverters is shown at (c). The circuits in Fig. 4.12-3 will be convenient for deriving the formula for $K_{k,k+1}$ in terms of L_{ak}, L_{ak+1}, and the prototype element values g_k and g_{k+1}. The open- and short-circuits are introduced merely to simplify the equations.

Referring to Fig. 4.12-3, in the circuit at (a),

$$Z_k = j\omega L_k + \frac{1}{j\omega C_{k+1}} \qquad (4.12\text{-}3)$$

Meanwhile in the circuit at (c)

$$Z'_k = j\omega L_{ak} + \frac{K^2_{k,k+1}}{j\omega L_{ak+1}} \qquad (4.12\text{-}4)$$

Now Z'_k must be identical to Z_k except for an impedance scale change of L_{ak}/L_k. Therefore

$$Z'_k = \frac{L_{ak}}{L_k} Z_k = j\omega L_{ak} + \frac{L_{ak}}{L_k} \frac{1}{j\omega C_{k+1}} . \qquad (4.12\text{-}5)$$

Equating the second terms in Eqs. (4.12-4) and (4.12-5) gives, after some rearrangement,

$$K_{k,k+1} = \sqrt{\frac{L_{ak} L_{ak+1}}{L_k C_{k+1}}} . \qquad (4.12\text{-}6)$$

Since $L_k = g_k$ and $C_{k+1} = g_{k+1}$, Eq. (4.12-6) is equivalent to the equation for $K_{k,k+1}$ given in Fig. 4.12-2(a). It is easily seen that by moving the positions of the open- and short-circuit points correspondingly, the same procedure would apply for calculation of the K's for all the inverters except those at the ends. Hence, Eq. (4.12-6) applies for $k = 1, 2, \ldots, n-1$.

Next consider Fig. 4.12-4. At (a) is shown the last two elements of a prototype circuit and at (b) is shown a corresponding form with a K inverter. In the circuit at (a)

$$Z_n = j\omega L_n + \frac{1}{G_{n+1}} \qquad (4.12\text{-}7)$$

while at (b)

$$Z'_n = j\omega L_{an} + \frac{K^2_{n,n+1}}{R_B} . \qquad (4.12\text{-}8)$$

Since Z'_n must equal Z_n within a scale factor L_{an}/L_n,

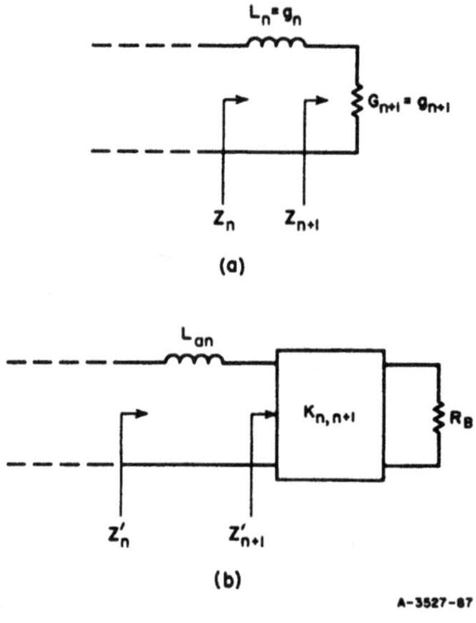

FIG. 4.12-4 ADDITIONAL CIRCUITS DISCUSSED IN SEC. 4.12
The end portion of a prototype circuit is shown at (a) while at (b) is shown the corresponding end portion of a circuit with K-inverters.

$$Z'_n = \frac{L_{an}}{L_n} Z_n = j\omega L_{an} + \frac{L_{an}}{L_n} \frac{1}{G_{n+1}} \quad . \tag{4.12-9}$$

Equating the second terms of Eqs. (4.12-8) and (4.12-9) leads to the result

$$K_{n,n+1} = \sqrt{\frac{L_{an} R_B}{L_n G_{n+1}}} \quad . \tag{4.12-10}$$

Substituting g_n and g_{n+1} for L_n and G_{n+1}, respectively, gives the equation for $K_{n,n+1}$ shown in Fig. 4.12-2.

The derivation of the equations for the $J_{k,k+1}$ parameters in Fig. 4.12-2(b) may be carried out in like manner on the admittance (*i.e.*, dual) basis.

SEC. 4.13, EFFECTS OF DISSIPATIVE ELEMENTS IN PROTOTYPES FOR LOW-PASS, BAND-PASS, OR HIGH-PASS FILTERS

Any practical microwave filter will have elements with finite Q's, and in many practical situations it is important to be able to estimate the effect of these finite element Q's on pass-band attenuation. When a filter has been designed from a low-pass prototype filter it is convenient to relate the microwave filter element Q's to dissipative elements in the prototype filter and then determine the effects of the dissipative elements on the prototype filter response. Then the increase in pass-band attenuation of the prototype filter due to the dissipative elements will be the same as the increase in pass-band attenuation (at the corresponding frequency) of the microwave filter due to the finite element Q's.

The element Q's referred to below are those of the elements of a low-pass filter at its cutoff frequency ω_1 and are defined as

$$Q_k = \frac{\omega_1 L_k}{R_k} \quad \text{or} \quad \frac{\omega_1 C_k}{G_k} \tag{4.13-1}$$

where R_k is the parasitic resistance of the inductance L_k, and G_k is the parasitic conductance of the capacitance C_k.* In the case of a band-pass

* Here, the unprimed L_k, R_k, C_k, G_k, and ω_1 values are meant to apply to any low-pass filter, whether it is a normalized prototype or not. Later in this section primes will be introduced to aid in distinguishing between the low-pass prototype parameters and those of the corresponding band-pass or high-pass filter.

filter which is designed from a low-pass prototype, if $(Q_{BP})_k$ is the mid-band unloaded Q of the kth resonator of the band-pass filter, then the corresponding Q of the kth reactive element of the prototype is

$$Q_k = w(Q_{BP})_k \qquad (4.13\text{-}2)$$

In this equation w is the fractional bandwidth of the band-pass filter as measured to its pass-band edges which correspond to the ω_1' pass-band edge of the low-pass prototype (see Chapter 8). The unloaded Q of the resonators can be estimated by use of the data in Chapter 5, or it can be determined by measurements as in Sec. 11.02.

In the case of a high-pass filter designed from a low-pass prototype, the element Q's of the prototype should be made to be the same as the Q's of the corresponding elements of the high-pass filter at its cutoff frequency.

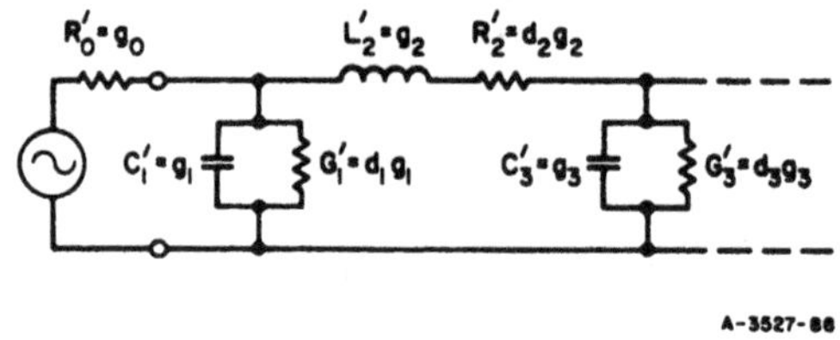

FIG. 4.13-1 LOW-PASS PROTOTYPE FILTER WITH DISSIPATIVE ELEMENTS ADDED

Figure 4.13-1 shows a portion of a low-pass prototype filter with parasitic loss elements introduced. Note that the parasitic loss element to go with reactive element g_k is designated as $d_k g_k$, where d_k will be referred to herein as a *dissipation factor*. Using this notation Eq. (4.13-1) becomes $Q_k = \omega_1' g_k/(d_k g_k) = \omega_1'/d_k$ where ω_1' is the cutoff frequency of the low-pass prototype. Thus,

$$d_k = \frac{\omega_1'}{Q_k} \qquad (4.13\text{-}3)$$

Then for a series branch of a prototype filter

$$Z_k = j\omega' L_k' + R_k' = (j\omega' + d_k)g_k \qquad (4.13\text{-}4)$$

and for a shunt branch

$$Y_k = j\omega' C_k' + G_k' = (j\omega' + d_k)g_k \qquad (4.13\text{-}5)$$

A special case of considerable practical interest is that where the Q's of all the elements are the same so that $d_k = d$ for $k = 1$ to n. Then, as can be seen from Eqs. (4.13-4) and (4.13-5), the effects of dissipation can be accounted for by simply replacing the frequency variable $j\omega'$ for the lossless circuit by $(j\omega' + d)$ to include the losses. For example, this substitution can be made directly in the transfer functions in Eqs. (4.07-1), (4.08-5) to (4.08-8) in order to compute the transfer characteristics with parasitic dissipation included. At DC the function $(j\omega' + d)$ becomes simply d, so that if

$$\frac{(E_2)_{avail}}{E_2} = P_n(p')\big|_{p'=j\omega'} = a_n(j\omega')^n + \ldots + a_1\, j\omega' + a_0 \qquad (4.13\text{-}6)$$

for a dissipationless prototype, the DC loss for a prototype with uniform dissipation d is for $\omega' = 0$

$$\left.\frac{(E_2)_{avail}}{E_2}\right|_{\omega'=0} = P_n(d) = a_n d^n + \ldots + a_1 d + a_0 \qquad (4.13\text{-}7)$$

where $(E_2)_{avail}/E_2$ is as defined in Sec. 2.10. Usually d is small so that only the last two terms of Eq. (4.13-7) are significant. Then it is easily shown that

$$(\Delta L_A)_0 = 20 \log_{10} [C_n d + 1] \text{ db} \qquad (4.13\text{-}8)$$

$$\approx 8.686\, C_n d$$

where $(\Delta L_A)_0$ is the db *increase* in attenuation at $\omega' = 0$ when d is finite, over the attenuation when $d = 0$ (*i.e.*, when there is no dissipation loss).* The coefficient $C_n = a_1/a_0$ where a_1 and a_0 are from polynomial $P_n(j\omega')$ in Eq. (4.13-6).

In the case of low-pass prototypes for band-pass filters, $(\Delta L_A)_0$ is also the increase in the *midband* loss of the corresponding band-pass filter as a result of finite resonator Q's. For high-pass filters designed

---

* For example, a dissipationless, 0.5-db ripple Tchebyscheff filter with $n = 4$ would have $L_A =$ 0.5 db for $\omega' = 0$. If uniform dissipation is introduced the attenuation for $\omega' = 0$ will become $L_A = 0.5 + (\Delta L_A)_0$ db.

from low-pass prototypes, $(\Delta L_A)_0$ relates to the attenuation as $\omega \to \infty$. Equation (4.13-8) applies both for prototypes such as those in Sec. 4.05 which for the case of no dissipation loss have points where L_A is zero, and also for the impedance-matching network prototypes in Secs. 4.09 and 4.10 which even for the case of no dissipation have non-zero L_A at all frequencies.

Table 4.13-1 is a tabulation of the coefficients C_n for prototype filters having maximally flat attenuation with their 3-db point at $\omega_1' = 1$. Figure 4.13-2 shows the C_n coefficients for Tchebyscheff filters plotted *vs* db pass-band ripple. In this case the equal-ripple band edge is $\omega_1' = 1$. Note that above about 0.3 db-ripple, the curves fall for n even and rise for n odd. This phenomenon is related to the fact that a Tchebyscheff prototype filter with n even has a ripple maximum at $\omega' = 0$, while a corresponding filter with n odd has a ripple minimum at that frequency. There is apparently a tendency for the effects of dissipation to be most pronounced at ripple minima.

Table 4.13-1

MAXIMALLY FLAT ATTENUATION FILTER COEFFICIENTS C_n FOR USE IN EQ. (4.13-8)

These coefficients are for filters with their 3-db point at $\omega_1' = 1$ and are equal to the group time delay in seconds as ω' approaches zero [see Eq. (4.08-2)]

n	C_n	n	C_n
1	1.00	9	5.76
2	1.41	10	6.39
3	2.00	11	7.03
4	2.61	12	7.66
5	3.24	13	8.30
6	3.86	14	8.93
7	4.49	15	9.57
8	5.13		

Bode[19] gives an equation for ΔL_A, the increase in attenuation due to uniform dissipation, as a function of the attenuation phase slope and the dissipation factor, d. Bode's equation may be expressed in the form

$$\Delta L_A \approx 8.686\, d \frac{d\phi}{d\omega'} \quad \text{db} \tag{4.13-9}$$

where

$$\phi = \arg \frac{(E_2)_{\text{avail}}}{E_2} \quad , \tag{4.13-10}$$

and in this case ΔL_A is the increase in attenuation at, ω', the frequency, at which $d\phi/d\omega'$ is evaluated.* Thus, this equation provides a convenient

* It can be seen from Eqs. (4.13-8) and (4.13-9) that the C_n coefficients in Table 4.13-1 and Fig. 4.13-2 are equal to the group time delay in seconds as ω' approaches zero.

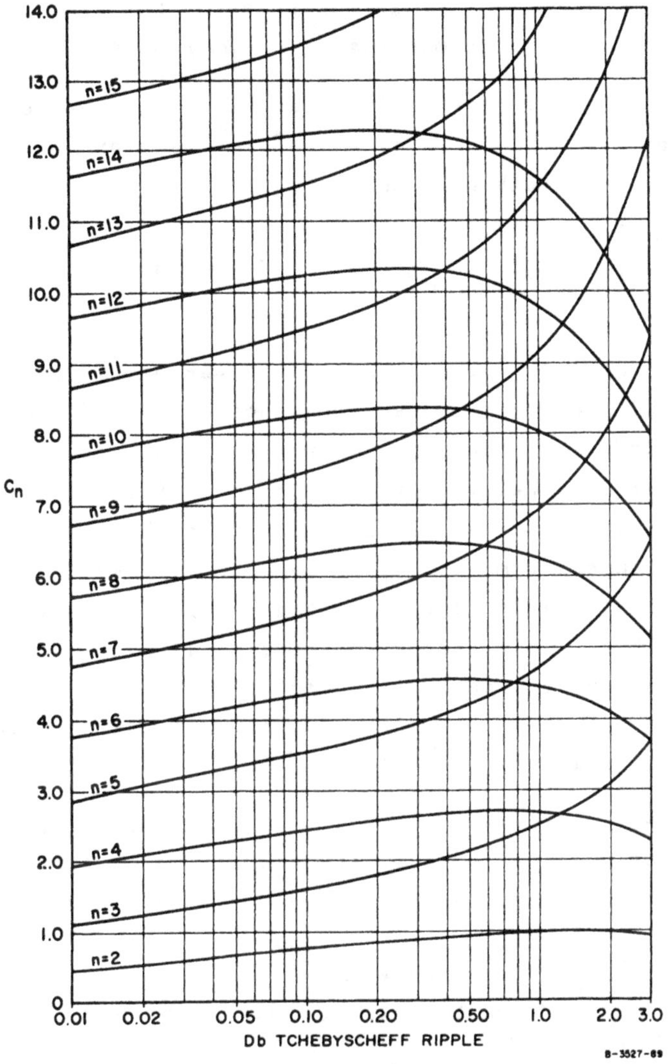

FIG. 4.13-2 PROTOTYPE TCHEBYSCHEFF FILTER COEFFICIENTS C_n vs. db TCHEBYSCHEFF RIPPLE, FOR PROTOTYPES WITH n REACTIVE ELEMENTS AND $\omega_1' = 1$

These coefficients are for use in Eq. (4.13-8). They are also the group time delay in seconds as ω' approaches zero (see Eq. 4.08-2).

means for estimating the effects of uniform dissipation at any frequency. Bode's discussion[19] indicates that for cases where all the inductances have a given Q, Q_L, and all of the capacitances have another Q, Q_c, good results can be obtained by computing d as $2\omega_1'/(Q_L + Q_c)$.

Cohn[20] has presented another formula which is convenient for estimating the effects of dissipation loss of low-pass prototypes for $\omega' = 0$. His formula may be expressed in the form

$$(\Delta L_A)_0 = 4.343 \sum_{k=1}^{n} d_k g_k \quad \text{db} \tag{4.13-11}$$

where the d_k are given by Eq. (4.13-3), and the prototype element values g_k are specifically assumed to have been normalized so that $g_0 = 1$ (as has been done for all of the prototypes discussed in this chapter). Note that this formula does not require that the dissipation be uniform. Equation (4.13-11) was derived by assuming that the load and source resistances are both one ohm, and that the effect of each R_k' or G_k' in Fig. 4.13-1 at $\omega' = 0$ is to act as voltage or current divider with respect to one ohm.[20] As a result, Eq. (4.13-11) can lead to appreciable error if the load and source resistances are sizeably different, though it generally gives very good results if the terminations are equal or at least not very greatly different.*

Table 4.13-2 compares the accuracy of Eqs. (4.13-8), (4.13-9), and (4.13-11) for various Tchebyscheff filters having uniform dissipation. Cases 1 to 3, which are for filters with $n = 4$ reactive elements, have

Table 4.13-2

COMPARISON OF ACCURACY OF EQS. (4.13-8), (4.13-9), AND (4.13-11) FOR COMPUTING $(\Delta L_A)_0$ FOR VARIOUS TCHEBYSCHEFF FILTERS HAVING UNIFORM DISSIPATION

CASE	n	db RIPPLE	Q	$(\Delta L_A)_0$ ACTUAL VALUE	$(\Delta L_A)_0$ BY EQ. (4.13-8)	$(\Delta L_A)_0$ BY EQ. (4.13-9)	$(\Delta L_A)_0$ BY EQ. (4.13-11)
1	4	0.5	100	0.236	0.232	--	0.264
2	4	2.0	100	0.223	0.214	--	0.346
3	4	2.0	10	2.39	1.95	--	3.46
4	5	0.5	100	0.364	0.357	0.35	0.365
5	5	0.5	10	3.55	3.05	3.5	3.65

* Equation (4.13-11) can be made to be more accurate for the case of unequal terminations by multiplying its right-hand side by $4R_0R_{n+1}/(R_0 + R_{n+1})^2$ where R_0 and R_{n+1} are the resistances of the terminations. This can be seen from the alternate point of view in Sec. 6.14.

unequal terminations; hence, Eq. (4.13-11) has relatively low accuracy if the pass-band ripples are large. Equation (4.13-8) gives reduced accuracy if the value of Q is very low. This happens as a result of using only the last two terms in Eq. (4.13-7). The actual value of $(\Delta L_A)_0$ was computed by using as many terms in Eq. (4.13-7) as was required in order to obtain high accuracy. The values computed using Eq. (4.13-9) were obtained by computing phase slope from Fig. 4.08-1. Note that the results are quite good. The $Q = 10$ values included in Table 4.13-2 are of practical interest since in the case of low-pass prototypes of band-pass filters the element Q's for the low-pass prototype can become quite low if the fractional bandwidth w of the band-pass filter is small [see Eq. (4.13-2)].

The above discussion treats the effects of parasitic dissipation at $\omega' = 0$, and the important question arises as to what the loss will be elsewhere in the pass band. Equation (4.13-9) provides convenient means for obtaining an approximate answer to this question. Since it says that ΔL_A at any frequency is proportional to the attenuation phase slope (*i.e.*, the group time delay) at that frequency, we can estimate ΔL_A across the pass band by examining the phase slope across that band. As seen from the examples in Fig. 4.08-1, the phase slope in typical cases is greatest near the cutoff frequency. In Fig. 4.08-1 the slope near cutoff is 2.66, 1.73, and 1.49 times the slope at $\omega' = 0$ for the cases of 0.5-db ripple, 0.01-db ripple, and maximally flat responses, respectively. Thus ΔL_A near cutoff will be greater than $(\Delta L_A)_0$ at $\omega' = 0$ by about these factors. These results are typical and are useful in obtaining an estimate of what to expect in practical situations.

SEC. 4.14, APPROXIMATE CALCULATION OF PROTOTYPE STOP-BAND ATTENUATION

Cohn[20] has derived a convenient formula for computing the attenuation of low-pass filters at frequencies well into their stop bands. This formula is derived using the assumption that the reactances of the series inductances are very large compared to the reactances of the shunt capacitors. When this condition holds, the voltage at one node of the filter may be computed with good accuracy from that at the preceding node using a simple voltage divider computation.[20] Cohn further simplifies his formula by use of the assumption that $(\omega^2 L_k C_{k+1} - 1) \approx \omega^2 L_k C_{k+1}$.

Cohn's formula, when put in the notation of the low-pass prototype filters in this chapter, is

$$L_A = 20 \log_{10} [(\omega')^n (g_1 g_2 g_3 \cdots g_n)]$$

$$- 10 \log_{10} \left(\frac{4}{g_0 g_{n+1}} \right) \quad \text{db} \qquad (4.14\text{-}1)$$

where g_0, g_1, ..., g_{n+1} are the prototype element values defined in Fig. 4.04-1(a), (b) and ω' is the prototype radian frequency variable. For this formula to have high accuracy, ω' should be a number of times as large as ω_1', the filter cutoff frequency.

As an example, consider a Tchebyscheff filter with $n = 4$ reactive elements and 0.2 db ripple. By Table 4.05-2(a), $g_0 = 1$, $g_1 = 1.3028$, $g_2 = 1.2844$, $g_3 = 1.9761$, $g_4 = 0.8468$, $g_5 = 1.5386$, and the cutoff frequency is $\omega_1' = 1$. By Eq. (4.14-1), to slide-rule accuracy

$$L_A = 20 \log_{10} [(\omega')^4 (4.29)] - 10 \log_{10} 6.15 \quad . \qquad (4.14\text{-}2)$$

Evaluating Eq. (4.14-2) for $\omega' = 3$ gives $L_A = 43.1$ db. By Fig. 4.03-6 we find that the actual attenuation is 42 db. Repeating the calculation for $\omega' = 2$ gives $L_A = 28.8$ db as compared to 26.5 db by Fig. 4.03-6. Thus it appears that even for values of ω'/ω_1' as small as 2, Eq. (4.14-1) gives fairly good results. The error was +2.3 db for $\omega' = 2$ and +1.1 db for $\omega' = 3$.

Equation (4.14-1) neglects the effects of dissipation in the circuit. This is valid as long as the dissipative elements in the prototype can be assumed to be arranged as are those in Fig. 4.13-1. This arrangement of dissipative elements is usually appropriate for prototypes for low-pass, band-pass, and high-pass filters. However, in the case of prototypes for band-stop filters, the different arrangement of dissipative elements discussed in Sec. 4.15 should be assumed. For that case Eq. (4.14-1) will be quite inaccurate in some parts of the stop band.

SEC. 4.15, PROTOTYPE REPRESENTATION OF DISSIPATION LOSS IN BAND-STOP FILTERS

In the case of band-stop filters, the effects of parasitic dissipation in the filter elements are usually more serious in the stop band than in the pass band. The stop band usually has one or more frequencies where, if the filter had no dissipation loss, the attenuation would be infinite. However, dissipation loss in the resonators will prevent the attenuation from going to infinity and in some cases may reduce the maximum stop-band attenuation to an unacceptably low value. If a band-stop filter is designed from a low-pass prototype, it is quite easy to compute the effects of finite resonator Q's on the maximum stop-band attenuation.

The solid lines in Fig. 4.15-1 shows a Tchebyscheff low-pass prototype response along with the response of a band-stop filter designed from

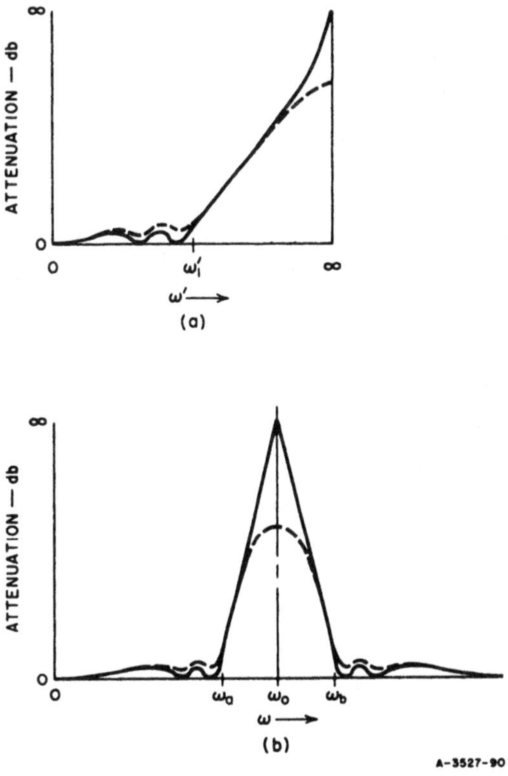

FIG. 4.15-1 A LOW-PASS PROTOTYPE RESPONSE IS SHOWN AT (a), AND THE CORRESPONDING BAND-STOP FILTER RESPONSE IS SHOWN AT (b)
The dashed lines show the effects of dissipation loss.

this prototype, both for the case of no incidental dissipation. For a typical band-stop filter the resonators are resonant at the center of the stop band (instead of at the center of the pass band as is the case for a typical band-pass filter), and as a result the loss effects are most severe at the center of the stop band. The dashed line in Fig. 4.15-1(b) shows how dissipation loss in the resonators will round off the attenuation characteristic of a band-stop filter. The dashed line in Fig. 4.15-1(a) shows the corresponding effect in a low-pass prototype filter.

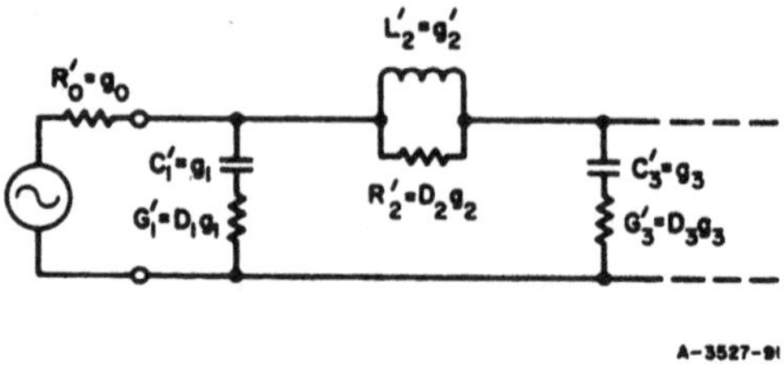

FIG. 4.15-2 LOW-PASS PROTOTYPE FILTER WITH DISSIPATIVE ELEMENTS ADDED AS REQUIRED FOR COMPUTING PEAK STOP-BAND ATTENUATION OF CORRESPONDING BAND-STOP FILTERS

It is easily seen that in order for resistor elements to affect the attenuation of a prototype filter as shown by the dashed line in Fig. 4.15-1, they should be introduced into the prototype circuit as shown in Fig. 4.15-2. Note that in this case as $\omega' \to \infty$, the reactive elements have negligible influence and the circuit operates in the same way as a ladder network of resistors. In Fig. 4.15-2 the Q of the kth reactive element is given by*

$$Q_k = \frac{R'_k}{\omega'_1 L'_k} \quad \text{or} \quad \frac{G'_k}{\omega'_1 C'_k} \tag{4.15-1}$$

$$= \frac{D_k}{\omega'_1} \tag{4.15-2}$$

where ω'_1 is the cutoff frequency in Fig. 4.15-1(a). The unloaded Q, $(Q_{BSF})_k$, of the kth resonator of the band-stop filter is related to Q_k of the prototype (at frequency ω'_1) by

$$Q_k = w(Q_{BSF})_k \tag{4.15-3}$$

* Note that these unusual definitions of Q result from the manner in which the dissipative elements are introduced in each branch of the filter.

where

$$w = \frac{\omega_b - \omega_a}{\omega_0} \tag{4.15-4}$$

and ω_a, ω_b, and ω_0 are as defined in Fig. 4.15-1(b). By Eq. (4.15-2),

$$D_k = \omega_1' Q_k \tag{4.15-5}$$

and as shown in Fig. 4.15-2,

$$R_k' \quad \text{or} \quad G_k' = D_k g_k \Big|_{k=1,2,\ldots,n} \tag{4.15-6}$$

where the g_k are the prototype filter elements as defined in Fig. 4.04-1.

As previously mentioned, when $\omega' \to \infty$ the reactive elements in Fig. 4.15-2 may be neglected, and the attenuation can be computed from the remaining network of resistors. In typical cases, resistances of the series branches will be very large compared to the resistances of the shunt branches, and Cohn's method for computing the stop-band attenuation of low-pass filters[20] can be adapted to cover this case also. The resulting equation is

$$(L_A)_\infty = 20 \log_{10} [(D_1 D_2 \ldots D_n)(g_1 g_2 \ldots g_n)] - 10 \log_{10} \left(\frac{4}{g_0 g_{n+1}}\right) \text{ db} \tag{4.15-7}$$

which is analogous to Eq. (4.14-1) for the reactive attenuation of a low-pass filter.

As an example, let us suppose that a band-stop filter is desired with a fractional stop-band width of $w = 0.02$ (referred to the 3 db points), and that maximally flat pass bands are desired. Let us assume further that the resonator Q's at the mid-stop-band frequency are 700 and that the maximum stop-band attenuation is to be computed. By Eq. (4.15-3) $Q_1 = Q_2 = 0.02\ (700) = 14$. By Table 4.05-1(a) the elements values of the desired $n = 2$ low-pass prototype are $g_0 = 1$, $g_1 = 1.414$, $g_2 = 1.414$, and $g_3 = 1$. Also, ω_1' which in this case is the 3-db band-edge frequency, is equal to unity. By Eq. (4.15-5), $D_1 = D_2 = 14$, and, to slide-rule accuracy, Eq. (4.15-7) gives $(L_A)_\infty = 45.8$ db. In comparison, using the method of Sec. 2.13 to compute the attenuation from the ladder of resistors gives $(L_A)_\infty = 46.7$ db.

As is suggested by the dashed lines in Fig. 4.15-1, the effects of dissipation in the pass band are for this case most severe at the pass-band edge, and they decrease to zero as the frequency moves away from the pass-band edge (within the pass band). The increase in loss due to dissipation at the band-edge frequency can be estimated by use of the formula*

$$(\Delta L_A)\big|_{\omega_1'} \approx 8.686 \sum_{k=1}^{n} \frac{\omega_1' g_k}{Q_k} \quad . \qquad (4.15\text{-}8)$$

This formula represents only an estimate, but should be reasonably accurate for cases such as when an $n = 5$, 0.1-db ripple prototype is used. For cases where very large Tchebyscheff ripples are used this equation will underestimate the loss; when very small ripples are used it will over-estimate the loss. For 0.1-db ripple, if n were reduced to 2 or 1, Eq. (4.15-8) would tend to overestimate the band-edge loss. For typical practical cases, Eq. (4.15-8) should never have an error as great as a factor of 2.

Equation (4.15-8) was obtained from Eq. (4.13-11) by the use of two approximations. The first is that for the arrangement of dissipative elements shown in Fig. 4.13-1, the added loss ΔL_A due to dissipation at the band edge ω_1' is roughly twice the value $(\Delta L_A)_0$ of the loss due to dissipation when $\omega' = 0$. This was shown by examples in Sec. 4.13 to be a reasonably good approximation for typical low-pass prototype filters, though it could be markedly larger if very large pass-band ripples are used. The second approximation assumes that a filter with dissipative elements as shown in Fig. 4.15-2 can be approximated at the frequency ω_1' by the corresponding circuit in Fig. 4.13-1. The reactive element values g_k are assumed to have been unchanged, and also the Q's of the individual reactive element are assumed to be unchanged; however, *the manner in which the dissipation is introduced has been changed.* This approximation is valid to the extent that

---

*This formula is based on Eq. (4.13-11) which assumes that the prototype element values have been normalized so that $g_0 = 1$.

$$\frac{\omega_1' g_k}{Q_k} + j g_k \omega_1' \approx \frac{Q_k \omega_1' g_k}{Q_k^2 + 1} + j \frac{\omega_1' g_k}{1 + \left(\frac{1}{Q_k}\right)^2} \tag{4.15-9}$$

represents a good approximation. It is readily seen that this is a good approximation even for Q's as low as 10. Thus to summarize the basis for Eq. (4.15-8)—the equation as it stands gives a rough estimate of the attenuation due to dissipation at band edge for the situation where the dissipative elements are introduced as shown in Fig. 4.13-1. We justify the use of this same equation for the case of dissipative elements arranged as in Fig. 4.15-2 on the basis of the approximation in Eq. (4.15-9). It shows that as long as the reactive elements are the same, and the element Q's are the same, and around 10 or higher, it doesn't make much difference which way the dissipative elements are connected as far as their effect on transmission loss is concerned.

REFERENCES

1. S. Darlington, "Synthesis of Reactance 4-Poles Which Produce Prescribed Insertion Loss Characteristics," *Jour. Math. and Phys.*, Vol. 18, pp. 257-353 (September 1939).
2. E. A. Guillemin, *Synthesis of Passive Networks* (John Wiley and Sons, Inc., New York, 1957).
3. M. E. Van Valkenburg, *Introduction to Modern Network Synthesis* (John Wiley and Sons, New York, 1960).
4. V. Belevitch, "Tchebyscheff Filters and Amplifier Networks," *Wireless Engineer*, Vol. 29, pp. 106-110 (April 1952).
5. H. J. Orchard, "Formula for Ladder Filters," *Wireless Engineer*, Vol. 30, pp. 3-5 (January 1953).
6. E. Green, "Synthesis of Ladder Networks to Give Butterworth or Chebyshev Response in the Pass Band," *Proc. IEE* (London) Part IV, Monograph No. 88 (1954).
7. E. Green, *Amplitude-Frequency Characteristics of Ladder Networks*, pp. 62-78, Marconi's Wireless Telegraph Co., Ltd., Chelmsford, Essex, England (1954).
8. L. Weinberg, "Network Design by Use of Modern Synthesis Techniques and Tables," *Proc. of Nat. Elec. Conf.*, Vol. 12 (1956).
9. L. Weinberg, "Additional Tables for Design of Optimum Ladder Networks," Parts I and II, *Journal of the Franklin Institute*, Vol. 264, pp. 7-23 and 127-138 (July and August 1957).
10. L. Storch, "Synthesis of Constant-Time-Delay Ladder Networks Using Bessel Polynomials," *Proc. IRE* **42**, pp. 1666-1675 (November 1954).
11. W. E. Thomson, "Networks with Maximally Flat Delay," *Wireless Engineer*, Vol. 29, pp. 255-263 (October 1952).
12. S. B. Cohn, "Phase-Shift and Time-Delay Response of Microwave Narrow-Band Filters," *The Microwave Journal*, Vol. 3, pp. 47-51 (October 1960).
13. M. J. Di Toro, "Phase and Amplitude Distortion in Linear Networks," *Proc. IRE* **36**, pp. 24-36 (January 1948).
14. R. M. Fano, "Theoretical Limitations on the Broadband Matching of Arbitrary Impedances," *J. Franklin Inst.*, Vol. 249, pp. 57-83 and 139-154 (January and February 1950).
15. B. F. Barton, "Design of Efficient Coupling Networks," Technical Report 44, Contract DA 36-039-SC-63203, Electronic Defense Group, University of Michigan, Ann Arbor, Michigan (March 1955).
16. H. J. Carlin, "Gain-Bandwidth Limitations on Equalizers and Matching Networks," *Proc. IRE* **42**, pp. 1676-1685 (November 1954).
17. G. L. Matthaei, "Synthesis of Tchebyscheff Impedance-Matching Networks, Filters, and Interstages," *IRE Trans. PGCT* **3**, pp. 162-172 (September 1956).
18. H. J. Carlin, "Synthesis Techniques for Gain-Bandwidth Optimization in Passive Transducers," *Proc. IRE* **48**, pp. 1705-1714 (October 1960).
19. H. W. Bode, *Network Analysis and Feedback Amplifier Design* pp. 216-222 (D. Van Nostrand Co., Inc., New York, 1945).
20. S. B. Cohn, "Dissipation Loss in Multiple-Coupled-Resonator Filters," *Proc. IRE* **47**, pp. 1342-1348 (August 1959).
21. W. J. Getsinger, "Prototypes for Use in Broadbanding Reflection Amplifiers," scheduled to appear in the November 1963 issue of the *IEEE Trans. PTGMTT*.

CHAPTER 5

PROPERTIES OF SOME COMMON MICROWAVE FILTER ELEMENTS

SEC. 5.01, INTRODUCTION

Previous chapters have summarized a number of important concepts necessary for the design of microwave filters and have outlined various procedures for later use in designing filters from the image viewpoint and from the insertion-loss viewpoint. In order to construct filters that will have measured characteristics as predicted by these theories, it is necessary to relate the design parameters to the dimensions and properties of the structures used in such filters. Much information of this type is available in the literature. The present chapter will attempt to summarize information for coaxial lines, strip lines, and waveguides that is most often needed in filter design. No pretense of completeness is made, since a complete compilation of such data would fill several volumes. It is hoped that the references included will direct the interested reader to sources of more detailed information on particular subjects.

SEC. 5.02, GENERAL PROPERTIES OF TEM-MODE TRANSMISSION LINES

Transmission lines composed of two conductors operating in the transverse electromagnetic (TEM) mode are very useful as elements of microwave filters. Lossless lines of this type have a characteristic or image impedance Z_0, which is independent of frequency f, and waves on these lines are propagated at a velocity, v, equal to the velocity of light in the dielectric filling the line. Defining R, L, G, and C as the resistance, inductance, conductance and capacitance per unit length for such a line, it is found that Z_0 and the propagation constant γ_t are given by

$$Z_0 = \frac{1}{Y_0} = \sqrt{\frac{R + j\omega L}{G + j\omega C}} = \sqrt{\frac{Z}{Y}} \text{ ohms} \qquad (5.02\text{-}1)$$

$$\gamma_t = \alpha_t + j\beta_t = \sqrt{(R + j\omega L)(G + j\omega C)} = \sqrt{ZY} \qquad (5.02\text{-}2)$$

where $\omega = 2\pi f$. When the line is lossless, α_t is zero and

$$\beta_t = \omega\sqrt{LC} \qquad \text{radians/unit length} \qquad (5.02\text{-}3)$$

$$v = \frac{\omega}{\beta_t} = \frac{1}{\sqrt{LC}} \qquad \text{distance/second} \qquad (5.02\text{-}4)$$

$$Z_0 = \sqrt{\frac{L}{C}} = \frac{1}{vC} = vL \text{ ohms} \quad . \qquad (5.02\text{-}5)$$

In practice a line will have some finite amount of attenuation

$$\alpha_t = \alpha_c + \alpha_d \qquad (5.02\text{-}6)$$

where α_c is the attenuation due to conductor loss and α_d the attenuation due to loss in the dielectric. For small attenuations *

$$\alpha_c = \frac{R}{2Z_0} = \frac{\beta_t}{2Q_c} \qquad \text{nepers} \qquad (5.02\text{-}7)$$

$$\alpha_d = \frac{G}{2Y_0} = \frac{\beta_t}{2Q_d} = \frac{\beta_t}{2\tan\delta} \qquad \text{nepers} \qquad (5.02\text{-}8)$$

where $Q_c = \omega L/R$, $Q_d = \omega C/G$, and $\tan\delta$ is the loss tangent of the dielectric material filling the line. The total Q of the transmission line used as a resonator is given by

$$\frac{1}{Q} = \frac{1}{Q_c} + \frac{1}{Q_d} \quad . \qquad (5.02\text{-}9)$$

These definitions are in agreement with those given in terms of the resonator reactance and susceptance slope parameters in Sec. 5.08. For a slightly lossy line the characteristic impedance and propagation constant become

* To convert nepers to decibels, multiply by 8.686.

$$\beta_t = \omega\sqrt{LC}\left[1 - \frac{1}{4Q_cQ_d} + \frac{1}{8Q_d^2} + \frac{1}{8Q_c^2}\right] \qquad (5.02\text{-}10)$$

$$Z_0 = \sqrt{\frac{L}{C}}\left[1 + \frac{j}{2}\left(\frac{1}{Q_d} - \frac{1}{Q_c}\right)\right] \quad . \qquad (5.02\text{-}11)$$

The TEM modes can also propagate on structures containing more than two conductors. Examples of such structures with two conductors contained within an outer shield are described in Sec. 5.05. Two principal modes can exist on such two-conductor structures: an even mode in which the currents in the two conductors flow in the same direction, and an odd mode in which the currents on the conductors flow in opposite directions. The velocity of propagation of each of these modes in the lossless case is equal to the velocity of light in the dielectric medium surrounding the conductors. However, the characteristic impedance of the even mode is different from that of the odd mode.

SEC. 5.03, SPECIAL PROPERTIES OF COAXIAL LINES

The characteristic impedance Z_0 of a coaxial line of outer diameter b and inner diameter d, filled with a dielectric material of relative dielectric constant ϵ_r, is

$$Z_0 = \frac{60}{\sqrt{\epsilon_r}} \ln \frac{b}{d} \quad \text{ohms} \quad . \qquad (5.03\text{-}1)$$

This expression is plotted in Fig. 5.03-1. The attenuation α_c of a copper coaxial line due to ohmic losses in the copper is

$$\alpha_c = 1.898 \times 10^{-4} \sqrt{\epsilon_r}\sqrt{f_{Gc}} \left(\frac{1 + b/d}{b \ln b/d}\right) \quad \text{db/unit length} \qquad (5.03\text{-}2)$$

where f_{Gc} is measured in gigacycles. (Here the copper is assumed to be very smooth and corrosion-free.) The attenuation is a minimum for b/d of 3.6 corresponding to $\sqrt{\epsilon_r}\, Z_0$ of 77 ohms.

The attenuation α_d of the coaxial line (or any other TEM line) due to losses in the dielectric is

$$\alpha_d = \frac{27.3\sqrt{\epsilon_r}\tan\delta}{\lambda} \quad \text{db/unit length} \qquad (5.03\text{-}3)$$

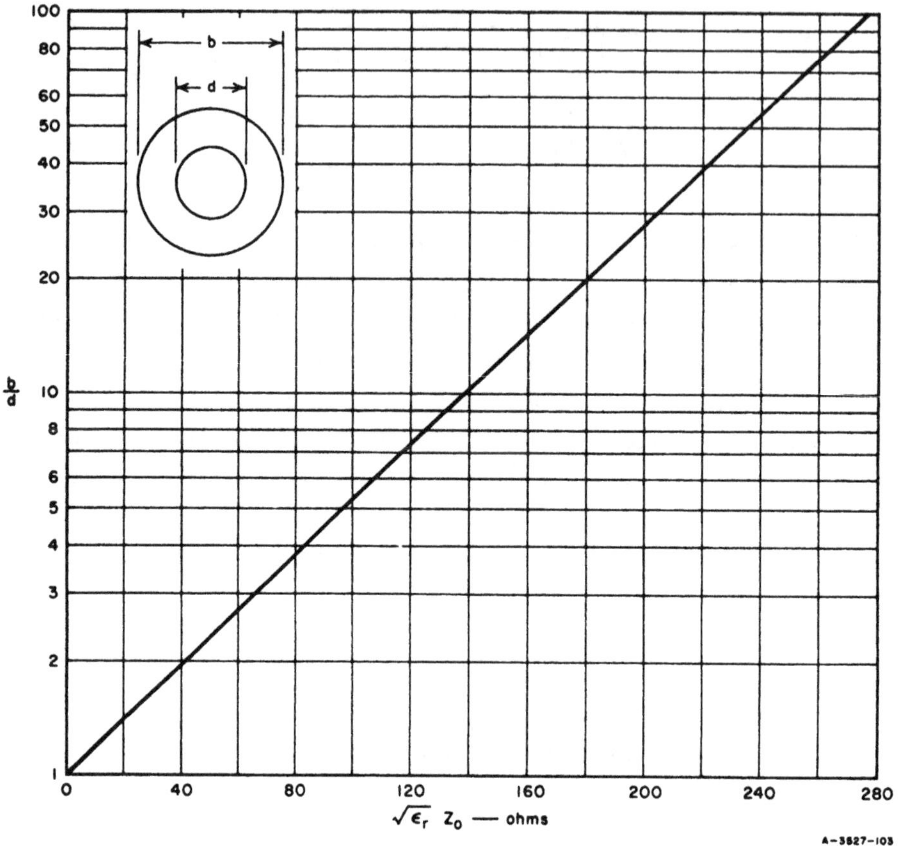

FIG. 5.03-1 COAXIAL-LINE CHARACTERISTIC IMPEDANCE

where tan δ is the loss tangent of the dielectric, and λ is the free-space wavelength. The total attenuation α_t is the sum of α_c and α_d. The attenuation of a coaxial line due to ohmic losses in the copper is shown in Fig. 5.03-2.

The Q of a dielectric-filled coaxial line may be expressed as

$$\frac{1}{Q} = \frac{1}{Q_c} + \frac{1}{Q_d} \qquad (5.03\text{-}4)$$

where $Q_c = \pi\sqrt{\epsilon_r}/\lambda\alpha_c$ depends only on the conductor loss and Q_d depends only on the dielectric loss. The Q_c of a dielectric-filled coaxial line is independent of ϵ_r and is given by the expression

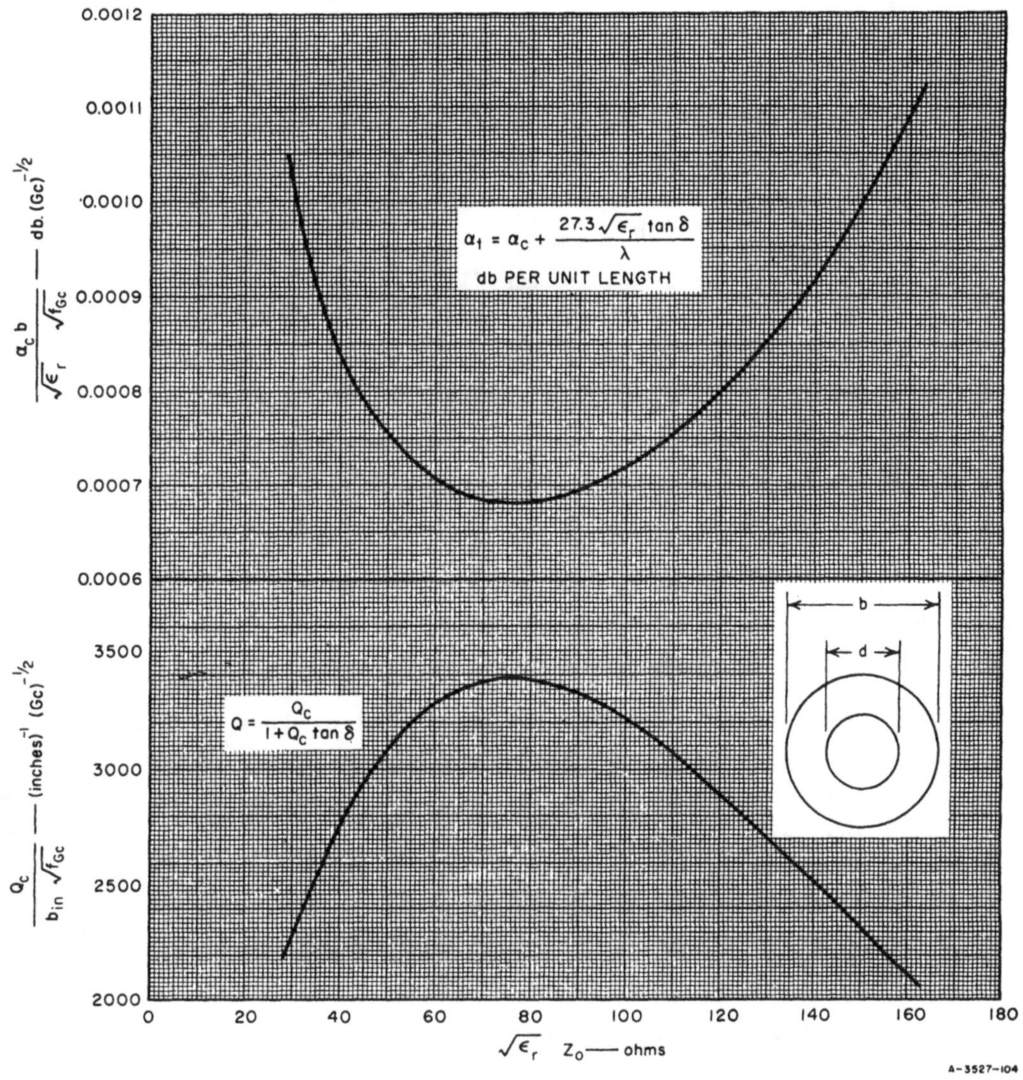

FIG. 5.03-2 COAXIAL-LINE ATTENUATION AND Q

$$Q_c = 1.215 \times 10^4 \sqrt{f_{Gc}} \frac{b \ln b/d}{[1 + b/d]} \tag{5.03-5}$$

where b and d are measured in inches. The value of Q_d for the coaxial line or any TEM line is

$$Q_d = \frac{1}{\tan \delta} \quad . \tag{5.03-6}$$

The values of Q_c for a copper coaxial line are plotted in Fig. 5.03-2.

Breakdown will occur in an air-filled coaxial line at atmospheric pressure when the maximum electric field E_m reaches a value of approximately 2.9×10^4 volts per cm. The average power P that can be transmitted on a matched coaxial line under these conditions is

$$P = \frac{E_m^2}{480} b^2 \frac{\ln b/d}{(b/d)^2} \quad \text{watts} \quad . \tag{5.03-7}$$

When the outer diameter b is fixed, the maximum power can be transmitted when b/d is 1.65, corresponding to Z_0 of 30 ohms.

The first higher-order TE mode in a coaxial line will propagate wher the average circumference of the line is approximately equal to the wavelength in the medium filling the line. The approximate cutoff frequency, f_c (in gigacycles), of this mode is

$$(f_c)_{Gc} = \frac{7.51}{\sqrt{\epsilon_r}} \frac{1}{(b + d)} \tag{5.03-8}$$

where b and d are measured in inches.

SEC. 5.04, SPECIAL PROPERTIES OF STRIP LINES

The characteristic impedance of strip line can be calculated by conformal mapping techniques; however, the resulting formulas are rather complex. Figure 5.04-1 shows the characteristic impedance, Z_0, of a common type of strip line with a rectangular center conductor,[1,2] for various values of $t/b \leq 0.25$, and $0.1 \leq w/b \leq 4.0$. The values shown are exact for $t/b = 0$ and are accurate to within about 1 percent for other

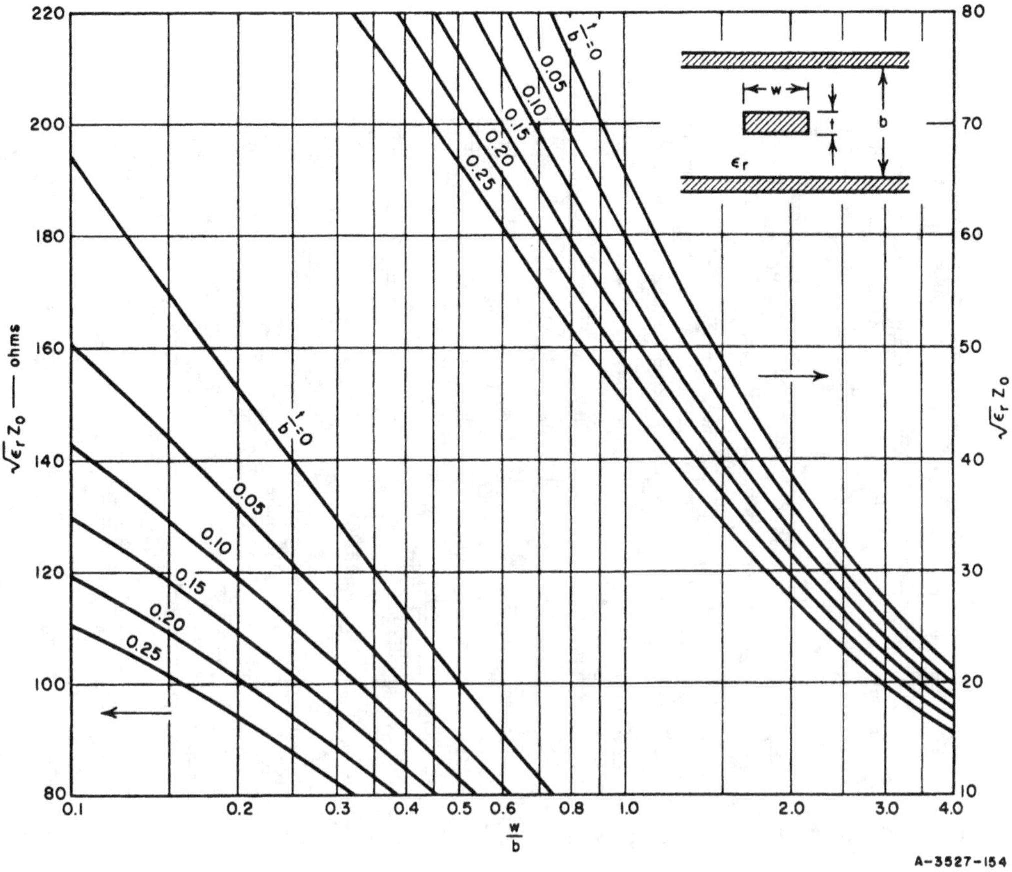

SOURCE: **Final Report, Contract DA 36-039 SC-63232, SRI; reprinted in *IRE Trans., PGMTT* (see Ref. 2, by S. B. Cohn).**

FIG. 5.04-1 GRAPH OF Z_0 vs. w/b FOR VARIOUS VALUES OF t/b

values of t/b. Figure 5.04-2 shows exact values of Z_0 for all values of t/b, and $w/b \leq 1.6$.[3]

The theoretical attenuation α_c due to ohmic losses in a copper strip line filled with a dielectric of relative dielectric constant ϵ_r, is shown in Fig. 5.04-3. The attenuation α_d due to the dielectric loss is given by Eq. (5.03-3). As in the case of the coaxial line, the total attenuation α_t is the sum of α_c and α_d.

The Q of a dielectric-filled strip line is given by Eq. (5.03-4). The Q_c of a dielectric-filled line is shown plotted in Fig. 5.04-4.[1,2] As in the case of the coaxial line, Q_d is the reciprocal of tan δ.

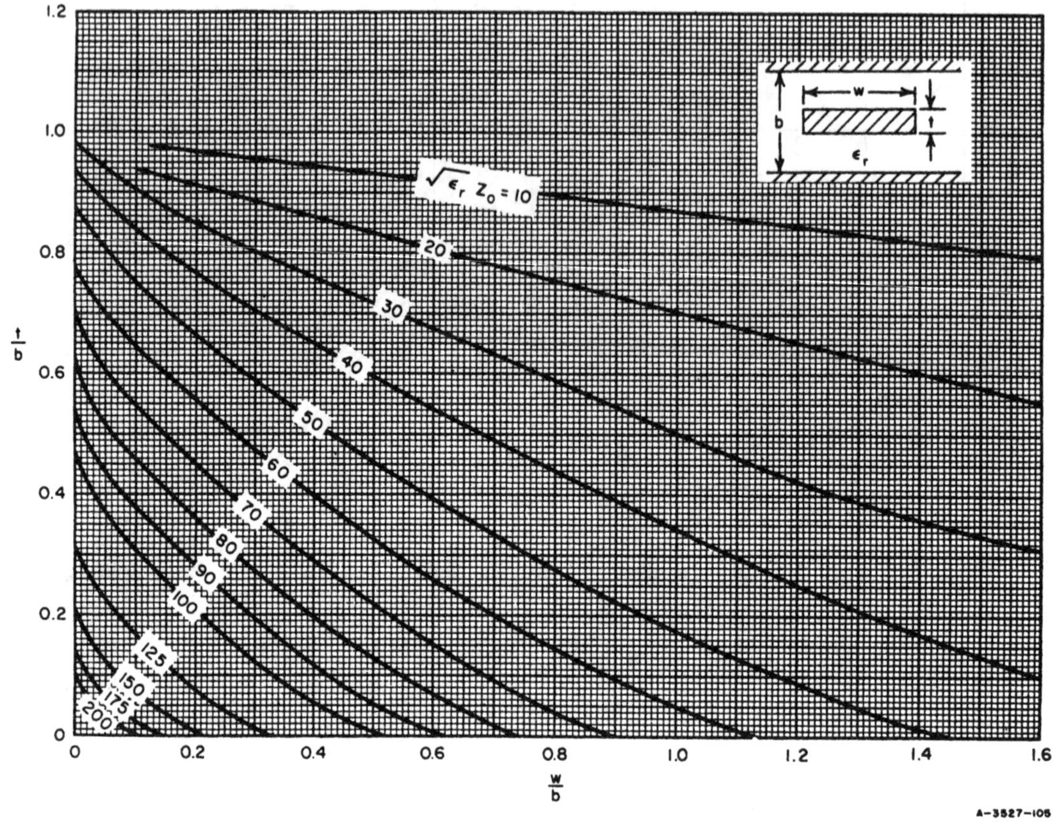

SOURCE: *IRE Trans. PGMTT* (see Ref. 3, by R. H. T. Bates).

FIG. 5.04-2 GRAPH OF Z_0 vs. w/b FOR VARIOUS VALUES OF t/b

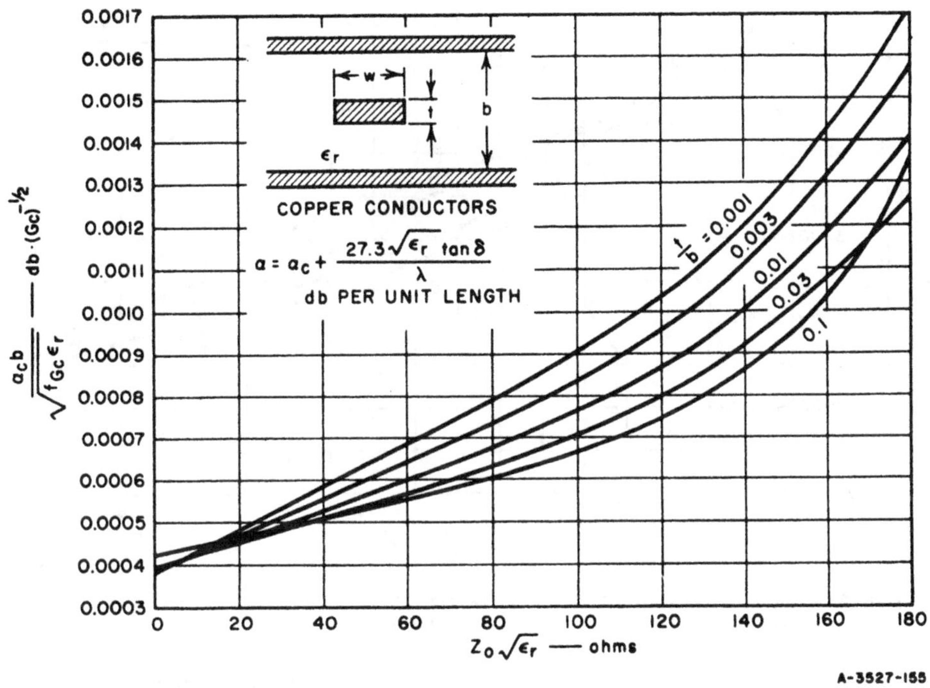

SOURCE: Final Report, Contract DA 36-039 SC-63232, SRI; reprinted in *IRE Trans., PGMTT* (see Ref. 2, by S. B. Cohn).

FIG. 5.04-3 THEORETICAL ATTENUATION OF COPPER-SHIELDED STRIP LINE IN A DIELECTRIC MEDIUM ϵ_r

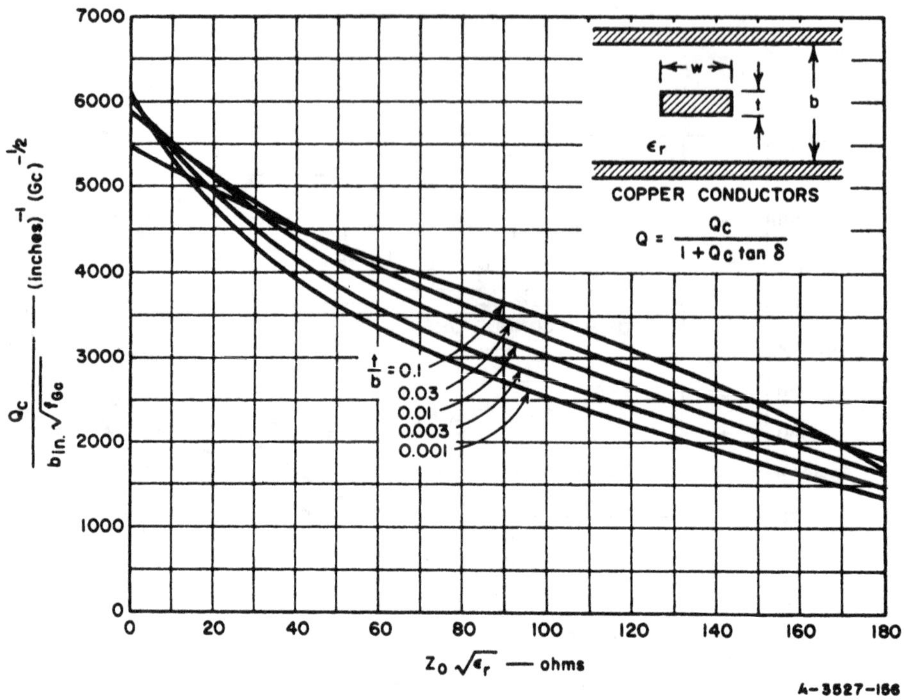

SOURCE: Final Report, Contract DA 36-039 SC-63232, SRI; reprinted in *IRE Trans., PGMTT* (see Ref. 2, by S. B. Cohn).

FIG. 5.04-4 THEORETICAL Q OF COPPER-SHIELDED STRIP LINE IN A DIELECTRIC MEDIUM ϵ_r

The average power, P (measured in kw), that can be transmitted along a matched strip line having an inner conductor with rounded corners is plotted in Fig. 5.04-5. In this figure the ground plane spacing b is measured in inches, and the breakdown strength of air is taken as 2.9×10^4 volts/cm. An approximate value of Z_0 can be obtained from Figs. 5.04-1 and 5.04-2.

The first higher-order mode that can exist in a strip line, in which the two ground planes have the same potential, has zero electric-field strength on the longitudinal plane containing the center line of the strip, and the electric field is oriented perpendicular to the strip and ground plane. The free-space cutoff wavelength, λ_c of this mode is

$$\lambda_c = \sqrt{\epsilon_r}\left[\frac{2w}{b} + \frac{4d}{b}\right]b$$

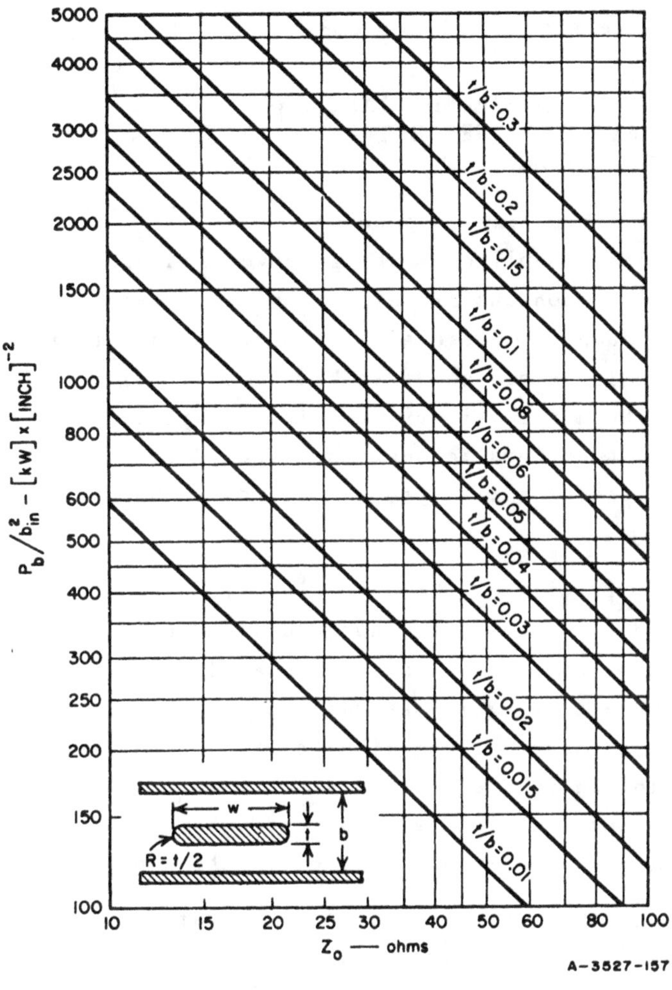

FIG. 5.04-5 THEORETICAL BREAKDOWN POWER OF AIR-DIELECTRIC ROUNDED-STRIP TRANSMISSION LINE

where d is a function of the cross section of the strip line. If $t/b \approx 0$ and $w/b \geq 0.35$, then $4d/b$ is a function of b/λ_c alone and is given in Table 5.04-1.

Table 5.04-1

THE QUANTITY $4d/b$ *vs.* b/λ_c FOR $w/b \geq 0.35$ AND $t/b \approx 0$

b/λ_c	$4d/b$
0.00	0.882
0.20	0.917
0.30	0.968
0.35	1.016
0.40	1.070
0.45	1.180
0.50	1.586

SEC. 5.05, PARALLEL-COUPLED LINES AND ARRAYS OF LINES BETWEEN GROUND PLANES

A number of strip-line components utilize the natural coupling existing between parallel conductors. Examples of such components are directional couplers, filters, baluns, and delay lines such as interdigital lines. A number of examples of parallel-coupled lines are shown in Fig. 5.05-1. The (a), (b), and (c) configurations shown are primarily useful in applications where weak coupling between the lines is desired. The (d), (e), (f), and (g) configurations are useful where strong coupling between the lines is desired.

The characteristics of these coupled lines can be specified in terms of Z_{oe} and Z_{oo}, their even and odd impedances, respectively. Z_{oe} is defined as the characteristic impedance of one line to ground when equal currents are flowing in the two lines. Z_{oo} is defined as the characteristic impedance of one line to ground when equal and opposite currents are flowing in the two lines. Figure 5.05-2 illustrates the electric field configuration over the cross section of the lines shown in Fig. 5.05-1(a) when they are excited in the even and odd modes.

Thin Strip Lines—The exact even-mode characteristic impedance of the infinitesimally thin strip configuration of Fig. 5.05-1(a) is[4]

$$Z_{oe} = \frac{30\pi}{\sqrt{\epsilon_r}} \cdot \frac{K(k_e')}{K(k_e)} \quad \text{ohms} \tag{5.05-1}$$

where

$$k_e = \tanh\left(\frac{\pi}{2} \cdot \frac{w}{b}\right) \cdot \tanh\left(\frac{\pi}{2} \cdot \frac{w+s}{2}\right) \tag{5.05-2}$$

$$k_e' = \sqrt{1 - k_e^2} \tag{5.05-3}$$

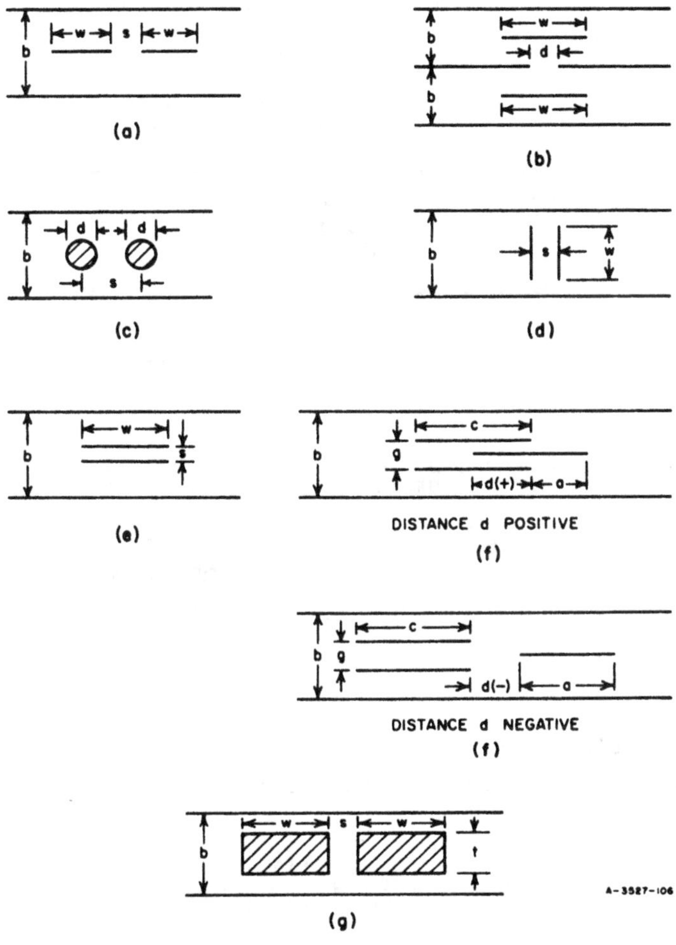

FIG. 5.05-1 CROSS SECTIONS OF VARIOUS COUPLED-TRANSMISSION-LINE CONFIGURATIONS

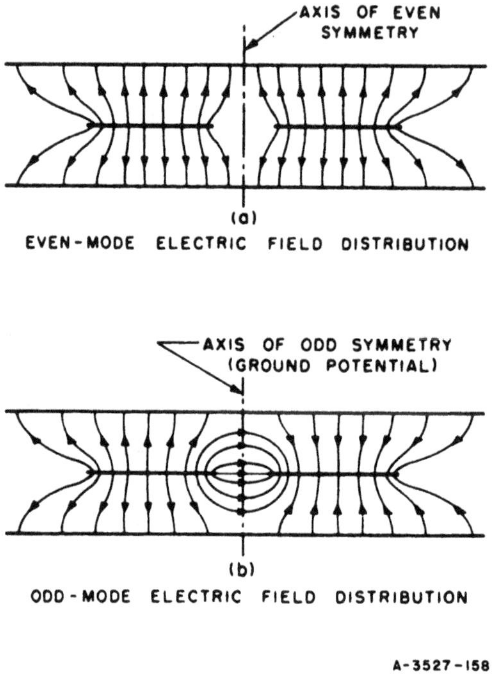

SOURCE: Final Report, Contract DA 36-039 SC-63232, SRI; reprinted in *IRE Trans., PGMTT* (see Ref. 4, by S. B. Cohn).

FIG. 5.05-2 FIELD DISTRIBUTIONS OF THE EVEN AND ODD MODES IN COUPLED STRIP LINE

and ϵ_r is the relative dielectric constant of the medium of propagation. The exact odd-mode impedance in the same case is[1]

$$Z_{oo} = \frac{30\pi}{\sqrt{\epsilon_r}} \cdot \frac{K(k_o')}{K(k_o)} \text{ ohms} \tag{5.05-4}$$

where

$$k_o = \tanh\left(\frac{\pi}{2} \cdot \frac{w}{b}\right) \cdot \coth \frac{\pi}{2} \cdot \left(\frac{w + s}{b}\right) \tag{5.05-5}$$

$$k_o' = \sqrt{1 - k_o^2} \tag{5.05-6}$$

and K is the complete elliptic integral of the first kind. Convenient tables of $K(k')/K(k)$ have been compiled by Oberhettinger and Magnus.[5] Nomographs giving the even- and odd-mode characteristic impedances are presented in Figs. 5.05-3(a) and (b).

Thin Lines Coupled Through a Slot—The thin-strip configuration shown in Fig. 5.05-1(b) with a thin wall separating the two lines has a value of $Z_{oo} \approx Z_0$, which is the characteristic impedance of an uncoupled line as given in Sec. 5.04. The even-mode characteristic impedance Z_{oe} is given approximately by

$$\frac{1}{Z_{oe}} = \frac{1}{2}\left[\frac{1}{Z_{oo}} + \frac{1}{Z}\right] \qquad (5.05\text{-}7)$$

where

$$Z = \frac{30\pi}{\sqrt{\epsilon_r}} \frac{K(k')}{K(k)} \qquad (5.05\text{-}8)$$

$$k = \sqrt{\frac{\cosh \frac{\pi w}{b} - 1}{\cosh \frac{\pi w}{b} + \cosh \frac{\pi d}{b}}} \qquad (5.05\text{-}9)$$

and

$$k' = \sqrt{1 - k^2} \quad . \qquad (5.05\text{-}10)$$

Round Wires—The even- and odd-mode characteristic impedances of round lines placed midway between ground planes as shown in Fig. 5.05-1(c) are given approximately by

$$Z_{oe} - Z_{oo} = \frac{120}{\sqrt{\epsilon_r}} \ln \coth \frac{\pi s}{2b} \qquad (5.05\text{-}11)$$

$$Z_{oe} + Z_{oo} = \frac{120}{\sqrt{\epsilon_r}} \ln \frac{4b}{\pi d} \quad . \qquad (5.05.12)$$

These should give good results, at least when $d/b < 0.55$ and $s/b > 2\, d/b$.

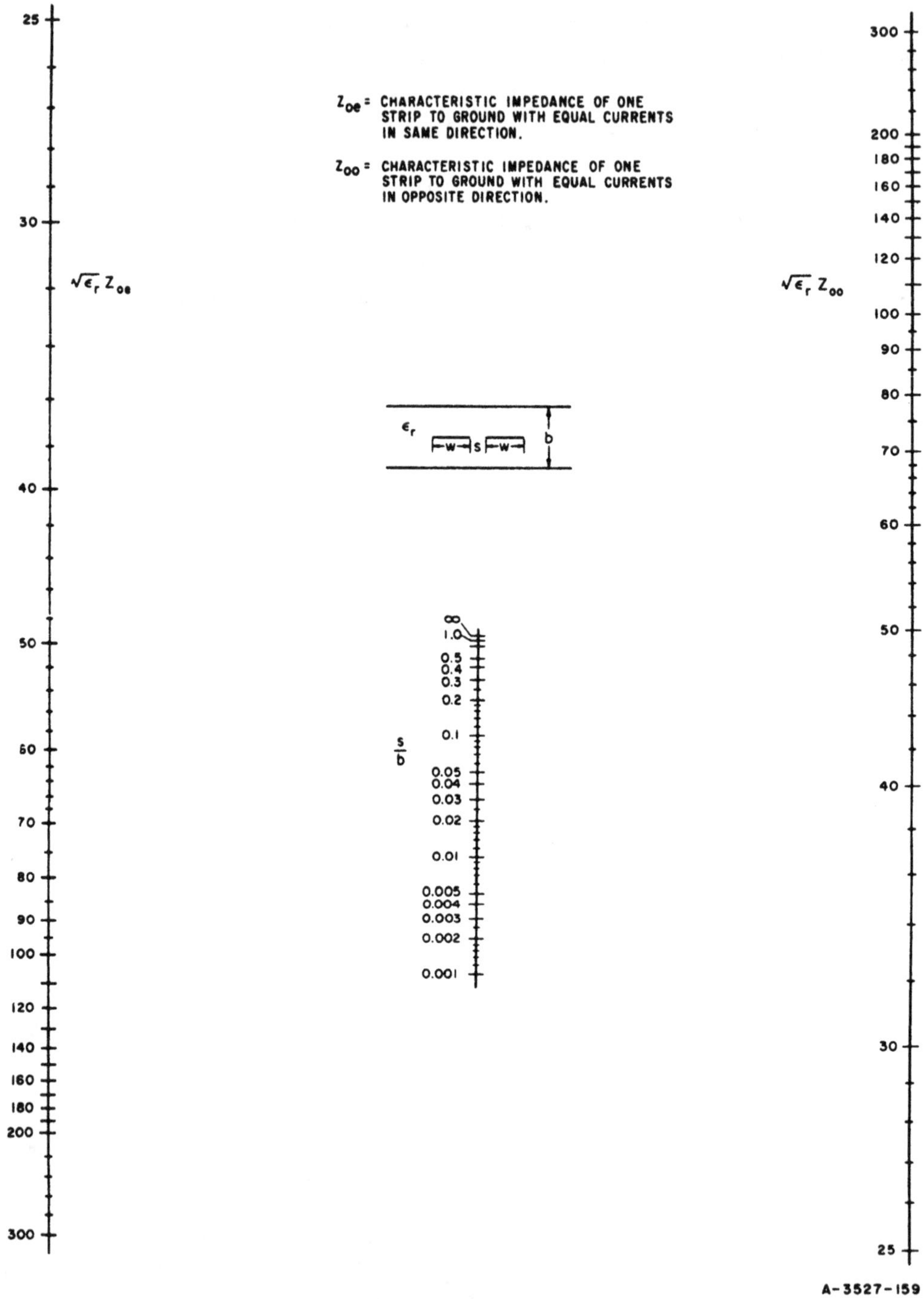

SOURCE: Final Report, Contract DA 36-039 SC-63232, SRI; reprinted in *IRE Trans., PGMTT* (see Ref. 4, by S. B. Cohn).

FIG. 5.05-3(a) NOMOGRAM GIVING s/b AS A FUNCTION OF Z_{oe} AND Z_{oo} IN COUPLED STRIP LINE

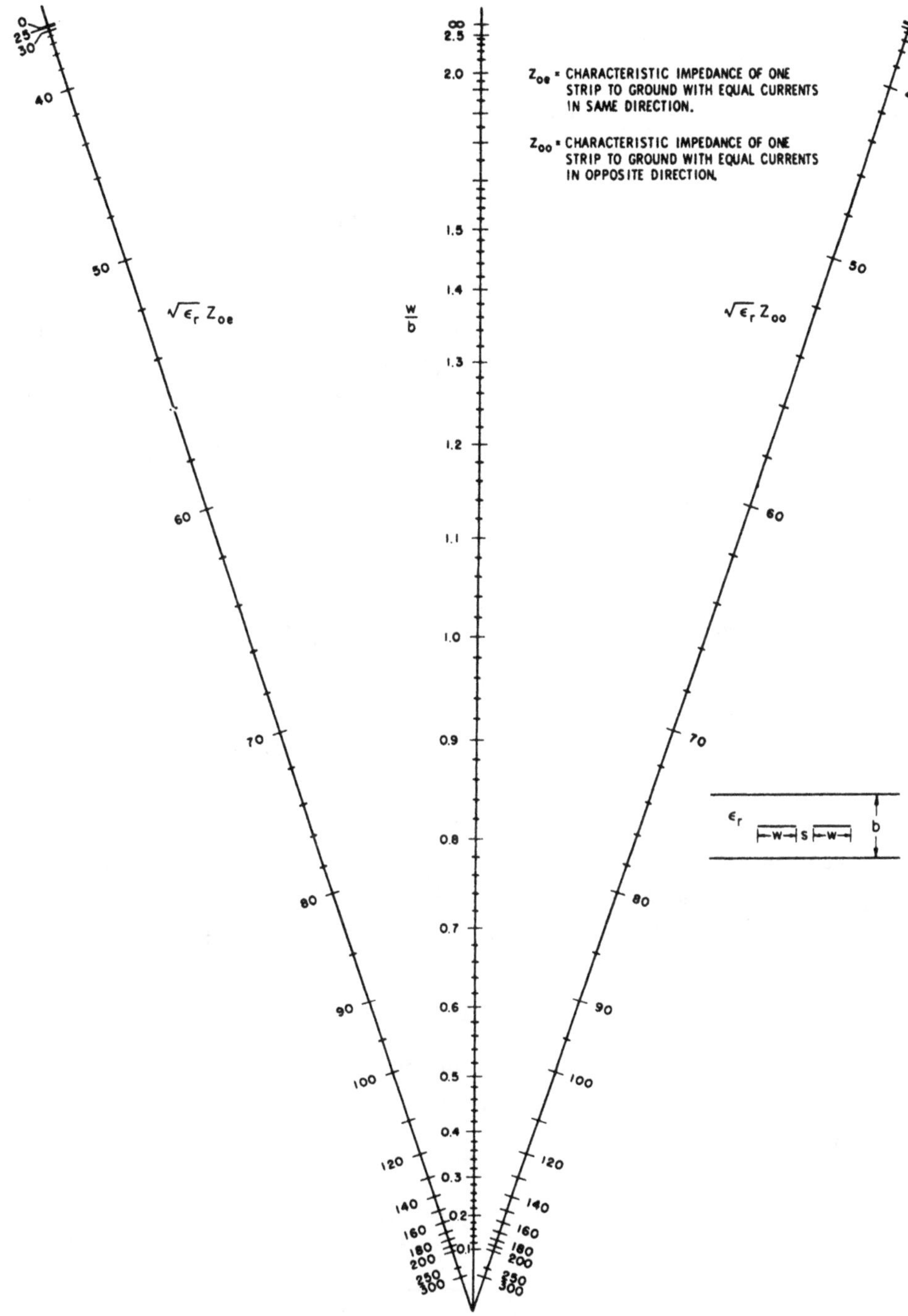

SOURCE: Final Report, Contract DA 36-039 SC-63232, SRI; reprinted in *IRE Trans., PGMTT* (see Ref. 4, by S. B. Cohn).

FIG. 5.05-3(b) NOMOGRAM GIVING w/b AS A FUNCTION OF Z_{oe} AND Z_{oo} IN COUPLED STRIP LINE

Thin Lines Vertical to the Ground Planes—The even- and odd-mode characteristic impedances of the thin coupled lines shown in Fig. 5.05-1(d) are given approximately by the formulas[6]

$$Z_{oe} = \frac{188.3}{\sqrt{\epsilon_r}} \frac{K(k)}{K(k')} \qquad (5.05\text{-}13)$$

$$Z_{oo} = \frac{296.1}{\sqrt{\epsilon_r}} \frac{1}{\frac{b}{s}\cos^{-1} k + \ln \frac{1}{k}} \quad . \qquad (5.05\text{-}14)$$

In these formulas k' is a parameter equal to $\sqrt{1 - k^2}$, and K is the complete elliptic integral of the first kind. The ratio w/b is given by

$$\frac{w}{b} = \frac{2}{\pi}\left\{\tan^{-1}\left[\frac{k'}{k}\sqrt{\frac{1 - \frac{k}{k'}\frac{s}{b}}{1 + \frac{k'}{k}\frac{s}{b}}}\right] - \frac{s}{b}\tanh^{-1}\sqrt{\frac{1 - \frac{k}{k'}\frac{s}{b}}{1 + \frac{k'}{k}\frac{s}{b}}}\right\} . \qquad (5.05\text{-}15)$$

The inverse cosine and tangent functions are evaluated in radians between 0 and $\pi/2$. To find the dimensions of the lines for particular values of Z_{oe} and Z_{oo}, one first determines the value of the k from Eq. (5.05-13) and the tables of $K(k)/K(k')$ *vs.* k in Ref. 5. Then b/s is determined from Eq. (5.05-14) and finally w/b is determined from Eq. (5.05-15). Equations (5.05-13) through (5.05-15) are accurate for all values of w/b and s/b, as long as w/s is greater than about 1.0.

Thin Lines Superimposed—The formulas for the even- and odd-mode characteristic impedances of the coupled lines shown in Fig. 5.05-1(e) reduce to fairly simple expressions when $(w/b)/(1 - s/b) \geq 0.35$.[6] It is found that

$$Z_{oe} = \frac{188.3/\sqrt{\epsilon_r}}{\frac{w/b}{1 - s/b} + \frac{C'_{fe}}{\epsilon}} \qquad (5.05\text{-}16)$$

$$Z_{oo} = \frac{188.3/\sqrt{\epsilon_r}}{\frac{w/b}{1 - s/b} + \frac{w}{s} + \frac{C'_{fo}}{\epsilon}} \quad . \qquad (5.05\text{-}17)$$

The capacitance C'_{fe} is the capacitance per unit length that must be added at each edge of each strip to the parallel plate capacitance, so that the total capacitance to ground for the even mode will be correct. C'_{fo} is the corresponding quantity for the odd mode and ϵ_r is the relative dielectric constant. The even- and odd-mode fringing capacitances are plotted in Fig. 5.05-4.

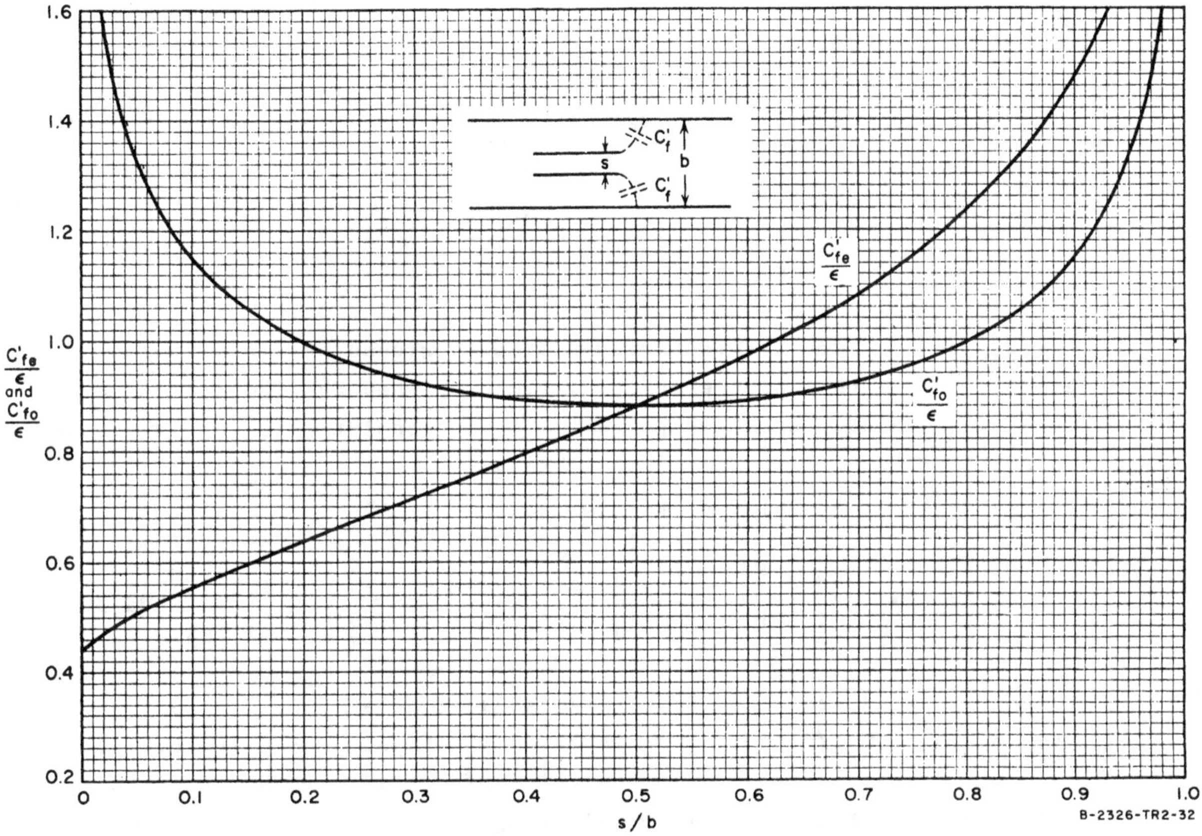

SOURCE: Final Report, Contract DA-36-039 SC-74862, SRI; reprinted in *IRE Trans., PGMTT* (see Ref. 5, by S. B. Cohn).

FIG. 5.05-4 EVEN- AND ODD-MODE FRINGING CAPACITANCES FOR BROADSIDE-COUPLED VERY THIN STRIPS PARALLEL TO THE GROUND PLANES

The even- and odd-mode characteristic impedances of the coupled lines shown in Figs. 5.05-1(a), (d), and (e) are modified slightly when the strips have a finite thickness. Correction terms that account for the effects of finite thickness have been derived by Cohn.[7]

Interleaved Thin Lines—The configuration of coupled strip lines illustrated in Fig. 5.05-1(f), in which the two lines of width c are always operated at the same potential, is particularly useful when it is desired to obtain tight coupling with thin strips that are supported by a homogeneous dielectric, of relative dielectric constant ϵ_r, that completely fills the region between the ground planes.[32] The dimensions of the strips for particular values of Z_{oe} and Z_{oo} can be determined with the aid of Figs. 5.05-5 through 5.05-8. For this purpose one needs the definitions that

$$\sqrt{\epsilon_r}\, Z_{oe} = \frac{376.6\epsilon}{C_{oe}} \qquad (5.05\text{-}18)$$

$$\sqrt{\epsilon_r}\, Z_{oo} = \frac{376.6\epsilon}{C_{oo}} \qquad (5.05\text{-}19)$$

where C_{oe} and C_{oo} are the total capacities to ground per unit length of the strips of width c or the strip of width a, when the lines are excited in the even and odd modes, respectively. The absolute dielectric constant ϵ is equal to $0.225\ \epsilon_r$ pf per inch. Using the values of Z_{oe} and Z_{oo} which are assumed to be known, one then computes $\Delta C/\epsilon$ from

$$\frac{\Delta C}{\epsilon} = \frac{188.3}{\sqrt{\epsilon_r}}\left[\frac{1}{Z_{oo}} - \frac{1}{Z_{oe}}\right] = \frac{1}{2}\left[\frac{C_{oo}}{\epsilon} - \frac{C_{oe}}{\epsilon}\right] \qquad (5.05\text{-}20)$$

Values of b and g are then selected and d/g is determined from Fig. 5.05-5. Next, values of $C'_{c\infty}/\epsilon$ and C'_{ce}/ϵ are read from Figs. 5.05-6 and 5.05-7. These quantities, together with the value of C_{oe}/ϵ from Eq. (5.05-18), are then substituted in Eq. (5.05-21) to give c/b:

$$c/b = \frac{1 - g/b}{2}\left[\frac{1}{2}C_{oe}/\epsilon - C'_{c\infty}/\epsilon - C'_{ce}/\epsilon\right] \quad . \qquad (5.05\text{-}21)$$

Finally, C'_{ae}/ϵ is found from Fig. 5.05-8 and substituted in Eq. (5.05-22) to give a/b:

$$\frac{a}{b} = \frac{1}{2}\left[\frac{1}{2}C_{oe}/\epsilon - C'_{ae}/\epsilon - 0.441\right] \quad . \qquad (5.05\text{-}22a)$$

Thus all the physical dimensions are determined.

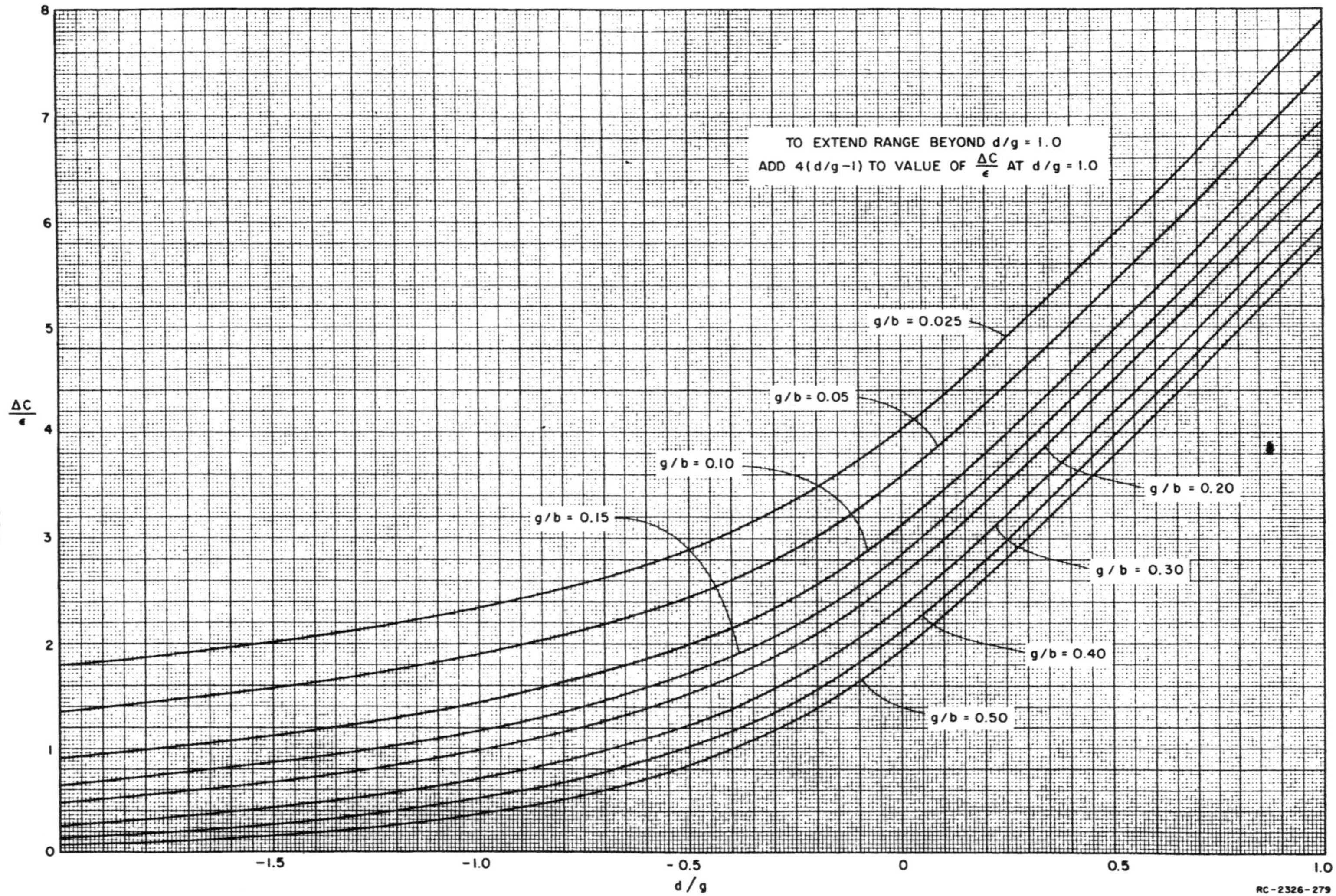

SOURCE: Final Report, Contract DA 36-039 SC-74862, SRI; reprinted in *IRE Trans., PGMTT* (see Ref. 32, by W. J. Getsinger).

FIG. 5.05-5 INTER-STRIP CAPACITANCE FOR CONFIGURATION IN FIG. 5.05-1(f)

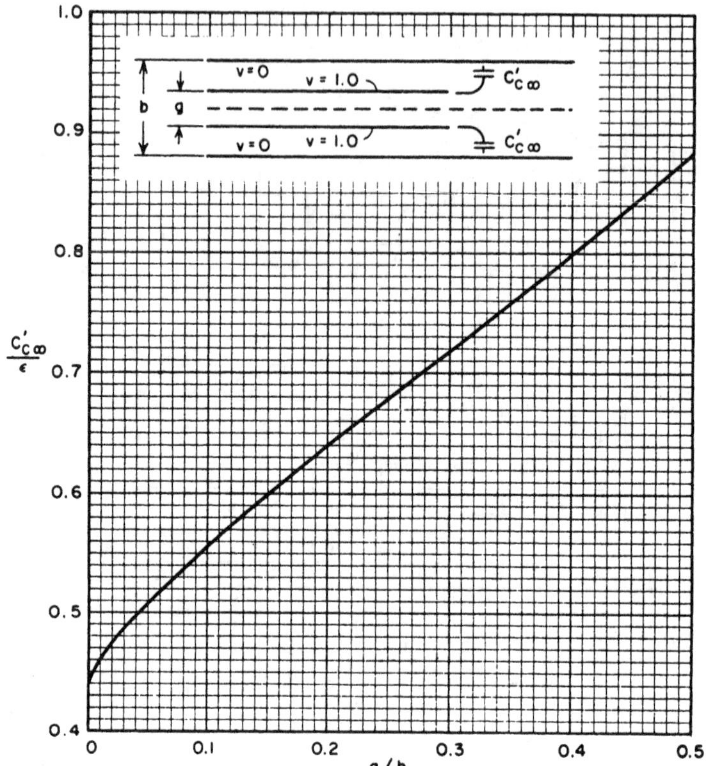

SOURCE: Final Report, Contract DA 36-039 SC-74862, SRI; reprinted in *IRE Trans., PGMTT* (see Ref. 32, by W. J. Getsinger).

FIG. 5.05-6 FRINGING CAPACITANCE OF OFFSET THIN STRIP IN FIG. 5.05-1(f)

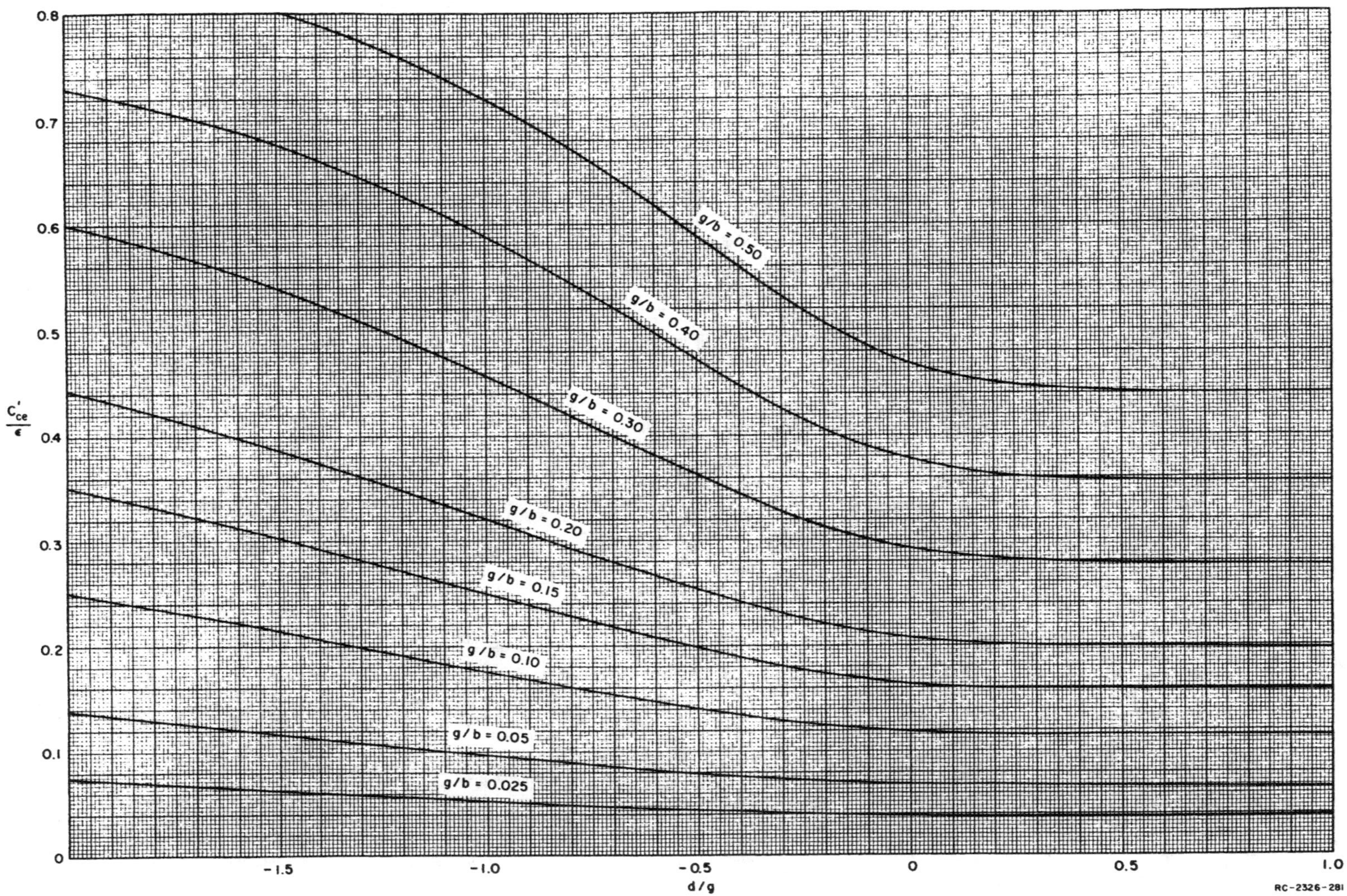

SOURCE: Final Report, Contract DA 36-039 SC-74862, SRI; reprinted in *IRE Trans., PGMTT* (see Ref. 32, by W. J. Getsinger).

FIG. 5.05-7 EVEN MODE, RIGHT-EDGE, FRINGING CAPACITANCE TO GROUND FOR EACH OF THE DOUBLE STRIPS OF WIDTH c IN FIG. 5.05-1(f)

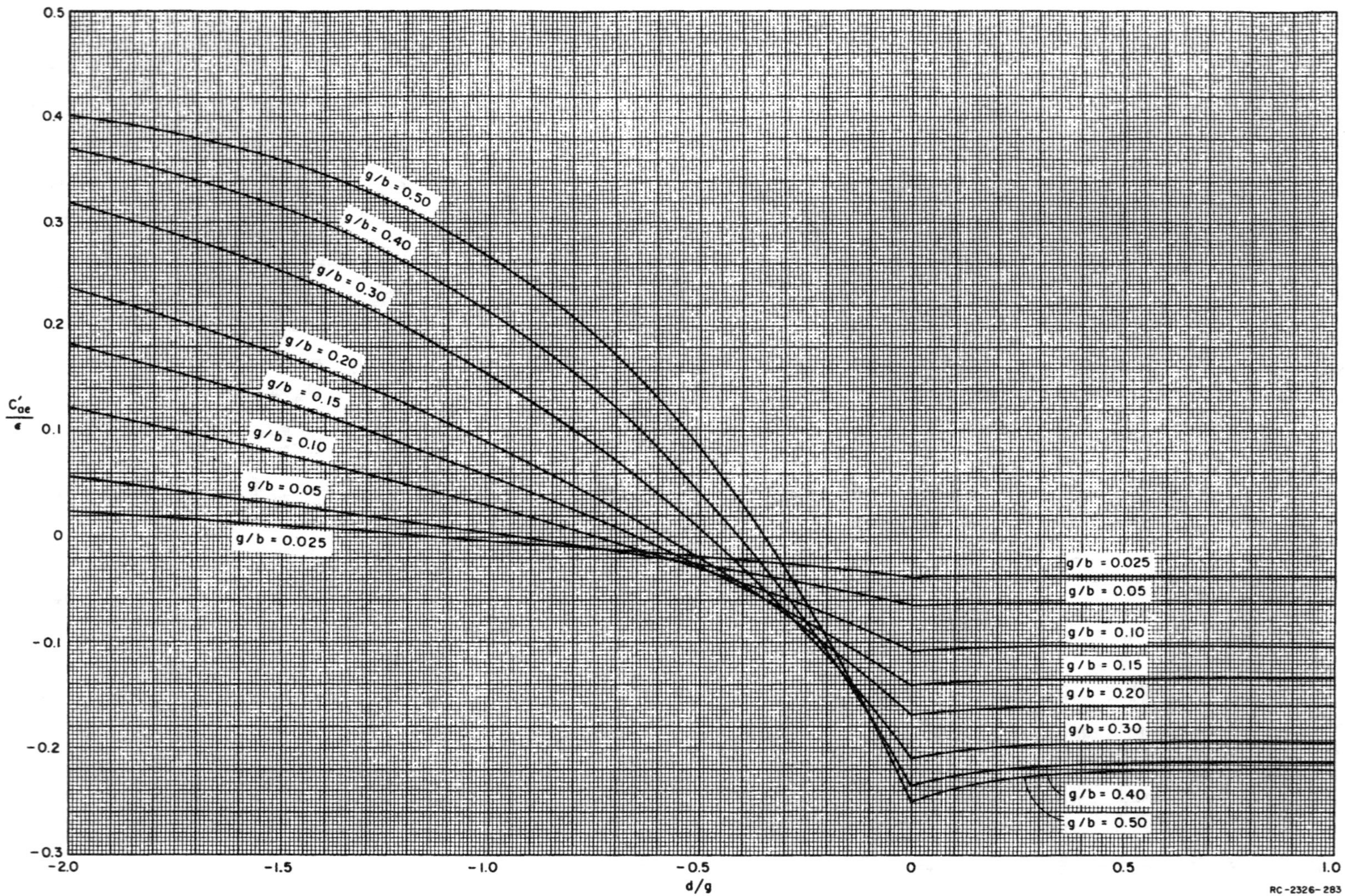

SOURCE: Final Report, Contract DA 36-039 SC-74862, SRI; reprinted in *IRE Trans., PGMTT* (see Ref. 32, by W. J. Getsinger).

FIG. 5.05-8 EVEN MODE, LEFT-EDGE, FRINGING CAPACITANCE TO GROUND FOR THE SINGLE STRIP OF WIDTH a OR a + d IN FIG. 5.05-1(f)

These formulas are exact in the limit of a and $c \gg b$ so that fringing fields at opposite edges of the strips do not interact. They are accurate to within 1.24 percent when $a/b \geq 0.35$ and $[(c/b)/(1 - g/b)] \geq 0.35$. If these conditions are not satisfied, it is possible to make approximate corrections based on increasing the parallel plate capacitance to compensate for the loss of fringing capacitance due to interaction of the fringing fields. If an initial value a_1/b is found to be less than 0.35, a new value, a_2/b, can be used where

$$\frac{a_2}{b} = \frac{0.07 + a_1/b}{1.20} \qquad (5.05\text{-}22b)$$

provided $0.1 < a_2/b < 0.35$. A similar formula for correcting an initial value c_1/b gives a new value c_2/b, as

$$\frac{c_2}{b} = \frac{[0.07(1 - g/b) + c_1/b]}{1.20} , \qquad (5.05\text{-}22c)$$

provided g/b is fairly small and $0.1 < (c_2/b)/(1 - g/b)$.

When the strip of width a is inserted so far between the strips of width c that $d/g > 1.0$ the even-mode values C'_{ae}/ϵ and C'_{ce}/ϵ, do not change from their values at $d/g = 1.0$. However the value of $\Delta C/\epsilon$ does change and it can be found simply by adding $4(d/g - 1)$ to the value of $\Delta C/\epsilon$ at $d/g = 1.0$. For spacing between the strips of width c greater than $g/b = 0.5$, or for a separation $d/g <$ -2.0, some of the configurations shown in Fig. 5.05-1(a), (b), or (c) are probably more suitable.

Thick Rectangular Bars—The thick rectangular bar configuration of coupled transmission lines, illustrated in Fig. 5.05-1(g) can also be conveniently used where tight coupling between lines is desired.[33] The dimensions of the strips for particular values of Z_{oe} and Z_{oo} can be determined with the aid of Figs. 5.05-9 and 5.05-10(a), (b). A convenient procedure for using the curves is as follows. First one determines $\Delta C/\epsilon$ from Eq. (5.05-20), using the specified values of Z_{oe} and Z_{oo}. Next a convenient value of t/b is selected and the value of s/b is determined from Fig. 5.05-9. The value of w/b is then determined from the equation

$$\frac{w}{b} = \frac{1}{2}\left(1 - \frac{t}{b}\right)\left[\frac{C_{oe}}{2\epsilon} - \frac{C'_{fe}}{\epsilon} - \frac{C'_f}{\epsilon}\right] . \qquad (5.05\text{-}23)$$

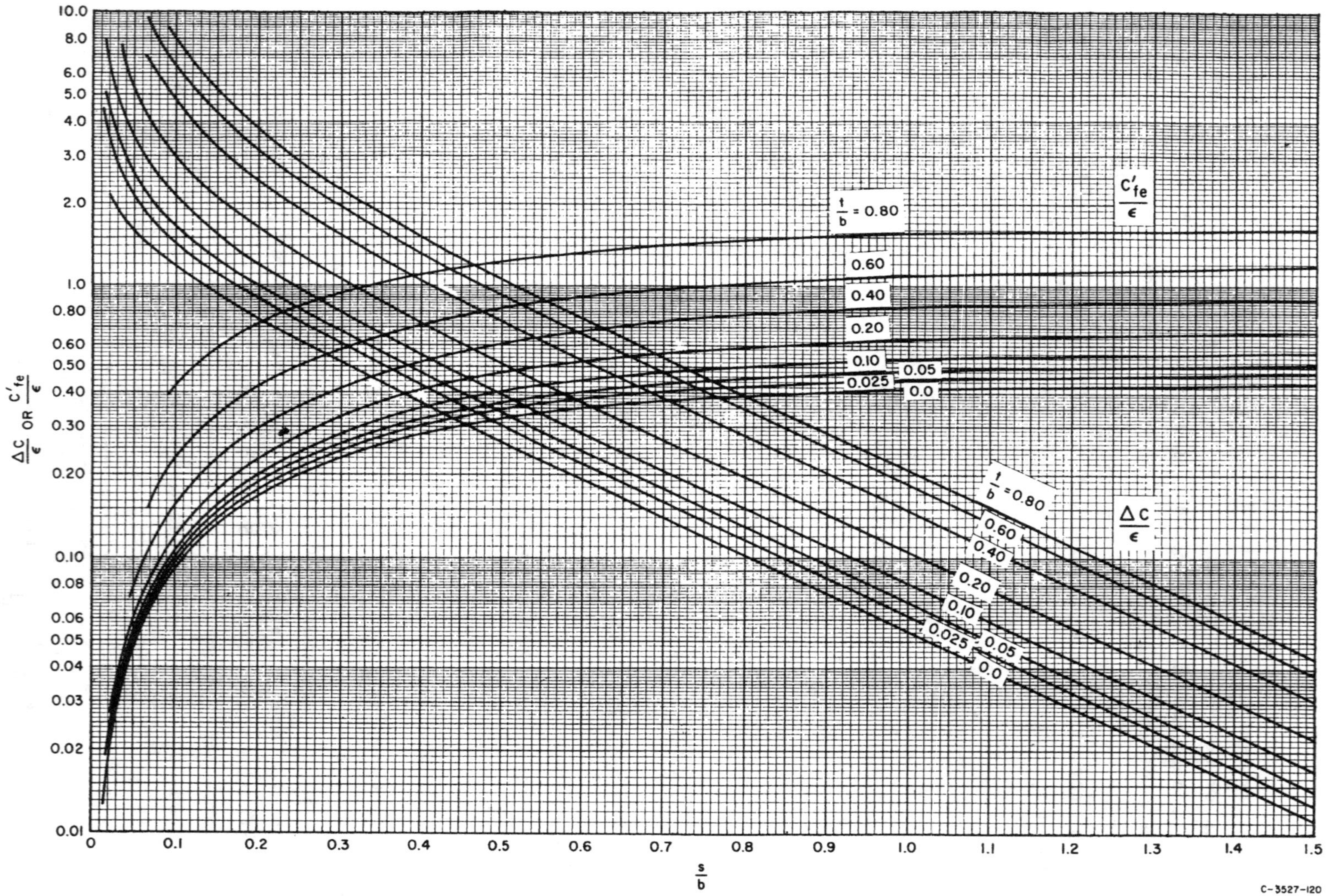

SOURCE: Quarterly Report 2, Contract DA 36-039 SC-87398, SRI; reprinted in *IRE Trans., PGMTT* (see Ref. 33, by W. J. Getsinger).

FIG. 5.05-9 NORMALIZED EVEN-MODE FRINGING CAPACITANCE C'_{fe}/ϵ AND INTERBAR CAPACITANCE $\Delta C/\epsilon$ FOR COUPLED RECTANGULAR BARS

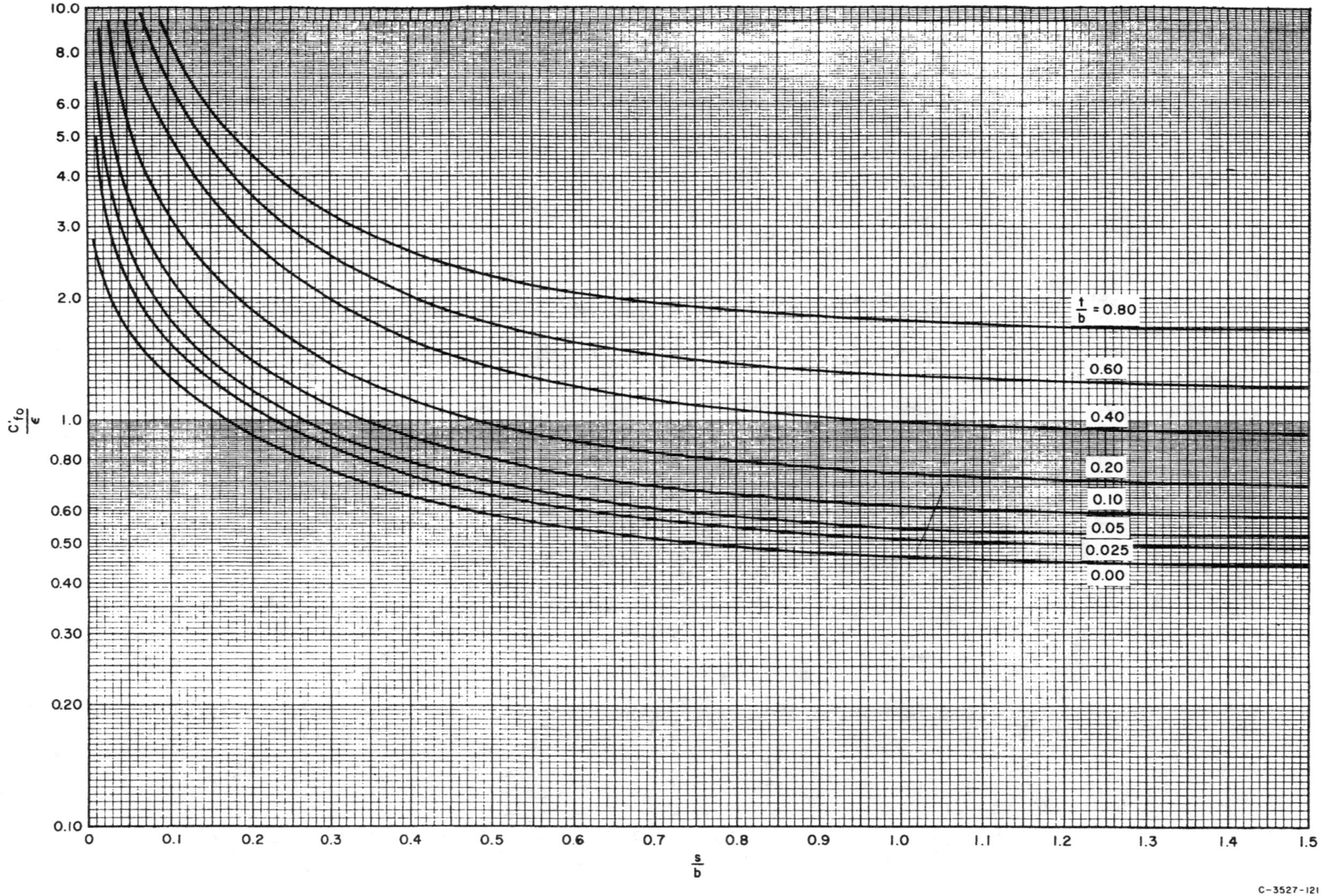

SOURCE: Quarterly Report 2, Contract DA 36-039 SC-87398, SRI; reprinted in *IRE Trans., PGMTT* (see Ref. 33, by W. J. Getsinger).

FIG. 5.05-10(a) NORMALIZED ODD-MODE FRINGING CAPACITANCE FOR RECTANGULAR BARS

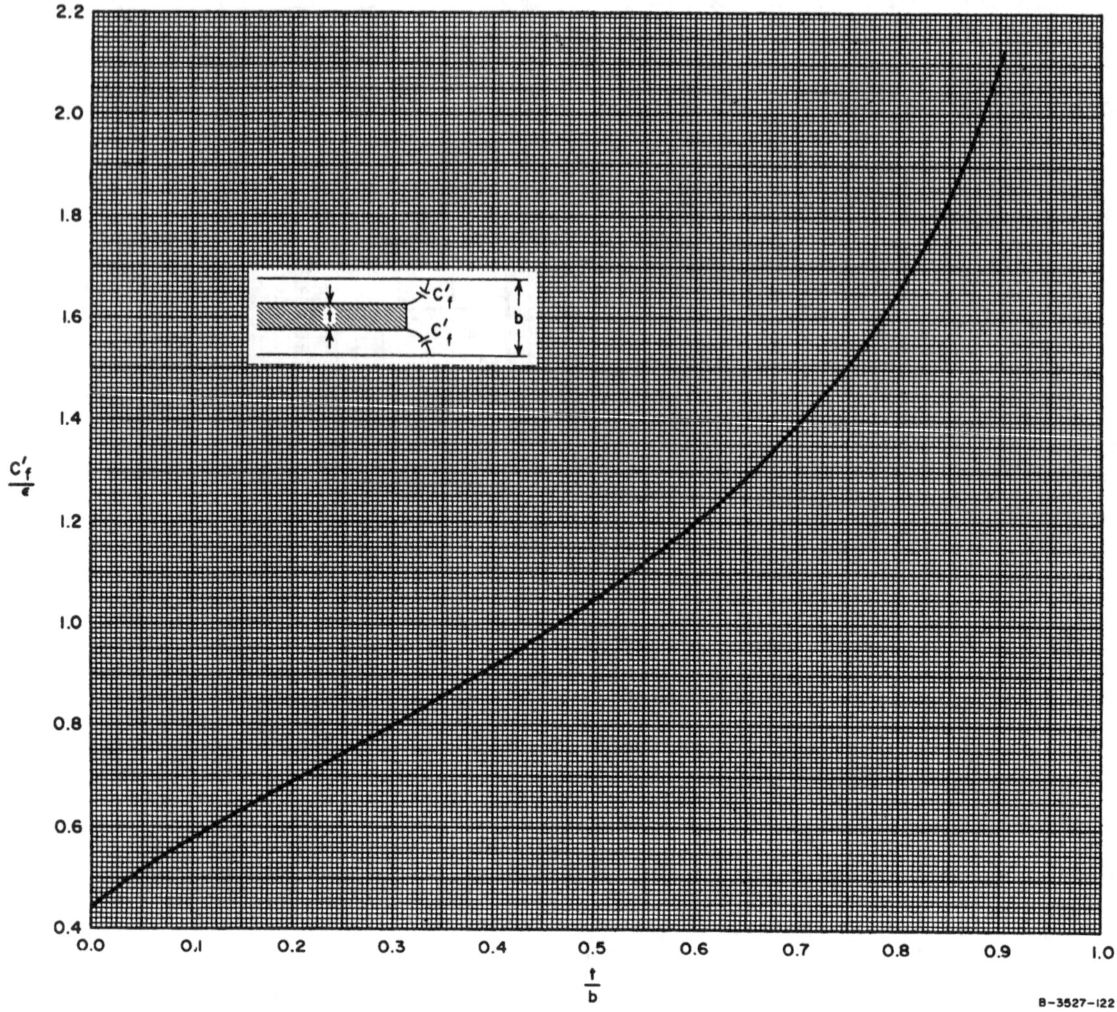

SOURCE: **Quarterly Report 2, Contract DA 36-039 SC-87398, SRI; reprinted in *IRE Trans., PGMTT* (see Ref. 33, by W. J. Getsinger).**

FIG. 5.05-10(b) NORMALIZED FRINGING CAPACITANCE FOR AN ISOLATED RECTANGULAR BAR

The value of C_{oe} to use is determined from the specified value of Z_{oe} using Eq. (5.05-18). The fringing capacitance C'_{fe} for the even mode can be read from Fig. 5.05-9, and C'_f can be determined from Fig. 5.05-10(b). The curves in Fig. 5.05-10(a) allow one to determine C'_{fo} directly.

The various fringing and parallel-plate capacitances used in the above discussion are illustrated in Fig. 5.05-11. Note that the odd-mode fringing capacitances C'_{fo} correspond to the fringing capacitances between the inner edges of the bars and a metallic wall halfway between the bars. It is seen that the total odd mode capacitance of a bar is

$$\frac{C_{oo}}{\epsilon} = 2\left[\frac{C_p}{\epsilon} + \frac{C'_{fo}}{\epsilon} + \frac{C'_f}{\epsilon}\right] \quad (5.05\text{-}24)$$

and the total even mode capacitance of a bar is

$$\frac{C_{oe}}{\epsilon} = 2\left[\frac{C_p}{\epsilon} + \frac{C'_{fe}}{\epsilon} + \frac{C'_f}{\epsilon}\right] \quad . \quad (5.05\text{-}25)$$

The normalized per-unit-length parallel plate capacitance $C_p/\epsilon = 2w/(b - t)$, and $\epsilon = 0.225\epsilon_r$ pf per inch.

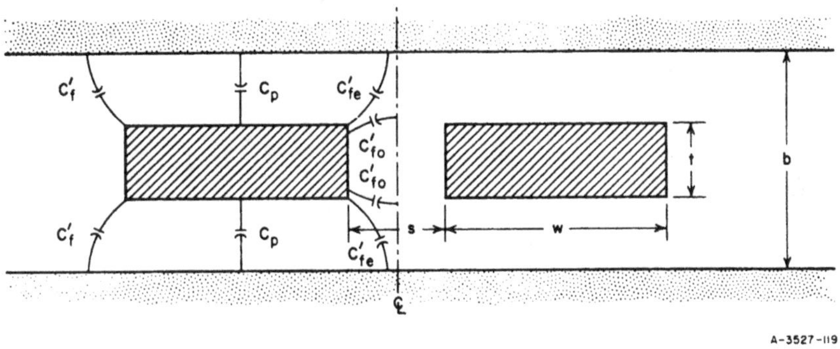

SOURCE: Quarterly Report 2, Contract DA 36-039 SC-87398, SRI; reprinted in *IRE Trans., PGMTT* (see Ref. 33, by W. J. Getsinger).

FIG. 5.05-11 COUPLED RECTANGULAR BARS CENTERED BETWEEN PARALLEL PLATES ILLUSTRATING THE VARIOUS FRINGING AND PARALLEL PLATE CAPACITIES

The even- and odd-mode fringing capacitances C'_{fe}/ϵ and C'_{fo}/ϵ were derived by conformal mapping techniques and are exact in limits of $[w/b/(1 - t/b)] \to \infty$. It is believed that when $[w/b/(1 - t/b)] > 0.35$ the interaction between the fringing fields is small enough so that the values of C_{oo}/ϵ and C_{oe}/ϵ determined from Eqs. (5.05-24) and (5.05-25) are reduced by a maximum of 1.24 percent of their true values.

In situations where an initial value, w/b is found from Eq. (5.05-23) to be less than $0.35\ [1 - (t/b)]$ so that the fringing fields interact, a new value of w'/b can be used where

$$\frac{w'}{b} = \frac{\left\{0.07\left[1 - \frac{t}{b}\right] + \frac{w}{b}\right\}}{1.20} \tag{5.05-26}$$

provided $0.1 < (w'/b)/[1 - (t/b)] < 0.35$.

Unsymmetrical Parallel-Coupled Lines—Figure 5.05-12 shows an unsymmetrical pair of parallel-coupled lines and various line capacitances. Note that C_a is the capacitance per unit length between Line a and ground, C_{ab} is the capacitance per unit length between Line a and Line b, while C_b is the capacitance per unit length between Line b and ground. When C_a is not equal to C_b, the two lines will have different odd- and even-mode admittances as is indicated by Eqs. (1) in Table 5.05-1. In terms of odd- and even-mode capacitances, for Line a

$$C^a_{oo} = C_a + 2C_{ab} , \qquad C^a_{oe} = C_a \tag{5.05-27}$$

while for Line b

$$C^b_{oo} = C_b + 2C_{ab} , \qquad C^b_{oe} = C_b \quad . \tag{5.05-28}$$

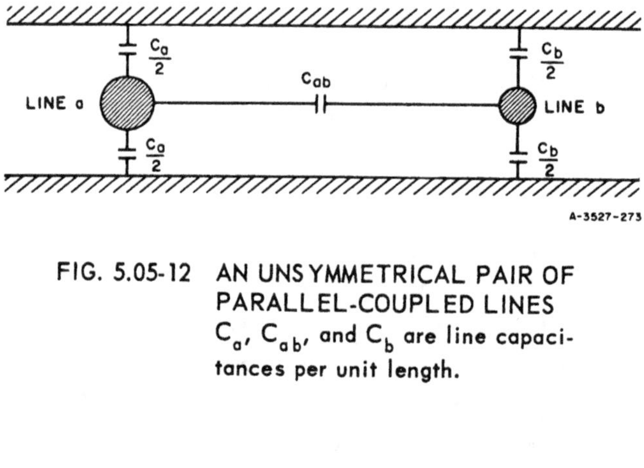

FIG. 5.05-12 AN UNSYMMETRICAL PAIR OF PARALLEL-COUPLED LINES
C_a, C_{ab}, and C_b are line capacitances per unit length.

For symmetrical parallel-coupled lines the odd-mode impedances are simply the reciprocals of the odd-mode admittances, and analogously for the even-mode impedances and admittances. However, as can be demonstrated from Eqs. (2) in Table 5.05-1, this is not the case for unsymmetrical parallel-coupled lines. For unsymmetrical lines, the odd- and even-mode impedances are not simply the reciprocals of the odd- and even-mode

Table 5.05-1

RELATIONS BETWEEN LINE ADMITTANCES, IMPEDANCES, AND CAPACITANCES PER UNIT LENGTH OF UNSYMMETRICAL PARALLEL-COUPLED LINES

v = velocity of light in media of propagation

= $1.18 \times 10^{10}/\sqrt{\epsilon_r}$ inches/sec.

η_0 = intrinsic impedance of free space = 376.7 ohms

ϵ = dielectric constant = $0.225\ \epsilon_r$ $\mu\mu$f/inch

$$\left.\begin{aligned} Y^a_{oe} &= vC_a\ , & Y^a_{oo} &= v(C_a + 2C_{ab}) \\ Y^b_{oe} &= vC_b\ , & Y^b_{oo} &= v(C_b + 2C_{ab}) \end{aligned}\right\} \qquad (1)$$

$$\left.\begin{aligned} Z^a_{oe} &= \frac{C_b + 2C_{ab}}{vF}\ , & Z^a_{oo} &= \frac{C_b}{vF} \\ Z^b_{oe} &= \frac{C_a + 2C_{ab}}{vF}\ , & Z^b_{oo} &= \frac{C_a}{vF} \end{aligned}\right\} \qquad (2)$$

where $F = C_a C_b + C_a C_{ab} + C_b C_{ab}$

$$\left.\begin{aligned} \frac{C_a}{\epsilon} &= \frac{\eta_0 Y^a_{oe}}{\sqrt{\epsilon_r}}\ , & \frac{C_{ab}}{\epsilon} &= \frac{\eta_0}{\sqrt{\epsilon_r}}\left(\frac{Y^a_{oo} - Y^a_{oe}}{2}\right) \\ \frac{C_b}{\epsilon} &= \frac{\eta_0 Y^b_{oe}}{\sqrt{\epsilon_r}}\ , & \frac{C_{ab}}{\epsilon} &= \frac{\eta_0}{\sqrt{\epsilon_r}}\left(\frac{Y^b_{oo} - Y^b_{oe}}{2}\right) \end{aligned}\right\} \qquad (3)$$

$$\left.\begin{aligned} \frac{C_a}{\epsilon} &= \frac{\eta_0 2Z^b_{oo}}{\sqrt{\epsilon_r}H}\ , & \frac{C_{ab}}{\epsilon} &= \frac{\eta_0}{\sqrt{\epsilon_r}}\left(\frac{Z^b_{oe} - Z^b_{oo}}{H}\right) \\ \frac{C_b}{\epsilon} &= \frac{\eta_0 2Z^a_{oo}}{\sqrt{\epsilon_r}\ H}\ , & \frac{C_{ab}}{\epsilon} &= \frac{\eta_0}{\sqrt{\epsilon_r}}\left(\frac{Z^a_{oe} - Z^a_{oo}}{H}\right) \end{aligned}\right\} \qquad (4)$$

where $H = Z^a_{oe}\ Z^b_{oo} + Z^b_{oe}\ Z^a_{oo}$.

admittances. The reason for this lies in the fact that when the odd- and even-mode *admittances* are computed the basic definition of these admittances assumes that the lines are being driven with *voltages of identical magnitude* with equal or opposite phase, while the *currents* in the lines may be of different magnitudes. When the odd- and even-mode *impedances* are computed, the basic definition of these impedances assumes that the lines are being driven by *currents of identical magnitude* with equal or opposite phases, while magnitudes of the *voltages* on the two lines may be different. These two different sets of boundary conditions can be seen to lead to different voltage-current ratios if the lines are unsymmetrical.

Some unsymmetrical parallel-coupled lines which are quite easy to design are shown in Fig. 5.05-13. Both bars have the same height, and both are assumed to be wide enough so that the interactions between the

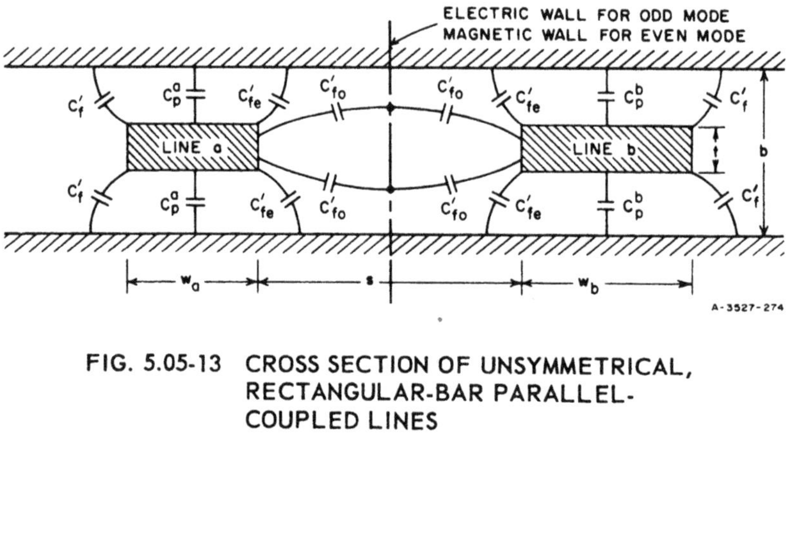

FIG. 5.05-13 CROSS SECTION OF UNSYMMETRICAL, RECTANGULAR-BAR PARALLEL-COUPLED LINES

fringing fields at the right and left sides of each bar are negligible, or at least small enough to be corrected for by use of Eq. (5.05-26). On this basis the fringing fields are the same for both bars, and their different capacitances C_a and C_b to ground are due entirely to different parallel-plate capacitances C_p^a and C_p^b. For the structure shown

$$\begin{aligned} C_a &= 2(C_p^a + C'_f + C'_{fe}) \\ C_{ab} &= (C'_{fo} - C'_{fe}) \qquad\qquad (5.05\text{-}29) \\ C_b &= 2(C_p^b + C'_f + C'_{fe}) \quad . \end{aligned}$$

To design a pair of lines such as those in Fig. 5.05-13 so as to have specified odd- and even-mode admittances or impedances, first use Eqs. (3) or (4) in Table 5.05-1 to compute C_a/ϵ, C_{ab}/ϵ, and C_b/ϵ. Select a convenient value for t/b, and noting that

$$\frac{\Delta C}{\epsilon} = \frac{C_{ab}}{\epsilon} \quad , \tag{5.05-30}$$

use Fig. 5.05-9 to determine s/b, and also C'_{fe}/ϵ. Using t/b and Fig. 5.05-10(b) determine C'_f/ϵ, and then compute

$$\frac{w_a}{b} = \frac{1}{2}\left(1 - \frac{t}{b}\right)\left[\frac{1}{2}\left(\frac{C_a}{\epsilon}\right) - \frac{C'_{fe}}{\epsilon} - \frac{C'_f}{\epsilon}\right] \tag{5.05-31}$$

$$\frac{w_b}{b} = \frac{1}{2}\left(1 - \frac{t}{b}\right)\left[\frac{1}{2}\left(\frac{C_b}{\epsilon}\right) - \frac{C'_{fe}}{\epsilon} - \frac{C'_f}{\epsilon}\right] \quad . \tag{5.05-32}$$

Knowing the ground-plane spacing b, the required bar widths w_a and w_b are then determined. This procedure also works for the thin-strip case where $t/b \approx 0$. If either w_a/b or w_b/b is less than $0.35[1 - t/b]$, Eq. (5.05-26) should be applied to obtain corrected values.

Arrays of Parallel-Coupled Lines—Figure 5.05-14 shows an array of parallel-coupled lines such as is used in the interdigital-line filters discussed in Chapt. 10. In the structure shown, all of the bars have the same t/b ratio and the other dimensions of the bars are easily obtained

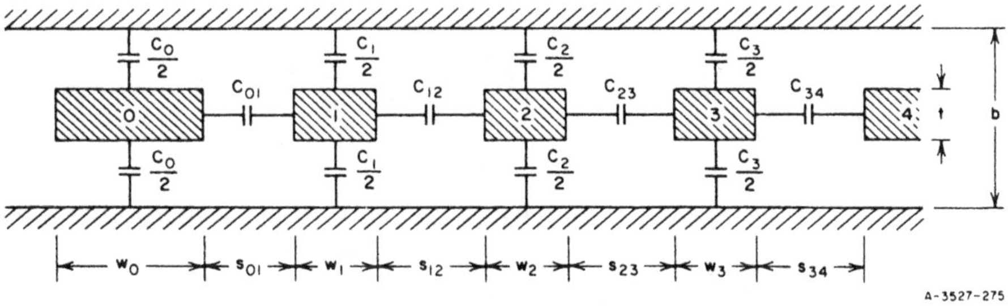

SOURCE: Quarterly Progress Report 4, Contract DA 36-039 SC-87398, SRI; reprinted in the *IRE Trans. PGMTT* (see Ref. 3 of Chapter 10, by G. L. Matthaei)

FIG. 5.05-14 CROSS SECTION OF AN ARRAY OF PARALLEL-COUPLED LINES BETWEEN GROUND PLANES

by generalizing the procedure described for designing the unsymmetrical parallel-coupled lines in Fig. 5.05-13. In the structure in Fig. 5.05-14 the electrical properties of the structure are characterized in terms of the self-capacitances C_k per unit length of each bar with respect to ground, and the mutual capacitances $C_{k,k+1}$ per unit length between adjacent bars k and $k+1$. This representation is not necessarily always highly accurate because there can conceivably be a significant amount of fringing capacitance in some cases between a given line element, and, for example, the line element beyond the nearest neighbor. However, at least for geometries such as that shown, experience has shown this representation to have satisfactory accuracy for applications such as interdigital filter design.

For design of the parallel-coupled array structures discussed in this book, equations will be given for the normalized self and mutual capacitances C_k/ϵ and $C_{k,k+1}/\epsilon$ per unit length for all the lines in the structure. Then the cross-sectional dimensions of the bars and spacings between them are determined as follows. First, choose values for t and b. Then, since

$$\frac{(\Delta C)_{k,k+1}}{\epsilon} = \frac{C_{k,k+1}}{\epsilon} \tag{5.05-33}$$

Fig. 5.05-9 can be used to determine $s_{k,k+1}/b$. In this manner, the spacings $s_{k,k+1}$ between all the bars are obtained. Also, using Fig. 5.05-9, the normalized fringing capacitances $(C'_{fe})_{k,k+1}/\epsilon$ associated with the gaps $s_{k,k+1}$ between bars are obtained. Then the normalized width of the kth bar is

$$\frac{w_k}{b} = \frac{1}{2}\left(1 - \frac{t}{b}\right)\left[\frac{1}{2}\left(\frac{C_k}{\epsilon}\right) - \frac{(C'_{fe})_{k-1,k}}{\epsilon} - \frac{(C'_{fe})_{k,k+1}}{\epsilon}\right] \quad . \tag{5.05-34}$$

In the case of the bar at the end of the array (the bar at the far left in Fig. 5.05-14), C'_{fe}/ϵ for the edge of the bar which has no neighbor must be replaced by C'_f/ϵ which is determined from Fig. 5.05-10(b). Thus, for example, for Bar 0 in Fig. 5.05-14,

$$\frac{w_0}{b} = \frac{1}{2}\left(1 - \frac{t}{b}\right)\left[\frac{1}{2}\left(\frac{C_0}{\epsilon}\right) - \frac{C'_f}{\epsilon} - \frac{(C'_{fe})_{01}}{\epsilon}\right] \quad . \tag{5.05-35}$$

If $w_k/b < 0.35[1 - t/b]$ for any of the bars, the width correction given in Eq. (5.05-26) should be applied to those bars where this condition exists.

SEC. 5.06, SPECIAL PROPERTIES OF WAVEGUIDES

A waveguide consisting of a single hollow conductor that can propagate electromagnetic energy above a certain cutoff frequency, f_c, is also a very useful element in microwave filters. A waveguide can propagate an infinite number of modes, which can be characterized as being either TE (transverse electric) or TM (transverse magnetic). The TE modes have a magnetic field but no electric field in the direction of propagation, while TM modes have an electric field but no magnetic field in the direction of propagation. Usually a waveguide is operated so that it propagates energy in a single mode, and under this condition it can be described as a transmission line with a propagation constant γ_t and a characteristic impedance Z_0. The propagation constant for a waveguide is uniquely defined. The characteristic impedance of a waveguide can be considered to be the wave impedance of the guide, Z_w (*i.e.*, the ratio of the transverse electric to the transverse magnetic field in the guide), multiplied by a constant. The value of the constant depends on what definition of characteristic impedance is employed (*i.e.*, voltage-current, voltage-power, or current-power). Thus it is seen that the characteristic impedance of a waveguide is not a unique quantity, as it is in the case of a TEM transmission line. However, this lack of uniqueness turns out to be unimportant in waveguide filter calculations because one can always normalize all waveguide equivalent circuit elements to the characteristic impedance of the guide.

In a lossless waveguide filled with dielectric of relative dielectric constant ϵ_r, the guide wavelength λ_g, free-space wavelength λ, wavelength in the dielectric λ_1, and cutoff wavelength λ_c, are related as

$$\frac{1}{\lambda_1^2} = \frac{\epsilon_r}{\lambda^2} = \frac{1}{\lambda_g^2} + \frac{1}{\lambda_c^2} \quad . \qquad (5.06\text{-}1)$$

The characteristic impedance that we shall assume for convenience to equal the wave impedance is

$$Z_0 = \frac{1}{Y_0} = \begin{cases} \dfrac{377}{\sqrt{\epsilon_r}} \lambda_g/\lambda_1 & \text{TE modes} \\[2ex] \dfrac{377}{\sqrt{\epsilon_r}} \lambda_1/\lambda_g & \text{TM modes} \end{cases} \quad . \tag{5.06-2}$$

The propagation phase constant β_t is

$$\beta_t = \frac{2\pi}{\lambda_g} \text{ radians/unit length} \quad . \tag{5.06-3}$$

The most common form of waveguide for use in microwave filters is a rectangular waveguide of width a and height b operating in a TE_{10} mode. TE_{m0} modes have cutoff wavelengths

$$\lambda_c = \frac{2a}{m} \quad . \tag{5.06-4}$$

The index m equals the number of half-waves of variation of the electric field across the width, a, of the guide. The cutoff frequency f_c (measured in gigacycles) is related to the cutoff wavelength in inches as

$$f_c = \frac{11.8}{\lambda_c} \sqrt{\epsilon_r} \quad . \tag{5.06-5}$$

The dominant mode, that is, the one with the lowest cutoff frequency, is the TE_{10} mode.

The dominant mode in circular waveguide of diameter D is the TE_{11} mode. The cutoff wavelength of the TE_{11} mode is $1.706D$.

The attenuation of these modes due to losses in the copper conductors are for TE_{m0} modes in rectangular guide

$$\alpha_c\,(TE_{m0}) = \frac{1.90 \times 10^{-4} \sqrt{\epsilon_r}\sqrt{f}}{b} \frac{\left[1 + \dfrac{2b}{a}\left(\dfrac{f_c}{f}\right)^2\right]}{\sqrt{1 - \left(\dfrac{f_c}{f}\right)^2}} \text{ db/unit length} \tag{5.06-6}$$

and for the TE_{11} mode in circular guide

$$\alpha_{c(TE_{11})} = \frac{3.80 \times 10^{-4} \sqrt{\epsilon_r}\sqrt{f}}{D} \frac{\left[\left(\frac{f_c}{f}\right)^2 + 0.420\right]}{\sqrt{1 - \left(\frac{f_c}{f}\right)^2}} \quad \text{db/unit length} \qquad (5.06\text{-}7)$$

where f is measured in gigacycles. These values of attenuation are plotted in Fig. 5.06-1.

The attenuation caused by losses in the dielectric in any waveguide mode is

$$\alpha_d = \frac{27.3 \tan \delta}{\lambda_1}\left(\frac{\lambda_g}{\lambda_1}\right) \quad \text{db/unit length} \qquad (5.06\text{-}8)$$

where $\tan \delta$ is the loss tangent of the dielectric. The unloaded Q, of a waveguide* is

$$\frac{1}{Q} = \frac{1}{Q_d} + \frac{1}{Q_c} \qquad (5.06\text{-}9)$$

where Q_d depends only on losses in the dielectric and is given by

$$Q_d = \frac{1}{\tan \delta} \qquad (5.06\text{-}10)$$

and Q_c is a function of the ohmic losses in the waveguide walls and the ratio f/f_c, and is given by

$$Q_c = \frac{\pi\lambda_g}{\lambda_1^2 \alpha_c} \quad . \qquad (5.06\text{-}11)$$

* Additional discussion relevant to the use of waveguides as resonators will be found in Sec. 5.08.

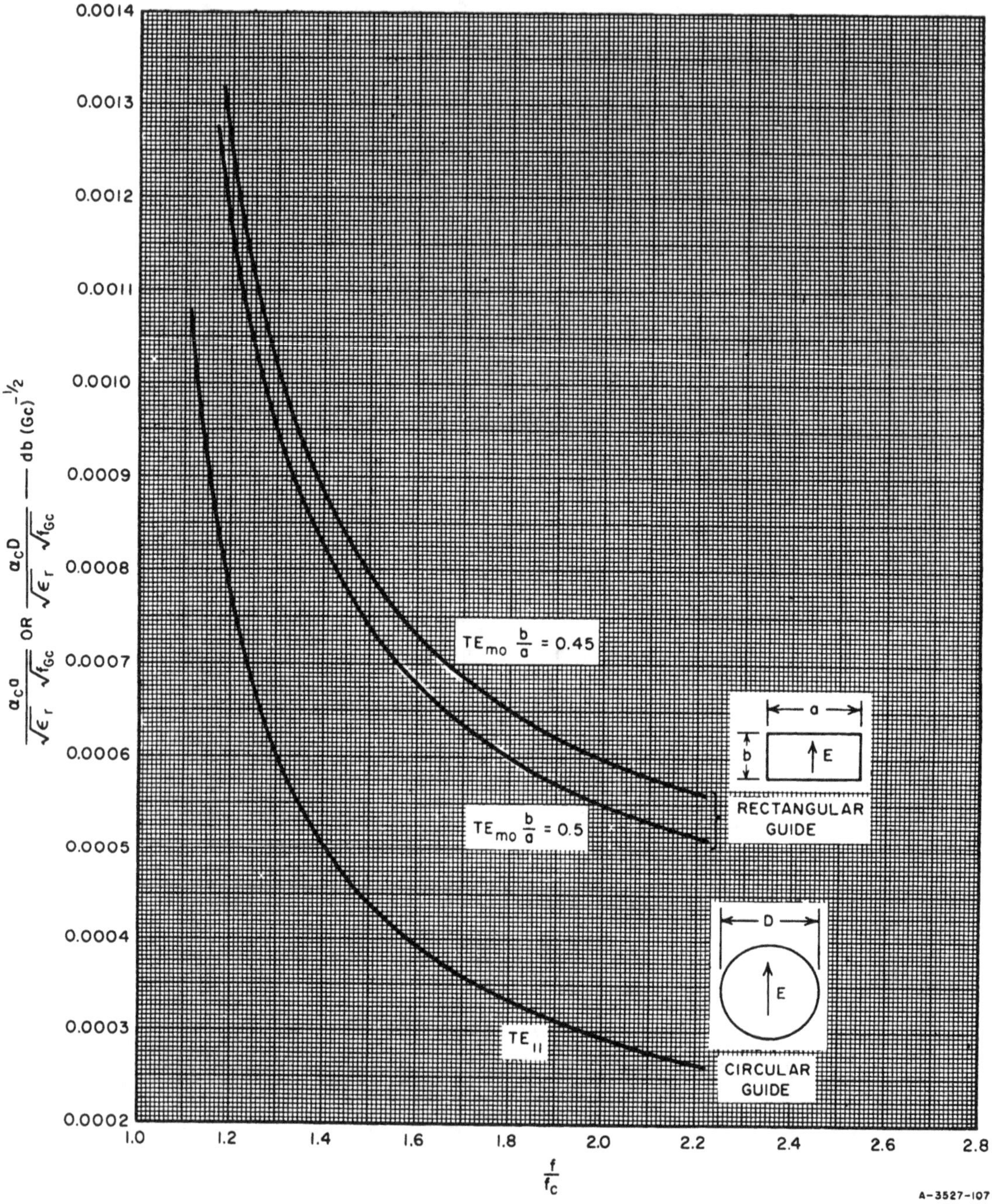

FIG. 5.06-1 WAVEGUIDE ATTENUATION DATA

For rectangular copper waveguides operating in the TE_{m0} mode, we have

$$Q_c(TE_{m0}) = \frac{1.212 \times 10^4 \, b\sqrt{f}}{1 + \dfrac{2b}{a}\left(\dfrac{f_c}{f}\right)^2} \qquad (5.06\text{-}12)$$

where a and b are measured in inches, and f in gigacycles. For a circular waveguide operating in the TE_{11} mode, we have

$$Q_c(TE_{11}) = \frac{0.606 \times 10^4 \, D\sqrt{f}}{0.420 + \left(\dfrac{f_c}{f}\right)^2} \qquad (5.06\text{-}13)$$

where D is measured in inches and f in gigacycles. These expressions for Q_c are plotted in Fig. 5.06-2.

The pulse-power capacity P_{max} of air-filled guides,* at atmospheric pressure, assuming a breakdown strength of 29 kv/cm, for the TE_{m0} mode in rectangular guide is

$$P_{max}(TE_{m0}) = 3.6 \, ab \, \frac{\lambda}{\lambda_g} \quad \text{megawatts} \quad , \qquad (5.06\text{-}14)$$

and for the TE_{11} mode in circular guide

$$P_{max}(TE_{11}) = 2.7 \, D^2 \, \frac{\lambda}{\lambda_g} \quad \text{megawatts} \qquad (5.06\text{-}15)$$

where the dimensions are in inches.

In a rectangular waveguide operating in the TE_{10} mode, with an aspect ratio b/a of 0.5 or 0.45, the next higher-order mode is the TE_{20} with cutoff wavelength $\lambda_c = a$. Next come the TE_{11} or TM_{11} modes each of which has the same cutoff wavelength, $\lambda_c = 2ab/\sqrt{a^2 + b^2}$. In the circular waveguide, the next higher-order mode is the TM_{01} mode, which has $\lambda_c = 1.305 \, D$.

---

* See Sec. 15 02 for further discussion of this subject

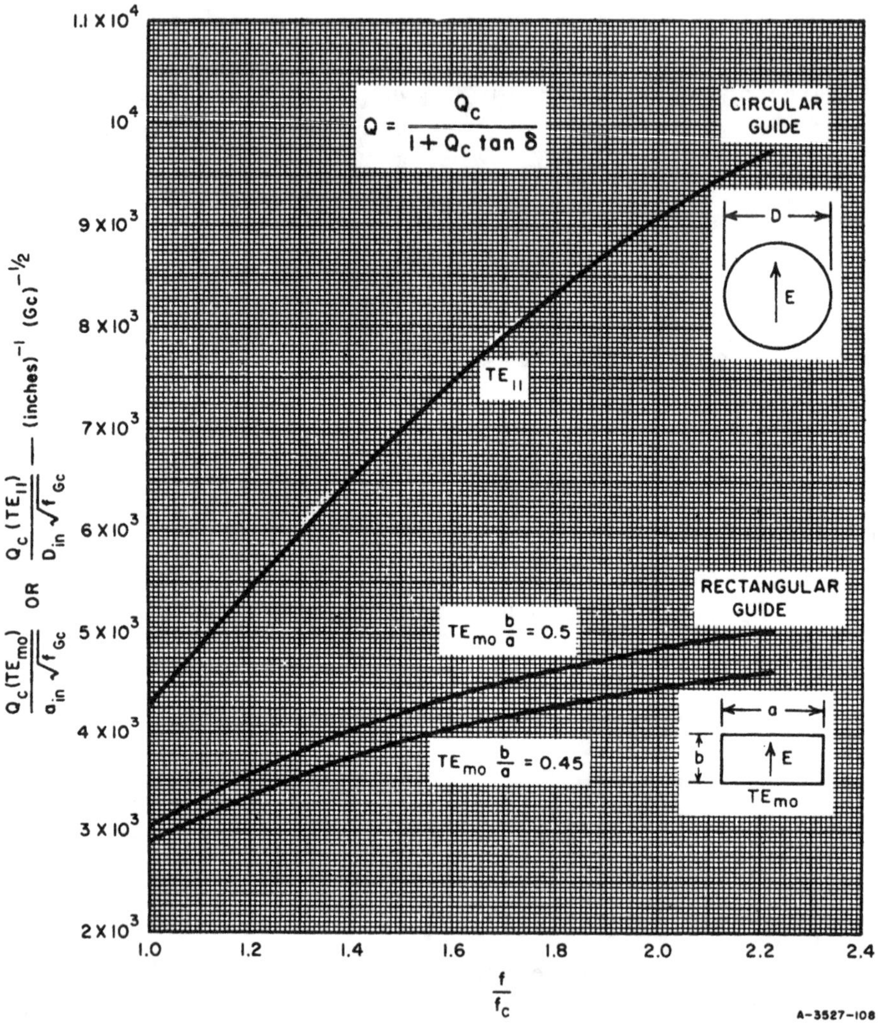

FIG. 5.06-2 WAVEGUIDE Q DATA

SEC. 5.07, COMMON TRANSMISSION LINE DISCONTINUITIES*

This section presents formulas and curves for some of the common discontinuities in transmission lines. Other more complete results are to be found in the literature.[8, 9, 10, 11, 12, 13]

Changes in Diameter of Coaxial Lines—When a change is made in the diameter of either the inner or outer conductor of a coaxial line, or in both conductors simultaneously, the equivalent circuits can be represented as shown in Fig. 5.07-1.[10, 11] The equivalent shunt capacity, C_d, for each of these cases is given in Fig. 5.07-2. These equivalent circuits apply when the operating frequency is appreciably below the cutoff frequency of the next higher-order propagating mode.

Changes in Width of Center Conductor of a Strip Line—The change in width of the center conductor of a strip line introduces an inductive reactance in series with the line.[12] In most situations this reactance is small and can be neglected. The approximate equivalent circuit for this situation is shown in Fig. 5.07-3.

Compensated Right-Angle Corner in Strip Line—A low-VSWR right-angle corner can be made in strip line if the outside edge of the strip is beveled. Figure 5.07-4 shows the dimensions of some matched right-angle corners for a plate-spacing-to-wavelength ratio, b/λ, of 0.0845. These data were obtained for a center strip conductor having negligible thickness; however, the data should apply with acceptable accuracy for strips of moderate thickness.

Fringing Capacitance for Semi-Infinite Plate Centered Between Parallel Ground Planes—The exact fringing capacitance, C'_f, from one corner of a semi-infinite plate centered between parallel ground planes is

$$C'_f = \frac{\epsilon}{\pi}\left\{\frac{2}{1-\frac{t}{b}}\, ln\left(\frac{1}{1-\frac{t}{b}}+1\right) - \left(\frac{1}{1-\frac{t}{b}}-1\right) ln\left[\frac{1}{\left(1-\frac{t}{d}\right)^2}-1\right]\right\} \quad \mu\mu\text{f/inch}$$

where $\epsilon = 0.225\,\epsilon_r$ micromicrofarads per inch and ϵ_r is the relative dielectric constant of the material between the semi-infinite plate and the ground planes. Fringing capacitance, C'_f, is plotted in Fig. 5.07-5.

* Data for some additional discontinuity configurations will be found in Secs. 8.05, 8.06, 8.08, 8.12, and 9.05.

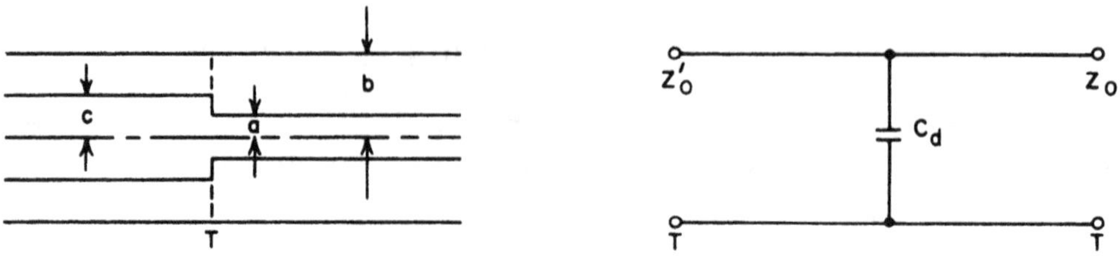

LONGITUDINAL SECTION EQUIVALENT CIRCUIT

$$\frac{Z_0'}{Z_0} = \frac{\ln \frac{b}{c}}{\ln \frac{b}{a}} \qquad C_d = 2\pi b C'_{d1}$$

(a) STEP IN INNER CONDUCTOR

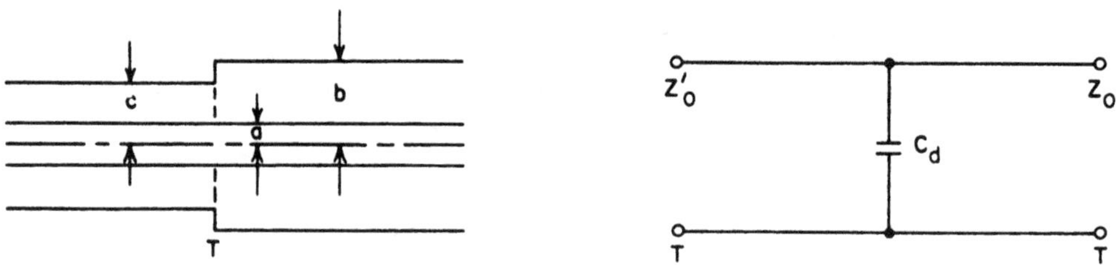

LONGITUDINAL SECTION EQUIVALENT CIRCUIT

$$\frac{Z_0'}{Z_0} = \frac{\ln \frac{c}{a}}{\ln \frac{b}{a}} \qquad C_d = 2\pi a C'_{d2}$$

(b) STEP IN OUTER CONDUCTOR

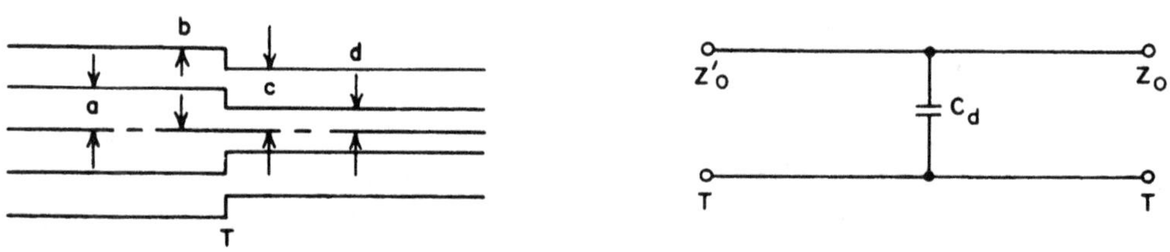

LONGITUDINAL SECTION EQUIVALENT CIRCUIT

$$\frac{Z_0'}{Z_0} = \frac{\ln \frac{b}{a}}{\ln \frac{c}{d}} \qquad C_d = 2\pi c C'_{d1} + 2\pi a C'_{d2}$$

(c) STEP IN INNER AND OUTER CONDUCTORS

A-3527-139

FIG. 5.07-1 COAXIAL-LINE DISCONTINUITIES

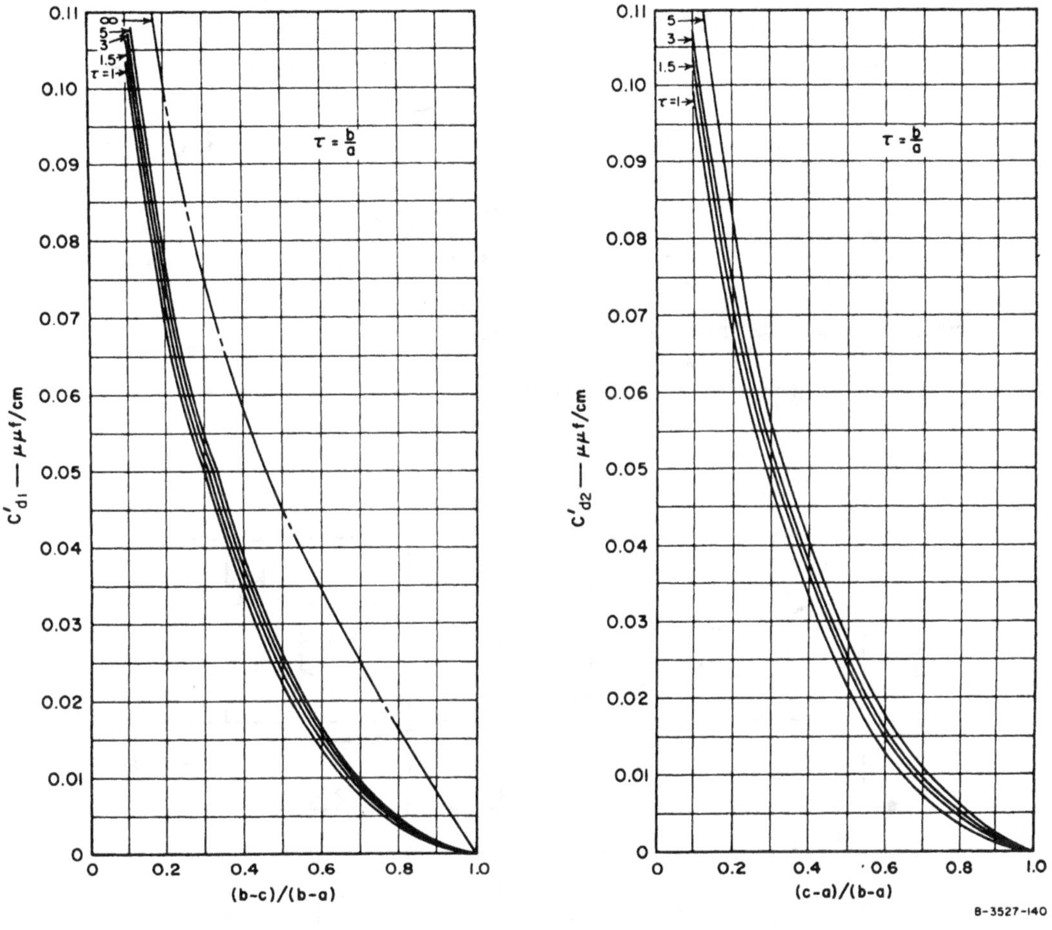

SOURCE: *Proc. IRE* (see Ref. 10 and 11 by J. R. Whinnery and H. W. Jamieson).

FIG. 5.07-2 COAXIAL-LINE-STEP FRINGING CAPACITIES

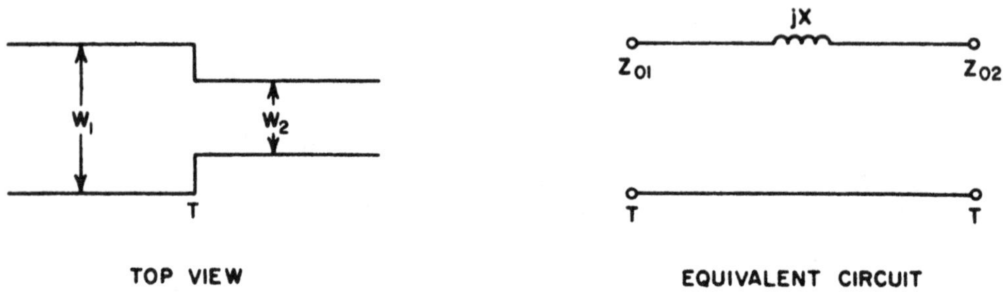

$$X = \frac{60\pi b}{\lambda} \; ln \; csc \left[\frac{\pi}{2}\frac{Z_{02}}{Z_{01}}\right]$$

A-3527-141

SOURCE: *IRE Trans., PGMTT* (see Ref. 12, by A. A. Oliner).

FIG. 5.07-3 STRIP-LINE STEP EQUIVALENT CIRCUIT

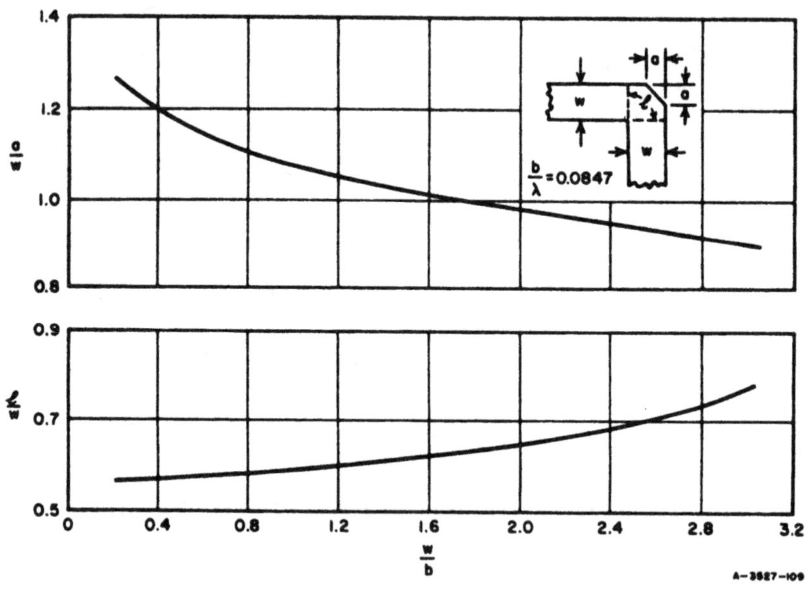

FIG. 5.07-4 MATCHED STRIP-LINE CORNER
The parameter ℓ is the effective length around the corner.

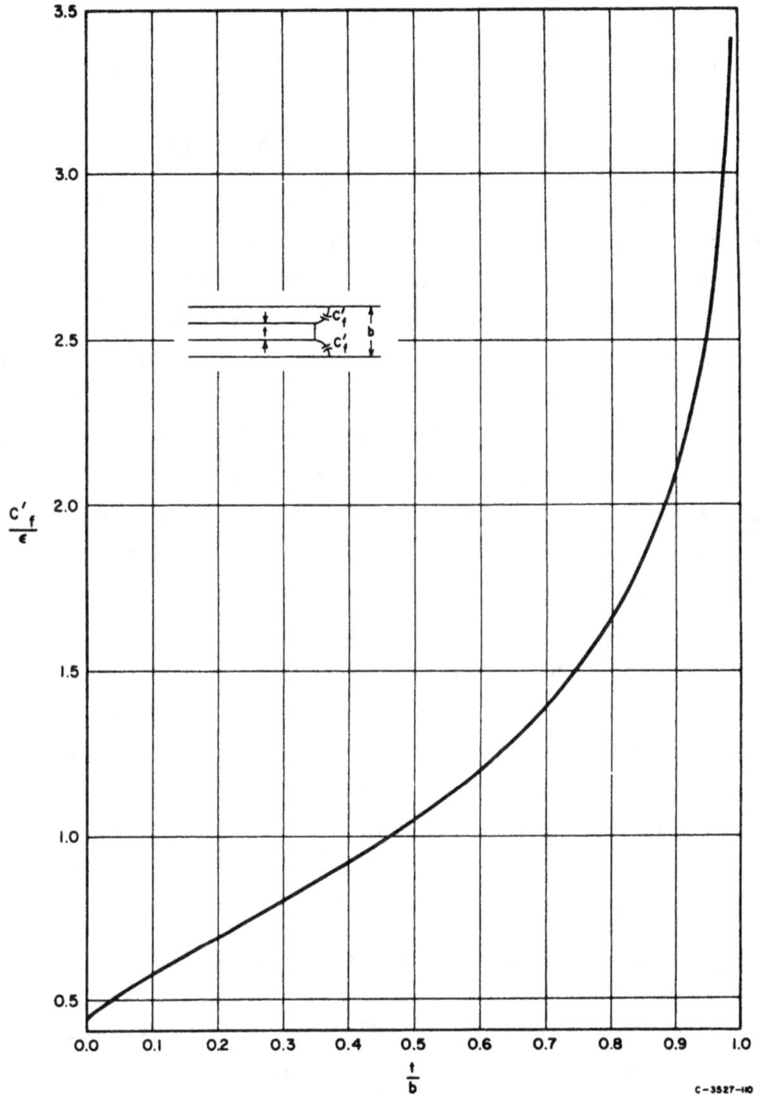

SOURCE: Final Report Contract DA 36-039 SC-63232, SRI; reprinted in *IRE Trans., PGMTT* (see Ref. 2, by S. B. Cohn).

FIG. 5.07-5 EXACT FRINGING CAPACITANCE FOR A SEMI-INFINITE PLATE CENTERED BETWEEN PARALLEL GROUND PLANES

Strip-Line T-Junctions—A symmetrical strip-line *T*-junction of the type illustrated in Fig. 5.07-6(a) can be represented by the equivalent circuit shown in Fig. 5.07-6(b). A short-circuit placed in turn in each of the three arms, at distances equal to multiples of one-half wavelength from the corresponding reference planes labeled P_1 and P_2, will block transmission between the other two arms of the junction.

Measured values obtained for the equivalent circuit parameters of sixteen different strip-line *T*-junctions are shown in Figs. 5.07-7, 5.07-8, and 5.07-9. The thickness, t, of the strips used in these measurements was 0.020 inch, while the ground-plane spacing was 0.500 inch. The widths of the strips having 35, 50, 75, and 100 ohms characteristic impedance were 1.050, 0.663, 0.405, and 0.210 inches, respectively. Measurements carried out in the frequency band extending from 2 to 5 Gc, corresponding to values of b/λ varying from 0.085 to 0.212. It was found that the reference plane positions were almost independent of frequency for all sixteen *T*-junctions, and therefore only the values corresponding to b/λ of 0.127 are shown in Fig. 5.07-7. It is seen from an inspection of Fig. 5.07-8 that A, the equivalent transformer turns ratio squared, is sensitive to frequency and has a value approximately equal to unity for b/λ very small, and decreases considerably for larger values of b/λ. The values of the discontinuity susceptance, B_d, vary considerably from one junction to another, and in some instances are quite frequency-sensitive. It is believed that B_d is essentially capacitive in nature. Thus positive values of B_d correspond to an excess of capacitance at the junction, while negative values correspond to a deficiency.

Although the data presented in Figs. 5.07-7, 5.07-8, and 5.07-9 are for *T*-junctions with air-filled cross section and with the ratio $t/b = 0.040$, these data may be applied to other cross sections. For instance, it is expected that these data should hold for any strip-thickness ratio, t/b, up to at least 0.125 if the same characteristic impedances are maintained.

In the case of a dielectric-filled section, $\epsilon_r > 1$, the data are expected to apply with good accuracy if one divides the characteristic impedances Z_{01} and Z_{02} by $\sqrt{\epsilon_r}$ and multiples b/λ and B_d/Y_o by $\sqrt{\epsilon_r}$.

Change in Height of a Rectangular Waveguide[8]—The equivalent circuit of the junction of two waveguides of different height but the same width, which are both operating in the TE_{10} mode can be represented as shown in

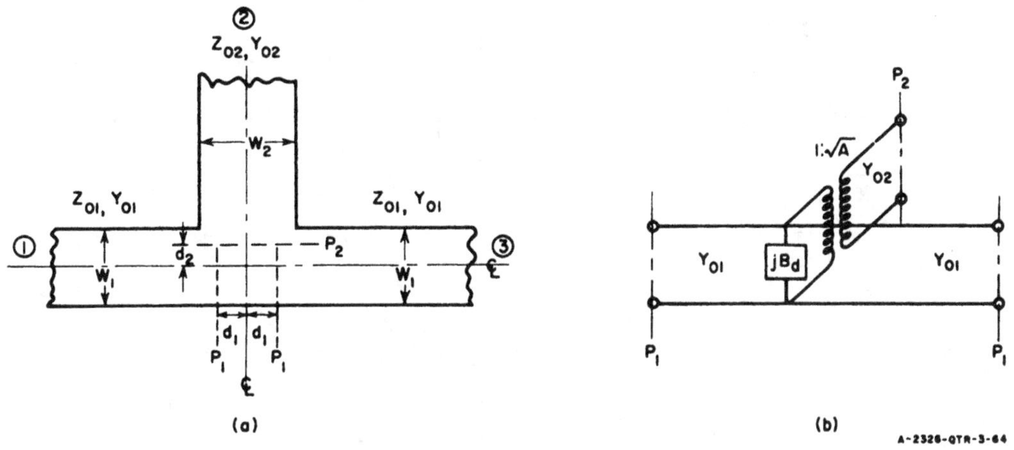

FIG. 5.07-6 EQUIVALENT CIRCUIT OF A STRIP-LINE T-JUNCTION

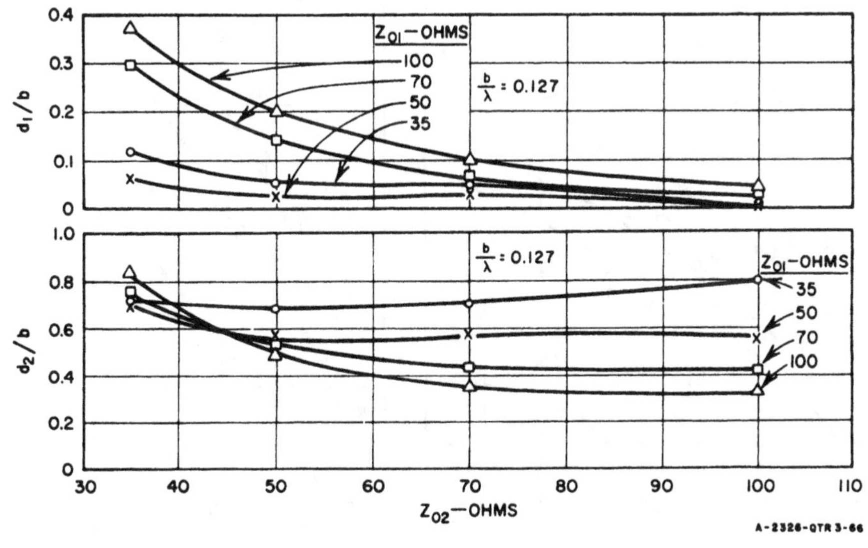

FIG. 5.07-7 REFERENCE-PLANE LOCATIONS vs. Z_{02}

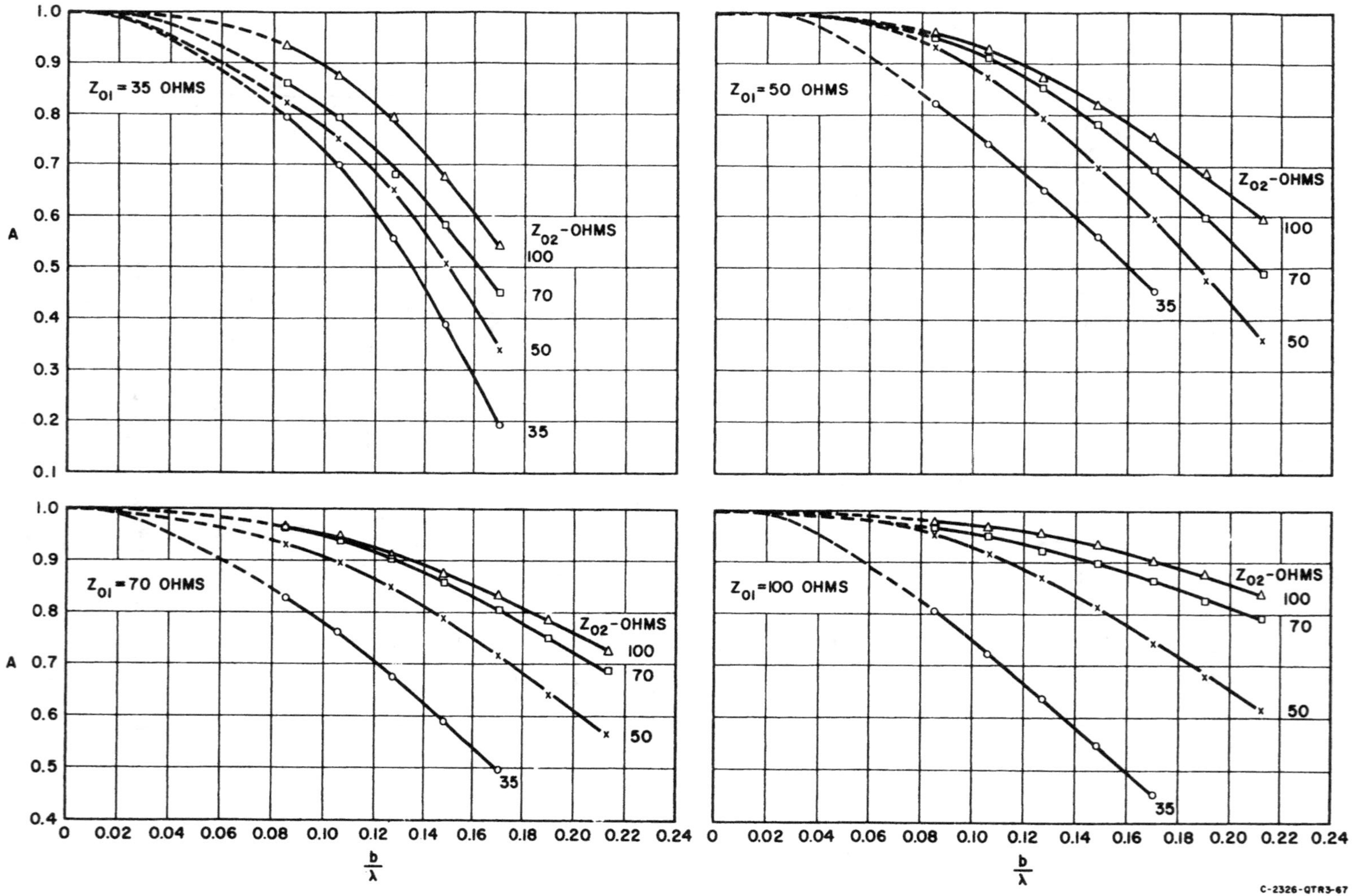

FIG. 5.07-8 MEASURED TRANSFORMER-TURNS RATIO SQUARED FOR SIXTEEN STRIP-LINE T-JUNCTIONS

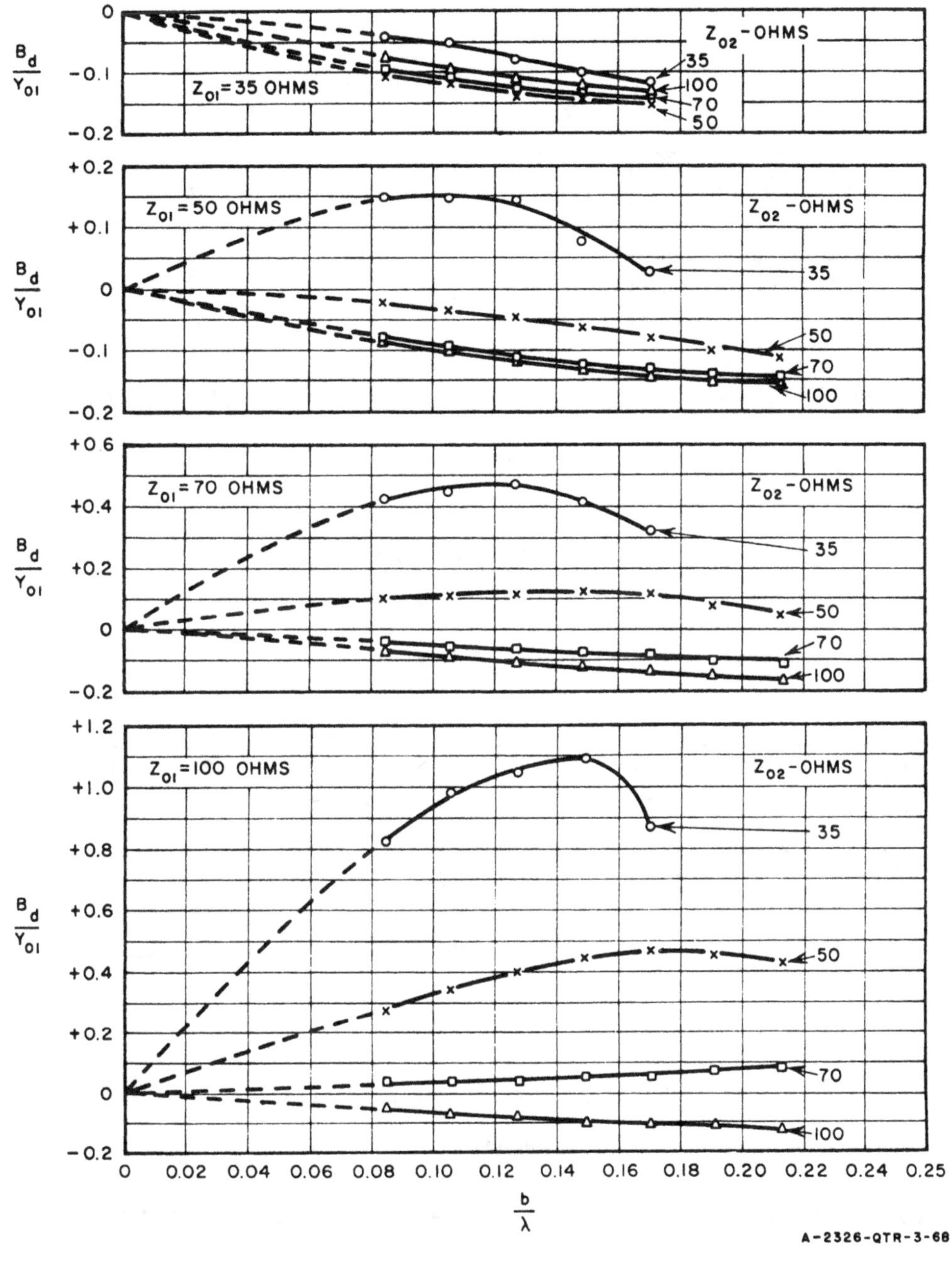

FIG. 5.07-9 MEASURED DISCONTINUITY SUSCEPTANCE OF SIXTEEN STRIP-LINE T-JUNCTIONS

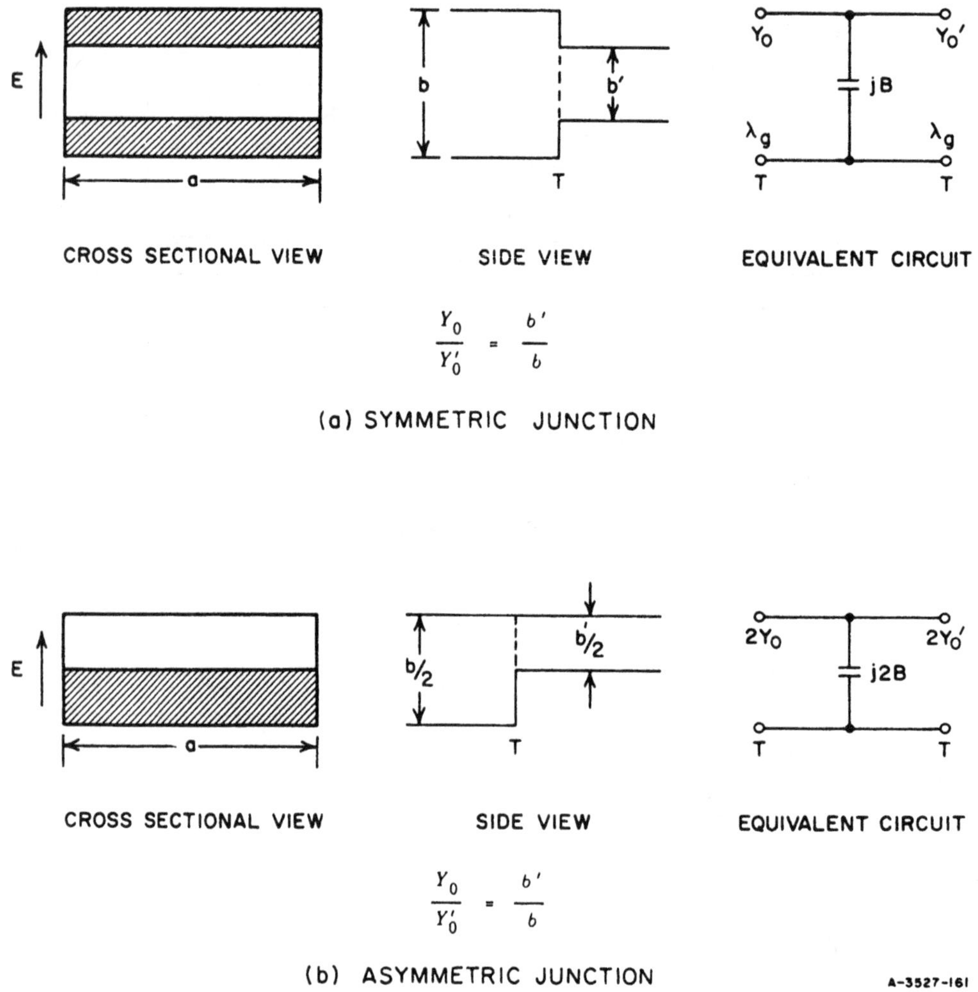

SOURCE: *Waveguide Handbook* (see Ref. 8, edited by N. Marcuvitz).

FIG. 5.07-10 EQUIVALENT CIRCUIT FOR CHANGE IN HEIGHT OF RECTANGULAR WAVEGUIDE

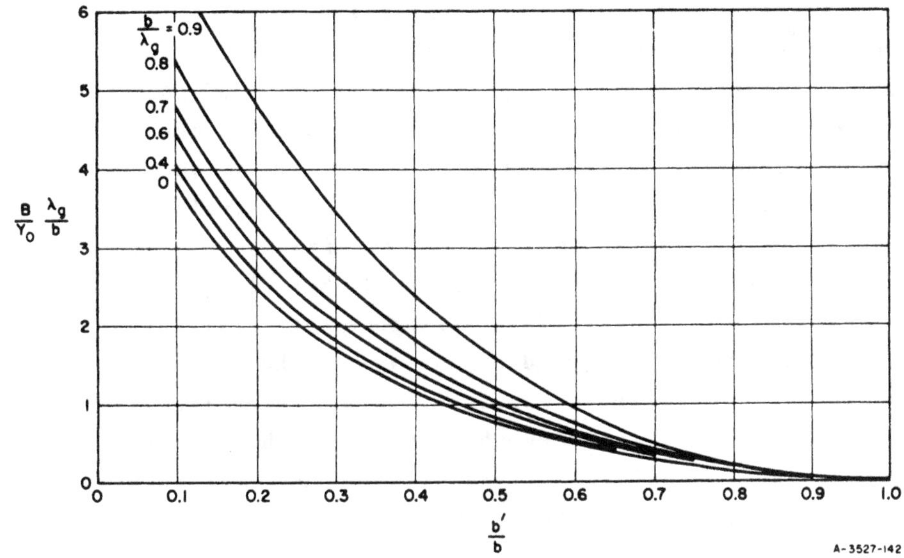

SOURCE: *Waveguide Handbook* (see Ref. 8, edited by N. Marcuvitz).

FIG. 5.07-11 SHUNT SUSCEPTANCE FOR CHANGE IN HEIGHT OF RECTANGULAR GUIDE

Fig. 5.07-10. The normalized susceptance $B\lambda_g/Y_0 b$ is plotted in Fig. 5.07-11 for various values of b/λ_g, and is accurate to about 1 percent for $b/\lambda_g \leq 1$.

SEC. 5.08, TRANSMISSION LINES AS RESONATORS

In many microwave filter designs, a length of transmission line terminated in either an open-circuit or a short-circuit is often used as a resonator. Figure 5.08-1 illustrates four resonators of this type, together with their lumped-constant equivalent circuits. It is to be noted that the resonators in Fig. 5.08-1(a) and 5.08-1(b) each have lengths which are multiples of one-half guide wavelength, and that the lumped-constant equivalent circuit of the transmission line which is short-circuited at one end is the dual of the equivalent circuit of the transmission line with an open-circuit termination. Similarly, the resonators in Fig. 5.08-1(c) and 5.08-1(d) have lengths which are odd multiples of one-quarter guide wavelength, and their lumped constant equivalent circuits are also duals of one another. The quantities α_t, λ_{g0} and λ_0 are the attenuation of the transmission line in nepers per unit length, the guide wavelength at the resonant frequency, and the plane-wave wavelength at the resonant frequency, respectively, in the dielectric medium filling the resonator.

The equivalence between the lumped constant circuits and the microwave circuits shown was established in the following fashion. The values of the resistance, R, and conductance, G, in the lumped-constant equivalent circuits were determined as the values of these quantities for the various lines at the resonance angular frequency, ω_0. The reactive elements in the lumped-constant equivalent circuits were determined by equating the slope parameters (defined below) of the lumped-element circuits to those of the transmission-line circuits which exhibited the same type of resonance. The general definition of the *reactance slope parameter* $\mathscr{X}$, which applies to circuits that exhibit a series type of resonance, is

$$\mathscr{X} = \frac{\omega_0}{2} \left.\frac{dX}{d\omega}\right|_{\omega 0} \quad \text{ohms} \qquad (5.08\text{-}1)$$

where X is the reactance portion of the input impedance to the circuit. The *susceptance slope parameter* $\mathscr{B}$, which applies to circuits that exhibit a parallel type of resonance, is

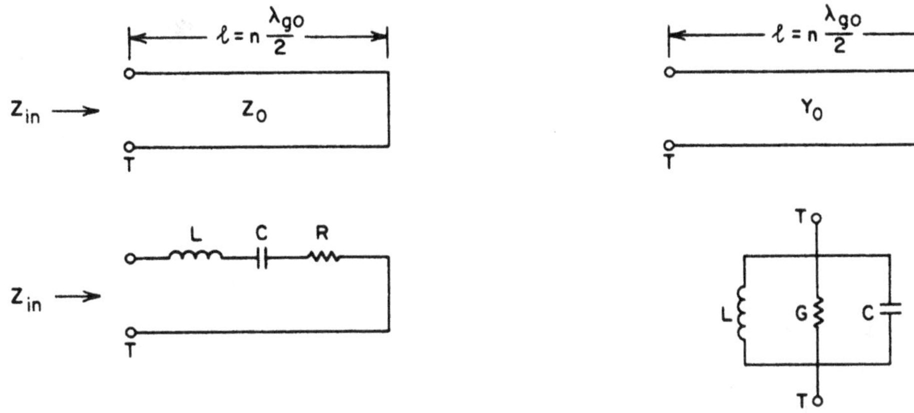

$$R \approx Z_0 \alpha_t \ell = \frac{n}{2} Z_0 \alpha_t \lambda_{g0}$$

$$x = \omega_0 L = \frac{1}{\omega_0 C} \approx \frac{n\pi Z_0}{2}\left(\frac{\lambda_{g0}}{\lambda_0}\right)^2$$

$$Z_{in} = R + jx\left(\frac{\omega}{\omega_0} - \frac{\omega_0}{\omega}\right)$$

$$Q = \frac{x}{R} \approx \frac{\pi\lambda_{g0}}{\alpha_t \lambda_0^2}$$

$$n = 1, 2, 3, \ldots$$

(a)

$$G \approx Y_0 \alpha_t \ell = \frac{n}{2} Y_0 \alpha_t \lambda_{g0}$$

$$b = \omega_0 C = \frac{1}{\omega_0 L} \approx \frac{n\pi Y_0}{2}\left(\frac{\lambda_{g0}}{\lambda_0}\right)^2$$

$$Y_{in} = G + jb\left(\frac{\omega}{\omega_0} - \frac{\omega_0}{\omega}\right)$$

$$Q = \frac{b}{G} \approx \frac{\pi\lambda_{g0}}{\alpha_t \lambda_0^2}$$

$$n = 1, 2, 3, \ldots$$

(b)

A-3527-152

FIG. 5.08-1 SOME TRANSMISSION LINE RESONATORS

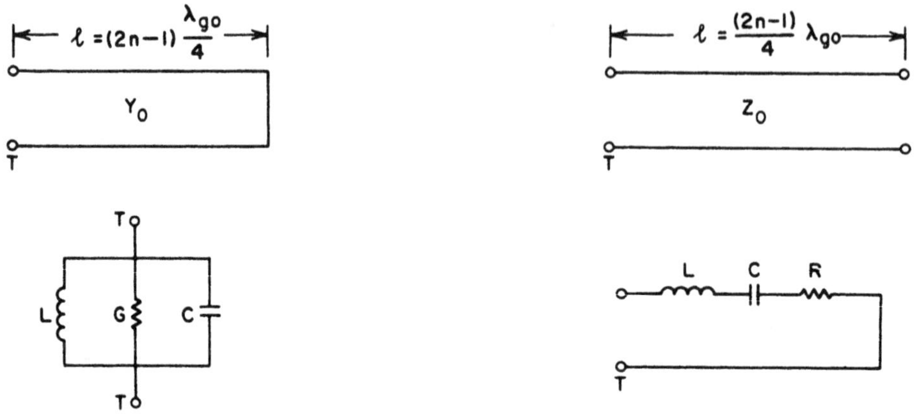

$$G \approx Y_0 \alpha_t \ell = \frac{(2n-1)}{4} Y_0 \alpha_t \lambda_{g0}$$

$$\mathscr{b} = \omega_0 C = \frac{1}{\omega_0 L} \approx \frac{(2n-1)}{4} \pi Y_0 \left(\frac{\lambda_{g0}}{\lambda_0}\right)^2$$

$$Y_{in} = G + j\mathscr{b}\left(\frac{\omega}{\omega_0} - \frac{\omega_0}{\omega}\right)$$

$$Q = \frac{\mathscr{b}}{G} \approx \frac{\pi \lambda_{g0}}{\alpha_t \lambda_0^2}$$

$$n = 1, 2, 3, \ldots$$

(c)

$$R \approx Z_0 \alpha_t \ell = \frac{(2n-1)}{4} Z_0 \alpha_t \lambda_{g0}$$

$$\mathscr{x} = \omega_0 L = \frac{1}{\omega_0 C} \approx \frac{(2n-1)}{4} \pi Z_0 \left(\frac{\lambda_{g0}}{\lambda_0}\right)^2$$

$$Z_{in} = R + j\mathscr{x}\left(\frac{\omega}{\omega_0} - \frac{\omega_0}{\omega}\right)$$

$$Q = \frac{\mathscr{x}}{R} \approx \frac{\pi \lambda_{g0}}{\alpha_t \lambda_0^2}$$

$$n = 1, 2, 3, \ldots$$

(d)

A-3527-153

FIG. 5.08-1 Concluded

$$\mathscr{b} = \frac{\omega_0}{2} \frac{dB}{d\omega_0}\bigg|_{\omega_0} \quad \text{mhos} \tag{5.08-2}$$

where B is the susceptance component of the input admittance of the circuit.

The above general definitions for slope parameters provide a convenient means for relating the resonance properties of any circuit to a simple lumped equivalent circuit such as those in Fig. 5.08-1. The reactance slope parameter $\mathscr{x}$ given by Eq. (5.08-1) is seen to be equal to $\omega_0 L = 1/(\omega_0 C)$ for the equivalent, series, lumped-element circuit, while the susceptance slope parameter $\mathscr{b}$ is equal to $\omega_0 C = 1/(\omega_0 L)$ for the equivalent, parallel, lumped-element circuit. Considerable use will be made of these parameters in later chapters dealing with band-pass and band-stop microwave filters.

It should be noted in Fig. 5.08-1 that the use of reactance or susceptance slope parameters also leads to convenient expressions for Q, and for the input impedance or admittance of the circuit in the vicinity of resonance. For narrow-band microwave applications, the approximate equivalence

$$\left(\frac{\omega}{\omega_0} - \frac{\omega_0}{\omega}\right) \approx 2\left(\frac{\omega - \omega_0}{\omega_0}\right) \tag{5.08-3}$$

is often convenient for use in the expressions for input impedance or admittance.

SEC. 5.09, COUPLED-STRIP-TRANSMISSION-LINE FILTER SECTIONS

The natural electromagnetic coupling that exists between parallel transmission lines can be used to advantage in the design of filters and directional couplers.[14, 15, 16, 17, 18, 19, 20] In this section, formulas are given for filter sections constructed of parallel-coupled lines of the types illustrated in Fig. 5.05-1. Several cases involving unsymmetrical parallel-coupled lines as in Figs. 5.05-12 and 5.05-13 are also considered.

The ten coupling arrangements that can be obtained from a pair of symmetrical, coupled transmission lines by placing open- or short-circuits on various terminal pairs, or by connecting ends of the lines together, are illustrated in Fig. 5.09-1. In this figure, schematic diagrams of single sections of each type are shown, together with their image parameters and either their open-circuit impedances or their short-circuit admittances. In addition, equivalent open-wire transmission-line circuits for eight of the coupled transmission line sections are shown beneath the corresponding schematic diagram.

In the schematic diagrams of the coupled-transmission-line sections in Fig. 5.09-1, the input and output ports are designated by small open circles. The image impedance seen looking into each of these ports is also indicated near each port. Open-circuited ports of the coupled lines are shown with no connection, while short-circuited ports are designated with the standard grounding symbol. In the equivalent transmission-line circuits shown beside the schematic diagrams, a two-wire line representation is used. In each case, the characteristic impedance or admittance of the lengths of transmission line is shown, together with the electrical length, θ. The equivalence between the parallel-coupled line sections and the non-parallel-coupled line sections shown is exact.

Figure 5.09-2 shows the same parallel-coupled sections as appear in Figs. 5.09-1(b), (c), (d), but for cases where the strip transmission lines have unsymmetrical cross sections.* The line capacitances C_a, C_{ab}, and C_b per unit length are as defined in Fig. 5.05-12. It is interesting to note that in the case of Fig. 5.09-2(a) the line capacitances per unit length for the left and right shunt stub in the equivalent open-wire representation are the same as the corresponding capacitances per unit length between Line *a* and ground, and Line *b* and ground, respectively. Meanwhile, the capacitance per unit length for the connecting line in the open-wire circuit is the same as the capacitance per unit length *between* Lines *a* and *b* of the parallel-coupled representation. In Fig. 5.09-2(b) the dual situation holds, where L_a and L_b are the self-inductances per unit length of Lines *a* and *b* in the parallel-coupled representation, while L_{ab} is the mutual inductance per unit length between the parallel-coupled lines. Since the line capacitances are more convenient to deal with, the line impedances of the equivalent open-wire circuit are also given in terms of C_a, C_{ab}, and C_b, for all three cases in Fig. 5.09-2. The quantity v indicated in Fig. 5.09-2 is the velocity of light in the medium of propagation.

---

* The results in Fig. 5.09-2 and also those in Figs. 5.09-3 and 5.09-4 were obtained by extension of the results in Refs. 19 and 20.

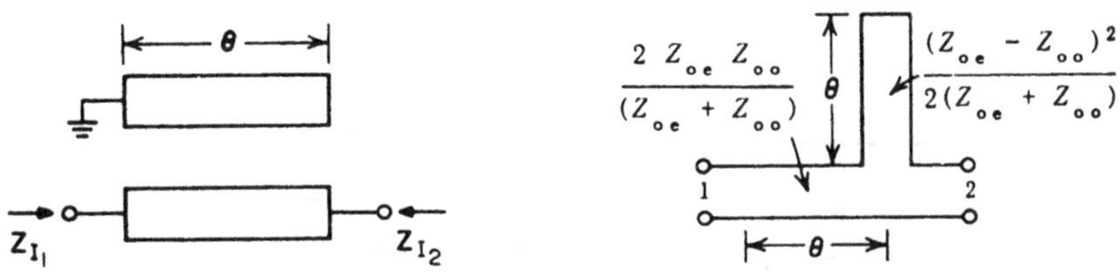

SCHEMATIC AND EQUIVALENT CIRCUIT

$$z_{11} - z_{12} = j\,\frac{2\,Z_{oe}\,Z_{oo}}{Z_{oe} + Z_{oo}}\tan\frac{\theta}{2}$$

$$z_{12} = -j\,\frac{2\,Z_{oe}\,Z_{oo}}{Z_{oe} + Z_{oo}}\csc\theta$$

$$z_{22} - z_{11} = j\,\frac{(Z_{oe} - Z_{oo})^2}{2(Z_{oe} + Z_{oo})}\tan\theta$$

TWO-PORT CIRCUIT PARAMETERS

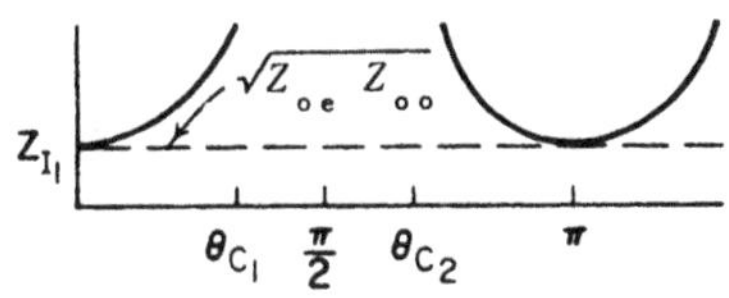

$$\cos\theta_{c_1} = -\cos\theta_{c_2} = \left[\frac{\dfrac{Z_{oe}}{Z_{oo}} - 1}{\dfrac{Z_{oe}}{Z_{oo}} + 1}\right]$$

$$\cosh(\alpha + j\beta) = \frac{[(Z_{oe} + Z_{oo})^2\cos^2\theta - (Z_{oe} - Z_{oo})^2]^{1/2}}{2\sqrt{Z_{oe}\,Z_{oo}}}$$

$$Z_{I_2} = \frac{Z_{oe}\,Z_{oo}}{Z_{I_1}}\,, \qquad Z_{I_1} = \frac{2\,Z_{oe}\,Z_{oo}\cos\theta}{[-(Z_{oe} - Z_{oo})^2 + (Z_{oe} + Z_{oo})^2\cos^2\theta]^{1/2}}$$

IMAGE PARAMETERS

(a) LOW PASS

A-3527-A162

SOURCE: Adapted from figures in Final Report, Contract DA 36-039 SC-64625, SRI; which were reprinted in *IRE Trans. PGMTT* (see Ref. 19 by E. M. T. Jones and J. T. Bolljahn).

FIG. 5.09-1 SOME PARALLEL-COUPLED TRANSMISSION-LINE FILTER SECTIONS

(continued on p. 220)

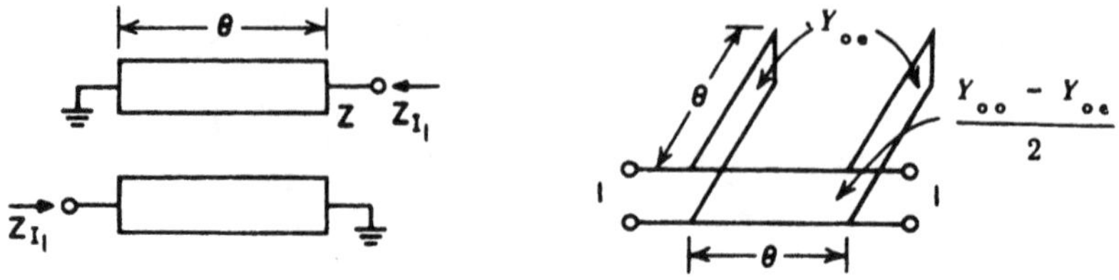

SCHEMATIC AND EQUIVALENT CIRCUIT

$$y_{11} = -j \frac{(Y_{oo} + Y_{oe})}{2} \cot \theta$$

$$y_{12} = -j \frac{(Y_{oo} - Y_{oe})}{2} \csc \theta$$

TWO-PORT CIRCUIT PARAMETERS

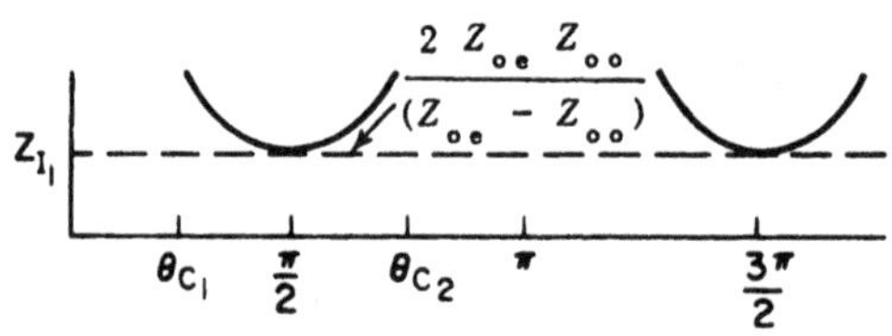

$$\cosh(\alpha + j\beta) = \left[\frac{\dfrac{Z_{oe}}{Z_{oo}} + 1}{\dfrac{Z_{oe}}{Z_{oo}} - 1}\right] \cos \theta$$

$$\cos \theta_{c_1} = -\cos \theta_{c_2} = \left[\frac{\dfrac{Z_{oe}}{Z_{oo}} - 1}{\dfrac{Z_{oe}}{Z_{oo}} + 1}\right]$$

$$Z_{I_1} = \frac{2\, Z_{oe}\, Z_{oo} \sin \theta}{[(Z_{oe} - Z_{oo})^2 - (Z_{oe} + Z_{oo})^2 \cos^2 \theta]^{1/2}}$$

IMAGE PARAMETERS

(b) BAND PASS

A-3527-B162

FIG. 5.09-1 Continued

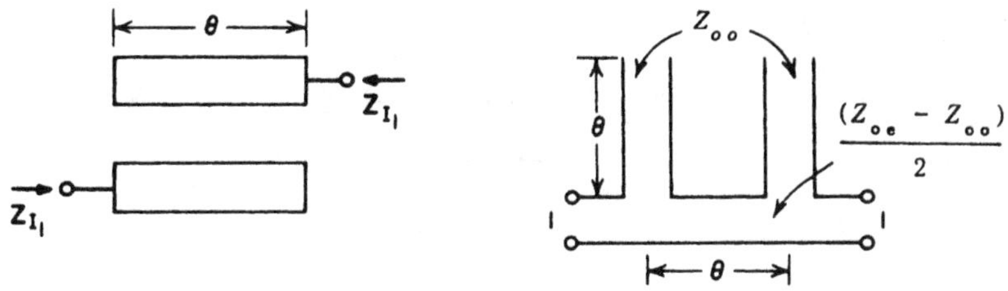

SCHEMATIC AND EQUIVALENT CIRCUIT

$$z_{11} = -j \frac{(Z_{oe} + Z_{oo})}{2} \cot \theta$$

$$z_{12} = -j \frac{(Z_{oe} - Z_{oo})}{2} \csc \theta$$

TWO-PORT CIRCUIT PARAMETERS

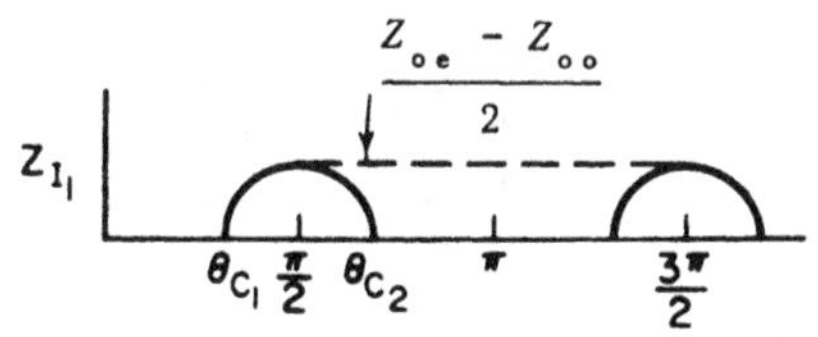

$$\cos \theta_{c_1} = -\cos \theta_{c_2} = \left[\frac{\dfrac{Z_{oe}}{Z_{oo}} - 1}{\dfrac{Z_{oe}}{Z_{oo}} + 1}\right]$$

$$Z_{I_1} = \frac{[(Z_{oe} - Z_{oo})^2 - (Z_{oe} + Z_{oo})^2 \cos^2 \theta]^{1/2}}{2 \sin \theta}$$

$$\cosh (\alpha + j\beta) = \left[\frac{\dfrac{Z_{oe}}{Z_{oo}} + 1}{\dfrac{Z_{oe}}{Z_{oo}} - 1}\right] \cos \theta$$

IMAGE PARAMETERS

(c) BAND PASS

A-3527-C162

FIG. 5.09-1 Continued

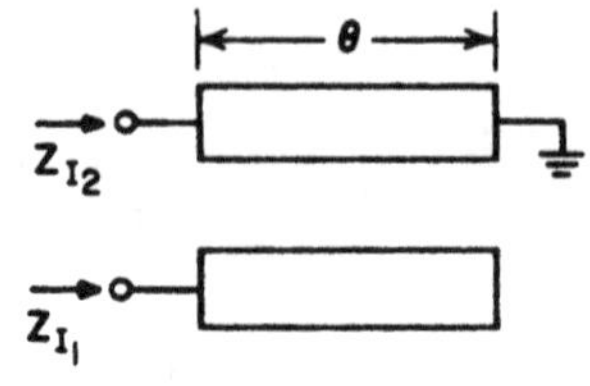

SCHEMATIC

$$z_{11} = -j\,(Z_{oe} + Z_{oo})\,\frac{\cot\theta}{2} + j\,\frac{(Z_{oe} - Z_{oo})^2}{Z_{oe} + Z_{oo}}\csc 2\theta$$

$$z_{12} = j\,(Z_{oe} - Z_{oo})\,\frac{\tan\theta}{2}$$

$$z_{22} = j\,(Z_{oe} + Z_{oo})\,\frac{\tan\theta}{2}$$

TWO-PORT CIRCUIT PARAMETERS

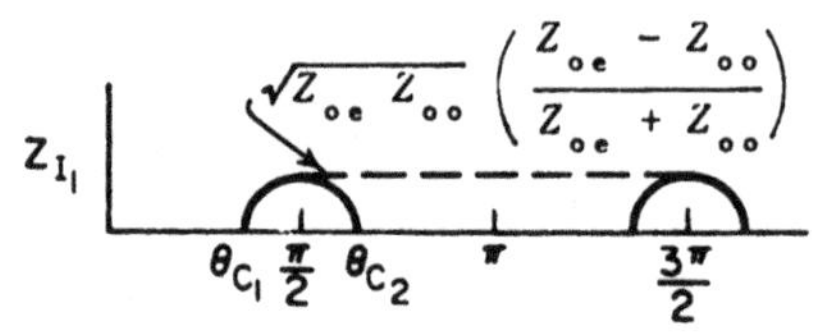

$$\cos\theta_{c_1} = -\cos\theta_{c_2} = \left[\frac{\dfrac{Z_{oe}}{Z_{oo}} - 1}{\dfrac{Z_{oe}}{Z_{oo}} + 1}\right]$$

$$Z_{I_1} = \frac{\sqrt{Z_{oe}Z_{oo}}\,[(Z_{oe} - Z_{oo})^2 - (Z_{oe} + Z_{oo})^2\cos^2\theta]^{1/2}}{\sin\theta\,(Z_{oe} + Z_{oo})}$$

$$Z_{I_2} = \frac{Z_{oe}Z_{oo}}{Z_{I_1}}$$

$$\cosh(\alpha + j\beta) = \frac{1}{\sin\theta}\left[1 - \left(\frac{Z_{oe} + Z_{oo}}{Z_{oe} - Z_{oo}}\right)^2\cos^2\theta\right]^{1/2}$$

IMAGE PARAMETERS

(d) BAND PASS A-3527-D162

FIG. 5.09-1 Continued

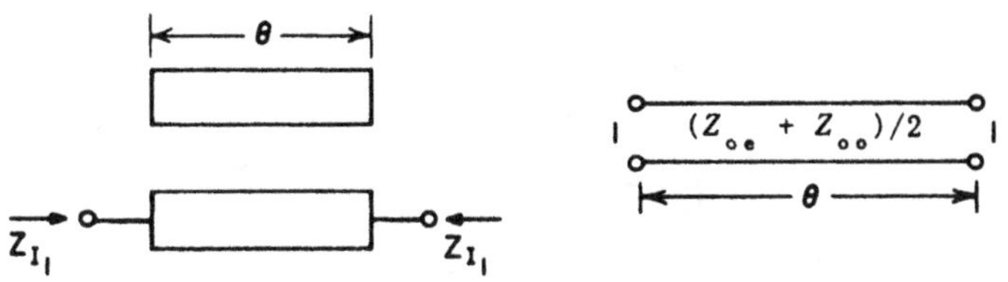

SCHEMATIC AND EQUIVALENT CIRCUIT

$$z_{11} - z_{12} = j \frac{(Z_{oe} + Z_{oo})}{2} \tan \frac{\theta}{2} \qquad z_{12} = -j \frac{(Z_{oe} + Z_{oo})}{2} \csc \theta$$

TWO-PORT CIRCUIT PARAMETERS

$$Z_{I_1} = \frac{Z_{oe} + Z_{oo}}{2} \qquad \beta = \theta$$

IMAGE PARAMETERS

(e) ALL PASS

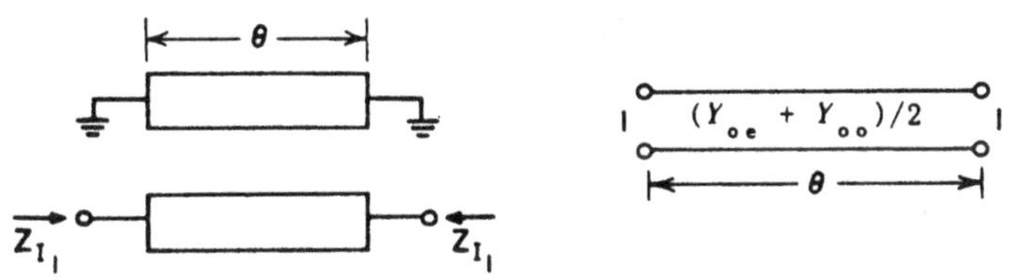

SCHEMATIC AND EQUIVALENT CIRCUIT

$$y_{11} - y_{12} = j \frac{(Y_{oe} + Y_{oo})}{2} \tan \frac{\theta}{2} \qquad y_{12} = -j \frac{(Y_{oe} + Y_{oo})}{2} \csc \theta$$

TWO-PORT CIRCUIT PARAMETERS

$$Z_{I_1} = \frac{2 Z_{oe} Z_{oo}}{Z_{oe} + Z_{oo}} \qquad \beta = \theta$$

IMAGE PARAMETERS

(f) ALL PASS

A-3527-E162

FIG. 5.09-1 Continued

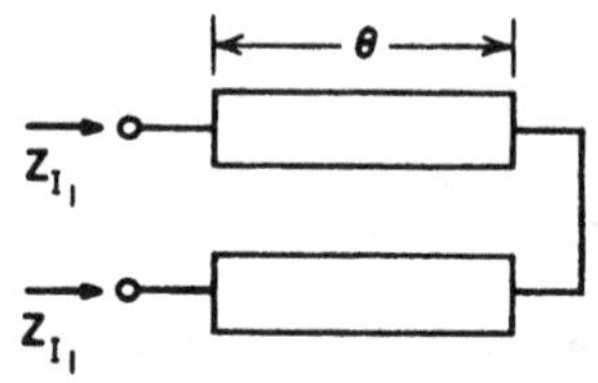

SCHEMATIC

$$z_{11} - z_{12} = jZ_{oo} \tan\theta$$

$$z_{12} = \frac{-j}{2}\left(Z_{oe}\cot\theta + Z_{oo}\tan\theta\right)$$

TWO-PORT CIRCUIT PARAMETERS

$$\cos\beta = \frac{\dfrac{Z_{oe}}{Z_{oo}} - \tan^2\theta}{\dfrac{Z_{oe}}{Z_{oo}} + \tan^2\theta} \quad , \quad \tan\beta = \frac{2\sqrt{Z_{oe}\,Z_{oo}}}{(Z_{oe}\cot\theta - Z_{oo}\tan\theta)} \qquad Z_{I_1} = \sqrt{Z_{oe}\,Z_{oo}}$$

IMAGE PARAMETERS

(g) ALL PASS

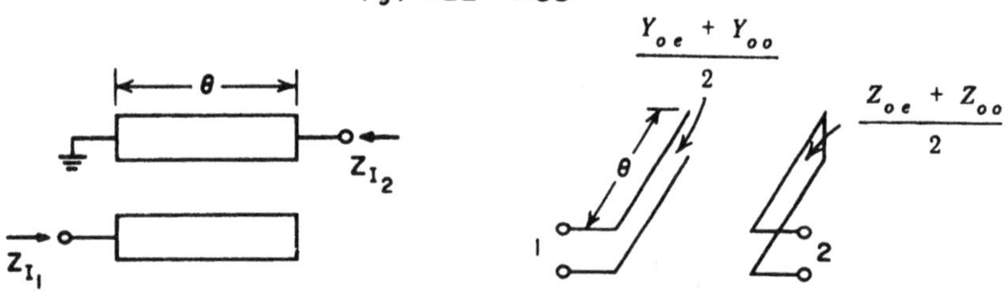

SCHEMATIC AND EQUIVALENT CIRCUIT

$$y_{11} = j\,\frac{(Y_{oo} + Y_{oe})}{2}\tan\theta \qquad z_{22} = j\,\frac{(Z_{oe} + Z_{oo})}{2}\tan\theta \qquad z_{12} = y_{12} = 0$$

TWO-PORT CIRCUIT PARAMETERS

$$Z_{I_1} = -j\,\frac{2\,Z_{oo}\,Z_{oe}}{Z_{oo} + Z_{oe}}\cot\theta \qquad Z_{I_2} = \frac{Z_{oe}\,Z_{oo}}{Z_{I_1}} \qquad \alpha = \infty$$

IMAGE PARAMETERS

(h) ALL STOP

A-3527-F162

FIG. 5.09-1 Continued

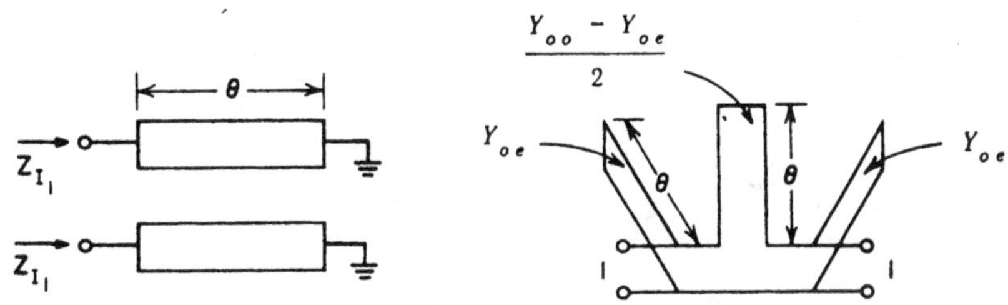

SCHEMATIC AND EQUIVALENT CIRCUIT

$$y_{11} - y_{12} = -j\, Y_{oe} \cot \theta \qquad y_{12} = -j\, \frac{(Y_{oo} - Y_{oe})}{2} \cot \theta$$

TWO-PORT CIRCUIT PARAMETERS

$$\cosh \alpha = \frac{Z_{oe} + Z_{oo}}{Z_{oe} - Z_{oo}} \qquad Z_{I_1} = j\sqrt{Z_{oe}\, Z_{oo}} \tan \theta$$

IMAGE PARAMETERS

(i) ALL STOP

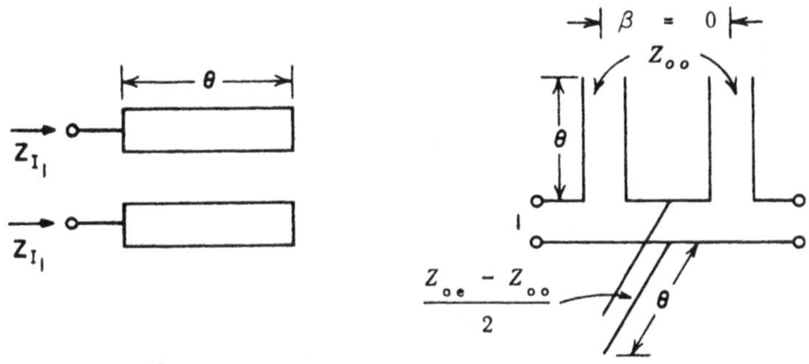

SCHEMATIC AND EQUIVALENT CIRCUIT

$$z_{11} - z_{12} = -j\, Z_{oo} \cot \theta \qquad z_{12} = -j\, \frac{(Z_{oe} - Z_{oo})}{2} \cot \theta$$

TWO-PORT CIRCUIT PARAMETERS

$$Z_{I_1} = -j\sqrt{Z_{oe}\, Z_{oo}} \cot \theta \qquad \cosh \alpha = \frac{Z_{oe} + Z_{oo}}{Z_{oe} - Z_{oo}}$$

IMAGE PARAMETERS

(j) ALL STOP A-3527-G162

FIG. 5.09-1 Concluded

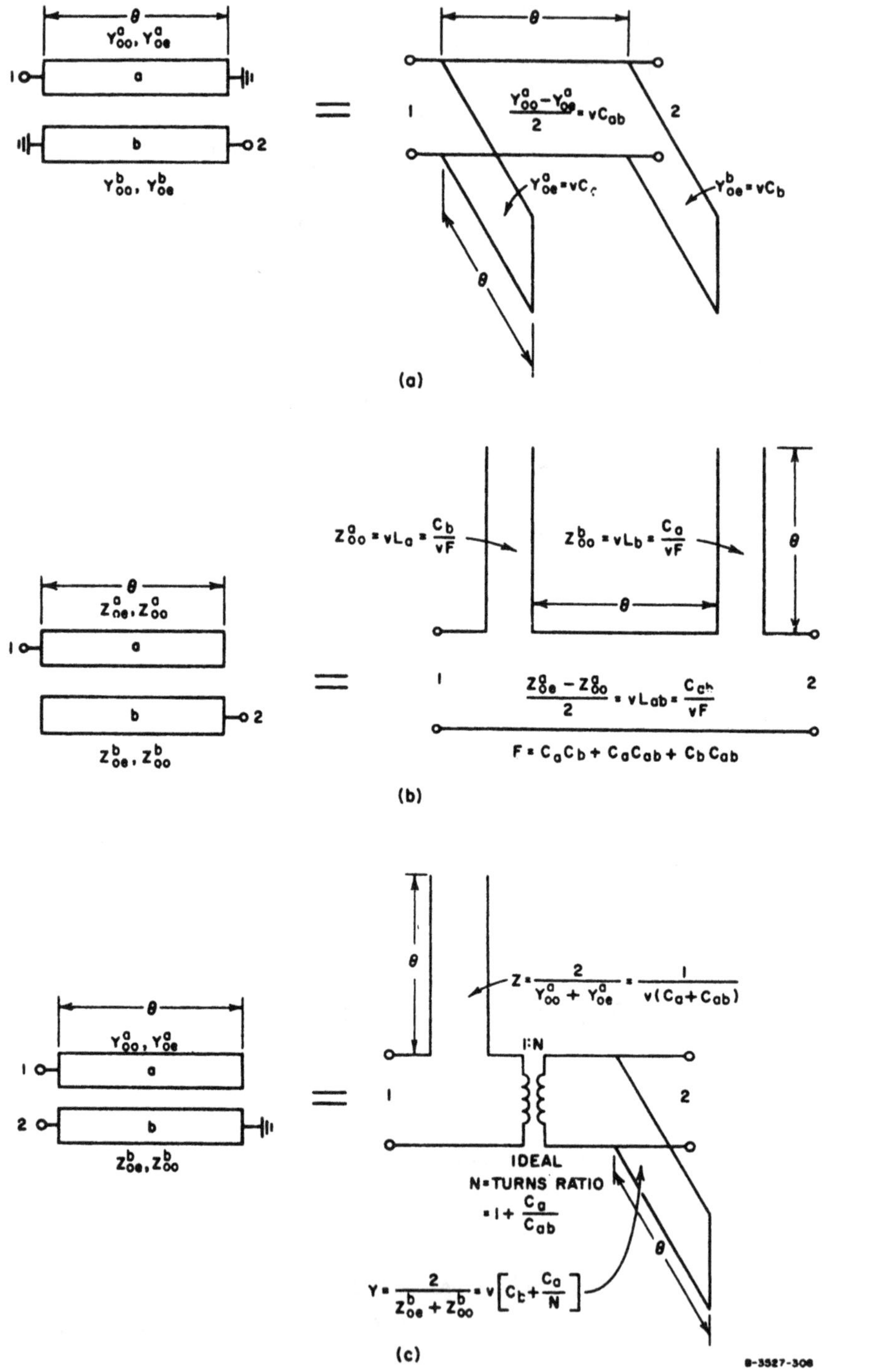

FIG. 5.09-2 SOME USEFUL UNSYMMETRICAL PARALLEL-COUPLED STRIP-LINE SECTIONS AND THEIR EQUIVALENT OPEN-WIRE LINE SECTIONS
Parameters C_a, C_{ab}, and C_b are line capacitances per unit length as defined in Fig. 5.05-12. v = velocity of propagation. All lines are of the same length.

SPECIAL CONSTRAINT: $Y^a_{oe}+Y^a_{oo}=2Y_A$

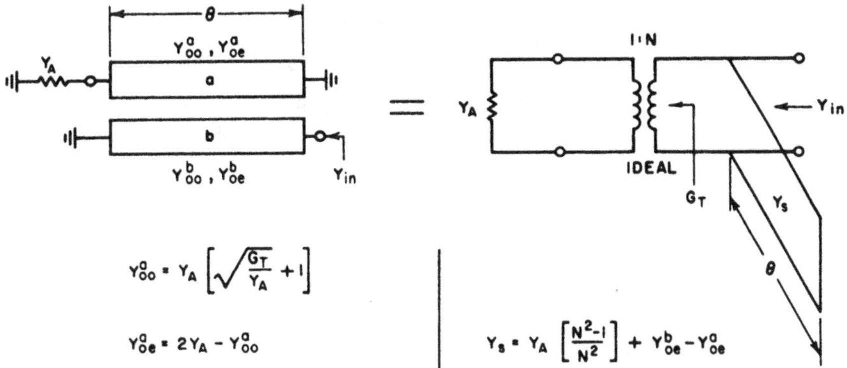

$$Y^a_{oo} = Y_A\left[\sqrt{\frac{G_T}{Y_A}}+1\right]$$

$$Y^a_{oe} = 2Y_A - Y^a_{oo}$$

$$Y^b_{oe} = Y_s - Y_A\left[1-\frac{G_T}{Y_A}\right]+Y^a_{oe}$$

$$Y^b_{oo} = Y^b_{oe} + Y^a_{oo} - Y^a_{oe}$$

$$Y_s = Y_A\left[\frac{N^2-1}{N^2}\right] + Y^b_{oe} - Y^a_{oe}$$

$$N = \text{TURNS RATIO} = \frac{2Y_A}{Y^a_{oo}-Y^a_{oe}}$$

$$G_T = \frac{Y_A}{N^2}$$

(a)

SPECIAL CONSTRAINT: $Z^a_{oe}+Z^a_{oo}=2Z_A$

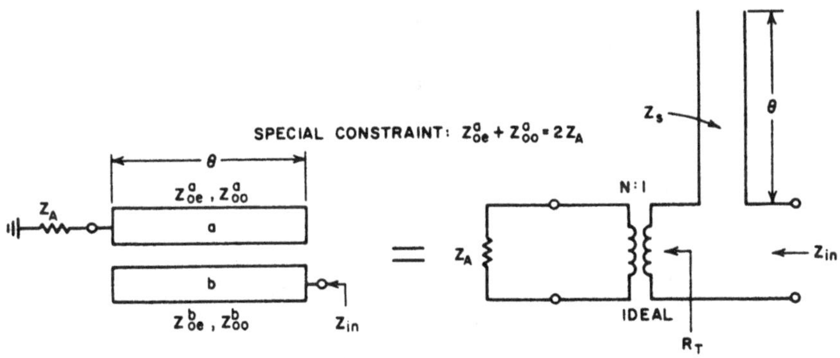

$$Z^a_{oe} = Z_A\left[\sqrt{\frac{R_T}{Z_A}}+1\right]$$

$$Z^a_{oo} = 2Z_A - Z^a_{oe}$$

$$Z^b_{oo} = Z_s - Z_A\left[1-\frac{R_T}{Z_A}\right]+Z^a_{oo}$$

$$Z^b_{oe} = Z^b_{oo} + Z^a_{oe} - Z^a_{oo}$$

$$Z_s = Z_A\left[\frac{N^2-1}{N^2}\right] + Z^b_{oo} - Z^a_{oo}$$

$$N = \text{TURNS RATIO} = \frac{2Z_A}{Z^a_{oe}-Z^a_{oo}}$$

$$R_T = \frac{Z_A}{N^2}$$

B-3527-475

(b)

FIG. 5.09-3 SOME PARALLEL-COUPLED STRIP-LINE AND OPEN-WIRE-LINE EQUIVALENCES WHICH APPLY UNDER SPECIAL CONSTRAINTS

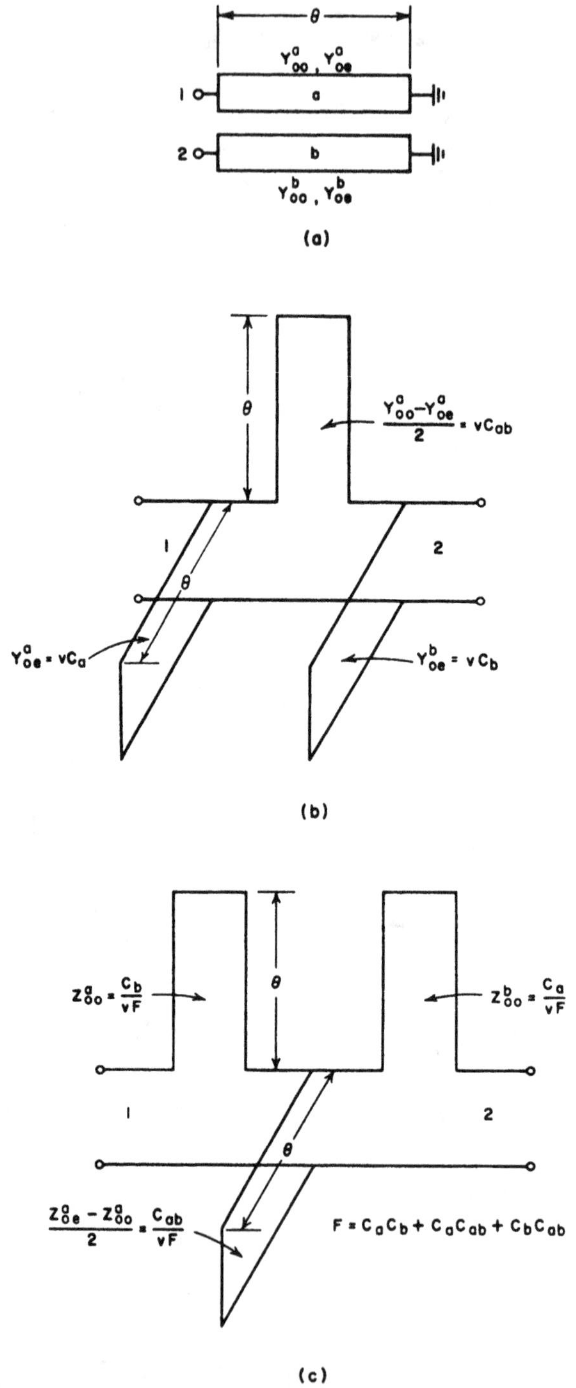

FIG. 5.09-4 A PARALLEL-COUPLED SECTION AND TWO OPEN-WIRE-LINE CIRCUITS WHICH ARE EXACTLY EQUIVALENT
The C_a, C_{ab}, and C_b are as indicated in Fig. 5.05-12.

In the cases of the circuits in Figs. 5.09-2(a), and (b), if the parallel-coupled sections are properly terminated, their equivalent open-wire line circuit simplifies in a very interesting and useful way. This is illustrated in Fig. 5.09-3(a) and (b). Note that when the indicated constraints are applied, the equivalent open-wire circuit reduces to simply an ideal transformer and a single stub. In spite of the constraint equations which are enforced in these circuits, there are still sufficient degrees of freedom so that for specified Y_s and G_T or Z_s and R_T, a wide range of Y_A or Z_A, respectively, can be accommodated. For this reason these two structures will prove quite useful for use with certain types of band-pass filters for the purpose of effectively realizing a series- or shunt-stub resonator, along with obtaining an impedance transformation which will accommodate some desired terminating impedance. In a somewhat more complex way, the circuit in Fig. 5.09-2(c) will also prove useful for similar purposes.

Figure 5.09-4 shows the parallel-coupled section in Fig. 5.09-1(i) generalized to cover the case where the two strip lines may be of different widths. At (a) is shown the structure under consideration, while at (b) and (c) are shown two open-wire line structures which are identically equivalent electrically to the strip-line structure at (a). As previously indicated, parallel-coupled structures of this sort are all-stop structures as they stand, but when properly used with lumped capacitances, they become the basis for the comb-line form of filter discussed in Sec. 8.13.

SEC. 5.10, IRIS-COUPLED WAVEGUIDE JUNCTIONS

Bethe[21, 22, 23, 24] has developed a general perturbation technique for calculating the scattering of power by small irises connecting one transmission line with another. The theory is applicable even though the two transmission lines have different cross sections and operate in different modes; however, it applies rigorously only to infinitesimally thin irises whose dimensions are small in terms of the operating wavelength. These irises should be located far from any corners, in a transmission-line wall whose radius of curvature is large in terms of wavelength. In practice it is found that the theory holds reasonably well even when the irises are located relatively close to sharp corners in transmission-line walls of fairly small radii of curvature. For irises of finite thickness, it is found that Bethe's theory is still applicable except that the transmission through the iris is reduced.[25] In many instances it is possible to use Cohn's frequency correction[25] where the iris dimensions are not negligibly small with respect to a wavelength.

Bethe's original derivations[21,22,23] appeared in a series of MIT Radiation Laboratory Reports, copies of which are quite difficult to obtain. Recently Collin[26] has derived some of Bethe's results using a different approach, and these results are readily available. Marcuvitz[27] recast much of Bethe's work and derived many equivalent circuits for iris-coupled transmission lines, many of which are presented in the *Waveguide Handbook*.[8] A paper by Oliner[28] contains some additional circuits for iris-coupled lines.

Bethe's calculation of the scattering of power by small irises actually consists of two distinct steps. The first step is the computation of the electric dipole moment, $\bar{p}$, and the magnetic dipole moment, $\bar{m}$, induced in the iris by the exciting fields. The next step is the calculation of the fields radiated by the electric and magnetic dipole moments.

Figure 5.10-1 illustrates two parallel-plane transmission lines connected by a small iris. The electric field, E_{on}, in the bottom line will couple through the iris in the manner shown in Fig. 5.10-1(a). To a first-order approximation, the distorted field within the iris can be considered to arise from two electric dipole moments, each of strength $\bar{p}$, induced in the iris by the exciting electric field E_{on} as shown in Fig. 5.10-1(b). The electric dipole moment in the upper line is parallel to E_{on}, while the electric dipole moment in the lower line is oppositely directed.

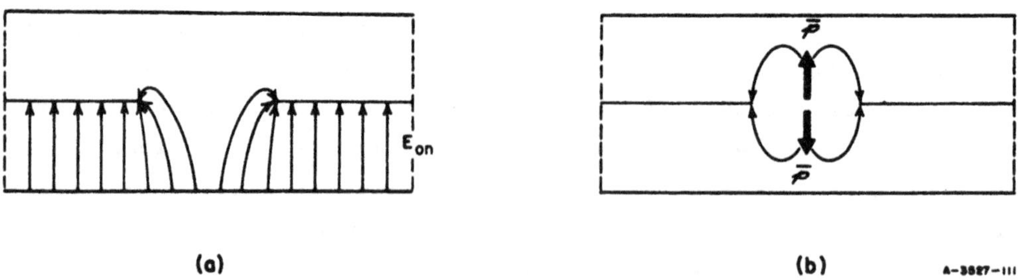

FIG. 5.10-1 ELECTRIC DIPOLE MOMENTS INDUCED IN AN IRIS BY AN ELECTRIC FIELD NORMAL TO THE PLANE OF THE IRIS

Figure 5.10-2 illustrates the magnetic field coupling through an iris connecting two parallel-plane transmission lines. Again the distorted magnetic field within the iris can be considered to arise from two magnetic dipole moments each of strength $\overline{m}$, induced in the iris by the exciting tangential magnetic field, H_{ot}. The magnetic dipole moment in the upper line is directed anti-parallel to H_{ot}, while that in the lower line is oppositely directed and parallel to H_{ot}.

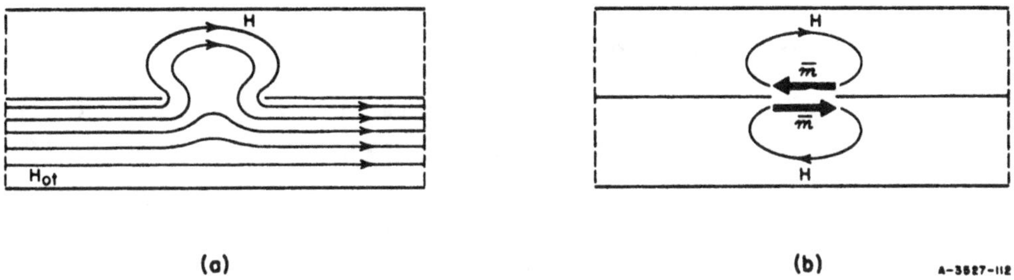

FIG. 5.10-2 MAGNETIC DIPOLE MOMENTS INDUCED IN AN IRIS BY A MAGNETIC FIELD TANGENTIAL TO THE PLANE OF THE IRIS

The strength of the electric dipole moment, $\overline{p}$, is proportional to the product of the electric polarizability P of the iris and the exciting field, E_{on}. Its value in mks units is

$$\overline{p} = \epsilon_0 P E_{on} \overline{n} \qquad (5.10\text{-}1)$$

where $\epsilon_0 = 8.854 \times 10^{-12}$ farads/meter, and $\overline{n}$ is a unit vector directed away from the iris on the side opposite from the exciting field.

The strength of the magnetic dipole moment is proportional to the product of the magnetic polarizability, M, of the iris and exciting tangential magnetic field H_{ot}. For the usual type of iris that has axes of symmetry, the magnetic dipole moment is, in mks units,

$$-\overline{m} = M_1 H_{ou} \overline{u} + M_2 H_{ov} \overline{v} \quad . \qquad (5.10\text{-}2)$$

In this expression the unit vectors $\overline{u}$ and $\overline{v}$ lie in the plane of the iris along the axes of symmetry, M_1 and M_2 are the magnetic polarizabilities, and H_{ou} and H_{ov} the exciting magnetic fields along the $\bar{u}$ and $\bar{v}$ axes, respectively.

The electric dipole moment, $\bar{p}$, set up in an iris by an exciting electric field, will radiate power into a given mode in the secondary waveguide only when the electric field of the mode to be excited has a component parallel to the dipole moment, $\bar{p}$. Similarly the magnetic dipole moment $\bar{m}$ set up in the aperture by an exciting magnetic field will radiate power into a given mode in the secondary waveguide only when the magnetic field of the mode to be excited has a component parallel to the magnetic dipole moment $\bar{m}$.

In order to be able to apply Bethe's theory, it is necessary to know the electric polarizability P and the magnetic polarizabilities M_1 and M_2 of the iris. Theoretical values of the polarizabilities can only be obtained for irises of simple shapes. For example, a *circular iris* of diameter d has a value of $M_1 = M_2 = d^3/6$ and $P = d^3/12$. A long, narrow iris of length l and width w has $P = M_2 = (\pi/16)\ lw^2$, if the exciting magnetic field is parallel to the narrow dimension of the slit (the $\bar{v}$ direction in this case), and the exciting electric field is perpendicular to the plane of the slit. The polarizabilities of elliptical irises have also been computed. In addition, the polarizabilities of irises of other shapes that are too difficult to calculate have been measured by Cohn[29, 30] in an electrolytic tank. The measured values of the polarizability of a number of irises are shown in Figs. 5.10-3 and 5.10-4(a),(b), together with the theoretical values for elliptical irises. Circular irises are the easiest to machine, but sometimes elongated irises are required in order to obtain adequate coupling between rectangular waveguides.

For many applications the equivalent-circuit representation of iris-coupled transmission lines is more convenient than the scattering representation. Figures 5.10-5 to 5.10-12 contain the equivalent-circuit representations of several two- and three-port waveguide junctions coupled by infinitesimally thin irises. Most of the information in the figures is self-explanatory. It is to be noted that in each case the reference planes for the equivalent circuits are at the center of gravity of the iris. The symbol K used in some circuits stands for an impedance inverter as defined in Sec. 4.12. Also included in each figure is the power transmission coefficient through the iris, expressed as the square of the magnitude of the scattering coefficient. (Sec. 2.12).

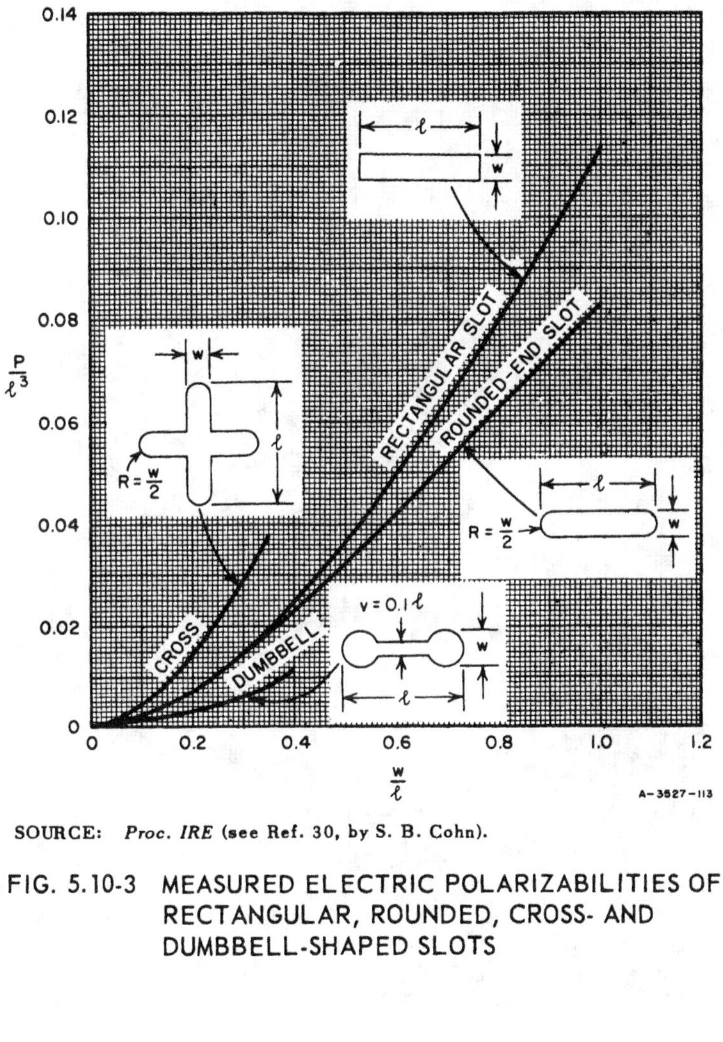

SOURCE: *Proc. IRE* (see Ref. 30, by S. B. Cohn).

FIG. 5.10-3 MEASURED ELECTRIC POLARIZABILITIES OF RECTANGULAR, ROUNDED, CROSS- AND DUMBBELL-SHAPED SLOTS

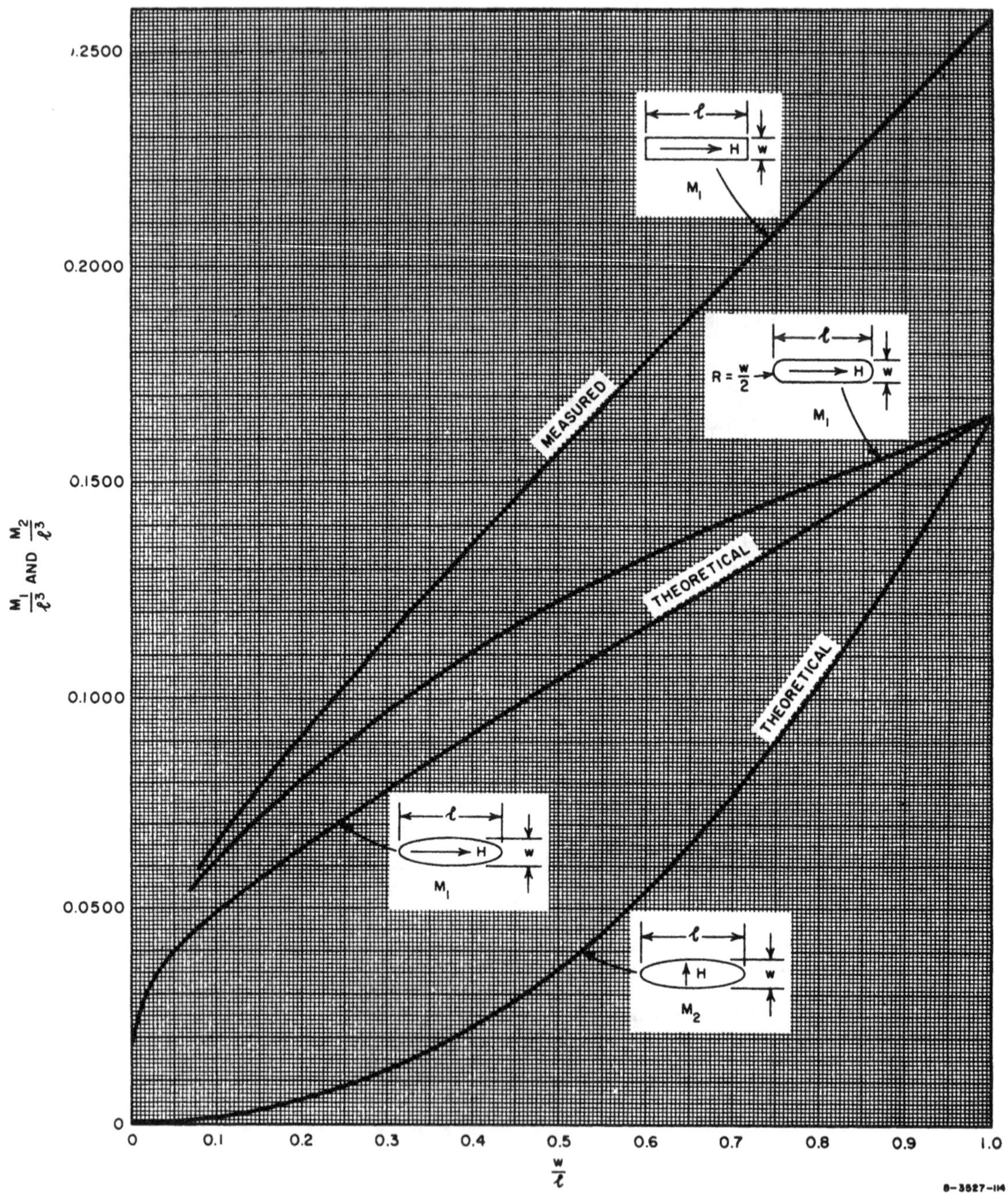

SOURCE: *Proc. IRE* (see Ref. 29, by S. B. Cohn).

FIG. 5.10-4(a) MAGNETIC POLARIZABILITIES OF RECTANGULAR, ROUNDED-END, AND ELLIPTICAL SLOTS

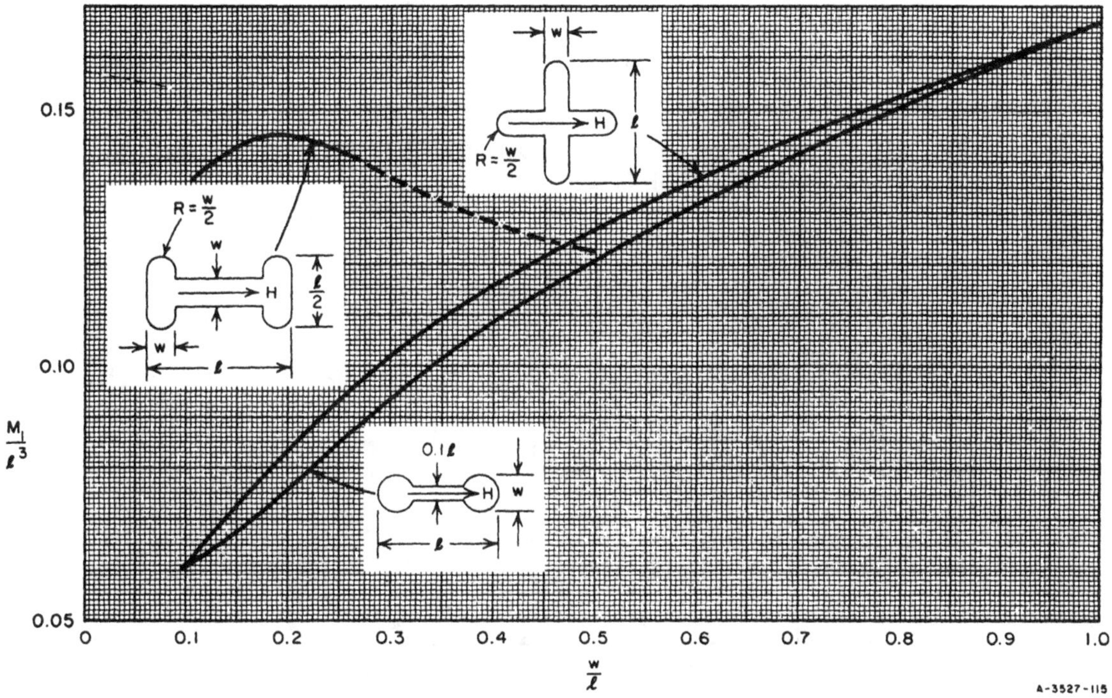

SOURCE: *Proc. IRE* (see Ref. 29, by S. B. Cohn).

FIG. 5.10-4(b) MAGNETIC POLARIZABILITIES OF H-, CROSS-, AND DUMBBELL-SHAPED APERTURES

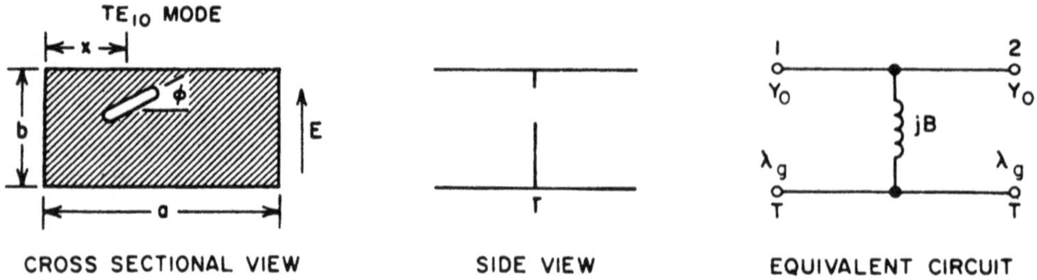

$$\frac{B}{Y_0} = -\frac{ab\lambda_g}{4\pi M \sin^2 \frac{\pi x}{a}} \qquad M = M_1 \cos^2 \phi + M_2 \sin^2 \phi \qquad \lambda_g = \frac{\lambda}{\sqrt{1 - \left(\frac{\lambda}{2a}\right)^2}}$$

$$|S_{12}|^2 \approx \frac{4Y_0^2}{|B|^2} = \frac{64\pi^2 M^2 \sin^4 \frac{\pi x}{a}}{a^2 b^2 \lambda_g^2}$$

A-3527-144

Adapted from the *Waveguide Handbook* (see Ref. 8 edited by N. Marcuvitz)

FIG. 5.10-5 IRIS CONNECTING RECTANGULAR WAVEGUIDES OF THE SAME CROSS SECTION

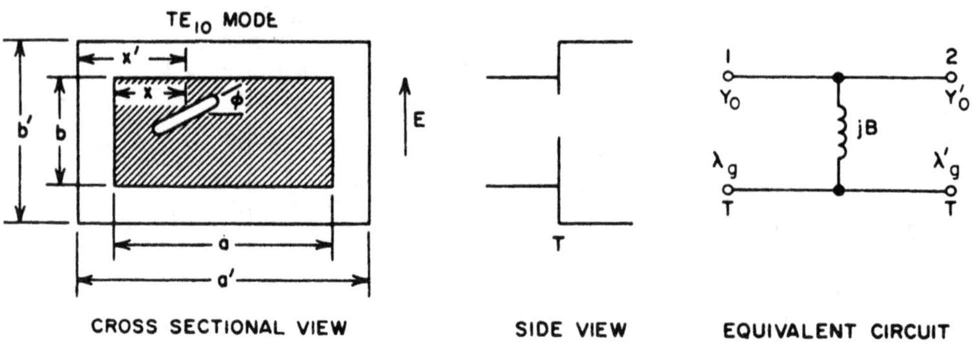

$$|S_{12}|^2 \approx \frac{4Y_0^2}{|B|^2} \cdot \frac{Y_0'}{Y_0} = \frac{64\pi^2 M^2 \sin^2 \frac{\pi x'}{a} \sin^2 \frac{\pi x}{a}}{aa'bb'\lambda_g \lambda_g'}$$

$$\lambda_g = \frac{\lambda}{\sqrt{1 - \left(\frac{\lambda}{2a}\right)^2}} \qquad M = M_1 \cos^2 \phi + M_2 \sin^2 \phi \qquad \lambda_g' = \frac{\lambda}{\sqrt{1 - \left(\frac{\lambda}{2a'}\right)^2}}$$

$$\frac{Y_0'}{Y_0} = \frac{\lambda_g ab}{\lambda_g' a'b'} \frac{\sin^2 \frac{\pi x'}{a'}}{\sin^2 \frac{\pi x}{a}} \qquad \frac{B}{Y_0} = -\frac{ab\lambda_g}{4\pi M \sin^2 \frac{\pi x}{a}}$$

A-3527-145

Adapted from the *Waveguide Handbook* (see Ref. 8 edited by N. Marcuvitz)

FIG. 5.10-6 IRIS CONNECTING RECTANGULAR WAVEGUIDES OF DIFFERENT CROSS SECTIONS

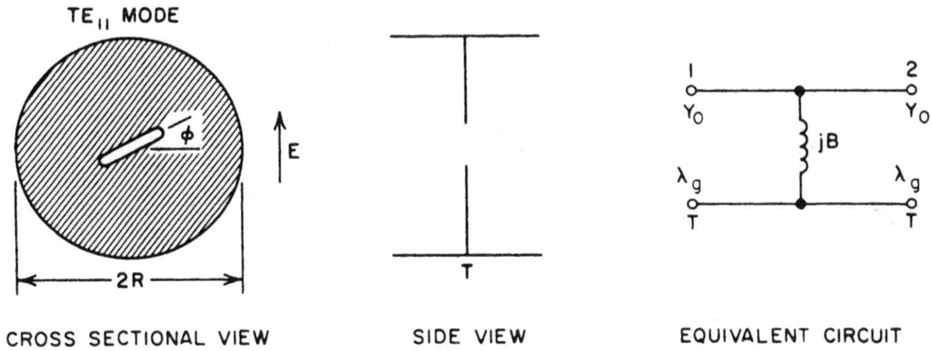

$$|S_{12}|^2 \approx \frac{4Y_0^2}{|B|^2} = \frac{64\pi^2 M^2}{9R^4\lambda_g^2} \qquad \lambda_g = \frac{\lambda}{\sqrt{1 - \left(\frac{\lambda}{3.41R}\right)^2}}$$

$$M = M_1 \cos^2\phi + M_2 \sin^2\phi \qquad \frac{B}{Y_0} = -\frac{0.955(\pi R^2)\lambda_g}{4\pi M}$$

Adapted from the *Waveguide Handbook* (see Ref. 8 edited by N. Marcuvitz) A-3527-146

FIG. 5.10-7 IRIS CONNECTING CIRCULAR WAVEGUIDES OF THE SAME RADIUS

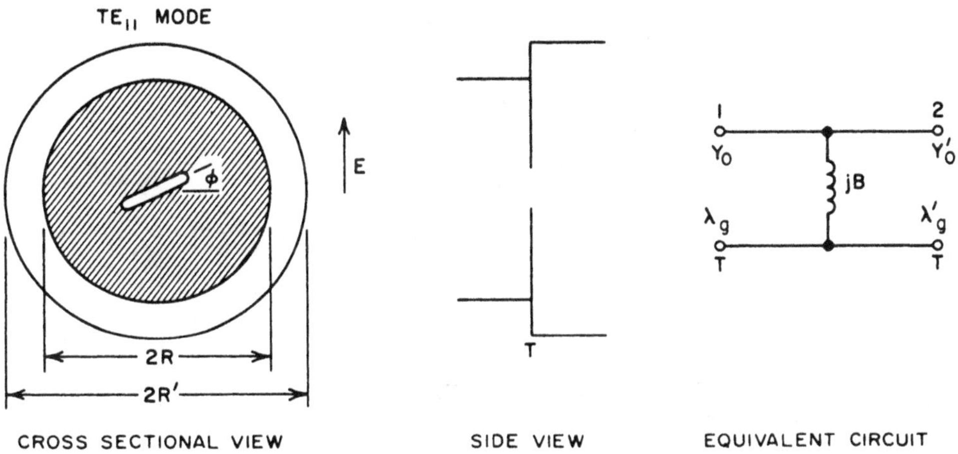

$$\lambda_g' = \frac{\lambda}{\sqrt{1 - \left(\frac{\lambda}{3.41R'}\right)^2}} \qquad \frac{Y_0'}{Y_0} = \frac{R^2\lambda_g}{R'^2\lambda_g'} \qquad \frac{B}{Y_0} = -\frac{0.955(\pi R^2)\lambda_g}{4\pi M}$$

$$\lambda_g = \frac{\lambda}{\sqrt{1 - \left(\frac{\lambda}{3.41R}\right)^2}} \qquad |S_{12}|^2 \approx \frac{4Y_0^2}{|B|^2}\cdot\frac{Y_0'}{Y_0} = \frac{64\pi^2 M^2}{9R^2R'^2\lambda_g\lambda_g'}$$

$$M = M_1 \cos^2\phi + M_2 \sin^2\phi$$

A-3527-143

Adapted from the *Waveguide Handbook* (see Ref. 8 edited by N. Marcuvitz)

FIG. 5.10-8 IRIS CONNECTING CIRCULAR WAVEGUIDES OF DIFFERENT RADII

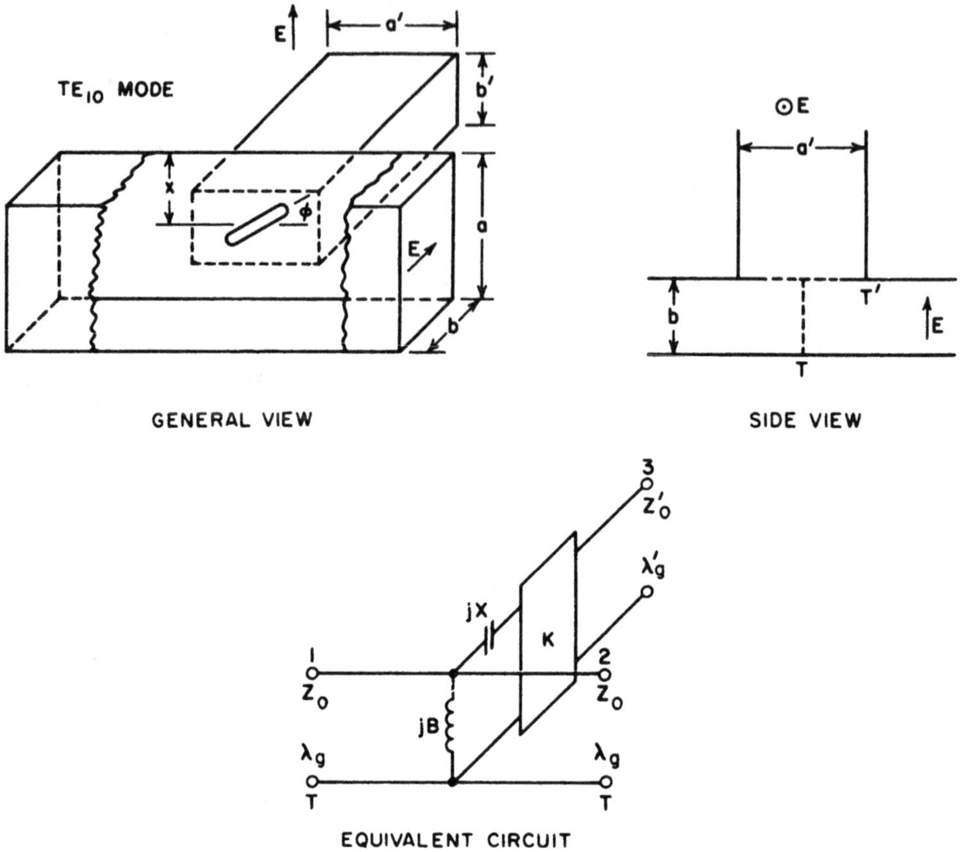

$$|S_{13}|^2 = |S_{23}|^2 = \frac{K^2}{Z_0^2} \cdot \frac{Z_0}{Z_0'} \cdot \frac{Z_0^2}{|X|^2} = \frac{4\pi^2 \lambda_g M^2}{\lambda_g' a' b' a^3 b} \cos^2 \frac{\pi x}{a}$$

$$\frac{X}{Z_0} = -\frac{a^3 b}{\pi \lambda_g M \cos^2 \frac{\pi x}{a}} \qquad \frac{B}{Y_0} = -\frac{4\pi\lambda_g P}{a\lambda^2 b} \sin^2 \frac{\pi x}{a} \qquad \frac{K}{Z_0} = \frac{2a}{\lambda_g \cos \frac{\pi x}{a}}$$

$$\frac{Z_0'}{Z_0} = \frac{\lambda_g' a' b'}{\lambda_g a b} . \qquad \lambda_g = \frac{\lambda}{\sqrt{1 - \left(\frac{\lambda}{2a}\right)^2}} \qquad \lambda_g' = \frac{\lambda}{\sqrt{1 - \left(\frac{\lambda}{2a'}\right)^2}}$$

$$M = M_1 \cos^2 \phi + M_2 \sin^2 \phi$$

A-3527-147

FIG. 5.10-9 IRIS-COUPLED SHUNT T-JUNCTION IN RECTANGULAR GUIDE E-PLANE

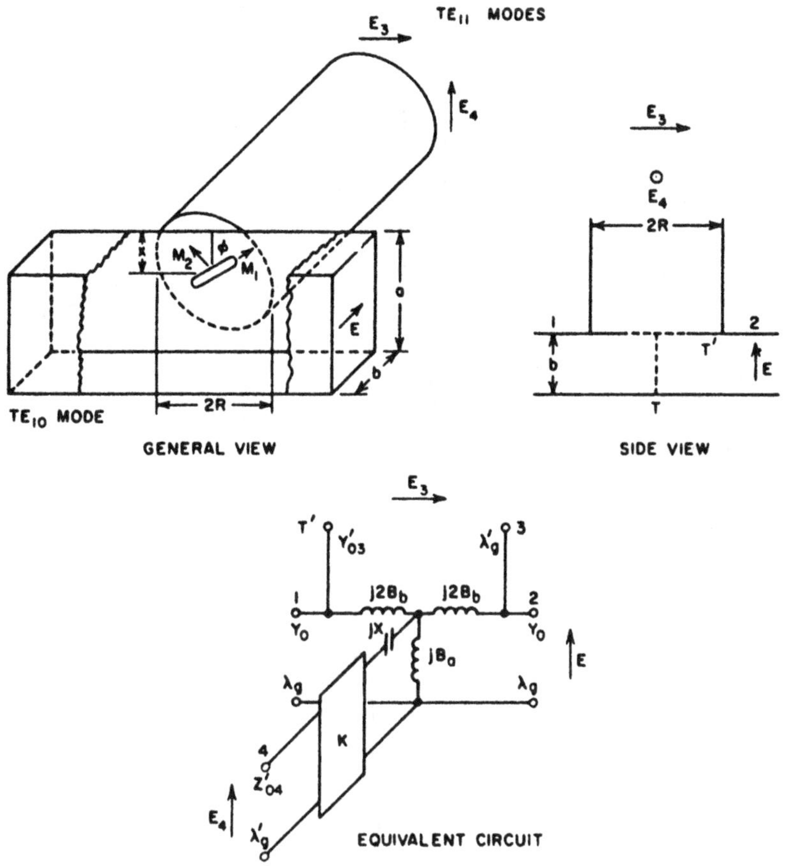

$$|S_{14}|^2 = |S_{24}|^2 \approx \frac{K^2}{Z_0^2} \cdot \frac{Z_0}{Z'_{04}} \cdot \frac{Z_0^2}{|X|^2} = \frac{4\pi^2 M'^2 \lambda_g \cos^2 \frac{\pi x}{a}}{3R^2 a^3 b \lambda'_g} \qquad \frac{K}{Z_0} = \frac{2a}{\lambda_g \cos \frac{\pi x}{a}}$$

$$|S_{13}|^2 = |S_{23}|^2 \approx \frac{Y'_{03}}{Y_0} \cdot \frac{Y_0^2}{|B_b|^2} = \frac{16\pi^2 M^2 \sin^2 \frac{\pi x}{a}}{3R^2 ab \lambda_g \lambda'_g} \qquad \frac{B_b}{Y_0} = -\frac{\lambda_g ab}{4\pi M \sin^2 \frac{\pi x}{a}}$$

$$\lambda'_g = \frac{\lambda}{\sqrt{1 - \left(\frac{\lambda}{3.41R}\right)^2}} \qquad \frac{B_a}{Y_0} = -\frac{4\pi\lambda_g P}{a\lambda^2 b} \sin^2 \frac{\pi x}{a} \qquad \lambda_g = \frac{\lambda}{\sqrt{1 - \left(\frac{\lambda}{2a}\right)^2}}$$

$$\frac{Y'_{03}}{Y_0} = \frac{ab\lambda_g}{3R^2 \lambda'_g \sin^2 \frac{\pi x}{a}} \qquad \frac{Z'_{04}}{Z_0} = \frac{3R^2 \lambda'_g}{ab\lambda_g} \qquad \frac{X}{Z_0} = -\frac{a^3 b}{\pi \lambda_g M' \cos^2 \frac{\pi x}{a}}$$

$$M' = M_1 \sin^2 \phi + M_2 \cos^2 \phi \qquad M = M_1 \cos^2 \phi + M_2 \sin^2 \phi$$

Adapted from the *Waveguide Handbook* (see Ref. 8 edited by N. Marcuvitz)

B-3527-148

FIG. 5.10-10 IRIS-COUPLED T-JUNCTION OF RECTANGULAR AND CIRCULAR GUIDES

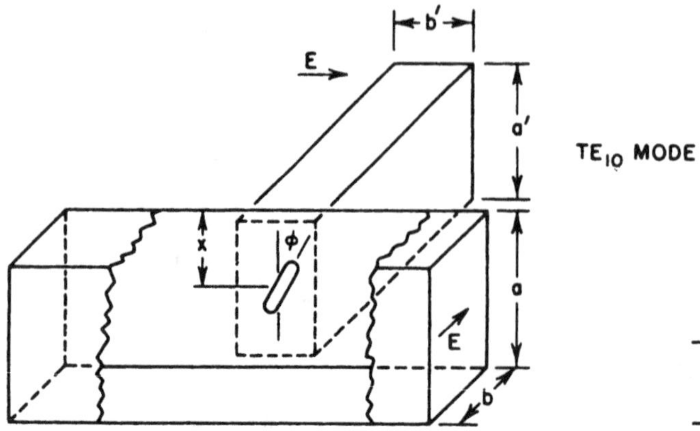

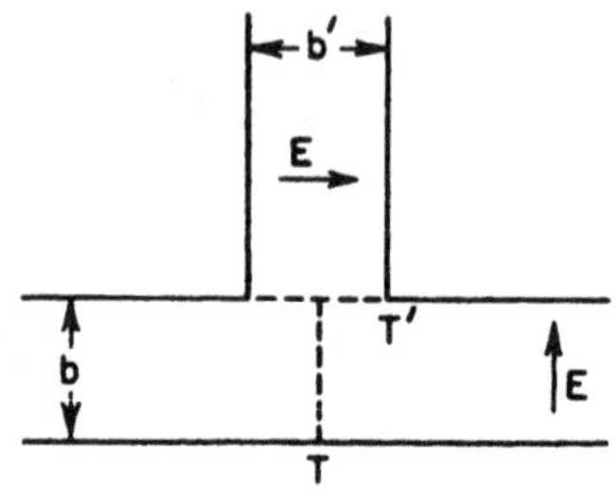

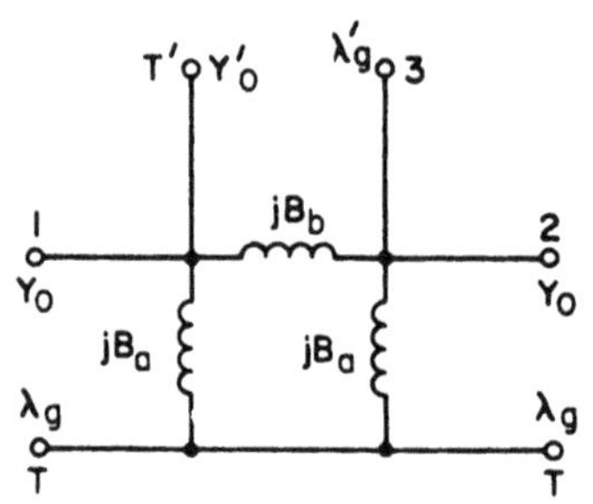

$$|S_{13}|^2 = |S_{23}|^2 = \frac{Y_0'}{Y_0} \cdot \frac{Y_0^2}{|B_b|^2} = \frac{16\pi^2 M^2 \sin^2 \frac{\pi x}{a}}{aa'bb'\lambda_g \lambda_g'}$$

$$\frac{B_b}{Y_0} = -\frac{\lambda_g ab}{4\pi M \sin^2 \frac{\pi x}{a}} \qquad \frac{B_a}{Y_0} = -\frac{2\pi\lambda_g P}{ab\lambda^2} \sin^2 \frac{\pi x}{a} + \frac{\pi M'\lambda_g}{2a^3 b} \cos^2 \frac{\pi x}{a}$$

$$\frac{Y_0'}{Y_0} = \frac{ab\lambda_g}{a'b'\lambda_g' \sin^2 \frac{\pi x}{a}} \qquad \lambda_g = \frac{\lambda}{\sqrt{1 - \left(\frac{\lambda}{2a}\right)^2}} \qquad \lambda_g' = \frac{\lambda}{\sqrt{1 - \left(\frac{\lambda}{2a'}\right)^2}}$$

$$M = M_1 \cos^2 \phi + M_2 \sin^2 \phi \qquad M' = M_1 \sin^2 \phi + M_2 \cos^2 \phi$$

Adapted from the *Waveguide Handbook* (see Ref. 8 edited by N. Marcuvitz) A-3527-149

FIG. 5.10-11 IRIS-COUPLED SERIES T-JUNCTION IN RECTANGULAR GUIDE, E-PLANE

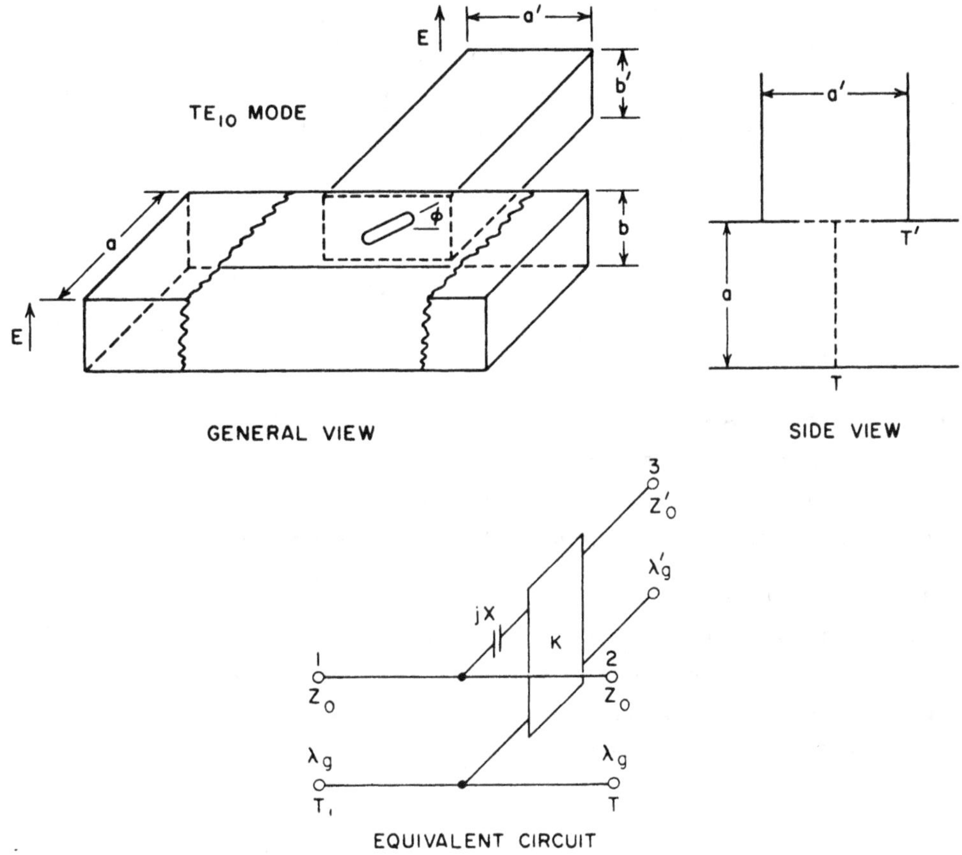

$$|S_{13}|^2 = |S_{23}|^2 \approx \frac{K^2}{Z_0^2} \cdot \frac{Z_0}{Z_0'} \cdot \frac{Z_0^2}{|X|^2} = \frac{4\pi^2 \lambda_g M^2}{\lambda_g' a' b' a^3 b}$$

$$\frac{X}{Z_0} = -\frac{a^3 b}{\pi \lambda_g M} \qquad M = M_1 \cos^2 \phi + M_2 \sin^2 \phi \qquad \frac{Z_0'}{Z_0} = \frac{\lambda_g' a' b'}{\lambda_g a b}$$

$$\lambda_g = \frac{\lambda}{\sqrt{1 - \left(\frac{\lambda}{2a}\right)^2}} \qquad \frac{K}{Z_0} = \frac{2a}{\lambda_g} \qquad \lambda_g' = \frac{\lambda}{\sqrt{1 - \left(\frac{\lambda}{2a'}\right)^2}}$$

A-3527-150

Adapted from the *Waveguide Handbook* (see Ref. 8 edited by N. Marcuvitz)

FIG. 5.10-12 IRIS-COUPLED SHUNT T-JUNCTION IN RECTANGULAR GUIDE, H-PLANE

When the irises are not small with respect to free-space wavelength, it is found that the equivalent circuits of Figs. 5.10-5 to 5.10-12 apply with good accuracy if the static magnetic polarizability M_1 given in Fig. 5.10-4 is replaced by the magnetic polarizability M_1'. The expression for M_1' is

$$M_1' = \frac{M_1}{1 - \left(\frac{\lambda_c}{\lambda}\right)^2} \qquad (5.10\text{-}3)$$

where λ_c is the free-space wavelength at the cutoff frequency for the lowest-order mode in a waveguide having the same cross section as the iris, and λ is the free-space wavelength at the frequency of operation. For long, thin irises of length l, λ_c is approximately equal to $2l$.

The finite thickness, t, of an iris reduces the transmission through it. It is found that the total attenuation α of a thick iris is predicted with reasonable approximation as the sum of the attenuation α_0 of a thin iris and the attenuation α_1 of a length of transmission line having a length equal to the iris thickness. Thus,

$$\alpha_0 = 10 \log_{10} \frac{1}{|S_{mn}|^2} \quad \text{db} \qquad (5.10\text{-}4)$$

and

$$\alpha_1 = \frac{54.6tA}{\lambda_c}\sqrt{1 - \left(\frac{\lambda_c}{\lambda}\right)^2} \quad \text{db} \qquad (5.10\text{-}5)$$

where A is an empirically determined constant approximately equal to one for a round hole.[25] For an elongated slot of length l in a wall t thick, A is about 3 if $t < 0.02\,l$, but A decreases in size as t becomes larger.[25]

The information in Eqs. (5.10-3), (5.10-4), and (5.10-5) can be combined to yield an equivalent polarizability M_{1t}' for a thick iris whose cross-sectional dimensions are not small in terms of a wavelength. The expression is

$$M'_{1t} = \frac{M_1}{1 - \left(\frac{\lambda_c}{\lambda}\right)^2} \, 10^{-\left[\frac{2.73tA}{\lambda_c}\sqrt{1-\left(\frac{\lambda_c}{\lambda}\right)^2}\right]} \quad . \qquad (5.10\text{-}6)$$

SEC. 5.11, RESONANT FREQUENCIES AND UNLOADED Q OF WAVEGUIDE RESONATORS

Two important characteristics of a waveguide resonator that are useful in the design of waveguide filters are the resonant frequency of the resonator and its unloaded Q, Q_u. This section presents curves and formulas yielding these quantities for completely closed cavities of the rectangular and cylindrical varieties. When a small coupling iris is cut in a cavity its resonant frequency and Q_u will be nearly the same as those of the unperturbed cavity.

Rectangular Waveguide Resonators—Rectangular resonators are probably used more often in waveguide filters than any other type. An example of such a resonator is illustrated in Fig. 5.11-1(a). The modes that can exist in this resonator are conveniently divided into two sets, the transverse electric TE-modes and the transverse-magnetic TM-modes. The TE-modes have no electric field components, E, along the z axis and the TM-modes have no magnetic field components, H, along the z axis. The two types of modes are further specified in terms of the integers l, m, and n. These are defined as

l = number of half-period variations of E and H along x
m = number of half-period variations of E and H along y
n = number of half-period variations of E and H along z.

For a given set of integers a mode is completely specified, and the modes are designated as either $TE_{l,m,n}$, or $TM_{l,m,n}$.

The resonant frequencies are given by the equation

$$f^2AB = 34.82 \left\{\frac{B}{A} l^2 + \frac{A}{B} m^2 + \frac{AB}{L^2} n^2\right\} \qquad (5.11\text{-}1)$$

where A, B, and L are measured in inches, and f is expressed in giga-cycles. Figure 5.11(a) also contains a mode chart in which f^2A^2 is plotted as a function of A^2/L^2 for all of the TE- and TM-modes having

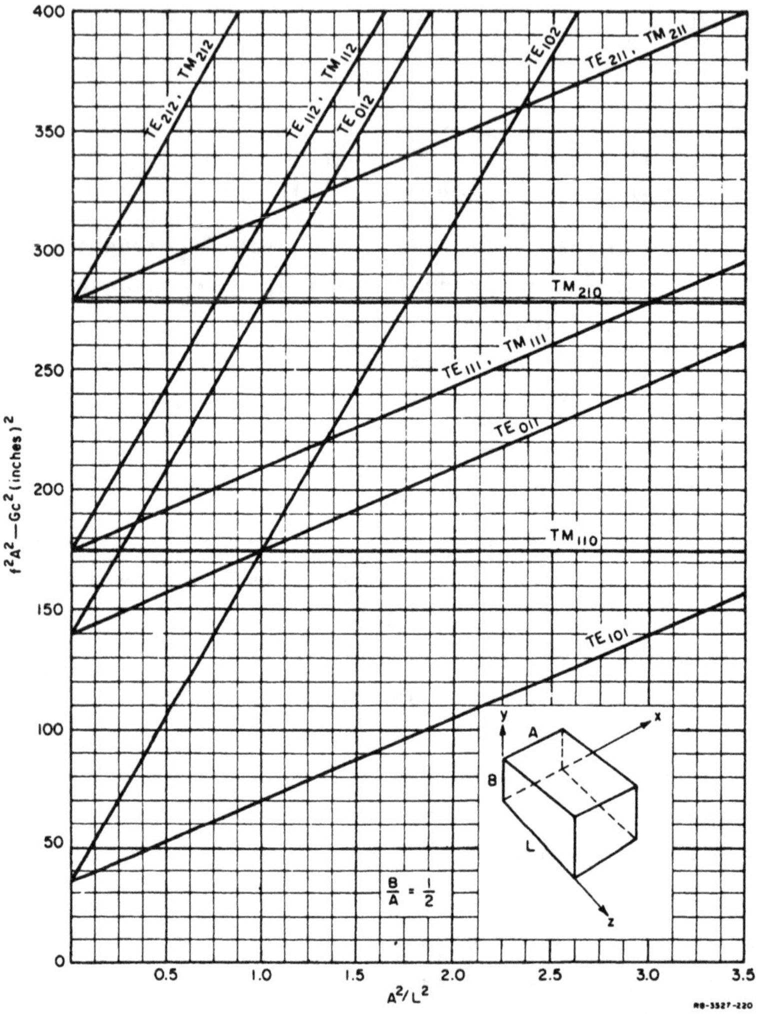

FIG. 5.11-1(a) MODE CHART FOR RECTANGULAR WAVEGUIDE RESONATOR WITH B/A = 1/2
The dimensions A, B, and L are measured in inches and frequency f is measured in gigacycles

$l, m, n \leq 2$ in a cavity in which $B/A = 1/2$. In this figure, all dimensions are in inches, and frequency is measured in gigacycles.

The unloaded Q of a cavity is most conveniently tabulated in the dimensionless form $Q_u \delta/\lambda$ where δ is the skin depth and λ is the free space wavelength. Table 5.11-1 presents values of δ/λ for various metals having polished, corrosion-free surfaces.

Table 5.11-1

VALUES OF δ/λ FOR VARIOUS METALS

The Values Given Are For Polished, Corrosion-Free Surfaces.
The Frequency f_{Gc} Is In Gigacycles.

Silver,	$\delta/\lambda = 6.76 \times 10^{-6} \sqrt{f_{Gc}}$
Copper,	$\delta/\lambda = 6.95 \times 10^{-6} \sqrt{f_{Gc}}$
Aluminum,	$\delta/\lambda = 8.70 \times 10^{-6} \sqrt{f_{Gc}}$
Brass,	$\delta/\lambda = 13.4 \times 10^{-6} \sqrt{f_{Gc}}$

For TE-modes we find that $Q_u(\delta/\lambda)$ is given by:[31]

$$Q_u \frac{\delta}{\lambda} = \frac{ABL}{4} \times \frac{(p^2 + q^2)\,(p^2 + q^2 + r^2)^{3/2}}{AL[p^2 r^2 + (p^2 + q^2)^2] + BL[q^2 r^2 + (p^2 + q^2)^2] + ABr^2(p^2 + q^2)}, \tag{5.11-2}$$

for (l and m) > 0;

$$Q_u \frac{\delta}{\lambda} = \frac{ABL}{2} \cdot \frac{(q^2 + r^2)^{3/2}}{q^2 L(B + 2A) + r^2 B(L + 2A)}, \quad \text{for } l = 0\,; \tag{5.11-3}$$

and

$$Q_u \frac{\delta}{\lambda} = \frac{ABL}{2} \cdot \frac{(p^2 + r^2)^{3/2}}{p^2 L(A + 2B) + r^2 A(L + 2B)}, \quad \text{for } m = 0 \tag{5.11-4}$$

where $p = l/A$, $q = m/B$, $r = n/L$. Figure 5.11-1(b) shows a chart of $Q_u(\delta/\lambda)$ *versus* A/L for various aspect ratios $k = A/B$ for the TE_{101} mode.

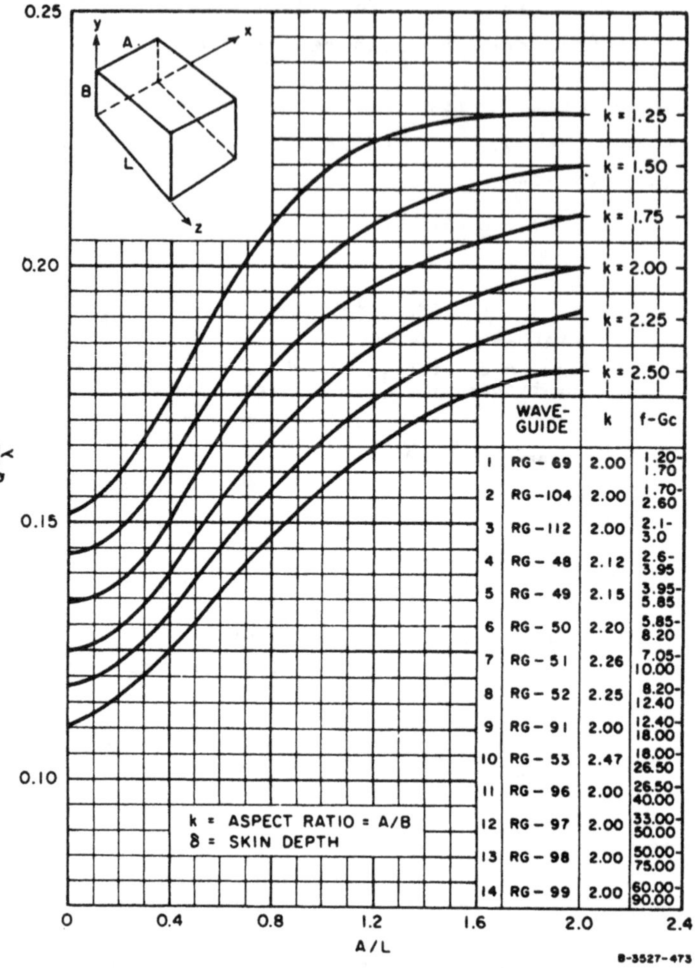

SOURCE: U. S. Army Signal Research and Development Laboratory, Ft. Monmouth, N.J.

FIG. 5.11-1(b) CHART FOR ESTIMATING THE UNLOADED Q OF TE_{101}-MODE RECTANGULAR WAVEGUIDE RESONATORS

For the TM-modes we find that Q_u is given by:[31]

$$Q_u \frac{\delta}{\lambda} = \frac{ABL}{4} \cdot \frac{(p^2 + q^2)\ (p^2 + q^2 + r^2)^{1/2}}{p^2 B(A + L) + q^2 A(B + L)} , \quad \text{for } n > 0 \ ; \qquad (5.11\text{-}5)$$

and

$$Q_u \frac{\delta}{\lambda} = \frac{ABL}{2} \cdot \frac{(p^2 + q^2)^{3/2}}{p^2 B(A + 2L) + q^2 A(B + 2L)} , \quad \text{for } n = 0 \ . \qquad (5.11\text{-}6)$$

Right-Circular-Cylinder Resonators—Cylindrical resonators of the type illustrated in Fig. 5.11-2 also have normal modes that can be characterized as TE-modes when there are no electric field components, E, along the z axis, and as TM-modes when there are no magnetic field components, H, along the z axis. The individual TE- and TM-modes are further identified by means of the three integers l, m, and n, which are defined as follows:

l = number of full-period variations of E_r with respect to θ

m = number of half-period variations of E_θ with respect to r

n = number of half-period variations of E_r with respect to z

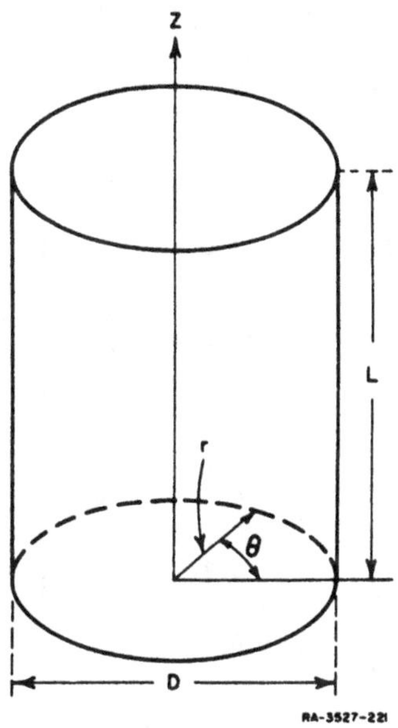

SOURCE: *Technique of Microwave Measurements*, see Ref. 31 by C. G. Montgomery

FIG. 5.11-2 RIGHT-CIRCULAR-CYLINDER RESONATOR

where E_r and E_θ are the field components in the r and θ directions. As in the case of the rectangular cavity modes the right circular cylinder modes are also designated as $TE_{l,m,n}$ or $TM_{l,m,n}$. The resonant frequencies of these modes are given by the expression[31]

$$f^2D^2 = 139.3\left\{\left(\frac{x_{l,m}}{\pi}\right)^2 + \left(\frac{nD}{2L}\right)^2\right\} \quad . \qquad (5.11\text{-}7)$$

In this expression f is measured in gigacycles, the dimensions D and L are measured in inches. The quantities $x_{l,m}$ are

$$\begin{aligned} x_{l,m} &= m\text{th root of } J'_l(x) = 0 \text{ for the TE-modes} \\ x_{l,m} &= m\text{th root of } J_l(x) = 0 \text{ for the TM-modes} \quad . \end{aligned} \tag{5.11-8}$$

Values of a few of these roots are given in Table 5.11-2.

Table 5.11-2

ROOTS OF $J_l(x)$ AND $J'_l(x)$

TE-mode	x_{lm}	TM-mode	x_{lm}
11*n*	1.841	01*n*	2.405
21*n*	3.054	11*n*	3.832
01*n*	3.832	21*n*	5.136
31*n*	4.201	02*n*	5.520
41*n*	5.318	31*n*	6.380
12*n*	5.332	12*n*	7.016
51*n*	6.415	41*n*	7.588
22*n*	6.706	22*n*	8.417
02*n*	7.016	03*n*	8.654
61*n*	7.501	51*n*	8.772
32*n*	8.016	32*n*	9.761
13*n*	8.536	61*n*	9.936
71*n*	8.578	13*n*	10.174
42*n*	9.283		
81*n*	9.648		
23*n*	9.970		
03*n*	10.174		

Source: *Technique of Microwave Measurements*, see Ref. 31, by C. G. Montgomery.

Figure 5.11-3 is a mode chart in which f^2D^2 is plotted as a function of D^2/L^2, for several of the lower-order TE- and TM-modes. In this figure all dimensions are in inches and frequency is measured in gigacycles.

Values of Q_u for right-circular-cylinder copper resonators are plotted for TE-modes in Figs. 5.11-4 and 5.11-5, and for TM-modes in Fig. 5.11-6.

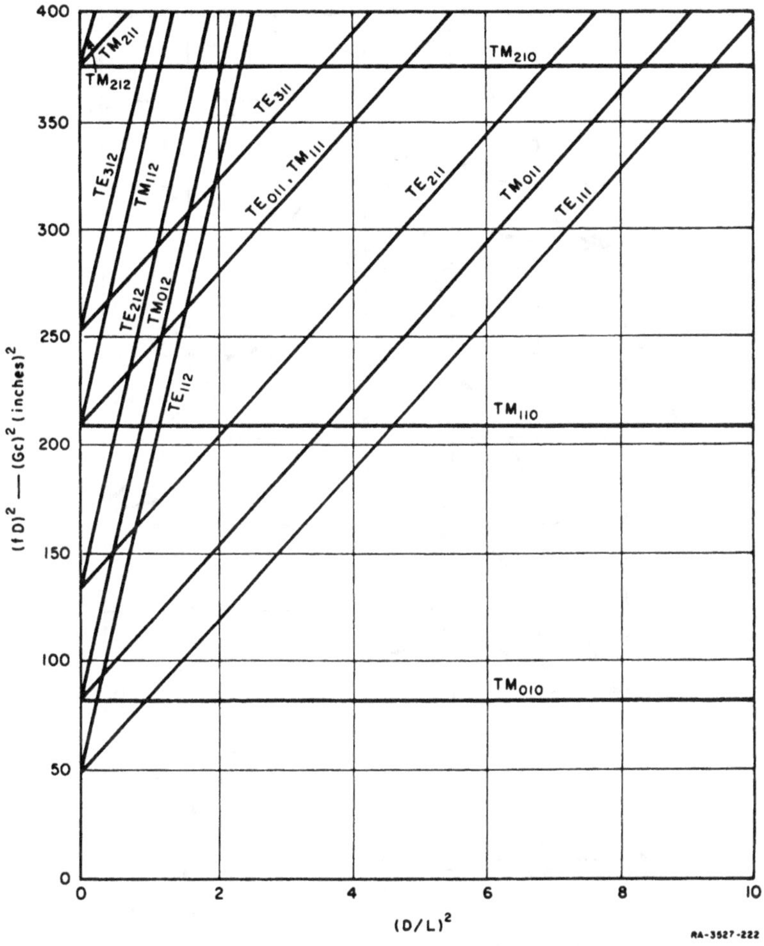

SOURCE: *Technique of Microwave Measurements*, see Ref. 31 by C. G. Montgomery

FIG. 5.11-3 MODE CHART FOR RIGHT-CIRCULAR-CYLINDER RESONATOR
The diameter D and length L are measured in inches and the frequency f is measured in gigacycles

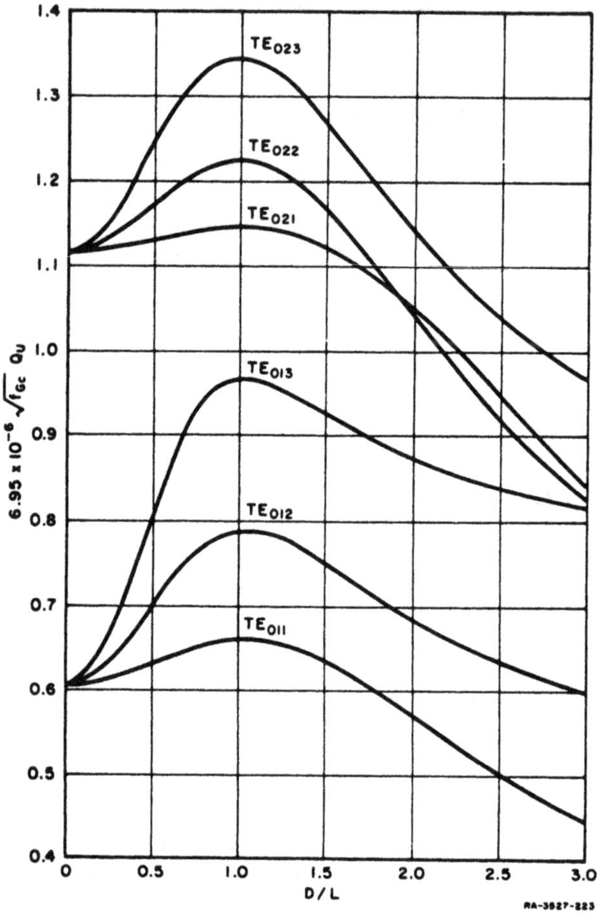

SOURCE: ***Technique of Microwave Measurements*****, see Ref. 31 by C. G. Montgomery**

FIG. 5.11-4 THEORETICAL UNLOADED Q OF SEVERAL TE_0-MODES IN A RIGHT-CIRCULAR-CYLINDER COPPER RESONATOR
Frequency is measured in gigacycles

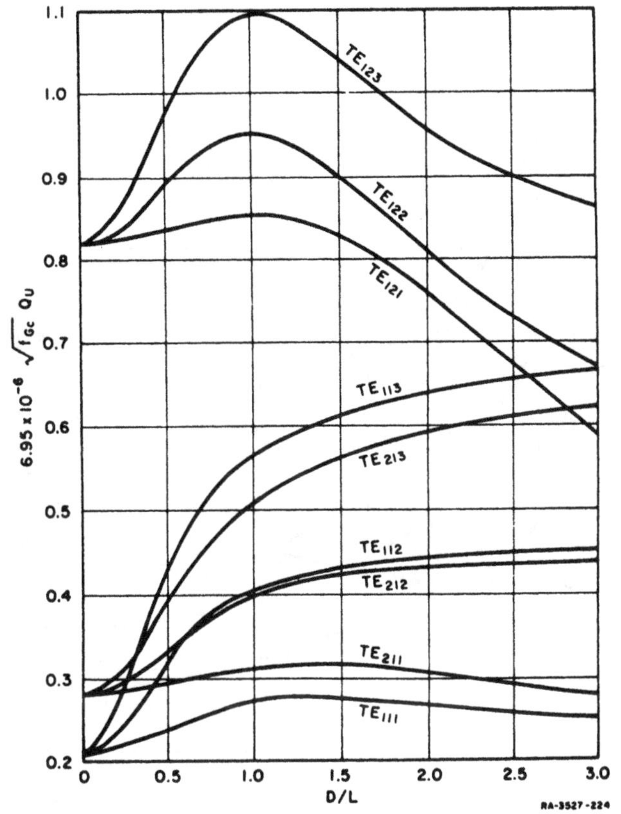

SOURCE: *Technique of Microwave Measurements*, see Ref. 31 by C. G. Montgomery

FIG. 5.11-5 THEORETICAL UNLOADED Q OF SEVERAL TE-MODES IN A RIGHT-CIRCULAR-CYLINDER COPPER RESONATOR
Frequency is measured in gigacycles

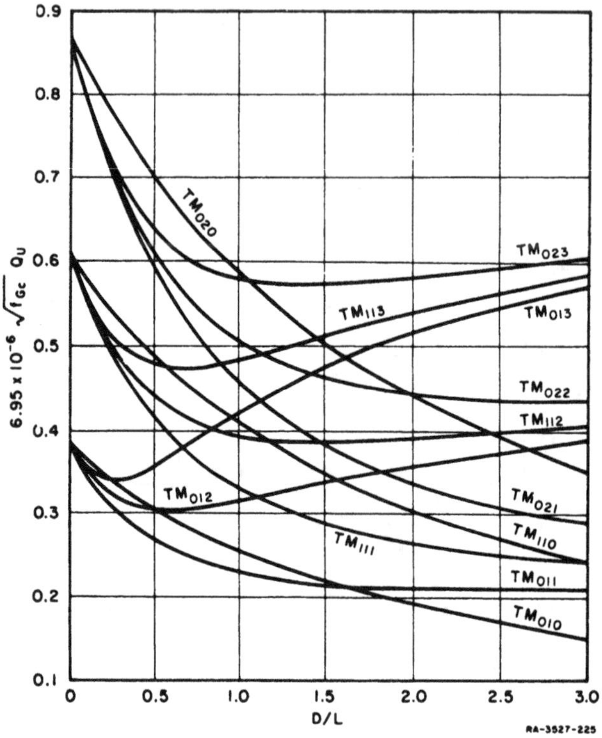

SOURCE: *Technique of Microwave Measurements*, see Ref. 31 by C. G. Montgomery

FIG. 5.11-6 THEORETICAL UNLOADED Q OF SEVERAL TM-MODES IN A RIGHT-CIRCULAR-CYLINDER RESONATOR
Frequency is measured in gigacycles

REFERENCES

1. S. B. Cohn, "Characteristic Impedance of the Shielded-Strip Transmission Line," *IRE Trans., PGMTT*-2, pp. 52-72 (July 1954).

2. S. B. Cohn, "Problems in Strip Transmission Lines," *IRE Trans., PGMTT*-3, 2, pp. 119-126 (March 1955).

3. R. H. T. Bates, "The Characteristic Impedance of Shielded Slab Line," *IRE Trans., PGMTT*-4, pp. 28-33 (January 1956).

4. S. B. Cohn, "Shielded Coupled-Strip Transmission Lines," *IRE Trans., PGMTT*-3, pp. 29-38 (October 1955).

5. F. Oberhettinger and W. Magnus, *Anwendung der Elliptischen Functionen in Physick and Technik*, (Springer-Verlag, Berlin, 1949).

6. S. B. Cohn, "Characteristic Impedances of Broadside-Coupled Strip Transmission Lines," *IRE* Trans., PGMTT-8, 6, pp. 633-637 (November 1960).

7. S. B. Cohn, "Thickness Corrections for Capacitive Obstacles and Strip Conductors," *IRE Trans., PGMTT*-8, 6, pp. 638-644 (November 1960).

8. N. Marcuvitz, *Waveguide Handbook*, MIT Radiation Laboratory Series, Vol. 10 (McGraw Hill Book Co., Inc., New York City, 1951).

9. T. Moreno, *Microwave Transmission Design Data* (Dover Publications Inc., New York City, 1958).

10. J. R. Whinnery and H. W. Jamieson, "Equivalent Circuits for Discontinuities in Transmission Lines," *Proc. IRE* 32, 2, pp. 98-114 (February 1944).

11. J. R. Whinnery, H. W. Jamieson and T. E. Robbins, "Coaxial-Line Discontinuities," *Proc. IRE* 32, 11, pp. 695-709 (November 1944).

12. A. A. Oliner, "Equivalent Circuits for Discontinuities in Balanced Strip Transmission Line," *IRE Trans., PGMTT*-3, 2, pp. 134-143 (March 1955).

13. H. M. Altschuler and A. A. Oliner, "Discontinuities in the Center Conductor of Symmetric Strip Transmission Line," *IRE Trans., PGMTT*-8, 3, pp. 328-339 (May 1960).

14. A. Alford, "Coupled Networks in Radio-Frequency Circuits," *Proc. IRE* 29, pp. 55-70 (February 1941).

15. J. J. Karakash and D. E. Mode, "A Coupled Coaxial Transmission-Line Band-Pass Filter," *Proc. IRE* 38, pp. 48-52 (January 1950).

16. W. L. Firestone, "Analysis of Transmission Line Directional Couplers," *Proc. IRE* 42, pp. 1529-1538 (October 1954).

17. B. M. Oliver, "Directional Electromagnetic Couplers," *Proc. IRE*, Vol. 42, pp. 1686-1692 (November 1954).

18. R. C. Knechtli, "Further Analysis of Transmission-Line Directional Couplers," *Proc. IRE* 43, pp. 867-869 (July 1955).

19. E. M. T. Jones and J. T. Bolljahn, "Coupled-Strip-Transmission-Line Filters and Directional Couplers," *IRE Trans., PGMTT*-4, 2, pp. 75-81 (April 1956).

20. H. Ozaki and J. Ishii, "Synthesis of a Class of Strip-Line Filters," *IRE Trans., PGCT*-5, pp. 104-109 (June 1958).

21. H. A. Bethe, "Lumped Constants for Small Irises," Report 43-22, M.I.T. Radiation Laboratory, Cambridge, Massachusetts (March 1943).

22. H. A. Bethe, "Theory of Side Windows in Waveguides," Report 43-27, M.I.T. Radiation Laboratory, Cambridge, Massachusetts (April 1943).

23. H. A. Bethe, "Formal Theory of Waveguides of Arbitrary Cross Section," Report 43-26, M.I.T. Radiation Laboratory, Cambridge, Massachusetts (March 1943).

24. H. A. Bethe, "Theory of Diffraction by Small Holes," *Phys. Rev.* Vol. 66, pp. 163-182 (1944).

25. S. B. Cohn, "Microwave Coupling by Large Apertures," *Proc. IRE* 40, pp. 696-699 (June 1952).

26. R. E. Collin, *Field Theory of Guided Waves,* Sec. 7.3 (McGraw Hill Book Co., Inc., New York City, 1960).

27. N. Marcuvitz, "Waveguide Circuit Theory: Coupling of Waveguides by Small Apertures," Report No. R-157-47, Microwave Research Institute, Polytechnic Institute of Brooklyn (1947) PIB-106.

28. A. A. Oliner, "Equivalent Circuits for Small Symmetrical Longitudinal Apertures and Obstacles," *IRE Trans., PGMTT*-8 1, pp. 72-80 (January 1960).

29. S. B. Cohn, "Determination of Aperture Parameters by Electrolytic Tank Measurements," *Proc. IRE* 39, pp. 1416-1421 (November 1951).

30. S. B. Cohn, "The Electric Polarizability of Apertures of Arbitrary Shape," *Proc. IRE* 40, pp. 1069-1071 (September 1952).

31. C. G. Montgomery, *Technique of Microwave Measurements*, Secs. 5.4 and 5.5 (McGraw-Hill Book Co., New York City, N.Y., 1947).

32. W. J. Getsinger, "A Coupled Strip-Line Configuration Using Printed-Circuit Construction that Allows Very Close Coupling," *IRE Trans., PGMTT*-9, pp. 535-544 (November 1961).

33. W. J. Getsinger, "Coupled Rectangular Bars Between Parallel Plates," *IRE Trans., PGMTT*-10, pp. 65-72 (January 1962).

CHAPTER 6

STEPPED-IMPEDANCE TRANSFORMERS AND FILTER PROTOTYPES

SEC. 6.01, INTRODUCTION

The objective of this chapter is to present design equations and numerical data for the design of quarter-wave transformers, with two applications in mind: the first application is as an impedance-matching device or, literally, transformer; the second is as a prototype circuit, which shall serve as the basis for the design of various band-pass and low-pass filters.

This chapter is organized into fifteen sections, with the following purpose and content:

- Section 6.01 is introductory. It also discusses applications, and gives a number of definitions.
- Sections 6.02 and 6.03 deal with the performance characteristics of quarter-wave transformers and half-wave filters. In these parts the designer will find what *can* be done, not how to do it.
- Sections 6.04 to 6.10 tell *how* to design quarter-wave transformers and half-wave filters. If simple, general design formulas were available, and solvable by nothing more complicated than a slide-rule, these sections would be much shorter.
- Section 6.04 gives exact formulas and tables of complete designs for Tchebyscheff and maximally flat transformers of up to four sections.
- Section 6.05 gives tables of designs for maximally flat (but not Tchebyscheff) transformers of up to eight sections.
- Section 6.06 gives a first-order theory for Tchebyscheff and maximally flat transformers of up to eight sections, with explicit formulas and numerical tables. It also gives a general first-order formula, and refers to existing numerical tables published elsewhere which are suitable for up to 39 sections, and for relatively wide (but not narrow) bandwidths.
- Section 6.07 presents a modified first-order theory, accurate for larger transformer ratios than can be designed by the (unmodified) first-order theory of Sec. 6.06.

- Section 6.08 deals with the discontinuity effects of non-ideal junctions, and first-order corrections to compensate for them.

- Sections 6.09 and 6.10 apply primarily to prototypes for filters, since they are concerned with large impedance steps. They become exact only in the limit as the output-to-input impedance ratio, R, tends to infinity. Simple formulas are given for any number of sections, and numerical tables on lumped-constant filters are referred to.

Note: Sections 6.09 and 6.10 complement Secs. 6.06 and 6.07, which give exact results only in the limit as R tends to zero. It is pointed out that the dividing line between "small R" and "large R" is in the order of $[2/(\text{quarter-wave transformer bandwidth})]^{2n}$, where n is the number of sections. This determines whether the first-order theory of Secs. 6.06 and 6.07, or the formulas of Secs. 6.09 and 6.10 are to be used. An example (Example 3 of Sec. 6.09) where R is in this borderline region, is solved by both the "small R" and the "large R" approximations, and both methods give tolerably good results for most purposes.

- Sections 6.11 and 6.12 deal with "inhomogeneous" transformers, which are not uniformly dispersive, since the cutoff wavelength changes at each step.

- Section 6.13 describes a particular transformer whose performance and over-all length are similar to those of a single-section quarter-wave transformer, but which requires only matching sections whose characteristic impedances are equal to the input and output impedances.

- Section 6.14 considers dissipation losses. It gives a general formula for the midband dissipation loss.

- Section 6.15 relates group delay to dissipation loss in the pass band, and presents numerical data in a set of universal curves.

Quarter-wave transformers have numerous applications besides being impedance transformers; an understanding of their behavior gives insight into many other physical situations not obviously connected with impedance transformations. The design equations and numerical tables have, moreover, been developed to the point where they can be used conveniently for the synthesis of circuits, many of which were previously difficult to design.

Circuits that can be designed using quarter-wave transformers as a prototype include: impedance transformers[1-6] (as in this chapter); reactance-coupled filters[7-9] (Chapt. 9); short-line low-pass filters (Sec. 7.06); branch-guide directional couplers[10] (Chapt. 13); as well as optical multi-layer filters and transformers,[11,12] and acoustical transformers.[13,14]

The attenuation functions considered here are all for maximally flat or Tchebyscheff response in the pass band. It is of interest to note that occasionally other response shapes may be desirable. Thus TEM-mode coupled-transmission-line directional couplers are analytically equivalent to quarter-wave transformers (Chapt. 13), but require functions with maximally flat or equal-ripple characteristics in the *stop* band. Other attenuation functions may be convenient for other applications, but will not be considered here.

As in the design of all microwave circuits, one must distinguish between the ideal circuits analyzed, and the actual circuits that have prompted the analysis and which are the desired end product. To bring this out explicitly, we shall start with a list of definitions:[15]

Homogeneous transformer—a transformer in which the ratios of internal wavelengths and characteristic impedances at different positions along the direction of propagation are independent of frequency.

Inhomogeneous transformer—a transformer in which the ratios of internal wavelengths and characteristic impedances at different positions along the direction of propagation may change with frequency.

Quarter-wave transformer—a cascade of sections of lossless, uniform* transmission lines or media, each section being one-quarter (internal) wavelength long at a common frequency.

* A uniform transmission line, medium, etc., is here defined as one in which the physical and electrical characteristics do not change with distance along the direction of propagation. This is a generalization of the IRE definition of *uniform waveguide* (see Ref. 16).

Note: Homogeneous and inhomogeneous quarter-wave transformers are now defined by a combination of the above definitions. For instance, an *inhomogeneous quarter-wave transformer* is a quarter-wave transformer in which the ratios of internal wavelengths and characteristic impedances taken between different sections, may change with frequency.

Ideal junction—the connection between two impedances or transmission lines, when the electrical effects of the connecting wires, or the junction discontinuities, can be neglected. (The junction effects may later be represented by equivalent reactances and transformers, or by positive and negative line lengths, etc.)

Ideal quarter-wave transformer—a quarter-wave transformer in which all of the junctions (of guides or media having different characteristic impedances) may be treated as ideal junctions.

Half-wave filter—a cascade of sections of lossless uniform transmission lines or media, each section being one-half (internal) wavelength long at a common frequency.

Synchronous tuning condition—a filter consisting of a series of discontinuities spaced along a transmission line is synchronously tuned if, at some fixed frequency in the pass band, the reflections from any pair of *successive* discontinuities are phased to give the maximum cancellation. (A quarter-wave transformer is a synchronously tuned circuit if its impedances form a monotone sequence. A half-wave filter is a synchronously tuned circuit if its impedances alternately increase and decrease at each step along its length.)

Synchronous frequency—the "fixed frequency" referred to in the previous definition will be called the *synchronous frequency*. (In the case of quarter-wave transformers, all sections are one-quarter wavelength long at the synchronous frequency; in the case of half-wave filters, all sections are one-half wavelength long at the synchronous frequency. Short-line, low-pass filters may also be derived from half-wave filters, with the synchronous frequency being thought of as zero frequency.)

The realization of transmission-line discontinuities by impedance steps is equivalent to their realization by means of ideal impedance inverters (Sec. 4.12). The main difference is that while impedance steps can be physically realized over a wide band of frequencies (at least for small steps), ideal impedance inverters can be approximated over only limited bandwidths. As far as using either circuit as a mathematical model, or prototype circuit, is concerned, they give equivalent results, as can be seen from Fig. 6.01-1.

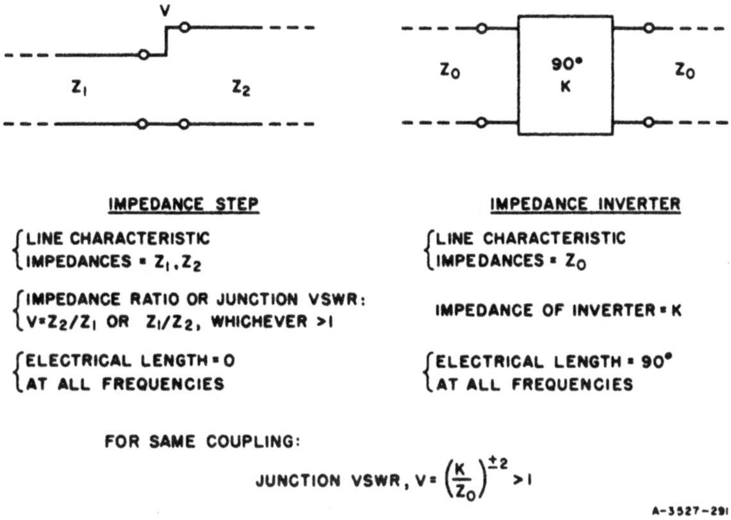

SOURCE: Quarterly Progress Report 4, Contract DA 36-039 SC-87398, SRI; reprinted in *IRE Trans. PGMTT* (See Ref. 36 by L. Young)

FIG. 6.01-1 CONNECTION BETWEEN IMPEDANCE STEP AND IMPEDANCE INVERTER

SEC. 6.02, THE PERFORMANCE OF HOMOGENEOUS QUARTER-WAVE TRANSFORMERS

This section summarizes the relationships between the pass-band and stop-band attenuation, the fractional bandwidth, w_q, and the number of sections or resonators, n. Although the expressions obtained hold exactly only for ideal quarter-wave transformers, they hold relatively accurately for real physical quarter-wave transformers and for certain filters, either without modification or after simple corrections have been applied to account for junction effects, etc.

A quarter-wave transformer is depicted in Fig. 6.02-1. Define the quarter-wave transformer fractional bandwidth, w_q, by

$$w_q = 2\left(\frac{\lambda_{g1} - \lambda_{g2}}{\lambda_{g1} + \lambda_{g2}}\right) \tag{6.02-1}$$

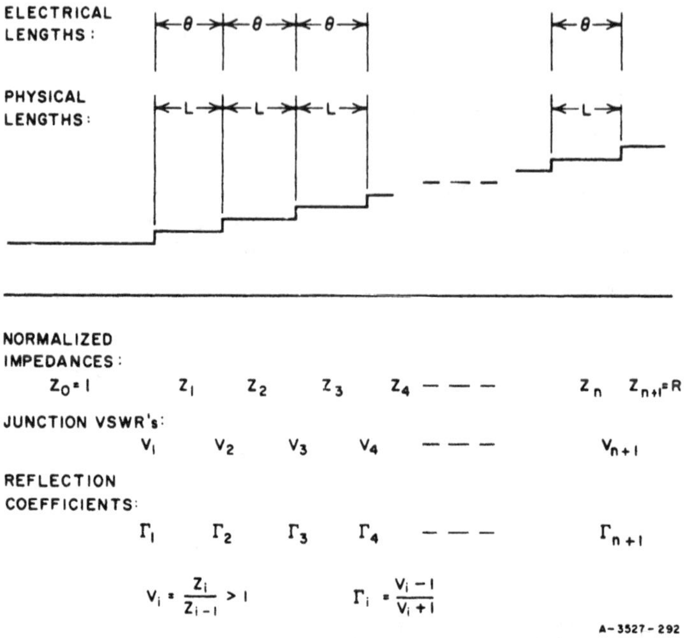

SOURCE: Quarterly Progress Report 4, Contract DA 36-039 SC-87398, SRI; reprinted in *IRE Trans. PGMTT* (See Ref. 36 by L. Young)

FIG. 6.02-1 QUARTER-WAVE TRANSFORMER NOTATION

where λ_{g1} and λ_{g2} are the longest and shortest guide wavelengths, respectively, in the pass band of the quarter-wave transformer. The length, L, of each section (Fig. 6.02-1) is nominally one-quarter wavelength at center frequency and is given by

$$L = \frac{\lambda_{g1}\lambda_{g2}}{2\,(\lambda_{g1} + \lambda_{g2})} = \frac{\lambda_{g0}}{4} \qquad (6.02\text{-}2)$$

where the center frequency is defined as that frequency at which the guide wavelength λ_g is equal to λ_{g0}.

When the transmission line is non-dispersive, the free-space wavelength λ may be used in Eqs. (6.02-1) and (6.02-2), which then become

$$w_q = 2\left(\frac{\lambda_1 - \lambda_2}{\lambda_1 + \lambda_2}\right) = 2\left(\frac{f_2 - f_1}{f_2 + f_1}\right) \qquad (6.02\text{-}3)$$

and

$$L = \frac{\lambda_1 \lambda_2}{2(\lambda_1 + \lambda_2)} = \frac{\lambda_0}{4} \qquad (6.02\text{-}4)$$

where f stands for frequency.

The transducer loss ratio (Sec. 2.11) is defined as the ratio of P_{avail}, the available generator power, to P_L, the power actually delivered to the load. The "excess loss," $\mathcal{E}$, is herein defined by

$$\mathcal{E} = \frac{P_{avail}}{P_L} - 1 \quad . \qquad (6.02\text{-}5)$$

For the maximally flat quarter-wave transformer of n sections and over-all impedance ratio R (Fig. 6.02-1) $\mathcal{E}$ is given by

$$\mathcal{E} = \frac{(R-1)^2}{4R} \cos^{2n} \theta = \mathcal{E}_a \cos^{2n} \theta \qquad (6.02\text{-}6)$$

where

$$\theta = \frac{\pi}{2} \frac{\lambda_{g0}}{\lambda_g} \quad , \qquad (6.02\text{-}7)$$

λ_{g0} being the guide wavelength at band center, when $\theta = \pi/2$; and where

$$\mathcal{E}_a = \frac{(R-1)^2}{4R} \qquad (6.02\text{-}8)$$

is the greatest excess loss possible. (It occurs when θ is an integral multiple of π, since the sections then are an integral number of half-wavelengths long.)

The 3-db fractional bandwidth of the maximally flat quarter-wave transformer is given by

$$w_{q,3db} = \frac{4}{\pi} \sin^{-1} \left[\frac{4R}{(R-1)^2} \right]^{1/2n} \qquad (6.02\text{-}9)$$

The fractional bandwidth of the maximally flat quarter-wave transformer between the points of x-db attenuation is given by

$$w_{q,xdb} = \frac{4}{\pi} \sin^{-1} \left\{ \frac{4R\,[\text{antilog}\,(x/10) - 1]}{(R-1)^2} \right\}^{1/2n} . \qquad (6.02\text{-}10)$$

For the Tchebyscheff transformer of fractional bandwidth w_q,

$$\left.\begin{aligned} \mathcal{E} &= \frac{(R-1)^2}{4R}\,\frac{T_n^2(\cos\theta/\mu_0)}{T_n^2(1/\mu_0)} \\ &= \mathcal{E}_r T_n^2(\cos\theta/\mu_0) \end{aligned}\right\} \tag{6.02-11}$$

where

$$\mu_0 = \sin\left(\frac{\pi w_q}{4}\right) , \tag{6.02-12}$$

T_n is a Tchebyscheff polynomial (of the first kind) of order n, and where the quantity

$$\mathcal{E}_r = \frac{(R-1)^2}{4R}\,\frac{1}{T_n^2(1/\mu_0)} = \frac{\mathcal{E}_a}{T_n^2(1/\mu_0)} \tag{6.02-13}$$

is the maximum excess loss in the pass band. [Compare also Eq. (6.02-18), below.] The shape of these response curves for maximally flat and Tchebyscheff quarter-wave transformers is shown in Fig. 6.02-2. Notice that the peak transducer loss ratio for any quarter-wave transformer is

$$\frac{P_{avail}}{P_L} = \mathcal{E}_a + 1 = \frac{(R+1)^2}{4R} \tag{6.02-14}$$

and is determined solely by the output-to-input impedance ratio, R.

For the maximally flat transformer, the 3-db fractional bandwidth, $w_{q,3db}$, is plotted against log R for $n = 2$ to $n = 15$ in Fig. 6.02-3. The attenuation given by Eq. (6.02-6) can also be determined from the corresponding lumped-constant, low-pass, prototype filter (Sec. 4.03). If ω' is the frequency variable of the maximally flat, lumped-constant, low-pass prototype, and ω_1' is its band edge, then

$$\frac{\omega'}{\omega_1'} = \frac{\cos\theta}{\mu_0} \tag{6.02-15}$$

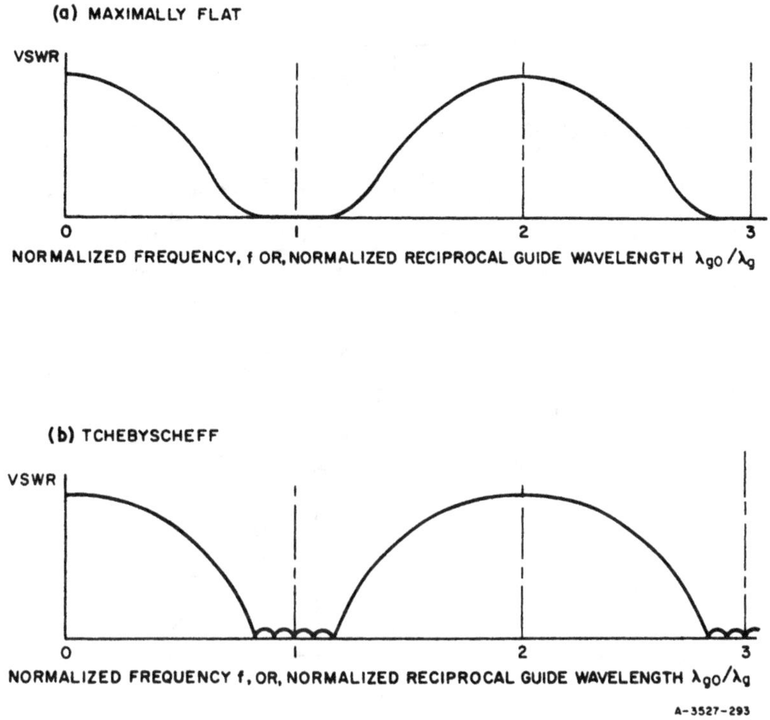

SOURCE: Quarterly Progress Report 4, Contract DA 36-039 SC-87398, SRI; reprinted in *IRE Trans. PGMTT* (See Ref. 36 by L. Young)

FIG. 6.02-2 QUARTER-WAVE TRANSFORMER CHARACTERISTICS

where μ_0 is defined by Eq. (6.02-12), and w_q (which occurs in the definition of μ_0) is the fractional bandwidth of the maximally flat quarter-wave transformer between points of the same attenuation as the attenuation of the maximally flat low-pass filter at $\omega' = \omega_1'$. This enables one to turn the graph of attenuation versus ω'/ω_1' in Fig. 4.03-2 into a graph of attenuation versus $\cos\theta$ of the quarter-wave transformer, using Eq. (6.02-15).

For the Tchebyscheff transformer,

$$\frac{\mathcal{E}_a}{\mathcal{E}_r} = T_n^2(1/\mu_0) = M(n, w_q) \tag{6.02-16}$$

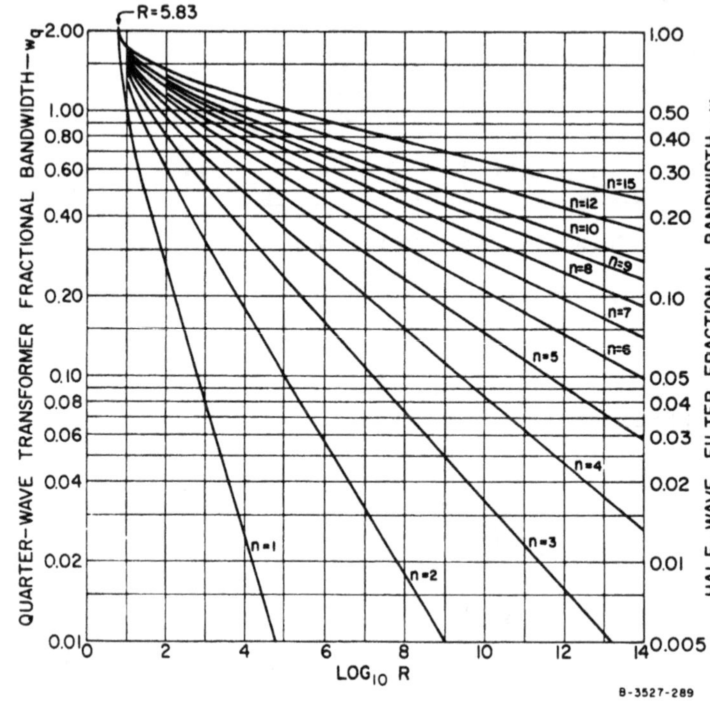

SOURCE: Quarterly Progress Report 4, Contract DA 36-039 SC-87398, SRI; reprinted in *IRE Trans. PGMTT* (See Ref. 36 by L. Young)

FIG. 6.02-3 3-db BANDWIDTHS OF MAXIMALLY FLAT TRANSFORMERS

where M is thus defined as a function of the number of sections, n, and the bandwidth, w_q. It shows how much the pass-band tolerance increases when it is desired to improve the peak rejection. The function M in Eq. (6.02-16) is given in Table 6.02-1 for all fractional bandwidths, w_q, in steps of 10 percent, for $n = 2$ to $n = 15$. The smallest fractional bandwidth in Table 6.02-1 is $w_q = 0.1$. For small bandwidths,

$$\frac{\mathcal{E}_a}{\mathcal{E}_r} = T_n^2(1/\mu_0) = M(n, w_q) \approx \frac{1}{4}\left(\frac{8}{\pi w_q}\right)^{2n} , \quad (w_q \text{ small}) . \tag{6.02-17}$$

Table 6.02-1

$$M(n, w_q) = T_n^2 \left[\frac{1}{\sin\,(\pi w_q/4)}\right]$$

n \ w_q	0.1	0.2	0.3	0.4	0.5	0.6	0.7	0.8	0.9	1.0
2	0.1049 * 6	0.6517 * 4	0.1274 * 4	0.3978 * 3	0.1601 * 3	0.7575 * 2	0.4001 * 2	0.2293 * 2	0.1400 * 2	0.9000 * 1
3	0.6795 * 8	0.1052 * 7	0.9094 * 5	0.1584 * 5	0.4036 * 4	0.1306 * 4	0.4972 * 3	0.2130 * 3	0.9966 * 2	0.5000 * 2
4	0.4402 *11	0.1699 * 9	0.6491 * 7	0.6313 * 6	0.1020 * 6	0.2265 * 5	0.6246 * 4	0.2013 * 4	0.7291 * 3	0.2890 * 3
5	0.2851 *14	0.2742 *11	0.4634 * 9	0.2517 * 8	0.2578 * 7	0.3930 * 6	0.7852 * 5	0.1906 * 5	0.5353 * 4	0.1682 * 4
6	0.1847 *17	0.4427 *13	0.3308 *11	0.1003 *10	0.6516 * 8	0.6819 * 7	0.9872 * 6	0.1806 * 6	0.3933 * 5	0.9801 * 4
7	0.1196 *20	0.7148 *15	0.2361 *13	0.3999 *11	0.1646 *10	0.1183 * 9	0.1241 * 8	0.1710 * 7	0.2890 * 6	0.5712 * 5
8	0.7751 *22	0.1154 *18	0.1685 *15	0.1594 *13	0.4162 *11	0.2052 *10	0.1560 * 9	0.1620 * 8	0.2123 * 7	0.3329 * 6
9	0.5021 *25	0.1863 *20	0.1203 *17	0.6355 *14	0.1052 *13	0.3561 *11	0.1961 *10	0.1535 * 9	0.1560 * 8	0.1940 * 7
10	0.3252 *28	0.3008 *22	0.8590 *18	0.2533 *16	0.2658 *14	0.6178 *12	0.2466 *11	0.1454 *10	0.1146 * 9	0.1131 * 8
11	0.2107 *31	0.4856 *24	0.6132 *20	0.1010 *18	0.6720 *15	0.1072 *14	0.3100 *12	0.1377 *11	0.8422 * 9	0.6592 * 8
12	0.1365 *34	0.7840 *26	0.4377 *22	0.4026 *19	0.1698 *17	0.1860 *15	0.3898 *13	0.1304 *12	0.6188 *10	0.3842 * 9
13	0.8842 *36	0.1266 *29	0.3124 *24	0.1605 *21	0.4292 *18	0.3227 *16	0.4901 *14	0.1235 *13	0.4547 *11	0.2239 *10
14	0.5728 *39	0.2044 *31	0.2230 *26	0.6397 *22	0.1084 *20	0.5598 *17	0.6161 *15	0.1170 *14	0.3340 *12	0.1305 *11
15	0.3710 *42	0.3299 *33	0.1592 *28	0.2550 *24	0.2742 *21	0.9712 *18	0.7746 *16	0.1108 *15	0.2454 *13	0.7607 *11

Table 6.02-1 (concluded)

n \ w_q	1.1	1.2	1.3	1.4	1.5	1.6	1.7	1.8	1.9	2.0
2	0.6046 * 1	0.4226 * 1	0.3066 * 1	0.2308 * 1	0.1804 * 1	0.1467 * 1	0.1243 * 1	0.1103 * 1	0.1024 * 1	1.0
3	0.2654 * 2	0.1479 * 2	0.8611 * 1	0.5234 * 1	0.3331 * 1	0.2236 * 1	0.1601 * 1	0.1241 * 1	0.1056 * 1	1.0
4	0.1230 * 3	0.5553 * 2	0.2634 * 2	0.1308 * 2	0.6802 * 1	0.3739 * 1	0.2213 * 1	0.1454 * 1	0.1102 * 1	1.0
5	0.5771 * 3	0.2125 * 3	0.8288 * 2	0.3398 * 2	0.1459 * 2	0.6610 * 1	0.3219 * 1	0.1762 * 1	0.1162 * 1	1.0
6	0.2713 * 4	0.8170 * 3	0.2631 * 3	0.8965 * 2	0.3206 * 2	0.1206 * 2	0.4853 * 1	0.2197 * 1	0.1239 * 1	1.0
7	0.1276 * 5	0.3145 * 4	0.8380 * 3	0.2379 * 3	0.7120 * 2	0.2239 * 2	0.7490 * 1	0.2802 * 1	0.1334 * 1	1.0
8	0.6006 * 5	0.1211 * 5	0.2671 * 4	0.6327 * 3	0.1588 * 3	0.4197 * 2	0.1174 * 2	0.3639 * 1	0.1450 * 1	1.0
9	0.2826 * 6	0.4666 * 5	0.8515 * 4	0.1684 * 4	0.3552 * 3	0.7907 * 2	0.1858 * 2	0.4790 * 1	0.1590 * 1	1.0
10	0.1329 * 7	0.1797 * 6	0.2715 * 5	0.4483 * 4	0.7950 * 3	0.1493 * 3	0.2959 * 2	0.6371 * 1	0.1756 * 1	1.0
11	0.6257 * 7	0.6923 * 6	0.8656 * 5	0.1194 * 5	0.1780 * 4	0.2825 * 3	0.4730 * 2	0.8542 * 1	0.1954 * 1	1.0
12	0.2944 * 8	0.2667 * 7	0.2760 * 6	0.3179 * 5	0.3986 * 4	0.5347 * 3	0.7581 * 2	0.1152 * 2	0.2187 * 1	1.0
13	0.1385 * 9	0.1027 * 8	0.8800 * 6	0.8465 * 5	0.8928 * 4	0.1012 * 4	0.1216 * 3	0.1560 * 2	0.2463 * 1	1.0
14	0.6518 * 9	0.3956 * 8	0.2806 * 7	0.2254 * 6	0.1999 * 5	0.1918 * 4	0.1954 * 3	0.2120 * 2	0.2787 * 1	1.0
15	0.3067 *10	0.1524 * 9	0.8947 * 7	0.6003 * 6	0.4478 * 5	0.3632 * 4	0.3142 * 3	0.2888 * 2	0.3167 * 1	1.0

*4 means "multiply by 10^4," and so on.

SOURCE: Quarterly Progress Report 4, Contract DA 36-039 SC-87398, SRI; reprinted in *IRE Trans. PGMTT* (see Ref. 36 by L. Young)

Equation (6.02-17) is accurate to better than about 1 percent for w_q less than 0.1.

The attenuation given by Eq. (6.02-11) for the Tchebyscheff quarter-wave transformer can also be determined from the graphs in Figs. 4.03-4 to 4.03-10 for the corresponding lumped-constant, low-pass, prototype filter [as already explained for the maximally flat case in connection with Eq. (6.02-15)] by using the same Eq. (6.02-15) except that now ω_1' is the Tchebyscheff (equal-ripple) band edge of the low-pass filter.

In the design of transformers as such, one is interested only in the pass-band performance for small R (usually less than 100), and this is expressed in terms of maximum VSWR rather than maximum attenuation. Tables 6.02-2 through 6.02-5 give directly the maximum VSWR inside the pass band for transformers with output-to-input impedance ratios, R, of less than 100, and fractional bandwidths, w_q, up to 120 percent, for transformers of n = 1, 2, 3, and 4 sections.[4] For all other cases, the maximum VSWR may be worked out from Table 6.02-1, using the relation

$$\mathcal{E}_r = \frac{(V_r - 1)^2}{4V_r} \qquad (6.02\text{-}18)$$

where V_r is the ripple VSWR (maximum VSWR in the pass band), together with Eqs. (6.02-8) and (6.02-16).

Example 1—Determine the minimum number of sections for a transformer of impedance ratio R = 100 to have a VSWR of less than 1.15 over a 100-percent bandwidth (w_q = 1.0).

From Eq. (6.02-18), for V_r = 1.15,

$$\mathcal{E}_r = 0.00489 \qquad (6.02\text{-}19)$$

and from Eq. (6.02-8), for R = 100,

$$\mathcal{E}_a = 24.5 \quad . \qquad (6.02\text{-}20)$$

Table 6.02-2

MAXIMUM VSWR FOR SINGLE-SECTION QUARTER-WAVE TRANSFORMERS

IMPEDANCE RATIO, R	BANDWIDTH, w_q					
	0.2	0.4	0.6	0.8	1.0	1.2
1.25	1.03	1.07	1.11	1.14	1.17	1.20
1.50	1.06	1.13	1.20	1.27	1.33	1.39
1.75	1.09	1.19	1.30	1.39	1.49	1.57
2.00	1.12	1.24	1.38	1.51	1.64	1.76
2.50	1.16	1.34	1.53	1.73	1.93	2.12
3.00	1.20	1.43	1.68	1.95	2.21	2.47
4.00	1.26	1.58	1.95	2.35	2.76	3.15
5.00	1.32	1.73	2.21	2.74	3.30	3.83
6.00	1.37	1.86	2.45	3.12	3.82	4.50
8.00	1.47	2.11	2.92	3.86	4.86	5.84
10.00	1.55	2.35	3.37	4.58	5.88	7.16
12.50	1.65	2.63	3.92	5.47	7.15	8.81
15.00	1.75	2.90	4.47	6.36	8.41	10.46
17.50	1.84	3.17	5.01	7.25	9.67	12.10
20.00	1.92	3.43	5.54	8.11	10.93	13.74
25.00	2.08	3.95	6.60	9.86	13.44	17.02
30.00	2.24	4.45	7.65	11.60	15.95	20.30
40.00	2.54	5.45	9.73	15.07	20.96	26.85
50.00	2.82	6.43	11.81	18.54	25.97	33.40
60.00	3.10	7.40	13.88	22.00	30.98	39.95
80.00	3.63	9.34	18.02	28.92	40.98	53.04
100.00	4.16	11.27	22.15	35.83	50.98	66.13

Table 6.02-3

MAXIMUM VSWR FOR TWO-SECTION TCHEBYSCHEFF QUARTER-WAVE TRANSFORMERS*

IMPEDANCE RATIO, R	BANDWIDTH, w_q					
	0.2	0.4	0.6	0.8	1.0	1.2
1.25	1.00	1.01	1.03	1.05	1.08	1.11
1.50	1.01	1.02	1.05	1.09	1.15	1.22
1.75	1.01	1.03	1.07	1.13	1.21	1.32
2.00	1.01	1.04	1.08	1.16	1.27	1.41
2.50	1.01	1.05	1.12	1.22	1.37	1.58
3.00	1.01	1.06	1.14	1.27	1.47	1.74
4.00	1.02	1.08	1.19	1.37	1.64	2.04
5.00	1.02	1.09	1.23	1.45	1.80	2.33
6.00	1.03	1.11	1.26	1.53	1.95	2.60
8.00	1.03	1.13	1.33	1.67	2.23	3.13
10.00	1.04	1.15	1.38	1.80	2.50	3.64
12.50	1.04	1.18	1.45	1.95	2.82	4.27
15.00	1.05	1.20	1.51	2.09	3.13	4.89
17.50	1.05	1.22	1.57	2.23	3.44	5.50
20.00	1.05	1.24	1.62	2.36	3.74	6.11
25.00	1.06	1.27	1.72	2.62	4.33	7.32
30.00	1.07	1.30	1.82	2.87	4.91	8.52
40.00	1.08	1.36	2.00	3.36	6.06	10.91
50.00	1.09	1.41	2.17	3.83	7.20	13.29
60.00	1.10	1.46	2.34	4.30	8.33	15.66
80.00	1.12	1.55	2.65	5.21	10.57	20.41
100.00	1.13	1.63	2.96	6.11	12.81	25.15

SOURCE: *IRE Trans. PGMTT* (see Ref. 4 by L. Young)

Table 6.02-4

MAXIMUM VSWR FOR THREE-SECTION TCHEBYSCHEFF QUARTER-WAVE TRANSFORMERS*

IMPEDANCE RATIO, R	BANDWIDTH, w_q					
	0.2	0.4	0.6	0.8	1.0	1.2
1.25	1.00	1.00	1.01	1.02	1.03	1.06
1.50	1.00	1.00	1.01	1.03	1.06	1.11
1.75	1.00	1.00	1.02	1.04	1.08	1.16
2.00	1.00	1.01	1.02	1.05	1.11	1.20
2.50	1.00	1.01	1.03	1.07	1.14	1.28
3.00	1.00	1.01	1.03	1.08	1.18	1.35
4.00	1.00	1.01	1.04	1.11	1.24	1.47
5.00	1.00	1.01	1.05	1.13	1.29	1.59
6.00	1.00	1.02	1.06	1.15	1.33	1.69
8.00	1.00	1.02	1.07	1.18	1.42	1.88
10.00	1.00	1.02	1.08	1.21	1.49	2.06
12.50	1.00	1.03	1.09	1.25	1.58	2.28
15.00	1.00	1.03	1.11	1.28	1.66	2.48
17.50	1.00	1.03	1.12	1.31	1.73	2.68
20.00	1.00	1.03	1.12	1.34	1.81	2.87
25.00	1.00	1.04	1.14	1.39	1.95	3.25
30.00	1.01	1.04	1.16	1.43	2.08	3.62
40.00	1.01	1.05	1.19	1.52	2.33	4.34
50.00	1.01	1.06	1.21	1.60	2.57	5.05
60.00	1.01	1.06	1.23	1.68	2.80	5.75
80.00	1.01	1.07	1.28	1.82	3.25	7.13
100.00	1.01	1.08	1.31	1.95	3.69	8.51

SOURCE: *IRE Trans. PGMTT* (see Ref. 4 by L. Young)

Table 6.02-5

MAXIMUM VSWR FOR FOUR-SECTION TCHEBYSCHEFF QUARTER-WAVE TRANSFORMERS*

IMPEDANCE RATIO, R	BANDWIDTH, w_q					
	0.2	0.4	0.6	0.8	1.0	1.2
1.25	1.00	1.00	1.00	1.00	1.01	1.03
1.50	1.00	1.00	1.00	1.01	1.02	1.06
1.75	1.00	1.00	1.00	1.01	1.03	1.08
2.00	1.00	1.00	1.00	1.02	1.04	1.10
2.50	1.00	1.00	1.01	1.02	1.06	1.14
3.00	1.00	1.00	1.01	1.03	1.07	1.17
4.00	1.00	1.00	1.01	1.03	1.09	1.22
5.00	1.00	1.00	1.01	1.04	1.11	1.27
6.00	1.00	1.00	1.01	1.05	1.13	1.31
8.00	1.00	1.00	1.02	1.06	1.16	1.39
10.00	1.00	1.00	1.02	1.07	1.18	1.46
12.50	1.00	1.00	1.02	1.08	1.21	1.54
15.00	1.00	1.00	1.02	1.08	1.24	1.62
17.50	1.00	1.00	1.03	1.09	1.26	1.69
20.00	1.00	1.01	1.03	1.10	1.28	1.76
25.00	1.00	1.01	1.03	1.11	1.33	1.88
30.00	1.00	1.01	1.04	1.13	1.36	2.01
40.00	1.00	1.01	1.04	1.15	1.43	2.24
50.00	1.00	1.01	1.05	1.17	1.50	2.46
60.00	1.00	1.01	1.05	1.18	1.56	2.67
80.00	1.00	1.01	1.06	1.22	1.67	3.08
100.00	1.00	1.01	1.07	1.25	1.78	3.48

SOURCE: *IRE Trans. PGMTT* (see Ref. 4 by L. Young)

* Corresponding Tchebyscheff transformer designs are presented on pp. 275-278.

Hence, Eq. (6.02-16) gives

$$M(n, w_q) = T_n^2(1/\mu_0) = \frac{\mathcal{E}_a}{\mathcal{E}_r} = 0.501 \times 10^4 \qquad (6.02\text{-}21)$$

From Table 6.02-1, in the column $w_q = 1.0$, it is seen that this value of $M(n, w_q)$ falls between $n = 5$ and $n = 6$. Therefore, the transformer must have at least six sections. (See also Example 1 of Sec. 6.07).

SEC. 6.03, THE PERFORMANCE OF HOMOGENEOUS HALF-WAVE FILTERS

The half-wave filter was defined in Sec. 6.01. It is shown in Fig. 6.03-1. Its fractional bandwidth w_h is defined [compare Eq. (6.02-1)] by

$$w_h = 2\left(\frac{\lambda_{g1} - \lambda_{g2}}{\lambda_{g1} + \lambda_{g2}}\right) \qquad (6.03\text{-}1)$$

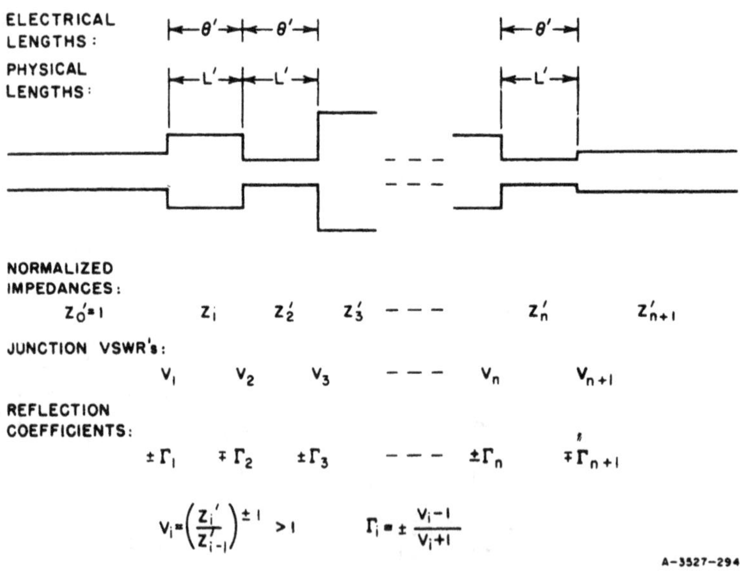

SOURCE: Quarterly Progress Report 4, Contract DA 36-039 SC-87398, SRI; reprinted in *IRE Trans. PGMTT* (See Ref. 36 by L. Young)

FIG. 6.03-1 HALF-WAVE FILTER NOTATION

and the length L' of each section [compare Eq. (6.02-2)] is

$$L' = \frac{\lambda_{g1}\lambda_{g2}}{\lambda_{g1} + \lambda_{g2}} = \frac{\lambda_{g0}}{2} \qquad (6.03.2)$$

where λ_{g1} and λ_{g2} are the longest and shortest wavelengths, respectively, in the pass band *of the half-wave filter*. This can be simplified for non-dispersive lines by dropping the suffix "g," as in Eqs. (6.02-3) and (6.02-4). A half-wave filter *with the same junction VSWRs* V_i (Figs. 6.02-1 and 6.03-1) as a quarter-wave transformer of bandwidth w_q has a bandwidth

$$w_h = \frac{w_q}{2} \qquad (6.03\text{-}3)$$

since its sections are twice as long and therefore twice as frequency-sensitive. The performance of a half-wave filter generally can be determined directly from the performance of the quarter-wave transformer with the same number of sections, n, and junction VSWRs V_i, by a linear scaling of the frequency axis by a scale-factor of 2. Compare Figs. 6.03-2 and 6.02-2. The quarter-wave transformer with the same n and V_i as the half-wave filter is herein called its *prototype* circuit.

In the case of the half-wave filter, R is the maximum VSWR, which is no longer the output-to-input impedance ratio, as for the quarter-wave transformer, but may generally be defined as the product of the junction VSWRs :

$$R = V_1 V_2 \ldots V_{n+1} \quad . \qquad (6.03\text{-}4)$$

This definition applies to both the quarter-wave transformer and the half-wave filter, as well as to filters whose prototype circuits they are. (In the latter case, the V_i are the individual discontinuity VSWRs as in Chapter 9.)

The equations corresponding to Eqs. (6.02-6) through (6.02-18) will now be restated, wherever they differ, for the half-wave filter.

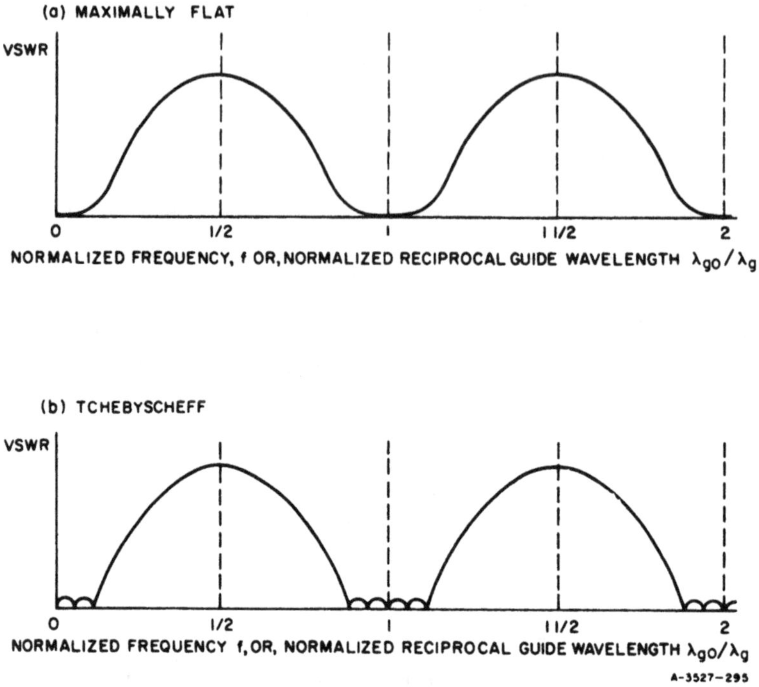

SOURCE: Quarterly Progress Report 4, Contract DA 36-039 SC-87398, SRI; reprinted in *IRE Trans. PGMTT* (See Ref. 36 by L. Young)

FIG. 6.03-2 HALF-WAVE FILTER CHARACTERISTICS

For the maximally flat half-wave filter of n sections,

$$\mathcal{E} = \frac{(R-1)^2}{4R} \sin^{2n} \theta' = \mathcal{E}_a \sin^{2n} \theta' \qquad (6.03\text{-}5)$$

where

$$\theta' = \pi \frac{\lambda_{g0}}{\lambda_g} = 2\theta \qquad (6.03\text{-}6)$$

instead of Eq. (6.02-7), so that $\theta' = \pi$ (instead of $\theta = \pi/2$) at band center. The 3-db bandwidth of the maximally flat half-wave filter is

$$w_{h,3\,db} = \frac{w_{q,3\,db}}{2} \qquad (6.03\text{-}7)$$

and the bandwidth between the points of x-db attenuation is

$$w_{h,x\,db} = \frac{w_{q,x\,db}}{2} \tag{6.03-8}$$

which can be obtained from Eqs. (6.02-9) and (6.02-10).

For the Tchebyscheff half-wave filter,

$$\left.\begin{aligned} \mathcal{E} &= \frac{(R-1)^2}{4R}\,\frac{T_n^2(\sin\theta'/\mu_0)}{T_n^2(1/\mu_0)} \\ &= \mathcal{E}_r T_n^2(\sin\theta'/\mu_0) \end{aligned}\right\} \tag{6.03-9}$$

where

$$\mu_0 = \sin\left(\frac{\pi w_q}{4}\right) = \sin\left(\frac{\pi w_h}{2}\right) \quad . \tag{6.03-10}$$

The quantities $\mathcal{E}_a$, $\mathcal{E}_r$, and the maximum transducer loss ratio are still given by Eqs. (6.02-8), (6.02-13), and (6.02-14). For maximally flat half-wave filters, the graph of Fig. 6.02-3 can again be used, but with the right-hand scale.

The lumped-constant, low-pass, prototype filter graphs in Figs. 4.03-2 and 4.03-4 to 4.03-10 may again be used for both the maximally flat and Tchebyscheff half-wave filters by substituting

$$\frac{\omega'}{\omega_1'} = \frac{\sin\theta'}{\mu_0} \tag{6.03-11}$$

for Eq. (6.02-15), where μ_0 is given by Eq. (6.03-10).

Equation (6.02-16) and Table 6.02-1 still apply, using Eq. (6.03-3) to convert between w_q and w_h.

Example 1—Find R for a half-wave filter of six sections having a Tchebyscheff fractional bandwidth of 60 percent with a pass-band ripple of 1 db.

Here, $w_h = 0.6$, or $w_q = 1.2$. From Eq. (6.02-13),

$$\text{antilog}\,(0.1) - 1 = \frac{(R-1)^2}{4R}\,\frac{1}{T_6^2(1/\mu_0)} \tag{6.03-12}$$

and from Table 6.02-1 for $w_q = 1.2$,

$$1.259 - 1 = \frac{(R-1)^2}{4R}\,\frac{1}{817} \ .$$

Hence, $R = 850$.

SEC. 6.04, EXACT TCHEBYSCHEFF AND MAXIMALLY FLAT SOLUTIONS FOR UP TO FOUR SECTIONS

Enough exact solutions will be presented to permit the solution of all intermediate cases by interpolation for Tchebyscheff and maximally flat transformers and filters having up to four sections.

The solutions were obtained from Collin's formulas.[2] With the notation of Fig. 6.02-1, they can be reduced to the expressions given below. The equations are first given for maximally flat transformers and then for Tchebyscheff transformers.

For maximally flat transformers with $n = 2$, 3, and 4:

$\underline{n = 2}$

$$\left.\begin{aligned} V_1 &= R^{1/4} \\ V_2 &= R^{1/2} \end{aligned}\right\} \tag{6.04-1}$$

$\underline{n = 3}$

$$\left.\begin{aligned} V_1^2 + 2R^{1/2}V_1 - \frac{2R^{1/2}}{V_1} - \frac{R}{V_1^2} &= 0 \\ V_2 &= R^{1/2}/V_1 \end{aligned}\right\} \tag{6.04-2}$$

$\underline{n = 4}$

$$\left.\begin{aligned} V_1 &= A_1 R^{1/8} \\ V_2 &= R^{1/4} \\ V_3 &= R^{1/4}/A_1^2 \end{aligned}\right\} \tag{6.04-3}$$

where

$$\left(\frac{1}{A_1^2} - A_1^2\right) = 2\left(\frac{R^{1/4} - 1}{R^{1/4} + 1}\right) . \qquad (6.04\text{-}3) \text{ (contd.)}$$

For Tchebyscheff Transformers with n = 2, 3, and 4:

$\underline{n = 2}$

$$V_1^2 = \sqrt{C^2 + R} + C$$

$$V_2 = R/V_1^2$$

where

$$C = \frac{(R - 1)\mu_0^2}{2(2 - \mu_0^2)} \qquad (6.04\text{-}4)$$

and μ_0 is given by Eq. (6.01-12).

$\underline{n = 3}$

$$V_1^2 + 2\sqrt{R}\, V_1 - \frac{2\sqrt{R}}{V_1} - \frac{R}{V_1^2} = \frac{3\mu_0^2(R - 1)}{4 - 3\mu_0^2}$$

$$V_2 = R^{1/2}/V_1 \qquad (6.04\text{-}5)$$

$\underline{n = 4}$

$$V_1 = \left\{R\left[B + \left(B^2 + \frac{A^2}{R}\right)^{1/2}\right]\right\}^{1/2}$$

$$V_2 = \frac{1}{A}$$

$$V_3 = \frac{A^2 R}{V_1^2} \qquad (6.04\text{-}6)$$

where

$$
\left.\begin{aligned}
A^2 &= \frac{1 - 1/R}{2t_1t_2} + \left[\frac{(1 - 1/R)^2}{4t_1^2t_2^2} + \frac{1}{R}\right]^{1/2} \\
B &= \frac{1}{2}\left(\frac{A}{A + 1}\right)^2 \left[(t_1 + t_2)\left(A^2 - \frac{1}{A^2R}\right) - 2A + \frac{2}{AR}\right] \\
&\text{and} \\
t_1 &= \frac{2\sqrt{2}}{(\sqrt{2} + 1)\mu_0^2} - 1 \\
t_2 &= \frac{2\sqrt{2}}{(\sqrt{2} - 1)\mu_0^2} - 1 \quad .
\end{aligned}\right\} \quad (6.04\text{-}6)\ (\text{contd.})
$$

A difference between typical quarter-wave transformers, and half-wave filters suitable for use as prototypes for microwave filters, is that, for the former, R is relatively small (usually less than 100) and only the pass-band performance is of interest; for the latter, R is relatively large, and the performance in both pass band and stop band is important. Two sets of tables are presented for $n = 2$, 3, and 4. The first set (Tables 6.04-1 to 6.04-4) cover R from 1 to 100. Since these tables are most likely to be used in the design of transformers, the impedances Z_1 and Z_2 (Fig. 6.02-1) are tabulated; the remaining impedances are obtained from the symmetry relation, which can be written (for any n)

$$Z_iZ_{n+1-i} = R \qquad (6.04\text{-}7)$$

(where the Z_i are normalized so that $Z_0 = 1$), or

$$V_i = V_{n+2-i} \qquad (6.04\text{-}8)$$

$$\Gamma_i = \pm\Gamma_{n+2-i} \quad . \qquad (6.04\text{-}9)$$

Table 6.04-1

Z_1 FOR TWO-SECTION TCHEBYSCHEFF QUARTER-WAVE TRANSFORMERS*

(For $w_q = 2.0$, $Z_1 = Z_2 = \sqrt{R}$)

IMPEDANCE RATIO, R	BANDWIDTH, w_q									
	0.0	0.2	0.4	0.6	0.8	1.0	1.2	1.4	1.6	1.8
1.00	1.00000	1.00000	1.00000	1.00000	1.00000	1.00000	1.00000	1.000	1.000	1.000
1.25	1.05737	1.05810	1.06034	1.06418	1.06979	1.07725	1.08650	1.096	1.107	1.115
1.50	1.10668	1.10808	1.11236	1.11973	1.13051	1.14495	1.16292	1.183	1.203	1.218
1.75	1.15016	1.15218	1.15837	1.16904	1.18469	1.20572	1.23199	1.261	1.291	1.314
2.00	1.18921	1.19181	1.19979	1.21360	1.23388	1.26122	1.29545	1.334	1.373	1.402
2.50	1.25743	1.26113	1.27247	1.29215	1.32117	1.36043	1.40979	1.466	1.522	1.564
3.00	1.31607	1.32079	1.33526	1.36042	1.39764	1.44816	1.51179	1.584	1.656	1.711
4.00	1.41421	1.42080	1.44105	1.47640	1.52892	1.60049	1.69074	1.793	1.894	1.971
5.00	1.49535	1.50366	1.52925	1.57405	1.64084	1.73205	1.84701	1.977	2.105	2.200
6.00	1.56508	1.57501	1.60563	1.65937	1.73970	1.84951	1.98768	2.143	2.295	2.407
8.00	1.68179	1.69473	1.73475	1.80527	1.91107	2.05579	2.23693	2.439	2.633	2.775
10.00	1.77828	1.79402	1.84281	1.92906	2.05879	2.23607	2.45663	2.700	2.931	3.100
12.50	1.88030	1.89934	1.95846	2.06334	2.22139	2.43686	2.70282	2.994	3.266	3.463
15.00	1.96799	1.99014	2.05909	2.18171	2.36672	2.61818	2.92611	3.259	3.568	3.791
17.50	2.04531	2.07045	2.14880	2.28850	2.49938	2.78500	3.13212	3.505	3.847	4.093
20.00	2.11474	2.14275	2.23019	2.38640	2.62224	2.94048	3.32447	3.733	4.107	4.374
25.00	2.23607	2.26955	2.37439	2.56229	2.84580	3.22539	3.67741	4.152	4.583	4.888
30.00	2.34035	2.37903	2.50046	2.71863	3.04734	3.48399	3.99798	4.533	5.013	5.353
40.00	2.51487	2.56334	2.71614	2.99167	3.40499	3.94578	4.57017	5.210	5.779	6.179
50.00	2.65915	2.71681	2.89921	3.22888	3.72073	4.35536	5.07697	5.808	6.454	6.907
60.00	2.78316	2.84956	3.06024	3.44157	4.00711	4.72769	5.53691	6.350	7.065	7.565
80.00	2.99070	3.07359	3.33788	3.81681	4.51833	5.39296	6.35680	7.314	8.150	8.733
100.00	3.16228	3.26067	3.57565	4.14625	4.97177	5.98279	7.08181	8.164	9.107	9.763

* Z_2 is given by

$$Z_2 = R/Z_1$$

The maximum pass-band VSWR's of these designs are tabulated on p. 267.

SOURCE: *IRE Trans. PGMTT* (see Ref. 4 by L. Young)

Table 6.04-2

Z_1 FOR THREE-SECTION TCHEBYSCHEFF QUARTER-WAVE TRANSFORMERS*

(For $w_q = 2.0$, $Z_1 = Z_2 = Z_3 = \sqrt{R}$)

IMPEDANCE RATIO, R	BANDWIDTH, w_q									
	0.0	0.2	0.4	0.6	0.8	1.0	1.2	1.4	1.6	1.8
1.00	1.00000	1.00000	1.00000	1.00000	1.00000	1.00000	1.00000	1.000	1.000	1.000
1.25	1.02829	1.02883	1.03051	1.03356	1.03839	1.04567	1.05636	1.071	1.091	1.109
1.50	1.05202	1.05303	1.05616	1.06186	1.07092	1.08465	1.10495	1.134	1.170	1.207
1.75	1.07255	1.07396	1.07839	1.08646	1.09933	1.11892	1.14805	1.189	1.243	1.298
2.00	1.09068	1.09247	1.09808	1.10830	1.12466	1.14966	1.18702	1.240	1.310	1.382
2.50	1.12177	1.12422	1.13192	1.14600	1.16862	1.20344	1.25594	1.332	1.434	1.535
3.00	1.14793	1.15096	1.16050	1.17799	1.20621	1.24988	1.31621	1.413	1.543	1.673
4.00	1.19071	1.19474	1.20746	1.23087	1.26891	1.32837	1.41972	1.556	1.736	1.917
5.00	1.22524	1.23013	1.24557	1.27412	1.32078	1.39428	1.50824	1.679	1.907	2.133
6.00	1.25439	1.26003	1.27790	1.31105	1.36551	1.45187	1.58676	1.790	2.060	2.329
8.00	1.30219	1.30916	1.33128	1.37253	1.44091	1.55057	1.72383	1.985	2.333	2.677
10.00	1.34089	1.34900	1.37482	1.42320	1.50397	1.63471	1.84304	2.159	2.577	2.984
12.50	1.38110	1.39048	1.42039	1.47674	1.57157	1.72651	1.97543	2.354	2.849	3.329
15.00	1.41512	1.42564	1.45924	1.52282	1.63055	1.80797	2.09480	2.532	3.098	3.640
17.50	1.44475	1.45630	1.49328	1.56355	1.68331	1.88193	2.20457	2.698	3.325	3.924
20.00	1.47108	1.48359	1.52371	1.60023	1.73135	1.95013	2.30687	2.848	3.541	4.191
25.00	1.51650	1.53075	1.57661	1.66464	1.81693	2.07364	2.49446	3.129	3.934	4.678
30.00	1.55498	1.57080	1.62184	1.72040	1.89229	2.18447	2.66499	3.384	4.288	5.124
40.00	1.61832	1.63691	1.69719	1.81471	2.02249	2.38028	2.97034	3.845	4.920	5.909
50.00	1.66978	1.69080	1.75924	1.89378	2.13434	2.55256	3.24219	4.249	5.480	6.600
60.00	1.71340	1.73661	1.81246	1.96266	2.23376	2.70860	3.49018	4.616	5.987	7.226
80.00	1.78522	1.81232	1.90144	2.08004	2.40750	2.98700	3.93524	5.286	6.896	8.338
100.00	1.84359	1.87411	1.97500	2.17928	2.55856	3.23420	4.33178	5.870	7.700	9.318

* Z_2 and Z_3 are given by

$$Z_2 = \sqrt{R}$$

$$Z_3 = R/Z_1$$

The maximum pass-band VSWR's of these designs are tabulated on p. 267.

SOURCE: *IRE Trans. PGMTT* (see Ref. 4 by L. Young)

Table 6.04-3

Z_1 FOR FOUR-SECTION TCHEBYSCHEFF QUARTER-WAVE TRANSFORMERS*

(For $w_q = 2.0$, $Z_1 = Z_2 = Z_3 = Z_4 = \sqrt{R}$)

IMPEDANCE RATIO, R	BANDWIDTH, w_q									
	0.0	0.2	0.4	0.6	0.8	1.0	1.2	1.4	1.6	1.8
1.00	1.00000	1.00000	1.00000	1.00000	1.00000	1.00000	1.00000	1.000	1.000	1.000
1.25	1.01405	1.01440	1.01553	1.01761	1.02106	1.02662	1.03560	1.050	1.073	1.102
1.50	1.02570	1.02635	1.02842	1.03227	1.03866	1.04898	1.06576	1.093	1.137	1.193
1.75	1.03568	1.03659	1.03949	1.04488	1.05385	1.06838	1.09214	1.131	1.194	1.277
2.00	1.04444	1.04558	1.04921	1.05598	1.06726	1.08559	1.11571	1.165	1.247	1.354
2.50	1.05933	1.06088	1.06577	1.07494	1.09026	1.11531	1.15681	1.226	1.342	1.495
3.00	1.07176	1.07364	1.07963	1.09086	1.10967	1.14059	1.19218	1.280	1.426	1.622
4.00	1.09190	1.09435	1.10216	1.11685	1.14159	1.18259	1.25182	1.371	1.574	1.847
5.00	1.10801	1.11093	1.12026	1.13784	1.16759	1.21721	1.30184	1.450	1.703	2.045
6.00	1.12153	1.12486	1.13549	1.15559	1.18974	1.24702	1.34555	1.520	1.820	2.225
8.00	1.14356	1.14758	1.16043	1.18482	1.22654	1.29722	1.42054	1.642	2.028	2.545
10.00	1.16129	1.16588	1.18060	1.20863	1.25683	1.33920	1.48458	1.749	2.213	2.828
12.50	1.17961	1.18483	1.20156	1.23353	1.28883	1.38421	1.55461	1.869	2.420	3.146
15.00	1.19506	1.20082	1.21931	1.25475	1.31638	1.42350	1.61690	1.977	2.609	3.433
17.50	1.20847	1.21471	1.23478	1.27335	1.34074	1.45869	1.67357	2.077	2.784	3.699
20.00	1.22035	1.22703	1.24854	1.28998	1.36269	1.49074	1.72593	2.170	2.948	3.946
25.00	1.24078	1.24824	1.27232	1.31891	1.40125	1.54791	1.82099	2.342	3.249	4.399
30.00	1.25803	1.26618	1.29251	1.34367	1.43467	1.59831	1.90654	2.498	3.524	4.809
40.00	1.28632	1.29564	1.32587	1.38498	1.49127	1.68552	2.05820	2.780	4.015	5.538
50.00	1.30920	1.31953	1.35308	1.41905	1.53879	1.76055	2.19214	3.031	4.451	6.182
60.00	1.32853	1.33974	1.37624	1.44833	1.58022	1.82732	2.31378	3.261	4.848	6.765
80.00	1.36025	1.37297	1.41455	1.49736	1.65091	1.94412	2.53156	3.674	5.556	7.801
100.00	1.38591	1.39992	1.44587	1.53798	1.71073	2.04579	2.72559	4.043	6.183	8.715

* See Footnote, Table 6.04-4

SOURCE: *IRE Trans. PGMTT* (see Ref. 4 by L. Young)

Table 6.04-4

Z_2 FOR FOUR-SECTION TCHEBYSCHEFF QUARTER-WAVE TRANSFORMERS*

(For $w_q = 2.0$, $Z_1 = Z_2 = Z_3 = Z_4 = \sqrt{R}$)

IMPEDANCE RATIO, R	BANDWIDTH, w_q									
	0.0	0.2	0.4	0.6	0.8	1.0	1.2	1.4	1.6	1.8
1.00	1.00000	1.00000	1.00000	1.00000	1.00000	1.00000	1.00000	1.000	1.000	1.000
1.25	1.07223	1.07260	1.07371	1.07559	1.07830	1.08195	1.08683	1.093	1.102	1.112
1.50	1.13512	1.13584	1.13799	1.14162	1.14685	1.15394	1.16342	1.176	1.193	1.214
1.75	1.19120	1.19224	1.19537	1.20065	1.20827	1.21861	1.23248	1.251	1.277	1.307
2.00	1.24206	1.24340	1.24745	1.25431	1.26420	1.27764	1.29572	1.320	1.354	1.393
2.50	1.33204	1.33396	1.33974	1.34954	1.36370	1.38300	1.40907	1.445	1.494	1.551
3.00	1.41051	1.41296	1.42036	1.43290	1.45105	1.47583	1.50943	1.556	1.620	1.694
4.00	1.54417	1.54760	1.55795	1.57553	1.60102	1.63596	1.68360	1.750	1.842	1.947
5.00	1.65686	1.66118	1.67423	1.69642	1.72864	1.77292	1.83358	1.918	2.037	2.170
6.00	1.75529	1.76043	1.77600	1.80248	1.84098	1.89401	1.96694	2.069	2.212	2.371
8.00	1.92323	1.92990	1.95009	1.98446	2.03453	2.10376	2.19954	2.335	2.524	2.730
10.00	2.06509	2.07315	2.09756	2.13915	2.19984	2.28397	2.40096	2.568	2.798	3.046
12.50	2.21803	2.22770	2.25698	2.30691	2.37988	2.48134	2.62317	2.826	3.105	3.399
15.00	2.35186	2.36303	2.39686	2.45455	2.53898	2.65667	2.82190	3.059	3.383	3.719
17.50	2.47169	2.48426	2.52237	2.58739	2.68264	2.81570	3.00321	3.273	3.639	4.014
20.00	2.58072	2.59463	2.63681	2.70880	2.81433	2.96208	3.17095	3.472	3.878	4.288
25.00	2.77447	2.79089	2.84069	2.92575	3.05065	3.22609	3.47548	3.836	4.315	4.789
30.00	2.94423	2.96299	3.01989	3.11712	3.26008	3.46148	3.74905	4.165	4.711	5.243
40.00	3.23492	3.25798	3.32792	3.44754	3.62377	3.87328	4.23198	4.750	5.415	6.049
50.00	3.48136	3.50835	3.59021	3.73029	3.93704	4.23091	4.65555	5.266	6.038	6.759
60.00	3.69752	3.72816	3.82111	3.98025	4.21547	4.55096	5.03760	5.734	6.601	7.401
80.00	4.06810	4.10544	4.21877	4.41293	4.70063	5.11329	5.71502	6.568	7.603	8.543
100.00	4.38263	4.42610	4.55802	4.78420	5.12003	5.60394	6.31175	7.304	8.487	9.548

* Z_1 is given in Table 6.04-3, Z_3 and Z_4 are given by

$$Z_3 = R/Z_2$$

$$Z_4 = R/Z_1$$

The maximum pass-band VSWR's of these designs are tabulated on 267.

SOURCE: *IRE Trans. PGMTT* (see Ref. 4 by L. Young)

The characteristic impedances, Z_i , are obtained from the junction VSWRs, V_i, using Fig. 6.02-1 for the quarter-wave transformer and Fig. 6.03-1 for the half-wave filter. It is convenient to normalize with respect to Z_0, and as a result, the values of Z_1, Z_2, ... given in the tables are for $Z_0 = 1$. The tables giving the Z_i all refer to quarter-wave transformers. To obtain the Z_i' of half-wave filters, obtain the V_i from Fig. 6.02-1, and use these V_i to obtain the Z_i' from Fig. 6.03-1. This gives the half-wave filter with the same attenuation characteristics as the quarter-wave transformer, but having a bandwidth $w_h = \frac{1}{2} w_q$. (Compare Figs. 6.02-2 and 6.03-2.)

The solutions of Eqs. (6.04-1) to (6.04-6) for larger values of R are presented in the second set of tables (Tables 6.04-5 to 6.04-8). They give the values of V_2 and V_3 for $n = 2$, 3, and 4. The remaining values of V are obtained from Eq. (6.04-8) and

$$V_1 V_2 \ldots V_{n+1} = R \qquad (6.04\text{-}10)$$

which, for even n, reduces to

$$(V_1 V_2 \ldots V_{n/2})^2 V_{(n/2)+1} = R \qquad (6.04\text{-}11)$$

and for odd n, reduces to

$$(V_1 V_2 \ldots V_{(n+1)/2})^2 = R \quad . \qquad (6.04\text{-}12)$$

Equations (6.04-7) to (6.04-12) hold for all values of n .

Tables 6.04-5 to 6.04-8 give the step VSWRs for R from 10 to ∞ in multiples of 10. Note that for Tchebyscheff transformers V_2, V_3, ... V_n and $V_1/(R)^{1/2} = V_{n+1}/(R)^{1/2}$ tend toward finite limits as R tends toward infinity, as can be seen from Eqs. 6.04-1 to 6.04-6 for n up to 4, by letting R tend toward infinity. (For limiting values as R tends toward infinity and $n > 4$, see Sec. 6.10.) The tables give fractional bandwidths, w_q, from 0 to 2.00 in steps of 0.20. [The greatest possible bandwidth is $w_q = 2.00$, by definition, as can be seen from Eq. (6.02-1).]

Table 6.04-5

V_2 FOR TWO-SECTION QUARTER-WAVE TRANSFORMERS

w_q \ Log R	0	1	2	3	4	5	6	7	8	9	10	∞
0	1.0	3.1622	10.0000	31.6227	100.0000	316.2278	*	*	*	*	*	∞
0.2	1.0	3.1070	9.4056	26.0349	55.6931	76.0577	80.2075	80.6742	80.7215	80.7263	80.7267	80.7267
0.4	1.0	2.9446	7.8214	15.2942	19.2101	19.8657	19.9363	19.9434	19.9441	19.9442	19.9443	19.9443
0.6	1.0	2.6872	5.8168	8.1357	8.6395	8.6971	8.7030	8.7036	8.7037	8.7037	8.7037	8.7037
0.8	1.0	2.3592	4.0455	4.6882	4.7783	4.7878	4.7887	4.7888	4.7889	4.7889	4.7889	4.7889
1.0	1.0	2.0000	2.7937	2.9763	2.9976	2.9997	3.0000	3.0000	3.0000	3.0000	3.0000	3.0000
1.2	1.0	1.6569	1.9939	2.0491	2.0550	2.0556	2.0557	2.0557	2.0557	2.0557	2.0557	2.0557
1.4	1.0	1.3708	1.5000	1.5172	1.5190	1.5192	1.5192	1.5192	1.5192	1.5192	1.5192	1.5192
1.6	1.0	1.1635	1.2055	1.2105	1.2110	1.2111	1.2111	1.2111	1.2111	1.2111	1.2111	1.2111
1.8	1.0	1.0405	1.0491	1.0500	1.0501	1.0502	1.0502	1.0502	1.0502	1.0502	1.0502	1.0502
2.0	1.0	1.0000	1.0000	1.0000	1.0000	1.0000	1.0000	1.0000	1.0000	1.0000	1.0000	1.0000

SOURCE: Quarterly Progress Report 4, Contract DA 36-039 SC-87398, SRI; reprinted in *IRE Trans. PGMTT* (see Ref. 36 by L. Young)

Table 6.04-6

V_2 FOR THREE-SECTION QUARTER-WAVE TRANSFORMERS

w_q \ Log R	0	1	2	3	4	5	6	7	8	9	10	∞
0	1.0	2.358	5.424	12.14	26.66	57.99	125.50	270.94	584.31	*	*	∞
0.2	1.0	2.344	5.337	11.68	24.46	47.69	79.82	101.7501	106.8067	107.3998	107.4610	107.4679
0.4	1.0	2.300	5.064	10.30	18.10	24.4154	26.1709	26.3912	26.4138	26.4161	26.4163	26.4164
0.6	1.0	2.222	4.5885	8.1080	10.6644	11.3276	11.4081	11.4163	11.4171	11.4172	11.4173	11.4173
0.8	1.0	2.103	3.9083	5.5671	6.1014	6.1728	6.1802	6.1809	6.1810	6.1810	6.1810	6.1810
1.0	1.0	1.9344	3.0919	3.6649	3.7630	3.7736	3.7747	3.7748	3.7748	3.7748	3.7748	3.7748
1.2	1.0	1.7158	2.3085	2.4686	2.4884	2.4904	2.4906	2.4907	2.4907	2.4907	2.4907	2.4907
1.4	1.0	1.4647	1.7022	1.7428	1.7472	1.7477	1.7477	1.7477	1.7477	1.7477	1.7477	1.7477
1.6	1.0	1.2269	1.2995	1.3089	1.3099	1.3100	1.3101	1.3101	1.3101	1.3101	1.3101	1.3101
1.8	1.0	1.0596	1.0731	1.0746	1.0747	1.0748	1.0748	1.0748	1.0748	1.0748	1.0748	1.0748
2.0	1.0	1.0000	1.0000	1.0000	1.0000	1.0000	1.0000	1.0000	1.0000	1.0000	1.0000	1.0000

SOURCE: Quarterly Progress Report 4, Contract DA 36-039 SC-87398, SRI; reprinted in *IRE Trans. PGMTT* (see Ref. 36 by L. Young)

Table 6.04-7

V_2 FOR FOUR-SECTION QUARTER-WAVE TRANSFORMERS

w_q \ Log R	0	1	2	3	4	5	6	7	8	9	10	∞
0	1.0	1.7782	3.1622	5.6234	10.0000	17.7827	31.6227	56.2341	100.0000	177.8279	316.2277	∞
0.2	1.0	1.7781	3.1616	5.6200	9.9808	17.6752	31.0220	52.9321	82.9105	106.5498	113.2186	114.1604
0.4	1.0	1.7766	3.1524	5.5677	9.6904	16.1089	23.4997	27.3837	28.0996	28.1787	28.1867	28.1877
0.6	1.0	1.7698	3.1107	5.3364	8.4944	11.2447	12.1344	12.2543	12.2667	12.2680	12.2681	12.2682
0.8	1.0	1.7503	2.9928	4.7320	6.1880	6.6331	6.6915	6.6975	6.6981	6.6982	6.6982	6.6982
1.0	1.0	1.7054	2.7392	3.7126	4.0665	4.1171	4.1225	4.1230	4.1230	4.1231	4.1231	4.1231
1.2	1.0	1.6172	2.3157	2.6617	2.7224	2.7290	2.7297	2.7298	2.7298	2.7298	2.7298	2.7298
1.4	1.0	1.4676	1.8065	1.8903	1.9004	1.9014	1.9016	1.9016	1.9016	1.9016	1.9016	1.9016
1.6	1.0	1.2645	1.3725	1.3886	1.3904	1.3905	1.3906	1.3906	1.3906	1.3906	1.3906	1.3906
1.8	1.0	1.0768	1.0956	1.0978	1.9080	1.0980	1.0980	1.0980	1.0980	1.0980	1.0980	1.0980
2.0	1.0	1.0000	1.0000	1.0000	1.0000	1.0000	1.0000	1.0000	1.0000	1.0000	1.0000	1.0000

SOURCE: Quarterly Progress Report 4, Contract DA 36-039 SC-87398, SRI; reprinted in *IRE Trans. PGMTT* (see Ref. 36 by L. Young)

Table 6.04-8

V_3 FOR FOUR-SECTION QUARTER-WAVE TRANSFORMERS

w_q \ Log R	0	1	2	3	4	5	6	7	8	9	10	∞
0	1.0	2.3448	5.2063	10.7833	21.1024	39.7366	73.0562	132.4190	238.0478	425.9232	760.0403	∞
0.2	1.0	2.3266	5.1045	10.3536	19.5265	34.4781	56.7950	85.7685	114.9996	132.4555	136.7239	137.7307
0.4	1.0	2.2728	4.8133	9.1941	15.6771	23.5245	30.2006	33.0054	33.4806	33.5323	33.5375	33.5382
0.6	1.0	2.1853	4.3689	7.6018	11.1871	13.5788	14.2420	14.3279	14.3367	14.3376	14.3377	14.3378
0.8	1.0	2.0664	3.8146	5.8599	7.2224	7.5865	7.6326	7.6374	7.6379	7.6380	7.6380	7.6380
1.0	1.0	1.9169	3.1842	4.1904	4.5121	4.5564	4.5610	4.5615	4.5615	4.5615	4.5615	4.5615
1.2	1.0	1.7347	2.5101	2.8558	2.9136	2.9199	2.9205	2.9206	2.9206	2.9206	2.9206	2.9206
1.4	1.0	1.5163	1.8739	1.9587	1.9688	1.9699	1.9700	1.9700	1.9700	1.9700	1.9700	1.9700
1.6	1.0	1.2767	1.3880	1.4045	1.4062	1.4064	1.4064	1.4064	1.4064	1.4064	1.4064	1.4064
1.8	1.0	1.0777	1.0967	1.0989	1.0991	1.0992	1.0992	1.0992	1.0992	1.0992	1.0992	1.0992
2.0	1.0	1.0000	1.0000	1.0000	1.0000	1.0000	1.0000	1.0000	1.0000	1.0000	1.0000	1.0000

SOURCE: Quarterly Progress Report 4, Contract DA 36-039 SC 87398, SRI; reprinted in *IRE Trans. PGMTT* (see Ref. 36 by L. Young)

When interpolating, it is generally sufficient to use only the two nearest values of V or Z. In that case, a linear interpolation on a log V or log Z against log R scale is preferable. Such interpolations, using only first differences, are most accurate for small R and for large R, and are least accurate in the neighborhood

$$R \sim \left(\frac{2}{w_q}\right)^{2(n-1)} \quad . \tag{6.04-13}$$

In this region, second- or higher-order differences may be used (or a graphical interpolation may be more convenient) to achieve greater accuracy.

Example 1—Design a quarter-wave transformer for $R = 2.5$, to have a VSWR less than 1.02 over a 20-percent bandwidth.

Here, $R = 2.5$ and $w_q = 0.2$. From Table 6.02-2, it can be seen that one section is not enough, but Table 6.02-3 indicates that two sections will do. From Table 6.04-1, we obtain $Z_1 = 1.261$, and from Eq. (6.04-7), $Z_2 = 1.982$.

Example 2—Find the step VSWRs V_1, V_2, V_3, and V_4 for a three-section quarter-wave transformer of 80-percent bandwidth and $R = 200$. Also, find the maximum pass-band VSWR.

Here, $n = 3$ and $w_q = 0.8$. For $R = 100$, from Table 6.04-6,

$$V_2 = 3.9083 \ ,$$

$$\therefore \log V_2 = 0.5920 \quad .$$

For $R = 1000$,

$$V_2 = 5.5671 \ ,$$

$$\therefore \log V_2 = 0.7456 \quad .$$

Now, for $R = 200$,

$$\log R = 2.301 \quad .$$

Interpolating linearly,

$$\log V_2 \approx 0.5920 + 0.301(0.7456 - 0.5920)$$

$$= 0.6382$$

$$\therefore \quad V_2 = 4.347 = V_3 \text{ also} \quad .$$

From Eq. (6.04-10) or (6.04-12),

$$(V_1 V_2)^2 = R$$

$$\therefore \quad V_1 = V_4 = 2.086 \quad .$$

The maximum pass-band VSWR, V_r, is found from Eqs. (6.02-8), (6.02-13), and Table 6.02-1, which give $\mathcal{E}_r = 0.23$, and then Eq. (6.02-18) determines the maximum pass-band VSWR, $V_r = 2.5$.

SEC. 6.05, EXACT MAXIMALLY FLAT SOLUTIONS FOR UP TO EIGHT SECTIONS

Enough exact solutions will be presented to permit the solution of all intermediate cases by interpolation, for maximally flat transformers with up to eight sections.

The solutions were obtained by Riblet's method.[3] This is a tedious procedure to carry out numerically; it requires high accuracy, especially for large values of R. In the limit as R becomes very large, approximate formulas adapted from the direct-coupled cavity filter point of view in Chapter 8 become quite accurate, and become exact in the limit, as R tends to infinity. This will be summarized in Sec. 6.09. For our present purposes, it is sufficient to point out that, for maximally flat transformers, the ratios

$$\left.\begin{aligned} A_1 &= A_{n+1} = V_1/R^{1/2n} \\ A_i &= V_i/R^{1/n}, \quad i \neq 1 \text{ or } n+1 \end{aligned}\right\} \qquad (6.05\text{-}1)$$

tend to finite limits as R tends to infinity (see Sec. 6.10).

Table 6.05-1 gives the impedances Z_1 to Z_4 (Fig. 6.02-1) of maximally flat quarter-wave transformers of 5, 6, 7, and 8 sections for values of R up to 100. The impedances of maximally flat transformers of 2, 3, and 4 sections were already given in Tables 6.04-1 to 6.04-4 (case of $w_q = 0$). The remaining impedances not given in these tables are determined from Eq. (6.04-7).

Table 6.05-2 gives the A_i defined in Eq. (6.05-1) for maximally flat transformers of from 3 to 8 sections for values of R from 1 to ∞ in multiples of 10. The A_i change relatively little over the infinite range of R, thus permitting very accurate interpolation. The V_i are then obtained from Eqs. (6.05-1), (6.04-8), and (6.04-10). The case $n = 2$ is not tabulated, since the formulas in Eq. (6.04-1) are so simple.

SEC. 6.06, APPROXIMATE DESIGN WHEN R IS SMALL

First-Order Theory—Exact numerical Tchebyscheff solutions for $n > 4$, corresponding to the maximally flat solutions up to $n = 8$ in Sec. 6.05 have not yet been computed. When the output-to-input impedance ratio, R, approaches unity, the reflection coefficients of the impedance steps approach zero, and a first-order theory is adequate. The first-order theory assumes that each discontinuity (impedance step) sets up a reflected wave of small amplitude, and that these reflected waves pass through the other small discontinuities without setting up further second-order reflections. This theory holds for "small R" as defined by

$$R < \left(\frac{2}{w_q}\right)^{n/2} \qquad (6.06\text{-}1)$$

and can be useful even when R approaches $(2/w_q)^n$, particularly for large bandwidths. [Compare with Eqs. (6.07-2) and (6.09-1).]

Table 6.05-1

IMPEDANCES OF MAXIMALLY FLAT TRANSFORMERS

R	$n = 5$		$n = 6$			$n = 7$			$n = 8$			
	Z_1	Z_2	Z_1	Z_2	Z_3	Z_1	Z_2	Z_3	Z_1	Z_2	Z_3	Z_4
1.5	1.01277	1.07904	1.00636	1.04540	1.14960	1.00318	1.02570	1.09628	1.00158	1.01438	1.06041	1.15872
2.0	1.02201	1.13908	1.01096	1.07904	1.26929	1.00547	1.04448	1.17039	1.00273	1.02481	1.10571	1.28658
2.5	1.02931	1.18816	1.01458	1.10608	1.37082	1.00727	1.05944	1.23157	1.00363	1.03307	1.14243	1.39558
3.0	1.03539	1.23002	1.01759	1.12884	1.45995	1.00878	1.07195	1.28415	1.00438	1.03997	1.17355	1.49162
3.5	1.04061	1.26672	1.02018	1.14861	1.53996	1.01007	1.08275	1.33055	1.00503	1.04590	1.20071	1.57813
4.0	1.04521	1.29954	1.02246	1.16613	1.61292	1.01121	1.09229	1.37227	1.00560	1.05114	1.22490	1.65722
4.5	1.04932	1.32931	1.02450	1.18191	1.68026	1.01223	1.10085	1.41030	1.00611	1.05583	1.24678	1.73039
5.0	1.05305	1.35663	1.02635	1.19631	1.74297	1.01315	1.10863	1.44534	1.00658	1.06009	1.26681	1.79870
6.0	1.05962	1.40549	1.02961	1.22186	1.85731	1.01479	1.12240	1.50837	1.00740	1.06762	1.30252	1.92356
7.0	1.06530	1.44845	1.03243	1.24413	1.96010	1.01620	1.13436	1.56414	1.00812	1.07414	1.33381	2.03617
8.0	1.07032	1.48696	1.03493	1.26395	2.05396	1.01746	1.14496	1.61440	1.00875	1.07992	1.36177	2.13926
9.0	1.07482	1.52196	1.03717	1.28186	2.14066	1.01859	1.15451	1.66032	1.00932	1.08513	1.38714	2.23474
10.0	1.07892	1.55413	1.03921	1.29822	2.22148	1.01962	1.16322	1.70270	1.00984	1.08987	1.41041	2.32393
15.0	1.09531	1.68600	1.04740	1.36450	2.56378	1.02375	1.19830	1.87818	1.01194	1.10895	1.50543	2.70350
20.0	1.10760	1.78804	1.05356	1.41497	2.84017	1.02688	1.22484	2.01581	1.01354	1.12335	1.57860	3.01198
25.0	1.11753	1.87251	1.05855	1.45628	3.07621	1.02942	1.24645	2.13089	1.01484	1.13507	1.63889	3.27666
30.0	1.12592	1.94524	1.06277	1.49152	3.28448	1.03158	1.26482	2.23080	1.01594	1.14502	1.69087	3.51111
35.0	1.13322	2.00950	1.06646	1.52243	3.47223	1.03347	1.28087	2.31965	1.01692	1.15371	1.73661	3.72308
40.0	1.13969	2.06729	1.06973	1.55006	3.64407	1.03515	1.29518	2.40004	1.01778	1.16146	1.77770	3.91762
45.0	1.14552	2.12000	1.07268	1.57510	3.80311	1.03667	1.30812	2.47372	1.01857	1.16845	1.81513	4.09813
50.0	1.15084	2.16856	1.07538	1.59807	3.95162	1.03805	1.31996	2.54192	1.01928	1.17485	1.84958	4.26701
60.0	1.16027	2.25588	1.08017	1.63911	4.22331	1.04052	1.34106	2.66530	1.02056	1.18624	1.91145	4.57684
70.0	1.16847	2.33312	1.08434	1.67513	4.46845	1.04268	1.35951	2.77519	1.02168	1.19620	1.96609	4.85724
80.0	1.17575	2.40267	1.08805	1.70736	4.69297	1.00460	1.37597	2.87473	1.02269	1.20507	2.01523	5.11474
90.0	1.18230	2.46613	1.09139	1.73661	4.90095	1.04634	1.39087	2.96605	1.02359	1.21310	2.06003	5.35379
100.0	1.18828	2.52464	1.09444	1.76343	5.09522	1.04793	1.40450	3.05064	1.02442	1.22043	2.10129	5.57761

SOURCE: Quarterly Progress Report 4, Contract DA 36-039 SC-87398, SRI; reprinted in *IRE Trans. PGMTT* (see Ref. 36 by L. Young)

Table 6.05-2

A_i OF MAXIMALLY FLAT TRANSFORMERS

$$\begin{pmatrix} A_1 = A_{n+1} = V_1/R^{1/2n} \\ A_i = V_i/R^{1/n}, \text{ when } i \neq 1,\ n+1 \end{pmatrix}$$

Log R	$n = 3$	$n = 4$*	$n = 5$		$n = 6$			$n = 7$			$n = 8$			
	A_1	A_1	A_1	A_2	A_1	A_2	A_3	A_1	A_2	A_3	A_1	A_2	A_3	A_4
0	1.0000	1.0000	1.0000	1.0000	1.0000	1.0000	1.0000	1.0000	1.0000	1.0000	1.0000	1.0000	1.0000	1.0000
1	0.9135	0.8708	0.8570	0.9088	0.8577	0.8510	1.1658	0.8649	0.8210	1.0534	0.8744	0.8093	0.9704	1.2355
2	0.8557	0.7793	0.7497	0.8458	0.7456	0.7478	1.3411	0.7541	0.6941	1.1250	0.7682	0.6699	0.9682	1.4926
3	0.8239	0.7221	0.6755	0.8084	0.6619	0.6837	1.5107	0.6664	0.6110	1.2147	0.6803	0.5736	0.9966	1.7432
4	0.8080	0.6883	0.6263	0.7873	0.6013	0.6451	1.6629	0.5987	0.5578	1.3131	0.6090	0.5084	1.0468	1.9665
5	0.8004	0.6689	0.5943	0.7753	0.5582	0.6217	1.7911	0.5473	0.5238	1.4103	0.5519	0.4645	1.1088	2.1524
6	0.7968	0.6579	0.5738	0.7684	0.5281	0.6073	1.8934	0.5087	0.5016	1.4992	0.5069	0.4348	1.1745	2.2997
7	0.7951	0.6516	0.5607	0.7643	0.5071	0.5983	1.9717	0.4801	0.4871	1.5760	0.4717	0.4144	1.2384	2.4125
8	0.7943	0.6481	0.5523	0.7618	0.4926	0.5924	2.0296	0.4590	0.4773	1.6394	0.4444	0.4003	1.2969	2.4976
9	--	0.6461	0.5471	0.7603	0.4827	0.5886	2.0716	0.4436	0.4707	1.6900	0.4234	0.3904	1.3481	2.5610
10	--	0.6450	0.5437	0.7594	0.4758	0.5861	2.1013	0.4324	0.4661	1.7293	0.4074	0.3834	1.3914	2.6078
11	--	--	0.5416	0.7588	0.4712	0.5845	2.1222	0.4242	0.4630	1.7593	0.3952	0.3784	1.4270	2.6423
12	--	--	0.5403	0.7584	0.4680	0.5833	2.1366	0.4183	0.4607	1.7817	0.3860	0.3747	1.4557	2.6681
∞	0.7937	0.6436	0.5380	0.7579	0.4612	0.5810	2.1684	0.4031	0.4553	1.8433	0.3578	0.3646	1.5538	2.7430

* For $n = 4$, $A_2 = 1.0000$

SOURCE: Quarterly Progress Report 4, Contract DA 36-039 SC-87398, SRI; reprinted in *IRE Trans. PGMTT* (see Ref. 36 by L. Young)

Denote the reflection coefficients of an n = section transformer or filter by

$$\Gamma_i \quad , \text{ where } i = 1, 2, \ldots, n + 1$$

to give a Tchebyscheff response of bandwidth, w_q. Let

$$c = \cos\left(\frac{\pi w_q}{4}\right) \tag{6.06-2}$$

The quantity c is related to μ_0 of Eq. (6.02-12) by

$$c^2 + \mu_0^2 = 1 \quad . \tag{6.06-3}$$

Then, for n-section Tchebyscheff transformers, the following ratio formulas relate the reflection coefficients up to n = 8.

For n = 2,

$$\Gamma_1:\Gamma_2 = 1:2c^2 \quad . \tag{6.06-4}$$

For n = 3,

$$\Gamma_1:\Gamma_2 = 1:3c^2 \quad . \tag{6.06-5}$$

For n = 4,

$$\Gamma_1:\Gamma_2:\Gamma_3 = 1:4c^2:2c^2(2 + c^2) \quad . \tag{6.06-6}$$

For n = 5,

$$\Gamma_1:\Gamma_2:\Gamma_3 = 1:5c^2:5c^2(1 + c^2) \quad . \tag{6.06-7}$$

For n = 6,

$$\Gamma_1:\Gamma_2:\Gamma_3:\Gamma_4 = 1:6c^2:3c^2(2 + 3c^2):2c^2(3 + 6c^2 + c^4) \quad . \tag{6.06-8}$$

For n = 7,

$$\Gamma_1:\Gamma_2:\Gamma_3:\Gamma_4 = 1:7c^2:7c^2(1 + 2c^2):7c^2(1 + 3c^2 + c^4) \quad . \tag{6.06-9}$$

For n = 8,

$$\Gamma_1:\Gamma_2:\Gamma_3:\Gamma_4:\Gamma_5 = 1:8c^2:4c^2(2 + 5c^2):8c^2(1 + 4c^2 + 2c^4): 2c^2(4 + 18c^2 + 12c^4 + c^6) \quad . \tag{6.06-10}$$

Table 6.06-1 tabulates the Γ_i/Γ_1 for all fractional bandwidths in steps of 20 percent in w_q, for transformers of up to eight sections. The Γs are obtained from the appropriate one of the above equations, or from Table 6.06-1, together with Eq. (6.04-9) and the specified value of R (see Example 1 of Sec. 6.06). When $w_q = 0$ (maximally flat case), the Γs reduce to the binomial coefficients. (A general formula for any n will be given below.)

Range of Validity of First-Order Theory—For a transformer of given bandwidth, as R increases from unity on up, the Γ_i all increase at the same rate according to the first-order theory, keeping the ratios Γ_i/Γ_1 constant. Eventually one of the Γ_i would exceed unity, resulting in a physically impossible situation, and showing that the first-order theory has been pushed too far. To extend the range of validity of the first-order theory, it has been found advantageous to substitute $\log V_i$ for Γ_i. This substitution,[17] which appears to be due to W. W. Hansen,[1] might be expected to work better, since, first, $\log V_i$ will do just as well as Γ_i when the Γ_i are small compared to unity, as then

$$\left.\begin{aligned} \log V_i &= \log \frac{1+\Gamma_i}{1-\Gamma_i} \\ &\approx \text{constant} \times \Gamma_i \end{aligned}\right\} \qquad (6.06\text{-}11)$$

and, second, $\log V_i$ can increase indefinitely with increasing $\log R$ and still be physically realizable.

The first-order theory generally gives good results in the pass band when $\log V_i$ is substituted for Γ_i, provided that R is "small" as defined by Eq. (6.06-1). (Compare end of Sec. 6.10.)

Example 1—Design a six-section quarter-wave transformer of 40-percent bandwidth for an impedance ratio of $R = 10$. [This transformer will have a VSWR less than 1.005 in the pass band, from Eqs. (6.02-8) and (6.02-18) and Table 6.02-1.]

Here $(2/w_q)^{n/2} = 125$, which is appreciably greater than $R = 10$. Therefore, we can proceed by the first-order theory. From Table 6.06-1,

Table 6.06-1

TABLE OF Γ_i/Γ_1

BAND-WIDTH, w_q	$n = 2$	$n = 3$	$n = 4$		$n = 5$		$n = 6$			$n = 7$			$n = 8$			
	$i = 2$	$i = 2$	$i = 2$	$i = 3$	$i = 2$	$i = 3$	$i = 2$	$i = 3$	$i = 4$	$i = 2$	$i = 3$	$i = 4$	$i = 2$	$i = 3$	$i = 4$	$i = 5$
0.0	2.0000	3.0000	4.0000	6.0000	5.0000	10.0000	6.0000	15.0000	20.0000	7.0000	21.0000	35.0000	8.0000	28.0000	56.0000	70.0000
0.2	1.9511	2.9266	3.9021	5.8054	4.8776	9.6359	5.8532	14.4181	19.1298	6.8287	20.1519	33.3120	7.8042	26.8373	53.1111	66.1559
0.4	1.8090	2.7135	3.6180	5.2543	4.5225	8.6132	5.4270	12.7903	16.7247	6.3316	17.7855	28.6925	7.2361	23.5988	45.2566	55.7879
0.6	1.5878	2.3817	3.1756	4.4361	3.9695	7.1208	4.7634	10.4357	13.3273	5.5572	14.3810	22.2954	6.3511	18.9564	34.5254	41.8439
0.8	1.3090	1.9635	2.6180	3.4748	3.2725	5.4144	3.9271	7.7825	9.6284	4.5816	10.5789	15.5402	5.2361	13.8037	23.4303	27.7539
1.0	1.0000	1.5000	2.0000	2.5000	2.5000	3.7500	3.0000	5.2500	6.2500	3.5000	7.0000	9.6250	4.0000	9.0000	14.0000	16.1250
1.2	0.6910	1.0365	1.3820	1.6207	1.7275	2.3243	2.0729	3.1472	3.5878	2.4184	4.0895	5.2138	2.7639	5.1512	7.2434	8.0793
1.4	0.4122	0.6183	0.8244	0.9094	1.0305	1.2429	1.2366	1.6190	1.7639	1.4428	2.0375	2.3961	1.6489	2.4985	3.1483	3.3919
1.6	0.1910	0.2865	0.3820	0.4002	0.4775	0.5231	0.5730	0.6550	0.6841	0.6684	0.7961	0.8660	0.7639	0.9463	1.0697	1.1133
1.8	0.0489	0.0734	0.0979	0.0991	0.1224	0.1254	0.1468	0.1522	0.1540	0.1713	0.1797	0.1840	0.1958	0.2078	0.2152	0.2177
2.0	0	0	0	0	0	0	0	0	0	0	0	0	0	0	0	0

SOURCE: Quarterly Progress Report 4, Contract DA 36-039 SC-87398, SRI; reprinted in *IRE Trans. PGMTT* (see Ref. 36 by L. Young)

$$\log V_1 : \log V_2 : \log V_3 : \log V_4 = 1:5.4270:12.7903:16.7247$$

$$\therefore \quad \frac{\log V_1}{\log R} = \frac{\log V_1}{\sum_{i=1}^{7} \log V_i} = \frac{1}{55.1593} = 0.01813 \quad .$$

Since $\log R = \log 10 = 1$,

$$\therefore V_1 = V_7 = \text{antilog } (0.01813) = 1.0426$$

$$V_2 = V_6 = \text{antilog } (5.4270 \times 0.01813) = 1.254$$

$$V_3 = V_5 = \text{antilog } (12.7903 \times 0.01813) = 1.705$$

and

$$V_4 = \text{antilog } (16.7247 \times 0.01813) = 2.010 \quad .$$

Hence

$$\begin{aligned} Z_1 &= V_1 = 1.0426 \\ Z_2 &= V_2 Z_1 = 1.308 \\ Z_3 &= V_3 Z_2 = 2.228 \\ Z_4 &= V_4 Z_3 = 4.485 \\ Z_5 &= V_5 Z_4 = 7.65 \\ Z_6 &= V_6 Z_5 = 9.60 \\ R = Z_7 &= V_7 Z_6 = 10.00 \end{aligned}$$

Relation to Dolph-Tchebyscheff Antenna Arrays—When R is small, numerical solutions of certain cases up to $n = 39$ may be obtained through the use of existing antenna tables. The first-order Tchebyscheff transformer problem is mathematically the same as Dolph's solution[18] of the linear array, and the correspondences shown in Table 6.06-2 may be set up.

Table 6.06-2

TRANSFORMER-ARRAY CORRESPONDENCES

TCHEBYSCHEFF TRANSFORMER	DOLPH-TCHEBYSCHEFF ARRAY
First-order theory	Optical diffraction theory
Synchronous tuning	Uniform phase (or linear phase taper)
Frequency	Angle in space
Transformer length	Array length
Pass band	Side-lobe region
Stop band	Main lobe
Reflection coefficient	Radiation field
Number of steps (n + 1)	Number of elements
$M(n, w_q)$	Side-lobe ratio
10 $\log_{10} M$	Side-lobe level in db
$\log V_i$	Element currents, I_i

SOURCE: Quarterly Progress Report 4, Contract DA 36-039 SC-87398, SRI; reprinted in *IRE Trans. PGMTT* (see Ref. 36 by L. Young)

The calulation of transformers from tables or graphs of array solutions is best illustrated by an example.

Example 2—Design a transformer of impedance ratio R = 5 to have a maximum VSWR, V_r, of less than 1.02 over a 140-percent bandwidth (w_q = 1.4).

It is first necessary to determine the minimum number of sections. This is easily done as in Example 1 of Sec. 6.02, using Table 6.02-1, and is determined to be n = 11.

Applying the test of Eq. (6.06-1)

$$\left(\frac{2}{w_q}\right)^{n/2} = 50$$

whereas R is only 5, and so we may expect the first-order theory to furnish an accurate design.

The most extensive tables of array solutions are contained in Ref. 19. (Some additional tables are given in Ref. 20.) We first work out M from Eqs. (6.02-8), (6.02-18), and (6.02-16), and find M = 8000. Hence the side-lobe level is

$$10 \log_{10} M = 39.0 \text{ db} \quad .$$

From Table II in Ref. 19, the currents of an $n + 1 = 12$ element array of side-lobe level 39 db are respectively proportional to 3.249, 6.894, 12.21, 18.00, 22.96, 25.82, 25.82, 22.96, 18.00, 12.21, 6.894, and 3.249. Their sum is 178.266. Since the currents are to be proportional to log V_i, and since $R = 5$, log $R = 0.69897$, we multiply these currents by $0.69897/178.266 = 0.003921$ to obtain the log V_i. Taking antilogarithms yields the V_i and, finally, multiplying yields the Z_i (as in Example 1). Thus Z_0 through R are respectively found to be 1.0, 1.0298, 1.09585, 1.2236, 1.4395, 1.7709, 2.2360, 2.8233, 3.4735, 4.0861, 4.5626, 4.8552, and 5.0000. The response of this transformer is plotted in Fig. 6.06-1, and is found to satisfy the specifications almost perfectly.

In antenna theory, one is usually not interested in side-lobe ratios in excess of 40 db; this is as far as the antenna tables take us. Only fairly large bandwidths can be calculated with this 40-db limit. For

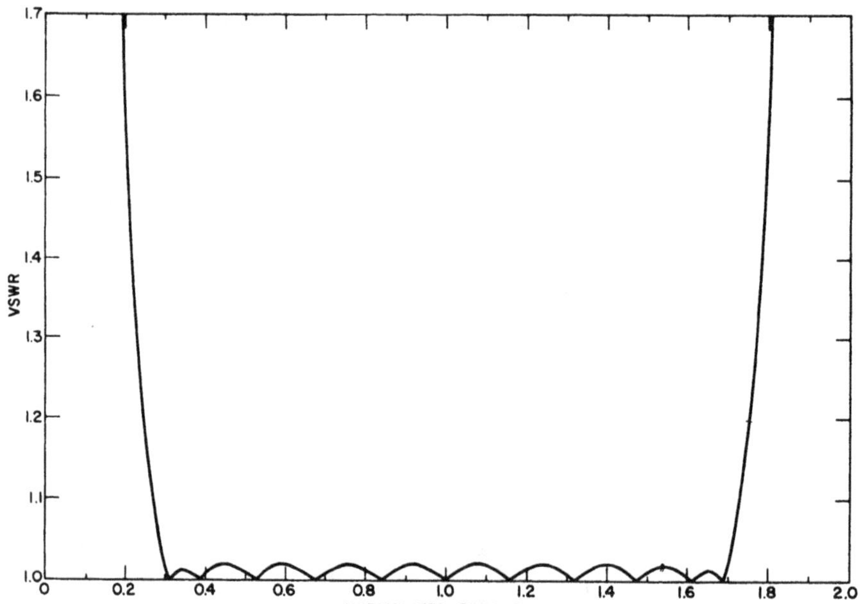

SOURCE: Quarterly Progress Report 4, Contract DA 36-039 SC-87398, SRI; reprinted in *IRE Trans. PGMTT* (See Ref. 36 by L. Young)

FIG. 6.06-1 ANALYZED PERFORMANCE OF TRANSFORMER DESIGNED IN EXAMPLE 2 OF SEC. 6.06

example, Table 6.02-1 shows that for $n = 2$ this limits us to $w_q > 0.18$; for $n = 4$, to $w_q > 0.67$; for $n = 8$, to $w_q > 1.21$; and for $n = 12$, to $w_q > 1.52$. A general formula for all cases has been given by G. J. Van der Maas,[21] which becomes, when adapted to the transformer,

$$\frac{\Gamma_i}{\Gamma_1} = \frac{n}{n+1-i} \sum_{r=0}^{i-2} \binom{n+1-i}{r+1} \binom{i-2}{r} c^{2(r+1)} \tag{6.06-12}$$

for $2 \le i \le (n/2) + 1$, where c is given by Eq. (6.06-2), and $\binom{a}{b}$ are the binomial coefficients

$$\binom{a}{b} = \frac{a!}{b!(a-b)!} \tag{6.06-13}$$

SEC. 6.07, APPROXIMATE DESIGN FOR UP TO MODERATELY LARGE R

Modified First-Order Theory—In Sec. 6.06 a first-order theory was presented which held for "small" values of R as defined by Eq. (6.06-1). In Sec. 6.09, there will be presented formulas that hold for "large" values of R as defined by Eq. (6.09-1). This leaves an intermediate region without explicit formulas. Since exact numerical solutions for maximally flat transformers of up to eight sections have been tabulated (Tables 6.05-1 and 6.05-2), these might be used in conjunction with either the "small R" or the "large R" theories to extend the one upward or the other downward in R, and so obtain more accurate solutions for Tchebyscheff transformers with R in this intermediate region. This idea is applied here to the first-order ("small R") theory only, as will be explained. It extends the range of the first-order theory from the upper limit given by Eq. (6.06-1) up to "moderately large" values of R as defined by

$$R < \left(\frac{2}{w_q}\right)^n \tag{6.07-1}$$

and gives acceptable results even up to the square of this limit,

$$R < \left(\frac{2}{w_q}\right)^{2n} \quad . \tag{6.07-2}$$

[Compare with Eqs. (6.06-1) and (6.09-1).] Of course, when R is less than specified by Eq. (6.06-1), there is no need to go beyond the simpler first-order theory of Sec. 6.06.

The first step in the proposed modification of the first-order theory is to form ratios of the Γ_i, which will be denoted by γ_i, with the property that

$$\left(\frac{\Gamma_i}{\sum_{i=1}^{n+1} \Gamma_i}\right)_{\substack{\text{Tchebyscheff}\\ \text{transformer}}} = \gamma_i \left(\frac{\Gamma_i}{\sum_{i=1}^{n+1} \Gamma_i}\right)_{\substack{\text{maximally flat}\\ \text{transformer}}} \tag{6.07-3a}$$

The γ_i are functions of n (the same n for both transformers) and w_q (the bandwidth of the desired Tchebyscheff transformer). The substitution of log V_i for Γ_i will again be used, and therefore $\sum_{i=1}^{n+1}\Gamma_i$ is replaced by log R, according to Eq. (6.04-10). If now we choose R to be the same for both the Tchebyscheff transformer and the corresponding maximally flat transformer, then Eq. (6.07-3) reduces to

$$(\log V_i)_{\substack{\text{Tchebyscheff}\\ \text{transformer}}} = \gamma_i (\log V_i)_{\substack{\text{maximally flat}\\ \text{transformer}}} \tag{6.07-3b}$$

The modification to the first-order theory now consists in using the *exact* log V_i of the maximally flat transformer where these are known (Tables 6.05-1 and 6.05-2). The γ_i could be obtained from Eq. (6.07-3) . and Table 6.06-1, but are tabulated for greater convenience in Table 6.07-1. The numbers in the first row of this table are, by definition, all unity. The application of this table is illustrated by an example given below.

Range of Validity of the Modified First-Order Theory—The analyzed performance of a first-order design, modified as explained above and to be illustrated in Example 1, agrees well with the predicted performance, provided that R satisfies Eq. (6.07-1) or at least Eq. (6.07-2). (In this regard, compare the end of Sec. 6.10.)

As a rough but useful guide, the first-order modification of the exact maximally flat design generally gives good results when the pass-band maximum VSWR is less than or equal to $(1 + w_q^2)$, where w_q is the equal-ripple quarter-wave transformer bandwidth [Eq. (6.02-1)]. By definition, it becomes exact when $w_q = 0$.

Example 1—In Example 1 of Sec. 6.02, it was shown that a quarter-wave transformer of impedance ratio $R = 100$, fractional bandwidth $w_q = 1.00$, and maximum pass-band VSWR of less than 1.15 must have at least six sections ($n = 6$). Calculate the normalized line impedances, Z_i, of this quarter-wave transformer. Predict the maximum pass-band VSWR, V_r. Then, also find the bandwidth, w_h, and normalized line impedances, Z_i', of the corresponding half-wave filter.

First, check that R is small enough for the transformer to be solved by a first-order theory. Using Eq. (6.06-1),

$$\left(\frac{2}{w_q}\right)^{n/2} = 2^3 = 8 \quad . \tag{6.07-4}$$

Therefore the unmodified first-order theory would not be expected to give good results, since $R = 100$ is considerably greater than 8. Using Eqs. (6.07-1) and (6.07-2),

$$\left.\begin{aligned} \left(\frac{2}{w_q}\right)^{n} &= 64 \\ \left(\frac{2}{w_q}\right)^{2n} &= 2048 \quad . \end{aligned}\right\} \tag{6.07-5}$$

Table 6.07-1

TABLE OF γ_i

BAND-WIDTH, w_q	$n = 2$		$n = 3$		$n = 4$			$n = 5$		
	$i = 1$	$i = 2$	$i = 1$	$i = 2$	$i = 1$	$i = 2$	$i = 3$	$i = 1$	$i = 2$	$i = 3$
0	1.00000	1.00000	1.00000	1.00000	1.00000	1.00000	1.00000	1.00000	1.00000	1.00000
0.2	1.01237	0.98762	1.01869	0.99376	1.02501	0.99992	0.99176	1.03135	1.00611	0.99380
0.4	1.05014	0.94985	1.07715	0.97428	1.10418	0.99873	0.96695	1.13188	1.02379	0.97491
0.6	1.11488	0.88511	1.18283	0.93905	1.25124	0.99336	0.92510	1.32337	1.05062	0.94234
0.8	1.20882	0.79117	1.34975	0.88341	1.49381	0.97770	0.86512	1.65171	1.08104	0.89430
1.0	1.33333	0.66666	1.60000	0.80000	1.88235	0.94117	0.78431	2.20689	1.10344	0.82758
1.2	1.48643	0.51356	1.96415	0.67861	2.50599	0.86581	0.67690	3.16718	1.09426	0.73614
1.4	1.65823	0.34176	2.47172	0.50942	3.51015	0.72344	0.53202	4.88788	1.00739	0.60751
1.6	1.82565	0.17434	3.10921	0.29692	5.05657	0.48290	0.33727	7.99760	0.76377	0.41835
1.8	1.95226	0.04773	3.72647	0.09117	6.97198	0.17063	0.11515	12.82256	0.31389	0.16079
2.0	2.0	0	4.0	0	8.0	0	0	16.0	0	0

Table 6.07-1 (*concluded*)

BAND-WIDTH, w_q	$n = 6$				$n = 7$				$n = 8$				
	$i = 1$	$i = 2$	$i = 3$	$i = 4$	$i = 1$	$i = 2$	$i = 3$	$i = 4$	$i = 1$	$i = 2$	$i = 3$	$i = 4$	$i = 5$
0	1.00000	1.00000	1.00000	1.00000	1.00000	1.00000	1.00000	1.00000	1.00000	1.00000	1.00000	1.00000	1.00000
0.2	1.03774	1.01235	0.99748	0.99258	1.04417	1.01861	1.00200	0.99381	1.05063	1.02492	1.00701	0.99643	0.99294
0.4	1.16027	1.04946	0.98935	0.97026	1.18937	1.07581	1.00731	0.97503	1.21921	1.10279	1.02757	0.98531	0.97167
0.6	1.39965	1.11118	0.97375	0.93268	1.48033	1.17521	1.01374	0.94298	1.56565	1.24295	1.05997	0.96526	0.93590
0.8	1.82608	1.19520	0.94743	0.87911	2.01888	1.32138	1.01702	0.89639	2.23202	1.46088	1.10036	0.93387	0.88496
1.0	2.58585	1.29292	0.90505	0.80808	3.02958	1.51479	1.00986	0.83313	3.54939	1.77469	1.14087	0.88734	0.81762
1.2	3.99301	1.37951	0.83778	0.71630	5.03077	1.73806	0.97968	0.74941	6.33721	2.18942	1.16586	0.81969	0.73143
1.4	6.75454	1.39211	0.72904	0.59571	9.30719	1.91834	0.90301	0.63717	12.81069	2.64044	1.14312	0.72021	0.62075
1.6	12.45111	1.18908	0.54369	0.42589	19.21633	1.83488	0.72848	0.47546	29.51655	2.81846	0.99755	0.56381	0.46943
1.8	23.25581	0.56899	0.23596	0.17906	41.69381	1.02030	0.35677	0.21919	74.08908	1.81333	0.54984	0.28471	0.23041
2.0	32.0	0	0	0	64.0	0	0	0	128.0	0	0	0	0

SOURCE: Quarterly Progress Report 4, Contract DA 36-039 SC-87398, SRI; reprinted in *IRE Trans. PGMTT* (see Ref. 36 by L. Young)

Therefore the modified first-order theory should work quite well, although we may expect noticeable but not excessive deviation from the desired performance since $R = 100$ is slightly greater than $(2/w_q)^n = 64$.

From Table 6.05-1 and Fig. 6.02-1, or from Table 6.05-2 and Eq. (6.05-1), it can be seen that a maximally flat transformer of six sections with $R = 100$ has

$$\left.\begin{aligned} V_1 &= V_7 = 1.094 \quad \therefore \log V_1 = 0.0391 \\ V_2 &= V_6 = 1.610 \quad \therefore \log V_2 = 0.2068 \\ V_3 &= V_5 = 2.892 \quad \therefore \log V_3 = 0.4612 \\ & V_4 = 3.851 \quad \therefore \log V_4 = 0.5856 \end{aligned}\right\} \quad (6.07\text{-}6)$$

The log VSWRs of the required 100-percent bandwidth transformer are now obtained, according to Eq. (6.07-3b), multiplying the log Vs in Eq. (6.07-6) by the appropriate values of γ in Table 6.07-1:

$$\left.\begin{aligned} \log V_1 &= 0.0391 \times 2.586 = 0.1011 \\ \log V_2 &= 0.2068 \times 1.293 = 0.2679 \\ \log V_3 &= 0.4612 \times 0.905 = 0.4170 \\ \log V_4 &= 0.5856 \times 0.808 = 0.4733 \end{aligned}\right\} \quad (6.07\text{-}7)$$

$$\left.\begin{aligned} \therefore V_1 &= V_7 = 1.262 \\ V_2 &= V_6 = 1.853 \\ V_3 &= V_5 = 2.612 \\ & V_4 = 2.974 \end{aligned}\right\} \quad (6.07\text{-}8)$$

Now this product $V_1 V_2 \ldots V_7$ equals 105.4, instead of 100. It is therefore necessary to scale the V_i slightly downward, so that their product reduces to exactly 100. The preferred procedure is to reduce V_1 and V_7 by a factor of $(100/105.4)^{1/12}$ while reducing $V_2, \ldots, V_6$ by a factor of $(100/105.4)^{1/6}$. [In general, if R' and R are respectively the trial and desired impedance ratios, then for an n-section transformer, the scaling factor is $(R/R')^{1/n}$ for V_2, V_3, ... V_n, and $(R/R')^{1/2n}$ for V_1 and V_{n+1}.] It can be shown [see Example 2 of Sec. 6.09 and Eq. (6.09-2)] that this type of scaling, where V_1 and V_{n+1} are scaled by the square root of the scaling factor for V_2, ..., V_n, has as its principal effect a slight increase in bandwidth while leaving the pass-band ripple almost unaffected. Since the approximate designs generally fall slightly short in bandwidth, while coming very close to, or even improving on, the specified pass-band ripple, this method of scaling is preferable. Subtracting 0.0038 from log V_1 and 0.0076 from the remaining log V_i in Eq. (6.07-7) gives the new V_i

$$\left.\begin{aligned} V_1 &= V_7 = 1.251 \\ V_2 &= V_6 = 1.821 \\ V_3 &= V_5 = 2.566 \\ V_4 &= 2.922 \end{aligned}\right\} \qquad (6.07\text{-}9)$$

and for the corresponding normalized line impedances of the quarter-wave transformer (Fig. 6.02-1),

$$\left.\begin{aligned} Z_0 & & &= 1.0 \\ Z_1 &= V_1 &&= 1.251 \\ Z_2 &= Z_1V_2 &&= 2.280 \\ Z_3 &= Z_2V_3 &&= 5.850 \\ Z_4 &= Z_3V_4 &&= 17.10 \\ Z_5 &= Z_4V_5 &&= 43.91 \\ Z_6 &= Z_5V_6 &&= 79.94 \\ R &= Z_6V_7 &&= 100.00 \end{aligned}\right\} \qquad (6.07\text{-}10)$$

We note in passing that the product of the VSWRs before reduction was 105.4 instead of the specified 100. If the discrepancy between these two numbers exceeds about 5 to 10 percent, the predicted performance will usually not be realized very closely. This provides an additional internal check on the accuracy of the design.

The maximum transducer attenuation and VSWR in the pass band predicted from Eq. (6.02-16) and Table 6.02-1 are

$$\left.\begin{aligned} \mathcal{E}_r &= 0.0025 \text{ , or } 0.011 \text{ db} \quad . \\ &\text{Therefore by Eq. (6.02-18) ,} \\ V_r &= 1.106 \quad . \end{aligned}\right\} \qquad (6.07\text{-}11)$$

The computed plot of V against normalized frequency, f, of this transformer (or against λ_{g0}/λ_g if the transformer is dispersive) is shown in Fig. 6.07-1. The bandwidth is 95 percent (compared to 100 percent predicted) for a maximum pass-band VSWR of 1.11.

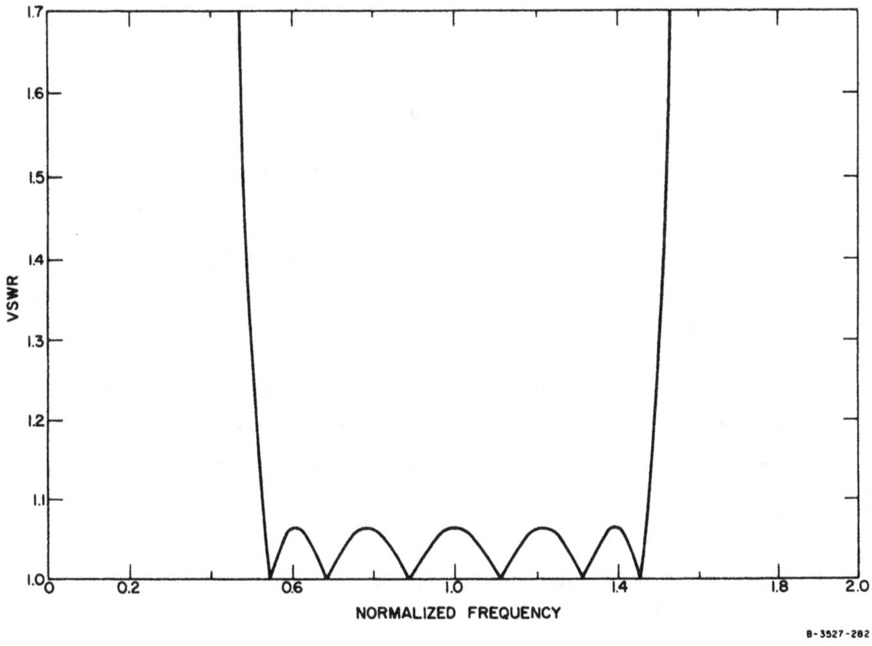

SOURCE: Quarterly Progress Report 4, Contract DA 36-039 SC-87398, SRI; reprinted in *IRE Trans. PGMTT* (See Ref. 36 by L. Young)

FIG. 6.07-1 ANALYZED PERFORMANCE OF TRANSFORMER DESIGNED IN EXAMPLE 1 OF SEC. 6.07

(Notice that the response has equal ripple heights with a maximum VSWR of 1.065 over an 86-percent bandwidth.)

The bandwidth w_h of the half-wave filter for a maximum VSWR of 1.11 will be just half the corresponding bandwidth of the quarter-wave transformer, namely, 47.5 percent (instead of the desired 50 percent). The normalized line impedances of the half-wave filter are (see Fig. 6.03-1):

$$\left.\begin{aligned}
Z'_0 & & &= 1.0 \quad \text{(input)} \\
Z'_1 &= V_1 & &= 1.251 \\
Z'_2 &= Z'_1/V_2 & &= 0.6865 \\
Z'_3 &= Z'_2\, V_3 & &= 1.764 \\
Z'_4 &= Z'_3/V_4 & &= 0.604 \\
Z'_5 &= Z'_4\, V_5 & &= 1.550 \\
Z'_6 &= Z'_5/V_6 & &= 0.850 \\
Z'_7 &= Z'_6\, V_7 & &= 1.065 \quad \text{(output)} \quad .
\end{aligned}\right\} \qquad (6.07\text{-}12)$$

It should be noticed that the output impedance, Z'_7, of the half-wave filter is also the VSWR of the filter or transformer at center frequency[9] (Fig. 6.07-1).

In this example it was not necessary to interpolate from the tables for the V_i or Z_i. When R is not given exactly in the tables, the interpolation procedure explained at the end of Sec. 6.04 should be followed.

SEC 6.08, CORRECTION FOR SMALL-STEP DISCONTINUITY CAPACITANCES[37]

A discontinuity in waveguide or coaxial-line cross-section cannot be represented by a change of impedance only--*i.e.*, practical junctions are non-ideal (see Sec. 6.01). The equivalent circuit for a small change in inner or outer diameter of a coaxial line can be represented by an ideal junction shunted by a capacitance,[22] and the same representation is possible for an E-plane step in rectangular waveguide.[23] This shunt capacitance has only a second-order effect on the magnitude of the junction VSWR, since it contributes a smaller component in quadrature with the (already small) reflection coefficient of the step. Its main effect is to move the reference planes with real Γ out of the plane of the junction. Since the spacing between adjacent and facing reference planes should be one-quarter wavelength at center frequency, the physical junctions should be moved the necessary amount to accomplish this. Formulas have been given by Cohn.[1] The procedure outlined here is equivalent to Cohn's formulas, but is in pictorial form, showing the displaced reference planes, and should make the numerical working of a problem a little easier. The necessary formulas are summarized in Fig. 6.08-1 which shows the new reference plane positions. The low-impedance end is shown on the left, the high-impedance end on the right. There are two reference planes with real Γ associated with each junction, one seen from the low-impedance side, and one seen from the high-impedance side (Fig. 6.08-1). When the two "terminal-pairs" of a junction are situated in the appropriate reference planes, it is equivalent to an ideal junction. The following results can be shown to hold generally when the step discontinuity can be represented by a shunt capacitance:

(1) The two reference planes associated with any junction are both in the higher impedance line (to the right of the junction in Fig. 6.08-1).

(2) The two reference planes associated with any junction are always in the order shown in Fig. 6.08-1--*i.e.*, the reference plane seen from the higher impedance line is nearer to the junction.

(3) As the step vanishes, both reference planes fall into the plane of the junction.

(4) The reference plane seen from the higher impedance line (the one nearer to the junction) is always within one-eighth of a wavelength of the junction. (The other reference plane is not so restricted.)

The spacing between junctions is then determined as shown in Fig. 6.08-1. It is seen that the 90-degree lengths overlap, and that the separation between junctions will therefore generally be *less* than one-quarter wavelength, although this does not necessarily always hold (*e.g.*, if $x_1 > X_2$).

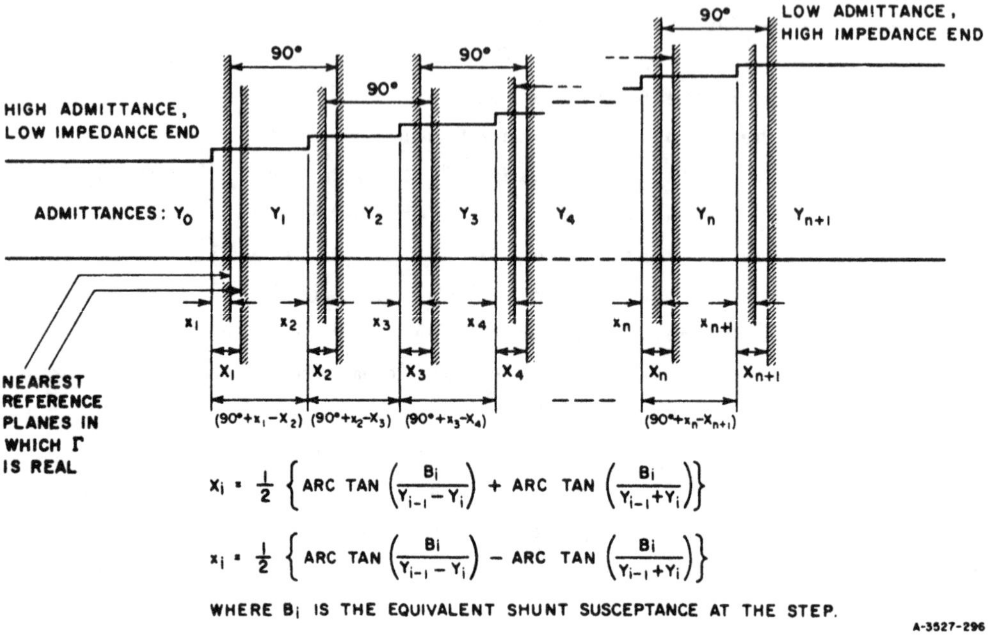

FIG. 6.08-1 LENGTH CORRECTIONS FOR DISCONTINUITY CAPACITANCES

Example 1—Design a transformer from 6.5- by 1.3-inch rectangular waveguide to 6.5- by 3.25-inch rectangular waveguide to have a VSWR less than 1.03 from at least 1180 to 1430 megacycles.

Here $R = 2.5$

$$\lambda_{g1} = 15.66 \text{ inches} , \lambda_{g2} = 10.68 \text{ inches} .$$

From Eq. (6.02-2),

$$\lambda_{g0} = 12.68 \text{ inches} , \text{ and } \frac{\lambda_{g0}}{4} = 3.17 \text{ inches} ,$$

while Eq. (6.02-1) gives $w_q = 0.38$. From Tables 6.02-3 and 6.02-4, it can be seen that at least three sections are needed. We shall select $w_q = 0.50$, which still meets the specification that the pass-band VSWR be less than 1.03 (see Table 6.02-4). From Table 6.04-2, the b dimensions of such a transformer are

$$b_0 = 1.300 \text{ inches}$$

$$b_1 = 1.479 \text{ inches}$$

$$b_2 = 2.057 \text{ inches}$$

$$b_3 = 2.857 \text{ inches}$$

$$b_4 = 3.250 \text{ inches}$$

Make all the steps symmetrical (as in Fig. 6.08-2), since in this case the length corrections would be appreciable if the steps were unsymmetrical.

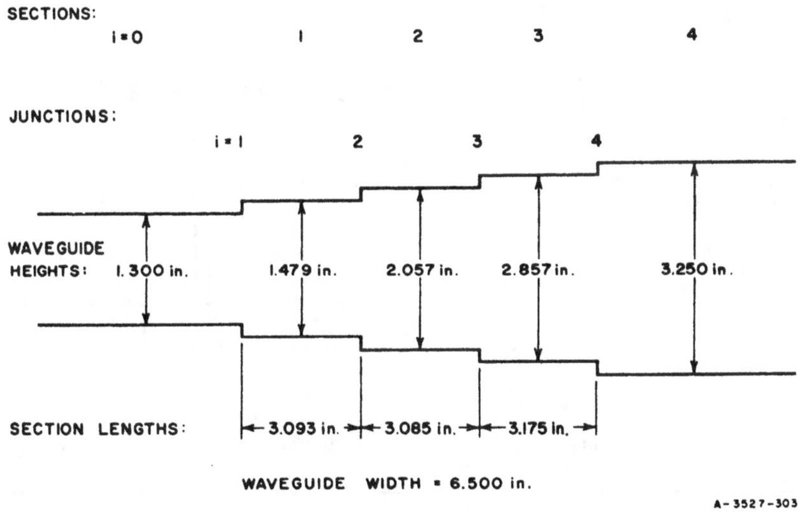

FIG. 6.08-2 SOLUTION TO EXAMPLE 1 OF SEC. 6.08 ILLUSTRATING LENGTH CORRECTIONS FOR DISCONTINUITY CAPACITANCES

Now make up a table as follows:

LINE	QUANTITY	SECTION OR JUNCTION NO. (see Fig. 6.08-2) 1	2	3	4
1.	b_i/λ_{g0}	0.117	0.162	0.225	0.256
2.	$b_{i-1}/b_i = Y_i/Y_{i-1}$	0.88	0.72	0.72	0.88
3.	$\left(\dfrac{B_i}{Y_i}\dfrac{\lambda_{g0}}{b_i}\right)$ (from Figs. 5.07-10 and -11)	0.06	0.26	0.26	0.06
4.	B_i/Y_{i-1} = Line 1 x Line 2 x Line 3	0.0062	0.0303	0.0421	0.0135
5.	$\dfrac{B_i/Y_{i-1}}{1 - Y_i/Y_{i-1}} = \dfrac{B_i}{Y_{i-1} - Y_i}$	0.052	0.108	0.150	0.113
6.	$\dfrac{B_i/Y_{i-1}}{1 + Y_i/Y_{i-1}} = \dfrac{B_i}{Y_{i-1} + Y_i}$	0.0033	0.0176	0.0245	0.0072
7.	X_i electrical degrees (from Fig. 6.08-1)	1.59	3.60	5.00	3.45
8.	x_i electrical degrees (from Fig. 6.08-1)	1.40	2.59	3.60	3.03
9.	$(X_{i+1} - x_i)$ electrical degrees	2.20	2.41	−0.15	
10.	$\dfrac{\lambda_{g0}}{360}(X_{i+1} - x_i)$ inches	0.077	0.085	−0.005	

The last line subtracted from 3.17 inches gives the section lengths. The first two sections are somewhat shorter than one-quarter wavelength, while the third section is slightly longer. The final dimensions are shown in Fig. 6.08-2.

SEC. 6.09, APPROXIMATE DESIGN WHEN R IS LARGE

Theory—Riblet's procedure,[3] while mathematically elegant and although it holds for all values of R, is computationally very tedious, and the accuracy required for large R can lead to difficulties even with a large digital computer. Collin's formulas[2] are more convenient (Sec. 6.04) but do not go beyond $n = 4$ (Tables 6.04-1 to 6.04-8). Riblet's procedure has been used to tabulate maximally flat transformers up to $n = 8$ (Tables 6.06-1 and 6.06-2). General solutions applicable only to "small R" have been given in Secs. 6.06 and 6.07, and are tabulated in Tables 6.06-1 and 6.07-1. In this part, convenient formulas will be given which become exact only when R is "large," as defined by

$$R \gg \left(\frac{2}{w_q}\right)^n \quad . \tag{6.09-1}$$

These solutions are suitable for most practical *filter* applications (but not for practical *transformer* applications). [Compare with Eqs. (6.06-1) and (6.07-2).]

For "large R" (or small w_q), stepped impedance transformers and filters may be designed from low-pass, lumped-constant, prototype filters (Chapter 4) whose elements are denoted by g_i ($i = 0, 1, \ldots, n+1$).* The transformer or filter step VSWRs are obtained from

$$\left.\begin{aligned} V_1 &= V_{n+1} = \frac{4}{\pi}\,\frac{g_0 g_1 \omega_1'}{w_q} \\ V_i &= \frac{16}{\pi^2}\,\frac{\omega_1'^2}{w_q^2}\,g_{i-1}g_i \ , \quad \text{when } 2 \le i \le n \\ &(V_i \text{ large, } w_q \text{ small}) \end{aligned}\right\} \tag{6.09-2}$$

where ω_1' is the radian cutoff frequency of the low-pass prototype and w_q is the quarter-wave transformer fractional bandwidth [given by Eq. (6.02-1) for Tchebyscheff transformers and Eqs. (6.02-9) or (6.02-10) for maximally flat transformers]. Again, the half-wave filter bandwidth, w_h, is equal to one-half w_q [Eq. (6.03-3)].

* Note: Here it is assumed that in the prototypes defined in Fig. 4.04-1, the circuit is symmetric or antimetric (see Sec. 4.05).

The V_i and Γ_i are symmetrical about the center in the sense of Eqs. (6.04-8) and (6.04-9), when the prototype is symmetrical or antimetrical as was assumed.

With Tables 4.05-1, 4.05-2, 4.06-1, 4.06-2, and 4.07-1, it is easy to use Eq. (6.09-1). One should, however, always verify that the approximations are valid, and this is explained next. Procedures to be used in borderline cases, and the accuracy to be expected, will be illustrated by examples.

Range of Validity—The criteria given in Eqs. (6.06-1) and (6.07-1) are reversed. The validity of the design formulas given in this part depends on R being large enough. It is found that the analyzed performance agrees well with the predicted performance (after adjusting R, if necessary, as in Examples 2 and 3 of this section) provided that Eq. (6.09-1) is satisfied; R should exceed $(2/w_q)^n$ by preferably a factor of about 10 or 100 or more. (Compare end of Sec. 6.10.) The ranges of validity for "small R" and "large R" overlap in the region between Eqs. (6.07-2) and (6.09-1), where both procedures hold only indifferently well. (See Example 3 of this section.)

For the maximally flat transformer, Eq. (6.09-1) still applies fairly well, when $w_{q,3db}$ is substituted for w_q.

As a rough but useful guide, the formulas of this section generally result in the predicted performance in the pass band when the pass-band maximum VSWR exceeds about $(1 + w_q^2)$. This rule must be considered indeterminate for the maximally flat case ($w_q = 0$), when the following rough generalization may be substituted: The formulas given in this section for maximally flat transformers or filters generally result in the predicted performance when the maximally flat quarter-wave transformer 3-db fractional bandwidth, $w_{q,3db}$, is less than about 0.40.*
The half-wave filter fractional bandwidth, $w_{h,3db}$, must, of course, be less than half of this, or 0.20.

After the filter has been designed, a good way to check on whether it is likely to perform as predicted is to multiply all the VSWRs, $V_1 V_2 \ldots V_{n+1}$, and to compare this product with R derived from the performance specifications using Table 6.02-1 and Eq. (6.02-13). If they

* Larger 3-db fractional bandwidths can be designed accurately for small n, for example up to about $w_{q,3db} = 0.60$ for $n = 2$.

agree within a factor of about 2, then after scaling each V so that their VSWR product finally equals R, good agreement with the desired performance may be expected.

Three examples will be worked out, illustrating a narrow-band and a wide-band design, and one case where Eq. (6.09-1) is no longer satisfied.

Example 1—Design a half-wave filter of 10-percent fractional bandwidth with a VSWR ripple of 1.10, and with at least 30-db attenuation 10 percent from center frequency.

Here $w_h = 0.1$, $\therefore w_q = 0.2$. A VSWR of 1.10 corresponds to an insertion loss of 0.01 db. From Eqs. (6.03-12) and (6.03-10), or (6.02-17) and (6.02-12),

$$\mu_0 = \sin\frac{\pi w_h}{2} = \sin 9^\circ = 0.1564 \quad .$$

At 10 percent from center frequency, by Eq. (6.03-11),

$$\frac{\omega'}{\omega_1'} = \frac{\sin\theta'}{\mu_0} = \frac{\sin 172^\circ}{0.1564} = 1.975 \quad .$$

From Fig. 4.03-4, a 5-section filter would give only 24.5 db at a frequency 10 percent from band center, but a six-section filter will give 35.5 db. Therefore, we must choose $n = 6$ to give at least 30-db attenuation 10 percent from center frequency.

The output-to-input impedance ratio of a six-section quarter-wave transformer of 20-percent fractional bandwidth and 0.01-db ripple is given by Table 6.02-1 and Eq. (6.02-13) and yields (with $\varepsilon_r = 0.0023$ corresponding to 0.01-db ripple)

$$R = 4.08 \times 10^{10} \quad . \tag{6.09-3}$$

Thus R exceeds $(2/w_q)^n$ by a factor of 4×10^4, which by Eq. (6.09-1) is ample, so that we can proceed with the design.

From Table 4.05-2(a), for n = 6 and 0.01-db ripple (corresponding to a maximum VSWR of 1.10), and from Eq. (6.09-2)

$$\left.\begin{aligned} V_1 &= V_7 = 4.98 \\ V_2 &= V_6 = 43.0 \\ V_3 &= V_5 = 92.8 \\ & \quad V_4 = 105.0 \end{aligned}\right\} \quad (6.09\text{-}4)$$

This yielded the response curve shown in Fig. 6.09-1, which is very close to the design specification in both the pass and stop bands. The half-wave filter line impedances are

$$\left.\begin{aligned} Z'_0 & & &= 1.0 \quad \text{(input)} \\ Z'_1 &= V_1 &&= 4.98 \\ Z'_2 &= Z'_1/V_2 &&= 0.1158 \\ Z'_3 &= Z'_2\,V_3 &&= 10.74 \\ Z'_4 &= Z'_3/V_4 &&= 0.1023 \\ Z'_5 &= Z'_4\,V_5 &&= 9.50 \\ Z'_6 &= Z'_5/V_6 &&= 0.221 \\ Z'_7 &= Z'_6\,V_7 &&= 1.10 \quad \text{(output)} \quad . \end{aligned}\right\} \quad (6.09\text{-}5)$$

Note that Z'_7 = 1.10 is also the VSWR at center frequency (Fig. 6.09-1).

The corresponding quarter-wave transformer has a fractional bandwidth of 20 percent; its line impedances are

$$\left.\begin{aligned} Z_0 & & &= 1.0 \quad \text{(input)} \\ Z_1 &= V_1 &&= 4.98 \\ Z_2 &= Z_1 V_2 &&= 2.14 \times 10^2 \\ Z_3 &= Z_2 V_3 &&= 1.987 \times 10^4 \\ Z_4 &= Z_3 V_4 &&= 2.084 \times 10^6 \\ Z_5 &= Z_4 V_5 &&= 1.9315 \times 10^9 \\ Z_6 &= Z_5 V_6 &&= 8.30 \times 10^9 \\ R = Z_7 &= Z_6 V_7 &&= 4.135 \times 10^{10} \quad \text{(output)} \quad . \end{aligned}\right\} \quad (6.09\text{-}6)$$

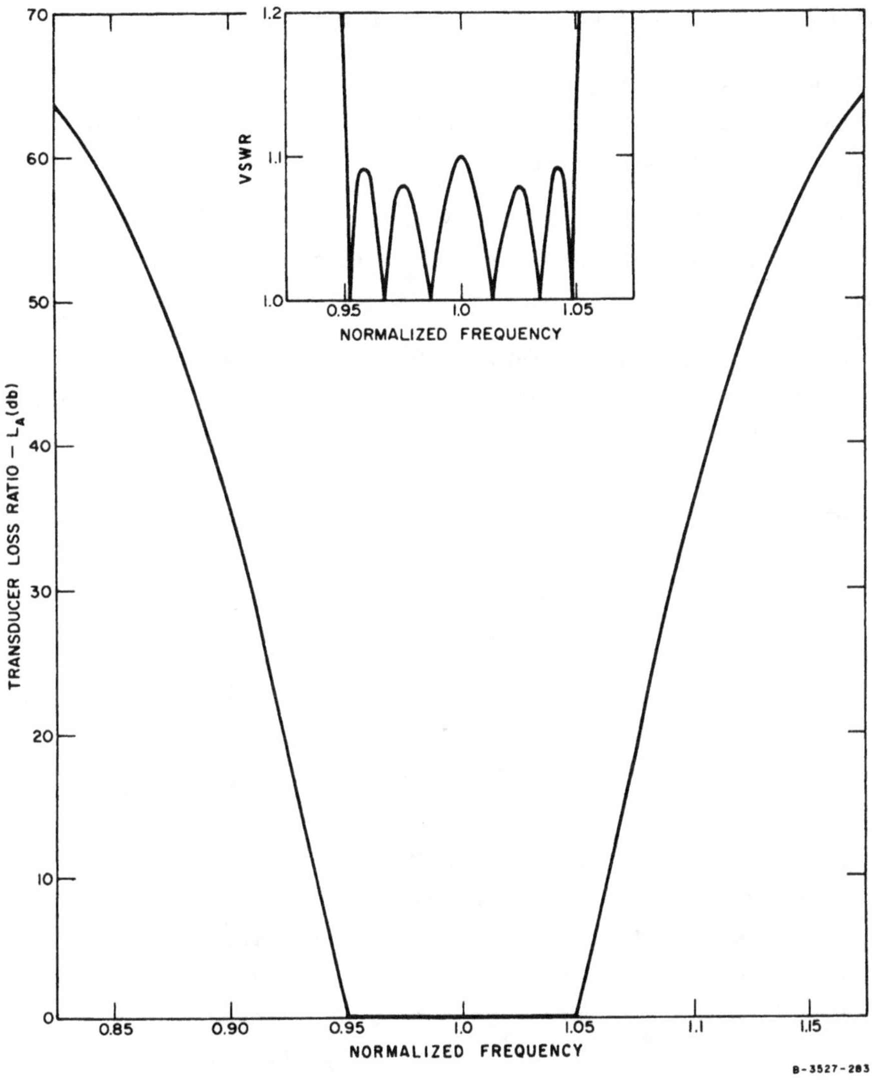

SOURCE: Quarterly Progress Report 4, Contract DA 36-039 SC-87398, SRI; reprinted in *IRE Trans. PGMTT* (See Ref. 36 by L. Young)

FIG. 6.09-1 ANALYZED PERFORMANCE OF HALF-WAVE FILTER DESIGNED IN EXAMPLE 1 OF SEC. 6.09

which is within about 1½ percent of R in Eq. (6.09-3). Therefore we would expect an accurate design, which is confirmed by Fig. 6.09-1. The attenuation of 35.5 db at $f = 1.1$ is also exactly as predicted.

Example 2—It is required to design a half-wave filter of 60-percent bandwidth with a 2-db pass-band ripple. The rejection 10 percent beyond the band edges shall be at least 20 db.

Here $w_h = 0.6$, $\therefore w_q = 1.2$. As in the previous example, it is determined that at least six sections will be required, and that the rejection 10 percent beyond the band edges should then be 22.4 db.

From Eq. (6.02-13) and Table 6.02-1 it can be seen that, for an exact design, R would be 1915; whereas $(2/w_q)^n$ is 22. Thus R exceeds $(2/w_q)^n$ by a factor of less than 100, and therefore, by Eq. (6.09-1), we would expect only a fairly accurate design with a noticeable deviation from the specified performance. The step VSWRs are found by Eq. (6.09-2) to be

$$\left.\begin{aligned} V_1 &= V_7 = 3.028 \\ V_2 &= V_6 = 2.91 \\ V_3 &= V_5 = 3.93 \\ &\quad\; V_4 = 4.06 \end{aligned}\right\} \qquad (6.09\text{-}7)$$

Their product is 4875, whereas from Eq. (6.02-13) and Table 6.02-1, R should be 1915. The V_i must therefore be reduced. As in Example 1 of Sec. 6.07, we shall scale the V_i so as to slightly increase the bandwidth, without affecting the pass-band ripple. Since from Eq. (6.09-2) V_1 and V_{n+1} are inversely proportional to w_q, whereas the other $(n - 1)$ junction VSWRs, namely $V_2, V_3, \ldots V_n$, are inversely proportional to the square of w_q, reduce V_1 and V_7 by a factor of

$$\left(\frac{1915}{4875}\right)^{1/2n} = \left(\frac{1915}{4875}\right)^{1/12} = 0.9251 \quad ,$$

and V_2 through V_6 by a factor of

$$\left(\frac{1915}{4875}\right)^{1/n} = \left(\frac{1915}{4875}\right)^{1/6} = 0.8559 \quad .$$

(Compare Example 1 of Sec. 6.07.) This reduces R from 4875 to 1915. Hence,

$$\left.\begin{aligned} V_1 &= V_7 = 2.803 \\ V_2 &= V_6 = 2.486 \\ V_3 &= V_5 = 3.360 \\ & V_4 = 3.470 \quad . \end{aligned}\right\} \qquad (6.09\text{-}8)$$

The half-wave filter line impedances are now

$$\left.\begin{aligned} Z'_0 &= 1.0 \quad \text{(input)} \\ Z'_1 &= 2.803 \\ Z'_2 &= 1.128 \\ Z'_3 &= 3.788 \\ Z'_4 &= 1.092 \\ Z'_5 &= 3.667 \\ Z'_6 &= 1.475 \\ Z'_7 &= 4.135 \quad \text{(output)} \quad . \end{aligned}\right\} \qquad (6.09\text{-}9)$$

Since the reduction of R, from 4875 to 1915, is a relatively large one, we may expect some measurable discrepancy between the predicted and the analyzed performance. The analyzed performances of the designs given by Eqs. (6.09-7) and (6.09-8), before and after correction for R, are shown in Fig. 6.09-2. For most practical purposes, the agreement after correction for R is quite acceptable. The bandwidth for 2-db insertion loss is 58 percent instead of 60 percent; the rejection is exactly as specified.

Discussion—The half-wave filter of Example 1 required large impedance steps, the largest being $V_4 = 105$. It would therefore be impractical to build it as a stepped-impedance filter; it serves, instead as a prototype for a reactance-coupled cavity filter (Sec. 9.04).

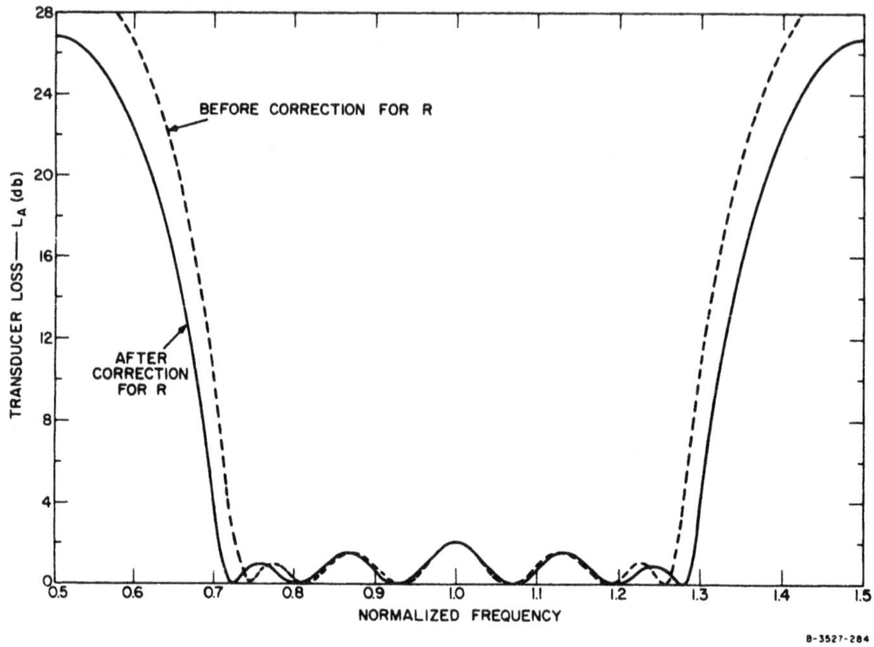

SOURCE: Quarterly Progress Report 4, Contract DA 36-039 SC-87398, SRI; reprinted in *IRE Trans. PGMTT* (See Ref. 36 by L. Young)

FIG. 6.09-2 ANALYZED PERFORMANCE OF TWO HALF-WAVE FILTERS DESIGNED IN EXAMPLE 2 OF SEC. 6.09

This is typical of narrow-band filters. The filter given in the second example, like many wide-band filters, may be built directly from Eq. (6.09-9) since the largest impedance step is $V_4 = 3.47$ and it could be constructed after making a correction for junction discontinuity capacitances (see Sec. 6.08). Such a filter would also be a low-pass filter (see Fig. 6.03-2). It would have identical pass bands at all harmonic frequencies, and it would attain its peak attenuation at one-half the center frequency (as well as at 1.5, 2.5, etc., times the center frequency, as shown in Fig. 6.03-2). The peak attenuation can be calculated from Eqs. (6.02-8) and (6.09-3). In Example 1 of Sec. 6.09 the peak attenuation is 100 db, but the impedance steps are too large to realize in practice. In Example 2 of Sec. 6.09 the impedance steps could be realized, but the peak attenuation is only 27 db. Half-wave filters are therefore more useful as prototypes for other filter-types which are easier to realize physically. If shunt inductances or series capacitances were used (in place of the impedance steps) to realize the V_i and to form a direct-coupled-cavity filter, then the attenuation below the pass band is increased and reaches infinity at zero frequency; the attenuation above the pass band is reduced, as compared with the symmetrical response of the half-wave filters (Figs. 6.09-1 and 6.09-2). The derivation of such filters from the quarter-wave transformer or half-wave filter prototypes will be presented in Chapter 9.

Example 3—This example illustrates a case when neither the first-order theory (Sec. 6.06) nor the method of this part are accurate, but both may give usable designs. These are compared to the exact design.

It is required to design the best quarter-wave transformer of four sections, with output-to-input impedance ratio $R = 31.6$, to cover a fractional bandwidth of 120 percent.

Here $n = 4$ and $w_q = 1.2$. From Eq. (6.02-13) and Table 6.02-1, the maximum VSWR in the pass band is 2.04. Proceeding as in the previous example, and after reducing the product $V_1 V_2 \ldots V_5$ to 31.6 (this required a relatively large reduction factor of 4), yields Design A shown in Table 6.09-1. Its computed VSWR is plotted in Fig. 6.09-3 (continuous line, Case A).

Table 6.09-1

THE THREE DESIGNS OF EXAMPLE 3

A—"Large R" Approximation.
B—"Small R" Approximation.
C—Exact Design.

	DESIGN		
	A	B	C
$V_1 = V_5$	1.656	1.780	1.936
$V_2 = V_4$	2.028	2.091	1.988
V_3	2.800	2.289	2.140

SOURCE: Quarterly Progress Report 4, Contract DA 36-039 SC-87398, SRI; reprinted in *IRE Trans. PGMTT* (see Ref. 36 by L. Young)

Since R exceeds $(2/w_q)^n$ by a factor of only 4 [see Eq. 6.09-1)], the first-order procedure of Sec. 6.07 may be more appropriate. This is also indicated by Eq. (6.07-2), which is satisfied, although Eq. (6.07-1) is not. Proceeding as in Example 1 of Sec. 6.07 yields Design B, shown in Table 6.09-1 and plotted in Fig. 6.09-3 (dash-dot line, Case B).

In this example, the exact design can also be obtained from Tables 6.04-3 and 6.04-4, by linear interpolation of log V against log R. This gives Design C shown in Table 6.09-1 and plotted in Fig. 6.09-3 (broken line, Case C).

Designs A and B both give less fractional bandwidth than the 120 percent asked for, and smaller VSWR peaks than the 2.04 allowed.

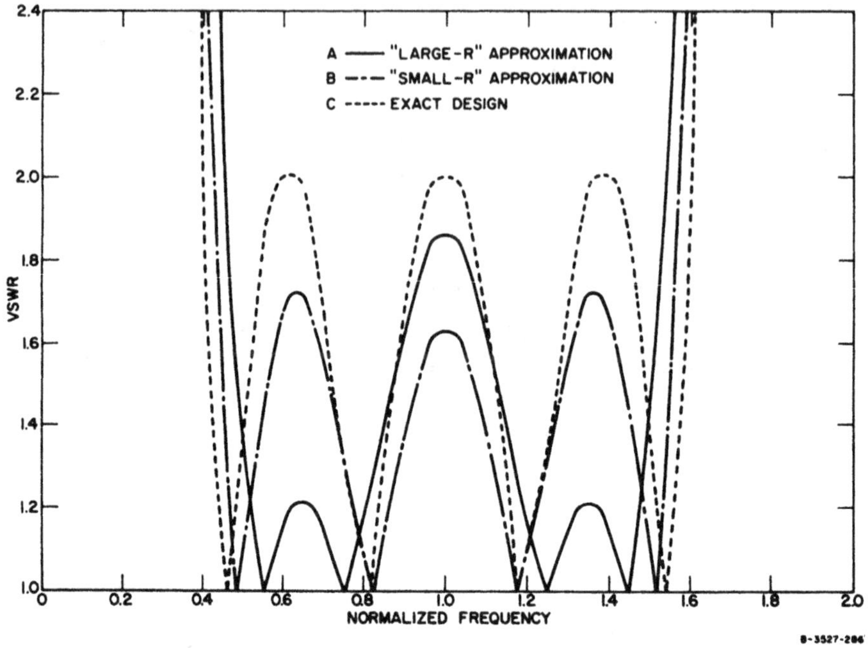

SOURCE: Quarterly Progress Report 4, Contract DA 36-039 SC-87398, SRI; reprinted in *IRE Trans. PGMTT* (See Ref. 36 by L. Young)

FIG. 6.09-3 ANALYZED PERFORMANCE OF THREE QUARTER-WAVE TRANSFORMERS DESIGNED IN EXAMPLE 3 OF SEC. 6.09

The fractional bandwidth (between $V = 2.04$ points) of Design A is 110 percent, and of Design B is 115 percent, and only the exact equal-ripple design, Design C, achieves exactly 120 percent. It is rather astonishing that two approximate designs, one based on the premise $R \approx 1$, and one on $R \to \infty$, should agree so well.

SEC. 6.10, ASYMPTOTIC BEHAVIOR AS R TENDS TO INFINITY

Formulas for direct-coupled cavity filters with reactive discontinuities are given in Chapter 8. These formulas become exact only in the limit as the bandwidth tends to zero. This is not the only restriction. The formulas in Secs. 8.05 and 8.06 for transmission-line filters, like the formulas in Eq. (6.09-2), hold only when Eq. (6.09-1) or its equivalent is satisfied. [Define the V_i as the VSWRs of the reactive discontinuities at center frequency; R is still given by Eq. (6.04-10); for w_q in Eq. (6.09-1), use twice the filter fractional bandwidth in reciprocal guide wavelength.] The variation of the V_i with bandwidth is correctly given by Eq. (6.09-2) for small bandwidths. These formulas can be adapted for design of both quarter-wave transformers and half-wave filters, as in Eq. (6.09-2), and hold even better in this case than when the discontinuities are reactive. [This might be expected since the line lengths between discontinuities for half-wave filters become exactly one-half wavelength at band-center, whereas they are only approximately 180 electrical degrees long in direct-coupled cavity filters (see Fig. 8.06-1)].

Using Eq. (6.09-2) and the formulas of Eqs. (4.05-1) and (4.05-2) for the prototype element values g_i ($i = 0, 1, 2, \ldots, n, n+1$), one can readily deduce some interesting and useful results for the V_i as R tends to infinity. One thus obtains, for the junction VSWRs of Tchebyscheff transformers and filters,

$$\left.\begin{aligned}\lim_{R\to\infty} V_i &= \left(\frac{8}{\pi w_q}\right)^2 \frac{\sin\left(\dfrac{2i-3}{2n}\pi\right)\sin\left(\dfrac{2i-1}{2n}\pi\right)}{\sin^2\left(\dfrac{i-1}{n}\pi\right)} \\ &= \left(\frac{4}{\pi w_h}\right)^2 \frac{\sin\left(\dfrac{2i-3}{2n}\pi\right)\sin\left(\dfrac{2i-1}{2n}\pi\right)}{\sin^2\left(\dfrac{i-1}{n}\pi\right)} \\ &= \left(\frac{4}{\pi w_h}\right)^2 \left[1 - \left(\frac{\sin\dfrac{\pi}{2n}}{\sin\left(\dfrac{i-1}{n}\pi\right)}\right)^2\right] \\ &\qquad (i = 2, 3, \ldots, n) \quad .\end{aligned}\right\} \qquad (6.10\text{-}1)$$

The quantity

$$w_h^2 \lim_{R\to\infty} (V_i) = \left(\frac{w_q}{2}\right)^2 \lim_{R\to\infty} (V_i) \qquad (6.10\text{-}2)$$

is tabulated in Table 6.10-1 for $i = 2, 3, \ldots, n$ and for $n = 2, 3, \ldots, 14$.

Table 6.10-1

TABLE OF $\left(\frac{w_q}{2}\right)^2 \lim_{R\to\infty} (V_i)$ FOR SMALL w_q

$[V_i = V_{n+2-i}]$

n	$i = 2$	$i = 3$	$i = 4$	$i = 5$	$i = 6$	$i = 7$	$i = 8$
2	0.81056						
3	1.08075						
4	1.14631	1.38372					
5	1.17306	1.44999					
6	1.18675	1.47634	1.51254				
7	1.19474	1.48981	1.53668				
8	1.19981	1.49773	1.54885	1.55943			
9	1.20325	1.50282	1.55596	1.57073			
10	1.20568	1.50631	1.56052	1.57727	1.58146		
11	1.20747	1.50880	1.56365	1.58145	1.58762		
12	1.20882	1.51066	1.56589	1.58431	1.59153	1.59351	
13	1.20987	1.51207	1.56757	1.58636	1.59419	1.59723	
14	1.21070	1.51318	1.56886	1.58789	1.59610	1.59975	1.60081

SOURCE: Quarterly Progress Report 4, Contract DA 36-039 SC-87398, SRI; reprinted in *IRE Trans. PGMTT* (see Ref. 36 by L. Young)

We notice that for Tchebyscheff transformers and filters, the $V_i (i \neq 1, n + 1)$ tend to finite limits, and thus $V_1 = V_{n+1}$ tend to a constant times $R^{1/2}$. We also see that

$$w_h^2 V_i < \frac{16}{\pi^2} = 1.62115 \quad (i = 2, 3, \ldots, n) \tag{6.10-3}$$

for all n, and tends to $16/\pi^2$ only in the limit $i \to n/2 \to \infty$.

For maximally flat transformers, the V_i all tend to infinity with R, but the quantities

$$\left. \begin{aligned} A_1 = A_{n+1} &= \frac{V_1}{R^{1/2n}} \\ A_i &= \frac{V_i}{R^{1/n}} \quad (i = 2, 3, \ldots, n) \end{aligned} \right\} \tag{6.10-4}$$

tend toward finite limits given by

$$\left.\begin{aligned}\lim_{R\to\infty} A_1 &= 2^{(n-1)/n}\,\sin\left(\frac{\pi}{2n}\right)\\ \lim_{R\to\infty} A_i &= 2^{2(n-1)/n}\,\sin\left(\frac{2i-1}{2n}\,\pi\right)\sin\left(\frac{2i-3}{2n}\,\pi\right)\\ &\qquad (i \neq 1\,,\ n+1)\end{aligned}\right\}\quad(6.10\text{-}5)$$

from which we see that

$$\left.\begin{aligned}V_1 = V_{n+1} &< \left(\frac{4^{n-1}}{R}\right)^{1/2n}\\ V_i &< \left(\frac{4^{n-1}}{R}\right)^{1/n}\quad (i \neq 1\,,\ n+1)\end{aligned}\right\}\quad(6.10\text{-}6)$$

for all n. They tend toward the values on the right hand side only in the limit $i \to n/2 \to \infty$.

To show how a typical V_i approaches its asymptotic value, the exact solution for V_2 when $n = 4$ is plotted in Fig. 6.10-1 for all fractional bandwidths w_q in steps of 0.20. It is seen that each curve consists of two almost linear regions with a sharp knee joining them. In the sloping region above the origin ("small R"), the approximations of Sec. 6.06 or 6.07 apply; in the horizontal region ("large R"), the approximations of Sec. 6.09 apply. These two sets of approximations probably hold as well as they do because the knee region is so small.

The exact asymptotic values of $w_h^2 V_i = (w_q/2)^2 V_i$ are plotted against w_q in Fig. 6.10-2. If Eq. (6.10-1) were exact instead of approximate, then all of the curves would be horizontal straight lines. As it is, Eq. (6.10-1) gives the correct value only on the $w_q = 0$ axis. As the bandwidth increases, $w_h^2 V_i$ departs from the value at $w = 0$ slowly at first, then reaches a minimum, and finally all curves pass through unity at $w_q = 2$ ($w_h = 1$). The values of $(w_q/2)^2 V_i$ at $w_q = 0$ up to $n = 8$ are also shown in Fig. 6.10-2. (They can be obtained more accurately from Table 6.10-1.) They all lie below the value $16/\pi^2 = 1.62115$, and

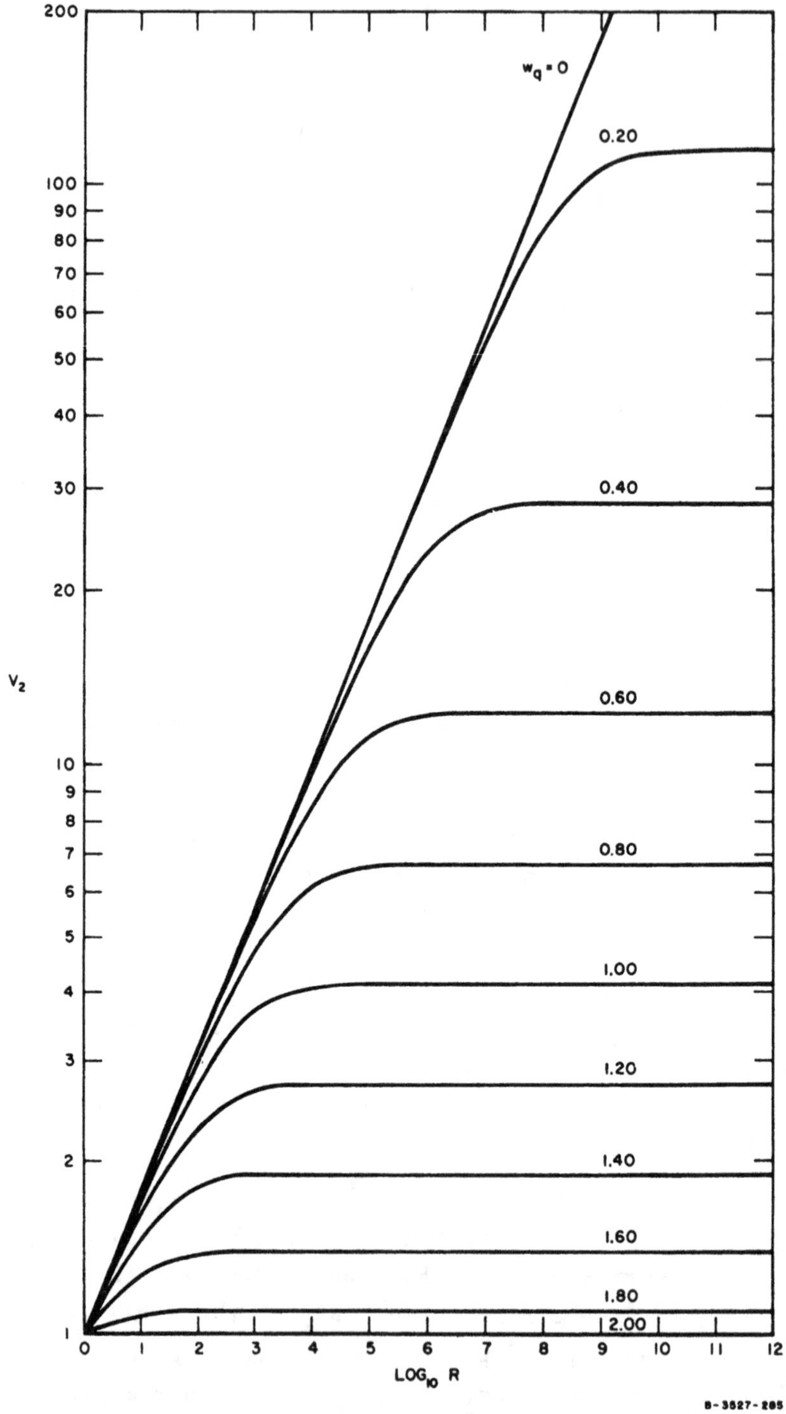

SOURCE: Quarterly Progress Report 4, Contract DA 36-039 SC-87398, SRI; reprinted in *IRE Trans. PGMTT* (See Ref. 36 by L. Young)

FIG. 6.10-1 V_2 VERSUS log R OF FOUR-SECTION TRANSFORMER FOR ALL FRACTIONAL BANDWIDTHS IN STEPS OF 0.20

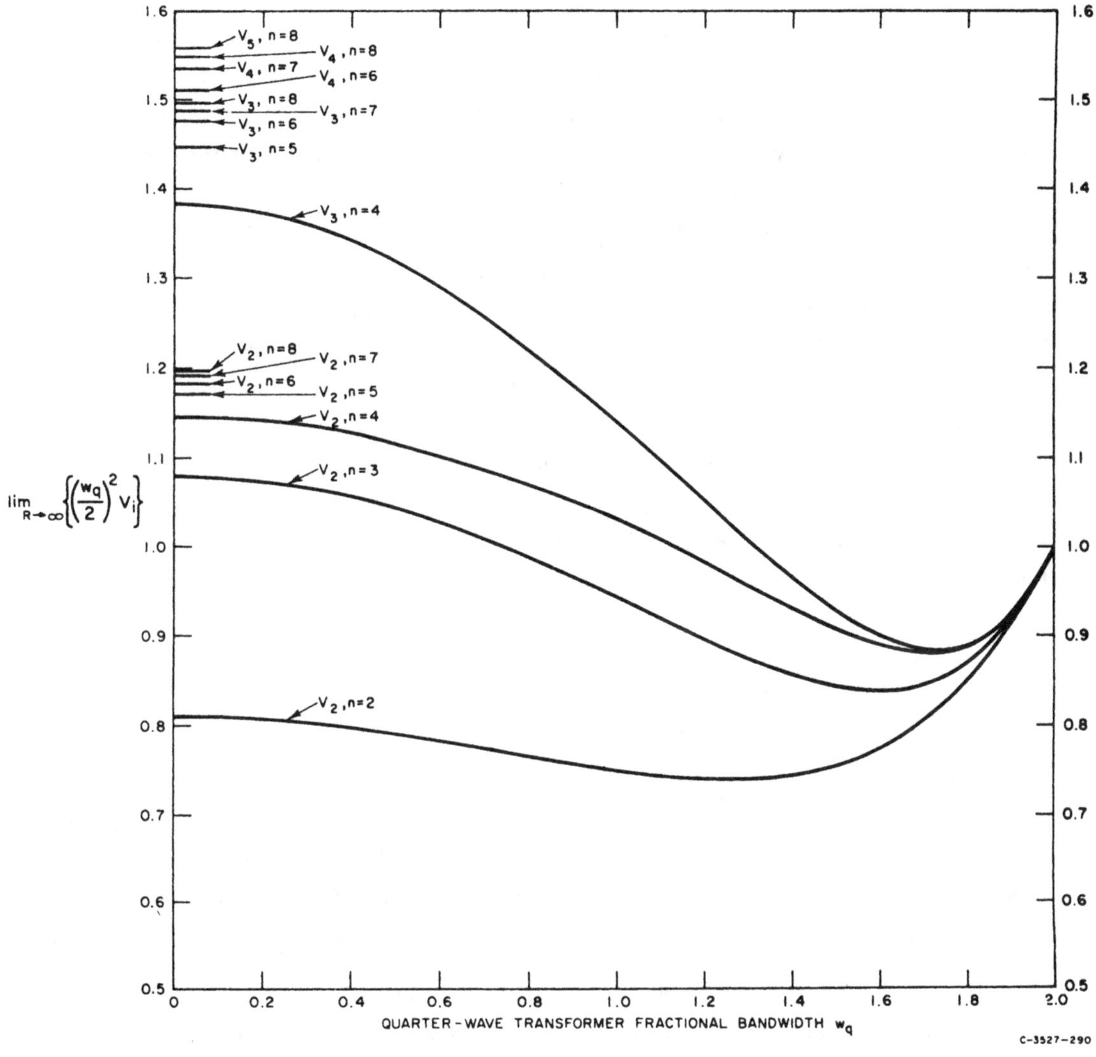

SOURCE: Quarterly Progress Report 4, Contract DA 36-039 SC-87398, SRI; reprinted in *IRE Trans. PGMTT* (see Ref. 36 by L. Young)

FIG. 6.10-2 $\lim_{R\to\infty} (w_q/2)^2 V_i$ PLOTTED AGAINST FRACTIONAL BANDWIDTH FOR TRANSFORMERS HAVING UP TO FOUR SECTIONS, AND SHOWN FOR SMALL w_q UP TO EIGHT SECTIONS

may be expected to exhibit the same sort of general behavior as do the curves up to $n = 4$, for which the exact solutions were obtained from Eqs. (6.04-4) to (6.04-6).

The asymptotic values of the V_i for $i = 2, 3, \ldots, n$, and for a given fractional bandwidth, are seen to be fairly independent of n, on examination of Eq. (6.10-1), Table 6.10-1, or Fig. 6.10-2. It follows that the same is true of $V_1/\sqrt{R} = V_{n+1}/\sqrt{R}$. Thus, as R increases indefinitely, so do V_1 and V_{n+1}; on the other hand for "small R," V_1^2 and V_{n+1}^2 are less than the other V_i (*not* squared) for small and moderately wide fractional bandwidths (up to about 100-percent bandwidths, by Table 6.06-1). If we assume that in the knee region (Fig. 6.10-1) $V_1^2 = V_{n+1}^2$ are of the order of the other V_i, then in the knee region R is of the order of $(V_i)^n$, for any $i \neq 1, n+1$. From Eq. (6.09-2), R is therefore inversely proportional to (const. $\times\, w_q)^{2n}$, and from the previous remarks this constant of proportionality is reasonably independent of n. Using Fig. 6.10-1 for example, the constant is very close to the value ½. This leads to the magnitude formulas of Eqs. (6.06-1), (6.07-1), (6.07-2), and (6.09-1), which have been confirmed by numerous sample solutions.

SEC. 6.11, INHOMOGENEOUS WAVEGUIDE QUARTER-WAVE TRANSFORMERS OF ONE SECTION

Inhomogeneous transformers were defined in Sec. 6.01. They come about, for instance, when rectangular waveguides having different 'a' dimensions are cascaded; or when rectangular waveguides are combined with ridged, circular, or other types of waveguide; or when the materials of an optical multi-layer are not uniformly dispersive.

At first, only ideal waveguide transformers will be considered. The junction effects in non-ideal transformers can be compensated by adjusting the lengths as in Sec. 6.08, except that the step discontinuity effects cannot usually be represented by a shunt capacitance alone. Only very limited information on waveguide junctions (other than E-plane steps) is available,[23] and for large steps the designer may have to make individual measurements on each junction.

The notation for an inhomogeneous quarter-wave transformer of one section is shown in Fig. 6.11-1.

To obtain zero reflection at center frequency (where the section length is one-quarter guide-wavelength) a sufficient condition is that

$$Z_1 = (Z_0 Z_2)^{1/2} \qquad (6.11\text{-}1)$$

where Z_0, Z_1, and Z_2 are the characteristic impedances of the input waveguide, the transformer section, and the output waveguide, respectively (Fig. 6.11-1). For a *homogeneous* transformer Eq. (6.11-1) determines the design completely, since the three cutoff wavelengths are the same ($\lambda_{c0} = \lambda_{c1} = \lambda_{c2}$); in the case of rectangular waveguide, the three wide dimensions are then equal ($a_0 = a_1 = a_2$). However, even when a homogeneous transformer is possible, that is, when $\lambda_{c0} = \lambda_{c2}$, we may prefer to make λ_{c1} different, and thus choose to make the transformer inhomogeneous. This gives an extra degree of freedom, which, it turns out, can always be used to: (1), lower the VSWR near center frequency, and simultaneously (2), shorten the transformer.

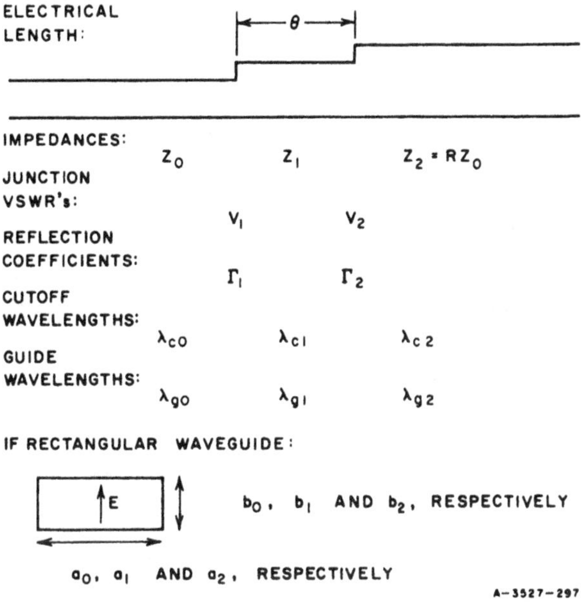

FIG. 6.11-1 INHOMOGENEOUS QUARTER-WAVE TRANSFORMER OF ONE SECTION

When λ_{c0} and λ_{c2} are not equal, an inhomogeneous transformer results of necessity. For a match at center frequency, Eq. (6.11-1) still holds, but there are an infinity of possible cutoff wavelengths, λ_{c1} (equal to $2a_1$ for rectangular waveguide). This general case will now be considered. (If a homogeneous transformer is required, then λ_{c0} can be set equal to λ_{c2} at any stage.)

It can be shown[5] that the excess loss [see Eq. (6.02-5)] is given by

$$\mathcal{E} = \frac{1}{T_1^2 T_2^2} [(\Gamma_2 - \Gamma_1)^2 + 4\Gamma_1\Gamma_2 \cos^2 \theta] \quad . \tag{6.11-2}$$

For no attenuation at center frequency ($\theta = \pi/2$), it is only necessary that $\Gamma_1 = \Gamma_2$, which is equivalent to Eq. (6.11-1). Minimizing the frequency variation of $\mathcal{E}$ at center frequency, leads for both TE and TM modes to:

$$\lambda_{g1}^2 = \frac{1}{2}\left[\frac{\lambda_{g0}^2 + \lambda_{g2}^2}{1 + \left(\frac{\pi}{4}\right)^2 \frac{(Z_2 - Z_0)^2}{Z_2 Z_0}}\right] \quad . \tag{6.11-3}$$

Note that

$$(\lambda_{g1\ opt.})^2 < \frac{1}{2}(\lambda_{g0}^2 + \lambda_{g2}^2) \tag{6.11-4}$$

and that further, if $\lambda_{c0} = \lambda_{c2}$,

$$\lambda_{c1\ opt.} > \lambda_{c0} = \lambda_{c2} \quad . \tag{6.11-5}$$

Therefore, one can always improve upon a homogeneous transformer ($\lambda_{c1} = \lambda_{c0} = \lambda_{c2}$). The computed VSWR against normalized wavelength of three transformers matching from a_0 = 0.900 in., b_0 = 0.050 in., to a_2 = 0.900 in., b_0 = 0.400 in. waveguide, at a center frequency of 7211 megacycles (λ_0 = 1.638 in.) is shown in Fig. 6.11-2 for transformer guide widths of a_1 = 0.900 in. (homogeneous), a_1 = 0.990 in., and a_1 = 1.90 in. (optimum). Beyond this value the performance deteriorates again. The performance changes very slowly around the optimum value.

It is seen that for the best inhomogeneous transformer (a_1 = 1.90 in.) the VSWR *vs.* frequency slope is slightly better than 45 percent of that for the homogeneous transformer. Moreover a_1 is so uncritical that it

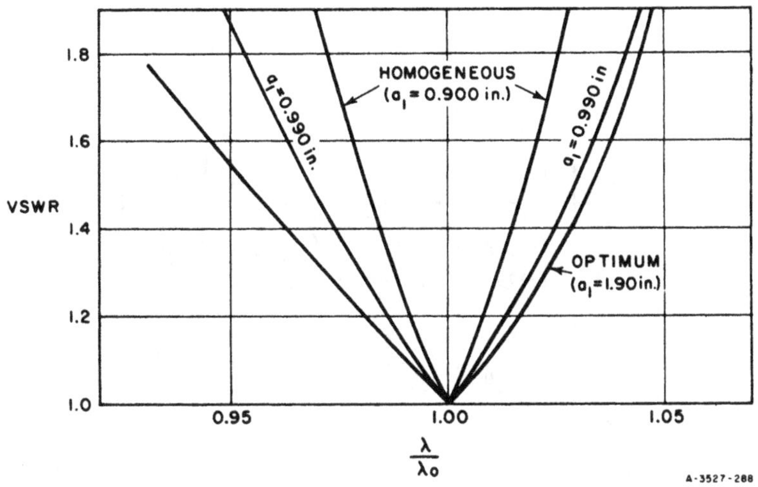

SOURCE: *IRE Trans. PGMTT* (See Ref. 5 by L. Young)

FIG. 6.11-2 VSWR AGAINST WAVELENGTH OF THREE QUARTER-WAVE TRANSFORMERS OF ONE SECTION, ALL FROM 0.900-INCH BY 0.050-INCH WAVEGUIDE TO 0.900-INCH BY 0.400-INCH WAVEGUIDE. CENTER FREQUENCY = 7211 Mc

may be reduced from 1.90 in. to 1.06 in. and the improvement remains better than 50 percent. This is very useful in practice, since a_1 cannot be made much greater than a_0 or a_2 without introducing higher-order modes or severe junction discontinuities.

The example selected above for numerical and experimental investigation has a higher transformer impedance ratio (R = 8), and operates considerably closer to cutoff (λ_0/λ_c = 0.91), than is common. In such a situation the greatest improvement can be obtained from optimizing a_1. In most cases (low R and low dispersion) the improvement obtained in making the transformer section less dispersive than that of a homogeneous transformer will only be slight. This technique, then, is most useful only for highly dispersive, high-impedance-ratio transformers.

Table 6.11-1 connects (λ/λ_c) with (λ_g/λ), and is useful in the solution of inhomogeneous transformer problems.

To compensate for the junction effects, we note that a non-ideal junction can always be represented by an ideal junction, but the non-ideal junction's reference planes (in which the junction reflection coefficient

Table 6.11-1

RELATIONS BETWEEN λ_g, λ_c, AND λ

$\left(\frac{\lambda}{\lambda_c}\right)$	$\left(\frac{\lambda_g}{\lambda}\right)$	$\left(\frac{\lambda_g}{\lambda}\right)^2$	$\left(\frac{\lambda_g}{\lambda_c}\right)$	$\left(\frac{\lambda_c}{\lambda_g}\right)$
0.50	1.1547	1.3333	0.5773	1.7320
0.51	1.1625	1.3515	0.5929	1.6866
0.52	1.1707	1.3706	0.6087	1.6426
0.53	1.1792	1.3906	0.6250	1.5999
0.54	1.1881	1.4116	0.6415	1.5586
0.55	1.1973	1.4336	0.6585	1.5184
0.56	1.2070	1.4568	0.6759	1.4794
0.57	1.2170	1.4812	0.6937	1.4414
0.58	1.2275	1.5069	0.7119	1.4045
0.59	1.2385	1.5339	0.7307	1.3684
0.60	1.2500	1.5625	0.7500	1.3333
0.61	1.2619	1.5926	0.7698	1.2990
0.62	1.2745	1.6244	0.7902	1.2654
0.63	1.2876	1.6580	0.8112	1.2326
0.64	1.3014	1.6937	0.8329	1.2005
0.65	1.3159	1.7316	0.8553	1.1691
0.66	1.3310	1.7717	0.8785	1.1382
0.67	1.3470	1.8145	0.9025	1.1080
0.68	1.3638	1.8601	0.9274	1.0782
0.69	1.3815	1.9087	0.9532	1.0489
0.70	1.4002	1.9607	0.9801	1.0202
0.71	1.4200	2.0165	1.0082	0.9918
0.72	1.4409	2.0764	1.0375	0.9638
0.73	1.4631	2.1408	1.0681	0.9362
0.74	1.4867	2.2104	1.1001	0.9089
0.75	1.5118	2.2857	1.1338	0.8819
0.76	1.5386	2.3674	1.1693	0.8551
0.77	1.5672	2.4563	1.2068	0.8286
0.78	1.5980	2.5536	1.2464	0.8022
0.79	1.6310	2.6602	1.2885	0.7760
0.80	1.6666	2.7777	1.3333	0.7500
0.81	1.7052	2.9078	1.3812	0.7239
0.82	1.7471	3.0525	1.4326	0.6980
0.83	1.7928	3.2144	1.4880	0.6720
0.84	1.8430	3.3967	1.5481	0.6459
0.85	1.8983	3.6036	1.6135	0.6197
0.86	1.9596	3.8402	1.6853	0.5933
0.87	2.0281	4.1135	1.7645	0.5667
0.88	2.1053	4.4326	1.8527	0.5397
0.89	2.1931	4.8100	1.9519	0.5123
0.90	2.2941	5.2631	2.0647	0.4843
0.91	2.4119	5.8173	2.1948	0.4556
0.92	2.5515	6.5104	2.3474	0.4259
0.93	2.7206	7.4019	2.5302	0.3952
0.94	2.9310	8.5910	2.7551	0.3629
0.95	3.2025	10.2564	3.0424	0.3286
0.96	3.5714	12.7551	3.4285	0.2916
0.97	4.1134	16.9204	3.9900	0.2506
0.98	5.0251	25.2525	4.9246	0.2030
0.99	7.0888	50.2512	7.0179	0.1424
1.00	∞	∞	∞	0

Γ is real) are no longer in the plane of the junction. This can be compensated for E-plane steps, as explained in Sec. 6.08. In compound junctions involving both E-plane and H-plane steps, if the junction discontinuities of these steps are small enough, they may be treated separately of each other using the junction data in Marcuvitz;[23] the two corrections are then superimposed. In most cases, fortunately, these two corrections tend to oppose each other; the shunt inductance effect of the H-plane step partly cancels the shunt capacitance effect of the E-plane step. When for a rectangular waveguide operating in the TE_{10} mode, both the width a and height b are to be increased together (or decreased together), the condition for resonance of the two reactive discontinuities coincides with the condition for equal characteristic impedances,

$$\left(\frac{b}{a}\frac{\lambda_g}{\lambda}\right)_{\text{Waveguide 1}} = \left(\frac{b}{a}\frac{\lambda_g}{\lambda}\right)_{\text{Waveguide 2}} \qquad (6.11\text{-}6)$$

according to Ref. 24, p. 170; when an increase in the 'a' dimension is accompanied by a decrease in the 'b' dimension (or vice versa), then an empirical equation showing when the reactive discontinuities resonate and so cancel is given in Ref. 25, but it is not known how accurate this empirical data is.

In addition to the phase perturbation introduced by the non-ideal junction, there may also be a noticeable effect on the magnitude of the reflection coefficient. (In the case of E-plane steps alone, the latter is usually negligible; see Sec. 6.08.) The increase in the magnitude of the reflection coefficient for H-plane steps in rectangular waveguide can be derived from the curves in Marcuvitz[23] (pp. 296-304). The junction VSWR is then greater than the impedance ratio of the two guides. For instance, in the example already quoted, the output-to-input impedance ratio, R, is equal to 8 with ideal junctions. However, because of the additional reflection due to junction susceptances, this goes up to an effective R of 9.6 (confirmed experimentally[5]).

As a general rule, for rectangular waveguides the change in the 'a' dimension of an H-plane step should be kept below about 10-20 percent if the junction effects are to be treated as first-order corrections to the ideal transformer theory. This is mainly to keep the reference plane from moving too far out of the junction plane (see Marcuvitz,[23] Fig. 5.24-2, p. 299, and Fig. 5.24-5, p. 303). Symmetrical junctions are to be preferred to asymmetrical junctions. Larger H-plane steps are permissible as the guide nears cutoff (smaller 'a' dimension).

SEC. 6.12, INHOMOGENEOUS WAVEGUIDE QUARTER-WAVE TRANSFORMERS OF TWO OR MORE SECTIONS

The condition that an ideal inhomogeneous transformer of two sections (Fig. 6.12-1) be *maximally flat* can be written for both TE and TM modes:

$$\left(\frac{Z_2}{Z_1}\right)^2 = R \tag{6.12-1}$$

$$\lambda_{g2}^2 - \lambda_{g1}^2 = \frac{1}{2}(\lambda_{g3}^2 - \lambda_{g0}^2) \tag{6.12-2}$$

$$\left(\frac{Z_1}{Z_0}\right)^2 = \frac{\lambda_{g1}^2 + \lambda_{g2}^2 R^{1/2}}{\lambda_{g1}^2 + \lambda_{g2}^2 R^{-1/2}} \tag{6.12-3}$$

with the notation of Fig. 6.12-1. Equations (6.12-1) to (6.12-3) are only three conditions for the four parameters λ_{g1}, λ_{g2}, Z_1, Z_2; or in rectangular waveguide, for a_1, a_2, b_1, b_2. Thus there are an infinity of maximally flat transformers of two sections (just as there was an infinity of matching transformers of one section), and some have flatter responses than others. An example is shown in Fig. 6.12-2, in which ideal junctions are assumed. The transformation in this case is between two rectangular waveguides, namely a_0 = 8 in., b_0 = 2 in., to be transformed to a_3 = 5 in., b_3 = 3 in., at a center frequency of 1300 megacycles. The various values of a_1 taken are shown in Fig. 6.12-2. There is probably an optimum (or "flattest maximally flat") transformer, but this has not been found. Instead, it is suggested that a_1 and a_2 be chosen to minimize junction discontinuities and keep the transformer as nearly ideal as possible.

Equation (6.12-3) is plotted in Fig. 6.12-3, with $(\lambda_{g2}/\lambda_{g1})^2$ running from 0.5 to 2.5, for R = 1, 2, ..., 9, 10.

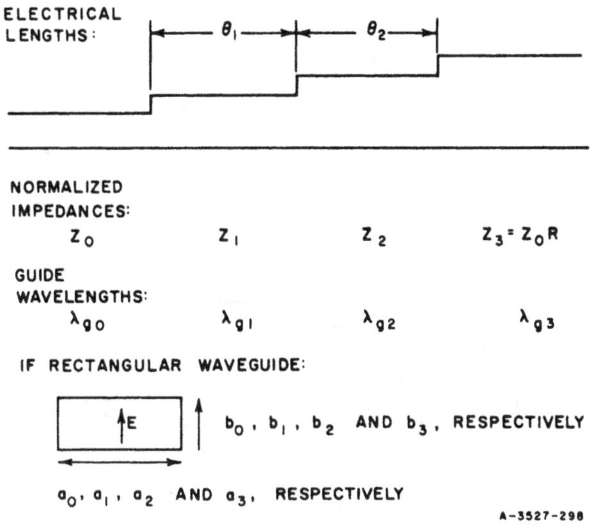

FIG. 6.12-1 INHOMOGENEOUS QUARTER-WAVE TRANSFORMER OF TWO SECTIONS

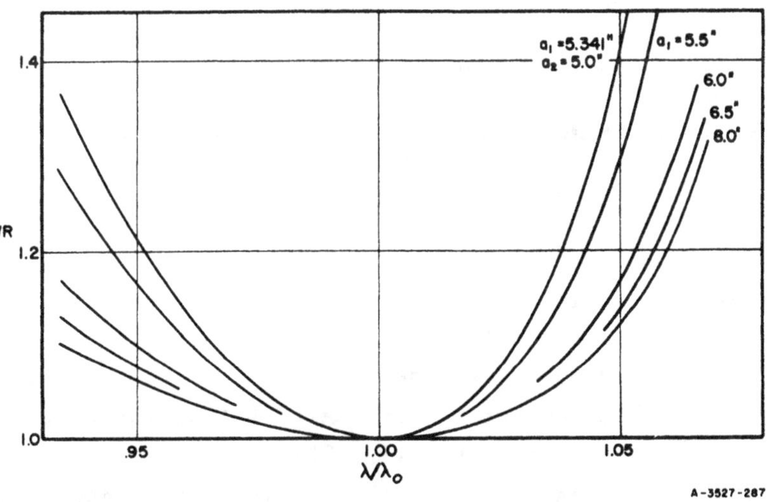

SOURCE: *IRE Trans. PGMTT* (see Ref. 6 by L. Young)

FIG. 6.12-2 VSWR AGAINST WAVELENGTH OF SEVERAL TWO-SECTION MAXIMALLY FLAT TRANSFORMERS, ALL FROM 8-INCH BY 2-INCH WAVEGUIDE TO 5-INCH BY 3-INCH WAVEGUIDE. CENTER FREQUENCY = 1300 Mc

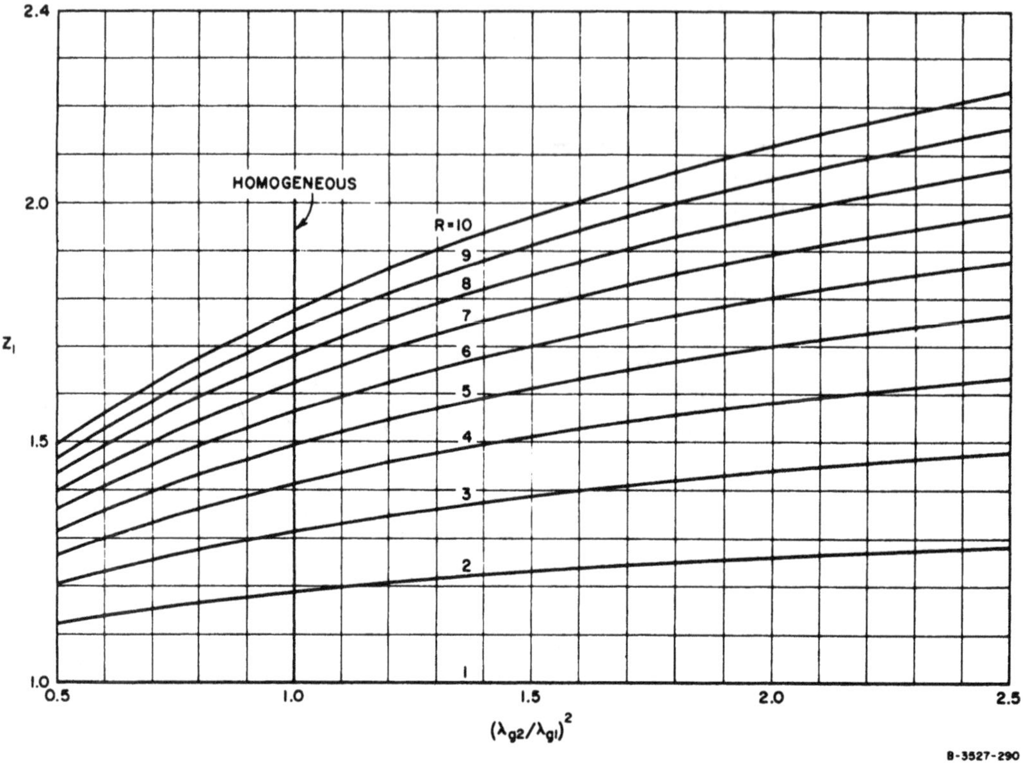

SOURCE: The Microwave Journal (see Ref. 15 by L. Young)

FIG. 6.12-3 DEPENDENCE OF Z_1 ON $(\lambda_{g2}/\lambda_{g1})^2$ TO OBTAIN A MAXIMALLY FLAT TRANSFORMER, FOR TEN VALUES OF R

In most practical applications the transformer should have minimum reflection over a finite frequency band, rather than have maximally flat frequency response. No exact method has yet been found to *broad-band* inhomogeneous transformers, but an approximate design procedure has worked out very well. This consists of first designing for maximally flat response, and then applying the inhomogeneous transformer theory as a multiplicative "correction" to the impedance ratios.[6]

Example 1—Design a transformer from 0.900- by 0.400-inch (WR-90, or RG-52/U or RG-67/U) to 0.750- by 0.400-inch waveguide to have a VSWR of better than 1.10 over a 13-percent frequency band. Here, λ_0 = 1.390 inches.

The reciprocal-guide-wavelength fractional bandwidth is approximately $(d\lambda_g/\lambda_g)/(d\lambda/\lambda) = (\lambda_g/\lambda)^2$ times the frequency fractional bandwidth of 0.13. The arithmetic mean of $(\lambda_g/\lambda)^2$ for the a = 0.900-inch and the a = 0.750-inch waveguides is $(2.47 + 7.04)/2 = 4.75$, so that the $(1/\lambda_g)$ bandwidth is approximately $4.75 \times 13 = 62$ percent. The characteristic impedance is proportional to $(b/a)\ (\lambda_g/\lambda)$, as in Eq. (6.11-6), and the output-to-input impedance ratio, R, is 2.027. A homogeneous transformer of R = 2.027, to have a VSWR of less than 1.10 over a 62-percent bandwidth, must have at least two sections, according to Table 6.02-3. Therefore choose n = 2.

Since the transformer is inhomogeneous, first design the maximally flat transformer. The choice of one waveguide 'a' dimension is arbitrary, so long as none of the steps exceeds about 10-20 percent. Selecting a_1 = 0.850 inch, Eq. (6.12-2) yields a_2 = 0.771 inch and then Eqs. (6.12-1) and (6.12-3), or Fig. 6.12-3, yield b_1 = 0.429 inch, b_2 = 0.417 inch. (Note that none of the H-plane steps exceed 10 percent.) The computed performance of this maximally flat transformer, assuming ideal waveguide junctions, is shown by the broken line in Fig. 6.12-4.

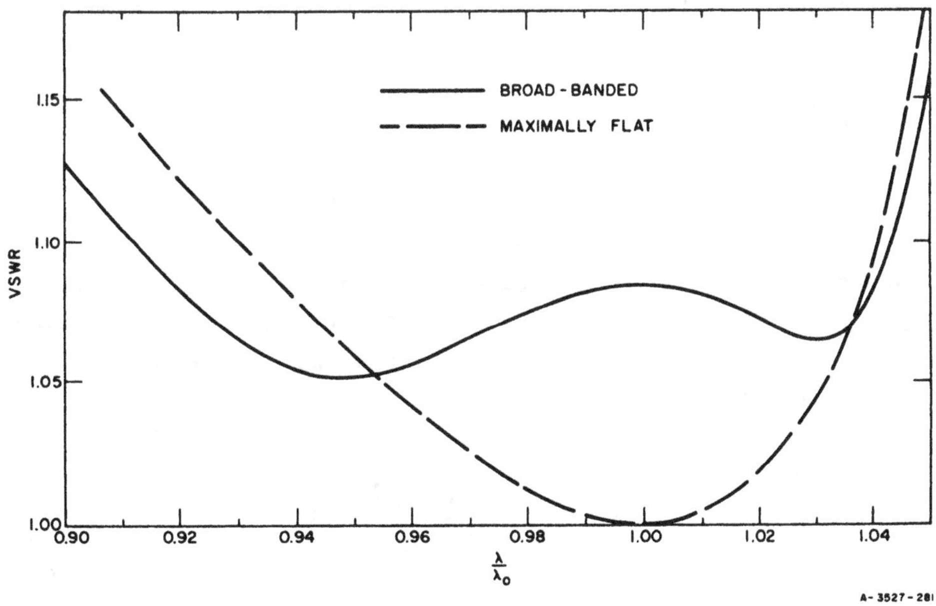

SOURCE: *IRE Trans. PGMTT* (see Ref. 6 by L. Young)

FIG. 6.12-4 VSWR AGAINST WAVELENGTH OF BROADBANDED AND MAXIMALLY FLAT TRANSFORMERS

To broadband this transformer (minimize its reflection over the specified 13 percent frequency band), we note from Table 6.04-1 that, for a two-section homogeneous transformer of $R = 2.027$ to be modified from maximally flat to 62 percent bandwidth, Z_1 increases about 2 percent, and Z_2 is reduced about 2 percent. Applying exactly the same "corrections" to b_1 and b_2 then yields $b_1 = 0.437$ inch and $b_2 = 0.409$ inch. The 'a' dimensions are not affected. The computed performance of this transformer is shown in Fig. 6.12-4 (solid line), and agrees very well with the predicted performance.

In the computations, the effects of having junctions that are non-ideal have not been allowed for. Before such a transformer is built, these effects should be estimated and first-order length corrections should be applied as indicated in Secs. 6.11 and 6.08.

Transformers having $R = 1$—It is sometimes required to change the 'a' dimension keeping the input and output impedances the same ($R = 1$). It may also sometimes be convenient to effect an inhomogeneous transformer by combining a homogeneous transformer (which accounts for all or most of the impedance change) with such an inhomogeneous transformer (which accounts for little or none of the impedance change but all of the change in the 'a' dimension). Such inhomogeneous transformers are sketched in Fig. 6.12-5. We set $R = 1$ in Eqs. (6.12-1) and (6.12-2) and obtain

$$Z_0 = Z_1 = Z_2 = Z_3 \quad . \tag{6.12-4}$$

The reflection coefficients at each junction are zero at center frequency, and we may add the requirement that the rates of change of the three reflection coefficients with frequency be in the ratio 1:2:1. This then leads to

$$\left.\begin{aligned} \lambda_{g1}^2 &= \frac{3\lambda_{g0}^2 + \lambda_{g3}^2}{4} \\ \lambda_{g2}^2 &= \frac{\lambda_{g0}^2 + 3\lambda_{g3}^2}{4} \quad . \end{aligned}\right\} \tag{6.12-5}$$

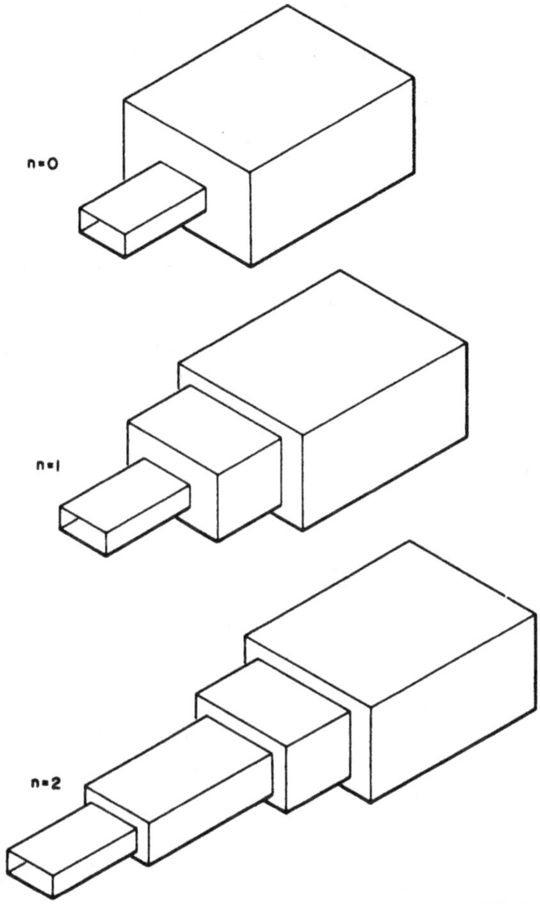

FIG. 6.12-5 INHOMOGENEOUS TRANSFORMERS WITH R = 1

Equations (6.12-2), (6.12-4), and (6.12-5) then determine all the waveguide dimensions.

Example 2—Find the 'a' dimensions of an ideal two-section quarter-wave transformer in rectangular waveguide from a_{in} = 1.372 inches to a_{out} = 1.09 inches to have $R = 1$ and to conform with Eqs. (6.12-2), (6.12-4), and (6.12-5). Here, λ_0 = 1.918 inches.

The solution is readily found to be a_1 = 1.226 inches and a_2 = 1.117 inches. In order for the impedances to be the same at center frequency, as required by Eq. (6.12-4), the 'b' dimensions have to be in the ratio $b_0:b_1:b_2:b_3$ = 1:0.777:0.582:0.526, since $Z \propto (b/a)\ (\lambda_g/\lambda)$. The performance of this transformer is shown in Fig. 6.12-6.

The performances of two other transformers are also shown in Fig. 6.12-6, both with the same input and output waveguide dimensions as in Example 2, given above, and both therefore also with $R = 1$. The optimum one-section transformer has $Z_2 = Z_1 = Z_0$, from Eq. (6.11-3), but requires $\lambda^2_{g1} = (\lambda^2_{g0} + \lambda^2_{g2})/2$, where suffix 2 now refers to the output. This yields a_1 = 1.157 inches. The third, and only V-shaped, characteristic in Fig. 6.12-6 results when the two waveguides are joined without benefit of intermediate transformer sections. The match at center frequency is ensured by the 'b' dimensions which are again chosen so that $R = 1$ at center frequency.

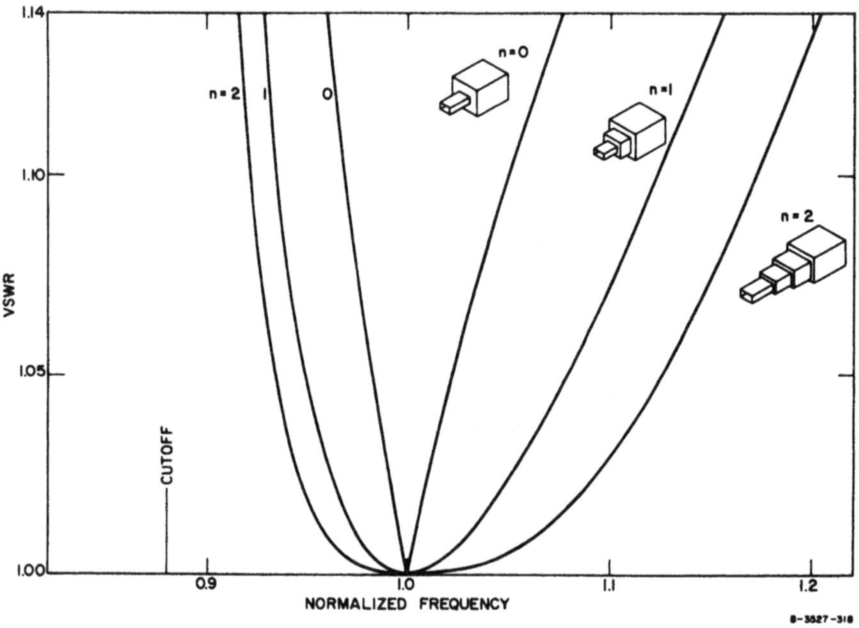

FIG. 6.12-6 PERFORMANCE OF THREE INHOMOGENEOUS TRANSFORMERS ALL WITH R = 1, HAVING NO INTERMEDIATE SECTION (n = 0), ONE SECTION (n = 1), AND TWO SECTIONS (n = 2), RESPECTIVELY

Transformers with more than two sections—No design equations have been discovered for $n > 2$. If a two-section transformer, as in Example 1 of Sec. 6.12, does not give adequate performance, there are two ways open to the designer: When the cutoff wavelengths λ_c of the input and output waveguides are only slightly different, the transformer may be designed as if it were homogeneous. In this case the λ_c of the intermediate sections may be assigned arbitrary values intermediate to the input and output values of λ_c; the impedances are selected from the tables for homogeneous transformers for a fractional bandwidth based on the guide wavelength, Eq. (6.02-1), of that waveguide which is nearest to being cutoff. Even though the most dispersive guide is thus selected for the homogeneous prototype, the frequency bandwidth of the inhomogeneous transformer will still come out less, and when the spread in λ_c is appreciable, considerably less. Thus, this method applies only to transformers that are nearly homogeneous in the first place.

The second method is to design the transformer in two parts: one an inhomogeneous transformer of two sections with $R = 1$, as in Example 2 of this section; the other a homogeneous transformer with the required R, preferably built in the least dispersive waveguide.

Example 3—Design a quarter-wave transformer in rectangular waveguide from $a_{in} = 1.372$ inches to $a_{out} = 1.09$ inches, when $R = 4$. Here, $\lambda_0 = 1.918$ inches.

Selecting a three-section homogeneous transformer of prototype bandwidth $w_q = 0.30$ and $R = 4$, in $a = 1.372$-inch waveguide, followed by the two-section inhomogeneous transformer of Example 2 of this section, gives

$$\begin{aligned}
a_0 &= 1.372 \text{ inches}, & Z_0 &= 1.0, \\
a_1 &= 1.372 \text{ inches}, & Z_1 &= 1.19992, \\
a_2 &= 1.372 \text{ inches}, & Z_2 &= 2.0, \\
a_3 &= 1.372 \text{ inches}, & Z_3 &= 3.33354, \\
a_4 &= 1.276 \text{ inches}, & Z_4 &= 4.0, \\
a_5 &= 1.117 \text{ inches}, & Z_5 &= 4.0, \\
a_6 &= 1.090 \text{ inches}, & Z_6 &= 4.0 .
\end{aligned}$$

The 'b' dimensions may again be obtained from $Z \propto (b/a)\ (\lambda_g/\lambda)$, as in Example 2 of this section. The performance of this five-section transformer is shown in Fig. 6.12-7. Its VSWR is less than 1.05 over a 20-percent frequency band, although it comes within 6 percent of cutoff at one end.

Where a low VSWR over a relatively wide pass band is important, and where there is room for four or five sections, the method of Example 3 of this section is generally the best.

SEC. 6.13, A NONSYNCHRONOUS TRANSFORMER

All of the quarter-wave transformers considered so far have been synchronously tuned (see Sec. 6.01); the impedance ratio at any junction has been less than the output-to-input impedance ratio, R. It is possible to obtain the same or better electrical performance with an ideal

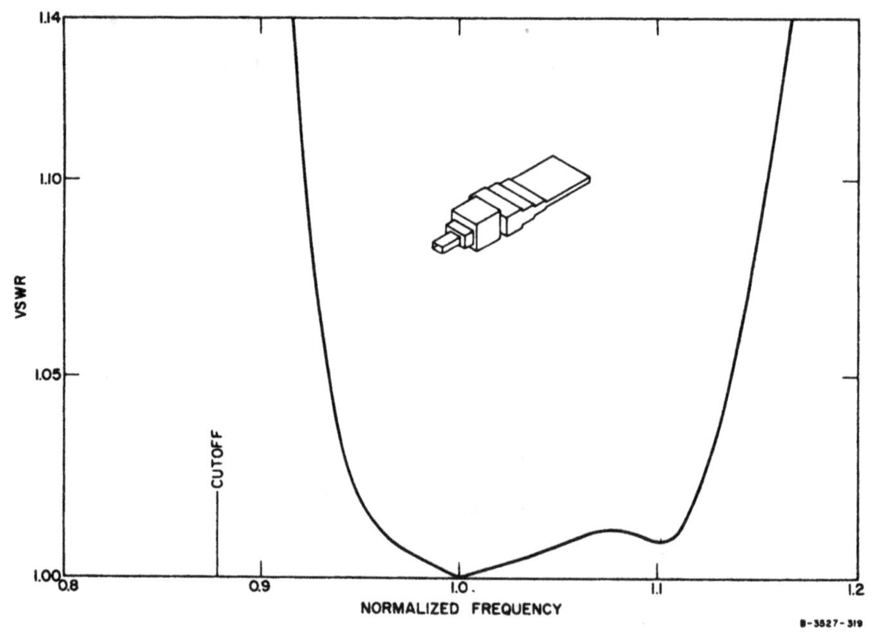

FIG. 6.12-7 PERFORMANCE OF A FIVE-SECTION INHOMOGENEOUS TRANSFORMER

nonsynchronous transformer of shorter length; however, the impedance ratios at the junctions generally exceed R by a large factor, and for more than two sections such "supermatched" transformers appear to be impractical. There is one case of a nonsynchronous transformer that is sometimes useful. It consists of two sections, whose respective impedances are equal to the output and input impedances, as shown in Fig. 6.13-1. The whole transformer is less than one-sixth wavelength long, and its performance is about the same as that of a single-section quarter-wave transformer. It can be shown[26] that the length of each section for a perfect match has to be equal to

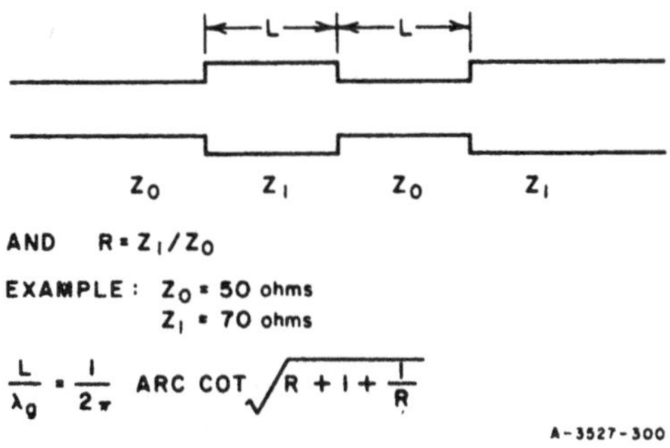

FIG. 6.13-1 A NONSYNCHRONOUS TRANSFORMER

$$L = \frac{1}{2\pi} \text{ arc cot } \left(R + 1 + \frac{1}{R}\right)^{1/2} \quad \text{wavelengths} \qquad (6.13\text{-}1)$$

which is always less than 30 electrical degrees, and becomes 30 degrees only in the limit as R approaches unity. It can be shown further that, for small R, the slope of the VSWR *vs.* frequency characteristic is greater than that for the corresponding quarter-wave transformer by a factor of $2/\sqrt{3}$ (about 15 percent greater); but then the new transformer is only two-thirds the over-all length ($\lambda_g/6$ compared to $\lambda_g/4$).

The main application of this transformer is in cases where it is difficult to come by, or manufacture, a line of arbitrary impedance. Thus if it is desired to match a 50-ohm cable to a 70-ohm cable, it is not necessary to look for a 59.1-ohm cable; instead, the matching sections can be one piece of 50-ohm and one piece of 70-ohm cable. Similarly, if it is desired to match one medium to another, as in an optical multilayer antireflection coating, this could be accomplished without looking for additional dielectric materials.

In Sec. 4.13 a formula was derived for the center-frequency increase in attenuation $(\Delta L_A)_0$ due to dissipation losses. Equation (4.13-11) applies to lumped-constant filters which are reflectionless at band center, and also includes those transmission-line filters which can be derived from the low-pass lumped-constant filters of Chapter 4 (see, for example, Sec. 6.09). If, however, the filter has not been derived from a lumped-constant prototype, then it is either impossible or inconvenient to use Eq. (4.13-11). What is required is a formula giving the dissipation loss in terms of the transmission-line filter parameters, such as the V_i instead of the g_i.

Define S_i as the VSWR at center frequency seen inside the *i*th filter cavity, or transformer section, when the output line is matched (Fig. 6.14-1). Here the numbering is such that $i = 1$ refers to the section or transmission-line cavity nearest the generator. Let

$$|\rho_i| = \frac{S_i - 1}{S_i + 1} \tag{6.14-1}$$

be the amplitude of the reflection coefficient in the *i*th cavity, corresponding to the VSWR S_i. Let[12,27]

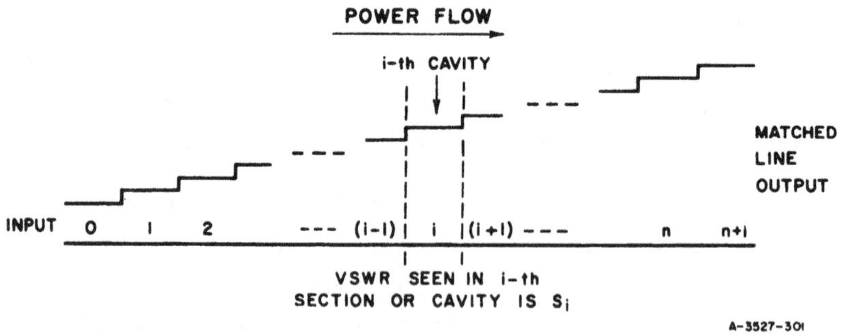

FIG. 6.14-1 VSWR INSIDE A FILTER OR TRANSFORMER

$$U_i = \frac{\text{Gross Power Flow}}{\text{Net Power Flow}}$$

$$= \frac{1 + |\rho_i|^2}{1 - |\rho_i|^2} \qquad (6.14\text{-}2)$$

$$= \frac{S_i^2 + 1}{2S_i} \quad .$$

The attenuation of transmission lines or dielectric media is usually denoted by α, but it is measured in various units for various purposes. Let

$$\left.\begin{aligned} \alpha_d &= \text{attenuation measured in decibels per unit length} \\ \alpha_n &= \text{attenuation measured in nepers per unit length} \\ \alpha_0 &= \text{absorption coefficient (used in optics}^{12}\text{)} \quad . \end{aligned}\right\} (6.14\text{-}3)$$

The absorption coefficient, α_0, is defined as the fraction of the incident power absorbed per unit length. Thus, if P_{inc} is the incident power (or irradiance) in the z-direction, then

$$\alpha_0 = -\frac{1}{P_{inc}} \cdot \frac{dP_{inc}}{dz} \quad . \qquad (6.14.4)$$

These three attenuation constants, α_d, α_n, and α_0, are related as follows:

$$\begin{aligned} \alpha_n &= \alpha_0/2 \text{ nepers} \\ \alpha_d &= (10 \log_{10} e)\alpha_0 \approx 4.343\alpha_0 \text{ decibels} \qquad (6.14\text{-}5) \\ &= (20 \log_{10} e)\alpha_n \approx 8.686\alpha_n \text{ decibels} \quad . \end{aligned}$$

Denote the length of the ith cavity or section by l_i. If each l_i is equal to an integral number of quarter-wavelengths, with impedance maxima and minima at the ends, as is the case with synchronously tuned, stepped-impedance filters and transformers at center frequency, then the dissipation loss (if small) is given by[12]

$$
\left.
\begin{aligned}
(\Delta L_A)_0 &= (1 - |\rho_0|^2) \sum_{i=1}^{n} \alpha_{di} l_i U_i \quad \text{decibels} \\
&= (1 - |\rho_0|^2) \sum_{i=1}^{n} \alpha_{ni} l_i U_i \quad \text{nepers} \\
&= (1 - |\rho_0|^2) \sum_{i=1}^{n} \alpha_{0i} l_i U_i \\
&\quad \text{as a fraction of the incident power}
\end{aligned}
\right\} \qquad (6.14\text{-}6)
$$

where $|\rho_0|$ is again the reflection coefficient amplitude at the input.

To calculate the dissipation loss from Eq. (6.14-6), the gross-to-net power flow ratio, U_i, has to be determined from Eq. (6.14-2). For half-wave filters this is particularly simple, since

$$
S_i = \left(\frac{Z'_i}{Z'_{n+1}} \right)^{\pm 1} > 1 \qquad (6.14\text{-}7)
$$

where Z'_i is the impedance of the line forming the ith cavity and Z'_{n+1} is the output impedance of the half-wave filter. The half-wave filter impedances, Z'_i, can be worked out as in Example 1 of Sec. 6.07, or Examples 1 and 2 of Sec. 6.09, or from Fig. 6.03-1. Since the filter or transformer is synchronously tuned,

$$
\left.
\begin{aligned}
S_n &= V_{n+1} \\
S_{n-1} &= \left(\frac{V_n}{S_n} \right)^{\pm 1} > 1
\end{aligned}
\right\} \qquad (6.14\text{-}8)
$$

$$
\left.
\begin{aligned}
S_i &= \left(\frac{V_{i+1}}{S_{i+1}}\right)^{\pm 1} > 1 \\
&\text{------} \\
S_1 &= \left(\frac{V_2}{S_2}\right)^{\pm 1} > 1 \\
\text{Input VSWR} = S_0 &= \left(\frac{V_1}{S_1}\right)^{\pm 1} > 1 \quad .
\end{aligned}
\right\} \qquad (6.14\text{-}8) \text{ (contd.)}
$$

The internal VSWR, S_i, for synchronous filters, can also be written in the form

$$
S_i = \left(\frac{V_{i+1} V_{i+3} V_{i+5} \cdots}{V_{i+2} V_{i+4} \cdots\cdots\cdot}\right)^{\pm 1} > 1 \quad . \qquad (6.14\text{-}9)
$$

The highest suffix of any V in this equation is $n + 1$.

Narrow-Band Filters—For narrow-band filters of large R (filters with large stop-band attenuation), Eq. (6.09-2) combined with the formulas[7] for the g_i (Sec. 4.03) shows that the V_i increase toward the center (compare Table 6.10-1 or Fig. 6.10-2). Therefore, the positive exponent must be taken in Eq. (6.14-9) and hence throughout Eq. (6.14-8). Then

$$
V_i = S_i S_{i-1} \quad (i = 1, 2, \ldots, n + 1) \quad . \qquad (6.14\text{-}10)
$$

Since the output is matched ($S_{n+1} = 1$), and from Eq. (6.04-10), the maximum possible VSWR (in the stop band) is

$$
R = S_0 (S_1 S_2 \ldots S_n)^2 \quad . \qquad (6.14\text{-}11)
$$

With the restriction of constant R, it can be shown[12] that when all the $\alpha_i l_i$ products are equal, Eq. (6.14-6) gives minimum dissipation loss when all the S_i are made equal. The internal V_i are then all equal to each other, and equal to the square of $V_1 = V_{n+1}$. Such a filter (called a "periodic filter") gives minimum band-center dissipation loss for a given R (*i.e.*, for a given maximum stop-band attenuation). (In optical terms, it gives maximum "contrast".) General formulas including filters of this type have been given by Mielenz[28] and by Abelès.[29]

Since the attenuation, α_n, and the unloaded Q, Q_u, are related by[30]

$$\alpha_n = \frac{1}{Q_u}\frac{\pi}{\lambda_g}\left(\frac{\lambda_g}{\lambda}\right)^2 , \qquad (6.14\text{-}12)$$

therefore $(\Delta L_A)_0$ can be expressed in terms of Q_u,

$$\left.\begin{aligned} (\Delta L_A)_0 &= (1 - |\rho_0|^2)\,\pi \sum_{i=1}^{n} \frac{1}{Q_{ui}}\frac{l_i}{\lambda_{gi}}\left(\frac{\lambda_{gi}}{\lambda}\right)^2 U_i \quad \text{nepers} \\ &= 27.28\,(1 - |\rho_0|^2) \sum_{i=1}^{n} \frac{1}{Q_{ui}}\frac{l_i}{\lambda_{gi}}\left(\frac{\lambda_{gi}}{\lambda}\right)^2 U_i \quad \text{decibels} \end{aligned}\right\} \qquad (6.14\text{-}13)$$

To relate this to Eq. (4.13-11), we must assume narrow-band filters with large R. As in Chapter 4 and Ref. 31, it is convenient to normalize the low-pass filter prototype elements to $g_0 = 1$. In Eq. (4.13-2) and in Ref. 31, w is the frequency fractional bandwidth, related to w_q or w_h (Secs. 6.02 and 6.03) of dispersive waveguide filters by[32]

$$w = w_q\left(\frac{\lambda}{\lambda_g}\right)^2 \quad \text{or} \quad w_h\left(\frac{\lambda}{\lambda_g}\right)^2 , \qquad (6.14\text{-}14)$$

whichever is appropriate. This can be shown to lead, for small w and large R, to

$$
\left.
\begin{aligned}
(\Delta L_A)_0 &\approx (1 - |\rho_0|^2) \frac{\omega_1'}{2w} \sum_{i=1}^{n} \frac{g_i}{Q_{ui}} && \text{nepers} \\
&= 4.343\,(1 - |\rho_0|^2) \frac{\omega_1'}{w} \sum_{i=1}^{n} \frac{g_i}{Q_{ui}} && \text{decibels}
\end{aligned}
\right\} \quad (6.14\text{-}15)
$$

It differs from Eq. (4.13-11) and Ref. 31 for the low-pass lumped-constant filter by an additional factor

$$
(1 - |\rho_0|^2) = 1/\text{antilog}\,[|(L_A)_0|/10] \quad . \qquad (6.14\text{-}16)
$$

If this factor is added to Eq. (4.13-11) or Eq. (1) in Ref. 31, they also become more accurate. [For instance, multiplying the last column in Table 4.13-2 by the factor in Eq. (6.14-16), approximates the exact values in the first column for $(L_A)_0$ more closely, reducing the error by an order of magnitude in every case.]

Equation (6.14-6) is the most accurate available formula for the dissipation loss at center frequency of a quarter-wave or half-wave filter, and can be applied to any such filter directly; Eq. (6.14-15) is the most accurate available formula for band-pass filters derived from the low-pass lumped-constant filter prototype of Chapter 4. Equation (6.14-6) or (6.14-15) determines the dissipation loss at the center of the pass band. The dissipation loss generally stays fairly constant over most of the pass band, rising to sharp peaks just outside both edges, as indicated in Fig. 6.14-2(a). When the total attenuation (reflection loss plus dissipation loss) is plotted against frequency, the appearance of the response curve in a typical case is as shown in Fig. 6.14-2(b); the two "dimples" are due to the two dissipation peaks shown in Fig. 6.14-2(a).

The two peaks of dissipation loss near the two band-edges may be attributed to a build-up in the internal fields and currents. Thus we would expect the power-handling capacity of the filter to be approximately

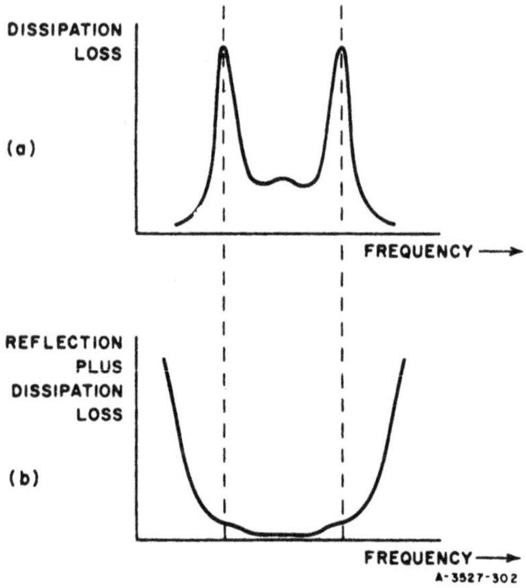

SOURCE: Jour. Opt. Soc. Am. (see Ref. 12 by L. Young)

FIG. 6.14-2 ATTENUATION CHARACTERISTICS OF FILTERS

inversely proportional to the dissipation loss, as the frequency changes. An increase in stored energy for a matched filter is in turn associated with a reduced group velocity,[32] or increased group delay. Thus we would expect the group delay through the filter to be approximately proportional to the dissipation loss, as the frequency changes. This has already been pointed out in Sec. 4.13. These questions are taken up further in Sec. 6.15.

Example 1—The parameters of a half-wave filter are: $Z_0' = 1$, $Z_1' = 245.5$, $Z_2' = 0.002425$, $Z_3' = 455.8$, $Z_4' = 0.0045$, $Z_5' = 1.106$ (corresponding to a 0.01-db pass-band ripple for a lossless filter of bandwidth $w_h = 0.00185$). Calculate the center-frequency dissipation loss if this filter is constructed in waveguide having an attenuation of 4.05 db/100 ft. Wavelength $= \lambda_0 = 1.437$ inches; waveguide width $= a = 1.015$ inches. (See also Examples 1 and 2, pages 344 and 352.)

The guide wavelength is

$$\lambda_{g0} = 2.034 \text{ inches}$$

and

$$(\lambda_{g0}/\lambda_0)^2 = 2.00 \quad .$$

The internal VSWRs are by Eq. (6.14-7),

$$S_1 = (Z_1'/Z_5') = 222.0$$

$$S_2 = (Z_5'/Z_2') = 455.8$$

$$S_3 = (Z_3'/Z_5') = 412.5$$

$$S_4 = (Z_5'/Z_4') = 245.5$$

$$S_5 = 1.0 \quad \text{(by definition)} \quad .$$

Summing these gives

$$\sum_{i=1}^{4} U_i \approx \frac{1}{2} \sum_{i=1}^{4} S_i = 667.9$$

Since the center-frequency input VSWR is equal to $Z_5' = 1.106$, therefore

$$|\rho_0|^2 = 0.0025 \quad .$$

Hence from the first of Eqs. (6.14-6),

$$(\Delta L_A)_0 = 0.9975 \times \frac{4.05}{100 \times 12} \times 1.017 \times 667.9 \text{ decibels}$$

$$= 2.29 \text{ db} \quad .$$

SEC. 6.15, GROUP DELAY

The slope of the phase-versus-frequency curve of a matched filter is a measure of the group delay through the filter. This has already been discussed in Sec. 4.08, and results for some typical low-pass filter prototypes with $n = 5$ elements are given in Figs. 4.08-1 and 4.08-2. In this section, group delay, dissipation loss, and power-handling capacity will be examined in terms of stepped-impedance filters, such as the quarter-wave transformer prototype.

It can be shown[33] that the group delay at center frequency f_0 through a homogeneous matched quarter-wave transformer is given by

$$f_0(t_d)_0 = \frac{1}{4}\left(\frac{\lambda_{g0}}{\lambda_0}\right)^2 \sum_{i=1}^{n} U_i \tag{6.15-1}$$

where t_d is the phase slope $d\phi/d\omega$ and may be interpreted as the group delay in the pass band. (The phase slope $t_d = d\phi/d\omega$ will, as usual, be referred to as the group delay also outside the pass band, although its physical meaning is not clear when the attenuation varies rapidly with frequency.)

The group delay of a half-wave filter is just twice that of its quarter-wave transformer prototype; in general, the group delay of any matched stepped-impedance filter at center frequency is given by[33]

$$f_0(t_d)_0 = \sum_{i=1}^{n} \left(\frac{\lambda_{gi}}{\lambda}\right)_0^2 \left(\frac{l_i}{\lambda_{gi}}\right)_0 U_i \quad . \tag{6.15-2}$$

Combining Eq. (6.15-2) with Eq. (6.14-6) when $\rho_0 = 0$ (filter matched at center frequency), and when the attenuation constants α and guide wavelengths λ_g are the same in each section, yields

$$\Delta L_A = \alpha\lambda_g(\lambda/\lambda_g)^2 f_0 t_d \tag{6.15-3}$$

where α may be measured in units of nepers per unit length (α_n), or in units of decibels per unit length (α_d), ΔL_A being measured accordingly in nepers or decibels.

Equation (6.15-3) can also be written*

$$\Delta L_A = \frac{\pi}{Q_u} f_0 t_d \qquad \text{nepers} \quad . \tag{6.15-4}$$

These equations have been proved for center frequency only. It can be argued from the connection between group velocity and stored energy[33] that the relations (6.15-3) and (6.15-4) between dissipation loss and group delay should hold fairly well over the entire pass band. For this reason the suffix 0 has been left out of Eqs. (6.15-3) and (6.15-4). This conclusion can also be reached through Eqs. (4.13-2), (4.13-3) and (4.13-9) in Chapter 4.

Example 1—Calculate the time delay $(t_d)_0$ at the center frequency of the filter in Example 1 of Sec. 6.14 from its center-frequency dissipation loss, $(\Delta L_A)_0$.

* To convert nepers to decibels, multiply by 8.686.

From Eq. (6.15-3),

$$f_0(t_d)_0 = \left(\frac{\lambda_g}{\lambda}\right)^2 \frac{(\Delta L_A)_0}{\alpha\lambda_g} \text{ cycles at center frequency}$$

$$= 2.00 \times \frac{2.29}{\left(\dfrac{4.05 \times 2.034}{100 \times 12}\right)} \text{ cycles at center frequency}$$

$$= 668 \text{ cycles at center frequency} \quad .$$

Since $\lambda_0 = 1.437$ inches, which corresponds to $f_0 = 8220$ Mc, therefore

$$(t_d)_0 = \frac{668}{8220} \text{ microseconds}$$

$$= 81.25 \text{ nanoseconds} \quad .$$

Universal Curves of Group Delay—Curves will be presented in Figs. 6.15-1 through 6.15-10 which apply to stepped-impedance transformers and filters of large R and small bandwidth (up to about $w_q = 0.4$). They were computed for specific cases (generally $R \approx 10^{2n}$ and $w_q = 0.20$), but are plotted in a normalized fashion and then apply generally for large R, small w. The response is plotted not directly against frequency, but against

$$x = \pm\sigma\left(\frac{f - f_0}{f_0}\right) \tag{6.15-5}$$

with σ given by

$$\sigma = pR^{1/2n} \tag{6.15-6}$$

where p is the length of each section measured in quarter-wavelengths. (Thus $p = 1$ for a quarter-wave transformer, and $p = 2$ for a half-wave

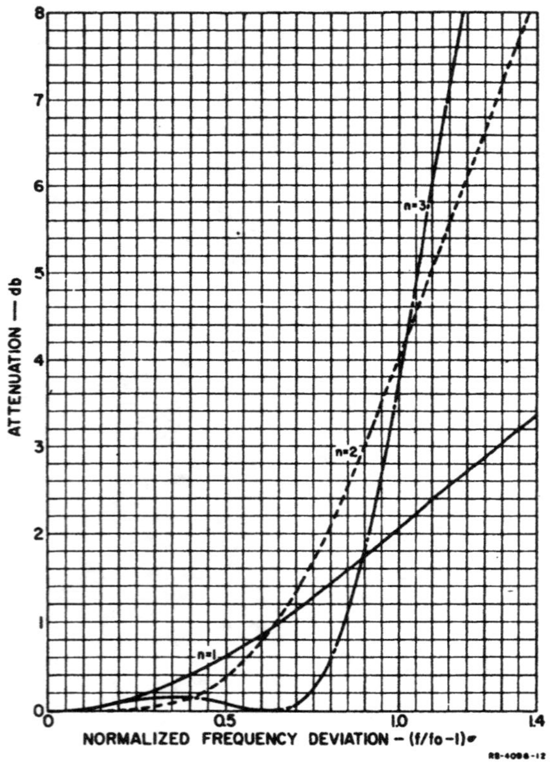

SOURCE: Quarterly Progress Report 1, Contract AF 30(602)-2734, SRI (see Ref. 33 by L. Young)

FIG. 6.15-1 ATTENUATION CHARACTERISTICS OF PERIODIC FILTERS OF n = 1, 2, AND 3 SECTIONS

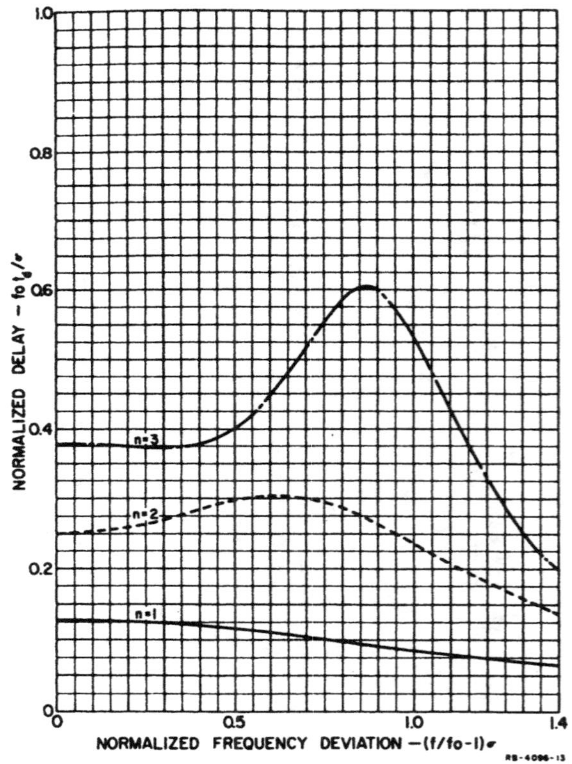

SOURCE: Quarterly Progress Report 1, Contract AF 30(602)-2734, SRI (see Ref. 33 by L. Young)

FIG. 6.15-2 GROUP DELAY OF PERIODIC FILTERS OF n = 1, 2, AND 3 SECTIONS

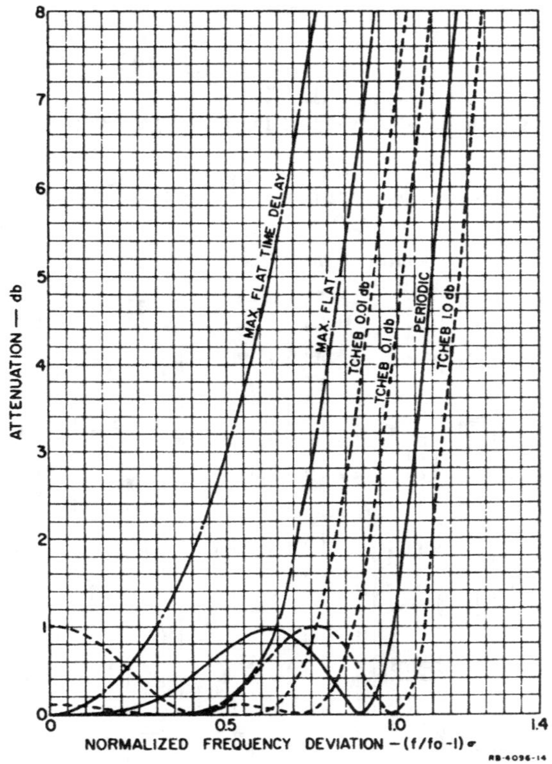

SOURCE: Quarterly Progress Report 1, Contract AF 30(602)-2734, SRI (see Ref. 33 by L. Young)

FIG. 6.15-3 ATTENUATION CHARACTERISTICS OF SIX FILTERS OF n = 4 SECTIONS

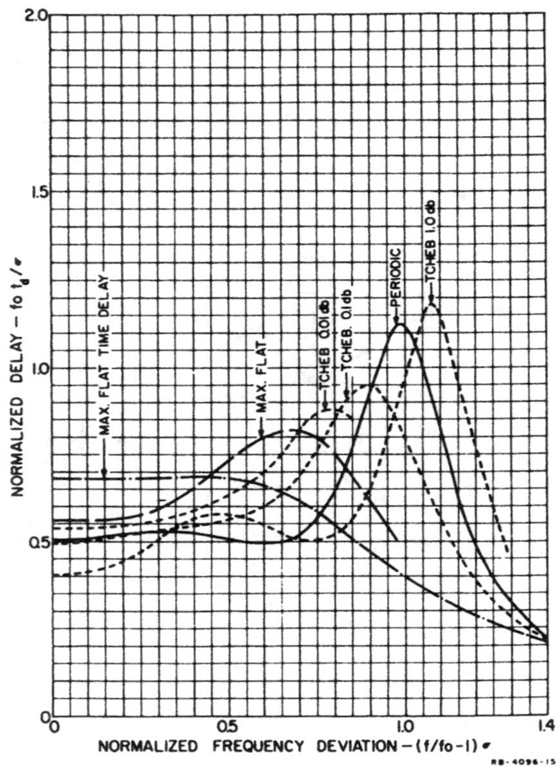

SOURCE: Quarterly Progress Report 1, Contract AF 30(602)-2734, SRI (see Ref. 33 by L. Young)

FIG. 6.15-4 GROUP DELAY CHARACTERISTICS OF SIX FILTERS OF n = 4 SECTIONS

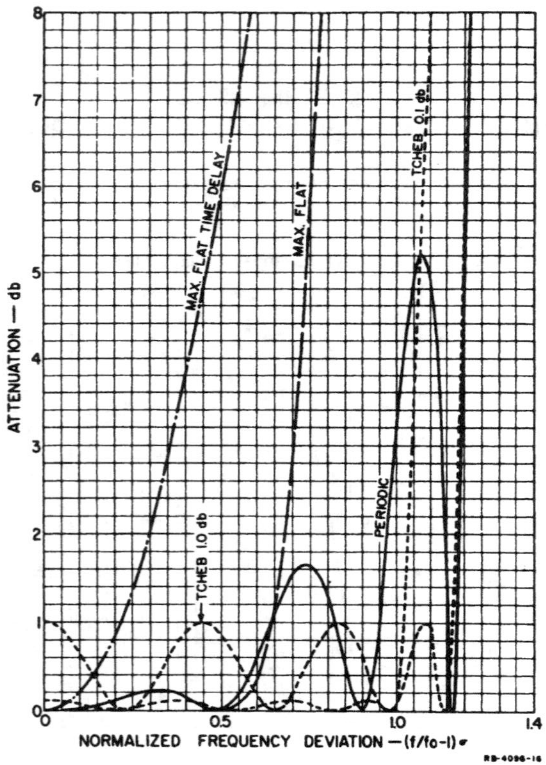

SOURCE: Quarterly Progress Report 1, Contract AF 30(602)-2734, SRI (see Ref. 33 by L. Young)

FIG. 6.15-5 ATTENUATION CHARACTERISTICS OF FIVE FILTERS OF n = 8 SECTIONS

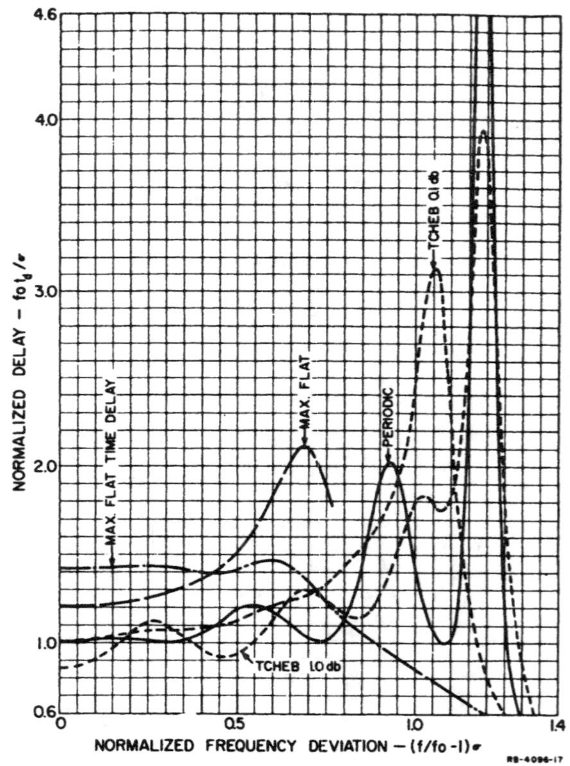

SOURCE: Quarterly Progress Report 1, Contract AF 30(602)-2734, SRI (see Ref. 33 by L. Young)

FIG. 6.15-6 GROUP DELAY CHARACTERISTICS OF FIVE FILTERS OF n = 8 SECTIONS

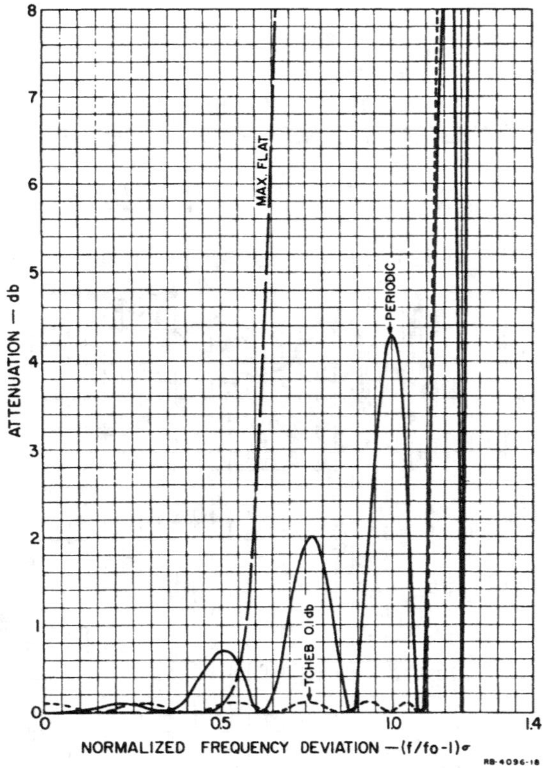

SOURCE: Quarterly Progress Report 1, Contract AF 30(602)-2734, SRI (see Ref. 33 by L. Young)

FIG. 6.15-7 ATTENUATION CHARACTERISTICS OF THREE FILTERS OF n = 12 SECTIONS

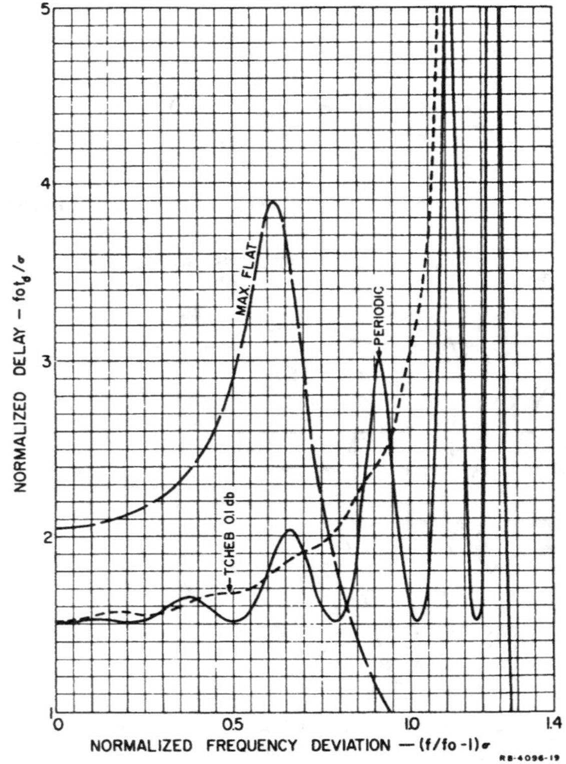

SOURCE: Quarterly Progress Report 1, Contract AF 30(602)-2734, SRI (see Ref. 33 by L. Young)

FIG. 6.15-8 GROUP DELAY CHARACTERISTICS OF THREE FILTERS OF n = 12 SECTIONS

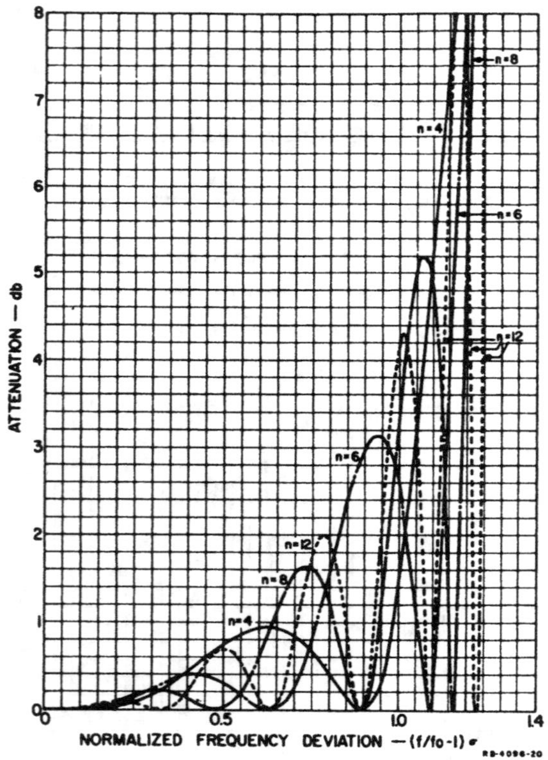

SOURCE: Quarterly Progress Report 1, Contract AF 30(602)-2734, SRI (see Ref. 33 by L. Young)

FIG. 6.15-9 ATTENUATION CHARACTERISTICS OF FOUR PERIODIC FILTERS WITH UP TO n = 12 SECTIONS

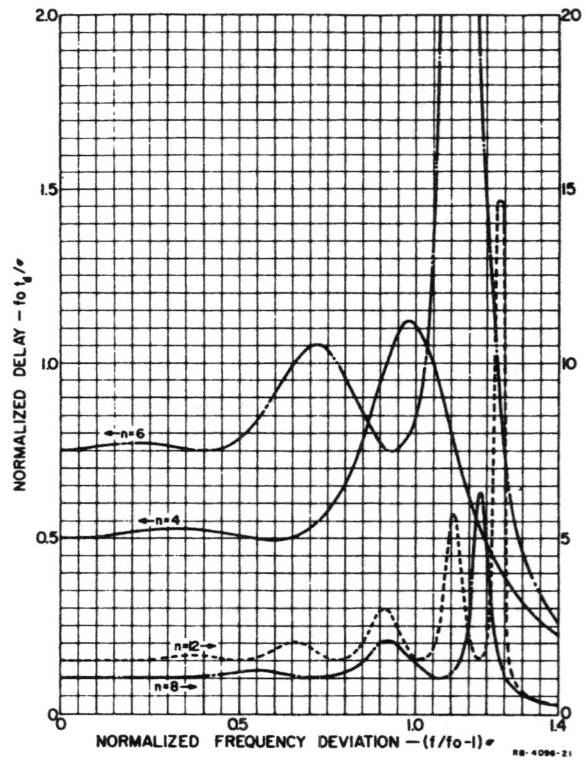

SOURCE: Quarterly Progress Report 1, Contract AF 30(602)-2734, SRI (see Ref. 33 by L. Young)

FIG. 6.15-10 GROUP DELAY CHARACTERISTICS OF FOUR PERIODIC FILTERS WITH UP TO n = 12 SECTIONS

filter.) For maximally flat filters, Eq. (6.15-6) with the aid of Sec. 6.02 reduces to

$$\sigma = \frac{4}{\pi}\left(\frac{4^{1/2n}}{w_{3\text{-db}}}\right) \tag{6.15-7}$$

where $w_{3\text{-db}}$ in Eq. (6.15-7) is the 3-db fractional bandwidth; while for Tchebyscheff transformers,

$$\sigma = \frac{8}{\pi}\left(\frac{\varepsilon_r^{\,1/2n}}{w}\right) . \tag{6.15-8}$$

Similarly it can be shown[33] for maximally flat time-delay filters, that

$$\left.\begin{aligned} \sigma &= 8\left[\frac{2(n!)}{(2n)!}\right]^{1/n} f_0(t_d)_0 \\ &= 4\left[\frac{2}{1.3.5.7.\ \ldots\ .(2n-1)}\right]^{1/n} f_0(t_d)_0 \end{aligned}\right\} \tag{6.15-9}$$

and that for equal-element filters (corresponding to periodic filters),

$$\sigma = \frac{4}{\pi}\,\frac{\omega_1'}{w}\,g \quad . \tag{6.15-10}$$

It can be deduced from Eq. (6.09-2) that the attenuation characteristics are independent of bandwidth or the value of R when plotted against x, defined by Eq. (6.15-5). Similarly, it follows from Eqs. (6.15-1) and (6.09-2) that the time delay should be plotted as

$$y = \frac{f_0 t_d}{\sigma} \tag{6.15-11}$$

so that it should become independent of bandwidth and the quantity R (still supposing small bandwidth, large R).

By using Eq. (6.15-7) through (6.15-10) to obtain σ, the curves in Figs. 6.15-1 through 6.15-10 can be used also for lumped-constant filters.

These curves are useful not only for predicting the group delay, but also for predicting the dissipation loss and (less accurately) the power-handling capacity in the pass band, when the values of these quantities at midband are already known [as, for instance, by Eq. (6.14-6) or (6.14-15)].

The following filter types are presented: maximally flat; Tchebyschef (0.01 db ripple, 0.1 db ripple and 1.0 db ripple); maximally flat time delay and periodic filters. The last-named are filters in which $V_1^2 = V_i = V_{n+1}^2$ for $i = 2, 3, \ldots, n$. [They correspond to low-pass prototype filters in which all the g_i ($i = 1, 2, \ldots, n$) in Fig. 4.04-1 are equal to one another. For large R and small bandwidth periodic filters give minimum band-center dissipation loss[12, 31] and greatest power-handling capacity[34] for a given selectivity.]

The figures go in pairs, the first plotting the attenuation characteristics, and the second the group delay. Figures 6.15-1 and 6.15-2 are for three periodic filters. The case $n = 1$ cannot be labelled, as it belongs to all types. The case $n = 2$ periodic is also maximally flat. The case $n = 3$ periodic is equivalent to a Tchebyscheff filter of about 0.15 db ripple.

Figures 6.15-3 to 6.15-8 are for $n = 4$, $n = 8$, and $n = 12$ sections, respectively, and include various conventional filter types. Figures 6.15-9 and 6.15-10 are for several periodic filters, showing how the characteristics change from $n = 4$ to $n = 12$ sections.

Example 2—Calculate the dissipation loss at band-edge of the filter in Example 1 of Sec. 6.14.

It was shown in that example that the band-center dissipation loss for that filter is 2.29 db. Since this is a Tchebyscheff 0.01-db ripple filter with $n = 4$, we see from Fig. 6.15-4 that the ratio* of band-edge to band-center dissipation loss is approximately 0.665/0.535 = 1.243. Therefore the band-edge dissipation loss is approximately 2.29 × 1.243 = 2.85 db.

The application of the universal curves to the power-handling capacity of filters is discussed in Section 15.03.

* Exact calculations made later showed that the ratio dissipation loss/group delay remained constant to within one-percent over the pass band.

REFERENCES

1. S. B. Cohn, "Optimum Design of Stepped Transmission-Line Transformers," *IRE Trans. PGMTT*-3, pp. 16-21 (April 1955).

2. R. E. Collin, "Theory and Design of Wide-Band Multisection Quarter-Wave Transformers," *Proc. IRE* 43, pp. 179-185 (February 1955).

3. H. J. Riblet, "General Synthesis of Quarter-Wave Impedance Transformers," *IRE Trans. PGMTT*-5, pp. 36-43 (January 1957).

4. Leo Young, "Tables for Cascaded Homogeneous Quarter-Wave Transformers," *IRE Trans. PGMTT*-7, pp. 233-237 (April 1959), and *PGMTT*-8, pp. 243-244 (March 1960).

5. Leo Young, "Optimum Quarter-Wave Transformers," *IRE Trans. PGMTT*-8, pp. 478-482 (September 1960).

6. Leo Young, "Inhomogeneous Quarter-Wave Transformers of Two Sections," *IRE Trans. PGMTT*-8, pp. 645-649 (November 1960).

7. S. B. Cohn, "Direct-Coupled-Resonator Filters," *Proc. IRE* 45, pp. 187-196 (February 1957).

8. G. L. Matthaei, "Direct-Coupled Band-Pass Filters with $\lambda/4$ Resonators," IRE National Convention Record, Part 1, pp. 98-111 (March 1958).

9. Leo Young, "The Quarter-Wave Transformers Prototype Circuit," *IRE Trans. PGMTT*-8, pp. 483-489 (September 1960).

10. Leo Young, "Synchronous Branch Guide Directional Couplers for Low and High Power Applications," *IRE Trans. PGMTT*-10, pp. 459-475 (November 1962).

11. Leo Young, "Synthesis of Multiple Antireflection Films over a Prescribed Frequency Band," *J. Opt. Soc. Am.*, 51, pp. 967-974 (September 1961).

12. Leo Young, "Prediction of Absorption Loss in Multilayer Interference Filters," *J. Opt. Soc. Am.*, 52, pp. 753-761 (July 1962).

13. J. E. Holte and R. F. Lambert, "Synthesis of Stepped Acoustic Transmission Systems," *J. Acoust. Soc. Am.*, 33, pp. 289-301 (March 1961).

14. Leo Young, "Stepped Waveguide Transformers and Filters," Letter in *J. Acoust. Soc. Am.*, 33, p. 1247 (September 1961).

15. Leo Young, "Inhomogeneous Quarter-Wave Transformers," *The Microwave Journal*, 5, pp. 84-89 (February 1962).

16. IRE Standards on Antennas and Waveguides, *Proc. IRE* 47, pp. 568-582 (1959). See especially, p. 581.

17. G. C. Southworth, *Principles and Application of Waveguide Transmission* (D. Van Nostrand Co., Inc., New York City, 1950).

18. C. L. Dolph, "A Current Distribution for Broadside Arrays Which Optimizes the Relationship Between Beam Width and Side-Lobe Level," *Proc. IRE* 34, pp. 335-348 (June 1946). See also Discussion in *Proc. IRE* 35, pp. 489-492 (May 1947).

19. L. B. Brown and G. A. Sharp, "Tchebyscheff Antenna Distribution, Beamwidth and Gain Tables," NAVORD Report 4629 (NOLC Report 383), Naval Ordnance Laboratory, Corona, California (28 February 1958).

20. M. L. Reuss, Jr., "Some Design Considerations Concerning Linear Arrays Having Dolph-Tchebyscheff Amplitude Distributions," NRL Report 5240, ASTIA Number AD 212 621 (12 February 1959).

21. G. J. Van der Maas, "A Simplified Calculation for Dolph-Tchebyscheff Arrays," *J. Appl. Phys.* 24, p. 1250 (September 1953).

22. J. R. Whinnery, H. W. Jamieson, and Theo Eloise Robbins, "Coaxial Line Discontinuities," *Proc. IRE* 32, pp. 695-709 (November 1944).

23. N. Marcuvitz, *Waveguide Handbook*, MIT Rad. Lab. Series, Vol. 10 (McGraw-Hill Book Co., Inc., New York City, 1951).

24. C. G. Montgomery, R. H. Dicke, and E. M. Purcell, *Principles of Microwave Circuits*, MIT Rad. Lab. Series Vol. 9 (McGraw-Hill Book Co., Inc., New York City, 1948).

25. E. A. Ohm, "A Broad-Band Microwave Circulator," *IRE Trans. PGMTT*-4, pp. 210-217 (October 1956).

26. B. Bramham, "A Convenient Transformer for Matching Coaxial Lines," *Electronic Engineering* 33, pp. 42-44 (January 1961).

27. G. L. Ragan, *Microwave Transmission Circuits*, MIT Rad. Lab. Series, Vol. 9, pp. 29-36 (McGraw-Hill Book Co., Inc., New York City, 1946).

28. K. D. Mielenz, "Use of Chebychev Polynomials in Thin Film Computations," *J. Res. Nat. Bur. Stand.*, 63A, pp. 297-300 (November-December 1959). [There is a misprint in Eq. (17b): the lower left element in the matrix should be $S_{n-1}a_{21}$.]

29. F. Abelès, "Sur l'élévation à la puissance *n* d'une matrice carrée a quatre éléments à l'aide des polynomes de Tchébychev,"Comptes Rendus 226, pp. 1872-1874 (1948) and "Transmission de la lumière à travers un systeme de lames minces alternées," Comptes Rendus, 226, pp. 1808-1810 (1948).

30. Leo Young, "Q-Factors of a Transmission Line Cavity," *IRE Trans PGCT*-4, pp. 3-5 (March 1957).

31. S. B. Cohn, "Dissipation Loss in Multiple-Coupled-Resonator Filters," *Proc. IRE* 47, pp. 1342-1348 (August 1959).

32. Leo Young, "Analysis of a Transmission Cavity Wavemeter," *IRE Trans. PGMTT*-8, pp. 436-439 (July 1960).

33. Leo Young, "Suppression of Spurious Frequencies," Sec. III, Quarterly Progress Report 1, SRI Project 4096, Contract AF 30(602)-2734, Stanford Research Institute, Menlo Park, California (July 1962). See also: Leo Young, "Group Delay and Transmission Loss in Transmission-Line Filters," *IEEE Trans. PTGMTT*-11, pp. 215-217 (May 1963).

34. S. B. Cohn, "Design Considerations for High-Power Microwave Filters," *IRE-Trans. PGMTT*-7, pp. 149-153 (January 1959).

NOTE: The material for this chapter is largely derived from:

35. Leo Young and G. L. Matthaei, "Microwave Filters and Coupling Structures," Quarterly Progress Report 4, SRI Project 3527, Contract DA 36-039 SC-87398, Stanford Research Institute, Menlo Park, California (January 1962).

Sections 6.01 through 6.07 and Sections 6.09 and 6.10 are mostly contained in:

36. Leo Young, "Stepped Impedance Transformers and Filter Prototypes," *IRE Trans. PGMTT*-10, pp. 339-359 (September 1962).

Experimental results on a seven-section quarter-wave transformer, and a comparison with a linear taper are given in:

37. Leo Young, "Practical Design of a Wide-Band Quarter-Wave Transformer in Waveguide," *The Microwave Journal*, 6, No. 10, pp. 76-79 (October 1963).

CHAPTER 7

LOW-PASS AND HIGH-PASS FILTERS USING SEMI-LUMPED ELEMENTS OR WAVEGUIDE CORRUGATIONS

SEC. 7.01, PROPERTIES OF THE FILTERS DISCUSSED IN THIS CHAPTER

Unlike most of the filter structures to be discussed in later chapters, the microwave filters treated in this chapter consist entirely of elements which are small compared to a quarter-wavelength (at pass-band frequencies). In the cases of the TEM-mode filters treated, the design is carried out so as to approximate an idealized lumped-element circuit as nearly as possible. In the cases of the corrugated and waffle-iron low-pass waveguide filters discussed, the corrugations are also small compared to a quarter-wavelength. Such filters are a waveguide equivalent of the common series-L, shunt-C, ladder type of low-pass filter, but due to the waveguide nature of the structure, it is more difficult to design them as a direct approximation of a lumped-element, low-pass filter. Thus, in this chapter the waveguide filters with corrugations are treated using the image method of design (Chapter 3).

In Sec. 7.02 will be found a discussion of how lumped elements may be approximated using structures which are practical to build for microwave applications. In later sections the design of filters in specific common types of construction are discussed, but using the principles in Sec. 7.02 the reader should be able to devise additional forms of construction as may be advantageous for special situations.

Figure 7.01-1(a) shows a coaxial form of low-pass filter which is very common. It consists of short sections of high-impedance line (of relatively thin rod or wire surrounded by air dielectric) which simulate series inductances, alternating with short sections of very-low-impedance line (each section consisting of a metal disk with a rim of dielectric) which simulate shunt capacitances. The filter shown in Fig. 7.01-1(a) has tapered lines at the ends which permit the enlarging of the coaxial region at the center of the filter so as to reduce dissipation loss. However, it is more common to build this type of filter with the outer conductor consisting of a uniform, cylindrical metal tube. The popularity

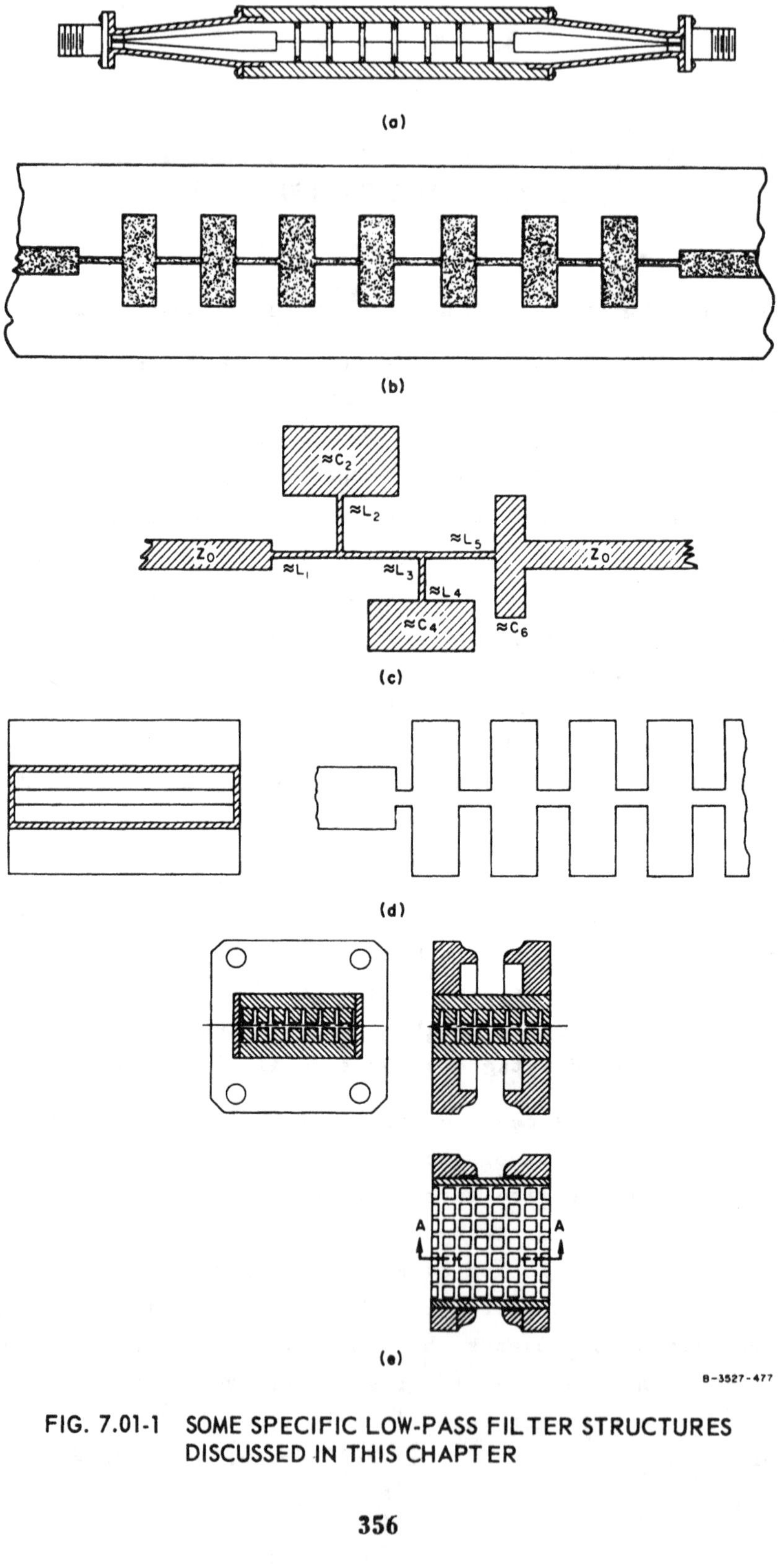

FIG. 7.01-1 SOME SPECIFIC LOW-PASS FILTER STRUCTURES DISCUSSED IN THIS CHAPTER

of this type of low-pass filter results from its simplicity of fabrication and its excellent performance capabilities. Its first spurious pass band occurs, typically, when the high-impedance lines are roughly a half-wavelength long. It is not difficult with this type of filter to obtain stop bands which are free of spurious responses up as far as five times the cutoff frequency of the filter. Filters of this type are commonly built with cutoff frequencies ranging from a few hundred megacycles up to around 10 Gc. A discussion of their design will be found in Sec. 7.03.

Figure 7.01-1(b) shows a printed-circuit, strip-line filter which is equivalent to the filter in Fig. 7.01-1(a) in most respects, but which has somewhat inferior performance characteristics. The great advantage of this type of filter is that it is unusually inexpensive and easy to fabricate. It usually consists primarily of two sheets of low-loss dielectric material with a photo-etched, copper-foil, center-conductor [shown in Fig. 7.01-1(b)] sandwiched in between, and with copper foil or metal plates on the outer surfaces of the dielectric pieces to serve as ground planes. When this type of circuit is used the dissipation loss is generally markedly higher than for the filter in Fig. 7.01-1(a) because of the presence of dielectric material throughout the circuit. Also, when this type of construction is used it is generally not possible to obtain as large a difference in impedance level between the high- and low-impedance line sections as is readily feasible in the construction shown in Fig. 7.01-1(a). As a result of this, the attenuation level at frequencies well into the stop band for filters constructed as shown in Fig. 7.01-1(b) is generally somewhat lower than that for filters constructed as shown in Fig. 7.01-1(a). Also, spurious responses in the stop band generally tend to occur at lower frequencies for the construction in Fig. 7.01-1(b). Filters using this latter construction can also be used in the 200-Mc to 10-Gc range. However, for the high portion of this range they must be quite small and they tend to have considerable dissipation loss. A discussion of the design of this type of filter will be found in Sec. 7.03.

Figure 7.01-1(c) shows another related type of printed-circuit low-pass filter. The symbols L_1, L_2, C_2, etc., indicate the type of element which different parts of the circuit approximate. Elements L_2 and C_2 in series approximate an L-C branch which will short-circuit transmission at its resonant frequency. Likewise for the part of the circuit which approximates L_4 and C_4. These branches then produce peaks of high attenuation at frequencies above the cutoff frequency and fairly close to it, and by so doing, they increase the sharpness of the cutoff characteristic. This type of filter is also easy to fabricate in photo-etched, printed-circuit construction, but has not been used as much as the type in Fig. 7.01-1(b), probably because it is somewhat more difficult to design accurately. This type of filter can also be designed in coaxial or coaxial split-block form so as to obtain improved performance, but such a filter would, of course, be markedly more costly to build. Discussion of the design of filters such as that in Fig. 7.01-1(c) will be found in Sec. 7.03.

The filter shown in Fig. 7.01-1(d) is a waveguide version of the filters in Figs. 7.01-1(a) and (b). In this case the low- and high-impedance sections of line are realized by raising and lowering the height of the guide, which has led to the name "corrugated waveguide filter" by which it is commonly known. It is a low-pass filter in its operation, but since the waveguide has a cutoff frequency, it cannot operate, of course, to DC as do most low-pass filters. This type of filter can be made to have very low pass-band loss because of its waveguide construction, and it can be expected to have a higher power rating than equivalent TEM-mode filters. However, this type of filter has disadvantages compared to, say, the coaxial filter in Fig. 7.01-1(a) because (1) it is larger and more costly to build, (2) the stop bands cannot readily be made to be free of spurious responses to as high a frequency even for the normal TE_{10} mode of propagation, and (3) there will be numerous spurious responses in the stop-band region for higher-order modes, which are easily excited at frequencies above the normal TE_{10} operating range of the waveguide. Due to the presence of the corrugations in the guide, modes having variations in the direction of the waveguide height will be cut off up to very high frequencies. Therefore, TE_{n0} modes will be the only ones that need be considered. If the waveguide is excited by a probe on its center line, the TE_{20}, TE_{40}, and other even-order modes will not be excited. In this case, the first higher-order mode that will be able to cause trouble is the TE_{30} mode which has a cutoff frequency three times that of the TE_{10} mode. In typical cases the TE_{30} mode might give a spurious response at about 2.5 times the center frequency of the first pass band. Thus, if the TE_{20} mode is not excited, or if a very wide stop band is not required, corrugated waveguide filters will frequently be quite satisfactory. The only limitations on their useful frequency range are those resulting from considerations of size and ease of manufacture. Filters of this type (or the waffle-iron filters discussed below) are probably the most practical forms of low-pass filters for frequencies of 10 Gc or higher. This type of filter is discussed in Sec. 7.04.

Figure 7.01-1(e) shows a waffle-iron filter which in many respects is equivalent to the corrugated waveguide filter in Fig. 7.01-1(d), but it includes a feature which reduces the problem of higher-order modes introducing spurious responses in the stop band. This feature consists of the fact that the low-impedance sections of the waveguide are slotted in the longitudinal direction so that no matter what the direction of the components of propagation in the waveguide are, they will see a low-pass filter type of structure, and be attenuated. Filters of this type have been constructed with stop bands which are free of spurious responses up to three times the cutoff frequency of the filter. The inclusion of longitudinal slots makes them somewhat more difficult to build than corrugated waveguide filters, but they are often worth the extra trouble. Their characteristics are the same as those of the corrugated waveguide filter, except for the improved stop band. This type of filter is discussed in Sec. 7.05.

Figure 7.01-2 shows a common type of high-pass filter using coaxial split-block construction. This type of filter is also designed so that its elements approximate lumped elements. In this case the short-circuited coaxial stubs represent shunt inductances, and the disks with Teflon spacers represent series capacitors. This type of filter has

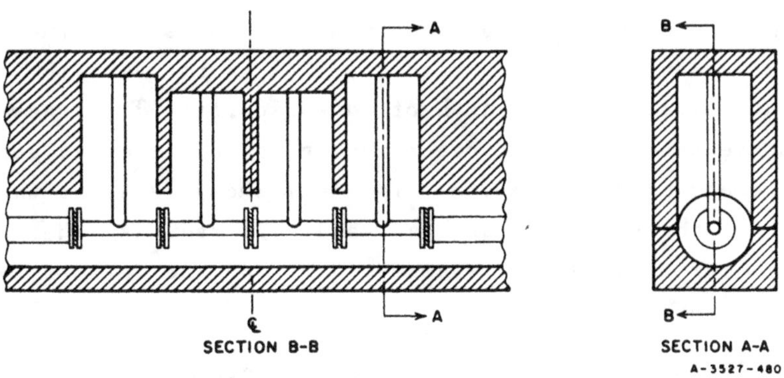

FIG. 7.01-2 A HIGH-PASS FILTER IN SPLIT-BLOCK COAXIAL CONSTRUCTION

excellent cutoff characteristics since for a design with n reactive elements there is an nth-order pole of attenuation (Sec. 2.04) at zero frequency. Typical filters of this sort have a low-attenuation, low-VSWR pass band extending up about an octave above the cutoff frequency, with relatively low attenuation extending up to considerably higher frequencies. The width of the pass band over which the filter will simulate the response of its idealized, lumped prototype depends on the frequency at which the elements no longer appear to be sufficiently like lumped elements. To achieve cutoffs at high microwave frequencies, structures of this type have to be very small, and they require fairly tight manufacturing tolerance. This makes them relatively difficult to construct for high microwave frequency applications. For this reason they are used most often for cutoffs in the lower microwave frequency range (200 to 2000 Mc) where their excellent performance and compactness has considerable advantage, but they are also sometimes miniaturized sufficiently to operate with cutoffs as high as 5 or 6 Gc. Usually at the higher microwave frequency ranges the need for high-pass filters is satisfied by using wideband band-pass filters (see Chapters 9 and 10). The type of high-pass filter in Fig. 7.01-2 has not been fabricated in equivalent printed-circuit form much because of the difficulties in obtaining good short-circuits on the inductive stubs in printed circuits, and in obtaining adequately large series capacitances.

SEC 7.02, APPROXIMATE MICROWAVE REALIZATION OF LUMPED ELEMENTS

A convenient way to realize relatively wide-band filters operating in the frequency range extending from about 100 Mc to 10,000 Mc is to construct them from short lengths of coaxial line or strip line, which approximate lumped-element circuits. Figure 7.02-1 illustrates the exact T- and π-equivalent circuits of a length of non-dispersive TEM transmission line. Also shown are the equivalent reactance and susceptance values of the networks when their physical length l is small enough so that the electrical length $\omega l/v$ of the line is less than about $\pi/4$ radians. Here we have used the symbol ω for the radian frequency and v for the velocity of propagation along the transmission line.

For applications where the line lengths are very short or where an extremely precise design is not required, it is often possible to represent a short length of line by a single reactive element. For example, inspection of Fig. 7.02-1 shows that a short length of high-Z_0 line terminated at both ends by a relatively low impedance has an effect equivalent to that

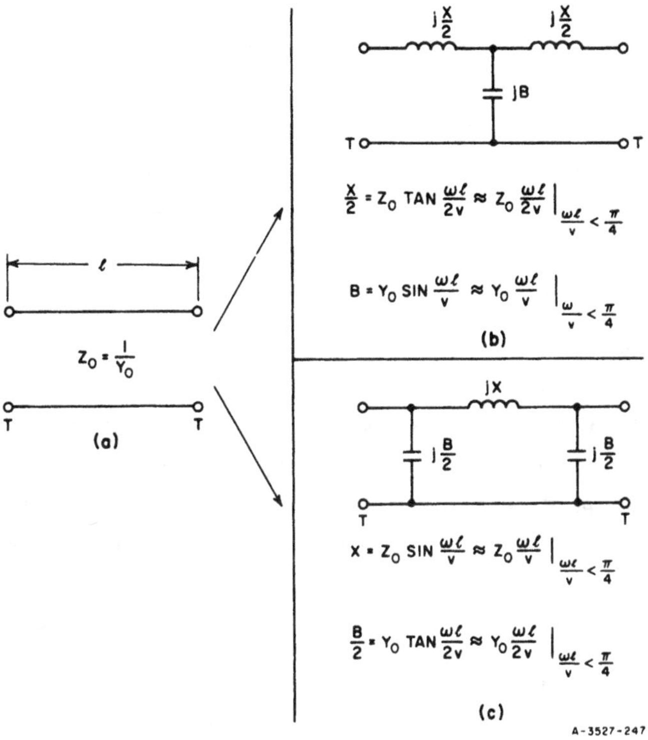

FIG. 7.02-1 TEM-LINE EQUIVALENT CIRCUITS

of a series inductance having a value of $L = Z_0 l/v$ henries. Similarly, a short length of low-Z_0 line terminated at either end by a relatively high impedance has an effect equivalent to that of a shunt capacitance $C = Y_0 l/v = l/Z_0 v$ farads. Such short sections of high-Z_0 line and low-Z_0 line are the most common ways of realizing series inductance and shunt capacitance, respectively, in TEM-mode microwave filter structures.

A lumped-element shunt inductance can be realized in TEM transmission line in several ways, as illustrated in Fig. 7.02-2(a). The most convenient way in most instances is to employ a short length of high-Z_0 line, short-circuited to ground at its far end, as shown in the strip-line example. For applications where a very compact shunt inductance is required, a short length of fine wire connected between the inner and outer conductors can be used, as is illustrated in the coaxial line example in Fig. 7.02-2(a).

Also, a lumped element series capacitance can be realized approximately in TEM transmission lines in a variety of ways, as illustrated in Fig. 7.02-2(b). Often the most convenient way is by means of a gap in the center conductor.[1] Where large values of series capacitance are required in a coaxial system a short length of low-Z_0, open-circuited line, in series with the center conductor can be used. Values of the series capacitance of overlapping strip lines are also shown in Fig. 7.02-2(b). Section 8.05 presents some further data on capacitive gaps.

A lumped-element, series-resonant, shunt circuit can be realized in strip line in the manner shown in Fig. 7.02-2(c). It is usually necessary when computing the capacitive reactance of the low-impedance (Z_{01}) line in Fig. 7.02-2(c) to include the fringing capacitance at the end of the Z_{01} line and at the step between lines. The end fringing capacitance can be accounted for as follows. First, compute the per-unit-length capacitance

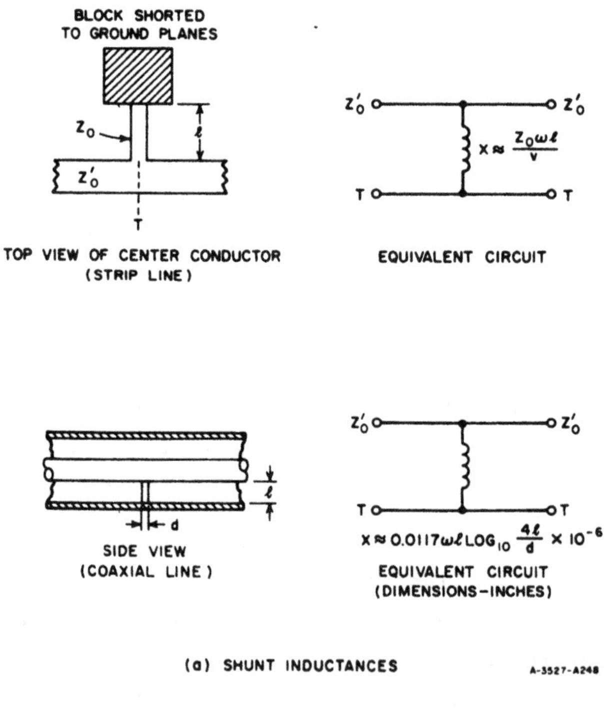

FIG. 7.02-2 SEMI-LUMPED-ELEMENT CIRCUITS IN TEM TRANSMISSION LINE

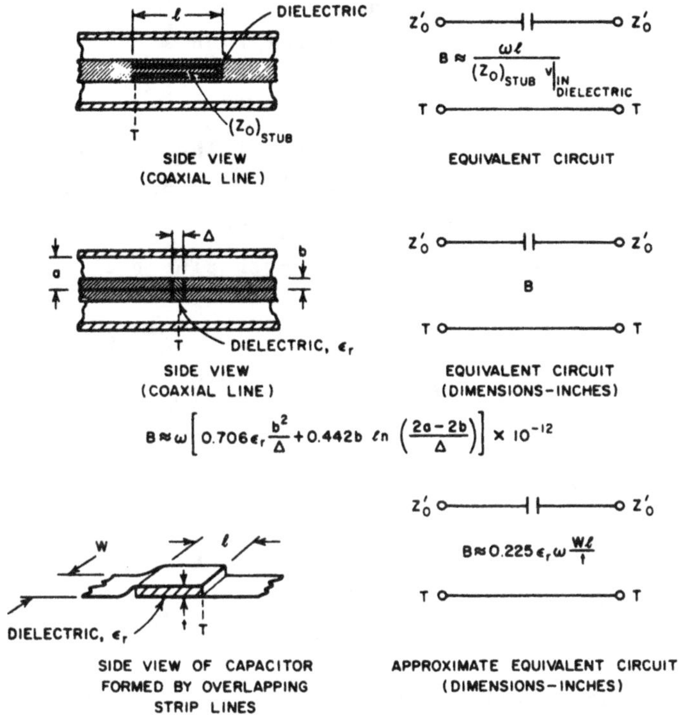

(b) SERIES CAPACITANCES

A-3527-B248

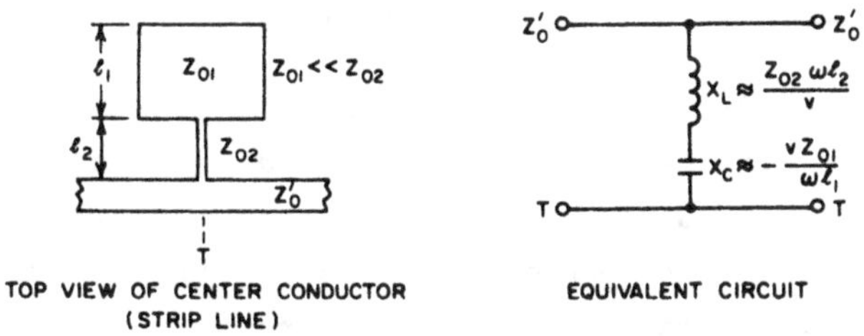

(c) SERIES-RESONANT SHUNT CIRCUIT

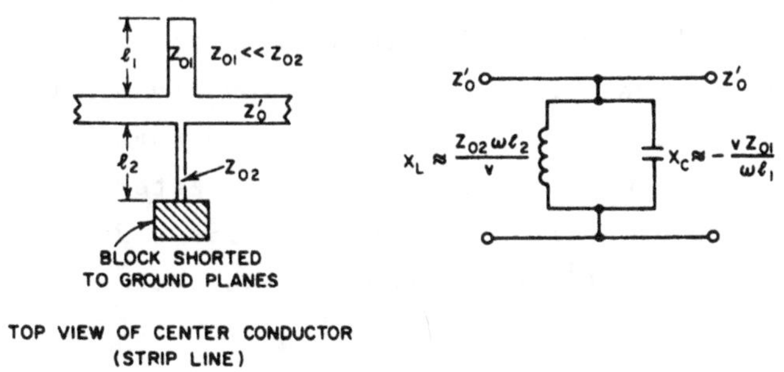

(d) PARALLEL-RESONANT SHUNT CIRCUIT

A-3527-C248

FIG. 7.02-2 Concluded

$$C_{01} = \frac{84.73\sqrt{\epsilon_r}}{Z_{01}} \ \mu\mu\text{f/inch} \tag{7.02-1}$$

for the Z_{01} line, where ϵ_r is the relative dielectric constant. Then the effect of the fringing capacitances at the ends of the line can be accounted for, approximately, by computing the total effective electrical length of the Z_{01} line as the measured length plus a length

$$\Delta l = \frac{0.450\, w\epsilon_r}{C_{01}} \left(\frac{C'_f}{\epsilon}\right) \ \text{inches} \tag{7.02-2}$$

added at each end. In Eqs. (7.02-1) and (7.02-2), w is the width of the strip in inches, and C'_f/ϵ is obtained from Fig. 5.07-5. A further refinement in the design of resonant elements such as that in Fig. 7.02-2(c) can be made by correcting for the junction inductance predicted by Fig. 5.07-3; however, this correction is usually quite small.

A lumped-element parallel-resonant shunt circuit can be realized in the manner shown in Fig. 7.02-2(d). Here too it is necessary, when computing the capacitive reactance of the low-impedance (Z_{01}) line, to include the fringing capacitance at the end of the open-circuited line.

The series-resonance and parallel-resonance characteristics of the lumped elements of Figs. 7.02-2(c) and 7.02-2(d) can also be approximated over limited frequency bands by means of quarter-wavelength lines, respectively, open-circuited or short-circuited at their far ends. Formulas for computing the characteristics of such lines are given in Fig. 5.08-1.

Series circuits having either the characteristics of lumped series-resonant circuits or lumped parallel-resonant circuits are very difficult to realize in semi-lumped-form TEM transmission lines. However, they can be approximated over limited frequency bands, in coaxial lines, by means of quarter wavelength stubs in series with the center conductor, that are either open-circuited or short-circuited at their ends, respectively. Such stubs are usually realized as lines within the center conductor in a manner similar to the first example in Fig. 7.02-2(b).

SEC. 7.03, LOW-PASS FILTERS USING SEMI-LUMPED ELEMENTS

The first step in the design of filters of this type is to select an appropriate lumped-element design (usually normalized), such as those in the tables of low-pass prototypes in Secs. 4.05 to 4.07. The choice of the type of the response (for example, the choice between a 0.1- or 0.5-db ripple Tchebyscheff response) will depend on the requirements of a specific application. Also, the number *n* of reactive elements will be determined by the rate of cutoff required for the filter. For Tchebyscheff and maximally flat series-L, shunt-C, ladder low-pass filters the required value of *n* is easily determined from the normalized attenuation curves in Sec. 4.03.

Having obtained a suitable lumped-element design, the next step is to find a microwave circuit which approximates it. Some examples will now be considered.

An Example of a Simple L-C Ladder Type of Low-Pass Filter—It is particularly advantageous to design low-pass filters in coaxial- or printed-circuit form using short lengths of transmission line that act as semi-lumped elements. In order to illustrate the design procedure for this type of filter the design of a 15-element filter is described in this section. The design specifications for this filter are 0.1-db equal-ripple insertion loss in the pass band extending from zero frequency to 1.971 Gc, and at least 35-db attenuation at 2.168 Gc. A photograph of the filter constructed from coaxial elements using the "split-block" coaxial line construction technique is shown in Fig. 7.03-1.

The form of the 15-element low-pass prototype chosen for this filter has a series inductance as the first element, as illustrated in the schematic of Fig. 7.03-2(a). At the time this filter was designed the element values in Table 4.05-2(b) were not available, but the element values for filters containing up to 10 elements as listed in Table 4.05-2(a) were available. Therefore, the 15-element prototype was approximated by using the nine-element prototype in Table 4.05-2(a), augmented by repeating three times each of the two middle elements of the nine-element filter. Comparison of these values with the more recently obtained exact values from Table 4.05-2(b) shows that the end elements of the filter are about 1.2 percent too small and that the error in the element values increases gradually toward the center of the filter so that the center element is about 4.2 percent too small. These errors are probably too

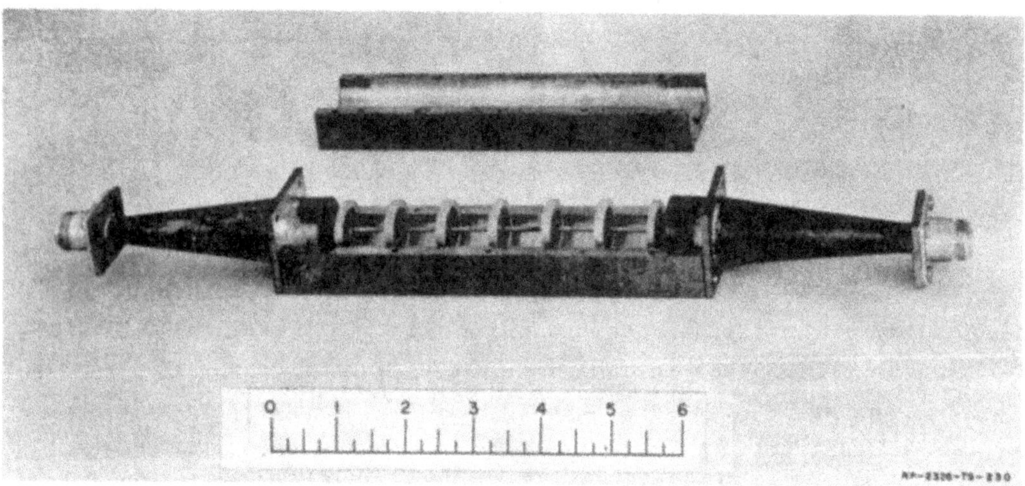

FIG. 7.03-1 A MICROWAVE LOW-PASS FILTER

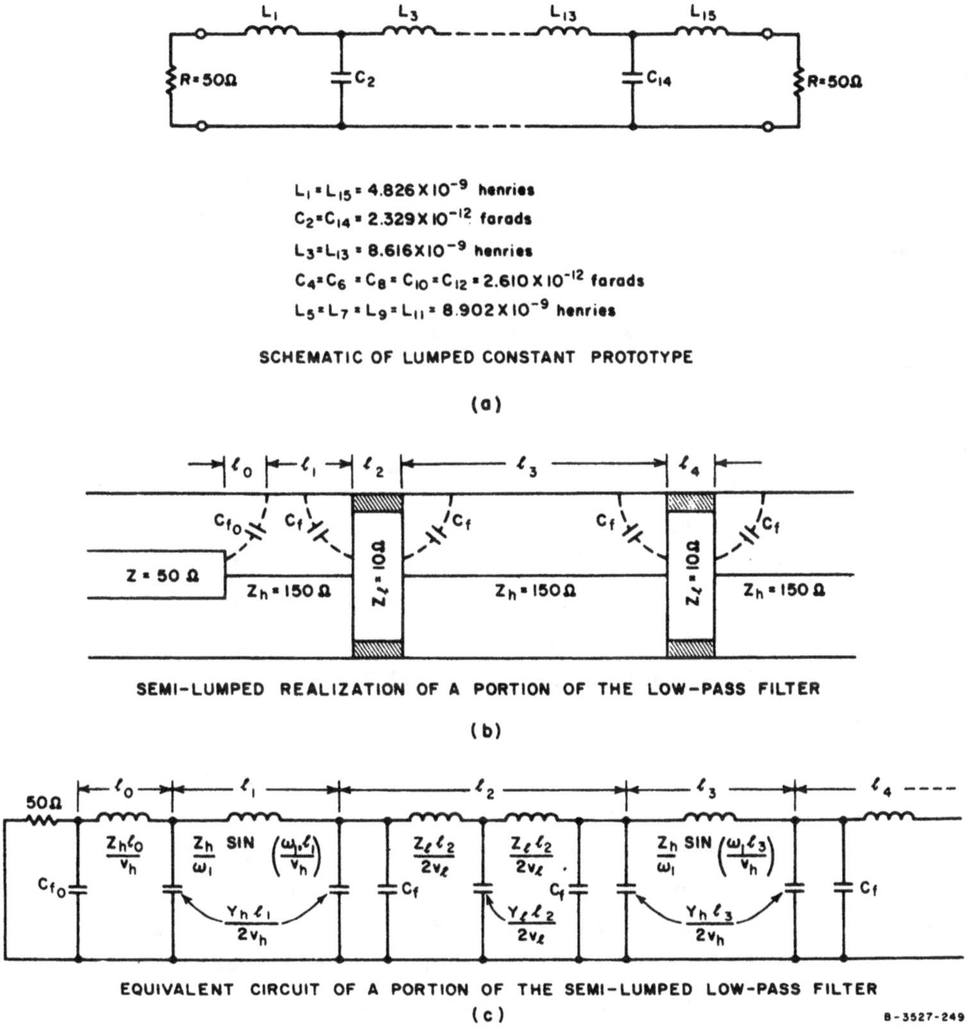

FIG. 7.03-2 STEPS IN THE REALIZATION OF A MICROWAVE LOW-PASS FILTER

small to be of significance in most applications. It should be noted that since tables going to $n = 15$ are now available, good designs for even larger n's can be obtained by augmenting $n = 14$ or $n = 15$ designs, in the above manner.

The schematic of the lumped-constant prototype used in the design of the actual filter is shown in Fig. 7.03-2(a). This filter is scaled to operate at a 50-ohm impedance level with an angular band-edge frequency ω_1 of 12.387×10^9 radians per second. The values of the inductances and capacitances used in the lumped-constant circuit are obtained from the low-pass prototype by means of Eqs. (4.04-3) and (4.04-4). That is, all inductances in the low-pass prototype are multiplied by $50/(12.387 \times 10^9)$ and all capacitances are multiplied by $1/(50 \times 12.387 \times 10^9)$. Sometimes, instead of working with inductance in henries and capacitance in farads, it is more convenient to work in terms of reactance and susceptance. Thus, a reactance $\omega_1' L_k'$ for the prototype becomes simply $\omega_1 L_k = (\omega_1' L_k')(R_0/R_0')$ for the actual filter, where R_0' is the resistance of one of the prototype terminations and R_0 is the corresponding resistance for the scaled filter. Also, the shunt susceptances $\omega_1' C_k'$ for the prototype become $\omega_1 C_k = (\omega_1' C_k')(R_0'/R_0)$ for the scaled filter. This latter approach will be utilized in the numerical procedures about to be outlined.

The semi-lumped realization of a portion of the filter is shown in Fig. 7.03-2(b). It is constructed of alternate sections of high-impedance ($Z_h = 150$ ohms) and low-impedance ($Z_l = 10$ ohms) coaxial line, chosen so that the lengths of the high-impedance line would be approximately one-eighth wavelength at the equal-ripple band-edge frequency of 1.971 Gc. The whole center conductor structure is held rigidly aligned by dielectric rings ($\epsilon_r = 2.54$) surrounding each of the low-impedance lengths of line. The inside diameter of the outer conductor was chosen to be 0.897 inch so that the 2.98-Gc cutoff frequency of the first higher-order mode* that can propagate in the low-impedance sections of the filter is well above the 1.971-Gc band-edge frequency of the filter. The values of the inductances and capacitances in the lumped-constant circuit, Fig. 7.03-2(a), are realized by adjusting the lengths of the high- and low-impedance lines respectively.

---

* As discussed in Sec. 5.03, the first higher mode can occur when $f \geq 7.51(b + d)/\sqrt{\epsilon_r}$, where f is in Gc and b and d are the outer and inner diameters in inches.

The exact equivalent circuit of the semi-lumped realization of the first three end elements of the filter are shown in Fig. 7.03-2(c). In this figure C_{f0} is the fringing capacity at the junction of the 50-ohm terminating line and the 150-ohm line representing the first element in the filter, as determined from Fig. 5.07-2. Similarly, C_f is the fringing capacitance at each junction between the 10-ohm and 150-ohm lines in the filter. It is also determined from Fig. 5.07-2, neglecting the effect on fringing due to the dielectric spacers in 10-ohm lines. The velocity of propagation v_h of a wave along the 150-ohm line is equal to the velocity of light in free space while the velocity of propagation v_l along the 10-ohm line is $v_h/\sqrt{\epsilon_r}$.

Some of the 150-ohm lines in this filter attain electrical lengths of approximately 50 electrical degrees at the band-edge frequency ω_1. For lines of this length it has been found that the pass-band bandwidth is most closely approximated if the reactances of the lumped-constant inductive elements at frequency ω_1 are matched to the exact inductive reactance of the transmission line elements at frequency ω_1 using the formulas in Fig. 7.02-1. The inductive reactance of the 10-ohm lines can also be included as a small negative correction to the lengths of the 150-ohm lines. Following this procedure we have

$$\omega_1 L_1 = Z_h \sin\left(\frac{\omega_1 l_1}{v_h}\right) + \frac{Z_l l_2 \omega_1}{2v_l} \quad \text{ohms} \tag{7.03-1}$$

$$\omega_1 L_3 = Z_h \sin\left(\frac{\omega_1 l_3}{v_h}\right) + \frac{Z_l l_2 \omega_1}{2v_l} + \frac{Z_l l_4 \omega_1}{2v_l}$$

etc.

The capacitance of each shunt element in the low-pass filter in Fig. 7.03-2(a) is realized as the sum of the capacitance of a short length of 10-ohm line, plus the fringing capacitances between the 10-ohm line and the adjacent 150-ohm lines, plus the equivalent 150-ohm-line capacitance as lumped at the ends of the adjacent 150-ohm lines. Thus, we can determine the lengths of the 10-ohm lines by means of the relations

$$\omega_1 C_2 = \frac{Y_l l_2 \omega_1}{v_l} + 2C_f \omega_1 + \frac{Y_h l_1 \omega_1}{2v_h} + \frac{Y_h l_3 \omega_1}{2v_h} \quad \text{mhos} \qquad (7.03\text{-}2)$$

$$\omega_1 C_4 = \frac{Y_l l_4 \omega_1}{v_l} + 2C_f \omega_1 + \frac{Y_h l_3 \omega_1}{2v_h} + \frac{Y_h l_5 \omega_1}{2v_h}$$

etc.

In Eqs. (7.03-1) above, the first term in each equation on the right is the major one, and the other terms on the right represent only small corrections. Thus, it is convenient to start the computations by neglecting all but the first term on the right in each of Eqs. (7.03-1), which makes it possible to solve immediately for preliminary values of the lengths l_1, l_3, l_5, etc., of the series-inductive elements. Having approximate values for l_1, l_3, l_5, etc., it is then possible to solve each of Eqs. (7.03-2) for the lengths l_2, l_4, l_6, etc., of the capacitive elements. Then, having values for l_2, l_4, l_6, etc., these values may then be used in the correction terms in Eqs. (7.03-1), and Eqs. (7.03-1) can then be solved to give improved values of the inductive element lengths l_1, l_3, l_5, etc.

The iterative process described above could be carried on to insert the improved values of l_1, l_3, l_5, etc., in Eq. (7.03-2) in order to recompute the lengths l_2, l_4, l_6, etc. However, this is unnecessary because the last two terms on the right in each of Eqs. (7.03-2) are only small correction terms themselves, and a small correction in them would have negligible effect on the computed lengths of the capacitor elements.

The reactance or susceptance form of Eqs. (7.03-1) and (7.03-2) is convenient because it gives numbers of moderate size and avoids the necessity of carrying multipliers such as 10^{-12}. The velocity of light is $v = 1.1803 \times 10^{10}/\sqrt{\epsilon_r}$ inches per second, so that the ratios ω_1/v_h and ω_1/v_h are of moderate size.

The effect of the discontinuity capacitances C_{f0} and $Y_h l_1/2v_h$ at the junction between the 50-ohm lines terminating the filter and the 150-ohm lines comprising the first inductive elements of the filter can be minimized by increasing the length of the 150-ohm lines by a small amount l_0 to simulate the series inductance and shunt capacitance of a short length of 50-ohm line. The necessary line length l_0 can be determined from the relation

$$Z_0 = 50 = \sqrt{\frac{\dfrac{Z_h l_0}{v_h}}{C_{f0} + \dfrac{Y_h l_1}{2v_h}}} = \sqrt{\frac{\text{Series Inductance}}{\text{Shunt Capacitance}}} \quad .$$

Solving for l_0 gives

$$l_0 = Z_0^2 \left[\frac{(\omega_1 C_{f0}) v_h}{Z_h \omega_1} + \frac{l_1}{2Z_h^2}\right] \quad . \tag{7.03-3}$$

Figure 7.03-3(a) shows the dimensions of the filter determined using the above procedures, while Fig. 7.03-3(b) shows the measured response of the filter. It is seen that the maximum pass-band ripple level as determined from VSWR measurements is about 0.12 db over most of the pass band while rising to 0.2 db near the edge of the pass band. It is believed that the discrepancy between the measured pass-band ripple

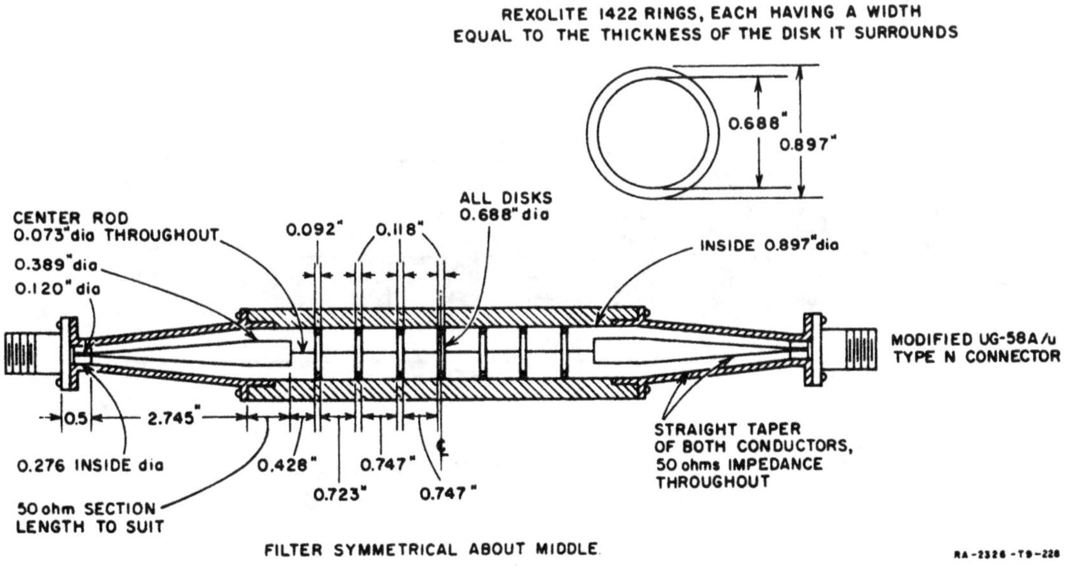

FIG. 7.03-3(a) DIMENSIONS OF THE FILTER IN FIG. 7.03-1

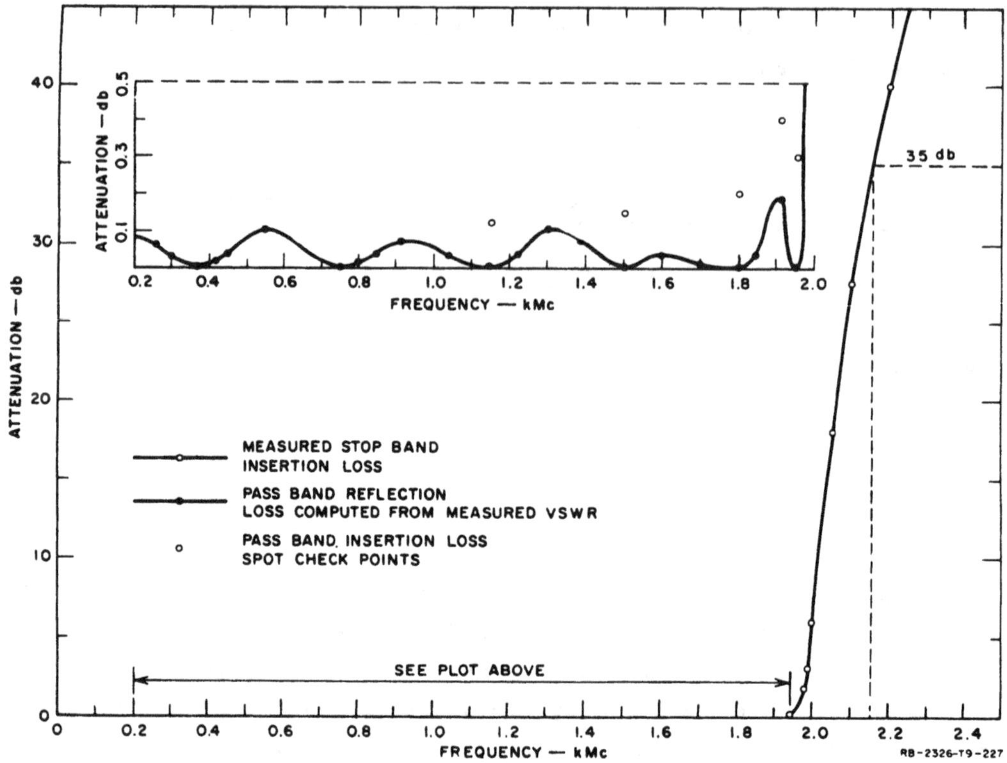

FIG. 7.03-3(b) MEASURED RESPONSE OF THE FILTER IN FIG. 7.03-1

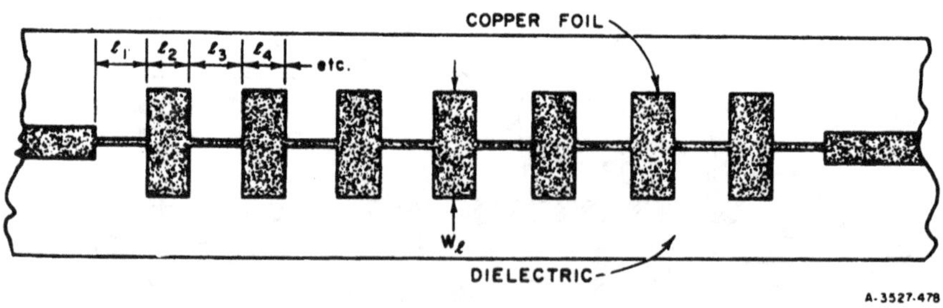

FIG. 7.03-3(c) A POSSIBLE PRINTED-CIRCUIT VERSION OF THE LOW-PASS FILTER IN FIG. 7.03-3(a)

level and the theoretical 0.1-db level is caused primarily by the fact that the approximate prototype low-pass filter was used rather than the exact prototype as given in Table 4.05-2(b). The actual pass-band attenuation of the filter, which includes the effect of dissipation loss in the filter, rises to approximately 0.35 db near the edge of the pass band. This behavior is typical and is explained by the fact that $d\phi/d\omega'$, the rate of change of phase shift through the low-pass prototype filter as a function of frequency, is more rapid near the pass-band edge, and this leads to increased attenuation as predicted by Eq. (4.13-9). A more complete discussion of this effect is contained in Sec. 4.13.

This filter was found to have some spurious responses in the vicinity of 7.7 to 8.5 Gc, caused by the fact that many of the 150-ohm lines in the filter were approximately a half-wavelength long at these frequencies. No other spurious responses were observed, however, at frequencies up through *X*-band. In situations where it is desired to suppress these spurious responses it is possible to vary the length and the diameter of the high-impedance lines to realize the proper values of series inductance, so that only a few of the lines will be a half-wavelength long at any frequency within the stop band.

The principles described above for approximate realization of low-pass filters of the form in Fig. 7.03-2(a) can also be used with other types of filter constructions. For example, Fig. 7.03-3(c) shows how the filter in Fig. 7.02-3(a) would look if realized in printed-circuit, strip-line construction. The shaded area is the copper foil circuit which is photo-etched on a sheet of dielectric material. In the assembled filter the photo-etched circuit is sandwiched between two slabs of dielectric, and copper foil or metal plates on the outside surfaces serve as the ground planes. The design procedure is the same as that described above, except that in this case the line impedances are determined using Fig. 5.04-1 or 5.04-2, and the fringing capacitance C_f in Eqs. (7.03-2) is determined using Fig. 5.07-5. It should be realized that C'_f in Fig. 5.07-5 is the capacitance per unit length from one edge of the conductor to one ground plane.* Thus, C_f in Eqs. (7.03-2) is $C_f = 2C'_f W_l$, where W_l is the width of the low-impedance line sections [Fig. 7.03-3(c)] The calculations then proceed exactly as described before. The relative

* If in computing C'_f from C'_f/ϵ in Fig. 5.07-5, $\epsilon = 0.255\epsilon_r \times 10^{-12}$ is used, then C'_f will have the units of farads/inch.

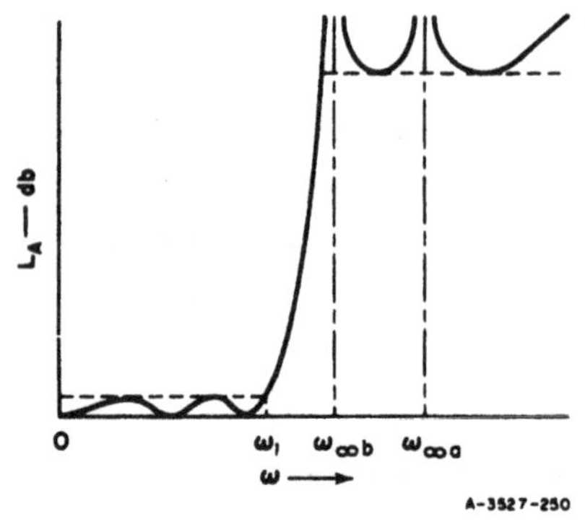

FIG. 7.03-4 TCHEBYSCHEFF FILTER CHARACTERISTIC WITH INFINITE ATTENUATION POINTS AT FINITE FREQUENCIES

advantages and disadvantages of printed-circuit *vs.* coaxial construction are discussed in Sec. 7.01.

Low-Pass Filters Designed from Prototypes Having Infinite Attenuation at Finite Frequencies—The prototype filters tabulated in Chapter 4 all have their frequencies of infinite attenuation (see Secs. 2.02 to 2.04) at $\omega = \infty$. The corresponding microwave filters, such as the one just discussed in this section, are of a form which is very practical to build and commonly used in microwave engineering. However, it is possible to design filters with an even sharper rate of cutoff for a given number of reactive elements, by using structures giving infinite attenuation at finite frequencies. Figure 7.03-4 shows a Tchebyscheff attenuation characteristic of this type, while Fig. 7.03-5 shows a filter structure which can give such a characteristic. Note that the filter structure has series-resonant branches connected in shunt, which short out transmission at the frequencies $\omega_{\infty a}$ and $\omega_{\infty b}$, and thus give the corresponding infinite attenuation points shown in Fig. 7.03-4. In addition this structure has a second-order pole of attenuation at $\omega = \infty$ since the $\omega_{\infty a}$ and $\omega_{\infty b}$ branches have no effect at that frequency, and the inductances L_1, L_3, and L_5 block transmission by having infinite

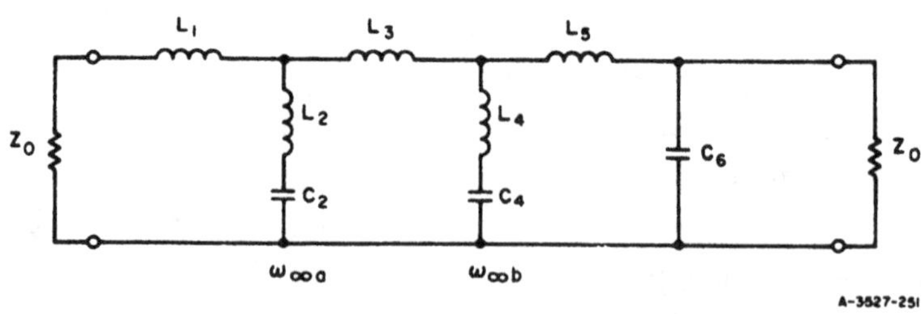

FIG. 7.03-5 A FILTER STRUCTURE WHICH IS POTENTIALLY CAPABLE OF REALIZING THE RESPONSE IN FIG. 7.03-4

series reactance, while C_6 shorts out transmission by having infinite shunt susceptance (see Sec. 2.04).

Filters of the form in Fig. 7.03-5 having Tchebyscheff responses such as that in Fig. 7.03-4 are mathematically very tedious to design. However, Saal and Ulbrich[2] have tabulated element values for many cases. If desired, of course, one may obtain designs of this same general class by use of the classical image approach discussed in Secs. 3.06 and 3.08. Such image designs are sufficiently accurate for many less critical applications.

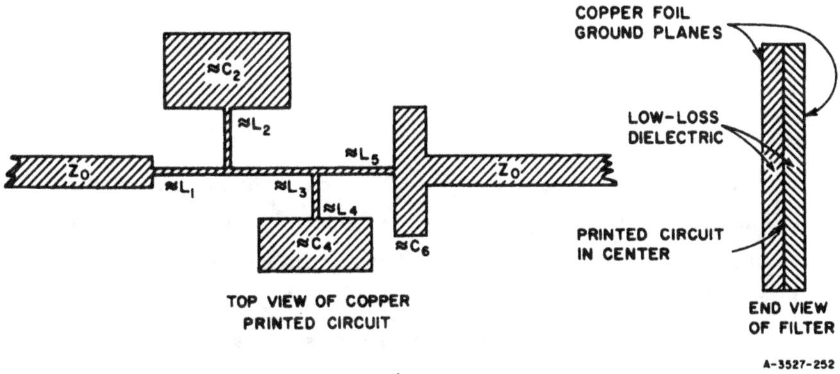

FIG. 7.03-6 A STRIP-LINE PRINTED-CIRCUIT FILTER WHICH CAN APPROXIMATE THE CIRCUIT IN FIG. 7.03-5

Figure 7.03-6 shows how the filter in Fig. 7.03-5 can be realized, approximately, in printed-circuit, strip-line construction. Using this construction, low-loss dielectric sheets are used, clad on one or both sides with thin copper foil. The circuit is photo-etched on one side of one sheet, and the printed circuit is then sandwiched between the first sheet of dielectric and a second sheet, as shown at the right in the figure. Often, the ground planes consist simply of the copper foil on the outer sides of the dielectric sheets.

The *L*'s and *C*'s shown in Fig. 7.03-6 indicate portions of the strip line circuit which approximate specific elements in Fig. 7.03-5. The various elements are seen to be approximated by use of short lengths of high- and low-impedance lines, and the actual dimensions of the line elements are computed as discussed in Sec. 7.02. In order to obtain best accuracy, the shunt capacitance of the inductive line elements should be compensated for in the design. By Fig. 7.02-1(c) the lengths of the inductive-line-elements can be computed by the equation

$$l_k = \frac{v}{\omega_1} \sin^{-1} \frac{\omega_1 L_k}{Z_k} ,$$

and the resulting equivalent capacitive susceptance *at each end* of the pi-equivalent circuit of inductive-line-element k is then

$$\omega_1 (C_\pi)_k = \frac{1}{Z_k} \tan \frac{\omega_1 l_k}{2v} \qquad (7.03\text{-}4)$$

where ω_1 is the cutoff frequency, Z_k is the characteristic impedance of inductive-line-element k, l_k is the length of the line element, and v is again the velocity of propagation. Now, for example, at the junction of the inductive line elements for L_1, L_2, and L_3 in Fig. 7.03-6 there is an unwanted total equivalent capacitive susceptance of $\omega_1 C_L = \omega_1 (C_\pi)_1 + \omega_1 (C_\pi)_2 + \omega_1 (C_\pi)_3$ due to the three inductance line elements. The unwanted susceptance $\omega_1 C_L$ can be compensated for by correcting the susceptance of the shunt branch formed by L_2 and C_2 so that

$$B_2 = \omega_1 C_L + B_2^c \qquad (7.03\text{-}5)$$

where B_2 is the susceptance at frequency ω_1 of the branch formed by L_2 and C_2 in Fig. 7.03-5, and B_2^c is the susceptance of a "compensated" shunt branch which has L_2 and C_2 altered to become L_2^c and C_2^c in order to compensate for the presence of C_L. Solving Eq. (7.03-5) for $\omega_1 C_2^c$ and $\omega_1 L_2^c$ gives

$$\omega_1 C_2^c = \omega_1 C_2 - \omega_1 C_L \left[1 - \left(\frac{\omega_1}{\omega_{\infty a}}\right)^2\right] \qquad (7.03\text{-}6)$$

$$\omega_1 L_2^c = \frac{1}{\omega_1 C_2^c} \left(\frac{\omega_1}{\omega_{\infty a}}\right)^2 \qquad (7.03\text{-}7)$$

where

$$\left(\frac{\omega_1}{\omega_{\infty a}}\right)^2 = \omega_1 L_2 \omega_1 C_2 \quad . \tag{7.03-8}$$

Then the shunt branch is redesigned using the compensated values L_2^c and C_2^c which should be only slightly different from the original values computed by neglecting the capacitance of the inductive elements.

In filters constructed as shown in Fig. 7.03-6 (or in filters of any analogous practical construction) the attenuation at the frequencies $\omega_{\infty a}$ and $\omega_{\infty b}$ (see Fig. 7.03-4) will be finite as a result of losses in the circuit. Nevertheless, the attenuation should reach high peaks at these frequencies, and the response should have the general form in Fig. 7.03-4, at least up to stop-band frequencies where the line elements are of the order of a quarter-wavelength long.

Example—One of the designs tabulated in Ref. 2 gives normalized element values for the circuit in Fig. 7.03-5 which are as follows:

Z_0'	=	1.000	L_4'	=	0.7413
L_1'	=	0.8214	C_4'	=	0.9077
L_2'	=	0.3892	L_5'	=	1.117
C_2'	=	1.084	C_6'	=	1.136
L_3'	=	1.188	ω_1'	=	1.000

This design has a maximum pass-band reflection coefficient of 0.20 (0.179 db attenuation) and a theoretical minimum stop-band attenuation of 38.1 db which is reached by a frequency $\omega' = 1.194\ \omega_1'$. As an example of how the design calculations for such a filter will go, calculations will be made to obtain the dimensions of the portions of the circuit in Fig. 7.03-6 which approximate elements L_1 to L_3. The impedance level is to be scaled so that $Z_0 = 50$ ohms, and so that the un-normalized cutoff frequency is $f_1 = 2$ Gc or $\omega_1 = (2\pi)2 \times 10^9 = 12.55 \times 10^9$ radians/sec.

A printed-circuit configuration with a ground-plane spacing of $b = 0.25$ inch using dielectric with $\epsilon_r = 2.7$ is assumed. Then, for the input and output line $\sqrt{\epsilon_r} Z_0 = 1.64\ (50) = 82$, and by Fig. 5.04-1, $W_0/b = 0.71$, and a width $W_0 = 0.71\ (0.25) = 0.178$ inch is required.

Now $v = 1.1803 \times 10^{10}/\sqrt{\epsilon_r}$ inches/sec

so

$$\frac{v}{\omega_1} = \frac{1.1803 \times 10^{10}}{(1.64)(12.55 \times 10^9)} = 0.573 \quad .$$

For inductor L_1, $\omega_1 L_1 = \omega_1' L_1' (Z_0/Z_0') = 1(0.8214)(50)/1 = 41.1$ ohms. Assuming a line impedance of $Z_1 = 118$ ohms, $\sqrt{\epsilon_r} Z_1 = 193$, and Fig. 5.04-1 calls for a line width of $W_1 = 0.025$ inch. Then the length of the L_1-inductive element is

$$l_1 = \frac{v}{\omega_1} \sin^{-1} \frac{\omega_1 L_1}{Z_1} = 0.573 \sin^{-1} \frac{41.1}{118} = 0.204 \quad \text{inch} \quad .$$

The effective, unwanted capacitive susceptance at each end of this inductive line is

$$\omega_1 (C_\pi)_1 = \frac{1}{2}\left(\frac{\omega_1}{v}\right) \frac{l_1}{Z_1} = \frac{0.204}{2(0.573)118} = 0.0015 \quad \text{mho} \quad .$$

After some experimentation it is found that in order to keep the line element which realizes L_2 from being extremely short, it is desirable to use a lower line impedance of $Z_2 = 90$ ohms, which gives a strip width of $W_2 = 0.055$ inch. Then $\omega_1 L_2 = \omega_1' L_2' (Z_0/Z_0') = 19.95$ and

$$l_2 = \frac{v}{\omega_1} \sin^{-1} \frac{\omega_1 L_2}{Z_2} = 0.573 \sin^{-1} \frac{19.95}{90} = 0.128 \quad \text{inch} \quad .$$

Even a lower value of Z_2 might be desirable in order to further lengthen l_2 so that the large capacitive piece realizing C_2 in Fig. 7.03-6 will be further removed from the L_1 and L_2 lines. However, we shall proceed with the sample calculations. The effective unwanted capacitance susceptance at each end of l_2 is

$$\omega_1 (C_\pi)_2 = \frac{1}{2} \frac{\omega_1}{v} \frac{l_2}{Z_2} = \frac{0.128}{2(0.573)90} = 0.0012 \quad \text{mho} \quad .$$

Similar calculations for L_3 give $l_3 = 0.302$ inch and $\omega_1(C_\pi)_3 = 0.0022$ mho, where Z_3 is taken to be 118 ohms as was Z_1. Then the net unwanted susceptance due to line capacitance at the junction of L_1, L_2, and L_3 is

$$\omega_1 C_L = \omega_1(C_\pi)_1 + \omega_1(C_\pi)_2 + \omega_1(C_\pi)_3 = 0.0049 \text{ mho} \quad .$$

Now $\omega_1 C_2 = \omega_1' C_2' (Z_0'/Z_0) = 1(1.084)/50 = 0.0217$ mhos. Then by Eq. (7.03-8)

$$\left(\frac{\omega_1}{\omega_{\infty a}}\right)^2 = 19.45(0.0217) = 0.422 \ ,$$

and by Eq. (7.03-6) the compensated value for $\omega_1 C_2$ is

$$\omega_1 C_2^c = 0.217 - 0.0049\,[1 - 0.422] = 0.0189 \text{ mho} \quad .$$

Now the compensated value for $\omega_1 L_2$ is

$$\omega_1 L_2^c = \left(\frac{\omega_1}{\omega_{\infty a}}\right)^2 \frac{1}{\omega_1 C_2^c} = 22.3 \text{ ohms} \quad .$$

Then the compensated value for the length l_2 of the line for L_2 is

$$l_2 = 0.573 \sin^{-1} \frac{22.3}{90} = 0.144 \text{ inch} \quad .$$

To realize C_2 we assume a line of impedance $Z_{C_2} = 30.5$ ohms which calls for a strip width of $W_{C_2} = 0.362$ inch. This strip should have a capacitive susceptance of $\omega_1 C_2^c - \omega_1(C_\pi)_2 = 0.0189 - 0.0012 = 0.0177$ mho. Neglecting end-fringing, this will be obtained by a strip of length

$$l_{C_2} = [\omega_1 C_2^c - \omega_1(C_\pi)_2]\, Z_{C_2} \frac{v}{\omega_1}$$

$$= 0.0177(30.5)(0.573) = 0.309 \text{ inch} \quad .$$

To correct for the fringing capacitance at the ends of this strip we first use Eq. (7.02-1) to obtain the line capacitance

$$C = \frac{84.73\sqrt{\epsilon_r}}{Z_{C_2}} = \frac{84.73(1.64)}{30.5} = 4.55 \ \mu\mu\text{f per inch}$$

Then by Fig. 5.07-5, $C'_f/\epsilon = 0.45$, and by Eq. (7.02-2) we need to subtract about

$$\Delta l = \frac{0.450W\epsilon_r}{C}\left(\frac{C'_f}{\epsilon}\right)$$

$$= \frac{0.450(0.362)}{4.55}(2.7)(0.45) = 0.0435 \quad \text{inch}$$

from each end of the capacitive strip, realizing C_2^c in order to correct for end-fringing. The corrected length of the strip is then $l_{C_2} - 2\Delta l = 0.222$ inch. This calculation ignores the additional fringing from the corners of the C_2 strip (Fig. 7.03-6), but there appear to be no satisfactory data for estimating the corner-fringing. The corner-fringing will be counter-balanced in some degree by the loss in capacitance due to the shielding effect of the line which realizes L_2.

In this manner the dimensions of the portions of the circuit in Fig. 7.03-6 which are to realize L_1, L_2, C_2, and L_3 in Fig. 7.03-5 are fixed. It would be possible to compensate the length of the line realizing L_1 so as to correct for the fringing capacitance at the junction between L_1 and Z_0 (Fig. 7.03-6), but in this case the correction would be very small and difficult to determine accurately.

SEC. 7.04, LOW-PASS CORRUGATED-WAVEGUIDE FILTER

A low-pass* corrugated-waveguide filter of the type illustrated schematically in Fig. 7.04-1 can be designed to have a wide, well-matched

---

* That is the filter is low-pass in nature except for the cutoff effect of the waveguide.

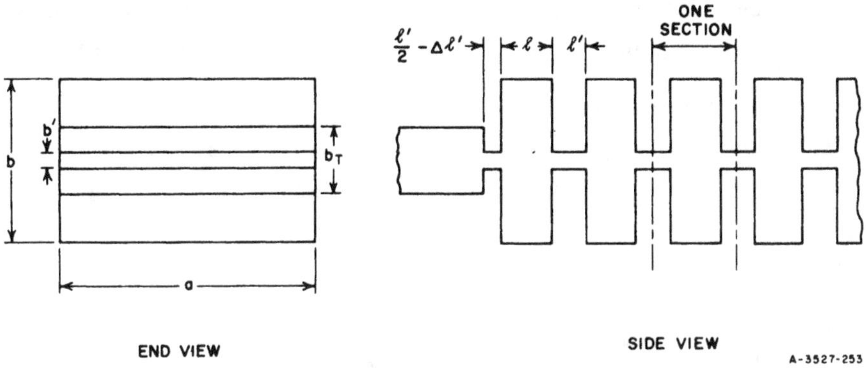

SOURCE: Proc. IRE (See Ref. 4 by S. B. Cohn)

FIG. 7.04-1 A LOW-PASS CORRUGATED WAVEGUIDE FILTER

pass band and a wide, high-attenuation stop band, for power propagating in the dominant TE_{10} mode. Because the corrugations are uniform across the width of the waveguide the characteristics of this filter depend only on the guide wavelength of the TE_{n0} modes propagating through the filter, and not on their frequency. Therefore, while this type of filter can be designed to have high attenuation over a particular frequency band for power propagating in the TE_{10} mode, it may offer little or no attenuation to power incident upon it in the TE_{20} or TE_{30} modes in this same frequency band, if the guide wavelengths of these modes falls within the range of guide wavelengths which will give a pass band in the filter response.

A technique for suppressing the propagation of the higher-order TE_{n0} modes, consisting of cutting longitudinal slots through the corrugations, thus making a "waffle-iron" filter, is described in Sec. 7.05. However, the procedure for designing the unslotted corrugated waveguide filter will be described here because this type of filter is useful in many applications, and an understanding of design techniques for it is helpful in understanding the design techniques for the waffle-iron filter.

The design of the corrugated waveguide filter presented here follows closely the image parameter method developed by Cohn.[3,4] When $b \leqq l$ the design of this filter can be carried out using the lumped-element prototype approach described in Sec. 7.03; however, the present design applies for unrestricted values of b. Values of l' are restricted, however, to be greater than about $b'/2$ so that the fringing fields at either end of the line sections of length l' will not interact with each other.

Figure 7.04-2 illustrates the image parameters of this type of filter as a function of frequency. The pass band extends from f_c, the cutoff frequency of the waveguide, to f_1, the upper cutoff frequency

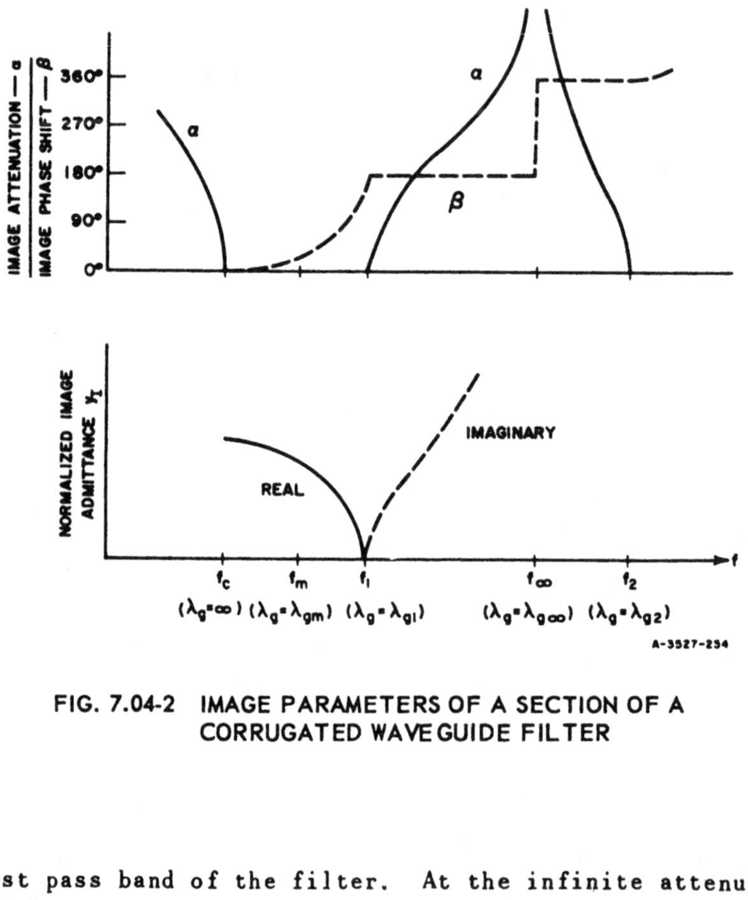

FIG. 7.04-2 IMAGE PARAMETERS OF A SECTION OF A CORRUGATED WAVEGUIDE FILTER

of the first pass band of the filter. At the infinite attenuation frequency, f_∞, the image phase shift per section changes abruptly from 180 to 360 degrees. The frequency f_2 is the lower cutoff frequency of the second pass band. The *normalized* image admittance y_I of the filter is maximum at f_c (where the guide wavelength $\lambda_g = \infty$) and zero at f_1 (where $\lambda_g = \lambda_{g1}$).

The equivalent circuit of a single half-section of the filter is illustrated in Fig. 7.04-3. *For convenience all admittances are normalized with respect to the waveguide characteristic admittance of the portions of*

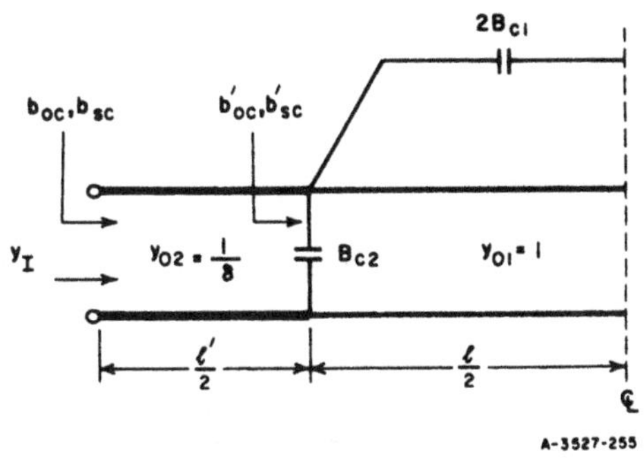

FIG. 7.04-3 NORMALIZED EQUIVALENT CIRCUIT OF A WAVEGUIDE CORRUGATED FILTER HALF-SECTION
y_{01} and y_{02} are normalized characteristic admittances and y_I is the normalized image admittance

the filter of height b and width a. Thus, the normalized characteristic admittance of the terminating lines are b/b_T, where b and b_T are defined in Fig. 7.04-1.

The half-section open- and short-circuit susceptances are given by

$$b_{oc} = \frac{1}{\delta} \tan\left[\frac{\pi l'}{\lambda_g} + \tan^{-1}(\delta b'_{oc})\right] \tag{7.04-1}$$

$$b_{sc} = \frac{1}{\delta} \tan\left[\frac{\pi l'}{\lambda_g} + \tan^{-1}(\delta b'_{sc})\right] \tag{7.04-2}$$

where

$$b'_{oc} = \tan\left(\frac{\pi l}{\lambda_g}\right) + B_{c2} \tag{7.04-3}$$

$$b'_{sc} = -\cot\left(\frac{\pi l}{\lambda_g}\right) + B_{c2} + 2B_{c1} \tag{7.04-4}$$

and

$$\delta = b'/b \quad .$$

The susceptances marked *oc* are evaluated with the ends of the wires on the right in Fig. 7.04-3 left open-circuited, while the susceptances marked *sc* are evaluated with the ends of the wires on the right all shorted together at the center line.

When $\delta \leqq 0.15$, the shunt susceptance B_{c2} is given accurately by the equation

$$B_{c2} \approx \frac{2b}{\lambda_g}\left\{ \ln\frac{1}{\delta} - 0.338 + \sum_{k=1}^{\infty}\left[\frac{\tanh\dfrac{k\pi lF}{b}}{F} - 1\right]\right\} - 0.09\,\frac{b}{\lambda_g} \qquad (7.04\text{-}5)$$

and the series susceptance B_{c1} has the value

$$B_{c1} \approx \frac{2b}{\lambda_g}\sum_{k=1}^{\infty}\frac{\operatorname{csch}\dfrac{2\pi k lF}{b}}{F} \qquad (7.04\text{-}6)$$

where

$$F = \sqrt{1 - \left(\frac{b}{k\lambda_g}\right)^2} \quad .$$

The normalized image admittance $y_I = \sqrt{y_{oc}y_{sc}}$ is

$$y_I = j\,\frac{\cot\dfrac{\theta'}{2}}{\delta}\sqrt{\frac{\left(b'_{oc} + \dfrac{\tan\dfrac{\theta'}{2}}{\delta}\right)\left(b'_{sc} + \dfrac{\tan\dfrac{\theta'}{2}}{\delta}\right)}{\left(b'_{sc} - \dfrac{\cot\dfrac{\theta'}{2}}{\delta}\right)\left(b'_{oc} - \dfrac{\cot\dfrac{\theta'}{2}}{\delta}\right)}} \qquad (7.04\text{-}7)$$

and the image propagation constant for a *full section* is

$$\gamma = \alpha + j\beta = 2\tanh^{-1}\sqrt{\frac{y_{oc}}{y_{sc}}}$$

or

$$\gamma = 2\tanh^{-1}\sqrt{\frac{\left(b'_{oc} + \dfrac{\tan\dfrac{\theta'}{2}}{\delta}\right)\left(b'_{sc} - \dfrac{\cot\dfrac{\theta'}{2}}{\delta}\right)}{\left(b'_{oc} - \dfrac{\cot\dfrac{\theta'}{2}}{\delta}\right)\left(b'_{sc} + \dfrac{\tan\dfrac{\theta'}{2}}{\delta}\right)}} \tag{7.04-8a}$$

where $\theta' = 2\pi l'/\lambda_g$ is the electrical length of the low-impedance lines of length l'.*

The attenuation per section of a corrugated filter can be computed by use of Eq. (7.04-8a) (for frequencies where the equivalent circuit in Fig. 7.04-3 applies). However, once the image cutoff frequency of the sections has been determined, with its corresponding guide wavelength λ_{g1}, the approximate formula

$$\alpha = 17.372\cosh^{-1}\frac{\lambda_{g1}}{\lambda_g} \quad \text{db/section} \tag{7.04-8b}$$

is convenient, where λ_g is the guide wavelength at a specified stop-band frequency. Equation (7.04-8b) is based on Eq. (3.06-7) which is for lumped-element filters. Thus, Eq. (7.04-8b) assumes that the corrugations are small compared to a wavelength. Note that a section of this filter is defined as the region from the center of one tooth of the corrugation to the center of the next tooth. The approximate total attenuation is, of course, α times the number of sections.

Equations (7.04-7) and (7.04-8a) can be interpreted most easily with the aid of Fig. 7.04-4, which shows a sketch of the quantities in these equations as a function of reciprocal guide wavelength. It is seen that the image cutoff frequency f_1 at which $y_I = 0$, is determined by the condition that

$$b'_{sc} + \frac{\tan\dfrac{\theta'}{2}}{\delta} = 0 \quad . \tag{7.04-9}$$

* The equations used here for y_I and γ are essentially the same as equations which can be found in Table 3.03-1. Their validity for the case in Fig. 7.04-3, where there are more than two terminals on the right, can be proved by use of Bartlett's Bisection Theorem.[5]

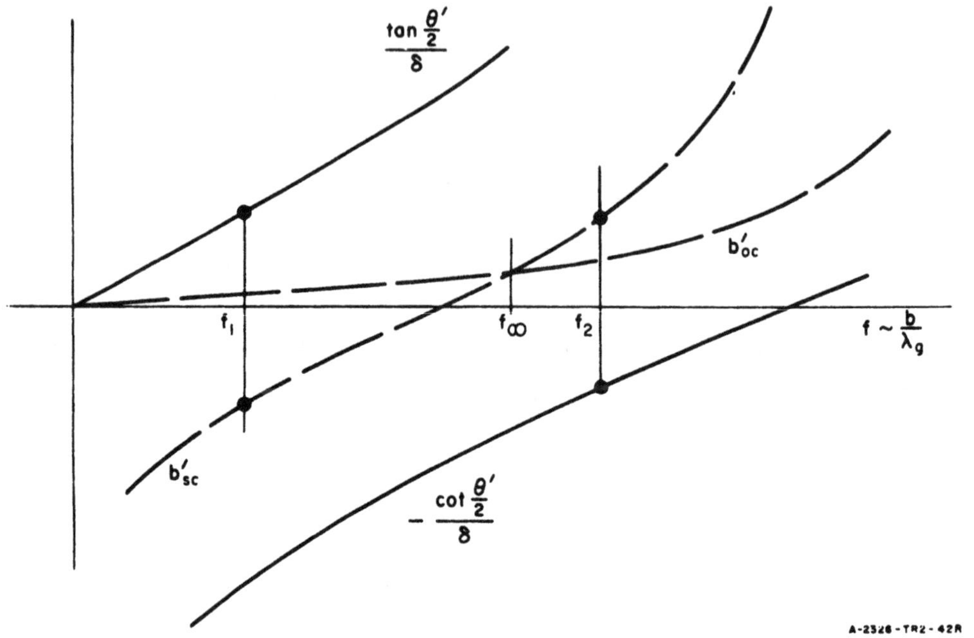

FIG. 7.04-4 GRAPH OF QUANTITIES WHICH DETERMINE CRITICAL FREQUENCIES IN CORRUGATED-WAVEGUIDE FILTER RESPONSE

The infinite attenuation frequency f_∞ is determined by the condition that

$$b'_{sc} = b'_{oc} \quad . \tag{7.04-10}$$

Finally, the image cutoff frequency f_2 at the upper edge of the first stop band is determined from the condition that

$$b'_{sc} - \frac{\cot \dfrac{\theta'}{2}}{\delta} = 0 \quad . \tag{7.04-11}$$

Design Procedure—One can design corrugated waveguide filters by means of Eqs. (7.04-1) to (7.04-11), using computed values of b'_{sc} and b'_{oc} or the values plotted by Cohn[3] for $l/b = 1/\pi$, $1/2\pi$, and $1/4\pi$. Alternatively one can use the values of b'_{sc} and b'_{oc} derived from the equivalent circuit of a waveguide E-plane T-junction as tabulated by Marcuvitz[6] for $l/b' \lessapprox 1.0$. However, it is generally easier to use the design graphs

(Figs. 7.04-5, 7.04-6, and 7.04-7) prepared by Cohn,[7] which are accurate to within a few percent for $\delta \leqq 0.20$.* In using these graphs the first step is to specify f_c, f_1, and f_∞. The width a is then fixed, since

$$a = \frac{5.9}{(f_c)_{Gc}} \tag{7.04-12}$$

where a is measured in inches and $(f_c)_{Gc}$ in gigacycles. Values of λ_{g1} and $\lambda_{g\infty}$ measured in inches are then calculated in the usual way from the relation

$$\lambda_{gn} = \frac{11.8}{\sqrt{(f_n)^2_{Gc} - (f_c)^2_{Gc}}} \tag{7.04-13}$$

using $n = 1$ and $n = \infty$.

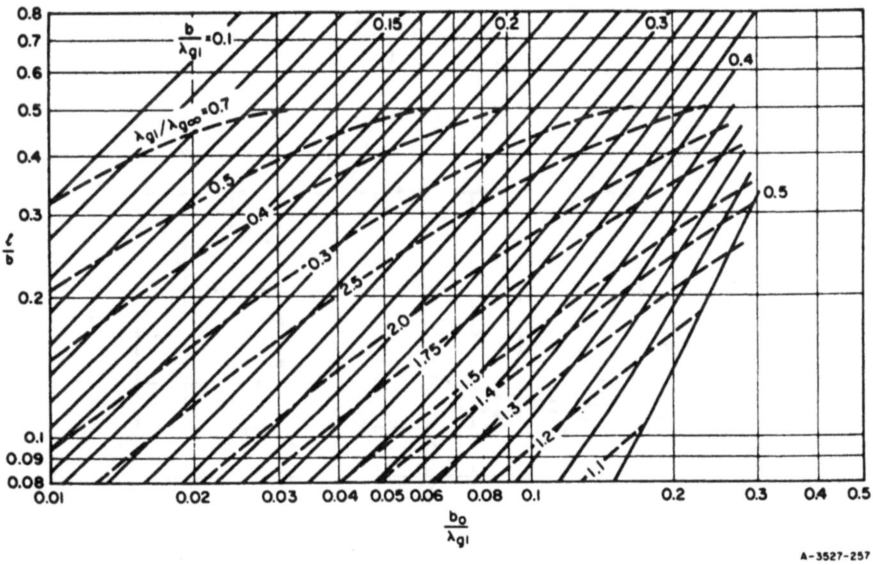

SOURCE: Proc. IRE (See Ref. 7 by S. B. Cohn)

FIG. 7.04-5 DESIGN GRAPH GIVING THE PARAMETER b USED IN SEC. 7.04

* It will be noted that use of Cohn's graphs bypasses the use of Eqs. (7.04-5) and (7.04-6), which were accurate for $\delta \leqq 0.15$.

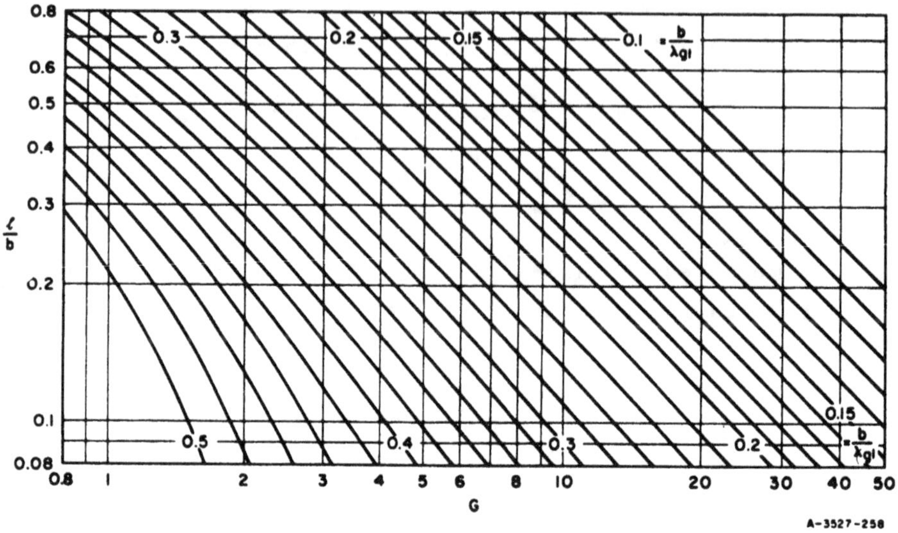

SOURCE: Proc. IRE (See Ref. 7 by S. B. Cohn)

FIG. 7.04-6 DESIGN GRAPH GIVING THE PARAMETER G USED IN SEC. 7.04

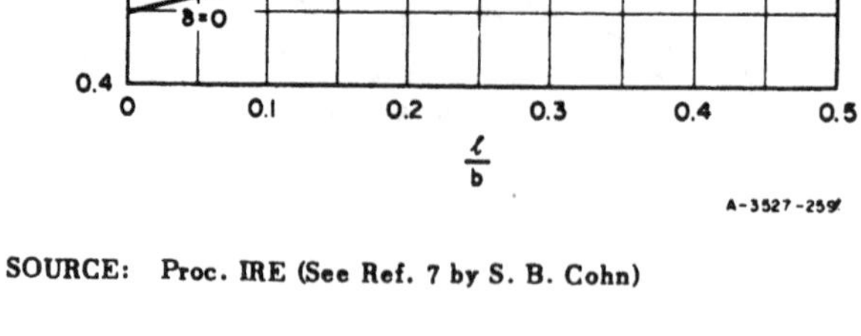

SOURCE: Proc. IRE (See Ref. 7 by S. B. Cohn)

FIG. 7.04-7 INFINITE-ATTENUATION-WAVELENGTH CURVE

The next step in the filter design is to choose a convenient value of l/b. Using this value of l/b and the value of $\lambda_{g1}/\lambda_{g\infty}$ one then enters Fig. 7.04-5 and determines b/λ_{g1} and b_0/λ_{g1}, thus fixing the values of b, b_0, and l. Here b_0 is the terminating guide height which will match the filter as λ_g approaches infinity. Then one determines the design parameter G from Fig. 7.04-6 in terms of l/b and b/λ_{g1}. Finally, one assumes a value of $\delta \leqq 0.20$ and calculates l' from the relation

$$\tan \frac{\pi l'}{\lambda_{g'}} = \pi\delta \frac{b}{\lambda_{g1}} \left[G - \frac{2}{\pi} \ln \frac{1}{\delta} + 0.215 \right] \quad . \tag{7.04-14}$$

If l'/b' is less than 0.5, a different value of δ should be used.

The image admittance in the pass band of the filter, normalized to a guide of height b, is given to very good approximation by

$$y_I = \frac{b}{b_0} \sqrt{1 - \left(\frac{\lambda_{g1}}{\lambda_g}\right)^2} \tag{7.04-15}$$

where λ_{g1} is the guide wavelength at frequency f_1. In order that a perfect match to the filter be achieved at some frequency f_m (for which $\lambda_g = \lambda_{gm}$), the height b_T of the terminating guide may be adjusted so that

$$b_T = \frac{b_0}{\sqrt{1 - \left(\frac{\lambda_{g1}}{\lambda_{gm}}\right)^2}} \quad . \tag{7.04-16}$$

If $b_0 \approx 0.7\, b_T$ a fairly good over-all match in the pass band is obtained. The amount of mismatch can be estimated by use of Eq. (7.04-15) and Fig. 3.07-2, where the abscissa of Fig. 3.07-2 is $a = y_I b_T/b$. A superior alternative for achieving a wide-band match is to use transforming end sections as described in Sec. 3.08. In this case, one sets $b_0 \approx b_T$, both for the internal sections and for the transforming end sections. However, the internal sections are designed to have a cutoff at λ_{g1}, while the transforming end sections are designed to have their cutoff at about $\lambda_{g1}/1.3$.

An explicit relation for $b/\lambda_{g\infty}$ in terms of l/b is also presented in Fig. 7.04-7, which is often useful in design work.

Unfortunately Cohn's[7] simplified design procedure does not enable one to specify f_2. However, it is generally found that f_2 is only about 20% higher in frequency than f_∞. Therefore, it is wise in any design situation to place f_∞ quite near the upper edge of the prescribed stop band.

The length $l'/2$ of the low-impedance line of height b', connecting to the terminating line of height b_T, must be reduced by an amount $\Delta l'$ to account for the discontinuity susceptance B of the junction. This is illustrated in Fig. 7.04-1. The amount of $\Delta l'$ that the line should be decreased in length is given by the expression

$$\Delta l' = \frac{b'}{2\pi}\left[\frac{\lambda_g}{b_T}\frac{B}{Y_0}\right] \tag{7.04-17}$$

where Y_0 is the characteristic admittance of the terminating line. The appropriate value of $[(\lambda_g/b_T)(B/Y_0)]$ is easily determined from Fig. 5.07-11.

Two examples of this procedure as applied to the design of waffle-iron filters will be presented in the next section.

SEC. 7.05, LOW-PASS WAFFLE-IRON FILTERS* HAVING VERY WIDE STOP BANDS

This section describes the design of low-pass corrugated waveguide filters containing longitudinal slots cut through the corrugations. These types of filter, known as waffle-iron filters, have wide, well matched pass bands and wide, high-attenuation stop bands which can be made to be free of spurious responses for all modes. Several specific designs will be discussed.

Figure 7.05-1 is a drawing of a waffle-iron filter, illustrating the metal islands, or bosses (from which it derives its name) lying between the longitudinal and transverse slots. In these filters it is essential that the center-to-center spacing of the bosses be no greater than a half of a free-space wavelength at the highest required stop-band frequency. Under these conditions the waffle-iron structure is essentially isotropic and has the same characteristics, at a given frequency, for TEM waves

---

* This type of filter was originated by S. B. Cohn. Additional data will be found in Sec. 15.05.

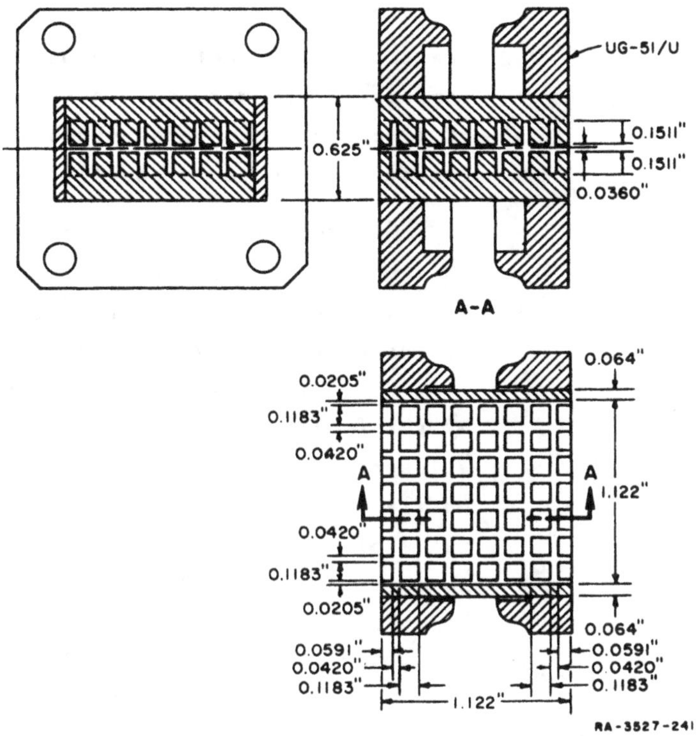

FIG. 7.05-1 DETAILS OF A TYPICAL WAFFLE-IRON FILTER

propagating through it in any direction. Thus, since any TE_{m0} mode can be resolved into TEM waves traveling in different directions through the filter it is seen that the properties of the waffle-iron filter for TE_{m0} modes are functions of frequency only. This is in contrast to the unslotted corrugated waveguide filters described in Sec. 7.04, whose response properties involve guide dimensions and mode numbers also, and are functions of guide wavelength.

Incident modes having horizontal components of electric field can excite slot modes that will propagate through the longitudinal slots in the filter at frequencies where the slot height b is greater than one half a free-space wavelength. Usually these modes are troublesome only at the highest stop-band frequencies. However, when unslotted step transformers are used to match the waffle-iron filters to waveguide of the standard

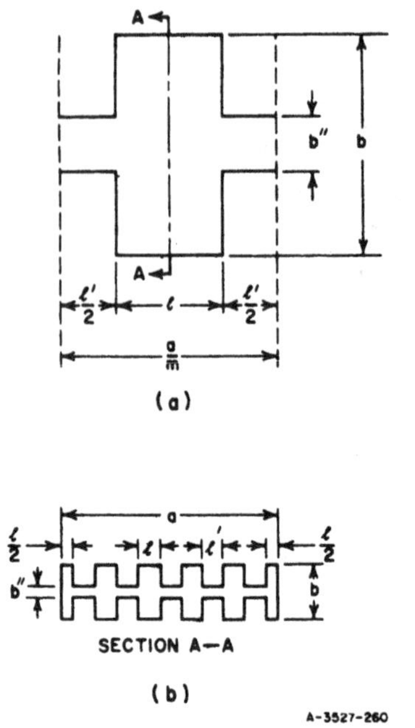

FIG. 7.05-2 A SINGLE-FILTER SECTION OF A WAFFLE-IRON FILTER

height, the reduced height of the stepped transformers effectively suppresses the incident modes with horizontal components of electric field which could otherwise excite slot modes in the filters.

Design Utilizing Cohn's Corrugated Filter Data—Waffle-iron filters can be designed approximately using the technique described in Sec. 7.04 *if the guide wavelength λ_g used there is everywhere replaced by the free-space wavelength λ_1.* As an illustration we will consider the design of Waffle-Iron-Filter-I, used with WR-650 waveguide of width a = 6.5 inches. We use the notation in Sec. 7.04 and that shown in Fig. 7.05-2, and choose f_1 = 2.02 Gc (λ_1 = 5.84 inches), f_∞ = 5.20 Gc (λ_∞ = 2.27 inches), so that λ_1/λ_∞ = 2.57. Letting l/b = 0.318 = $1/\pi$, we find from Fig. 7.04-5 that b_0/λ_1 = 0.077, and b/λ_1 = 0.275, so that b_0 = 0.450 inch, b = 1.607 inches and l = 0.511 inch. Referring to Fig. 7.04-6 we find the design parameter G to be 3.85. Now, we make the assumption that we want five bosses across the a = 6.5-inch width of the filter so that $l + l'$ = 6.5/5 = 1.3 inches and l' = 0.789 inch. From various trial designs, it has been found that for a 3:1 stop-band width, five bosses is about optimum in terms of giving convenient dimensions. For narrower stop-band widths more bosses can be used. Substituting the values of G and l' into Eq. (7.04-16) we find that $\delta = b'/b$ = 0.176 and since b = 1.607 inches, b' = 0.282 inch.

The presence of the longitudinal slots through the filter has the effect of decreasing the capacitance per unit length of the low-impedance lines. This decrease in capacitance can be compensated for by decreasing the dimension b' for an unslotted filter to b''. The ratio b''/b' is given approximately by

$$\frac{b''}{b'} = \frac{l'}{l + l'} + \frac{2}{\pi}\,\frac{l}{(l + l')}\left[\tan^{-1}\left(\frac{b''}{b'}\,\frac{b'}{l}\right) + \frac{\ln\sqrt{1 + (lb'/b'b'')^2}}{\dfrac{l}{b'}\,\dfrac{b'}{b''}}\right] \quad . \qquad (7.05\text{-}1)$$

Solving the above equation gives $b''/b' \approx 0.81$. However, the filter described here has the edges of the bosses rounded with a 0.0625-inch radius to increase its power-handling capacity, and this rounding further decreases the capacitance of the low-impedance lines. Therefore, b''/b' was chosen to be 0.75, yielding a value of $b'' = 0.210$ inch.

The height of the unslotted terminating guide b_T necessary to match this filter at some pass-band design frequency f_m is related to b_0, the height to give a match when $\lambda_g \to \infty$, by

$$b_T \approx \frac{b_0}{\sqrt{1 - \left(\frac{f_m}{f_1}\right)^2}} \quad . \tag{7.05-2}$$

In order to maintain a reasonably good match across the band, f_m should not be too close to f_1; typically, $f_m \leqq 0.7 f_1$ is desirable. For the best wide-band match, matching end sections should be used, as will be described in a following example. Using $f_m = 1.3$ Gc, Eq. (7.05-2) predicts $b_T = 0.555$ inch. Step transformers were used at each end to match this guide of height b_T to standard guide.

The attenuation per section in the stop-band region just above the pass band can be estimated by use of Eq. (7.04-8b), with λ_g and λ_{g1} replaced by λ and λ_1.

Figure 7.05-3 shows the measured insertion loss of this filter in the stop band. It is seen to be everywhere greater than 60 db from 2.2 to 5.7 Gc. The VSWR in the 1.25-to-1.40-Gc required pass band of this filter was less than 1.08 and the attenuation was less than 0.1 db. As will be discussed at the end of this section, a broader band of good impedance match could have been obtained if the filter had been constructed to start and end in the center of a boss [*i.e.*, at plane *A-A* in Fig. 7.05-2(a)] instead of in the center of a row of teeth [*i.e.*, in the plane of one of the dotted lines in Fig. 7.05-2(a)].

Design Using the T-Junction Equivalent Circuit of Marcuvitz—Though the method described above is usually easier, waffle-iron filters can also be designed using the equivalent circuit of a waveguide *T*-junction as given by Marcuvitz[6] when $l/b' \leqq 1$, for arbitrary values of $\delta = b'/b$,

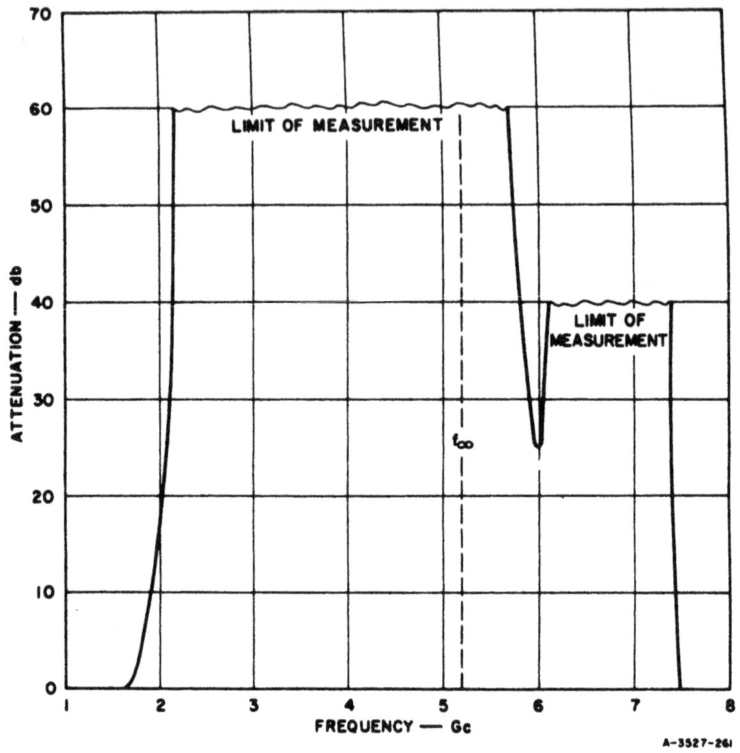

SOURCE: Technical Note 2, Contract AF 30(602)-2392, SRI (See Ref. 8 by Eugene Sharp)

FIG. 7.05-3 MEASURED INSERTION LOSS OF WAFFLE-IRON FILTER I

so long as l'/b' is greater than about 0.5. Cohn's graphs apply only when $l/b' > 1$, so if $l/b' < 1$, the use of Marcuvitz's data is the most convenient. In order to illustrate this procedure we will now describe the design of Waffle-Iron-Filter-II, used with WR-112 waveguide of width a = 1.122 inch. It has a pass band extending from 7.1 to 8.6 Gc and a stop band with greater than 40-db attenuation extending from 14 to 26 Gc. This filter could also be designed by the technique described above but the alternate procedure is presented here for completeness.

Figure 7.05-4 illustrates the bottom half of a single section of the waffle-iron filter together with its equivalent circuit. The part of the equivalent circuit representing the junction of the series stub with the main transmission line of characteristic impedance Z_0 is taken from Marcuvitz's Fig. 6.1-2. (The parameter labeled b/λ_g on Marcuvitz's curves[6] in his Figs. 6.1-4 to 6.1-14 should in reality be $2b/\lambda_g$.) The normalized image impedance of a filter section is

$$\frac{Z_I}{Z_0} = \frac{1}{Z_0}\sqrt{Z_{sc}Z_{oc}} = \frac{1}{2}\sqrt{\frac{1 + \cot\dfrac{\phi}{2}\left(\dfrac{X}{Z_0} + n^2\dfrac{Z_0'}{Z_0}\tan\phi'\right)}{1 - \tan\dfrac{\phi}{2}\left(\dfrac{X}{Z_0} + n^2\dfrac{Z_0'}{Z_0}\tan\phi'\right)}} \tag{7.05-3}$$

while the image attenuation constant $\gamma = \alpha + j\beta$ per section is related to the bisected section open- and short-circuit impedance Z_{oc} and Z_{sc} by

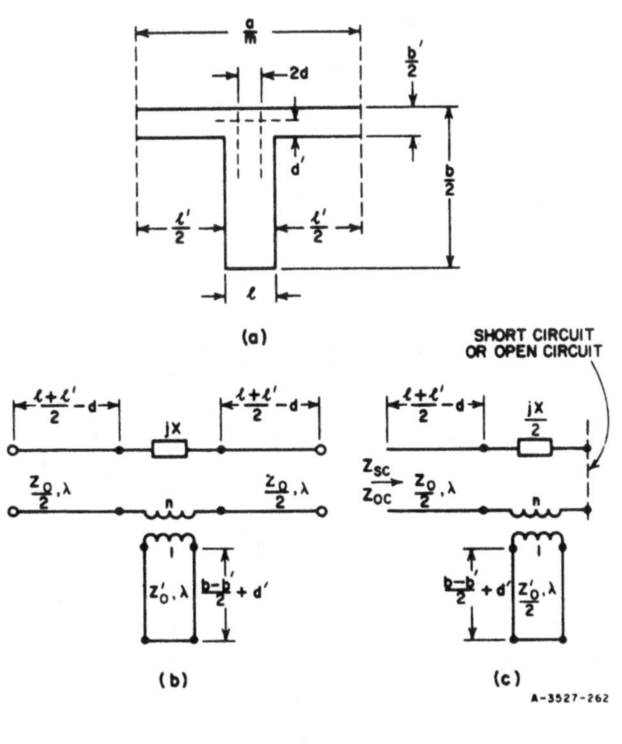

FIG. 7.05-4 FULL-FILTER SECTION — CROSS SECTION OF WAFFLE-IRON FILTER AND EQUIVALENT CIRCUIT
At (c) the equivalent circuit has been bisected

$$\tanh \frac{(\alpha + j\beta)}{2} = \sqrt{\frac{Z_{sc}}{Z_{oc}}} = \sqrt{\frac{\dfrac{X}{Z_0} + n^2 \dfrac{Z_0'}{Z_0} \tan \phi' + \tan \dfrac{\phi}{2}}{\dfrac{X}{Z_0} + n^2 \dfrac{Z_0'}{Z_0} \tan \phi' - \cot \dfrac{\phi}{2}}} \qquad (7.05\text{-}4)$$

where

$$\frac{\phi}{2} = \frac{2\pi}{\lambda}\left[\frac{l + l'}{2} - d\right] , \qquad (7.05\text{-}5)$$

$$\phi' = \frac{2\pi}{\lambda}\left[\frac{b - b'}{2} + d'\right] , \qquad (7.05\text{-}6)$$

and the remaining parameters are as indicated in Fig. 7.05-4. In applying Eqs. (7.05-3) and (7.05-4) it has been found that $\phi/\phi' \approx 1$ is nearly optimum. Values of $\phi/\phi' \approx 2$ are to be avoided because they cause the filter to have a narrow spurious pass band near the infinite attenuation frequency f_∞.

The design of this filter proceeded by a trial and error technique using Eq. (7.05-4) to determine the dimensions to yield approximately equal attenuations at 14 and 26 Gc. In this design the curves for the equivalent-circuit element values for series T-junctions in Marcuvitz[6] were extrapolated to yield equivalent-circuit parameters for $l'/b \approx 1.17$, and λ_g was replaced by λ. The choice of dimensions was restricted to some extent in order to have an integral number, m, of bosses across the width of the guide. The value of m was chosen to be 7. The calculated attenuation per section was calculated to be 7.6 db at 14 Gc and 8.8 db at 26 Gc. The total number of sections along the length of the filter was chosen to be 7 in order to meet the design specifications. Reference to Eq. (7.05-1) showed that b'' was within 5 percent of b', so $b'' = b'$ was used. The final dimensions of the filter obtained by this method are those shown in Fig. 7.05-1.

The normalized image impedance Z_I/Z_0 of the filter was computed from Eq. (7.05-3) to be 2.24 at 7.9 Gc. Thus, it is expected that the height b_T of the terminating guide should be

$$b_T = b'' \frac{Z_I}{Z_0} \qquad (7.05\text{-}7)$$

or 0.036 × 2.24 = 0.080 inch. Experimentally, it was determined that the optimum value for b_T is 0.070 inch at 7.9 Gc.

This filter was connected to standard WR-112 waveguide by means of smooth tapered transitions which had a VSWR of less than 1.06 over the frequency band from 7.1 to 8.6 Gc, when they were placed back-to-back. The measured insertion loss of the filter and transitions in the stop band was less than 0.4 db from 6.7 to 9.1 Gc while the VSWR was less than 1.1 from 7 to 8.6 Gc. The measured stop-band attenuation of the filter is shown in Fig. 7.05-5, and it is seen to agree quite closely with the theoretical analysis.

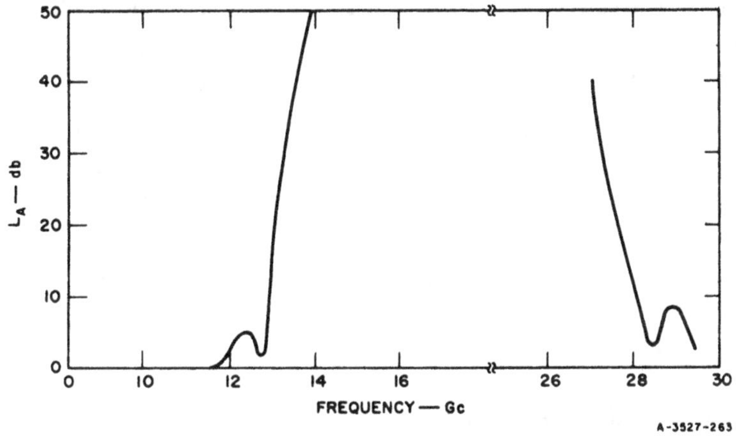

FIG. 7.05-5 STOP-BAND ATTENUATION OF WAFFLE-IRON FILTER II

No spurious responses were measured on either of the above described filters in the stop band when they were terminated by centered waveguides. However, if the terminating waveguides are misaligned at each end of the filter, it is found that spurious transmissions can occur when $\lambda < 2b$. These spurious responses are caused by power propagating through the longitudinal slots in the filter in a mode having a horizontal component of electric field. Thus, it is seen to be essential to accurately align the waveguides terminating waffle-iron filters if maximum stop-band width is desired.

A Third Example with Special End-Sections to Improve Impedance Match—As a final example, the design of a low-pass waffle-iron filter naving integral longitudinally slotted step transformers will be described. This filter is designed to be terminated at either end with WR-51 waveguide. The pass band of the filter extends from 15 to 21 Gc and the stop band which has greater than 40-db attenuation, extends from 30 to 63 Gc. A photograph of this filter is shown in Fig. 7.05-6, illustrating the split-block construction, chosen so that the four parts of the filter would be easy to machine.

The longitudinal slots in the stepped transformers necessitate that the design of this filter be different than those described previously. This occurs because these slots allow modes incident on the transformers such as the TE_{11} or TM_{11} to set up the previously described slot modes, having horizontal electric fields, which propagate through the filter when $b \gtrless \lambda/2$. Thus, it is necessary in the design of this filter to choose $b \leqq \lambda/2$ at the highest stop-band frequency of 63 Gc. In the design presented here, $b = 0.0803$ and $f_1 = 24.6$ Gc ($\lambda_1 = 0.480$ inch). It was decided to use 5 bosses across the width of the guide with $l = 0.0397$ inch and $l' = 0.0623$ inch. Referring to Fig. 7.04-5 we find $b_0 = 0.021$ inch, and from Fig. 7.04-6 we find the design parameter $G = 7$. Substituting in Eq. (7.04-16) we find $\delta = 0.139$ or $b' = 0.0113$ inch. We find the reduction in gap height due to the presence of the longitudinal slots from Eq. (7.05-1), which predicts $b''/b' \approx 0.77$ or $b'' = 0.0087$ inch.

The height b_T of a parallel-plate terminating guide that will give a match at 18 Gc is determined from Eq. (7.04-16) to be 0.031 inch. The actual height of the longitudinally slotted lines used in this design is $b_T = 0.030$ inch.

In order to further improve the match of this filter over the operating band, transforming end sections were used at either end having the same values of b, b'', and l, but with l' reduced from 0.0623 inch to 0.040 inch. This reduction in the value of l' causes the end sections to have a low-frequency image admittance about 14 percent lower than that of the middle sections and an image cutoff frequency about 14 percent higher than that of the middle sections. Figure 7.05-7 shows a sketch of the image admittance of the middle and end sections of the filter normalized to the admittance of a parallel-plate guide of height $b = 0.0803$ inch. The image phase shift of the end sections is 90 degrees at 21 Gc (the upper edge of the operating band) and not greatly different from 90 degrees over the rest of the operating band. The approximate admittance level of the

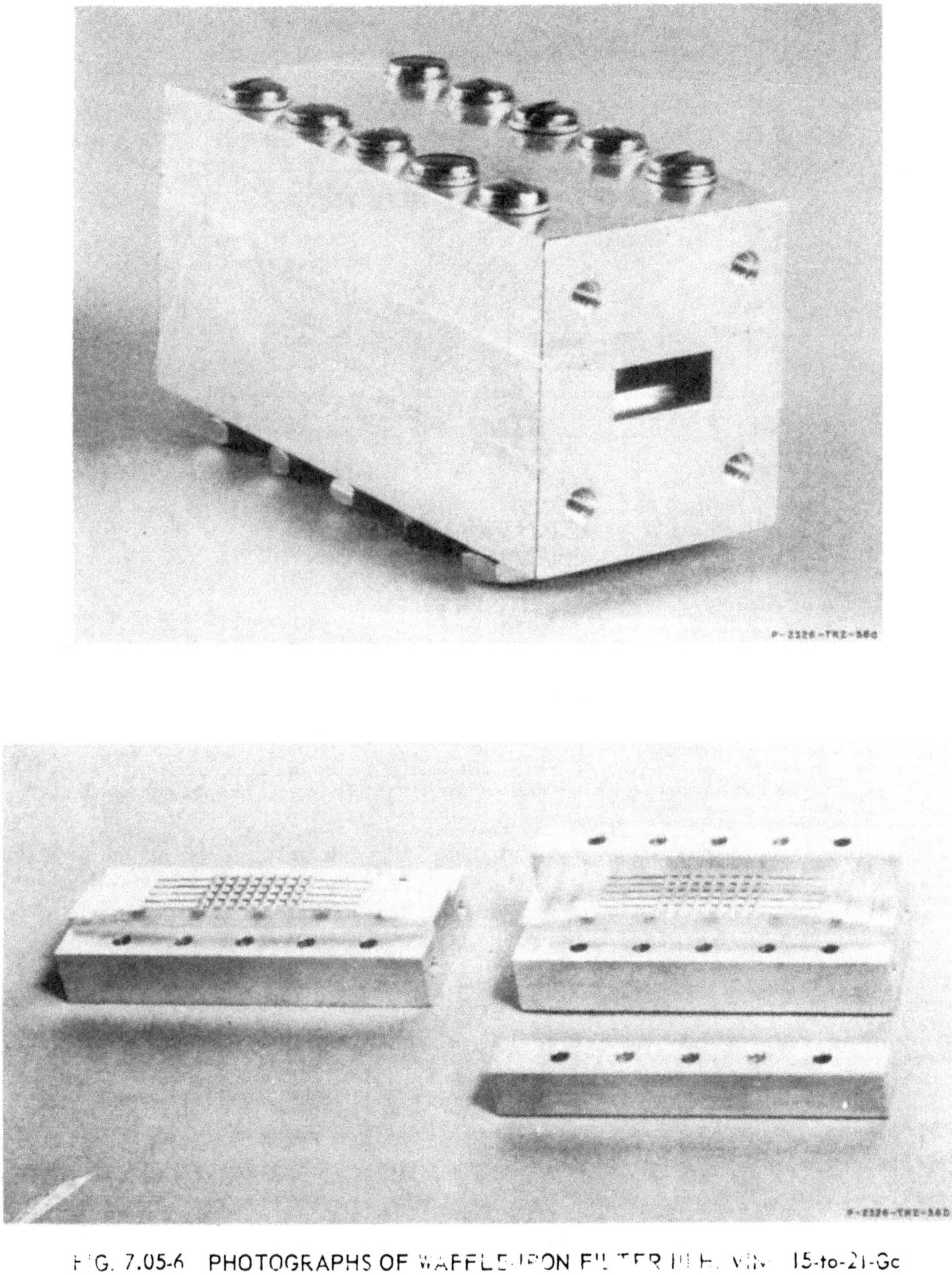

FIG. 7.05-6 PHOTOGRAPHS OF WAFFLE-IRON FILTER [illegible] 15-to-21-Gc PASS BAND AND 30-[illegible] STOP [illegible]

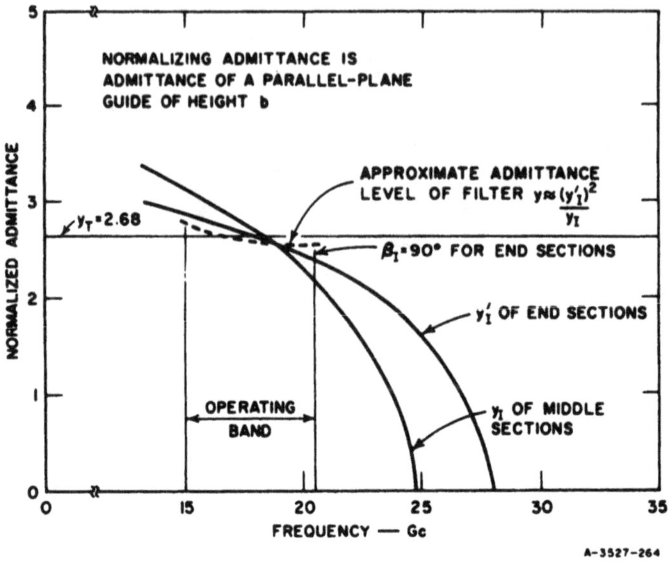

FIG. 7.05-7 SKETCH OF NORMALIZED IMAGE ADMITTANCE vs. FREQUENCY OF MIDDLE- AND END-SECTIONS OF WAFFLE-IRON FILTER III

filter is transformed to closely approximate the normalized terminating admittance $y_T = 2.68$ over the operating band, as indicated in the figure. A more general discussion of this matching technique is presented in Sec. 3.08.

The discontinuity capacity at the junction between each end section and the terminating line was compensated for by reducing the length of each end section by 0.004 inch as predicted by Eq. (7.04-17).

Quarter-Wave Transformers with Longitudinal Slots—Quarter-wave transformers, some of whose sections contained longitudinal slots, were designed for Waffle-Iron Filter III using the methods presented in Chapter 6. If there were no longitudinal slots in any of the steps of the transformers the appropriate transformation ratio to use in the design of the transformers would be the ratio of the height of the terminating guide, which is 0.255 inch, to the height of the guide which properly terminates the filter, which in this case is 0.030 inch. Thus, the transformation ratio would be 0.255/0.030 = 8.5.

If the filters and the step transformers are made from the same piece of material it is difficult to machine longitudinal slots in the main body of the filter without machining them in the step transformers at the end also. However, this difficulty can be avoided if the step transformers are made as inserts or as removable sections. Alternately, the step transformers can be designed to include longitudinal slots.

The presence of the longitudinal slots would tend to increase the transformation ratio about 8 percent since the impedance of a slotted transformer step is slightly lower than that of an unslotted step. The procedure used to calculate the impedance of a slotted waveguide is explained in detail later in this section. Qualitatively, however, it can be seen that the impedance of a slotted waveguide tends to be increased because the capacity between the top and bottom of the waveguide is reduced. On the other hand, the slots also reduce the guide wavelength which tends to *decrease* the waveguide impedance. Ordinarily it is found that the net result of these two competing effects is that the impedance of a longitudinally slotted waveguide is less than that for an unslotted waveguide.

The present design was carried out including the presence of the slots; however, it is believed that in future designs they may well be neglected in the design calculations.* The ratio of guide wavelengths at the lower and upper edge of the operating band of the transformers was chosen to be 2.50, which allowed ample margin to cover the 2.17 ratio of the guide wavelengths at the lower and upper edges of the operating band of the filter. The maximum theoretical pass-band VSWR is 1.023, and five $\lambda_g/4$ steps were used.

The procedure used to account for the presence of the longitudinal slots in the step transformers is as follows:

One assumes that the impedance Z_{0_g} of the longitudinally slotted guide is

$$Z_{0_g} = Z_0(\infty) \Bigg/ \sqrt{1 - \left(\frac{\lambda}{\lambda_c}\right)^2} \tag{7.05-8}$$

* Calculations have shown that at least in some cases the corrections for the presence of the slots is quite small.

where $Z_0(\infty)$ is the impedance of the slotted waveguide at infinite frequency and λ_c is the cutoff wavelength of the slotted waveguide. Both $Z_0(\infty)$ and λ/λ_c are functions of the guide height h_i, which is taken as the independent variable for the purpose of plotting curves of these quantities. [If Fig. 7.05-2(b) is interpreted as a cross section of the longitudinally slotted transformers, h_i corresponds to b''.]

First $Z_0(\infty)$ is calculated for several values of $h < b$ [where b is again as indicated in Fig. 7.05-2(b)] by considering TEM propagation in the longitudinal direction. Since the line is uniform in the direction of propagation

$$Z_0(\infty) = \frac{84.73\ 10^{-12}}{C_0} \quad \text{ohms} \tag{7.05-9}$$

where C_0 is the capacitance in farads per inch of length for waveguide a inches wide. The capacitance C_0 can be expressed as

$$C_0 = C_{pp} + C_d \quad . \tag{7.05-10}$$

Here the total parallel-plate capacitance C_{pp} of the longitudinal ridges of the waveguide of width a is given approximately by

$$C_{pp} = 0.225 \times 10^{-12} \left(\frac{l'}{l' + l}\right) \frac{a}{h_i} \quad \text{farads/inch} \quad . \tag{7.05-11}$$

The total discontinuity capacitance C_d of the $2m$ step discontinuities across the width of the guide is given approximately by

$$C_d = (2m)\frac{4}{\pi}\frac{l}{h_i} \times 0.225 \times 10^{-12} \left[\tan^{-1}\frac{h_i}{l} + \frac{ln\sqrt{1 + l^2/h_i^2}}{l/h_i}\right] \quad \text{farads/inch} \quad . \tag{7.05-12}$$

The cutoff wavelength, λ_c, of a rectangular waveguide with longitudinal slots is then calculated from the condition of transverse resonance for the values of h_i used above. For this calculation it is necessary to consider the change in inductance as well as the change in capacitance for waves propagating in a direction perpendicular to the longitudinal slots, back and forth across the guide of width a.

We will use static values of capacitance and inductance, and to be specific, consider that the waves propagating back and forth across the width of the guide are bounded by magnetic walls transverse to the longitudinal axis of the guide and spaced a distance w inches apart. The capacitance per slice w wide, per inch of guide width (transverse to the longitudinal axis of the guide), is

$$\frac{C_0 w}{a} \quad \text{farads/inch} \quad . \tag{7.05-13}$$

The inductance per inch of the same slice is approximately

$$L_0 = 0.032 \times 10^{-6} \frac{(lb + l'b')}{w(l + l')} \quad \text{henries/inch} \tag{7.05-14}$$

where all dimensions are in inches. A new phase velocity in the transverse direction is then calculated to be

$$v_p = \frac{1}{\sqrt{L_0 C_0 \frac{w}{a}}} \quad \text{inches/second} \quad . \tag{7.05-15}$$

The new cutoff wavelength is now

$$\lambda_c = 2a \left(\frac{v}{v_p}\right) \quad \text{inches} \tag{7.05-16}$$

where v is the plane-wave velocity of light in air—*i.e.*, 1.1803×10^{10} inches/second.

A graph of Z_{0g} *vs.* h is then made using Eq. (7.05-8), and from this graph the guide height, h_i, is obtained for each Z_i of the stepped transformer, and also for the optimum filter terminating impedance, all as previously calculated. Finally, new values of step length are calculated at the middle of the pass band for each slotted step using the values of λ_g computed from the new values of λ_c by means of the relation

$$\lambda_g = \frac{\lambda}{\sqrt{1 - \left(\frac{\lambda}{\lambda_c}\right)^2}} \quad . \qquad (7.05\text{-}17)$$

Figure 7.05-8 shows a dimensioned drawing of the filter. The lengths of the terminating guides at each end of the filter were experimentally adjusted on a lower-frequency scale model of this filter for best pass-band match. By this procedure a maximum pass-band VSWR of 1.4, and a maximum pass-band attenuation of 0.7 db was achieved. The stop-band attenuation of this filter as determined on the scale model is shown in Fig. 7.05-9. The circled points within the stop band represent spurious transmission through the filter when artificially generated higher-order modes are incident upon it. These higher-order modes were generated by twisting and displacing the terminating waveguides. The freedom from spurious responses over most of the stop band in Fig. 7.05-9, even when higher-order modes were deliberately excited, shows that this waffle-iron filter does effectively reflect all modes incident upon it in its stop band.

A Simple Technique for Further Improving the Pass-Band Impedance Match[9-11]—In the preceding examples step transformers were used to match standard waveguide into waveguide of the proper height needed to give a reasonably good match into the waffle-iron filter structure. In Waffle-Iron Filter III, besides a step transformer, additional end sections designed by the methods of Sec. 3.08 were used to further improve the impedance match. As this material is being prepared for press an additional design insight has been obtained, and is described in the following paragraphs. This insight can improve pass-band performance even more, when used in conjunction with the previously mentioned techniques.

Waffle-iron filters starting with half-capacitances (half-teeth) at either end, as used in the examples so far, are limited in the bandwidth of their pass band. The reason for this is the change of image impedance with frequency. This variation is shown in Fig. 3.06-1 for Z_{IT} and $Z_{I\pi}$. The waffle-iron with half-teeth presents an image impedance $Z_{I\pi}$, whose value increases with frequency. (The image admittance then decreases with frequency, as indicated in Fig. 7.05-7.) However, the characteristic impedance Z_0 of rectangular waveguide decreases with frequency as

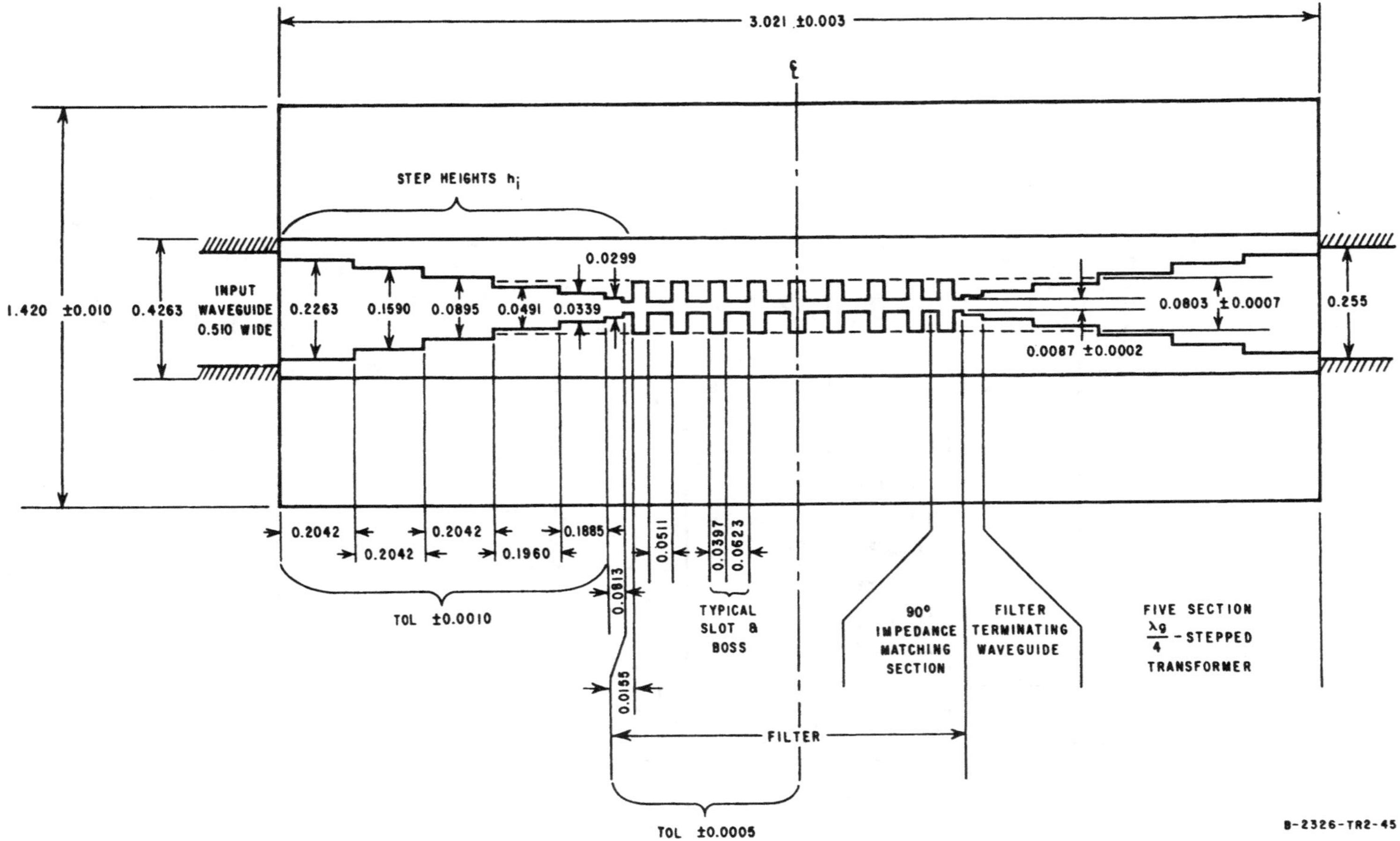

FIG. 7.05-8 SKETCH OF WAFFLE-IRON FILTER III GIVING DIMENSIONS

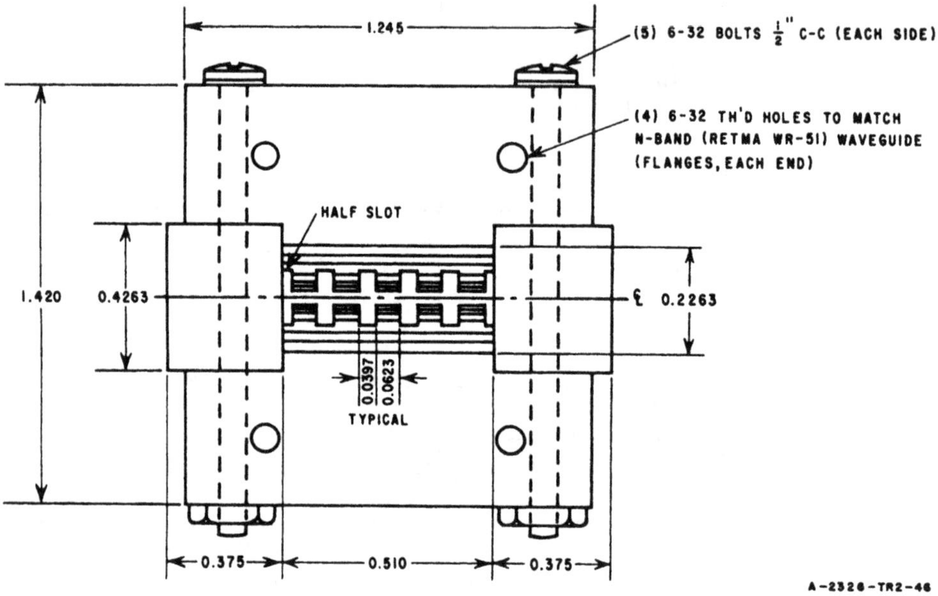

FIG. 7.05-8 Concluded

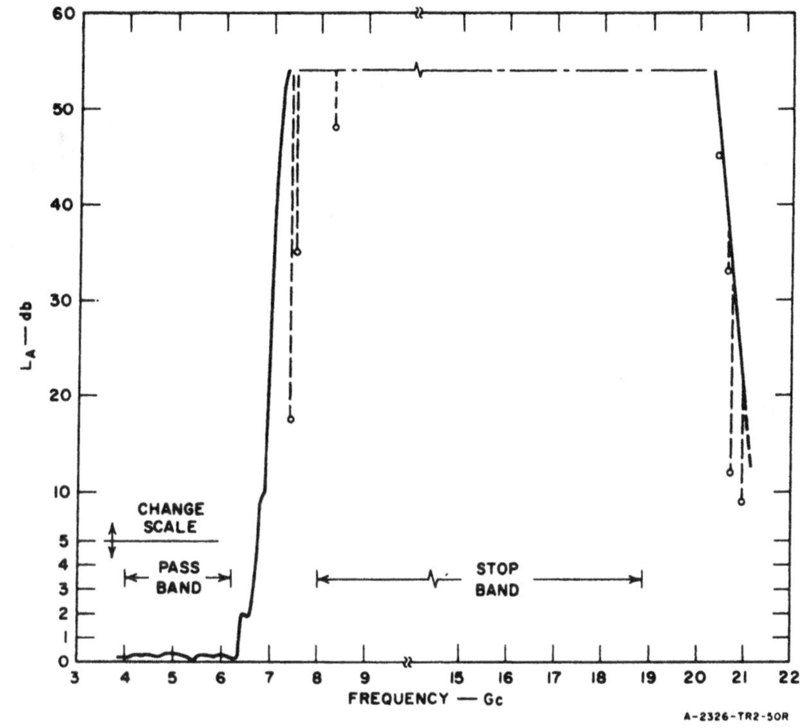

FIG. 7.05-9 MEASURED PERFORMANCE OF SCALE MODEL OF WAFFLE-IRON FILTER III SHOWING EFFECT OF ARTIFICIALLY GENERATED HIGHER MODES
The scale factor was 3.66

indicated by $Z_0 \approx \sqrt{1 - (f_c/f)^2}$ where f_c is the cutoff frequency of the waveguide. Thus, while it is possible to match the image impedance $Z_{I\pi}$ of the filter to the characteristic impedance Z_0 of the waveguide at one frequency, $Z_{I\pi}$ and Z_0 diverge rapidly with frequency, resulting in a relatively narrow pass band.

By terminating the filter with a half T-section, the image impedance Z_{IT} (Fig. 3.06-1) runs parallel to the wave guide impedance Z_0 over a substantial frequency band; then by matching Z_{IT} to Z_0 at one frequency, they stay close together over a relatively wide frequency band. Such a filter[9,10] has been built and is shown in Fig. 7.05-10. This L-band, five-section filter has circular (instead of square) teeth to improve the power-handling capacity by an estimated factor of 1.4. The dimensions of this filter, using the notation of Fig. 7.05-2, were: b = 1.610 inches,

SOURCE: Quarterly Progress Report 1, Contract AF 30(602)-2734
(See Ref. 9 by Leo Young)

FIG. 7.05-10 EXPLODED VIEW OF WAFFLE-IRON FILTER WITH ROUND TEETH AND HALF-INDUCTANCES AT THE ENDS

b'' = 0.210 inch, a = 6.500 inches, center-to-center spacing = 1.300 inches, tooth diameter = 0.893 inch, edge radius of the rounded teeth is R = 0.063 inch. This filter is in fact based on the Waffle-Iron Filter I design, whose stop-band performance is shown in Fig. 7.05-3.

The new filter (Fig. 7.05-10) had a stop-band performance which almost duplicates Fig. 7.05-3 (after allowance is made for the fact that it has five rather than ten sections), showing that neither the tooth shape (round, not square), nor the end half-sections (half-T, not half-π) affect the stop-band performance.

In the pass band, the filter (Fig. 7.05-10) was measured first with 6.500-inch-by-0.375-inch waveguide connected on both sides. The VSWR was less than 1.15 from 1200 to 1640 megacycles. (It was below 1.08 from 1250 to 1460 megacycles). The same filter was then measured connected to 6.500-inch-by-0.350-inch waveguide, and its VSWR remained below 1.20 from 1100 to 1670 megacycles (as compared to 1225 to 1450 megacycles for 1.2 VSWR or less with Waffle-Iron Filter I). Thus the VSWR remains low over almost the whole of L-band.

The estimated power-handling capacity of the filter in Fig. 7.05-10 is over two megawatts in air at atmospheric pressure. This power-handling capacity was later quadrupled by paralleling four such filters (Sec. 15.05).

SEC. 7.06, LOW-PASS FILTERS FROM QUARTER-WAVE TRANSFORMER PROTOTYPES

This section is concerned with the high-impedance, low-impedance short-line filter, which is the most common type of microwave low-pass filter, and which has been treated in Sec. 7.03 in terms of an approximately lumped-constant structure (Fig. 7.03-1). Such an approximation depends on:

(1) The line lengths being short compared to the shortest pass-band wavelength

(2) The high impedances being very high and the low ones very low—*i.e.*, the impedance steps should be large.

There is then a close correspondence between the high-impedance lines of the actual filter and series inductances of the lumped-constant prototype, on the one hand, and the low-impedance lines and shunt capacitances, on the other.

There is another way of deriving such a transmission-line low-pass filter, which is exact when:

(1) All line lengths are equal (and not necessarily vanishingly short)

(2) When the step discontinuity capacities are negligible.

When either of these, or both, are not satisfied, approximations have to be made, as in the design from the lumped-constant prototype. Which one of the two prototypes is more appropriate depends on which of the two sets of conditions (1) and (2) above are more nearly satisfied. Whereas the lumped-constant prototype (Sec. 7.03) is usually the more appropriate design procedure, the method outlined in this section gives additional insight, especially into the stop-band behavior, and into the spurious pass bands beyond.

This second way of deriving the short-line low-pass filter can best be understood with reference to Fig. 7.06-1. In Fig. 7.06-1(a) is shown a quarter-wave transformer (Sec. 6.02) with its response curve. Each section is a quarter-wave long at a frequency inside the first pass band,

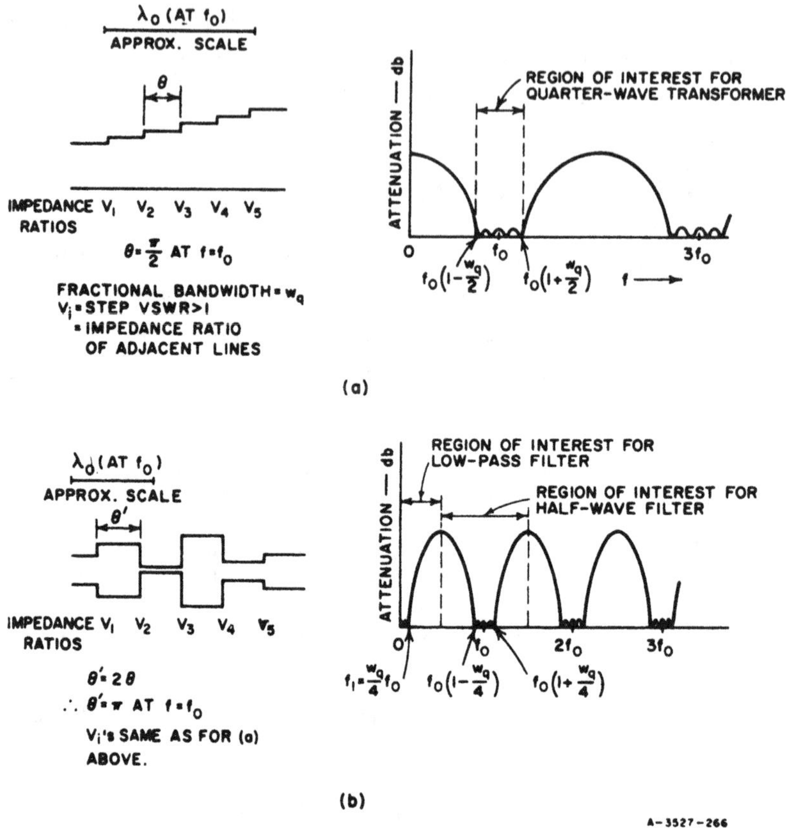

FIG. 7.06-1 CONNECTION BETWEEN QUARTER-WAVE TRANSFORMERS (a) AND CORRESPONDING LOW-PASS FILTERS (b)

called the band center, f_0. The "low-pass filter" is sketched in Fig. 7.06-1(b). Its physical characteristics differ from the quarter-wave transformer in that the impedance steps are alternately up and down, instead of forming a monotone sequence; it is essentially the same structure as the "half-wave filter" of Sec. 6.03. Each section is a half-wave long at a frequency f_0 at the center of the first band-pass pass band. However, note that there is also a low-pass pass band from $f = 0$ to f_1, and that the stop band above f_1 is a number of times as wide as the low-pass pass band. The fractional bandwidth of the spurious pass band at f_0 for the low-pass filter has half the fractional pass-band bandwidth, w_q, of the quarter-wave transformer. The VSWRs V_i of the corresponding steps in the step-transformer and in the low-pass filter are the same for both structures, the VSWRs here being defined as equal to the ratio (taken so as to be greater than one) of the impedances of adjacent lines.

Low-pass filters are generally made of non-dispersive lines (such as strip lines or coaxial lines), and such lines will be assumed here. If wave-guides or other dispersive lines are used, it is only necessary to replace normalized frequency f/f_0 by normalized reciprocal guide wavelength λ_{g0}/λ_g. Since the low-pass filter sections are a half-wavelength long at $f = f_0$, the over-all length of a low-pass filter of n sections is at most $nw_q/8$ wavelengths at any frequency in the (low-pass) pass band, this being its length at the low-pass band-edge, $f_1 = w_q f_0/4$. Note that the smaller w_q for the step-transformer is chosen to be, the larger the size of the stop band above f_1 will be for the low-pass filter, relative to the size of the low-pass pass band.

Exact solutions for Tchebyscheff quarter-wave transformers and half-wave filters have been tabulated up to $n = 4$ (Sec. 6.04); and for maximally flat filters up to $n = 8$ (Sec. 6.05); all other cases have as yet to be solved by approximate methods, such as are given in Secs. 6.06 to 6.09.

The low-pass filter (as designed by this method) yields equal line lengths for the high- and low-impedance lines. When the impedance steps, V_i, are not too large (as in the wide-band examples of Sec. 6.09), then the approach described in this section can be quite useful.* Corrections for the discontinuity capacitances can be made as in Sec. 6.08. If large impedance steps are used, as is usually desirable, the discontinuity effects become dominant over the transmission-line effects, and it is usually more straightforward to use lumped-element prototypes as was done for the first example in Sec. 7.03.

SEC. 7.07, HIGH-PASS FILTERS USING SEMI-LUMPED ELEMENTS

High-pass filters, having cutoff frequencies up to around 1.5 or possibly 2.0 Gc can be easily constructed from semi-lumped elements. At frequencies above 1.5 or 2.0 Gc the dimensions of semi-lumped high-pass

---

* It should be noted that small impedance steps imply a relatively limited amount of attenuation. Thus, small steps will be desired only in certain special situations.

filters become so small that it is usually easier to use other types of structures. The wide-band band-pass filters discussed in Chapters 9 and 10 are good candidates for many such applications.

In order to illustrate the design of a semi-lumped-element high-pass filter we will first describe the general technique for designing a lumped-element high-pass filter from a lumped-element low-pass prototype circuit. Next we will use this technique to determine the dimensions of a split-block, coaxial-line high-pass microwave filter using semi-lumped elements.

Lumped-Element High-Pass Filters from Low-Pass Prototype Filters—The frequency response of a lumped-element high-pass filter can be related to that of a corresponding low-pass prototype filter such as that shown in Fig. 4.04-1(b) by means of the frequency transformation

$$\omega' = -\frac{\omega_1 \omega_1'}{\omega} \quad . \tag{7.07-1}$$

In this equation ω' and ω are the angular frequency variables of the low- and high-pass filters respectively while ω_1' and ω_1 are the corresponding band-edge frequencies of these filters. It is seen that this transformation has the effect of interchanging the origin of the frequency axis with the point at infinity and the positive frequency axis with the negative frequency axis. Figure 7.07-1 shows a sketch of the response, for positive frequencies, of a nine-element low-pass prototype filter together with the response of the analogous lumped-element high-pass filter obtained by means of the transformation in Eq. (7.07-1).

Equation (7.07-1) also shows that any inductive reactance $\omega' L'$ in the low-pass prototype filter is transformed to a capacitive reactance $-\omega_1 \omega_1' L'/\omega = -1/(\omega C)$ in the high-pass filter, and any capacitive susceptance $\omega' C'$ in the low-pass prototype filter is transformed into an inductive susceptance $-\omega_1 \omega_1' C'/\omega = -1/(\omega L)$ in the high-pass filter.

Thus, any inductance L' in the low-pass prototype filter is replaced in the high-pass filter by a capacitance

$$C = \frac{1}{\omega_1 \omega_1' L'} \quad . \tag{7.07-2}$$

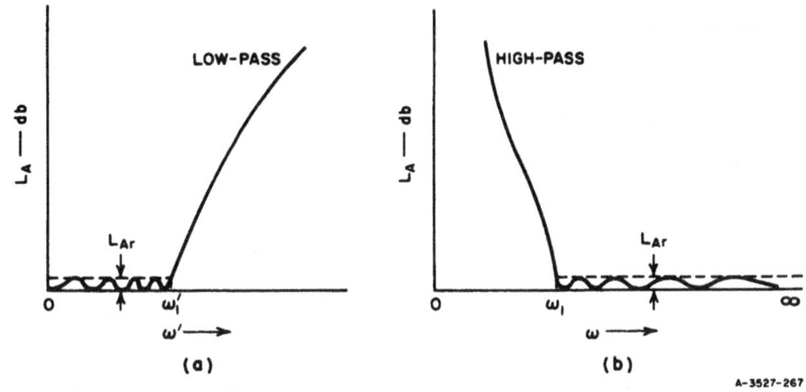

FIG. 7.07-1 FREQUENCY RESPONSE OF A LOW-PASS PROTOTYPE AND OF A CORRESPONDING HIGH-PASS FILTER

Likewise any capacitance C' in the low-pass prototype is replaced in the high-pass filter by an inductance

$$L = \frac{1}{\omega_1 \omega_1' C'} \quad . \tag{7.07-3}$$

Figure 7.07-2 illustrates the generalized equivalent circuit of a high-pass filter obtained from the low-pass prototype in Fig. 4.04-1(b) by these methods. A dual filter with an identical response can be obtained by applying Eqs. (7.07-2) and (7.07-3) to the dual low-pass prototype in Fig. 4.04-1(a). The impedance level of the high-pass filter may be scaled as discussed in Sec. 4.04.

Design of a Semi-Lumped-Element High-Pass Filter—In order to illustrate the technique for designing a semi-lumped-element high-pass filter we will consider the design of a nine-element high-pass filter with a pass-band ripple L_{Ar} of 0.1 db, a cutoff frequency of 1 Gc ($\omega_1 = 2\pi \times 10^9$), that will operate between 50-ohm terminations. The first step in the design is to determine the appropriate values of the low-pass prototype elements from Table 4.05-2(a). It should be noted that elements in this table are normalized so that the band-edge frequency $\omega_1' = 1$ and the termination element $g_0 = 1$. The values of the inductances and capacitances for the high-pass filter operating between 1-ohm terminations are then

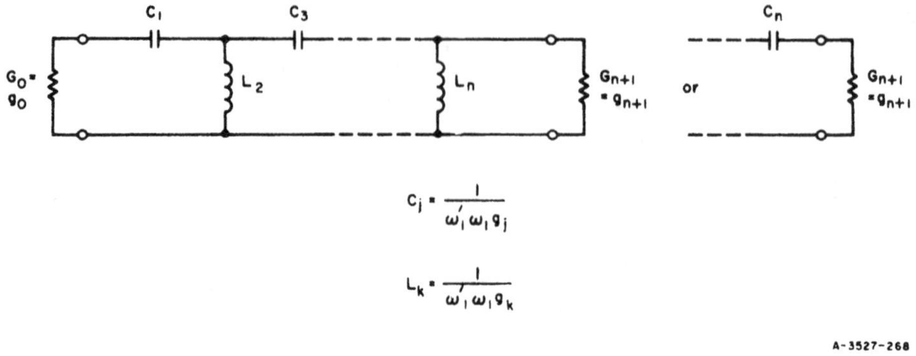

FIG. 7.07-2 HIGH-PASS FILTER CORRESPONDING TO THE LOW-PASS PROTOTYPE IN FIG. 4.04-1(b)
Frequencies ω_1' and ω_1 are defined in Fig. 7.07-1. A dual form of this filter corresponding to the low-pass filter in Fig. 4.04-1(a) is also possible

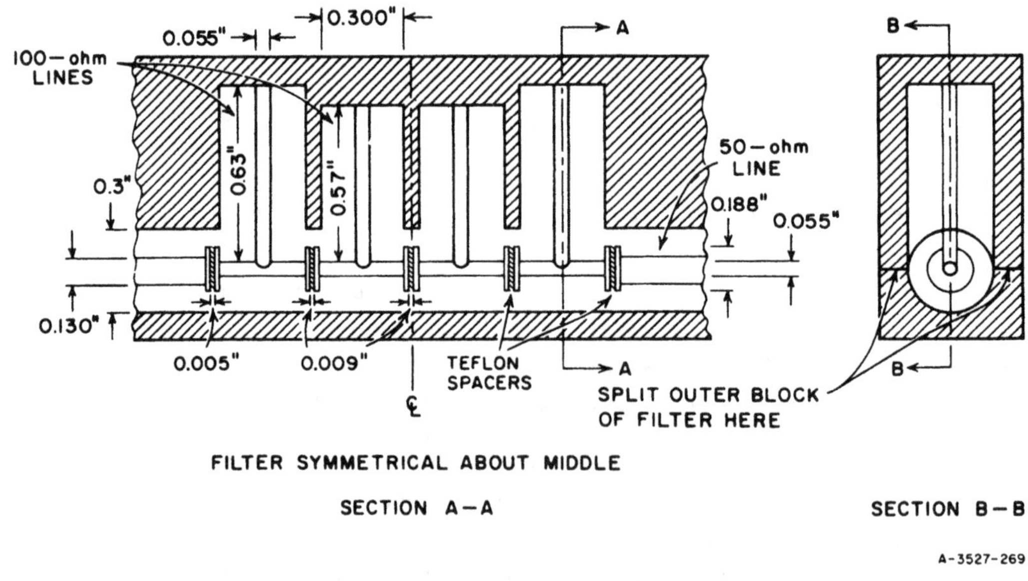

FIG. 7.07-3 DRAWING OF COAXIAL LINE HIGH-PASS FILTER CONSTRUCTED FROM SEMI-LUMPED ELEMENTS USING SPLIT-BLOCK CONSTRUCTION

determined using the formulas in Fig. 7.07-2, upon setting $\omega_1' = 1$, $\omega_1 = 2\pi \times 10^9$, and using the g_k values selected from Table 4.05-2(a). In order to convert the above design to one that will operate at a 50-ohm impedance level it is necessary to divide all the capacitance and conductance values obtained by 50 and to multiply all the inductance values obtained by 50. When this procedure is carried through we find that $C_1 = C_9 = 2.66$ $\mu\mu$f, $L_2 = L_8 = 5.51$ mμh, $C_3 = C_7 = 1.49$ $\mu\mu$f, $L_4 = L_6 = 4.92$ mμh, and $C_5 = 1.44$ $\mu\mu$f.

A sketch showing a possible realization of such a filter in coaxial line, using split-block construction, is shown in Fig. 7.07-3. Here it is seen that the series capacitors are realized by means of small metal disks utilizing Teflon ($\epsilon_r = 2.1$) as dielectric spacers. The shunt inductances are realized by short lengths of $Z_0 = 100$-ohm line short-circuited at the far end. In determining the radius r of the metal disks, and the separation s between them, it is assumed that the parallel-plate capacitance is much greater than the fringing capacitance, so that the capacitance C of any capacitor is approximately

$$C \approx \epsilon_r \, 0.225 \frac{\pi r^2}{s} \qquad \mu\mu\text{f} \tag{7.07-4}$$

where all dimensions are measured in inches. The lengths l of the short-circuited lines were determined by means of the formula

$$L = 0.0847 \, Z_0 l \qquad \text{m}\mu\text{h} \tag{7.07-5}$$

where Z_0 is measured in ohms and l is measured in inches. Equation (7.07-4) is adapted from one in Fig. 7.02-2(b), while Eq. (7.07-5) is adapted from one in Fig. 7.02-1(a).

The dimensions presented in Fig. 7.07-3 must be regarded as tentative, because a filter having these particular dimensions has not been built and tested. However, the electrical length of each of the lines in the filter is very short—even the longest short-circuited lines forming the shunt inductors have an electrical length of only 19.2 degrees at 1 Gc. Therefore, it is expected that this semi-lumped-constant filter will have very close to the predicted performance from low frequencies up to at least 2.35 Gc, where two of the short-circuited lines are an eighth-wavelength long and have about 11 percent higher reactance than the idealized lumped-constant design. Above this frequency some increase in pass-band attenuation will probably be noticed (perhaps one or two db) but not a really large increase. At about 5 Gc when the short-circuited lines behave as open circuits, the remaining filter structure formed from the series capacitors and the short lengths of series lines has a pass band, so that the attenuation should be low even at this frequency. However, somewhere between 5 Gc and 9 Gc (where the short-circuited lines are about 180 degrees long) the attenuation will begin to rise very rapidly.

SEC. 7.08, LOW-PASS AND HIGH-PASS IMPEDANCE-MATCHING NETWORKS

Some microwave loads which can be approximated by an inductance and a resistance in series, or by a capacitance and a conductance in parallel, can be given a satisfactory broadband impedance match by use of low-pass matching networks. Having L and R, or C and G to represent the load, the decrement

$$\delta = \frac{R}{\omega_1 L} \quad \text{or} \quad \frac{G}{\omega_1 C} \tag{7.08-1}$$

is computed, where ω_1 is the pass-band cutoff frequency above which a good impedance match is no longer required. Though the prototype filter to be used in designing the matching network may have a considerably different impedance level and cutoff frequency ω_1', it must have the same decrement δ. Thus, having computed δ from the given microwave load elements and required cutoff frequency ω_1, an appropriate impedance-matching-network prototype filter can be selected from the computed value of δ and the charts of prototype element values in Sec. 4.09. Having selected a satisfactory prototype filter, the impedance-matching network can be designed by scaling the prototype in frequency and impedance level and by using the semi-lumped-element realization techniques discussed in Sec. 7.03. As was illustrated in Fig. 4.09-1, the microwave load to be matched provides the microwave circuit elements corresponding to the prototype elements g_0 and g_1, the microwave impedance-matching network corresponds to the prototype elements g_2 through g_n, and the microwave driving-source resistance or conductance corresponds to g_{n+1}.

Though low-pass microwave impedance-matching structures are quite practical for some applications, they do, nevertheless, have some inherent disadvantages compared to the band-pass impedance-matching networks discussed in Secs. 11.08 to 11.10. One of these disadvantages is that a good impedance match all the way from dc up to microwave frequencies is rarely really necessary. As was discussed in Sec. 1.03, allowing energy to be transmitted in frequency bands where energy transmission is not needed will detract from the efficiency of transmission in the band where good transmission is really needed. Thus if the decrement computed using Eq. (7.08-1) is found to be so small that Fig. 4.09-3 indicates an unacceptable amount of pass-band attenuation, the possibility of using a band-pass matching network instead should be considered. If a band-pass transmission characteristic is usable, better performance can be obtained.

Another disadvantage of low-pass impedance-matching networks is that the designer is not free to choose the driving source resistance. For a given R-L or G-C load circuit and a given cutoff frequency ω_1, the charts in Sec. 4.09 will lead to matching networks which must use the driving source resistances (or conductances) specified by the charts, if the predicted performance is to be obtained. In many microwave applications, adjustments of the driving-source impedance level will not be convenient. In such cases the use of band-pass impedance-matching networks is again recommended since in the case of band-pass filters, impedance-level transformations are easily achieved in the design of the filter, without affecting the transmission characteristic.

High-pass impedance-matching networks have basically the same disadvantages as low-pass impedance-matching networks. Nevertheless they are of practical importance for some applications. Loads which can be approximated by a capacitance and resistance in series, or by an inductance and conductance in parallel can be given a high-pass impedance match by using the methods of this book. In this case the decrement is computed by use of the formula

$$\delta = \omega_1 CR \quad \text{or} \quad \omega_1 LG \tag{7.08-2}$$

where in this case ω_1 is the cutoff frequency for the desired high-pass matching characteristic. Knowing δ, the $(L_A)_{max}$ values for various numbers of matching elements are checked and a prototype is then selected, as discussed in Sec. 4.09. [Again, if the values of $(L_A)_{max}$ for the computed value of δ are too large, the possibility of band-pass matching should be considered.] The low-pass prototype is then transformed to a high-pass filter as discussed in Sec. 7.07, and its frequency scale and impedance level are adjusted so as to conform to the required ω_1 value and the specified microwave load. If the cutoff frequency ω_1 is not too high, it should be practical to realize the microwave impedance-matching structure by use of the semi-lumped-element high-pass filter techniques discussed in Sec. 7.07.

SEC. 7.09, LOW-PASS TIME-DELAY NETWORKS

Most of the primary considerations in the design of low-pass time-delay networks have been previously discussed in Secs. 1.05, 4.07, and 4.08. The maximally flat time-delay networks tabulated in Sec. 4.07 were seen to give extremely flat time-delay* characteristics, but at the expense of having an attenuation characteristic which varies considerably in the operating band. Maximally flat time-delay networks also are unsymmetrical, which makes their fabrication more difficult. In Sec. 4.08 it was noted that Tchebyscheff filters with small pass-band ripple should make excellent time-delay networks for many practical applications. As was discussed in Sec. 1.05, the amount of time delay can be increased considerably for a given circuit complexity by using, where possible, a band-pass rather than a low-pass structure for the delay network (see Secs. 1.05 and 11.11). High-pass delay networks are also conceivable, but they would not give much delay, except, possibly, near cutoff.

Example—As an example of the initial steps in the design of a low-pass time-delay network, let us suppose that a time delay of about 7.2 nanoseconds is required from frequencies of a few megacycles up to 200 Mc. From considerations such as those discussed in Sec. 4.08, let us further suppose that it has been decided to use a 0.1-db ripple Tchebyscheff filter with a cutoff of f_1 = 250 Mc, as the delay network. From Eq. (4.08-3), the low-frequency time delay of a corresponding normalized prototype filter with a cutoff of $\omega_1' = 1$ radian/sec is

$$t'_{do} = t_{do}\frac{\omega_1}{\omega_1'} = \frac{7.2(10^{-9})2\pi(0.25)10^9}{1} = 11.3 \text{ seconds}$$

---

* Here time delay is assumed to imply group time delay (Sec. 1.05).

By Eq. (4.08-2) and Fig. 4.13-2, this nominal time delay will be achieved by a 0.10-db ripple filter having n = 13 reactive elements. Hence, an n = 13, L_{ar} = 0.10 db prototype should be selected from Table 4.05-2(b). The actual microwave filter is then designed from the prototype as discussed in Sec. 7.03. If desired, this filter could be designed to be a few inches long, while it would take approximately 7 feet of air-filled coaxial line to give the same time delay.

REFERENCES

1. N. Marcuvitz, *Waveguide Handbook,* p. 178 (McGraw Hill Book Company, New York, N.Y., 1951).
2. R. Saal and E. Ulbrich, "On the Design of Filters by Synthesis," *Trans. IRE,* PGCT-5, pp. 284-327 (December 1958). The same tables and many more will be found in the book, R. Saal, *Der Entwurf von Filtern mit Hilfe des Kataloges normierter Tiefpasse,* Telefunken GMBH, Backnang, Wurttemburg, Germany (1961).
3. S. B. Cohn, "A Theroretical and Experimental Study of a Waveguide Filter Structure," Cruft Laboratory Report 39, ONR Contract N50 RI-76, Harvard University (April 1948).
4. S. B. Cohn, "Analysis of a Wide-Band Waveguide Filter," *Proc. IRE 37,* pp. 651-656 (June 1949).
5. E. A. Guillemin, *Communication Networks, Vol. 2,* p. 439 (John Wiley and Sons, New York, N.Y., 1935).
6. N. Marcuvitz, *op. cit.,* p. 336-350.
7. S. B. Cohn, "Design Relations for the Wide-Band Waveguide Filter," *Proc. IRE* 38, pp. 799-803 (July 1950).
8. Eugene Sharp, "A High-Power Wide-Band Waffle-Iron Filter," Tech. Note 2, SRI Project 3478, Contract AF 30(602)-2392, Stanford Research Institute, Menlo Park, California (January 1962). RADC-TDR-62-183. Also, *IEEE Trans. PTGMTT*-11, pp. 111-116 (March 1963).
9. Leo Young, "Suppression of Spurious Frequencies," Quarterly Progress Report 1, SRI Project 4096, Contract AF 30(602)-2734, Stanford Research Institute, Menlo Park, California (July 1962).
10. L. Young and B. M. Schiffman, "New and Improved Types of Waffle-Iron Filters," *Proc. IEE* (London)-110, pp. 1191-1198 (July 1963).
11. Leo Young, "Postscript to Two Papers on Waffle-Iron Filters," *IEEE Trans. PTGMTT*-11, pp. 555-557 (November 1963).

CHAPTER 8

BAND-PASS FILTERS
(A GENERAL SUMMARY OF BAND-PASS FILTERS, AND A VERSATILE DESIGN TECHNIQUE FOR FILTERS WITH NARROW OR MODERATE BANDWIDTHS)

SEC. 8.01, A SUMMARY OF THE PROPERTIES OF THE BAND-PASS OR PSEUDO HIGH-PASS FILTERS TREATED IN CHAPTERS 8, 9, AND 10

This chapter is the first of a sequence of four chapters concerning band-pass filter design. Chapters 8, 9, and 10 deal with the design theory and specific types of microwave filters, while Chapter 11 discusses various experimental and theoretical techniques which are generally helpful in the practical development of many kinds of band-pass filters and impedance-matching networks. This present chapter (Chapter 8) utilizes a design point of view which is very versatile but involves narrow-band approximations which limit its usefulness to designs having fractional bandwidths typically around 0.20 or less. The design procedure utilized in Chapter 9 makes use of step transformers as prototypes for filters, and the procedures given there are useful for either narrow or wide bandwidths. Chapter 10 uses yet another viewpoint for design, and the method described there is also useful for either narrow or wide bandwidths. The procedures in Chapter 9 are most advantageous for filters consisting of transmission lines with lumped discontinuities placed at intervals, while the methods in Chapter 10 are most advantageous when used for filters consisting of lines and stubs or of parallel-coupled resonators.

In this chapter the general design point of view is first described in a qualitative way, then design equations and other data for specific types of filters are presented, and finally the background details of how the design equations for specific filters were derived are presented. Chapters 9 and 10 also follow this pattern as far as is possible.

It is recognized that some designers may have little interest in filter design theory, and that they may only wish to pick out one design for one given job. To help meet this need, Table 8.01-1 has been prepared. It summarizes the more significant properties of the various types of filters discussed in Chapters 8, 9, and 10, and tells the reader in which sections design data for a given type of filter can be found.

Table 8.01-1

SUMMARY OF BAND-PASS AND PSEUDO HIGH-PASS FILTERS IN CHAPTERS 8, 9, AND 10

Symbols

ω_0 = pass-band center frequency

ω_{SPB} = center frequency of second pass band

$(L_A)_{USB}$ = peak attenuation (in db) in upper stop band (between ω_0 and ω_{SPB})

L_{Ar} = peak attenuation (in db) in pass band

w = fractional bandwidth

λ_0 = wavelength at ω_0

λ_g = guide wavelength

$\lambda_{g0}, \lambda_{g1}, \lambda_{g2}$ = guide wavelengths at ω_0 and at lower and upper pass-band-edge frequencies

$w_\lambda = \dfrac{\lambda_{g1} - \lambda_{g2}}{\lambda_{g0}}$ = guide-wavelength fractional bandwidth

STRIP-LINE (OR COAXIAL) AND SEMI-LUMPED-ELEMENT FILTERS

Typical Resonator or Section	Filter Properties
1 ★ 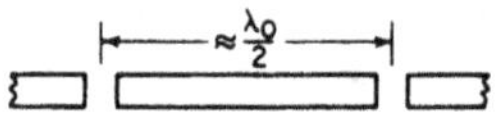 STRIP LINE	$\omega_{SPB} \approx 2\omega_0$. $(L_A)_{USB}$ decreases with increasing w. $(L_A)_{USB}$ is usually sizeable for w = 0.20 or less, but it is usually only 5 or 10 db for w = 0.70. Has first-order pole of attenuation at ω = 0. Dielectric support required for resonators. Coupling gaps may become quite small for w much larger than 0.10, which presents tolerance considerations. See Sec. 8.05 for designs with w about 0.20 or less. See Chapter 9 for designs having larger w, or for designs with very small L_{Ar} (0.01 db, for example), or for designs for high-pass applications. Coaxial filters of this type are widely used as pseudo high-pass filters.
2 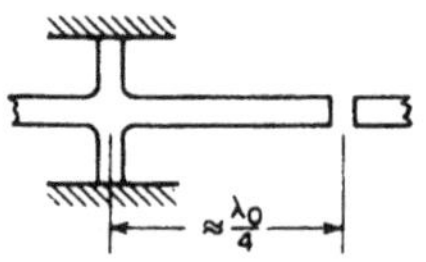 STRIP LINE	$\omega_{SPB} \approx 3\omega_0$. $(L_A)_{USB}$ decreases with increasing w, but for given w and ω_0, $(L_A)_{USB}$ will be larger than for Filter 1 above. Has multiple-order pole of attenuation at ω = 0. Inductive stubs can provide mechanical support for resonator structure so that dielectric is not required. For given w and ω_0 capacitive coupling gaps are larger than for Filter 1 above. See Sec. 8.08 for designs with $w \lesssim 0.30$. See Chapter 9 for designs having larger w, or for designs with very small L_{Ar} (0.01 db, for example), or for designs for high-pass applications.
3 ★ 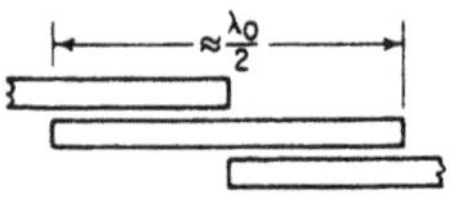 STRIP LINE	$\omega_{SPB} = 3\omega_0$. Has first-order pole of attenuation at ω = 0 and at $\omega = 2\omega_0$. However, is prone to have narrow spurious pass bands near $2\omega_0$ due to slightest mistuning. Dielectric support material required. Very attractive structure for printed circuit fabrication, when $w \lesssim 0.15$. See Sec. 8.09-1 for $w \lesssim 0.15$. See Sec. 10.02 for designs having larger w, or for designs for high-pass applications.

Table 8.01-1 *Continued*

STRIP-LINE (OR COAXIAL) AND SEMI-LUMPED ELEMENT FILTERS	
Typical Resonator or Section	**Filter Properties**
4 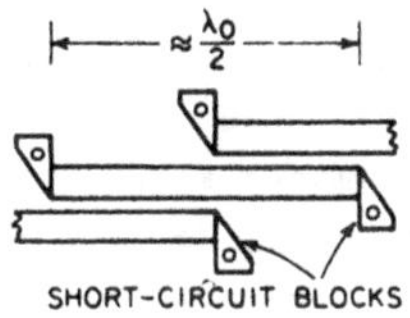 STRIP LINE	$\omega_{SPB} = 3\omega_0$. Has first-order pole of attenuation at $\omega = 0$ and at $\omega = 2\omega_0$. However, is prone to narrow spurious pass bands near $2\omega_0$ due to slightest mistuning. Short-circuit blocks provide mechanical support for resonators. Suitable for values of w from around 0.01 to 0.70 or more. See Sec. 10.02.
5 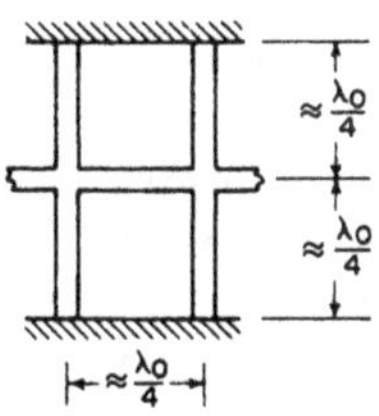 STRIP LINE	$\omega_{SPB} = 3\omega_0$. Has first-order pole of attenuation at $\omega = 0$ and at $\omega = 2\omega_0$. However, is prone to narrow spurious pass bands near $2\omega_0$ due to slightest mistuning. Short-circuits at ends of stubs provide mechanical support for structure. Suitable for values of w from around 0.40 to 0.70 or more. See Sec. 10.03. Also see Sec. 10.05 for case where series stubs are added at ends to give poles of attenuation at additional frequencies.
6 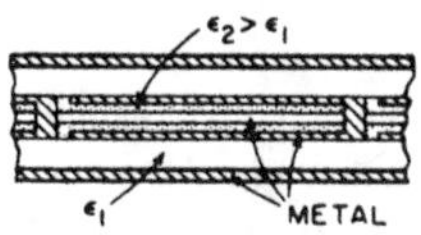 COAXIAL	Structure in coaxial form with series stubs fabricated within center conductor of main line. $\omega_{SPB} = 3\omega_0$. Has first-order pole of attenuation at $\omega = 0$ and at $\omega = 2\omega_0$. However, is prone to narrow spurious pass bands near $2\omega_0$ due to slightest mistuning. Structure requires dielectric support material. Suitable for values of w around 0.60 or more. See Sec. 10.03.
7 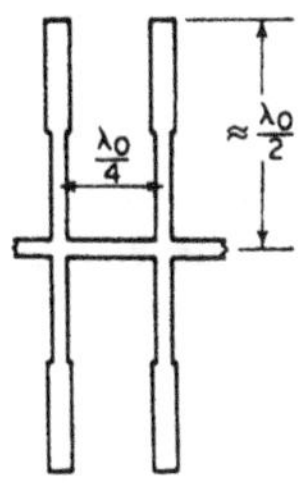 STRIP LINE	$\omega_{SPB} = 2\omega_0$, and also has a pass band around $\omega = 0$. Has poles of attenuation above and below ω_0 at frequencies ω_∞ and $(2\omega_0 - \omega_\infty)$, where ω_∞ may be specified. Requires dielectric material for support. Can conveniently be fabricated by printed circuit means. Little restriction on w if ω_∞ can be chosen appropriately. See Sec. 10.04. (continued on p. 424)

Table 8.01 *Continued*

STRIP-LINE (OR COAXIAL) AND SEMI-LUMPED-ELEMENT FILTERS	
Typical Resonator or Section	**Filter Properties**
8 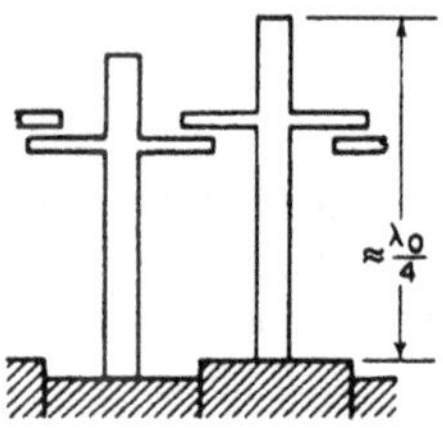 STRIP LINE	ω_{SPB} can be made to be as high as $5\omega_0$ or more. Has multiple-order poles of attenuation at $\omega = 0$. Short-circuited ends of resonators provide mechanical support so that dielectric material is not required. Structure is quite compact. See Sec. 8.12 for design data suitable for designs with $w \lessapprox 0.10$.
9 ★ 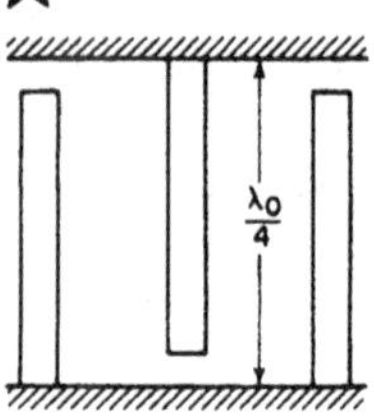 STRIP LINE	Interdigital Filter. $\omega_{SPB} = 3\omega_0$. Has multiple-order poles of attenuation at $\omega = 0$ and $\omega = 2\omega_0$. Can be fabricated without using dielectric support material. Spacings between resonator elements are relatively large which relaxes mechanical tolerances. Structure is very compact. See Secs. 10.06 and 10.07 for equations for designs with w ranging from small values up to large values around 0.70 or more.
10 ★ 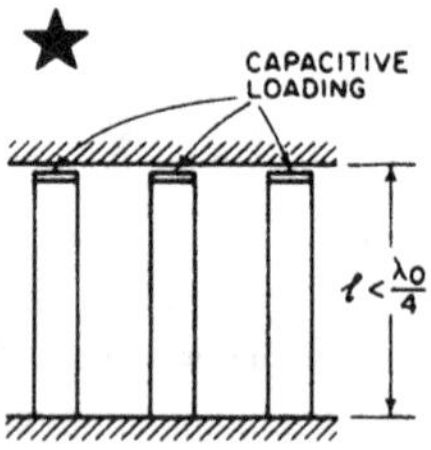 STRIP LINE	Comb-line filter. Resonator length l depends on amount of capacitive loading used. $\omega_{SPB} \gtrapprox \omega_0 \lambda_0/(2l)$ so filter can be designed for very broad upper stop band. Poles of attenuation at $\omega = 0$ and $\omega = \omega_0 \lambda_0/(4l)$. Extremely compact structure which can be fabricated without dielectric support material. Unloaded Q's of resonators somewhat less than those for Filter 9 for some strip-line cross-section. See Sec. 8.13 for designs having w up to about 0.15.
11 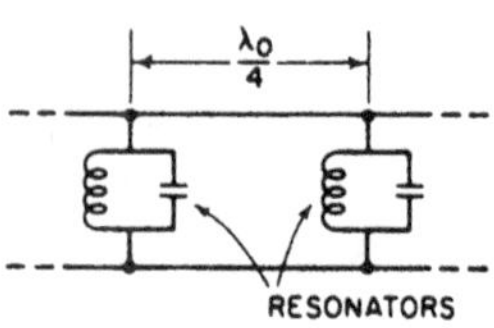 LUMPED ELEMENTS	Filter with quarter-wave-coupled resonators. Resonators may be cavities, resonant irises, or lumped-element resonators. See Sec. 8.08 for design data useful when w is around 0.05 or less.

Table 8.01-1 *Concluded*

Typical Resonator or Section	Filter Properties
STRIP-LINE (OR COAXIAL) AND SEMI-LUMPED-ELEMENT FILTERS	
12 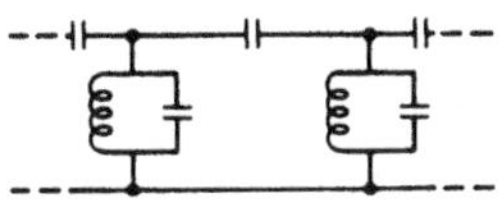 LUMPED ELEMENTS	Lumped-element circuit for use as a guide for design of semi-lumped-element microwave filters. See Sec. 8.11 for designs with $w \lesssim 0.20$.
13 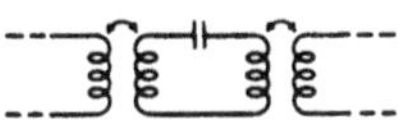 LUMPED ELEMENTS	Lumped-element circuit for use as a guide for design of semi-lumped-element microwave filters. See Sec. 8.11 for designs with $w \lesssim 0.20$.
WAVEGUIDE AND CAVITY FILTERS	
14 ★ 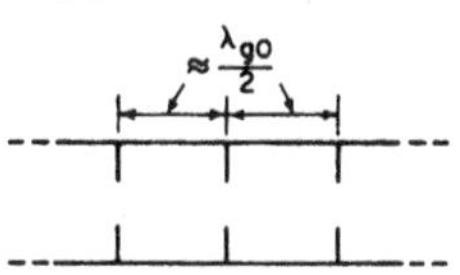 WAVEGUIDE	ω_{SPB} occurs when λ_g is about $\lambda_{g0}/2$; however, when higher-order modes can propagate, the upper stop band and second pass band may be disrupted. $(L_A)_{\mathrm{USB}}$ decreases with increasing w_λ. Waveguide resonators give relatively low dissipation loss for given w_λ. See Secs. 8.06 and 8.07 for designs with w_λ about 0.20 or less. See Chapter 9 for designs having larger w_λ, or for designs with very small L_{Ar} (0.01 db, for example), or for designs for high-pass applications.
15 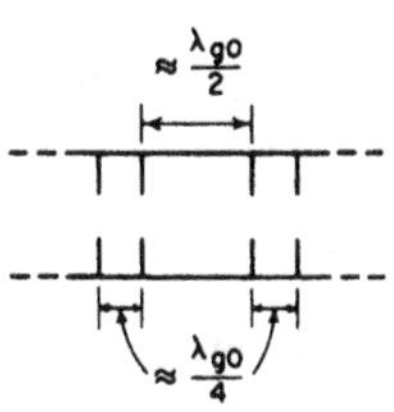 WAVEGUIDE	Use of $\lambda_{g0}/4$ couplings gives irises which are all nearly the same. If a disassembly joint is placed in the middle of each $\lambda_{g0}/4$ coupling region, resonators may be easily tested individually. ω_{SPB} occurs when λ_g is about $\lambda_{g0}/2$; however, when higher-order modes can propagate the upper stop band and second pass band may be disrupted. $(L_A)_{\mathrm{USB}}$ decreases with increasing w_λ. Waveguide resonators give relatively low dissipation loss for given w_λ. Satisfactory for designs having w_λ about 0.05 or less. See Sec. 8.08.

The filters whose properties are summarized in Table 8.01-1 are suitable for a wide range of applications. Some are suitable for either narrow- or wide-band band-pass filter applications. Also, since it is difficult, if not impossible, to build a microwave high-pass filter with good pass-band performance up to many times the cutoff frequency, pseudo high-pass filters, which are simply wideband band-pass filters, provide some of the most practical means for fabricating filters for microwave high-pass applications. Thus, many of the filters in Table 8.01-1 should also be considered as potential microwave high-pass filters.

Although most of the filters in Table 8.01-1 are pictured in strip-line form, many of them could be fabricated equally well in coaxial form or in split-block coaxial form (Fig. 10.05-3). One of the filter properties which is of interest in selecting a particular type of band-pass filter structure is the frequency at which the second pass band will be centered. In Table 8.01-1, this frequency is designated as ω_{SPB}, and it is typically two or three times ω_0, the center frequency of the first pass band. However, in the case of Filter 8 in Table 8.01-1, ω_{SPB} can be made to be as much as five or more times ω_0. Filter 10 is also capable of very broad stop bands.

All of the filters in Table 8.01-1 have at least one frequency, ω, where they have infinite attenuation (or where they would have infinite attenuation if it were not for the effects of dissipation loss). These infinite attenuation points, known as poles of attenuation (see Sec. 2.04), may be of first order or of multiple order; the higher the order of the pole of attenuation, the more rapidly the attenuation will rise as ω approaches the frequency of the pole. Thus, the presence of first-order or multiple-order poles of attenuation at frequencies ω are noted in Table 8.01-1 as a guide towards indicating what the relative strength of the stop band will be in various frequency ranges. Four of the filters in Table 8.01-1 (Filters 1, 2, 14, and 15) have no poles of attenuation in the stop-band region above the pass-band center ω_0, and the attenuation between the first and second pass-bands levels off at a value of $(L_A)_{USB}$ decibels. As is mentioned in Table 8.01-1, the values of $(L_A)_{USB}$ will in such cases be influenced by the fractional bandwidth w of the filter. Also, it should be noted that the filters which have a first-order pole of attenuation in the stop band above ω_0 may be liable to spurious responses close to this pole if there is any mistuning.

Another consideration in choosing a type of filter for a given job is the unloaded Q's obtainable with the resonator structures under consideration. Waveguide or cavity resonators will, of course, give the best unloaded Q's, and hence will result in filters with minimum insertion loss for a given fractional bandwidth. However, waveguide resonators have the disadvantages of being relatively bulky and of being useful over only a limited frequency range because of the possibility of higher-order modes. Thus, where wide pass bands or wide stop bands are required, strip-line, coaxial, or semi-lumped-element filters are usually preferable. If strip-line or coaxial constructuions are used, the presence of dielectric material, which may be required for mechanical support of the structure, will tend to further decrease the resonator Q's obtainable. For this reason, it is in many cases noted in Table 8.01-1 whether or not the specific structure can be fabricated without the use of dielectric support material.

The filter structures marked with stars in Table 8.01-1 are filter types which represent attractive compromise choices for many applications. However, they are by no means necessarily the best choices in all respects, and special considerations may dictate the use of some of the unstarred types of filters listed in the table.

Filter 1 in Table 8.01-1 was starred because, in coaxial form, it provides a very rugged and convenient way for manufacturing pseudo high-pass filters. Commercial coaxial high-pass filters are most commonly of this form.

Filter 3 in Table 8.01-1 has been starred because it is extremely easy to design and fabricate in printed-circuit construction when the fractional bandwidth is around 0.15 or less. However, its stop-band characteristics and its resonator Q's are inferior to those that can be obtained with some of the other types of strip-line or coaxial filters in the table.

Filter 9 was starred because it is easy to design for anywhere from small to large fractional bandwidths, it is compact, and it has strong stop bands on both sides of ω_0.

Filter 10 was starred because of its compactness and ease of design, and because it is capable of a very broad upper stop band.

Filter 14 was starred because it is the simplest and most commonly used type of waveguide filter. Within the single-mode frequency range of the waveguide, such filters generally give excellent performance.

SEC. 8.02, GENERAL PRINCIPLES OF COUPLED-RESONATOR FILTERS*

In this section we will discuss the operation of coupled-resonator filters in qualitative terms. For the benefit of those readers who are concerned

* The point of view used herein is that due to S. B. Cohn.[1] However, herein his point of view has been restated in more general terms, and it has been applied to additional types of filter structures not treated by Cohn. Some other points of view and earlier contributions are listed in References 2 to 8.

primarily with practical design, rather than with theory, this qualitative discussion will be followed by design data for specific types of filters. Details of the derivation of the design equations will be found in Sec. 8.14.

In the design procedures of this chapter, the lumped-element prototype filter designs discussed and tabulated in Chapter 4 will be used to achieve band-pass filter designs having approximately the same Tchebyscheff or maximally flat response properties. Thus, using a lumped-element prototype having a response such as the Tchebyscheff response shown in Fig. 8.02-1(a), the corresponding band-pass filter response will also be Tchebyscheff as shown in Fig. 8.02-1(b). As suggested in Fig. 8.02-1(b), the multiple resonances inherent in transmission-line or cavity resonators generally give band-pass microwave filters additional pass bands at higher frequencies.

Figure 8.02-2(a) shows a typical low-pass prototype design, and Fig. 8.02-2(b) shows a corresponding band-pass filter design, which can be obtained directly from the prototype by a low-pass to band-pass transformation to be discussed in Sec. 8.04. In the equations for the band-pass filter element values, the g_j are the prototype filter element values, ω' and ω'_1 are for the prototype filter response as indicated in Fig. 8.02-1(a) for a typical Tchebyscheff case, and ω, ω_0, ω_1, and ω_2 apply to the corresponding band-pass filter response as indicated in Fig. 8.02-1(b). Of course, the filter in Fig. 8.02-2(b) would not have the higher frequency pass bands suggested in Fig. 8.02-1(b) because it is composed of lumped elements.

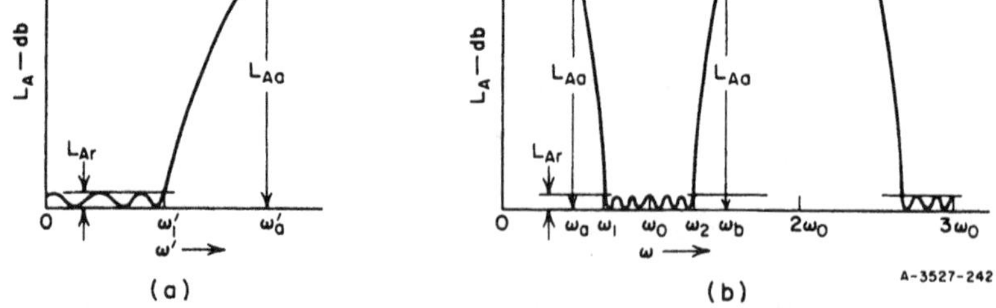

FIG. 8.02-1 LOW-PASS PROTOTYPE RESPONSE AND CORRESPONDING BAND-PASS FILTER RESPONSE

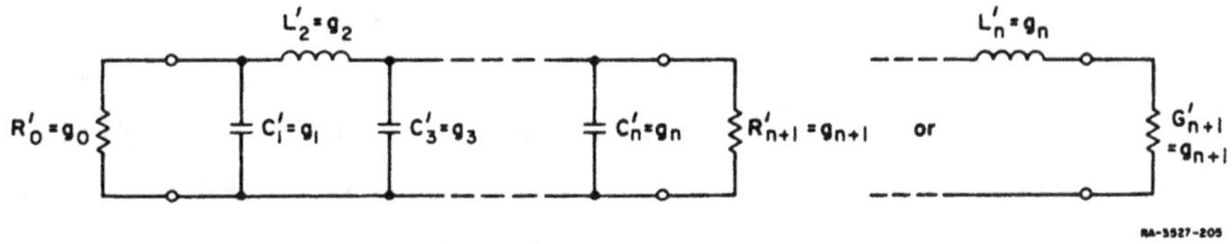

FIG. 8.02-2(a) A LOW-PASS PROTOTYPE FILTER

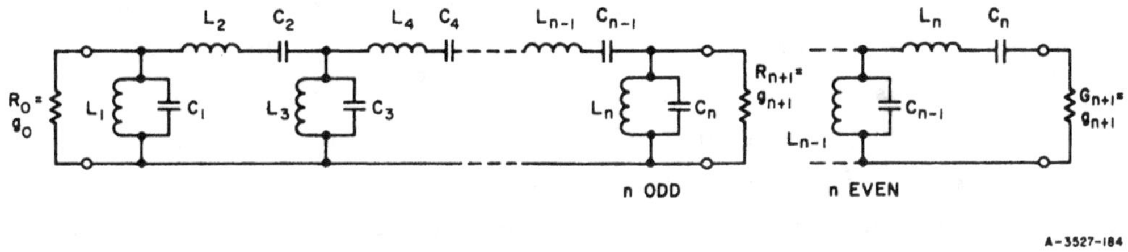

FOR SHUNT RESONATORS:

$$\mathscr{b}_j = \omega_0 C_j = \frac{1}{\omega_0 L_j} = \frac{\omega_1' g_j}{w} = \text{susceptance slope parameter} \qquad (1)$$

FOR SERIES RESONATORS:

$$\mathscr{x}_k = \omega_0 L_k = \frac{1}{\omega_0 C_k} = \frac{\omega_1' g_k}{w} = \text{reactance slope parameter} \qquad (2)$$

$$w = \frac{\omega_2 - \omega_1}{\omega_0}$$

$$\omega_0 = \sqrt{\omega_1 \omega_2}$$

FIG. 8.02-2(b) BAND-PASS FILTERS AND THEIR RELATION TO LOW-PASS PROTOTYPES
Frequencie s ω_1', ω_1, and ω_2 are defined in Fig. 8.02-1, ang g_0, g_1, ..., g_{n+1} are defined in Fig. 8.02-2(a)

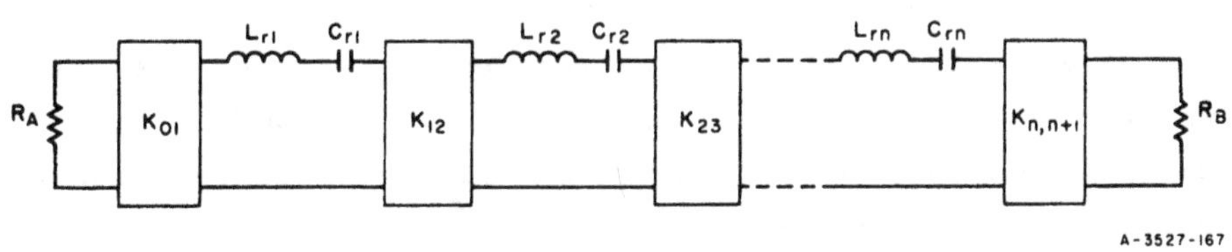

NOTE: Adapted from Final Report, Contract DA-36-039 SC-64625, SRI; reprinted in *Proc. IRE* (see Ref. 1 by S. B. Cohn).

FIG. 8.02-2(c) THE BAND-PASS FILTER IN FIG. 8.02-2(b) CONVERTED TO USE ONLY SERIES RESONATORS AND IMPEDANCE INVERTERS

The filter structure in Fig. 8.02-2(b) consists of series resonators alternating with shunt resonators, an arrangement which is difficult to achieve in a practical microwave structure. In a microwave filter, it is much more practical to use a structure which approximates the circuit in Fig. 8.02-2(c), or its dual. In this structure all of the resonators are of the same type, and an effect like alternating series and shunt resonators is achieved by the introduction of "impedance inverters," which were defined in Sec. 4.12, and are indicated by the boxes in Fig. 8.02-2(c). The band-pass filter in Fig. 8.02-2(c) can be designed from a low-pass prototype as in Fig. 8.02-2(a) by first converting the prototype to the equivalent low-pass prototype form in Fig. 4.12-2(a) which uses only series inductances and impedance inverters in the filter structure. Then a low-pass to band-pass transformation can be applied to the circuit in Fig. 4.12-2(a) to yield the band-pass circuit in Fig. 8.02-2(c). Practical means for approximate realization of impedance inverters will be discussed in Sec. 8.03 following.

Since lumped-circuit elements are difficult to construct at microwave frequencies, it is usually desirable to realize the resonators in distributed-element forms rather than the lumped-element forms in Figs. 8.02-2(b), (c). As a basis for establishing the resonance properties of resonators regardless of their form it is convenient to specify their *resonant frequency* ω_0 and their *slope parameter*. For any resonator exhibiting a series-type resonance (case of zero reactance at ω_0) the *reactance slope parameter*

$$\mathscr{X} = \frac{\omega_0}{2} \frac{dX}{d\omega}\bigg|_{\omega_0} \quad \text{ohms} \tag{8.02-1}$$

applies, where X is the reactance of the resonator. For a simple series L-C resonator, Eq. (8.02-1) reduces to $\mathscr{X} = \omega_0 L = 1/(\omega_0 C)$. For any resonator exhibiting a shunt-type resonator (case of zero *susceptance* at ω_0) the *susceptance slope parameter*

$$\mathscr{B} = \frac{\omega_0}{2} \frac{dB}{d\omega}\bigg|_{\omega_0} \quad \text{mhos} \tag{8.02-2}$$

applies where B is the susceptance of the resonator. For a shunt L-C resonator, Eq. (8.02-2) reduces to $\mathscr{b} = \omega_0 C = 1/(\omega_0 L)$. Note that in Fig. 8.02-2(b) the properties of the lumped resonators have been defined in terms of susceptance and reactance slope parameters. The slope parameters of certain transmission-line resonators were discussed in Sec. 5.08 and are summarized in Fig. 5.08-1. Any resonator having a series-type resonance with a reactance slope parameter $\mathscr{x}$ and series resistance R has a Q of

$$Q = \frac{\mathscr{x}}{R} \quad . \tag{8.02-3}$$

Likewise, any resonator having a shunt-type resonance with a susceptance slope parameter $\mathscr{b}$ and a shunt conductance G has a Q of

$$Q = \frac{\mathscr{b}}{G} \quad . \tag{8.02-4}$$

Figure 8.02-3(a) shows a generalized circuit for a band-pass filter having impedance inverters and series-type resonator characteristics as indicated by the resonator-reactance curve in Fig. 8.02-3(b). Let us suppose that a band-pass filter characteristic is desired like that in Fig. 8.02-1(b), and the filter is to be designed from a low-pass prototype having a response like that in Fig. 8.02-1(a) and having prototype parameters g_0, g_1, ..., g_{n+1}, and ω_1'. The resonator slope parameters $\mathscr{x}_1$, $\mathscr{x}_2$, ..., $\mathscr{x}_n$ for the band-pass filter may be selected arbitrarily to be of any size corresponding to convenient resonator designs. Likewise, the terminations R_A, R_B, and the fractional bandwidth w may be specified as desired. The desired shape of response is then insured by specifying the impedance-inverter parameters K_{01}, K_{12}, ..., $K_{n,n+1}$ as required by Eqs. (2) to (4) in Fig. 8.02-3. If the resonators of the filter in Fig. 8.02-3(a) were each comprised of a lumped L and C, and if the impedance inverters were not frequency sensitive, the equations in Fig. 8.02-3 would be exact regardless of the fractional bandwidth w of the filter. However, since the inverters used in practical cases are frequency sensitive (see Sec. 8.03), and since the resonators used will generally not be lumped, in practical cases the equations in Fig. 8.02-3 represent approximations which are best for narrow bandwidths. However, in some cases good results can be obtained for bandwidths as great as

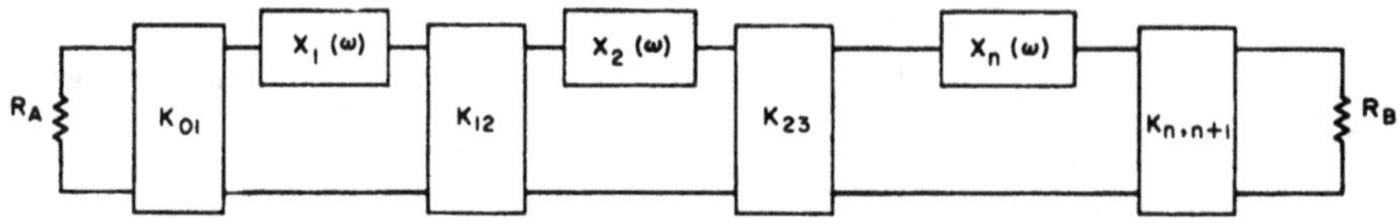

(a) A GENERALIZED, BAND-PASS FILTER CIRCUIT USING IMPEDANCE INVERTERS

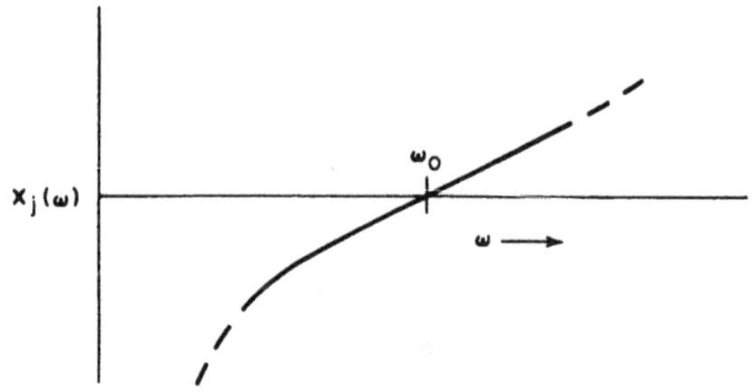

B-3527-185

(b) REACTANCE OF jth RESONATOR

$$\mathscr{X}_j = \frac{\omega_0}{2} \left. \frac{dX_j(\omega)}{d\omega} \right|_{\omega=\omega_0} \quad \text{ohms} \tag{1}$$

$$= \text{Reactance Slope Parameter}$$

$$K_{01} = \sqrt{\frac{R_A \mathscr{X}_1 w}{g_0 g_1 \omega_1'}} \tag{2}$$

$$\left. K_{j,j+1} \right|_{j=1 \text{ to } n-1} = \frac{w}{\omega_1'} \sqrt{\frac{\mathscr{X}_j \mathscr{X}_{j+1}}{g_j g_{j+1}}} \tag{3}$$

$$K_{n,n+1} = \sqrt{\frac{R_B \mathscr{X}_n w}{\omega_1' g_n g_{n+1}}} \tag{4}$$

$$w = \begin{array}{c}\text{fractional}\\ \text{bandwidth}\end{array} = \text{ or } \approx \frac{\omega_2 - \omega_1}{\omega_0} \tag{5}$$

where ω_1', ω_0, ω_1, and ω_2 are defined in Fig. 8.02-1, and $g_0, g_1, \ldots, g_{n+1}$ are as defined in Sec. 4.04 and Fig. 8.02-2(a).

For Experimental Determination of Couplings (As Discussed in Chapter 11)

External Q's are:

$$(Q_e)_A = \frac{\mathscr{X}_1}{(K_{01}^2/R_A)} = \frac{g_0 g_1 \omega_1'}{w} \tag{6}$$

$$(Q_e)_B = \frac{\mathscr{X}_n}{(K_{n,n+1}^2/R_B)} = \frac{\omega_1' g_n g_{n+1}}{w} \tag{7}$$

Coupling coefficients are:

$$\left. k_{j,j+1} \right|_{j=1 \text{ to } n-1} = \frac{K_{j,j+1}}{\sqrt{\mathscr{X}_j \mathscr{X}_{j+1}}} = \frac{w}{\omega_1' \sqrt{g_j g_{j+1}}} \tag{8}$$

FIG. 8.02-3 GENERALIZED EQUATIONS FOR DESIGN OF BAND-PASS FILTERS FROM LOW-PASS PROTOTYPES
Case of filters with resonators having series-type resonances. The K-inverters represent the couplings

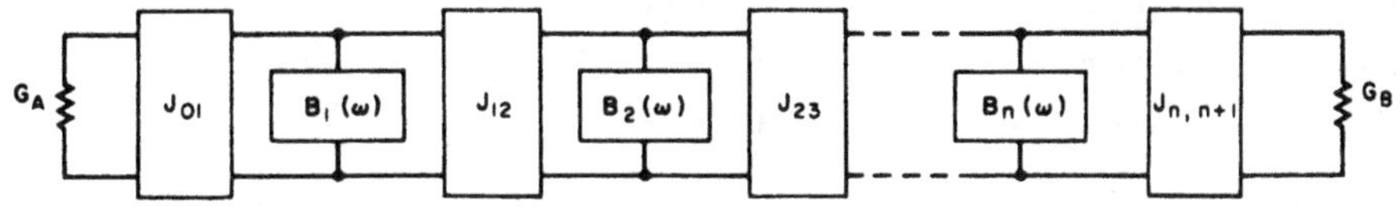

(a) A GENERALIZED, BAND-PASS FILTER CIRCUIT USING ADMITTANCE INVERTERS

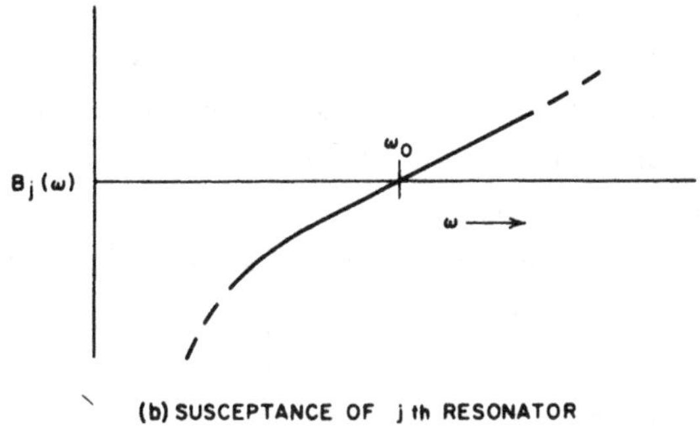

(b) SUSCEPTANCE OF j th RESONATOR

A-3527-186

$$b_j = \frac{\omega_0}{2} \left. \frac{dB_j(\omega)}{d\omega} \right|_{\omega=\omega_0} \quad \text{mhos} \tag{1}$$

$$= \text{Susceptance Slope Parameter}$$

$$J_{01} = \sqrt{\frac{G_A b_1 w}{g_0 g_1 \omega_1'}} \tag{2}$$

$$\left. J_{j,j+1} \right|_{j=1 \text{ to } n-1} = \frac{w}{\omega_1'} \sqrt{\frac{b_j b_{j+1}}{g_j g_{j+1}}} \tag{3}$$

$$J_{n,n+1} = \sqrt{\frac{G_B b_n w}{\omega_1' g_n g_{n+1}}} \tag{4}$$

$$w = \begin{array}{c}\text{fractional} \\ \text{bandwidth}\end{array} \begin{array}{c} = \\ \text{or} \approx \end{array} \frac{\omega_2 - \omega_1}{\omega_0} \tag{5}$$

where ω_1', ω_0, ω_1 and ω_2 are defined in Fig. 8.02-1, and g_0, g_1, ..., g_{n+1} are as defined in Sec. 4.04 and Fig. 8.02-2(a).

For Experimental Determination of Couplings (As Discussed in Chapter 11)

External Q's are:

$$(Q_e)_A = \frac{b_1}{(J_{01}^2/G_A)} = \frac{g_0 g_1 \omega_1'}{w} \tag{6}$$

$$(Q_e)_B = \frac{b_n}{(J_{n,n+1}^2/G_B)} = \frac{\omega_1' g_n g_{n+1}}{w} \tag{7}$$

Coupling coefficient are:

$$\left. k_{j,j+1} \right|_{j=1 \text{ to } n-1} = \frac{J_{j,j+1}}{\sqrt{b_j b_{j+1}}} = \frac{w}{\omega_1' \sqrt{g_j g_{j+1}}} \tag{8}$$

FIG. 8.02-4 GENERALIZED EQUATIONS FOR DESIGN OF BAND-PASS FILTERS FROM LOW-PASS PROTOTYPES

Case of filters having resonators with only shunt-type resonances. The J-inverters represent the couplings

20 percent when half-wavelength resonators are used, and when quarter-wavelength resonators are used, good results can be obtained in some cases for bandwidths approaching 40 percent.

Equations (6) to (8) in Fig. 8.02-3 are forms which are particularly convenient when the resonator couplings are to be adjusted by experimental procedures discussed in Chapter 11. The external Q, $(Q_e)_A$, is the Q of Resonator 1 coupled by the Inverter K_{01} to the termination R_A. The external Q, $(Q_e)_B$ is the corresponding Q of resonator n coupled by $K_{n,n+1}$ to R_B. The expression for the coupling coefficients $k_{j,j+1}$ is a generalization of the usual definition of coupling coefficient. For lumped-element resonators with inductive couplings $k_{j,j+1} = M_{j,j+1}/\sqrt{L_j L_{j+1}}$ where L_j and L_{j+1} are self inductances and $M_{j,j+1}$ is the mutual inductance. By specifying the coupling coefficients between resonators and the external Q's of the end resonators as indicated in Eqs. (6) to (8) in Fig. 8.02-3, the response of the filter is fixed. Equations (2) to (4) and Eqs. (6) to (8) are equivalent.

The band-pass filter in Fig. 8.02-4(a) uses *admittance inverters* and *shunt-type* resonator characteristics as indicated by the resonator-*susceptance* curve in Fig. 8.02-4(b). Admittance inverters are in principle the same as impedance inverters, but for convenience they are here characterized by an admittance parameter, $J_{j,j+1}$, instead of an impedance parameter, $K_{j,j+1}$ (see Sec. 4.12). The equations in Fig. 8.02-4 are duals of those in Fig. 8.02-3, and the same general principles discussed in the preceding paragraphs apply.

In the discussions to follow K-inverter impedance parameters will be used whenever the resonators have a series-type resonance, and J-inverter admittance parameters will be used whenever the resonators have a shunt-type resonance.

SEC. 8.03, PRACTICAL REALIZATION OF K- AND J-INVERTERS

One of the simplest forms of inverters is a quarter-wavelength of transmission line. Observe that such a line obeys the basic impedance-inverter definition in Fig. 4.12-1(a), and that it will have an inverter parameter of $K = Z_0$ ohms where Z_0 is the characteristic impedance of the line. Of course, a quarter-wavelength of line will also serve as an admittance inverter as can be seen from Fig. 4.12-1(b), and the admittance inverter parameter will be $J = Y_0$ where Y_0 is the characteristic admittance of the line.

Although its inverter properties are relatively narrow-band in nature, a quarter-wavelength line can be used satisfactorily as an impedance or admittance inverter in narrow-band filters. Thus, if we have six identical cavity resonators, and if we connect them by lines which are a quarter-wavelength long at frequency ω_0, then by properly adjusting the coupling at each cavity it is possible to achieve a six resonator Tchebyscheff response such as that in Fig. 8.02-1(b). Note that if the resonators all exhibit, say, series-type resonances, and if they were connected together directly without the impedance inverters, they would simply operate like a *single* series resonator with a slope parameter equal to the sum of the slope parameters of the individual resonators. Some sort of inverters between the resonators are essential in order to obtain a multiple-resonator response if all of the resonators are of the same type, *i.e.*, if all exhibit a series-type resonance or all exhibit a shunt-type resonance.

Besides a quarter-wavelength line, there are numerous other circuits which operate as inverters. All necessarily give an image phase (see Sec. 3.02) of some odd multiple of ±90 degrees, and many have good inverting properties over a much wider bandwidth than does a quarter-wavelength line. Figure 8.03-1 shows four inverting circuits which are of special interest for use as K-inverters (*i.e.*, inverters to be used with series-type resonators). Those shown in Figs. 8.03-(a),(b) are particularly useful in circuits where the negative L or C can be absorbed into adjacent positive series elements of the same type so as to give a resulting circuit having all positive elements. The inverters shown in Figs. 8.03-1(c),(d) are particularly useful in circuits where the line of positive or negative electrical length ϕ shown in the figures can be added to or subtracted from adjacent lines of the same impedance. The circuits shown at (a) and (c) have an over-all image phase shift of −90 degrees, while those at (b) and (d) have an over-all image phase shift of +90 degrees. The impedance-inverter parameter K indicated in the figure is equal to the image impedance (see Sec. 3.02) of the inverter network and is analogous to the characteristic impedance of a transmission line. The networks in Fig. 8.03-1 are much more broadband inverters than is a quarter-wavelength line.*

* In the cases of Figs. 8.03-1(c),(d), this statement assumes that $|X/Z_0| \ll 1$ which is usually the case in the practical application of these circuits.

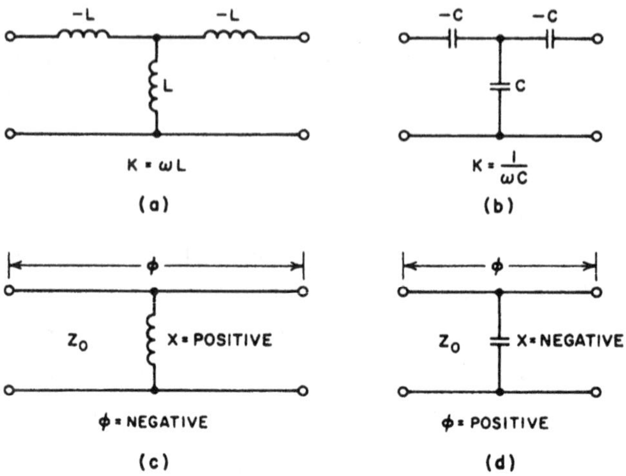

For both cases (c) and (d)

$$K = Z_0 \tan\left|\frac{\phi}{2}\right| \quad \text{ohms}$$

$$\phi = -\tan^{-1}\frac{2X}{Z_0} \quad \text{radians}$$

$$\left|\frac{X}{Z_0}\right| = \frac{\frac{K}{Z_0}}{1 - \left(\frac{K}{Z_0}\right)^2}$$

A-3527-187

FIG. 8.03-1 SOME CIRCUITS WHICH ARE PARTICULARLY USEFUL AS K-INVERTERS (Inverters To Be Used with Series-Type Resonators)

Figure 8.03-2 shows four inverting circuits which are of special interest for use as J-inverters (*i.e.*, inverters to be used with shunt-type resonators). These circuits will be seen to be the duals of those in Fig. 8.03-1, and the inverter parameters J are the image admittances of the inverter networks.

Figure 8.03-3 shows two more circuits which operate as inverters. These circuits are useful for computing the impedance-inverting properties of certain types of discontinuities in transmission lines. Examples will be cited in Secs. 8.05 and 8.06. Figure 8.03-4 shows yet another form of inverter composed of transmission lines of positive and negative

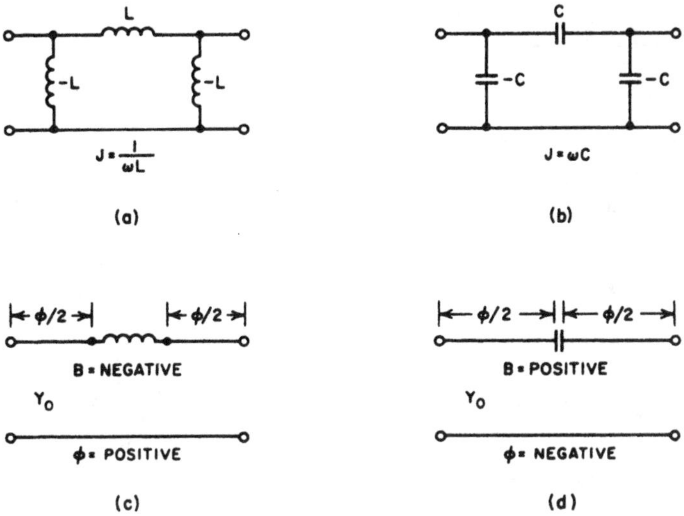

For both cases (c) and (d):

$$J = Y_0 \tan \left|\frac{\phi}{2}\right| \qquad \text{mhos}$$

$$\phi = -\tan^{-1} \frac{2B}{Y_0} \qquad \text{radians}$$

$$\left|\frac{B}{Y_0}\right| = \frac{\dfrac{J}{Y_0}}{1 - \left(\dfrac{J}{Y_0}\right)^2}$$

A-3527-188

FIG. 8.03-2 SOME CIRCUITS WHICH ARE PARTICULARLY USEFUL AS J-INVERTERS (Inverters to be Used with Shunt-Type Resonators)

characteristic admittance. The negative admittances are in practice absorbed into adjacent lines of positive admittance.

Numerous other circuits will operate as impedance or admittance inverters, the requirements being that their image impedance be real in the frequency band of operation, and that their image phase be some odd multiple of $\pm\pi/2$. For any symmetrical inverter, these conditions will be satisfied if

$$(X_{1/2})_{oc} = -(X_{1/2})_{sc} \tag{8.03-1}$$

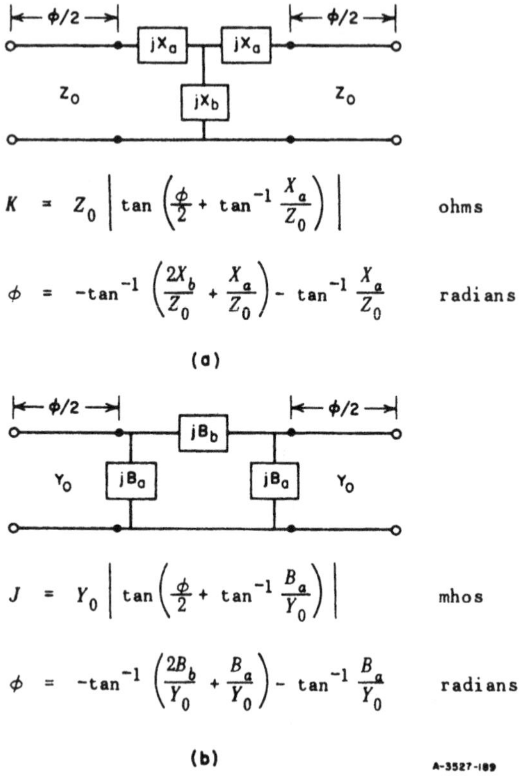

$$K = Z_0 \left| \tan\left(\frac{\phi}{2} + \tan^{-1}\frac{X_a}{Z_0}\right) \right| \qquad \text{ohms}$$

$$\phi = -\tan^{-1}\left(\frac{2X_b}{Z_0} + \frac{X_a}{Z_0}\right) - \tan^{-1}\frac{X_a}{Z_0} \qquad \text{radians}$$

$$J = Y_0 \left| \tan\left(\frac{\phi}{2} + \tan^{-1}\frac{B_a}{Y_0}\right) \right| \qquad \text{mhos}$$

$$\phi = -\tan^{-1}\left(\frac{2B_b}{Y_0} + \frac{B_a}{Y_0}\right) - \tan^{-1}\frac{B_a}{Y_0} \qquad \text{radians}$$

FIG. 8.03-3 TWO CIRCUITS WHICH ARE USEFUL FOR REPRESENTING THE INVERTER PROPERTIES OF CERTAIN DISCONTINUITIES IN TRANSMISSION LINES

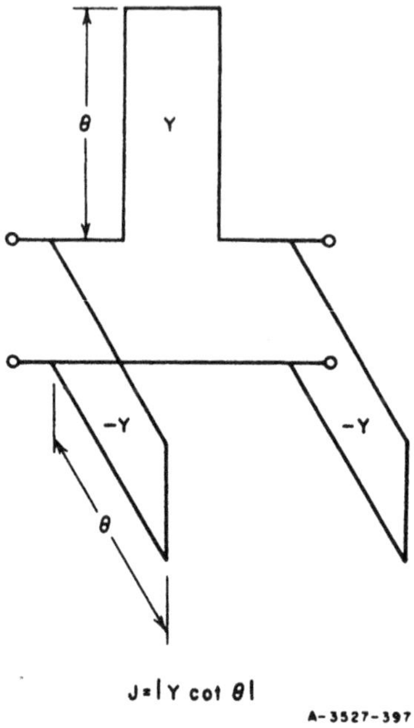

FIG. 8.03-4 AN ADMITTANCE INVERTER FORMED FROM STUBS OF ELECTRICAL LENGTH θ

where $(X_{1/2})_{oc}$ is the input reactance of the circuit when cut in half and the cut wires are left open-circuited, while $(X_{1/2})_{sc}$ is the corresponding reactance when the cut wires are shorted together.

SEC. 8.04, USE OF LOW-PASS TO BAND-PASS MAPPINGS

The response of a low-pass prototype circuit such as either of those in Fig. 4.04-1 can be related *exactly* to the response of a corresponding band-pass filter as shown in Fig. 8.02-2(b) by a well known *low-pass* to *band-pass mapping*

$$\frac{\omega'}{\omega_1'} = \frac{1}{w}\left(\frac{\omega}{\omega_0} - \frac{\omega_0}{\omega}\right) \qquad (8.04\text{-}1)$$

where w is the fractional bandwidth

$$w = \frac{\omega_2 - \omega_1}{\omega_0} , \tag{8.04-2}$$

$$\omega_0 = \sqrt{\omega_2 \omega_1} , \tag{8.04-3}$$

and ω' and ω_1' refer to the low-pass filter response as indicated in Fig. 8.02-1(a) while ω, ω_0, ω_1, and ω_2 refer to the corresponding band-pass filter response as shown in Fig. 8.02-1(b). Mappings of this sort are particularly useful in determining the number of resonators needed to meet given attenuation requirements. For example, suppose that an audio-frequency filter of the form in Fig. 8.02-2(b) was desired with a 1.0-db Tchebyscheff ripple from $f_1 = 2$ kc to $f_2 = 4$ kc and with at least 50-db attenuation at 1.5 kc. It is then desired to know how many resonators will be required to do the job. Using the mapping Eq. (8.04-1)

$$w = \frac{\omega_2 - \omega_1}{\sqrt{\omega_2 \omega_1}} = \frac{f_2 - f_1}{\sqrt{f_2 f_1}} = \frac{4 - 2}{\sqrt{(4)(2)}}$$

$$= 0.707 .$$

Now

$$\frac{\omega'}{\omega_1'} = \frac{1}{w}\left(\frac{\omega}{\omega_0} - \frac{\omega_0}{\omega}\right) = \frac{1}{w}\left(\frac{f}{f_0} - \frac{f_0}{f}\right) ,$$

$f_0 = \sqrt{f_2 f_1} = 2.825$ kc, and we wish, 50-db attenuation or more at $f = 1.5$ kc. Then the low-pass prototype must have at least 50-db attenuation for

$$\frac{\omega'}{\omega_1'} = \frac{1}{0.707}\left(\frac{1.5}{2.825} - \frac{2.825}{1.5}\right) = -1.914 .$$

The minus sign in the above result occurs because, mathematically, the portion of the band-pass filter response below ω_0 in Fig. 8.02-1(b) maps to negative values of the low-pass filter frequency variable ω', while, mathematically, the low-pass filter response in Fig. 8.02-1(a) for negative values of ω' is a mirror image of its response for positive values of ω. For our present purposes we may ignore the minus sign. The chart shown in Fig. 4.03-8 shows the cutoff characteristics of filters with 1.0-db Tchebyscheff ripple. Using this chart we see that an $n = 6$ reactive element prototype will give 54.5 db attenuation for $|\omega'/\omega'_1| = 1.914$ (*i.e.*, $|\omega'/\omega'_1| - 1 = 0.914$) as required, and $n = 5$ elements will give only 43 db attenuation. Thus, the corresponding band-pass filter with $f_1 = 2$ kc and $f_2 = 4$ kc will require $n = 6$ resonators in order to meet the attenuation requirement at $f = 1.5$ kc.

The various microwave filter structures about to be discussed approximate the performance of the filter in Fig. 8.02-2(b) very well for narrow bandwidths, but their rates of cutoff will differ noticeably from that for the filter in Fig. 8.02-2(b) when the bandwidth becomes appreciable (more than five percent or so). However, in most cases in this chapter, approximate mappings will be suggested which are more accurate than Eq. (8.04-1) for the given structure. In many cases the suggested mappings give very accurate results for filters with bandwidths as great as 20 percent or somewhat more. Though the mapping functions will be somewhat different from Eq. (8.04-1), they are used in exactly the same way for determining the required number of resonators for a given application.

SEC. 8.05, CAPACITIVE-GAP-COUPLED TRANSMISSION LINE FILTERS

Figure 8.05-1 presents design relations for coupled-resonator filters consisting of transmission-line resonators which are approximately a half-wavelength long at the midband frequency ω_0, and which have series-capacitance coupling between resonators. In this case the inverters are of the form in Fig. 8.03-2(d). These inverters tend to reflect high impedance levels to the ends of each of the half-wavelength resonators, and it can be shown that this causes the resonators to exhibit a shunt-type resonance (see Sec. 8.14). Thus, the filters under consideration operate like the shunt-resonator type of filter whose general design equations were shown in Fig. 8.02-4.

If the capacitive gaps operate like purely series capacitances, then the susceptance of the capacitive couplings can be computed by use of Eqs. (1) to (4) in Fig. 8.05-1, and the electrical distance between the series capacitance discontinuities is obtained by Eq. (5). However, in

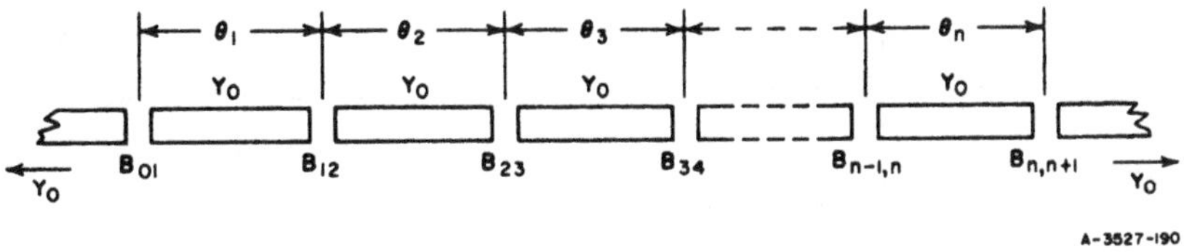

$$\frac{J_{01}}{Y_0} = \sqrt{\frac{\pi}{2}\,\frac{w}{g_0 g_1 \omega_1'}} \qquad (1) \qquad \left.\frac{J_{j,j+1}}{Y_0}\right|_{j=1 \text{ to } n-1} = \frac{\pi w}{2\omega_1'}\,\frac{1}{\sqrt{g_j g_{j+1}}} \qquad (2)$$

$$\frac{J_{n,n+1}}{Y_0} = \sqrt{\frac{\pi w}{2 g_n g_{n+1} \omega_1'}} \qquad (3)$$

where $g_0, g_1, \ldots, g_n$ are as defined in Fig. 4.04-1, ω' is defined in Fig. 8.02-1(a), and w is the fractional bandwidth defined below. The $J_{j,j+1}$ are admittance inverter parameters and Y_0 is the characteristic admittance of the line.

Assuming the capacitive gaps act as perfect, series-capacitance discontinuities of susceptance $B_{j,j+1}$ as in Fig. 8.03-2(d)

$$\frac{B_{j,j+1}}{Y_0} = \frac{\dfrac{J_{j,j+1}}{Y_0}}{1 - \left(\dfrac{J_{j,j+1}}{Y_0}\right)^2} \qquad (4)$$

and

$$\theta_j = \pi - \frac{1}{2}\left[\tan^{-1}\left(\frac{2B_{j-1,j}}{Y_0}\right) + \tan^{-1}\left(\frac{2B_{j,j+1}}{Y_0}\right)\right] \quad \text{radians} \qquad (5)$$

where the $B_{j,j+1}$ and θ_j are evaluated at ω_0.

For the construction in Figs. 8.05-3(a),(b); determine the gap spacings Δ from the $J_{j,j+1}/Y_0$ values and Figs. 8.05-3(a),(b); determine the $\phi_{j,j+1}$ values from the Δ's and Fig. 8.05-3(c); then

$$\left.\theta_j\right|_{j=1 \text{ to } n} = \pi + \frac{1}{2}\left[\phi_{j-1,j} + \phi_{j,j+1}\right] \quad . \qquad (6)$$

where the $\phi_{j,j+1}$ will usually be negative.

To map low-pass prototype filter response to corresponding band-pass filter response use the approximation

$$\frac{\omega'}{\omega_1'} = \frac{2}{w}\left(\frac{\omega - \omega_0}{\omega}\right) \qquad (7)$$

where

$$w = 2\left(\frac{\omega_2 - \omega_1}{\omega_2 + \omega_1}\right) \quad , \qquad (8) \qquad \omega_0 = \left(\frac{2\omega_2\omega_1}{\omega_2 + \omega_1}\right) \quad , \qquad (9)$$

where ω' and ω_1' are as defined in Fig. 8.02-1(a); and ω, ω_0, ω_1, and ω_2 are defined in Fig. 8.02-1(b).

FIG. 8.05-1 DESIGN EQUATIONS FOR CAPACITIVE-GAP-COUPLED TRANSMISSION-LINE FILTERS

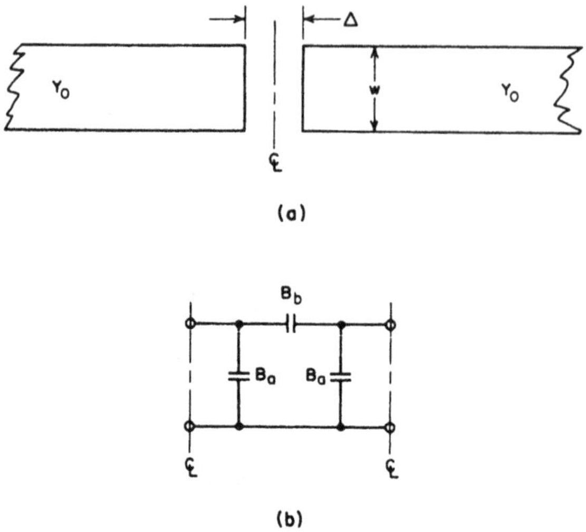

$$\frac{B_a}{Y_0} \approx -\frac{2b}{\lambda}\, ln\left[\cosh\left(\frac{\pi\Delta}{b}\right)\right] \qquad (1)$$

$$\frac{B_b}{Y_0} \approx \frac{b}{\lambda}\, ln\left[\coth\frac{\pi\Delta}{2b}\right] \qquad (2)$$

A-3527-191

FIG. 8.05-2 GAP EQUIVALENT CIRCUIT, AND OLINER'S EQUATIONS[9,10] FOR CAPACITIVE-GAP SUSCEPTANCES FOR THIN STRIP LINE
Parameter b is the ground-plane spacing, and λ is the wavelength in media of propagation, in same units. Equations are most accurate for w/b ≈ 1.2 or more and t/b ≈ 0, where t is the strip thickness.

many practical situations the capacitive gaps between resonators will be so large that they cannot be treated as simple series capacitances. Consider, for example, the capacitive gap in a strip transmission line shown in Fig. 8.05-2(a). If the length of each resonator is defined as extending from the *centerline* of one capacitive gap to the *centerline* of the next gap (as is done in Fig. 8.05-1), then an equivalent circuit for the gap, *as referred to the centerline of the gap,* will include series capacitance and some *negative* shunt capacitance to account for the fact that the gap reduces the shunt capacitance in the vicinity of the centerline. Figure 8.05-2(b) shows such an equivalent circuit for the gap, and also shows some equations due to Oliner[9] which give approximate values for the susceptances, for the case of strip line of nearly zero thickness. (Altschuler and Oliner[10] point out that these equations are reasonably accurate if w/b is fairly large as is the case for a 50-ohm strip line having nearly zero thickness and air dielectric. However, if w/b is small the error is considerable.) Having reasonably accurate values for the susceptances in Fig. 8.05-2(b), the corresponding admittance inverter parameters for a given gap size can be computed by use of Fig. 8.03-3(b). The gap sizes must be chosen to give the $J_{j,j+1}/Y_0$ values called for by Eqs. (1) to (3) in Fig. 8.05-1, and the corresponding values of ϕ obtained from Fig. 8.03-3(b) are then used with Eq. (6) in Fig. 8.05-1 in order to obtain the proper electrical distance between the centerlines of the coupling gaps. It should be noted that all susceptances and electrical distances are to be evaluated at the midband frequency ω_0.

Figures 8.05-3(a) to (c) present data for capacitive-gap filters which were obtained by experimental procedures[11] (see Chapter 11). These data are for the particular rectangular-bar strip line construction shown in Fig. 8.05-3(a). Figures 8.05-3(a),(b) give data to be used for determining the proper gap spacing Δ in inches to give a specified J/Y_0 value, while Fig. 8.05-3(c) is for use in determining the proper negative line length to be associated with the inverter. A simple numerical example will clarify the use of these charts.

Suppose that a filter is desired with a 0.5-db ripple Tchebyscheff pass band from $f_1 = 3.0$ Gc to $f_2 = 3.20$ Gc and that 30-db attenuation

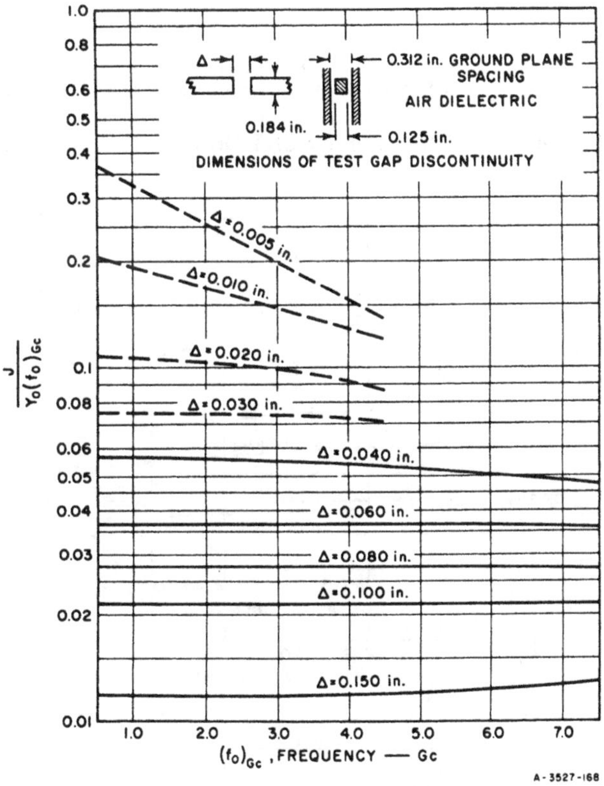

SOURCE: Reference 11, by G. L. Matthaei. (By courtesy of the Ramo-Wooldridge Div. of the Thompson-Ramo-Wooldridge Corp.)

FIG. 8.05-3(a) $J/[Y_0(f_0)_{Gc}]$ vs. $(f_0)_{Gc}$ FOR CAPACITIVE-GAP J-INVERTERS IN BAR TRANSMISSION-LINE CONSTRUCTION
The characteristic impedance of the transmission lines is $Z_0 = 1/Y_0 = 50$ ohms and $(f_0)_{Gc}$ is the band center frequency of the filter in Gc

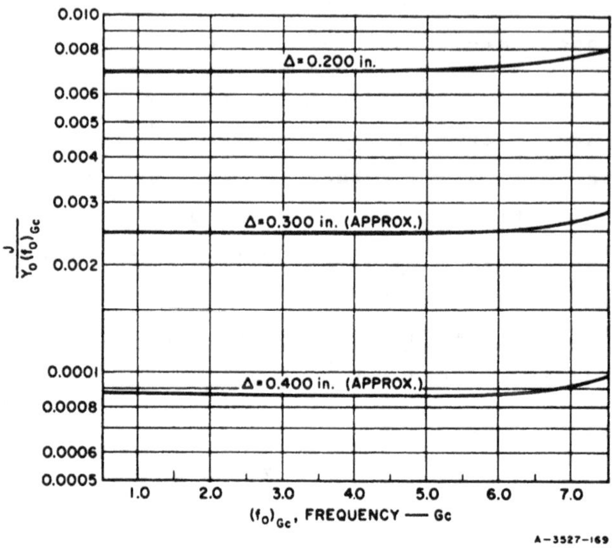

SOURCE: Reference 11, by G. L. Matthaei. (By courtesy of the Ramo-Wooldridge Div. of the Thompson-Ramo-Wooldridge Corp.)

FIG. 8.05-3(b) CONTINUATION OF FIG. 8.05-3(a)

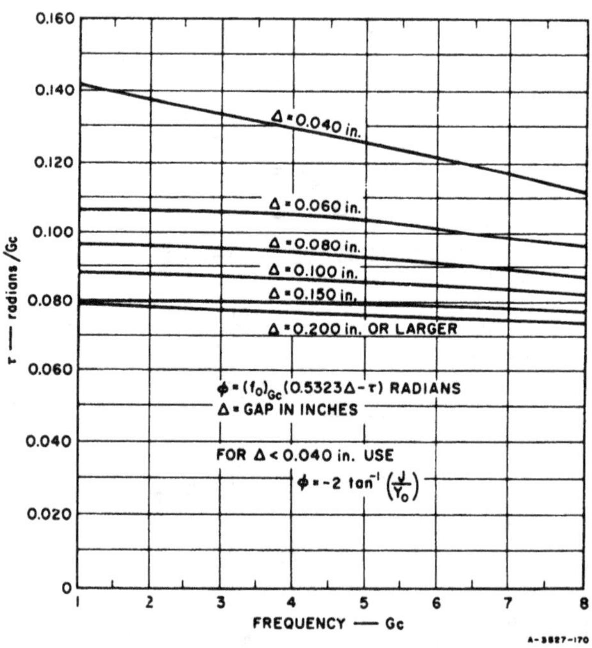

SOURCE: Reference 11, by G. L. Matthaei. (By courtesy of the Ramo-Wooldridge Div. of the Thompson-Ramo-Wooldridge Corp.)

FIG. 8.05-3(c) PARAMETER ϕ FOR THE J-INVERTER DISCONTINUITY IN FIG. 8.05-3(a) IN TERMS OF AN AUXILIARY PARAMETER τ

is required at f_a = 2.50 Gc and at f_b = 3.50 Gc. By Eqs. (7) to (9) in Fig. 8.05-1

$$w = 2\left(\frac{\omega_2 - \omega_1}{\omega_2 + \omega_1}\right) = 2\left(\frac{f_2 - f_1}{f_2 + f_1}\right) = 0.0645 \quad ,$$

$$f_0 = \frac{2 f_2 f_1}{f_2 + f_1} = 3.10 \text{ Gc} \quad ,$$

$$\frac{\omega'}{\omega'_1} = \frac{2}{w}\left(\frac{f - f_0}{f}\right)$$

which for f_a = 2.5 Gc gives ω'/ω'_1 = −7.45, while for f_b = 3.5 Gc it gives ω'/ω'_1 = 3.55. Since ω'/ω'_1 has the smaller magnitude for f = 3.5 Gc, the restriction at that frequency controls the design. Using Fig. 4.03-7 and the procedure discussed in Sec. 8.04, for L_{Ar} = 0.5 db we find that for a three-resonator design, L_A should be about 35 db at 3.5 Gc, and it should be about 55 db at 2.5 Gc. Thus, three resonators will be sufficient.

By Table 4.05-2(a), the element values for an n = 3 reactive element 0.5 db-ripple Tchebyscheff prototype are g_0 = 1, g_1 = 1.5963, g_2 = 1.0967, g_3 = 1.5963, and g_4 = 1.000. By Eqs. (1) to (3) in Fig. 8.05-1, $J_{01}/Y_0 = J_{34}/Y_0$ = 0.252, $J_{12}/Y_0 = J_{23}/Y_0$ = 0.0769.* Since f_0 = 3.1 Gc, $J_{01}/Y_0(f_0)_{Gc}$ = 0.252/3.1 = 0.0813 and $J_{12}/Y_0(f_0)_{Gc}$ = 0.0769/3.1 = 0.0248. Using Fig. 8.05-3(a) a plot of Δ *vs.* $J/Y_0(f_0)_{Gc}$ for f_0 = 3.1 Gc is made for purposes of interpolation, and from this plot the required gaps are found to be $\Delta_{01} = \Delta_{34}$ = 0.027 inch, and $\Delta_{12} = \Delta_{23}$ = 0.090 inch.

Using Fig. 8.05-3(c) for determining the $\phi_{j,j+1}$, since Δ_{01} < 0.040" we use

$$\phi_{01} = \phi_{34} = -2\tan^{-1}\left(\frac{J_{01}}{Y_0}\right) = -2\tan^{-1}(0.252) = -0.494 \quad \text{radian} \ .$$

For the Δ_{12} = 0.090-inch gap we use the chart to get τ = 0.090 radian/Gc for f_0 = 3.1 Gc. Then

* Filters designed using Fig. 8.05-1 and any symmetrical or antimetrical prototype such as those in Tables 4.05-1(a),(b), or 4.05-2(a),(b) will always be symmetrical.

$$\phi_{12} = \phi_{23} = (f_0)_{Gc}(0.5323\,\Delta - \tau) = -0.130 \quad \text{radian}$$

By Eq. (6) in Fig. 8.05-1

$$\theta_1 = \theta_3 = \pi + \frac{1}{2}[-0.494 - 0.130] = 2.830 \quad \text{radians}$$

and

$$\theta_2 = \pi + \frac{1}{2}[-0.130 - 0.130] = 3.012 \quad \text{radians}$$

For propagation in air, λ = 3.810 inches at 3.10 Gc, and the distances between the *centerlines* of the capacitive gaps is $l_1 = l_3 = \theta_1\lambda/2\pi$ = 1.715 inches for Resonators 1 and 3, and $l_2 = \theta_2\lambda/2\pi$ = 1.825 inches for Resonator 2. The resonator bars may be supported by Polyfoam or by thin horizontal slabs of dielectric passing through the sides of the bars. Of course, some correction in resonator bar dimensions will be required to compensate for the effect of the dielectric supporting material on the velocity of propagation and line impedance. In order to tune the filter precisely tuning screws may be used as described in Sec. 11.05, or alternately the resonant frequency of the various resonators may be checked by testing them individually or in pairs as is also described in Secs. 11.03 to 11.05.

Figure 8.05-4(a) shows a filter constructed using the design charts in Fig. 8.05-3(a) to (c). This is a four-resonator filter designed for a 1.0-percent bandwidth maximally flat response centered at f_0 = 6.120 kMc. In this filter the resonators are supported by 0.062-inch-thick Rexolite 2200 dielectric slabs which pass through the sides of the resonator bars, the slabs being held by clamp strips at the sides of the filter. The four bars in the interior of the filter are resonators while the bar at each end is a 50-ohm input or output line. The resonant frequencies of the resonators were checked by testing them in pairs as discussed in Sec. 11.04. These tests indicated small errors in the lengths of the resonator bars, and the required corrections were made. Figure 8.05-4(b) shows the resulting measured response obtained after the filter was assembled without tuning screws.[11]

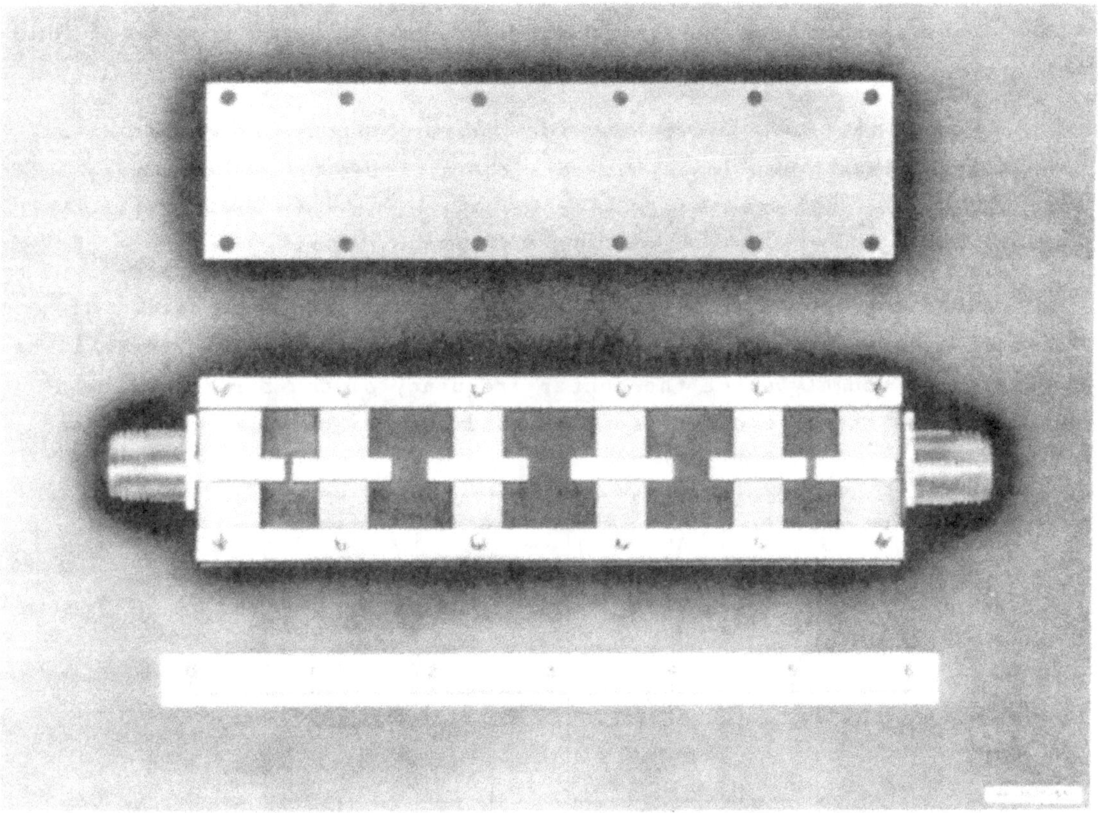

SOURCE: Reference 11, by G. L. Matthaei. (By courtesy of the Ramo-Wooldridge Div. of the Thompson-Ramo-Wooldridge Corp.)

FIG. 8.05-4(a) A FILTER WITH 0.9% BANDWIDTH CENTERED AT 6.120 Gc AS SHOWN IN FIG. 8.05-4(b)
As is seen from the photograph, this filter uses four, $\lambda_0/2$ resonators in bar construction

Figure 8.05-5 shows the measured response of a six-resonator filter in similar construction.[12] This filter was designed for 1-db Tchebyscheff ripple and 20 percent bandwidth. The x's show the measured data while the circles show points mapped from the low-pass prototype using the mapping in Eqs. (7) to (9) of Fig. 8.05-1. As can be seen, even for bandwidths as large as 20 percent the design procedure and the mapping give good accuracy. However, the bandwidth for which this procedure is accurate depends somewhat on the pass-band ripple tolerance. For ripples as small as 0.01 db, this design procedure will not meet the design objectives for as large bandwidths as it will when the ripples are, say, 0.5 or 1.0 db. For bandwidths of around 15 percent or more and very small pass-band ripples, the procedures in Chapter 9 are recommended for this type of filter.

Observe that, the wider bandwidth filter response in Fig. 8.05-5 shows less dissipation loss than does the narrow-band response in Fig. 8.05-4(b). The unloaded Q for resonators in this construction has been found to be typically about 1000 to 1300 at S band.

Other considerations in the practical design and application of filters of this type are that the second pass band of the filter will be centered at roughly twice the center frequency of the first pass band, and that the attenuation in the stop band between the first and second

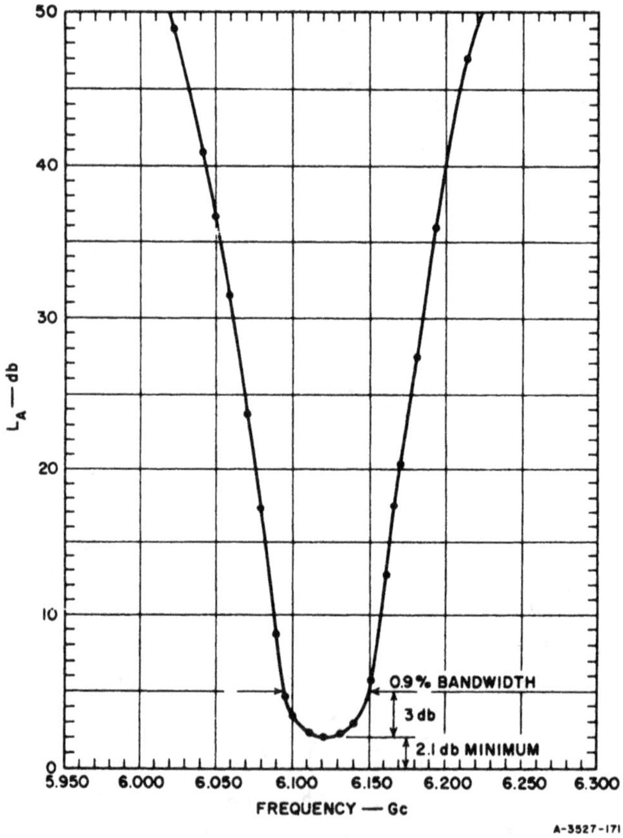

SOURCE: Reference 11, by G. L. Matthaei. (By courtesy of the Ramo-Wooldridge Div. of the Thompson-Ramo-Wooldridge Corp.)

FIG. 8.05-4(b) THE ATTENUATION CHARACTERISTIC OF THE FILTER IN FIG. 8.05-4(a)

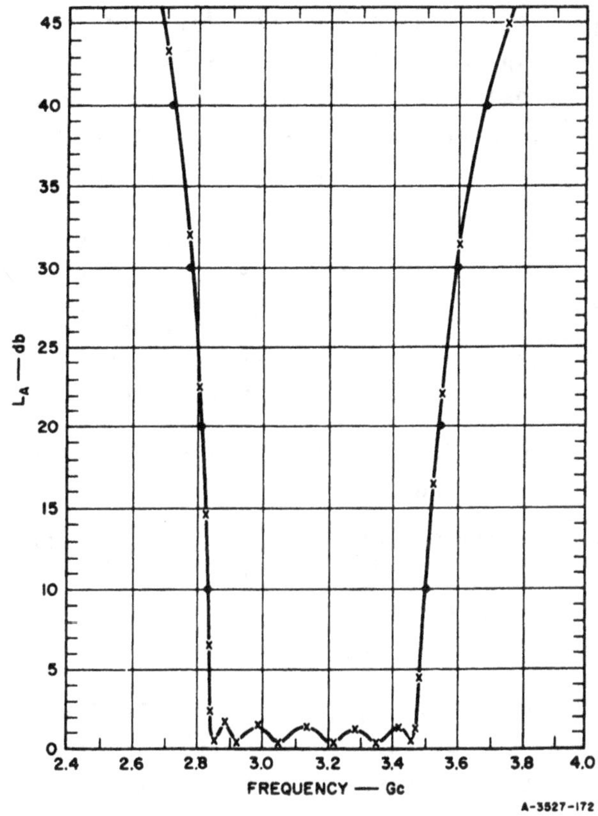

SOURCE: Reference 12, by G. L. Matthaei. (By courtesy of the Ramo-Wooldridge Div. of the Thompson-Ramo-Wooldridge Corp.)

FIG. 8.05-5 THE ATTENUATION CHARACTERISTIC OF A 6-RESONATOR FILTER
The x's indicated measured attenuation while the circles are theoretical points calculated using the mapping in Eqs. (7) to (9) of Fig. 8.05-1

pass band will level off to some peak finite value of $(L_A)_{USB}$ decibels, which occurs at about $\omega = 3\omega_0/2$. The size of this maximum attenuation in the upper stop band can be estimated by use of the formula

$$(L_A)_{USB} \approx 20 \log_{10}\left[\frac{1}{\left(\dfrac{B_{01}}{Y_0}\right)\left(\dfrac{B_{12}}{Y_0}\right)\cdots\left(\dfrac{B_{n,n+1}}{Y_0}\right)}\right] - (n+1)3.53 - 6.02 \quad \text{db} \tag{8.05-1}$$

where the $B_{j,j+1}/Y_0$ are computed from the $J_{j,j+1}/Y_0$ by use of Eq. (4) in Fig. 8.05-1. The stop band below the pass band has a first-order pole of attenuation (Sec. 2.04) at $\omega = 0$. Thus, in the case of the lower stop band the attenuation continues to grow as the frequency goes lower, and the attenuation approaches an infinite value as ω approaches zero.

If sizeable attenuation in the upper stop band is required for a given application, $(L_A)_{USB}$ should be computed. The attenuation predicted by Eqs. (7) to (9) in Fig. 8.05-1 for upper-stop-band frequencies near the pass band, will be reasonably accurate only so long as the computed attenuation values are around 20 db or more below $(L_A)_{USB}$.

In the case of the three-resonator numerical example discussed above, $J_{01}/Y_0 = J_{34}/Y_0 = 0.252$, and $J_{12}/Y_0 = J_{23}/Y_0 = 0.0769$. By Eq. (4) in Fig. 8.05-1, $B_{01}/Y_0 = B_{34}/Y_0 = 0.269$, and $B_{12}/Y_0 = B_{23}/Y_0 = 0.077$. Then by Eq. (8.05-1), $(L_A)_{USB} \approx 54$ db. Thus, the 35-db value computed for 3.5 Gc by use of the mapping should be reasonably accurate since the 35-db value is about 19 db below $(L_A)_{USB}$.

It will be found that $(L_A)_{USB}$ *decreases* rapidly as the fractional bandwidth increases, but at the same time $(L_A)_{USB}$ *increases* rapidly as the number of resonators is increased. Thus, if $(L_A)_{USB}$ is found to be too small, it can be increased by adding more resonators.

SEC. 8.06, SHUNT-INDUCTANCE-COUPLED, WAVEGUIDE FILTERS

The waveguide filter in Fig. 8.06-1 is in most respects the dual of the capacitive-gap coupled filter in Fig. 8.05-1. In this case, the inverters are of the type in Fig. 8.03-1(c) and the structure operates like the filter with series resonators shown in Fig. 8.02-3. The low-pass to band-pass transformation in Eqs. (6) to (8) in Fig. 8.06-1 for the waveguide filter is the same as that in Eqs. (7) to (9) for the capacitive-gap coupled filter if both transformations are expressed in terms of guide wavelength. However, since the guide wavelength for waveguide varies with frequency in a different way from the guide wavelength in a TEM-mode structure, the frequency responses will be somewhat different for the two types of filters. In particular, for a given range of guide wavelength, the waveguide-type of filter will have narrower frequency bandwidth because of the more rapid change in guide wavelength for non-TEM modes of propagation.

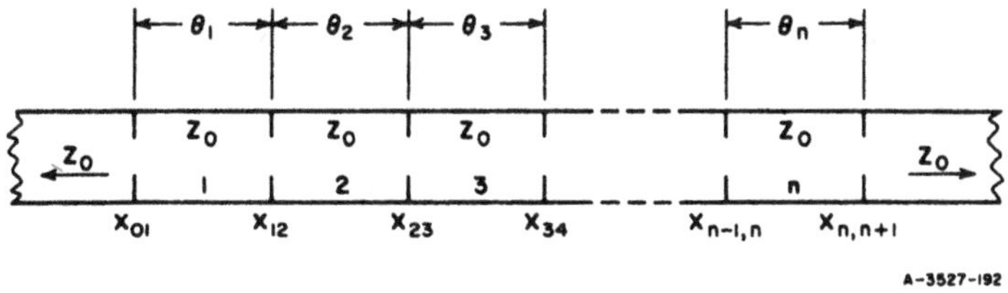

$$\frac{K_{01}}{Z_0} = \sqrt{\frac{\pi}{2} \frac{w_\lambda}{g_0 g_1 \omega_1'}} \qquad (1)$$

$$\left.\frac{K_{j,j+1}}{Z_0}\right|_{j=1 \text{ to } n-1} = \frac{\pi w_\lambda}{2\omega'} \frac{1}{\sqrt{g_j g_{j+1}}} \qquad (2)$$

$$\frac{K_{n,n+1}}{Z_0} = \sqrt{\frac{\pi}{2} \frac{w_\lambda}{g_n g_{n+1} \omega_1'}} \qquad (3)$$

where g_0, g_1, ..., g_{n+1} are as defined in Fig. 4.04-1, ω_1' is defined in Fig. 8.02-1(a), and w_λ is the guide-wavelength fractional bandwidth defined below. The $K_{j,j+1}$ are impedance inverter parameters and Z_0 is the guide impedance.

For purely lumped-inductance discontinuities having shunt reactance $X_{j,j+1}$,

$$\frac{X_{j,j+1}}{Z_0} = \frac{\dfrac{K_{j,j+1}}{Z_0}}{1 - \left(\dfrac{K_{j,j+1}}{Z_0}\right)^2} \qquad (4)$$

and

$$\theta_j = \pi - \frac{1}{2}\left[\tan^{-1}\left(\frac{2X_{j-1,j}}{Z_0}\right) + \tan^{-1}\left(\frac{2X_{j,j+1}}{Z_0}\right)\right] \quad \text{radians} . \qquad (5a)$$

For discontinuities with more complicated equivalent circuits use Fig. 8.03-3 and

$$\theta_j = \pi + \frac{1}{2}\left[\phi_{j-1,j} + \phi_{j,j+1}\right] \quad \text{radians} , \qquad (5b)$$

where the ϕ's will usually be negative.

(continued on p. 452)

FIG. 8.06-1 DESIGN EQUATIONS FOR SHUNT-INDUCTANCE-COUPLED WAVEGUIDE FILTERS

To map low-pass filter response to corresponding band-pass filter response use

$$\frac{\omega'}{\omega_1'} = \frac{2}{w_\lambda}\left(\frac{\lambda_{g0} - \lambda_g}{\lambda_{g0}}\right) \tag{6}$$

where

$$w_\lambda = \left[\frac{\lambda_{g1} - \lambda_{g2}}{\lambda_{g0}}\right] \approx \left(\frac{\lambda_{g0}}{\lambda_0}\right)^2 \left(\frac{\omega_2 - \omega_1}{\omega_0}\right) \quad , \tag{7}$$

$$\lambda_{g0} = \frac{\lambda_{g1} + \lambda_{g2}}{2} \quad , \tag{8}$$

λ_{g0}, λ_{g1}, λ_{g2}, and λ_g are the guide wavelengths at frequencies ω_0, ω_1, ω_2, and ω as defined in Fig. 8.02-1(b); ω' and ω_1' are as defined in Fig. 8.02-1(a); and λ_0 is the wavelength of a plane wave at frequency ω_0 in the medium of the guide.

FIG. 8.06-1 Concluded

Assuming that the waveguide propagates the TE_{10} mode of propagation and that all higher-order modes are cut off, the procedure for using the equations in Fig. 8.06-1 is very similar to that for the equations in Fig. 8.05-1. Figures 8.06-2 and 8.06-3(a),(b) present inductive iris and inductive post coupling discontinuity data from Marcuvitz.[13] The reactances plotted relate to the equivalent circuit in Fig. 8.06-4. Since for a very thin iris, $X_a \approx 0$, Eqs. (4) and (5) in Fig. 8.06-1 which assume a simple, shunt, lumped-inductance discontinuity may be used. For the

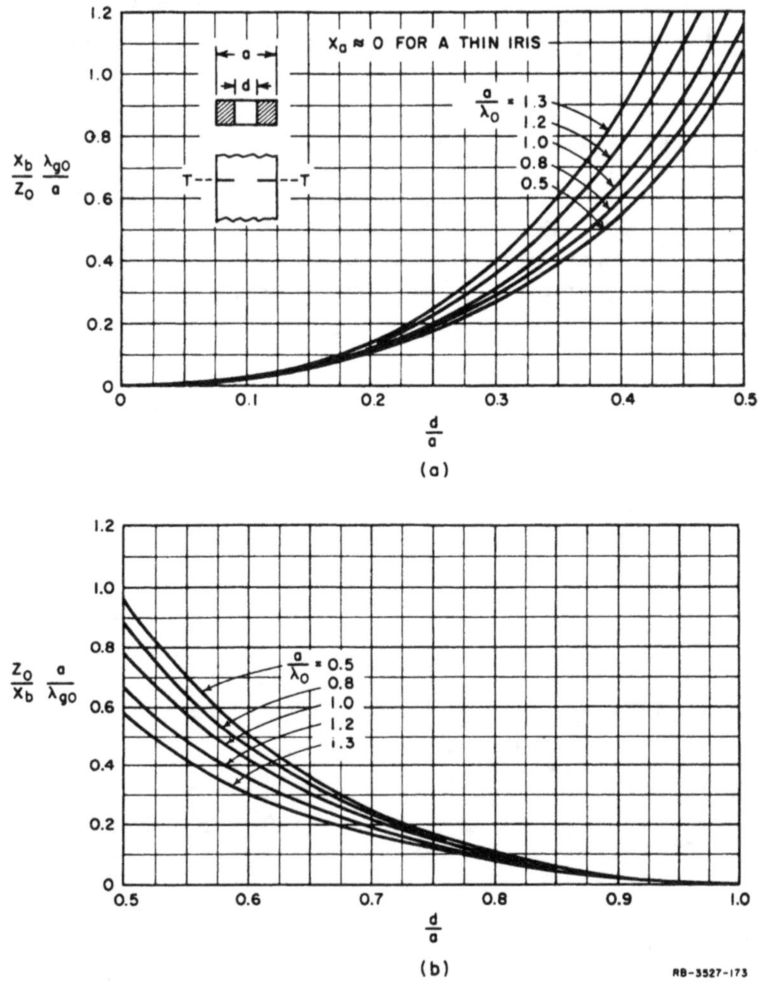

SOURCE: *Waveguide Handbook* edited by N. Marcuvitz.[13]

FIG. 8.06-2 SHUNT REACTANCE OF SYMMETRICAL INDUCTIVE WINDOW IN RECTANGULAR GUIDE
Fig. 8.06-4 shows the equivalent circuit for this discontinuity

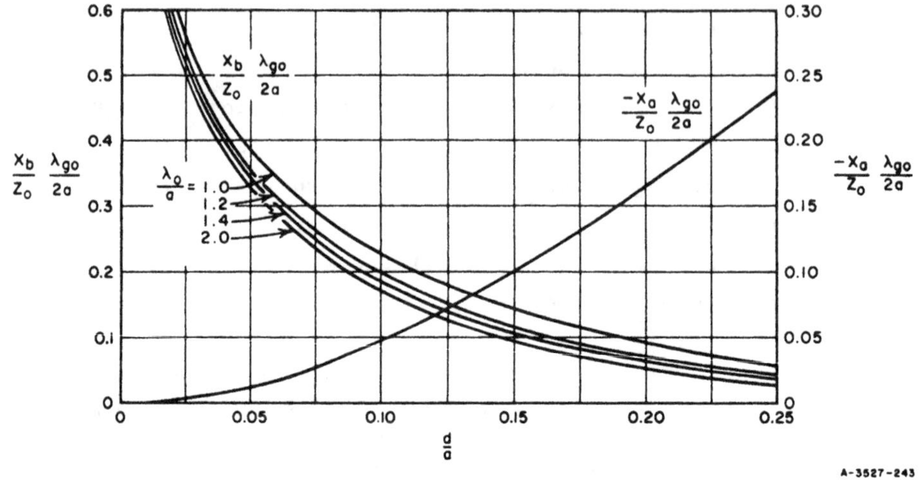

SOURCE: *Waveguide Handbook* edited by N. Marcuvitz.[13]

FIG. 8.06-3(a) CIRCUIT PARAMETERS OF CENTERED INDUCTIVE POST IN RECTANGULAR GUIDE
The guide wavelength at midband is λ_{g0} while λ_0 is the corresponding free-space wavelength

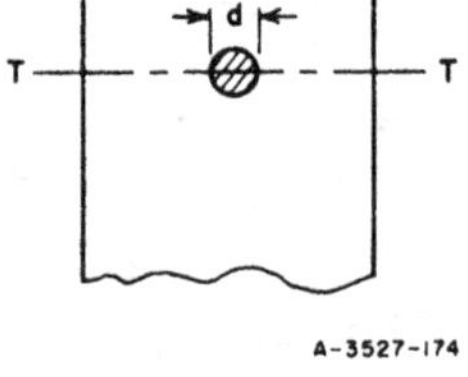

FIG. 8.06-3(b) DEFINITION OF THE DIMENSIONS IN FIG. 8.06-3(a)

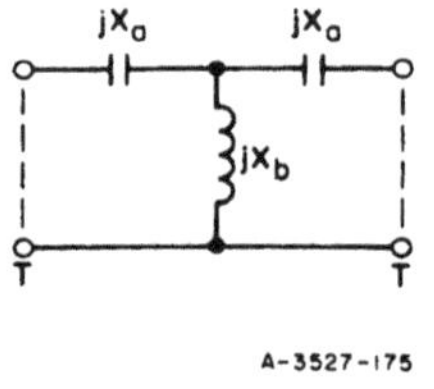

FIG. 8.06-4 EQUIVALENT CIRCUIT FOR THE SHUNT-INDUCTIVE DISCONTINUITIES IN FIG. 8.06-2 AND 8.06-3(a), (b)
Note that X_a 0 for case of Fig. 8.06-2

case of the inductive post (or a thick iris), X_a is not negligible and it should be accounted for in the design process. This can be done for the case of inductive posts by first computing the required normalized inverter parameter values $K_{j,j+1}/Z_0$ by use of Eqs. (1) to (3) in Fig. 8.06-1. Then, using the data in Fig. 8.06-3(a) along with Fig. 8.03-3(a), a plot is made of K/Z_0 and ϕ *vs.* d/a, for the desired midband guide wavelength λ_{g0}, corresponding plane-wave wavelength λ_0, and waveguide width a. From this chart the post diameters d which will give the normalized impedance inverting parameters $K_{j,j+1}/Z_0$ can be determined, and also the corresponding values of $\phi_{j,j+1}$. Then, paralleling the analogous case for Fig. 8.05-1, the electrical distance between the centers of the posts at each end of Resonator j is

$$\theta_j = \pi + \frac{1}{2}[\phi_{j-1,j} + \phi_{j,j+1}] \quad \text{radians} \quad . \tag{8.06-1}$$

Except possibly for the case of large posts, the $\phi_{j,j+1}$ should be negative. The distance between post centers for Resonator j is then

$$l_j = \frac{\lambda_{g0}\theta_j}{2\pi} \quad . \tag{8.06-2}$$

This design procedure should give good accuracy for designs having guide-wavelength fractional bandwidths w_λ [see Eq. (7) in Fig. 8.06-1] of 20 percent,[1] with diminishing accuracy for larger bandwidths.

Analogously to the strip-line filter in Sec. 8.05, this waveguide filter will have for TE_{10}-mode propagation a second pass band centered approximately at the frequency for which $\lambda_g = \lambda_{g0}/2$. This frequency would be somewhat less than $2\omega_0$ because of the manner in which λ_g and the $X_{j,j+1}$ vary with frequency. Also, the attenuation between the first and second pass bands for TE_{10}-mode propagation will level off with a peak value of $(L_A)_{USB}$, which can be estimated by use of the equation

$$(L_A)_{USB} \approx 20 \log_{10}\left[\frac{1}{\left(\frac{X_{01}}{Z_0}\right)\left(\frac{X_{12}}{Z_0}\right)\cdots\left(\frac{X_{n,n+1}}{Z_0}\right)}\right] - (n+1)3.53 - 6.02 \quad \text{db} \tag{8.06-3}$$

where the $X_{j,j+1}/Z_0$ are computed from the $K_{j,j+1}/Z_0$ by use of Eq. (4) in Fig. 8.06-1. Equation (8.06-1) is the dual of Eq. (8.05-1), and some further ramifications concerning its use are discussed at the end of Sec. 8.05.

As for the type of filter in Sec. 8.05, the waveguide filter in Fig. 8.06-1 will have monotonically increasing attenuation for frequencies varying from the pass-band frequency downward. Thus, the attenuation in the lower stop band rises to an infinite value at $\omega = 0$, due to the attenuating effects of the irises, and due to the cutoff of the waveguide.

It should be noted that the discussion above assumes that only the TE_{10} mode is present. If other modes are also present (as is likely to happen for frequencies which are around 1.5 or more times ω_0), the performance can be greatly disrupted. This disruption arises because higher-order modes have different guide wavelengths than that for the TE_{10} mode. As a result the pass and stop bands for energy in the higher modes will occur at quite different frequencies than for the TE_{10} mode. Thus, the possible effects of higher-order modes should be kept in mind when this or any other type of waveguide filter is to be used.

In order to clarify the differences between strip-line and waveguide filter design, a waveguide filter design example will now be considered which is closely related to the strip-line filter example in Sec. 8.05. Let us suppose that a pass band with 0.5-db Tchebyscheff ripple is desired from $f_1 = 3.047$ to $f_2 = 3.157$ Gc, and that at least 30-db attenuation is required at the frequencies $f_a = 2.786$ Gc and $f_b = 3.326$ Gc. Let us suppose that WR-284 waveguide is to be used. The design calculations are those outlined in Table 8.06-1.

In Part (a) of Table 8.06-1, guide wavelengths are computed that correspond to the various frequencies of importance. In Part (b), w_λ and $w = (f_2 - f_1)/f_0$ are computed, and it should be noted that w_λ, the guide-wavelength fractional bandwidth, is nearly twice as large as the frequency fractional bandwidth w. Also, normalized prototype frequencies ω'/ω'_1 are computed corresponding to f_a and f_b for the waveguide filter, and the attenuation is predicted by use of the chart in Fig. 4.03-7. It will be noted that $w_\lambda = 0.0645$ for this example, which corresponds exactly to $w = 0.0645$ for the example in Sec. 8.05. Also, the ratios $\lambda_{g0}/\lambda_{ga} = 5.130/6.361 = 0.806$ and $\lambda_{g0}/\lambda_{gb} = 5.130/4.544 = 1.129$ correspond exactly

Table 8.06-1

OUTLINE OF A WAVEGUIDE FILTER DESIGN CALCULATION

Part (a)

Assume *WR*-284 guide. Width $= a = 2.840$ inches

Height $= b = 1.340$ inches

$$\lambda_g = \frac{1}{\sqrt{(0.08472f)^2 - \left(\frac{1}{2a}\right)^2}} \text{ inches} \qquad (1)$$

where a is in inches and f is in Gc.

$$f_1 = 3.047 \text{ Gc} \,, \quad \lambda_{g1} = 5.296 \text{ inches}$$

$$f_2 = 3.157 \text{ Gc} \,, \quad \lambda_{g2} = 4.965 \text{ inches}$$

$$\lambda_{g0} = \frac{\lambda_{g1} + \lambda_{g2}}{2} = 5.130 \text{ inches } (f_0 = 3.100 \text{ Gc})$$

$$\lambda_0 = (\text{Plane wavelength at } f_0) = 3.807 \text{ inches}$$

$$f_a = 2.786 \text{ Gc} \,, \quad \lambda_{ga} = 6.361 \text{ inches}$$

$$f_b = 3.326 \text{ Gc} \qquad \lambda_{gb} = 4.544 \text{ inches}$$

Part (b)

$$w_\lambda = \frac{\lambda_{g1} - \lambda_{g2}}{\lambda_{g0}} = 0.0645 \,, \quad w = \frac{f_2 - f_1}{f_0} = 0.0355$$

Alternately:

$$w_\lambda \approx \left(\frac{\lambda_{g0}}{\lambda_0}\right)^2 w = \left(\frac{5.130}{3.807}\right)^2 (0.0355) = 0.0645$$

$$\frac{\omega'}{\omega_1'} = \frac{2}{w_\lambda}\left(\frac{\lambda_{g0} - \lambda_g}{\lambda_{g0}}\right) \qquad (2)$$

For $f = f_a = 2.786$ Gc , $\lambda_g = \lambda_{ga}$ and $\frac{\omega'}{\omega_1'} = -7.45$

For $f = f_b = 3.326$ Gc , $\lambda_g = \lambda_{gb}$ and $\frac{\omega'}{\omega_1'} = 3.55$

By Fig. 4.03-7, for a 0.5-db ripple $n = 3$ design:

For $f = f_a \left(|\omega'/\omega_1'| = 7.5\right)$, $L_A = 55$ db.

For $f = f_b \left(|\omega'/\omega_1'| = 3.55\right)$, $L_A = 35$ db.

(continued on p. 458)

Table 8.06-1 *Continued*

Part (c)

For $n = 3$, 0.5-db ripple Tchebyscheff prototype, by Table 4.05-2(*a*): $g_0 = 1$, $g_1 = 1.5963$, $g_2 = 1.0967$, $g_3 = 1.5963$, $g_4 = 1.0000$, and $\omega_1' = 1$.

$$\frac{K_{01}}{Z_0} = \sqrt{\frac{\pi}{2}\frac{w_\lambda}{g_0 g_1 \omega_1'}} = \frac{K_{34}}{Z_0} = 0.252$$

$$\frac{K_{12}}{Z_0} = \frac{\pi w_\lambda}{2\omega_1'}\frac{1}{\sqrt{g_1 g_2}} = \frac{K_{23}}{Z_0} = 0.0769$$

$$\frac{X_{j,j+1}}{Z_0} = \frac{\dfrac{K_{j,j+1}}{Z_0}}{1-\left(\dfrac{K_{j,j+1}}{Z_0}\right)^2} \qquad (3)$$

$$\frac{X_{01}}{Z_0} = \frac{X_{34}}{Z_0} = 0.269$$

$$\frac{X_{12}}{Z_0} = \frac{X_{23}}{Z_0} = 0.0774$$

$$\frac{X_{01}}{Z_0}\frac{\lambda_{g0}}{a} = \frac{X_{34}}{Z_0}\frac{\lambda_{g0}}{a} = \frac{0.269(5.130)}{2.840} = 0.486$$

$$\frac{X_{12}}{Z_0}\frac{\lambda_{g0}}{a} = \frac{X_{23}}{Z_0}\frac{\lambda_{g0}}{a} = \frac{0.0774(5.130)}{2.840} = 0.140$$

By Fig. 8.06-2, with $a/\lambda_0 = 2.840/3.807 = 0.746$:

For X_{01} and X_{34}, $d/a = 0.37$ and $d = 1.050$ inches

For X_{12} and X_{23}, $d/a = 0.22$ and $d = 0.625$ inch

Part (d)

$$\theta_j = \pi - \frac{1}{2}\left[\tan^{-1}\frac{2X_{j-1,j}}{Z_0} + \tan^{-1}\frac{2X_{j,j+1}}{Z_0}\right] \qquad (4)$$

$\theta_1 = \theta_3 = 2.819$ radians, $\theta_2 = 2.989$ radians

the spacing between irises is:

$$l_1 = l_3 = \frac{\theta_1\lambda_{g0}}{2\pi} = 2.302 \text{ inches}$$

$$l_2 = \frac{\theta_2\lambda_{g0}}{2\pi} = 2.441 \text{ inches}$$

to the $f_a/f_0 = 2.5/3.10 = 0.806$ and $f_b/f_0 = 3.5/3.10 = 1.129$ ratios for the example of Sec. 8.05. The attenuations are seen to be the same for these corresponding ratios. In fact, using λ_{g0}/λ_g as a normalized frequency variable, the response of the waveguide filter would be identical to that of the strip-line filter example in Sec. 8.05, plotted *vs.* f/f_0. But note that the waveguide filter response plotted as a function of frequency will be quite different. As is seen from the calculations, an $n = 3$ design gives an adequate rate of cutoff, and over 30-db attenuation at both f_a and f_b.

In Part (c) of the table the dimensions of the coupling irises are determined with the aid of the chart in Fig. 8.06-2, and in Part (d) the spacings between irises are determined. The iris data in Fig. 8.06-2 are for thin irises, and if the iris is, say, 0.020-inch thick, the error due to thickness should not be serious for most purposes, since the main effect will be on the resonant frequency of the cavities. There are presently no data available which give an accurate thickness correction for irises of the form in Fig. 8.06-2 with holes as large as are to be used in this filter. A suggested procedure is to measure the resonator lengths l_1, l_2, and l_3 from the *centerline* of one iris to the *centerline* of the next. This should make the resonant frequencies of the resonators a trifle high, so that they can be tuned down to the correct frequency using tuning screws and the alternating short- and open-circuit method discussed in Sec. 11.05. If a precision design without tuning screws is desired, the single- or double-resonator test procedures described in Secs. 11.03 to 11.05 are recommended for precision determination of the iris sizes and resonator tunings.

The peak attenuation $(L_A)_{USB}$ between the first and second pass bands will be about 54 db just as for the example in Sec. 8.05. However, it should be recalled that this holds only if the TE_{10} mode alone is present.

SEC. 8.07, NARROW-BAND CAVITY RESONATOR FILTERS COUPLED BY SMALL IRISES

The design of cavity resonator filters coupled by small irises can be carried out in a general fashion by means of Bethe's small-aperture theory (see Sec. 5.10). For most of the filters discussed in this chapter, it will be convenient to carry out the design in terms of the resonator slope parameters $\mathscr{X}_j$ or $\mathscr{B}_j$ and the inverter parameters $K_{j,j+1}$ or $J_{j,j+1}$.

However, in this section it will be more convenient to use the entirely equivalent approach which deals in terms of the external Q's, $(Q_e)_A$ and $(Q_e)_B$ of each end resonator loaded by its adjacent termination, and the coupling coefficients $k_{j,j+1}$ for the coupling between adjacent resonators. These matters were introduced in Sec. 8.02, and equations for the external Q's and coupling coefficients are given in Eqs. (6) to (8) of Figs. 8.02-3 and 8.02-4.

Figure 8.07-1 presents formulas for the external Q's of a rectangular cavity coupled to a terminated waveguide in any of various ways. In the equations and in the discussion below λ is the free-space wavelength, λ_g and λ_{g1} are the guide wavelengths

$$\lambda_g = \frac{\lambda}{\sqrt{1 - \left(\frac{\lambda}{2a}\right)^2}} \quad \text{and} \quad \lambda_{g1} = \frac{\lambda}{\sqrt{1 - \left(\frac{\lambda}{2a_1}\right)^2}} \quad , \qquad (8.07\text{-}1)$$

s is the number of half guide-wavelengths in the l_1 dimension of the cavity, M_1 is the magnetic polarizability of the iris, and the quantities a, b, a_1, b_1, and l_1 are dimensions defined in the figures. Having computed the required values of $(Q_e)_A$ and $(Q_e)_B$ from Eqs. (6) and (7) of Figs. 8.02-3 or 8.02-4, the appropriate equation in Fig. 8.07-1 can be used to solve for the required magnetic polarizability M_1. Then, by use of Figs. 5.10-4(a),(b), the dimensions of the coupling iris can be obtained. It should be noted that M_1 has dimensions of (length)^3, which is consistent with the equations in Fig. 8.07-1, and with the normalization of the ordinates in Figs. 5.10-4(a),(b).

Figure 8.07-2 shows formulas for the coupling coefficient k for two rectangular resonators coupled by a small iris in either the end or side wall. The significance of the other parameters in the equations is the same as for Fig. 8.07-1. The required coupling coefficient values for the couplings between the various adjacent resonators of a filter can be computed by use of Eq. (8) of Fig. 8.02-3 or Fig. 8.02-4. Then, by use of the appropriate formula in Fig. 8.07-2, the magnetic polarizability M_1 of the various coupling irises can be solved for. As for the end irises, the dimensions of the internal irises can be determined with the aid of Fig. 5.10-4(a),(b).

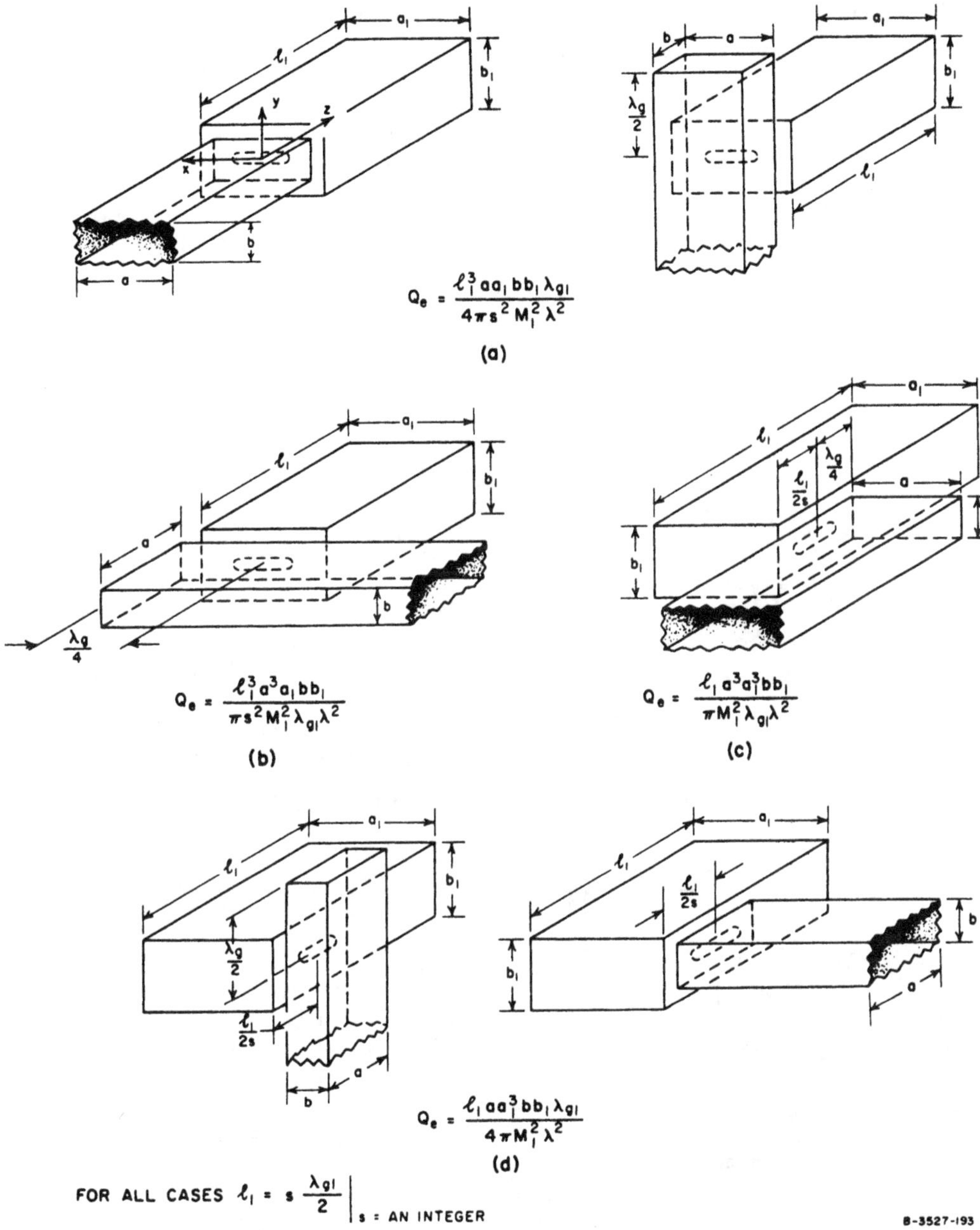

FIG. 8.07-1 EXTERNAL Q, Q_e, OF A RECTANGULAR CAVITY COUPLED TO A TERMINATED WAVEGUIDE BY A SMALL IRIS IN VARIOUS WAYS

For narrow-band filters such as those discussed in this section, the low-pass to band-pass mapping

$$\frac{\omega'}{\omega_1'} = \frac{2}{w}\left(\frac{f - f_0}{f_0}\right) \qquad (8.07\text{-}2)$$

where

$$w = \frac{\omega_2 - \omega_1}{\omega_0}$$

and

$$\omega_0 = \frac{\omega_2 + \omega_1}{2}$$

should give satisfactory accuracy.

As an example of the use of this method we consider the design of a three-cavity direct-coupled filter having a 0.01-db pass-band ripple to operate at a center frequency of 10 Gc in WR-90 waveguide (a = 0.900 inch, b = 0.400 inch). We choose the bandwidth to be 50 Mc (w = 0.005) and choose $l_1 = \lambda_{g1}/2 = 0.7815$ (s = 1). The elements of the low-pass prototype

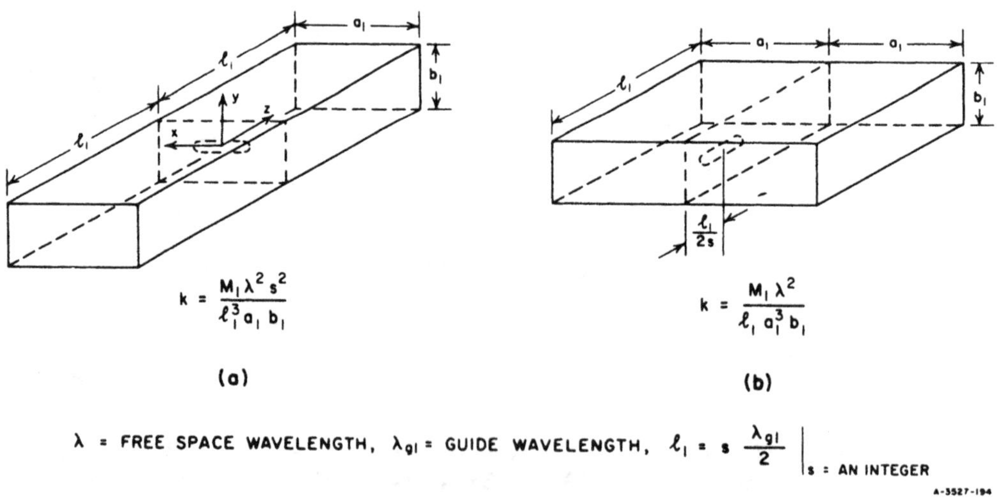

FIG. 8.07-2 COUPLING COEFFICIENT k FOR RECTANGULAR CAVITIES COUPLED BY A SMALL IRIS IN THE END WALL OR SIDE WALL

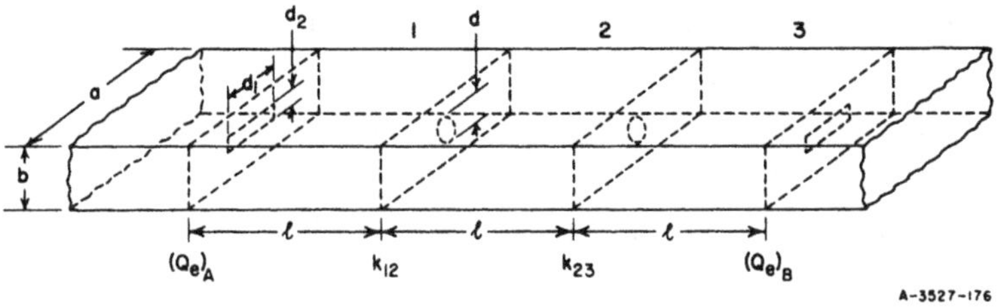

FIG. 8.07-3 REALIZATION OF NARROW-BAND DIRECT-COUPLED FILTER USING SMALL IRISES

filter are determined from Table 4.05-2(a) to be $g_0 = g_4 = 1.000$, $g_1 = g_3 = 0.6291$, and $g_2 = 0.9702$. Figure 8.07-3 illustrates the realization of this filter. We determine from Fig. 8.02-3 that $(Q_e)_A = (Q_e)_B = g_0 g_1\ \omega_1'/w = 125.8$ and that $k_{12} = k_{23} = w/(\omega_1'\sqrt{g_1 g_2}) = 0.0064$. Using Figs. 8.07-1(a) and 8.07-2(a) we find the polarizabilities M_1 for the external and internal apertures to be 6.62×10^{-3} and 0.79×10^{-3} respectively. For the rectangular end irises we choose $d_2/d_1 = 0.5$ (see Fig. 8.07-3). Referring to Fig. 5.10-4(a), we find from the curve for rectangular irises, an initial value of $d_2 = 0.344$ inch. However, d_2 is an appreciable fraction of $\lambda = 1.18$ inches, so that we use Eq. (5.10-3) to determine an approximate correction and find as final values $d_2 = 0.31$ inch and $d_1 = 0.155$ inch. For the circular middle irises we find $d = (6M_1)^{1/3} = 0.168$ inch (see Sec. 5.10). If the thickness of the irises is 0.005 inch or less, the thickness correction of Eq. (5.10-5) is negligible. However, for greater thickness this correction should be applied.

The presence of the apertures will have the effect of lowering the resonant frequencies of the resonators slightly from what they were before the apertures were added. If desired, a small correction in the lengths of the resonators in Fig. 8.07-3 could be made by applying Eq. (5) of Fig. 8.06-1. For this example the normalized reactances $X_{j,j+1}/Z_0$ can be obtained from Fig. 5.10-5, which for the centered irises in Fig. 8.07-3 gives

$$\left.\frac{X_{j,j+1}}{Z_0}\right|_{j=0\ \text{to}\ 3} = \frac{4\pi(M_1)_{j,j+1}}{ab\lambda_{g1}} , \qquad (8.07\text{-}3)$$

where X_{01}/Z_0 and X_{34}/Z_0 are for the irises at the ends.

The design method of this section is based on Bethe's small-aperture theory and is very versatile. However, it does rely on the assumption that the coupling irises are relatively small, which also implies that the fractional bandwidth w of the filter is small (say, of the order of 0.01 or less). Some discussion of the derivation of the equations in Figs. 8.07-1 and 8.07-2 will be found in Sec. 8.14.

SEC. 8.08, FILTERS USING TWO-PORT, QUARTER-WAVELENGTH RESONATORS

The filters discussed in Sec. 8.05 use J-inverters of the type in Fig. 8.03-2(d) along with half-wavelength resonators, and their design equations can be derived from Fig. 8.02-4 as will be outlined in Sec. 8.14. The filters discussed in Sec. 8.06 use K-inverters of the type in Fig. 8.03-1(c) along with half-wavelength resonators, and their design equations can be derived from Fig. 8.02-3. If quarter-wavelength resonators are used in an analogous way, they themselves have an inverting effect so that if at one end of each resonator they behave like the series resonators in Fig. 8.02-3, at their other ends they will operate like the shunt resonators in Fig. 8.02-4. In this manner it can be shown that filters can be constructed using two-port, quarter-wavelength resonators if they are coupled alternately by K- and J-inverters.[14]

Though other types of construction and other types of K- and J-inverters may also be used, Fig. 8.08-1 gives design data for a TEM-mode type of filter using quarter-wavelength resonators with capacitive-gap J-inverters, and shunt inductance K-inverters. Except for the use of two different kinds of inverters and other minor differences which result from the fact that the resonators are a quarter-wavelength rather than a half-wavelength long, the design procedure is much the same as for the preceding cases. Using the strip-line construction shown in Fig. 8.05-3(a), the J-inverter capacitive-gap spacing and the electrical length ϕ can be determined by use of the data in Figs. 8.05-3(a),(b), and (c). Figures 8.08-2(a) to 8.08-4(b) show data for inductive-stub K-inverters. Note that the ordinates on these graphs are normalized with respect to frequency in Gc, and that due to the junction effect the ϕ values are not always negative in this case.

Figure 8.08-5(a) shows a filter with six quarter-wavelength resonators designed using the charts just discussed.[11] The construction is

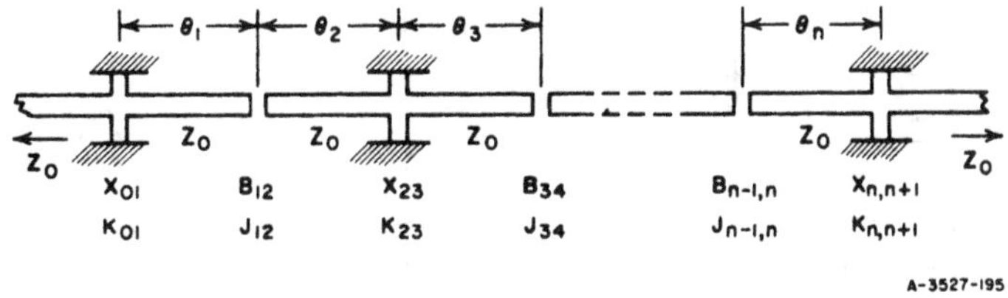

$$\frac{K_{01}}{Z_0} = \sqrt{\frac{\pi w}{4g_0 g_1 \omega_1'}} \qquad (1)$$

$$\left.\frac{J_{j,j+1}}{Y_0} \text{ or } \frac{K_{j,j+1}}{Z_0}\right|_{j=1 \text{ to } n-1} = \frac{\pi w}{4\omega_1'} \frac{1}{\sqrt{g_j g_{j+1}}} \qquad (2)$$

$$\frac{K_{n,n+1}}{Z_0} = \sqrt{\frac{\pi w}{4g_n g_{n+1} \omega_1'}} \qquad (3)$$

where $g_0, g_1, \ldots, g_n$ are as defined in Fig. 4.04-1, ω_1' is defined in Fig. 8.02-1(a), and w is the fractional bandwidth defined below. In this structure, impedance inverters (with parameters $K_{j,j+1}$) alternate with admittance inverters (with parameters $J_{j,j+1}$), and $Z_0 = 1/Y_0$ is the characteristic impedance of the line between inverters.

Using $K_{j,j+1}$ inverters of the form in Fig. 8.03-1(c) and $J_{j,j+1}$ inverters of the form in Fig. 8.03-2(d), the $X_{j,j+1}$, $B_{j,j+1}$, and $\phi_{j,j+1}$ values can be computed from the equations in those figures. Then

$$\theta_j = \frac{\pi}{2} + \frac{1}{2}\left[\phi_{j-1,j} + \phi_{j,j+1}\right] \quad \text{radians} \qquad (4)$$

where the $\phi_{k,k+1}$ are negative.

Using the construction shown in Figs. 8.08-5(a), the gap spacings Δ and the ϕ values for the $J_{j,j+1}$ inverters may be determined by Figs. 8.05-3(a), (b), (c). The stub lengths and ϕ values for the $K_{j,j+1}$ inverters may be determined by Figs. 8.08-2(a) to 8.08-4(b).

To map low-pass prototype filter response to corresponding band-pass filter response use the approximation

$$\omega' = \frac{2\omega_1'}{w}\left(\frac{\omega - \omega_0}{\omega}\right) , \qquad (5)$$

where

$$w = 2\left(\frac{\omega_2 - \omega_1}{\omega_2 + \omega_1}\right) , \qquad (6)$$

and

$$\omega_0 = \frac{2\omega_2\omega_1}{\omega_2 + \omega_1} . \qquad (7)$$

FIG. 8.08-1 DESIGN EQUATIONS FOR FILTERS WITH TWO-PORT, QUARTER-WAVELENGTH RESONATORS

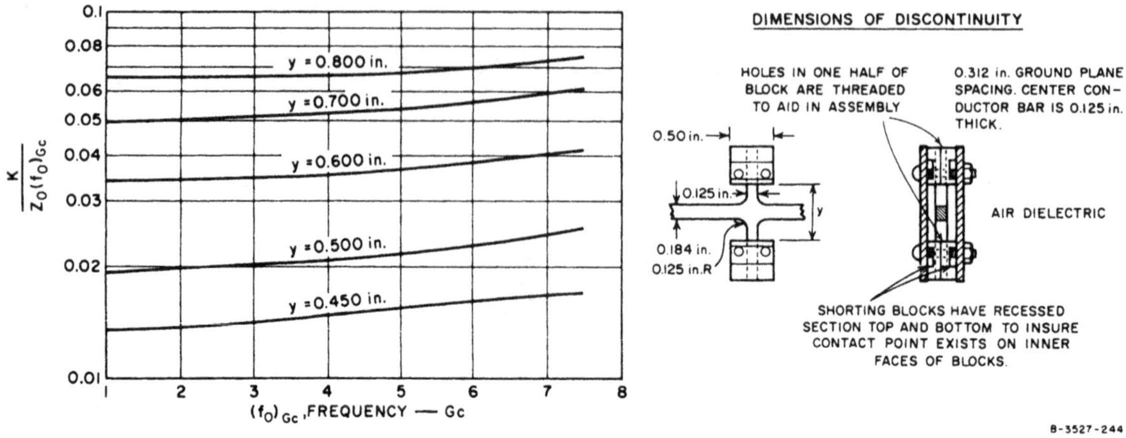

SOURCE: Reference 11, by G. L. Matthaei. (By courtesy of the Ramo-Wooldridge Div. of the Thompson-Ramo Wooldridge Corp.)

FIG. 8.08-2(a) $K/[Z_0(f_0)_{Gc}]$ vs. $(f_0)_{Gc}$ FOR DOUBLE-STUB, SHUNT INDUCTANCE K-INVERTERS IN BAR TRANSMISSION-LINE CONSTRUCTION
The characteristic impedance of the resonator transmission line is $Z_0 = 1/Y_0 = 50$ ohms, and $(f_0)_{Gc}$ is the resonant frequency in Gc

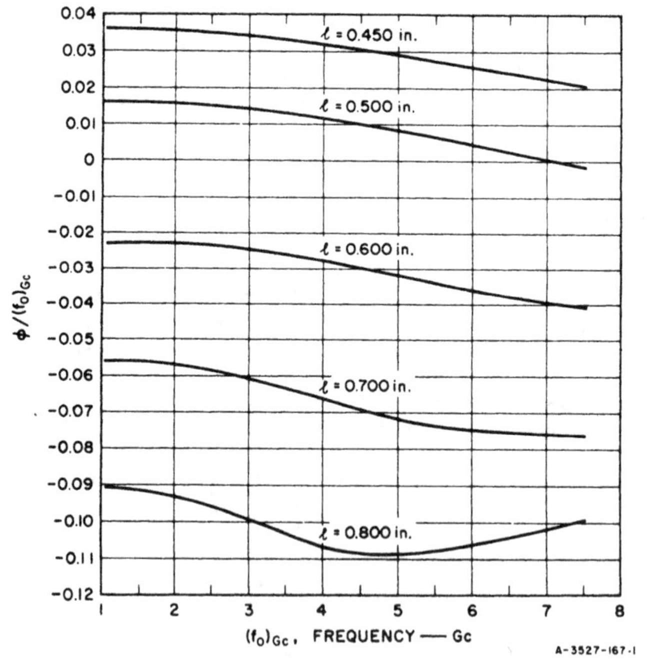

SOURCE: Reference 11, by G. L. Matthaei. (By courtesy of the Ramo-Wooldridge Div. of the Thompson-Ramo-Wooldridge Corp.)

FIG. 8.08-2(b) $\phi/(f_0)_{Gc}$ vs. $(f_0)_{Gc}$ vs. $(f_0)_{Gc}$ FOR THE DOUBLE-STUB K-INVERTER IN FIG. 8.08-2(a)

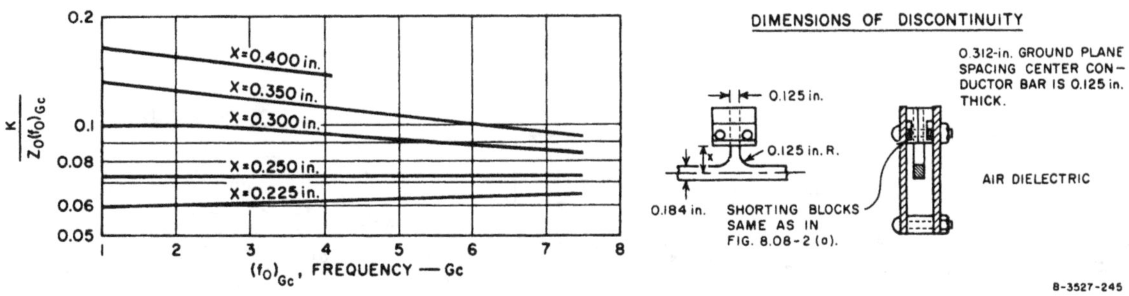

SOURCE: Reference 11, by G. L. Matthaei. (By courtesy of the Ramo-Wooldridge Div. of the Thompson-Ramo-Wooldridge Corp.)

FIG. 8.08-3(a) $K/[Z_0(f_0)_{Gc}]$ vs. $(f_0)_{Gc}$ FOR A SINGLE-STUB, SHUNT INDUCTANCE K-INVERTER IN BAR TRANSMISSION-LINE CONSTRUCTION
The characteristic impedance of the resonator transmission line is $Z_0 = 1/Y_0 = 50$ ohms, and $(f_0)_{Gc}$ is the resonant frequency in Gc

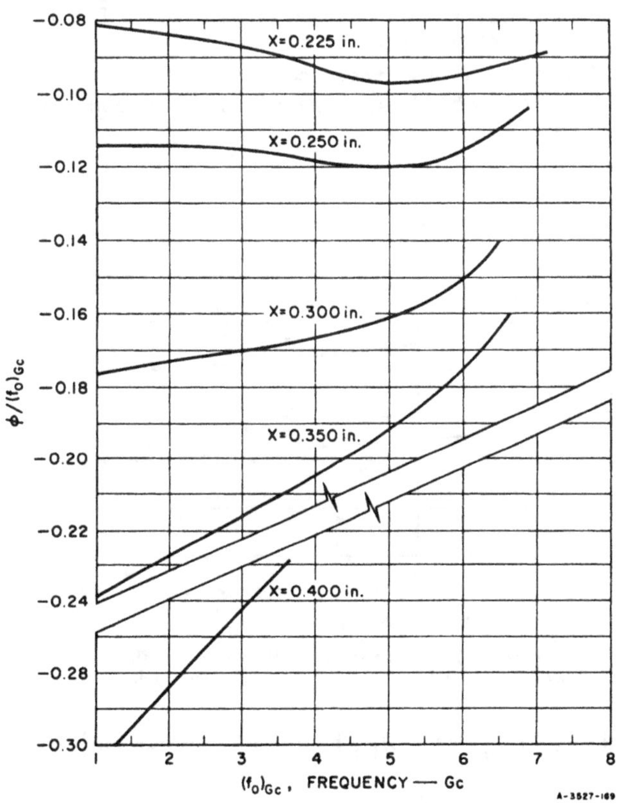

SOURCE: Reference 11, by G. L. Matthaei. (By courtesy of the Ramo-Wooldridge Div. of the Thompson-Ramo-Wooldridge Corp.)

FIG. 8.08-3(b) $\phi(f_0)_{Gc}$ vs. $(f_0)_{Gc}$ FOR THE SINGLE-STUB K-INVERTER IN FIG. 8.08-3(a)

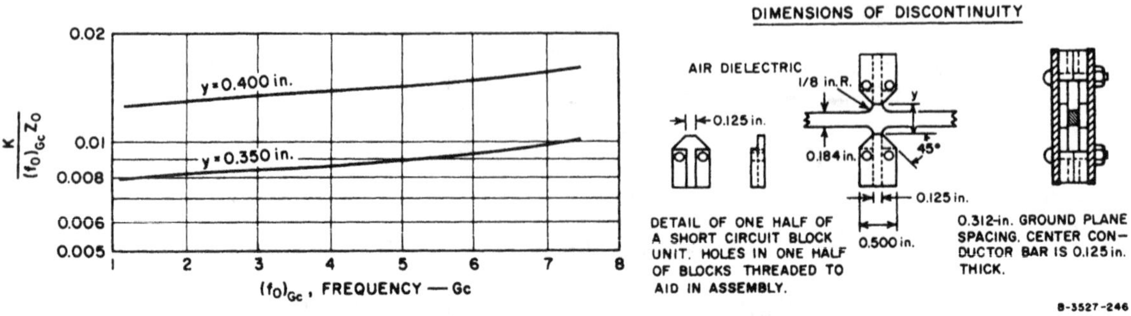

SOURCE: Reference 11, by G. L. Matthaei. (By courtesy of the Ramo-Wooldridge Div. of the Thompson-Ramo Wooldridge Corp.)

FIG. 8.08-4(a) $K/[Z_0(f_0)_{Gc}]$ vs. $(f_0)_{Gc}$ FOR A SHUNT INDUCTANCE K-INVERTER DESIGNED TO PERMIT RELATIVELY LOOSE COUPLINGS IN BAR TRANSMISSION LINE
The characteristic impedance of the resonator transmission line is $Z_0 = 1/Y_0 = 50$ ohms, and $(f_0)_{Gc}$ is the resonant frequency in Gc

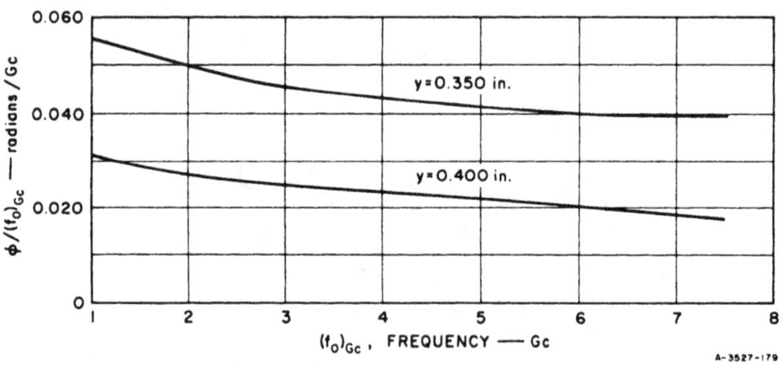

SOURCE: Reference 11, by G. L. Matthaei. (By courtesy of the Ramo-Wooldridge Div. of the Thompson-Ramo-Wooldridge Corp.)

FIG. 8.08-4(b) $\phi/(f_0)_{Gc}$ vs. $(f_0)_{Gc}$ FOR THE K-INVERTER IN FIG. 8.08-4(a)

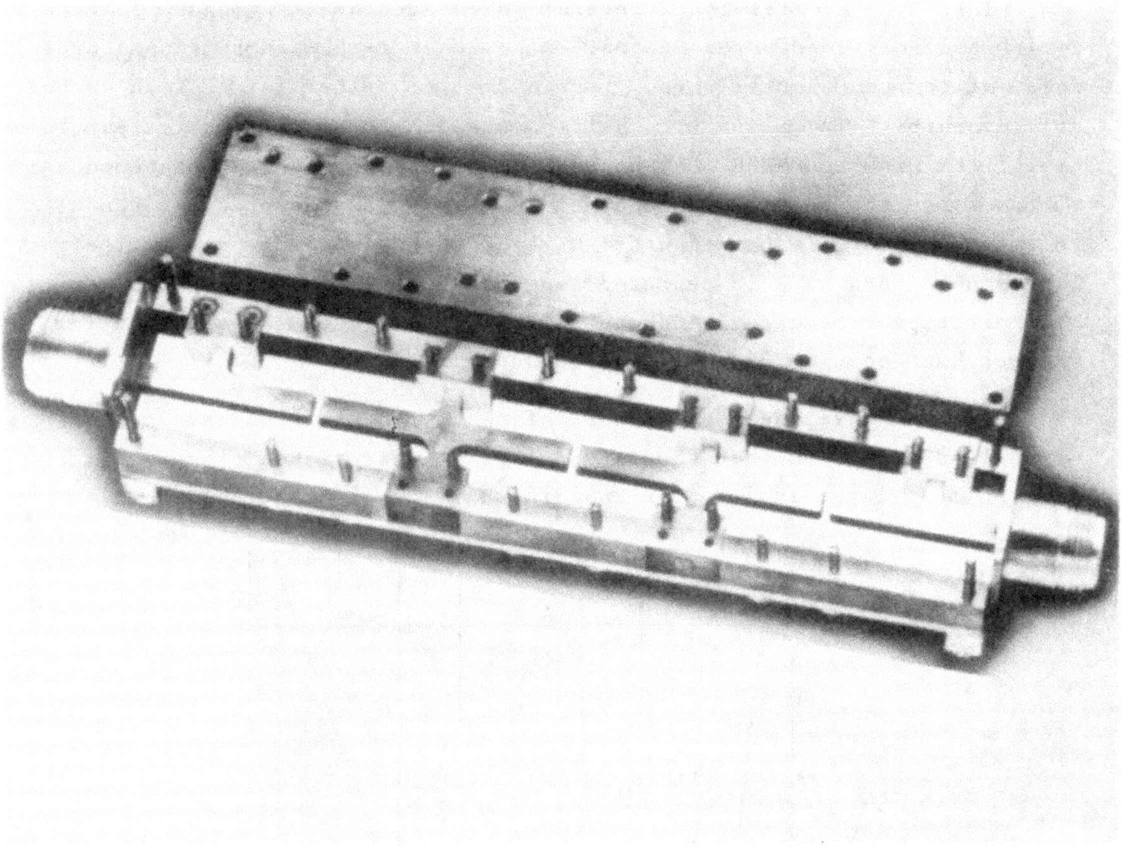

SOURCE: Reference 11, by G. L. Matthaei. (By courtesy of the Ramo-Wooldridge Div. of the Thompson-Ramo-Wooldridge Corp.)

FIG. 8.08-5(a) A FILTER WITH SIX, $\lambda_0/4$ RESONATORS IN BAR CONSTRUCTION
The response is shown in Fig. 8.08-5(b)

quite rugged, and no dielectric support material is required. The resonators in this filter were tested in pairs by the methods described in Secs. 11.04 and 11.05 to insure that their tuning was correct. The design pass band was from 2.6 to 3.4 Gc, and as can be seen from Fig. 8.08-5(b), this was achieved with good accuracy. The mapping defined in Eqs. (5) to (7) in Fig. 8.08-1 is not quite as accurate, however, for this type of filter as for the type in Fig. 8.05-1. In this case, the predicted attenuation at 2.4 Gc is about 40 db, which is only about 2 db more than was measured; however, the predicted attenuation at 3.7 Gc is about 37 db as against a measured attenuation of 32 db.

This type of filter has several advantages over analogous filters using half-wavelength resonators.[14] The quarter-wavelength resonators are, of course, shorter which gives a smaller filter for a given number of resonators. A filter with half-wavelength resonators equivalent to the filter in Fig. 8.08-5(a) would have a second pass band centered at about twice the center frequency of the first pass band, or at about 6 kMc. However, in this quarter-wavelength-resonator type of filter, the second pass band is centered roughly three times the band center of the first pass band, or at about 9 kMc in this case. This particular filter has about 61.5 db attenuation at 6 kMc.

Quarter-wavelength resonators of the type described have an additional advantage in that their reactance or susceptance slope parameters are half as large as for corresponding half-wavelength resonators.

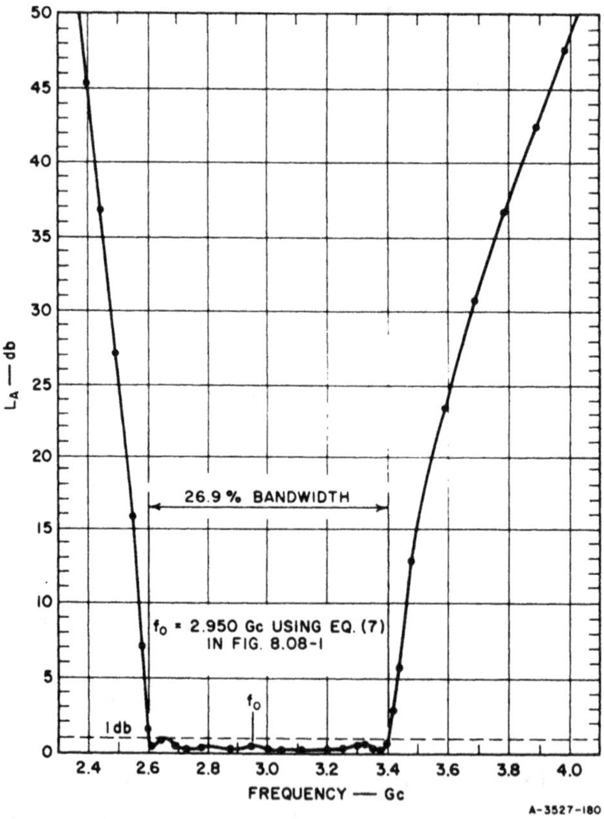

SOURCE: Reference 11, by G. L. Matthaei. (By courtesy of the Ramo-Wooldridge Div. of the Thompson-Ramo-Wooldridge Corp.)

FIG. 8.08-5(b) THE ATTENUATION CHARACTERISTIC OF THE FILTER IN FIG. 8.08-5(a)

Because of this, for a given bandwidth and pass-band shape the couplings are considerably looser for the quarter-wavelength than for the half-wavelength resonator types of filters. This calls for larger capacitive gaps so that tolerances are less of a problem, and it also results in considerably higher maximum attenuation $(L_A)_{USB}$ in the stop band above the pass band. Also, because of the shorter resonators and looser couplings the circuit is more nearly lumped, and as a result, the design equations in Fig. 8.08-1 will be found to give filters with specified pass-band characteristics accurately for greater bandwidths. They should give good results for many filters having bandwidths as large as 30 percent. As in the preceding cases, the equations are more accurate for larger bandwidths if the pass-band ripple tolerance is 0.5 to 1.0 db than if a very small tolerance such as 0.01 db is called for.

For this type of filter, the maximum attenuation between the first and second pass band is always finite (just as for the filters in Secs. 8.05, 8.06, and 8.07), but in this case, the attenuation levels off to a maximum value near $\omega = 2\omega_0$. This maximum upper-stop-band attenuation can be estimated by use of the formula

$$(L_A)_{USB} \approx 20 \log_{10} \left[\frac{1}{\left(\frac{X_{01}}{Z_0}\right)\left(\frac{B_{12}}{Y_0}\right) \cdots \left(\frac{B_{n-1,n}}{Y_0}\right)\left(\frac{X_{n,n+1}}{Z_0}\right)} \right] - (n+1)6.02 - 6.02 \quad \text{db} \tag{8.08-1}$$

where

$$\frac{X_{j,j+1}}{Z_0} = \frac{\dfrac{K_{j,j+1}}{Z_0}}{1 - \left(\dfrac{K_{j,j+1}}{Z_0}\right)^2} \quad , \tag{8.08-2}$$

$$\frac{B_{j,j+1}}{Y_0} = \frac{\dfrac{J_{j,j+1}}{Y_0}}{1 - \left(\dfrac{J_{j,j+1}}{Y_0}\right)^2} \quad , \tag{8.08-3}$$

and the $K_{j,j+1}/Z_0$ and $J_{j,j+1}/Y_0$ are computed by use of Eq. (1) to (3) in Fig. 8.08-1. An n-resonator filter of this type will have an $(n + 1)$-order pole of attenuation (Sec. 2.04) at $\omega = 0$. For that reason, this type of filter will have a very fast rate of cutoff below the pass band, as can be seen in the case of the response in Fig. 8.08-5(b).

SEC. 8.09, FILTERS WITH PARALLEL-COUPLED STRIP-LINE RESONATORS

Figure 8.09-1 presents design equations (which are a modified form of equations due to Cohn[15]) for filters using half-wavelength strip-line resonators, positioned so that adjacent resonators parallel each other along half of their length. This parallel arrangement gives relatively large coupling for a given spacing between resonator strips, and thus, this construction is particularly convenient for printed-circuit filters up to about 10 or 15 percent bandwidth.[15] For larger bandwidths the resonators can be constructed from bars having a rectangular cross section (which permits tighter coupling), and for that case the wide-band filter equations in Chapter 10 are recommended.

The use of the equations in Fig. 8.09-1 is best illustrated by use of an example. Let us suppose that a low VSWR in the pass band is desired so that a 0.01-db ripple, Tchebyscheff prototype is to be used in the design. The desired fractional bandwidth is assumed to be $w = 0.10$, and the design center frequency is to be $f_0 = 1207$ Mc. We shall assume that 25-db attenuation is required at $f = 1100$ Mc. Then, by mapping Eqs. (6) to (8) in Fig. 8.09-1 for $f = 1100$ Mc,

$$\frac{\omega'}{\omega_1'} = \frac{2}{w}\left(\frac{\omega - \omega_0}{\omega_0}\right) = \frac{2}{w}\left(\frac{f - f_0}{f_0}\right)$$

$$= \frac{2}{0.10}\left(\frac{1100 - 1207}{1207}\right) = -1.77 \quad .$$

By Fig. 4.03-4 it is found that an $n = 6$ design has 29 db attenuation for $|\omega'/\omega_1'| - 1 = 0.77$ while an $n = 5$ design has $L_A = 18.5$ db. Thus, $n = 6$ is required. By Table 4.05-2(a), the desired $n = 6$ prototype parameters are $g_0 = 1$, $g_1 = 0.7813$, $g_2 = 1.3600$, $g_3 = 1.6896$, $g_4 = 1.5350$, $g_5 = 1.4970$, $g_6 = 0.7098$, $g_7 = 1.1007$, and $\omega_1' = 1$.

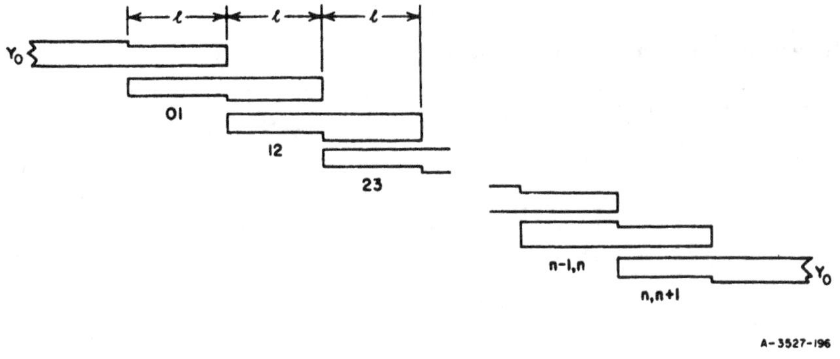

$$\frac{J_{01}}{Y_0} = \sqrt{\frac{\pi w}{2 g_0 g_1}} \qquad (1)$$

$$\left.\frac{J_{j,j+1}}{Y_0}\right|_{j=1 \text{ to } n-1} = \frac{\pi w}{2\omega_1'} \frac{1}{\sqrt{g_j g_{j+1}}} \qquad (2)$$

$$\frac{J_{n,n+1}}{Y_0} = \sqrt{\frac{\pi w}{2 g_n g_{n+1}}} \qquad (3)$$

where $g_0, g_1, \ldots, g_{n+1}$ are as defined in Fig. 4.04-1, ω_1' is as defined in Fig. 8.02-1(a), and w is the fractional bandwidth defined below. The $J_{j,j+1}$ are admittance inverter parameters and Y_0 is the characteristic admittance of the terminating lines. The even- and odd-mode impedances of the strips are

$$\left.(Z_{oe})_{j,j+1}\right|_{j=0 \text{ to } n} = \frac{1}{Y_0}\left[1 + \frac{J_{j,j+1}}{Y_0} + \left(\frac{J_{j,j+1}}{Y_0}\right)^2\right] \qquad (4)$$

$$\left.(Z_{oo})_{j,j+1}\right|_{j=0 \text{ to } n} = \frac{1}{Y_0}\left[1 - \frac{J_{j,j+1}}{Y_0} + \left(\frac{J_{j,j+1}}{Y_0}\right)^2\right] \qquad (5)$$

and the strip dimensions can be determined by use of Sec. 5.05.

To map the low-pass prototype response to the band-pass filter response use the approximation

$$\frac{\omega'}{\omega_1'} = \frac{2}{w}\left(\frac{\omega - \omega_0}{\omega_0}\right) \qquad (6)$$

where

$$w = \frac{\omega_2 - \omega_1}{\omega_0} \qquad (7)$$

$$\omega_0 = \frac{\omega_2 + \omega_1}{2} \qquad (8)$$

and ω_1 and ω_2 are as defined in Fig. 8.02-1(b).

FIG. 8.09-1 DESIGN EQUATIONS FOR FILTERS WITH PARALLEL-COUPLED RESONATORS

Table 8.09-1

DESIGN PARAMETERS FOR EXPERIMENTAL PARALLEL-COUPLED STRIP-LINE-RESONATOR FILTER

j	$J_{j,j+1}/Y_0$	$(Z_{oe})_{j,j+1}$	$(Z_{oo})_{j,j+1}$	$W_{j,j+1}$	$s_{j,j+1}$	$d_{j,j+1}$
0	0.449	82.5 ohms	37.6 ohms	0.236 inch	0.021 inch	0.073 inch
1	0.1529	58.8 ohms	43.5 ohms	0.346 inch	0.110 inch	0.084 inch
2	0.1038	55.7 ohms	45.3 ohms	0.360 inch	0.158 inch	0.085 inch
3	0.0976	55.4 ohms	45.6 ohms	0.361 inch	0.163 inch	0.085 inch

SOURCE: Final Report, Contract DA 36-039 SC-64625, SRI; reprinted in *IRE Trans. PGMTT* (see Ref. 15 by S. B. Cohn).

Table 8.09-1 shows the $J_{j,j+1}/Y_0$, $(Z_{oe})_{j,j+1}$, and $(Z_{oo})_{j,j+1}$ values as computed from the equations in Fig. 8.09-1. This filter was constructed[15] using polystyrene dielectric with a relative dielectric constant ϵ_r of 2.55. Using a 0.5-inch ground plane spacing and copper-foil resonators of negligible thickness, by use of Figs. 5.05-3(a),(b) the dimensions of the strip widths $W_{j,j+1}$, and the gaps $s_{j,j+1}$, were obtained and they are as is also shown in Table 8.09-1. The significance of these dimensions is further illustrated in Fig. 8.09-2.

The dimensions $d_{j,j+1}$ indicated in Table 8.09-1 and Fig. 8.09-2 are resonator length corrections to account for the fringing capacitance from

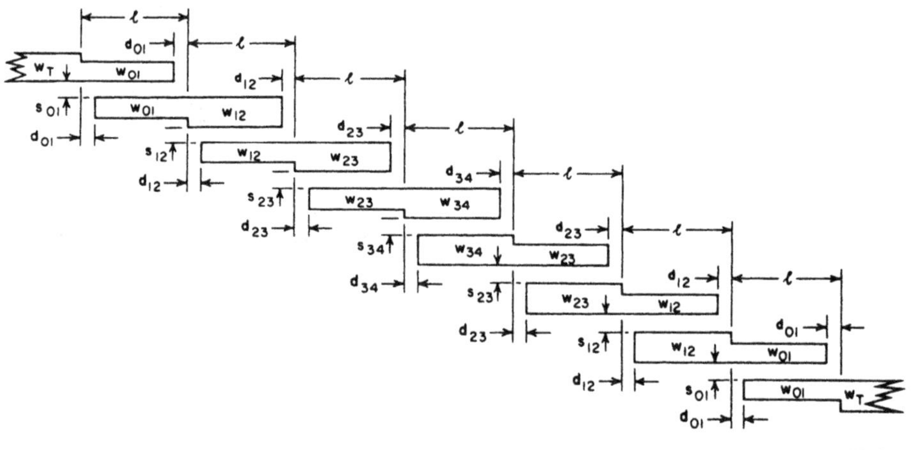

SOURCE: Final report, Contract DA 36-039 SC-64625, SRI; reprinted in *IRE Trans., PGMTT* (see Ref. 15 by S. B. Cohn).

FIG. 8.09-2 LAYOUT OF PARALLEL-COUPLED-RESONATOR FILTER

the end of each strip. The basic length l indicated in the Fig. 8.09-2 is a quarter-wavelength at frequency ω_0 in the medium of propagation, while the actual strip lengths are shortened by the amount $d_{j,j+1}$. Although Table 8.09-1 indicates some variation in the $d_{j,j+1}$ values, Cohn[15] has found that a constant correction of $d_{j,j+1} = 0.165b$ (where b is the ground-plane spacing) is apparently satisfactory.

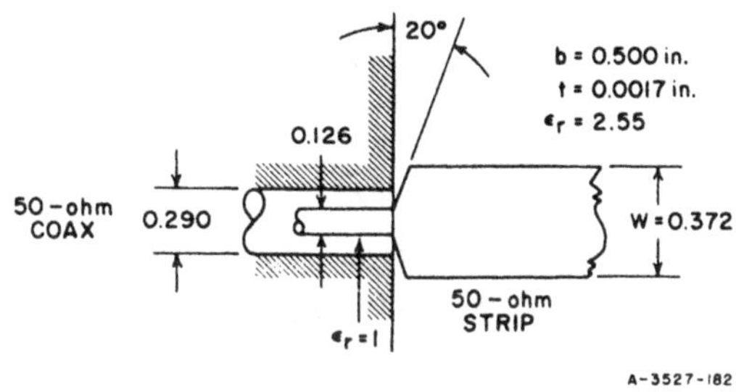

FIG. 8.09-3 COAXIAL-LINE TO STRIP-LINE JUNCTION

As a result of the filter being designed from an antimetric prototype filter (see Sec. 4.05), the resulting parallel-coupled microwave filter has symmetry about its center. For that reason only the dimensions of half the filter are shown in Table 8.09-1. The input and output lines are of 50 ohms impedance which requires that they be $W_T = 0.372$ inch wide as determined from Fig. 5.04-1 with $b = 0.50$, $t = 0$, and $\epsilon_r = 2.55$. Figure 8.09-3 shows the manner in which the input and output strips were beveled to give a low-reflection transition from the printed-circuit strip line to coaxial line.

Figure 8.09-4 shows a photograph of the completed printed-circuit filter with its upper half removed. The circles in Fig. 8.09-5 show measured attenuation values while the solid curve shows the theoretical attenuation as computed from the low-pass prototype attenuation with the aid of the mappings in Eqs. (6) to (8) of Fig. 8.09-1. As can be seen from the figure, the agreement is very good. Of course, as a result of dissipation loss, the pass-band attenuation is considerably above the 0.01 db theoretical value for a lossless filter. Working back from the measured attenuation using Eqs. (4.13-2), (4.13-8), and Fig. 4.13-2, the Q of the resonators in this filter is estimated to be roughly 600.

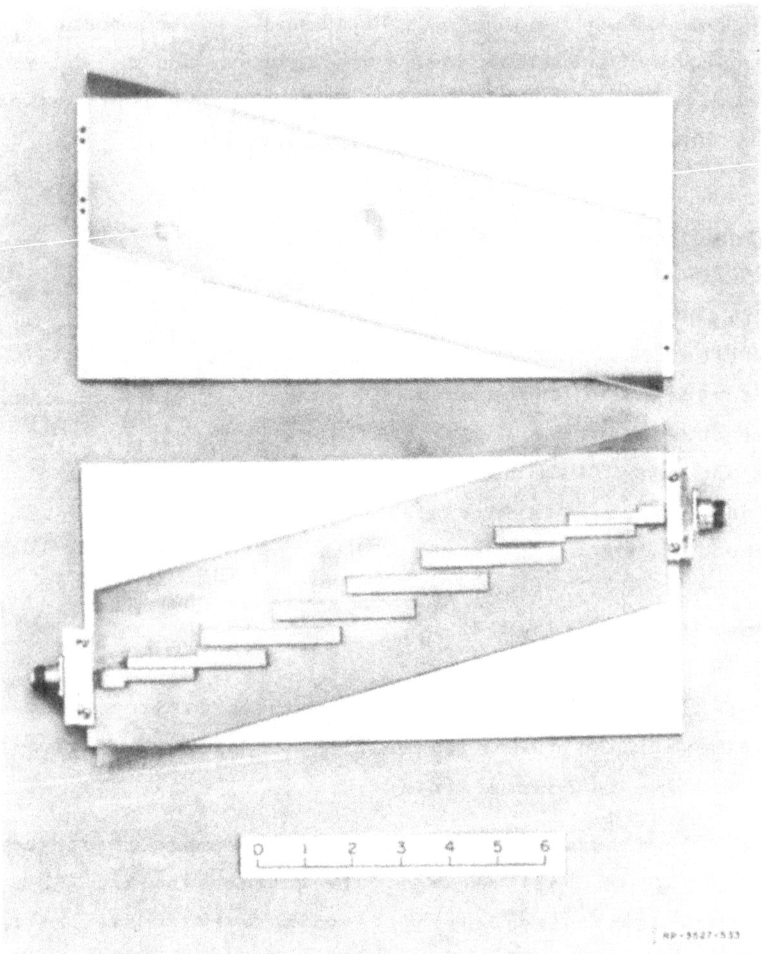

SOURCE: Final Report, Contract DA 36-039 SC-64625, SRI; reprinted in *IRE Trans.*, *PGMTT* (see Ref. 15 by S. B. Cohn).

FIG. 8.09-4 PHOTOGRAPH OF THE EXPERIMENTAL PARALLEL-COUPLED FILTER WITH ITS COVER PLATE REMOVED

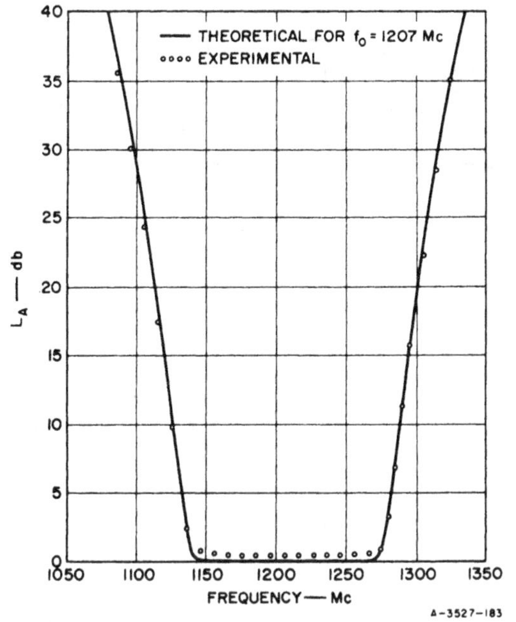

SOURCE: Final report, Contract DA 36-039 SC-64625, SRI; reprinted in *IRE Trans., PGMTT* (see Ref. 15 by S. B. Cohn).

FIG. 8.09-5 THEORETICAL AND MEASURED ATTENUATION FOR THE FILTER IN FIG. 8.09-4

SEC. 8.10, FILTERS WITH QUARTER-WAVELENGTH COUPLINGS

As has been previously mentioned, quarter-wavelength lines can be used satisfactorily as K- or J-inverters in narrow-band filters (*i.e.*, filters with bandwidths of the order of a few percent or less). Figure 8.10-1 shows a filter with quarter-wavelength lines for inverters and presents the appropriate design equations. The $\pi/4$ and $\pi/2$ terms in the equations for the normalized resonator susceptance slope parameters $\mathscr{b}_j/Y_0$ represent correction terms for the added selectivity introduced by the quarter-wavelength lines.[4] The particular structure shown gives perfect transmission at the midband frequency ω_0, hence, it is only applicable for achieving responses that have this property (*i.e.*, no reflection loss at ω_0). Therefore, the data given is valid for use with the maximally flat low-pass prototypes in Tables 4.05-1(a),(b), having any value of n, but only for Tchebyscheff prototypes in Tables 4.05-2(a),(b), having an *odd* number of reactive elements n.*

* Tchebyscheff responses corresponding to n even can also be achieved with this type of filter if the coupling lines are allowed to have Y_0 values different from that of the terminations. In Fig. 8.10-1, the lines are all the same, for simplicity.

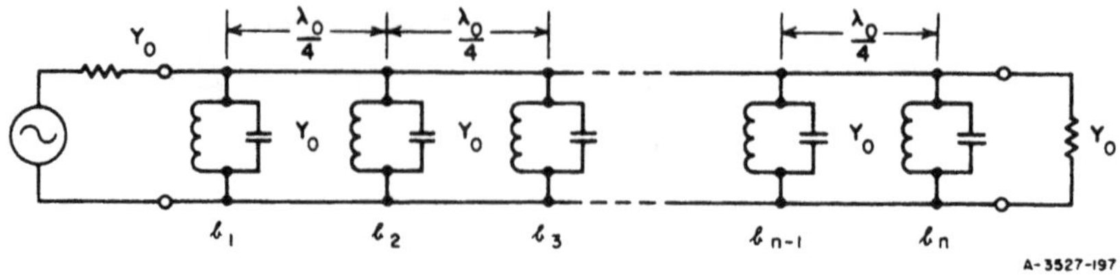

For maximally flat filters with n even or odd, or Tchebyscheff filters with n odd:

$$\frac{\mathscr{b}_1}{Y_0} = \frac{\omega_1' g_1 g_0}{w} - \frac{\pi}{4} \tag{1}$$

$$\left.\frac{\mathscr{b}_j}{Y_0}\right|_{j=\text{even}<n} = \frac{\omega_1' g_j}{w g_0} - \frac{\pi}{2} \tag{2}$$

$$\left.\frac{\mathscr{b}_j}{Y_0}\right|_{j=\text{odd}<n} = \frac{\omega_1' g_j g_0}{w} - \frac{\pi}{2} \tag{3}$$

$$\frac{\mathscr{b}_n}{Y_0} = \frac{\mathscr{b}_1}{Y_0} \tag{4}$$

where the $g_0, g_1, \ldots, g_{n+1}$ are as defined in Fig. 4.04-1, ω_1' is defined in Fig. 8.02-1(a), the $\mathscr{b}_j$ are susceptance slope parameters defined in Fig. 8.02-4, w is as defined below, λ_0 is the propagation wavelength at the midband frequency ω_0, and Y_0 is the admittance of the transmission line connecting the resonators.

To map low-pass filter response to corresponding band-pass filter response use (for narrow-band designs):

$$\frac{\omega'}{\omega_1'} = \frac{2}{w}\left(\frac{\omega - \omega_0}{\omega_0}\right) \tag{5}$$

where

$$w = \left(\frac{\omega_2 - \omega_1}{\omega_0}\right) \quad , \tag{6}$$

$$\omega_0 = \left(\frac{\omega_2 + \omega_1}{2}\right) \quad , \tag{7}$$

and ω_0, ω_1, and ω_2 are as defined in Fig. 8.02-1(b).

FIG. 8.10-1 DESIGN EQUATIONS FOR FILTERS WITH QUARTER-WAVELENGTH COUPLINGS

The resonators for filters of this type can be formed from semi-lumped elements, cavities with loop couplings, resonant irises,[5] and other means. One common way of realizing the desired resonators is illustrated in Fig. 8.10-2.[4] In this case, the resonator used is a half-wavelength resonator with a K-inverter at each end, as indicated at (b) in the figure. The K-inverters tend to reflect low-impedance levels to the ends of the half-wavelength line section which, it can be shown, will make it operate like a series resonator (see Sec. 8.14). However, this series resonance operation when viewed from the outside through the K-inverters looks like a shunt resonance equivalent to that of the shunt-tuned circuit shown in Fig. 8.10-2(a). Using waveguide and inductive irises of shunt reactance X_j to realize the K_j inverters, the resulting quarter-wavelength-coupled waveguide filter takes the form shown in Fig. 8.10-3. Note that the half-wavelength resonators are corrected for the electrical length ϕ_j associated with the K-inverters, and that the quarter-wavelength coupling lengths should be corrected in a similar way.

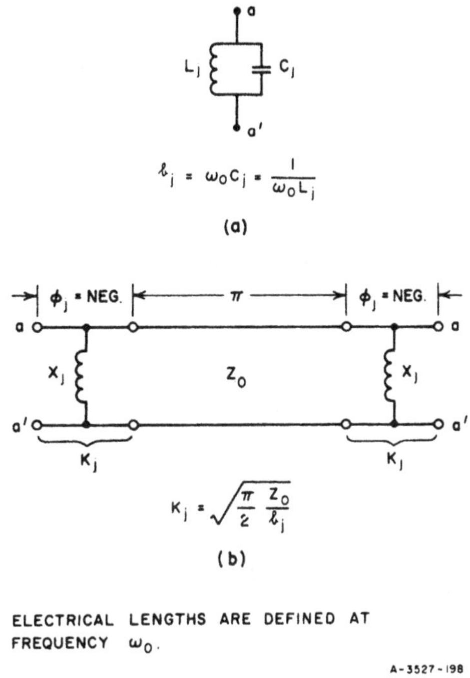

FIG. 8.10-2 REPLACEMENT OF A SHUNT RESONATOR BY A HALF-WAVELENGTH RESONATOR WITH TWO K-INVERTERS
The K-Inverters shown are of the type defined in Fig. 8.03-1(c)

The main advantage of this type of filter appears to be that the resonators are easily tested individually. If a waveguide joint is placed in the center of each quarter-wavelength coupling, the filter can easily be disassembled and each resonator checked by itself. Each resonator should, of course, resonate at ω_0, and if Fig. 8.10-1 calls for a susceptance slope parameter of $\mathscr{b}_j$ for the jth resonator, then if the resonator is connected to matched source and load waveguides of the same dimensions (and characteristic admittance Y_0), the resonator should exhibit a doubly loaded Q of

$$Q_j = \frac{\mathcal{b}_j}{2Y_0} = \frac{f_0}{f_b - f_a} \qquad (8.10\text{-}1)$$

where f_b and f_a are the 3-db points of the transmission through the resonator.

To summarize the procedure for the design of a quarter-wavelength coupled waveguide filter of the type in Fig. 8.10-3, the number of resonators and the value for w_λ and the required number n of resonators should be determined by use of Eqs. (6) to (8) in Fig. 8.06-1, as discussed in Sec. 8.04.* Then w_λ should be used to replace w in Eqs. (1) to (4) in Fig. 8.10-1 (since guide-wavelength variation controls the bandwidth in this case) and the normalized susceptance slope parameters $\mathcal{b}_j/Y_0$ should be determined using the desired lumped-element prototype parameters. By Fig. 8.10-2

$$\frac{K_j}{Z_0} = \sqrt{\frac{\pi}{2}\frac{Y_0}{\mathcal{b}_j}} \qquad (8.10\text{-}2)$$

since $Z_0 = 1/Y_0$. Having values for the K_j/Z_0, the dimensions of the discontinuities and their corresponding ϕ_j values can be determined as previously discussed in Sec. 8.06. The radian electrical spacings θ_j and

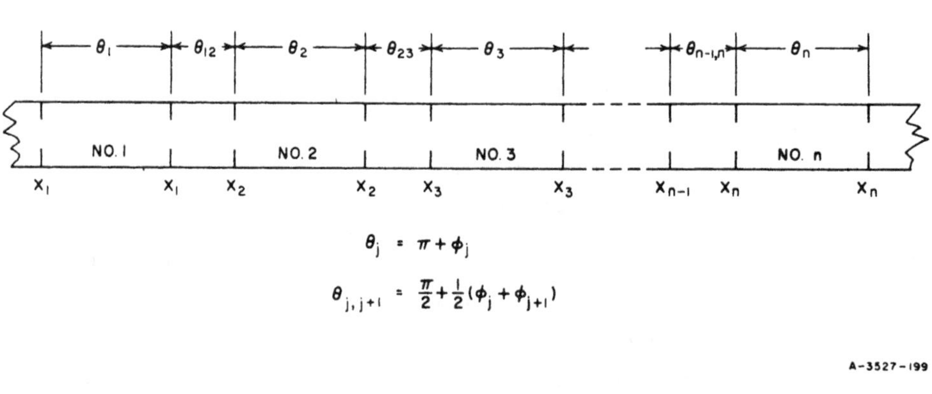

FIG. 8.10-3 WAVEGUIDE FILTER USING SHUNT-INDUCTIVE IRISES AND QUARTER-WAVELENGTH COUPLINGS
The ϕ_j are as indicated in Figs. 8.10-2 and 8.03-1(c)

* See Sec. 8.14 for discussion of the use of λ_{g0}/λ_g as a frequency parameter in design of waveguide filters.

$\theta_{j,j+1}$ (with respect to guide wavelength) of the discontinuities are then determined as indicated in Fig. 8.10-3.

SEC. 8.11, LUMPED-ELEMENT, COUPLED-RESONATOR FILTERS

At the lower microwave frequencies it may be possible to use semi-lumped elements, and analysis in terms of the lumped-element structures in Fig. 8.11-1 or 8.11-2 may be helpful. The structure in Fig. 8.11-1 approximates that in Fig. 8.02-4 using lumped, shunt resonators $B_j(\omega)$ and lumped J-inverters of the form in Fig. 8.03-2(b). In Fig. 8.11-1 the capacitances C_{rj} are the effective capacitances for determining the resonant frequency and susceptance slope parameters of the resonators. But, the actual shunt-capacitor elements used are smaller than the C_{rj}, as indicated in Eqs. (8) to (10). This is because the negative shunt capacitance of the J-inverters must be subtracted from the positive resonator capacitance to give the net shunt capacitance actually inserted in the circuit. The end coupling capacitances C_{01} and $C_{n,n+1}$ are treated in a somewhat different manner, as discussed in Sec. 8.14, in order to prevent having to deal with a negative shunt capacitance next to the terminations G_A and G_B. Note that G_A, G_B, and the C_{rj} may be given any values desired.

The circuit in Fig. 8.11-2(b) is the exact dual of that in Fig. 8.11-1 if L_{r0} and L_{rn+1} are chosen to equal M_{01} and $M_{n,n+1}$, respectively, which will make L_0 and L_{n+1} zero. The equations are slightly modified from those in Fig. 8.11-1, however, in order to also accommodate the circuit form in Fig. 8.11-2(a).

The low-pass to band-pass mappings shown in Figs. 8.11-1 and 8.11-2 are accurate for narrow bandwidths only; however, Cohn[1] has shown that the approximate mapping

$$\frac{\omega'}{\omega_1'} = \left| \frac{2 - \dfrac{\omega_0}{\omega} - \dfrac{1}{2 - \dfrac{\omega_0}{\omega}}}{2 - \dfrac{\omega_0}{\omega_2} - \dfrac{1}{2 - \dfrac{\omega_0}{\omega_2}}} \right| \tag{8.11-1}$$

where

$$\omega_0 = \omega_1 + \omega_2 - \sqrt{(\omega_2 - \omega_1)^2 + \omega_1\omega_2} \tag{8.11-2}$$

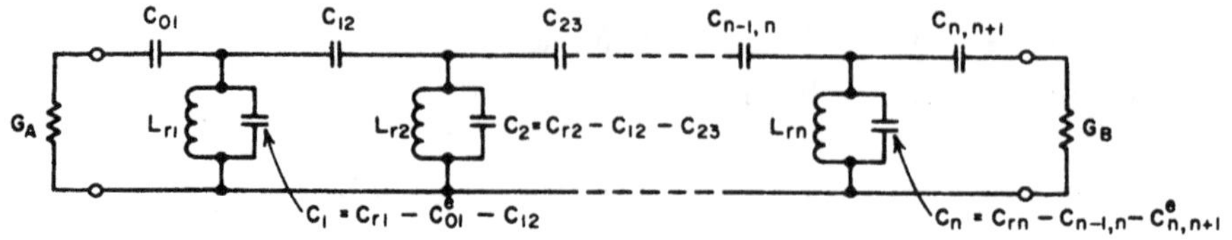

For definitions of the g_j, ω_1', ω_0, ω_1, ω_2, and the $J_{j,j+1}$; see Figs. 4.04-1, 8.02-1(a),(b), 8.02-4, and 8.03-2(b).

Choose values for G_A, C_{r1}, C_{r2}, ..., C_{rn}, and G_B. Then:

$$L_{rj}\Big|_{j=1 \text{ to } n} = \frac{1}{C_{r_j}\omega_0^2} \tag{1}$$

$$J_{01} = \sqrt{\frac{G_A \omega_0 C_{r1} w}{g_0 g_1 \omega_1'}} \tag{2}$$

$$J_{j,j+1}\Big|_{j=1 \text{ to } n-1} = \frac{w\omega_0}{\omega_1'}\sqrt{\frac{C_{rj}C_{rj+1}}{g_j g_{j+1}}} \tag{3}$$

$$J_{n,n+1} = \sqrt{\frac{G_B \omega_0 C_{rn} w}{g_n g_{n+1} \omega_1'}} \quad , \tag{4}$$

where w is defined below.

The coupling capacitances are:

$$C_{01} = \frac{J_{01}}{\omega_0\sqrt{1 - \left(\frac{J_{01}}{G_A}\right)^2}} \tag{5}$$

$$C_{j,j+1}\Big|_{j=1 \text{ to } n-1} = \frac{J_{j,j+1}}{\omega_0} \tag{6}$$

$$C_{n,n+1} = \frac{J_{n,n+1}}{\omega_0\sqrt{1 - \left(\frac{J_{n,n+1}}{G_B}\right)^2}} \quad . \tag{7}$$

FIG. 8.11-1 DESIGN FORMULAS FOR CAPACITIVELY COUPLED LUMPED-ELEMENT FILTERS

The net shunt capacitances are:

$$C_1 = C_{r1} - C_{01}^e - C_{12} \tag{8}$$

$$C_j \Big|_{j=2 \text{ to } n-1} = C_{rj} - C_{j-1,j} - C_{j,j+1} \tag{9}$$

$$C_n = C_{rn} - C_{n-1,n} - C_{n,n+1}^e \tag{10}$$

where the $C_{j,j+1}$ are given by Eqs. (5) to (7) and

$$C_{01}^e = \frac{C_{01}}{1 + \left(\dfrac{\omega_0 C_{01}}{G_A}\right)^2} \tag{11}$$

$$C_{n,n+1}^e = \frac{C_{n,n+1}}{1 + \left(\dfrac{\omega_0 C_{n,n+1}}{G_B}\right)^2} \quad . \tag{12}$$

For mapping low-pass prototype response approximately to band-pass response use, if $\omega_2/\omega_1 \leqq 1.05$,

$$\omega' = \frac{\omega_1'}{w}\left(\frac{\omega}{\omega_0} - \frac{\omega_0}{\omega}\right) \tag{13}$$

where

$$\omega_0 = \sqrt{\omega_1 \omega_2} \tag{14}$$

$$w = \frac{\omega_2 - \omega_1}{\omega_0} \quad . \tag{15}$$

For $\omega_2/\omega_1 > 1.05$ see text for a suitable mapping and definition of w and ω_0

FIG. 8.11-1 Concluded

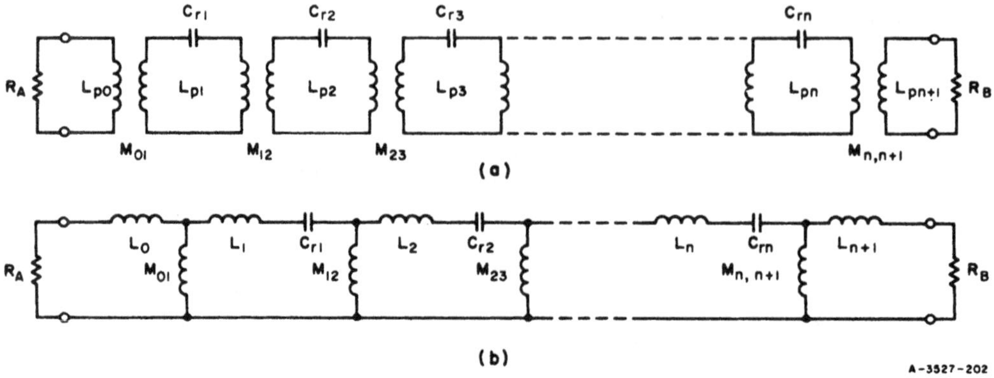

For definitions of the g_j, ω_1', ω_0, ω_1, ω_2, and the $K_{j,j+1}$; see Figs. 4.04-1, 8.02-1(a),(b), and 8.03-1(a).

Choose values for R_A, R_B, L_{r0}, L_{r1}, ..., L_{rn}, L_{rn+1} where the L_{rj} are related to the L_{pj} as indicated in Eqs. (15) to (19) below.

$$C_{rj}\Big|_{j=1\text{ to }n} = \frac{1}{L_{rj}\omega_0^2} \tag{1}$$

$$K_{01} = \sqrt{\frac{R_A\omega_0 L_{r1} w}{g_0 g_1 \omega_1'}} \tag{2}$$

$$K_{j,j+1}\Big|_{j=1\text{ to }n-1} = \frac{w\,\omega_0}{\omega_1'}\sqrt{\frac{L_{rj}L_{rj+1}}{g_j g_{j+1}}} \tag{3}$$

$$K_{n,n+1} = \sqrt{\frac{R_B\omega_0 L_{rn} w}{g_n g_{n+1}\omega_1'}} \tag{4}$$

where w is as defined below.

The mutual couplings are:

$$M_{01} = \frac{K_{01}}{\omega_0}\sqrt{1 + \left(\frac{\omega_0 L_{r0}}{R_A}\right)^2} \tag{5}$$

$$M_{j,j+1}\Big|_{j=1\text{ to }n-1} = \frac{K_{j,j+1}}{\omega_0} \tag{6}$$

$$M_{n,n+1} = \frac{K_{n,n+1}}{\omega_0}\sqrt{1 + \left(\frac{\omega_0 L_{rn+1}}{R_B}\right)^2} \quad . \tag{7}$$

The series inductances drawn at (b) above are

$$L_0 = L_{r0} - M_{01} \tag{8}$$

FIG. 8.11-2 DESIGN FORMULAS FOR INDUCTIVELY COUPLED, LUMPED-ELEMENT FILTERS

$$L_1 = L_{r1} - M^e_{01} - M_{12} \tag{9}$$

$$L_j\Big|_{j=2\text{ to }n-1} = L_{rj} - M_{j-1,j} - M_{j,j+1} \tag{10}$$

$$L_n = L_{rn} - M_{n-1,n} - M^e_{n,n+1} \tag{11}$$

$$L_{n+1} = L_{rn+1} - M_{n,n+1} \tag{12}$$

where

$$M^e_{01} = \frac{M_{01} + \dfrac{(L_{r0} - M_{01})\omega_0^2 M_{01} L_{p0}}{R_A^2}}{1 + \left(\dfrac{\omega_0 L_{r0}}{R_A}\right)^2} \tag{13}$$

$$M^e_{n,n+1} = \frac{M_{n,n+1} + \dfrac{(L_{rn+1} - M_{n,n+1})\omega_0^2 M_{n,n+1} L_{pn+1}}{R_B^2}}{1 + \left(\dfrac{\omega_0 L_{rn+1}}{R_B}\right)^2} \quad . \tag{14}$$

For form shown at (a) above, the L_{pj} are the total loop inductances and

$$L_{p0} = L_{r0} \tag{15}$$

$$L_{p1} = L_{r1} + M_{01} - M^e_{01} \tag{16}$$

$$L_{pj}\Big|_{j=2\text{ to }n-1} = L_{rj} \tag{17}$$

$$L_{pn} = L_{rn} + M_{n,n+1} - M^e_{n,n+1} \tag{18}$$

$$L_{pn+1} = L_{rn+1} \tag{19}$$

For mapping low-pass response approximately to band-pass response, if $\omega_2/\omega_1 \leqq 1.05$ use

$$\omega' = \frac{\omega_1'}{w}\left(\frac{\omega}{\omega_0} - \frac{\omega_0}{\omega}\right) \tag{20}$$

where

$$\omega_0 = \sqrt{\omega_1 \omega_2} \tag{21}$$

$$w = \frac{\omega_2 - \omega_1}{\omega_0} \tag{22}$$

For $\omega_2/\omega_1 > 1.05$, see text for a suitable mapping and definition of w and ω_0.

FIG. 8.11-2 Concluded

gives good results to bandwidths around 20 percent. A definition* of w for use in such cases is[1]

$$w = \left(\frac{\omega_0}{\omega_1} - \frac{\omega_0}{\omega_2}\right) \quad . \tag{8.11-3}$$

SEC. 8.12, BAND-PASS FILTERS WITH WIDE STOP BANDS[16]

All of the filter structures discussed so far that involve transmission lines tend to have additional pass bands at frequencies which are multiples of their first pass-band frequencies, or at least at frequencies which are odd multiples of their pass-band frequency. Figure 8.12-1 shows a filter structure which when properly designed can be made to be free of higher-order pass bands up to quite high frequencies. The shunt capacitances C_j^t in Fig. 8.12-1 are not necessary to the operation of the device, but are stray capacitances that will usually be associated with the coupling capacitances $C_{j,j+1}$. At the pass-band center frequency of the filter, each resonator line is somewhat less than a quarter-wavelength long, as measured from its short-circuited end to its open-circuited end. (They would all be exactly a quarter-wavelength long, if it were not for the capacitive loading due to the C_j^t and the $C_{j,j+1}$.) As seen from the connection points at which the resonator lines are attached, at midband the short-circuited portion of each line looks like a shunt inductance, while the open-circuited portion looks like a shunt capacitance, so the circuit is very similar to that in Fig. 8.11-1.

The circuit in Fig. 8.12-1 will tend to have additional pass bands when the length of the transmissions line resonators is roughly an odd multiple of a quarter-wavelength long. However, it can be seen that such pass bands can be suppressed if, when a line is resonant, the length from the short-circuited end of the line to the connection point is exactly one-half wavelength or a multiple thereof, while the electrical distance from the open-circuited end to the connection point is exactly an odd multiple of one-quarter wavelength. Under these conditions the connection point of such a resonator is at a voltage null, and the resonance looks like a series resonance which short-circuits the signal to ground, instead

---

* The definition of w used here differs from the w' that Cohn[1] uses for this case, by a factor of ω_0/ω_1'. This fact is consistent with the equations used herein and gives the same end result. The w defined here is fractional bandwidth, while Cohn's w' is not.

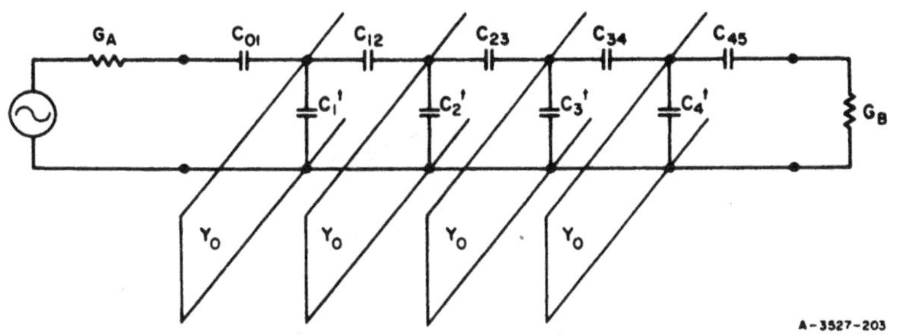

For definitions of the g_j, ω_1', ω_0, ω_1, ω_2, and the $J_{j,j+1}$; see Figs. 4.04-1, 8.02-1(a),(b), 8.02-4, and 8.03-2(b).

Choose values for G_A, G_B, and Y_0 and estimate:

$$B_1^J = \left(\frac{\omega_0 C_{01}}{1+\left(\frac{\omega_0 C_{01}}{G_A}\right)^2} + \omega_0 C_1^t + \omega_0 C_{12}\right) \tag{1}$$

$$B_j^J\Big|_{j=2 \text{ to } n-1} = \left(\omega_0 C_{j-1,j} + \omega_0 C_j^t + \omega_0 C_{j,j+1}\right) \tag{2}$$

$$B_n^J = \left(\omega_0 C_{n-1,n} + \omega_0 C_n^t + \frac{\omega_0 C_{n,n+1}}{1+\left(\frac{\omega_0 C_{n,n+1}}{G_B}\right)^2}\right) \ . \tag{3}$$

Obtain slope parameters $\mathscr{b}_j$ from the B_j^J, Y_0, and Fig. 8.12-2 or Fig. 8.12-3 or Eq. (8.12-4).

$$J_{01} = \sqrt{\frac{G_A \mathscr{b}_1 w}{g_0 g_1 \omega_1'}} \tag{4}$$

$$J_{j,j+1}\Big|_{j=1 \text{ to } n-1} = \frac{w}{\omega_1'}\sqrt{\frac{\mathscr{b}_j \mathscr{b}_{j+1}}{g_j g_{j+1}}} \tag{5}$$

$$J_{n,n+1} = \sqrt{\frac{G_B \mathscr{b}_n w}{g_n g_{n+1} \omega_1'}} \tag{6}$$

(continued on p. 488)

FIG. 8.12-1 DATA FOR BAND-PASS FILTERS WITH WIDE STOP BANDS

where w is given by (11) below.

$$\omega_0 C_{01} = \frac{J_{01}}{\sqrt{1 - \left(\frac{J_{01}}{G_A}\right)^2}} \tag{7}$$

$$\omega_0 C_{j,j+1}\Big|_{j=1 \text{ to } n-1} = J_{j,j+1} \tag{8}$$

$$\omega_0 C_{n,n+1} = \frac{J_{n,n+1}}{\sqrt{1 - \left(\frac{J_{n,n+1}}{G_B}\right)^2}} \quad . \tag{9}$$

For mapping low-pass prototype response approximately to band-pass response use

$$\omega' = \frac{2\omega_1'}{w}\left(\frac{\omega - \omega_0}{\omega}\right) \tag{10}$$

where

$$w = 2\left(\frac{\omega_2 - \omega_1}{\omega_2 + \omega_1}\right) \tag{11}$$

and

$$\omega_0 = \frac{2\omega_2\omega_1}{\omega_2 + \omega_1} \tag{12}$$

FIG. 8.12-1 Concluded

of a shunt resonance which passes the signal. Since for this higher resonance the connection point has zero voltage, the C_j^t and the $C_{j,j+1}$ have no effect on the higher resonant frequency. By designing the various resonators to suppress different pass bands, it should be possible to make the stop band extend very far without any spurious pass bands.

The B_j^J in Eqs. (1) to (3) in Fig. 8.12-1 are susceptances which account for effects of the C_j^t and $C_{j,j+1}$ on the tuning of the resonators and on their susceptance slope parameters at the midband frequency ω_0. The total susceptance of the jth resonator is then

$$B_j(\omega) = Y_0 \tan\left(\frac{\theta_{aj}\omega}{\omega_0}\right) - Y_0 \cot\left(\frac{\theta_{bj}\omega}{\omega_0}\right) + \frac{\omega}{\omega_0} B_j^J \qquad (8.12\text{-}1)$$

where Y_0 is the characteristic admittance of the resonator line, θ_{aj} is the electrical length of the open-circuited portion of the resonator line at frequency ω_0, and θ_{bj} is the electrical length of the short-circuited portion at the same frequency. At frequency ω_0 we require that $B_j(\omega_0) = 0$ which calls for

$$\frac{B_j^J}{Y_0} = \cot \theta_{bj} - \tan \theta_{aj} \quad . \qquad (8.12\text{-}2)$$

In order to short-circuit pass bands at $3\omega_0$ or $5\omega_0$, etc., it is only necessary that $\theta_{aj} = \theta_{bj}/2$, or $\theta_{aj} = \theta_{bj}/4$, etc., respectively, as previously discussed. Having related θ_{aj} and θ_{bj}, one may solve Eq. (8.12-2) for the total electrical length required at frequency ω_0 in order to give resonance in the presence of the susceptance B_j^J. If l_j is the resonator length, then

$$\frac{l_j}{\lambda_0/4} = \frac{\theta_{aj} + \theta_{bj}}{\pi/2} \qquad (8.12\text{-}3)$$

where λ_0 is the wavelength in the medium of propagation at the frequency ω_0. Applying Eq. (1) of Fig. 8.02-4 to Eq. (8.12-1) gives, for the susceptance slope parameter $\mathscr{b}_j$ normalized with respect to Y_0,

$$\frac{\mathscr{b}_j}{Y_0} = \frac{1}{2}\left[\frac{\theta_{aj}}{\cos^2 \theta_{aj}} + \frac{\theta_{bj}}{\sin^2 \theta_{bj}} + \frac{B_j^J}{Y_0}\right] \quad . \tag{8.12-4}$$

Figure 8.12-2 shows a plot of $l_j/(\lambda_0/4)$ and $\mathscr{b}_j/Y_0$ *vs.* B_j^J/Y_0 for resonators which are to suppress transmission at the $3\omega_0$ pass band. Figure 8.12-3 shows corresponding data for resonators designed to suppress the pass band in the vicinity of $5\omega_0$.

When using the design data in Figs. 8.12-1 to 8.12-3, some iteration in the design calculations will be necessary if high accuracy is desired.

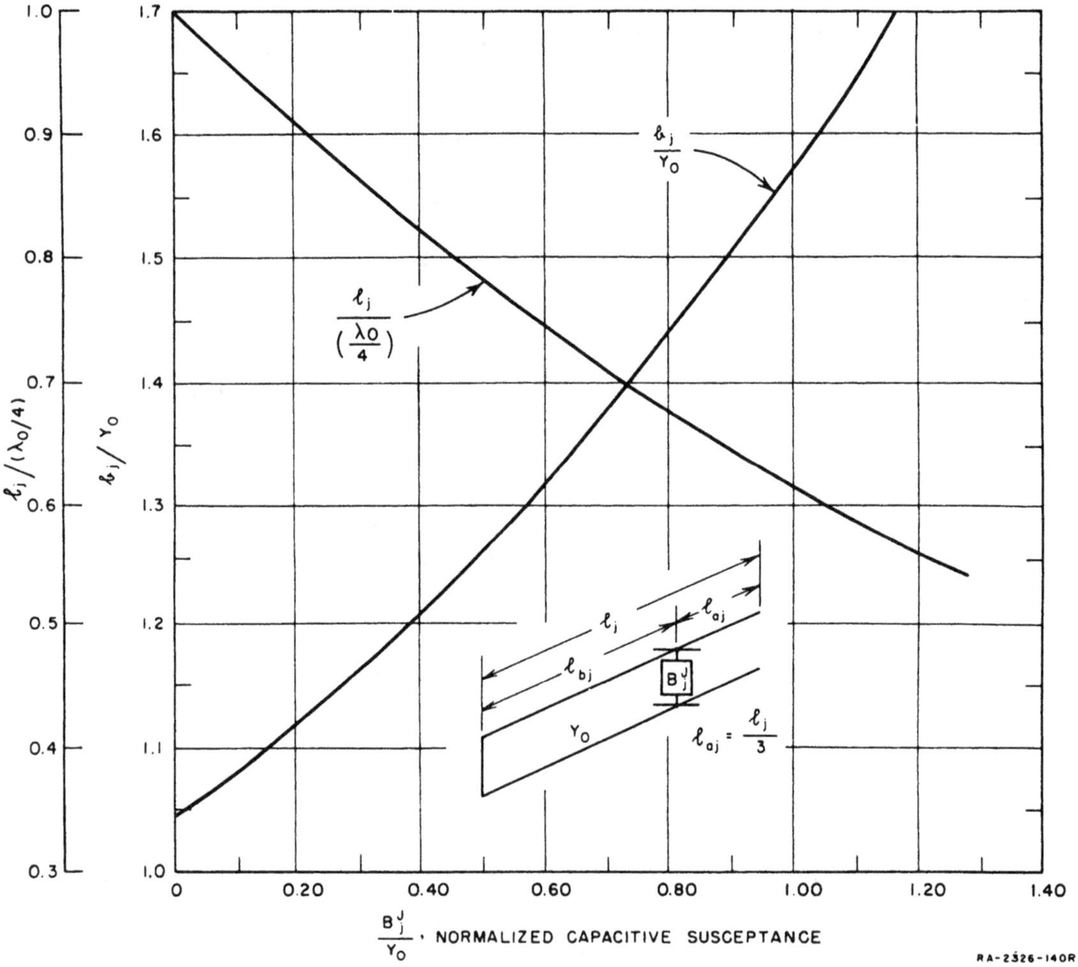

FIG. 8.12-2 CHART FOR DESIGN OF RESONATORS TO SUPPRESS THE SPURIOUS PASS BAND IN THE VICINITY OF $3\omega_0$

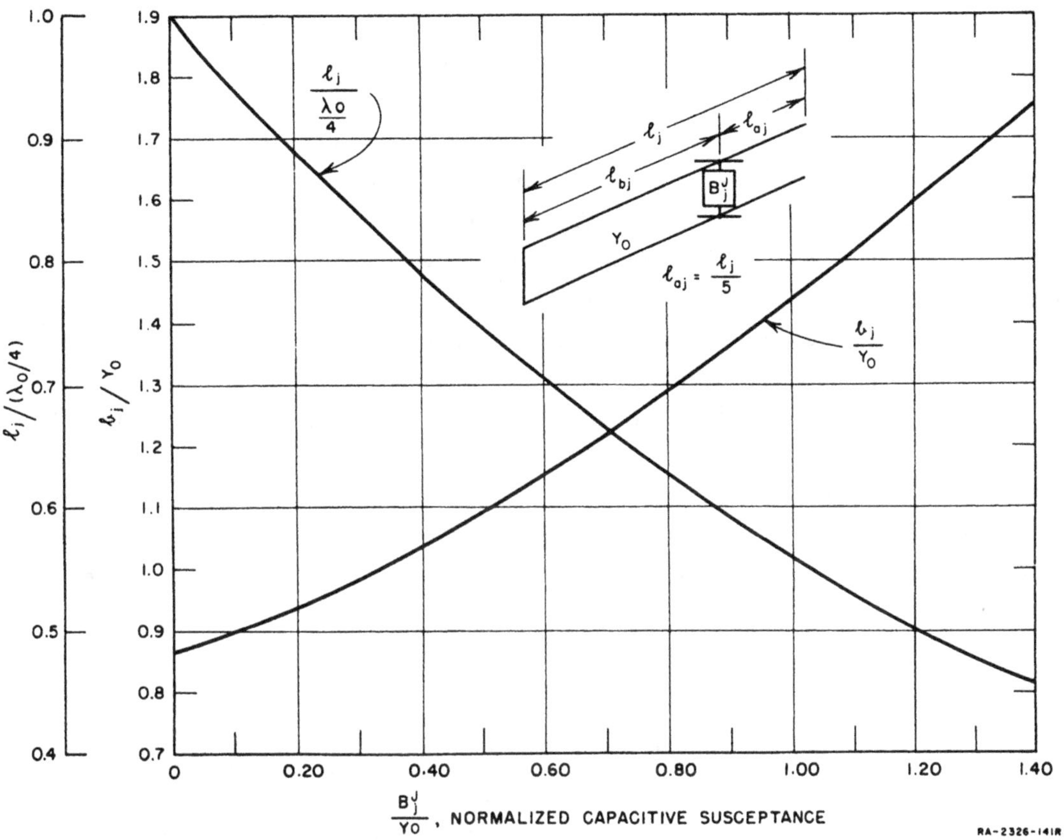

FIG. 8.12-3 CHART FOR DESIGN OF RESONATORS TO SUPPRESS THE SPURIOUS PASS BAND IN THE VICINITY OF $5\omega_0$

This is because the B_j^J must be known in order to compute the coupling capacitances $C_{j,j+1}$ (and usually the C_j^t) accurately, while in turn the $C_{j,j+1}$ and C_j^t must be known in order to determine the B_j^J accurately. However, since the B_j^J generally have a relatively minor influence on the coupling capacitance values $C_{j,j+1}$ required, the calculations converge quickly and are not difficult. First the B_j^J are estimated and corresponding values of the $C_{j,j+1}$ and C_j^t are obtained. Then improved values for the B_j^J are computed, and from them improved values for the $C_{j,j+1}$ and $l_j/(\lambda_0/4)$ are obtained. These latter values should be sufficiently accurate.

Figure 8.12-4 shows a possible form of construction for the filters under consideration. The resonators are in 50-ohm (Y_0 = 0.020 mhos) rectangular-bar strip-transmission-line form, with small coupling tabs between the resonator bars. The spacing between resonators has been shown to give adequate isolation between resonators as evidenced by tests on trial, two-resonator and four-resonator designs.[16] Figure 8.12-5(a) shows a plot of estimated coupling capacitance $C_{j,j+1}$ *vs.* gap spacing y for various amounts of coupling tab overlap x. The similar data in Fig. 8.12-5(b) are for the shunt capacitance to ground $C^s_{j,j+1}$ of an individual tab in the $j,j+1$th coupling. Using the data in Fig. 8.12-5(b), the junction capacitance C^t_j for the jth junction is

$$C^t_j = C^s_{j-1,j} + C^s_{j,j+1} + C^+ \qquad (8.12\text{-}5)$$

where C^+ introduces an additional junction shunt susceptance like that for the T-junctions in Sec. 5.07. Calculations from measurements on the two-resonator filter mentioned above suggest that C^+ should be taken as

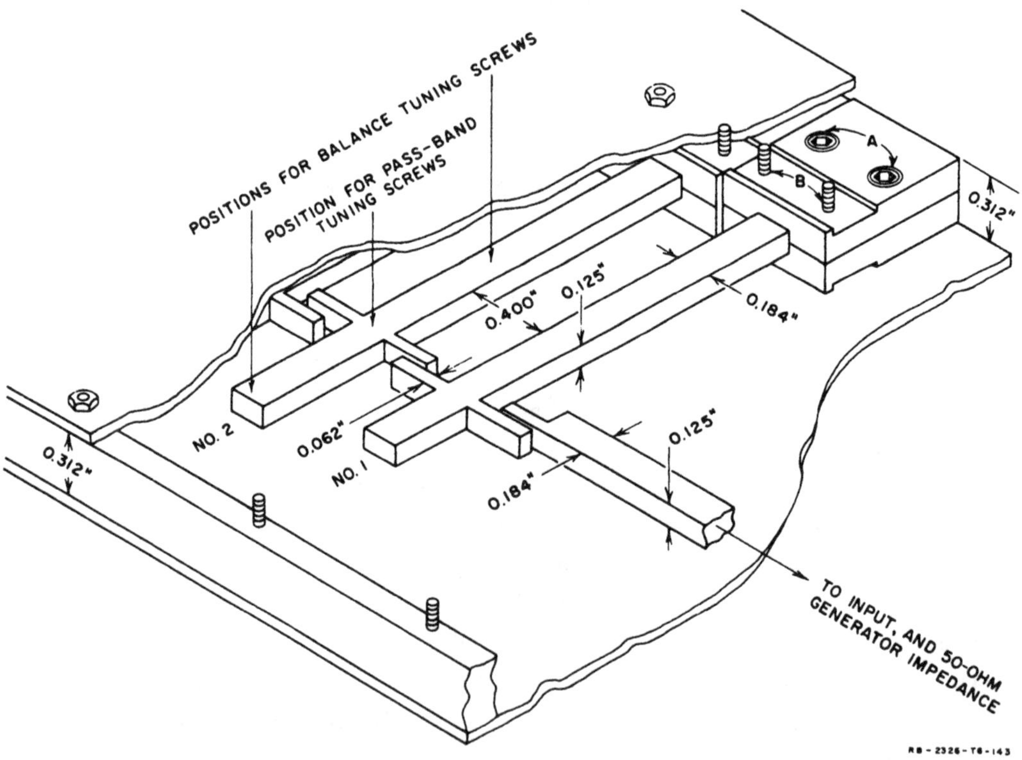

FIG. 8.12-4 A BAR TRANSMISSION LINE CONSTRUCTION FOR THE FILTER IN FIG. 8.12-1

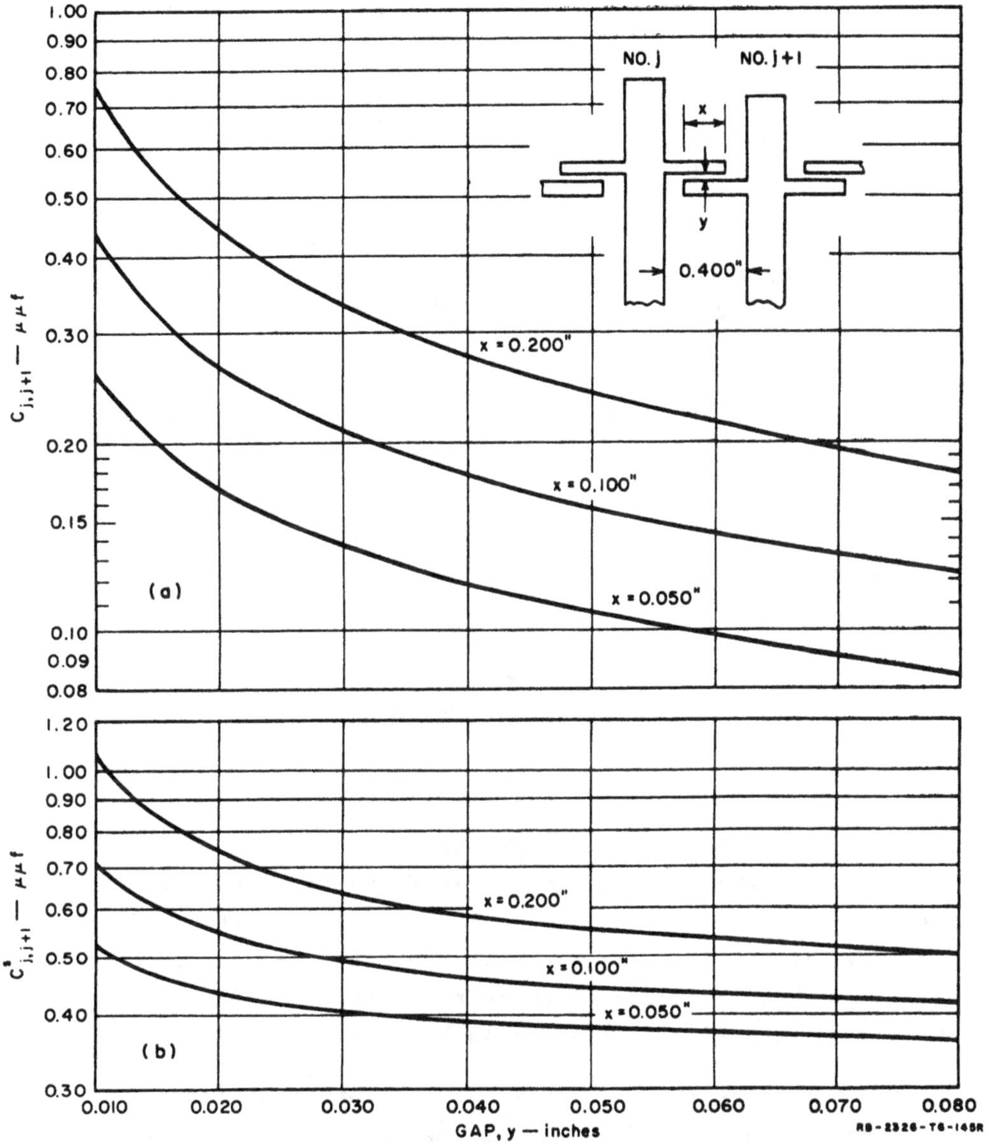

FIG. 8.12-5 CHARTS OF ESTIMATED VALUES OF THE CAPACITANCES ASSOCIATED WITH THE COUPLINGS FOR THE CONSTRUCTION IN FIG. 8.12-4

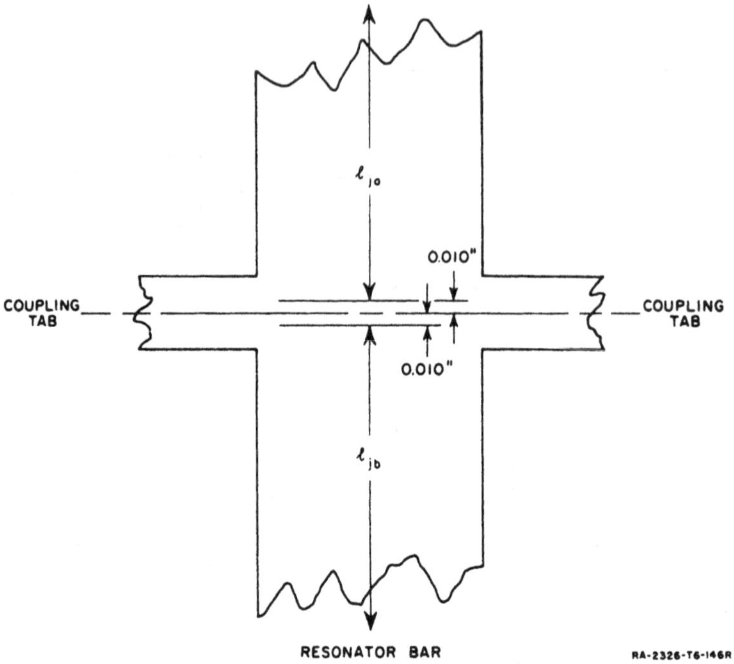

FIG. 8.12-6 DEFINITION OF THE JUNCTION REFERENCES PLANES FOR THE CONSTRUCTION IN FIG. 8.12-4

about −0.10 $\mu\mu$f.* Approximate reference planes for fixing the lengths of the open- and short-circuited sides of the resonator are shown in Fig. 8.12-6. In fixing the length of the open-circuited end, allowance must be made for the fringing capacitance from the end of the bar. It is estimated that, in order to correct for this capacitance, the length l_{aj} (see Figs. 8.12-2, -3, -6) should be reduced by about 0.055 inch.

The two-resonator filter built in the construction in Fig. 8.12-4 was intended to suppress the $3\omega_0$ pass band, but at first did not do so. The reason was that the open- and short-circuited sides of the resonators did not reflect short-circuits to the connection points at exactly the same frequencies, as they must for high attenuation. To correct this, "balance" tuning screws were added at two points on each resonator indicated by the arrows in Fig. 8.12-4. In addition, pass-band tuning screws were placed directly over the coupling-tab junction of each resonator.

* The negative sign merely indicates that with the *junction reference planes being used*, some capacitance must be subtracted in order to represent the junction.

The balance screws were adjusted first to give high attenuation in the vicinity of $3\omega_0$ and then the pass-band tuning screws were adjusted using the procedure discussed in Sec. 11.05. Since the pass-band tuning screws are at a voltage null point for the resonance in the vicinity of $3\omega_0$, the adjustment of the pass-band tuning screws will not affect the balance tuning adjustment of the resonators. However, it should be noted that the balance adjustment must be made before the pass-band tuning adjustment since the setting of the balance tuning screws will affect the pass-band tuning.

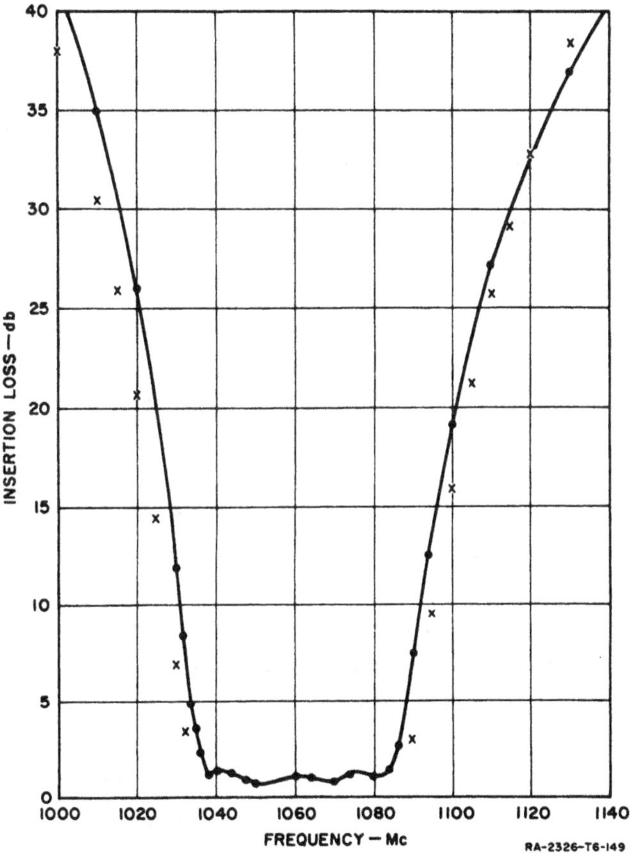

FIG. 8.12-7 THE MEASURED RESPONSE OF A FOUR-RESONATOR FILTER OF THE FORM IN FIG. 8.12-4
The solid line is the measured response while the x's represent attenuation values mapped from the low-pass prototype using Eq. (10) in Fig. 8.12-1

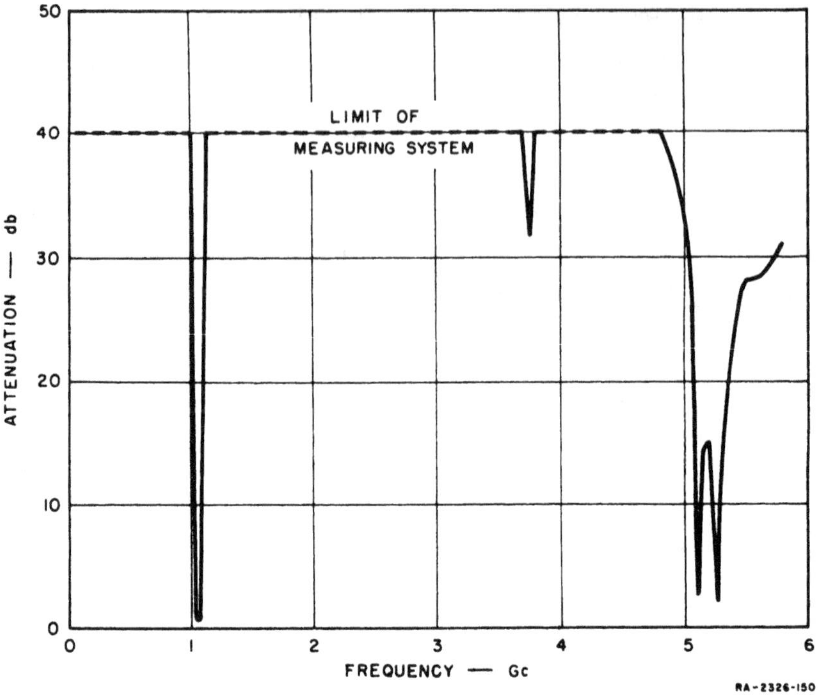

FIG. 8.12-8 THE STOP-BAND RESPONSE OF A FOUR-RESONATOR FILTER OF THE FORM IN FIG. 8.12-4

Figures 8.12-7 and 8.12-8 show the measured response of a four-resonator filter constructed in the form in Fig. 8.12-4 using the design data discussed above. As can be seen from Fig. 8.12-7, the bandwidth is about 10 percent narrower than called for by the points mapped from the low-pass prototype (which are indicated by x's). This is probably due largely to error in the estimated coupling capacitances in Fig. 8.12-5. If desired, this possible source of error can be compensated for by using values of w which are 10 percent larger than actually required. The approximate mapping used is seen to be less accurate on the high side of the response in Fig. 8.12-7 than on the low side for this type of filter.

The four-resonator filter discussed above was designed using one pair of resonators to suppress the $3\omega_0$ resonance and a second pair to suppress the $5\omega_0$ resonance. Since the two sets of resonators had their higher resonances at somewhat different frequencies it was hoped that balance tuning would be unnecessary. This was practically true for the $3\omega_0$ resonance since high attenuation was attained without balance tuning of the resonators intended to suppress that resonance. However, there was a small dip in attenuation at about 3.8 kMc (see Fig. 8.12-8) which probably could easily have been removed by balance tuning.

The pass band near $5\omega_0$ would not disappear in this case no matter how the balance screws were adjusted on the resonators meant to suppress that pass band. Some experimentation with the device suggested that this was due to a resonance effect in the coupling tabs, which was greatly aggravated by the fact that the resonators involved were the end resonators (which have relatively large coupling capacitances). This difficulty can probably be avoided by putting the resonators to suppress pass bands near $5\omega_0$ or higher in the interior of the filter and putting the resonators to suppress the pass band near $3\omega_0$ at the ends of the filter. Also, keeping the coupling tabs as short as possible should help.

SEC. 8.13, COMB-LINE, BAND-PASS FILTERS

Figure 8.13-1(a) shows a comb-line band-pass filter in strip-line form and Fig. 8.13-1(b) presents design equations for this type of filter. The resonators consist of line elements which are short-circuited at one end, with a lumped capacitance C_j^s between the other end of each resonator line element and ground. In Fig. 8.13-1(a) Lines 1 to n, along with their associated lumped capacitances C_1^s to C_n^s comprise resonators, while Lines 0 and $n+1$ are not resonators but simply part of impedance-transforming

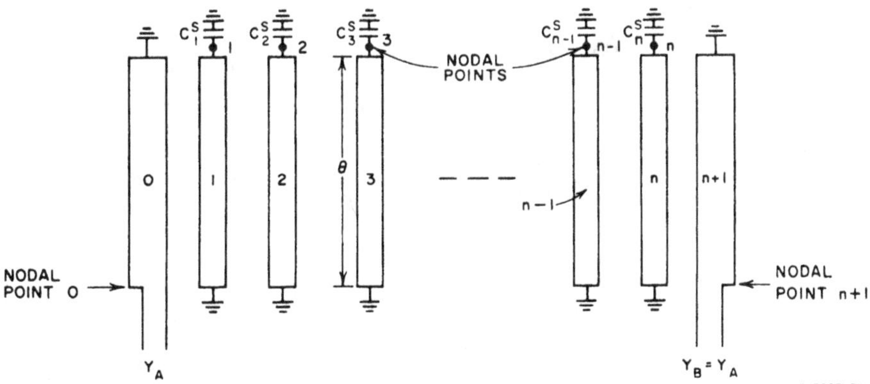

SOURCE: Quarterly Progress Report 5, Contract DA 36-039 SC-87398, SRI; reprinted in *The Microwave Journal* (see Ref. 18 by G. L. Matthaei)

FIG. 8.13-1(a) A COMB-LINE, BAND-PASS FILTER
The nodal points are defined for use in the design equation derivations discussed in Sec. 8.14

Choose the normalized characteristic admittances Y_{aj}/Y_A so as to give good resonator unloaded Q's. (See text.) Then compute:

$$\left.\frac{\mathscr{b}_j}{Y_A}\right|_{j=1\text{ to }n} = \frac{Y_{aj}}{Y_A}\left(\frac{\cot\theta_0 + \theta_0\csc^2\theta_0}{2}\right) \qquad (1)$$

where θ_0 is the electrical length of the resonator elements at the mid-band frequency ω_0.

Compute:

$$\frac{G_{T1}}{Y_A} = \frac{w\dfrac{\mathscr{b}_1}{Y_A}}{g_0 g_1 \omega_1'} \qquad (2)$$

$$\left.\frac{J_{j,j+1}}{Y_A}\right|_{j=1\text{ to }n-1} = \frac{w}{\omega_1'}\sqrt{\frac{(\mathscr{b}_j/Y_A)(\mathscr{b}_{j+1}/Y_A)}{g_j g_{j+1}}} \qquad (3)$$

$$\frac{G_{Tn}}{Y_A} = \frac{w\left(\dfrac{\mathscr{b}_n}{Y_A}\right)}{g_n g_{n+1}\omega_1'} \qquad (4)$$

where w is the fractional bandwidth defined below.

The normalized capacitances per unit length between each line and ground are

$$\frac{C_0}{\epsilon} = \frac{376.7\, Y_A}{\sqrt{\epsilon_r}}\left(1 - \sqrt{\frac{G_{T1}}{Y_A}}\right)$$

$$\frac{C_1}{\epsilon} = \frac{376.7\, Y_A}{\sqrt{\epsilon_r}}\left(\frac{Y_{a1}}{Y_A} - 1 + \frac{G_{T1}}{Y_A} - \frac{J_{12}}{Y_A}\tan\theta_0\right) + \frac{C_0}{\epsilon}$$

$$\left.\frac{C_j}{\epsilon}\right|_{j=2\text{ to }n-1} = \frac{376.7\, Y_A}{\sqrt{\epsilon_r}}\left(\frac{Y_{aj}}{Y_A} - \frac{J_{j-1,j}}{Y_A}\tan\theta_0 - \frac{J_{j,j+1}}{Y_A}\tan\theta_0\right) \qquad (5)$$

$$\frac{C_n}{\epsilon} = \frac{376.7\, Y_A}{\sqrt{\epsilon_r}}\left(\frac{Y_{an}}{Y_A} - 1 + \frac{G_{Tn}}{Y_A} - \frac{J_{n-1,n}}{Y_A}\tan\theta_0\right) + \frac{C_{n+1}}{\epsilon}$$

$$\frac{C_{n+1}}{\epsilon} = \frac{376.7\, Y_A}{\sqrt{\epsilon_r}}\left(1 - \sqrt{\frac{G_{Tn}}{Y_A}}\right)$$

FIG. 8.13-1(b) DESIGN EQUATIONS FOR COMB-LINE FILTERS

where ϵ is the absolute dielectric constant of the medium of propagation, and ϵ_r is the relative dielectric constant.

The normalized mutual capacitances per unit length between adjacent lines are:

$$\frac{C_{01}}{\epsilon} = \frac{376.7\ Y_A}{\sqrt{\epsilon_r}} - \frac{C_0}{\epsilon}$$

$$\left.\frac{C_{j,j+1}}{\epsilon}\right|_{j=1 \text{ to } n-1} = \frac{376.7\ Y_A}{\sqrt{\epsilon_r}} \left(\frac{J_{j,j+1}}{Y_A} \tan\theta_0\right) \tag{6}$$

$$\frac{C_{n,n+1}}{\epsilon} = \frac{376.7\ Y_A}{\sqrt{\epsilon_r}} - \frac{C_{n+1}}{\epsilon} \quad .$$

The lumped capacitances C_j^s are:

$$\left. C_j^s \right|_{j=1 \text{ to } n} = Y_A \left(\frac{Y_{aj}}{Y_A}\right) \frac{\cot\theta_0}{\omega_0} \quad . \tag{7}$$

A suggested low-pass to band-pass transformation is

$$\frac{\omega'}{\omega_1'} = \frac{2}{w}\left(\frac{\omega - \omega_0}{\omega_0}\right) \tag{8}$$

where

$$w = \frac{\omega_2 - \omega_1}{\omega_0} \tag{9}$$

and

$$\omega_0 = \frac{\omega_2 + \omega_1}{2} \quad . \quad . \tag{10}$$

SOURCE: Quarterly Progress Report 5, Contract DA 36-039 SC-87398, SRI; reprinted in *The Microwave Journal* (see Ref. 18 by G. L. Matthaei)

FIG. 8.13-1(b) Concluded

sections at the ends. Coupling between resonators is achieved in this type of filter by way of the fringing fields between resonator lines. With the lumped capacitors C_j^s present, the resonator lines will be less than $\lambda_0/4$ long at resonance (where λ_0 is the wavelength in the medium of propagation at midband), and the coupling between resonators is predominantly magnetic in nature. Interestingly enough, if the capacitors C_j^s were not present, the resonator lines would be a full $\lambda_0/4$ long at resonance, *and the structure would have no pass band!*[17] This is so because, without some kind of reactive loading at the ends of the resonator line elements, the magnetic and electric coupling effects cancel each other out, and the comb-line structure becomes an all-stop structure.*

For the reasons described above, it is usually desirable to make the capacitances C_j^s in this type of filter sufficiently large that the resonator lines will be $\lambda_0/8$ or less, long at resonance. Besides having efficient coupling between resonators (with sizeable spacings between adjacent resonator lines), the resulting filter will be quite small. In this type of filter, the second pass band occurs when the resonator line elements are somewhat over a half-wavelength long, so if the resonator lines are $\lambda_0/8$ long at the primary pass band, the second pass band will be centered at somewhat over four times the frequency of the center of the first pass band. If the resonator line elements are made to be less than $\lambda_0/8$ long at the primary pass band, the second pass band will be even further removed. Thus, like the filter in Sec. 8.12, comb-line filters also lend themselves to achieving very broad stop bands above their primary pass bands.

Since the coupling between the resonators is distributed in nature, it is convenient to work out the design of the resonator lines in terms of their capacitance to ground C_j per unit length, and the mutual capacitances $C_{j,j+1}$ per unit length between neighboring lines j and $j+1$. These capacitances are illustrated in the cross-sectional view of the line elements shown in Fig. 8.13-2. Fringing capacitance effects beyond nearest neighbors will be neglected. Figure 8.13-2 also defines various dimensions for the case where the resonator lines are to be constructed in rectangular-bar strip line. Using the design formulas in Fig. 8.13-1(b), the distributed line capacitances will be computed in normalized form to

---

* However, if every other unloaded, $\lambda_0/4$ resonator were turned end for end so that the structure had open- and short-circuited ends alternating, the band-stop structure would become a band-pass structure. The resulting configuration is that of the interdigital filters discussed in Secs. 10.06 and 10.07.

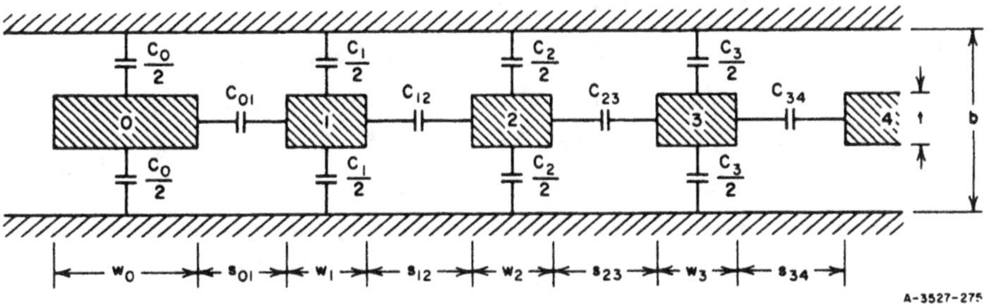

SOURCE: Quarterly Progress Report 5, Contract DA 36-039 SC-87398, SRI; reprinted in *The Microwave Journal* (see Ref. 18 by G. L. Matthaei)

FIG. 8.13-2 DEFINITIONS OF THE LINE CAPACITANCES AND SOME OF THE DIMENSIONS INVOLVED IN COMB-LINE FILTER DESIGN

give C_j/ϵ and $C_{j,j+1}/\epsilon$ values, where ϵ is the absolute dielectric constant of the medium of propagation. Then by use of the charts and formulas in Sec. 5.05 the corresponding rectangular-bar line dimensions w_j and $s_{j,j+1}$ in Fig. 8.13-2 can be determined for specified t and b.

To carry out the design of a comb-line filter by use of Fig. 8.13-1(b), the low-pass prototype filter parameters g_0, g_1, ..., g_{n+1} and ω_1' are selected in the usual manner (Secs. 8.02 and 8.04). The low-pass to band-pass mapping indicated in Eqs. (8) to (10) is a commonly used, simplified, narrow-band mapping, but unfortunately it is not outstandingly accurate for this type of filter when the bandwidth is as large as 10 percent or so. From the trial design described below, the largest error is seen to occur on the high side of the pass band where the narrow-band mapping does not predict as large a rate of cutoff as actually occurs. The reason that the actual rate of cutoff tends to be unusually large on the high-frequency side of the pass band is that the structure has infinite attenuation (theoretically) at the frequency for which the resonator lines are a quarter-wavelength long. Thus, the steepness of the attenuation characteristic on the high side will depend to some extent upon the choice of θ_0, the electrical length of the resonator lines at the pass-band center frequency. Although the simplified mapping in Eqs. (8) to (10) of Fig. 8.13-1(b) cannot account for these more subtle effects in the response of this type of filter, it is sufficiently accurate to serve as a useful guide in estimating the number of resonators required for a given application.

Next the terminating line admittance Y_A, the midband electrical length θ_0 of the resonator lines, the fractional bandwidth w, and the normalized line admittances Y_{aj}/Y_A must all be specified. As indicated above, it is usually desirable to make $\theta_0 = \pi/4$ radians or less. The choice of the resonator line admittances Y_{aj} fixes the admittance level within the filter, and this is important in that it influences the unloaded Q's that the resonators will have. At the time of this writing the line characteristic admittances to give optimum unloaded Q's for structures of this type have not been determined. However, choosing the Y_{aj} in Eq. (1) of Fig. 8.13-1(b) to correspond to about 0.0143 mho (*i.e.*, about 70 ohms), appears to be a reasonable choice. [The admittance Y_{aj} in Fig. 8.13-1(b) is interpreted physically as the admittance of Line j with the adjacent Lines $j-1$ and $j+1$ grounded.] The remainder of the calculations proceed in a straightforward manner as presented in the figure. As mentioned above, having the C_j/ϵ and $C_{j,j+1}/\epsilon$, the required line dimensions are obtained from the data in Sec. 5.05.

Table 8.13-1 summarizes various parameters used and computed in the design of a trial four-resonator, comb-line filter designed for a fractional bandwidth of $w = 0.10$, and 0.1-db Tchebyscheff ripple. Due to a misprint in the table of prototype-filter element values which were used for the design of this filter, the g_1 element value is, unfortunately, off by about 10 percent. However, a computed response for this filter revealed that this error should not have any sizeable effect on the shape of the response. In this design $\theta_0 = \pi/4$ radians so that the resonator lines are $\lambda_0/8$ long at the midband frequency, which was to be 1.5 Gc.

Table 8.13-1

VARIOUS PARAMETERS WHICH WERE SPECIFIED OR COMPUTED IN THE DESIGN OF THE TRIAL, FOUR-RESONATOR, COMB-LINE FILTER

j	$\frac{J_{j,j+1}}{Y_A}$	$\frac{C_{j,j+1}}{\epsilon}$	$s_{j,j+1}$ (inches)	j	$\frac{C_j}{\epsilon}$	w_j (inches)
0 and 4		2.130	0.116	0 and 5	5.404	0.362
1 and 3	0.0730	0.550	0.337	1 and 4	3.022	0.152
2	0.0572	0.431	0.381	2 and 3	4.119	0.190

$g_0 = 1$	$g_3 = 1.7703$	$w = 0.10$	$\theta_0 = \pi/4$ radian
$g_1 = 1.0880$*	$g_4 = 0.8180$	$Y_A = 0.020$ mho	$b = 0.625$ inch
$g_2 = 1.3061$	$g_5 = 1.3554$	$\mathscr{b}_j/Y_A = 0.870$ } $j=1$ to 4	$t = 0.188$ inch
	$\omega_1' = 1$	$Y_{aj}/Y_A = 0.677$ } $j=1$ to 4	

* This value should have been $g_1 = 1.1088$ for a true 0.1-db ripple prototype.

SOURCE: Quarterly Progress Report 5, Contract DA 36-039 SC-87398, SRI; reprinted in *The Microwave Journal* (see Ref. 18 by G. L. Matthaei)

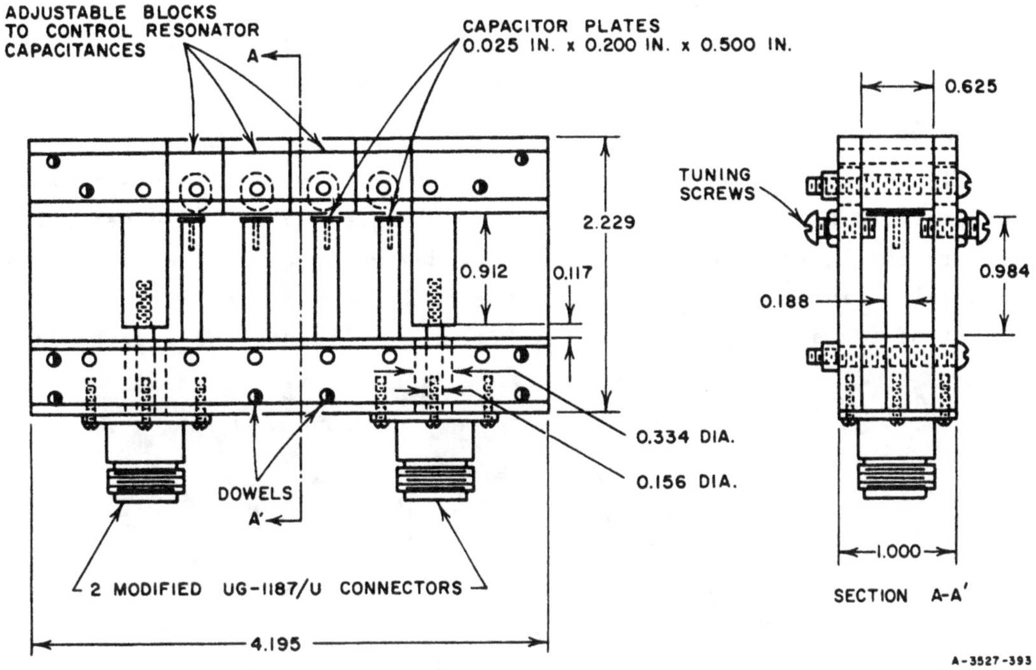

SOURCE: Quarterly Progress Report 5, Contract DA 36-039 SC-87398, SRI; reprinted in *The Microwave Journal* (see Ref. 18 by G. L. Matthaei)

FIG. 8.13-3 DRAWING OF THE TRIAL, FOUR-RESONATOR, COMB-LINE FILTER
Additional dimensions of electrical importance are given in Table 8.13-1

Note that $Y_{aj}/Y_A = 0.677$ which with $Y_A = 0.020$ mho makes $Y_{aj} = 0.0135$ mho, or $1/Y_{aj} = 74$ ohms. The electrically important dimensions of this filter are summarized in Table 8.13-1 along with Figs. 8.13-2 and 3. Figure 8.13-4 shows the completed filter with its cover plate removed.

The filter was tuned using a slotted line and the alternating short-circuit and open-circuit procedure described in Sec. 11.05. To adjust the capacitance of an individual resonator, first its sliding block (shown in Fig. 8.13-3) was adjusted to give slightly less than the required resonator capacitance, and then the tuning screws on the resonator were used to bring the resonator to the exact desired frequency. In this case the bandwidth was sufficiently large so that the alternating short-circuit and open-circuit procedure did not give entirely satisfactory results as evidenced by some lack of symmetry in the pass-band response. However, it was found that this could be easily corrected by readjusting the tuning screws on the end resonators* while using a sweep-generator and recording-reflectometer

* Since the end resonators have adjacent couplings which are quite different from those of the interior resonators, it is usually the end resonators that cause tuning difficulties when using the alternating short-circuit and open-circuit procedure.

set-up. After the tuning was completed, the measured input VSWR was as shown in Fig. 8.13-5 and the measured attenuation as shown in Fig. 8.13-6.

The VSWR characteristic in Fig. 8.13-5 corresponds to roughly a 0.2-db Tchebyscheff ripple rather than a 0.1-db ripple. The discrepancy is believed to be due to the fact that coupling effects beyond nearest-neighbor lines have been neglected in the design procedure in Fig. 8.13-1. If a smaller ripple were necessary, this could be achieved by small adjustment of the spacings s_{01} and s_{45} between the input line and the first

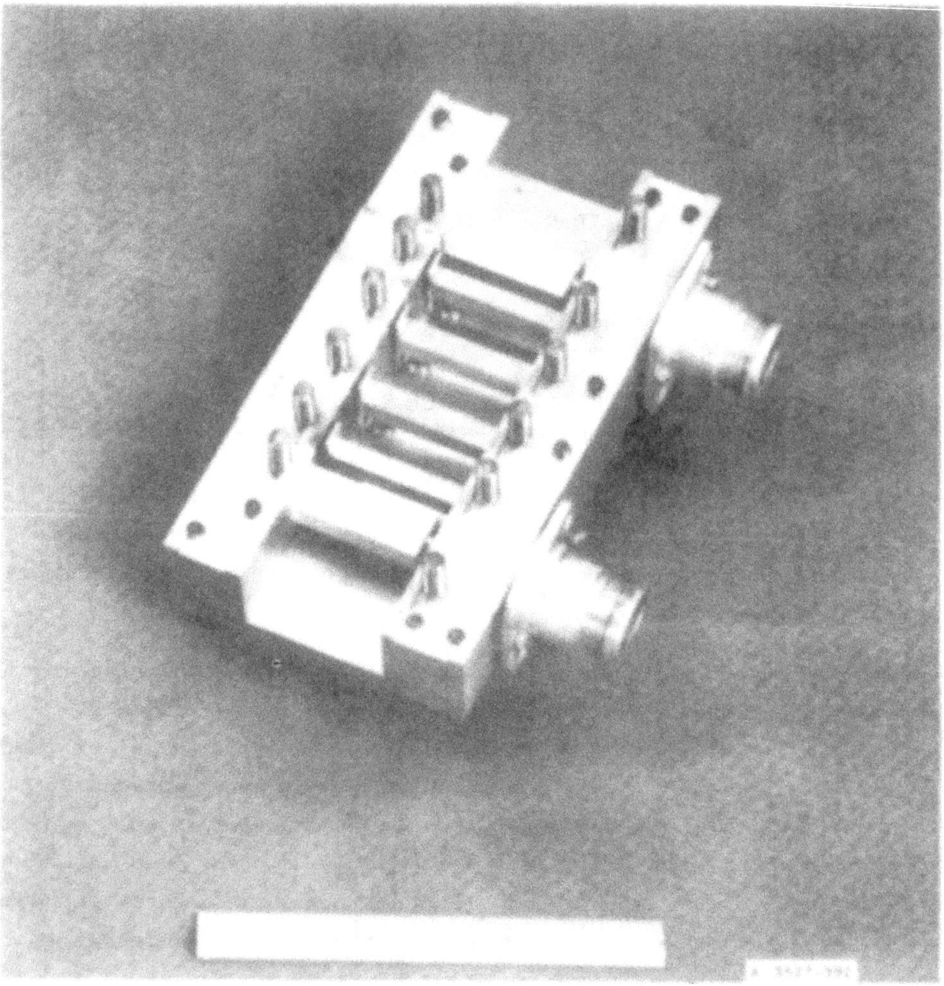

SOURCE: Quarterly Progress Report 5, Contract DA 36-039 SC-87398, SRI; reprinted in *The Microwave Journal* (see Ref. 18 by G. L. Matthaei)

FIG. 8.13-4 A FOUR-RESONATOR COMB-LINE FILTER WITH ITS COVER PLATE REMOVED

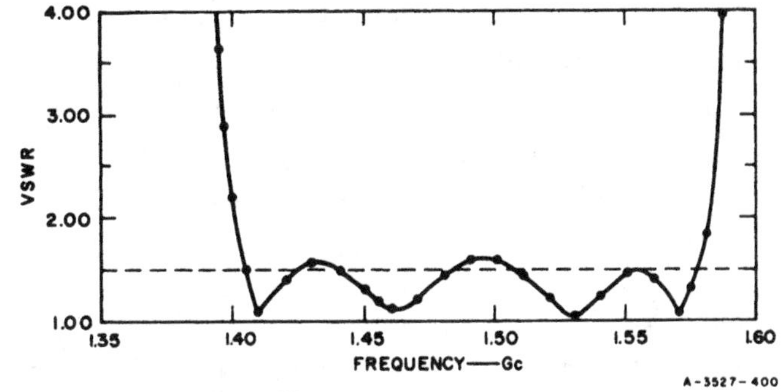

SOURCE: Quarterly Progress Report 5, Contract DA 36-039 SC-87398, SRI; reprinted in *The Microwave Journal* (see Ref. 18 by G. L. Matthaei)

FIG. 8.13-5 MEASURED VSWR OF THE FILTER IN FIG. 8.13-4

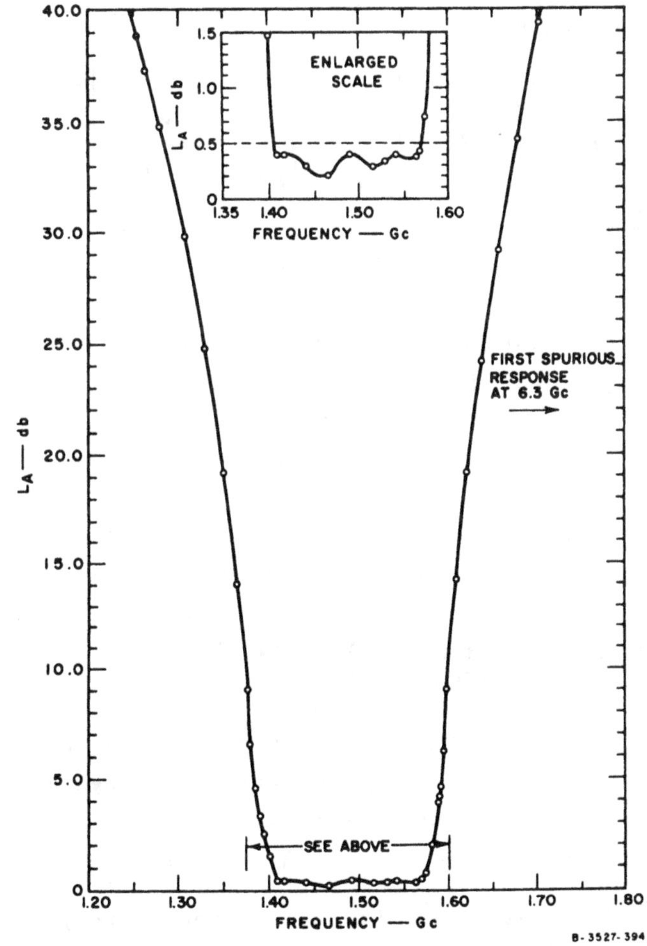

SOURCE: Quarterly Progress Report 5, Contract DA 36-039 SC-87398, SRI; reprinted in *The Microwave Journal* (see Ref. 18 by G. L. Matthaei)

FIG. 8.13-6 MEASURED ATTENUATION OF THE FILTER IN FIG. 8.13-4

resonator, and between Resonator 4 and the output line. A similar phenomenon occurred in the interdigital line filter example discussed in Sec. 10.06. In that case the size of the ripples was easily reduced by decreasing the sizes of end-gap spacings s_{01} and $S_{n,n+1}$. In the case of Fig. 8.13-5, the ripples were not considered to be sufficiently oversized to warrant expenditure of time on additional adjustments.

Table 8.13-2

COMPARISON OF ATTENUATION VALUES OBTAINED BY MAPPING AND BY MEASUREMENT

MAPPING CONDITIONS	f (Gc)	L_A BY MAPPING (db)	L_A MEASURED (db)
A Original Specifications, w = 0.10, 0.10-db Tchebyscheff Ripple, f_0 = 1.491 Gc	1.25	41.5	39
	1.70	36.5	39
B Measured Specifications, w = 0.116, 0.20-db Tchebyscheff Ripple, f_0 = 1.491 Gc	1.25	39.5	39
	1.70	34.0	39

From the VSWR characteristic in Fig. 8.13-5 the measured fractional bandwidth at the equal-VSWR-ripple level is found to be w = 0.116 instead of the specified w = 0.100. This somewhat oversize bandwidth may also be due to coupling effects beyond nearest neighbor line elements, which were neglected in the derivation of the design equations in Fig. 8.13-1(b). Table 8.13-2 compares attenuation values computed by use of the mapping Eqs. (8) to (10) of Fig. 8.13-1(b) as compared to the actual measured values. Conditions A are for the original specifications while Conditions B are for the w = 0.116 fractional bandwidth and approximately 0.2-db ripple indicated by the VSWR characteristic in Fig. 8.13-5. Note that in either case the attenuation predicted by the mapping for f = 1.25 Gc (f below f_0) has come out close to being correct, while the attenuation predicted by the mapping for f = 1.70 Gc (f above f_0) is somewhat low, for reasons previously discussed.

SEC. 8.14, CONCERNING THE DERIVATION OF SOME OF THE PRECEDING EQUATIONS

For convenience in using the preceding sections for practical filter design, some background theoretical matters have been delayed until this section. Let us first note how the design equations for the general, coupled-series-resonator case in Fig. 8.02-3 are derived.

In Sec. 4.12 it was shown that the lumped-prototype circuit in Fig. 8.02-2(a) can be converted to the form in Fig. 4.12-2(a) (where R_A, R_B, and the L_{aj} may be chosen arbitrarily), and the same transmission response will result. This low-pass circuit may be transformed to a corresponding lumped-element band-pass circuit by use of the transformation

$$\omega' = \frac{\omega_1'}{w}\left(\frac{\omega}{\omega_0} - \frac{\omega_0}{\omega}\right) \tag{8.14-1}$$

where

$$w = \frac{\omega_2 - \omega_1}{\omega_0} \quad , \tag{8.14-2}$$

$$\omega_0 = \sqrt{\omega_2\omega_1} \quad , \tag{8.14-3}$$

and ω', ω_1', ω, ω_0, ω_1, and ω_2 are as indicated in Figs. 8.02-1(a), (b) for the case of Tchebyscheff filters. Then the series reactances $\omega' L_{aj}$ in Fig. 4.12-2(a) transform as follows:

$$\omega' L_{aj} = \frac{\omega_1'}{w}\left(\frac{\omega}{\omega_0} - \frac{\omega_0}{\omega}\right) L_{aj} \tag{8.14-4}$$

$$= L_{rj}\omega - \frac{1}{C_{rj}\omega} \tag{8.14-5}$$

where

$$L_{rj} = \frac{\omega_1' L_{aj}}{w\omega_0} \quad \text{and} \quad C_{rj} = \frac{w}{\omega_1' L_{aj}\omega_0} \tag{8.14-6}$$

This reasoning may then be used to convert the low-pass circuit in Fig. 4.12-2(a) directly into the band-pass circuit in Fig. 8.02-2(c). To derive the corresponding general equations in Fig. 8.02-3 we can first use the function

$$X_j(\omega) = L_{rj}\omega - \frac{1}{C_{rj}\omega} \tag{8.14-7}$$

for the resonator reactances in Fig. 8.02-2(c) in order to compute the resonator slope parameters

$$\mathscr{X}_j = \frac{\omega_0}{2} \frac{dX_j(\omega)}{d\omega}\bigg|_{\omega=\omega_0} = L_{rj}\omega_0 = \frac{1}{C_{rj}\omega_0} \quad . \tag{8.14-8}$$

Then by Eqs. (8.14-6) and (8.14-8)

$$L_{aj} = \frac{\mathscr{X}_j w}{\omega_1'} \quad . \tag{8.14-9}$$

Substitution of this result in the equations in Fig. 4.12-2(a) yields Eqs. (2) to (4) in Fig. 8.02-3.

Equations (6) and (7) in Fig. 8.02-3 can be derived by use of Eq. (8.14-8), Fig. 4.12-1, and the fact that the external Q of each end resonator is simply $\omega_0 L_{r1}$ or $\omega_0 L_{rn}$ divided by the resistive loading reflected through the adjacent impedance inverter. The basis for Eq. (8) in Fig. 8.02-3 can be seen by replacing the idealized impedance inverters in Fig. 8.02-2(c) by inverters of the form in Fig. 8.03-1(a), yielding a circuit similar to that in Fig. 8.11-2(b) with the equivalent transformer-coupled form shown in Fig. 8.11-2(a). Then the coupling coefficients of the interior resonators of the filter are

$$k_{j,j+1}\bigg|_{j=1 \text{ to } n-1} = \frac{\omega_0 M_{j,j+1}}{\sqrt{\omega_0 L_{pj}\omega_0 L_{pj+1}}} \quad . \tag{8.14-10}$$

Equation (8) in Fig. 8.02-3 will be seen to be a generalized expression for this same quantity. For example, for Fig. 8.11-2, $K_{j,j+1} = \omega_0 M_{j,j+1}$, and the $\mathscr{X}_j = \omega_0 L_{pj}$. If these quantities are substitutes in Eq. (8.14-10), Eq. (8) of Fig. 8.02-3 will result.

The derivations of the equations in Fig. 8.02-4 follow from Fig. 4.12-2(b) in exactly the same manner, but on the dual basis. The equations for the K- or J-inverter parameters for the various filter structures discussed in this chapter are obtained largely by evaluation of the reactance or susceptance slope parameters $\mathscr{X}$ or $\mathscr{b}$ for the particular resonator structure under consideration, and then inserting these quantities in the equations in Fig. 8.02-3 or 8.02-4. *Thus the derivations of the design equations for the various types of filters discussed in this chapter rest largely on the general design equations in Figs. 8.02-3 and 8.02-4.*

The Capacitively-Coupled Filters of Sec. 8.05—Let us now derive the resonator, susceptance slope parameters for the capacitive-gap-coupled transmission-line filter in Fig. 8.05-1. In this case, the resonator lines are roughly a half-wavelength long in the pass band of the filter, and if Z_L is the impedance connected to one end of a resonator line the impedance looking in at the other end will be

$$Z_{in} = Z_0 \left[\frac{Z_L + jZ_0 \tan \frac{\pi\omega}{\omega_0}}{Z_0 + jZ_L \tan \frac{\pi\omega}{\omega_0}} \right] \tag{8.14-11}$$

$$\approx Z_0 \left[\frac{Z_L + jZ_0 \pi \left(\frac{\omega - \omega_0}{\omega_0} \right)}{Z_0 + jZ_L \pi \left(\frac{\omega - \omega_0}{\omega_0} \right)} \right]_{\omega \text{ near } \omega_0} . \tag{8.14-12}$$

Filters of the form in Fig. 8.05-1 which have narrow or moderate bandwidth will have relatively small coupling capacitances. It can be shown that because of this each resonator will see relatively large impedances at each end. Applying this condition to Eq. (8.14-12), $|Z_L| >> Z_0$, and at least for frequencies near ω_0 Eq. (8.14-12) reduces to

$$Z_{in} \approx \frac{1}{Y_L + jB(\omega)} , \tag{8.14-13}$$

where

$$B(\omega) = Y_0 \pi \left(\frac{\omega - \omega_0}{\omega_0} \right) , \tag{8.14-14}$$

$$Y_L = 1/Z_L \quad \text{and} \quad Y_0 = 1/Z_0 . \tag{8.14-15}$$

Thus, Z_{in} looking into the line looks like the load admittance Y_L in parallel with a resonator susceptance function $B(\omega)$. Applying Eq. (1) of Fig. 8.02-4, to Eq. (8.14-14) for the jth transmission line resonator gives, for the susceptance slope parameter

$$\mathscr{b}_j = \frac{\pi}{2} Y_0 . \tag{8.14-16}$$

Since all of the lines in Fig. 8.05-1 have the same characteristic admittance Y_0, all of the $\mathscr{b}_j$ are the same in this case. Inserting Eq. (8.14-16) in Eqs. (2) to (4) in Fig. 8.02-3 yields Eqs. (1) to (3) of Fig. 8.05-1. It is interesting to note that filters of the type in Fig. 8.05-1 can also be constructed using resonators which are nominally n half-wavelengths long at the desired pass-band center frequency ω_0. In that case the susceptance slope parameters become

$$\mathscr{b}_j = \frac{n\pi}{2} Y_0 \quad . \tag{8.14-17}$$

The Waveguide Filters in Sec. 8.06—The waveguide filter in Fig. 8.06-1 with shunt-inductance couplings is the dual of the capacitively-coupled filter in Fig. 8.05-1 except for one important factor. This factor is that the additional frequency effect due to the dispersive variation of the guide wavelength λ_g in the waveguide must also be accounted for. It can be shown that the response of the waveguide filter in Fig. 8.06-1 will have the same form as that of an equivalent strip line filter as in Fig. 8.05-1 if the waveguide filter response is plotted with $1/\lambda_g$ as a frequency variable instead of ω. Thus, the equations in Fig. 8.06-1 are simply the duals of those in Fig. 8.05-1 with frequency ratios ω/ω_0, ω_1/ω_0 and ω_2/ω_0 replaced by corresponding guide-wavelength ratios λ_{g0}/λ_g, $\lambda_{g0}/\lambda_{g1}$, and $\lambda_{g0}/\lambda_{g2}$, where λ_{g0} is the guide wavelength at midband. The half-wavelength resonators in this case have a series-type resonance with slope parameter

$$\mathscr{x}_j = \frac{\pi}{2} Z_0 \quad . \tag{8.14-18a}$$

Equation (8.14-18a) applies to waveguide resonators *only if the frequency variable is in terms of reciprocal guide-wavelength* (or λ_{g0}/λ_g); however, it applies to TEM-mode resonators on *either* a frequency or reciprocal-guide-wavelength basis. If radian frequency ω is to be used as the frequency variable of a waveguide filter, the slope parameter must be computed including the additional effects of λ_g as a function of frequency. Using ω as the frequency variable, the slope parameter

$$\mathscr{x}_j = \frac{\pi}{2} Z_0 \left(\frac{\lambda_{g0}}{\lambda_0}\right)^2 \tag{8.14-18b}$$

discussed in Sec. 5.08 must be used. In an actual filter design the difference between the slope parameters given by Eqs. (8.14-18a) and (8.14-18b) is compensated for by the fact that the fractional bandwidth w in terms of frequency will be different from the fractional bandwidth w_λ in terms of guide wavelength by the factor $(\lambda_{g0}/\lambda_0)^2$, at least for narrow-band cases. [See Eq. (7) of Fig. 8.06-1.] The reciprocal guide wavelength approach appears to be the most natural for most waveguide cases, though either may be used.

Insertion of Eq. (8.13-18a), $R_A = R_B = Z_0$, and w_λ (in place of w) in Eqs. (2) to (4) of Fig. 8.02-3 gives Eqs. (1) to (3) of Fig. 8.06-1.

The Narrow-Band, Cavity Filters of Sec. 8.07—As an example of the derivation of the equations in Sec. 8.07, consider the case of Fig. 8.07-1(a) which shows a cavity connected to a rectangular waveguide propagating the TE_{10} mode by a small iris with magnetic polarizability M_1 (see Sec. 5.10). The fields within the cavity in MKS units are

$$E_{y1} = jH_1 \sqrt{\frac{\mu_0}{\epsilon_0}} \frac{2l_1}{s\lambda} \cos\frac{\pi x}{a_1} \sin\frac{s\pi z}{2l_1}$$

$$H_{x1} = H_1 \cos\frac{\pi x}{a_1} \cos\frac{s\pi z}{2l_1} \qquad (8.14\text{-}19)$$

$$H_{z1} = \frac{H_1 l_1}{sa_1} \sin\frac{\pi x}{a_1} \sin\frac{s\pi z}{2l_1} \quad .$$

In these equations $\sqrt{\mu_0/\epsilon_0} = 376.6$ ohms (the intrinsic impedance of free space), λ is free space wavelength and s is the number of field variation along the length, l_1, of the cavity. The normal mode fields in the waveguide are

$$E_y = H \sqrt{\frac{\mu_0}{\epsilon_0}} \frac{\lambda_g}{\lambda} \cos\frac{\pi x}{a} e^{j[\omega t + (2\pi/\lambda_g)z]}$$

$$H_x = H \cos\frac{\pi x}{a} e^{j[\omega t + (2\pi/\lambda_g)z]} \qquad (8.14\text{-}20)$$

$$H_z = -jH \frac{\lambda_g}{2a} \sin \frac{\pi x}{a} e^{j[\omega t + (2\pi/\lambda_g) z]} \qquad (8.14\text{-}20) \text{ Cont.}$$

where λ_g is given by Eq. (8.07-1). We define Q_e as

$$Q_e = \frac{\omega W}{P_L} \qquad (8.14\text{-}21)$$

where $\omega = 2\pi f$ is the angular resonance frequency, W is stored energy within the cavity and P_L is the average power lost through the iris to the terminating guide.

The stored energy within the cavity is

$$W = \frac{1}{2} \epsilon_0 \iiint |E_{y1}|^2 \, dx \, dy \, dz = \frac{\mu_0 H_1^2 a_1 b_1 l_1^3}{2 s^2 \lambda^2} \qquad (8.14\text{-}22)$$

where we have used Eq. (8.13-19).

The power lost through the iris is

$$P_L = \frac{|A_a|^2 S_a}{2} \qquad (8.14\text{-}23)$$

where A_a, the amplitude of the normal mode fields excited in the terminating guide, is given by

$$A_a = \frac{\omega \mu_0 M_1 H H_1}{S_a} \quad . \qquad (8.14\text{-}24)$$

The amplitude of the tangential normal-mode magnetic field in the terminating waveguide at the center of gravity of the window is H, and H_1 is the amplitude of the tangential magnetic field in the cavity at the center of gravity of the window. The quantity S_a is the peak power of the normal mode in the rectangular waveguide or

$$S_a = \iint a \sqrt{\frac{\mu_0}{\epsilon_0}} \frac{\lambda_g}{\lambda} H^2 \cos^2 \frac{\pi x}{a} \, dx \, dy$$

$$= \sqrt{\frac{\mu_0}{\epsilon_0}} \frac{\lambda_g}{\lambda} H^2 \frac{ab}{2} \quad . \qquad (8.14\text{-}25)$$

Substituting Eqs. (8.14-24) and (8.14-25) into Eq. (8.14-23) we find

$$P_L = \sqrt{\frac{\mu_0}{\epsilon_0}} \frac{4\pi^2 M_1^2 H_1^2}{ab\lambda_g \lambda} \quad . \tag{8.14-26}$$

When Eq. (8.14-26) and Eq. (8.14-22) are substituted in Eq. (8.14-21) we find

$$Q_e = \frac{a_1 b_1 ab l_1^3 \lambda_g}{4\pi s^2 M_1^2 \lambda^2} \tag{8.14-27}$$

as given in Fig. 8.07-1(a).

When two resonant cavities are connected together by a small iris as shown in Fig. 8.07-2(a) they will have two natural resonant frequencies ω_r and $\omega_r - \Delta\omega$. When the tangential magnetic fields are pointing in the same direction on either side of the iris the cavities will oscillate at frequency ω_r, which is the natural resonant frequency of a cavity with no iris. When the tangential magnetic fields are pointing in opposite directions on either side of the window, the natural resonant frequency is $\omega_r - \Delta\omega$. When $\Delta\omega$ is small the coupling coefficient k can be defined as

$$k = \frac{\Delta\omega}{\omega_r} = \frac{M_1 \mu_0 H_1^2}{\epsilon_0 \iiint |E_{y1}|^2 \, dx \, dy \, dz} \quad . \tag{8.14-28}$$

Substituting Eq. (8.13-19) into Eq. (8.13-28) we find

$$k = \frac{M_1 \lambda^2 s^2}{l_1^3 a_1 b_1} \quad . \tag{8.14-29}$$

as for Fig. 8.07-2(a).

The Quarter-Wavelength-Resonator Filter of Sec. 8.08—As discussed in Sec. 8.08, the filter structure in Fig. 8.08-1 looks like the filter type in Fig. 8.02-3 when observed from its K-inverters, but looks like the filter type in Fig. 8.02-4 when observed from its J-inverters. Thus, at one end of each quarter-wavelength resonator a reactance slope parameter applies, while at the other end a susceptance slope parameter applies.

By analysis similar to that in Eqs. (8.14-11) to (8.14-16) it can be shown that for quarter-wavelength resonators exhibiting series resonance

$$\mathscr{X}_j = \frac{\pi}{4} Z_0 \quad . \tag{8.14-30}$$

and when exhibiting shunt resonance

$$\mathscr{B}_j = \frac{\pi}{4} Y_0 \quad . \tag{8.14-31}$$

Insertion of these equations in the appropriate equations in Figs. 8.02-3 and 8.02-4 gives Eqs. (1) to (3) of Fig. 8.08-1.

The Parallel-Coupled Filters of Sec. 8.09—The equations presented in Fig. 8.09-1 can be derived by showing that for narrow or moderate bandwidths each of the parallel-coupled sections $j, j+1$ of length l in Fig. 8.09-1 is equivalent to a J-inverter with a length of line on each side, the lines being a quarter-wavelength long at frequency ω_0. A complete derivation of the equations in Fig. 8.09-1 (in somewhat different form) can be found in Ref. 15.

The Quarter-Wavelength-Coupled Filters of Sec. 8.10—The design equations (1) to (4) in Fig. 8.10-1 can be derived from those in Fig. 8.02-4 by setting G_A, G_B, and the inverter parameters $J_{j,j+1}$ all equal to Y_0, and then solving for the $\mathscr{B}_j/Y_0$. As previously discussed in Sec. 8.10, the $\pi/4$ and $\pi/2$ terms were introduced in these equations to account for the added selectivity introduced by the quarter-wavelength lines.[4] The correction is $\pi/4$ for the end resonators which have only one, quarter-wavelength line adjacent to them, and is twice as large for the interior resonators which have a quarter-wavelength line on each side. Note that this $\pi/4$ correction per quarter-wavelength line corresponds to the $\mathscr{B}_j/Y_0$ values for the quarter-wavelength resonators discussed in connection with Eq. (8.14-31).

The Lumped-Element Filters of Sec. 8.11—The resonator susceptance slope parameters for the capacitively-coupled, lumped-element filter in Fig. 8.11-1 are simply

$$\mathscr{B}_j = \omega_0 C_{rj} \tag{8.14-32}$$

and these values inserted in Eqs. (2) to (4) of Fig. 8.02-4 yield Eqs. (2) to (4) in Fig. 8.11-1. The J-inverters in this case are of the form in Fig. 8.03-2(b). The negative shunt capacitances required for these inverters are lumped with the resonator capacitances C_{rj} to yield the somewhat smaller net shunt capacitance actually used in constructing the filter. However, in the case of the inverters between the end resonators and the terminations, this procedure does not work since there is no way of absorbing the negative capacitance that would appear across the resistor termination. This difficulty in analysis can be avoided by analyzing the end couplings in a somewhat different way.

Looking from Resonator 1 in Fig. 8.11-1 out toward C_{01} and G_A in series, the admittance is

$$Y = \frac{1}{\dfrac{1}{jB_{01}} + \dfrac{1}{G_A}} = \frac{\left(\dfrac{B_{01}^2}{G_A}\right)}{1 + \left(\dfrac{B_{01}}{G_A}\right)^2} + \frac{jB_{01}}{1 + \left(\dfrac{B_{01}}{G_A}\right)^2} \tag{8.14-33}$$

where $B_{01} = \omega_0 C_{01}$. Meanwhile, looking left from Resonator 1 in Fig. 8.02-4 into the J_{01} inverter the conductance

$$G = \frac{J_{01}^2}{G_A} \tag{8.14-34}$$

is seen. Equating G in Eq. (8.14-34) to the real part of Y in Eq. (8.14-33) and solving for C_{01} gives Eq. (5) in Fig. 8.11-1, and ensures that the conductance loading on Resonator 1 will be the same as that called for by the general equations in Fig. 8.02-4. The imaginary part of Y in Eq. (8.14-33) can be dealt with satisfactorily by replacing it by a shunt capacitive susceptance $\omega_0 C_{01}^e$ of the same size which then leads to Eq. (11) in Fig. 8.11-1. Since C_{01}^e effectively increases the shunt capacitance of Resonator 1, this amount should be subtracted from C_{r1} as indicated in Eq. (8) in Fig. 8.11-1 when computing the net shunt capacitance to be used in constructing Resonator 1. Of course, the same reasoning applies for design of the $C_{n,n+1}$ coupling at the other end of the filter.

It should be noted why the procedure discussed above is necessary for the lumped-element circuit in Fig. 8.11-1 when it was not necessary for the circuits with transmission-line resonators in Figs. 8.05-1, 8.06-1, and 8.08-1. In these latter cases, the inverters used involved a negative length of line (rather than a negative capacitance or inductance). Since this negative length of line had a characteristic impedance equal to the impedance of the terminations, the negative line lengths can be regarded as simply reducing the length of the matched lines connected to the generator and load impedances. These terminating lines are matched, so that their length has no effect on the attenuation characteristics of the filter. Thus, in this manner, the inverter negative lines adjacent to the terminations can, effectively, be absorbed into the generator and load terminations.

The derivation of the design equations for the filter structure in Fig. 8.11-2 follows reasoning essentially the dual of that for Fig. 8.11-1. The end couplings in this case are slightly more complex, but the same general approach applies.

The Wide-Stop-Band Filters of Sec. 8.12—The derivation of the design equations for the filter in Fig. 8.12-1 is very similar to that for the filter in Fig. 8.11-1, except for certain aspects previously discussed in Sec. 8.12. Note that in this case it is not feasible to reduce the net resonator shunt capacitance as was done in Eqs. (8) to (10) of Fig. 8.11-1. Therefore, for purposes of computing resonant frequencies and susceptance slope parameters, the negative shunt capacitances associated with the J-inverters in the interior of the filter must be compensated for by a corresponding increase in positive shunt capacitance to be attributed to the resonators. This is why it was necessary to introduce the positive shunt susceptances equal to $\omega_0 C_{j,j+1}$ in Eqs. (1) to (3) of Fig. 8.12-1.

The Comb-Line Filters of Sec. 8.13—The comb-line filter design equations in Fig. 8.13-1(b) were derived by use of the approximate, open-wire-line representation in Fig. 8.14-1. In Fig. 8.14-1 nodal points are indicated which correspond to nodal points also indicated in Fig. 8.13-1(a). The circuitry in Fig. 8.14-1 between Nodal Points 0 and 1 was specified with the aid of the equivalences in Figs. 5.09-2(a) so as to correspond to the parallel-coupled strip lines in Fig. 8.13-1(a) between Nodal Points 0 and 1. The circuitry between Nodal Points 1 and 2, 2 and 3, ..., $n - 1$ and n, was specified with the aid of the equivalences in Figs. 5.09-4(a) and (b) so as to correspond to the pairs of parallel-coupled sections formed by the strip lines connected to Nodal Points 1

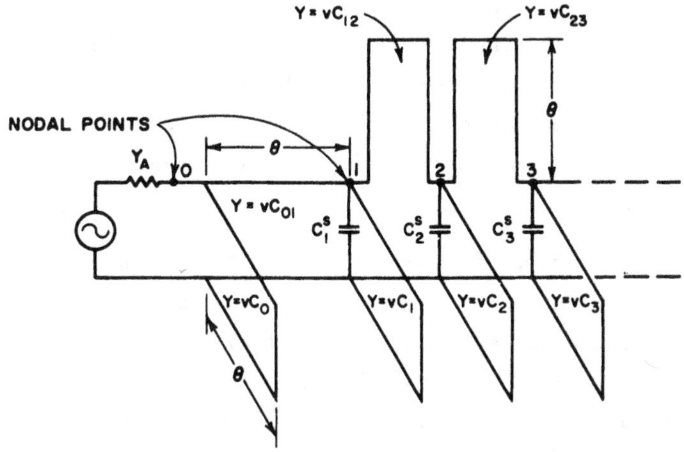

SOURCE: Quarterly Progress Report 5, Contract DA 36-039 SC-87398, SRI; reprinted in *The Microwave Journal* (see Ref. 18 by G. L. Matthaei)

FIG. 8.14-1 AN APPROXIMATE EQUIVALENT CIRCUIT OF THE COMB-LINE FILTER IN FIG. 8.13-1(a)
The capacitances C_i and $C_{i,i+1}$ are the self and mutual capacitances per unit length of the lines in Fig. 8.13-2, while v is the velocity of propagation

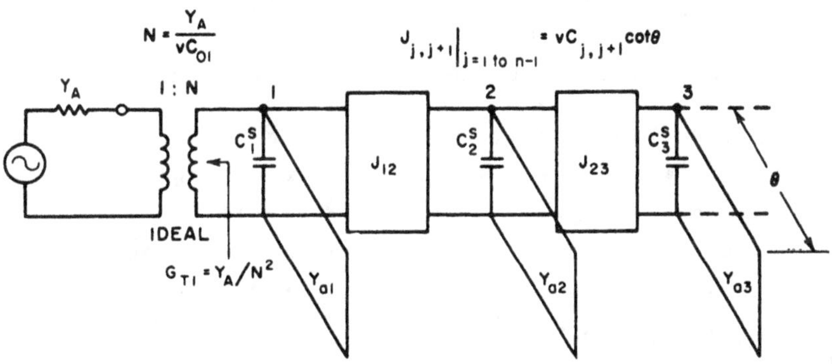

$$Y_{a1} = Y_A \left[1 - \left(\frac{vC_{01}}{Y_A}\right)^2\right] + v\left(C_1 + C_{12} - C_0\right)$$

$$Y_{aj}\Big|_{=2 \text{ to } n-1} = v\left(C_j + C_{j-1,j} + C_{j,j+1}\right)$$

A-3527-396

SOURCE: Quarterly Progress Report 5, Contract DA 36-039 SC-87398, SRI; reprinted in *The Microwave Journal* (see Ref. 18 by G. L. Matthaei)

FIG. 8.14-2 THE CIRCUIT IN FIG. 8.14-1 REORGANIZED TO INCLUDE ADMITTANCE INVERTERS
The constraint $vC_0 = Y_A - vC_{01}$ has also been applied

and 2, 2 and 3, etc. Note that the line admittances in Fig. 8.14-1 are defined in terms of the line capacitances per unit length C_j and $C_{j,j+1}$ as defined in Fig. 8.13-2, times the velocity of propagation (which gives the dimensions of admittance). This representation of a comb-line filter is approximate, and neglects the effects of fringing capacitances beyond nearest neighbors.[17]

The design equations in Fig. 8.13-1(b) are based on the general equations in Fig. 8.02-4. In order to modify the circuit in Fig. 8.14-1 to a form such that the data in Fig. 8.02-4 can be easily applied, the series stubs between Nodal Points 1 and 2, 2 and 3, etc., in Fig. 8.14-1 were incorporated into J-inverters of the form in Fig. 8.03-4, which gave the result shown in Fig. 8.14-2. Since each of the inverters $J_{j,j+1}$ consists of a pi configuration of a series stub of characteristic admittance Y, and two shunt stubs of characteristic admittance $-Y$, it was necessary to increase the characteristic admittances of the actual shunt stubs on each side in order to compensate for the negative admittances ascribed to the inverters. This is why the shunt stubs 2 to $n - 1$ in Fig. 8.14-2 now have the admittances $Y_{aj} = v\,(C_j + C_{j-1,j} + C_{j,j+1})$ instead of just $v\,C_j$. The portion of the circuit in Fig. 8.14-1 between Nodal Points 0 and 1 has been converted to the form shown in Fig. 8.14-2 by use of a simplifying constraint which brings about the properties summarized in Fig. 5.09-3(a).

When applying the general relations in Fig. 8.02-4 to the circuit in Fig. 8.14-2 to derive design equations for comb-line filters, the admittance-inverter parameters $J_{j,j+1}$ are, of course, evaluated at midband, and the resonator slope parameters are computed from the resonator circuits consisting of the lines of admittance Y_{aj} shunted by the lumped capacitances C_j^s. The terminating admittance G_{T1} in Fig. 8.14-2 is specified so that $G_{T1} = J_{01}^2/Y_A$, where the value of J_{01} is as given in Fig. 8.02-4.

REFERENCES

1. S. B. Cohn, "Direct-Coupled-Resonator Filters," *Proc. IRE,* Vol. 45, pp. 187-196 (February 1957).

2. W. L. Pritchard, "Quarter Wave Coupled Filters," *J. Appl. Phys.,* Vol. 18, pp. 862-872 (October 1947).

3. R. M. Fano and A. W. Lawson, "Microwave Filters Using Quarter-Wave Couplings," *Proc. IRE,* Vol. 35, pp. 1318-1323 (November 1947).

4. W. W. Mumford, "Maximally Flat Filters in Waveguide," *Bell System Tech. J.,* Vol. 27, pp. 648-714 (October 1948).

5. G. L. Ragan, *Microwave Transmission Circuits,* McGraw-Hill Book Co., Inc., New York City, 1948. Chapters 9 and 10, by R. M. Fano and A. W. Lawson.

6. G. C. Southworth, *Principles and Applications of Waveguide Transmission,* D. Van Nostrand Co., Inc., New York City, 1950, pp. 285-293.

7. J. Reed, "Low *Q* Microwave Filters," *Proc. IRE,* Vol. 38, pp. 793-796 (July 1950).

8. H. J. Riblet, "A Unified Discussion of High-*Q* waveguide Filter Design Theory," *IRE Trans. PGMTT,* Vol. MTT-6, pp. 359-368 (October 1958).

9. A. A. Oliner, "Equivalent Circuits for Discontinuities in Balanced Strip Transmission Line," *IRE Trans. PGMTT,* Vol. MTT-3, pp. 134-143 (March 1955).

10. H. M. Altschuler and A. A. Oliner, "Discontinuities in the Center Conductor of Symmetric Strip Transmission Line," *IRE Trans. PGMTT,* Vol. MTT-8, pp. 328-339 (May 1960).

11. G. L. Matthaei, "Final Report on Microwave Filter and Ferrite Device Research," Report ERL-123, Electronics Research Laboratory, Ramo-Wooldridge Division of T-R-W Corp., Canoga Park, California (20 August 1958).

12. G. L. Matthaei, "Theory, Design, and Special Applications of Direct-Coupled Strip Transmission Line, Band-Pass Filters," Report ERL-115, Electronics Research Laboratory, Ramo-Wooldridge Division of T-R-W Corp., Canoga Park, California (18 December 1957).

13. N. Marcuvitz, *Waveguide Handbook,* McGraw-Hill Book Co., New York City, 1951, Chapter 5.

14. G. L. Matthaei, "Direct-Coupled-Band-Pass Filters with $\lambda_0/4$ Resonators," *1958 IRE National Convention Record,* Part 1, pp. 98-111.

15. S. B. Cohn, "Parallel-Coupled Transmission-Line-Resonator Filters," *IRE Trans. PGMTT,* Vol. MTT-6, pp. 223-231 (April 1958).

16. G. L. Matthaei, *et al.,* "Design Criteria for Microwave Filters and Coupling Structures," Final Report, Chapter 16, SRI Project 2326, Contract DA 36-039 SC-74862, Stanford Research Institute, Menlo Park, California (January 1961).

17. J. T. Bolljahn and G. L. Matthaei, "A Study of the Phase and Filter Properties of Arrays of Parallel Conductors Between Ground Planes," *Proc. IRE,* Vol. 50, pp. 299-311 (March 1962).

18. G. L. Matthaei, "Comb-Line Band-Pass Filters of Narrow or Moderate Bandwidth," *The Microwave Journal,* Vol. 6, pp. 82-91 (August 1963).

CHAPTER 9

BAND-PASS FILTERS, CONTINUED (WIDE-BAND AND NARROW-BAND BAND-PASS FILTERS CONSISTING OF TRANSMISSION LINES WITH REACTIVE DISCONTINUITIES)

SEC. 9.01, INTRODUCTION

The band-pass filter design techniques discussed in this chapter are based on the quarter-wave transformer prototype circuit (Chapter 6). They apply to band-pass filters with transmission line resonators alternating between coupling elements which are series capacitances or shunt inductances. The design bandwidths may range from narrow-band on up to such wide bandwidths that the filters can be used for microwave high-pass applications. Filters of these general types were also discussed in Secs. 8.05 to 8.08, using the design view point of Chapter 8, which is applicable to narrow and moderate bandwidths. The design view point of this chapter was developed to obtain a design method which would hold for wider bandwidths, and for smaller pass-band Tchebyscheff ripples, as well.

Section 9.02 introduces the quarter-wave transformer prototype circuit, and Sec. 9.03 gives basic design formulas for synchronously tuned filters.

Section 9.04 treats narrow-band filters from the view point of this chapter, showing the connection with the lumped-constant low-pass prototype used in Chapter 8. It has been found that the design technique of Sec. 9.04 and Chapter 8 for narrow-band filters generally works well up to fractional bandwidths of about 20 percent or more (compare Sec. 8.01), provided that the pass-band ripple is not too small: the ripple VSWR should exceed about $1 + (2w)^2$, where w is the fractional bandwidth of the narrow-band filter, if it is to be derivable from a lumped-constant low-pass prototype.

The remainder of this chapter, from Sec. 9.05 on, is concerned mainly with the design of wide-band and pseudo-high-pass filters, for which the method of the quarter-wave transformer prototype is principally intended.

Section 9.05 deals with the realization of the reactive discontinuities (for filters of any bandwidth). In Sec. 9.06 some standardized designs are given which can be adapted for many high-pass applications, and experimental results are given in Sec. 9.07.

The basic theory, design data, and examples will be found in Secs. 9.08 through 9.11. Finally, Section 9.12 deals briefly with reactance-coupled quarter-wave filters.

In this chapter the frequency is introduced everywhere as the normalized frequency, usually denoted by f/f_0, the ratio of the frequency f to the synchronous frequency f_0. For waveguide filters the "normalized frequency" is to be understood to refer to the quantity λ_{g0}/λ_g, the ratio of the guide wavelength λ_{g0} at the frequency of synchronous tuning, to the guide wavelength λ_g. (For example, an experimental waveguide filter is described in Sec. 9.07.)

SEC. 9.02, FILTERS WITH IMPEDANCE STEPS AND IMPEDANCE INVERTERS

Stepped-impedance filters (quarter-wave transformers and half-wave filters) have been treated in Chapter 6. This section points out their equivalence to filters with impedance inverters, and, serves as an introduction to the design of wide-band reactance-coupled half-wave filters.

An impedance (or admittance) step [Fig. 9.02-1(a)] can always be replaced by an impedance (or admittance) inverter [Fig. 9.02-1(b) and (c)] without affecting the filter response curve, provided that the input and output ports are properly terminated. Thus the two types of circuit in Fig. 9.02-1 are entirely equivalent as a starting point for the design of filters. The impedance-inverter (or admittance-inverter) point of view [Fig. 9.02-1(b) and (c)] was the more natural one to adopt in Chapter 8 to convert the lumped-constant low-pass prototype of Chapter 4 into a transmission-line filter; whereas in this chapter a stepped-impedance-filter point of view is more convenient to utilize directly the design data of Chapter 6.

The stepped-impedance filter is turned into a reactance-coupled filter by replacing each impedance step with a reactance having the same discontinuity-VSWR and spacing the reactances to obtain synchronous tuning

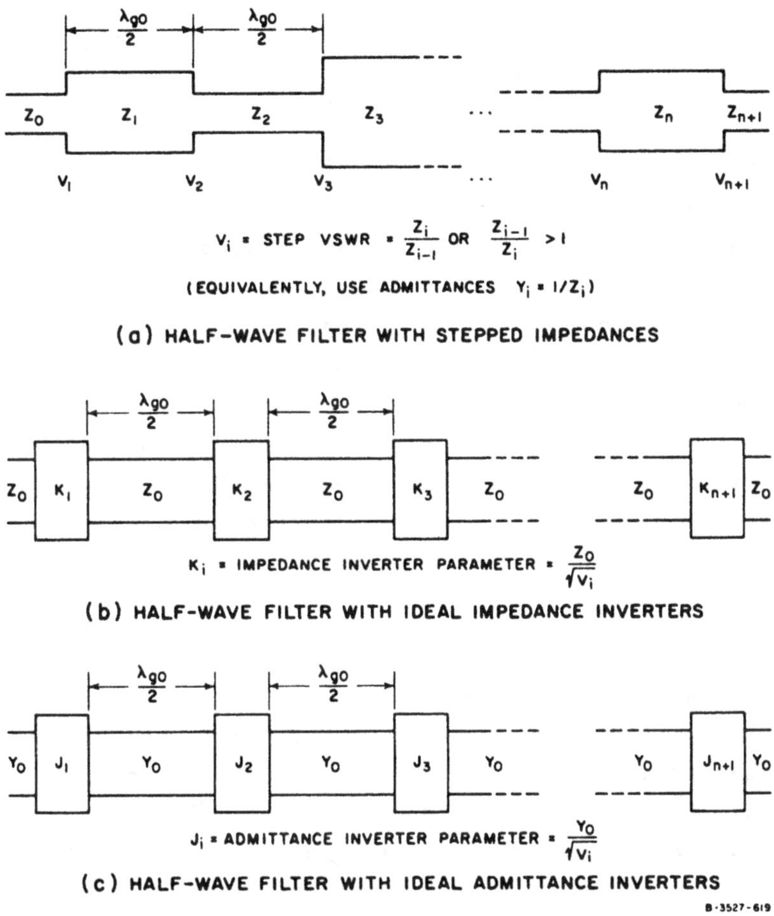

SOURCE: Quarterly Progress Report 5, Contract DA 36-039 SC-87398, SRI; reprinted in the *IEEE Trans. PTGMTT* (see Ref. 18 by Leo Young)

FIG. 9.02-1 A STEPPED-IMPEDANCE HALF-WAVE FILTER, AND EQUIVALENT FILTERS USING IMPEDANCE OR ADMITTANCE INVERTERS

(Sec. 9.03). The step-VSWRs will generally be obtained from Chapter 6. For narrow-band filters Eq. (6.09-2) may be used. This is equivalent (through Fig. 9.07-1) to the formulas in Fig. 8.05-1 for the normalized inverter parameters $K_{j,j+1}/Z_0$. Other equivalent circuits for impedance inverters suitable for narrow-band filter design are given in Sec. 8.03, but we shall be concerned in this chapter only with the shunt-inductance of Fig. 8.03-1(c) and the series-capacitance of Fig. 8.03-2(d).

One important difference in approach between this chapter and Chapter 8 is that in this chapter the starting point or prototype circuit is one of the circuits in Fig. 9.02-1 (whose synthesis is precisely controlled), whereas in Chapter 8 the exact synthesis is pushed back one stage to the lumped-constant prototype circuit of Chapter 4. For example, the performance of the circuits shown in Fig. 9.02-1(b), (c), having ideal inverters, would not give exactly the prescribed response if designed by the methods of Chapter 8 (although the approximations would be very close for narrow or moderate bandwidths). However, the circuits in Fig. 9.02-1(b) (c) have transmission characteristics identical to those of the half-wave filter in Fig. 9.02-1(a).

The other important difference over the previous chapter is that the frequency-behavior of the reactive discontinuities (shunt-inductances or series-capacitances) is examined in detail. The behavior of the discontinuities leads to increasing distortion of the filter response (*e.g.*, pass-band bandwidth and stop-band attenuation), as the frequency spread is increased. This type of consideration can be left out in the design of narrow-band filters thereby simplifying the design process considerably. However, it is important to predict the distortion for filters having large bandwidths.

A quarter-wave transformer and the notation associated with it is shown in Fig. 9.02-2. The characteristics of maximally flat and Tchebyscheff quarter-wave transformers are sketched in Fig. 9.02-3. Closely related to the quarter-wave transformer is the stepped-impedance half-wave filter (Sec. 6.03) sketched in Fig. 9.02-4. Its characteristics, shown in Fig. 9.02-5 are similar to those of the quarter-wave transformer, shown in Fig. 9.02-3. If the impedance steps or junction VSWRs $V_i (i = 1, 2, \ldots, n + 1)$ of a quarter-wave transformer and a stepped-impedance half-wave filter are the same, then the characteristics of the latter can be obtained from those of the former by a linear change of

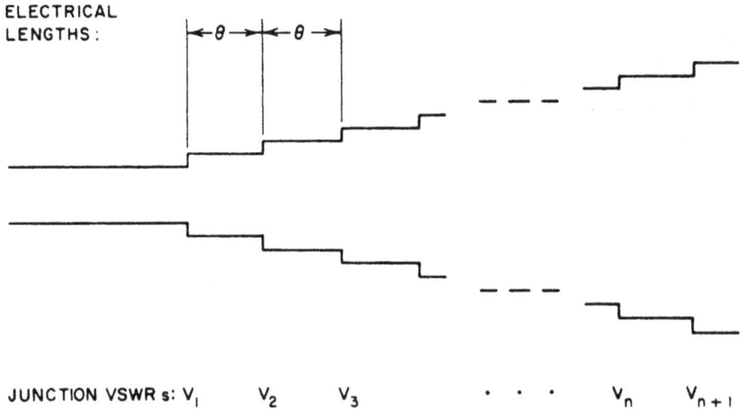

PRODUCT, $V_1 V_2 \cdots V_{n+1}$ = R = OUTPUT / INPUT IMPEDANCE RATIO.

ALL ELECTRICAL LENGTHS θ ARE 90 DEGREES AT THE SYNCHRONOUS FREQUENCY, WHICH IS ALSO THE CENTER FREQUENCY.

A-3527-364

SOURCE: Quarterly Progress Report 5, Contract DA 36-039 SC-87398, SRI; reprinted in the *IEEE Trans. PTGMTT* (see Ref. 18 by Leo Young)

FIG. 9.02-2 QUARTER-WAVE TRANSFORMER USED AS PROTOTYPE CIRCUIT

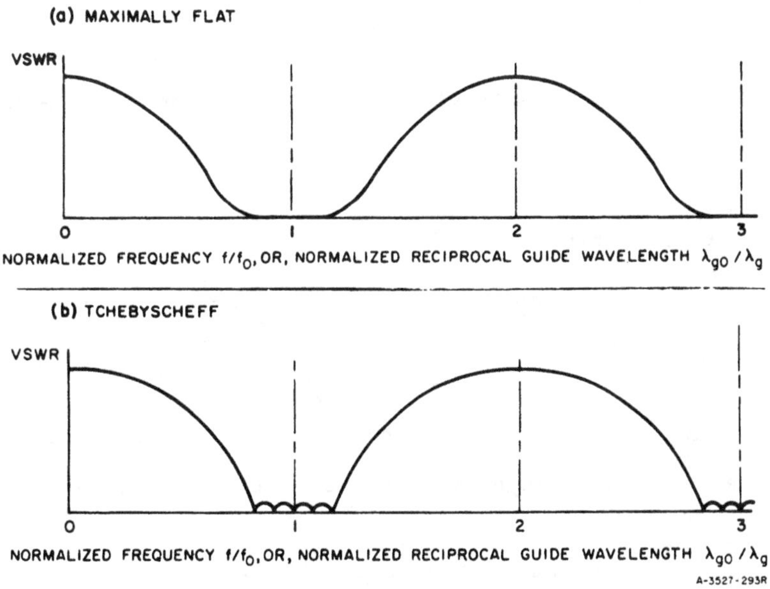

SOURCE: Quarterly Progress Report 5, Contract DA 36-039 SC-87398, SRI; reprinted in the *IEEE Trans. PTGMTT* (see Ref. 18 by Leo Young)

FIG. 9.02-3 QUARTER-WAVE TRANSFORMER CHARACTERISTICS

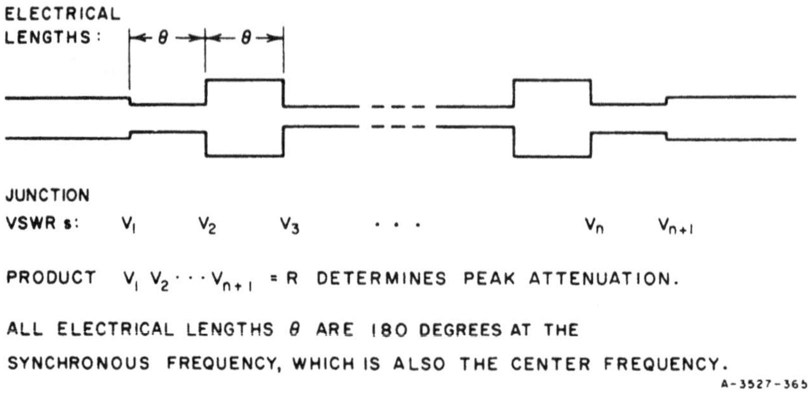

SOURCE: Quarterly Progress Report 5, Contract DA 36-039 SC-87398, SRI; reprinted in the *IEEE Trans. PTGMTT* (see Ref. 18 by Leo Young)

FIG. 9.02-4 STEPPED HALF-WAVE FILTER USED AS PROTOTYPE CIRCUIT

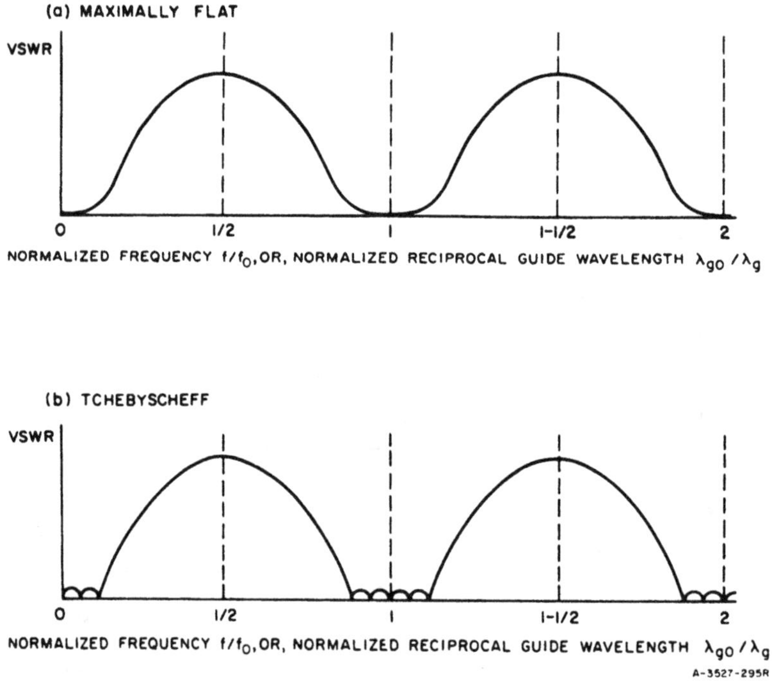

SOURCE: Quarterly Progress Report 5, Contract DA 36-039 SC-87398, SRI; reprinted in the *IEEE Trans. PTGMTT* (see Ref. 18 by Leo Young)

FIG. 9.02-5 STEPPED HALF-WAVE FILTER CHARACTERISTICS

scale on the frequency axis; the stepped-impedance half-wave filter bandwidth becomes one-half the quarter-wave transformer bandwidth.

The parameter R is again defined as the product of the discontinuity VSWRs [compare Eq. (6.03-4)]:

$$R = V_1 V_2 \ldots V_{n+1} \quad . \tag{9.02-1}$$

If the fractional bandwidth w is less than about 20-percent, and if, by Eq. (6.09-1),

$$R >> \left(\frac{1}{w}\right)^n \tag{9.02-2}$$

then the filter may be considered narrow-band; this case will be treated in Sec. 9.04.

The quarter-wave transformer prototype circuit is suitable for designing reactance-coupled filters up to very large bandwidths. As a result, it is subject to certain limitations that do not complicate narrow-band design-procedures. It is well to understand these differences at the outset. Basically they arise from the fact that it is not possible to convert the specified performance of the filter into the performance of the appropriate prototype transformer over large frequency bands by means of simple equations or tabulated functions. Instead, the frequency variations of the reactive couplings have been used to modify the known response functions of quarter-wave transformers to predict the performance of the derived filters (Fig. 9.02-1) over large frequency ranges. Thus it is possible from the graphs to quickly calculate the principal filter characteristics from the transformer characteristics, but not the other way around, as would be more desirable. In the case of wide-band designs where the variation of reactive coupling across the pass band is appreciable, it is necessary first to guess what prototype should be used, and then to match the predicted filter performance against the specified filter performance; if they are not close enough, the process must be repeated with another prototype. What makes this method feasible and practical is the speed with which, by means of the design graphs, this prediction can be made. Most of these design graphs will be presented in Sec. 9.08.

SEC. 9.03, SYNCHRONOUSLY TUNED REACTANCE-COUPLED HALF-WAVE FILTERS

Band-pass filters of the two configurations shown in Fig. 9.03-1 are of considerable practical importance since they are easily realized in practice. These two circuits are duals of each other: the first, shown in Fig. 9.03-1(a), consists of a number of series capacitances alternating with a number of transmission-line sections; the second, shown in Fig. 9.03-1(b), consists of a number of shunt inductances alternating with a number of transmission-line sections. Both filters shown in Fig. 9.03-1 will be called reactance-coupled *half-wave* filters, in the sense that all line lengths between reactances approach one-half wavelength (or a multiple thereof) as the couplings become weak. Each line length between discontinuities constitutes a resonator, so that the filters in Fig. 9.03-1 have n resonators. Notice that the series elements in Fig. 9.03-1(a) are stipulated to be capacitances, that is, their susceptances are supposed to be positive and proportional to frequency.

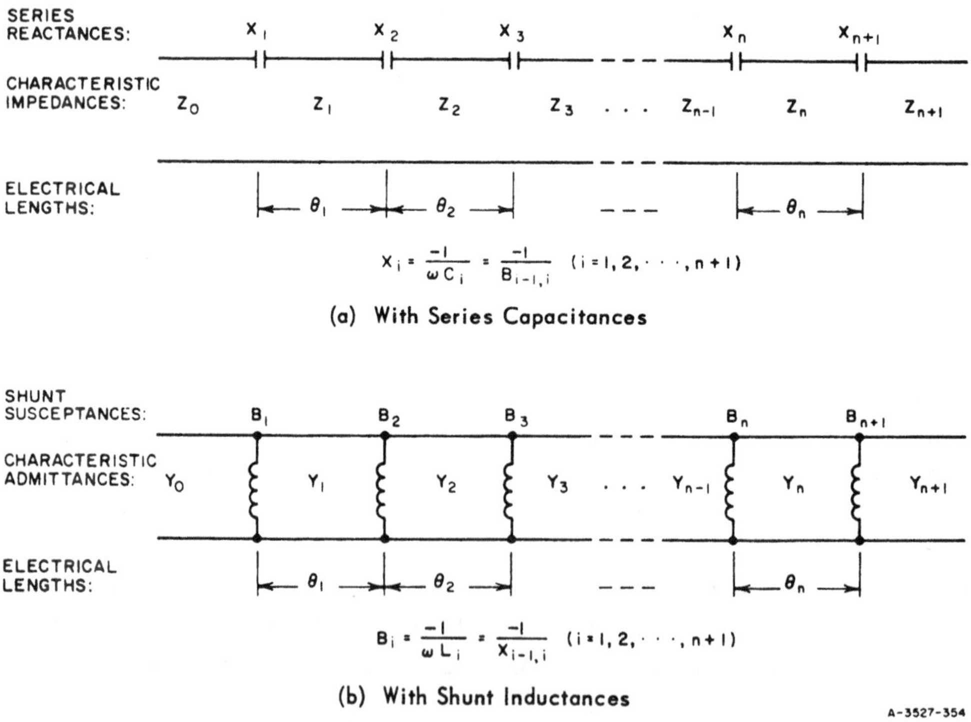

SOURCE: Quarterly Progress Report 5, Contract DA 36-039 SC-87398, SRI; reprinted in the *IEEE Trans. PTGMTT* (see Ref. 18 by Leo Young)

FIG. 9.03-1 REACTANCE-COUPLED HALF-WAVE FILTERS

Similarly, the shunt elements in Fig. 9.03-1(b) are stipulated to be inductances, that is, their reactances are supposed to be positive and proportional to frequency. (If the transmission line is dispersive, these statements are to be modified by replacing frequency by reciprocal guide wavelength.)

All the filters described in this chapter are synchronously tuned as defined in Sec. 6.01; that is, all discontinuities are so spaced that the reflections from any two adjacent discontinuities are phased to give maximum cancellation at a fixed frequency (the synchronous frequency) in the pass band. At the synchronous frequency the filter is interchangeable with a stepped-impedance half-wave filter (Sec. 6.03), and the ith reactive discontinuity has the same discontinuity VSWR, V_i, as the ith impedance step.[1,2] Impedances are shown for the series-reactance coupled filter in Fig. 9.03-1(a), and admittances for the shunt-susceptance coupled filter in Fig. 9.03-1(b); then, let

$$h_i = \frac{Z_i}{Z_{i-1}} \quad \text{or} \quad \frac{Y_i}{Y_{i-1}} \qquad (9.03\text{-}1)$$

and

$$u_i = \frac{X_i}{Z_{i-1}} \quad \text{or} \quad \frac{B_i}{Y_{i-1}} \quad . \qquad (9.03\text{-}2)$$

Most frequently $h_i = 1$, since usually this is mechanically the most convenient. Sometimes it may be advantageous for electrical or mechnical reasons to make some of the characteristic impedance ratios h_i different from unity. For instance, it may be desirable to combine the filter with an impedance transformer instead of cascading a filter with a separate transformer; also, in some cases the filter performance can be improved appreciably when the values of h_i are selected carefully, as in Sec. 9.11.

The u_i of the reactance-coupled filter are obtained from the V_i of the stepped-impedance filter (Chapter 6), and the h_i, from

$$u_i = \sqrt{\left(V_i + \frac{1}{V_i}\right)h_i - (1 + h_i^2)} \quad . \qquad (9.03\text{-}3)$$

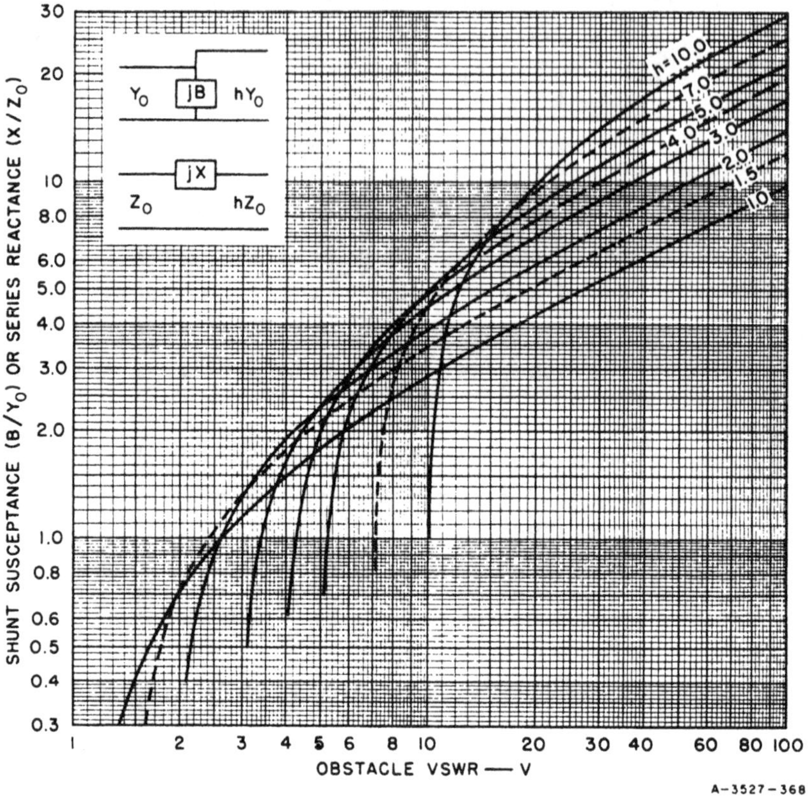

SOURCE: Quarterly Progress Report 5, Contract DA 36-039 SC-87398, SRI; reprinted in the *IEEE Trans. PTGMTT* (see Ref. 18 by Leo Young)

FIG. 9.03-2 SHUNT SUSCEPTANCE (or Series Reactance) AS A FUNCTION OF DISCONTINUITY VSWR FOR SEVERAL CHARACTERISTIC ADMITTANCE (or Impedance) RATIOS, h

The graph of Fig. 9.03-2 gives some solutions of this equation. Generally it will be most convenient to select $h_i = 1$ (that is, all Z_i or Y_i equal), and then Eq. (9.03-3) simplifies to

$$u_i = \sqrt{V_i} - \frac{1}{\sqrt{V_i}} \quad . \tag{9.03-4}$$

The spacings θ_i in Fig. 9.03-1 are determined as follows:[1,2] A single discontinuity of a series-reactance coupled filter is shown in Fig. 9.03-3. (A similar notation, but with Y for Z and B for X, applies to a shunt susceptance coupled filter.) It represents the ith discontinuity of the filter (Fig. 9.03-1). If the reflection coefficients of this discontinuity in the two reference planes shown are to be pure imaginary quantities,[1,2] then one has to set

$$\psi_i' = \frac{1}{2} \text{ arc tan} \left(\frac{u_i^2 + h_i^2 - 1}{2u_i} \right) \qquad (9.03\text{-}5)$$

$$\psi_i'' = \frac{1}{2} \text{ arc tan} \left(\frac{u_i^2 + 1 - h_i^2}{2h_i u_i} \right) \quad . \qquad (9.03\text{-}6)$$

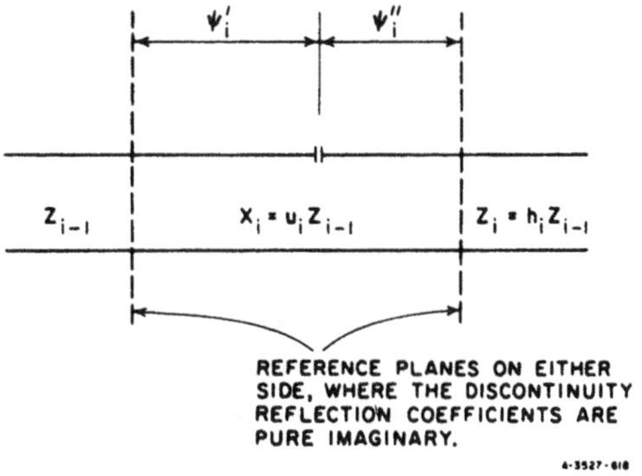

SOURCE: Quarterly Progress Report 5, Contract DA 36-039 SC-87398, SRI; reprinted in the *IEEE Trans. PTGMTT* (see Ref. 18 by Leo Young)

FIG. 9.03-3 SERIES-REACTANCE COUPLING OF TWO LINES WITH DIFFERENT CHARACTERISTIC IMPEDANCES

The spacings θ_i in Fig. 9.03-1 are now given (in radians) by

$$\theta_i = \frac{\pi}{2} + \psi_i'' + \psi_{i+1}' \quad . \qquad (9.03\text{-}7)$$

When $h_i = 1$, these equations reduce to

$$\psi_i' = \psi_i'' = \psi_i = \frac{1}{2} \text{ arc tan} \left(\frac{u_i}{2} \right) \qquad (9.03\text{-}8)$$

and then*

$$\left. \begin{aligned} \theta_i &= \frac{\pi}{2} + \psi_i + \psi_{i+1} \\ &= \frac{\pi}{2} + \frac{1}{2} \left[\text{arc tan} \left(\frac{u_i}{2} \right) + \text{arc tan} \left(\frac{u_{i+1}}{2} \right) \right] \\ &= \pi - \frac{1}{2} \left[\text{arc tan} \left(\frac{2}{u_i} \right) + \text{arc tan} \left(\frac{2}{u_{i+1}} \right) \right] . \end{aligned} \right\} \qquad (9.03\text{-}9)$$

* These equations are equivalent to Eqs. (5) of Figs. 8.05-1 and 8.06-1.

SEC. 9.04, NARROW-BAND HALF-WAVE FILTERS

The main application of this chapter is to wide-band filters. However, since the design of narrow-band filters is simpler, it will be convenient to use the narrow-band case to illustrate the method in its simplest form.

When the impedance-steps of a narrow-band (Sec. 9.01) stepped-impedance half-wave filter (Sec. 6.03) are replaced by reactances having the same discontinuity-VSWRs, and the filter is again synchronously tuned (Sec. 9.03), then there is little change in the characteristics of the filter in and near the pass-band region. All the formulas necessary to carry out this conversion have been given in Sec. 9.03. It is not necessary to maintain uniform line impedance (all Z_i or Y_i the same), but it is usually convenient to do so.

For narrow-band filters, both quarter-wave transformers and lumped-constant low-pass filters will serve as a prototype, and the conversion from either prototype into the actual filter is equally convenient. The choice of prototype depends on two factors:

(1) Which prototype results in a filter that meets the design specifications more closely, and

(2) Which prototype design is more readily available.

The quarter-wave transformer is better as regards Point (1), but the difference in accuracy is usually negligible for *narrow-band* filters; the lumped-constant low-pass filter, on the other hand, is generally more convenient as regards (2). The reason for this is that explicit formulas exist for the lumped-constant low-pass filter of n elements (Chapter 4), whereas the numerical design of transformers demands great arithmetical accuracy, and becomes convenient only for those cases where the solutions have been tabulated (Chapter 6).

A lumped-constant low-pass filter (Chapter 4) can serve as a prototype circuit for a narrow-band stepped-impedance half-wave filter. Equations (6.09-2) with the substitution $w_q = 2w$, reduce to

$$\left.\begin{aligned} V_1 &= V_{n+1} = \frac{2}{\pi} g_0 \frac{g_1\omega_1'}{w} \\ V_i &= \frac{4}{\pi^2} \frac{\omega_1'^2}{w^2} g_{i-1}g_i, \qquad \text{when } 2 \le i \le n \end{aligned}\right\} \qquad (9.04\text{-}1)$$

where w is the fractional bandwidth of the narrow-band half-wave filter. The reactances are then obtained from Eq. (9.03-3) or (9.03-4) and the spacings from Eq. (9.03-5) through (9.03-9). The low-pass prototype filter is here assumed to be either symmetric or antimetric (Sec. 4.05), and element values for maximally flat and Tchebyscheff prototypes of this type can be found in Sec. 4.05. The parameter ω_1' is the cutoff frequency of the low-pass prototype filter.

Example—It is desired to design a reactance-coupled half-wave filter to have a pass-band VSWR of better than 1.10 over a 10-percent bandwidth, and to have at least 25 db of attenuation at a frequency 10 percent above band center (*i.e.*, twice as far out as the desired band-edge).

This filter can be considered narrow-band, and may be based on a low-pass prototype circuit (Sec. 4.05), since the ripple VSWR of 1.10 exceeds the quantity $1 + (2w)^2 = 1.04$, as mentioned in Sec. 9.01.

We must next determine the minimum number of resonators with which these specifications can be met. Selecting a quarter-wave transformer of fractional bandwidth $w_q = 0.20$, since $w = 0.10$, and with $V_r = 1.10$, the attenuation at twice the band-edge frequency-increment (see the first example in Sec. 6.09) is 24.5 db for $n = 5$ sections and 35.5 db for $n = 6$ sections. Since the filter attenuation at the corresponding frequency above the pass-band will be somewhat less than it was, $n = 5$ is certainly not enough resonators. We then tentatively select $n = 6$. It will be shown in Sec. 9.08 that the attenuation in the stop band of the narrow-band reactance-coupled half-wave filter of Fig. 9.03-1 differs from the attenuation of the narrow-band stepped-impedance half-wave filter of Sec. 6.09 by approximately

$$\Delta L_A = 20\ (n + 1) \log_{10}\left(\frac{f_0}{f}\right) \quad \text{db} \qquad (9.04\text{-}2)$$

where f/f_0 is the normalized frequency (the ratio of the frequency f to the center frequency f_0). This ΔL_A has to be added to the attenuation of the stepped-impedance filter to give the attenuation of the reactance-coupled filter.

Let us, for instance, calculate the attenuation of the filter at $f/f_0 = 1.10$. Using Eq. (9.04-2), with $n = 6$,

$$\Delta L_A = -20 \times 7 \times \log_{10} (1.1) = -5.8 \text{ db} \qquad (9.04\text{-}3)$$

which shows that the filter attenuation is 5.8 db less than the attenuation of the half-wave stepped filter at $f = 1.1\ f_0$, that is, $35.5 - 5.8 = 29.7$ db. This exceeds the 25-db attenuation specified, which confirms our choice of $n = 6$. The discontinuity VSWRs are then given by Eq. (6.09-4). Taking the shunt-inductance-coupled filter of Fig. 9.03-1, with all Y_i equal to Y_0, yields

$$\left.\begin{aligned} \frac{B_1}{Y_0} &= \frac{B_7}{Y_0} = -1.780 \\ \frac{B_2}{Y_0} &= \frac{B_6}{Y_0} = -6.405 \\ \frac{B_3}{Y_0} &= \frac{B_5}{Y_0} = -9.544 \\ \frac{B_4}{Y_0} &= -10.154 \end{aligned}\right\} \qquad (9.04\text{-}4)$$

and from Eq. (9.03-9),

$$\left.\begin{aligned} \theta_1 &= \theta_6 = 147.16 \text{ degrees} \\ \theta_2 &= \theta_5 = 165.41 \text{ degrees} \\ \theta_3 &= \theta_4 = 168.51 \text{ degrees} . \end{aligned}\right\} \qquad (9.04\text{-}5)$$

This filter was analyzed and its computed response is shown in Fig. 9.04-1 (solid line) together with the computed response of the stepped-impedance half-wave filter (broken line). (The stepped-impedance half-wave filter has the same characteristics as the quarter-wave transformer, except for a linear change of scale, by a factor of 2 along the frequency axis.) It is seen that the band edges of the filter and its stepped-impedance half-wave filter prototype very nearly coincide, and that the peak ripples in the two pass bands are nearly the same height.

The VSWR ripple in the pass band is very close to 1.10. The quarter-wave transformer design is itself only approximate, and the ripple heights (broken line, top of Fig. 9.04-1) are not exactly the same, since the transformer was designed from a lumped-constant low-pass prototype (first example in Sec. 6.09). The causes of imperfection in the filter response in Fig. 9.04-1 may in this case be ascribed partly to (1) the imperfect quarter-wave transformer response, since the transformer was derived by an approximation from a lumped-constant circuit, and partly to (2) the further approximation involved in deriving the filter with its unequal spacings and frequency-sensitive couplings from the transformer.

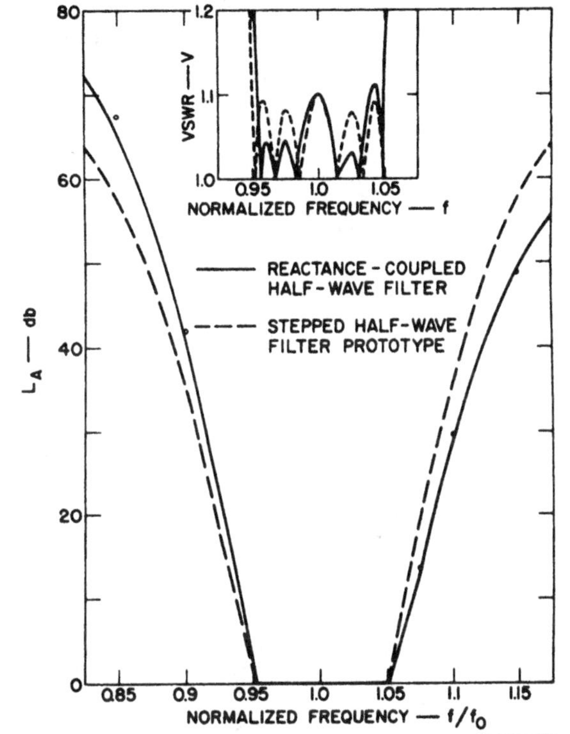

SOURCE:
Quarterly Progress Report 5, Contract DA 36-039 SC-87398, SRI; reprinted in the *IEEE Trans. PTGMTT* (see Ref. 18 by Leo Young)

FIG. 9.04-1 CHARACTERISTICS OF A SIX-RESONATOR FILTER AND ITS STEPPED-IMPEDANCE HALF-WAVE FILTER PROTOTYPE

With regard to the imperfect quarter-wave transformer response, there is clearly little room for improvement, as can be seen from Fig. 9.04-1. As for the further approximations involved, one could adjust the line characteristic impedances to improve the performance (as is explained in Sec. 9.11), but this would also result in only a very small improvement. These adjustments were not considered further in the present example.

The attenuation of the filter at f/f_0 = 1.1 had been predicted from Eq. (9.04-2) to be 29.7 db. This gives one of the circle points in Fig. 9.04-1, and falls very close to the curve computed by analysis of the filter (solid line); other points predicted using Eq. (9.04-2) also fall very close to this computed curve.

Series Capacitances[4]--Series-capacitive coupling can be realized easily in coaxial line or strip transmission line by breaking the inner conductor as shown in Fig. 9.05-1. The gap-spacing is controlled to produce the desired capacitance. When air dielectric is used, filters coupled by series capacitances should have lower dissipation losses than

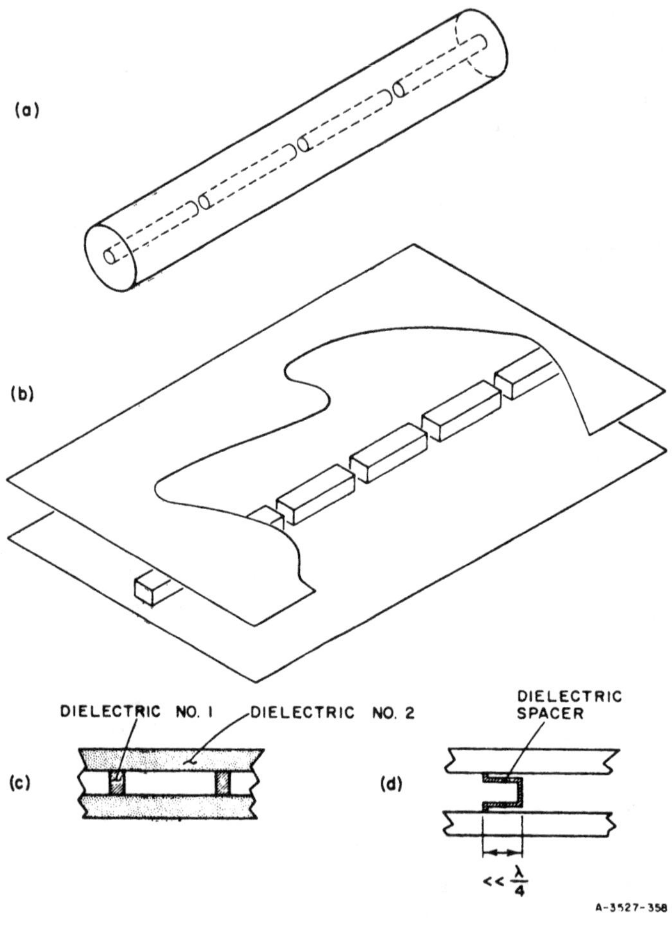

SOURCE: Quarterly Progress Report 5, Contract DA 36-039, SC 87398, SRI; reprinted in the Microwave Journal (See Ref. 4 by Leo Young).

FIG. 9.05-1 REALIZATION OF SERIES-CAPACITANCE REACTANCE-COUPLED HALF-WAVE FILTERS
(a) Filter in Coaxial Line
(b) Filter in Strip Transmission Line
(c) and (d) Series Capacitance in Coaxial Line

filters using shunt inductive posts or irises (considered below), because of their ohmic losses.

The gap spacing that will produce the desired coupling can be determined either experimentally or theoretically. Experimentally one can proceed as follows: The desired VSWR of each discontinuity is, by definition, equal to the junction VSWR of the corresponding step of the prototype transformer. The VSWR-versus-gap-spacing curve can be determined experimentally by direct measurement, or by measuring attenuation, or by using two identical gaps to obtain resonance and measuring the 3 db bandwidth. The curve can then be plotted for the particular transmission line at the synchronous frequency of the pass band. The desired gap spacings are read off from this curve.

In general it would be difficult to calculate the capacitance of a gap for arbitrary cross-sectional shape. However good approximations for the circular inner conductor inside a circular outer conductor (Fig. 9.05-1) can be obtained in the following two ways.

The first approximation is as follows: The inside diameter of the outer tube is b inches; the diameter of the concentric inner conductor is a inches; and the gap spacing is s inches, as shown in Fig. 9.05-2.

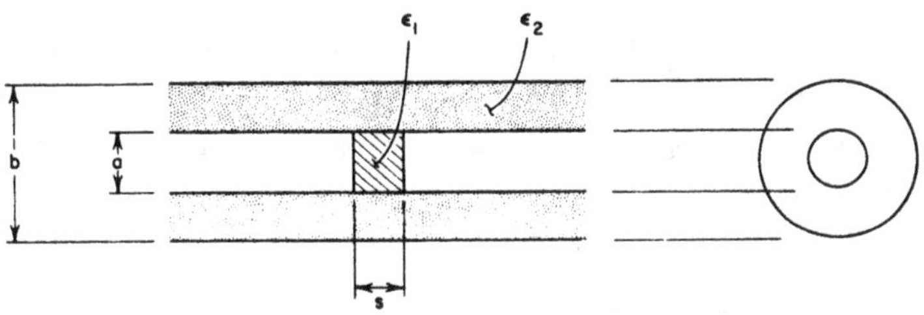

a = INNER CONDUCTOR DIAMETER

b = OUTER CONDUCTOR DIAMETER

s = GAP SPACING

ϵ_1 = DIELECTRIC CONSTANT IN GAP

ϵ_2 = DIELECTRIC CONSTANT IN COAXIAL LINE

A-3527-359

SOURCE: Quarterly Progress Report 5, Contract DA 36-039, SC 87398, SRI; reprinted in the Microwave Journal (See Ref. 4 by Leo Young).

FIG. 9.05-2 NOTATION FOR SERIES CAPACITIVE GAP IN COAXIAL LINE

The following formula is adapted from an approximate formula given by Marcuvitz[5] which is valid when $(b - a) < \lambda$, $s << \lambda$, and $s << (b - a)$, where λ is the free-space wavelength. Then

$$C = C_p + C_f \qquad (9.05\text{-}1)$$

where

$$C_p = \frac{0.1764\ \epsilon_{r1} a^2}{s} \text{ picofarads} \qquad (9.05\text{-}2)$$

is the parallel-plate capacitance,

$$C_f = 0.2245\ \epsilon_{r2} a\ ln \left(\frac{b - a}{s}\right) \text{ picofarads} \qquad (9.05\text{-}3)$$

is the fringing capacitance, and ϵ_{r1} and ϵ_{r2} are the relative dielectric constants of the materials (Fig. 9.05-2). Equations (9.05-2) and (9.05-3) are given also in mks units in Ref. 4.

Second, a more accurate, but still approximate, estimate of the capacitance in Fig. 9.05-2 can be obtained[6] by Fig. 5.05-9: use the curve for $t/b = 0$ and determine the quantity $\Delta C/\epsilon$ using the same s/b (Fig. 9.05-2) on the abscissa. In Fig. 5.05-9 ϵ is our present ϵ_2, which is the absolute dielectric constant for Dielectric 2 in Fig. 9.05-2. Then the capacitance is given by Eqs. (9.05-1) and (9.05-2) but with Eq. (9.05-3) replaced by

$$C_f = 0.353\ \epsilon_{r2} a \left(\frac{\Delta C}{\epsilon}\right)_G \text{ picofarads} \qquad (9.05\text{-}4)$$

where $(\Delta C/\epsilon)_G$ is the quantity $\Delta C/\epsilon$ in Fig. 5.05-9. Both Eqs. (9.05-3) and (9.05-4) tend to underestimate the true capacitance. For $d/b < 0.1$, the two approximate formulas for the total capacitance C agree to within 5 percent.

Numerical data for a rectangular strip transmission line with a strip cross section 0.184 inch by 0.125 inch is given in Fig. 8.05-3.

Example—Find the capacitance of a gap in a coaxial line whose dimensions (Fig. 9.05-2) are $b = 0.575$ inch, $a = 0.250$ inch, $s = 0.025$ inch.

The relative dielectric constant in the gap is $\epsilon_{r1} = 2.0$ and the relative dielectric constant between the conductors is $\epsilon_{r2} = 2.5$. Then find the normalized reactance at a frequency of 1.3 Gc. From Eq. (9.05-2),

$$C_p = 0.882 \text{ picofarad} \tag{9.05-5}$$

and from Eq. (9.05-3)

$$C_f = 0.371 \text{ picofarad} \quad . \tag{9.05-6}$$

Thus the total capacitance is

$$C = C_p + C_f = 1.253 \text{ picofarads} \quad . \tag{9.05-7}$$

The line impedance is

$$Z_0 = \frac{60}{\sqrt{\epsilon_{r2}}} \, ln \, \frac{b}{a} = 31.6 \text{ ohms} \tag{9.05-8}$$

and therefore the normalized reactance at 1.3 Gc is

$$\left.\begin{aligned} \frac{X}{Z_0} &= \frac{10^{12}}{2\pi \times 1300 \times 10^6 \times 1.253 \times 31.6} \\ &= 3.08 \quad . \end{aligned}\right\} \tag{9.05-9}$$

In this case, by Eqs. (9.05-5) and (9.05-6), about three-tenths of the total capacitance is due to the fringing fields; as the gap is reduced, both C_p and C_f increase, but an increasingly higher proportion of the total capacitance is due to C_p.

Filters with series-capacitance couplings were also treated in Sec. 8.05, and have been discussed by Cohn,[7] Ragan,[8] and Torgow.[9] The sleeve-like coupling shown in Fig. 9.05-1(d) is a short, open-circuited stub rather than a lumped capacitance. When its length is increased to one-quarter wavelength at center frequency, and the connecting lines are also made one-quarter wavelength long at center frequency, then a different type of filter, although still with somewhat similar electrical characteristics, results from this change. This type of filter has been treated by Matthaei,[10] and constructional details have also been given by Bostick.[11]

Shunt Inductances--Shunt-inductive couplings can easily be realized in coaxial line, strip transmission line, and waveguide. Some of the common structures are shown in Fig. 9.05-3. Numerical data on some of them have been given in Sec. 8.06, and a few further references will be mentioned here.

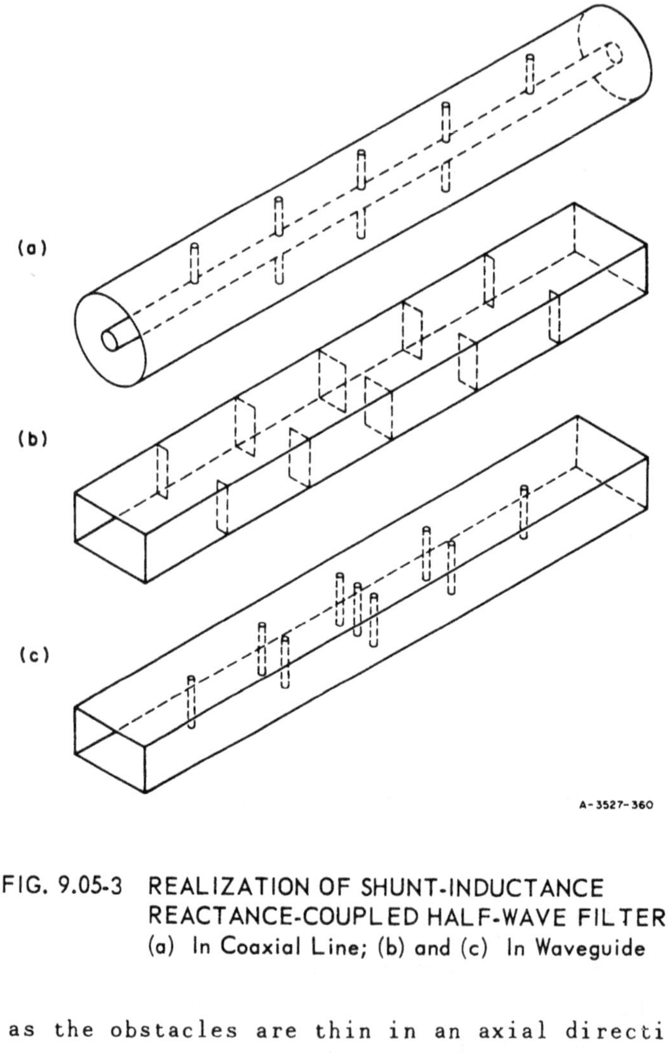

FIG. 9.05-3 REALIZATION OF SHUNT-INDUCTANCE REACTANCE-COUPLED HALF-WAVE FILTER
(a) In Coaxial Line; (b) and (c) In Waveguide

As long as the obstacles are thin in an axial direction, their reactances are nearly proportional to frequency, or reciprocal guide wavelength in dispersive waveguide. In addition, short-circuit stubs also behave approximately as inductances when their lengths are well below one-quarter wavelength. In this case, however, the stub equivalent circuit is more complicated than a simple shunt inductance, since the *T*-junction reference plane positions and the coupling ratio all change with frequency, and the stub can then be approximated by a single inductance only over a limited frequency band.

Numerical data are available for a single post, for double posts, and for four posts in coaxial line,[12,13] as shown in Figs. 9.05-3(a) and 9.05-4. Data on shunt-inductive irises [Fig. 9.05-3(b)] and a single center post in waveguide are graphed in Figs. 8.06-2 and 8.06-3. As shown in Fig. 9.05-3(c), one can also use double symmetrical posts,[14] equally spaced triple posts,[15] and even more than three equally spaced posts[16] in waveguide. Data on strip-line shunt-inductive stubs are given in Figs. 8.08-2(a), (b) to 8.08-4(a), (b).

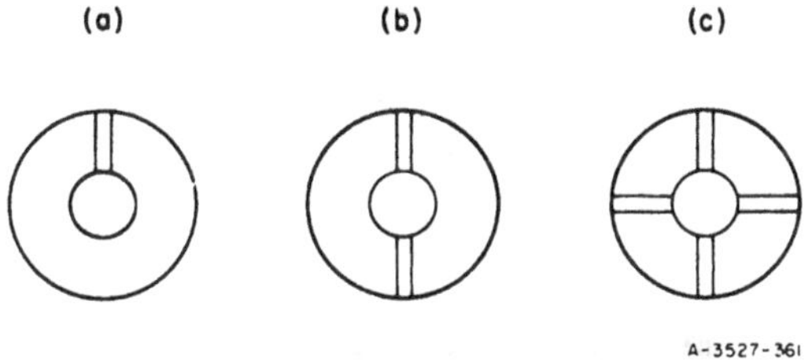

FIG. 9.05-4 CROSS-SECTIONAL VIEW OF ROUND SHUNT-INDUCTIVE POSTS IN COAXIAL LINE
(a) Single Post, (b) Two Posts, (c) Four Posts

As long as all of these discontinuities can be represented by shunt inductances—that is to say by positive reactances whose reactance values are proportional to frequency or to reciprocal guide wavelength—the design data reported here apply up to very large bandwidths.

SEC. 9.06, SOME STANDARDIZED PSEUDO-HIGH-PASS FILTER DESIGNS

A problem frequently encountered is the design of a high-pass filter with capacitive couplings of the type shown in Fig. 9.03-1. Actually, this filter is not high-pass but band-pass; however, for large bandwidths the upper stop band becomes vestigial, because the series reactances diminish rapidly with increasing frequency. Such a filter may therefore be called *pseudo-high-pass*. In the past it has usually been designed on an image-impedance basis (See Chapter 3), often leading to increasing ripple amplitude in the pass band near the cutoff frequency (lower band edge). An equal-ripple characteristic would generally be preferable; a systematic design procedure for equal-ripple wide-band filters is given in this chapter.

The fractional bandwidth, w, of nondispersive filters is defined as in Chapter 6, by

$$w = 2\left(\frac{f_2 - f_1}{f_2 + f_1}\right) \tag{9.06-1}$$

$$= 2\left(\frac{\lambda_1 - \lambda_2}{\lambda_1 + \lambda_2}\right) \tag{9.06-2}$$

where f_1 and f_2 are the lower and upper pass-band edge frequencies, and λ_1, and λ_2 are the corresponding wavelengths. For dispersive filters, guide wavelength has to be substituted in Eq. (9.06-2), and then

$$w = 2\left(\frac{\lambda_{g1} - \lambda_{g2}}{\lambda_{g1} + \lambda_{g2}}\right) \quad . \tag{9.06-3}$$

The design of an eight-resonator (*i.e*, $n = 8$) filter with 0.1-db pass-band ripple and fractional bandwidth 0.85 is treated in the last example of Sec. 9.09. (A fractional bandwidth of 0.85 corresponds to $f_2/f_1 = 2.5$.) Its normalized series reactances X_i/Z_0 (or shunt susceptances B_i/Y_0) were found to be −0.2998, −0.4495, −0.613, −0.700, −0.725, −0.700, 0.613, −0.4495 and −0.2998. Since the three central elements were so nearly alike, they were each set equal to 0.710 for the present purpose. The responses of several filters with normalized series reactances (or shunt susceptances) −0.2998, −0.4495, −0.613, −0.710, −0.710, ..., −0.710, −0.613, −0.4495, −0.2998 were then analyzed numerically on a digital computer. Since all Z_i are the same, then by Eq. (9.03-9) the line lengths are 100.60, 104.85, 108.29, 109.54, ..., 109.54, 108.29, 104.85, and 100.60 electrical degrees. When eight or more resonator lines are used, involving three or more identical central elements, the filter may be thought of as a periodic structure with three (instead of the usual one) impedance-matching resonators at each end. The performance of five such filters with eight, nine, ten, twelve, and fifteen resonators is shown in Figs. 9.06-1 and 9.06-2. It is seen that the computed performance closely meets the predicted one: only in the fifteen-resonator filter (which contains ten identical coupling reactances) are there any ripples appreciably greater than 0.1 db, and even then there are only two large ripples, one just inside each band edge (Fig. 9.06-1).

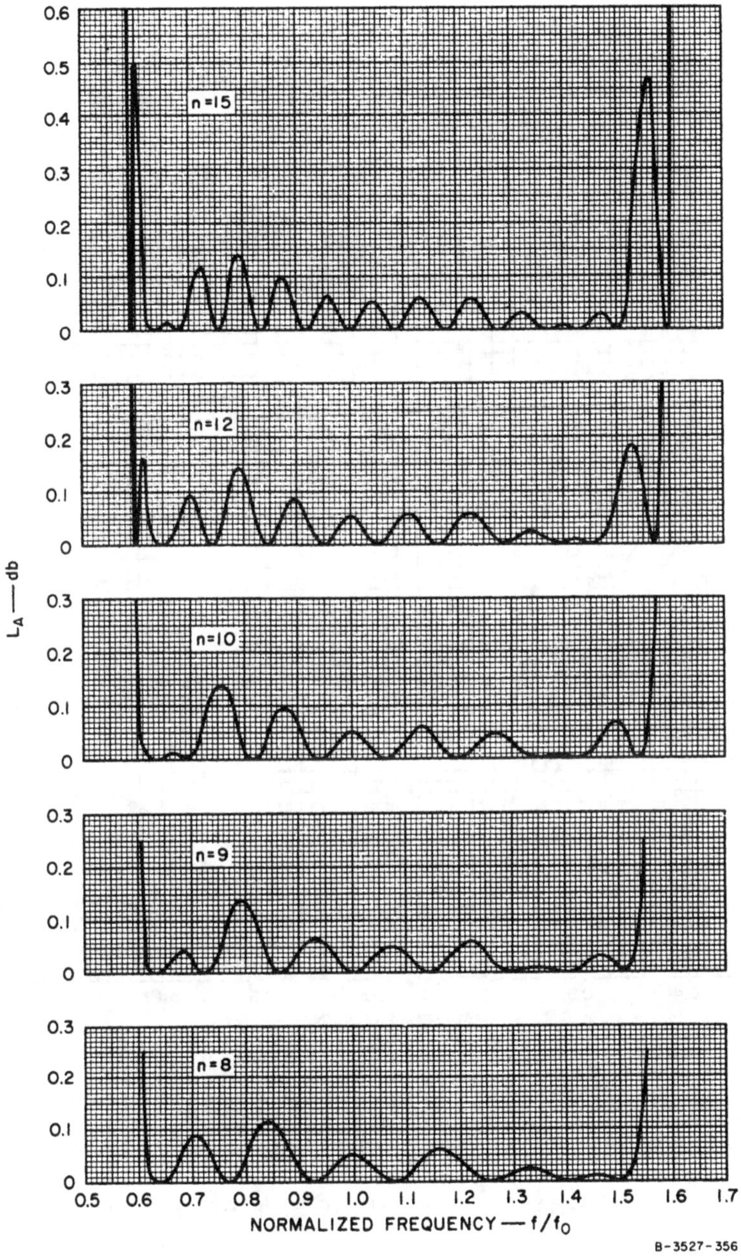

SOURCE: Quarterly Progress Report 5, Contract DA 36-039, SC 87398, SRI; reprinted in the Microwave Journal (See Ref. 17 by Leo Young and B. M. Schiffman).

FIG. 9.06-1 PASS-BAND RESPONSE OF FIVE PSEUDO-HIGH-PASS FILTERS

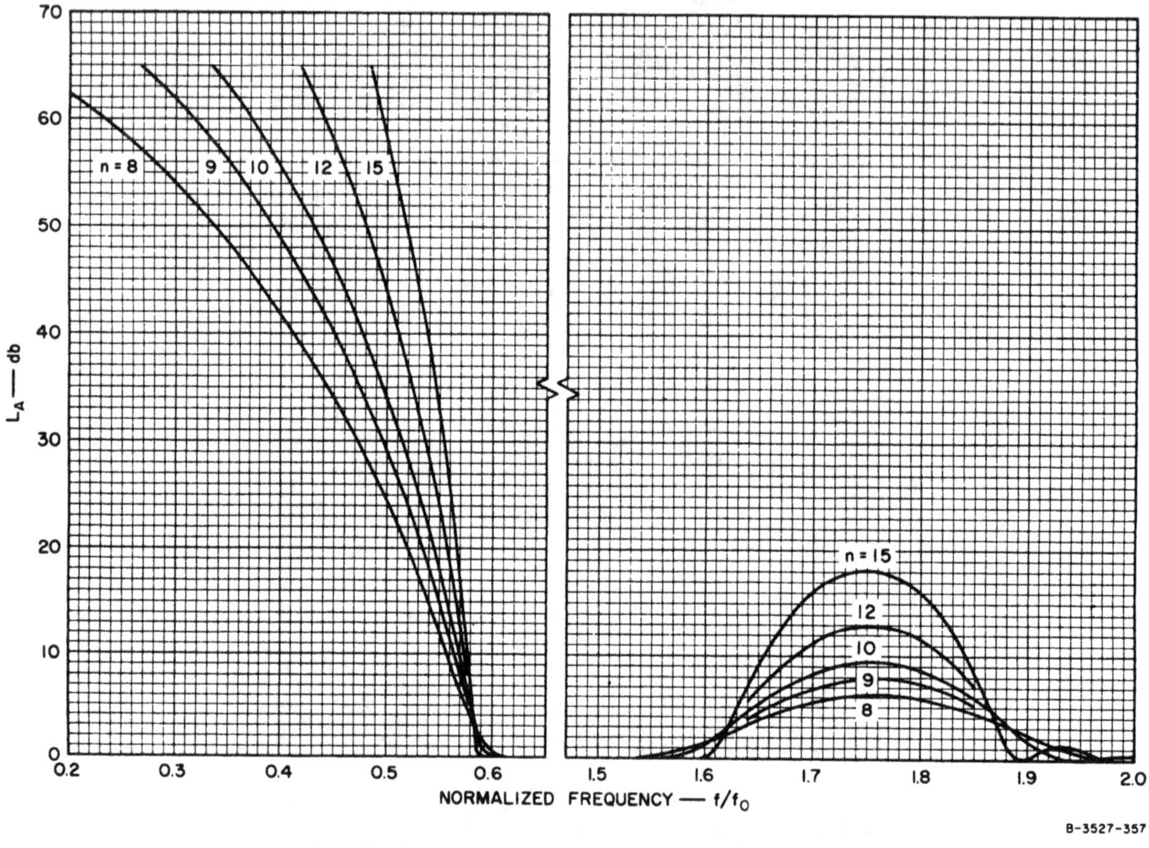

SOURCE: Quarterly Progress Report 5, Contract DA 36-039, SC 87398, SRI; reprinted in the Microwave Journal (See Ref. 17 by Leo Young and B. M. Schiffman).

FIG. 9.06-2 STOP-BAND RESPONSE OF FIVE PSEUDO-HIGH-PASS FILTERS

The rate of cutoff (skirt selectivity) improves as the number of resonators is increased, as can be seen from Fig. 9.06-2, increasing by approximately the same number of decibels per resonator added. Since the upper stop band is not particularly useful, these filters will find application as high-pass rather than as band-pass filters.

SEC. 9.07, AN EXPERIMENTAL WIDE-BAND WAVEGUIDE FILTER

In order to test the performance of the pseudo-high-pass filter described in Sec. 9.06, a six-cavity filter derived from the eight-cavity design of Sec. 9.09 was constructed in waveguide. Because all of the central cavities have been removed, the six-cavity design is the smallest possible filter of the group of periodic structures with three matching resonators on each end. The end three shunt susceptances, the single shunt susceptance in the center, and the line lengths are unchanged.

A photograph of the waveguide filter is shown in Fig. 9.07-1. For this design, WR-137 waveguide (1.372 inches × 0.622 inch I.D.) was used and the lower cutoff frequency was chosen to be 5.4 Gc, which is well above the waveguide cutoff frequency of 4.3 Gc. The upper band edge is 9.0 Gc and the design center frequency is 6.8 Gc. The shunt susceptances are symmetrical irises 0.020 inch thick designed with the aid of graphs in Fig. 8.06-2(b).

An empirical adjustment was made to allow for the iris thickness, namely, the aperture d (Fig. 8.06-2) was increased by 0.020 inch (an amount equal to the iris thickness). It is known that the shunt susceptance of a waveguide iris is not an exact linear function of guide wavelength; the iris inductances cannot therefore be treated as constants independent of frequency. Since the most important single frequency of a high-pass filter is the lower cutoff frequency, the iris susceptances were made to coincide with the values of the design susceptances at 5.4 Gc (the lower band-edge or cutoff frequency) rather than at 6.8 Gc (the band center). The response of the filter was then computed using the curves in Fig. 8.06-2(b) to obtain the numerical values of the shunt susceptances at each frequency. It was found that the computed difference in performance between the physically more realistic filter [based on Fig. 8.06-2(b)] and the one assuming constant iris inductance (*i.e.*, susceptance proportional to guide wavelength) could hardly have been detected experimentally. We therefore conclude that the irises might equally well have been designed to have the correct values at the band center.

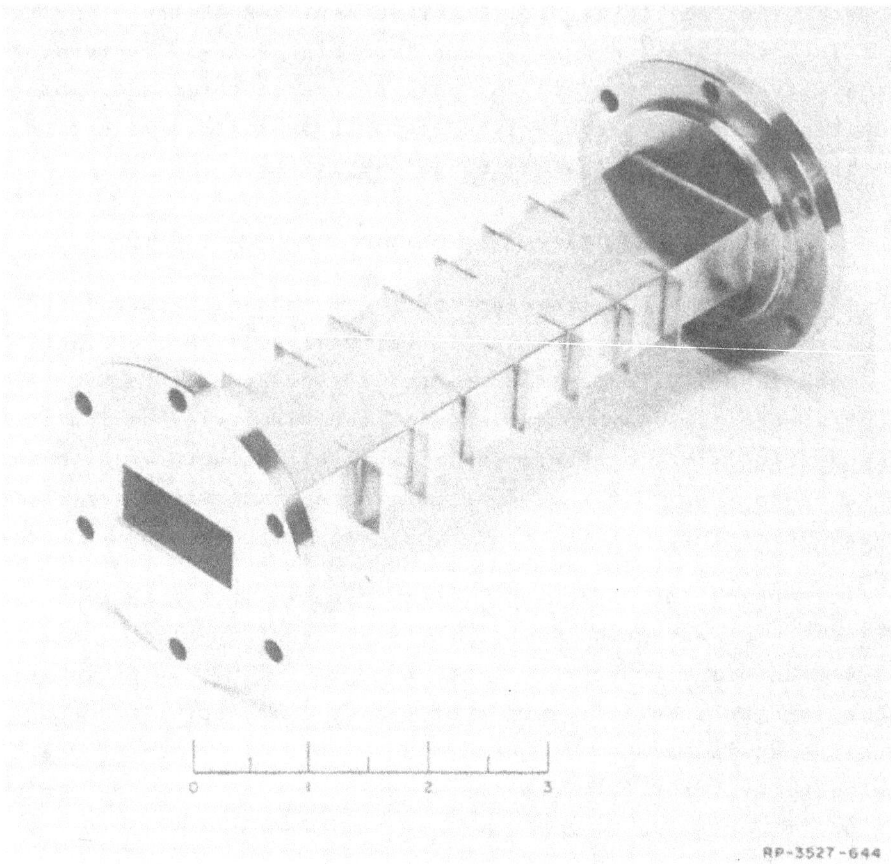

FIG. 9.07-1 EXPERIMENTAL WAVEGUIDE FILTER FOR C-BAND

The widths of the iris-openings for the experimental six-cavity filter (starting at one end and going towards the center) are 1.026, 0.958, 0.898, and 0.870 inch, respectively. The cavity lengths (again starting from one end) are 0.626, 0.653, and 0.674 inch, respectively. Insertion loss and VSWR measurements were made, and the results were as shown in Fig. 9.07-2. While the general shape of the computed response curve* was followed remarkably well by the measured points, a slight shift toward the higher-frequency region is evident. This may be due in part to the fact that cavity lengths were measured between center lines of the 0.020-inch irises with no allowance for iris thickness.

* The computed curve shown was programmed for constant inductance, *i.e.*, susceptance directly proportional to guide wavelength.

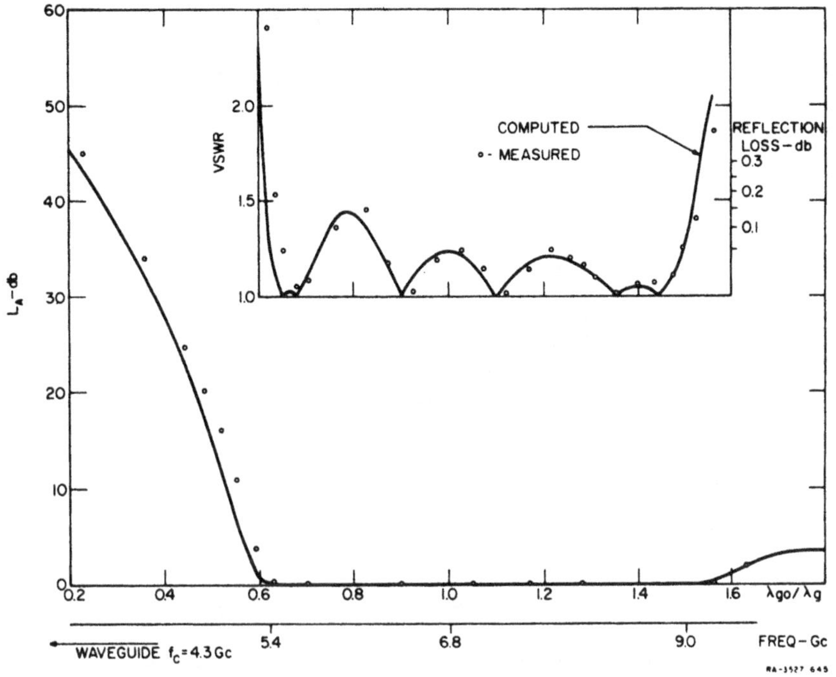

COMPUTED AND MEASURED PASS-BAND RESPONSE OF
6-RESONATOR EXPERIMENTAL FILTER FOR C-BAND

SOURCE: *The Microwave Journal* (see Ref. 17 by L. Young and B. M. Schiffman)

FIG. 9.07-2 COMPUTED AND MEASURED PASS-BAND RESPONSE OF SIX-RESONATOR EXPERIMENTAL FILTER FOR C-BAND

SEC. 9.08, DESIGN FOR SPECIFIED BAND EDGES AND STOP-BAND ATTENUATION

Typical characteristics of a quarter-wave transformer and the reactively coupled filter derived from it are shown in Fig. 9.08-1. The transformer prototype has a symmetrical response (broken line) on a frequency scale. Denoting its band edges by f_1' and f_2' (Fig. 9.08-1), the frequency of synchronous operation (Sec. 9.03) is also the mean or center frequency,

$$f_0 = \frac{f_1' + f_2'}{2} \quad . \tag{9.08-1}$$

The response is symmetrical about f_0. When the impedance steps of the transformer (Chapter 6) are replaced by series capacitances or shunt inductances, then the new response is as indicated by the solid line in Fig. 9.08-1. The following general changes should be noted:

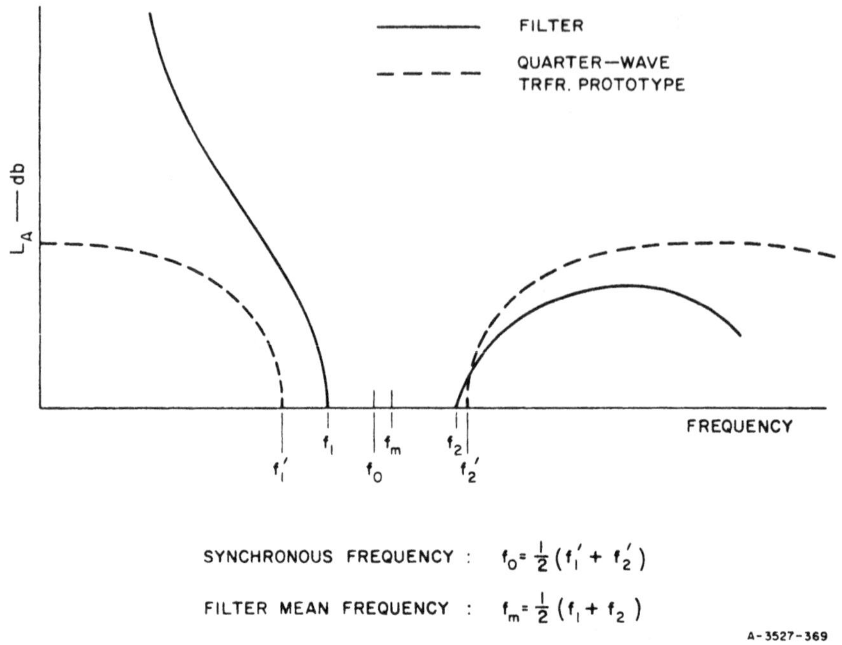

SOURCE: Quarterly Progress Report 5, Contract DA 36-039 SC-87398, SRI; reprinted in the *IEEE Trans. PTGMTT* (see Ref. 18 by Leo Young)

FIG. 9.08-1 GENERAL CHARACTERISTICS OF REACTANCE-COUPLED HALF-WAVE FILTER AND QUARTER-WAVE TRANSFORMER WITH SAME DISCONTINUITY VSWRs AND SAME SYNCHRONOUS FREQUENCY

(1) The bandwidth has contracted. (For small bandwidths this is the only change of major concern.)

(2) The lower band edge has contracted (f_1' to f_1) more than the upper band edge (f_2' to f_2). If both the transformer and the filter have the same synchronous frequency f_0, then the new mean frequency (defined as the arithmetic mean of f_1 and f_2)

$$f_m = \frac{f_1 + f_2}{2} \tag{9.08-2}$$

is greater than f_0, the frequency of synchronous tuning. Also, the two curves in the upper stop band cross each other, and the response is not symmetrical about f_m.

(3) The ripple amplitude inside the pass band, for a Tchebyscheff filter, has not changed appreciably. (This is not indicated in Fig. 9.08-1.)

Bandwidth Contraction--We shall define the fractional bandwidth w of the filter in the usual way by

$$w = \frac{f_2 - f_1}{f_m} \quad . \tag{9.08-3}$$

The fractional bandwidth w_q of the transformer is

$$w_q = \frac{f_2' - f_1'}{f_0} \quad . \tag{9.08-4}$$

The bandwidth contraction factor β is then defined by

$$\beta = \frac{w}{w_q} \tag{9.08-5}$$

and can be obtained from the graphs given in Fig. 9.08-2 as a function of R, the ratio (greater than one) of the output impedance to the input impedance of the transformer prototype. For narrow bandwidths, the pass band is nearly symmetrical on a frequency scale, and so the bandwidth also determines the band edges. (For narrow-band filters, β will be close to 0.5, as in the example in Sec. 9.04.) For wide-band filters, the band-width contraction does not give the whole story, and one has to consider the movement of the two band edges separately, as will now be shown.

Pass-Band Distortion--It will be shown in Sec. 9.10 that one would expect the response to be approximately symmetrical when plotted not against frequency, but against the quantity

$$x = \frac{\Delta f}{(f/f_0)^{\alpha}} = \frac{f - f_0}{(f/f_0)^{\alpha}} \tag{9.08-6}$$

as shown in Fig. 9.08-3, and that for highly selective filters (filters corresponding to large transformer output-to-input impedance ratio, R), the exponent α is given approximately by

$$\alpha \approx 1 + \frac{1}{n} \quad (\text{large } R) \tag{9.08-7}$$

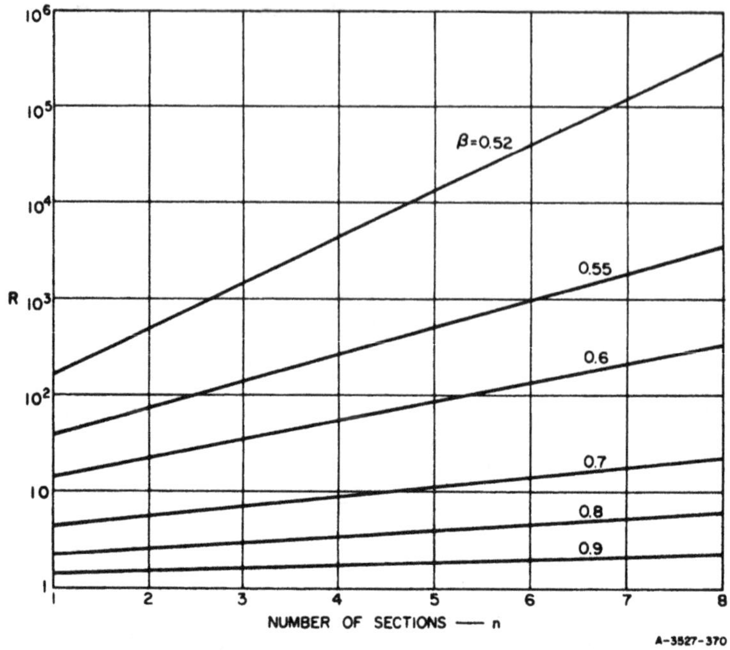

SOURCE: Quarterly Progress Report 5, Contract DA 36-039 SC-87398, SRI; reprinted in the *IEEE Trans. PTGMTT* (see Ref. 18 by Leo Young)

FIG. 9.08-2 BANDWIDTH CONTRACTION FACTOR β AS A FUNCTION OF n (Number of Resonators) AND R (Discontinuity-VSWR Product)

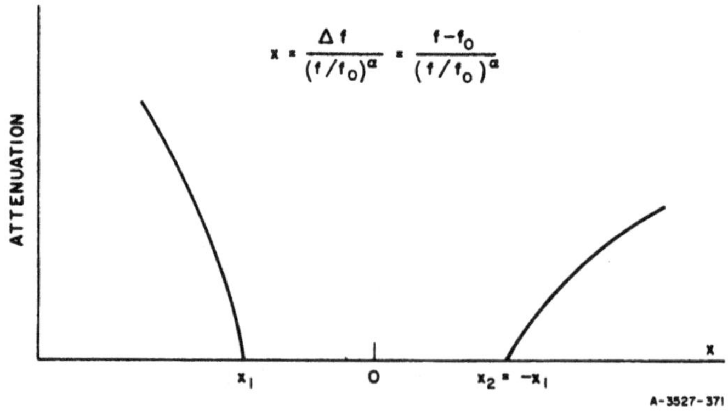

SOURCE: Quarterly Progress Report 5, Contract DA 36-039 SC-87398, SRI; reprinted in the *IEEE Trans. PTGMTT* (see Ref. 18 by Leo Young)

FIG. 9.08-3 A METHOD OF TRANSFORMING THE FREQUENCY VARIABLE TO OBTAIN AN APPROXIMATELY SYMMETRICAL FILTER CHARACTERISTIC

where n is the number of transformer sections or filter resonators. (The case $\alpha \approx 1$, corresponding to large n, leads to a symmetrical response on a wavelength scale, as previously noted by Cohn,[3] using different arguments.)

When R approaches unity, α will approach zero for synchronous filters for all n, regardless of the frequency dependence of the couplings. Thus any curve in Fig. 9.08-4 must pass through the origin. Similarly, Eq. (9.08-7) supplies the asymptotes for the graph of Fig. 9.08-4. Equation (9.08-6) can be made exact for the two band-edge frequencies, f_1 and f_2, by defining

$$\alpha = \frac{\log (\Delta f_2/\Delta f_1)}{\log (f_2/f_1)} \tag{9.08-8}$$

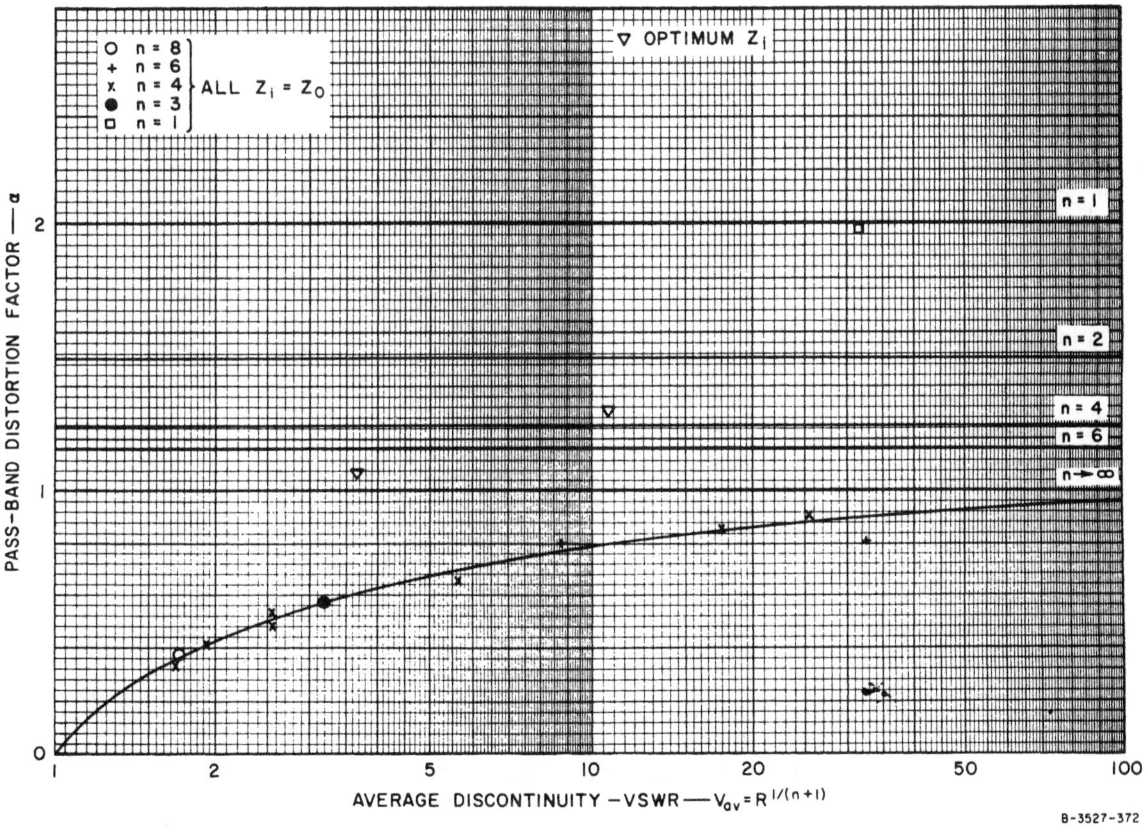

SOURCE: Quarterly Progress Report 5, Contract DA 36-039 SC-87398, SRI; reprinted in the *IEEE Trans. PTGMTT* (see Ref. 18 by Leo Young)

FIG. 9.08-4 PASS-BAND DISTORTION FACTOR, α, VERSUS AVERAGE DISCONTINUITY VSWR, V_{av}, SHOWING THE SOLUTIONS FOR FOURTEEN PARTICULAR CASES JOINED BY A SMOOTH CURVE

where

$$\left.\begin{aligned} \Delta f_2 &= f_2 - f_0 \\ \Delta f_1 &= f_0 - f_1 \end{aligned}\right\} \quad . \tag{9.08-9}$$

Equation (9.08-8) will henceforth be used as the definition of α. The parameter α was thus calculated for fourteen widely different filters whose response curves had been computed, having from $n = 3$ to 8 resonators plus one with $n = 1$ resonator, and for bandwidths varying from narrow (10 percent) through medium to wide (85 percent). These fourteen points are plotted in Fig. 9.08-4 against the average discontinuity VSWR,

$$V_{av} = R^{1/(n+1)} \quad . \tag{9.08-10}$$

It is seen that eleven points can be joined by a smooth curve running through or very close to them. The three exceptions are explained as follows: One is for $n = 1$ (see Sec. 9.10), and then by Eq. (9.08-7) one would expect $\alpha \approx 2$, which is indeed the case. The other two points, shown by triangles, correspond to non-uniform-impedance filters, to be dealt with in Sec. 9.11. It may be concluded that the curve in Fig. 9.08-4 can generally be used to obtain the pass-band distortion factor α for filters with uniform line impedances (all Z_i equal to Z_0 in Fig. 9.03-1), and having more than about $n = 3$ resonators.

It can be shown from Eqs. (9.08-8) and (9.08-9) that the frequency displacement $f_m - f_0$ is given by

$$\left.\begin{aligned} \delta &= \frac{f_m - f_0}{f_m} = \frac{\Delta f_2 - \Delta f_1}{2f_m} = \left(\frac{A-1}{A+1}\right)\frac{w}{2} \\ &\text{where} \\ \log A &= \log\,(\Delta f_2/\Delta f_1) \\ &= \alpha \log\,(f_2/f_1) \\ &= \alpha \log\left(\frac{2+w}{2-w}\right) . \end{aligned}\right\} \tag{9.08-11}$$

This equation is *exact* when Eq. (9.08-8) is regarded as a definition as it is here. Equation (9.08-11) is plotted in Fig. 9.08-5, showing the relative mean-to-synchronous frequency displacement, $(f_m - f_0)/f_m$ as a function of the fractional bandwidth, w, for several values of the parameter α. [When $\alpha = 0$, the displacement $(f_m - f_0)$ is zero.]

This completes the discussion of the effect on the pass-band edges of changing the discontinuities from impedance steps to reactive elements. We shall now show how the stop-band attenuation is affected by this change.

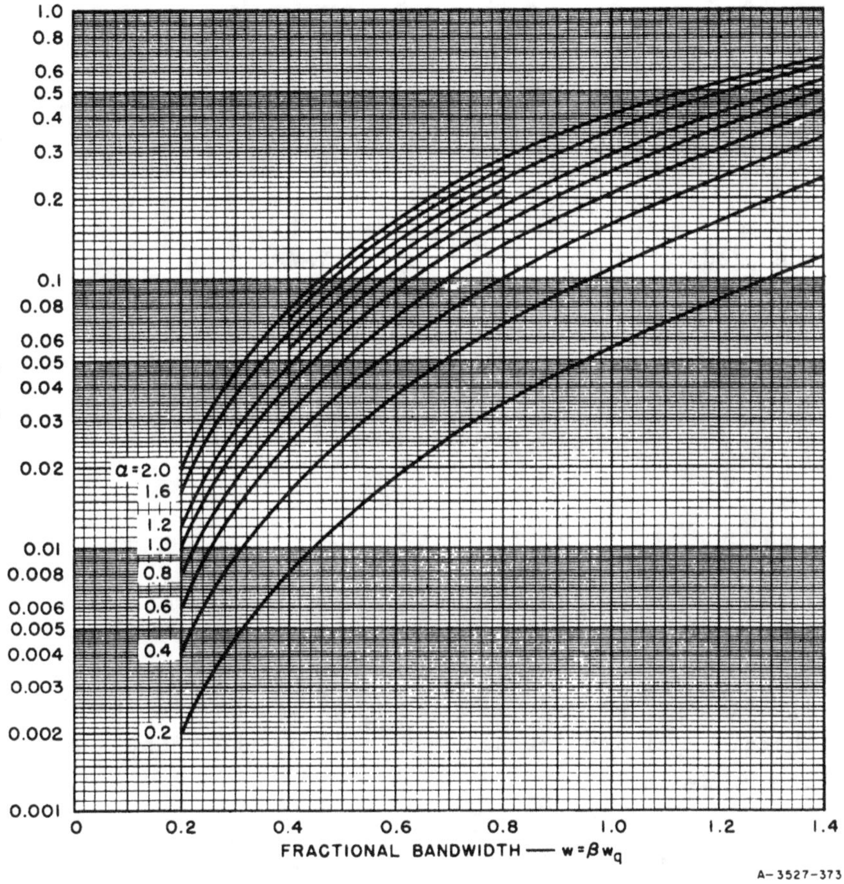

SOURCE: Quarterly Progress Report 5, Contract DA 36-039 SC-87398, SRI; reprinted in the *IEEE Trans. PTGMTT* (see Ref. 18 by Leo Young)

FIG. 9.08-5 RELATIVE DISPLACEMENT OF MEAN FREQUENCY FROM SYNCHRONOUS FREQUENCY, δ, AS A FUNCTION OF FRACTIONAL BANDWIDTH FOR SEVERAL VALUES OF α

Stop-Band Attenuation--A simple procedure will be developed for predicting the skirt of the filter response. The approximations made are such that this prediction holds closely over most of the rising portion of the skirt, but will be relatively less accurate very close to the band edge, as well as past the first attenuation maximum above the pass band; these are not serious limitations in practice. (The accuracy obtainable will be illustrated by several examples in Secs. 9.09 and 9.11.)

The excess loss, $\mathcal{E}$, of a stepped-impedance half-wave filter is (see Sec. 6.03)

$$\mathcal{E} = \frac{P_{\text{available}}}{P_{\text{load}}} - 1 = \frac{(R-1)^2}{4R}\,\frac{T_n^2(\sin\theta/\mu_0)}{T_n^2(1/\mu_0)} \tag{9.08-12}$$

where R is the product of the discontinuity VSWRs

$$R = V_1 V_2 V_3 \ldots V_{n+1} \tag{9.08-13}$$

Here, T_n is a Tchebyscheff polynomial of order n and μ_0 is a constant (Sec. 6.02),

$$\mu_0 = \sin\left(\frac{\pi w_q}{4}\right) \quad . \tag{9.08-14}$$

The response of the reactive-element filter is also given by Eq. (9.08-12) except that R is no longer constant, since the V_i become functions of frequency, as a result of the changing susceptances or reactances. Therefore at any frequency f [and for the shunt-susceptance filter of Fig. 9.03-1(b)]

$$V_i(f) = \frac{[4 + (B_i/Y_0)^2(f_0/f)^2]^{1/2} + (B_i/Y_0)(f_0/f)}{[4 + (B_i/Y_0)^2(f_0/f)^2]^{1/2} - (B_i/Y_0)(f_0/f)} \tag{9.08-15}$$

when all the line impedances are equal. [For the series-reactance filter of Fig. 9.02-1(a), substitute (X_i/Z_0) for (B_i/Y_0).] For large enough V_i and B_i, Eq. (9.08-15) reduces approximately to

$$V_i(f) \approx (B_i f_0/f)^2 \quad . \tag{9.08-16}$$

This equation is accurate to within 20 percent for $|B| > 3$; 8 percent for $|B| > 5$; 2 percent for $|B| > 7$; and 1 percent for $|B| > 8$. For smaller $|B|$, Eq. (9.08-15) should be used. The numerical solution of Eq. (9.08-15) for $f = f_0$ is the curve marked $h = 1$ in Fig. 9.03-2.

The attenuation of the filter on both skirts of the response curve may be estimated simply and fairly accurately from the known attenuation of the transformer prototype. Using Eq. (9.08-16), R becomes a function of frequency such that approximately

$$R \propto (f_0/f)^{2(n+1)} \tag{9.08-17}$$

and by Eq. (9.08-12) the attenuation will be multiplied by the same factor when R is large. [More accurately Eq. (9.08-15) rather than Eq. (9.08-16) should be used when some of the V_i are small.] Thus to estimate the filter attenuation at a specified frequency not too close to the band edge, we may first find the transformer attenuation in decibels at the corresponding frequency and then add $20(n + 1) \log_{10} (f_0/f)$ decibels, as already stated in Eq. (9.04-2).

By the *corresponding frequency*, we here mean that frequency on the quarter-wave transformer characteristic, f', (Fig. 9.08-1) which is obtained from a linear scaling

$$\frac{\Delta f'}{\Delta f_1'} = \frac{\Delta f}{\Delta f_2} \tag{9.08-18}$$

or

$$\frac{\Delta f'}{\Delta f_2'} = \frac{\Delta f}{\Delta f_2} \tag{9.08-19}$$

depending on whether the frequency f is below the lower band edge, f_1, or above the upper band edge, f_2 (Fig. 9.08-1).

The stop-band attenuation of the filter can thus be predicted fairly accurately from the prototype transformer characteristic. More often the reverse problem has to be solved. Thus the quantities specified may include the stop-band attenuation of the *filter* at some frequency, besides (for instance) the pass-band ripple and bandwidth; it is then required to find the minimum number of resonators, n, to meet these specifications.

This problem can be solved explicitly only for the prototype circuit (Chapter 6). To find the number of resonators, n, for the reactively coupled half-wave filter to meet a specified pass-band ripple, bandwidth, and skirt selectivity (stop-band attenuation) requires trial solutions in which numbers are assumed for n until the filter meets the specifications. Where Eqs. (9.08-17) and (9.04-2) are valid, this can be worked out quickly, as illustrated in the example of Sec. 9.04. Otherwise Eqs. (9.08-13) and (9.08-15) should be used; the numerical solution is facilitated by the graph in Fig. 9.03-2. Usually it is not necessary to solve for all the V_i, but to solve only for one average discontinuity VSWR, V_{av}, given by Eq. (9.08-10), which saves time in making the calculations; this method is used in the last example of Sec. 9.09.

This completes the necessary background material required for the selection of transformer prototypes which will lead to filters of specified characteristics. The design procedure will now be summarized.

Summary of Design Procedure--The design procedure to be followed then consists of the following steps:

(1) From the filter specifications select a quarter-wave transformer prototype that may be expected to yield a filter with nearly the desired performance. (The selected transformer will have the same pass-band ripple as specified for the filter.)

(2) Determine β from Fig. 9.08-2, and so estimate $w = \beta w_q$. If w is not as specified, repeat with another transformer with different bandwidth w_q until this specification is met.

(3) Determine α from Fig. 9.08-4, and then $\delta = (f_m - f_0)/f_m$ from Fig. 9.08-5. If $(f_m - f_0)$ is small enough to be neglected (as will generally be the case for filters below about 10-percent bandwidth), omit Steps 4 and 5.

(4) If $(f_m - f_0)$ is significant, find f_0 from

$$f_0 = (1 - \delta)f_m \quad . \qquad (9.08\text{-}20)$$

This is the synchronous frequency, which is also the center frequency of the transformer.

(5) The upper and lower band-edges, f_2 and f_1, are next found from

$$\left.\begin{aligned} f_2 &= f_m\left(1 + \frac{w}{2}\right) \\ &= f_0\left(\frac{1 + w/2}{1 - \delta}\right) \\ \text{and} \qquad & \\ f_1 &= f_m\left(1 - \frac{w}{2}\right) \\ &= f_0\left(\frac{1 - w/2}{1 - \delta}\right) \ . \end{aligned}\right\} \tag{9.08-21}$$

(6) The values of the reactances or susceptances and their spacings are given in Eqs. (9.03-1) through (9.03-9), and must be determined at the synchronous frequency f_0.

SEC. 9.09, EXAMPLES OF FILTERS HAVING MEDIUM AND LARGE BANDWIDTHS

In this section, two further examples will be given, illustrating the design of a medium-bandwidth (20 percent), and a large-bandwidth (85 percent) filter (all of the type shown in Fig. 9.03-1). Their predicted and analyzed performances will be compared to show how accurate the method may be expected to be.

A 20-Percent-Bandwidth Filter--It is required to design a filter with four resonators to have a pass-band VSWR of better than 1.10 over a 20-percent bandwidth.

Thus $n = 4$, $w = 0.20$, $V_r = 1.10$.

Here $V_r = 1.10$ is less than $1 + (2w)^2 = 1.16$, but not very much less; reference to Sec. 9.01 suggests that this is a borderline case, for which the low-pass prototype will not work too well, but is worth trying. Using Eqs. (9.04-1) and (9.03-4), one obtains*

* These results could also be obtained using Eqs. (1) to (5) of Fig. 8.06-1 with $w_\lambda = w$ and $X_{j,j+1}/Z_0 = Y_0/B_{j+1}$.

$$\left.\begin{aligned}\frac{B_1}{Y_0} &= \frac{B_5}{Y_0} = -0.842\\ \frac{B_2}{Y_0} &= \frac{B_4}{Y_0} = -2.607\\ &\frac{B_3}{Y_0} = -3.758\end{aligned}\right\} \qquad (9.09\text{-}1)$$

with

$$\left.\begin{aligned}\theta_1 &= \theta_4 = 127.67 \text{ degrees}\\ \theta_2 &= \theta_3 = 147.24 \text{ degrees}\end{aligned}\right\} \qquad (9.09\text{-}2)$$

which also corresponds to a quarter-wave transformer or half-wave filter that would have

$$\left.\begin{aligned}V_1 &= V_5 = 2.27\\ V_2 &= V_4 = 8.67\\ &V_3 = 16.07\end{aligned}\right\} \qquad (9.09\text{-}3)$$

The product $R = V_1 V_2 \ldots V_5$ is equal to 6215 which is only about ten times $(1/w)^n = (1/0.2)^4 = 625$, which confirms that this is a borderline case. [See Eq. (9.02-2) and Sec. 6.09.]

The analyzed performance curves of the filter defined by Eqs. (9.09-1) and (9.09-2), and the stepped-impedance half-wave filter defined by Eq. (9.09-3), are plotted in Fig. 9.09-1. Neither characteristic meets the specifications very closely, because the narrow-band condition, Eq. (9.09-2), is not satisfied well enough.

Let us re-design the prototype quarter-wave transformer, or stepped-impedance half-wave filter, and derive the reactance-coupled filter from the quarter-wave transformer prototype. Selecting $n = 4$, $w_q = 0.40$, $V_r = 1.10$, which by Table 6.02-1 gives $R = 5625$, yields

$$\left.\begin{aligned}V_1 &= V_5 = 2.398\\ V_2 &= V_4 = 8.45\\ &V_3 = 13.71\end{aligned}\right\} \qquad (9.09\text{-}4)$$

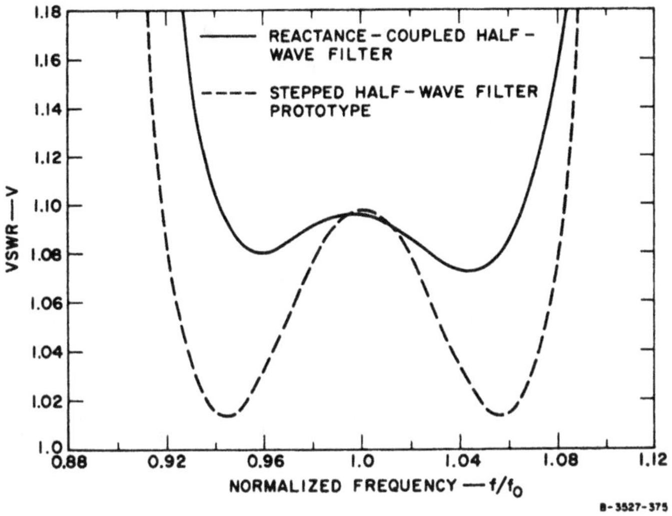

SOURCE: Quarterly Progress Report 5, Contract DA 36-039 SC-87398, SRI; reprinted in the *IEEE Trans. PTGMTT* (see Ref. 18 by Leo Young)

FIG. 9.09-1 CHARACTERISTICS OF A FOUR-RESONATOR FILTER AND ITS STEPPED HALF-WAVE FILTER PROTOTYPE, BOTH BASED ON AN EQUAL-RIPPLE LOW-PASS LUMPED-CONSTANT PROTOTYPE

These VSWRs do not seem to differ greatly from those in Eq. (9.09-3), yet they will be enough to turn the broken-line characteristic in Fig. 9.09-1 into an equi-ripple prototype response and greatly improved filter response.

From Fig. 9.08-2 for $n = 4$ and $R = 5625$, one obtains $\beta = 0.52$, so that we would expect the reactance-coupled filter bandwidth to be $w = \beta w_q = 0.52 \times 0.40 = 0.208$. From Fig. 9.08-4, for

$$V_{av} = R^{(1/n+1)} = (5625)^{1/5} = 5.65 \qquad (9.09\text{-}5)$$

we read off $\alpha = 0.65$. Hence, from Fig. 9.08-5, $\delta = 0.0064$. Then, from Eq. (9.08-21)

$$\left.\begin{aligned} f_2 &= 1.110\ f_0 \\ f_1 &= 0.902\ f_0 \end{aligned}\right\} \qquad (9.09\text{-}6)$$

where f_0 is the synchronous frequency.

For the reactance-coupled filter derived from Eqs. (9.09-4) using Eqs. (9.03-4) and (9.03-9) (which assume that $h_i = 1$)

$$\left.\begin{aligned} \frac{B_1}{Y_0} &= \frac{B_5}{Y_0} = -0.902 \\ \frac{B_2}{Y_0} &= \frac{B_4}{Y_0} = -2.563 \\ \frac{B_3}{Y_0} &= -3.436 \end{aligned}\right\} \qquad (9.09\text{-}7)$$

and therefore

$$\left.\begin{aligned} \theta_1 &= \theta_4 = 128.15 \text{ degrees} \\ \theta_2 &= \theta_3 = 145.92 \text{ degrees} . \end{aligned}\right\} \qquad (9.09\text{-}8)$$

The analyzed response of the stepped-impedance half-wave filter prototype corresponding to the exact Tchebyscheff transformer design Eq. (9.09-4) is shown by the broken line in Fig. 9.09-2, and the reactance-coupled filter response is shown by the solid line in that figure. This is an appreciable improvement on the performance of the preceding filter and transformer design based on the first procedure using the lumped-constant low-pass prototype. The analyzed performance of the filter shows that $f_1 = 0.909\ f_0$ (compare $0.902\ f_0$ predicted) and $f_2 = 1.103\ f_0$ (compare $1.110\ f_0$ predicted). The fractional bandwidth w is 0.193 (compare 0.208 predicted), and the relative mean-to-synchronous frequency displacement $\delta = (f_m - f_0)/f_m$ is 0.006 (just as predicted).

It is clear, comparing the solid and broken lines of Fig. 9.09-2, that there is still room for improvement. The main discrepancy between predicted and analyzed performance is in the bandwidth, which is 1.5 percent less than predicted. The reason for this is the difference in the frequency sensitivities of the resonator lengths; this difference is typical of filters in which some of the discontinuity VSWRs are in the neighborhood of 2.0 and others differ appreciably from the value 2.0 [see Eq. (9.09-4)]. The reason for this will be explained in Sec. 9.10.

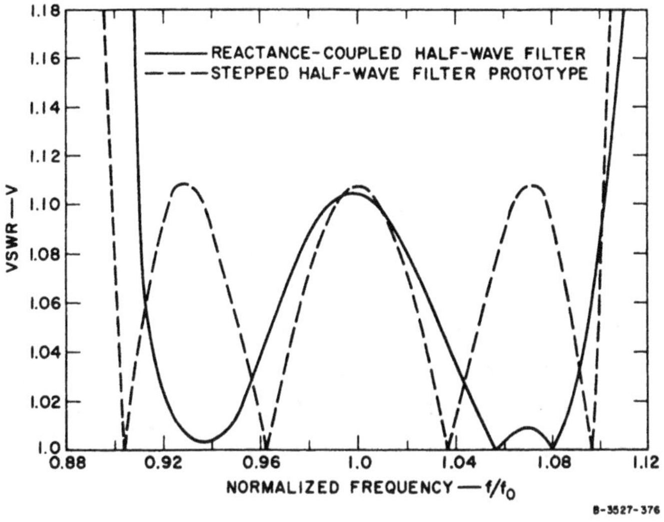

SOURCE: Quarterly Progress Report 5, Contract DA 36-039 SC-87398, SRI; reprinted in the *IEEE Trans. PTGMTT* (see Ref. 18 by Leo Young)

FIG. 9.09-2 CHARACTERISTICS OF A FOUR-RESONATOR FILTER AND ITS EQUAL-RIPPLE STEPPED HALF-WAVE FILTER PROTOTYPE

This example will then be continued in the first example of Sec. 9.11, where it will be shown that the line-length frequency-sensitivities can be equalized by optimizing the line impedances (instead of setting them all equal to each other). This generally leads to a very nearly equal-ripple characteristic with slightly more than the predicted bandwidth.

An 85-Percent-Bandwidth Filter—A pseudo-high-pass filter of eight sections is to be designed to have a pass-band frequency ratio, f_2/f_1, of approximately 2.5:1, and a pass-band attenuation (reflection-loss) of less than 0.1 db.

Since $f_2/f_1 = 2.5$,

$$w = 2\left(\frac{f_2 - f_1}{f_2 + f_1}\right) \approx 0.85 \quad . \tag{9.09-9}$$

We design a quarter-wave transformer prototype by the modified first-order theory of Sec. 6.07, specifying $w_q = 1.40$, and a 0.2 db pass-band attenuation ripple. (This approximate method always gives slightly less bandwidth, and slightly less ripple, than specified.) The modified first-order theory gives

$$\left.\begin{aligned} V_1 &= V_9 = 1.348 \\ V_2 &= V_8 = 1.561 \\ V_3 &= V_7 = 1.829 \\ V_4 &= V_6 = 1.985 \\ V_5 &= 2.034 \end{aligned}\right\} \qquad (9.09\text{-}10)$$

The computed response is shown by the broken line in Fig. 9.09-3, and it is seen that it has a pass-band attenuation of less than 0.12 db over a 135-percent bandwidth (w_q = 1.35). The quarter-wave transformer output-to-input impedance ratio, R, is

$$R = V_1 V_2 \ldots V_9 = 118.4 \qquad (9.09\text{-}11)$$

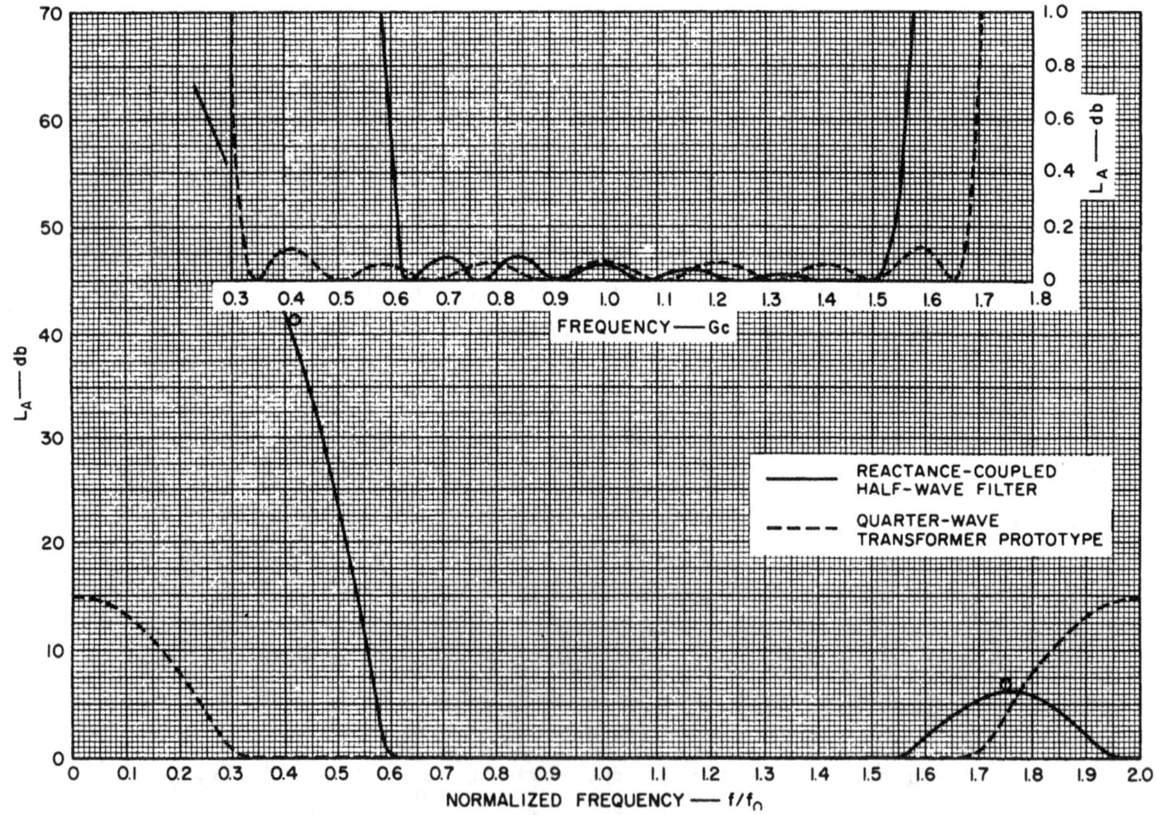

SOURCE: Quarterly Progress Report 5, Contract DA 36-039 SC-87398, SRI; reprinted in the *IEEE Trans. PTGMTT* (see Ref. 18 by Leo Young)

FIG. 9.09-3 CHARACTERISTICS OF AN EIGHT-RESONATOR FILTER AND ITS QUARTER-WAVE TRANSFORMER PROTOTYPE

From Fig. 9.08-2, the bandwidth contraction factor is $\beta = 0.63$, and the expected fractional bandwidth of the filter is therefore $w = \beta w_q = 0.85$, which is the specified bandwidth.

We also find from Fig. 9.08-4 that for $V_{av} = (118.4)^{1/9} = 1.70$, $\alpha = 0.36$. Then, from Fig. 9.08-5, one obtains $\delta = 0.06$. Therefore, by Eq. (9.08-21), we shall expect

$$\left.\begin{aligned} f_2 &= 1.52\, f_0 \\ f_1 &= 0.61\, f_0 \; . \end{aligned}\right\} \qquad (9.09\text{-}12)$$

For the reactance-coupled filter derived from Eq. (9.09-10) by use of the equations in Sec. 9.03 with $h_i = 1$,

$$\left.\begin{aligned} \frac{B_1}{Y_0} &= \frac{B_9}{Y_0} = -0.2998 \\ \frac{B_2}{Y_0} &= \frac{B_8}{Y_0} = -0.4495 \\ \frac{B_3}{Y_0} &= \frac{B_7}{Y_0} = -0.613 \\ \frac{B_4}{Y_0} &= \frac{B_6}{Y_0} = -0.700 \\ \frac{B_5}{Y_0} &= -0.725 \end{aligned}\right\} \qquad (9.09\text{-}13)$$

and

$$\left.\begin{aligned} \theta_1 &= \theta_8 = 100.60 \text{ degrees} \\ \theta_2 &= \theta_7 = 104.85 \text{ degrees} \\ \theta_3 &= \theta_6 = 108.17 \text{ degrees} \\ \theta_4 &= \theta_5 = 109.61 \text{ degrees} \; . \end{aligned}\right\} \qquad (9.09\text{-}14)$$

The response of this filter was analyzed and is shown by the solid line in Fig. 9.09-3. It is seen that the attenuation in the pass band is everywhere less than 0.1 db, the fractional bandwidth w is 0.85, the band edges are $f_2 = 1.53$ and $f_1 = 0.62$, and the relative mean-to-synchronous frequency displacement is 0.075; all of these are very close to the predicted values.

The stop-band attenuation was worked out at two frequencies, as explained in Sec. 9.08; Fig. 9.03-2, based on Eq. (9.08-15), was utilized in this calculation. These two points are shown by the small rings in Fig. 9.09-3, and fall very close to the curve obtained by analysis on a digital computer.

This filter was used to obtain the "standardized" pseudo-high-pass filters of Sec. 9.06. The three central elements, which are nearly equal (0.700, 0.725, 0.700), were averaged and each set equal to 0.710. This element was then repeated periodically as explained in Sec. 9.06, with the results shown in Figs. 9.06-1 and 9.06-2.

SEC. 9.10, DERIVATION OF THE DATA FOR BANDWIDTH CONTRACTION AND PASS-BAND DISTORTION

The basic ideas on the conversion of the quarter-wave transformer prototype into a filter with reactive elements have already been explained. The design procedure and numerical data were presented—mostly without proof. We now proceed to fill in the details of the over-all picture presented thus far.

Bandwidth Contraction--The frequency-sensitivity[1,2] (and hence bandwidth) of the reactance-coupled filters of Fig. 9.03-1 is strongly influenced by the angles ψ' and ψ'', in Fig. 9.03-3, which correspond to the electrical distances between the coupling reactance and the two reference planes with pure imaginary reflection coefficient on either side of it. Both reference planes move closer to the reactance as the frequency increases, partly (1), because a given *electrical* separation shrinks in *physical* length as the frequency increases; and partly (2) because the *electrical* lengths ψ' and ψ'' do not remain constant, but decrease with increasing frequency for shunt inductances (or series capacitances), since their susceptance (or reactance) values decrease with frequency. The movement of the reference planes is measured quantitatively by two parameters d' and d'' defined by

$$d' = \frac{4}{\pi}\left[\psi' - \frac{d\psi'}{d(f/f_0)}\right]_{f=f_0} \tag{9.10-1}$$

$$d'' = \frac{4}{\pi}\left[\psi'' - \frac{d\psi''}{d(f/f_0)}\right]_{f=f_0} \tag{9.10-2}$$

where the first term in the square brackets corresponds to Cause (1), and the second term to Cause (2). The parameters d' and d'' measure the rate of change with frequency of the reference planes in Fig. 9.03-3 as compared to the rate of change of a 45-degree line length. The spacings θ_i (Fig. 9.03-1) between reactances are given at band center by Eq. (9.03-7). The spacings are thus always longer electrically than 90-degrees, and accordingly increase with frequency faster than does a quarter-wave line length. The bandwidth of the filter is therefore always less than the bandwidth of its quarter-wave transformer prototype by a factor β. The bandwidth contraction factor associated with the *i*th resonator or line-section, β_i, is given by

$$\frac{1}{\beta_i} = \frac{2}{\pi}\left[\frac{d\theta_i}{d(f/f_0)}\right]_{f=f_0} = 1 + \frac{d''_i + d'_{i+1}}{2} \tag{9.10-3}$$

If all the β_i, defined in Eq. (9.10-3) were the same for a particular filter, then its bandwidth would be

$$w = \beta w_q \quad , \tag{9.10-4}$$

where w_q is the quarter-wave transformer bandwidth. Usually the β_i are not all equal; the smallest of the β_i should then be used for β in the above equation since the most frequency-sensitive resonator tends to determine the filter bandwidth.

To cover both the series-reactance-coupled and the shunt-susceptance-coupled filter in Fig. 9.03-1, we shall use the word *immittance* when we mean impedance for the former [Fig. 9.03-1(a)] or admittance for the latter [Fig. 9.03-1(b)]. When the line immittances are all equal, then $d'_i = d''_i$, but when the line immittances are not all equal, the d'_i and d''_i

are not equal: The larger d_i is associated with the ψ_i in the line with lower immittance, and is given by the solid lines in Fig. 9.10-1; conversely, the smaller d_i is associated with the ψ_i in the line with higher immittance, and is given by the broken lines in Fig. 9.10-1. The curves in Fig. 9.10-1 were worked out for infinitesimal bandwidths, following Eqs. (9.10-1) and (9.10-2). The curves in Fig. 9.10-2 were worked out for several finite bandwidths, replacing the differential terms in Eqs. (9.10-1) and (9.10-2) by finite increment ratios; only filters with uniform line immittances ($h_i = 1$) are shown in Fig. 9.10-2. Figure 9.08-2 was then worked out[1] with the aid of the curve $h = 1$ in Fig. 9.10-1, which is the same as the curve $w = 0$ in Fig. 9.10-2.

It has been found that Fig. 9.08-2 has predicted bandwidths closely for all the filters analyzed. The accuracy is least for filters that have a considerable spread among their β_i. According to Eq. (9.10-3) and Fig. 9.10-1 or 9.10-2, this occurs when the discontinuity VSWRs in one filter range into and out of the neighborhood of 2.0, where d can change appreciably in either direction (see Figs. 9.10-1 or 9.10-2). In that case it may be worthwhile to optimize the line impedances, as in Sec. 9.11.

Pass-Band Distortion—The distinction has already been made between the synchronous frequency, f_0, and the arithmetic mean frequency, f_m, which is always greater than f_0, since the portion of the pass band above the synchronous frequency is greater than the portion below. This phenomenon is due to the declining discontinuity VSWRs with increasing frequency when series capacitances or shunt inductances are used, and may be put on a quantitative footing as follows.

The excess loss has already been cited in Eq. (9.08-12). Consider now the case of large R, so that the approximation Eq. (9.09-16) holds. The largest term of the Tchebyscheff polynomial well inside the stop-band is the highest power of $(\sin \theta/\mu_0)$, and then Eq. (9.08-12) reduces to

$$\mathcal{E} = \frac{P_{\text{avail}}}{P_{\text{load}}} - 1 \approx \text{const.} \; \frac{\sin^{2n}\theta}{(f/f_0)^{2(n+1)}} \tag{9.10-5}$$

$$\approx \text{const.} \left[\frac{\Delta f}{(f/f_0)^{1+1/n}}\right]^{2n} \tag{9.10-6}$$

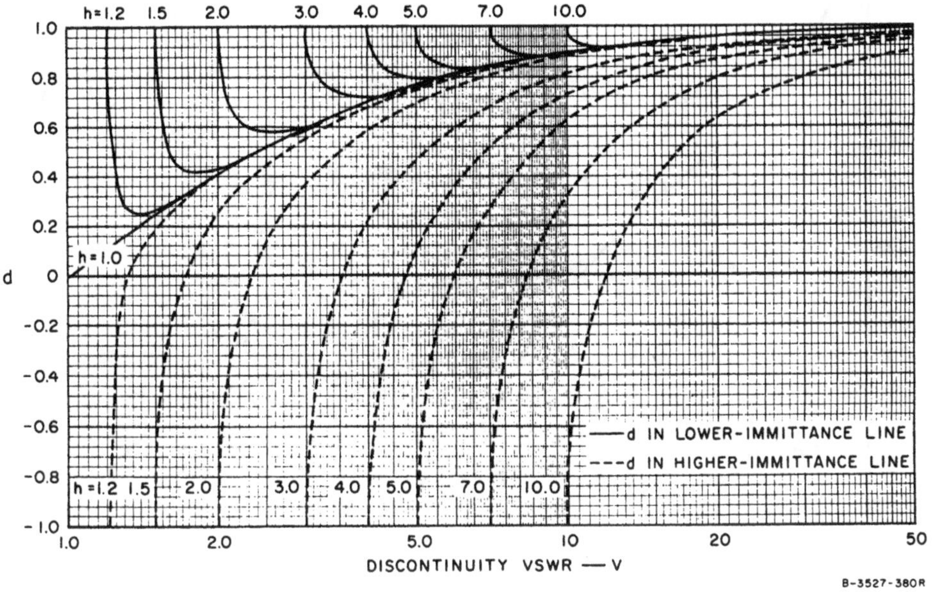

FIG. 9.10-1 CHART DETERMINING FREQUENCY SENSITIVITIES OF INDIVIDUAL RESONATORS OVER SMALL FREQUENCY BANDS

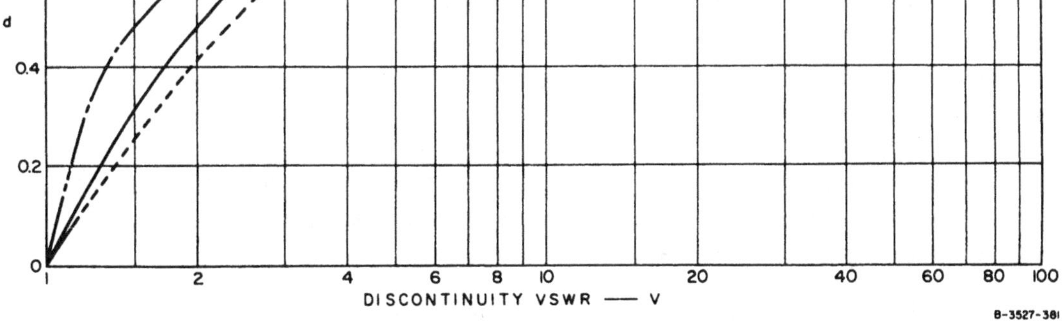

FIG. 9.10-2 CHART DETERMINING FREQUENCY SENSITIVITIES OF INDIVIDUAL RESONATORS OVER VARIOUS FRACTIONAL BANDWIDTHS, w, WHEN THE LINE IMPEDANCES ARE CONSTANT

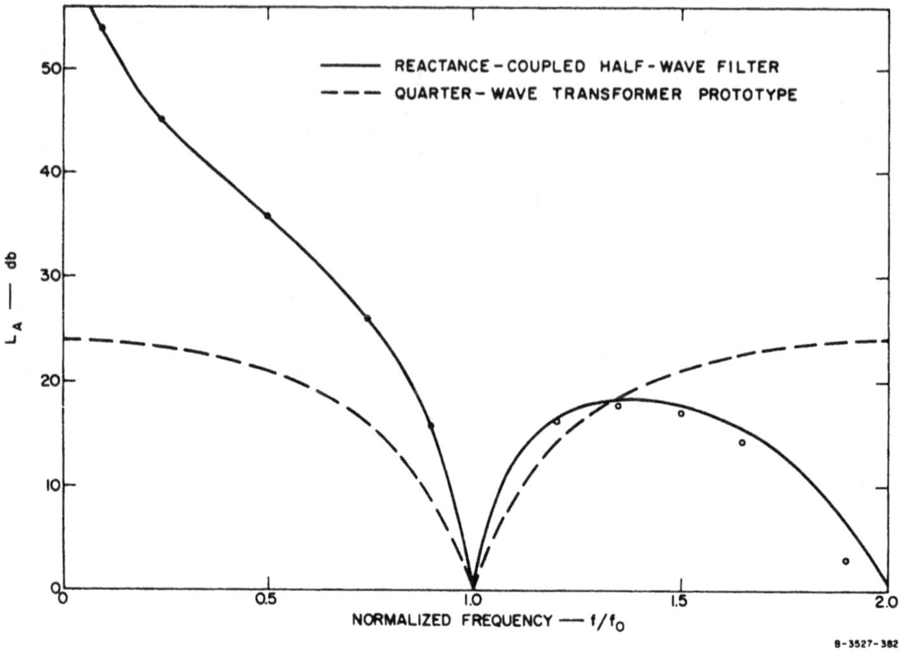

SOURCE: Quarterly Progress Report 5, Contract DA 36-039 SC-87398, SRI; reprinted in the *IEEE Trans. PTGMTT* (see Ref. 18 by Leo Young)

FIG. 9.10-3 CHARACTERISTICS OF A SINGLE-SECTION FILTER AND ITS QUARTER-WAVE TRANSFORMER PROTOTYPE

where $\Delta f = f - f_0$. This proves the result stated in Eqs. (9.08-6) and (9.08-7); for large n and large R, the exponent of (f/f_0) reduces to unity, leading to a more symmetrical response on a wavelength (rather than frequency) scale. As a counter-example, a single-resonator filter ($n = 1$) was analyzed and α calculated from Eq. (9.08-8). The response of this filter ($n = 1$, $R = 1000$) is plotted in Fig. 9.10-3, and it was found that (using the 14-db band edges for convenience) $\alpha = 1.97$, which is close to $1 + 1/n = 2.0$ as required for $n = 1$. This point is the square one in Fig. 9.08-4.

The circle points in Fig. 9.10-3 were calculated using the approximate Eq. (9.04-2) in conjunction with the prototype characteristic. This method is seen once again to give excellent results.

The choice of mapping or frequency distortion by a function of the form $\Delta f/(f/f_0)^{\alpha}$, was based on the above considerations, and is further developed in Sec. 9.08.

It was pointed out in Sec. 9.10 that different line sections of a single filter yield different bandwidth contraction factors β_i, because the quantities d_i', d_i'' vary from resonator to resonator. So far we have only considered examples of filters with uniform line impedances, where all Z_i are equal to Z_0. In deriving the discontinuity parameters, the discontinuity VSWR is always set equal to the corresponding step-VSWR of the prototype transformer; this VSWR can be obtained in the filter by an infinity of combinations of reactances with impedance ratios, since the *two* parameters h and u (Fig. 9.03-3) produce the desired discontinuity VSWR. Thus, if V_i is given and h_i is selected, X_i or B_i is determined from Eqs. (9.03-2) and (9.03-3). The problem is now to select h_i, V_i being given, so that all the β_i are the same. This can easily be done with the aid of Fig. 9.10-1, and is best illustrated by an example.

A 20 Percent-Bandwidth Filter with Optimized Line Impedances—It is required to improve the performance of the filter defined by Eqs. (9.09-7) and (9.09-8) and shown by the solid line in Fig. 9.09-2.

We see from Eq. (9.09-4) that the V_i range in numerical value from about 2 to nearly 14. Thus the different resonators have appreciably different β_i, and we might expect a noticeable deviation from an equal-ripple response, as already pointed out for this situation.

Here we have a four-resonator filter. The two central resonators, Nos. 2 and 3, are each flanked by discontinuity VSWRs of 8.45 and 13.71, according to Eq. (9.09-4). Keeping the characteristic admittances of the lines forming the four resonators the same, we find from Fig. 9.10-1 that $d_2'' = 0.88$ (corresponding to $h = 1$, $V = 8.45$) and $d_3' = 0.93$ (corresponding to $h = 1$, $V = 13.71$), so that for the two central resonators,

$$\frac{d_2'' + d_3'}{2} = \frac{d_3'' + d_4'}{2} = 0.905 \quad . \tag{9.11-1}$$

If we kept the input and output admittances the same also, so that $h = 1$ at both the first and last discontinuities, then for $V = 2.398$ [Eq. (9.09-4)] we would have $d_1'' = 0.50$, which is considerably different from the other d. Since $d_2' = d_2'' = 0.88$, this would yield $(d_1'' + d_2')/2 = 0.69$ for the outside resonators, which differs appreciably from 0.905 for the central resonators. Hence the relatively poor response shape in

Fig. 9.09-2. To obtain a value of $(d_1'' + d_2')/2$ equal to 0.905, as for the central resonators, requires $d_1'' = d_3' = 0.93$. We then find the value of h from Fig. 9.10-1. Finding the intersection of the horizontal line for $d = 0.93$ with the vertical line for $V = 2.398$ gives $h = 2.38$. One then obtains the following filter parameters:

$$\frac{Y_1}{Y_0} = \frac{Y_2}{Y_0} = \frac{Y_3}{Y_0} = \frac{Y_4}{Y_0} = \frac{1}{2.38} = 0.4202 \tag{9.11-2}$$

$$\left.\begin{aligned} \frac{B_1}{Y_0} &= \frac{B_5}{Y_0} = -0.7895 \\ \frac{B_2}{Y_1} &= \frac{B_4}{Y_1} = -2.564 \\ &\ \ \frac{B_3}{Y_1} = -3.433 \end{aligned}\right\} \tag{9.11-3}$$

$$\left.\begin{aligned} \theta_1 &= \theta_4 = 158.74 \quad \text{degrees} \\ \theta_2 &= \theta_3 = 145.92 \quad \text{degrees} . \end{aligned}\right\} \tag{9.11-4}$$

The predicted bandwidth is

$$w = \beta w_q = 0.40/1.905 = 0.210 \quad . \tag{9.11-5}$$

The appearance of such a filter with shunt-inductive irises in waveguide, or with series-capacitive gaps in strip line, is shown in Fig. 9.11-1.

The analyzed performance of this filter is shown by the solid line (marked C) in Fig. 9.11-2. The original design obtained from a lumped-constant low-pass prototype through Eqs. (9.04-1) and (9.03-4) is shown for comparison (Curve A); the performance of the filter based on the same transformer prototype as Curve C, but with *uniform* line impedances (all $h_i = 1$: see Sec. 9.09) is also shown (Curve B). It is seen that the new design, after optimizing the line impedances, gives an almost equal-ripple response. Its bandwidth is 21.8 percent, slightly more than the 21.0 percent predicted.

The distortion factor, α, worked out from Eq. (9.08-8) amounts to 1.33, and is shown by the upper triangle-point in Fig. 9.08-4. It does not fall in line with the points calculated for the constant-line-impedance filters. Most of the improvement in bandwidth is due to an increase in the upper band-edge frequency (Fig. 9.11-2) which has the effect of increasing α. A possible explanation is that V_1 and V_5 are largely determined by the impedance step, which is independent of frequency, whereas the other VSWRs (V_2, V_3, and V_4) are determined by reactances which decrease as the frequency rises. This corresponds, on the high-frequency side, to having the filter turn into a wider-band design, thus pushing the upper band edge even further up. The reverse holds

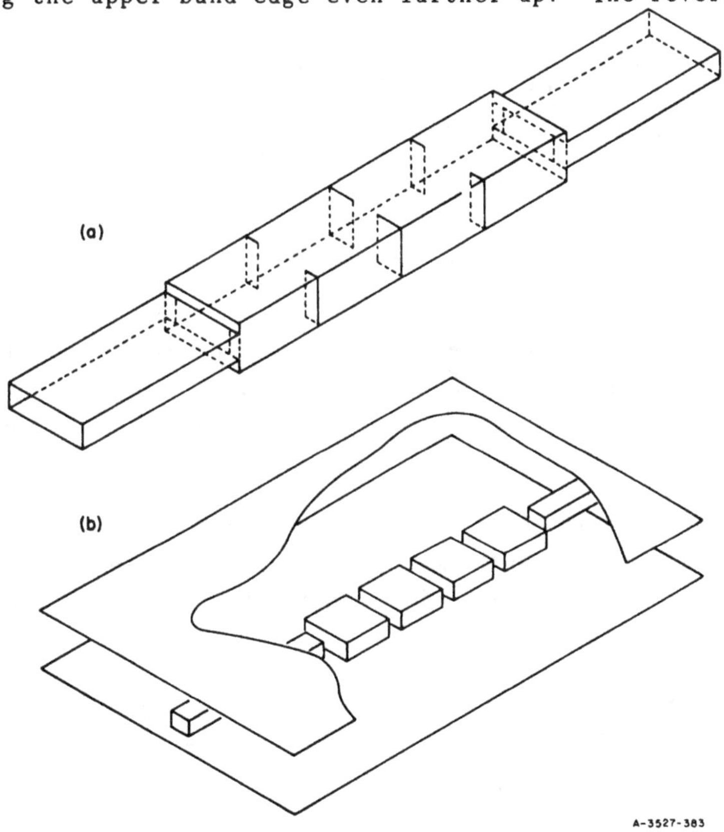

SOURCE: Quarterly Progress Report 5, Contract DA 36-039 SC-87398, SRI; reprinted in the *IEEE Trans. PTGMTT* (see Ref. 18 by Leo Young)

FIG. 9.11-1 FILTERS IN WHICH THE LINE IMPEDANCES CHANGE

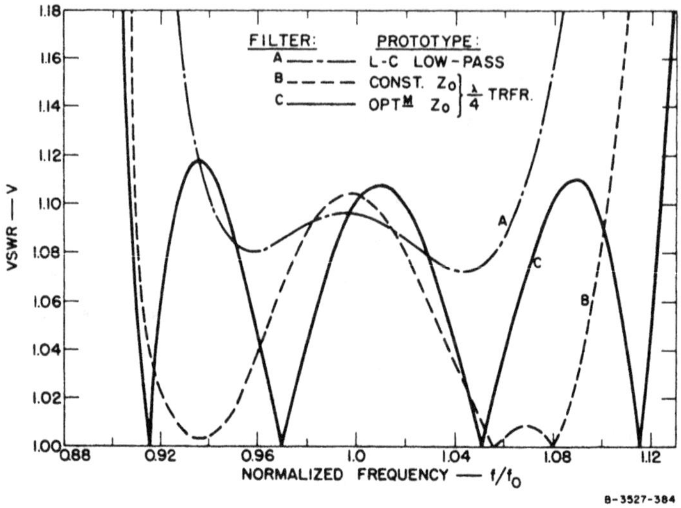

SOURCE: Quarterly Progress Report 5, Contract DA 36-039 SC-87398, SRI; reprinted in the *IEEE Trans. PTGMTT* (see Ref. 18 by Leo Young)

FIG. 9.11-2 CHARACTERISTICS IN THE PASS BAND OF THREE FILTERS DESIGNED TO THE SAME SPECIFICATIONS

(the V_i increase) below band center, which here partly cancels the improvement in bandwidth due to making all β_i equal to one another, and so the lower band edge moves less. Thus α increases by Eq. (9.08-8).

The pass- and stop-band responses of the two filters based on the same quarter-wave transformer prototype are shown in Fig. 9.11-3, along with the response of the transformer. The circle-points were calculated for the uniform-line impedance filter curve (B in Figs. 9.11-2 and -3) by the method described in Sec. 9.08.

A 30 Percent-Bandwidth Filter—The following example is worked the same way as the previous one, and only the results are given. It is based on the following prototype Tchebyscheff transformer:

$$\left.\begin{aligned} n &= 4 \\ R &= 100 \\ w_q &= 0.6 \\ \text{Ripple VSWR} &= 1.07 \end{aligned}\right\} \qquad (9.11\text{-}6)$$

giving

$$\left.\begin{aligned} V_1 &= V_5 = 1.538 \\ V_2 &= V_4 = 3.111 \\ V_3 &= 4.368 \ . \end{aligned}\right\} \qquad (9.11\text{-}7)$$

Transformed into a filter with uniform line impedances, the reactance-parameters and line-lengths are

$$\left.\begin{aligned} \frac{B_1}{Y_0} &= \frac{B_5}{Y_0} = -0.4335 \\ \frac{B_2}{Y_0} &= \frac{B_4}{Y_0} = -1.1971 \\ \frac{B_3}{Y_0} &= -1.6115 \ . \end{aligned}\right\} \qquad (9.11\text{-}8)$$

$$\left.\begin{aligned} \theta_1 &= \theta_4 = 111.57 \ \text{degrees} \\ \theta_2 &= \theta_3 = 124.88 \ \text{degrees} \ . \end{aligned}\right\} \qquad (9.11\text{-}9)$$

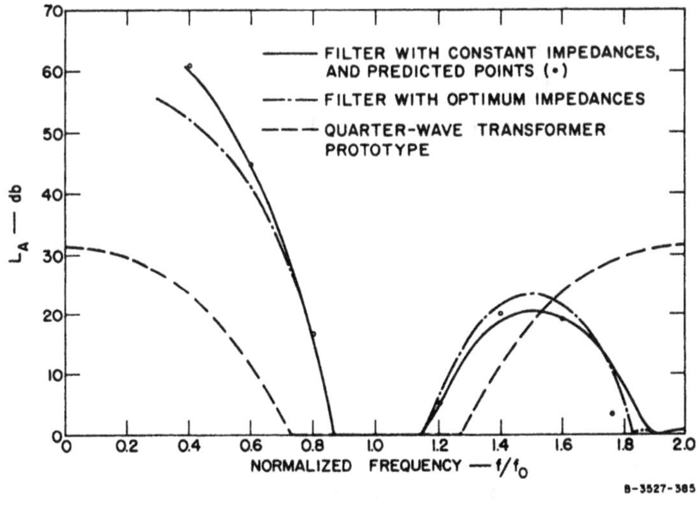

SOURCE: Quarterly Progress Report 5, Contract DA 36-039 SC-87398, SRI; reprinted in the *IEEE Trans. PTGMTT* (see Ref. 18 by Leo Young)

FIG. 9.11-3 CHARACTERISTICS IN THE STOP BAND OF A QUARTER-WAVE TRANSFORMER PROTOTYPE AND OF THE TWO FILTERS (B and C in Fig. 9.11-2) DERIVED FROM THE TRANSFORMER

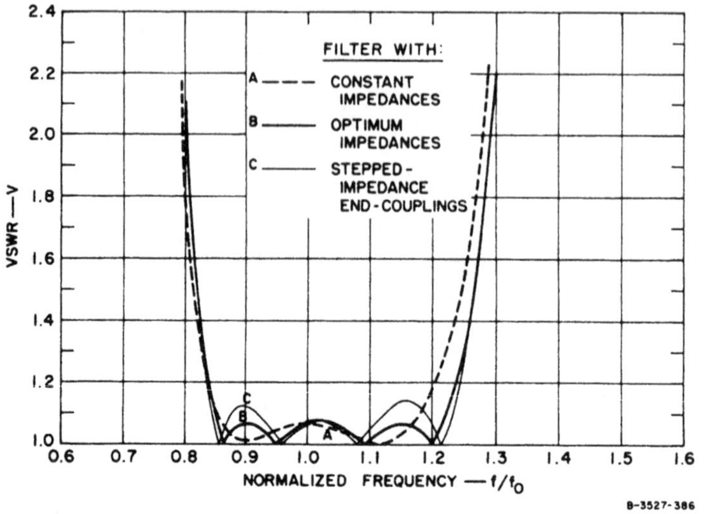

SOURCE: Quarterly Progress Report 5, Contract DA 36-039 SC-87398, SRI; reprinted in the *IEEE Trans. PTGMTT* (see Ref. 18 by Leo Young)

FIG. 9.11-4 PASS-BAND CHARACTERISTICS OF THREE FILTERS DESIGNED FOR 30-PERCENT BANDWIDTH

This example has been selected because of the appreciable spread of V_i about the numerical value 2.

With these values of V_i and all $Z_i = Z_0$, Fig. 9.10-1 shows a considerable variation in d, from about 0.25 to 0.75, and we would expect these different frequency sensitivities to result in a poor response shape. This is borne out in Fig. 9.11-4, in which Curve A is the filter response analyzed on a digital computer. It has a bandwidth of 30.7 percent, compared to 34.7 percent predicted.

To optimize the line impedances, Fig. 9.10-1 determines $h = 1.5$ for the end-couplings, and one obtains

$$\frac{Y_1}{Y_0} = \frac{Y_2}{Y_0} = \frac{Y_3}{Y_0} = \frac{Y_4}{Y_0} = \frac{1}{1.5} = 0.6667 \qquad (9.11\text{-}10)$$

$$\left.\begin{aligned} \frac{B_1}{Y_0} = \frac{B_5}{Y_0} &= -0.1198 \\ \frac{B_2}{Y_1} = \frac{B_4}{Y_1} &= -1.1971 \\ \frac{B_3}{Y_1} &= -1.6115 \end{aligned}\right\} \qquad (9.11\text{-}11)$$

$$\left.\begin{aligned}\theta_1 &= \theta_4 = 142.62 \quad \text{degrees} \\ \theta_2 &= \theta_3 = 124.88 \quad \text{degrees} \;.\end{aligned}\right\} \qquad (9.11\text{-}12)$$

The physical appearance of this filter would again be as indicated in Fig. 9.11-1. The response of the filter was analyzed and is plotted as Curve B in Fig. 9.11-4. Again there is very nearly an equal-ripple response, and the bandwidth is 36.2 percent, which is slightly more than the 35.8 percent predicted. Most of the improvement in bandwidth occurs above the band-center. A possible explanation for this effect was offered in the previous example. The distortion factor α here equals 1.07 and is shown by the lower triangle-point in Fig. 9.08-4.

Most of the end-coupling VSWR of 1.538 is due to the impedance ratio of 1.50, and only a small part is due to the normalized susceptance of 0.1198. Since most of $V_1 = V_5 = 1.538$ is due to the 1.5:1 impedance step, it is of practical interest to investigate what happens to the performance when the reactances B_1 and B_5 are left out, and the impedance ratio is increased to 1.538:1 to achieve the desired V_1 and V_5. The result is

$$\frac{Y_1}{Y_0} = \frac{Y_2}{Y_0} = \frac{Y_3}{Y_0} = \frac{Y_4}{Y_0} = \frac{1}{1.538} = 0.6502 \qquad (9.11\text{-}13)$$

$$\left.\begin{aligned}\frac{B_1}{Y_0} &= \frac{B_5}{Y_0} = 0 \\ \frac{B_2}{Y_1} &= \frac{B_4}{Y_1} = -1.1971 \\ \frac{B_3}{Y_1} &= -1.6115\end{aligned}\right\} \qquad (9.11\text{-}14)$$

$$\left.\begin{aligned}\theta_1 &= \theta_4 = 150.45 \quad \text{degrees} \\ \theta_2 &= \theta_3 = 124.88 \quad \text{degrees} \;.\end{aligned}\right\} \qquad (9.11\text{-}15)$$

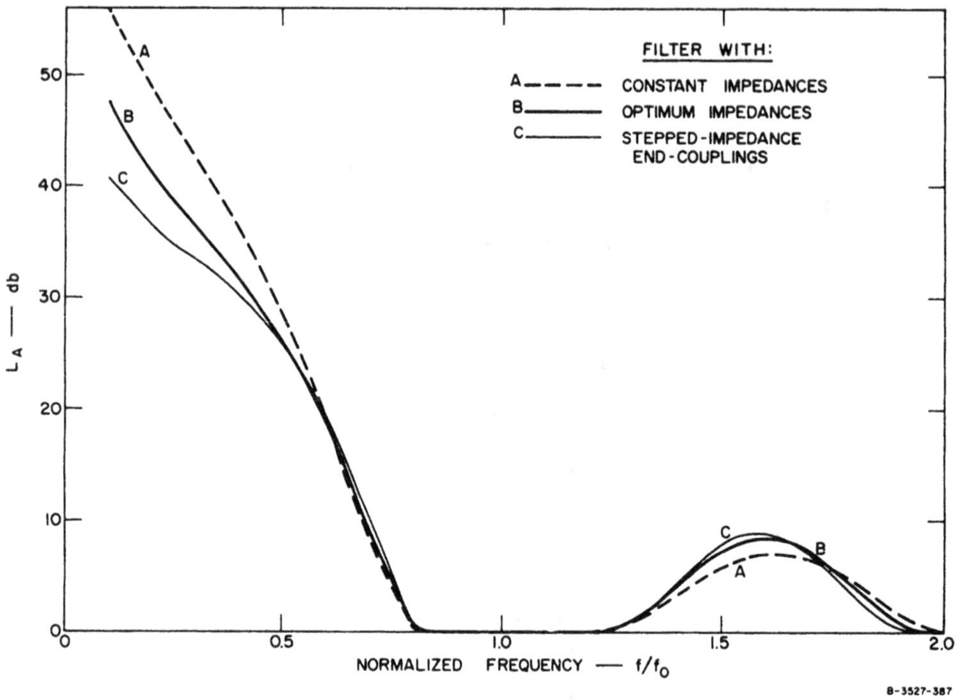

SOURCE: Quarterly Progress Report 5, Contract DA 36-039 SC-87398, SRI; reprinted in the *IEEE Trans. PTGMTT* (see Ref. 18 by Leo Young)

FIG. 9.11-5 STOP-BAND CHARACTERISTICS OF THREE FILTERS DESIGNED FOR 30-PERCENT BANDWIDTH

The analyzed performance of this filter is shown in Fig. 9.11-4 by Curve C. This filter has passed beyond the optimum performance; the peak reflection has almost doubled in the pass band, and the ripples are no longer equal. Even so, this performance is better than the first design with uniform line impedances (Curve A).

The pass- and stop-band characteristics of all three filters are shown in Fig. 9.11-5, and are in the expected relationships to each other, since the end couplings of Design A have the most capacitance, and those of Design C have none.

SEC. 9.12, REACTANCE-COUPLED QUARTER-WAVE FILTERS

The circuit of a reactance-coupled quarter-wave filter is shown in Fig. 9.12-1. This circuit was described in Sec. 8.08, where it was designed from a lumped-constant low-pass prototype. The method of Chapter 8 holds up to about 20 percent bandwidths for reactance-coupled half-wave filters, and up to about 40 percent bandwidths for reactance-coupled quarter-wave filters, which generally have about twice the fractional bandwidth when the discontinuity VSWRs are the same.

The spacings between elements at the synchronous frequency are exactly one-quarter wavelength less than for a reactance-coupled half-wave filter (Fig. 9.03-1) having the same discontinuity VSWRs, and can thus be determined with the aid of Eq. (9.03-7) or Eq. (9.03-9). [The series reactances and shunt susceptances are still determined from the quarter-wave transformer prototype step-VSWRs using Eqs. (9.03-3) or (9.03-4).] The quantities d' and d'' can be determined from Fig. 9.10-1, but the new bandwidth contraction or expansion factor β' is given by

$$\frac{1}{\beta'_i} = \frac{d''_i + d'_{i+1}}{2} \tag{9.12-1}$$

instead of Eq. (9.10-3). The bandwidth contraction factor β' for the quarter-wave reactance-coupled filter can also be determined from Fig. 9.08-2 by substituting

$$\beta' = \frac{1}{\frac{1}{\beta} - 1} = \frac{\beta}{1 - \beta} \tag{9.12-2}$$

and may thus be expected to be greater than unity. This is in keeping with the fact that all discontinuities are separated by distances *less* than one-quarter wavelength at the synchronous frequency, and the bandwidth of the filter may therefore be expected to be greater than the bandwidth of its prototype quarter-wave transformer.

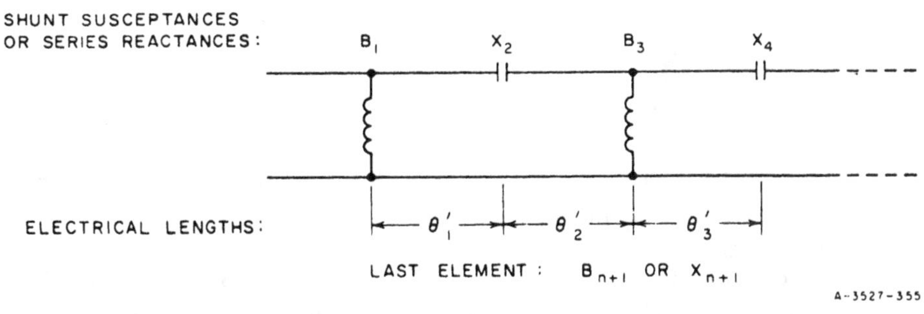

FIG. 9.12-1 REACTANCE-COUPLED QUARTER-WAVE FILTER

However, since a common 90-degree line-length (at the synchronous frequency) has been removed from each cavity, the remaining line lengths differ relatively more among themselves than before, so that one should expect less than the bandwidths predicted by Eq. (9.12-2) and Fig. 9.08-2. This will be so especially for strong couplings, when the small series reactances or shunt susceptances lead to close spacings (much less than 90 electrical degrees) between discontinuities. Then a small absolute difference in resonator lengths may amount to a large relative difference. Prediction by the previous method then becomes less and less accurate as the couplings become stronger (which usually means, as the bandwidth increases).

We shall now convert two half-wave reactance-coupled filters (one narrow-band and one wide-band) into quarter-wave reactance-coupled filters. The parameters of the filters are obtained on making the substitutions

$$\left.\begin{aligned} \frac{B_1}{Y_0} &\rightarrow \frac{B_1}{Y_0} \\ \frac{B_2}{Y_0} &\rightarrow \frac{X_2}{Z_0} \\ \frac{B_3}{Y_0} &\rightarrow \frac{B_3}{Y_0} \\ \frac{B_4}{Y_0} &\rightarrow \frac{X_4}{Z_0} \qquad \text{etc.} \end{aligned}\right\} \tag{9.12-3}$$

and

$$\theta_i - \frac{\pi}{2} \rightarrow \theta_i' \quad , \tag{9.12-4}$$

(that is, all separations are reduced by 90 degrees). The numerical values will therefore not be repeated.

A 20 Percent Bandwidth Filter—It is desired to design a quarter-wave reactance-coupled filter to have a pass-band VSWR of better than 1.10 over a 20-percent bandwidth, and to have at least 22 db of attenuation at a frequency 20 percent above band center (*i.e.*, twice as far out as the desired band edge).

A quarter-wave filter of sufficiently small bandwidth should have twice the fractional bandwidth of a half-wave filter with the same discontinuity VSWRs. Therefore the reactance-coupled quarter-wave filter specified above and the reactance-coupled half-wave filter of Sec. 9.04 will be derivable from the same prototype transformer, provided that the skirt selectivity requires the same number of resonators. The attenuation twice as far above band center as the band edge was specified as 25 db in the example of Sec. 9.04, but is only 22 db in the present example. We must expect a lower attenuation in the present example since by Eq. (9.08-17) R is reduced by a factor $(1.2)^{14}$ instead of $(1.1)^{14}$ as before. Therefore we may again try $n = 6$ resonators.

The pass-band response of this filter is shown in Fig. 9.12-2. Its bandwidth is almost twice (19.1 percent as compared with 9.7 percent) the bandwidth of the half-wave reactance-coupled resonator filter, which is also shown for comparison. This was expected. The deterioration in the response shape can be attributed to the relatively greater differences in the frequency sensitivities of the spacings.

The attenuation at $f/f_0 = 1.2$ may be calculated using Eq. (9.04-5),

$$(f/f_0)^{2(n+1)} = (1.2)^{14} = 12.8 \quad , \qquad (9.12\text{-}5)$$

which shows that the attenuation is 11.1 db less than the attenuation of the quarter-wave transformer at $(f/f_0) = 1.2$, or the attenuation of the stepped-impedance half-wave filter at $f/f_0 = 1.1$, which are both 35.5 db as in Sec. 9.04. Therefore the attenuation is predicted to be 35.5 − 11.1 = 24.4 db at $(f/f_0) = 1.2$, as shown by the circle in Fig. 9.12-3. It lies almost right on the computed curve.

The stop-band response up to the fifth harmonic is also shown in Fig. 9.12-3. Notice that the first spurious response occurs at the third (rather than at the second) harmonic; the next occurs at the fifth; and so on.

A High-Pass Filter—This filter is based on the last example in Sec. 9.09, which had an 85-percent bandwidth. It is beyond the point where any reliable predictions can be made. The response of this filter is shown in Fig. 9.12-4. Compare this figure with Fig. 9.09-3, for the reactance-coupled half-wave filter. The element spacings are now all between 10 and 20 degrees, making this almost a lumped-constant high-pass

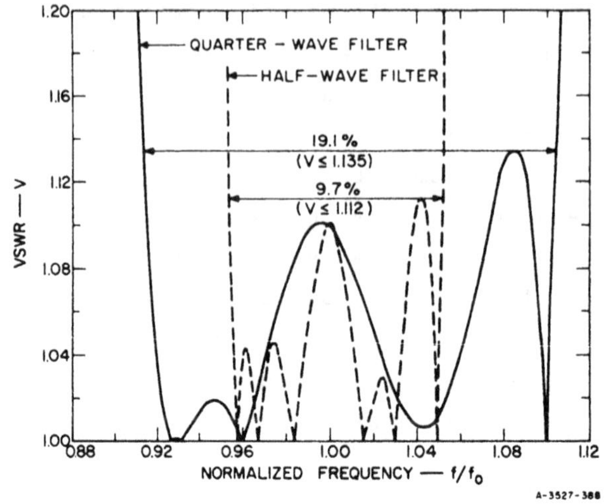

FIG. 9.12-2 PASS-BAND CHARACTERISTICS OF REACTANCE-COUPLED QUARTER-WAVE AND HALF-WAVE FILTERS

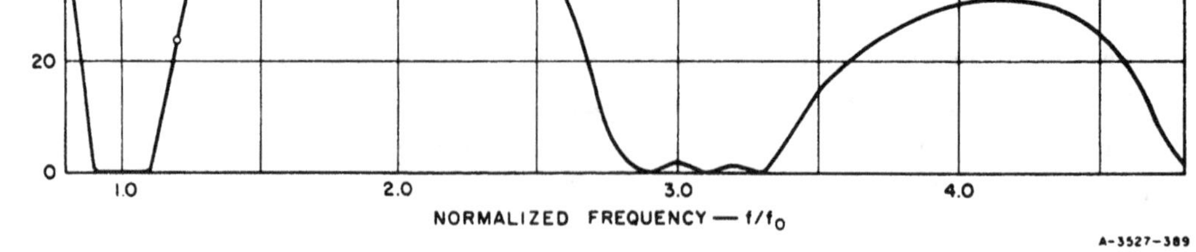

FIG. 9.12-3 STOP-BAND CHARACTERISTICS OF REACTANCE-COUPLED QUARTER-WAVE FILTER

filter. In fact, the general behavior of this filter at the low-frequency end can be predicted most simply by treating it as a lumped-constant filter. Its lower cutoff frequency, based on the formula $1/\sqrt{LC}$, is some sort of a geometric mean of the numbers in Eq. (9.09-13), as a fraction of the synchronous frequency. The actual cutoff frequency from Fig. 9.12-4 as a fraction of the synchronous frequency lies between 0.4 and 0.5, whereas the unweighted geometric mean of the numbers in Eq. (9.09-13) is 0.51.

As the frequency increases, the first stop band should occur when the element spacings are about 90 degrees. With spacings of from 10 to 20 degrees at the synchronous frequency, this stop band is expected to be centered at about 6 times the synchronous frequency. The peak attenuation will then be determined by multiplying all the discontinuity VSWRs at this frequency, which yields an input VSWR of 2.6. The peak attenuation should therefore fall just short of 1 db. All this appears to be borne out by Fig. 9.12-4, as far as it is plotted.

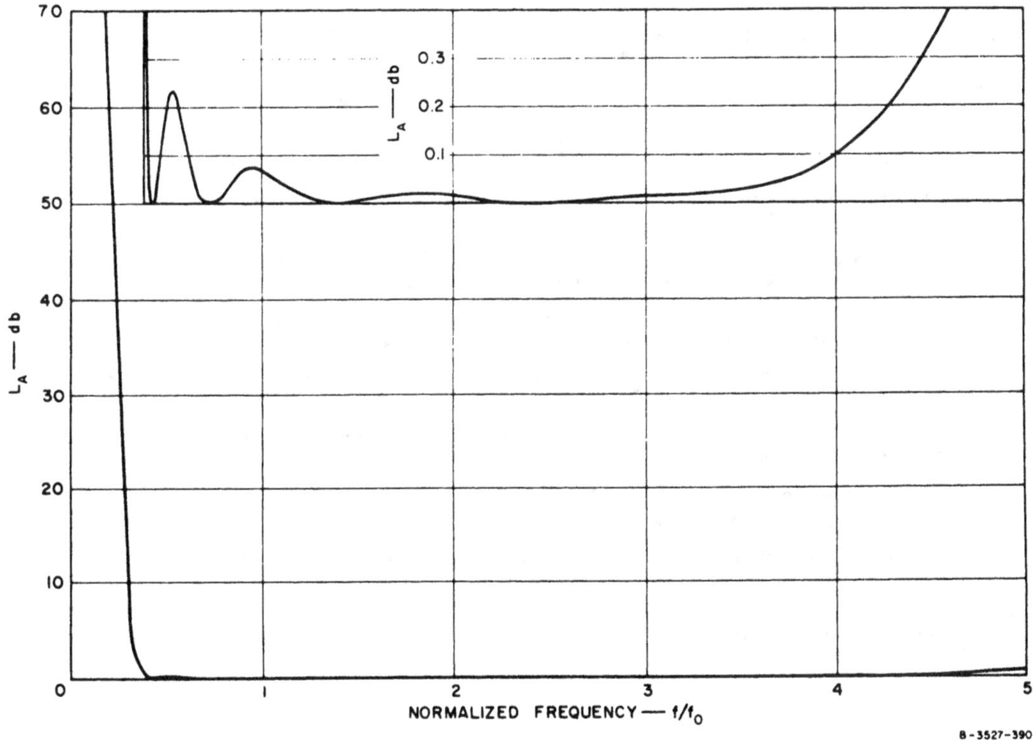

FIG. 9.12-4 CHARACTERISTICS OF A QUARTER-WAVE REACTANCE-COUPLED PSEUDO-HIGH-PASS FILTER

REFERENCES

1. G. L. Matthaei and Leo Young, "Microwave Filters and Coupling Structures, "Quarterly Progress Report 5, SRI Project 3527, Contract DA 36-039 SC-87398, Stanford Research Institute, Menlo Park, California (April 1962).

2. Leo Young, "The Quarter-Wave Transformer Prototype Circuit," *IRE Trans. PGMTT*-8, pp. 483-489 (September 1960).

3. S. B. Cohn, "Direct-Coupled-Resonator Filters," *Proc. IRE* 45, pp. 187-196 (February 1957). (See also Correspondence, p. 880, June 1957, for a discussion on the frequency dependence of the coupling elements.)

4. Leo Young, "The Practical Realization of Series-Capactive Couplings for Microwave Filters," *The Microwave Journal* 5, No. 12, pp. 79-81 (December 1962).

5. N. Marcuvitz, *Waveguide Handbook*, MIT Rad. Lab. Series Vol. 10, p. 178, (McGraw-Hill Book Co., Inc., New York City, 1951).

6. W. J. Getsinger, "Coupled Rectangular Bars Between Parallel Plates," *IRE Trans. PGMTT*-10, pp. 65-72 (January 1962).

7. Harvard University, Radio Research Laboratory, *Very High Frequency Techniques*, Vol. 2, Chapter 26 and 27. (McGraw-Hill Book Company, Inc., New York City, 1947).

8. G. L. Ragan, *Microwave Transmission*, MIT Rad. Lab. Series Vol. 9, p. 641 (McGraw-Hill Book Co., Inc., New York City, 1948).

9. E. N. Torgow, "Microwave Filters," *Electro-Technology* 67, pp. 90-96 (April 1961).

10. G. L. Matthaei, "Design of Wide-Band (and Narrow-Band) Band-Pass Microwave Filters on the Insertion Loss Basis," *IRE Trans. PGMTT*-8, pp. 580-593 (November 1960).

11. Glyn Bostick, "Design Procedure for Coaxial, High-Pass Filters," *Electronic Design*, 8, pp. 66-69 (13 April 1960).

12. H. Smith, "Design of Symmetrical Bandpass Filters," *Electronic Design* (Microwaves Section), Vol. 10, pp. 40-43 (12 April 1962). (Note: These data apply only to 50-ohm line.)

13. C. E. Muehe, "The Inductive Susceptance of Round Metal Posts Mounted in Coaxial Line," Group Report 46-32, MIT Lincoln Laboratory (5 November 1958).

14. H. Gruenberg, "Symmetrically Placed Inductive Posts in Rectangular Waveguide," *Canadian Journal of Physics* 30, pp. 211-217 (1952).

15. G. Craven and L. Lewin, "Design of Microwave Filters with Quarter-Wave Couplings," *Proc. IEE*, (London), Vol. 103B, pp. 173-177 (March 1956).

16. J. C. Simon and G. Broussaud, "Les Filtres Passe-Bande, en hyperfréquence," *Annales de Radioélectricité*," Vol. 3, pp. 3-19 (1953).

17. Leo Young and B. M. Schiffman, "A Useful High-Pass Filter Design," *The Microwave Journal* 6, No. 2, pp. 78-80, (February 1963).

18. Leo Young, "Direct Coupled Cavity Filters for Wide and Narrow Bandwidths," *IEEE Trans. PTGMTT*-11, pp. 162-178 (May 1963).

CHAPTER 10

BAND-PASS FILTERS, CONTINUED (BAND-PASS AND PSEUDO HIGH-PASS FILTERS CONSISTING OF LINES AND STUBS, OR PARALLEL-COUPLED ARRAYS OF LINES)

SEC. 10.01, CONCERNING THE FILTERS AND DESIGN PROCEDURES IN THIS CHAPTER

A summary of the various filter structures described in this chapter, along with their various advantages and disadvantages as compared with the other types of band-pass filters in Chapters 8, 9, and 10 will be found in Sec. 8.01. Wide-band filters of some of the forms discussed in this chapter are also useful for microwave high-pass applications.

The design procedures in this chapter make use of the lumped-element low-pass prototype filters in Chapter 4, as did the procedures in Chapter 8. Though the procedures in Chapter 8 are relatively simple and quite versatile, they involve fixing the various filter parameters at the midband frequency, and as a result the design equations obtained are accurate only for filters of narrow or moderate bandwidth. The design procedures in this chapter fix various filter parameters at band-edge frequencies as well as at midband. For this reason, good results are insured when these procedures are used for either narrow- or wide-bandwidth designs. Though the method used for deriving the equations in this chapter is very general in its potential application, in most cases of wide-band filters consisting of lines with lumped-reactance couplings (such as the filters in Chapter 9), the design viewpoint of this chapter leads to simultaneous equations that are a combination of transcendental and algebraic functions, which are very tedious to solve. In such cases design from step-transformer prototypes, as described in Chapter 9, is easier. However, in many cases of filters consisting of lines and stubs or parallel arrays of lines, the methods of this chapter give equations that are very easy to use.

Using a low-pass prototype filter having a response as shown in Fig. 10.01(a), along with the methods of this chapter, a band-pass filter response approximately like that in Fig. 10.01(b) will be obtained.

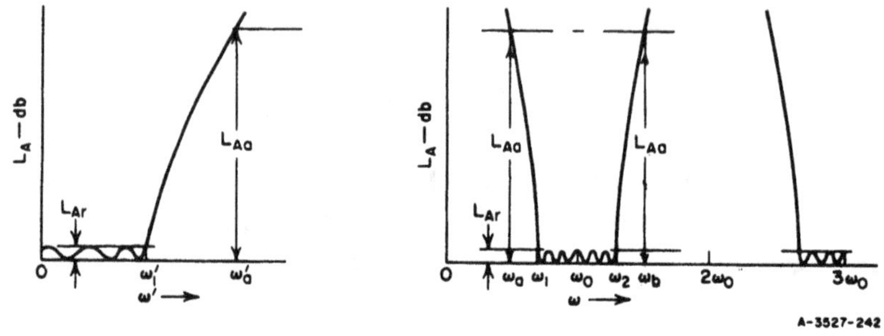

SOURCE: Final Report, Contract DA 36-039 SC-74862, SRI; reprinted in *IRE Trans. PGMTT* (see Ref. 1 by G. L. Matthaei)

FIG. 10.01-1 LOW-PASS PROTOTYPE RESPONSE AND CORRESPONDING BAND-PASS FILTER RESPONSE

Most of the filters in this chapter have additional pass bands centered at $3\omega_0$, $5\omega_0$, etc., though some (which will be noted) tend to have spurious pass bands also at $2\omega_0$, $4\omega_0$, etc., if they are not perfectly tuned. The fractional bandwidth

$$w = \frac{(\omega_2 - \omega_1)}{\omega_0} \tag{10.01-1}$$

of the band-pass filter pass band may be fixed by the designer as desired. As was done in Chapter 8, in order to estimate the rate of cutoff to be expected from the band-pass filter, approximate low-pass-to-band-pass transformations of the form

$$\frac{\omega'}{\omega_1'} = F(w, \omega/\omega_0) \tag{10.01-2}$$

will be utilized, where the primed frequencies refer to Fig. 10.01-1(a) and the unprimed frequencies to Fig. 10.01-1(b). The specific function $F(w,\omega/\omega_0)$ to be used will vary with the different filter structures considered. For two frequencies, ω' and ω, which satisfy such a mapping, the attenuation is the same for both the prototype and the band-pass filter. Hence, the low-pass prototype attenuation characteristics in Fig. 4.03-2 and Figs. 4.03-4 through 4.03-10 can be mapped into corresponding band-pass attenuation characteristics by use of such mappings. Readers who are unfamiliar with these procedures should find the example in Sec. 8.04 helpful.

SEC. 10.02, PARALLEL-COUPLED FILTERS WITH $\lambda_0/2$ RESONATORS

Figure 10.02-1(a) shows a strip-line filter having n parallel-coupled resonators which are short-circuited at both ends and which are a half-wavelength long at midband. The filter may be viewed as being pieced together from $n + 1$ parallel-coupled sections $S_{k,k+1}$ which are a quarter-wavelength long in the medium of propagation at the midband frequency ω_0. The filter in Fig. 10.02-1(b), which uses open-circuits at the ends of its resonators, is the dual of the filter in Fig. 10.02-1(a). Both types of filter can have identical transmission characteristics, the main bases for choice between them being related to their method of fabrication.

Figure 10.02-2 shows a possible way of fabricating the filter in Fig. 10.01-1(a). In this structure the resonators consist of rectangular bars which are supported by the short-circuit blocks at their ends. This construction requires no dielectric material (hence it eliminates dielectric loss), and can easily achieve the tight coupling between resonator elements that is required for wide-band filters. The required bar dimensions can be obtained from the odd- and even-mode admittances calculated in this section with the aid of Figs. 5.05-9 through 5.05-13 and their accompanying discussion.

The filter in Fig. 10.02-1(b) is of the form discussed in Sec. 8.09. However, the design equations presented in Sec. 8.09 are not accurate for large bandwidths, while the equations discussed in this section give good accuracy for either narrow or wide bandwidths.

If the printed-circuit form discussed in Sec. 8.09 is used for design of wide-band filters, it will be found that unreasonably small gaps will be required between resonator elements. One way of avoiding this problem while still using printed-circuit construction is to use the interleaved construction shown in Fig. 10.02-3. In this construction alternate resonator strips are printed on two parallel strips of dielectric so that the resonators can be interleaved to achieve tight coupling. Since the structure is symmetrical about a centerline midway between the ground planes, no ground-plane modes will be excited. Interleaved resonators of this type can be designed from the odd- and even-mode impedances computed in this section with the aid of the information in Figs. 5.05-4 through 5.05-8 and their accompanying discussion. Since the propagation is in

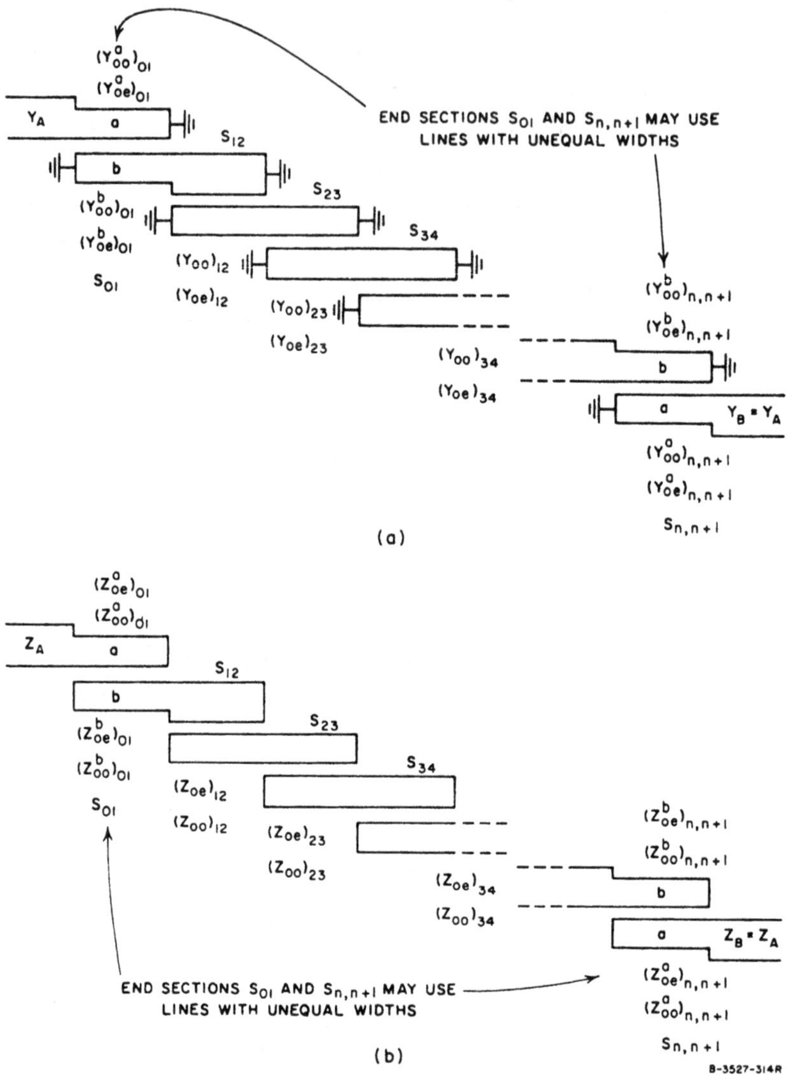

FIG. 10.02-1 TWO FORMS OF PARALLEL-COUPLED FILTERS WITH $\lambda_0/2$ RESONATORS
Each section $S_{k,k+1}$ is $\lambda_0/4$ long where λ_0 is the wavelength at frequency ω_0

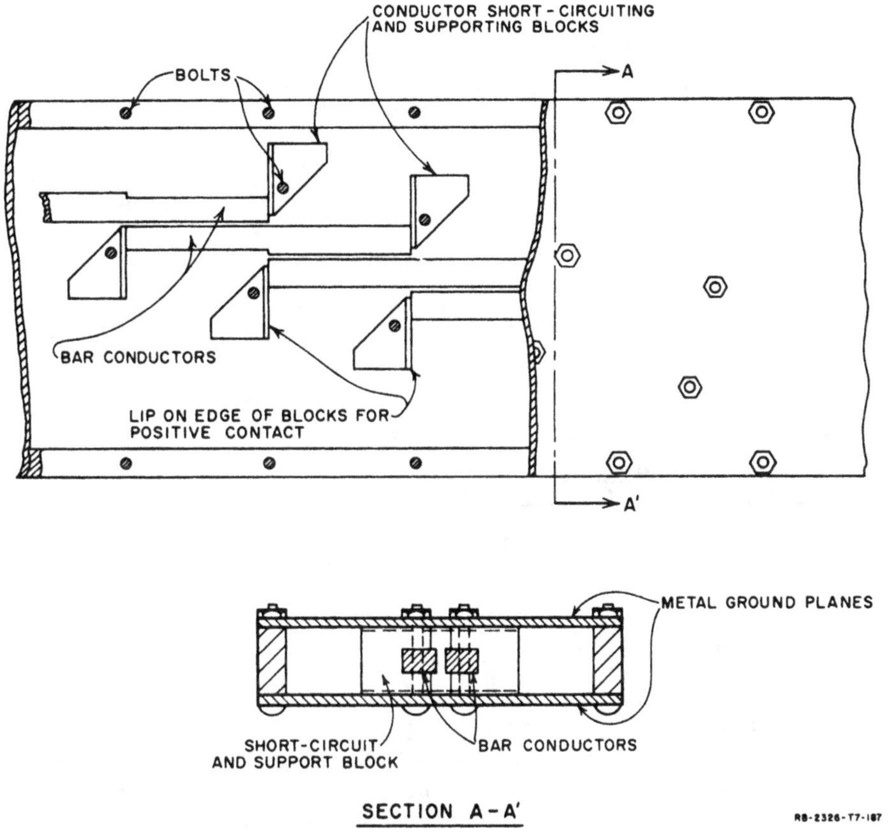

SOURCE: Final Report, Contract DA 36-039 SC-74862, SRI; reprinted in *IRE Trans. PGMTT* (see Ref. 1 by G. L. Matthaei)

FIG. 10.02-2 POSSIBLE MEANS FOR FABRICATING WIDE-BAND FILTERS OF THE TYPE IN FIG. 10.02-1(a) IN BAR TRANSMISSION-LINE CONSTRUCTION
The short-circuiting blocks support the bar conductors so that no dielectric material is required

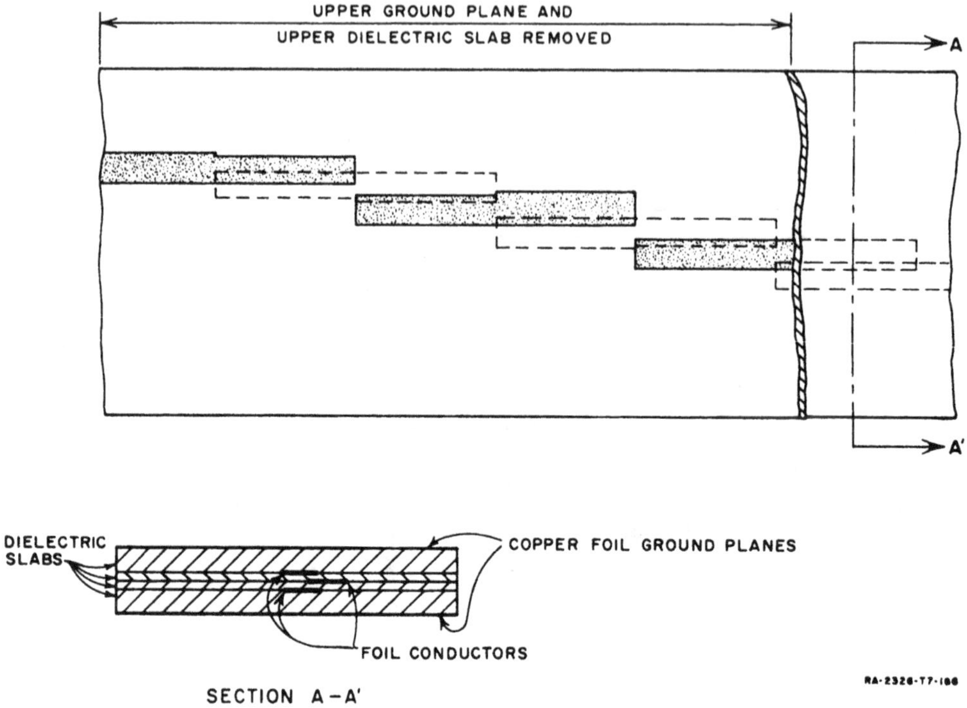

SOURCE: Final Report, Contract DA 36-039 SC-74862, SRI; reprinted in *IRE Trans. PGMTT* (see Ref. 1 by G. L. Matthaei)

FIG. 10.02-3 POSSIBLE MEANS FOR FABRICATING WIDE-BAND FILTERS OF THE TYPE IN FIG. 10.02-1(b) USING PRINTED CIRCUIT TECHNIQUES
In order to achieve tight coupling with reasonably large conductor spacings, alternate conductor strips are made to be double so that conductor strips can be interleaved

dielectric, the relative dielectric constant, ϵ_r, must be taken into account, of course, in computing the lengths and widths of the resonator elements. Each section having admittances $(Y_{oo})_{k,k+1}$ and $(Y_{oe})_{k,k+1}$ is a quarter wavelength long in the medium of propagation at frequency ω_0, and the $n + 1$ sections joined together operate as n half-wavelength resonators (when designed by the methods described in this section).

A convenient and moderately accurate low-pass-to-band-pass mapping (Sec. 10.01) which can be used to estimate the attenuation of the filters in Fig. 10.02-1 from their low-pass prototypes is

$$\frac{\omega'}{\omega_1'} = \frac{2}{w}\left(\frac{\omega - \omega_0}{\omega_0}\right) \qquad (10.02\text{-}1)$$

where

$$w = \frac{\omega_2 - \omega_1}{\omega_0} \qquad (10.02\text{-}2)$$

$$\omega_0 = \frac{\omega_2 + \omega_1}{2} \qquad (10.02\text{-}3)$$

and ω', ω_1', ω, ω_1, and ω_2 are as indicated in Figs. 10.01-1(a)(b). A more accurate mapping is

$$\frac{\omega'}{\omega_1'} = \frac{-\cos\left(\frac{\pi}{2}\frac{\omega}{\omega_0}\right)\sqrt[n]{\left|\sin\left(\frac{\pi\omega_1}{2\omega_0}\right)\right|}}{\cos\left(\frac{\pi}{2}\frac{\omega_1}{\omega_0}\right)\sqrt[n]{\left|\sin\left(\frac{\pi}{2}\frac{\omega}{\omega_0}\right)\right|}} \qquad (10.02\text{-}4)$$

where

$$\frac{\omega_1}{\omega_0} = 1 - \frac{w}{2} \qquad (10.02\text{-}5)$$

and n is the number of reactive elements in the low-pass prototype. Examples indicating the relative accuracy of these two mappings will be shown later in this section.

After the low-pass to band-pass transformation has been used to estimate the number n of resonators required to achieve a desired rate of cutoff for the filter, low-pass prototype element values g_0, g_1, g_2, ..., g_{n+1} are obtained (see Chapter 4), along with the prototype cutoff frequency, ω_1'. The fractional bandwidth w and the terminating conductances, $Y_A = Y_B$, having been specified, the odd- and even-mode admittances for the various sections of the filter can be computed in a straightforward fashion by use of the equations in Table 10.02-1. In the table the parameter h is a dimensionless admittance scale factor which can be chosen arbitrarily to adjust the admittance level within the interior of the filter as may be desired, without affecting the filter response.

Table 10.02-1

DESIGN EQUATIONS FOR PARALLEL-COUPLED FILTER IN FIG. 10.02-1(a)

Use mapping Eq. (10.02-1) or (10.02-4) to select low-pass prototype with required value of n. There will be $n + 1$ parallel-coupled sections for an n-reactive-element prototype.

END SECTIONS 0,1 AND $n, n+1$

For $k = 0$ and $k = n$ compute:

$$\frac{J_{k,k+1}}{Y_A} = \frac{1}{\sqrt{g_k g_{k+1} \omega_1'}} \ , \qquad (Y_{oo}^a)_{k,k+1} = Y_A\left(\frac{J_{k,k+1}}{Y_A}\sqrt{h} + 1\right)$$

$$(Y_{oe}^a)_{k,k+1} = 2Y_A - (Y_{oo}^a)_{k,k+1} \ , \qquad \theta_1 = \frac{\pi\omega_1}{2\omega_0} = \frac{\pi}{2}\left(1 - \frac{w}{2}\right)$$

$$(Y_{oe}^b)_{k,k+1} = (Y_{oe}^a)_{k,k+1} + hY_A\left[\frac{\tan\theta_1}{2} + \left(\frac{J_{k,k+1}}{Y_A}\right)^2\right] - Y_A$$

$$(Y_{oo}^b)_{k,k+1} = (Y_{oe}^b)_{k,k+1} + (Y_{oo}^a)_{k,k+1} - (Y_{oe}^a)_{k,k+1}$$

The parameter h is a dimensionless scale factor which may be chosen arbitrarily so as to give a convenient admittance level in the filter (see text).

INTERIOR SECTIONS 1,2 TO $n-1, n$

For $k = 1$ to $k = n - 1$ compute:

$$\frac{J_{k,k+1}}{Y_A} = \frac{1}{\omega_1'\sqrt{g_k g_{k+1}}} \ , \qquad N_{k,k+1} = \sqrt{\left(\frac{J_{k,k+1}}{Y_A}\right)^2 + \frac{\tan^2\theta_1}{4}}$$

$$(Y_{oo})_{k,k+1} = hY_A\left(N_{k,k+1} + \frac{J_{k,k+1}}{Y_A}\right)$$

$$(Y_{oe})_{k,k+1} = hY_A\left(N_{k,k+1} - \frac{J_{k,k+1}}{Y_A}\right)$$

A value for h which is usually satisfactory (and which makes at least the parallel-coupled section, S_{01}, of the filter have strips of equal width) is

$$h = \frac{1}{\dfrac{\tan\theta_1}{2} + \left(\dfrac{J_{01}}{Y_A}\right)^2} \tag{10.02-6}$$

where θ_1 and J_{01}/Y_A are given in Table 10.02-1. When the symmetric or antimetric prototypes such as those in Tables 4.05-1(a), (b) and 4.05-2(a), (b) are used, the strip widths in both the end sections, S_{01} and $S_{n,n+1}$, will all be the same, if Eq. (10.02-6) is used. Other values of h may be chosen to obtain other more convenient admittance levels (and dimensions) for the resonators or to optimize the unloaded Q of the resonator elements. (At present the choice of resonator dimensions to obtain optimum unloaded Q for parallel-coupled resonators of this sort has not been determined.) As previously mentioned, after the odd- and even-mode admittances for the $n + 1$ sections of the filter are obtained, the dimensions of the various lines are obtained by use of the data in Sec. 5.05.

Table 10.02-2 summarizes the odd- and even-mode admittances for three designs obtained by use of Table 10.02-1 along with Eq. (10.02-6). All three were designed from a Tchebyscheff low-pass prototype filter

Table 10.02-2

SUMMARY OF THE ODD-MODE AND EVEN-MODE ADMITTANCE VALUES FOR THE FILTERS OF FIGS. 10.02-4 TO 10.02-6

They were designed by use of Table 10.02-1 and realized in the form in Fig. 10.02-1(a). Equation (10.02-6) was applied so that $(Y_{oo})^a_{01} = (Y_{oo})^b_{01}$ and $(Y_{oe})^a_{01} = (Y_{oe})^b_{01}$.

	FIG. 10.02-4 (5% Bandwidth)	FIG. 10.02-5 (30% Bandwidth)	FIG. 10.02-6 (2 to 1 Bandwidth)
$(Y_{oo})_{01} = (Y_{oo})_{67}$	1.251	1.540	1.716
$(Y_{oo})_{12} = (Y_{oo})_{56}$	0.996	1.023	1.142
$(Y_{oo})_{23} = (Y_{oo})_{45}$	0.981	0.937	0.954
$(Y_{oo})_{34}$	0.980	0.927	0.933
$(Y_{oe})_{01} = (Y_{oe})_{67}$	0.749	0.460	0.284
$(Y_{oe})_{12} = (Y_{oe})_{56}$	0.881	0.491	0.208
$(Y_{oe})_{23} = (Y_{oe})_{45}$	0.895	0.536	0.250
$(Y_{oe})_{34}$	0.896	0.542	0.255

All values normalized so that $Y_A = 1$.

SOURCE: Final Report, Contract DA 36-039 SC-74862, SRI; reprinted in *IRE Trans. PGMTT* (see Ref. 1 by G. L. Matthaei).

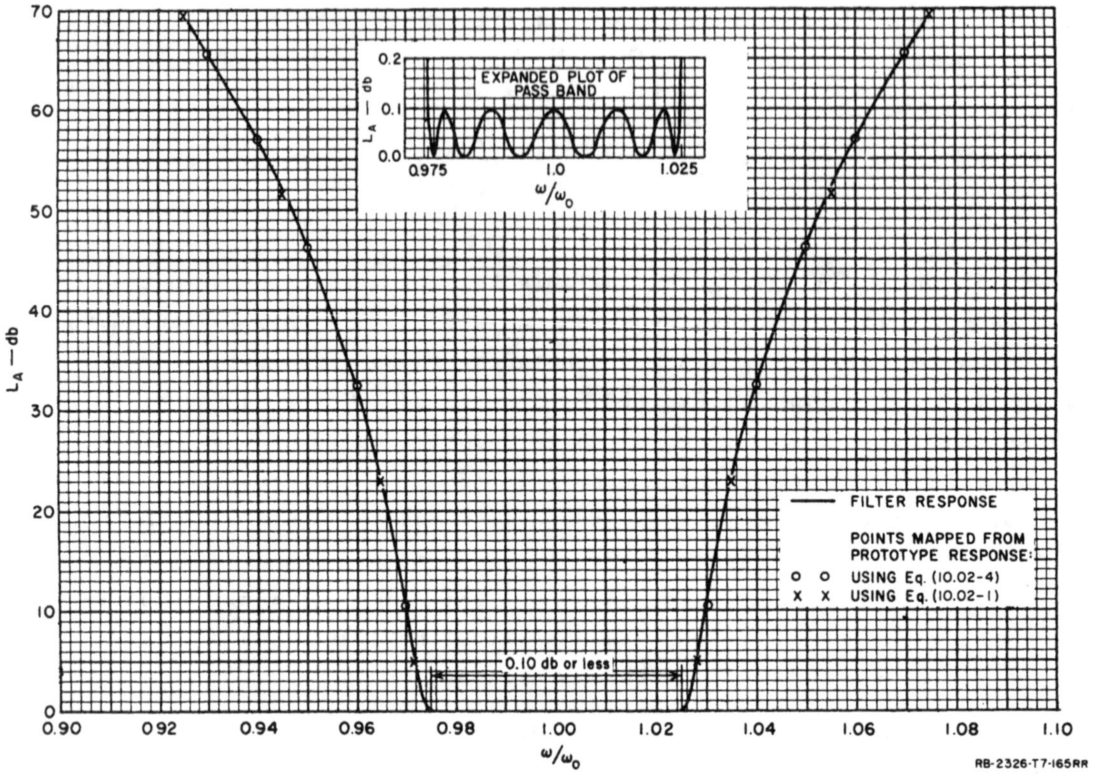

SOURCE: Final Report, Contract DA 36-039 SC-74862, SRI; reprinted in *IRE Trans. PGMTT* (see Ref. 1 by G. L. Matthaei)

FIG. 10.02-4 COMPUTED RESPONSE OF FILTERS DESIGNED AS IN TABLE 10.02-1 TO HAVE 5 PERCENT BANDWIDTH
Design value for ω_1/ω_0 was 0.975. Prototype had 0.10-db Tchebyscheff pass-band ripple with n = 6 reactive elements.

having 0.1-db ripple and n = 6 reactive elements [the element values being obtained from Table 4.05-2(a)]. As is seen from Table 10.02-2, the designs are for 5-percent, 30-percent, and an octave bandwidth; and the admittances have been normalized so that Y_A = 1. Figures 10.02-4, 10.02-5, and 10.02-6 show the computed responses of these designs. Note that even the octave-bandwidth design comes close to having the specified 0.1-db ripple in the pass band, the main error being a slight shrinkage in bandwidth. Points mapped from the low-pass prototype by use of the mapping of Eqs. (10.02-1) and (10.02-4) are also shown. Note that for the 5- and 30-percent-bandwidth designs the simple mapping in Eq. (10.02-1) gives quite good results even up to quite high attenuations. However, in the case of the octave-bandwidth design (Fig. 10.02-6), the more complicated Eq. (10.02-4) gives markedly better accuracy up around 60 db or above.

Design equations for the form of filter in Fig. 10.02-1(b) are obtained by applying duality to the equations in Table 10.02-1. Thus, the admittance quantities are replaced by impedance quantities as indicated below.

$$
\begin{aligned}
Y_A = Y_B &\longrightarrow Z_A = Z_B \\
J_{k,k+1} &\longrightarrow K_{k,k+1} \\
(Y_{oo})_{k,k+1} &\longrightarrow (Z_{oe})_{k,k+1} \\
(Y_{oe})_{k,k+1} &\longrightarrow (Z_{oo})_{k,k+1}
\end{aligned}
\tag{10.02-7}
$$

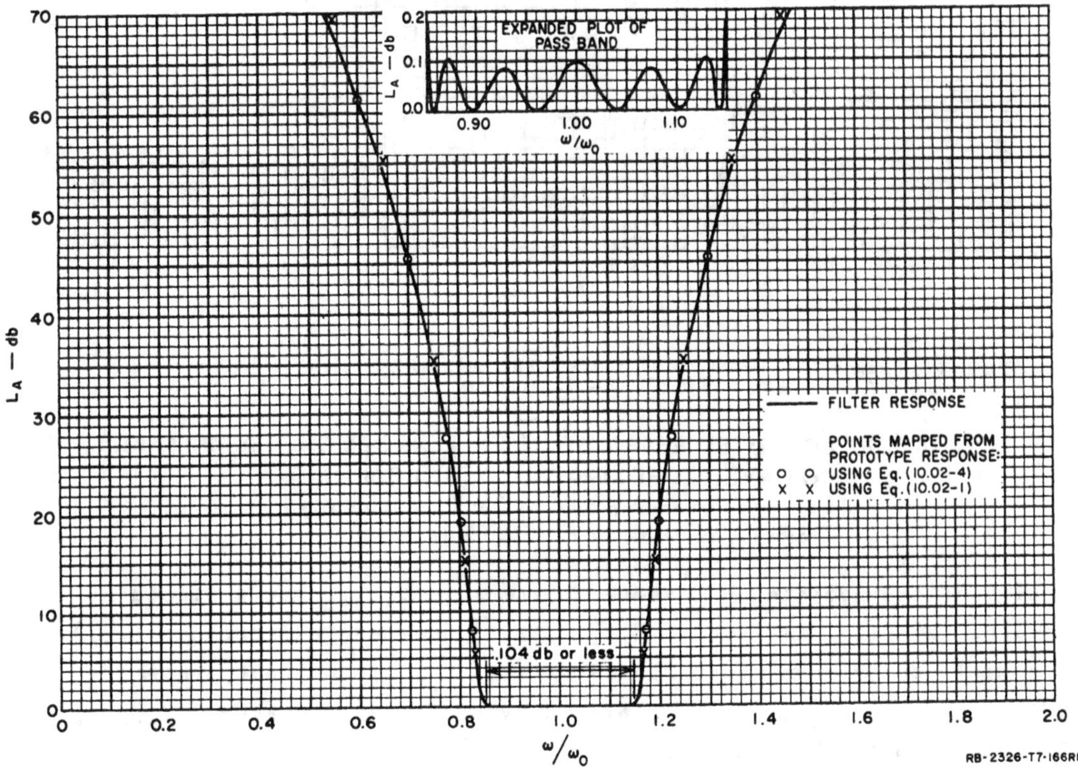

SOURCE: Final Report, Contract DA 36-039 SC-74862, SRI; reprinted in *IRE Trans. PGMTT* (see Ref. 1 by G. L. Matthaei)

FIG. 10.02-5 COMPUTED RESPONSE OF FILTERS DESIGNED AS IN TABLE 10.02-1 TO HAVE 30 PERCENT BANDWIDTH
Design value of ω_1/ω_0 was 0.850. Prototype same as for Fig. 10.02-4

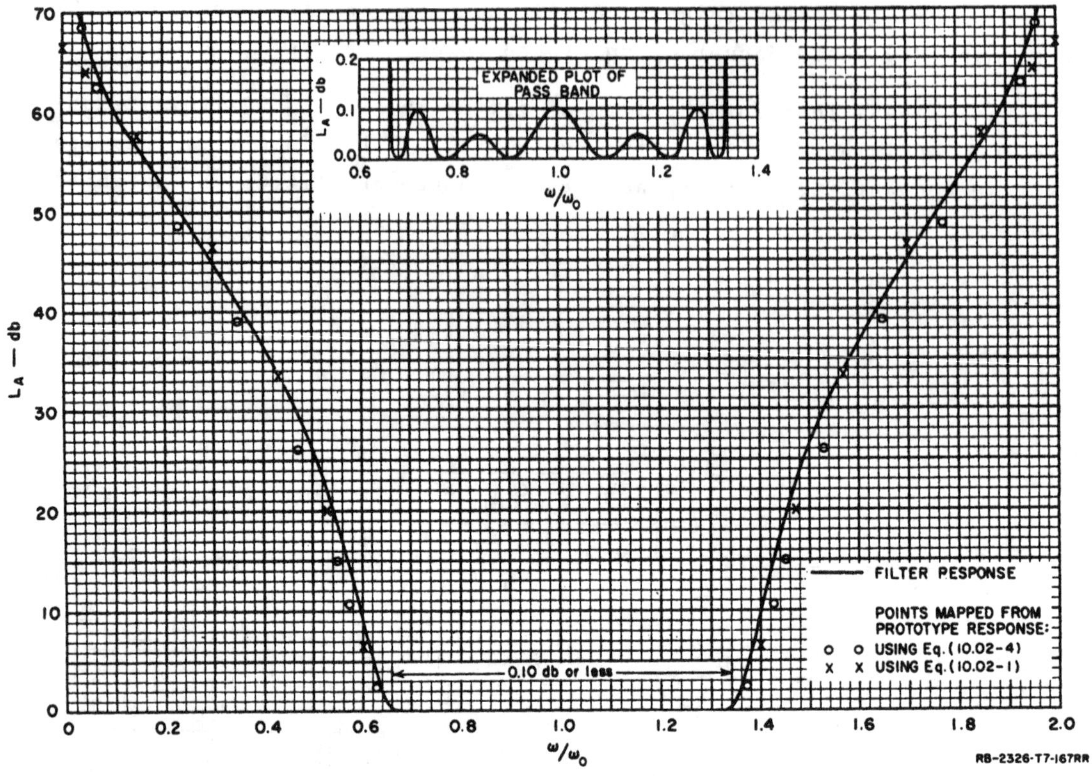

SOURCE: Final Report, Contract DA 36-039 SC-74862, SRI; reprinted in *IRE Trans. PGMTT* (see Ref. 1 by G. L. Matthaei)

FIG. 10.02-6 COMPUTED RESPONSE OF FILTERS DESIGNED AS IN TABLE 10.02-1 TO HAVE APPROXIMATELY 2 TO 1 BANDWIDTH
Design value for ω_1/ω_0 was 0.650, which calls for $\omega_2/\omega_1 = 2.077$. Prototype same as for Fig. 10.02-4

Note that the admittance inverter parameters $J_{k,k+1}$ are replaced by impedance inverter parameters, $K_{k,k+1}$; that *odd*-mode admittances $(Y_{oo})_{k,k+1}$ are replaced by *even*-mode impedances $(Z_{oe})_{k,k+1}$, and that *even*-mode admittances $(Y_{oe})_{k,k+1}$, are replaced by *odd*-mode impedances $(Z_{oo})_{k,k+1}$. The dimensionless scale factor, h, is used as before, except that in this case it scales the impedance level instead of the admittance level.

The filters in Figs. 10.02-1(a), (b) have their second pass band centered at $3\omega_0$, and in theory they have infinite attenuation at $2\omega_0$. However, the resonators are actually resonant at $2\omega_0$ (since the resonator strips are a wavelength long at that frequency), and the theoretical pole of attenuation arises only because the coupling regions between adjacent resonators are a half-wavelength long. (The coupling is maximum when the coupling gaps are $\lambda/4$ long and zero when they are $\lambda/2$ long.) However, for this to work out in fact, for the attenuation to remain high at and near ω_0 all of the coupling regions throughout the filter must be exactly $\lambda/2$ long at exactly the same frequency, which must also be exactly the frequency for which all the resonators are resonant. Since this is almost impossible to achieve in practice, there are almost always narrow spurious responses in the vicinity of $2\omega_0$ for these types of filters.

SEC. 10.03, FILTERS WITH SHUNT OR SERIES $\lambda_0/4$ STUBS

By use of equivalences which were summarized in Figs. 5.09-2(a), (b), it can be shown that parallel-coupled strip-line filters of the form in Fig. 10.02-1(a) are electrically exactly equivalent to stub filters of the form in Fig. 10.03-1. Likewise, parallel-coupled strip-line filters of the form in Fig. 10.02-1(b) are exactly equivalent to stub filters of the form in Fig. 10.03-2. It would be possible to work out a parallel-coupled filter design by the procedures in Sec. 10.02 and then convert it to either the form in Fig. 10.03-1 or that in Fig. 10.03-2; however, parallel-coupled filter designs that have reasonable impedance levels and dimensions generally have quite unreasonable impedance levels when converted to a stub-filter form. Similarly, practical stub filter designs generally do not convert to practical parallel-coupled designs. For this reason the design equations in this section are based on a somewhat different design procedure than are those in Sec. 10.02. This procedure gives reasonable impedance levels in typical stub filters, and also makes complete use of all of the natural modes of vibration of the

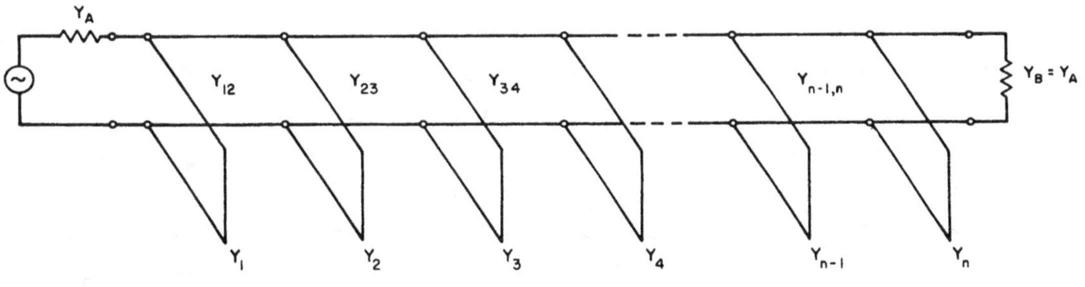

SOURCE: Final Report, Contract DA 36-039 SC-74862, SRI; reprinted in *IRE Trans. PGMTT* (see Ref. 1 by G. L. Matthaei)

FIG. 10.03-1 A BAND-PASS FILTER USING $\lambda_0/4$ CONNECTING LINES

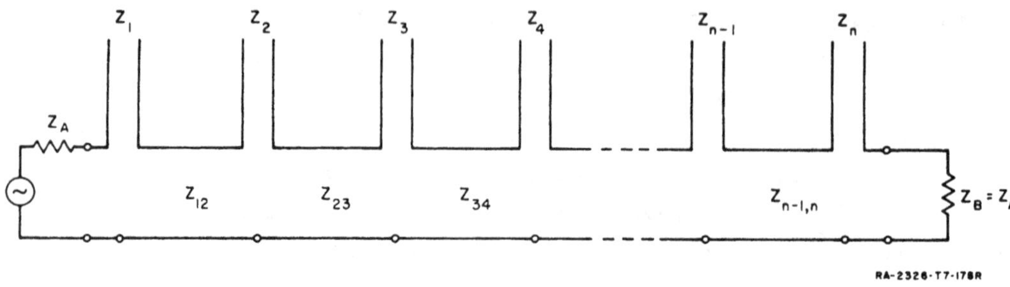

SOURCE: Final Report, Contract DA 36-039 SC-74862, SRI; reprinted in *IRE Trans. PGMTT* (see Ref. 1 by G. L. Matthaei)

FIG. 10.03-2 A BAND-PASS FILTER USING $\lambda_0/4$ SERIES STUBS AND $\lambda_0/4$ CONNECTING LINES
This filter is the dual of that in Fig. 10.03-1 and can have identical response characteristics

circuit (which the procedure in Sec. 10.02 does not, as a result of the manner in which the end sections are designed).

Both the filter in Fig. 10.03-1 and that in Fig. 10.03-2 are composed of stubs that are $\lambda_0/4$ long with connecting lines that are also $\lambda_0/4$ long, where λ_0 is the wavelength in the medium of propagation at the midband frequency ω_0. These two types of filters can be made to have identical transmission properties. However, the form in Fig. 10.03-1 is the one that is most commonly used since the series stubs of the filter in Fig. 10.03-2 are difficult to realize in a shielded structure. Since the filter in Fig. 10.03-1 is the most important for practical applications, it will be discussed first.

In order to determine how many reactive elements are required in the low-pass prototype in order to give a desired rate of cutoff, the low-pass to band-pass mapping in Eq. (10.02-1) or Eq. (10.02-4) should be used. Having the low-pass prototype parameters g_0, g_1, g_2, ..., g_{n+1} and ω_1', and having specified the fractional bandwidth, w, and the terminating conductances $Y_A = Y_B$, the designer can compute the characteristic admittances of the stubs and connecting lines in a straightforward fashion from the equations in Table 10.03-1. In the equations there is a dimensionless constant d which may be chosen for some adjustment of the admittance level in the interior of the filter. In the case of the equations in Table 10.02-1 the choice of the admittance scale factor, h, should have no effect on the transmission characteristics of the filter; however, in the case of the equations in Table 10.03-1, the choice of d

Table 10.03-1

DESIGN EQUATIONS FOR FILTERS WITH SHUNT $\lambda_0/4$ STUBS AND $\lambda_0/4$ CONNECTING LINES AS SHOWN IN FIG. 10.03-1

Use mapping Eq. (10.02-1) or (10.02-4) to select a low-pass prototype with the required value of n. The filter will have n shunt stubs for an n-reactive-element prototype.

Compute:

$$\theta_1 = \frac{\pi}{2}\frac{\omega_1}{\omega_0} = \frac{\pi}{2}\left(1 - \frac{w}{2}\right) ,$$

$$\frac{J_{12}}{Y_A} = g_0\sqrt{\frac{C_a}{g_2}} , \quad \left.\frac{J_{k,k+1}}{Y_A}\right|_{k=2 \text{ to } n-2} = \frac{g_0 C_a}{\sqrt{g_k g_{k+1}}} ,$$

$$\frac{J_{n-1,n}}{Y_A} = g_0\sqrt{\frac{C_a g_{n+1}}{g_0 g_{n-1}}} , \quad \text{where } C_a = 2dg_1$$

and d is a dimensionless constant (typically chosen to be 1.0) which can be chosen so as to give a convenient admittance level in the interior of the filter. (See text)

$$\left.N_{k,k+1}\right|_{k=1 \text{ to } n-1} = \sqrt{\left(\frac{J_{k,k+1}}{Y_A}\right)^2 + \left(\frac{g_0\omega_1' C_a \tan\theta_1}{2}\right)^2} .$$

The characteristic admittances of the shunt stubs are:

$$Y_1 = g_0 Y_A \omega_1'(1-d)g_1 \tan\theta_1 + Y_A\left(N_{12} - \frac{J_{12}}{Y_A}\right)$$

$$\left.Y_k\right|_{k=2 \text{ to } n-1} = Y_A\left(N_{k-1,k} + N_{k,k+1} - \frac{J_{k-1,k}}{Y_A} - \frac{J_{k,k+1}}{Y_A}\right)$$

$$Y_n = Y_A\omega_1'(g_n g_{n+1} - dg_0 g_1)\tan\theta_1 + Y_A\left(N_{n-1,n} - \frac{J_{n-1,n}}{Y_A}\right) .$$

The characteristic admittances of the connecting lines are:

$$\left.Y_{k,k+1}\right|_{k=1 \text{ to } n-1} = Y_A\left(\frac{J_{k,k+1}}{Y_A}\right) .$$

All stubs and connecting lines are $\lambda_0/4$ long, where λ_0 is the wavelength in the medium of propagation at the mid-band frequency ω_0.

Table 10.03-2

NORMALIZED LINE ADMITTANCES FOR THE TRIAL FILTER DESIGN WHOSE RESPONSE IS SHOWN IN FIG. 10.03-3

Filter designed using Table 10.03-1 from a 0.10 db ripple, n = 8, Tchebyscheff prototype using ω_1/ω_0 = 0.650 and d = 1.0

Y_1	$= Y_8$	= 1.042	Y_3	$= Y_6$	= 2.049
Y_{12}	$= Y_{78}$	= 1.288	Y_{34}	$= Y_{56}$	= 1.292
Y_2	$= Y_7$	= 2.050	Y_4	$= Y_5$	= 2.087
Y_{23}	$= Y_{67}$	= 1.364	Y_{45}	= 1.277	

All values normalized so $Y_A = Y_B = 1$

SOURCE: Final Report, Contract DA 36-039 SC-74862, SRI; reprinted in *IRE Trans. PGMTT* (see Ref. 1 by G. L. Matthaei).

may have some small effect on the approximations upon which the equations are based. To date, only values of d of 0.5 and 1.0 have been tested in this design procedure, but it is probable that considerably different values in the range $0 < d \leq 1$ would give satisfactory results.

Table 10.03-2 summarizes the line admittances for a design obtained using an n = 8-reactive element Tchebyscheff prototype with 0.10-db ripple [Table 4.05-2(a)]. The admittance level is normalized so that $Y_A = Y_B = 1$, and ω_1/ω_0 = 0.650 which calls for slightly over an octave bandwidth. Figure 10.03-3 shows the computed response of this design. Note that though the ripples at the edges of the pass band are undersized, the pass-band performance is quite close to what was specified. The x's and circles on the graph show points mapped from the low-pass prototype response by use of Eqs. (10.02-1) and (10.02-4). Note that in this case the more complicated Eq. (10.02-4) gives much better accuracy than does Eq. (10.02-1), for attenuation above 30 db.

Note in Table 10.03-2 that the admittances of the end stubs are about half of that for the stubs in the interior of the filter. For this reason it is sometimes convenient to build this type of filter with double stubs in the interior of the filter and with single stubs at the ends, as is illustrated in Fig. 10.03-4. Table 10.03-3 summarizes the line impedances for an n = 13 design which has the form in Fig. 10.03-4. This filter was constructed and tested in rectangular-bar strip-line form and its important dimensions are summarized in Figs. 10.03-5(a), (b), and (c). The filter was designed for a band-center frequency of $\omega_0/2\pi$ = 3.6 Gc, and it was necessary to take account of the junction effect where the stubs and main-line meet, in order to properly determine the lengths of the stubs and the connecting lines. It was assumed that the junction effect for a plus-junction must be similar to that of the T-junctions in Figs. 5.07-6 to 5.07-9. From tests on the filter, the junction equivalent circuit and reference planes were estimated to be about as indicated in Fig. 10.03-6.

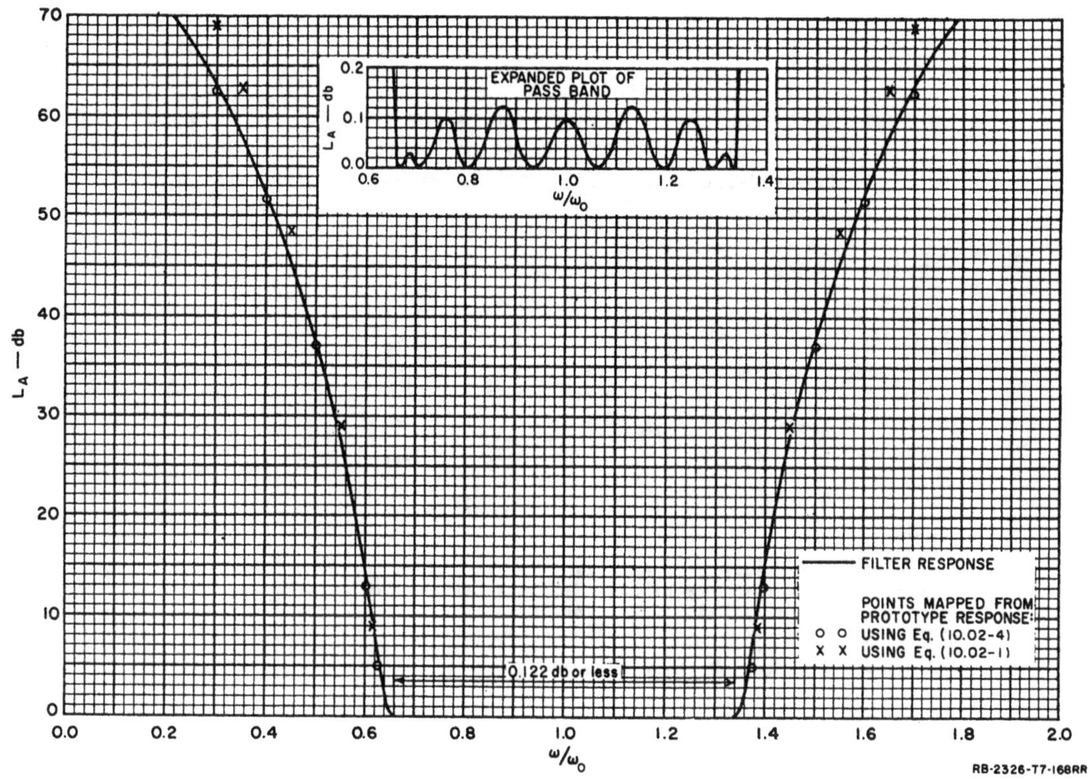

SOURCE: Final Report, Contract DA 36-039 SC-74862, SRI; reprinted in *IRE Trans. PGMTT* (see Ref. 1 by G. L. Matthaei)

FIG. 10.03-3 COMPUTED RESPONSE OF A FILTER DESIGNED AS IN TABLE 10.03-1 TO HAVE APPROXIMATELY 2 TO 1 BANDWIDTH
Design value for ω_1/ω_0 was 0.650. Prototype had 0.10-db Tchebyscheff pass-band ripple with n = 8 reactive elements

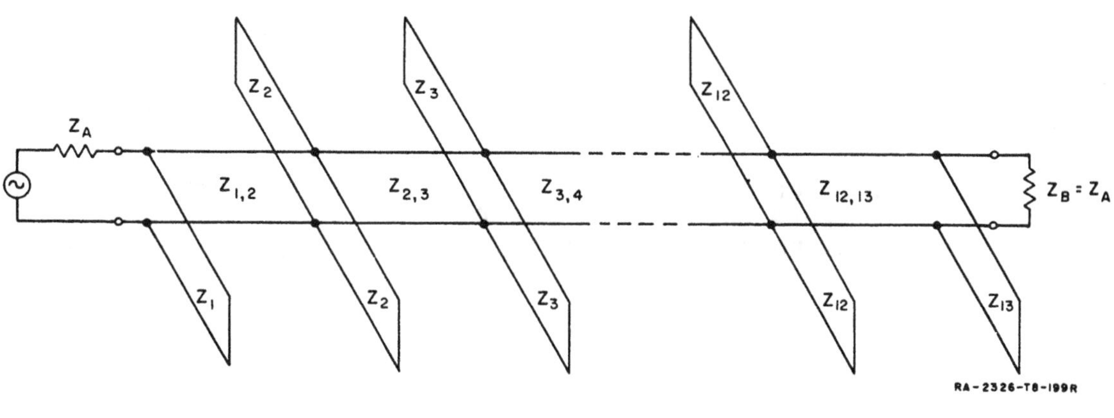

FIG. 10.03-4 OPEN-WIRE-LINE REPRESENTATION OF THE FILTER IN FIG. 10.03-5

Table 10.03-3

LINE IMPEDANCES FOR A BAND-PASS FILTER OF THE FORM IN FIG. 10.03-4 HAVING n = 13, A BAND-EDGE RATIO OF f_2/f_1 = 2.175, AND APPROXIMATELY 0.1-db TCHEBYSCHEFF PASS-BAND RIPPLE

$Z_0 = 50$ ohms	$Z_4 = Z_{10} = 52.4$ ohms
$Z_1 = Z_{13} = 52.8$	$Z_{4,5} = Z_{9,10} = 39.6$
$Z_{1,2} = Z_{12,13} = 38.8$	$Z_5 = Z_9 = 52.1$
$Z_2 = Z_{12} = 53.7$	$Z_{5,6} = Z_{8,9} = 39.8$
$Z_{2,3} = Z_{11,12} = 36.7$	$Z_6 = Z_8 = 52.0$
$Z_3 = Z_{11} = 53.7$	$Z_{6,7} = Z_{7,8} = 39.8$
$Z_{3,4} = Z_{10,11} = 38.9$	$Z_7 = 52.0$

These line impedances were computed from an n = 10, 0.10-db ripple prototype since at the time of this filter's design, tables of element values for n larger than 10 were not available. The design was augmented to n = 13 by adding additional lines and stubs at the center of the filter, the added lines and stubs having impedances the same as those at the center of the n = 10 design.

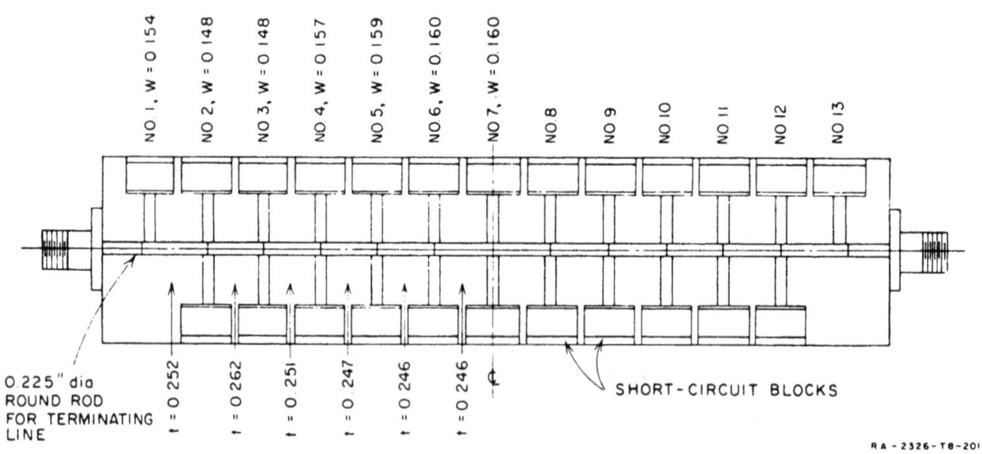

FIG. 10.03-5(a) LAYOUT OF STRIP TRANSMISSION LINE BAND-PASS FILTER WITH UPPER GROUND PLANE REMOVED
See Fig. 10.03-5(b) for definitions of t and w as used here

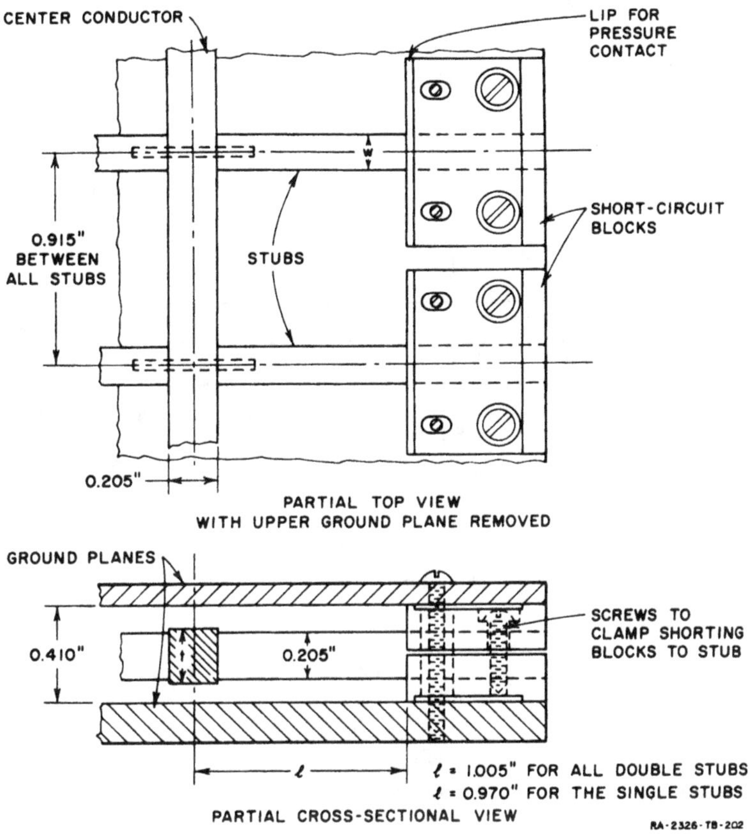

FIG. 10.03-5(b) SOME CONSTRUCTION DETAILS OF THE BAND-PASS FILTER IN FIG. 10.03-5(a)

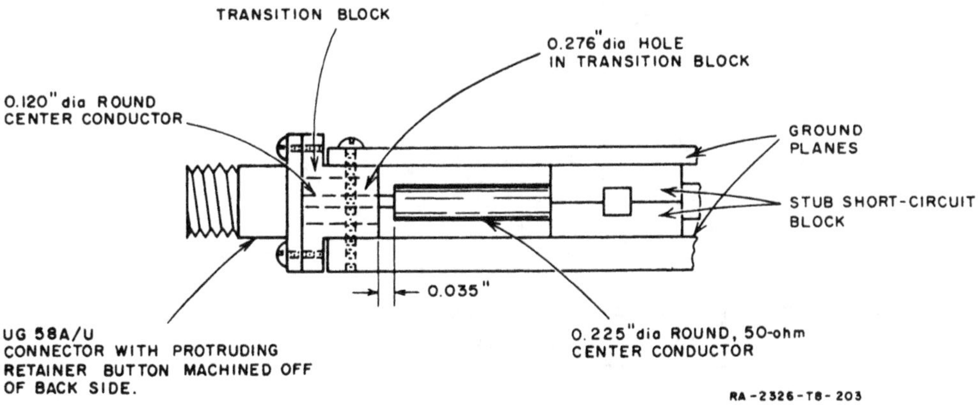

FIG. 10.03-5(c) DETAILS OF TRANSITION FROM FILTER TO TYPE-N CONNECTOR

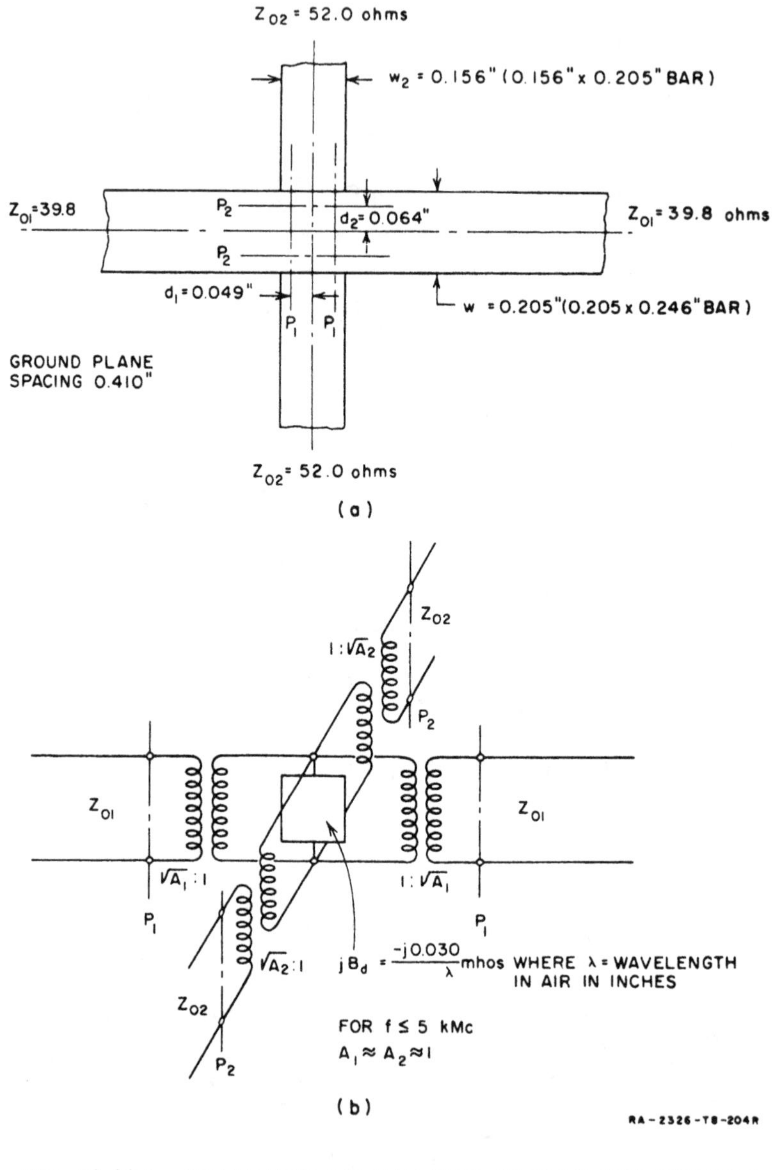

FIG. 10.03-6 ESTIMATED EQUIVALENT CIRCUIT FOR TYPICAL PLUS-JUNCTION IN THE FILTER OF FIGS. 10.03-5(a), (b), (c)

The $\lambda_0/4$ length for the stubs and the connecting lines was figured from the reference planes P_1 and P_2 indicated. The capacitive junction susceptance B_d is compensated for by reducing the lengths of each of the two side stubs by the amount

$$\Delta l \approx \frac{\lambda_0 B_d}{4\pi Y_s} \quad , \tag{10.03-1}$$

where Y_s is the characteristic admittance of each of the two stubs. (This equation effectively removes an amount of stub on each side so that each stub plus half of the small junction susceptance B_d will still be resonant at ω_0.) In the case of the T-junctions for the single stubs at the ends of the filter, the reference plane for determining the stub length was closer to the centerline of the main line so that the single, end stubs were made to be about 0.035 inch shorter than the double stubs in the interior of the filter.

Figure 10.03-7 shows the measured performance of this filter. The band-edge ratio, $\omega_2/\omega_1 = 2.21$, compared favorably with the 2.17 design value. In general, the performance is seen to be in excellent agreement with the design objective.

Design formulas for the filter in Fig. 10.03-2 can be obtained directly from the formulas in Table 10.03-1 by application of duality. Thus, admittance quantities in Table 10.03-1 are simply replaced by corresponding impedance quantities as listed below:

$$\begin{aligned} Y_A = Y_B &\longrightarrow Z_A = Z_B \\ J_{k,k+1} &\longrightarrow K_{k,k+1} \\ Y_k &\longrightarrow Z_k \\ Y_{k,k+1} &\longrightarrow Z_{k,k+1} \\ C_a &\longrightarrow L_a \end{aligned} \tag{10.03-2}$$

It would be possible to build a filter of the type in Fig. 10.03-2 shielded coaxial form by constructing the series stubs as stubs within the center conductor of the line, as shown in Fig. 10.03-8.

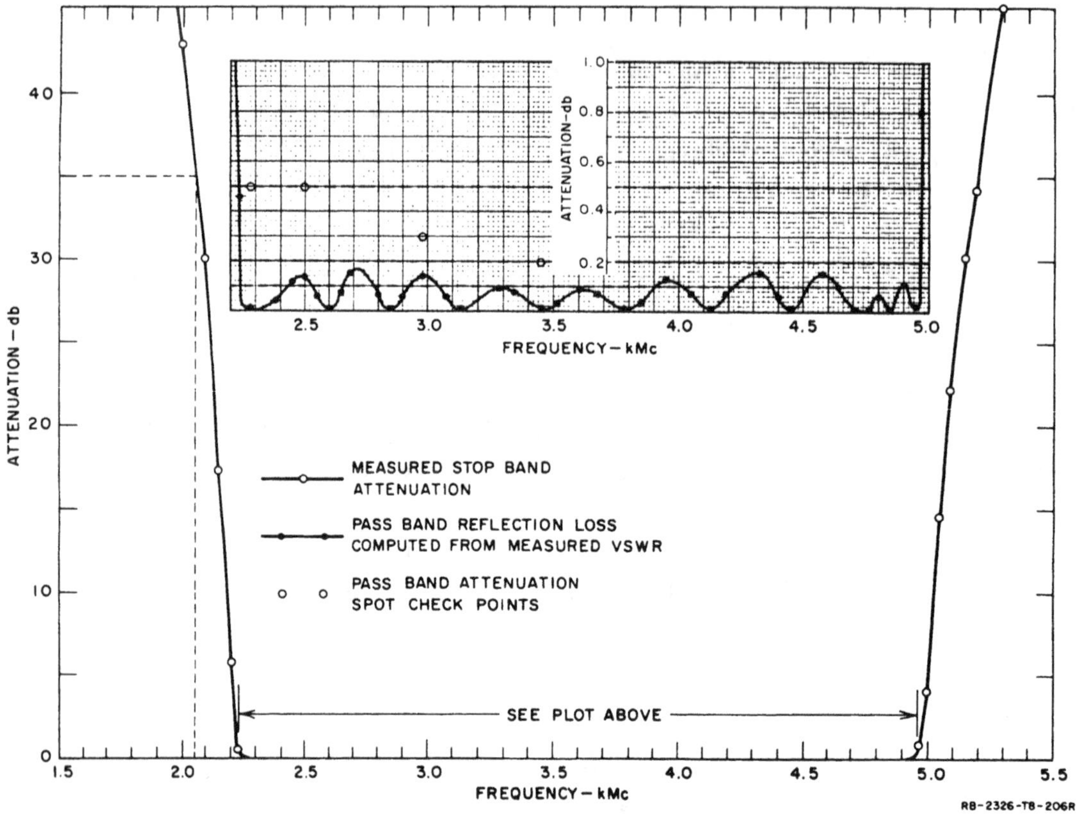

SOURCE: Final Report, Contract DA 36-039 SC-74862, SRI; reprinted in *IRE Trans. PGMTT* (see Ref. 1 by G. L. Matthaei)

FIG. 10.03-7 MEASURED RESPONSE OF THE FILTER SHOWN IN FIGS. 10.03-5(a), (b), (c)

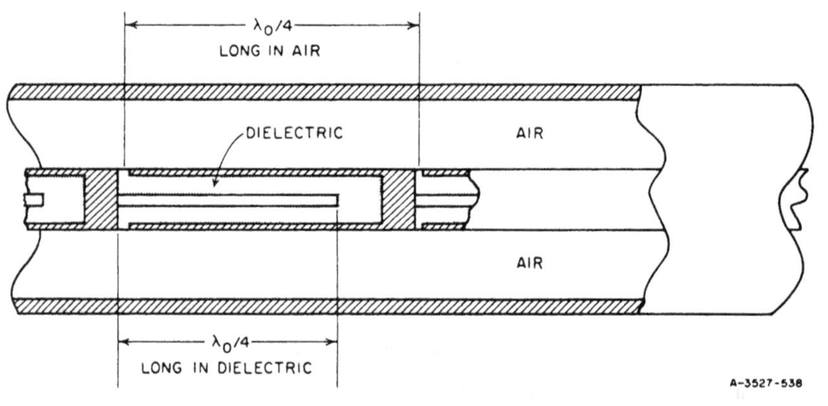

FIG. 10.03-8 POSSIBLE MEANS FOR REALIZING THE FILTER IN FIG. 10.03-2 IN COAXIAL FORM

Just as with the parallel-coupled filters discussed in Sec. 10.02, the filters in Figs. 10.03-1 and 10.03-2 have their second pass band centered by $3\omega_0$, but they are very prone to have narrow spurious responses in the vicinity of $2\omega_0$ if there is the slightest mistuning. Filters of the forms in Figs. 10.03-1 and 10.03-2 are candidates for use primarily as wide-band filters, because if narrow-band filters are designed in this form, their stubs will have unreasonably low impedance levels in the case of Fig. 10.03-1, and unreasonably high impedance levels for the case of Fig. 10.03-2.

SEC. 10.04, FILTERS WITH $\lambda_0/2$ STUBS AND $\lambda_0/4$ CONNECTING LINES

The filter in Fig. 10.04-1, which uses open-circuited $\lambda_0/2$ stubs spaced $\lambda_0/4$ apart, can be made to have pass-band characteristics similar to those of the filter in Fig. 10.03-1, which uses short-circuited $\lambda_0/4$ stubs spaced $\lambda_0/4$ apart (where λ_0 is the wavelength at the pass-band center frequency ω_0). However, the filter in Fig. 10.04-1 has quite different stop-band characteristics. If each $\lambda_0/2$ stub is of the same characteristic admittance throughout its length, then the stop-band will have infinite attenuation at the frequencies $\omega_0/2$ and $3\omega_0/2$. If the

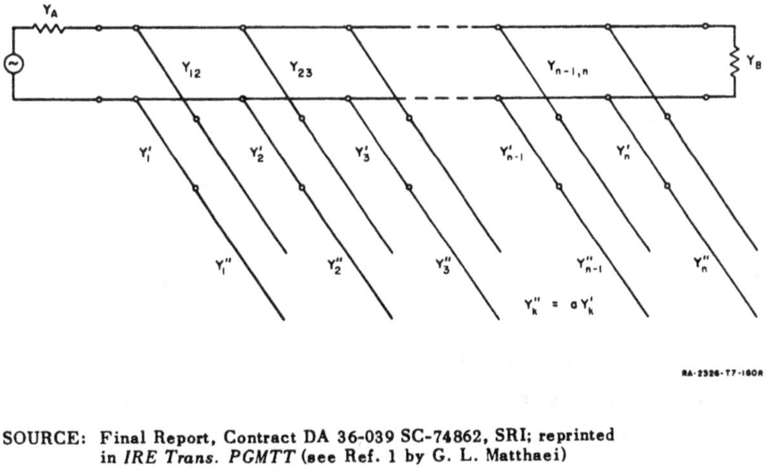

SOURCE: Final Report, Contract DA 36-039 SC-74862, SRI; reprinted in *IRE Trans. PGMTT* (see Ref. 1 by G. L. Matthaei)

FIG. 10.04-1 BAND-PASS FILTER WITH HALF-WAVELENGTH SHUNT STUBS AND QUARTER-WAVELENGTH CONNECTING LINES
The reference wavelength is the propagation wavelength at the midband frequency, ω_0

outer half of each stub is made to have an admittance that is a constant times the admittance of the inner half of the stub, as indicated in Fig. 10.04-1, then the frequencies of infinite attenuation can be made to occur at frequencies other than $\omega_0/2$ and $3\omega_0/2$. This type of filter will have additional pass bands in the vicinity of $\omega = 0$ and $\omega = 2\omega_0$, and at other corresponding periodic frequencies.

Filters in the form shown in Fig. 10.04-1 can be readily designed by a modified use of Table 10.03-1. The design is carried out first to give a filter in the form in Fig. 10.03-1 with the desired pass-band characteristic and bandwidth. Then each shunt, quarter-wavelength, short-circuited stub of characteristic admittance Y_k is replaced as shown in Fig. 10.04-1 by a shunt, half-wavelength, open-circuited stub having an inner quarter-wavelength portion with a characteristic admittance

$$Y'_k = \frac{Y_k (a \tan^2 \theta_1 - 1)}{(a + 1) \tan^2 \theta_1} \tag{10.04-1}$$

and an outer quarter-wavelength portion with a characteristic admittance

$$Y''_k = aY'_k \quad . \tag{10.04-2}$$

The parameter a is fixed by

$$a = \cot^2 \left(\frac{\pi\omega_\infty}{2\omega_0}\right)\Bigg|_{(\omega_\infty/\omega_0)<(\omega_1/\omega_0)} \tag{10.04-3}$$

where $\theta_1 = \pi\omega_1/2\omega_0$, and ω_∞ is a frequency at which the shunt lines present short circuits to the main line and cause infinite attenuation.* The principle upon which the above substitution is made is that Eqs. (10.04-1) through (10.04-3) are constrained to yield half-wavelength open-circuited stubs which have exactly the same susceptances at the band-edge frequency ω_1 as did the quarter-wavelength short-circuited stubs that they replace, while both kinds of stubs have zero admittance at ω_0.

To test out this procedure, a filter was designed as in Table 10.03-1 to give 30-percent bandwidth ($\omega_1/\omega_0 = 0.850$) using a 0.10-db Tchebyscheff

* Of course, in an actual filter the attenuation will be finite due to the losses in the stubs preventing the stub impedances from going to exactly zero.

prototype with $n = 8$. Then, choosing $\omega_\infty/\omega_0 = 0.500$, which gives $a = 1$, the quarter-wavelength stubs were replaced by half-wavelength stubs as described above, and the resulting computed response is shown in Fig. 10.04-2. Note that the pass band is almost exactly as prescribed, and that there are low-attenuation regions in the vicinity of $\omega = 0$ and $\omega = 2\omega_0$, which are to be expected. The element values for this filter are shown in Table 10.04-1.

The 2-to-1-bandwidth filter design (Fig. 10.03-1 and Table 10.03-2) was also converted to this form using $\omega_\infty/\omega_0 = 0.500$, and its response was computed. The features of the pass band looked much the same as those in the expanded plot in Fig. 10.03-3, while the stop bands consisted of

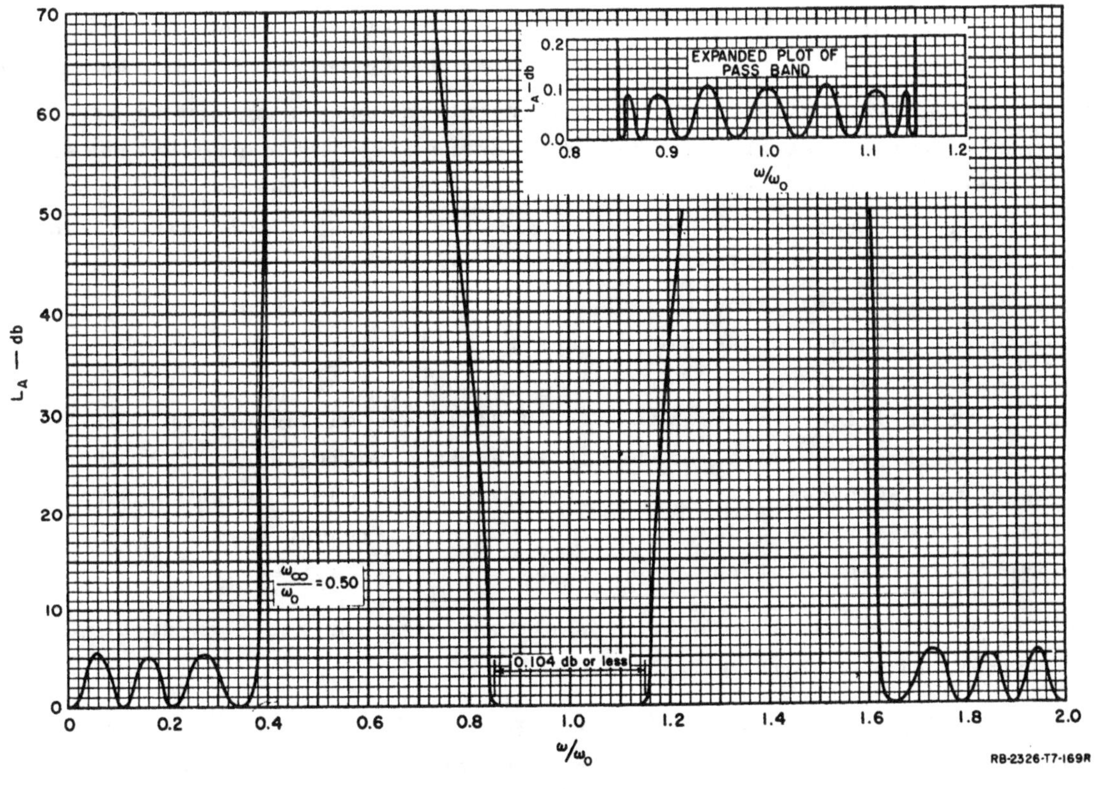

SOURCE: Final Report, Contract DA 36-039 SC-74862, SRI; reprinted in *IRE Trans. PGMTT* (see Ref. 1 by G. L. Matthaei)

FIG. 10.04-2 COMPUTED RESPONSE OF A 30-PERCENT-BANDWIDTH BAND-PASS FILTER DESIGNED IN THE FORM IN FIG. 10.04-1
Design value for $\omega_1/\omega_0 = 0.850$. Prototype had 0.10-db Tchebyscheff ripple with $n = 8$ reactive elements

Table 10.04-1

ELEMENT VALUES FOR THE FILTER OF FIG. 10.04-2 REALIZED AS SHOWN IN FIG. 10.04-1

Filter designed from a 0.10 db ripple, $n = 8$, Tchebyscheff prototype using $\omega_1/\omega_0 = 0.850$ and $\omega_\infty/\omega_0 = 0.500$. This, then, calls for $a = 1$ so that $Y'_k = Y''_k$ throughout.

$Y'_1 = Y'_8$	= 1.806		$Y'_3 = Y'_6$	= 3.584
$Y_{12} = Y_{78}$	= 1.288		$Y_{34} = Y_{56}$	= 1.292
$Y'_2 = Y'_7$	= 3.585		$Y'_4 = Y'_5$	= 3.614
$Y_{23} = Y_{67}$	= 1.364		Y_{45}	= 1.277

All values normalized so that $Y_A = 1$.

SOURCE: Final Report, Contract DA 36-039 SC-74862, SRI; reprinted in *IRE Trans. PGMTT* (see Ref. 1 by G. L. Matthaei)

very sharp attenuation spikes surrounding $\omega/\omega_0 = 0.500$, in a manner similar to that in Fig. 10.04-2, except that the attenuation bands were much narrower.

Filters of the form in Fig. 10.04-1 should be particularly useful where the pass bands around $\omega = 0$ and $\omega = 2\omega_0$ are not objectionable, and where there is a relatively narrow band of signals to be rejected. By the proper choice of ω_∞, the infinite attenuation point can be so placed as to give maximum effectiveness against the unwanted signals. Although using the same ω_∞ for all of the stubs should give the best pass-band response, it may be permissible to stagger the ω_∞ points of the stubs slightly to achieve broader regions of high attenuation. Filters of the form in Fig. 10.04-1 are practical for bandwidths narrower than those of filters of the form in Fig. 10.03-1 because of the larger susceptance slope of half-wavelength stubs for a given characteristic admittance. For example, in the case of Fig. 10.04-1, the shunt stubs for this filter as shown in Fig. 10.04-1 have characteristic admittance $Y'_k = Y''_k$ which are 0.471 times the characteristic admittances of the shunt stubs of the analogous filter in the form in Fig. 10.03-1 from which it was designed. Thus narrower bandwidths can be achieved without having the characteristic admittances of the shunt stubs become excessive.

No accurate low-pass-to-band-pass transformation has been developed for filters of the form in Fig. 10.04-1.

Since filters of the form in Fig. 10.04-1 do not involve any short-circuit connections, they are very easy to fabricate in printed-circuit strip-line form as suggested in Fig. 10.04-3.

SEC. 10.05, FILTERS USING BOTH SERIES AND SHUNT STUBS

The filter in Fig. 10.05-1 makes use of short-circuited $\lambda_0/4$ stubs spaced $\lambda_0/4$ apart, which makes it similar to the filter in Fig. 10.03-1. However, the filter in Fig. 10.05-1 has, in addition, a $\lambda_0/2$ *series*

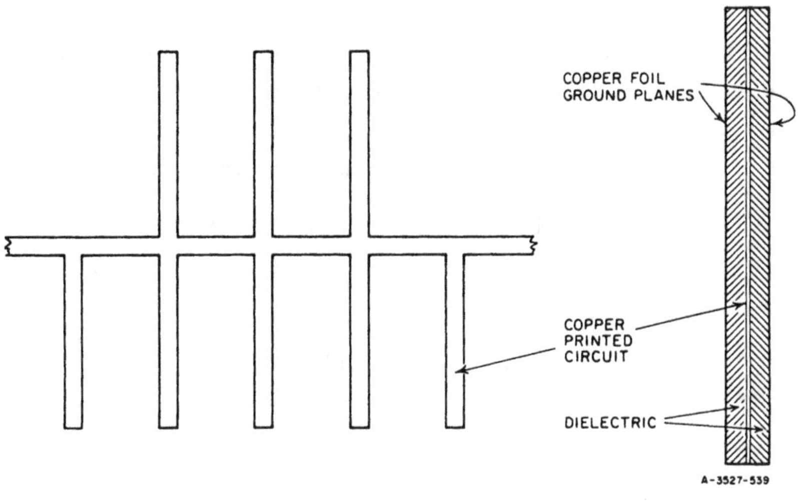

FIG. 10.04-3 A POSSIBLE PRACTICAL STRIP-LINE PRINTED-CIRCUIT VERSION OF THE FILTER IN FIG. 10.04-1

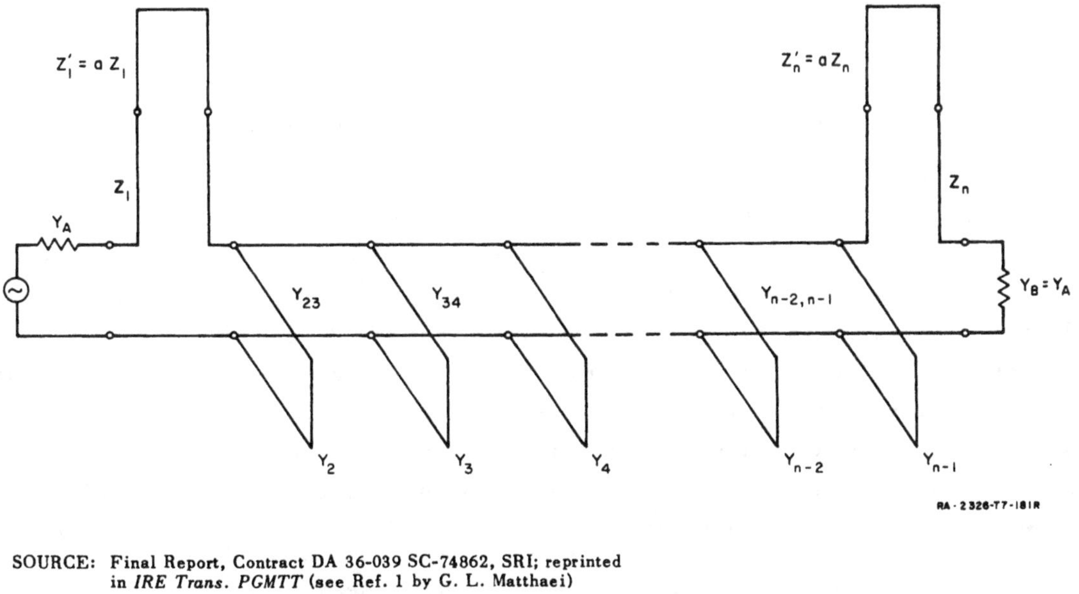

SOURCE: Final Report, Contract DA 36-039 SC-74862, SRI; reprinted in *IRE Trans. PGMTT* (see Ref. 1 by G. L. Matthaei)

FIG. 10.05-1 BAND-PASS FILTER WITH QUARTER-WAVELENGTH SHUNT STUBS, QUARTER-WAVELENGTH CONNECTING LINES, AND HALF-WAVELENGTH SERIES STUBS AT THE ENDS
The reference wavelength is that at the midband frequency, ω_0

short-circuited stub at each end. These $\lambda_0/2$ stubs yield frequencies ω where "infinite" attenuation* occurs close to the pass band, similar to those of the filter in Fig. 10.04-2, but in this case the attenuation remains high at $\omega = 0$ and $\omega = 2\omega_0$ (except for possible narrow spurious responses at $2\omega_0$ which can result from any mistuning).

In this case a reasonably accurate low-pass-to-band-pass transformation (Sec. 10.01) is

$$\frac{\omega'}{\omega_1'} = \frac{F_n\left(\dfrac{\omega}{\omega_0}\right)}{F_n\left(\dfrac{\omega_1}{\omega_0}\right)} , \tag{10.05-1}$$

where

$$F_n\left(\frac{\omega}{\omega_0}\right) = \frac{-\cos\left(\dfrac{\pi\omega}{2\omega_0}\right)}{\sqrt[n]{\left|\sin\left(\dfrac{\pi\omega}{2\omega_0}\right)\right|\left[\sin\dfrac{\pi}{2}\;\dfrac{(\omega-\omega_\infty)}{\omega_0}\right]^2\left[\sin\dfrac{\pi}{2}\;\dfrac{(\omega-2\omega_0+\omega_\infty)}{\omega_0}\right]^2}} , \tag{10.05-2}$$

$\omega_1/\omega_0 = 1 - w/2$, and ω_∞ is a stop-band frequency where infinite attenuation is desired.

After selection of a low-pass prototype with element values g_0, g_1, g_2, ..., g_{n+1} and cutoff frequency ω_1', and after specification of Y_A, w (or ω_1/ω_0), and ω_∞/ω_0, the design can be carried out by making the calculations indicated in Table 10.05-1. As in Table 10.03-1, the dimensionless parameter d can be used to give some degree of freedom in establishing the impedance level in the interior of the filter. The choice of d will have some minor influence on the approximations involved in the design process, but values of d in the range $0 < d \leq 1$ should be usable (to date only the value $d = 0.5$ has been used in trial designs).

Table 10.05-2 shows the results of a trial design computed using an $n = 8$ reactive element and a 0.10-db ripple Tchebyscheff prototype, and using $\omega_1/\omega_0 = 0.650$, $\omega_\infty/\omega_0 = 0.500$, $Y_A = 1$, and $d = 0.50$. Figure 10.05-2

---

* Of course, as a result of dissipation in the circuit the attenuation will always be finite for frequencies ω, but L_A will typically go very high at frequencies where the $\lambda_0/2$ stubs are anti-resonant.

Table 10.05-1

DESIGN EQUATIONS FOR FILTERS OF THE FORM IN FIG. 10.05-1

Use mapping Eqs. (10.05-1) and (10.05-2) to select low-pass prototype with the required value of n.

Compute:

$$\theta_1 = \frac{\pi}{2}\frac{\omega_1}{\omega_0} = \frac{\pi}{2}\left(1 - \frac{w}{2}\right) , \quad C_a = 2dg_2$$

where $d \leq 1$ is a dimensionless constant (typically one-half or somewhat larger) which may be chosen to give a desired impedance level in the interior of the filter.

$$\frac{J_{23}}{Y_A} = \frac{\sqrt{g_2 C_a}}{g_0\sqrt{g_2 g_3}} , \quad \left.\frac{J_{k,k+1}}{Y_A}\right|_{k=3 \text{ to } n-3} = \frac{C_a}{g_0\sqrt{g_k g_{k+1}}}$$

$$\frac{J_{n-2,n-1}}{Y_A} = \frac{1}{g_0}\sqrt{\frac{C_a g_0}{g_{n-2}g_{n+1}}}$$

$$\left.N_{k,k+1}\right|_{k=2 \text{ to } n-2} = \sqrt{\left(\frac{J_{k,k+1}}{Y_A}\right)^2 + \left(\frac{\omega_1' C_a \tan\theta_1}{2g_0}\right)^2}$$

$$\theta_\infty = \frac{\pi}{2}\frac{\omega_\infty}{\omega_0} , \quad a = \cot^2\theta_\infty$$

where ω_∞ is a frequency of infinite attenuation as indicated in the example in Fig. 10.05-2.

Referring to Fig. 10.05-1, for the stubs:

$$Z_1 = \frac{[a(\tan\theta_1)^2 - 1]\omega_1' g_0 g_1}{Y_A(a+1)\tan\theta_1} , \quad Z_1' = aZ_1$$

$$Y_2 = \frac{Y_A\omega_1'(1-d)g_2}{g_0}\tan\theta_1 + Y_A\left(N_{23} - \frac{J_{23}}{Y_A}\right)$$

$$\left.Y_k\right|_{k=3 \text{ to } n-2} = Y_A\left(N_{k-1,k} + N_{k+1,k} - \frac{J_{k-1,k}}{Y_A} - \frac{J_{k,k+1}}{Y_A}\right)$$

$$Y_{n-1} = \frac{Y_A\omega_1'(g_{n-1}g_0 - dg_2 g_{n+1})\tan\theta_1}{g_0 g_{n+1}} + N_{n-2,n-1} - \frac{J_{n-2,n-1}}{Y_A}$$

$$Z_n = \frac{[a(\tan\theta_1)^2 - 1]\omega_1' g_n g_{n+1}}{Y_A(a+1)\tan\theta_1} , \quad Z_n' = aZ_n .$$

For the connecting lines:

$$\left.Y_{k,k+1}\right|_{k=2 \text{ to } n-2} = \left(\frac{J_{k,k+1}}{Y_A}\right)Y_A .$$

shows the computed response of this filter (indicated by the solid lines), while the circles indicate points mapped from the low-pass prototype response using the mapping in Eqs. (10.05-1) and (10.05-2). Note that the 0.1-db point on the left side of the pass band is very nearly at $\omega/\omega_0 = 0.650$ as specified, and that the response in general is, for most

Table 10.05-2

ELEMENT VALUES FOR THE FILTER OF FIG. 10.05-2 REALIZED AS SHOWN IN FIG. 10.05-1

Filter designed using Table 10.05-1 from a 0.10-db ripple, $n = 8$, Tchebyscheff prototype using $\omega_1/\omega_0 = 0.650$, $\omega_\infty/\omega_0 = 0.500$, and $d = 0.5$.

Z_1	=	Z_8	=	0.606	Y_3	=	Y_6	=	1.235
Z_1'	=	Z_8'	=	0.606	Y_{34}	=	Y_{56}	=	0.779
Y_2	=	Y_7	=	1.779	Y_4	=	Y_5	=	1.258
Y_{23}	=	Y_{67}	=	0.823	Y_{45}	=	0.770		

Values normalized so that $Y_A = 1$.

SOURCE: Final Report, Contract DA 36-039 SC-74862, SRI; reprinted in *IRE Trans. PGMTT* (see Ref. 1 by G. L. Matthaei).

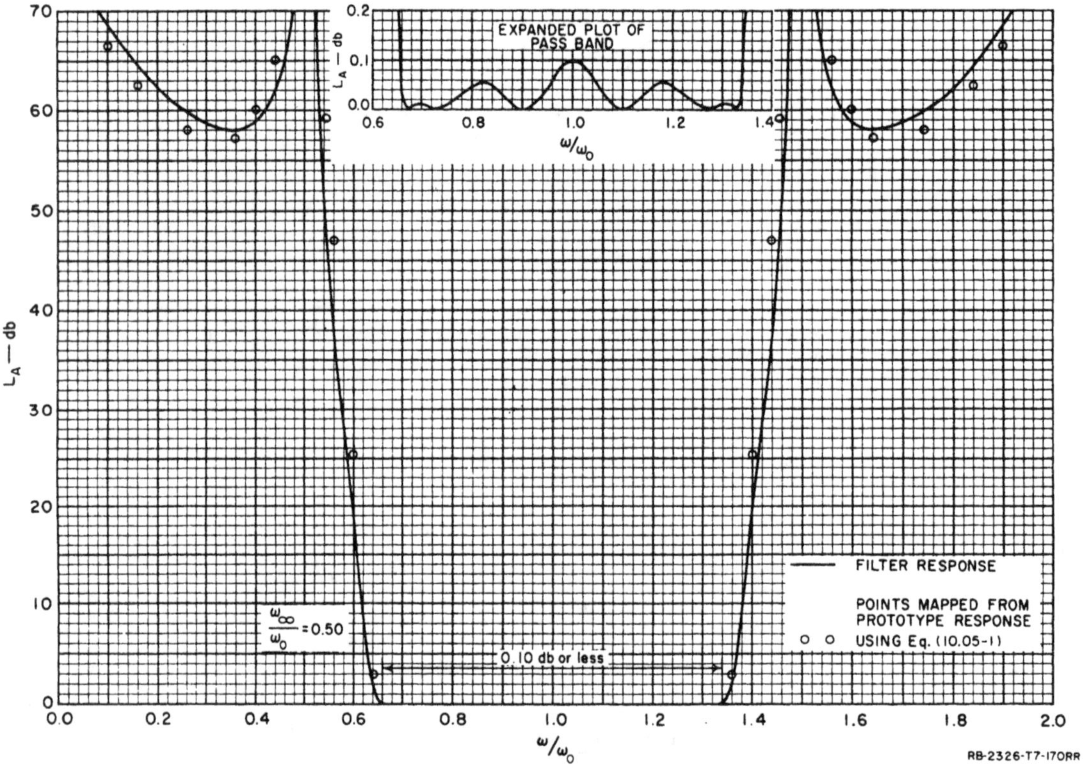

SOURCE: Final Report, Contract DA 36-039 SC-74862, SRI; reprinted in *IRE Trans. PGMTT* (see Ref. 1 by G. L. Matthaei)

FIG. 10.05-2 COMPUTED RESPONSE OF A FILTER AS IN FIG. 10.05-1 WITH APPROXIMATELY 2 TO 1 BANDWIDTH

Design value for ω_1/ω_0 was 0.650. Prototype had 0.10-db Tchebyscheff ripple with n = 8 reactive elements. Parameters d and ω_∞/ω_0 were both chosen as 0.500

engineering purposes, a satisfactory realization of the specified performance.

Figure 10.05-3 shows a possible way for constructing filters of the form in Fig. 10.05-1. The filter shown is in so-called split-block coaxial construction. The round center conductors are within cylindrical cavities machined from a solid split block. Note that the $\lambda_0/2$ series

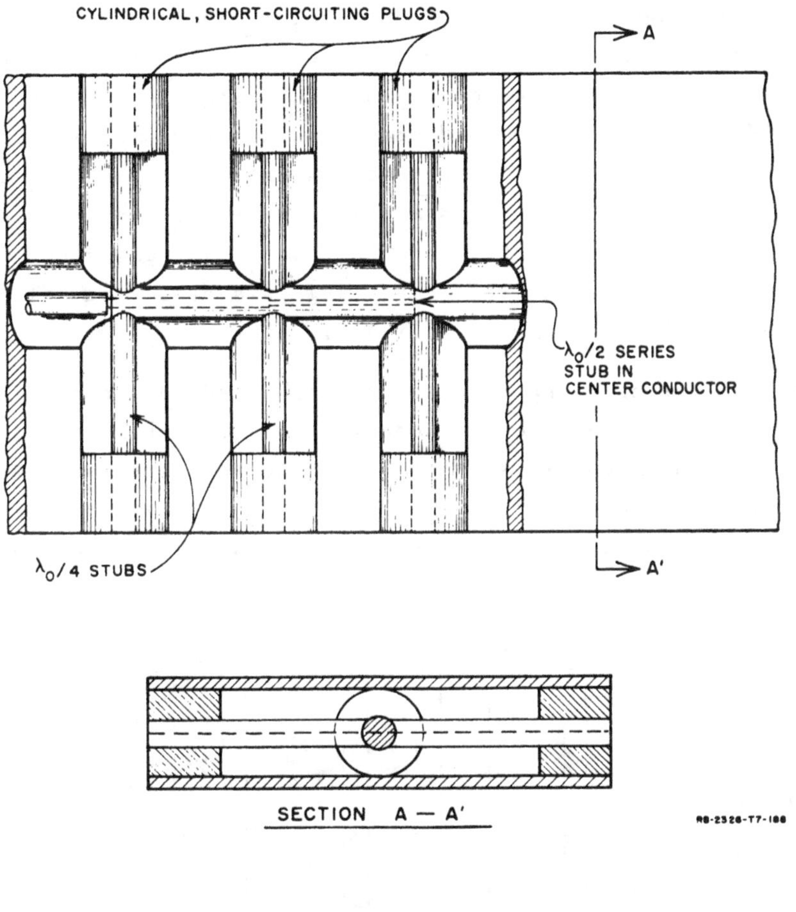

FIG. 10.05-3 POSSIBLE WAY FOR FABRICATING WIDE-BAND FILTERS OF THE TYPE IN FIG. 10.05-1 IN SPLIT-BLOCK CONSTRUCTION

The shunt, quarter-wavelength, short-circuited stubs are realized in parallel pairs so that the characteristic admittance of each stub will be cut in half, and so that the structure will be self-supporting. The series, half-wavelength, short-circuited stubs are inside the center conductor

stub at the input of the filter is realized as a coaxial stub within the main line of the filter.

SEC. 10.06, INTERDIGITAL-LINE FILTERS OF NARROW OR MODERATE BANDWIDTH

Figure 10.06-1 shows one type of interdigital filter to be discussed. The structure, as shown, consists of TEM-mode strip-line resonators between parallel ground planes. Each resonator element is a quarter-wavelength long at the midband frequency and is short-circuited at one end and open-circuited at the other end. Coupling is achieved by way of the fields fringing between adjacent resonator elements. Using the design procedure described in this section, Lines 1 to n in Fig. 10.06-1 serve as resonators. Lines 0 and $n + 1$, however, operate as impedance-transforming sections and not as resonators. Thus, using the procedures of this section, an n-reactive-element low-pass prototype will lead to an interdigital filter with $n + 2$ line elements.

If all of the coupling effects are accounted for, the mathematics that describe the performance of such interdigital filters as those discussed in this and the next section become quite unwieldy.[2] Since synthesizing a structure to have a prescribed response is a much more difficult problem than analyzing a given structure, and since an exact analysis of

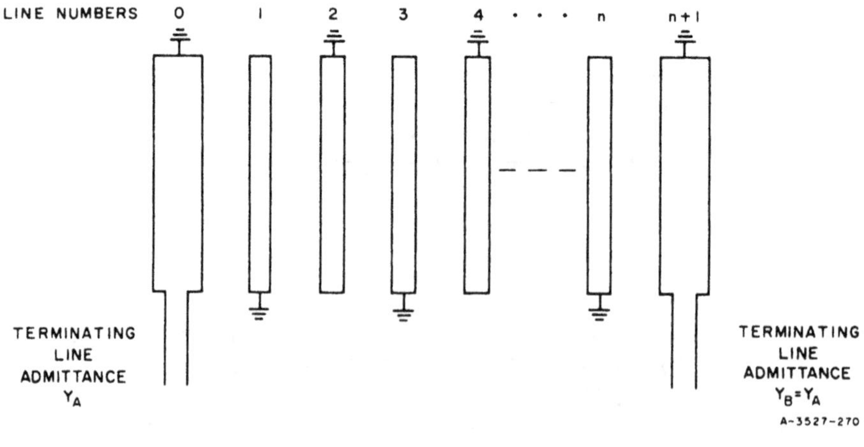

SOURCE: Quarterly Progress Report 4, Contract DA 36-039 SC-87398, SRI; reprinted in *IRE Trans. PGMTT* (see Ref. 3 by G. L. Matthaei)

FIG. 10.06-1 INTERDIGITAL FILTER WITH SHORT-CIRCUITED LINES AT THE ENDS

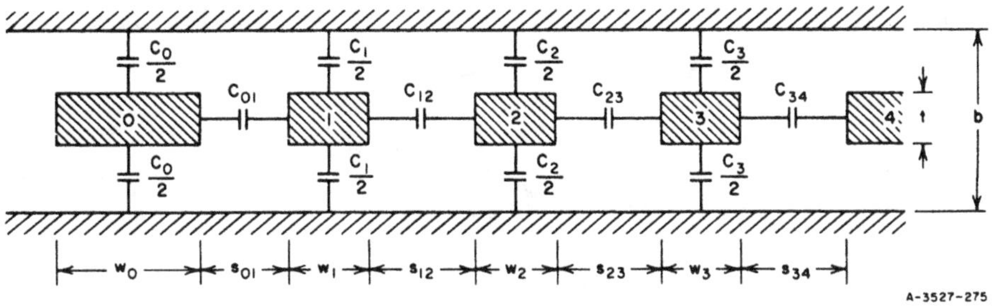

SOURCE: Quarterly Progress Report 4, Contract DA 36-039 SC-87398, SRI; reprinted in *IRE Trans. PGMTT* (see Ref. 3 by G. L. Matthaei)

FIG. 10.06-2 CROSS SECTION OF AN ARRAY OF PARALLEL-COUPLED LINES BETWEEN GROUND PLANES

such a structure is itself very tedious, the prospects of obtaining a usable exact synthesis procedure appear to be dim. Thus, the synthesis procedure given here involves several additional simplifying approximations (beyond those used for the procedures in Secs. 10.02 through 10.05) that permit straightforward, easy-to-use design calculations. Although the design formulas are approximate, the results of trial designs indicate that they are sufficiently accurate for most practical applications.

Figure 10.06-2 shows an array of parallel-coupled lines such as is used in an interdigital filter. The electrical properties of the structure are characterized in terms of the self-capacitances, C_k, per unit length of each bar with respect to ground, and the mutual capacitances, $C_{k,k+1}$, per unit length between adjacent bars k and $k + 1$. This representation is not always highly accurate; it is conceivable that a significant amount of fringing capacitance could exist between a given line element and, say, the line element beyond the nearest neighbor. However, at least for geometries such as that shown, experience has shown this representation to have satisfactory accuracy.

For designs of the interdigital-filter structures discussed herein, equations will be given for the normalized self and mutual capacitances, C_k/ϵ and $C_{k,k+1}/\epsilon$, for all the lines in the structure; where ϵ is the dielectric constant of the medium of propagation. Having these normalized capacitances the designer can obtain the dimensions of the bars, using the data in Sec. 5.05.

A convenient and reasonably accurate low-pass-to-band-pass transformation (Sec. 10.01) to use for estimating the attenuation characteristics of interdigital filters is

$$\frac{\omega'}{\omega_1'} = \frac{2}{w}\left(\frac{\omega - \omega_0}{\omega_0}\right) \tag{10.06-1}$$

where

$$w = \frac{\omega_2 - \omega_1}{\omega_0} \tag{10.06-2}$$

$$\omega_0 = \frac{\omega_2 + \omega_1}{2} \tag{10.06-3}$$

and ω', ω_1', ω_2', ω, ω_1, and ω_2 are as indicated in Figs. 10.01-1(a), (b).

Table 10.06-1 shows approximate design equations for interdigital filters of the form shown in Fig. 10.06-1. This type of design is most practical for filters having narrow or moderate bandwidth. Although no special investigation of this point has been made, it appears probable that one should consider the possibility of using the design equations in Sec. 10.07 when the bandwidth is of the order of 30 percent or more, instead of those in Table 10.06-1. Both sets of design equations are valid, however, for either narrow or wide bandwidths. The main drawback in applying the design procedure in this section to filters of wide bandwidth is that the gaps between Lines 0 and 1 and between Lines n and $n+1$ (see Fig. 10.06-1) tend to become inconveniently small when the bandwidth is large, and the widths of Bars 1 and n tend to become very small.

To use Table 10.06-1 for the design of an interdigital filter, first use Eqs. (10.06-1) through (10.06-3) and the charts in Sec. 4.03 to estimate the number, n, of reactive elements required in the low-pass prototype in order to give the desired rate of cutoff with the desired pass-band characteristics. When the prototype cutoff frequency ω_1' and element values g_0, g_1, ..., g_{n+1} have been obtained from the tables in Chapter 4, the design computations can begin. It is suggested that the filter fractional bandwidth, w, be specified to be 6 or 7 percent larger than is actually desired, since from the trial design described later it appears that there will be some shrinkage in bandwidth due to the approximate nature of the

Table 10.06-1

DESIGN EQUATIONS FOR INTERDIGITAL FILTERS OF THE FORM IN FIG. 10.06-1

Use mapping in Eqs. (10.06-1) to (10.06-3) to select a low-pass prototype with the required value of n. The input and output lines in this filter do not count as resonators, so that there are $n + 2$ lines for an n-reactive-element prototype.

Compute:

$$\theta_1 = \frac{\pi}{2}\frac{\omega_1}{\omega_0} = \frac{\pi}{2}\left(1 - \frac{w}{2}\right)$$

$$\frac{J_{01}}{Y_A} = \frac{1}{\sqrt{g_0 g_1 \omega_1'}} \quad , \quad \left.\frac{J_{k,k+1}}{Y_A}\right|_{k=1 \text{ to } n-1} = \frac{1}{\omega_1'\sqrt{g_k g_{k+1}}}$$

$$\frac{J_{n,n+1}}{Y_A} = \frac{1}{\sqrt{g_n g_{n+1}\omega_1'}}$$

$$\left.N_{k,k+1}\right|_{k=1 \text{ to } n-1} = \sqrt{\left(\frac{J_{k,k+1}}{Y_A}\right)^2 + \frac{\tan^2 \theta_1}{4}}$$

$$M_1 = Y_A\left(\frac{J_{01}}{Y_A}\sqrt{h} + 1\right) \quad , \quad M_n = Y_A\left(\frac{J_{n,n+1}}{Y_A}\sqrt{h} + 1\right)$$

where h is a dimensionless admittance scale factor to be specified arbitrarily so as to give a convenient admittance level in the filter. (See text.)

The normalized self capacitances C_k/ϵ per unit length for the line elements are:

$$\frac{C_0}{\epsilon} = \frac{376.7}{\sqrt{\epsilon_r}}\left[2Y_A - M_1\right]$$

$$\frac{C_1}{\epsilon} = \frac{376.7}{\sqrt{\epsilon_r}}\left\{Y_A - M_1 + hY_A\left[\frac{\tan \theta_1}{2} + \left(\frac{J_{01}}{Y_A}\right)^2 + N_{12} - \frac{J_{12}}{Y_A}\right]\right\}$$

(continued on p. 618)

Table 10.06-1 *concluded*

$$\left.\frac{C_k}{\epsilon}\right|_{k=2 \text{ to } n-1} = \frac{376.7}{\sqrt{\epsilon_r}}\, hY_A\left(N_{k-1,k} + N_{k,k+1} - \frac{J_{k-1,k}}{Y_A} - \frac{J_{k,k+1}}{Y_A}\right)$$

$$\frac{C_n}{\epsilon} = \frac{376.7}{\sqrt{\epsilon_r}}\left\{Y_A - M_n + hY_A\left[\frac{\tan\theta_1}{2} + \left(\frac{J_{n,n+1}}{Y_A}\right)^2 + N_{n-1,n} - \frac{J_{n-1,n}}{Y_A}\right]\right\}$$

$$\frac{C_{n+1}}{\epsilon} = \frac{376.7}{\sqrt{\epsilon_r}}\left[2Y_A - M_n\right]$$

where ϵ is the dielectric constant and ϵ_r is the relative dielectric constant in the medium of propagation.

The normalized mutual capacitances $C_{k,k+1}/\epsilon$ per unit length between adjacent line elements are:

$$\frac{C_{01}}{\epsilon} = \frac{376.7}{\sqrt{\epsilon_r}}\left[M_1 - Y_A\right] \quad ,$$

$$\left.\frac{C_{k,k+1}}{\epsilon}\right|_{k=1 \text{ to } n-1} = \frac{376.7\; hY_A}{\sqrt{\epsilon_r}}\left(\frac{J_{k,k+1}}{Y_A}\right) \quad ,$$

$$\frac{C_{n,n+1}}{\epsilon} = \frac{376.7}{\sqrt{\epsilon_r}}\left[M_n - Y_A\right] \quad .$$

SOURCE: Quarterly Progress Report 4, Contract DA 36-039 SC-87398, SRI; reprinted in *IRE Trans. PGMTT* (see Ref. 3 by G. L. Matthaei).

design equations. However, the desired value of w should be used in the mapping Eqs. (10.06-1) through (10.06-3) for determining n. Note that $Y_A = Y_B$ is the characteristic admittance of the terminating lines.

After all of the J/Y_A and N parameters in Table 10.06-1 have been computed, the admittance scale factor, h, must be fixed. One of the prime considerations in the choice of h is that the line dimensions must be such that the resonators will have a high unloaded Q. The dimensions that give optimum resonator Q's in such structures as interdigital filters are not known. However, it is known that for air-filled, coaxial-line resonators the optimum Q will result when the line impedance is about 76 ohms, and various approximate studies suggest that the optimum impedance for thick, rectangular-bar, strip-line resonators such as those in Fig. 10.06-2 is not greatly different. Thus, it is suggested that in this case h be chosen to make the quantity

$$\left.\frac{2C_{k-1,k}}{\epsilon} + \frac{C_k}{\epsilon} + \frac{2C_{k,k+1}}{\epsilon}\right|_{\substack{k=n/2 \text{ for } n \text{ even} \\ =(n+1)/2 \text{ for } n \text{ odd}}} = \text{(around 5.4)} \tag{10.06-4}$$

if air dielectric is used. If the quantity in Eq. (10.06-4) is set equal to 5.4 for the case of $\epsilon_r = 1$, this corresponds to making the line impedance 70 ohms for the resonator lines in the center of the filter, under the conditions that the adjacent lines are being excited with the same amplitude of voltage but with opposite phase (this is a generalized odd-mode admittance condition). A value for h, having been established, the remainder of the calculations follow in a straightforward manner. After the normalized capacitances, C_k/ϵ and $C_{k,k+1}/\epsilon$, have been computed, the line dimensions are determined as discussed in Sec. 5.05 [by use of Eqs. (5.05-33) through (5.05-35) and the accompanying charts].

A trial design was worked out using an $n = 6$ reactive-element Tchebyscheff prototype with $L_{Ar} = 0.10$ db. The prototype parameters were $g_0 = 1$, $g_1 = 1.1681$, $g_2 = 1.4039$, $g_3 = 2.0562$, $g_4 = 1.5170$, $g_5 = 1.9029$, $g_6 = 0.8618$, $g_7 = 1.3554$, and $\omega_1' = 1$. The design was worked out for a fractional bandwidth of $w = 0.10$ centered at $f_0 = 1.5$ Gc. Table 10.06-2 summarizes the various parameters used or computed in the calculations. The parameter h was chosen so that Eq. (10.06-4) would yield 5.4. The resulting circuit has symmetry in its dimensions because the Tchebyscheff prototype is antimetric (*i.e.*, one half of the network is reciprocal to the other half as discussed in Sec. 4.05).

Table 10.06-2

TABULATION OF QUANTITIES IN TABLE 10.06-1 AND IN FIG. 10.06-2 FOR A 10-PERCENT BANDWIDTH TRIAL DESIGN WITH $n = 6$

k	$J_{k,k+1}/Y_A$	$N_{k,k+1}$	$C_{k,k+1}/\epsilon$	$s_{k,k+1}$ (inches)	k	C_k/ϵ	w_k (inches)
0 and 6	0.9253		1.582	0.159*	0 and 7	5.950	0.405†
1 and 5	0.7809	6.401	0.301	0.419	1 and 6	3.390	0.152
2 and 4	0.5886	6.381	0.226	0.512	2 and 5	4.420	0.183
3	0.5662	6.379	0.218	0.520	3 and 4	4.496	0.183

$w = 0.10$ $\quad Y_A = 0.020$ mho
$\theta_1 = 1.492$ $\quad M_1 = M_6 = 0.02420$
$h = 0.05143$ $\quad b = 0.625$ inch
$\epsilon_r = 1$ $\quad t = 0.187$ inch

* Changed to 0.127 inch after laboratory tests.
† Changed to 0.437 inch after laboratory tests.
SOURCE: Quarterly Progress Report 4, Contract DA 36-039 SC-87398, SRI; reprinted in *IRE Trans. PGMTT* (see Ref. 3 by G. L. Matthaei).

Figure 10.06-3 shows a photograph of the completed filter, while Fig. 10.06-4 shows those dimensions of the filter not summarized in Table 10.06-2. The short-circuiting side walls of the structure are spaced exactly a quarter-wavelength apart at the midband frequency $f_0 = 1.5$ Gc ($\lambda_0/4 = 1.968$ inches). Because of the capacitance between the open-circuited ends of the resonator elements and the side walls, it was necessary to foreshorten the resonators so as to maintain their resonant frequency at 1.5 Gc. No very satisfactory means for accounting for all of the fringing capacitances at the open-circuit ends of the resonators has been devised, but some rough estimates were made using fringing capacitance data in Sec. 5.05 and various approximations. The estimated foreshortening for the resonators was 0.216 inch, but laboratory tests showed this to be excessive since the pass-band center was 1.56 instead of 1.50 Gc. Although the filter pass band can always be lowered in frequency by use of tuning screws, if the resonators had been foreshortened about 0.160 inch instead of 0.216 inch, the pass-band center frequency would probably have been about right.

Although the structure included tuning screws, no effort was made to lower the band-center frequency to 1.50 Gc. However, it was found that since Resonators 1 and 6 have different fringing-capacitance

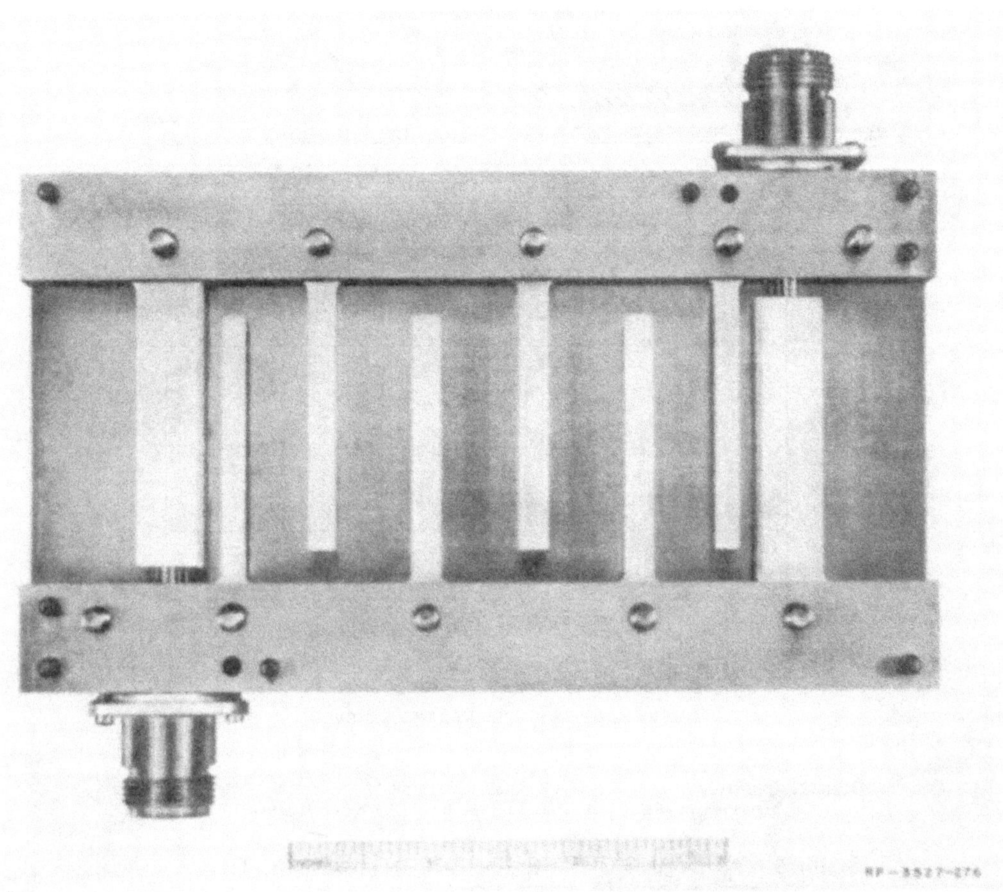

SOURCE: Quarterly Progress Report 4, Contract DA 36-039 SC-87398, SRI; reprinted in *IRE Trans. PGMTT* (see Ref. 3 by G. L. Matthaei)

FIG. 10.06-3 A 10-PERCENT BANDWIDTH INTERDIGITAL FILTER WITH ITS COVER PLATE REMOVED

conditions at their open-circuit ends than do the other resonators, it was necessary to increase the capacitance at their open-circuit ends by inserting the tuning screws. Before this was done, the pass-band response was not symmetrical (this is indicative of mistuning of some of the resonators with respect to the others).

When the filter was first tested, the pass-band VSWR reached peaks of 2.2, which is somewhat high since 0.1-db Tchebyscheff ripples call for VSWR peaks of only 1.36. Such conditions can usually be corrected by altering the couplings between the terminations and the first resonator on each end. Thus, a 0.032-inch-thick brass shim was added to the

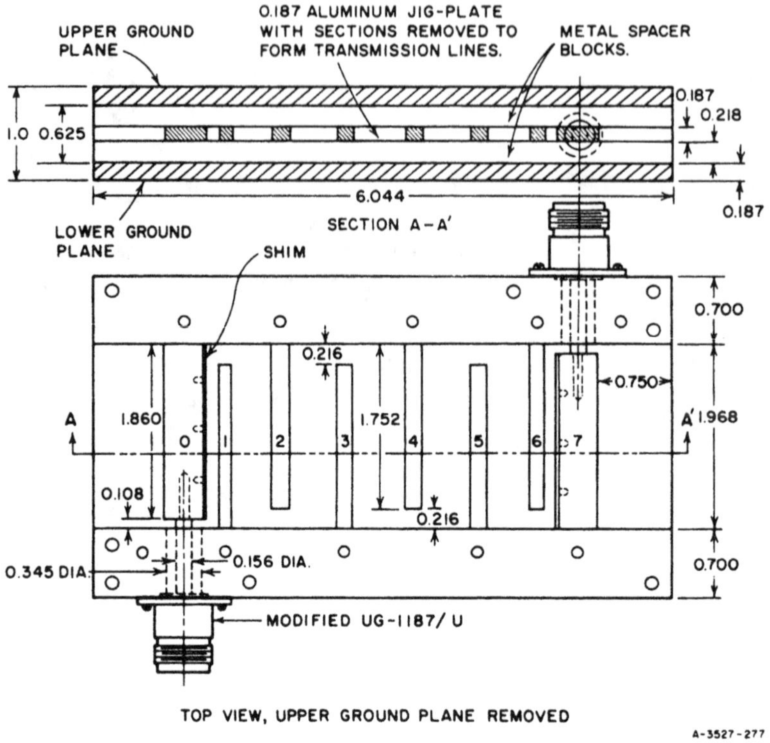

SOURCE: Quarterly Progress Report 4, Contract DA 36-039 SC-87398, SRI; reprinted in *IRE Trans. PGMTT* (see Ref. 3 by G. L. Matthaei)

FIG. 10.06-4 DRAWING OF THE 10-PERCENT BANDWIDTH INTERDIGITAL FILTER
Part of the dimensions are as specified by Table 10.06-2 along with Fig. 10.06-2

input and output lines (Lines 0 and 7) to reduce the adjacent gaps from $s_{01} = s_{67} = 0.159$ inch to 0.127 inch. This reduced the input VSWR to 1.30 or less across the band. It appears desirable that, in working out the design of trial models of interdigital filters as described herein, some provision should be made for experimental adjustment of the size of coupling gaps at the ends, if the pass-band VSWR is to be closely controlled.

Figure 10.06-5 shows the measured attenuation characteristic of this filter, while Fig. 10.06-6 shows the measured VSWR. The measured fractional bandwidth is slightly less than the design value ($w = 0.0935$ instead of 0.100). Using $w = 0.0935$ and $f_0 = \omega_0/(2\pi) = 1.563$ Gc, the measured attenuation was compared with that estimated by use of the mapping Eqs. (10.06-1) through (10.06-3) along with Fig. 4.03-5. The measured attenuation in the stop bands was found to be somewhat less than that predicted for a Tchebyscheff filter with L_{Ar} = 0.1-db ripple. However, the pass-band VSWR in Fig. 10.06-6 is, for the most part, much less than the 1.36 peak value corresponding to a 0.1-db Tchebyscheff ripple, and some of the VSWR amplitude shown may be due to the connectors and slight mistuning.* The attenuation for a L_{Ar} = 0.01-db ripple filter

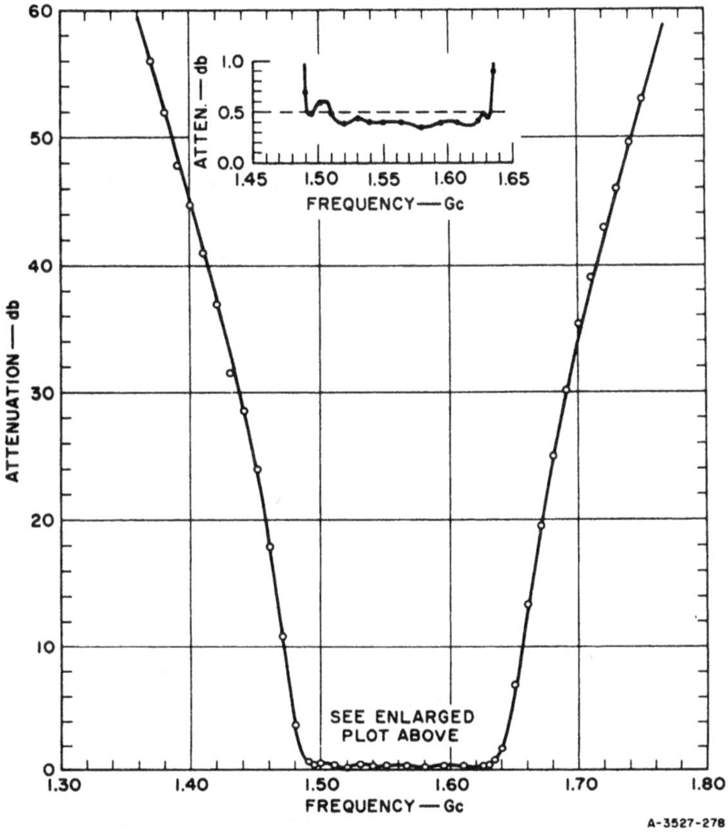

SOURCE: Quarterly Progress Report 4, Contract DA 36-039 SC-87398, SRI; reprinted in *IRE Trans. PGMTT* (see Ref. 3 by G. L. Matthaei)

FIG. 10.06-5 MEASURED ATTENUATION CHARACTERISTIC OF THE FILTER IN FIG. 10.06-3

---

* The pass-band VSWR is the most sensitive index of the correlation of the actual design as compared to the pass-band characteristic of the prototype. The pass-band attenuation as determined from through transmission measurements includes the additional attenuation due to dissipation loss, which may be markedly greater than 0.1 db depending on the Q's of the resonators.

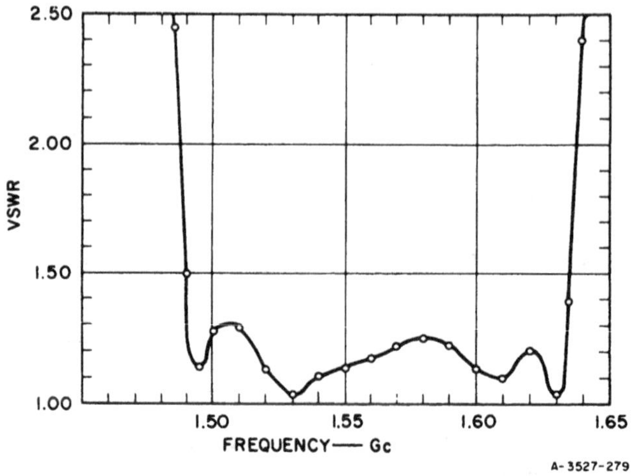

SOURCE: Quarterly Progress Report 4, Contract DA 36-039 SC-87398, SRI; reprinted in *IRE Trans. PGMTT* (see Ref. 3 by G. L. Matthaei)

FIG. 10.06-6 MEASURED VSWR OF THE FILTER IN FIG. 10.06-3

filter (VSWR = 1.1 peak) was also estimated, and the results are summarized in Table 10.06-3. Note that in all cases the measured stop-band attenuation is less than that computed for a 0.10-db ripple filter, but more than that for a 0.01-db ripple filter. Thus, the given approximate mapping procedure appears to be reasonably consistent with the measured results.

Table 10.06-3
COMPARISON OF MEASURED ATTENUATION IN FIG. 10.06-5 WITH ATTENUATION PREDICTED USING MAPPING EQS. (10.06-1) TO (10.06-3)
Computed values are for 0.01 and 0.10 db ripple, $n = 6$, $w = 0.0935$, and $f_0 = \omega_0/(2\pi) = 1.563$ Gc

f, Gc	L_A, db FOR $L_{Ar} = 0.01$	L_A, db MEASURED	L_A, db FOR $L_{Ar} = 0.10$
1.440	25	29	35.5
1.686	25	28	35.5
1.380	49.5	52	59.5
1.746	49.5	52	59.5

SOURCE: Quarterly Progress Report 4, Contract DA 36-039 SC-87398, SRI; reprinted in *IRE Trans. PGMTT* (see Ref. 3 by G. L. Matthaei.

The second pass band for interdigital filters is centered at $3\omega_0$ (where ω_0 is the center of the first pass band). There are multiple-order poles of attenuation (Sec. 2.04) at $\omega = 0$, $2\omega_0$, $4\omega_0$, etc. so the stop bands are very strong. Unlike all the filters discussed earlier in this chapter, interdigital filters cannot possibly have any spurious responses near $2\omega_0$, $4\omega_0$, etc., no matter how poorly they may be tuned.

Calculating from the measured pass-band attenuation of the trial design, it is estimated that the unloaded Q's of the resonators in this filter are about 1100. Using copper for the structure instead of aluminum would theoretically give a value about 25 percent higher for the unloaded Q's. It is possible that a different impedance level within the filter might also give higher Q's for a given ground plane spacing. Polishing the resonator bars and the ground planes would also help to raise the resonator Q's.

The lines in the trial interdigital filter (Figs. 10.06-3 and 10.06-4) were fabricated by machining Lines 0, 2, 4, and 6 in comb form from a single piece of jig plate. Lines 1, 3, 5, and 7 were cut out similarly from a second piece of jig plate. (Actually, both comb structures were machined at once, back-to-back.) Then, interleaving the two comb structures between ground planes gave the desired interdigital structure.

Figure 10.06-7 shows an alternative form of interdigital filter structure that should be even less expensive to fabricate. In this case, the interdigital line structure is photo-etched on a copper-clad dielectric card, and the dielectric material removed from the region between the copper-foil lines. In order to provide good support for the lines, the

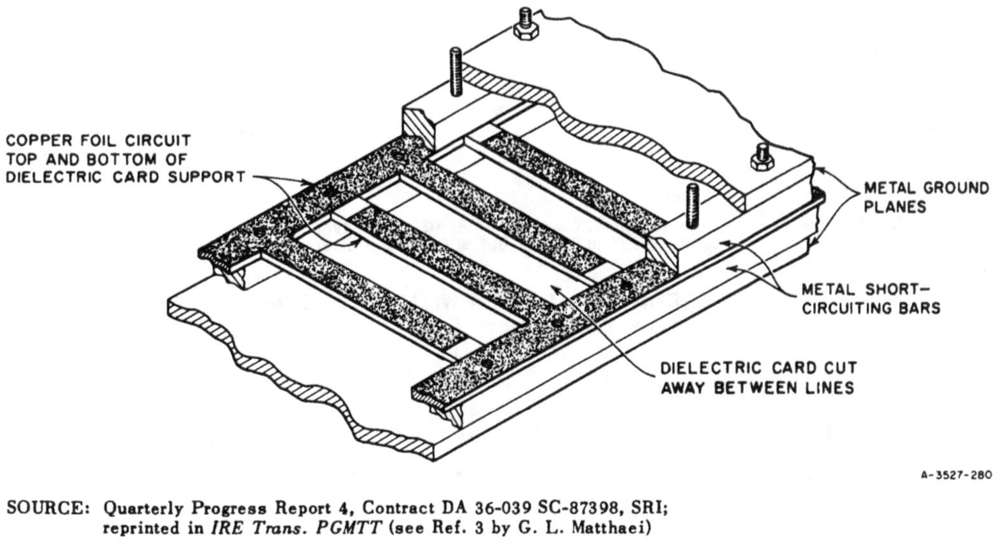

SOURCE: Quarterly Progress Report 4, Contract DA 36-039 SC-87398, SRI; reprinted in *IRE Trans. PGMTT* (see Ref. 3 by G. L. Matthaei)

FIG. 10.06-7 A PROPOSED LOW-COST CONSTRUCTION FOR INTERDIGITAL-LINE FILTERS

dielectric is not removed at the open-circuit ends of the lines, however. With the use of this structure, the propagation is largely in air, which should permit good performance; also, the dielectric-card line structures should be quite inexpensive to mass produce.

Round rods between ground planes also provide an attractive form for fabricating interdigital filters.[2] However, no data are yet available for accurate determination of the rod diameters and spacings from specified line capacitances.*

SEC. 10.07, INTERDIGITAL-LINE FILTERS HAVING WIDE BANDWIDTHS

In this section the interdigital band-pass filter discussion in Sec. 10.06 is extended to cover the filter structure shown in Fig. 10.07-1. Note that the structure in Fig. 10.07-1 differs from that in Fig. 10.06-1 in that the terminating lines are connected to open-circuited rather than

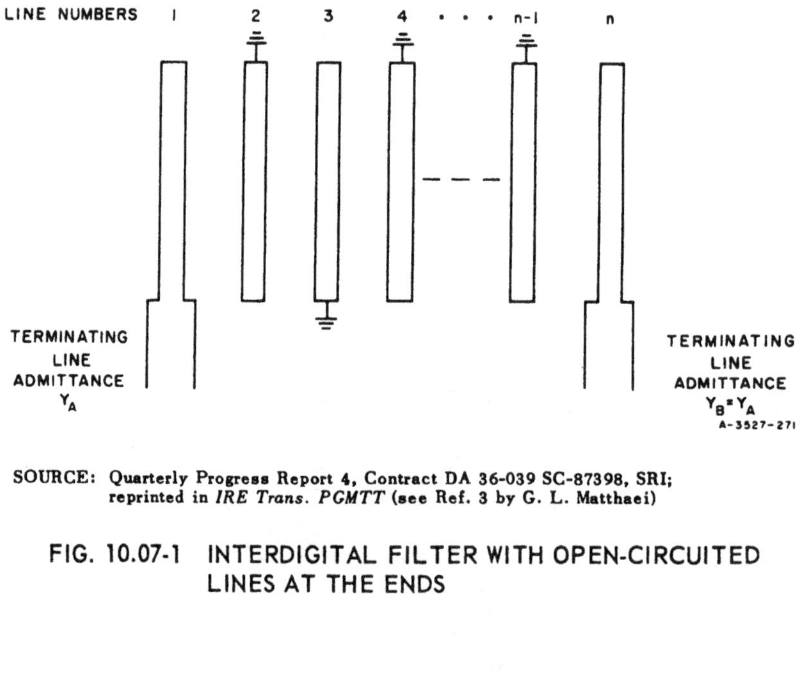

SOURCE: Quarterly Progress Report 4, Contract DA 36-039 SC-87398, SRI; reprinted in *IRE Trans. PGMTT* (see Ref. 3 by G. L. Matthaei)

FIG. 10.07-1 INTERDIGITAL FILTER WITH OPEN-CIRCUITED LINES AT THE ENDS

short-circuited line elements. In the case of the structure in Fig. 10.07-1 when it is designed by the methods of this section, all of the line elements (including Lines 1 and n) serve as resonators. Thus, when the procedure of this section is used, an n-reactive-element, low-pass prototype will lead to an interdigital filter with n interdigital line elements.

* Bolljahn and Matthaei (Ref. 2) give an approximate method for design of structures consisting of rods that are all the same size and have the same spacings. A procedure for accurate design where varying diameters and spacings are required has not as yet been obtained.

Table 10.07-1 presents approximate design equations for filters of the form in Fig. 10.07-1. This type of filter is most practical for designs having moderate or wide bandwidths (*i.e.*, perhaps bandwidths of around 30 percent or more), although the design procedure given is valid for either narrow- or wide-band filters. The main drawback in applying the procedure in Table 10.07-1 to narrow-band filters is that Lines 1 and n will tend to attain extremely high impedance in such designs.

Except for the somewhat different design equations, the procedure for the design of filters of the form shown in Fig. 10.07-1 is much the same as that described for filters in Sec. 10.06. It is suggested that for use in Table 10.07-1, w be made about 8 percent larger than the actual desired fractional bandwidth, in order to allow for some bandwidth shrinkage. Equation (10.06-4) is applicable when selecting a value for the admittance scale factor, h.

A trial design was worked out using an $n = 8$ reactive-element Tchebyscheff prototype with $L_{Ar} = 0.10$ db. The prototype parameters were $g_0 = 1$, $g_1 = 1.1897$, $g_2 = 1.4346$, $g_3 = 2.1199$, $g_4 = 1.6010$, $g_5 = 2.1699$, $g_6 = 1.5640$, $g_7 = 1.9444$, $g_8 = 0.8778$, $g_9 = 1.3554$, and $\omega_1' = 1$. The design was carried out for a fractional bandwidth of $w = 0.70$ centered at 1.50 Gc, and the parameter h was chosen to make Eq. (10.06-4) equal to 5.86. (This gives a generalized odd-mode impedance of 64.5 ohms for the middle lines.) Table 10.07-2 summarizes some of the quantities computed in the course of the design of this filter.

Figure 10.07-2 shows the completed filter, while the drawing in Fig. 10.07-3 shows additional construction details and additional dimensions not summarized in Table 10.07-2. The filter was fabricated in much the same manner as the filter described in Sec. 10.06, except that the resonator lines were foreshortened by 0.150 inch. The relatively small cross-sectional dimensions of the resonator elements, unfortunately, made 0.150 inch excessive, so that the measured band-center frequency was 1.55 Gc instead of 1.50 Gc. It is probable that foreshortening the line elements about 0.125 inch would have been about right.

When this filter was first tested, the VSWR was quite low across the band (about 1.2 or less) except at band center where the VSWR peaked to 1.8. This situation was altered by increasing the s_{12} and s_{78} gaps from 0.087 inch to 0.092 inch, which caused the VSWR peaks across the band to be more nearly even and to be 1.55 or less.

Table 10.07-1

DESIGN EQUATIONS FOR INTERDIGITAL FILTERS OF THE FORM IN FIG. 10.07-1

Use mapping in Eqs. (10.06-1) to (10.06-3) to select low-pass prototype with the required value of n. The input and output lines in this filter count as resonators, so that there are n line elements for an n-reactive-element prototype.

Compute:

$$\theta_1 = \frac{\pi}{2}\frac{\omega_1}{\omega_0} = \frac{\pi}{2}\left(1 - \frac{w}{2}\right)$$

$$\left.\frac{J_{k,k+1}}{Y_A}\right|_{k=2 \text{ to } n-3} = \frac{g_2}{g_0\sqrt{g_k g_{k+1}}} \;, \qquad \frac{J_{n-2,n-1}}{Y_A} = \frac{1}{g_0}\sqrt{\frac{g_2 g_0}{g_{n-2} g_{n+1}}}$$

$$\left.N_{k,k+1}\right|_{k=2 \text{ to } n-2} = \sqrt{\left(\frac{J_{k,k+1}}{Y_A}\right)^2 + \left(\frac{\omega_1' g_2 \tan\theta_1}{2g_0}\right)^2}$$

$$\frac{Z_1}{Z_A} = \omega_1' g_0 g_1 \tan\theta_1$$

$$\frac{Y_2}{Y_A} = \frac{\omega_1' g_2}{2g_0}\tan\theta_1 + N_{23} - \frac{J_{23}}{Y_A}$$

$$\left.\frac{Y_k}{Y_A}\right|_{k=3 \text{ to } n-2} = N_{k-1,k} + N_{k,k+1} - \frac{J_{k-1,k}}{Y_A} - \frac{J_{k,k+1}}{Y_A}$$

$$\frac{Y_{n-1}}{Y_A} = \frac{\omega_1'(2g_0 g_{n-1} - g_2 g_{n+1})\tan\theta_1}{2g_0 g_{n+1}} + N_{n-2,n-1} - \frac{J_{n-2,n-1}}{Y_A}$$

$$\frac{Z_n}{Z_A} = \omega_1' g_n g_{n+1} \tan\theta_1$$

The normalized self-capacitances, C_k/ϵ, per unit length for the line elements are:

$$\frac{C_1}{\epsilon} = \frac{376.7}{\sqrt{\epsilon_r}} Y_A \frac{(1 - \sqrt{h})}{(Z_1/Z_A)}$$

Table 10.07-1 *concluded*

$$\frac{C_2}{\epsilon} = \frac{376.7}{\sqrt{\epsilon_r}} Y_A h \left(\frac{Y_2}{Y_A}\right) - \sqrt{h}\,\frac{C_1}{\epsilon}$$

$$\left.\frac{C_k}{\epsilon}\right|_{k=3 \text{ to } n-2} = \frac{376.7}{\sqrt{\epsilon_r}} Y_A h \left(\frac{Y_k}{Y_A}\right)$$

$$\frac{C_{n-1}}{\epsilon} = \frac{376.7}{\sqrt{\epsilon_r}} Y_A h \left(\frac{Y_{n-1}}{Y_A}\right) - \sqrt{h}\,\frac{C_n}{\epsilon}$$

$$\frac{C_n}{\epsilon} = \frac{376.7}{\sqrt{\epsilon_r}} Y_A \frac{(1 - \sqrt{h})}{(Z_n/Z_A)}$$

where ϵ is the dielectric constant, ϵ_r is the relative dielectric constant in the medium of propagation, and h is a dimensionless admittance scale factor whose value should be chosen to give a convenient admittance level in the filter. (See Text.)

The normalized mutual capacitances $C_{k,k+1}/\epsilon$ per unit length between adjacent line elements are:

$$\frac{C_{12}}{\epsilon} = \frac{376.7}{\sqrt{\epsilon_r}} Y_A \frac{\sqrt{h}}{(Z_1/Z_A)}$$

$$\left.\frac{C_{k,k+1}}{\epsilon}\right|_{k=2 \text{ to } n-2} = \frac{376.7}{\sqrt{\epsilon_r}} Y_A h \left(\frac{J_{k,k+1}}{Y_A}\right)$$

$$\frac{C_{n-1,n}}{\epsilon} = \frac{376.7}{\sqrt{\epsilon_r}} Y_A \frac{\sqrt{h}}{(Z_n/Z_A)}$$

SOURCE: Quarterly Progress Report 4, Contract DA 36-039 SC-87398, SRI; reprinted in *IRE Trans. PGMTT* (see Ref. 3 by G. L. Matthaei).

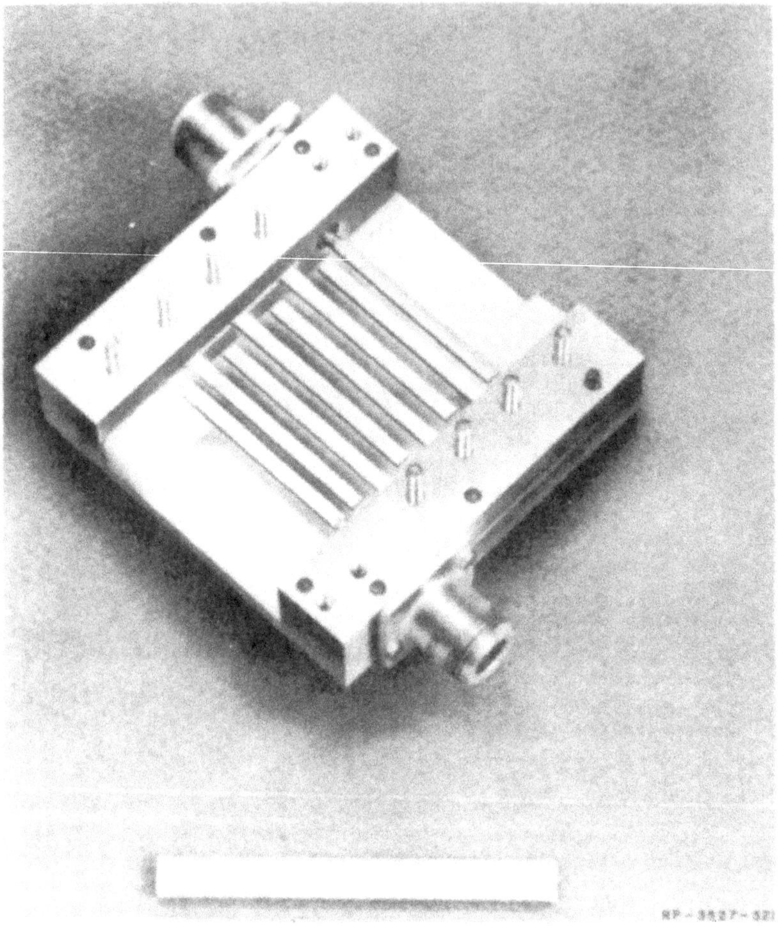

SOURCE: Quarterly Progress Report 4, Contract DA 36-039 SC-87398, SRI; reprinted in *IRE Trans. PGMTT* (see Ref. 3 by G. L. Matthaei)

FIG. 10.07-2 OCTAVE BANDWIDTH INTERDIGITAL FILTER WITH COVER PLATE REMOVED

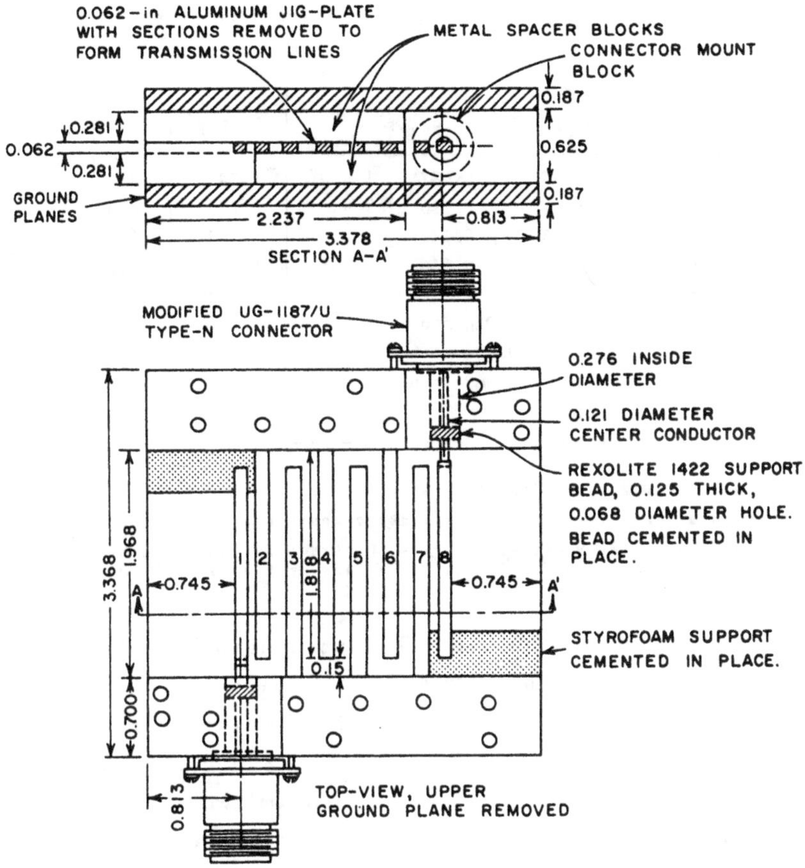

SOURCE: Quarterly Progress Report 4, Contract DA 36-039 SC-87398, SRI; reprinted in *IRE Trans. PGMTT* (see Ref. 3 by G. L. Matthaei)

FIG. 10.07-3 DRAWING OF THE INTERDIGITAL FILTER IN FIG. 10.07-2
Other dimensions not shown are as defined by Fig. 10.06-2 and Table 10.07-2

Table 10.07-2

TABULATION OF SOME OF THE PARAMETERS COMPUTED IN THE DESIGN OF THE TRIAL OCTAVE-BANDWIDTH INTERDIGITAL FILTER

The dimensions are as defined in Fig. 10.06-2

k	$J_{k,k+1}/Y_A$	$C_{k,k+1}/\epsilon$	$s_{k,k+1}$ (inches)	k	$\frac{Z_k}{Z_A}$ or $\frac{Y_k}{Y_A}$	C_k/ϵ	w_k (inches)
1 and 7		1.647	0.087*	1 and 8	1.941	2.235	0.126†
2 and 6	0.823	1.115	0.136	2 and 7	1.779	1.463	0.121†
3 and 5	0.779	1.056	0.143	3 and 6	1.235	1.675	0.126†
4	0.770	1.044	0.146	4 and 5	1.258	1.706	0.127†
w = 0.70		Y_A = 0.020 mho					
θ_1 = 1.021		b = 0.625 inch					
h = 0.18		t = 0.063 inch					
ϵ_r = 1							

* Changed to 0.092 inch after laboratory tests.

† Computed using Eq. (5.05-26) for width correction.

SOURCE: Quarterly Progress Report 4, Contract DA 36-039 SC-87398, SRI; reprinted in *IRE Trans. PGMTT* (see Ref. 3 by G. L. Matthaei.

Figure 10.07-4 shows the measured attenuation of this filter, while Fig. 10.07-5 shows its measured VSWR. The fractional bandwidth is 0.645 instead of the specified 0.700 value, which indicates a shrinkage of bandwidth of about 8 percent, as a result of the various approximations involved in the design equations. The attenuation characteristics in Fig. 10.07-4 was checked against the attenuation computed using the mapping Eqs. (10.06-1) through (10.06-3) with w = 0.645, and $f_0 = \omega_0/(2\pi)$ = 1.55 Gc, along with the n = 8 curve in Fig. 4.03-5. The resulting computed values are listed in Table 10.07-3, along with the corresponding measured values of attenuation. The agreement can be seen to be quite good.

Table 10.07-3

COMPARISON OF THE MEASURED ATTENUATION IN FIG. 10.07-4 WITH ATTENUATION PREDICTED USING MAPPING EQS. (10.06-1) TO (10.06-3)

Computed values are for 0.10-db ripple, n = 8, w = 0.645, and $f_0 = \omega_0/(2\pi)$ = 1.55 Gc

f, Gc	L_A, db FOR L_{Ar} = 0.10	L_A, db MEASURED
0.90	32	31
2.23	32	33
0.70	56	55
2.43	56	57

SOURCE: Quarterly Progress Report 4, Contract DA 36-039 SC-87398, SRI; reprinted in *IRE Trans. PGMTT* (see Ref. 3 by G. L. Matthaei).

SEC. 10.08, DERIVATION OF THE DESIGN EQUATIONS FOR PARALLEL-COUPLED AND STUB FILTERS

The first step in deriving the design equations in Tables 10.02-1 through

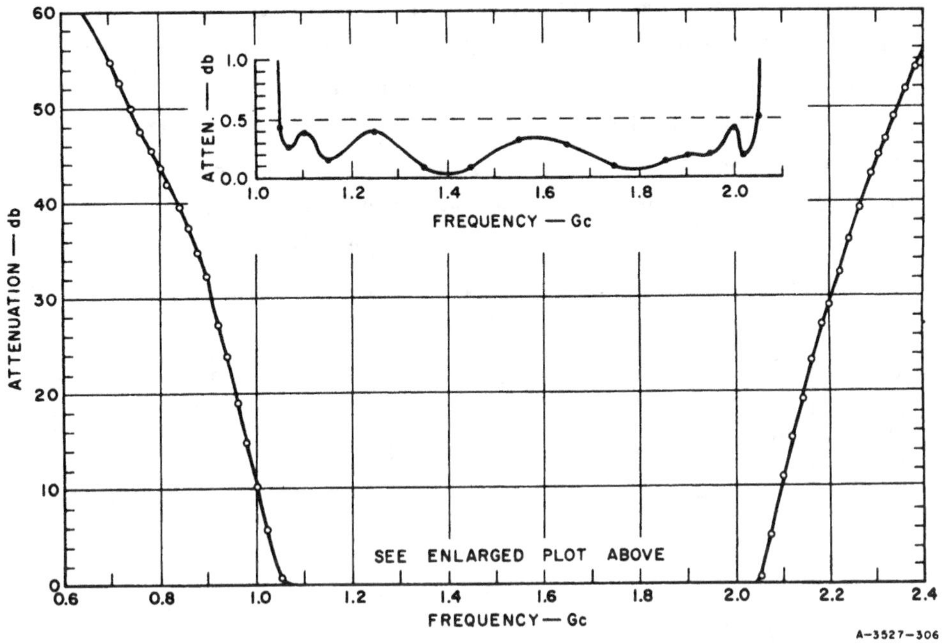

SOURCE: Quarterly Progress Report 4, Contract DA 36-039 SC-87398, SRI; reprinted in *IRE Trans. PGMTT* (see Ref. 3 by G. L. Matthaei)

FIG. 10.07-4 MEASURED ATTENUATION CHARACTERISTIC OF THE FILTER IN FIG. 10.07-2

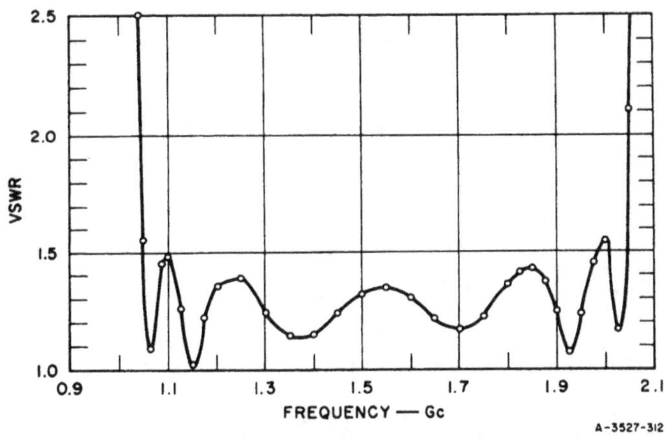

SOURCE: Quarterly Progress Report 4, Contract DA 36-039 SC-87398, SRI; reprinted in *IRE Trans. PGMTT* (see Ref. 3 by G. L. Matthaei)

FIG. 10.07-5 MEASURED VSWR OF THE FILTER IN FIG. 10.07-2

Table 10.05-1 is to convert the low-pass prototype (Sec. 4.04) into the modified form discussed in Sec. 4.12, which uses only one kind of reactive element along with impedance or admittance inverters as shown in Fig. 10.08-1, where the impedance and admittance inverters are assumed to be frequency-independent and to have the properties summarized in Fig. 10.08-2. It will be recalled that the low-pass prototype element values g_0, g_1, g_2, ..., g_{n+1} having been specified for the circuit in Fig. 10.08-1(a), the elements R_A, L_{a1}, L_{a2}, ..., L_{an}, R_B may be chosen

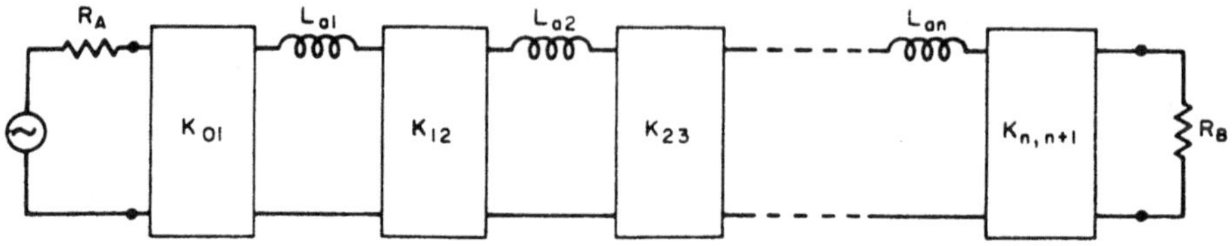

$$K_{01} = \sqrt{\frac{R_A L_{a1}}{g_0\, g_1}}\,, \qquad K_{k,k+1}\Big|_{k=1 \text{ to } n-1} = \sqrt{\frac{L_{ak}\, L_{a(k+1)}}{g_k\, g_{k+1}}}\,, \qquad K_{n,n+1} = \sqrt{\frac{L_{an}\, R_B}{g_n\, g_{n+1}}}$$

(a) MODIFIED PROTOTYPE USING IMPEDANCE INVERTERS

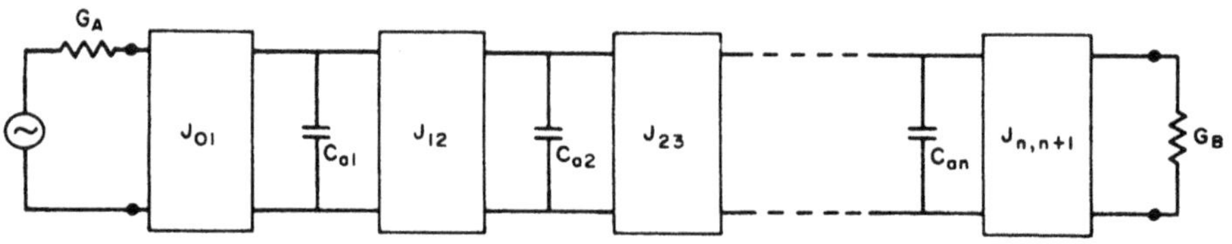

$$J_{01} = \sqrt{\frac{G_A C_{a1}}{g_0\, g_1}}\,, \qquad J_{k,k+1}\Big|_{k=1 \text{ to } n-1} = \sqrt{\frac{C_{ak}\, C_{a(k+1)}}{g_k\, g_{k+1}}}\,, \qquad J_{n,n+1} = \sqrt{\frac{C_{an}\, G_B}{g_n \cdot g_{n+1}}}$$

(b) MODIFIED PROTOTYPE USING ADMITTANCE INVERTERS

RB-2326-T7-172

SOURCE: Final Report, Contract DA 36-039 SC-74862, SRI; reprinted in *IRE Trans. PGMTT* (see Ref. 1 by G. L. Matthaei)

FIG. 10.08-1 LOW-PASS PROTOTYPES MODIFIED TO INCLUDE IMPEDANCE INVERTERS OR ADMITTANCE INVERTERS
The g_0, g_1, ..., g_n, g_{n+1} are obtained from the original prototype as in Fig. 4.04-1, while the R_A, L_{a1}, ..., L_{an}, and R_B or the G_A, C_{a1}, ..., C_{an}, and G_B may be chosen as desired

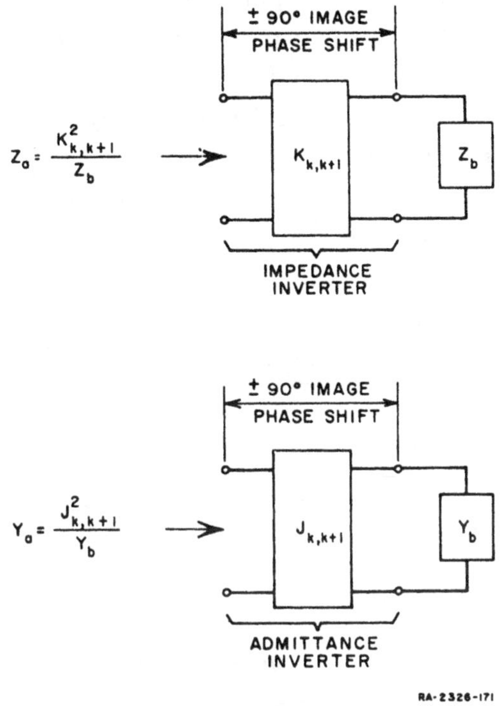

SOURCE: Final Report, Contract DA 36-039 SC-74862, SRI; reprinted in *IRE Trans. PGMTT* (see Ref. 1 by G. L. Matthaei)

FIG. 10.08-2 DEFINITION OF IMPEDANCE INVERTERS AND ADMITTANCE INVERTERS

as desired. Then the circuit with impedance inverters and series inductances will have exactly the same response as the original *L-C* ladder prototype if the impedance inverter parameters are specified as indicated by the equations in Fig. 10.08-1(a). An analogous situation also applies for the dual circuit in Fig. 10.08-1(b).

The derivations of the design equations in this chapter are based on the use of the converted prototypes in Fig. 10.08-1 using idealized, frequency-independent impedance or admittance inverters. The procedure will be to break up the modified prototype into symmetrical sections, and then to relate the *image properties* (Chapter 3) of the modified prototype sections to the image properties of corresponding sections of the actual band-pass microwave filter structure.

Procedure for Deriving the Equations in Table 10.02-1—The design equations in Table 10.02-1 are based on the modified prototype shown at (b) in Fig. 10.08-1, while Fig. 10.08-3 shows the manner in which the element values are specified, and the manner in which the prototype is broken into sections. The image admittance, $Y^i_{k,k+1}(\omega')$, and phase, $\beta_{k,k+1}$ (in the pass band) for each of the prototype interior sections (S'_{12} to $S'_{n-1,n}$) are readily shown to be

$$Y^i_{k,k+1}(\omega') = J_{k,k+1}\sqrt{1 - \left[\frac{\omega'(C_a/2)}{J_{k,k+1}}\right]^2} \qquad (10.08\text{-}1)$$

and

$$\beta_{k,k+1}\Bigg|_{\omega' \leq \frac{J_{k,k+1}}{C_a/2}} = \sin^{-1}\left[\frac{\omega'(C_a/2)}{J_{k,k+1}}\right] \mp \frac{\pi}{2} \qquad (10.08\text{-}2)$$

where, in this case, $C_a = G_A/\omega'_1$ and ω'_1 is the cutoff frequency for the low-pass prototype. The choice of $\mp\pi/2$ in Eq. (10.08-2) depends on whether the inverter is taken to have ±90-degree phase shift. The equations in Fig. 5.09-1(b) can be adapted to show that the image admittance and pass-band image phase for a parallel-coupled section, $S_{k,k+1}$ such as those in the filter in Fig. 10.02-1(a) is

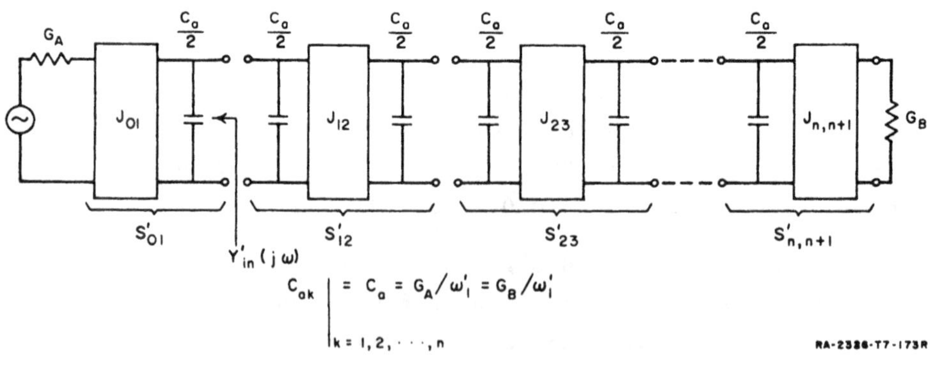

SOURCE: Final Report, Contract DA 36-039 SC-74862, SRI; reprinted in *IRE Trans. PGMTT* (see Ref. 1 by G. L. Matthaei)

FIG. 10.08-3 MODIFIED PROTOTYPE FOR DERIVING THE DESIGN EQUATIONS IN TABLE 10.02-1

$$Y_I = \frac{\sqrt{(Y_{oo} - Y_{oe})^2 - (Y_{oo} + Y_{oe})^2 \cos^2 \theta}}{2 \sin \theta} \qquad (10.08\text{-}3)$$

and

$$\beta = \cos^{-1}\left[\left(\frac{Y_{oo} + Y_{oe}}{Y_{oo} - Y_{oe}}\right) \cos \theta\right] \qquad (10.08\text{-}4)$$

where $\theta = \pi\omega/2\omega_0$ and Y_{oo} and Y_{oe} are the odd- and even-mode line admittances, respectively. The parameters of the parallel-coupled sections S_{12} to $S_{n-1,n}$ in Fig. 10.02-1(a) are related to sections S'_{12} to $S'_{n-1,n}$ of the prototype by forcing the following correspondences between the two structures:

(1) The image phase of the parallel-coupled sections when $\omega = \omega_0$ must be the same as the image phase of the prototype sections when $\omega' = 0$.

(2) The image admittances of the parallel-coupled sections when $\omega = \omega_0$ must be the same (within a scale factor h)* as the image admittances of the corresponding prototype sections when $\omega' = 0$.

(3) The image admittance of the parallel-coupled sections when $\omega = \omega_1$ must be the same (within a scale factor h)* as the image admittance of the corresponding prototype sections when $\omega' = \omega'_1$.

(10.08-5)

Correspondence (1) is fulfilled in this case by choosing the + sign in Eq. (10.08-2). Equating Eqs. (10.08-1) and (10.08-3) and evaluating each side at the appropriate frequencies indicated above, two equations are obtained from which the equations for interior sections in Table 10.02-1 may be derived (with the help of the information in Figs. 10.08-1 and 10.08-3) by solving for Y_{oo} and Y_{oe}.

The end sections, S_{01} and $S_{n,n+1}$, must be treated as a special case. If $Y_{in}(j\omega)$ is defined as the admittance seen looking in the right end of the parallel-coupled section S_{01} in Fig. 10.02-1(a), with the left end

---

* Taking $G_A = G_B = Y_A = Y_B$.

connected to the input line of admittance Y_A, the following correspondences are forced with respect to $Y'_{in}(j\omega')$ indicated in Fig. 10.08-3:

(1) Re $Y_{in}(j\omega_0)$ = Re $Y_{in}(j\omega_1)$ for the parallel-coupled terminating circuit, just as Re $Y'_{in}(j0)$ = Re $Y'_{in}(-j\omega'_1)$ for the terminating circuit of the prototype.

(10.08-6)

(2) $Im\ Y_{in}(j\omega_1)$/Re $Y_{in}(j\omega_1)$ must equal B'/G' = $Im\ Y'_{in}(-j\omega'_1)$/ Re $Y'_{in}(-j\omega'_1)$ computed from the prototype.

In order to obtain additional degrees of freedom for adjusting the admittance level within the interior of the filter, the two parallel-coupled strips for the end sections S_{01} and $S_{n,n+1}$ were allowed to be of unequal width and the special constraint conditions summarized in Fig. 5.09-3(a) were used in computing $Y_{in}(j\omega)$ for the actual filter. [The constraint $(Y^a_{oe})_{01} + (Y^a_{oo})_{01} = 2Y_A$ insures that Correspondence 1 in Eq. (10.08-6) will be satisfied.] From Fig. 10.08-3 it is easily seen that

$$Y'_{in}(j\omega') = \frac{J^2_{01}}{G_A} + j\frac{\omega' C_a}{2} \quad . \tag{10.08-7}$$

The equations in Table 10.02-1 for design of the end sections S_{01} and $S_{n,n+1}$ of the parallel-coupled filter were then obtained by using $Y_{in}(j\omega)$ computed using Fig. 5.09-3(a), and $Y'_{in}(j\omega')$ from Eq. (10.08-7) in the correspondences (10.08-6) above. It should be noted that the admittance scale factor $h = 1/\sqrt{N}$, where N is the turns ratio in Fig. 5.09-3(a).

Procedure for Deriving the Equations in Table 10.03-1—The equations in Table 10.03-1 were derived using much the same point of view as discussed above for Table 10.02-1, except that the somewhat different modified prototype in Fig. 10.08-4 was used. Note that in this case the J_{01} and $J_{n,n+1}$ admittance inverters are eliminated, and that the circuit has been split into a cascade of symmetrical sections. Note that the symmetrical sections in Fig. 10.08-4 are the same as those in Fig. 10.08-3, and that their image admittance and phase are given by Eqs. (10.08-1) and (10.08-2). Now the stub filter in Fig. 10.03-1 can be pieced together as a cascade of symmetrical stub sections as shown in Fig. 10.08-5. Also, note from Fig. 5.09-1(b) that these stub sections are exactly equivalent to the parallel-coupled sections in the filter in Fig. 10.02-1(a). Due

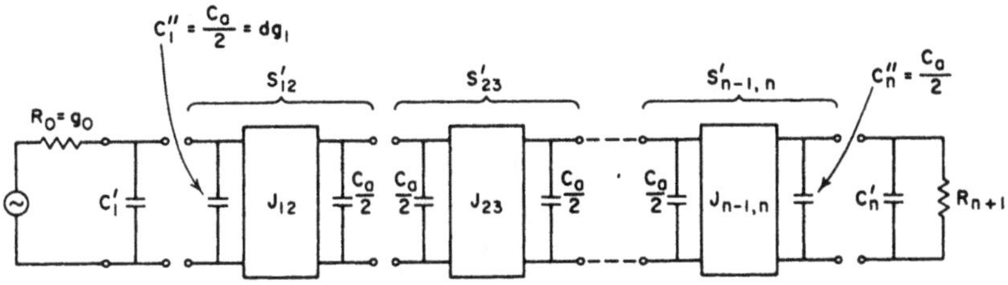

$$C_1' = g_1(1-d) \qquad C_o = 2dg_1 \qquad C_n' = \frac{g_n g_{n+1} - dg_1 R_{n+1}}{R_{n+1}}$$

$$C_1 = g_1 = C_1' + C_1'' \qquad C_n = g_n \frac{g_{n+1}}{R_{n+1}} = C_n' + C_n''$$

$$J_{12} = \sqrt{\frac{C_1 C_o}{g_1 g_2}} \;, \quad J_{k,k+1}\Big|_{k=2 \text{ TO } n-1} = \frac{C_o}{\sqrt{g_k g_{k+1}}} \;, \quad J_{n-1,n} = \sqrt{\frac{C_o C_n}{g_{n-1} g_n}}$$

A-3527-535

FIG. 10.08-4 MODIFIED PROTOTYPE FOR USE IN DERIVING THE DESIGN EQUATIONS IN TABLE 10.03-1
The parameter d may be chosen arbitrarily within the range $0 < d \leq 1$

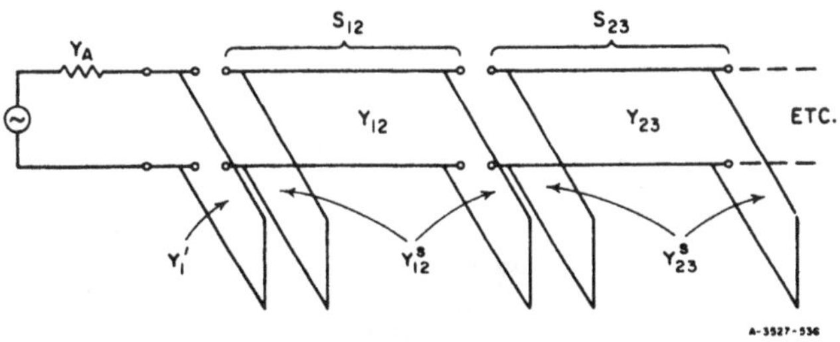

FIG. 10.08-5 THE FILTER IN FIG. 10.03-1 BROKEN INTO SYMMETRICAL SECTIONS
The sections of this filter are designed using the prototype sections in Fig. 10.08-4 as a guide

to this equivalence, Eqs. (10.08-3) and (10.08-4) also apply for stub sections by substituting

$$(Y_{oe})_{k,k+1} = Y^s_{k,k+1} \qquad (10.08\text{-}8)$$

and

$$(Y_{oo})_{k,k+1} = Y^s_{k,k+1} + 2Y_{k,k+1} \qquad (10.08\text{-}9)$$

where $Y_{k,k+1}$ is the characteristic admittance of the $\lambda_0/4$ connecting line and $Y^s_{k,k+1}$ is the characteristic admittance of the short-circuited $\lambda_0/4$ stubs in section $S_{k,k+1}$ of the filter (Fig. 10.08-5).

The sections $S'_{k,k+1}$ of the modified prototype in Fig. 10.08-4 were related to the corresponding sections $S_{k,k+1}$ (Fig. 10.08-5) of the band-pass filter by first setting $R_{n+1} = g_0$ (for Fig. 10.08-4) and then applying the correspondences (10.08-5) above. (Of course, in the statements of the correspondences "stub filter sections" should be used to replace "parallel-coupled sections.") The correspondences were first worked out using the admittance level of the prototype in Fig. 10.08-4; later the admittance level was altered by multiplying all admittances by $Y_A g_0$. Relating the prototype and band-pass filter sections made it possible to obtain equations for the line admittances in the band-pass filter sections, and then the final stub admittances for the filter in Fig. 10.03-1 were computed as the sums of the admittances of adjacent stubs, *i.e.*, for the kth stub

$$Y_k\Big|_{k=2 \text{ to } n-1} = Y^s_{k-1,k} + Y^s_{k,k+1} \quad . \qquad (10.08\text{-}10)$$

The stub Y'_1 in Fig. 10.08-5 is related to C'_1 in Fig. 10.08-4 by the relation

$$\omega'_1 C'_1 R_0 = \frac{Y'_1 \operatorname{ctn} \theta_1}{Y_A} \qquad (10.08\text{-}11)$$

which forces the susceptance of the Y'_1 stub at the band-edge frequency, ω_1, to be the same as the susceptance of C'_1 of the modified prototype at the prototype band-edge frequency, ω'_1, (except for a possible admittance scale change in the microwave filter). Then the end stub Y_1 has the total admittance

$$Y_1 = Y_1' + Y_{12}'' \quad . \tag{10.08-12}$$

The other end stub, Y_n, is treated in similar fashion.

Procedure for Deriving the Equations in Table 10.05-1—The stub filter in Fig. 10.05-1 is much the same as that in Fig. 10.03-1, except that series $\lambda_0/2$ short-circuited stubs have been added at the ends of the filter. In order to accommodate the series stubs, the modified prototype in Fig. 10.08-6 is used, where it should be noted that there are now series inductances at both ends of the filter. The series stub Z_1 of the filter in Fig. 10.05-1 is related to the series element L_1 of the filter in Fig. 10.08-6. This relation forces the reactance of the series stub at the band-edge frequency ω_1 for the band-pass filter to be the same as the reactance $\omega_1' L_1$ of the prototype at the prototype band-edge frequency, ω_1' (within a possible impedance scale change for the microwave filter). The same was done for the series stub Z_n at the other end of the filter. In all other respects the derivation of the equations in Table 10.05-1 is much the same as the derivation for the equations for the stub filter in Fig. 10.03-1, as discussed above.

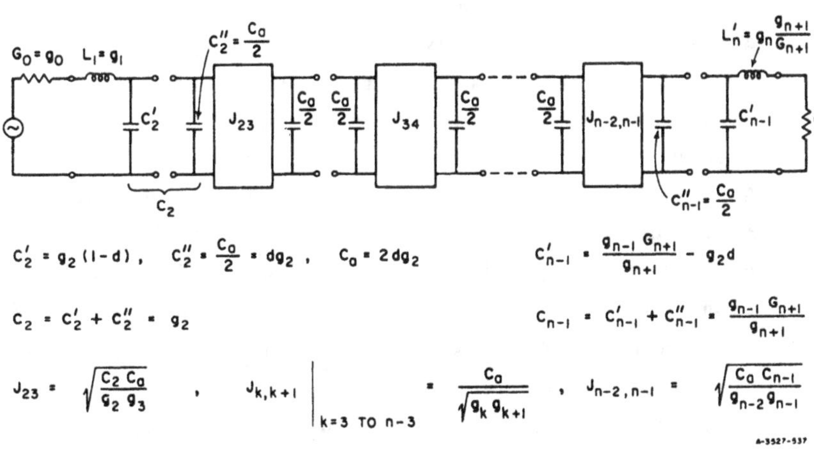

FIG. 10.08-6 MODIFIED PROTOTYPE FOR USE IN DERIVING THE EQUATIONS IN TABLE 10.05-1
The parameter d may be chosen arbitrarily within the range $0 < d \leq 1$

SEC. 10.09, DERIVATION OF THE DESIGN EQUATIONS FOR INTERDIGITAL-LINE FILTERS

The derivation of design equations for filters of the form in Fig. 10.09-1 was explained in Sec. 10.08. Filters of this type consist of parallel-coupled resonators which are $\lambda_0/2$ long at the midband frequency. There is a $\lambda_0/4$, short-circuited input and output coupling line at each end of this filter; these are designed to serve only as part of an admittance transforming section. An interdigital filter of the form in Fig. 10.06-1 is obtained from the filter in Fig. 10.09-1 if each $\lambda_0/2$ line is cut in the middle and folded double to give the structure in Fig. 10.09-2. It can be seen from Fig. 10.09-3 that this operation has little effect on the currents and voltages on the lines, at least at midband. Figure 10.09-3(a) shows the voltages and currents on a short-circuited $\lambda_0/2$ resonator, while Fig. 10.09-3(b) shows the voltages and currents after the resonator has been cut and folded. Note that the voltages and currents on the *a* and *b* portions of the resonator are the same in either case.

The circuits in Figs. 10.09-1 and 10.09-2 are clearly not electrically the same. First, if the structure in Fig. 10.09-2 has significant fringing capacitances extending beyond nearest-neighbor line elements, the coupling mechanism becomes much more complicated than is implied by the simple folding process. Second, the circuit in Fig. 10.09-1 can be shown to have only a first-order pole of attenuation (Sec. 2.04) at $\omega = 0,\ 2\omega_0,\ 4\omega_0$, etc., while the circuit in Fig. 10.09-2 has high-order poles of attenuation at these frequencies. For this reason it was at first believed that folding a filter as in Fig. 10.09-2 would have very little effect on its response for frequencies near ω_0 (provided that fringing beyond nearest neighbors is negligible), but that the error might be considerable at frequencies well removed from ω_0 (which would imply that the folding shown in Fig. 10.09-2 might considerably disturb the response of a wide-band filter). To check this point, the image cutoff frequency predicted by the approximation in Figs. 10.09-1 and 10.09-2 was compared with a previous interdigital-line exact analysis[2] for the case where there is no fringing beyond nearest neighbors. Surprisingly enough, the image bandwidths were practically the same (to slide rule accuracy) by either theory, even for bandwidths as great as an octave. This unexpected result indicates that the folding process

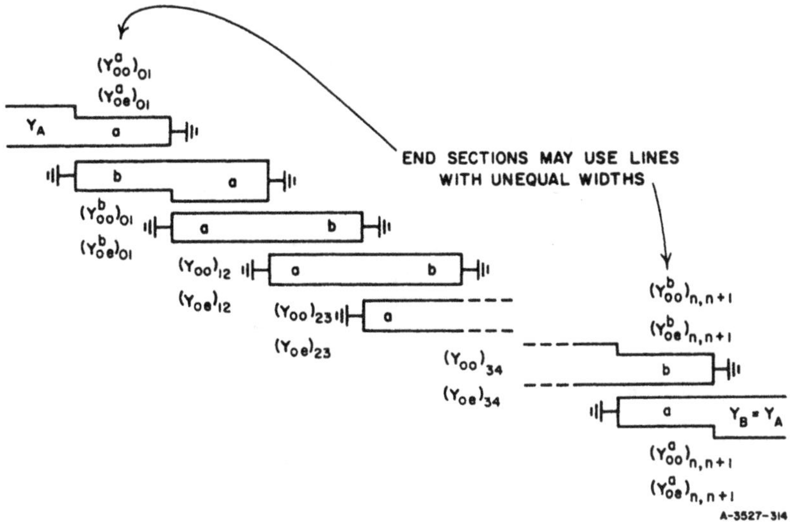

SOURCE: Quarterly Progress Report 4, Contract DA 36-039 SC-87398, SRI; reprinted in *IRE Trans. PGMTT* (see Ref. 3 by G. L. Matthaei)

FIG. 10.09-1 A PARALLEL-COUPLED STRIP-LINE FILTER WITH $\lambda_0/2$ RESONATORS

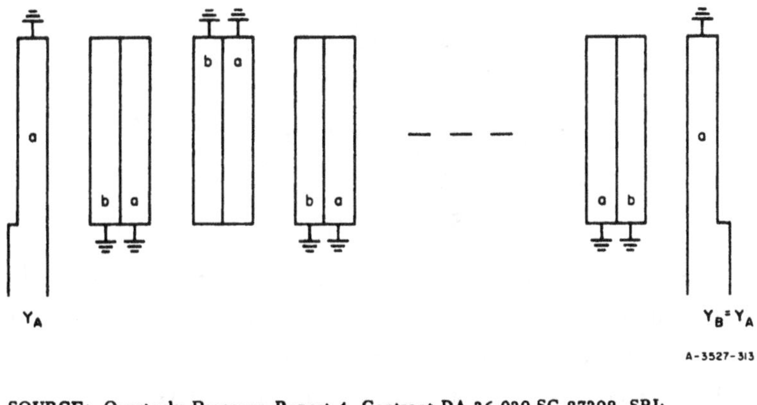

SOURCE: Quarterly Progress Report 4, Contract DA 36-039 SC-87398, SRI; reprinted in *IRE Trans. PGMTT* (see Ref. 3 by G. L. Matthaei)

FIG. 10.09-2 AN INTERDIGITAL FILTER FORMED FROM THE FILTER IN FIG. 10.09-1
Each $\lambda_0/2$ resonator has been cut in two in the middle and then folded double

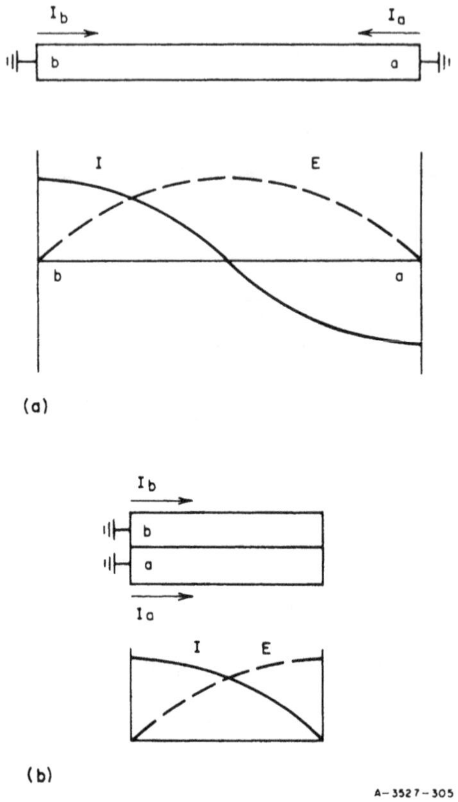

SOURCE: Quarterly Progress Report 4, Contract DA 36-039 SC-87398, SRI; reprinted in *IRE Trans. PGMTT* (see Ref. 3 by G. L. Matthaei)

FIG. 10.09-3 EFFECT OF FOLDING A $\lambda_0/2$ RESONATOR TO MAKE A $\lambda_0/4$ RESONATOR

in Fig. 10.09-2 should not greatly disturb the response of filters of the form in Fig. 10.09-1, even if the bandwidth is quite wide (provided that fringing capacitances beyond nearest-neighbor line elements are negligible). Experimental results show that in typical cases the fringing capacitance beyond nearest neighbor has no serious effect.

The design equations in Table 10.06-1 were obtained directly from those in Table 10.02-1 (which are for a filter of the form in Fig. 10.09-1), along with the "folding" approximation in Fig. 10.09-2. Since the equations for the filter in Fig. 10.09-1 were shown to be valid from narrow bandwidths to at least bandwidths of the order of an octave, and since the folding approximation appears to be reasonably good to such bandwidths, the design equations in Table 10.06-1 should, in principle, be good for large as well as for narrow bandwidths. However, the physical dimensions of wide-band filters of this type are not as desirable as those of the type shown in Fig. 10.07-1.

In order to derive design equations for interdigital filters with open-circuited terminating lines (Fig. 10.07-1), design equations were first derived for the type of filter shown in Fig. 10.09-4. This filter is nearly the same as the parallel-coupled filter in Fig. 10.09-1, except for the manner in which the terminating lines are coupled in. It is readily seen that if the folding process in Fig. 10.09-2 is applied to to the filter in Fig. 10.09-4, an interdigital filter with open-circuited input lines will result. It might seem at first that design equations for interdigital filters with open-circuited terminating lines could have been obtained by folding a parallel-coupled filter that had resonators open-circuited at the ends [Fig. 10.02-1(b)]. However, it will be seen that this cannot work, since the voltages at opposite ends of a $\lambda/2$ open-circuited resonator have opposite polarity.

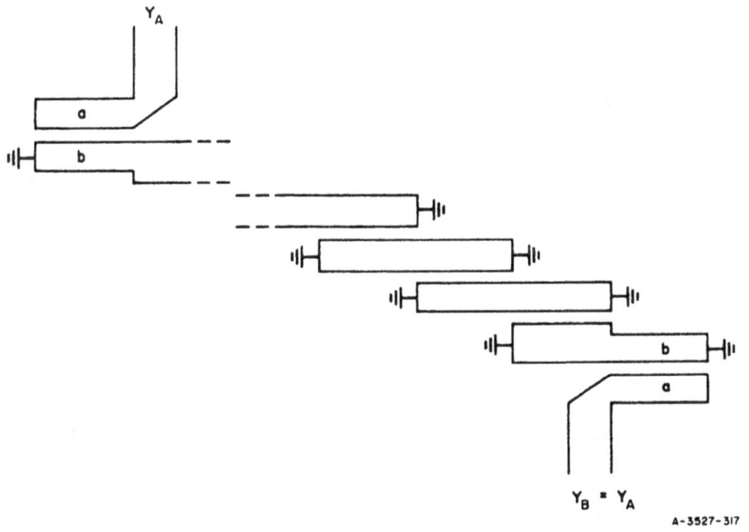

SOURCE: Quarterly Progress Report 4, Contract DA 36-039 SC-87398, SRI; reprinted in *IRE Trans. PGMTT* (see Ref. 3 by G. L. Matthaei)

FIG. 10.09-4 A PARALLEL-COUPLED FILTER WITH $\lambda_0/2$ SHORT-CIRCUITED RESONATORS AND OPEN-CIRCUITED TERMINATING LINES

Design equations for the filter in Fig. 10.09-4 were derived in a manner much like that used for deriving the equations for the filter in Fig. 10.09-1, except that the parallel-coupled line information summarized in Fig. 5.09-2(c) was used in designing the end sections. Note that this section also has impedance-transforming properties so that the admittance scale factor $h = 1/\sqrt{N}$, where N is the turns ratio of the ideal transformer in Fig. 5.09-2(c). Note that, in the derivation of the equations for filters of the form in Fig. 10.09-1, the mathematical constraint and equivalent circuit in Fig. 5.09-3(a) were used. The constraint given there causes one natural mode (Sec. 2.03) of each end section to be stifled. When end sections as in Fig. 5.09-2(c) are used, all natural modes are fully utilized.

When the strip-line and open-wire line equivalences in Figs. 5.09-2(a) and 5.09-2(c) are used, it will be seen that the strip-line circuit in Fig. 10.09-4 is electrically identical to the open-wire line circuit in Fig. 10.09-5. The filter circuit in Fig. 10.09-5 is very similar to the filter in Fig. 10.05-1, for which design equations were presented in Table 10.05-1. The filters in Fig. 10.09-5 and in Fig. 10.05-1 become identical if we set $a = \infty$ in Fig. 10.05-1 and Table 10.05-1, and if we introduce an ideal transformer at each end of the filter while altering the impedance level within the filter to make the impedances looking into the ends of the filter the same as before the transformers were introduced. Design equations for the filter in Fig. 10.09-5 were obtained in this manner,

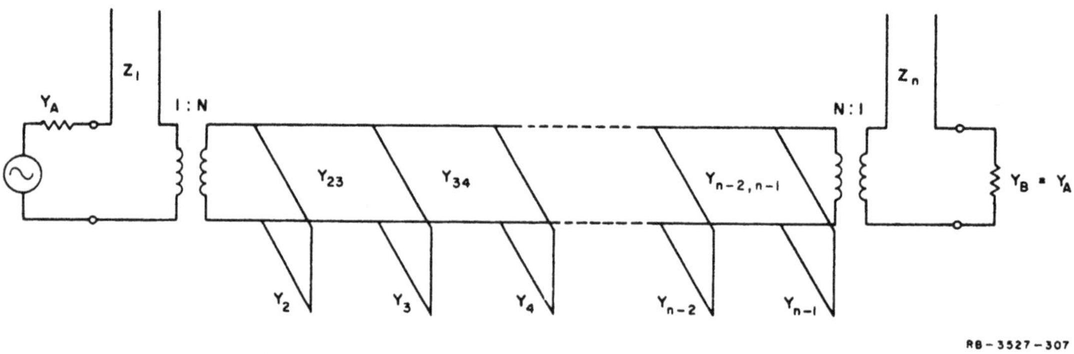

SOURCE: Quarterly Progress Report 4, Contract DA 36-039 SC-87398, SRI; reprinted in *IRE Trans. PGMTT* (see Ref. 3 by G. L. Matthaei)

FIG. 10.09-5 AN OPEN-WIRE-LINE EQUIVALENT CIRCUIT OF THE FILTER IN FIG. 10.09-4
All stubs and connecting lines are $\lambda_0/4$ long at midband

and from these results equations for the equivalent filter in Fig. 10.09-4 were obtained. Then the equations for the corresponding interdigital filter resulted from applying to the filter in Fig. 10.09-4 the folding approximation illustrated in Fig. 10.09-2. The reader will be interested to note that the $J_{k,k+1}$ in Table 10.07-1 correspond to characteristic admittances $Y_{k,k+1}$ of the connecting lines in Fig. 10.09-5, while Z_1, Y_2, Y_3, ..., Y_{n-1}, Z_n in Table 10.07-1 correspond to the characteristic impedances or admittances of the stubs in Fig. 10.09-5, for the limiting case where the transformer turns ratio is $N = 1$. For $N > 1$ the admittances are scaled by the factor h.

SEC. 10.10, SELECTION OF MAPPING FUNCTIONS

The plots presented herin show that when the function in Eq. (10.02-1) is used as indicated in Fig. 10.02-4 or 10.02-5 to map the response of a low-pass prototype, it will predict quite accurately the response of band-pass filters of the forms in Figs. 10.02-1 or 10.03-1 having narrow or moderate bandwidth. Although the function in Eq. (10.02-1) is very useful, it should not be expected to give high accuracy for wide-band cases because it is not periodic (which the filter responses in Sec. 10.02 and 10.03 are), nor does it go to infinity for $\omega = 0$, $2\omega_0$, $4\omega_0$, etc., which is necessary in order to predict the infinite attenuation frequencies (Sec. 2.04) in the response of the band-pass filter structure. It might at first seem that the function

$$F_n\left(\frac{\omega}{\omega_0}\right) = -\cot\left(\frac{\pi\omega}{2\omega_0}\right) \qquad (10.10\text{-}1)$$

would solve this problem nicely, since (1) it is periodic as desired, (2) it varies similarly to Eq. (10.02-1) in the vicinity of ω_0, and (3) it has poles at the desired frequencies, $\omega = 0$, $2\omega_0$, $4\omega_0$, etc. However, if the structures in Figs. 10.02-1 and 10.03-1 are analyzed, it will be seen that no matter what value of n is used, the poles of attenuation at $\omega = 0$, $2\omega_0$, $4\omega_0$, etc., are always first-order poles.* Meanwhile, an n-reactive-element prototype as in Fig. 4.04-1 (which will have an nth-order pole at $\omega' = \infty$) will map so as to give nth order poles at $\omega = 0$,

* For example, for the filter form in Fig. 10.03-1, as $\omega \to 0$ the effect of all of the shunt stubs can be reduced to that of a single, shunt, zero-impedance branch which would produce a first-order pole of attenuation at $\omega = 0$. (One way in which higher-order poles can be generated is to produce shunt, zero-impedance branches alternating with series branches having infinite impedance. See Sec. 2.04.)

$2\omega_0$, etc., if the function in Eq. (10.10-1) is used. This important source of error is corrected in the case of Eq. (10.02-4) by replacing cot $(\pi\omega/2\omega_0)$ by cos $(\pi\omega/2\omega_0)/|\sin(\pi\omega/2\omega_0)|$, and then taking the nth root of the denominator. In this manner the poles generated by the zeros of $|\sin(\pi\omega/2\omega_0)|$ become of $1/n$ order, which causes the nth-order pole at $\omega' = \infty$ for the prototype response to map into first-order poles of the band-pass filter response at the desired frequencies.

In the case of the circuit in Fig. 10.05-1, the poles of attenuation at $\omega = 0$, $2\omega_0$, $4\omega_0$, etc., will again always be of first order regardless of the value of n used. However, the series stubs at each end produce second-order poles at the frequency ω_∞ and at other corresponding points in the periodic response.* Thus, the

$$\sqrt[n]{|\sin(\pi\omega/2\omega_0)|}$$

factor in the denominator of Eq. (10.05-2) assures that the nth-order poles at $\omega' = \infty$ in the prototype response will always map to first-order poles at $\omega = 0$, $2\omega_0$, etc., for the band-pass filter response. In addition, the factor

$$\sqrt[n]{\left[\sin\frac{\pi}{2}\left(\frac{\omega - \omega_\infty}{\omega_0}\right)\right]^2\left[\sin\frac{\pi}{2}\left(\frac{\omega - 2\omega_0 + \omega_\infty}{\omega_0}\right)\right]^2}$$

is introduced to cause the nth-order pole at infinity in the prototype response to map to second-order poles at ω_∞ (and other periodic points) for the band-pass filter response. In this manner, all of the proper poles of attenuation are introduced with their proper order.

These principles can also be applied to the structure in Fig. 10.04-1, but this structure presents some new difficulties. It can be seen that this structure will develop nth-order poles of attenuation at ω_∞ and corresponding periodic points. However, the half-wavelength stubs also introduce additional natural modes of oscillation which create, in addition

---

* This can be seen as follows: For $\omega = \omega_\infty$, each of the series stubs represents an infinite-impedance series branch. For this single frequency, the interior part of the filter can be replaced by an equivalent T-section with a finite shunt impedance. Thus, the structure can be reduced (for the frequency ω_∞) to two, series, infinite-impedance branches separated by a finite, shunt-impedance branch. This can be seen to result in a second-order pole of attenuation. (If the impedance of the equivalent shunt branch had been zero, the pole of attenuation would have been raised to third order.)

to the desired pass band, a low-pass pass band (and corresponding periodic pass bands) as shown in the response in Fig. 10.04-2. This additional low-pass pass band approaches ω_∞ quite closely, with the result that, although the pole at ω_∞ is of relatively high order, its effectiveness is weakened by the close proximity of this low-pass pass band. The function

$$F_n\left(\frac{\omega}{\omega_0}\right) = \tan\left(\frac{\pi\omega}{\omega_0}\right) \tag{10.10-2}$$

for the case of $\omega_\infty/\omega_0 = 0.50$ would map the prototype response to give a low-pass pass band, an nth-order pole at ω_∞, and the desired pass band centered at ω_0. However, it would not properly predict how close the low-pass pass band comes to ω_∞, nor could it account for the oversize attenuation ripples which occur in this band (see Fig. 10.04-2). As a result, the function in Eq. (10.10-2) predicts an overly optimistic rate of cutoff at the edges of the pass band centered at ω_0. It is probable that a useful approximation could be obtained by using a mapping function such as that in Eq. (10.10-2) with additional factors added which create zeros in $F_n(\omega/\omega_0)$, close to, but somewhat off of, the $j\omega$ axis (regarded from the complex-frequency point of view as discussed in Secs. 2.03 and 2.04). These zeros could then be located to extend the low-pass pass band upwards toward ω_∞, which should give the proper effect.

The mapping in Eqs. (10.06-1) through (10.06-3) was found to predict the responses of the trial, interdigital filters reasonably well, so no further study was made of mapping for use in the design of interdigital filters.

REFERENCES

1. Matthaei, G. L., "Design of Wide-Band (and Narrow-Band) Band-Pass Microwave Filters on the Insertion Loss Basis," *IRE Trans. PGMTT*-8, pp. 580-593 (November 1960).

2. Bolljahn, J. T., and Matthaei, G. L., "A Study of the Phase and Filter Properties of Arrays of Parallel Conductors Between Ground Planes," *Proc. IRE,* **50**, pp. 299-311 (March 1962).

3. Matthaei, G. L., "Interdigital Band-Pass Filters," *IRE Trans. PGMTT*-10, pp. 479-491 (November 1962).

CHAPTER 11

SPECIAL PROCEDURES TO AID IN THE PRACTICAL DEVELOPMENT OF COUPLED-RESONATOR BAND-PASS FILTERS, IMPEDANCE-MATCHING NETWORKS, AND TIME-DELAY NETWORKS

SEC. 11.01, INTRODUCTION

The preceding three chapters have dealt with design information of different sorts for various specific types of band-pass filters, and also with general theory by which similar design information can be prepared for additional types of band-pass filters. This chapter continues the discussion of band-pass filters by treating various general techniques which are of considerable help in the practical development of filters. That is to say, this chapter introduces additional information of help in reducing theory to practice.

Sections 11.02 to 11.05 deal with laboratory procedures for determining the Q's of resonators, for adjusting their couplings to correspond to the couplings called for by the theory, and for the tuning adjustment of completed filters. Sections 11.06 and 11.07 discuss the effects of resonator losses and present special design information for applications where minimizing the midband loss of a band-pass filter is important. Sections 11.08 and 11.09 present supplementary information to aid in using the filter design methods of Chapters 8 and 10 for design of impedance-matching networks, and Sec. 11.10 explains how these same procedures can be used for design of coupling networks for negative-resistance devices. Section 11.11 is included to further clarify how the methods of Chapters 8 to 10 can be used for the design of band-pass filter networks with specified nominal time delay.

SEC. 11.02, MEASUREMENT OF Q_u, Q_e, AND Q_L OF A SINGLY LOADED RESONATOR

Figure 11.02(a) shows a resonator with impedance-inverter couplings K_1 and K_2. The inverter K_2 is open-circuited on its right, so that it reflects a short-circuit at its left side. As a result, the circuit in

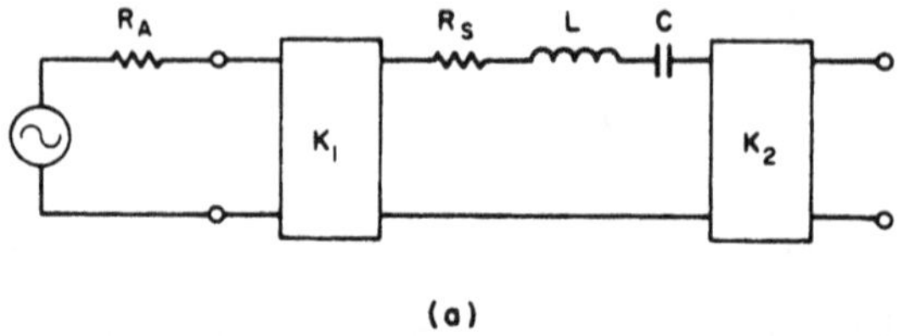

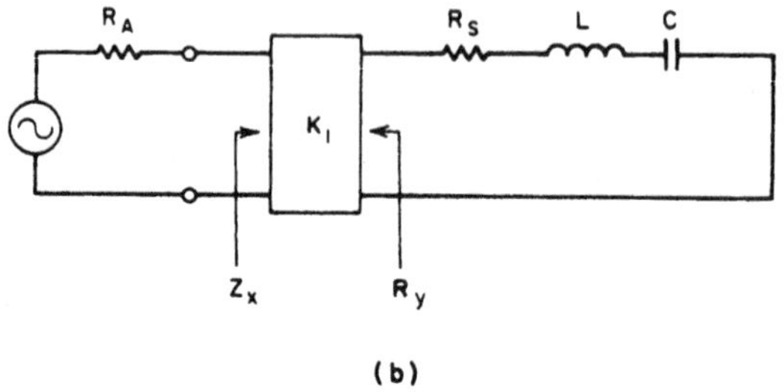

FIG. 11.02-1 TWO EQUIVALENT SINGLY LOADED RESONATORS

Fig. 11.02-1(a) may be replaced with the circuit in Fig. 11.02-1(b). The resonator as shown has a reactance slope parameter (Sec. 8.02) of

$$\mathscr{X} = \omega_0 L \tag{11.02-1}$$

and an *unloaded* Q of

$$Q_u = \frac{\mathscr{X}}{R_s} \quad . \tag{11.02-2}$$

The inverter K_1 reflects an impedance of $R_y = K_1^2/R_A$ to the resonator which gives it a *loaded* Q of

$$Q_L = \frac{\mathscr{X}}{R_y + R_s} = \frac{\mathscr{X}}{\dfrac{K_1^2}{R_A} + R_s} \quad . \tag{11.02-3}$$

The *external* Q of the resonator is defined as the Q with $R_s = 0$ (*i.e.*, with $Q_u = \infty$), so that the resistive loading of the resonator is due only to R_y. Thus, the external Q is

$$Q_e = \frac{\alpha}{R_y} = \frac{\alpha}{\left(\frac{K_1^2}{R_A}\right)} \quad . \tag{11.02-4}$$

Let us suppose that the resonator shown in Fig. 11.02-1(a), and its equivalent form in Fig. 11.02-1(b), symbolize a resonator which with its adjacent coupling discontinuities forms a resonator of a waveguide filter such as that in Fig. 8.06-1, or forms a resonator of a small-aperture-coupled cavity filter such as is discussed in Sec. 8.07. Let us further suppose that the resonator under consideration is the first resonator of the filter and its desired external Q has been computed by Eq. (6) of Fig. 8.02-3. The problem at hand then is to make measurements on the resonator to see if its external Q is as required by the calculations from the low-pass prototype element values.

Measurement of the values of Q_u, Q_L, and Q_e of a singly loaded resonator can be made by use of a slotted line of characteristic impedance $Z_0 = R_A$, along with procedures about to be explained. Since the resonator has some internal loss represented by the resistor R_s in Fig. 11.02-1, the VSWR at the resonant frequency f_0 will be finite. A plot of the VSWR in the vicinity of resonance will have a shape similar to that in Fig. 11.02-2.

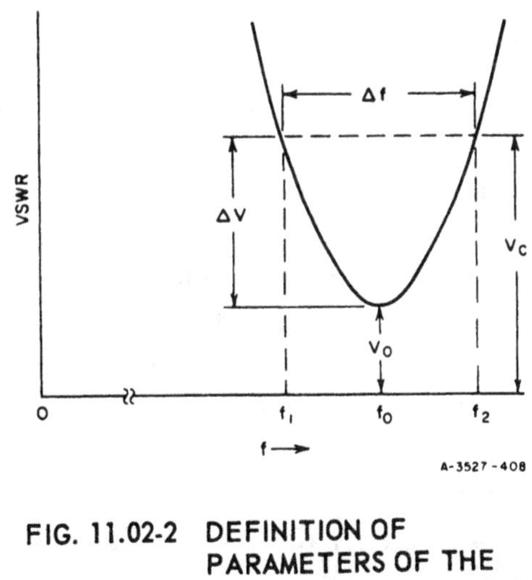

FIG. 11.02-2 DEFINITION OF PARAMETERS OF THE VSWR CHARACTERISTIC FOR A SINGLY LOADED RESONATOR

If in Fig. 11.02-1(b)

$$\left.\frac{Z_x}{R_A}\right|_{f=f_0} = \frac{\frac{K_1^2}{R_s}}{R_A} > 1 \quad , \tag{11.02-5}$$

the resonator is said to be *overcoupled.* This condition will be evidenced by the fact that the locations of the voltage minima on the slotted line at the resonant frequency f_0 (minimum VSWR frequency) will be about a quarter-wavelength away from their locations when the frequency is appreciably off resonance.* When this condition holds, the VSWR at resonance is

$$V_0 = \frac{\frac{K_1^2}{R_s}}{R_A} . \qquad (11.02\text{-}6)$$

For the so-called *undercoupled* case, the VSWR at resonance is

$$V_0 = \frac{R_A}{\frac{K_1^2}{R_s}} > 1 . \qquad (11.02\text{-}7)$$

This situation will be evidenced by the fact that the voltage minima on the slotted line will be in approximately the same positions at a frequency appreciably off of resonance as they are at the resonant frequency.† If at resonance the VSWR is $V_0 = 1$, the resonator is said to be *critically coupled.*

The procedure for measuring the various Q's of a resonator is then to first measure the cavity's VSWR in the vicinity of resonance, and then make a plot such as that in Fig. 11.02-2. At the same time, by noting how the voltage minima on the slotted line shift as the frequency deviates from the resonant frequency it should be determined whether the resonator is over-coupled, or undercoupled (or if $V_0 = 1$ it is critically coupled). Then by use of Fig. 11.02-3(a) for the overcoupled case, or Fig. 11.02-3(b) for the undercoupled case, a parameter ΔV is picked from the chart for the

* This assumes that the width of the resonance is quite narrow, so that there is little change in the electrical length of the slotted line over the frequency range of interest. The shift in the voltage minima results from the reflection coefficient at the input coupling of the cavity being 180° different in phase at resonance from its phase well off resonance.

† The assumptions mentioned in the immediately preceding footnote also apply here. In this undercoupled case the reflection coefficient between the slotted line and the resonator input coupling has the same phase at resonance, as it does well off of resonance. As a result, the voltage minima will assume approximately the same positions on the slotted line at the resonant frequency as they will at frequencies well off resonance, provided the frequency has not been shifted so much as to greatly change the electrical length of the slotted line. If at the resonant and off-resonance frequencies the electrical distance between the resonator and the region of interest on the slotted line changes appreciably compared to a quarter wavelength, this fact should be taken into account when determining whether a resonator is overcoupled or undercoupled.

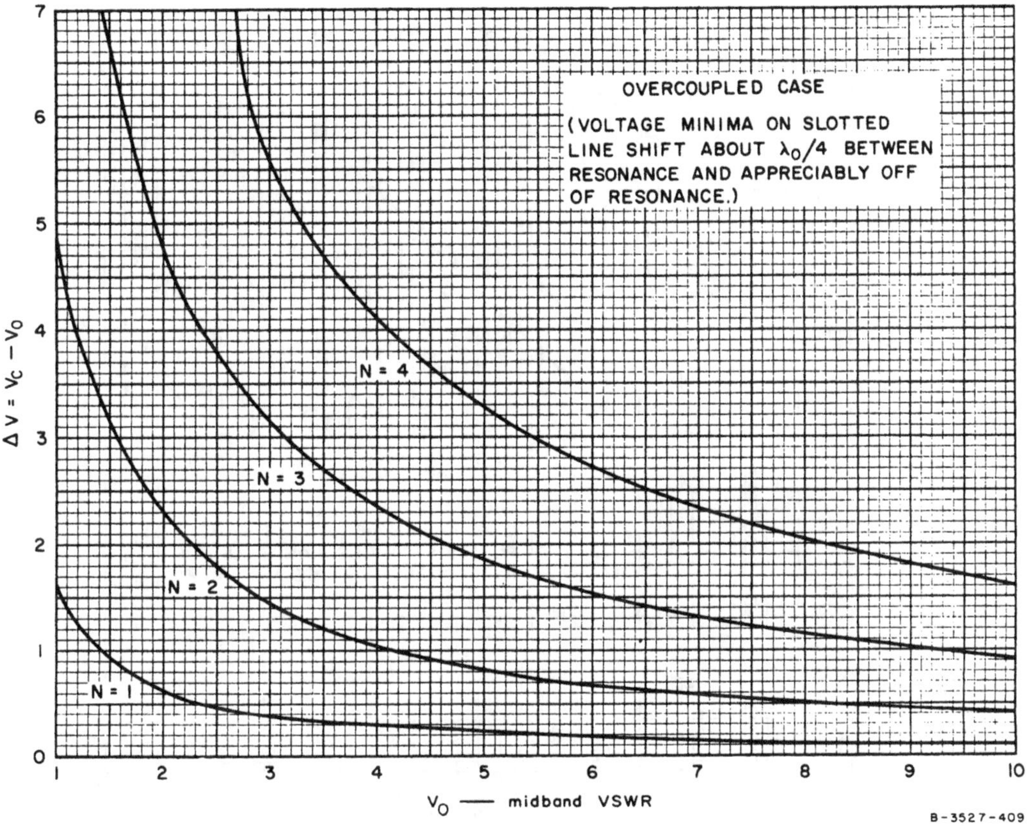

FIG. 11.02-3(a) CHART FOR USE IN DETERMINING Q_u OF AN OVER-COUPLED RESONATOR

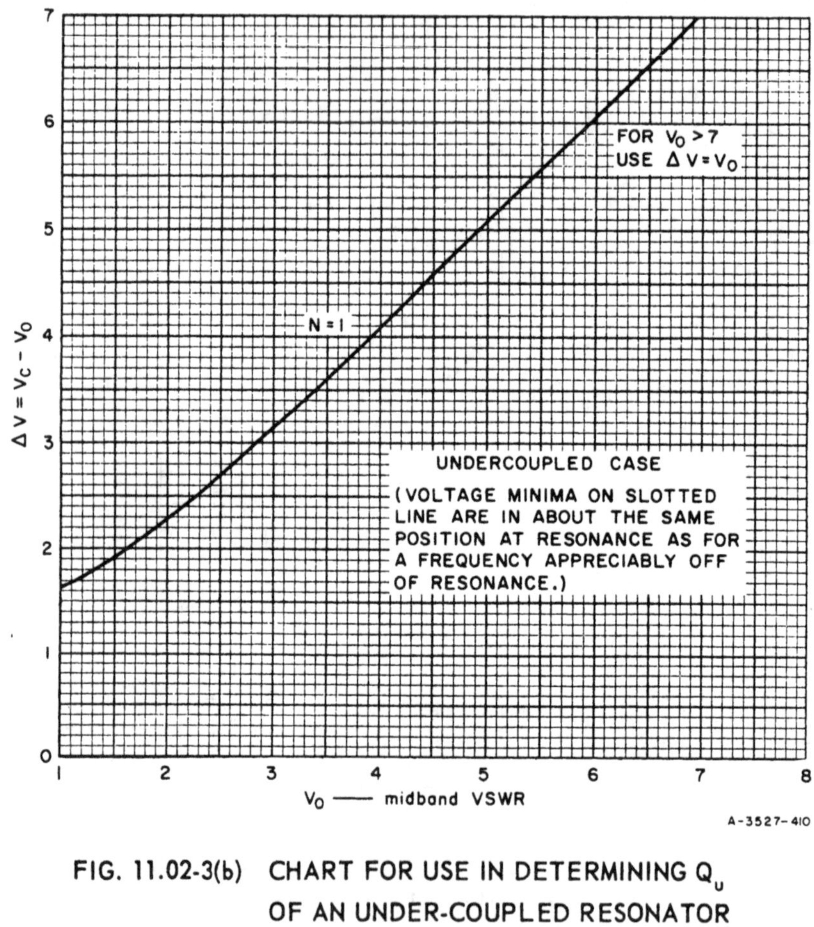

FIG. 11.02-3(b) CHART FOR USE IN DETERMINING Q_u OF AN UNDER-COUPLED RESONATOR

given midband VSWR value V_0 and for the value of N for the particular curve chosen. The choice of N is arbitrary, but it should be chosen large enough so that $V_C = V_0 + \Delta V$ will be appreciably different from V_0. In the undercoupled case V_0 and V_C are sufficiently different when $N = 1$, so that only the $N = 1$ case is shown in Fig. 11.02-3(b).*

When ΔV has been chosen,

$$V_C = V_0 + \Delta V \tag{11.02-8}$$

is computed. Next, as indicated in Fig. 11.02-2, the bandwidth Δf at the points where the VSWR equals V_C is determined. Then the unloaded Q of the resonator is computed by the formula

$$Q_u = \frac{N f_0}{\Delta f} \quad . \tag{11.02-9}$$

If the resonator is *overcoupled*, then the external Q is

$$Q_e = \frac{Q_u}{V_0} \quad , \tag{11.02-10}$$

and the loaded Q is

$$Q_L = \frac{Q_u}{V_0 + 1} \quad . \tag{11.02-11}$$

If the resonator is *undercoupled*, then the external Q is

$$Q_e = V_0 Q_u \quad , \tag{11.02-12}$$

and the loaded Q is

$$Q_L = \frac{Q_u V_0}{V_0 + 1} \quad . \tag{11.02-13}$$

---

* The significance of the value of N is as follows. The half-power point for the unloaded Q of a resonator occurs when the resonator reactance X equals the resonator resistance R_s. The curves in Fig. 11.02-3(a) correspond to the resonator reactance being equal to $X = NR_s$. For $N = 1$, Δf is the half-power bandwidth. For N other than one, $\Delta f \approx N$ times (half-power bandwidth).

If the resonator is critically coupled, the equations for both the over-coupled and the undercoupled cases will work.

To further clarify the use of these equations and charts, suppose that $V_0 = 7$ and it is found that the resonator is overcoupled. In order to make V_C significantly different from V_0, it is desirable to use the $N = 4$ curve in Fig. 11.02-3(a), and this value of N should be used in Eq. (11.02-8). If V_0 were even larger, it might be desirable to doubly load the resonator as described in the next section. Alternatively, measurements can be made on the singly loaded resonator by using the phase method described by Ginzton.[1]

The discussion so far has been phrased in terms of series-type resonators with couplings which simulate K-inverters. The same analysis on a dual basis applies to shunt-type resonators (which will have couplings that operate like J-inverters). As far as the laboratory procedures and the calculation of Q_u, Q_e, and Q_L from the laboratory data are concerned, there is no difference whatsoever.

The methods described above are very useful in determining if the couplings from the end resonators of a filter to their terminations are correct. After the external Q values are measured, they can be compared with the values computed from the lumped-element prototype elements by use of Eqs. (6) and (7) of Fig. 8.02-3 or 8.02-4. Equations (6) and (7) of Fig. 8.02-3 or 8.02-4, along with the laboratory procedures described in this section, are applicable to all of the filter types discussed in Chapter 8, as well as to numerous other possible forms of coupled-resonator filters. By these procedures the end couplings can be checked and adjusted to be correct to give a filter response corresponding to that of the low-pass prototype. The procedures of this section are also useful for checking the unloaded Q of resonators so that the over-all filter pass-band loss can be predicted.

If the reactance or susceptance slope parameter $\mathscr{x}$ or $\mathscr{b}$ of the resonators is known, this same procedure can be used for checking the couplings of all of the resonators of a filter. If a resonator with its adjacent couplings is removed from a filter and tested under singly loaded conditions as in Fig. 11.02-1(a), then it is easily shown that

$$\frac{K_1}{R_A} = \sqrt{\frac{\mathscr{x}}{R_A Q_e}} \quad . \qquad (11.02\text{-}14)$$

In the case of a filter with shunt-type resonance the dual equation

$$\frac{J_1}{G_A} = \sqrt{\frac{\mathscr{b}}{G_A Q_e}} \tag{11.02-15}$$

applies where G_A is the generator conductance, J_1 is the admittance-inverter parameter, and $\mathscr{b}$ is the susceptance slope parameter.

As an example, consider the case of filters of the form in Fig. 8.05-1. By Eq. (8.13-16) the susceptance slope parameter of these resonators is $\mathscr{b} = (\pi/2)Y_0$. Taking $G_A = Y_0$, Eq. (11.02-15) becomes

$$\frac{J_1}{Y_0} = \sqrt{\frac{\pi}{2Q_e}} \quad . \tag{11.02-16}$$

Thus, by measuring the Q_e of such a strip-line resonator connected as in Fig. 11.02-4, it is possible to determine the J/Y_0 value associated with a given size of capacitive coupling gap. Using a test unit such as that shown in Fig. 11.02-4, the proper coupling gaps to give the J/Y_0 values called for by Eqs. (1) to (3) of Fig. 8.05-1 can be determined. Analogous procedures will, of course, also work for other types of filters.

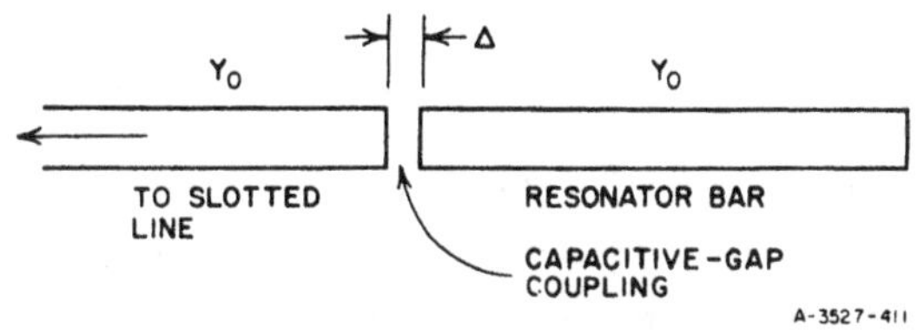

FIG. 11.02-4 A POSSIBLE ARRANGEMENT FOR EXPERIMENTALLY DETERMINING THE RESONATOR COUPLING GAPS FOR A STRIP-LINE FILTER OF THE FORM IN FIG. 8.05-1

The procedures described above concern themselves primarily with experimentally adjusting the resonator couplings to the proper values. The matter of obtaining the exactly correct resonant frequency will be treated in Sec. 11.05. Fortunately, tuning adjustments on resonators in typical cases has little effect on their couplings. The procedures of this section are most accurate when the Q's involved are relatively large. However, in some cases they can be helpful even when quite low Q's are involved (such as, say, 20 or so).

SEC. 11.03, TESTS ON SINGLE RESONATORS WITH LOADING AT BOTH ENDS

In some cases there is considerable advantage in testing a resonator under doubly loaded conditions as shown in Fig. 11.03-1, rather than under singly loaded conditions as shown in Fig. 11.02-1(a). The VSWRs to be measured will usually not be as high, and in some cases the resonant frequency of a single resonator will be exactly the same under doubly loaded conditions as it will be in a multiple-resonator filter. Thus, in such cases, both the couplings and the resonant frequency of resonators for a multiple-resonator filter can be checked by this procedure. The details concerning tuning will be discussed in Sec. 11.05. In order to check couplings other than the end couplings of a filter, it is necessary with this procedure to know the resonator slope parameters.

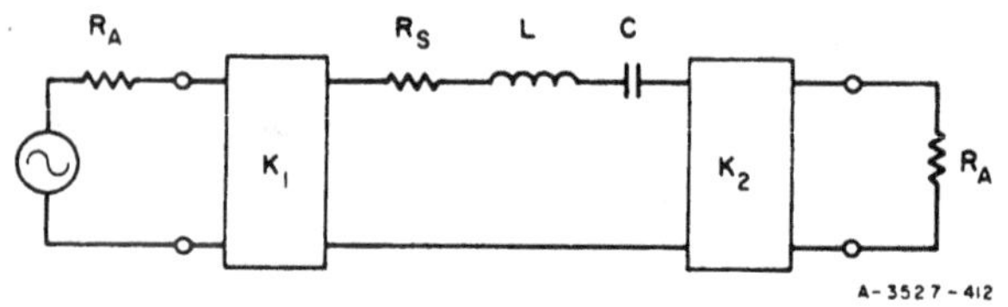

FIG. 11.03-1 A DOUBLY LOADED SINGLE RESONATOR

With respect to Fig. 11.03-1, the external Q, $(Q_e)_1$, will be defined as the Q when the circuit is loaded only by R_A on the left (*i.e.*, R_A on the right is removed so as to leave an open circuit, and $R_s = 0$). Similarly, $(Q_e)_2$ is the external Q when the circuit is loaded only by R_A on the right. Note that if K_1 and K_2 are different, $(Q_e)_1$ and $(Q_e)_2$ will be different. The unloaded Q, Q_u, of the resonator is its Q when both of the R_A terminations on the left and right are removed, and the resonator's only resistive loading is that due to its internal loss (represented by R_s in Fig. 11.03-1).

The loaded Q, Q_L, of the doubly loaded resonator is

$$Q_L = \frac{1}{\dfrac{1}{(Q_e)_1} + \dfrac{1}{Q_u} + \dfrac{1}{(Q_e)_2}} \tag{11.03-1}$$

$$= \frac{f_0}{(\Delta f)_{3\,\mathrm{db}}} \tag{11.03-2}$$

where f_0 is the resonant frequency of the resonator and $(\Delta f_0)_{3\,db}$ is here the bandwidth for which the attenuation for transmission through the resonator is up 3 db from that at resonance.

The attenuation through the resonator at resonance is

$$(L_A)_0 = 10 \log_{10} \left[\frac{(Q_e)_1 (Q_e)_2}{4Q_L^2} \right] \quad . \tag{11.03-3}$$

The definitions of $(Q_e)_1$, $(Q_e)_2$, Q_u, and Q_L mentioned above apply analogously to any resonator regardless of whether it is of the series- or shunt-resonance type. Equations (11.03-1) to (11.03-3) also apply regardless of the form of the resonator.

It is possible to check the couplings of a resonator by computing theoretical values for Q_L and $(L_A)_0$ and then by attenuation measurements compare the measured and computed $(\Delta f)_{3\,db}$ [which by Eq. (11.03-2) gives Q_L], and $(L_A)_0$. However, usually VSWR measurements are easier.

If the measurement procedures of Sec. 11.02 are applied by making VSWR measurements at the left side in Fig. 11.03-1, $(Q_e)_1$ can be determined along with an apparent unloaded Q which is equal to

$$(Q_u)_1 = \frac{1}{\dfrac{1}{Q_u} + \dfrac{1}{(Q_e)_2}} \quad . \tag{11.03-4}$$

Similarly, if VSWR tests are made from the right side in Fig. 11.03-1, $(Q_e)_2$ and an apparent unloaded Q

$$(Q_u)_2 = \frac{1}{\dfrac{1}{Q_u} + \dfrac{1}{(Q_e)_1}} \tag{11.03-5}$$

are obtained. The VSWR seen from the left at resonance will be

$$(V_0)_1\Big|_{\substack{\text{overcoupled}\\\text{case}}} \quad \text{or} \quad \frac{1}{(V_0)_1}\Big|_{\substack{\text{undercoupled}\\\text{case}}} = \frac{(Q_e)_2}{(Q_e)_1}\left[\frac{Q_u}{Q_u + (Q_e)_2}\right] \quad . \qquad (11.03\text{-}6)$$

The VSWR seen from the right at resonance will be

$$(V_0)_2\Big|_{\substack{\text{overcoupled}\\\text{case}}} \quad \text{or} \quad \frac{1}{(V_0)_2}\Big|_{\substack{\text{undercoupled}\\\text{case}}} = \frac{(Q_e)_1}{(Q_e)_2}\left[\frac{Q_u}{Q_u + (Q_e)_1}\right] \quad . \qquad (11.03\text{-}7)$$

Since $(Q_e)_1$ and $(Q_e)_2$ are obtained directly from the measurements as described above,it is possible to compute Q_u by use of Eq. (11.03-6) or (11.03-7). However, using single loading as described in Sec. 11.02 will usually give better accuracy for that purpose. If Q_u is known to be large compared to $(Q_e)_1$ and $(Q_e)_2$, it will be convenient to measure, say, $(Q_e)_1$ and then compute $(Q_e)_2$ from Eq. (11.03-6) using $Q_u = \infty$.

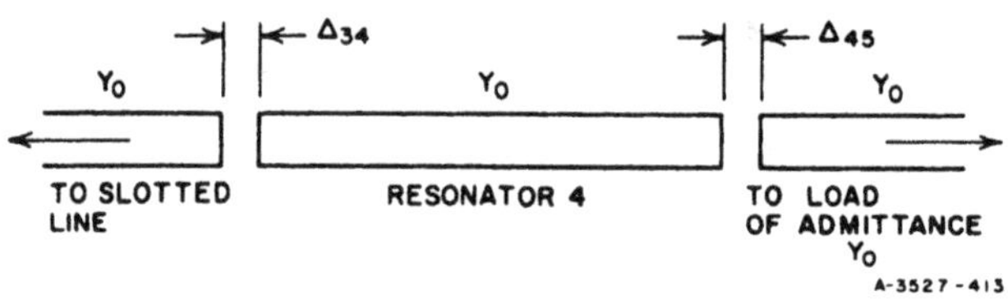

FIG. 11.03-2 AN ARRANGEMENT FOR TESTS ON A STRIP-LINE RESONATOR USING LOADING AT BOTH ENDS

Figure 11.03-2 shows how a resonator from a filter such as that in Fig. 8.05-1 can be tested using the procedure described above. The resonator shown is assumed to be Resonator 4, of a, say, six-resonator filter, and it is desired to see if the capacitive gaps Δ_{34} and Δ_{45} are correct to correspond to the J_{34}/Y_0 and J_{45}/Y_0 values computed by Eq. (2) of Fig. 8.05-1. By measurements as described above, external Q's $(Q_e)_{34}$ and $(Q_e)_{45}$ for loading at the left and right ends, respectively, of the resonator as set up in Fig. 11.03-2 are obtained, and then the corresponding values of J_{34}/Y_0 and J_{45}/Y_0 can be computed by use of Eq. (11.02-16) in the form*

$$\frac{J_{j,j+1}}{Y_0} = \sqrt{\frac{\pi}{2(Q_e)_{j,j+1}}} \quad . \tag{11.03-8}$$

If the $J_{j,j+1}/Y_0$ values are incorrect, the gap spacings can be altered to produce the correct values. The resonator in Fig. 11.03-2 would have exactly the same resonant frequency when operated as shown as it would when operated in a multiple-resonator filter as shown in Fig. 8.05-1. For that reason, after the resonant frequency of the resonator has been checked with the proper coupling gaps, the length of the resonator bar can be corrected if necessary to give the desired resonant frequency. (See Sec. 11.05 for discussion of tuning corrections for other types of filters.)

This procedure is particularly handy for the bar-strip-line resonators discussed in Sec. 8.05, because the individual resonators can be easily tested separately and then later inserted together in the complete multiple-resonator filter. In cases such as the analogous waveguide filters in Sec. 8.06, it may be desirable to build a test resonator with the coupling irises mounted in waveguide coupling flanges. In this way the irises being checked can easily be removed and their dimensions altered as indicated by the tests.

SEC. 11.04, TESTS ON SYMMETRICAL PAIRS OF RESONATORS†

Most microwave filters are symmetrical. Then, for each resonator with given couplings at one end of the filter there is another identical resonator with identical couplings at the other end of the filter. It is often feasible to check the couplings within a filter (and sometimes also to check the tuning of the resonators precisely) by disassembling the filter , connecting the pairs of identical resonators together, and testing them a pair at a time. A special advantage of this procedure

---

* As indicated in Sec. 11.02, this equation is based on the resonator slope parameters being $\mathcal{B} = (\pi/2)\, Y_0$ as is the case for filters of the form discussed in Sec. 8.05.

† Dishal has discussed a similar two-resonator technique in somewhat different terms. See Ref. 2.

over the procedures described in Secs. 11.02 and 11.03 is that the coefficient of coupling k [see Eqs. (8) of Fig. 8.02-3 and 8.02-4] for the coupling between resonators can be determined without specific knowledge of the slope parameters of the resonators. (See discussion in Sec. 11.02.) Also, in most cases, two sharp points of good transmission will be obtained, so that fewer measurements are required in order to get the desired information.

As an example, consider the symmetrical pair of cavity resonators shown in Fig. 11.04-1. In the discussion to follow, Q_u is the unloaded Q of either of the resonators by itself, Q_e is the external Q of either one of the resonators loaded by its adjacent termination (with Q_u equal to infinity), and k is the coefficient of coupling between the two resonators.

Cohn and Shimizu[3] have shown that the attenuation of a symmetrical pair of resonators is*

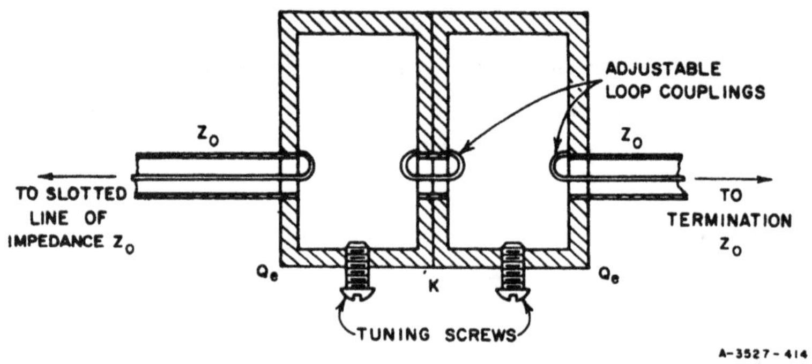

FIG. 11.04-1 SYMMETRICAL PAIR OF CAVITY RESONATORS WITH LOOP COUPLINGS

*As is implied by the use of $u = (f - f_0)/f_0$ as a frequency variable, this equation involves approximations which are most accurate for narrow-band cases.

$$L_A = 10 \log_{10} \left\{ \left[\frac{\left(1 + \frac{Q_e}{Q_u}\right)^2}{2kQ_e} + \frac{kQ_e}{2} \right]^2 \right.$$

$$\left. + 2 \left[\frac{\left(1 + \frac{Q_e}{Q_u}\right)^2}{k^2} - Q_e^2 \right] u^2 + \frac{4Q_e^2}{k^2} u^4 \right\} \quad \text{db}$$

(11.04-1)

where

$$u = \frac{f - f_0}{f_0}$$

f = frequency

f_0 = midband frequency

At midband the attenuation is seen to be

$$(L_A)_0 = 20 \log_{10} \left[\frac{\left(1 + \frac{Q_e}{Q_u}\right)^2}{2kQ_e} + \frac{kQ_e}{2} \right] \quad \text{db} \quad .$$

(11.04-2)

If the condition

$$k > \left(\frac{1}{Q_e} + \frac{1}{Q_u} \right) \qquad (11.04\text{-}3)$$

is satisfied the response will have a hump in the middle as shown in Fig. 11.04-2, and the resonators are said to be *overcoupled*. If

$$k = \left(\frac{1}{Q_e} + \frac{1}{Q_u} \right) \qquad (11.04\text{-}4)$$

the resonators are said to be *critically coupled*, while if

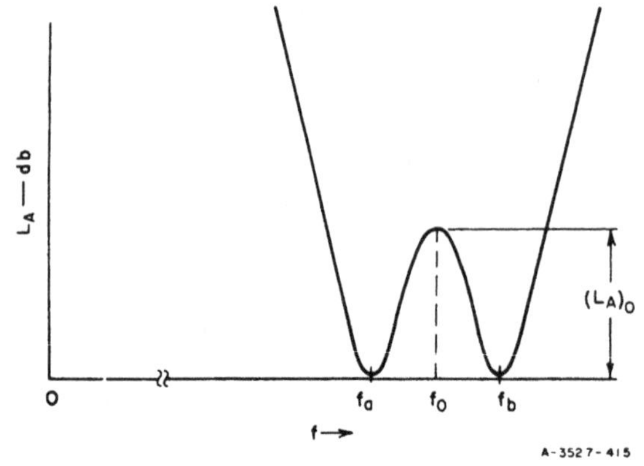

FIG. 11.04-2 TYPICAL RESPONSE OF AN "OVERCOUPLED" SYMMETRICAL PAIR OF RESONATORS

$$k < \frac{1}{Q_e} + \frac{1}{Q_u} \qquad (11.04\text{-}5)$$

the resonators are said to be *undercoupled*. In the critically coupled and undercoupled cases the response has no hump in the middle, and the midband loss increases as the resonators become more undercoupled. It should be noted that this use of the terms overcoupled, critically coupled, and undercoupled is entirely different from the use of these terms in discussing single resonators (see Sec. 11.02).

A possible way of making tests on a pair of resonators such as those in Fig. 11.04-1 is to first make single-resonator tests on, say, the resonator on the left using the methods of Sec. 11.02, with the resonator on the right grossly mistuned by running its tuning screw well in. This would make the second resonator have negligible effect on the resonance of the first resonator. From these tests, values for Q_e and Q_u can be obtained. Then, if the resonators are tuned to the same frequency (see Sec. 11.05) and the midband attenuation $(L_A)_0$ is measured, the coupling coefficient k can be obtained by solving Eq. (11.04-2) for k.

In most cases when testing pairs of resonators from a multiple-resonator filter, the response will be greatly overcoupled. In such cases, rather than measure $(L_A)_0$, it may be more convenient to measure the frequencies f_a and f_b in Fig. 11.04-2 by finding the points of minimum VSWR. Then it can be shown that

$$k = \sqrt{w_m^2 + \left(\frac{1}{Q_e} + \frac{1}{Q_u}\right)^2} \qquad (11.04\text{-}6)$$

where

$$w_m = \frac{f_b - f_a}{f_0} \qquad (11.04\text{-}7)$$

Note that if Q_e and Q_u are both sizeable compared to $1/w_m$, rather large percentage errors in Q_e or in Q_u will cause little error in k.

The strip-line impedance- and admittance-inverter data in Figs. 8.05-3(a), (b), (c), and 8.08-2(a) to 8.08-4(b) were obtained by laboratory tests using symmetrical, two-resonator test sections with adjustable coupling discontinuities between resonators. The couplings at the ends were held fixed and were made to be quite loose so that Q_e would be quite large and so that the response would be greatly overcoupled (which made the low-VSWR points at f_a and f_b in Fig. 11.04-2 very sharp and distinct). After the coupling coefficient between resonators had been determined using the procedures described above, it was possible to determine the inverter parameters since

$$k_{j,j+1} = \frac{J_{j,j+1}}{\sqrt{\mathscr{b}_j \mathscr{b}_{j+1}}} \quad \text{or} \quad \frac{K_{j,j+1}}{\sqrt{\mathscr{x}_j \mathscr{x}_{j+1}}} \tag{11.04-8}$$

as was discussed in Sec. 8.02, and since the resonator slope parameters in this case are $\mathscr{b}_j = d(\pi/4)Y_0$ or $\mathscr{x}_j = d(\pi/4)Z_0$, where d is an integer equal to the nominal number of quarter-wavelengths in the resonators (see Sec. 8.14).

Note that determining the inverter parameters from the coupling coefficients requires knowledge of the resonator slope parameters. The inverter description of a coupling is usually more useful for purposes of analysis because specification of the inverter parameter gives more information than does specification of a coupling coefficient. However, fixing the coupling coefficients $k_{j,j+1}$ between resonators and the external Q's of the resonators at the ends of a filter as called for by Eqs. (6) and (8) of Fig. 8.02-3 or 8.02-4 is adequate to fix the response of the filter as prescribed (*i.e.*, at least if the filter is of narrow or moderate bandwidth, and if the resonators are all properly tuned to the same frequency). Thus, the procedures of this section along with those of Sec. 11.02 are sufficient to properly adjust the the couplings of a filter of narrow or moderate bandwidth, even though the resonators may be of some arbitrary form for which the resonator slope parameters are unknown. If the filter is of such a form that the resonators can easily be removed, it may be convenient to remove the resonators two at a time and test pairs of identical resonators (assuming that the filter is symmetrical). If the resonators are not easily removed, it may be desirable to construct a separate, symmetrical, two-resonator test filter designed so that the couplings can easily be altered. This test filter can then be used to check out the designs of all of the couplings.

SEC. 11.05, TUNING OF MULTIPLE-RESONATOR BAND-PASS FILTERS

The word *tuning* as used herein refers to the process of adjusting all of the resonators to resonate at the same midband frequency ω_0. In general, synchronously tuned band-pass filters (of the sorts discussed in Chapters 8, 9, and 10) which are properly tuned will have a response that is symmetrical about the midband frequency ω_0, except for some possible skew as a result of the variation of the couplings with frequency [see, for example, the response in Fig. 8.08-5(b)]. In contrast, if a band-pass filter of this type has the resonators all properly tuned, but the couplings are not correct, the response will be nearly symmetrical, but it will otherwise have an improper shape (for example, pass-band ripples of the wrong size, or improper bandwidth).

*Alternating Short-Circuit and Open-Circuit Procedure**—A procedure which is frequently very useful for tuning synchronously tuned filters can be understood with the aid of Fig. 11.05-1. This figure shows a

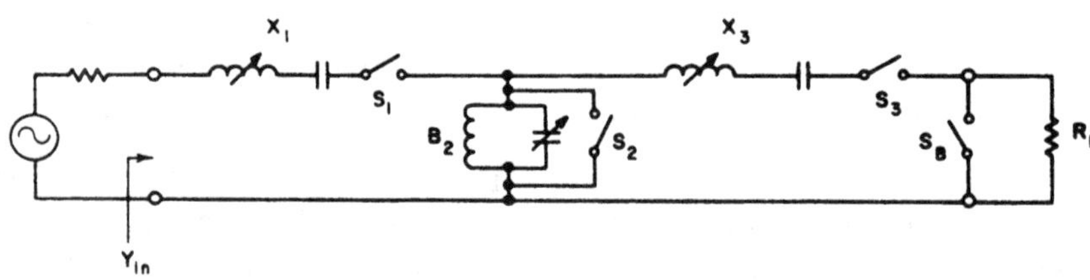

FIG. 11.05-1 A THREE-RESONATOR BAND-PASS FILTER
The switches are used to open-circuit the series resonators and to short-circuit the shunt resonator during the tuning process

*This procedure is the same as that described by Dishal in Ref. 2.

band-pass filter having two series and one shunt resonator, and switches are provided to open-circuit the series resonators and to short-circuit the shunt resonator. To start with, the series resonators are open-circuited by opening switches S_1 and S_3 while the shunt resonator is short-circuited by closing S_2. Then the admittance Y_{in} will be zero.

Next, a signal at the desired midband frequency ω_0 is applied, switch S_1 is closed, and Resonator 1 is tuned. Since Resonator 2 is still short-circuited, Resonator 1 will be tuned when $Y_{in} = \infty$. After Resonator 1 has been tuned it will present zero reactance and Resonator 2 can be tuned by opening S_2 and tuning Resonator 2 until $Y_{in} = 0$ (which occurs since S_2 is still open). Next, Resonator 3 is tuned (with S_B still closed to short-circuit the load) until $Y_{in} = \infty$. Then S_B is opened and the tuning process is complete.

Note that in the above process the series resonators were tuned to yield a *short-circuit* at the input while the shunt resonators were tuned to yield an *open-circuit* at the input.

Now let us consider the entirely equivalent filter in Fig. 11.05-2 which uses only shunt resonators separated by J-inverters. If the J-inverters are frequency invariant, this filter can have exactly the same Y_{in} and transmission characteristic as the filter in Fig. 11.05-1. However, since the resonators are in this case all in shunt, the tuning process is started with all of the resonators and the load on the right short-circuited. The tuning process then goes much the same as before and can be stated in general as follows:

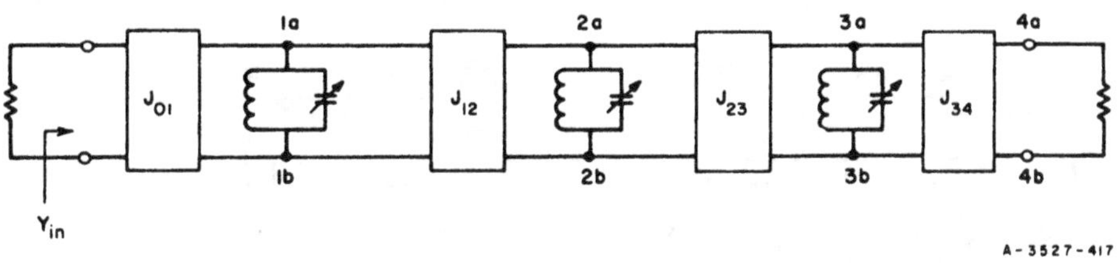

FIG. 11.05-2 DIRECT-COUPLED BAND-PASS FILTER WITH SHUNT RESONATORS
The J-inverters represent the couplings

(1) All of the shunt-resonators are short-circuited and an *open-circuit* is observed at the input (*i.e.*, $Y_{in} = 0$).

(2) The short-circuit is removed from Resonator 1, which is then tuned until a *short-circuit* is observed at the input at frequency ω_0.

(3) The short-circuit is removed from Resonator 2, which is then adjusted until an *open-circuit* is observed at the input at frequency ω_0.

(4) This process is repeated, tuning alternately for a short-circuit at the input when tuning one resonator, and for an open-circuit at the input when tuning the next. After the last resonator is tuned, the short-circuit is removed from the output.

If the filter uses series resonators and K-inverters, the procedure is essentially the same, except that all of the resonators must be open-circuited to start with, and for this condition the input will appear as a short-circuit.

As an example of how this procedure applies to microwave filters, let us consider capacitively coupled strip-line filters of the form discussed in Sec. 8.05. Tuning screws are added as shown in Fig. 11.05-3. To start the tuning process, the screws are screwed in all the way to short-out the resonators and the output line. The filter is connected to a slotted line and the position of a voltage minimum for frequency ω_0 is observed. Then the screws on the first resonator are backed-out and adjusted until the voltage minimum on the slotted line moves exactly a quarter wavelength. Next the second resonator is tuned to bring the minimum back to its original position. As consecutive resonators are tuned, the voltage minimum continues to move back and forth between the two points separated by $\lambda_0/4$.

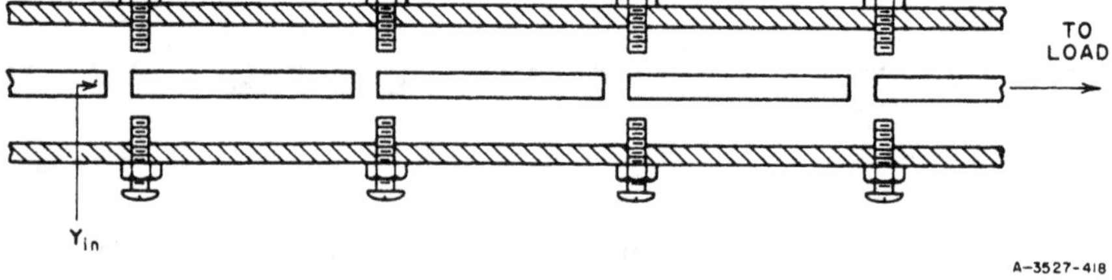

FIG. 11.05-3 A CAPACITIVELY COUPLED STRIP-LINE FILTER WITH TUNING SCREWS

In the case of the waveguide filters in Sec. 8.06, which have resonators with series-type resonances and K-inverters, the procedure is essentially the same except for that the tuning screws are located in the middle of the $\lambda/2$ resonators instead of at the ends. (This reflects an open-circuit to the K-inverters when the screws are all the way in.) Also, the screw on the output should be placed a quarter-wavelength away from the output iris.

In practical situations, the accuracy of the procedure described above depends very much on how close the idealized circuit in Fig. 11.05-2 (or its dual) is approximated. That is, in Fig. 11.05-2 short-circuiting a resonator will also apply a short-circuit directly across the terminals of the adjacent J-inverters. However, in the circuit in Fig. 11.05-3, the inverters are of the form in Fig. 8.03-2(d). These inverters include a negative length of line which in the actual filter structure is absorbed into adjacent positive line length of the same impedance. For this reason, the actual terminals of the inverter are not physically accessible, and the tuning screws cannot short-out the resonators at the exactly correct spot for perfect tuning. For narrow-band filters, however, the coupling susceptances will be very small, the negative length of line in the inverter will be very short, and the exact locations of the short-circuits on the resonator bars will not be critical. In such cases, the above tuning procedure works well. Similar considerations arise in tuning other types of direct-coupled filters.

In direct-coupled filters, the resonators that are most likely to give trouble when using this procedure are the first and last resonator, since their couplings generally differ more radically from those of the other resonators. Thus, if the bandwidth of the filter is, say, around 10 percent and this tuning procedure does not yield a suitably symmetrical response, it may be possible to correct this by experimental adjustment of the tuning of the end resonators alone. This is often quite easy to do if a sweeping signal generator is available.

Tuning of Resonators Singly or in Pairs—For precision tuning of filters having more than around 10-percent bandwidth, the best method appears to be to test and adjust the resonators individually or in identical pairs as was discussed in Secs. 11.03 and 11.04. Even with such procedures, however, difficulties can occur if the resonator couplings are at all tight. Filters such as those in Secs. 8.11, and 8.12 which use lumped-element couplings representable by the inverters in Fig. 8.03-1(a) and 8.03-2(b), will tune somewhat differently when connected to another resonator than they will when connected to resistor terminations. In the case of lumped-element shunt resonators with series-capacitance couplings, the coupling capacitors may be regarded as being part of inverters of the form in Fig. 8.03-2(b). In such cases the negative capacitance of the inverter can be absorbed into the shunt capacitances of the adjacent resonators; but when coupling a resonator to a resistor termination, the negative shunt capacitance next to the resistor cannot be absorbed. For this reason the tuning effect due to capacitor coupling between two lumped-element shunt resonators is different from that between a resonator and a resistor. These matters were previously discussed with regard to Eqs. (8.14-32) to (8.14-34). If the couplings are relatively tight, a correction can be applied by tuning the individual resonators (when tested individually while connected to resistor terminations) to a slightly different frequency calculated by use of the principles discussed in connection with Eqs. (8.14-32) and (8.14-34). If the couplings are relatively loose (case of narrow bandwidth) the correction required will be small and possibly negligible.

In the cases of the filter types in Figs. 8.05-1, 8.06-1, 8.07-3, 8.08-1, and 8.10-3, the difficulty described in the immediately preceding paragraph does not occur. These filters all consist of uniform transmission lines with appropriately spaced discontinuities. The inverters in these cases are of the forms in Figs. 8.03-1(c), 8.03-2(d), or 8.03-3. The negative line lengths ϕ involved in these inverters are of the same characteristic impedance as the resonator lines and the termination transmission lines. Thus, these lines can be absorbed equally well into the resonators or into the matched terminating lines, and if a resonator of characteristic impedance Z_0 is removed from the interior of a multiple-resonator filter and tested between matched terminating lines of the same characteristic impedance Z_0, *there will be no tuning error regardless of whether the couplings are loose or tight.* For example, the tuning of the resonators of filters of the form in Fig. 8.05-4(a) can be checked by testing each resonator individually as shown in Fig. 11.03-2, using exactly the same coupling gaps at each end as will be used when each resonator is installed in the multiple-resonator filter. The resonators in the filter in Fig. 8.05-4(a) were actually tested in symmetrical pairs as discussed in Sec. 11.04. Either the single- or double-resonator procedure gives high accuracy. After the resonators were tested, the lengths of the bars were corrected to give precisely the desired tuning frequency. Using this procedure, no tuning screws are required, and in many cases the tolerances required are reasonable enough that it should be practical to mass-produce additional filters of the same design without any tuning screws or adjustments.

Narrow-band parallel-coupled filters of the form in Fig. 8.09-1 can be tuned using the slotted-line procedure described with reference to Fig. 11.05-2. Figure 11.05-4 shows suggested locations for the tuning screws for parallel-coupled filters. The screws on the output line section $n,n+1$ should be backed-out flush with the ground planes after Resonator n has been tuned. It should usually not be necessary to tune relatively wide-band filters of this sort, since the accuracy of the synchronous tuning becomes less critical as the bandwidth becomes greater, and this type of filter tends to be relatively free of errors in tuning of any one resonator with respect to the others. If the pass band is not centered at the correct frequency, the lengths of all the resonators should be altered by a fixed amount computed to give the desired shift in pass-band center frequency.

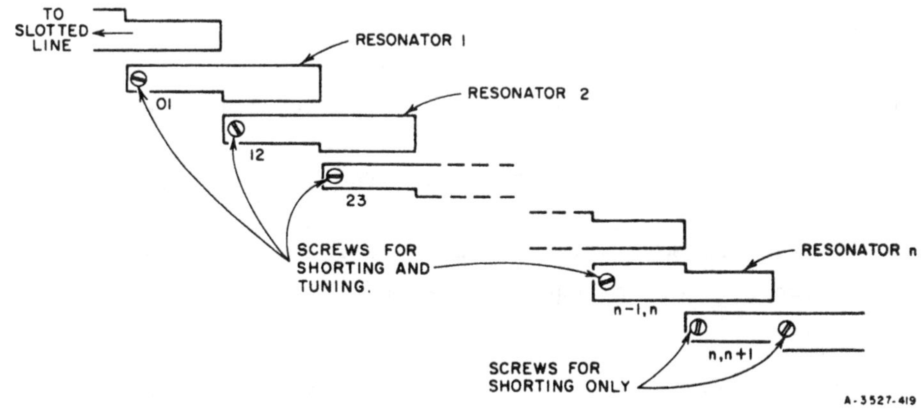

FIG. 11.05-4 ARRANGEMENT OF SCREWS FOR TUNING OF A NARROW-BAND PARALLEL-COUPLED FILTER

SEC. 11.06, CALCULATION OF THE MIDBAND DISSIPATION LOSS OF BAND-PASS FILTERS

The information necessary for computing the midband loss of band-pass filters was previously presented in Sec. 4.13, but an example in this chapter may be desirable. Let us suppose that a 10-percent-bandwidth, six-resonator, band-pass filter with a 0.10-db-ripple Tchebyscheff response is desired, and it is estimated that the resonator Q's will be about 1000 for the type of construction which is proposed. An estimate of the midband attenuation due to dissipation loss in the filter is desired. For this case, Fig. 4.13-2 is convenient, and from it for $n = 6$ and 0.10-db Tchebyscheff ripple we obtain $C_6 = 4.3$. By Eq. (4.13-2) the corresponding Q of the elements in the analogous low-pass prototype filter at the low-pass cutoff frequency ω_1' is

$$Q = wQ_{BP} = (0.10)(1000) = 100$$

where w is the band-pass filter fractional bandwidth; here $Q_{BP} = Q_u$ indicates the unloaded Q of the resonators of the band-pass filter. Then by Eq. (4.13-3) the dissipation factor is

$$d = \frac{\omega_1'}{Q} = \frac{1}{100} = 0.01$$

where, for the data in Fig. 4.13-2, the prototypes have been normalized so that $\omega_1' = 1$. Then by Eq. (4.13-8) the increase in attenuation at midband as a result of dissipation loss will be approximately

$$(\Delta L_A)_0 \approx 8.686 C_n d = 8.686(4.3)(0.01) = 0.37 \quad \text{db}.$$

As was discussed in Sec. 4.13, the attenuation due to dissipation at the pass-band edge frequencies can be expected to be around two to three times this amount, or from around 0.74 to 1.1 db.

It should be noted that it is sometimes useful to make the above calculations in the reverse sequence in order to estimate the unloaded Q's of the resonators of a band-pass filter from the measured increase in filter midband attenuation resulting from dissipation loss.

From Sec. 4.13 we obtain

$$Q_u = Q_{BP} = \frac{8.686 C_n \omega_1'}{w(\Delta L_A)_0} \quad . \tag{11.06-1}$$

Estimates of the unloaded Q values to use in computing band-pass filter attenuation can be obtained in a number of ways. One is to use data obtained by computing from the measured midband loss of existing filters as suggested in the preceding paragraph. Another is to make laboratory tests on a resonator of the type to be used—utilizing the techniques described in Sec. 11.02 to determine the unloaded Q. Another is to compute the unloaded Q by use of the information given in Chapter 5 for coaxial-line, strip-line, and waveguide structures. It should be remembered, however, that the values of Q which are theoretically possible are rarely achieved in practice. This is due to surface roughness of the metal, corrosion, and due to additional losses in the coupling elements which are difficult to accurately predict. In some strip-line filter structures, the Q's realized are typically about half of the theoretical value. In the case of waveguide filters, the agreement between theory and practice appears to be somewhat better, but in general some allowance should be made for the fact that the Q's in typical operating filters will be less than the theoretical optimum.

Some further matters involving filter dissipation loss will be treated in the next section.

SEC. 11.07, DESIGN OF NARROW-BAND FILTERS FOR MINIMUM MIDBAND LOSS AND SPECIFIED HIGH ATTENUATION AT SOME NEARBY FREQUENCY

In various practical situations such as the design of preselectors for superheterodyne receivers, narrow-band filters are desired which have as little loss as possible at band center, with some specified high attenuation at some nearby frequency. For example, in a superheterodyne receiver using a 30-Mc IF frequency, a preselector filter with high attenuation 60 Mc away from the pass-band center may be desired in order to suppress the image response. For microwave receivers of this sort the signal bandwidth is usually very small compared to the carrier frequencies involved. Thus, the design objectives for the preselector filter focus on obtaining minimum midband loss,

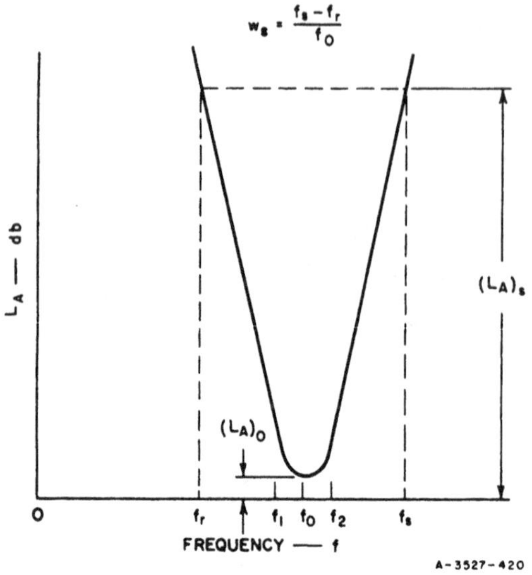

FIG. 11.07-1 DEFINITION OF TERMS USED IN SEC. 11.07

and specified high attenuation at some nearby frequency. Adequate pass-band width for the narrow signal spectrum to be received is virtually assured.

Figure 11.07-1 defines most of the symbols to be used in the following discussion. For the purposes of this analysis it will be assumed that the filters under discussion are of sufficiently narrow bandwidth so that the simplified low-pass to band-pass mapping

$$\frac{\omega'}{\omega_1'} = \frac{2}{w}\left(\frac{f - f_0}{f_0}\right)$$

where

$$w = \frac{f_2 - f_1}{f_0} \qquad (11.07\text{-}1)$$

and

$$f_0 = \frac{f_1 + f_2}{2}$$

will be valid regardless of the physical form of the resonators and couplings to be used. In Eq. (11.07-1), w is the fractional pass-band width at the pass-band edges corresponding to ω_1' of the prototype. Of more importance for this application is the fractional bandwidth w_s for which the attenuation is specified to have reached a specified high level $(L_A)_s$, as indicated in Fig. 11.07-1. The attenuation $(L_A)_0$ at midband is, of course, to be minimized.

Many low-pass prototype designs will lead to band-pass filter designs which will have $(L_A)_0 = 0$ *if the resonators of the band-pass filter are lossless.* However, when the inevitable resonator losses are included $(L_A)_0$ will necessarily be non-zero. Under these conditions filters which all have the same w_s for a specified $(L_A)_s$, and which would all have $(L_A)_0 = 0$ for the case of no losses, may have considerably different $(L_A)_0$ values when losses are included.

Schiffman[4] has developed design data for filters having two or three resonators which will yield minimum midband loss for specified w_s and $(L_A)_s$ as mentioned above.* Cohn[5] has studied the general case for an arbitrary number of resonators and has found that designs with very nearly minimum $(L_A)_0$ will be obtained if the filter is designed from a low-pass prototype having the parameters

$$g_0 = g_1 = g_2 = \ldots = g_n = g_{n+1} = 1$$

and (11.07-2)

$$\omega_1' = 1 \quad .$$

The significance of these parameters is the same as in Sec. 4.04.

By use of Eqs. (11.07-2), (4.13-3), and (4.13-11), the midband loss for filters designed from the prototype element values in Eq. (11.07-2) is approximately

$$(L_A)_0 \approx \frac{4.343n}{wQ_u} \quad \text{db} \tag{11.07-3}$$

where Q_u is the unloaded Q of the resonators. Here $(L_A)_0$ is practically the same as $(\Delta L_A)_0$ in Eq. (4.13-11), since this type of filter is perfectly matched at midband when no losses are present, and very nearly matched at midband when losses are included. The fractional bandwidth w in Eq. (11.07-3) is the pass-band fractional bandwidth corresponding to ω_1' for the prototype, and is the one to be used in the design equations of Chapters 8 to 11. However, the fractional bandwidth w_s is the one which is most useful for specifying the performance requirements for this application. By use of Eqs. (4.14-1), (11.07-1), and (11.07-2), w and w_s are found to be related by the approximate formula

* Taub and Bogner,[10] and Fubini and Guillemin[11] have studied true maximally flat and Tchebyscheff filters which have dissipation. It will be seen that, a true maximally flat or Tchebyscheff response in a filter having dissipation requires increased midband attenuation over that of the equal-element filters discussed in this section [*i.e.*, for the same $(L_A)_s$, w_s and resonator Q's].

$$\frac{w_s}{w} \approx \text{antilog}_{10} \frac{(L_A)_s + 6.02}{20n} \qquad (11.07\text{-}4)$$

where $(L_A)_s$ is in decibels. Thus, by Eqs. (11.07-3) and (11.07-4)

$$(L_A)_0 \approx \frac{4.343n \, \text{antilog}_{10} \left[\frac{(L_A)_s + 6.02}{20n}\right]}{w_s Q_u} \quad \text{db.} \qquad (11.07\text{-}5)$$

Figure 11.07-2 shows plots of data obtained from Eq. (11.07-5) for various values of $(L_A)_s$.* Note that $w_s Q_u (L_A)_0$ is plotted *vs.* the number, n, of resonators. Thus, for given w_s and Q_u the height of the curves is proportional to the midband loss that will occur. Observe that the optimum number of resonators needed to achieve minimum $(L_A)_0$ depends on the value of $(L_A)_s$ which has been specified. For $(L_A)_s = 30$ db the optimum number of resonators is four, while for $(L_A)_s = 70$ db, the optimum number of resonators is eight.

Equation (11.07-5) upon which Fig. 11.07-2 is based, involves a number of approximations such that the answers given will improve in accuracy as $(L_A)_s$ increases in size and as the product $Q_u w_s$ increases in size.* As a check on the accuracy of this equation, a typical trial design with moderate values of $(L_A)_s$ and $Q_u w_s$ will be considered. Let us suppose that $w_s = 0.03$ is required for $(L_A)_s = 40$ db, and that $Q_u = 1000$. Figure 11.07-2 shows that an $n = 4$ resonator design will be quite close to optimum. From the chart, for $n = 4$, $w_s Q_u (L_A)_0 = 65$. Thus, the estimated midband loss is

$$(L_A)_0 = \frac{65}{w_s Q_u} = \frac{65}{0.03(1000)} = 2.16 \quad \text{db.}$$

To check this result a lumped-element, $n = 4$, trial design of the form in Fig. 8.02-2(a) was worked out using $g_0 = g_1 = g_2 = g_3 = g_4 = g_5 = 1$, $\omega_1' = 1$, and $w = 0.007978$ [which was obtained from Eq. (11.07-4)]. The computed response of this design is shown in Fig. 11.07-3 for the cases of $Q_u = \infty$ (*i.e.*, not losses), and $Q_u = 1000$.

---

* Examples due to Mr. L. A. Robinson indicate that these curves are reasonably accurate from $n = 1$ to where they flatten out, but, because of the approximations involved, they do not rise as fast as they should towards the right.

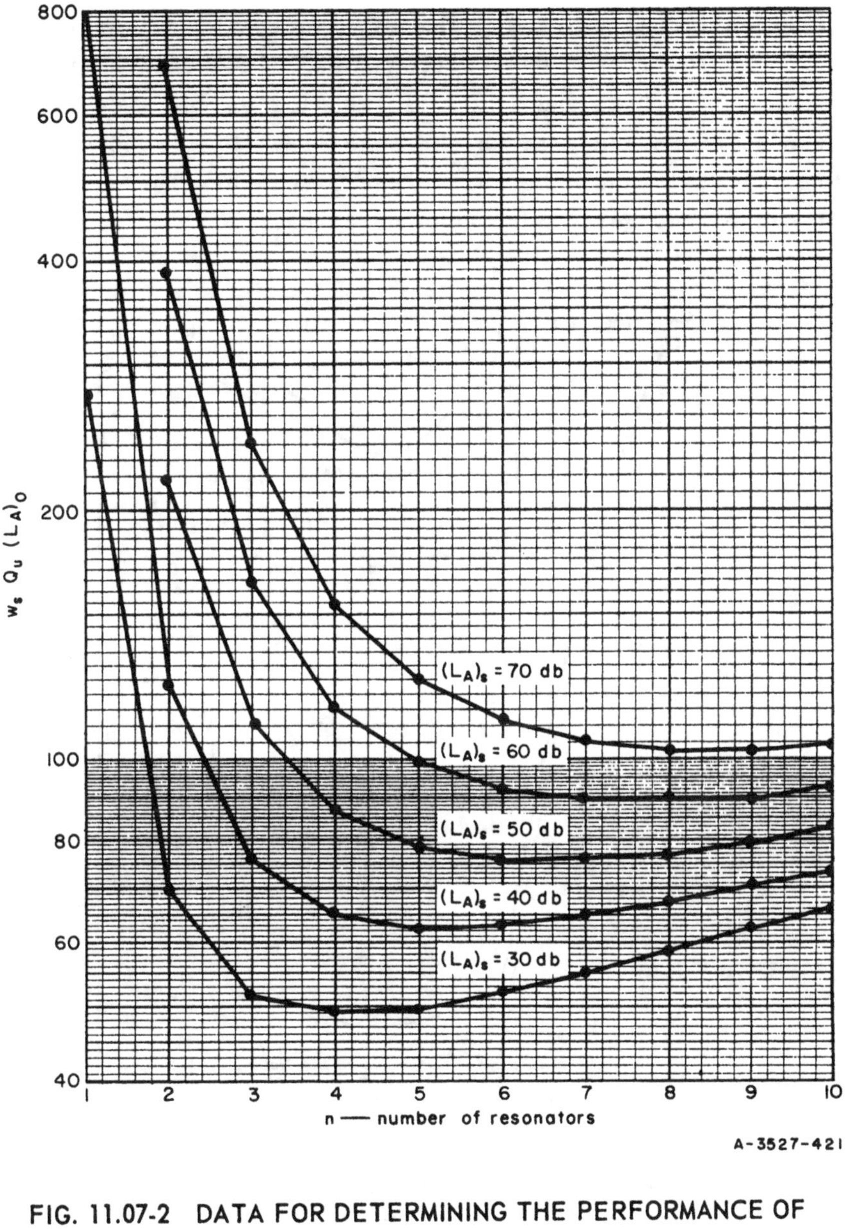

FIG. 11.07-2 DATA FOR DETERMINING THE PERFORMANCE OF BAND-PASS FILTERS DESIGNED FROM EQUAL-ELEMENT PROTOTYPES

Figure 11.07-3 shows the data in Fig. 11.07-2 to be satisfactory for this case. Note that the midband loss when Q_u = 1000 is for practical purposes exactly as predicted. The only noticeable error is that the attenuation at the w_s = 0.03 fractional bandwidth points (*i.e.*, the points were ω/ω_0 = 0.85 and 1.15) is 1 db low on one side and 1.5 db low on the other as compared to the specified $(L_A)_s$ = 40-db value. Thus, Fig. 11.07-2 should be sufficiently accurate for most practical applications of the sort under consideration.

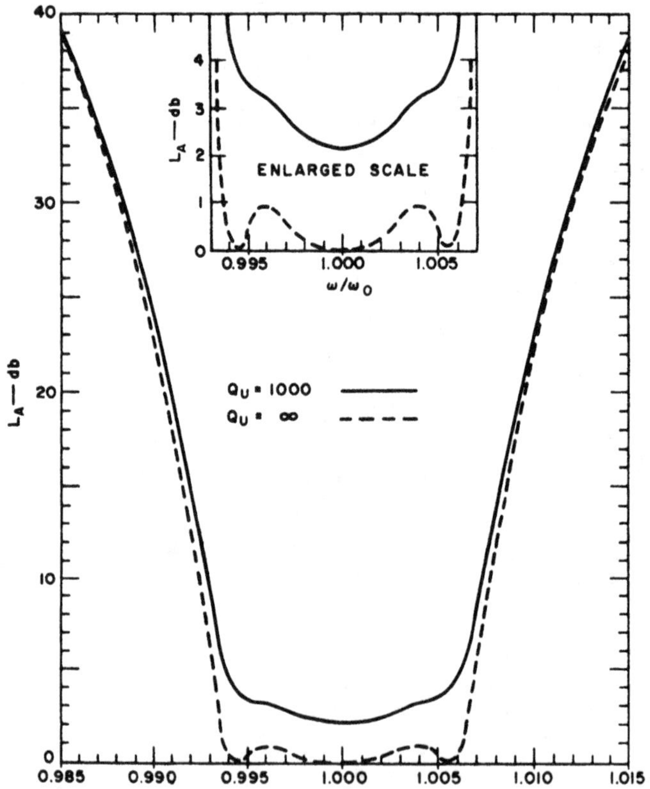

FIG. 11.07-3 RESPONSES WITH AND WITHOUT DISSIPATION LOSS OF A 4-RESONATOR FILTER DESIGNED FROM AN EQUAL-ELEMENT PROTOTYPE

The dashed line in Fig. 11.07-3 shows the attenuation of the filter when the resonators are lossless. The pass-band is seen to have ripples nearly 1 db in amplitude. Observe that when losses are included, the attenuation due to dissipation increases the loss at the band edges so much as to nearly obscure the ripples completely. This is typical of the effects of sizeable dissipation loss, regardless of the choice of low-pass prototype filter.

SEC. 11.08, DESIGN OF BAND-PASS IMPEDANCE-MATCHING NETWORKS USING THE METHODS OF CHAPTER 8

Very efficient impedance-matching networks can be designed for many applications by use of the filter-design techniques discussed in Chapter 8, along with the impedance-matching network low-pass prototypes discussed in Secs. 4.09 and 4.10. These procedures assume that the load to be matched can be approximated over the frequency band of interest by a simple R-L-C resonant circuit. This may seem like a serious restriction, but actually a very large number of microwave impedance-matching problems involve loads which approximate this situation satisfactorily over the frequency range for which a good impedance match is required.

The bandwidths for which the procedures described in this section will yield good impedance-matching network designs will depend partly on how closely the load to be matched resembles a simple R-L-C circuit within the frequency range of interest. However, since the procedures about to be described are based on the methods of Chapter 8, which involve approximations of a narrow-band sort, accurate results for bandwidths much over 20 percent can rarely be expected. Designs of 5 or 10 percent bandwidth which give nearly the theoretically optimum match for the given load and number of resonators should be relatively easy. The procedures described in Sec. 11.09, following, are recommended for the more wide-band cases since Sec. 11,09 is based on the methods of Chapter 10 which apply for very wide bandwidths as well as for narrow bandwidths.

Determination of the Load Parameters and the Low-Pass Prototype— Before a satisfactory network to give a good impedance match over a prescribed frequency range can be designed, it is necessary to establish parameters that define the load. The following procedure is recommended:

(1) Make impedance measurements of the load across the desired frequency range for which a good impedance match is desired. Plot the resulting impedance data on a Smith chart, or in some other convenient form. Also plot the corresponding admittance vs. frequency characteristic. If the real part of the impedance is more nearly constant than the real part of the admittance, the load is best suited for approximate representation as a *series* R-L-C circuit. If the real part of the admittance is the most nearly constant over the frequency range of interest, the load is best suited to approximate representation as a *parallel* R-L-C circuit.

(2) If the real part of the impedance is the most constant, add a reactance element in *series* so as to bring the load to series resonance at f_0, the midband frequency of the frequency range over which a good impedance match is desired. If the real part of the admittance is the most nearly constant, add a reactance element in *shunt* with the load so as to bring the load to parallel resonance at f_0.

(3) Using the procedures in Sec. 11.02, determine Q_A, the Q of the resonated load. If the load is series-resonant, also determine R_A, the resistance of the load at resonance. If the load is parallel-resonant, determine G_A, the conductance of the load at resonance.

After the load circuit has been brought to resonance at f_0, and a measured value has been found for its Q, Q_A, and for its midband resistance R_A, or conductance G_A, the load is defined for the purposes of this design procedure.

With regard to the steps outlined above, several additional points should be noted. In many cases the nature of the load circuit will be well enough understood so that the impedance and admittance plots in Step 1 above will not be necessary. When adding a reactance element to bring the load to series or parallel resonance at f_0, the additional element should be as nearly lumped as possible, and it should be as close to the load as possible, if best performance is desired. Using a resonating element of a size that approaches a quarter-wavelength or more will *increase* the Q of the resonant load and decrease the quality of the impedance match that will be possible over the specified band. Using a lumped resonating discontinuity at an appreciable electrical distance from the load has a similar deleterious effect. However, in some practical situations where the resulting degradation in performance can be tolerated, sizeable matching elements or the locating of a resonating discontinuity at some distance from the load may be desirable as a matter of practical convenience.

In order to select an appropriate low-pass prototype for the band-pass matching network, the decrement

$$\delta = \frac{1}{wQ_A} \tag{11.08-1}$$

must be computed where

$$w = \frac{f_2 - f_1}{f_0} \tag{11.08-2}$$

$$f_0 = \frac{f_2 + f_1}{2} \tag{11.08-3}$$

and where f_1 and f_2 are, respectively, the lower and upper edges of the frequency band over which a good match is desired. Knowing δ, the performance which is possible can be predicted as described in Secs. 4.09 and 4.10, and a suitable low-pass prototype can be determined as described in those sections.

It should be recalled that, for any load which has reactive elements as well as resistive elements in it, it is not possible to obtain a perfect match across a finite frequency band no matter how complicated an impedance-matching network is used (see Sec. 1.03). Thus, for any given load there are absolute restrictions on how good a match can be obtained across a given band of interest. Generally, adding one resonator (in addition to the resonator formed by the load) which calls for an $n = 2$ reactive element low-pass prototype circuit will give a very large improvement in performance. Going to a design worked out using an optimum $n = 3$ or $n = 4$ prototype will give still further improvements, but with rapidly diminishing returns for each increase in n. It would rarely be worthwhile to go to designs with n greater than 3 or 4; *i.e.*, to use more than two or three resonators in addition to the resonator formed by the resonated load.

When designing waveguide impedance-matching networks, it will frequently be convenient to use reciprocal guide wavelength as a frequency variable, as was done in the design data discussed in Sec. 8.06. On that basis, the decrement becomes

$$\delta_\lambda = \frac{1}{w_\lambda\left(\dfrac{\lambda_{g0}}{\lambda_{ga} - \lambda_{gb}}\right)} \approx \frac{1}{w_\lambda Q_A \left(\dfrac{\lambda_0}{\lambda_{g0}}\right)^2} \qquad (11.08\text{-}4)$$

where

$$w_\lambda = \frac{\lambda_{g1} - \lambda_{g2}}{\lambda_{g0}} \qquad (11.08\text{-}5)$$

$$\lambda_{g0} = \frac{\lambda_{g1} + \lambda_{g2}}{2} \qquad (11.08\text{-}6)$$

where λ_{g1} and λ_{g2} are the guide wavelengths at the edges of the frequency band over which a good match is desired, λ_{g0} is the guide wavelength at midband, λ_{ga} and λ_{gb} are the guide wavelengths at the frequencies of the half-power points of the resonated load, λ_0 is the plane-wave wavelength at the midband frequency in the same medium of propagation as exists within the guide, and Q_A is the Q of the resonated load as determined by the methods in Sec. 11.02, or by other equivalent methods. Having a value for the decrement δ_λ, the low-pass prototype to be used is determined by use of the data in Secs. 4.09 and 4.10, using δ_λ in place of δ.

Adaptation of the Data in Chapter 8 for Impedance-Matching Network Design—The methods for band-pass filter design discussed in Chapter 8 are readily adapted for the design of impedance-matching networks. After the load has been resonated as described above, the load provides the first resonator of the filter. From that point on, the remainder of the filter (which comprises the actual impedance-matching network) can be of any of the forms discussed in Chapter 8. Figure 11.08-1 presents generalized information for the design of impedance-matching networks for series-resonated loads with the matching network consisting of series resonators coupled by K-inverters. This generalized information supplements that in Fig. 8.02-3, and it can be applied to the design of impedance-matching networks using a wide variety of resonator structures as discussed in Chapter 8. Figure 11.08-2 shows analogous data for the case of a parallel-resonant load with the matching network consisting of shunt resonators and J-inverters. These data supplement the data in Fig. 8.02-4.

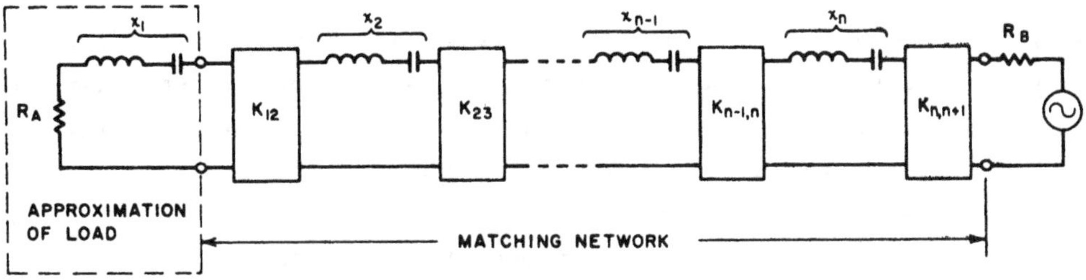

$$K_{12} = \frac{1}{\omega_1'}\sqrt{\frac{w R_A x_2}{g_1 g_2 \delta}} \;,\quad K_{k,k+1}\Big|_{k=2 \text{ TO } n} = \frac{w}{\omega_1'}\sqrt{\frac{x_k x_{k+1}}{g_k g_{k+1}}} \;,\quad K_{n,n+1} = \sqrt{\frac{R_B x_n w}{g_n g_{n+1} \omega_1'}}$$

WHERE R_B AND THE RESONATOR SLOPE PARAMETERS $x_2, x_3, \cdots, x_n$ MAY BE CHOSEN ARBITRARILY.

A-3527-423

FIG. 11.08-1 COUPLED-RESONATOR MATCHING NETWORK FOR SERIES-RESONANT LOADS
This data supplements that in Fig. 8.02-3

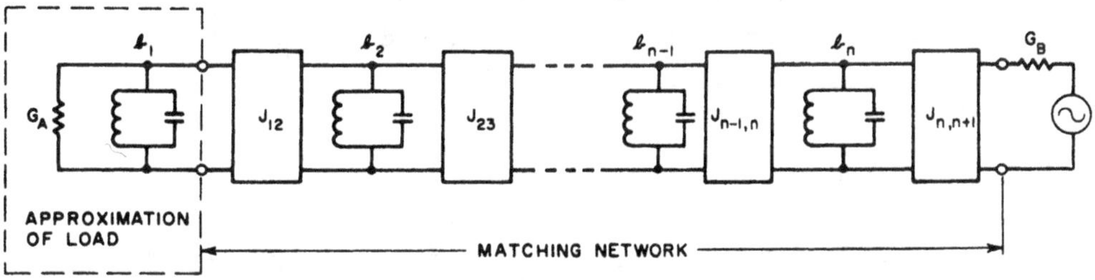

$$J_{12} = \frac{1}{\omega_1'}\sqrt{\frac{w G_A b_2}{g_1 g_2 \delta}} \;,\quad J_{k,k+1}\Big|_{k=2 \text{ TO } n} = \frac{w}{\omega_1'}\sqrt{\frac{b_k b_{k+1}}{g_k g_{k+1}}} \;,\quad J_{n,n+1} = \sqrt{\frac{G_B b_n w}{g_n g_{n+1} \omega_1'}}$$

WHERE G_B AND THE RESONATOR SLOPE PARAMETERS $b_2, b_3, \cdots, b_n$ MAY BE CHOSEN ARBITRARILY

A-3527-424

FIG. 11.08-2 COUPLED-RESONATOR MATCHING NETWORK FOR SHUNT-RESONANT LOADS
This data supplements that in Fig. 8.02-4

To further make clear how the generalized information in Figs. 11.08-1 and 11.08-2 is used for the practical design of impedance-matching networks, the data given will be restated as they apply to several of the specific filter structures discussed in Chapter 8. Figure 11.08-3 shows how the data apply to the design of strip-line impedance-matching filters with $\lambda_0/2$ resonators of the form in Sec. 8.05. Note that in this case the load must exhibit shunt resonance, and Resonator 2 is capacitively coupled to the load. When the load has been resonated and an appropriate low-pass prototype has been established as discussed above, and using the equations for the $J_{k,k+1}/Y_0$ as given in Fig. 11.08-3, the remainder of the design of the circuit is as discussed in Sec. 8.05. The capacitive coupling B_{12} between Resonator 2 and the load will have a small detuning effect on the resonated load. As a result, it will tend to shift the resonant frequency of the load by about

$$\Delta f = \frac{-f_0 B_{12}}{2 G_A Q_A} \quad . \tag{11.08-7}$$

To compensate for this, the load should be retuned after the matching network is installed, or it should be pretuned to $f_0 + |\Delta f|$ before the matching network is installed.

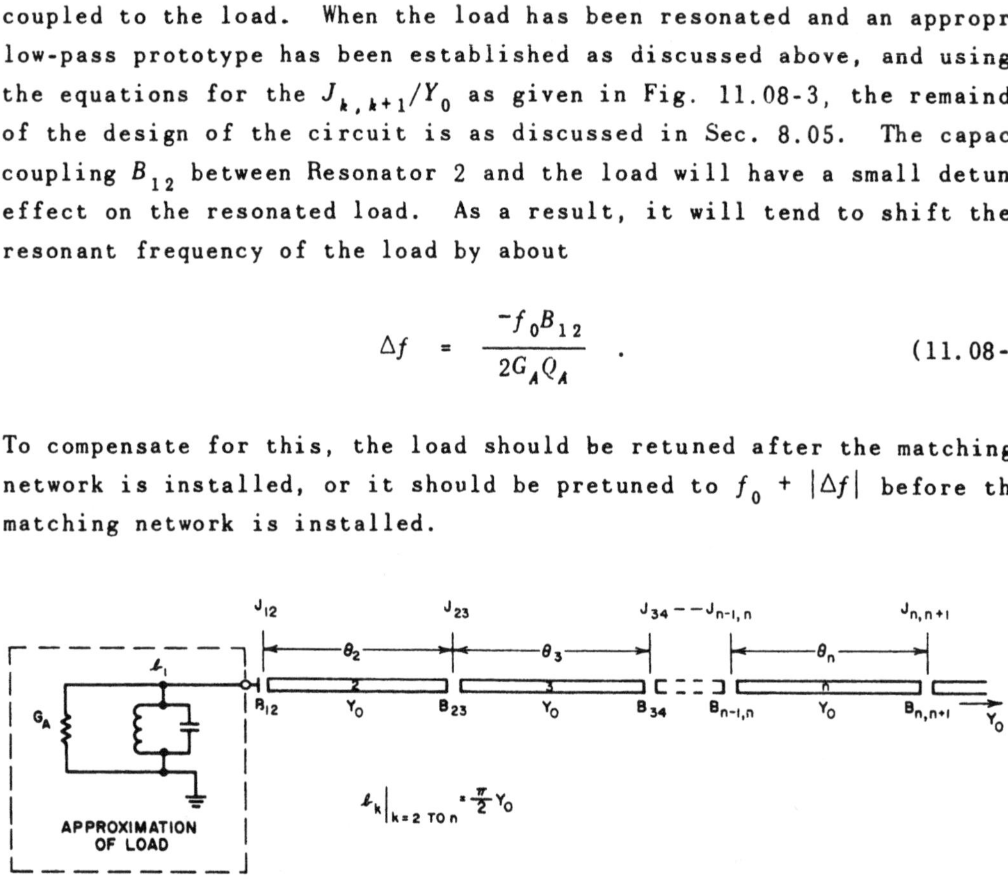

$$\frac{J_{12}}{Y_0} = \frac{1}{\omega_1'}\sqrt{\frac{\pi w G_A}{2 Y_0 g_1 g_2 \delta}} \; , \quad \left.\frac{J_{k,k+1}}{Y_0}\right|_{k=2 \text{ TO } n-1} = \frac{\pi w}{2\omega_1'}\,\frac{1}{\sqrt{g_k g_{k+1}}} \; , \quad \frac{J_{n,n+1}}{Y_0} = \sqrt{\frac{\pi w}{2 g_n g_{n+1} \omega_1'}}$$

A-3527-425

FIG. 11.08-3 SUPPLEMENTARY DATA FOR USE WITH THE MATERIAL IN SEC. 8.05 FOR DESIGN OF IMPEDANCE-MATCHING NETWORKS

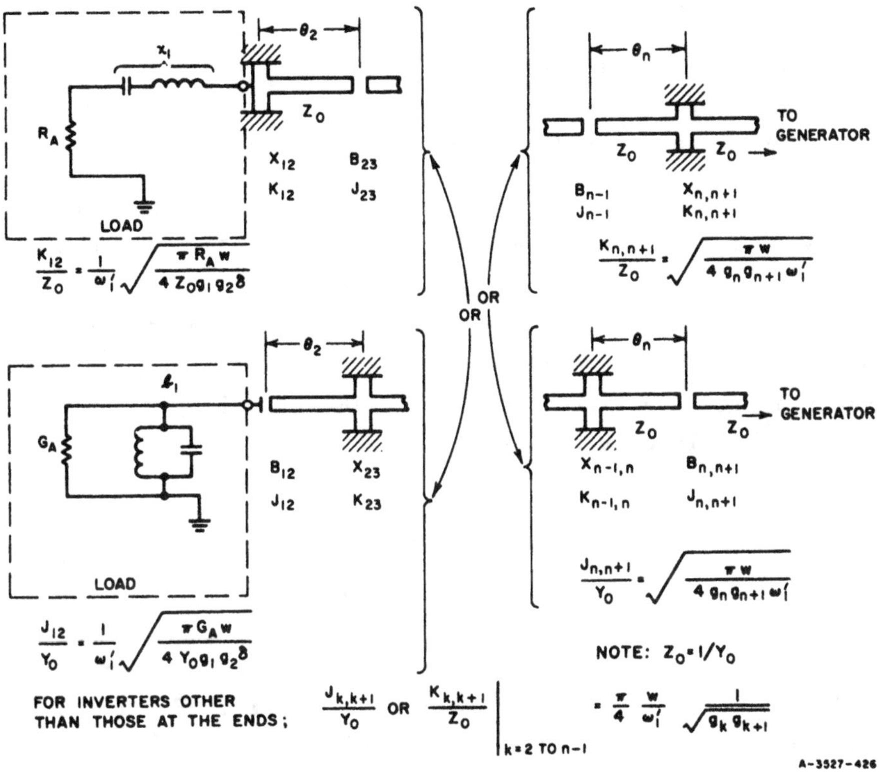

FIG. 11.08-4 SUPPLEMENTARY DESIGN DATA FOR USE WITH THE MATERIAL IN SEC. 8.08 FOR THE DESIGN OF IMPEDANCE-MATCHING NETWORKS

Figure 11.08-4 shows analogous data for design of impedance-matching networks whose resonators are in the quarter-wavelength, strip-line form discussed in Sec. 8.08. In this case the load may be either series- or shunt-resonated. If the load is series resonated, then it should be followed by a *K*-inverter consisting of a small shunt stub such as those discussed in Sec. 8.08. The shunt reactance X_{12} of this stub will have a small detuning effect on the load and will tend to shift its resonant frequency by about

$$\Delta f = \frac{-f_0 X_{12}}{2R_A Q_A} \quad . \tag{11.08-8}$$

This effect should be compensated for by retuning the load, or by precompensating the tuning of the load. If the load is parallel-resonated, then the load should be followed by a capacitive-gap *J*-inverter as is also shown in Fig. 11.08-4, and the detuning effect of the coupling susceptance

B_{12} is as given in Eq. (11.08-7). Figure 11.08-4 gives equations for the J- and K-inverters for design of the matching network, and the remainder of the design process is as described in Sec. 8.08.

Figure 11.08-5 presents data for the case where the matching network is to be a waveguide filter as discussed in Sec. 8.06. In this case the reference plane for the load must be referred to a point in the waveguide where the load will appear to be series-resonant. The load is coupled to the next resonator by a K-inverter consisting primarily of a shunt-inductive iris with shunt reactance X_{12}. This coupling reactance will tend to detune the load by an amount

$$\Delta f = \frac{-f_0(X_{12}/Z_0)}{2(R_A/Z_0)Q_A} \tag{11.08-9}$$

which can be compensated for either by retuning the load, or by moving the iris X_{12} slightly toward the load. Note that in Eq. (11.08-9) and in the equations in Fig. 11.08-5, the impedances are all normalized to the waveguide impedance, so that no difficulty arises from the ambiguity inherent in trying to define absolute values of waveguide impedance. Having computed the inverter parameters as given in Fig. 11.08-5, the

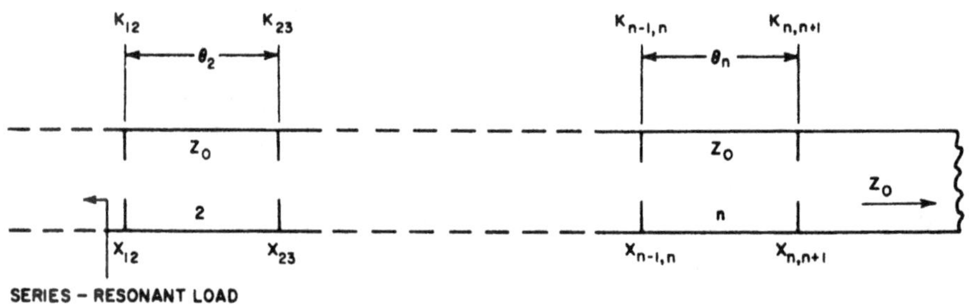

$$\frac{K_{12}}{Z_0} = \frac{1}{\omega_1'}\sqrt{\frac{\pi w_\lambda}{2g_1 g_2 \theta_\lambda}\left(\frac{R_A}{Z_0}\right)}, \quad \left.\frac{K_{k,k+1}}{Z_0}\right|_{k=2\text{ TO }n-1} = \frac{\pi w_\lambda}{2\omega_1'}\frac{1}{\sqrt{g_k g_{k+1}}}, \quad \frac{K_{n,n+1}}{Z_0} = \sqrt{\frac{\pi w_\lambda}{2g_n g_{n+1}\omega_1'}}$$

A-3527-427

FIG. 11.08-5 SUPPLEMENTARY DATA FOR USE WITH THE MATERIAL IN SEC. 8.06 FOR THE DESIGN OF WAVEGUIDE IMPEDANCE-MATCHING NETWORKS

remainder of the design process is as described in Sec. 8.06. Of course, in this case, the decrement δ_λ for use in determining the low-pass prototype and for use in the equation for K_{12}/Z_0, is defined as in Eq. (11.08-4).

The equations in Figs. 11.08-3 to 11.08-5 were obtained by simply inserting the appropriate slope parameter values in the generalized equations in Figs. 11.08-1 and 11.08-2. In a similar fashion design equations can be obtained for a wide variety of impedance-matching structures. The best structure to use in a given situation may vary widely, depending on such considerations as the impedance level of the load as compared to that of the source, the allowable size, the fractional bandwidth required, etc. As mentioned above, the resonated load should be kept in as compact a form as possible if best results are desired. Also, for best results the inverter coupling to the load should be of nearly lumped-element form. A quarter-wavelength of line could be used in relatively narrow-band cases as an inverter for coupling to the load, but the performance achieved would not be as good as when a lumped- or nearly lumped-element coupling is used. This is because the quarter-wavelength line itself has selectivity effects which would add to those of the resonated load and thus make impedance matching over a specified band more difficult. However, in some relatively non-critical situations where the resulting depreciation in performance can be tolerated, a quarter-wavelength coupling to the load may be a desirable practical compromise.

Experimental Adjustment of Coupled-Resonator Impedance-Matching Networks—In cases where one resonator can be added to the circuit at a time, and where the couplings between resonators and their tunings are readily adjustable, while the resonator slope parameters $\mathscr{x}_2$, $\mathscr{x}_3$, ..., $\mathscr{x}_n$ or $\mathscr{b}_2$, $\mathscr{b}_3$, ..., $\mathscr{b}_n$ are known, it may be convenient to determine the impedance-matching network design by a combination of theoretical and experimental procedures. Let us suppose that a load has been resonated in series as described above, and that its Q_A, its R_A, its decrement δ, and a desired $n = 2$ low-pass prototype have been determined. Assuming that the slope parameter $\mathscr{x}_2$ for Resonator 2 is known, the inverter parameters K_{12} and K_{23} are then computed by use of Fig. 11.08-1, and it is desired to adjust the resonator couplings to achieve the couplings indicated by these inverter parameters.

Figure 11.08-6 illustrates a procedure for experimental adjustment of such a filter. At (a) is shown the resonated load connected directly to the desired driving generator. The VSWR between the generator and load is then V_1 at resonance as given in the figure. Next, the coupling reactance associated with the inverter K_{12} is added, and it is adjusted

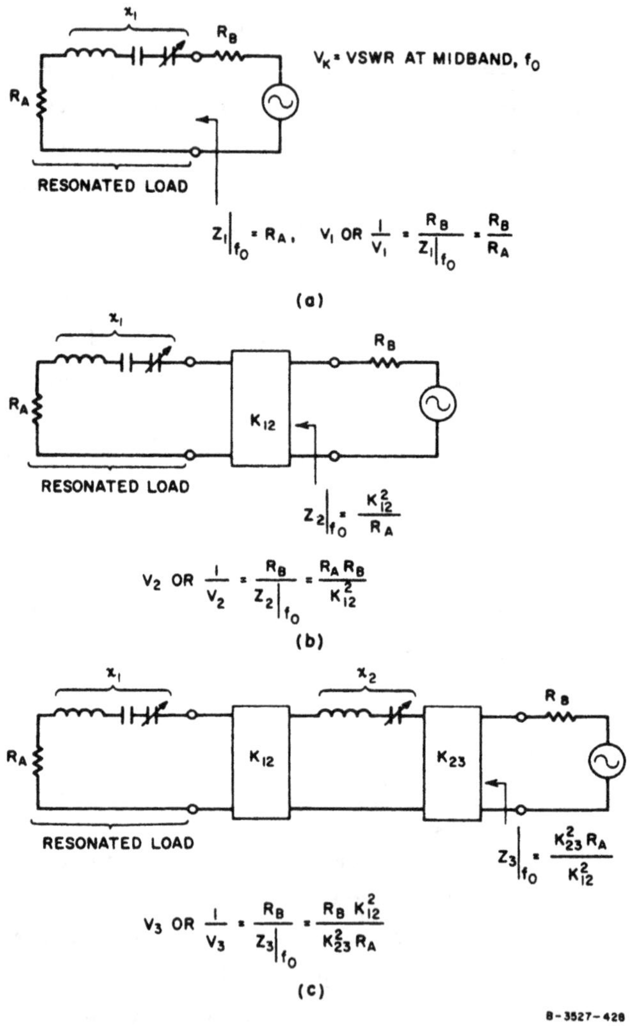

FIG. 11.08-6 PROCEDURE FOR EXPERIMENTAL ADJUSTMENT OF AN IMPEDANCE MATCHING NETWORK

until the VSWR at resonance is V_2, as indicated at (b) in the figure. Assuming that a slotted line is being used for the tests, the positions of the voltage minima must be checked to be sure that this V_2 corresponds to the desired $Z_2(\omega_0)$ and not its reciprocal (with respect to the slotted line impedance). Adding the coupling reactance associated with K_{12} will probably throw the load slightly off of the desired resonant frequency, so after the desired impedance is obtained at resonance, the load should be retuned. As long as V_2 is always measured at the frequency of resonance (no matter whether it is exactly the desired f_0 or not), the small correction required later in the load tuning should have negligible effect on the required coupling reactance.

Next, Resonator 2 along with the coupling reactance associated with the inverter K_{23} is added to the circuit as shown in Fig. 11.08-6(c). In this case, it is necessary both to bring Resonator 2 to the proper resonant frequency and to adjust the coupling reactance associated with K_{23} in order to achieve the desired impedance-matching effect. The tuning of Resonator 1 (*i.e.*, the resonated load) can be assumed to be very nearly correct so that any mistuning evident in the VSWR characteristic of the circuit must be due mainly to mistuning of Resonator 2. If Resonator 2 is correctly tuned to f_0, the VSWR characteristic will have either a maximum or a minimum value at f_0, and the characteristic will be symmetrical about f_0 (except possibly for some distortion of the frequency scale as a result of the fact that the coupling between resonators varies with frequency). Thus, the tuning of Resonator 2, and the coupling associated with the inverter K_{23} should be adjusted until a VSWR characteristic symmetrical about f_0 is obtained with a VSWR at f_0 equal to the value V_3 corresponding to the impedance $Z_3|_{f_0}$, computed from the data at (c) in Fig. 11.08-6. Once again, the locations of the voltage minima on the slotted line should be checked to be sure that this V_3 corresponds to $Z_3|_{f_0}$ and not its reciprocal with respect to the slotted line impedance.

This would complete the adjustment of an $n = 2$ matching network. The same procedure can be applied to the adjustment of impedance-matching networks of any complexity by repetition of part of the same steps. Of course, if the resonators are in shunt, the same procedure applies, but the analysis is on the dual basis.

A Practical Example and Some Additional Experimental Techniques—A practical example will now be considered. The circuit to be discussed was adjusted by experimental procedures as described above, but some additional techniques were involved that will also be explained. The main difference was that only the Q of the load was known, and its midband impedance was not known since all of the measurements on the load were made through a coupling reactance of unknown value. As will be seen, it was, nevertheless, still possible to carry out the required adjustments.

Figure 11.08-7 shows the circuit under consideration. It is a tunable up-converter in strip-line form.[6,7] A pumped, variable-capacitance diode provides both gain and frequency conversion. This device has a narrow-band, lower-sideband output circuit centered at 4037 Mc; hence, the response of the over-all circuit is narrow band. However, by varying the pump frequency from 4801 Mc to 5165 Mc the input frequency which

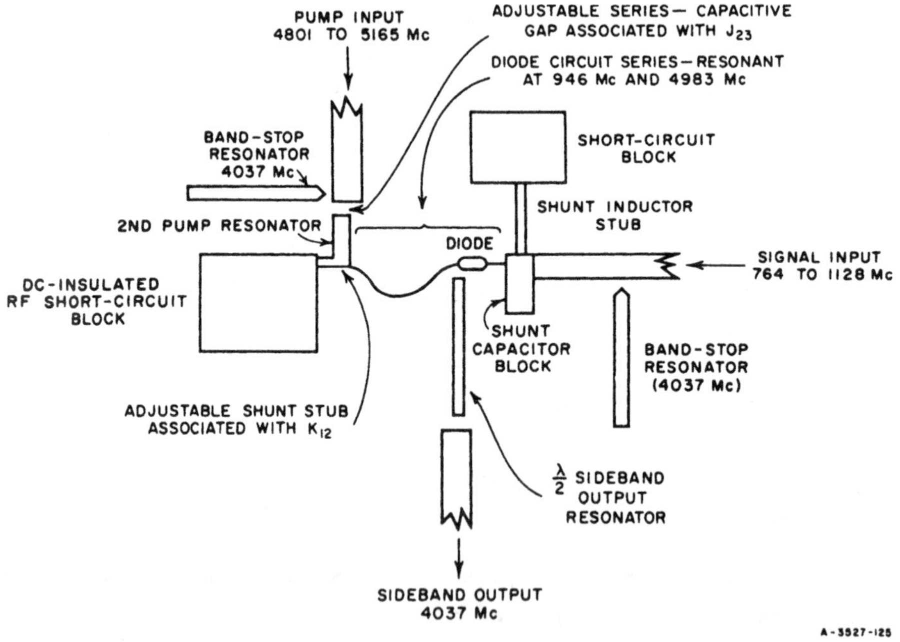

SOURCE: Quarterly Progress Report 2, Contract DA 36-039 SC-87398, SRI; reprinted in *Proc. IRE* (see Ref. 7 by G. L. Matthaei)

FIG. 11.08-7 SIMPLIFIED DRAWING OF A STRIP-TRANSMISSION-LINE TUNABLE UP-CONVERTER

would be accepted and amplified varies from 764 to 1128 Mc. In order to achieve the desired performance, impedance-matching filters are required at both the signal input and pump input ports. Herein, we will review the steps that were taken to broadband the pump input channel.

When the pump circuit of this device was adjusted, the portions of the circuit that were electrically of most importance were the diode plus its 0.020-inch diameter, 207-ohm wire lead (which together constituted the resonant load in this case), and the second pump resonator (see Fig. 11.08-7). The other portions of the circuit were greatly decoupled at the frequency range of interest. The second pump resonator was of the quarter-wavelength type in Fig. 11.08-4 having shunt-inductive coupling to the series-resonant diode circuit (which served as the first pump resonator), and having series-capacitance coupling to the pump input line. The second pump resonator shown in Fig. 11.08-7 had different cross-sectional proportions than the pump input line, but both were of 50 ohms impedance.

Figure 11.08-8 shows an approximate equivalent circuit for the pump channel of this device at frequencies around 5000 Mc. Though the load circuit (*i.e.*, the diode circuit) is actually quite complex, it behaves like a simple series-resonant load in the frequency range of interest. The impedance inverter K_{12} and the admittance inverter J_{23} relate to coupling discontinuities as indicated in Fig. 11.08-7.* Note that in this circuit the pump power is all ultimately delivered to the diode parasitic loss resistor R_s. Since R_s is quite small compared to the reactances involved, the impedance-matching problem is relatively severe.

For the circuit in Fig. 11.08-7, the shunt stub associated with K_{12} was set to a length estimated to be about right when the device was first assembled. To adjust the circuit, the capacitive gap associated with J_{23} was closed so that there was, in effect, a 50-ohm line right up to the K_{12} discontinuity. (This was equivalent to removing the J_{23} box in Fig. 11.08-8 to leave a 50-ohm line right up to K_{12}.) Then with a slotted line connected to the pump input port of the device, VSWR tests were made to determine the Q, Q_A, of the

---

*Inverters K_{12} and J_{23} are approximately of the forms in Figs. 8.03-1(c) and 8.03-2(d), respectively. Note that though the second pump resonator is of the quarter-wavelength type, its physical length will be considerably less than a quarter wavelength due to the negative lengths of line associated with the inverters. See, for example Eq. (4) of Fig. 8.08-1.

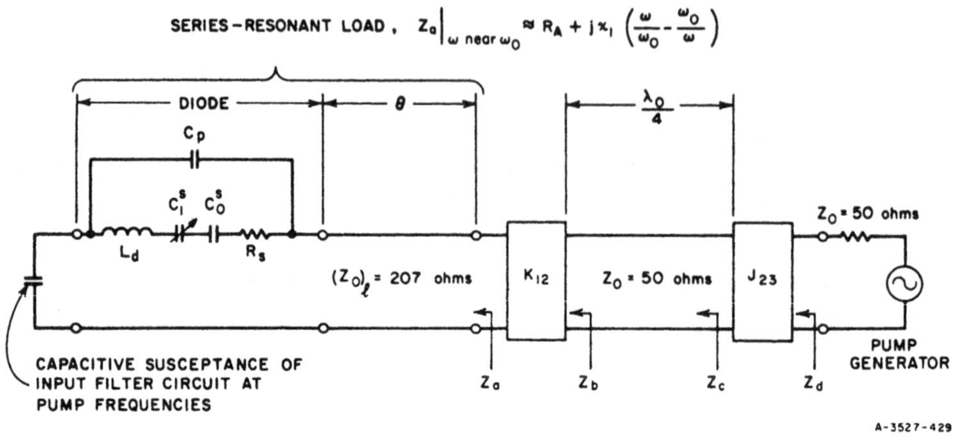

FIG. 11.08-8 APPROXIMATE EQUIVALENT CIRCUIT FOR THE PUMP INPUT CHANNEL OF THE CIRCUIT IN FIG. 11.08-7 FOR FREQUENCIES IN THE VICINITY OF 5000 Mc

diode-circuit at its pump-frequency resonance. Using the methods described in Sec. 11.02 the Q_A of the pump circuit was found to be 77. (Note that Q_A for the diode circuit corresponds to the unloaded Q as measured by techniques of Sec. 11.02, and that this Q is independent of the adjustment of the K_{12} shunt stub.)

The desired fractional bandwidth was specified to be w = 0.082, so that the decrement of the load was

$$\delta = \frac{1}{wQ_A} = \frac{1}{0.082(77)} = 0.16 \quad .$$

According to Figs. 4.09-3 and 4.09-4, an $n = 2$ resonator design with $\delta = 0.16$ calls for a maximum transducer loss of 3.3 db with 0.84-db ripple in the operating band. Since this was satisfactory for the application under consideration, a larger value of n was not considered. The corresponding optimum low-pass prototype element parameters were then from Fig. 4.09-6 found to be $g_0 = 1$, $g_1 = 6.30$, $g_2 = 0.157$, $g_3 = 6.30$, and $\omega_1' = 1$.

By Fig. 11.08-4,

$$K_{12} = \frac{Z_0}{\omega_1'} \sqrt{\frac{\pi R_A w}{4Z_0 g_1 g_2 \delta}} \qquad (11.08\text{-}10)$$

and

$$J_{23} = Y_0 \sqrt{\frac{\pi w}{4g_2 g_3 \omega_1'}} \quad , \qquad (11.08\text{-}11)$$

where $Y_0 = 1/Z_0$. All of the parameters needed to evaluate J_{23} were known, but K_{12} could not be evaluated because R_A was not known. (As previously mentioned, when Q_A was measured the K_{12} stub was set at some length estimated to be about right, but the length of the stub was not necessarily correct. It was not necessary to know K_{12} in order to determine Q_A by the methods of Sec. 11.02. However, the K_{12} parameter of the coupling stub would have to be known accurately in order to determine R_A.) This difficulty was bypassed by eliminating both K_{12} and R_A from the calculations in the following manner.

In Fig. 11.08-8, the impedance Z_b is at midband

$$Z_b \Big|_{f_0} = \frac{K_{12}^2}{R_A} \quad . \qquad (11.08\text{-}12)$$

Then from Eqs. (11.08-10) and (11.08-12)

$$Z_b \Big|_{f_0} = \frac{Z_0 \pi w}{(\omega_1')^2 4 g_1 g_2 \delta} \qquad (11.08\text{-}13)$$

$$= \frac{50\pi(0.082)}{(1)^2 4(6.30)(0.157)(0.16)} = 20.4 \text{ ohms.}$$

Thus, with the J_{23} gap closed so as to give a 50-ohm line up to the K_{12} stub, the length of the K_{12} stub was adjusted so that the VSWR seen looking in the pump input port at resonance was

$$V_b = \frac{Z_0}{Z_b\Big|_{f_0}} \tag{11.08-14}$$

$$= \frac{50}{20.4} = 2.45$$

while the vicinity of the K_{12} stub was a voltage minimum of the standing-wave pattern *vs.* length.

Next the J_{23} gap in Fig. 11.08-7 was opened and adjusted so as to give a VSWR of $V_d = 6.30$ at midband. For each trial adjustment of the gap spacing, tuning screws near the capacitive-gap end of the second pump resonator were adjusted to give a VSWR response *vs.* frequency which was approximately symmetrical about f_0. The $V_d = 6.30$ value can be computed as follows.

In Fig. 11.08-8 at midband

$$Z_c\Big|_{f_0} = \frac{Z_0^2}{Z_b\Big|_{f_0}} \quad . \tag{11.08-15}$$

Since J_{23} is an admittance inverter (Sec. 8.03)

$$Z_d\Big|_{f_0} = \frac{1}{Z_c\Big|_{f_0} J_{23}^2} = \frac{Z_b\Big|_{f_0}}{Z_0^2 J_{23}^2} \quad . \tag{11.08-16}$$

Then by Eqs. (11.08-11), (11.08-13), and (11.08-16),

$$Z_d\Big|_{f_0} = Z_0 g_0 g_3 \tag{11.08-17}$$

$$= 50(1)(6.30) = 315 \text{ ohms.}$$

Thus, the VSWR seen on the 50-ohm pump input line at midband should be

$$Z_d = \frac{Z_d\big|_{f_0}}{Z_0} \qquad (11.08\text{-}18)$$

$$= \frac{315}{50} = 6.30$$

where the vicinity of the J_{23} gap should be a voltage maximum of the standing-wave pattern *vs.* length.

This completed the adjustment of the pump impedance-matching circuit, and the resulting transducer attenuation* characteristic as computed from measured VSWR is shown in Fig. 11.08-9. The measured fractional bandwidth is 0.0875 as compared to the 0.082 design objective, the 3.2-db peak pass-band loss compares favorably with the 3.3 db predicted value, while the pass-band ripple is around 0.4 db as compared with the 0.84-db predicted value. The measured performance is somewhat superior to the predicted performance because when the data in Fig. 11.08-9 was taken the lower-sideband output resonator had been added to the circuit

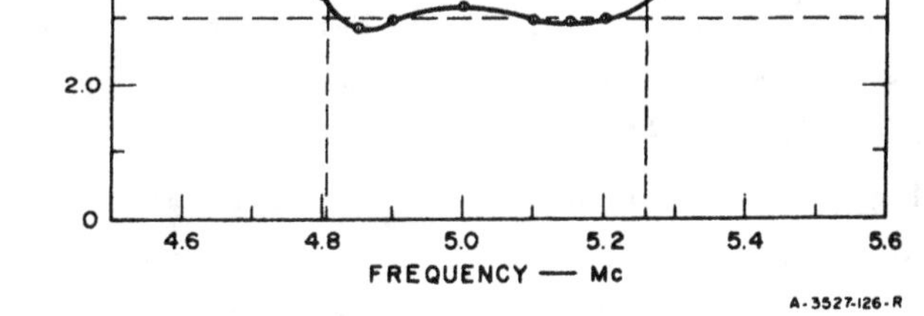

FIG. 11.08-9 TRANSDUCER ATTENUATION AT PUMP INPUT PORT AS COMPUTED FROM MEASURED VSWR

---

* See Sec. 2.11 for a discussion of this term.

which caused the effective Q_A of the diode circuit to drop a little from the Q_A = 77 value which was originally obtained without the output resonator.*

It is interesting to consider just how much has been accomplished by the addition of the second pump resonator as a matching network. Let us again assume that Q_A = 77 and that the desired fractional bandwidth is w = 0.082 so that δ = 0.16. If only the first pump resonator is used (*i.e.*, just the resonant diode circuit along with the K_{12} adjustable coupling stub) and if the K_{12} coupling stub were adjusted so as to give a perfect match to the 50-ohm pump input line at midband, then *the reflection loss at the band edges would be about 10.9 db.* If the K_{12} coupling stub were adjusted in accordance with an optimum n = 1 design, as determined using the prototype data in Sec. 4.09, the loss at the band edges would be $(L_A)_{max}$ = 5.6 db while the loss at midband would be $(L_A)_{min}$ = 3.3 db. Adding an additional resonator to give an n = 2 design reduces the pass-band loss limits to $(L_A)_{max}$ = 3.3 db and $(L_A)_{min}$ = 2.5 db. An n = 4 design would give $(L_A)_{max}$ = 2.5 db and $(L_A)_{min}$ = 2.3 db, while an $n = \infty$ design would give $(L_A)_{max} = (L_A)_{min}$ = 2.0 db. It is seen that going from an n = 1 to an n = 2 design gives a very large improvement, and that the incremental improvement for increasing values of n decreases very rapidly.

Other Forms of Coupled-Resonator Matching Networks—It would also be possible to construct a matching network in the form shown in Fig. 11.08-10, which has no inverter between the load and the adjacent matching-network resonator. This, however, is usually not very practical except in the cases of relatively wide-band matching networks. The slope parameter of Resonator 2 cannot be chosen arbitrarily in this case, and for narrow bandwidths $\mathscr{b}_2$ will often become so large as to be difficult to realize with practical circuitry. The dual case where the first resonator of the matching network is in series would be especially difficult.

The design procedures in Sec. 11.09 can give designs which are more or less equivalent to that in Fig. 11.08-10, but a different point of view is used. As indicated above, such designs become most practical for relatively large bandwidths, say, around 30 percent or more.

---

* This drop in diode circuit Q was due to the shunt capacitance introduced by the output resonator causing the diode internal resistance to be somewhat more tightly coupled.

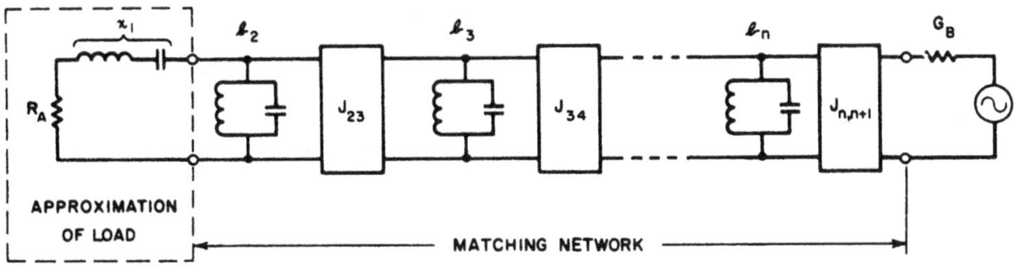

$$\mathscr{B}_2 = \frac{\omega_1' g_2}{w g_0 R_A}\,, \quad J_{k,k+1}\Big|_{k=2\text{ TO }n-1} = \frac{w}{\omega_1'}\sqrt{\frac{\mathscr{B}_k \mathscr{B}_{k+1}}{g_k\, g_{k+1}}}\,, \quad J_{n,n+1} = \sqrt{\frac{G_B\, \mathscr{B}_n\, w}{g_n\, g_{n+1}\, \omega_1'}}$$

WHERE $\mathscr{B}_3, \mathscr{B}_4, \cdots, \mathscr{B}_n$, AND G_B MAY BE CHOSEN ARBITRARILY

A-3527-430

FIG. 11.08-10 AN ADDITIONAL GENERAL FORM OF COUPLED-RESONATOR-FILTER IMPEDANCE-MATCHING NETWORK
The dual of this circuit applies when the load is shunt resonant

SEC. 11.09, DESIGN OF BAND-PASS IMPEDANCE-MATCHING NETWORKS BY THE METHODS OF CHAPTER 10

In this section, discussion of the design of impedance-matching networks will be continued so as to cover the design of matching networks of relatively large bandwidths. The underlying fundamentals are much the same as those discussed in Sec. 11.08, but the procedures are modified to use the design point of view of Chapter 10.

In the methods of this section the load is first brought to series- or shunt-resonance at midband f_0, just as was discussed in Sec. 11.08. However, since relatively large bandwidths are involved in this case it may be desirable to use something other than the Q of the load as a basis for computing the load decrement δ. This is because the Q of the load will typically be relatively low for cases where a wide-band match is required, and also because it may be desirable to examine the impedance or admittance characteristics of the load more thoroughly than is implied by a Q measurement alone.

Figure 11.09-1 shows a suggested procedure for computing the required load parameters δ and R_A or G_A. The impedance or admittance characteristics of the resonated load are measured across the frequency band of interest, and then the required parameters are readily determined from the measured data as indicated in the figures. Note that the frequencies f_1 and f_2

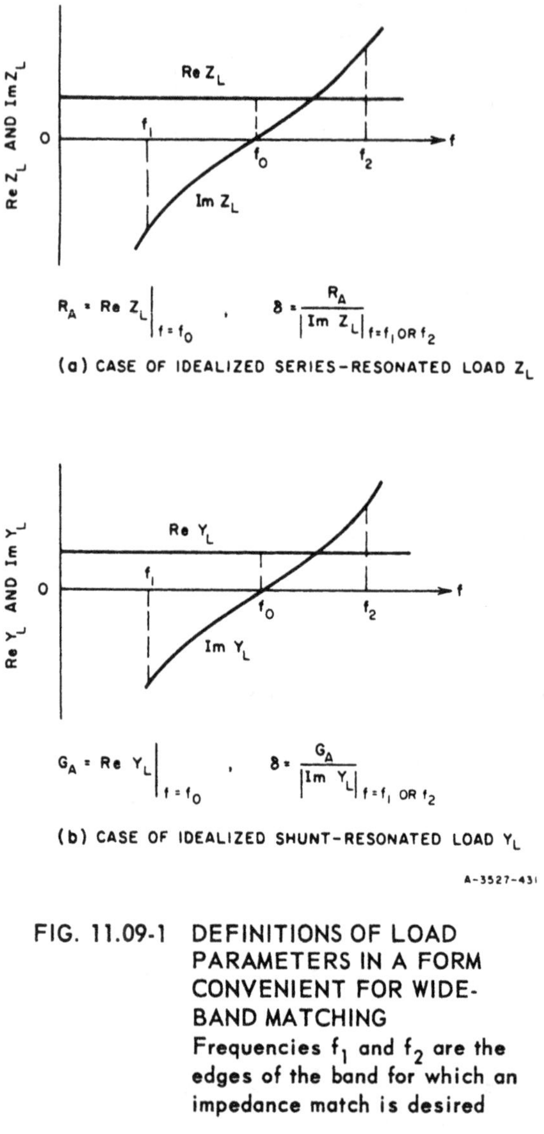

FIG. 11.09-1 DEFINITIONS OF LOAD PARAMETERS IN A FORM CONVENIENT FOR WIDE-BAND MATCHING
Frequencies f_1 and f_2 are the edges of the band for which an impedance match is desired

indicated in the figures are the edges of the frequency band over which a good impedance match is required. In the figure the characteristics are idealized in that the real parts are constant with frequency and the imaginary parts have odd symmetry about f_0 so that the magnitude of the imaginary part of the impedance or admittance is the same at f_1 as it is at f_2. Note that on this basis the decrement δ is simply the ratio of the real part at f_0 to the *magnitude* of the imaginary part at either f_1 or f_2. In practical cases where the load impedance or admittance characteristic may deviate appreciably from these idealized symmetrical characteristics, average values should be taken using the definitions in terms of the idealized cases as a guide. An example of a non-idealized case will be discussed later in this section.

After the parameters R_A or G_A, and the decrement δ have been determined for the load, an appropriate low-pass prototype filter should be selected to have the required δ value, as discussed in Sec. 4.09 and 4.10. The specified value of R_A or G_A, the prototype filter parameters, the specified internal resistance of the generator which is to drive the load, the specified fractional bandwidth w, and the specified center frequency f_0, then control the design of the matching network.

Figure 11.09-2 shows a form of impedance-matching network which can be designed by the methods of Chapter 10. The load as shown is of a form which would have an idealized symmetrical impedance characteristic of the form in Fig. 11.09-1(a); however, as an example will soon show, matching networks of the sort in the figure can be useful for considerably more complex loads with impedance characteristics which deviate considerably from the idealized symmetrical form. The matching network in Fig. 11.09-2 could be constructed in strip-line form such as that in the filter in Figs. 10.03-5(a), (b), and (c). Also, if desired, the matching network could be converted into parallel-coupled strip-line form. This could be accomplished by splitting each shunt stub Y_k (except for Y_2 and Y_n) into two parallel stubs Y'_k and Y''_k such that $Y'_k + Y''_k = Y_k$. Then the filter can be broken into open-wire-line sections such as that on the right side of Fig. 5.09-2(a), which are exactly equivalent to strip-line parallel-coupled sections such as that on the left side of Fig. 5.09-2(a). The parallel-coupled line section admittances are obtained by use of the relations in Fig. 5.09-2(a), and the line dimensions can be obtained from the line admittances by use of the data in Sec. 5.05. Whether the stub or parallel-coupled form is preferable would depend mainly on what dimensions were practical. The stub form would probably be convenient in most cases, though the parallel-coupled form has additional adjustability since it would not be difficult to

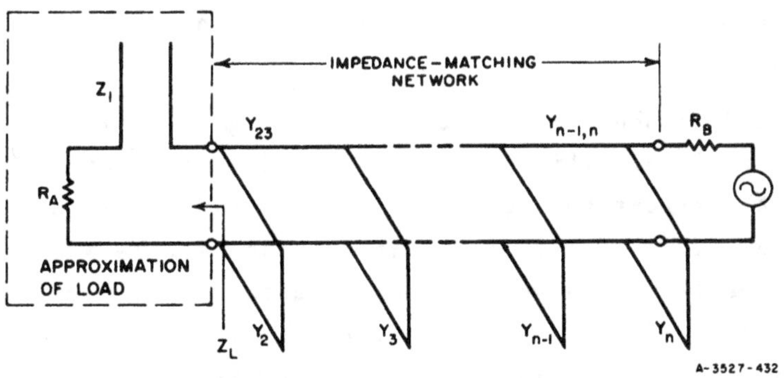

FIG. 11.09-2 IMPEDANCE-MATCHING NETWORK FORMED FROM STUBS AND CONNECTING LINES
The series stub Z_1, the shunt stubs Y_k, and the connecting lines $Y_{k,k+1}$ are all a quarter-wavelength long at midband

provide for experimental adjustment of the spacing between the parallel-coupled elements. This would be equivalent to adjusting the line admittances $Y_{k,k+1}$ in Fig. 11.09-2. Experimental adjustment of these line admittances might be desirable in cases where the load impedance is quite complicated and deviates greatly from the idealized symmetrical case in Fig. 11.09-1(a), so that some cut-and-try procedures are desirable.

Table 11.09-1 presents design equations for impedance-matching networks of the form in Fig. 11.09-2. These design equations were derived in essentially the same way as were the design equations for the filters in Secs. 10.03 and 10.05. (The derivations of the design equations in Secs. 10.03 and 10.05 were discussed in Sec. 10.08.) The main difference is that the more general modified low-pass prototype in Fig. 11.09-3 was used for the equations in Table 11.09-1. Comparing this prototype with the one in Fig. 10.08-6, note that the prototype in Fig. 11.09-3 does not use a series coil on the right, and provision has been made so that the termination R_{n+1} on the right can be specified arbitrarily. A multiplying factor s has been included which is introduced so as to alter the impedance level as gradually as possible from one side of the prototype to the other, by making $C_{k+1} = sC_k$ for $k = 2$ to $n - 2$. The relations between the prototype and the stub filter in Fig. 11.09-2 are much the same as for the cases discussed in Sec. 10.08. The parameter d in Table 11.09-1 can be chosen $0 < d$ so as to adjust the admittance level within the filter. Typical convenient values are $d = 0.5$ to 1.

Figure 11.09-4 shows an impedance-matching network for use with a shunt-resonated load. This circuit is the dual of that in Fig. 11.09-2, except that the filter has been converted to parallel-coupled form by use of the relations in Fig. 5.09-2(b). Since series stubs are difficult to realize in a shielded construction, the parallel-coupled form of this filter will usually be the practical one to use. Table 11.09-2 presents design equations for this type of matching network.*

---

*These design equations were obtained from those in Table 11.09-1 by duality. Using duality gave a matching network consisting of series stubs and connecting lines. Then this circuit was converted to parallel-coupled form using the relations in Fig. 5.09-2(b).

Table 11.09-1

DESIGN EQUATIONS FOR IMPEDANCE-MATCHING NETWORKS OF THE FORM IN FIG. 11.09-2 FOR SERIES-RESONATED LOADS

Determine the low-pass prototype parameters from the load decrement δ and the data in Secs. 4.09 and 4.10. The generator resistance R_B may be chosen arbitrarily, but R_A is a parameter of the load (see text). The following equations apply for $n \geq 3$.

Compute:

$$s = \sqrt[n-2]{\frac{R_A}{R_B}}$$

$$C_2 = g_2 \quad , \quad C_k \Big|_{\substack{k=3 \text{ to } n-1, \text{ if } n>3 \\ (\text{If } n=3 \text{ omit this equation.})}} = 2 d g_2 s^{k-2}$$

$$C_n \Big|_{n \geq 3} = g_0 g_n g_{n+1} \frac{R_A}{R_B} \quad ,$$

where d may be chosen arbitrarily within limit $0 < d$ (see text).

$$C_2' = g_2(1 - d) \quad , \quad C_2'' = d g_2$$

$$C_k' \Big|_{k=3 \text{ to } n} = C_{k-1}'' \quad , \quad C_k'' = C_k - C_k' \quad .$$

Letting

$$G_A = \frac{1}{R_A} \quad ,$$

$$\frac{J_{k,k+1}}{G_A} \Big|_{k=2 \text{ to } n-1} = \frac{1}{g_0} \sqrt{\frac{C_k C_{k+1}}{g_k g_{k+1}}}$$

$$N_{k,k+1} \Big|_{k=2 \text{ to } n-1} = \sqrt{\left(\frac{J_{k,k+1}}{G_A}\right)^2 + \left(\frac{C_k'' \tan \theta_1}{g_0}\right)^2}$$

where

$$\theta_1 = \frac{\pi}{2}\left(1 - \frac{w}{2}\right)$$

and $w = (f_2 - f_1)/f_0$ is the desired fractional bandwidth.

(continued on p. 704)

The characteristic admittances of the shunt stubs are:

$$Y_2 = \frac{G_A}{g_0} \omega_1' C_2' \tan \theta_1 + G_A \left(N_{23} - \frac{J_{23}}{G_A} \right)$$

$$Y_k \Big|_{k=3 \text{ to } n-1, \text{ if } n>3} = G_A \left(N_{k-1,k} + N_{k,k+1} - \frac{J_{k-1,k}}{G_A} - \frac{J_{k,k+1}}{G_A} \right)$$

$$Y_n \Big|_{n \geq 3} = \frac{G_A}{g_0} \omega_1' C_n'' \tan \theta_1 + G_A \left(N_{n-1,n} - \frac{J_{n-1,n}}{G_A} \right) .$$

The characteristic admittances of the connecting lines are:

$$Y_{k,k+1} \Big|_{k=2 \text{ to } n-1} = G_A \left(\frac{J_{k,k+1}}{G_A} \right) .$$

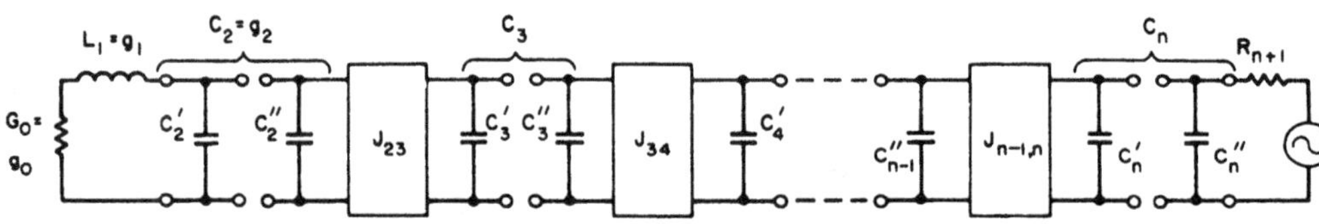

$$s = \sqrt[n-2]{\frac{1}{G_0 R_{n+1}}}$$ WHERE R_{n+1} IS ARBITRARY , $n \geq 3$,

$$C_2 = g_2 \ , \ C_k \Big|_{k=3 \text{ TO } n-1} = 2 d g_2 \ s^{k-2} \ , \ C_n = \frac{g_n g_{n+1}}{R_{n+1}} \ ,$$

$$C_2' = g_2 (1-d) \ , \ C_2'' = d g_2 \ , \ C_k' \Big|_{k=3 \text{ TO } n} = C_{k-1}'' \ , \ C_k'' = C_k - C_k' \ ,$$

WHERE d MAY BE CHOSEN ARBITRARILY WITHIN LIMIT 0<d.

$$J_{k,k+1} \Big|_{k=2 \text{ TO } n-1} = \sqrt{\frac{C_k C_{k+1}}{g_k g_{k+1}}}$$

A-3527-434

FIG. 11.09-3 MODIFIED LOW-PASS PROTOTYPE USED IN DERIVING THE EQUATIONS IN TABLE 11.09-1

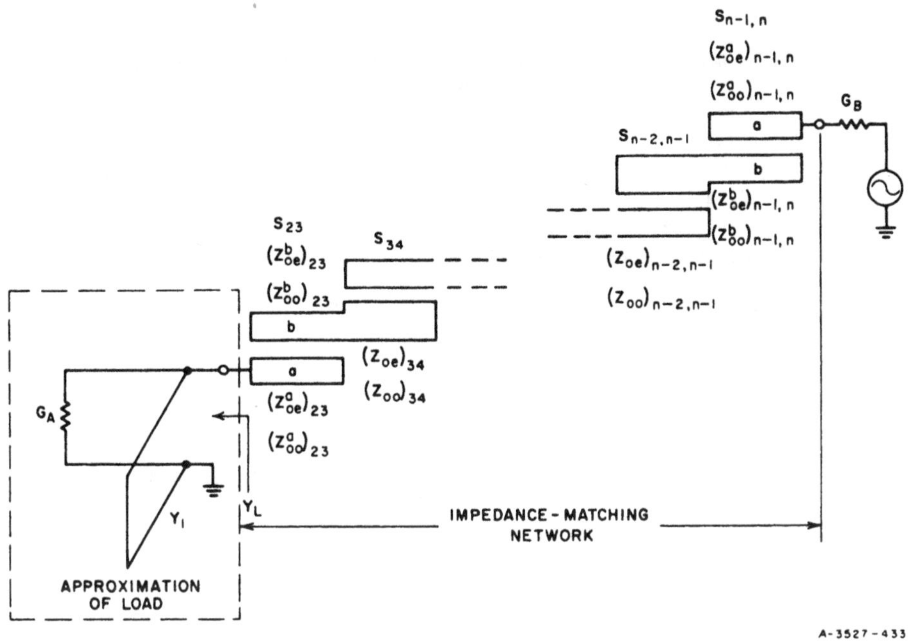

FIG. 11.09-4 IMPEDANCE-MATCHING NETWORK IN PARALLEL-COUPLED FORM FOR MATCHING SHUNT-RESONATED LOAD
The stub Y_1 and the parallel-coupled sections $S_{k,k+1}$ are all a quarter-wavelength long at midband

Let us now consider some examples of impedance-matching network design. Figure 11.09-5 shows a lumped-element load network which has been brought to resonance at f_0, and the figure also shows the resulting computed impedance characteristic. Let us assume that a good match is desired from $f_1/f_0 = 0.85$ to $f_2/f_0 = 1.15$, which calls for fractional bandwidth of $w = 0.30$. From Figs. 11.09-1(a) and 11.09-5, $R_A = ReZ_L|_{f_0} = 33.33$ ohms. However, $|ImZ_L|_{f_1} = 110$ ohms, while $|ImZ_L|_{f_2} = 95$ ohms, so there is some confusion about the calculations of δ. As a guess, an in-between value of 100 ohms was used for some trial calculations so that*

$$\delta = \frac{33.33}{100} \approx 0.3 \quad .$$

Table 11.09-3 summarizes the parameters and resulting line admittances for several $n = 3$ designs having different values of R_B or different choices for the parameter d. Note that Designs 1 and 2 both use $R_B = 100$ ohms, but Design 1 uses $d = 1$ while Design 2 uses $d = 0.5$.

* A useful significant figure was inadvertently dropped in these design calculations.

Table 11.09-2

DESIGN EQUATIONS FOR IMPEDANCE-MATCHING NETWORKS OF THE FORM IN FIG. 11.09-4 FOR SHUNT-RESONATED LOADS

Determine the low-pass prototype parameters from the load decrement δ and the data in Secs. 4.09 and 4.10. The generator conductance G_B may be chosen arbitrarily, but G_A is a parameter of the load (see text). The following equations cover cases where $n \geqq 3$.

Compute:

$$s = \sqrt[n-2]{\frac{G_A}{G_B}}$$

$$L_2 = g_2 \quad , \quad L_k \Bigg|_{\substack{k=3 \text{ to } n-1, \text{ if } n>3 \\ (\text{If } n=3 \text{ omit this equation.})}} = 2dg_2 s^{k-2}$$

$$L_n \Big|_{n \geqq 3} = g_0 g_n g_{n+1} \frac{G_A}{G_B} \quad ,$$

where d may be chosen arbitrarily within limit $0 < d$ (see text).

$$L_2' = g_2(1 - d) \quad , \quad L_2'' = dg_2$$

$$L_k' \Big|_{k=3 \text{ to } n} = L_{k-1}'' \quad , \quad L_k'' = L_k - L_k' \quad .$$

Letting

$$R_A = \frac{1}{G_A} \quad ,$$

$$\frac{K_{k,k+1}}{R_A} \Bigg|_{k=2 \text{ to } n-1} = \frac{1}{g_0} \sqrt{\frac{L_k L_{k+1}}{g_k g_{k+1}}}$$

$$N_{k,k+1} \Bigg|_{k=2 \text{ to } n-1} = \sqrt{\left(\frac{K_{k,k+1}}{R_A}\right)^2 + \left(\frac{L_k'' \tan \theta_1}{g_0}\right)^2}$$

where

$$\theta_1 = \frac{\pi}{2}\left(1 - \frac{w}{2}\right)$$

and $w = (f_2 - f_1)/f_0$ is the desired fractional bandwidth.

Table 11.09-2 *concluded*

The even- and odd-mode impedances of the parallel-coupled lines are:

If $n = 3$

$$(Z^a_{oe})_{23} = \frac{R_A \omega_1' L_2'}{g_0} \tan \theta_1 + R_A \left(N_{23} + \frac{K_{23}}{R_A}\right)$$

$$(Z^a_{oo})_{23} = \frac{R_A \omega_1' L_2'}{g_0} \tan \theta_1 + R_A \left(N_{23} - \frac{K_{23}}{R_A}\right)$$

$$(Z^b_{oe})_{23} = \frac{R_A \omega_1' L_3''}{g_0} \tan \theta_1 + R_A \left(N_{23} + \frac{K_{23}}{R_A}\right)$$

$$(Z^b_{oo})_{23} = \frac{R_A \omega_1' L_3''}{g_0} \tan \theta_1 + R_A \left(N_{23} - \frac{K_{23}}{R_A}\right) .$$

If $n \geq 4$,

$$(Z^a_{oe})_{23} = \frac{R_A}{g_0} \omega_1' L_2' \tan \theta_1 + R_A \left(N_{23} + \frac{K_{23}}{R_A}\right)$$

$$(Z^a_{oo})_{23} = \frac{R_A}{g_0} \omega_1' L_2' \tan \theta_1 + R_A \left(N_{23} - \frac{K_{23}}{R_A}\right)$$

$$(Z^b_{oe})_{23} = R_A \left(N_{23} + \frac{K_{23}}{R_A}\right)$$

$$(Z^b_{oo})_{23} = R_A \left(N_{23} - \frac{K_{23}}{R_A}\right)$$

$$(Z_{oe})_{k,k+1} \Bigg|_{\substack{k=3 \text{ to } n-2 \\ (\text{Omit if } n < 5)}} = R_A \left(N_{k,k+1} + \frac{K_{k,k+1}}{R_A}\right)$$

$$(Z_{oo})_{k,k+1} \Bigg|_{\substack{k=3 \text{ to } n-2 \\ (\text{Omit if } n < 5)}} = R_A \left(N_{k,k+1} - \frac{K_{k,k+1}}{R_A}\right)$$

$$(Z^a_{oe})_{n-1,n} = \frac{R_A}{g_0} \omega_1' L_n'' \tan \theta_1 + R_A \left(N_{n-1,n} + \frac{K_{n-1,n}}{R_A}\right)$$

$$(Z^a_{oo})_{n-1,n} = \frac{R_A}{g_0} \omega_1' L_n'' \tan \theta_1 + R_A \left(N_{n-1,n} - \frac{K_{n-1,n}}{R_A}\right)$$

$$(Z^b_{oe})_{n-1,n} = R_A \left(N_{n-1,n} + \frac{K_{n-1,n}}{R_A}\right)$$

$$(Z^b_{oo})_{n-1,n} = R_A \left(N_{n-1,n} - \frac{K_{n-1,n}}{R_A}\right)$$

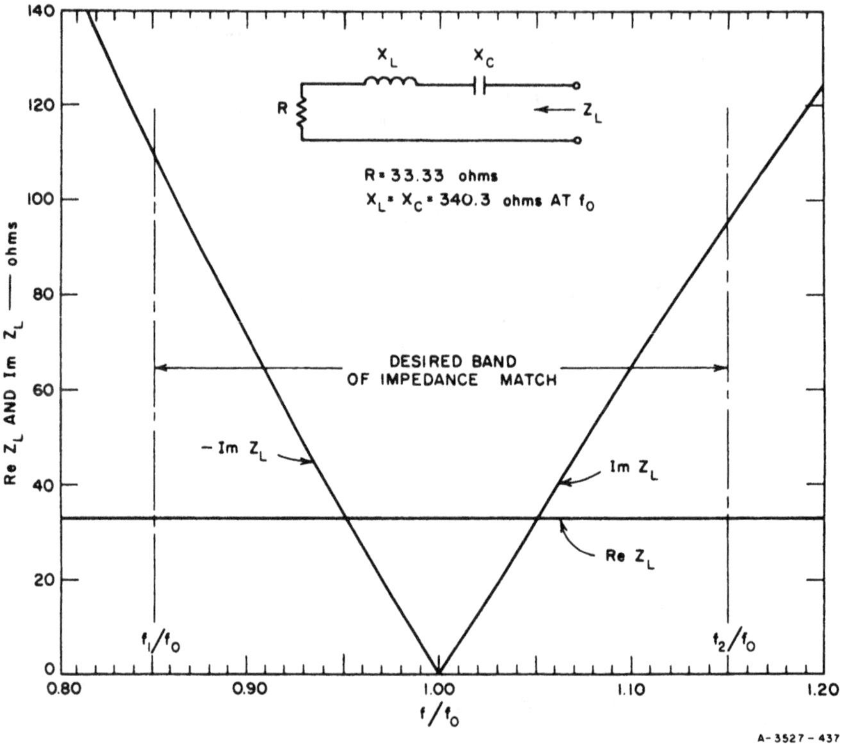

FIG. 11.09-5 A HYPOTHETICAL LOAD AND ITS COMPUTED IMPEDANCE

Table 11.09-3

IMPEDANCE-MATCHING NETWORK DESIGNS 1 to 4 FOR THE LOAD IN FIG. 11.09-5
The impedance-matching network is of the form in Fig. 11.09-2

For all designs below: $w = 0.30$, $R_A = 33.33$ ohms, $\delta = 0.30$; and for $n = 3$ Sec. 4.09 gives $g_0 = 1$, $g_1 = 3.30$, $g_2 = 0.52$, $g_3 = 2.10$, $g_4 = 0.40$ $\omega_1' = 1$, $(L_A)_{max} = 1.2$ db, and $(L_A)_{min} = 0.9$ db.

Design	d	R_B (ohms)	Y_2 (mhos)	Y_{23} (mhos)	Y_3 (mhos)
1	1.0	100	0.05494	0.01095	0.02495
2	0.5	100	0.05582	0.01095	0.02583
3	1.0	20	0.04495	0.02449	0.1549
4	1.0	500	0.06026	0.004899	0.002283

Within two significant figures the line admittances Y_2, Y_{23}, and Y_3 of Designs 1 and 2 are very nearly the same. This is to be expected since the choice of d should, in theory, not affect the design unless there are more than two resonators in the matching-network portion of the circuit. (In this case, the matching-network portion has only two resonators.)

Figure 11.09-6 shows the computed response of the load in Fig. 11.09-5 connected to matching-network Design 1 of Table 11.09-3. The dashed curve is for the load as shown in Fig. 11.09-5 without any retuning. Note that the attenuation (*i.e.*, transducer attenuation*) curve tilts a little to the right because of the dissymmetry of the reactance curve of the load. The solid curve in Fig. 11.09-6 shows the result of retuning the load by increasing the midband inductive reactance X_L from 340.3 ohms to 348.3 ohms. Note that the effect is to correct (actually overcorrect) the tilt in the response. By Figs. 4.09-3 and 4.09-4 the

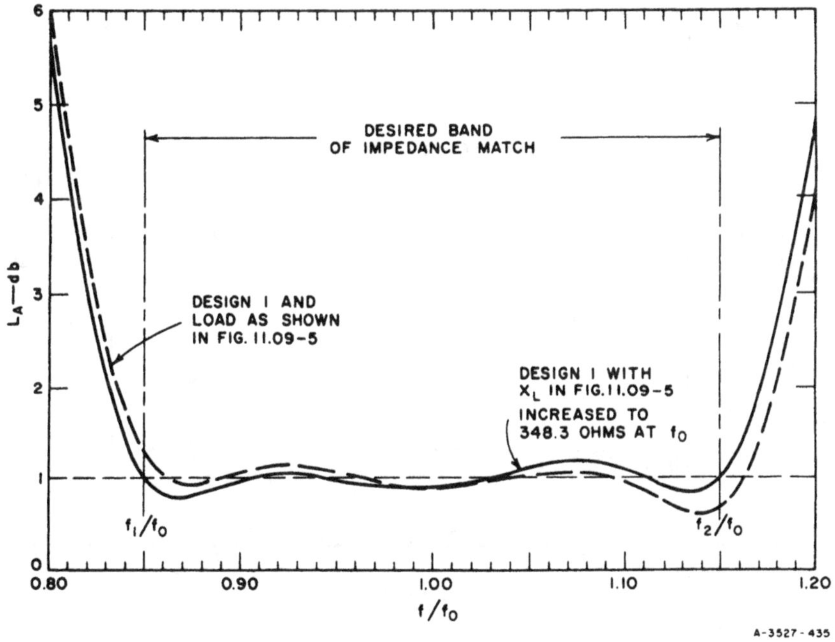

FIG. 11.09-6 TRANSDUCER ATTENUATION INTO THE LOAD IN FIG. 11.09-5 USING MATCHING-NETWORK DESIGN 1

---

* See Sec. 2.11 for a discussion of this term.

peak pass-band attenuation into the load should be 1.2 db for this case, while the minimum pass-band attenuation should be about 0.9 db. As can be seen from Fig. 11.09-6, the pass-band ripple characteristic comes quite close to meeting these values.

Figure 11.09-7 shows the computed attenuation characteristics for the same retuned load as described above (*i.e.*, with X_L = 348.3 ohms), but using matching-network Designs 3 and 4. Note that for Design 3 the generator resistance is R_B = 20 ohms, while for Design 4, R_B = 500 ohms. Thus, these two designs cover a range of choice of generator impedance of 25 to 1. As can be seen from the figure, the choice of R_B had no serious effect on the performance.

Figure 11.09-7 also shows the performance when no matching network is used but R_B is chosen to match the load impedance at resonance. Note

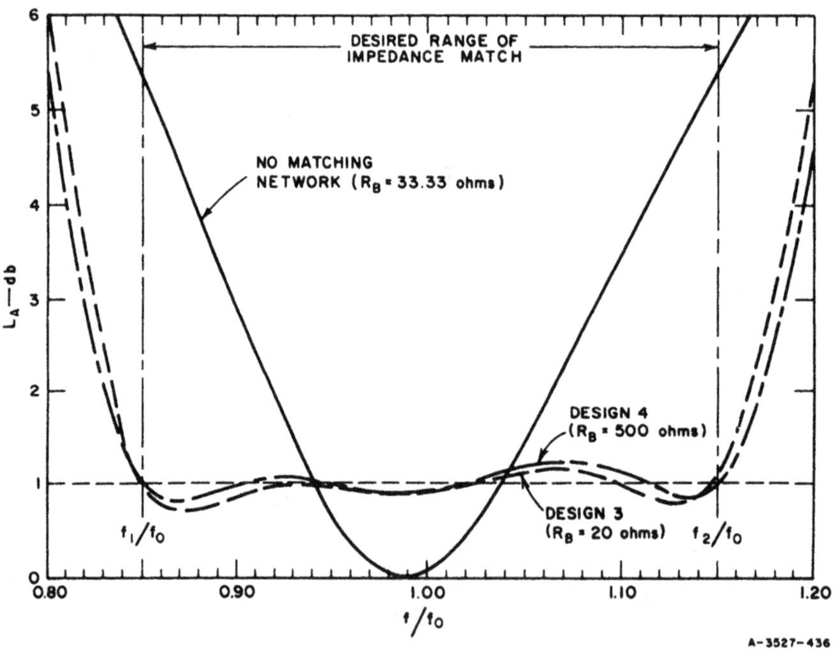

FIG. 11.09-7 TRANSDUCER ATTENUATION INTO THE LOAD IN FIG. 11.09-5 USING MATCHING-NETWORK DESIGNS 3 AND 4, AND ALSO FOR CASE OF NO MATCHING-NETWORK

In all cases shown the load is as in Fig. 11.09-5 except that X_L has been increased to X_L = 348.3 ohms at f_0

that the attenuation at the band edges in this case is 5.4 db as compared with around 1.2-db maximum attenuation across the 30 percent band when an n = 3 matching network is used. By Fig. 4.09-3, for a load with δ = 0.30, the best possible pass-band performance even with an infinite number of impedance-matching resonators would have a pass-band attenuation of 0.72 db.

Figure 11.09-8 shows another series-resonated load whose impedance characteristic deviates quite markedly from the idealized symmetrical characteristic in Fig. 11.09-1(a). Suppose that we wish to match over a 40-percent band, ReZ_L varies over a 2 to 1 range in the band of interest, while the reactance characteristic is also unsymmetrical. In this case $R_A = ReZ_L|_{f_0} = 20$ ohms, while $|ImZ_L|_{f_1} = 80$ ohms and $|ImZ_L|_{f_2} = 67$ ohms. Averaging these latter two values gives 73.5 ohms.

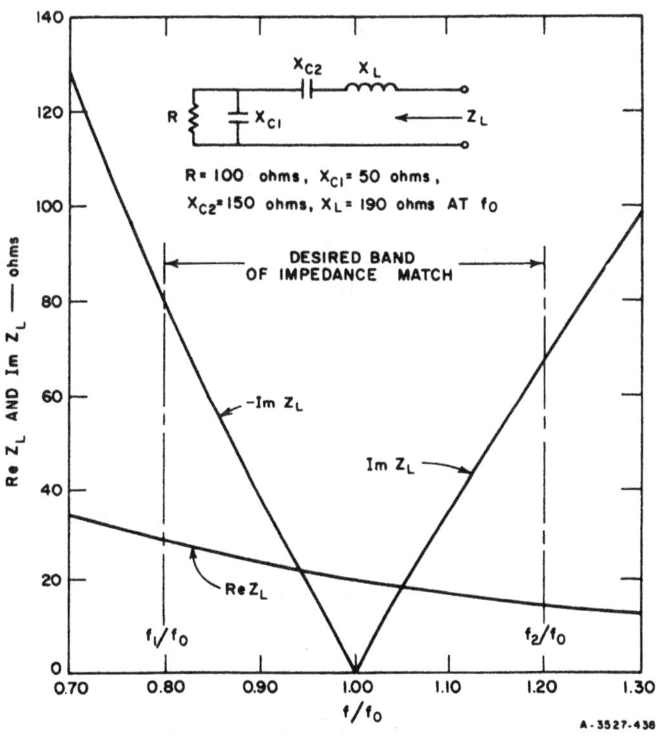

FIG. 11.09-8 A SERIES-RESONATED LOAD WHICH DEVIATES MARKEDLY FROM THE IDEALIZED CASE IN FIG. 11.09-1(a)

Thus, an estimated decrement value is

$$\delta = \frac{20}{73.5} = 0.272 \quad .$$

Table 11.09-4 summarizes the parameters of an $n = 3$ matching-network design for this load.

Table 11.09-4

IMPEDANCE-MATCHING NETWORK DESIGN 5 FOR THE LOAD IN FIG. 11.09-8

The impedance-matching network is of the form in Fig. 11.09-2

Design parameters: $w = 0.40$, $R_A = 20$ ohms, $\delta = 0.272$; and for $n = 3$ Sec. 4.09 gives $g_0 = 1.00$, $g_1 = 3.70$, $g_2 = 0.48$, $g_3 = 2.35$, $g_4 = 0.36$, $\omega_1' = 1$, $(L_A)_{max} = 1.4$ db, and $(L_A)_{min} = 1.07$ db.

Design	d	R_B (ohms)	Y_2 (mhos)	Y_{23} (mhos)	Y_3 (mhos)
5	1	100	0.06281	0.01342	0.01498

Figure 11.09-9 shows the computed transducer attenuation for the matching-network Design 5 in Table 11.09-4, used with the load in Fig. 11.09-8. The solid curve is for the original matching network with the stubs a quarter-wave-length long at frequency f_0. The dashed curve for Design 5 was computed with the Y_2 and Y_3 stubs lengthened by 2.5 percent in an effort to correct the skewedness of the response curve. As can be seen, this achieved the desired result but with a small loss in bandwidth. Note that the pass-band ripple maxima and minima are reasonably consistent with the $(L_A)_{max} = 1.4$ db and $(L_A)_{min} = 1.07$ db values predicted for $\delta = 0.272$ using Figs. 4.09-2 and 4.09-3. Figure 11.09-9 also shows the attenuation characteristic when no matching network is used but the generator resistance is chosen so as to give a match at resonance. Note that on this basis the attenuation is 5.9 db at the lower band edge and about 6.85 db at the upper band edge.

The above examples show that though the procedures for matching-network design based on the methods of Chapter 10 are approximate, they can give good results in close agreement with the theory. Also, the techniques described in this section are seen to be useful even if the load characteristics deviate appreciably from the idealized, symmetrical characteristics in Fig. 11.09-1. If the load impedance characteristics

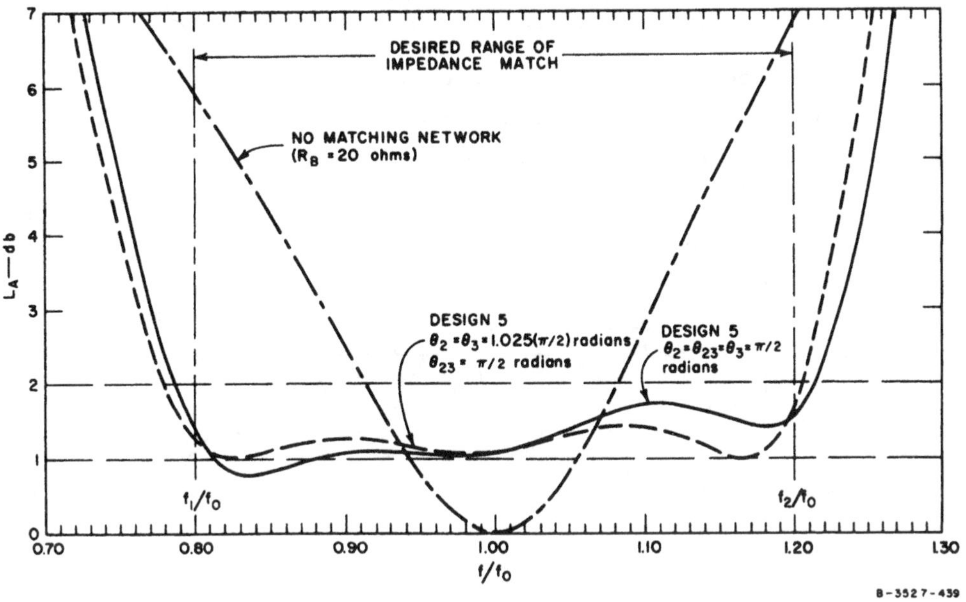

FIG. 11.09-9 TRANSDUCER ATTENUATION INTO THE LOAD IN FIG. 11.09-8 WITH AND WITHOUT MATCHING-NETWORK DESIGN 5
The parameters θ_2, θ_{23}, and θ_3 given are the electrical lengths of the lines Y_2, Y_{23}, and Y_3 of the matching network at frequency f_0

are skewed, this can often be compensated for by minor tuning adjustments on the load or on the resonant elements of the matching network.

In this section design data for only two specific types of wideband matching networks have been given. Though most of the other structures in Chapter 10 could also be used, the ones given in this section were chosen as particularly useful examples. If desired, data for other forms of matching networks can be obtained by the methods of Chapter 10 along with a modified low-pass prototype such as that in Fig. 11.09-3. Other forms of networks may be desirable in special situations where they might give more practical dimensions for the impedance levels and space limitations involved.

SEC. 11.10, DESIGN OF WIDE-BAND COUPLING STRUCTURES FOR NEGATIVE-RESISTANCE AMPLIFIERS

The same procedures as discussed in the preceding two sections can also be used for the design of coupling networks for negative-resistance devices. Consider, for example, the negative-resistance amplifier shown in Fig. 11.10-1. The required negative resistance is assumed to have been

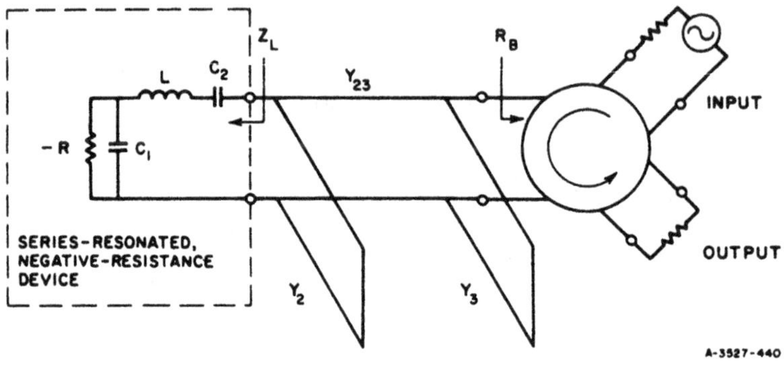

FIG. 11.10-1 EXAMPLE OF THE USE OF A FILTER STRUCTURE FOR BROAD-BANDING A NEGATIVE-RESISTANCE AMPLIFIER
The negative-resistance is assumed to have been achieved using a device such as a tunnel diode which has parasitic equivalent reactances as well as negative resistance

achieved by use of a device (such as a tunnel diode) which has parasitic reactive elements in its equivalent circuit as well as a negative resistance. A broad-banding network is then desirable in order to compensate for the effects of these reactive elements so as to give as good performance as possible over the required operating band.

In Sec. 1.04 it was explained that a negative resistance amplifier with a prescribed response characteristic can be obtained by first designing a corresponding filter having positive resistances at both ends. Then when one of the terminating resistors is replaced by a circulator having the same input resistance, while the other terminating resistor is replaced by its negative, a prescribed gain characteristic can be obtained at the circulator. Using this approach, Getsinger[12] has prepared extensive tables of low-pass-prototype negative-resistance amplifier designs, and his tables are reproduced in Sec. 4.11.

The following procedure is suggested for the design of coupling networks for band-pass negative-resistance amplifiers of this sort:

(1) The negative-resistance device should be mounted as it will be in the finished amplifier, including any voltage-bias leads, etc. Next determine whether it is most appropriate to resonate the device in series or in shunt.* (If necessary, both the impedance and admittance characteristics of the device should be plotted versus frequency, to aid in this decision). Then bring the device to resonance at the desired midband frequency f_0 by adding an inductance or a capacitance in series or in shunt, whichever is most appropriate.

(2) If the device is resonated in series, the impedance characteristic of the device shoult be plotted. If the device was resonated in shunt, its admittance characteristic should be obtained by impedance or admittance measurements on the device, by calculation if the equivalent circuit of the device is known, or by operating the device as a single-resonator amplifier as suggested by Getsinger.[12]

(3a) If it is desired to have a design with a given fractional bandwidth w, the designer will have to accept the gain limitations that go with this given bandwidth when using the given negative-resistance device. In this case the decrement of the negative-resistance load (*i.e.*, of the resonated negative-resistance device) can be determined as in Fig. 11.09-1 and the accompanying discussion. Having δ (a negative number in this case),

$$g_1 = -\frac{1}{\delta} \qquad (11.10\text{-}1)$$

is computed, and a suitable normalized low-pass prototype, having this value of g_1 (or larger), is selected from the Tables in Sec. 4.11.

(3b) If it is desired to have a given maximum and minimum gain in the operating band, the designer will have to accept the bandwidth limitations tha go with these gain parameters when using the given negative-resistance device. In this case it is convenient to determine a negative Q for the negative-resistance load using the formula

$$Q_A = \frac{-f_0}{f_b - f_a} \qquad (11.10\text{-}2)$$

* See the discussion in Sec. 11.08 on the determination of the load parameters. The principles are the same in this case except that the real part of the load impedance (and admittance) is negative.

where f_0 is the frequency of resonance, and f_b and f_a are, respectively, the frequencies above and below resonance where $[ReZ_L] = [ImZ_1]$ for the case of series resonance as in Fig. 11.10-1, or where $[ReY_L] = [ImY_L]$ for the case of shunt resonance. Then from Sec. 4.11, a normalized prototype design is selected that has the desired minimum gain $(G_t)_{min}$ and pass-band ripple G_{tr}. From the g_1 value of the normalized prototype compute

$$\delta = -\frac{1}{g_1} \quad , \tag{11.10-3}$$

and by Eqs. (11.08-1) and (11.10-2), the fractional bandwidth is

$$w = \frac{1}{\delta Q_A} \quad . \tag{11.10-4}$$

(4) If the resonator is resonated in series, determine the nominal (*i.e.*, average) real part of the input impedance, $ReZ_L \approx R_A$. If the resonator is resonated in shunt, determine the nominal value of the real part of the input admittance, $ReY_L \approx G_A$. Also determine the corresponding value R_B or G_B for the input resistance or conductance of the circulator to be used for the circuit.

(5) Using the low-pass prototype parameters g_0, g_1, ..., g_{n+1}, and $\omega_1' = 1$, along with w, δ, R_A or G_A, and R_B or G_B, temporarily drop the minus signs on δ, g_0, and the R_A or G_A value, and carry out the design of the network using the matching-network design techniques and equations in Secs. 11.08 and 11.09, exactly as for a passive matching network.

(6) Replace the positive-resistance load R_A or G_A and its adjacent resonator in the circuit obtained from Sec. 11.08 or 11.09 by the resonated negative-resistance device. Replace the R_B or G_B termination at the other end by a circulator having the same input resistance or conductance.

As an example of this procedure, let us imagine that the R = 100-ohm resistor in the circuit in Fig. 11.09-8 is replaced by a negative resistance of −100 ohms, and the resulting circuit is then the equivalent circuit of some form of resonated negative-resistance device. (If X_{c2} is removed, the circuit is similar to the equivalent circuit of a tunnel diode. But for a tunnel diode the shunt capacitor has a relatively large value so that resonating the diode in shunt would be preferable, while in this case resonating the circuit in series is best.) The curves in Fig. 11.09-8 then give the impedance characteristic of the series-resonated device as required by Step 2 above, except that with R negative instead of positive, the ReZ_L curve will be negative instead of positive.

Let us suppose that it is desired to obtain a minimum gain $(G_t)_{min}$ of 10 db, and to have a 1-db Tchebyscheff ripple G_{tr}. Then Step (3b) above should be used. From Fig. 11.09-8, $[ReZ_L] = [ImZ_L]$ is satisfied at $f_a/f_0 = 0.941$ and at $f_b/b_0 = 1.050$. Then by Eq. (11.10-2),

$$Q_A = \frac{-1}{1.050 - 0.941} = -9.2$$

Next we must select an appropriate low-pass prototype from Tables 4.11-1 through 4.11-3. From Table 4.11-1, for an $n = 2$ design having $(G_t)_{min}$ = 10 db and $G_{tr} = 1$ db, $g_1 = 1.5047$, and by Eqs. (11.10-3) and (11.10-4) the fractional bandwidth obtainable is $w = 0.163$. Similarly by Table 4.11-2 (which is for $n = 3$), $g_1 = 2.0122$ and w is 0.219; while by Table 4.11-3 (which is for $n = 4$), $g_1 = 2.2452$ and w is 0.244. Note that the improvement when going from $n = 2$ to $n = 3$ is more substantial than when going from $n = 3$ to $n = 4$. Let us suppose that the $w = 0.219$ bandwidth for $n = 3$ is adequate. Then from Table 4.11-2, $g_0 = -1$, $g_1 = 2.0122$, $g_3 = 0.6802$, $g_3 = 1.2303$, $g_4 = 0.5603$, and $\omega_1' = 1$.

Following Step 4 above we find from Fig. 11.09-8 that the nominal value of ReZ_L in the band of interest is $ReZ_L \approx R_A = -20$ ohms. Further, let us suppose that the circulator to be used has an input impedance of $R_B = 50$ ohms.

Finally, let us suppose that the stub form of filter in Fig. 11.09-2 is chosen, so then the design equations in Table 11.09-1 will be used. Carrying out Steps 5 and 6 above, the resulting design is as shown in Fig. 11.10-1, and the line admittances are $Y_2 = 0.1735$ mho, $Y_{23} = 0.0237$ mho, and $Y_3 = 0.0571$ mho. The value of d for use in Table 11.09-1 was arbitrarily selected as *one* (see discussion in Sec. 11.09).

The dashed curve marked G_t in Fig. 11.10-2 shows the computed gain for the amplifier design discussed above. Note that though the midband gain is exactly 11 db as it should be, and though the Tchebyscheff ripple is about $G_{tr} = 1$ db as it should be, the response is tilted to one side. This is due to the unsymmetrical impedance characteristic Z_L of the resonated negative-resistance device. Such tilting of the response curve was also noted in the impedance-matching network response

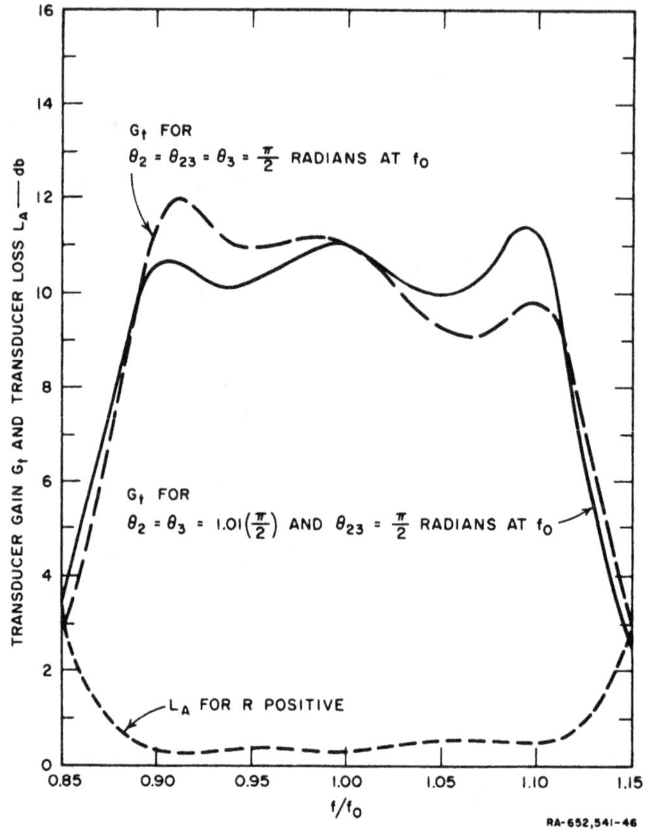

FIG. 11.10-2 COMPUTED RESPONSES FOR AN AMPLIFIER OF THE FORM IN FIG. 11.10-1
θ_2, θ_{23}, and θ_3 are the electrical lengths of lines Y_2, Y_{23}, and Y_3, respectively, at midband f_0

in Fig. 11.09-9, but was corrected by slightly altering the lengths of the shunt stubs. The same technique should be useful for leveling up the gain curve of the negative-resistance amplifier under discussion.

The dashed G_t curve in Fig. 11.10-2 was computed with the stubs Y_2 and Y_3, and the connecting line Y_{23} in Fig. 11.10-1 all a quarter-wavelength long at the midband frequency f_0. The solid-line G_t curve shows the result of "tuning" the stubs slightly by increasing their lengths by one percent, while leaving the length of the connecting line Y_{23} unchanged. It is seen that this adjustment yields a response that that has very nearly the $(G_t)_{min}$ = 10 db and G_{tr} = 1 db values specified, and the fractional bandwidth is in excellent agreement with the w = 0.219 theoretical value. Tuning adjustments of this type are particularly easy to do in an actual amplifier if a swept signal generator and an oscilloscope to display the response are available.

The lower dashed curve marked L_A in Fig. 11.10-2 shows the transducer loss of the circuit for transmission from the generator in Fig. 11.10-1 to the resistor at the left side of the circuit, for the case where the negative resistance $-R$ has been replaced by a positive resistance R of the same magnitude. The dashed G_t and L_A curves are related as indicated in Fig. 4.11-1, and by Eqs. (4.11-3) and (4.11-4).

Additional discussion of the design of negative-resistance will be found in Ref. 8. In the case of parametric amplifiers the design process is somewhat complicated by the fact that the impedance seen at the idler frequency is reflected through the time-varying capacitance so as to give an impedance component at the input frequency.[9,12,13,14] Nevertheless, the techniques described herein can still serve as a useful design guide for the design of such devices.[12,13]

SEC. 11.11, BAND-PASS TIME-DELAY FILTERS

Although all of the information required for the design of band-pass time-delay filters has been presented previously, it appears desirable to review some of the band-pass delay-filter design procedures. Some general concepts related to time-delay filters were discussed in Sec. 1.05; element values and related data were given in Sec. 4.07 for low-pass prototype filters having maximally flat time delay, and in Sec. 4.08 the time-delay characteristics of filters with Tchebyscheff and maximally flat attenuation characteristics were compared with those of maximally flat time-delay networks. It was noted in Sec. 4.08 that for many cases a Tchebyscheff filter having small pass-band ripple (say 0.1 db or less) may make a more desirable type of time-delay network than will a maximally flat time-delay network. This is because although the maximally flat time-delay network has a more nearly constant time-delay characteristic, its attenuation may vary considerably over the range of interest. Thus, in cases where both the attenuation and delay characteristics are important, Tchebyscheff filters with small pass-band ripple may make an excellent compromise.

Let us suppose that a band-pass time-delay filter is desired to give 0.016 microsecond delay in an operating band from 1000 to 1100 Mc, and that from the discussion in Sec. 4.08 it is decided to use a Tchebyscheff low-pass prototype for the delay filter having 0.01-db pass-band ripple. By Eq. (4.08-4) the midband time delay t_{d0} for the band-pass filter is

$$t_{d0} = \left(\frac{2\omega_1'}{\omega_2 - \omega_1}\right) t_{d0}' \qquad \text{seconds} \tag{11.11-1}$$

where ω_1 and ω_2 are the radian band-edge frequencies of the band-pass filter corresponding to the band-edge frequency ω_1' of the low-pass prototype, and t_{d0}' is the time delay in seconds of the low-pass prototype as $\omega' \to 0$. For the problem at hand,

$$t_{d0}' = \left(\frac{\omega_2 - \omega_1}{2\omega_1'}\right) t_{d0} \qquad \text{seconds} \tag{11.11-2}$$

$$= \frac{6.28(1100-1000)10^6}{2(1)} (0.016 \times 10^{-6})$$

$$= 5.02 \text{ seconds},$$

where $\omega_1' = 1$ is assumed for the low-pass prototype. By Eq. (4.08-2),

$$t_{d0}' = C_n \tag{11.11-3}$$

where by Fig. 4.13-2, for a Tchebyscheff filter with 0.01-db ripple, and a band-edge frequency of $\omega_1' = 1$ radian/sec, for $n = 7$, $C_7 = 4.7$ seconds, while for $n = 8$, $C_8 = 5.7$ seconds. Letting $n = 8$ so $t_{d0}' = C_8 = 5.7$ seconds, the desired delay can be obtained by enlarging the bandwidth slightly so that

$$\omega_2 - \omega_1 = \frac{2\omega_1' t_{d0}'}{t_{d0}} \tag{11.11-4}$$

$$= \frac{2(1)5.7}{0.016(10^{-6})} = 7.14 \times 10^8 \text{ rad/sec.}$$

$$f_2 - f_1 = \frac{7.14 \times 10^8}{6.28} = 113 \text{ Mc/sec.}$$

Thus, a band-pass filter designed from a 0.01-db-ripple, $n = 8$ Tchebyscheff prototype [whose element values are tabulated in Table 4.05-2(a)] so as to give a 113-Mc pass-band centered at 1050 Mc should have the desired midband time delay.*

When using Eq. (11.11-1), it should be realized that it is based on the assumption that the low-pass prototype phase characteristic maps linearly into the band-pass filter phase characteristic, so that the frequency mapping

$$\omega' = \frac{2\omega_1'}{w}\left(\frac{\omega - \omega_0}{\omega_0}\right) \tag{11.11-5}$$

where

$$w = \frac{\omega_2 - \omega_1}{\omega_0}$$

and

$$\omega_0 = \frac{\omega_2 + \omega_1}{2}$$

is valid. The mapping Eq. (11.11-5) is generally accurate for narrow-band designs (*i.e.*, for designs with fractional bandwidths of, say, around $w = 0.05$ or less), and depending on the type of filter it may have satisfactory accuracy to quite large bandwidths. However, it should be realized that to the degree that the band-pass filter pass-band characteristics *vs.* frequency are not a linear mapping of the low-pass filter pass-band characteristics, the shapes of the low-pass and band-pass time-delay characteristics will differ since time delay is the derivative of the phase with respect to frequency.

Another factor that can alter the time-delay characteristic of the band-pass filter from that of its prototype is the dissipation loss. In general, the dissipation loss in a filter tends to cause the natural frequencies of vibration of a circuit to move toward the left side of the complex frequency plane away from the $j\omega$ axis (see Secs. 2.02 to 2.04). This will cause the phase and time-delay characteristics to be somewhat different than they would be if the filter circuit had no dissipation loss. For cases where the attenuation due to dissipation is not large (say a few tenths of a db or less), the alteration of the

---

* This assumes that the time delay due to the physical length of the microwave filter can be neglected, as is usually true.

time-delay characteristic due to this cause is probably small. The part of the response where attenuation will be most evident will be near the band edges, and there it should tend to smooth the time-delay characteristic, which should not be harmful.

REFERENCES

1. E. L. Ginzton, *Microwave Measurements*, pp. 417-424 (McGraw-Hill Book Company, Inc., New York City, 1957).

2. M. Dishal, "Alignment and Adjustment of Synchronously Tuned Multiple-Resonator-Circuit Filters," *Proc. IRE* **39**, pp. 1448-1455 (November 1951).

3. S. B. Cohn and J. Shimizu, "Strip Transmission Lines and Components," Quarterly Progress Report 2, SRI Project 1114, Contract DA36-039 SC-63232, Stanford Research Institute, Menlo Park, California (May 1955).

4. S. B. Cohn, J. F. Cline, B. M. Schiffman, and P. M. Sherk, "Design Data for Antenna-Multicoupler Systems," Scientific Report 1, SRI Project 2183, Contract AF 19(604)-2247, Stanford Research Institute, Menlo Park, California (August 1957). Portions of this material are contained in the paper: J. F. Cline and B. M. Schiffman, "Tunable Passive Multi-couplers Employing Minimum-Loss Filters," *IRE Trans. PGMTT*-**7**, p. 121-127 (January 1959).

5. S. B. Cohn, "Dissipation Loss in Multiple-Coupled-Resonator Filters," *Proc. IRE* **47**, pp. 1342-1348 (August 1959).

6. G. L. Matthaei, "Design Theory of Up-Converters for Use as Electronically Tunable Filters," *IRE Trans. PGMTT*-**9**, pp. 425-435 (September 1961).

7. G. L. Matthaei, "An Electronically Tunable Up-Converter," *Proc. IRE* **49**, pp. 1703-1704 (November 1961).

8. E. S. Kuh and J. D. Patterson, "Design Theory of Optimum Negative-Resistance Amplifiers," *Proc. IRE* **49**, pp. 1043-1050 (June 1961).

9. G. L. Matthaei, "A Study of the Optimum Design of Wide-Band Parametric Amplifiers and Up-Converters," *IRE Trans. PGMTT*-**9**, pp. 23-38 (January 1961).

10. J. J. Taub and B. F. Bogner, "Design of Three-Resonator Dissipative Band-Pass Filters Having Minimum Insertion Loss," *Proc. IRE* 45, pp. 681-687 (May 1957).

11. E. G. Fubini and E. A. Guillemin, "Minimum Insertion Loss Filters," *Proc. IRE* 47, pp. 37-41 (January 1959).

12. W. J. Getsinger, "Prototypes for Use in Broadbanding Reflection Amplifiers," *IEEE Trans. PTGMTT-11*, pp. 486-497 (November 1963).

13. E. S. Kuh and M. Fukada, "Optimum Synthesis of Wide-Band Parametric Amplifiers and Converters," *IRE Trans. PGCT*-8, pp. 410-415 (December 1961).

14. W. J. Getsinger and G. L. Matthaei, "Some Aspects of the Design of Wide-Band Up-Converters and Nondegenerate Parametric Amplifiers," *IEEE Trans. PTGMTT*-12, pp. 77-87 (January 1964).

CHAPTER 12

BAND-STOP FILTERS

SEC. 12.01, INTRODUCTION

In most microwave systems the signal frequency has to be transmitted and guided from one place to another with a minimum of attenuation, while unwanted frequencies are strongly attenuated by band-pass filters which pass the signal frequency with low attenuation. The common types of band-pass filter provide adequate protection for most applications. However, should some interfering frequency be particularly strong, special measures may have to be taken to suppress it; or when a limited number of frequencies are generated in a frequency generator system, high attenuation may be needed only at certain frequencies. In such cases, a band-pass filter, which discriminates against wide ranges of frequencies outside the pass band, will not be as efficient as one or more band-stop filters which discriminate against specific unwanted frequencies.

Figure 12.01-1 shows two types of band-stop filters to be discussed in this chapter. Both of these filters are useful for situations where relatively narrow stop bands are desired. The structure in Fig. 12.01-1(a) is most suitable for designs having stop-band widths of, say, around 20 percent or less, while the same statement would also apply for the structure in Fig. 12.01-1(b) if reciprocal guide wavelength (*i.e.*, $1/\lambda_g$) is used as the frequency variable. The structure in Fig. 12.01-1(a) can be fabricated in either strip line or coaxial line, and it uses capacitively coupled, band-stop resonator stubs which are spaced a quarter wavelength apart at the stop-band center frequency. The waveguide filter structure in Fig. 12.01-1(b) is similar in principle, but uses band-stop resonator cavities which are coupled to the main waveguide by inductive irises. In order to avoid interaction between the fringing fields at the various resonator irises, the resonators are spaced three quarter guide wavelengths apart.

Figure 12.01-2 shows another type of band-stop filter discussed in this chapter. The structure shown is of strip-line form, though the stubs are often most conveniently made from wire rods. This type of

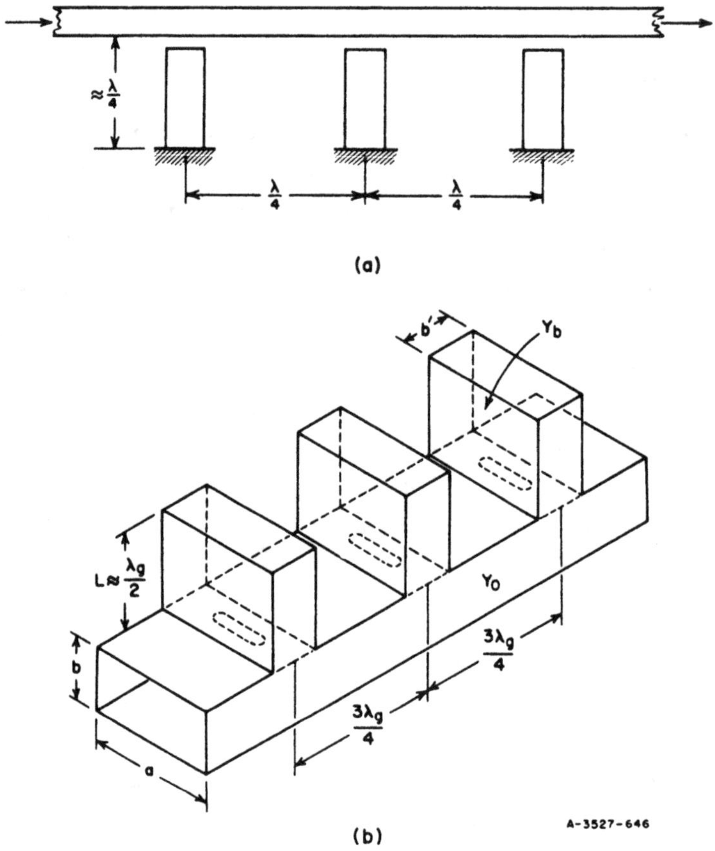

SOURCE: Quarterly Report 3, Contract DA 36-039 SC-87398, SRI; reprinted in *IRE Trans. PGMTT* (see Ref. 1 by Young, Matthaei, and Jones)

FIG. 12.01-1 STRIP-LINE AND WAVEGUIDE STRUCTURES FOR BAND-STOP FILTERS WITH NARROW STOP BANDS

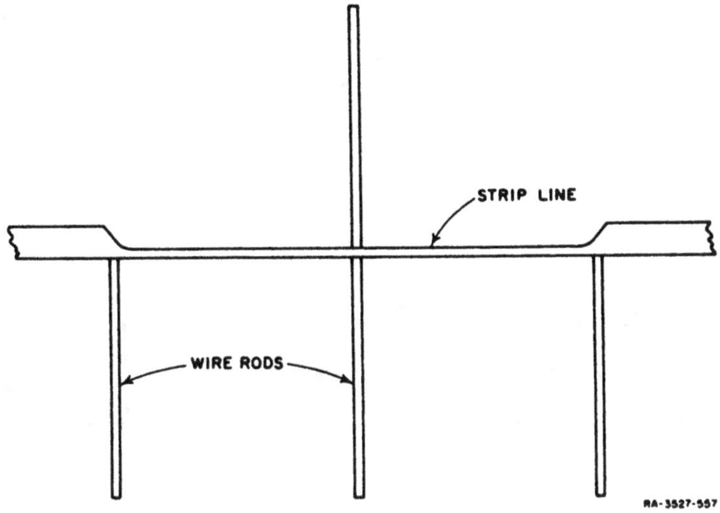

FIG. 12.01-2 A BAND-STOP, STRIP-LINE FILTER STRUCTURE HAVING A BROAD STOP BAND

filter can be designed to give quite wide stop bands, with well controlled Tchebyscheff or maximally flat pass bands, or pass bands of other shapes between the stop bands. The stop bands are centered at frequencies where the open-circuited stubs are a quarter-wavelength long or an odd multiple of a quarter-wavelength long. The structure in Fig. 12.01-2 could in principle also be used for filters having narrow stop bands; however, as the stop-band width becomes narrower, the stub impedance required may become very high. In order to avoid unreasonable impedances, the structures in Fig. 12.01-1 become preferable for narrow-stop-band or relatively narrow-stop-band cases.

SEC. 12.02, LUMPED-ELEMENT BAND-STOP FILTERS FROM LOW-PASS PROTOTYPES

A low-pass filter prototype such as those discussed in Sec. 4.05 can be transformed by suitable frequency transformations into either a band-pass, a high-pass, or a band-stop filter. The transition from low-pass to band-stop characteristic can be effected by the transformation

$$\frac{1}{\omega'} = \frac{1}{w\omega_1'}\left(\frac{\omega}{\omega_0} - \frac{\omega_0}{\omega}\right) \qquad (12.02\text{-}1)$$

where ω' stands for the radian frequency of the low-pass prototype filter and ω for that of the band-stop filter. The remaining quantities in Eq. (12.02-1) are defined in Figs. 12.02-1 and 12.02-2.

It can be seen from Eq. (12.02-1) and Figs. 12.02-1 and 12.02-2 that the prototype- and band-stop-filter frequencies in Table 12.02-1 correspond.

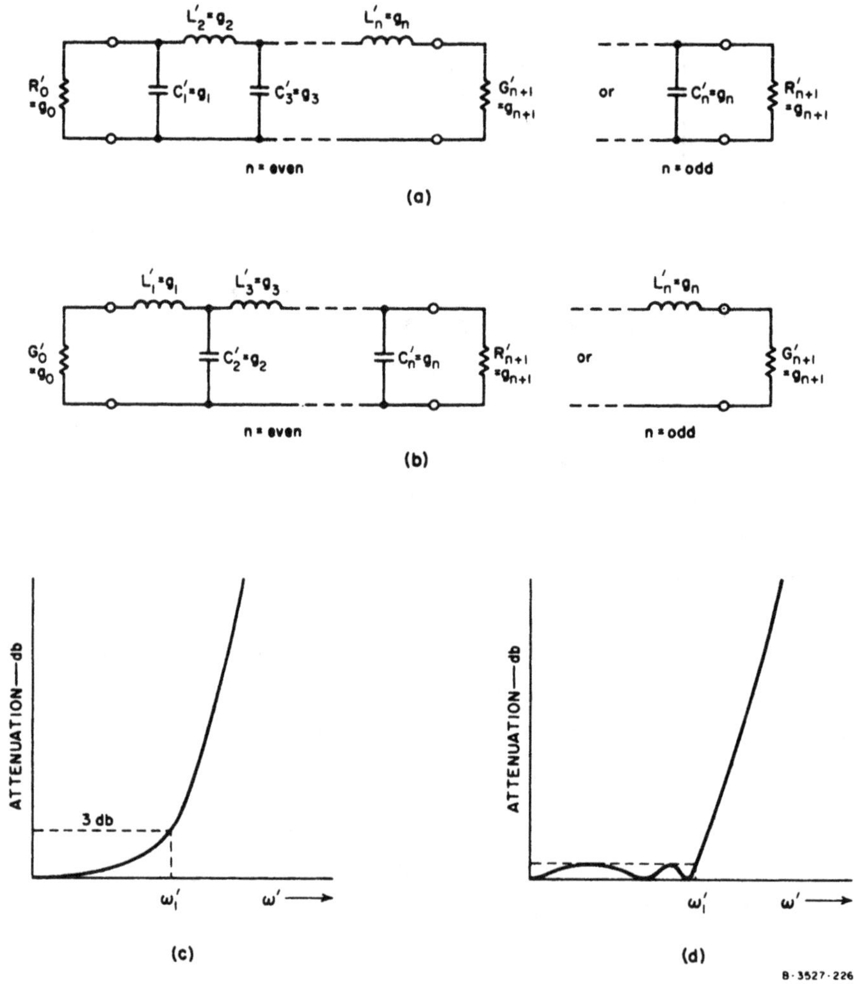

SOURCE: Quarterly Report 3, Contract DA 36-039 SC-87398, SRI; reprinted in *IRE Trans. PGMTT* (see Ref. 1 by Young, Matthaei, and Jones)

FIG. 12.02-1 LOW-PASS PROTOTYPE FILTER: (a) AND (b), FOUR BASIC CIRCUIT TYPES, DEFINING THE PARAMETERS $g_0, g_1, \ldots, g_{n+1}$; (c) AND (d), MAXIMALLY FLAT AND EQUI-RIPPLE CHARACTERISTICS, DEFINING THE BAND-EDGE ω_1'

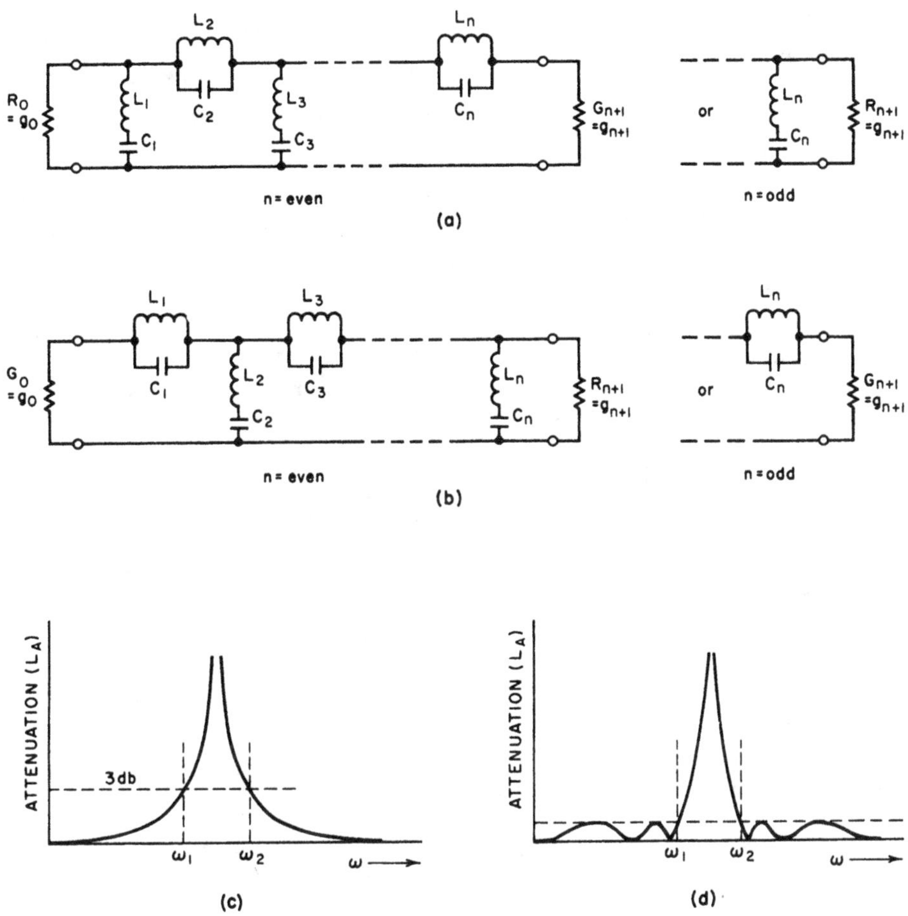

WHERE $\omega_0 = (\omega_1 \omega_2)^{1/2}$ and $w = \dfrac{\omega_2 - \omega_1}{\omega_0}$

FOR SHUNT BRANCHES, $x_i = \omega_0 L_i = \dfrac{1}{\omega_0 C_i} = \dfrac{1}{w \omega_1' g_i}$

FOR SERIES BRANCHES, $b_j = \omega_0 C_j = \dfrac{1}{\omega_0 L_j} = \dfrac{1}{w \omega_1' g_j}$

B-3527-227

SOURCE: Quarterly Report 3, Contract DA 36-039 SC-87398, SRI; reprinted in *IRE Trans. PGMTT* (see Ref. 1 by Young, Matthaei, and Jones)

FIG. 12.02-2 BAND-STOP FILTER: (a) AND (b), FOUR BASIC CIRCUIT TYPES DERIVABLE FROM THEIR FOUR LOW-PASS COUNTERPARTS IN FIG. 12.02-1; (c) AND (d), MAXIMALLY FLAT AND EQUI-RIPPLE CHARACTERISTICS DEFINING CENTER FREQUENCY ω_0, AND FRACTIONAL BANDWIDTH w

Table 12.02-1

RELATION BETWEEN VARIOUS FREQUENCIES IN THE PROTOTYPE- AND BAND-STOP FILTER RESPONSES

ω'	ω
0	0 or ∞
ω_1'	$\omega_{1,2} = \omega_0 \left[\left(1 + \frac{w^2}{4}\right)^{1/2} \mp \frac{w}{2}\right]$ $\approx \omega_0 \left(1 \mp \frac{w}{2}\right)$, when $w \ll 1$
∞	ω_0

To obtain the L_i and C_i in Fig. 12.02-2 in terms of the g_i in Fig. 12.02-1, it is simpler (but not essential) to derive Fig. 12.02-2(a) from Fig. 12.02-1(a) and Fig. 12.02-2(b), from Fig. 12.02-1(b) so that series impedances, go over to other series impedances, and shunt admittances to other shunt admittances. Multiplying both sides of Eq. (12.02-1) by $1/g_i$, one obtains the desired relation. For shunt branches, equating reactances,

$$\omega L_i - \frac{1}{\omega C_i} = \frac{1}{g_i \omega'} = \frac{1}{w\omega_1' g_i}\left(\frac{\omega}{\omega_0} - \frac{\omega_0}{\omega}\right) \tag{12.02-2}$$

which reduces to

$$\omega_0 L_i = \frac{1}{\omega_0 C_i} = \frac{1}{w\omega_1' g_i} \quad . \tag{12.02-3}$$

For series branches, equating susceptances,

$$\omega C_j - \frac{1}{\omega L_j} = \frac{1}{g_j \omega'} = \frac{1}{w\omega_1' g_j}\left(\frac{\omega}{\omega_0} - \frac{\omega_0}{\omega}\right) \tag{12.02-4}$$

which reduces to

$$\omega_0 C_j = \frac{1}{\omega_0 L_j} = \frac{1}{w\omega_1' g_j} \quad . \tag{12.02-5}$$

The *reactance slope parameter* x of a resonator reactance $X = \omega L - 1/\omega C$ is

$$x \equiv \frac{\omega_0}{2} \frac{dX}{d\omega}\bigg|_{\omega=\omega_0} = \omega_0 L = \frac{1}{\omega_0 C} \tag{12.02-6}$$

and similarly the *susceptance slope parameter* $\mathscr{b}$ of a resonator susceptance $B = \omega C - 1/\omega L$ is

$$\mathscr{b} \equiv \frac{\omega_0}{2} \frac{dB}{d\omega}\bigg|_{\omega=\omega_0} = \omega_0 C = \frac{1}{\omega_0 L} \quad . \tag{12.02-7}$$

Now Eq. (12.02-3) for a shunt branch becomes

$$x_i = \frac{1}{w\omega_1' g_i} \tag{12.02-8}$$

and Eq. (12.02-5) for a series branch becomes

$$\mathscr{b}_j = \frac{1}{w\omega_1' g_j} \quad . \tag{12.02-9}$$

The circuits shown in Fig. 12.02-2 have the same impedance levels as the prototypes in Fig. 12.02-1. To change to another impedance level, every R and every L should be multiplied by the impedance scale factor while every G and every C should be divided by it.

SEC. 12.03, THE EFFECTS OF DISSIPATION LOSS ON BAND-STOP FILTER PERFORMANCE

The discussion in this section will be phrased in terms of the lumped-element band-stop filters in Sec. 12.02; however, this discussion also applies to the microwave-filter structures such as those in Figs. 12.01-1 and 12.02-2.

The solid lines in Fig. 12.03-1 show the attenuation characteristic of a typical band-stop filter of the forms in Fig. 12.02-2 for the situation where the resonators have no dissipation loss. However, in practical filters there will inevitably be dissipation loss, and the dashed lines in Fig. 12.03-1

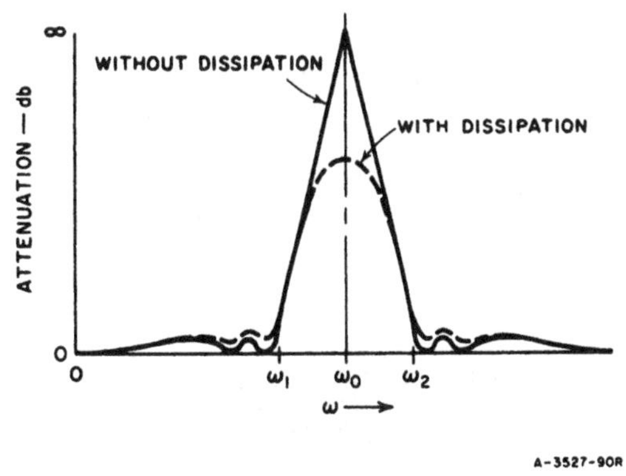

FIG. 12.03-1 THE EFFECTS OF DISSIPATION LOSS IN A BAND-STOP FILTER

illustrate its effects on filter performance. Note that the attenuation no longer goes to infinity at ω_0, but instead levels off at some finite value. Also note that near the edges of the pass bands of the filter there is an appreciable increase in the pass-band attenuation as a result of the dissipation. However, this increase in pass-band attenuation due to dissipation decreases rapidly as the frequency moves away from the stop band.

Figure 12.03-2 shows a filter of the type in Fig. 12.02-2 with resistors added to account for the dissipation loss of the resonators. If

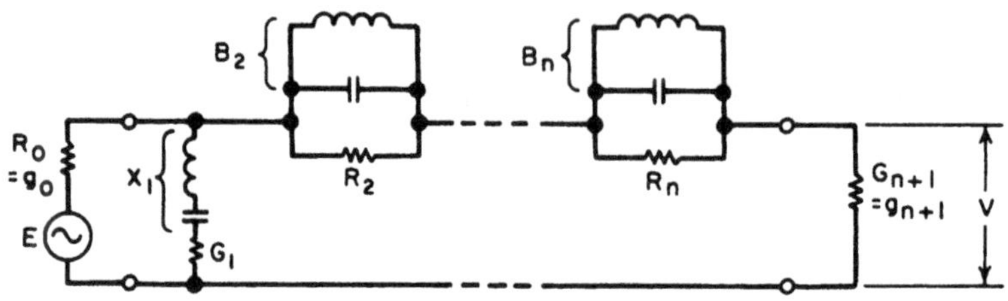

(1) ATTENUATION AT ω_0:

$$(L_A)_{\omega_0} \approx 20 \sum_{i=1}^{n} \log_{10} (g_i D_i) + 10 \log_{10} \left(\frac{g_0 g_{n+1}}{4}\right) \quad \text{db}$$

$$\text{where } D_i = w\omega_1' Q_{ui}$$

(2) RETURN LOSS AT ω_0:

$$10 \log_{10} \frac{1}{|\Gamma|^2} \approx \frac{17.37}{w\omega_1' g_0 g_1 Q_{u1}} \quad \text{db}$$

(3) DISSIPATION LOSS AT ω_1 AND ω_2:

$$\Delta L_A \approx 8.686 \frac{\omega_1'}{w} \left[\frac{g_1 g_0}{Q_{u1}} + \frac{g_2/g_0}{Q_{u2}} + \frac{g_3 g_0}{Q_{u3}} + \frac{g_4/g_0}{Q_{u4}} + \ldots, \right.$$

$$\left. \ldots + \frac{g_n/g_0}{Q_{un}} \right]_{\substack{\text{for} \\ n=\text{even}}} \quad \text{or} \quad \ldots + \left. \frac{g_n g_0}{Q_{un}} \right]_{\substack{\text{for} \\ n=\text{odd}}} \quad \text{db}$$

RA-3527-233R

SOURCE: Quarterly Report 3, Contract DA 36-039 SC-87398, SRI; reprinted in *IRE Trans. PGMTT* (see Ref. 1 by Young, Matthaei, and Jones)

FIG. 12.03-2 A BAND-STOP FILTER WITH DISSIPATION, AND FORMULAS FOR COMPUTING THE EFFECTS OF THE DISSIPATION

the unloaded Q's of the resonators are $Q_{u1}, Q_{u2}, \ldots$, etc., then for the shunt branches of the circuit

$$G_k = \frac{Q_{uk}}{\mathscr{X}_k} \qquad (12.03\text{-}1)$$

where $\mathscr{X}_k$ is the reactance slope parameter of the resonator, while for the series branches of the circuit

$$R_k = \frac{Q_{uk}}{\mathscr{B}_k} \qquad (12.03\text{-}2)$$

where $\mathscr{B}_k$ is the susceptance slope parameter of the resonator.

Figure 12.03-2 also presents a formula for computing the attenuation of the filter at the peak attenuation point ω_0. This formula was derived from the low-pass prototype circuit as discussed in Sec. 4.15. A formula is also given in Fig. 12.03-2 for the return loss at frequency ω_0. This is of interest if a band-stop filter is required to provide a short-circuit (or open-circuit) to an adjacent circuit at frequency ω_0. A formula is also given for estimating the increase in attenuation due to dissipation at the pass-band edge frequencies ω_1 and ω_2. This latter formula is based on several rather rough approximations which were discussed in Sec. 4.15. Though it is not highly accurate, this formula should be helpful in estimating the order of the dissipation loss at the pass-band edges. (It should give the ΔL_A in db at ω_1 and ω_2 within a factor of 2 or less.)

SEC. 12.04, AN APPROXIMATE DESIGN PROCEDURE FOR MICROWAVE FILTERS WITH VERY NARROW STOP BANDS

In this section a design approach for microwave filters with very narrow stop bands will be outlined. This approach is very simple and general in its application. However, in Sec. 12.09 will be found another approach which will give superior accuracy, especially if the stop-band width is appreciable. However, the design data there leads to designs which are somewhat more difficult to build because of transmission-line steps (which are not required when using the procedure of this section). Both design approaches are included for completeness.

To realize a band-stop filter in transmission line, it is more convenient to use only shunt branches, or only series branches. The circuit of Fig. 12.02-2(a) can be converted into a circuit with only series-resonant shunt branches by means of impedance inverters (Sec. 4.12) which here are approximated by 90-degree line lengths, as shown in Fig. 12.04-1. The reactance slope parameters of this circuit in terms of the low-pass prototype parameters $g_0, g_1, \ldots, g_{n+1}$, and ω_1' are also given in Fig. 12.04-1. These formulas can be deduced with the help of Figs. 12.02-1 and 12.02-2, and the impedance-inverting property of quarter-wave lines.

In the circuit of Fig. 12.04-1(a), the output impedance and the input impedance have both been set equal to Z_0. In the case of maximally flat filters, or Tchebyscheff filters with n odd, the whole sequence of impedance inverters can be uniform, with the 90-degree lines all having impedances $Z_1 = Z_0$. However, for Tchebyscheff filters with n even, the low-pass prototype is not symmetrical, and the simplest way to obtain a symmetrical band-stop filter of the type shown in Fig. 12.04-1(a) is to set the impedances of all the 90-degree lines equal to Z_1, which ceases to be equal to Z_0, and is given by

$$\left(\frac{Z_1}{Z_0}\right)^2 = \frac{1}{g_0 g_{n+1}} \qquad (12.04\text{-}1)$$

as stated in Fig. 12.04-1(a).

If the slope parameters determined by Fig. 12.04-1(a) are either too small or too large to realize conveniently, they may be adjusted up or down by controlling the impedances of the 90-degree lines. The formulas for the general case are given in Fig. 12.04-1(b). They reduce to the case of Fig. 12.04-1(a) when $Z_1 = Z_2 = \ldots = Z_{n-1}$. It should be noted that if the Z_i are chosen to be very unequal, then greater reflections will usually result somewhere in the pass band than would occur with a uniform impedance level $Z_i = Z_0$. This is because the 90-degree line sections between resonators are frequency-sensitive and approximate idealized inverters only over a limited frequency range. If the line sections in Fig. 12.04-1 were replaced by ideal inverters (Sec. 4.12) then the responses of the circuits in Fig. 12.04-1 would be identical to those of corresponding circuits of the form in Fig. 12.02-2 at all frequencies.

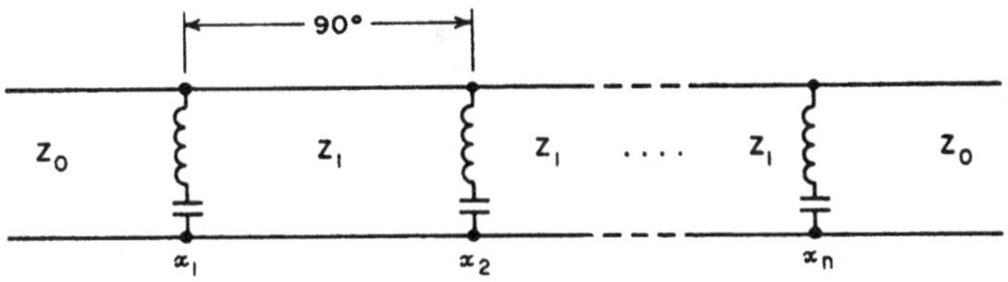

$$\frac{x_1}{Z_0} = \frac{1}{\omega_1' g_0 g_1 w}$$

$$\left.\frac{x_i}{Z_0}\right|_{i=\text{even}} = \left(\frac{Z_1}{Z_0}\right)^2 \frac{g_0}{\omega_1' g_i w}$$

$$\left.\frac{x_i}{Z_0}\right|_{i=\text{odd}} = \frac{1}{\omega_1' g_0 g_i w}$$

$$\left(\frac{Z_1}{Z_0}\right)^2 = \frac{1}{g_0 g_{n+1}}$$

If $n = \text{odd}$, $Z_1 = Z_0$

(a)

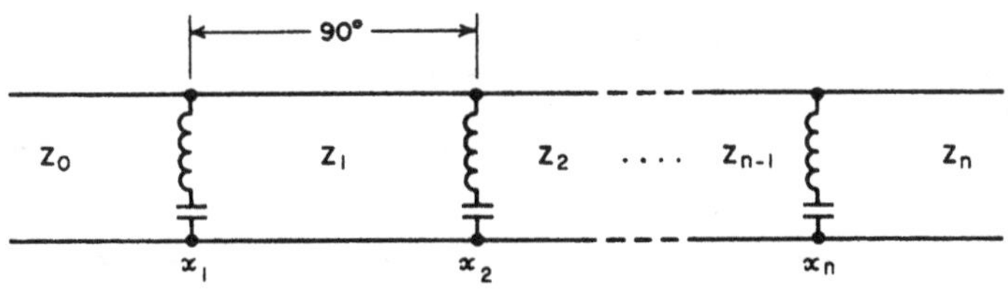

$$\frac{x_1}{Z_0} = \frac{1}{\omega_1' g_0 g_1 w}$$

$$\left.\frac{x_i}{Z_0}\right|_{i=\text{even}} = \left(\frac{Z_1 Z_3 \ldots\ldots Z_{i-1}}{Z_0 Z_2 \ldots\ldots Z_{i-2}}\right)^2 \frac{g_0}{\omega_1' g_i w}$$

$$\left.\frac{x_i}{Z_0}\right|_{\substack{i=\text{odd} \\ i\neq 1}} = \left(\frac{Z_2 Z_4 \ldots\ldots Z_{i-1}}{Z_1 Z_3 \ldots\ldots Z_{i-2}}\right)^2 \frac{1}{\omega_1' g_0 g_i w}$$

If $n = \text{even}$, $\dfrac{Z_0}{Z_n} = \left(\dfrac{Z_0 Z_2 \ldots\ldots Z_{n-2}}{Z_1 Z_3 \ldots\ldots Z_{n-1}}\right)^2 \dfrac{1}{g_0 g_{n+1}}$, or

If $n = \text{odd}$, $\dfrac{Z_0}{Z_n} = \left(\dfrac{Z_1 Z_3 \ldots\ldots Z_{n-2}}{Z_2 Z_4 \ldots\ldots Z_{n-1}}\right)^2 \dfrac{g_0}{g_{n+1}}$

(b)

B-3527-228

SOURCE: Quarterly Report 3, Contract DA 36-039 SC-87398, SRI; reprinted in *IRE Trans. PGMTT* (see Ref. 1 by Young, Matthaei, and Jones)

FIG. 12.04-1 BAND-STOP FILTER WITH SHUNT BRANCHES AND QUARTER-WAVE COUPLINGS:
(a) EQUAL INVERTER IMPEDANCES, Z_1;
(b) GENERAL CASE OF UNEQUAL INVERTER IMPEDANCES

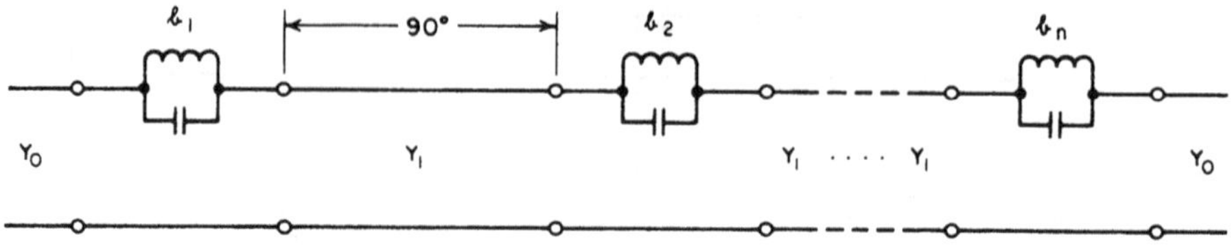

$$\frac{\mathscr{b}_1}{Y_0} = \frac{1}{\omega_1' g_0 g_1 w}$$

$$\left.\frac{\mathscr{b}_i}{Y_0}\right|_{i=\text{even}} = \left(\frac{Y_1}{Y_0}\right)^2 \frac{g_0}{\omega_1' g_i w}$$

$$\left.\frac{\mathscr{b}_i}{Y_0}\right|_{i=\text{odd}} = \frac{1}{\omega_1' g_0 g_i w}$$

$$\text{If } n = \text{even}, \quad \left(\frac{Y_1}{Y_0}\right)^2 = \frac{1}{g_0 g_{n+1}}$$

$$\text{If } n = \text{odd}, \quad Y_1 = Y_0$$

(a)

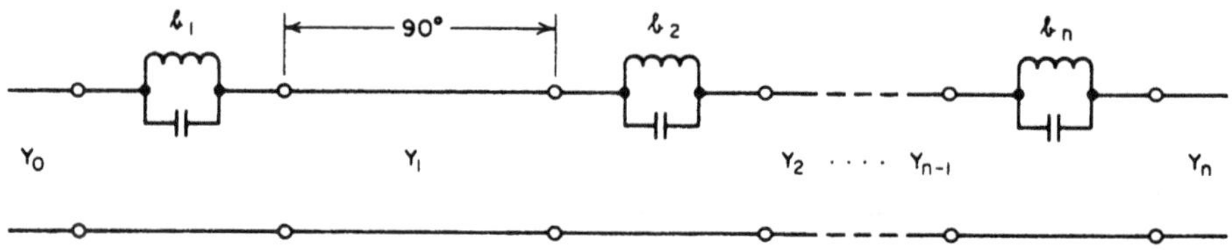

$$\frac{\mathscr{b}_1}{Y_0} = \frac{1}{\omega_1' g_0 g_1 w}$$

$$\left.\frac{\mathscr{b}_i}{Y_0}\right|_{i=\text{even}} = \left(\frac{Y_1 Y_3 \cdots Y_{i-1}}{Y_0 Y_2 \cdots Y_{i-2}}\right)^2 \frac{g_0}{\omega_1' g_i w}$$

$$\left.\frac{\mathscr{b}_i}{Y_0}\right|_{\substack{i=\text{odd}\\ i\neq 1}} = \left(\frac{Y_2 Y_4 \cdots Y_{i-1}}{Y_1 Y_3 \cdots Y_{i-2}}\right)^2 \frac{1}{\omega_1' g_0 g_i w}$$

$$\text{If } n = \text{even}, \quad \frac{Y_0}{Y_n} = \left(\frac{Y_0 Y_2 \cdots Y_{n-2}}{Y_1 Y_3 \cdots Y_{n-1}}\right)^2 \frac{1}{g_0 g_{n+1}}, \text{ or}$$

$$\text{If } n = \text{odd}, \quad \frac{Y_0}{Y_n} = \left(\frac{Y_1 Y_3 \cdots Y_{n-2}}{Y_2 Y_4 \cdots Y_{n-1}}\right)^2 \frac{g_0}{g_{n+1}}$$

(b)

B-3527-229

SOURCE: Quarterly Report 3, Contract DA 36-039 SC-87398, SRI; reprinted in *IRE Trans. PGMTT* (see Ref. 1 by Young, Matthaei, and Jones)

FIG. 12.04-2 BAND-STOP FILTER WITH SERIES BRANCHES AND QUARTER-WAVE COUPLINGS: (a) EQUAL INVERTER ADMITTANCES, Y_1; (b) GENERAL CASE OF UNEQUAL INVERTER ADMITTANCES

Figure 12.04-2 shows dual circuits to those in Fig. 12.04-1. The circuits in Fig. 12.04-2 use series-connected resonators which exhibit high reactance at resonance. The transmission properties of the filters in Fig. 12.04-2 are identical to those of Fig. 12.04-1.

We shall now consider the practical realization of resonators for microwave filters of the forms in Figs. 12.04-1 and 12.04-2.

SEC. 12.05, PRACTICAL REALIZATION OF BAND-STOP RESONATORS FOR NARROW-STOP-BAND FILTERS

The design procedures in Secs. 12.04 and 12.09 give the reactance-slope parameters $\mathscr{x}_i$ or susceptance-slope parameters $\mathscr{b}_i$ of the resonators of a narrow-stop-band filter and the impedances or admittances of the connecting lines. The practical design of microwave band-stop resonators so as to have prescribed slope parameters will now be considered.

In strip transmission line or coaxial line, the circuit of Fig. 12.04-1 may be realized by using non-contacting stubs, as shown in Fig. 12.01-1(a), with the gaps between the stubs and the main line forming the capacitances required in Fig. 12.04-1, and the stubs approximating the inductances. The stubs may be short-circuited and just under 90 degrees long, as shown in Fig. 12.01-1(a) and indicated in Fig. 12.05-1(a); or they may be open-circuited and just under 180 degrees long, as indicated in Fig. 12.05-1(b). Figures 12.05-1(c) and (d) show the corresponding dual cases. The inductively coupled resonator in Fig. 12.05-1(d) which is slightly less than a half wavelength long at resonance, is the type of resonator used in the waveguide band-stop filter in Fig. 12.01-1(b). The case of Fig. 12.05-1(a) will now be treated specifically, and then the cases of Figs. 12.05-1(b), (c), and (d) can be treated by extension of this first case.

The transmission lines will be supposed to be non-dispersive. (If waveguide or other dispersive line is used, replace normalized frequency ω/ω_0 by normalized reciprocal guide wavelength λ_{g0}/λ_g, where λ_g and λ_{g0} are the guide wavelengths at frequencies ω and ω_0, respectively.) Forcing the resonator reactance to be zero when $\omega = \omega_0$ and $\phi = \phi_0$ requires that

$$Z_b \tan \phi_0 = \frac{1}{\omega_0 C_b} \quad . \qquad (12.05\text{-}1)$$

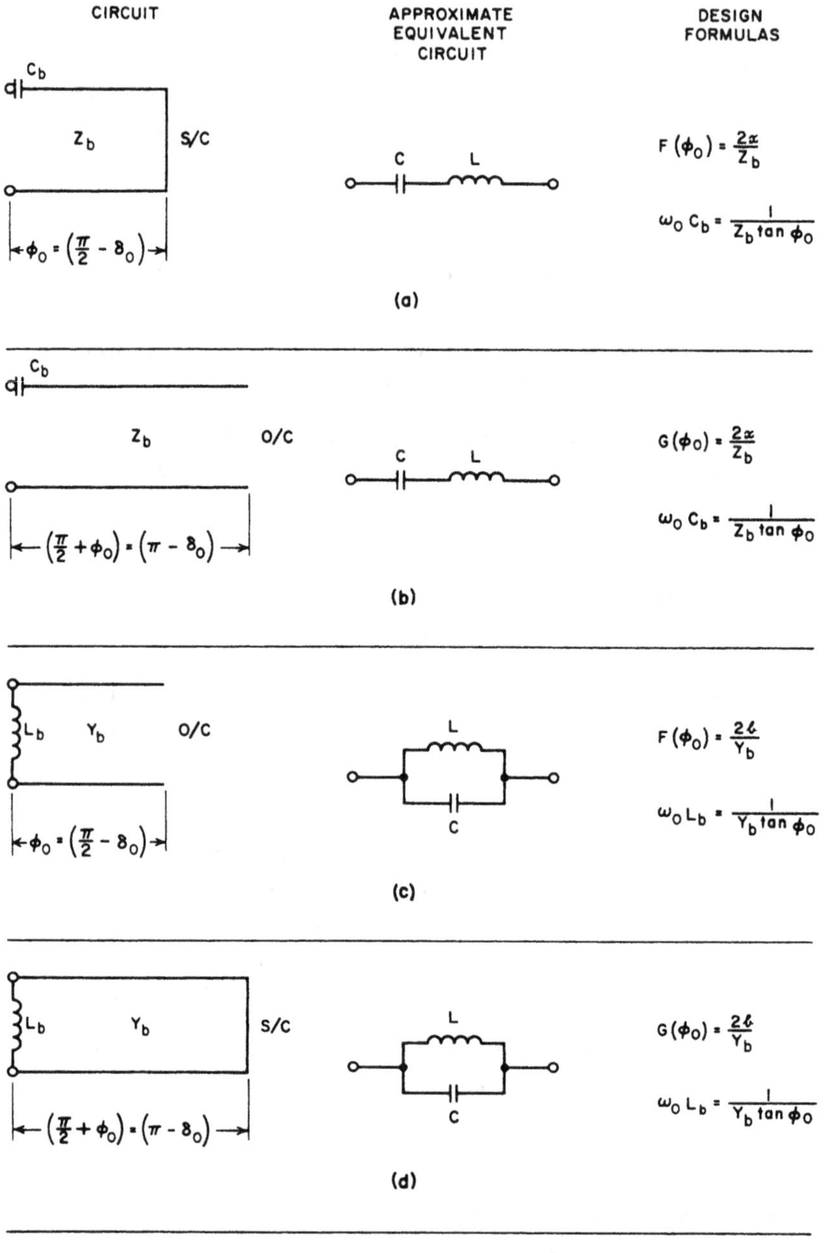

WHERE $F(\phi) = \phi \sec^2\phi + \tan\phi$ (TABLE 2)

$$G(\phi) = 2F(\phi) + \frac{1}{\pi}(2\delta - \sin 2\delta)$$

B-3527-231

SOURCE: Quarterly Report 3, Contract DA 36-039 SC-87398, SRI; reprinted in *IRE Trans. PGMTT* (see Ref. 1 by Young, Matthaei, and Jones)

FIG. 12.05-1 REALIZATION OF RESONANT CIRCUITS IN TRANSMISSION LINE: (a) AND (b), SERIES RESONANT CIRCUIT, SUITABLE FOR STRIP OR COAXIAL LINE; (c) AND (d), SHUNT RESONANT CIRCUIT, SUITABLE FOR WAVEGUIDE

Since ϕ is proportional to ω,

$$\frac{d\phi}{\phi} = \frac{d\omega}{\omega} \quad , \tag{12.05-2}$$

and the reactance slope parameter is

$$\mathscr{x} = \frac{\omega_0}{2}\frac{d}{d\omega}\left(Z_b \tan\phi - \frac{1}{\omega C}\right)\Bigg|_{\omega=\omega_0,\phi=\phi_0} \tag{12.05-3}$$

$$= \frac{Z_b}{2}\left(\phi_0 \sec^2\phi_0 + \tan\phi_0\right) = \frac{Z_b}{2}F(\phi_0) \tag{12.05-4}$$

where the function $F(\phi)$ is defined by

$$F(\phi) = \phi \sec^2\phi + \tan\phi \quad . \tag{12.05-5}$$

This function can be determined numerically from Table 12.05-1.

To determine the three parameters C_b, Z_b, and ϕ_0, one of them may be selected arbitrarily, for instance Z_b. The slope parameter $\mathscr{x}$ or $\mathscr{b}$ is determined from Fig. 12.04-1 or 12.04-2, or from Sec. 12.09. For the circuit of Fig. 12.05-1(a), ϕ_0 is determined from the assumed value for Z_b by use of Eq. (12.05-4) and Table 12.05-1, and finally Eq. (12.05-1) yields C_b. These formulas and the ones corresponding to all four cases are summarized in Fig. 12.05-1.

For a given ϕ_0 and Z_b or Y_b, the slope parameter of the near-180-degree line is almost twice as great as the slope parameter of the near-90-degree line. More precisely, for the near-180-degree lines [Cases (b) and (d) in Fig. 12.05-1] one has to substitute

$$G(\phi) = \left(\frac{\pi}{2} + \phi\right)\sec^2\phi + \tan\phi \tag{12.05-6}$$

(ϕ in radians) for $F(\phi)$, which was used with the near-90-degree lines [Cases (a) and (c) in Fig. 12.05-1]. This can be written more conveniently as

$$G(\phi) = 2F(\phi) + \frac{2\delta - \sin 2\delta}{\pi} \tag{12.05-7}$$

Table 12.05-1

TABLE OF $F(\phi) = \phi \sec^2 \phi + \tan \phi$

ϕ°	$F(\phi)$	ϕ°	$F(\phi)$	ϕ°	$F(\phi)$	ϕ°	$F(\phi)$	ϕ°	$F(\phi)$	ϕ°	$F(\phi)$
89.80	128907.01	83.80	134.59	77.80	35.03	71.80	15.88	65.80	9.05	59.00	5.54
89.60	32227.14	83.60	126.34	77.60	33.92	71.60	15.54	65.60	8.91	58.00	5.20
89.40	14324.32	83.40	118.82	77.40	32.86	71.40	15.22	65.40	8.77	57.00	4.89
89.20	8057.53	83.20	111.96	77.20	31.85	71.20	14.90	65.20	8.63	56.00	4.60
89.00	5157.05	83.00	105.68	77.00	30.88	71.00	14.59	65.00	8.49	55.00	4.34
88.80	3581.46	82.80	99.91	76.80	29.96	70.80	14.29	64.80	8.36	54.00	4.10
88.60	2631.38	82.60	94.60	76.60	29.09	70.60	14.00	64.60	8.23	53.00	3.88
88.40	2014.78	82.40	89.71	76.40	28.24	70.40	13.72	64.40	8.10	52.00	3.67
88.20	1592.03	82.20	85.19	76.20	27.44	70.20	13.45	64.20	7.98	51.00	3.48
88.00	1289.64	82.00	81.00	76.00	26.67	70.00	13.19	64.00	7.86	50.00	3.30
87.80	1065.91	81.80	77.11	75.80	25.93	69.80	12.93	63.80	7.74	49.00	3.13
87.60	895.73	81.60	73.50	75.60	25.22	69.60	12.68	63.60	7.62	48.00	2.98
87.40	763.30	81.40	70.14	75.40	24.55	69.40	12.44	63.40	7.51	47.00	2.83
87.20	658.21	81.20	67.01	75.20	23.89	69.20	12.21	63.20	7.40	46.00	2.69
87.00	573.44	81.00	64.08	75.00	23.27	69.00	11.98	63.00	7.29	45.00	2.57
86.80	504.05	80.80	61.34	74.80	22.67	68.80	11.76	62.80	7.19	44.00	2.44
86.60	446.55	80.60	58.77	74.60	22.09	68.60	11.54	62.60	7.08	43.00	2.33
86.40	398.36	80.40	56.36	74.40	21.53	68.40	11.33	62.40	6.98	42.00	2.22
86.20	357.58	80.20	54.10	74.20	21.00	68.20	11.13	62.20	6.88	41.00	2.12
86.00	322.76	80.00	51.97	74.00	20.48	68.00	10.93	62.00	6.79	40.00	2.02
85.80	292.79	79.80	49.97	73.80	19.99	67.80	10.73	61.80	6.69	39.00	1.93
85.60	266.82	79.60	48.08	73.60	19.51	67.60	10.55	61.60	6.60	38.00	1.84
85.40	244.16	79.40	46.29	73.40	19.05	67.40	10.36	61.40	6.51	37.00	1.76
85.20	224.27	79.20	44.61	73.20	18.60	67.20	10.18	61.20	6.42	36.00	1.68
85.00	206.73	79.00	43.01	73.00	18.17	67.00	10.01	61.00	6.33	35.00	1.61
84.80	191.16	78.80	41.50	72.80	17.76	66.80	9.84	60.80	6.24	34.00	1.53
84.60	177.29	78.60	40.07	72.60	17.36	66.60	9.68	60.60	6.16	33.00	1.46
84.40	164.89	78.40	38.71	72.40	16.97	66.40	9.51	60.40	6.08	32.00	1.40
84.20	153.74	78.20	37.42	72.20	16.59	66.20	9.36	60.20	6.00	31.00	1.33
84.00	143.69	78.00	36.19	72.00	16.23	66.00	9.20	60.00	5.92	30.00	1.27

Note:

$F(\phi) = (1/2)G(\phi)$

SOURCE: Quarterly Report 3, Contract DA 36-039 SC-87398, SRI; reprinted in *IRE Trans. PGMTT* (see Ref. 1 by Young, Matthaei, and Jones)

where

$$\delta = \frac{\pi}{2} - \phi \tag{12.05-8}$$

is the amount by which the line length falls short of $\pi/2$ or π radians, respectively. The last term in Eq. (12.05-7), namely $(2\delta - \sin 2\delta)/\pi$, is usually very small, so that $G(\phi)$ is very nearly twice $F(\phi)$. For instance, when $\phi = \pi/3 = 60$ degrees, the error in $G(\phi)$ due to neglecting this term in Eq. (12.05-7) is less than ½ percent, while for $\phi = \pi/4 = 45$ degrees, the error is 3.4 percent. One may therefore still use Table 12.05-1 to solve for ϕ_0, looking up $F(\phi) \approx (1/2)G(\phi)$, and subsequently making a small correction if ϕ_0 is smaller than about 60 degrees.

SEC. 12.06, EXPERIMENTAL ADJUSTMENT OF THE COUPLINGS AND TUNING OF BAND-STOP RESONATORS

In constructing resonators of the forms in Fig. 12.05, usually the coupling capacitances C_b are realized by capacitive gaps, while the coupling inductances L_b usually apply to waveguide filters where the coupling inductance is achieved by an inductive iris. Since it is often difficult to compute the appropriate gap or iris size with as much accuracy as is desired, it is frequently convenient to experimentally adjust these couplings to their proper value. This is done most easily by testing one resonator at a time with the other resonators removed (in the case of a strip-line filter with capacitive gaps), or with the coupling irises of the other resonators covered with aluminum tape (in the case of a waveguide filter). Using this procedure the measured 3-db bandwidth for each resonator can be compared with a computed value, and the coupling altered until the computed value of 3-db bandwidth is obtained. Since in many cases the filter will be symmetrical, it will usually be only necessary to make this test on half of the resonators. Once the proper coupling gaps or iris sizes have been determined, they are simply duplicated for the other half of the filter.

The 3-db fractional bandwidths of the four types of resonators in Fig. 12.05-1 are given in Fig. 12.06-1. Case (a) in Fig. 12.06-1 may, for instance, be derived by noting that the insertion loss of a reactance X in shunt with two resistances $R_0 = R_1$ is 3 db when the reactance has the value $X = (1/2)R_1$. By Eq. (12.05-4)

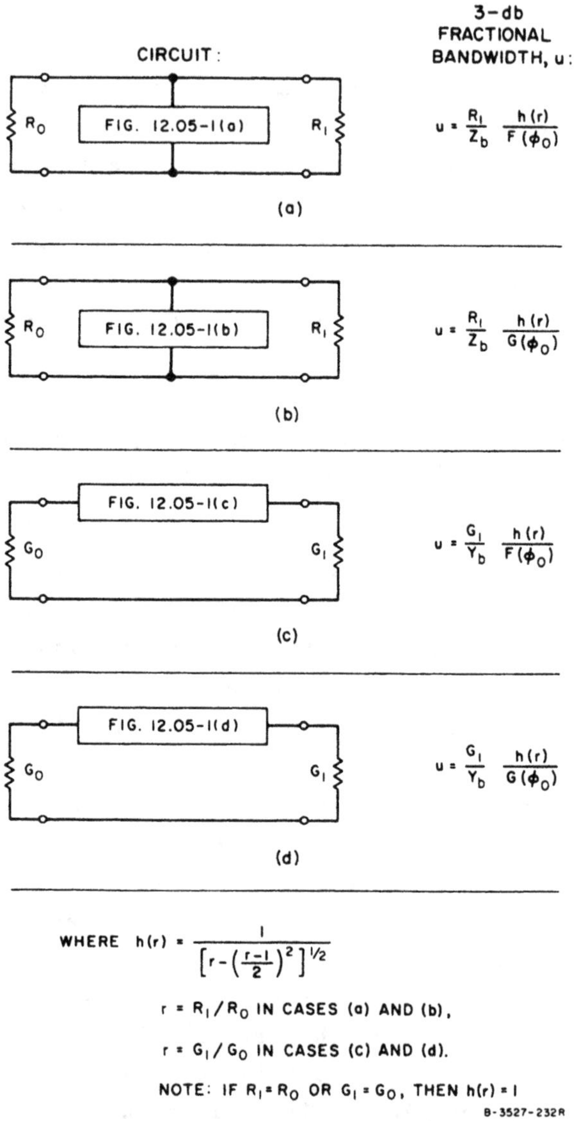

SOURCE: Quarterly Report 3, Contract DA 36-039 SC-87398, SRI; reprinted in *IRE Trans. PGMTT* (see Ref. 1 by Young, Matthaei, and Jones)

FIG. 12.06-1 THREE-db FRACTIONAL BANDWIDTHS OF THE RESONATORS IN FIG. 12.05-1 PLACED ACROSS THE JUNCTION OF TWO TRANSMISSION LINES

$$\frac{Z_b}{2} F(\phi_0) = \alpha = \frac{\omega_0}{2} \frac{dX}{d\omega}\bigg|_{\omega=\omega_0} \quad . \tag{12.06-1}$$

Since we are only considering small bandwidths, we may replace $2d\omega/\omega_0$ by the 3-db bandwidth, u, and dX by $(1/2)R_1$. Then

$$u = \frac{R_1}{Z_b} \frac{1}{F(\phi_0)} \quad . \tag{12.06-2}$$

This equation would apply to the case of a resonator of the form in Fig. 12.05-1(a) connected across a uniform transmission line of characteristic impedance equal to $Z_0 = R_0 = R_1$ which is terminated at both ends in its characteristic impedance.

The equations in Fig. 12.06-1(a) and 12.06-1(b) were derived in a similar but more general fashion. Note that these equations provide for unequal loading at the left and right sides of the resonators. This is of interest for cases where the main transmission line may have impedance steps in it. In narrow-stop-band filters the loading resistances R_0 and R_1 can be estimated by computing the impedances seen looking left and right from the resonator in question with all other resonators removed. These impedances will be purely real at ω_0, but, of course, they will vary with frequency if impedance steps are present. However, in the case of narrow-stop-band filters, it should in most cases be permissible to regard these impedances as constant resistances across the range of interest. This is because the frequency band involved is small, and in practical situations any impedance steps involved will usually also be small. In many practical cases the main-line impedances will all be equal so that the factor $h(r)$ in Fig. 12.06-1 is equal to one.

The discussion above also applies to the dual cases in Fig. 12.06-1(c) and 12.06-1(d), where the equations are in terms of admittances instead of impedances. We may note in passing that if $r = R_1/R_0$ or G_1/G_0 is greater than $3 + 2\sqrt{2} = 5.8284$ or less than $3 - 2\sqrt{2} = 0.1716$, then the 3-db bandwidth does not exist, since the intrinsic mismatch of the junction causes a 3-db reflection loss and any reactance at the junction can only serve to increase it. A 10-db or similar bandwidth would then have to be used for the experimental adjustment. However, such cases are not likely to arise in practice and will not be considered further.

In the case of filters with narrow stop bands, the tuning of the resonators must be quite precise if the response is to be as specified. Tuning is easily accomplished using tuning screws located in a high-voltage region of each resonator. One resonator should be tuned at a time while the other resonators are detuned or decoupled by adding foreign bodies near the other resonators (in the case of a strip-line filter) or by covering the coupling irises of the other resonators with aluminum tape (in the case of waveguide filters). With the other resonators detuned or decoupled, a signal at the mid-stop-band frequency is fed through the main line of the filter, and then the resonator in question is tuned to give maximum attenuation of the signal (around 30 or 40 db in many practical cases). The main advantage in incapacitating the other resonators while any given resonator is being tuned is that this technique helps to keep the signal level sufficiently high so that a distinct minimum can be observed when each resonator is tuned. If more than one resonator is resonant at a time the signal level may become so low as to be undetectable.

SEC. 12.07, EXAMPLE OF A STRIP-LINE, NARROW-STOP-BAND FILTER DESIGN

Let us suppose that a filter is desired with the following specifications:

Frequency of infinite attenuation, f_0: 4.0 Gc

Pass-band ripple: 0.5 db

Fractional stop-band bandwidth to equal-ripple (0.5-db) points: $w = 0.05$

Minimum attenuation over 2-percent stop band: 20 db.

In order to estimate the number of resonators required, Eq. (12.02-1) is used in the form

$$\left|\frac{\omega'}{\omega'_1}\right| = \left|\frac{w}{\dfrac{\omega}{\omega_0} - \dfrac{\omega_0}{\omega}}\right| \qquad (12.07\text{-}1)$$

which with $w = 0.05$ and $\omega/\omega_0 = 1.01$ (corresponding to the prescribed upper 20-db point) yields $|\omega'/\omega'_1| = 2.5$. From Fig. 4.03-7, which applies to Tchebyscheff prototypes with 0.5-db ripple, we find that for $|\omega'/\omega'_1| = 2.5$

the attenuation is 12.5 db for $n = 2$ and 26 db for $n = 3$. Thus a design with $n = 3$ resonators will be required. Since the peak attenuation of this filter will surely be much greater than 20 db, the calculation of peak attenuation at ω_0 (discussed in Sec. 12.03) need not be made in order to fix the design. The design equations in Sec. 12.04 will be used in working out this example; however, the somewhat more accurate equations in Sec. 12.09 could also be used.

From Table 4.05-2(a), the low-pass prototype parameters for 0.5-db ripple and $n = 3$ are

$$\begin{aligned} g_0 &= g_4 = 1.0 \\ g_1 &= g_3 = 1.5963 \\ g_2 &= 1.0967 \end{aligned} \qquad (12.07\text{-}2)$$

and $\omega_1' = 1$. The filter takes the form shown in Fig. 12.01-1(a). From Fig. 12.04-1(a) with $Z_0 = Z_1$,

$$\left.\begin{aligned} \frac{\alpha_1}{Z_0} &= \frac{\alpha_3}{Z_0} = 12.528 \\ \frac{\alpha_2}{Z_0} &= 18.232 \quad . \end{aligned}\right\} \qquad (12.07\text{-}3)$$

The physical embodiment of the filter is shown in Fig. 12.07-1.

The characteristic impedance of the main line, Z_0, was made 50 ohms, and the line consisted of a solid strip conductor 0.184 inch wide and 0.125 inch high, with a ground-plane spacing of 0.312 inch. The branch stubs were each made of square cross section, 0.125 inch by 0.125 inch, which gave the impedances $Z_{b1} = Z_{b2} = Z_{b3} = 59.4$ ohms (by Fig. 5.04-2). Therefore from Fig. 12.05-1(a),

$$\left.\begin{aligned} F(\phi_{01}) &= F(\phi_{03}) = \frac{2\alpha_1}{Z_0} \cdot \frac{Z_0}{Z_{b1}} = 21.1 \\ \text{and} \qquad F(\phi_{02}) &= \frac{2\alpha_2}{Z_0} \cdot \frac{Z_0}{Z_{b2}} = 30.7 \end{aligned}\right\} \qquad (12.07\text{-}4)$$

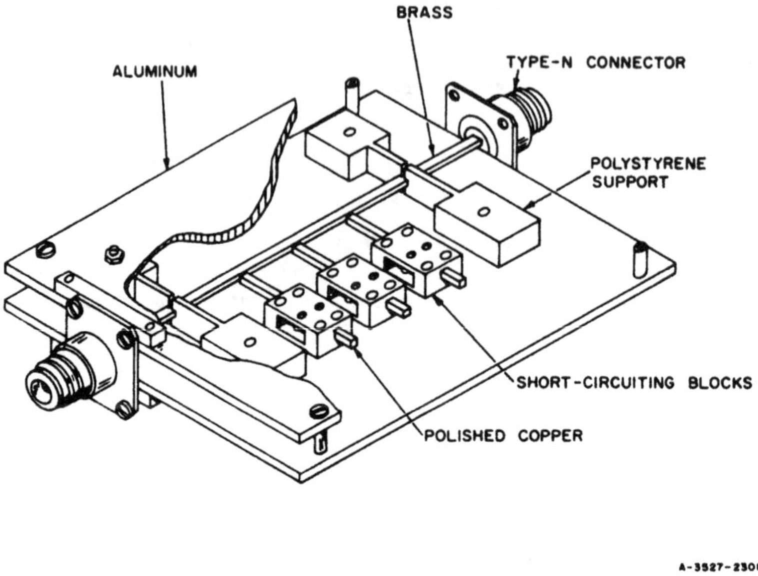

SOURCE: Quarterly Report 3, Contract DA 36-039 SC-87398, SRI; reprinted in *IRE Trans. PGMTT* (see Ref. 1 by Young, Matthaei, and Jones)

FIG. 12.07-1 A STRIP-LINE BAND-STOP FILTER WITH THREE RESONATORS

which can be solved by Table 12.05-1 to give

$$\left.\begin{aligned} \phi_{01} = \phi_{03} &= 74.2^\circ = 1.295 \quad \text{radians} \\ \phi_{02} &= 77.0^\circ = 1.344 \quad \text{radians} \quad . \end{aligned}\right\} \qquad (12.07\text{-}5)$$

Again from Fig. 12.05-1(a) we obtain

$$\left.\begin{aligned} \omega_0 C_{b_1} = \omega_0 C_{b_3} &= 0.004764 \quad \text{mho} \\ \omega_0 C_{b_2} &= 0.003888 \quad \text{mho} \end{aligned}\right\} \qquad (12.07\text{-}6)$$

which at 4.0 Gc yields

$$\left.\begin{aligned} C_{b_1} = C_{b_3} &= 0.1893 \quad \text{pf} \\ C_{b_2} &= 0.1546 \quad \text{pf} \quad . \end{aligned}\right\} \qquad (12.07\text{-}7)$$

With 0.125-by-0.125-inch stubs, such capacitances are obtained with a gap of the order of 0.031 inch, which is a quite suitable value, being small compared to a wavelength but still large enough to be achieved accurately without special measures such as using dielectric in the gap, which would increase the dissipation loss.

The length of the stubs is determined by Eq. (12.07-5) and the wavelength, which is 2.950 inches at 4.0 Gc. It was at first assumed that the reference plane was in the gap face opposite the main line. The capacitive gap of each stub was adjusted to give peak attenuation at 4.0 Gc and the 3-db bandwidth of the stop band of stubs 1 and 2 were then measured individually. The 3-db fractional bandwidths u_i are given by Fig. 12.06-1(a):

$$\left.\begin{aligned} u_1 = u_3 &= \frac{Z_0}{Z_{b_1}} \frac{1}{F(\phi_{01})} = 0.040 \\ u_2 &= \frac{Z_0}{Z_{b_2}} \frac{1}{F(\phi_{02})} = 0.0274 \quad . \end{aligned}\right\} \tag{12.07-8}$$

At first, the measured bandwidths were slightly too narrow, showing the coupling to be too loose. The coupling gaps of the stubs were reduced until the required 3-db bandwidths were obtained. Then the lengths of the stubs were corrected to give the desired 4-Gc resonant frequency. The final dimensions were: outer stub lengths 0.605 inch, gaps 0.0305 inch; middle stub length 0.629 inch, gap 0.045 inch. The three stubs were then placed along the line as shown in Fig. 12.07-1. They are nominally 90 degrees apart between reference planes at 4.0 Gc, but the spacing is not critical, and they were simply placed a quarter-wavelength apart between centers (0.738 inch), and not further adjusted.

It is important that there should be no interaction, or mutual coupling, between stubs: with a ground-plane spacing of 0.312 inch, the attenuation of the TE_{10} mode (electric vector parallel to the ground planes) from stub to stub is more than 50 db at 4.0 Gc, which should be adequate; it is also necessary to maintain proper centering so as not to excite the parallel-plate mode.

The response of this filter was computed on a digital computer and is plotted in Fig. 12.07-2. The points shown are the measured results on the experimental filter. The circles represent the measured reflection loss of the filter, which may be compared with the computed curve, while the solid dots represent the total measured attenuation, including the effect of dissipation. The computed reflection loss reaches a ripple height of 0.7 db below, and 0.8 db above 4.0 Gc (instead of 0.5 db, which was called for by the low-pass prototype). The measured reflection loss reaches a peak of 1.2 db below, and 0.7 db above 4.0 Gc.

The somewhat oversize pass-band ripples in this filter are due mainly to the frequency sensitivity of the $\lambda_0/4$ coupling lines between resonators

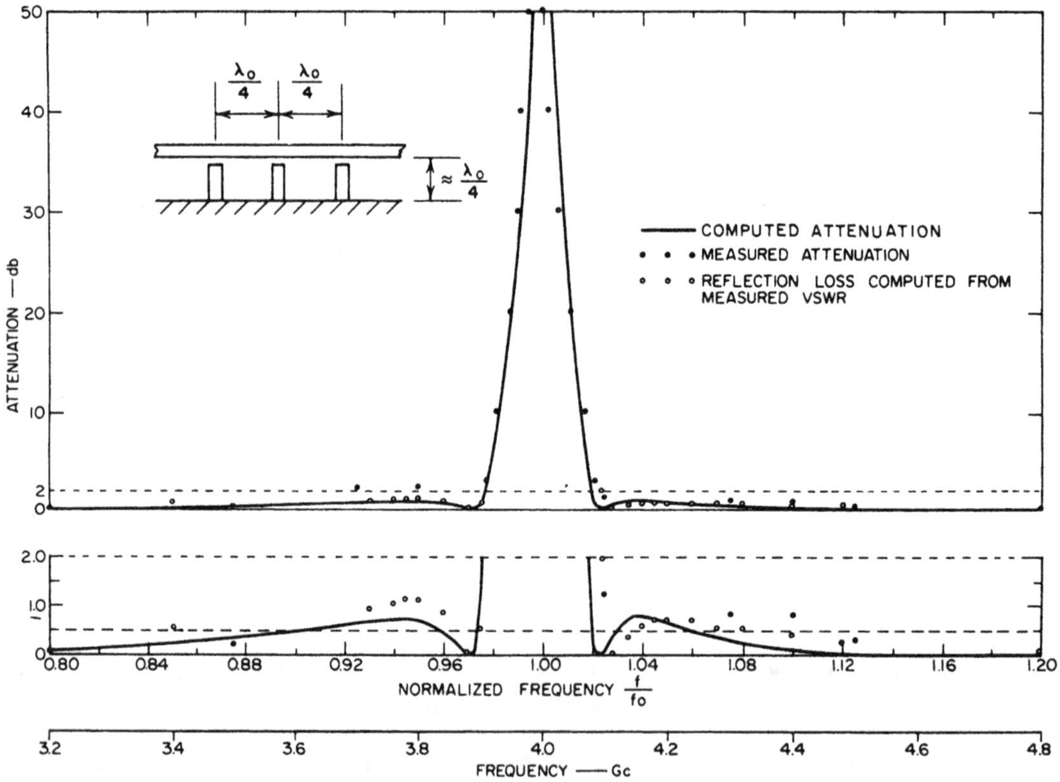

SOURCE: Quarterly Report 3, Contract DA 36-039 SC-87398, SRI; reprinted in *IRE Trans. PGMTT* (see Ref. 1 by Young, Matthaei, and Jones)

FIG. 12.07-2 COMPUTED AND MEASURED RESPONSE OF A STRIP-LINE BAND-STOP FILTER WITH $\lambda_0/4$ SPACINGS BETWEEN RESONATORS

If the design method described in Sec. 12.09 is used, selectivity of the coupling lines can be taken into account in the design process so that the pass-band ripples should be very nearly as prescribed. A possible drawback of the procedure in Sec. 12.09 is that it necessarily calls for steps in the impedance level of the main transmission line. However, in the case of a strip-line filter, this should cause little difficulty. If this same filter design were to be obtained by the methods of Sec. 12.09, the main-line impedances and the resonator slope parameters would be obtained by the equations of that section. From that point on, the design procedure would be the same as described above.

The peak attenuation of the three-resonator filter was too great to measure, but the peak attenuation of each branch separately was measured. This gave 32 db for each of the two outer branches, and 28 db for the middle branch alone. Working back from the data in Fig. 12.03-2, this leads to unloaded Q values of 1000. The theoretical unloaded Q of such a pure all-copper line at this frequency is 2600. In this case the strip conductors of the resonators were made of polished copper and the ground planes were aluminum. Other losses were presumably introduced by the current concentrations caused by fringing fields at the gap, and by the short-circuit clamps. The value of $Q_u = 1000$ is consistent with experimental results generally obtained. The peak attenuation of all three branches together should by the first formula of Fig. 12.03-2 reach a value of 104 db, which, however, was beyond the measurement range of the available test equipment. There are, of course, additional stop bands at frequencies where the stubs are roughly an odd multiple of $\lambda_0/4$ long.

According to the formula for the dissipation loss ΔL_A in Fig. 12.03-2, the dissipation loss should be about 0.74 db at the equal-ripple band edges ω_1 and ω_2. The measurements indicated dissipation loss of roughly that size at the pass-band edges.

The bandwidth of the stop band to the 0.5-db points, both measured and computed, is very nearly 5.0 percent, the design goal. The attenuation over a 2-percent band exceeds 25.4 db by computation. (Compare the 20-db design specification, and the 25.8 db predicted for a three-branch filter.) The measured points in the stop band fall close to the computed curve, as can be seen from Fig. 12.07-2. For frequencies moving away from the stop-band edges into the pass bands, the dissipation loss will drop quite rapidly.

SEC. 12.08, NARROW-STOP-BAND WAVEGUIDE FILTER DESIGN CONSIDERATIONS, AND AN EXAMPLE

A waveguide band-stop filter is most conveniently realized using resonators connected in series and spaced an odd multiple of a quarter wavelength apart along the waveguide. The equivalent circuit of such a filter is as shown in Fig. 12.04-2, while Fig. 12.01-1(b) shows a sketch of a waveguide realization of such a filter. It is seen that the resonators in Fig. 12.01-1(b) are spaced at intervals of three-quarters of a guide wavelength. It is believed that placing the resonators at one-quarter guide-wavelength intervals is not practical, because strong interaction results between the resonators by way of the fields about the coupling apertures. This interaction in an experimental band-stop filter, with quarter-guide-wavelength-spaced resonators, caused the stop-band response to have three peaks of high attenuation, with relatively low attenuation valleys in between, instead of the single high attenuation peak that is desired.

The appropriate normalized frequency variable to use with dispersive lines, as in a waveguide filter, is the normalized reciprocal guide wavelength, λ_{g0}/λ_g. Thus, a waveguide filter and a strip-line filter, each designed from the same low-pass prototype, can have identical responses if the waveguide filter response is plotted on a λ_{g0}/λ_g scale and the strip-line filter response on a ω/ω_0 scale. On this basis, the formula, which for the strip-line filter was

$$w = \frac{\omega_2 - \omega_1}{\omega_0} = \frac{\omega_2}{\omega_0} - \frac{\omega_1}{\omega_0} \quad \text{(TEM mode)} \tag{12.08-1}$$

where ω_0, ω_1, and ω_2 are as defined in Fig. 12.02-2, is now replaced for a waveguide filter by

$$w_\lambda = \frac{\lambda_{g0}}{\lambda_{g2}} - \frac{\lambda_{g0}}{\lambda_{g1}} \quad \text{(waveguide)} \tag{12.08-2}$$

where λ_{g0}, λ_{g1}, and λ_{g2} are analogously defined as the guide wavelengths at mid-stop-band, and at the lower and upper edges of the stop band, on a reciprocal guide-wavelength scale. If both responses are plotted on a frequency scale, however, the response of the waveguide filter derived from the same prototype will be considerably narrower. For small stop-band bandwidths, the fractional width of the waveguide filter stop band on a frequency scale will be approximately $(\lambda_0/\lambda_{g0})^2$ times w_λ, which is the fractional bandwidth on a reciprocal guide-wavelength scale given by Eq. (12.08-2). (λ_0 is the free-space wavelength at the center of the stop band.) Thus in designing for a frequency fractional bandwidth w, the bandwidth w_λ in the formulas is set to be equal to $(\lambda_{g0}/\lambda_0)^2 w$ for the narrow bandwidths considered here.

In the waveguide filter each resonator formed from a waveguide of characteristic admittance Y_b, which has a length slightly less than one-half guide wavelength, is connected to the main waveguide of characteristic admittance Y_0 by means of a small elongated coupling iris. Each coupling iris has a length, l, which is less than one-half free-space wavelength, and can be represented to a good approximation as an inductance in series with the main waveguide. The equivalent circuit of the resonator and coupling iris combination is shown in Fig. 12.05-1(d).

The susceptance, B, of each coupling iris can be determined approximately in terms of the magnetic polarizability, M', of the iris. From Fig. 5.10-11 the expression is found to be

$$\frac{B}{Y_b} = -\frac{\lambda_g a b'}{4\pi M_1'} \tag{12.08-3}$$

where a and b' are as defined in Fig. 12.01-1(b). For irises cut in walls of infinitesimal thickness, t, having a length l which is much less than one-half free-space wavelength, the magnetic polarizability $M_1' = M_1$ is given in Fig. 5.10-4(a). In the usual situation when t is not infinitesimal and l is not much less than one-half free-space wavelength, one must use the magnetic polarizability M_1', which is related to M_1 by the approximate empirical relation

$$M_1' = \frac{M_1}{1 - \left(\dfrac{2l}{\lambda_0}\right)^2} \; 10^{-\left(\frac{1.36t}{l}\right)\sqrt{\left(1 - \frac{2l}{\lambda_0}\right)^2}} \tag{12.08-4}$$

where λ_0 is again the free-space wavelength at center frequency corresponding to $\omega = \omega_0$, $\phi = \phi_0$, and $\lambda_g = \lambda_{g0}$. The resonant condition for each resonator becomes

$$Y_0 \tan \phi_0 = \frac{Y_b a b'}{4\pi M_1'} \lambda_{g0} \qquad (12.08\text{-}5)$$

where

$$\phi_0 = \frac{2\pi L}{\lambda_{g0}} - \frac{\pi}{2} \quad , \qquad (12.08\text{-}6)$$

and L is the length of the resonator.

The susceptance slope parameter, $\mathscr{b}$, for each resonator as viewed from the main transmission line becomes

$$\mathscr{b} = \frac{Y_b}{2} G(\phi) \approx Y_b F(\phi) \qquad (12.08\text{-}7)$$

where $F(\phi)$ is defined in Eq. (12.05-5) and $G(\phi)$ is defined in Eq. (12.05-7), and M_1' has been assumed to be frequency invariant.

Example of a Waveguide Filter Design—As an example of the use of the above technique, we consider here the design of a 3-resonator waveguide band-stop filter using the same low-pass prototype circuit used for the strip-line filter described in Sec. 12.07. The filter has a design center frequency $f_0 = 10$ Gc and the resonators and main transmission line are fabricated from WR-90 waveguide. The fractional bandwidth of the strip-line filter was $w = 0.05$ on a frequency basis [Eq. (12.08-1)]. Here, we use $w_\lambda = 0.05$ on a reciprocal guide-wavelength basis [Eq. (12.08-2)]. The susceptance slope parameters for the two end resonators and the middle resonator can be determined from Fig. 12.04-2(a) with $Y_1 = Y_0$. They are, respectively,

$$\frac{\mathscr{b}_1}{Y_0} = \frac{1}{\omega_1' g_0 g_1 w_\lambda} = \frac{\mathscr{b}_3}{Y_0} = 12.528$$

$$\frac{\mathscr{b}_2}{Y_0} = \frac{g_0}{\omega_1' g_2 w_\lambda} = 18.232 \qquad (12.08\text{-}8)$$

which correspond exactly to those previously computed for the strip-line filter. The stub characteristic admittances Y_b were chosen to be equal to Y_0. Therefore,

$$G(\phi_{01}) = G(\phi_{03}) = \frac{2\mathscr{b}_1}{Y_0} = 25.056$$

$$G(\phi_{02}) = \frac{2\mathscr{b}_2}{Y_0} = 36.464 \quad . \tag{12.08-9}$$

Referring to Table 12.05-1, we find

$$\phi_{01} + 90^\circ = \phi_{03} + 90^\circ = 159.5^\circ$$

$$\phi_{02} + 90^\circ = 163.0^\circ \quad . \tag{12.08-10}$$

At a frequency of 10 Gc, λ_{g0} in WR-90 waveguide is 1.5631 inches. Therefore the lengths $L_1 = L_3$ of the two end resonators and length L_2 of the middle resonator are

$$L_1 = L_3 = 0.693 \quad \text{inch}$$

$$L_2 = 0.709 \quad \text{inch} \quad . \tag{12.08-11}$$

The magnetic polarizability, $(M_1')_1 = (M_1')_3$, of the irises in the end cavities, and the magnetic polarizability, $(M_1')_2$, of the iris in the middle cavity can be determined from Eq. (12.08-5) to be

$$(M_1')_1 = (M_1')_3 = 0.0167 \quad \text{inch}^3$$

$$(M_1')_2 = 0.0137 \quad \text{inch}^3 \quad . \tag{12.08-12}$$

It was decided to use elongated irises with round ends, as for the second curve from the top in Fig. 5.10-4(a). The iris sizes were determined by first fixing the iris widths* as $w_1 = w_2 = w_3 = 0.125$ inch, and then computing successive approximations using Fig. 5.10-4(a) and Eq. (12.08-4). Since it was planned that the iris sizes would be checked experimentally as discussed in Sec. 12.06, the iris thickness t in Eq. (12.08-4) was assumed to be zero (which should cause the irises to come out slightly undersize). A suggested way for making these calculations is to convert Eq. (12.08-4) to the form

$$(M_1)_{k=1\text{ to }3} = (M_1')_k \left[1 - \left(\frac{2l_k}{\lambda_0}\right)^2\right] 10^{\left(\frac{1.36t}{l_k}\right)\sqrt{1-\left(\frac{2l_k}{\lambda_0}\right)^2}} \quad . \qquad (12.08\text{-}13)$$

Then using Fig. 5.10-4(a) and the desired polarizabilities $(M_1')_k$, preliminary values of the iris lengths l_k are obtained. Inserting these preliminary values of l_k in Eq. (12.08-13), compensated values $(M_1)_k$ for the polarizabilities are obtained which are then used with Fig. 5.10-4(a) to obtain compensated values for the iris lengths l_k. If desired, this procedure can be repeated in order to converge to greater mathematical accuracy; however, since Eq. (12.08-13) is in itself a rather rough approximation, the value of very high mathematical accuracy is rather doubtful.

By the above procedure, initial iris dimensions of $l_1 = l_3 = 0.430$ inch and $l_2 = 0.415$ inch for the iris lengths with $w_1 = w_2 = w_3 = 0.125$ inch were obtained. After tests made by measuring the 3-db bandwidth of Resonators 1 and 2 separately (by the methods discussed in Sec. 12.06), final values of $l_1 = l_3 = 0.481$ inch and $l_2 = 0.458$ inch for the iris lengths were obtained. The first version of this filter had $\lambda_0/4$ spacings between resonators, but this was found to be unsatisfactory due to interaction between the fringing fields around the irises of the various resonators (which resulted in considerable disruption of the stop-band performance). Thus a second version was constructed with $3\lambda_0/4$ spacings between resonators, and Fig. 12.08-1 shows the dimensions of this filter.

---

* The slot width w [Fig. 5.10-4(a)] is represented by the same symbol as the fractional bandwidth. This is the usual notation for both quantities. Since the slot calculation is only incidental in this context, it was felt to be less confusing to retain the w for slot width, than to change it.

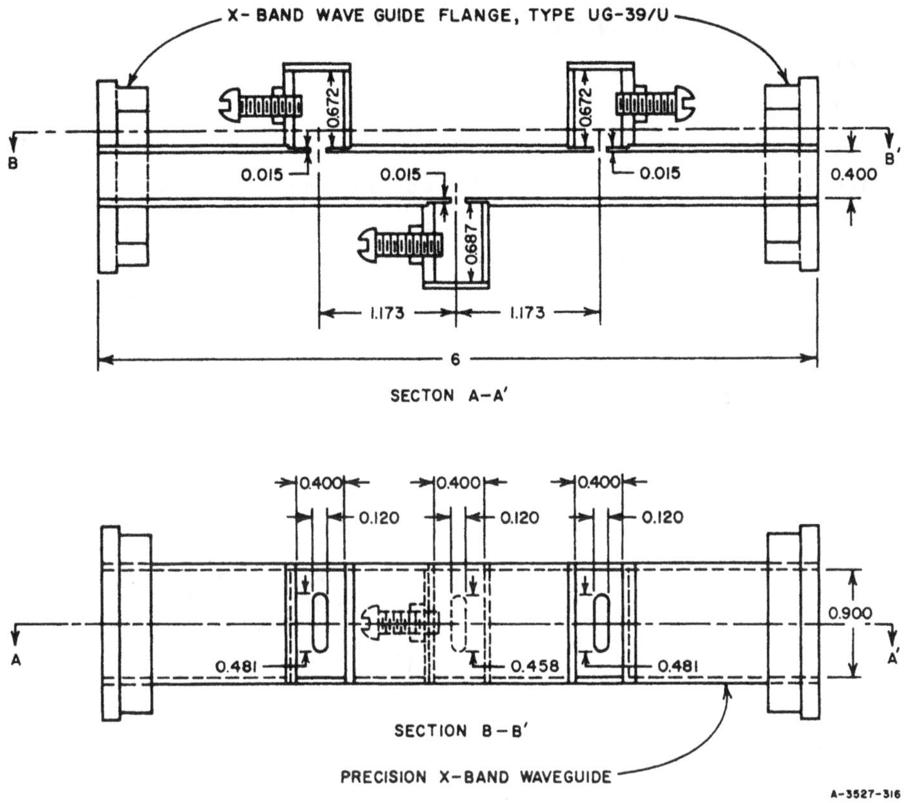

SOURCE: Quarterly Report 3, Contract DA 36-039 SC-87398, SRI; reprinted in *IRE Trans. PGMTT* (see Ref. 1 by Young, Matthaei, and Jones)

FIG. 12.08-1 DIMENSIONS OF A WAVEGUIDE BAND-STOP FILTER

The filter was tuned using the technique discussed in Sec. 12.06. Figure 12.08-2 shows points from the resulting measured response of the filter as compared with the theoretical response of this filter design as determined using a digital computer. As can be seen from the figure, the agreement is excellent.

Though the measured performance of this filter is in very good agreement with its computed performance, it will be noted from Fig. 12.08-2 that the computed performance indicates pass-band ripples of about 1.2 db instead of 0.5 db as was called for by the lumped-element prototype. These oversize ripples have been found to be due to the frequency sensitivity of the lines between resonators. If the $3\lambda_0/4$ coupling lines between resonators were replaced by ideal impedance inverters the

Tchebyscheff ripples would be exactly 0.5 db. Computer calculations show that if $\lambda_0/4$ coupling lines are used, the ripples should peak at approximately 0.7 db. In Fig. 12.08-2 the 1.2-db ripple peaks are due to the still greater selectivity of the $3\lambda_0/4$ coupling lines used in this design. Use of the design procedure discussed in Sec. 12.09 would make it possible to take into account the selectivity of the coupling lines so that the ripples should come out very closely as prescribed.* However, the price that must be paid for this is that the

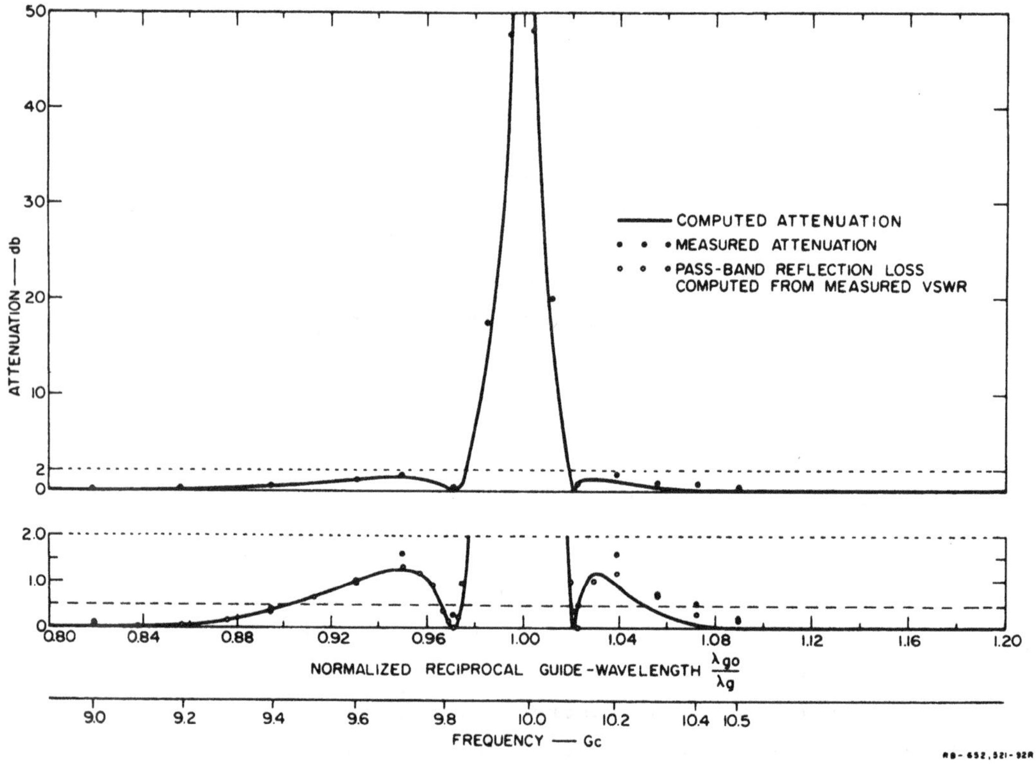

SOURCE: Quarterly Report 3, Contract DA 36-039 SC-87398, SRI; reprinted in *IRE Trans. PGMTT* (see Ref. 1 by Young, Matthaei, and Jones)

FIG. 12.08-2 COMPUTED AND MEASURED RESPONSE OF THE WAVEGUIDE BAND-STOP FILTER IN FIG. 12.08-1

* The design procedure described in Sec. 12.09 is exact if the filter is realized using only transmission lines and stubs. However, in the case of narrow-stop-band filters the stub impedances become unreasonable so that it is necessary to replace them with reactively coupled resonators. This substitution, of course, introduces an approximation.

procedure in Sec. 12.09 necessarily requires that there be steps in impedance along the main line of the filter. This introduces some added complication for manufacturing, but the difficulties introduced should not be great. Once the main line impedances and the resonator slope parameters are determined from the data in Sec. 12.09, the design process is the same as that described above.

SEC. 12.09, A DESIGN PROCEDURE WHICH IS ACCURATE FOR FILTERS WITH WIDE (AND NARROW) STOP-BAND WIDTHS*

In this section an exact design procedure will be discussed which is useful for the design of band-stop filters with either wide or narrow stop bands. The design equations given apply to filters with $\lambda_0/4$ stubs separated by connecting lines which are either $\lambda_0/4$ or $3\lambda_0/4$ long, where λ_0 is the wavelength at the mid-stop-band frequency. An example of this type of filter is shown in Fig. 12.01-2. In theory this type of filter can have any stop-band width; however, in practice the impedance of the stubs becomes unreasonable if the stop-band width is very narrow. In that case it would be desirable to replace the open-circuited stub resonators in the filter in Fig. 12.01-2 with capacitively coupled short-circuited stub resonators [as shown in Fig. 12.01-1(a)] having the same resonator slope parameters. This introduces some approximation into the design, but permits practical impedances. As has been mentioned previously, the main difference between a practical narrow-stop-band filter design obtained by the methods of this section and one obtained by the methods of Sec. 12.04 is that the method of this section calls for steps in transmission line, while usually none are necessary by the method of Sec. 12.04. However, the method of this section will give higher design accuracy even for narrow-stop-band cases where the approximation mentioned above has been introduced.

The methods of this section permit exact design of filters of the form in Fig. 12.09-1(a) or their duals (which consist of $\lambda_0/4$ short-circuited stubs in series with $\lambda_0/4$ connecting lines). Filters of these types which have been designed by the methods of this section have attenuation characteristics which are related to the attenuation characteristics of low-pass prototype filters of the form in Figs. 12.02-1 by the mapping†

* Reference 6 contains the material in this section plus additional design data for some structures studied after this was written.

† It can be shown that other methods of design[2,3] can lead to circuits of the form in Fig. 12.09-1(a) which have attenuation characteristics that are not mappings of the attenuation characteristics of low-pass prototypes of the form in Figs. 12.02-1(a), (b).

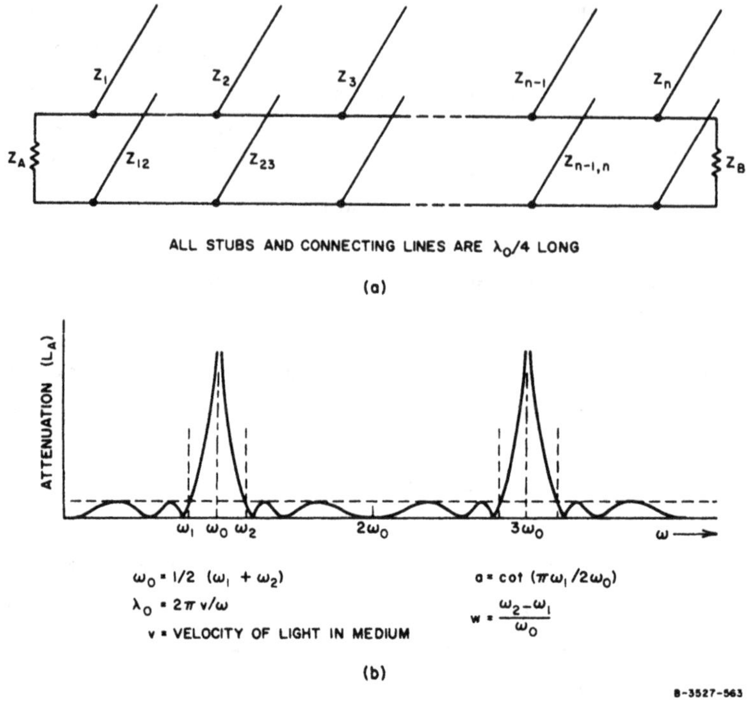

SOURCE: Quarterly Progress Report 7, Contract DA 36-039 SC-87398, SRI; reprinted in the *IEEE Trans. PTGMTT* (see Ref. 6 by B. M. Schiffman and G. L. Matthaei)

FIG. 12.09-1 BANDSTOP FILTER: (a) n-STUB TRANSMISSION-LINE FILTER DERIVED FROM n-ELEMENT LOW-PASS PROTOTYPE; (b) EQUI-RIPPLE CHARACTERISTIC DEFINING CENTER FREQUENCY ω_0, PARAMETER a, AND STOP BAND FRACTIONAL BANDWIDTH w

$$\omega' = \omega_1' a \tan\left(\frac{\pi}{2}\frac{\omega}{\omega_0}\right) \tag{12.09-1}$$

where

$$a = \cot\left(\frac{\pi}{2}\frac{\omega_1}{\omega_0}\right) \tag{12.09-2}$$

$$\omega_0 = \frac{\omega_1 + \omega_2}{2} \tag{12.09-3}$$

ω' and ω_1' are frequency points in the low-pass filter response [such as that in Fig. 12.02-1(d)], and ω, ω_0, ω_1, and ω_2 are frequency points in the corresponding band-stop response in Fig. 12.09-1(b). Note that the band-stop filter response in Fig. 12.09-1(b) has arithmetic symmetry, and that the infinite attenuation point at ω_0 is repeated at frequencies which are odd multiples of ω_0. At these frequencies the stubs in the filter in Fig. 12.09-1(a) are an odd multiple of $\lambda_0/4$ long, so that they short out the main line and cause infinite attenuation (theoretically). In the dual case which uses series-connected stubs, the stubs present open-circuits to the main line at these frequencies. Using the methods of this section, if the low-pass prototype has, say, a Tchebyscheff response with 0.1-db ripple, then the band-stop filter response should have *exactly* 0.1-db Tchebyscheff ripple, with the entire response exactly as predicted by the mapping Eqs. (12.09-1) to (12.09-3). (This statement, of course, ignores problems such as junction effects which occur in the practical construction of the filter.)

Table 12.09-1 presents design equations for filters of the form in Fig. 12.09-1(a). After the designer has selected a low-pass prototype filter (which gives him the parameters g_0, g_1, ..., g_{n+1}, and ω_1'); and has specified ω_1, ω_2, and the source impedance Z_A; he can then compute all of the line impedances. Table 12.09-1 treats filters of the form in Fig. 12.09-1(a) having $n = 1$ up to $n = 5$ stubs. The same equations also apply to the dual cases having short-circuited series stubs if all of the impedances in the equations are replaced by corresponding admittances. However, the case for which series stubs are of the most practical interest is the case of waveguide band-stop filters, and as was noted in Sec. 12.08, it will usually be desirable to use $3\lambda_0/4$ spacings between stubs for waveguide band-stop filters. Figure 12.09-2(a) shows a band-stop filter with $3\lambda_0/4$ spacings between $\lambda_0/4$, short-circuited, series stubs, while Fig. 12.09-2(b) shows the dual case having $\lambda_0/4$, open-circuited, shunt stubs. Table 12.09-2 presents design equations for filters of the form in Fig. 12.09-2(a) having $n = 2$ or $n = 3$ stubs.

As has previously been mentioned, the equations in Tables 12.09-1 and 12.09-2 are exact regardless of the stop-band width. However, in the case of Table 12.09-1 and filter structures of the form in Fig. 12.09-1(a), the impedances Z_j of the shunt stubs become so high as to be impractical, in the case of narrow-stop-band designs. This difficulty can be gotten around by replacing each open-circuited shunt stub, which has a resonator reactance slope parameter of

$$\mathscr{X}_j = \frac{\pi}{4} Z_j \tag{12.09-4}$$

Table 12.09-1

EXACT EQUATIONS FOR BAND-STOP FILTERS
WITH $\lambda_0/4$ SPACING BETWEEN STUBS

The filter structure is as shown in Fig. 12.09-1(a). For the dual case having short-circuited series stubs, replace all impedances in the equations below by corresponding admittances.

n = number of stubs

Z_A and Z_B = terminating impedances

Z_j (j = 1 to n) = impedances of open-circuited shunt stubs

$Z_{j-1,j}$ (j = 2 to n) = connecting line impedances

g_j = values of the elements of the low-pass prototype network as defined in Fig. 12.02-1.

$\Lambda = \omega_1' a$

where

ω_1' = low-pass prototype cutoff frequency

and

a = bandwidth parameter defined in Eq. (12.09-2)

(In all cases the left terminating impedance Z_A is arbitrary).

Case of $n = 1$:

$$Z_1 = \frac{Z_A}{\Lambda g_0 g_1} \; ; \qquad Z_B = \frac{Z_A g_2}{g_0}$$

Case of $n = 2$:

$$Z_1 = Z_A\left(1 + \frac{1}{\Lambda g_0 g_1}\right) ; \qquad Z_{12} = Z_A\left(1 + \Lambda g_0 g_1\right)$$

$$Z_2 = \frac{Z_A g_0}{\Lambda g_2} \; ; \qquad Z_B = Z_A g_0 g_3$$

Case of $n = 3$:

Z_1, Z_{12} and Z_2—same formulas as case $n = 2$

Table 12.09-1 *concluded*

$$Z_3 = \frac{Z_A g_0}{g_4}\left(1 + \frac{1}{\Lambda g_3 g_4}\right) \quad ; \qquad Z_{23} = \frac{Z_A g_0}{g_4}(1 + \Lambda g_3 g_4)$$

$$Z_B = \frac{Z_A g_0}{g_4}$$

Case of $n = 4$:

$$Z_1 = Z_A\left(2 + \frac{1}{\Lambda g_0 g_1}\right) \quad ; \qquad Z_{12} = Z_A\left(\frac{1 + 2\Lambda g_0 g_1}{1 + \Lambda g_0 g_1}\right)$$

$$Z_2 = Z_A\left(\frac{1}{1 + \Lambda g_0 g_1} + \frac{g_0}{\Lambda g_2(1 + \Lambda g_0 g_1)^2}\right) \quad ; \qquad Z_{23} = \frac{Z_A}{g_0}\left(\Lambda g_2 + \frac{g_0}{1 + \Lambda g_0 g_1}\right)$$

$$Z_3 = \frac{Z_A}{\Lambda g_0 g_1} \quad ; \qquad Z_{34} = \frac{Z_A}{g_0 g_5}(1 + \Lambda g_4 g_5)$$

$$Z_4 = \frac{Z_A}{g_0 g_5}\left(1 + \frac{1}{\Lambda g_4 g_5}\right) \quad ; \qquad Z_B = \frac{Z_A}{g_0 g_5}$$

Case of $n = 5$:

Z_1, Z_{12}, Z_2, Z_{23}, Z_3—same formulas as case $n = 4$

$$Z_4 = \frac{Z_A}{g_0}\left(\frac{1}{1 + \Lambda g_5 g_6} + \frac{g_6}{\Lambda g_4(1 + g_4 g_5)^2}\right) \quad ; \qquad Z_{34} = \frac{Z_A}{g_0}\left(\Lambda g_4 + \frac{g_6}{1 + \Lambda g_5 g_6}\right)$$

$$Z_5 = \frac{Z_A g_6}{g_0}\left(2 + \frac{1}{\Lambda g_5 g_6}\right) \quad ; \qquad Z_{45} = \frac{Z_A g_6}{g_0}\left(\frac{1 + 2\Lambda g_5 g_6}{1 + \Lambda g_5 g_6}\right)$$

$$Z_B = \frac{Z_A g_6}{g_0}$$

SOURCE: Quarterly Progress Report 7, Contract DA 36-039 SC-87398, SRI; reprinted in the *IEEE Trans. PTGMTT* (see Ref. 6 by B. M. Schiffman and G. L. Matthaei)

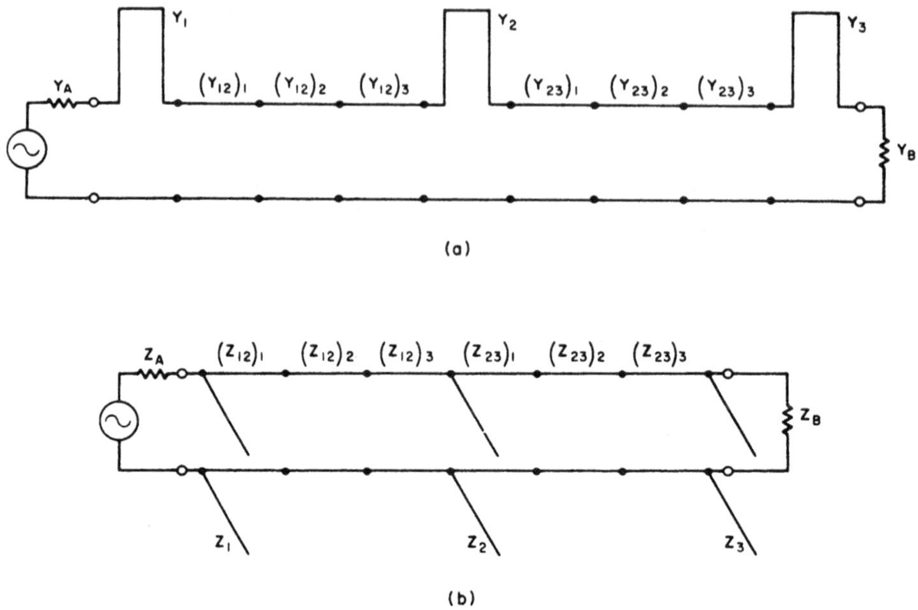

SOURCE: Quarterly Progress Report 7, Contract DA 36-039 SC-87398, SRI; reprinted in the *IEEE Trans. PTGMTT* (see Ref. 6 by B. M. Schiffman and G. L. Matthaei)

FIG. 12.09-2 BAND-STOP FILTERS WITH 3 $\lambda_0/4$ SPACING BETWEEN $\lambda_0/4$ STUBS

[see Fig. 5.08-1(d)], by a capacitively coupled short-circuited stub resonator as shown in Fig. 12.05-1(a), having the same resonator slope parameter. Then the design process becomes the same as that in the example of Sec. 12.07. The filter might take the strip-line form in Fig. 12.07-1, except that in this case the main line of the filter would have some steps in its dimensions since the connecting line impedances $Z_{j,j+1}$ are generally somewhat different from the terminating impedances. The computed performance of a filter design of this type obtained using the equations of this section is discussed in Sec. 12.10.

As previously mentioned, the case in Fig. 12.09-2(a) is of interest primarily for waveguide band-stop filters. Since the useful bandwidth of waveguides is itself rather limited, it is probable that band-stop filters with narrow stop-bands will be of most interest for the waveguide case. For narrow-stop-band designs the series-stub admittances Y_j will become so large as to make them difficult to construct (and especially difficult to

Table 12.09-2

EXACT EQUATIONS FOR BAND-STOP FILTERS WITH $3\lambda_0/4$ SPACING BETWEEN STUBS OR RESONATOR IRISES

The filter structure is of the form in Fig. 12.09-2(a). For the dual case in Fig. 12.09-2(b), replace all admittances in the equations below with corresponding impedances.

n = number of stubs

Y_A, Y_B = terminating admittances

Y_j (j = 1 to n) = admittances of short-circuited series stubs

$(Y_{j-1,j})_k$ = admittance of kth (k = 1, 2 or 3) connecting line from the left, between stubs $j-1$ and j

g_j = values of the elements of the low-pass prototype network as defined in Fig. 12.02-1

$\Lambda = \omega_1' \alpha$ where ω_1' = low-pass cutoff frequency and α = bandwidth parameter defined in Eq. (12.09-2)

(In all cases the left terminating impedance is arbitrary)

Case of n = 2:

$$Y_1 = Y_A\left(1 + \frac{1}{\Lambda g_0 g_1}\right); \qquad (Y_{12})_1 = Y_A(1 + \Lambda g_0 g_1)$$

$$Y_2 = Y_A \frac{g_0}{\Lambda g_2}(1 + 2\Lambda g_2 g_3); \qquad (Y_{12})_2 = Y_A g_0 g_3\left(\frac{1}{1 + \Lambda g_2 g_3}\right)$$

$$Y_B = Y_A g_0 g_3; \qquad (Y_{12})_3 = Y_A g_0 g_3\left(\frac{1 + 2\Lambda g_2 g_3}{1 + \Lambda g_2 g_3}\right)$$

Case of n = 3:

$$Y_1 = Y_A\left(3 + \frac{1}{\Lambda g_0 g_1}\right) \qquad (Y_{12})_1 = Y_A\left(\frac{1 + 3\Lambda g_0 g_1}{1 + 2\Lambda g_0 g_1}\right)$$

$$(Y_{12})_2 = Y_A\left(\frac{1 + \Lambda g_0 g_1}{1 + 2\Lambda g_0 g_1}\right)$$

$$Y_2 = Y_A \frac{g_0}{\Lambda g_2} \qquad (Y_{12})_3 = Y_A(1 + \Lambda g_0 g_1)$$

$$(Y_{23})_1 = Y_A \frac{g_0}{g_4}(1 + \Lambda g_3 g_4)$$

$$Y_3 = Y_A \frac{g_0}{g_4}\left(3 + \frac{1}{\Lambda g_3 g_4}\right) \qquad (Y_{23})_2 = Y_A \frac{g_0}{g_4}\left(\frac{1 + \Lambda g_3 g_4}{1 + 2\Lambda g_3 g_4}\right)$$

$$Y_B = Y_A \frac{g_0}{g_4} \qquad (Y_{23})_3 = Y_A \frac{g_0}{g_4}\left(\frac{1 + 3\Lambda g_3 g_4}{1 + 2\Lambda g_3 g_4}\right)$$

SOURCE: Quarterly Progress Report 7, Contract DA 36-039 SC-87398, SRI; reprinted in the *IEEE Trans. PTGMTT* (see Ref. 6 by B. M. Schiffman and G. L. Matthaei)

construct with good unloaded Q's). Now the series stubs have a resonator susceptance slope parameter of

$$\mathscr{b}_j = \frac{\pi}{4} Y_j \qquad (12.09\text{-}5)$$

and for narrow-stop band cases it is convenient to realize these resonator slope parameters using inductively coupled $\lambda_0/2$ short-circuited stub resonators of the form in Fig. 12.05-1(d). From this point on, the design process is the same as that in the example in Sec. 12.08. The completed filter might look like that in Fig. 12.08-1 except that the design equations of this section call for steps in the impedance along the main waveguide of the filter.

Derivation of the Design Equations in this Section—Exact methods for transmission-line filter design such as the methods of Ozaki and Ishii[2,3] and of Jones[4] have existed for some time. However, though these methods are mathematically elegant, they are computationally so tedious that they have found little application in actual practice. The case described herein is an exception in that it is an exact method of design, yet is also very simple to use for the design of practical filters. The fundamental principle of this design method was first suggested by Ozaki and Ishii,[3] while the design equations in this section were worked out by B. M. Schiffman.[5]

The design procedure of this section hinges on Kuroda's identity.[2,3] This identity in transmission-line form is as shown in Fig. 12.09-3. Note that this identity says that a circuit consisting of an open-circuited shunt stub and a connecting line, which are both of the same length, has an exact equivalent circuit consisting of a short-circuited series stub with a connecting line at the opposite side.

Figure 12.09-4 traces out the way in which Kuroda's identity is used to relate a band-stop filter of the type in Fig. 12.09-1(a) to a low-pass prototype filter. Figure 12.09-4(a) shows a low-pass prototype filter for the case of $n = 3$ reactive elements. Applying the mapping Eq. (12.09-1) to the shunt susceptances and series reactances of this filter gives

$$\omega' C'_j = C'_j \omega'_1 a \tan\left(\frac{\pi}{2}\frac{\omega}{\omega_0}\right) \qquad (12.09\text{-}6)$$

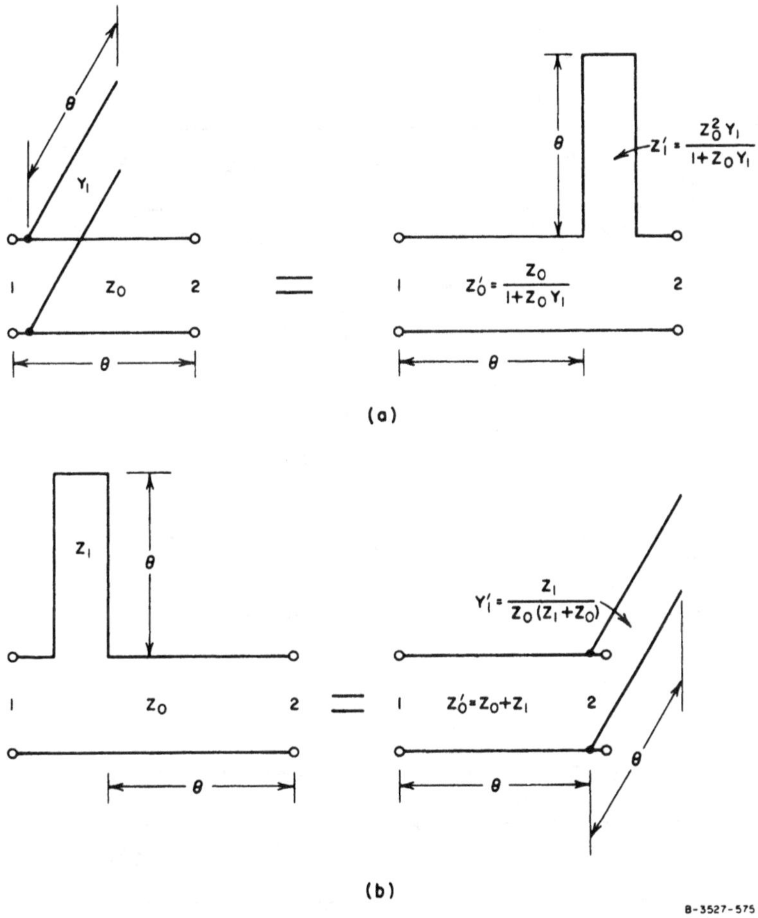

SOURCE: Quarterly Progress Report 7, Contract DA 36-039 SC-87398, SRI; reprinted in the *IEEE Trans. PTGMTT* (see Ref. 6 by B. M. Schiffman and G. L. Matthaei)

FIG. 12.09-3 KURODA'S IDENTITY IN TRANSMISSION LINE FORM

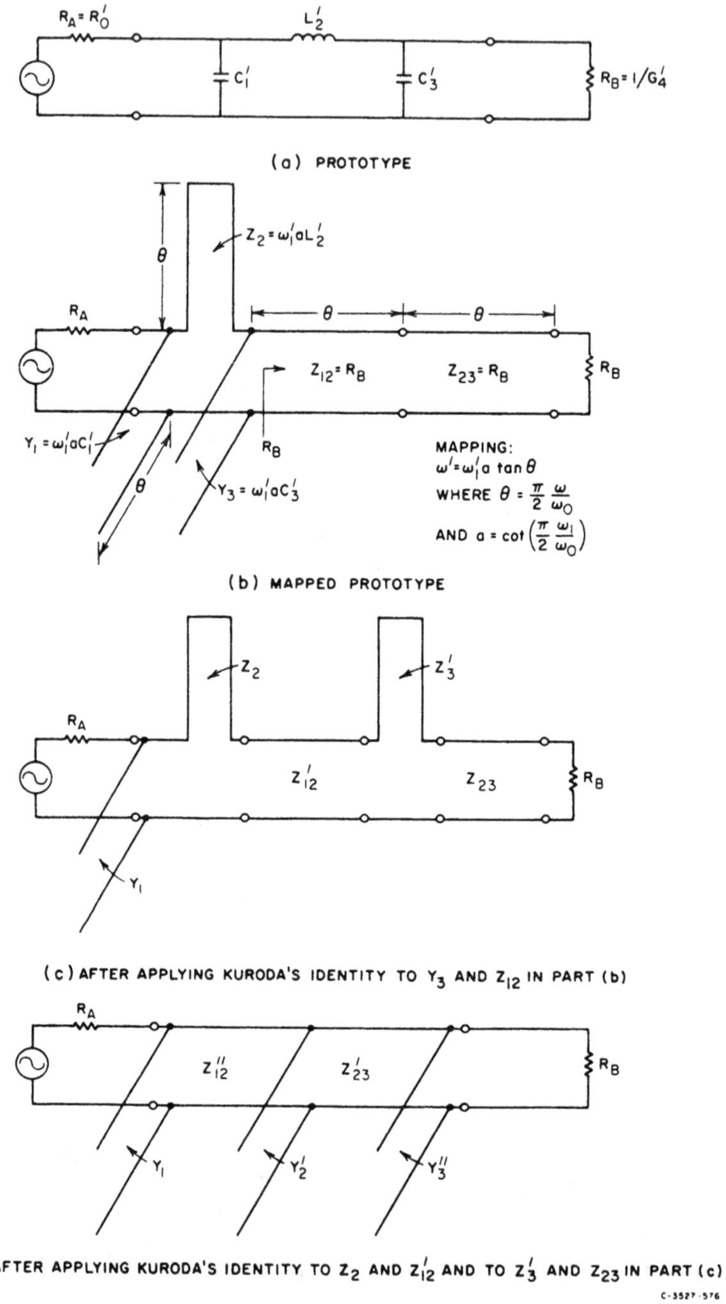

SOURCE: Quarterly Progress Report 7, Contract DA 36-039 SC-87398, SRI; reprinted in the *IEEE Trans. PTGMTT* (see Ref. 6 by B. M. Schiffman and G. L. Matthaei)

FIG. 12.09-4 STAGES IN THE TRANSFORMATION OF A LOW-PASS PROTOTYPE FILTER INTO A BAND-STOP TRANSMISSION-LINE FILTER

$$\omega' L'_j = L'_j \omega'_1 \, a \tan\left(\frac{\pi}{2}\frac{\omega}{\omega_0}\right) \quad . \tag{12.09-7}$$

Note that the right side of Eq. (12.09-6) corresponds to the susceptance of an open-circuited stub having a characteristic admittance

$$Y_j = C'_j \omega'_1 a \quad , \tag{12.09-8}$$

the stub being $\lambda_0/4$ long at ω_0. Similarly, the right side of Eq. (12.09-7) corresponds to the reactance of a short-circuited stub of characteristic impedance

$$Z_j = L'_j \omega'_1 a \quad , \tag{12.09-9}$$

the stub being $\lambda_0/4$ long at frequency ω_0. Thus, the shunt capacitors in the low-pass prototype become open-circuited shunt stubs in the mapped filter, while the series inductance in the prototype becomes a short-circuited series stub in the mapped filter.

Note that in the mapped filter in Fig. 12.09-4(b), the terminations seen by the reactive part of the filter are still R_A on the left and R_B on the right. However, on the right, two additional line sections of impedance $Z_{12} = Z_{23} = R_B$ have been added. Since their characteristic impedance matches that of the termination, they have no effect on the attenuation characteristic of the circuit, their only effect on the response being to give some added phase shift. The circuit in Fig. 12.09-4(b) then has a response which is the desired exact mapping of the low-pass prototype response. The only trouble with the filter in Fig. 12.09-4(b) is that it contains a series stub which is difficult to construct in a shielded TEM-mode microwave structure.

The series stub in Fig. 12.09-4(b) can be eliminated by application of Kuroda's identity (Fig. 12.09-3). Applying Kuroda's identity to stub Y_3 and line Z_{12} in Fig. 12.09-4(b) gives the circuit in Fig. 12.09-4(c). Then applying Kuroda's identity simultaneously to stub Z_2 and line Z'_{12}, and to stub Z'_3 and line Z_{23} in Fig. 12.09-4(c) gives the circuit in Fig. 12.09-4(d). Note that the circuit in Fig. 12.09-4(d) has exactly the same input impedance and over-all transmission properties as the circuit in Fig. 12.09-4(b), while the circuit in Fig. 12.09-4(d) has no series stubs.

The equations in Tables 12.09-1 and 12.09-2 were derived by use of repeated applications of the procedures described above. For reasons of convenience, the equations in the tables use a somewhat different notation than does the example in Fig. 12.09-4; however, the principles used are the same. The equations in the tables in this section also provide for a shift in impedance level from that of the low-pass prototype.

SEC. 12.10, SOME EXAMPLES ILLUSTRATING THE PERFORMANCE OBTAINABLE USING THE EQUATIONS IN SEC. 12.09

In this section both a wide-stop-band filter example and a narrow-stop-band filter example will be discussed, both being designed by the methods of Sec. 12.09. In the case of the narrow-stop-band design, the high-impedance shunt stubs will be replaced by capacitively coupled stubs of moderate impedance, as was discussed in the preceding section.

A Wide-Stop-Band Example—Let us assume that a design is desired having 0.1-db pass-band Tchebyscheff ripple with the band edges at $f_1 = 1.12$ Gc and $f_2 = 2.08$ Gc. This puts the stop-band center at $f_0 = (f_1 + f_2)/2 = 1.60$ Gc. Let us further suppose that at least 30-db attenuation is required at the frequencies $f = 1.600 \pm 0.115$ Gc. Now $\omega_1/\omega_0 = f_1/f_0 = 1.12/1.60 = 0.70$; while at the lower 30-db point, $\omega/\omega_0 = f/f_0 = (1.600-0.115)/1.500 = 0.9283$. By Eq. (12.09-2), $a = 0.5095$; and by Eq. (12.09-1), the low-pass prototype should have at least 30-db attenuation for $\omega'/\omega_1' = 4.5$. By Fig. 4.03-5 we find that an $n = 3$ prototype will have 34.5-db attenuation for $\omega'/\omega_1' = 4.5$, while an $n = 2$ design will have only 15.5-db attenuation. Hence, an $n = 3$ design must be used.

From Table 4.05-2(a) the low-pass prototype parameters $g_0 = 1$, $g_1 = 1.0315$, $g_2 = 1.1474$, $g_3 = 1.0315$, $g_4 = 1.0000$, and $\omega_1' = 1$ were obtained. Using Table 12.09-1 with $Z_A = 50$ ohms, the line impedances were then computed to be $Z_1 = Z_3 = 145.1$ ohms, $Z_{12} = Z_{23} = 76.3$ ohms, $Z_2 = 85.5$ ohms, and $Z_B = 50$ ohms. The filter was constructed in strip-line form, and its dimensions are shown in Fig. 12.10-1. Note that the $Z_2 = 85.5$-ohm stub has been realized as two 176-ohm stubs in parallel. This was done so that narrower stubs could be used which should have less junction effect.*

* As this material is going to press some evidence has been obtained on another filter structure, which indicates that at least under some circumstances it may be unwise to replace a single stub by double stubs in parallel. Some experimental results suggest that there may be interaction between the two stubs in parallel so that regardless of their tuning they will always give two separate resonances where they are both expected to resonate at the same frequency. Also, between the two resonances the attenuation of the pair of stubs may drop very low. This possible behavior of double stubs needs further study, and is mentioned here to alert the reader of what could be a pitfall.

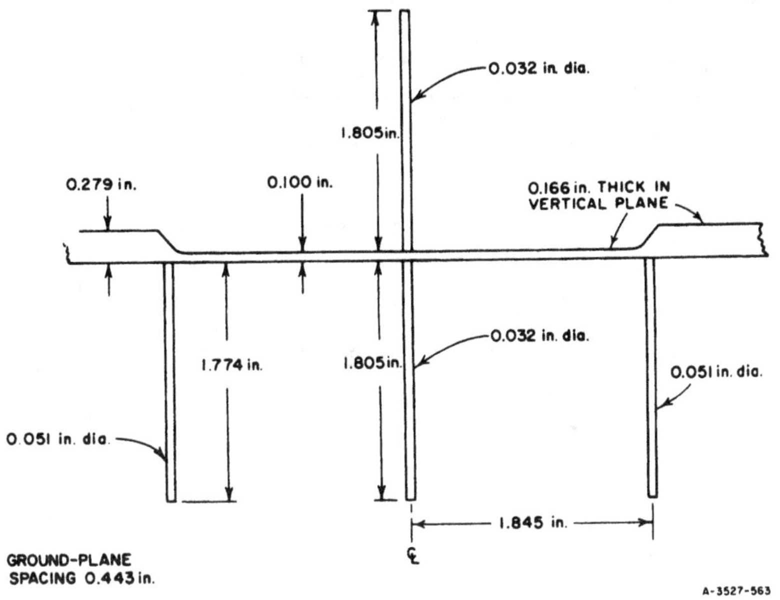

SOURCE: Quarterly Progress Report 7, Contract DA 36-039 SC-87398, SRI; reprinted in the *IEEE Trans. PTGMTT* (see Ref. 6 by B. M. Schiffman and G. L. Matthaei)

FIG. 12.10-1 A STRIP-LINE, WIDE-STOP-BAND FILTER

The stubs in this case are realized using round wire supported by Polyfoam. The dimensions of the round wires were determined by use of the approximate formula

$$d = \frac{4b}{\pi \operatorname{antilog}_{10}\left(\frac{\sqrt{\epsilon_r}Z}{138}\right)} \tag{12.10-1}$$

where d is the rod diameter, b is the ground plane spacing, ϵ_r is the relative dielectric constant, and Z is the desired line impedance.

Figure 12.10-2 shows the measured and computed performance of the filter in Fig. 12.10-1. Note that the computed performance is in perfect agreement with the specifications, as it should be. The measured performance of the filter is also in excellent agreement with the specifications, except that the filter is tuned slightly high. This could have been corrected by the addition of tuning screws.

A Narrow-Stop-Band Filter Example—Figure 12.10-3(a) shows another 3-stub band-stop filter design worked out using Table 12.09-1 and the same low-pass prototype as was used for the design described above. However, in this case the pass-band edges were defined so as to give a stop-band

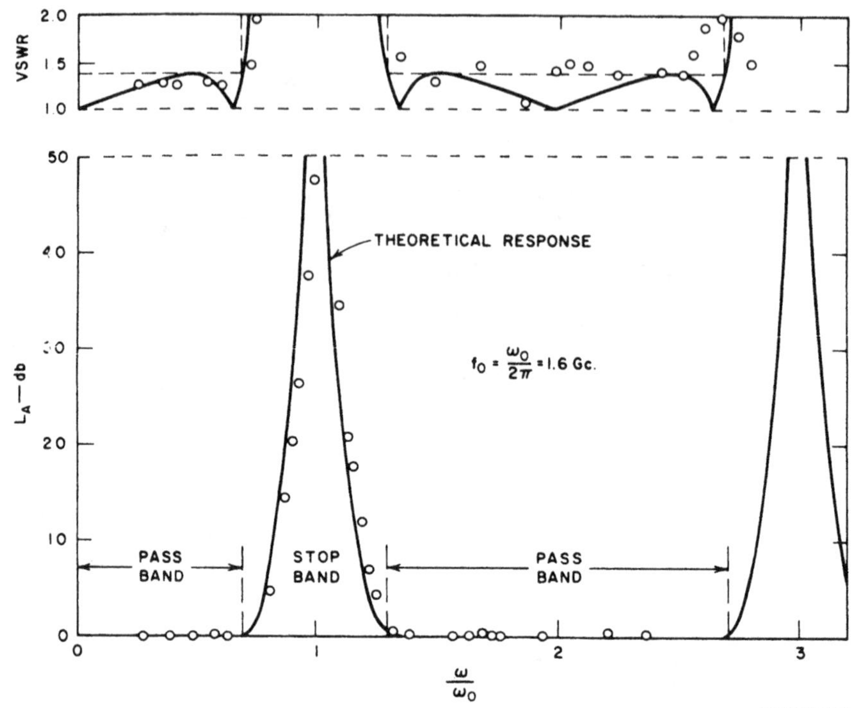

SOURCE: Quarterly Progress Report 7, Contract DA 36-039 SC-87398, SRI; reprinted in the *IEEE Trans. PTGMTT* (see Ref. 6 by B. M. Schiffman and G. L. Matthaei)

FIG. 12.10-2 COMPUTED AND MEASURED PERFORMANCE OF THE WIDE-STOP-BAND FILTER IN FIG. 12.10-1

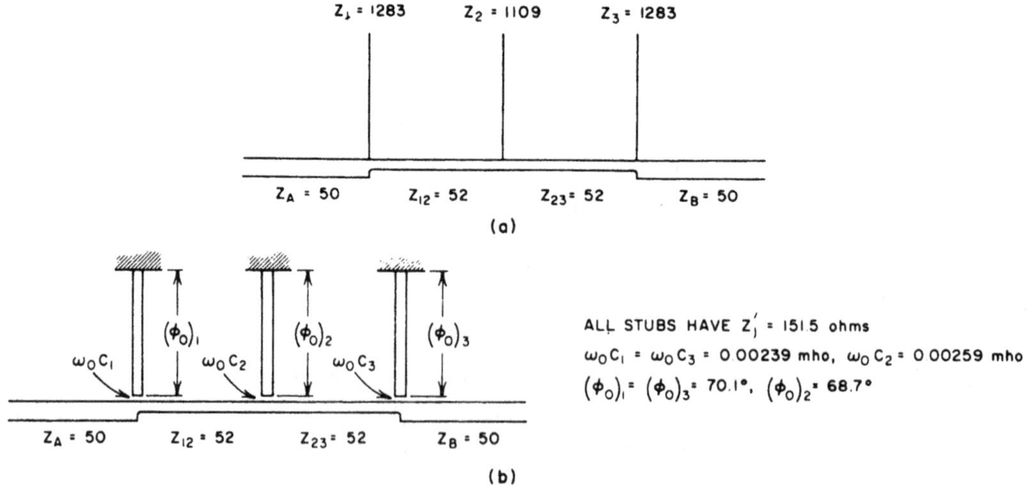

FIG. 12.10-3 NARROW-STOP-BAND FILTERS

fractional bandwidth of $w = 0.05$ as measured to the equal-ripple pass-band edges. As can be seen from the figure, in this case the line impedances $Z_{12} = Z_{23}$ are very nearly the same as the terminating impedances; however, the stub impedances are in excess of 1,000 ohms. To eliminate these impractical stub impedances, the open-circuited stub resonators were replaced by capacitively coupled short-circuited stub resonators as described in Sec. 12.09. The stub impedances were arbitrarily set at 151.5 ohms, but probably a lower impedance would have been a better choice from the standpoint of obtaining maximum resonator unloaded Q's. Figure 12.10-3(b) shows the design using capacitively coupled resonators.

Figure 12.10-4 shows the computed responses of these two designs. Note that the exact design [in Fig. 12.10-3(a)] has the desired 0.1-db pass-band ripple as specified. Though the approximate design with capacitively coupled stubs deviates a little from the specifications, it comes very close to the response over the frequency range shown. However, as can be seen from Fig. 12.10-5, which shows the same responses with enlarged scale over a larger frequency range, at higher frequencies the response of the approximate design deviates a good deal from that of the exact design. This is due to the variation of the coupling susceptances with frequency. Though either the method of Sec. 12.04 or the method of Sec. 12.09 will involve some error, narrow-band designs worked out by starting with the equations in Sec. 12.09 can be expected to give more accurate results.

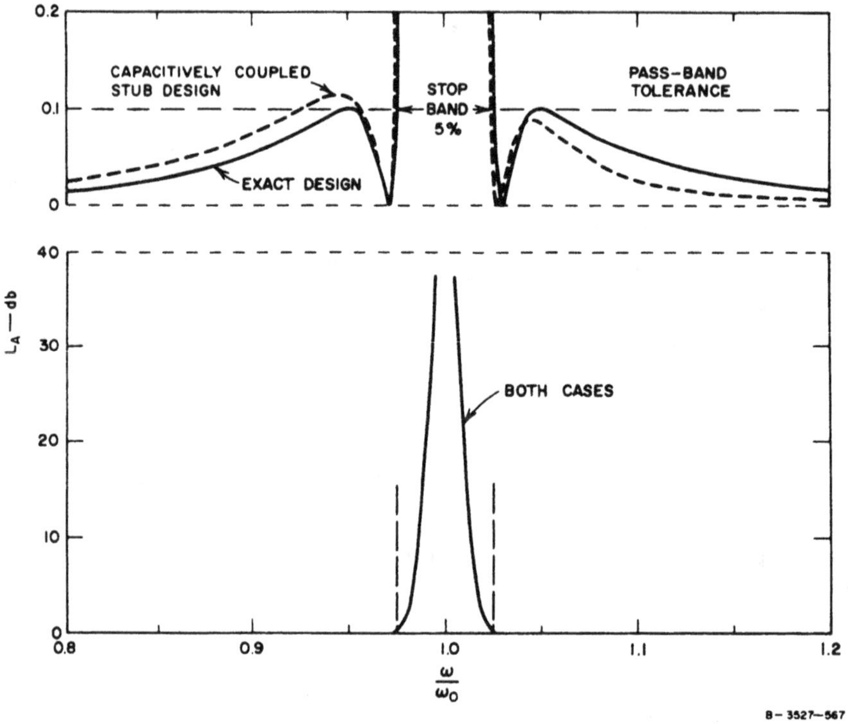

FIG. 12.10-4 COMPUTED RESPONSE OF THE FILTERS IN FIG. 12.10-3

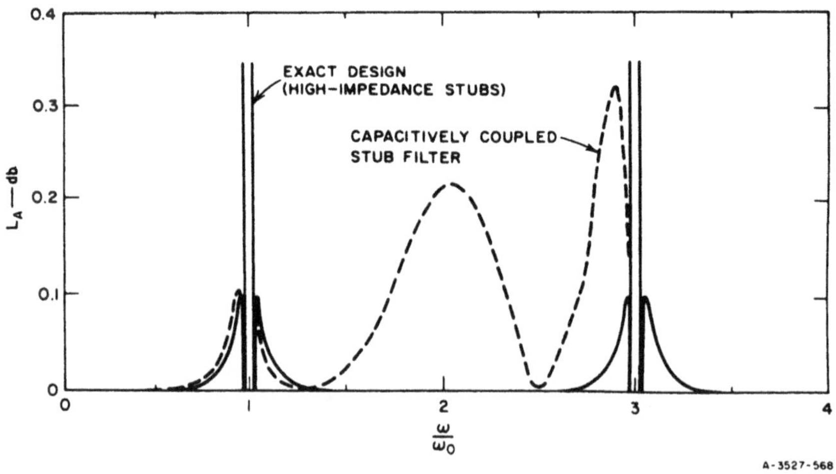

FIG. 12.10-5 THE RESPONSES IN FIG. 12.10-4 WITH AN ENLARGED SCALE

REFERENCES

1. Leo Young, G. L. Matthaei, and E. M. T. Jones, "Microwave Bandstop Filters with Narrow Stop Bands," *IRE Trans., PGMTT*-10, pp. 416-427 (November 1962).
2. H. Ozaki and J. Ishii, "Synthesis of Transmission-Line Networks and Design of UHF Filters," *IRE Trans., PGCT*-2, pp. 325-336 (December 1955).
3. H. Ozaki and J. Ishii, "Synthesis of a Class of Strip-Line Filters," *IRE Trans., PGCT*-5, pp. 104-109 (June 1958).
4. E. M. T. Jones, "Synthesis of Wide-Band Microwave Filters to Have Prescribed Insertion Loss," *IRE Convention Record,* 1956 National Convention, Part 5, pp. 119-128.
5. B. M. Schiffman, P. S. Carter, Jr., and G. L. Matthaei, "Microwave Filters and Coupling Structures," Quarterly Progress Report 7, Sec. II, SRI Project 3527, Contract DA 36-039 SC-87398, Stanford Research Institute, Menlo Park, California (October 1962).
6. B. M. Schiffman and G. L. Matthaei, "Exact Design of Band-Stop Microwave Filters," *IEEE Trans. PTGMTT-12*, pp. 6-15 (January 1964).

CHAPTER 13

TEM-MODE, COUPLED-TRANSMISSION-LINE DIRECTIONAL COUPLERS, AND BRANCH-LINE DIRECTIONAL COUPLERS

SEC. 13.01, INTRODUCTION

Though there are many possible kinds of directional couplers, the discussion in this book will be confined to only two types. They are TEM-mode couplers consisting of parallel, coupled transmission lines[1-7] (Secs. 13.01 to 13.08), and branch-line couplers which may be of either TEM-mode or waveguide form[8-11] (Secs. 13.09 to 13.14). TEM-mode, coupled-transmission-line couplers are included because they relate very closely to some of the directional filters discussed in Chapter 14, they are a very widely used form of coupler, and they are a type of coupler of which the authors have special knowledge. The branch-line couplers are included because, as is discussed in Chapter 15, they are useful as part of high-power filter systems, and because they also are a form of coupler of which the authors happen to have special knowledge in connection with high-power filter work. No effort will be made to treat the numerous other useful forms of couplers which appear at this time to be less closely related to the topics in this book.

Figure 13.01-1 illustrates schematically a quarter-wavelength coupler and a three-quarter-wavelength coupler, together with their frequency responses. These couplers, like all couplers analyzed in this chapter, have end-to-end symmetry. The coupled signal travels in the direction opposite to that of the input signal and therefore these couplers are often referred to as "backward couplers." The electrical length, θ, of each coupled section in the two couplers is 90 degrees at midband. The variation of coupling with frequency in the single-section, quarter-wavelength coupler is approximately sinusoidal. The coupling variation with frequency is much less in the 3-section coupler, which is formed by cascading three, quarter-wavelength couplers. It can be made to be maximally-flat or equal-ripple by adjusting the couplings of the three individual couplers. Even greater bandwidths can be obtained by cascading more than three couplers.

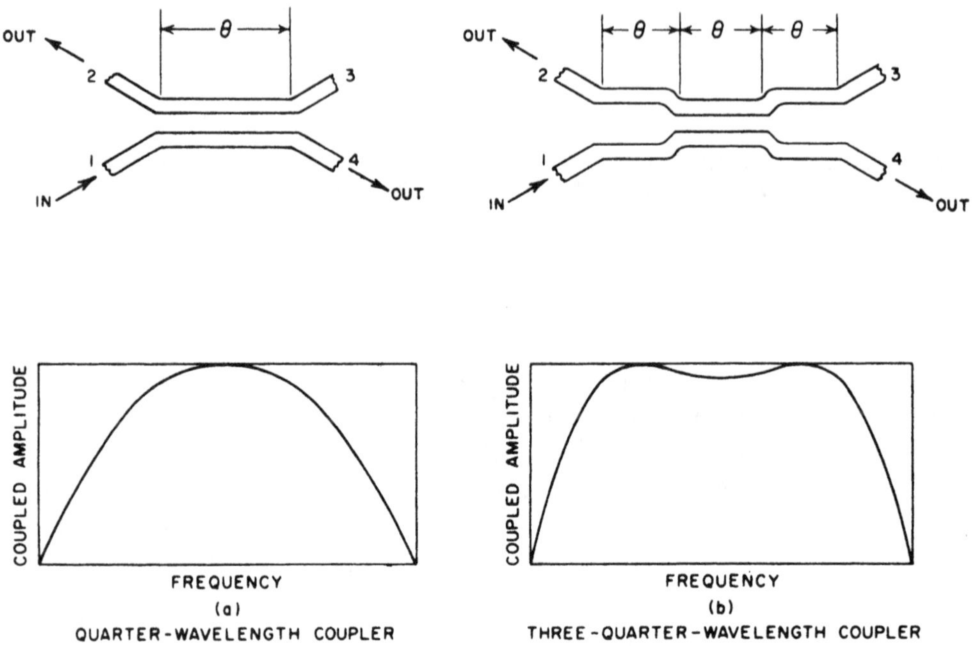

(a)
QUARTER-WAVELENGTH COUPLER

(b)
THREE-QUARTER-WAVELENGTH COUPLER

SOURCE: Final Report, Contract DA-36-039 SC-63232, SRI; reprinted in *IRE Trans. PGMTT* (see Ref. 5 by J. K. Shimizu and E. M. T. Jones)

FIG. 13.01-1 SKETCH SHOWING TYPICAL CONFIGURATIONS AND FREQUENCY RESPONSES FOR TEM-MODE, COUPLED-TRANSMISSION-LINE DIRECTIONAL COUPLERS OF ONE AND THREE SECTIONS

A variety of physical configurations for the coupled TEM lines can be used in these couplers, as is illustrated in Fig. 13.01-2. The configurations (a), (b) and (c) are most suitable for couplers having weak coupling such as 20 db, 30 db, etc. The configurations (d), (e), (f), (g) and (h) are most suitable for couplers having tight coupling such as 3 db. Intermediate values of coupling can usually most easily be obtained with configurations (d), (e), (f), and (g), although configuration (a) is often useful. Configuration (e) has a disadvantage in that the individual lines are unsymmetrically located with respect to the ground planes; hence the connections at the ends tend to excite ground-plane modes. This can be prevented, however, if the structure is closed in at the sides so that only the desired TEM modes can propagate. Configuration (f) avoids this difficulty by making one of the lines double, while the other line interleaves the double line in order to give tight coupling. In this manner the structure is made to be electrically balanced with respect to the ground planes, and ground-plane modes will not be excited. Configuration (h) uses conductors *A* and *B* of circular cross section, surrounded by a conductor *C*. The electrical potential of *C* is floating with respect to the potentials of the ground planes *D* and of conductors *A* and *B*. The addition of the floating shield *C* has the effect of greatly increasing the coupling between lines *A* and *B*.

The actual configuration to use in a particular application depends on a number of factors. However, configuration (a) is most frequently used for weak coupling while configuration (e) is often used for tight coupling. Both of these configurations can be fabricated by printed-circuit techniques. Configuration (c), utilizing round conductors, is particularly good for weak-coupling, high-power applications. When configuration (e) is fabricated using thick strips with rounded edges, it is

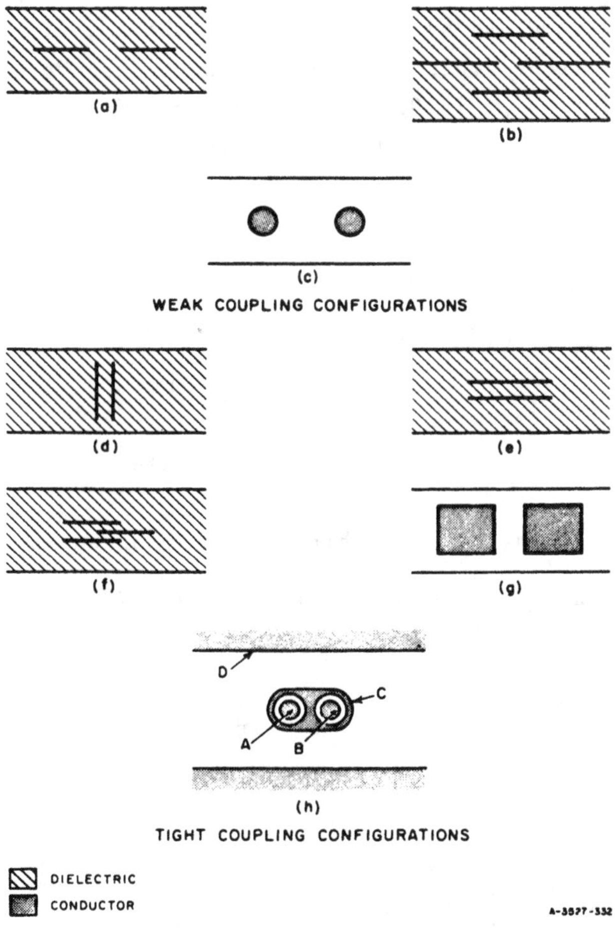

FIG. 13.01-2 CROSS SECTIONS OF TYPICAL TEM-MODE, COUPLED-TRANSMISSION-LINES

suitable for strong-coupling, high-power applications. Reference to the detailed designs presented in Secs. 13.05 and 13.06 may also help the reader decide on the most appropriate configuration for his particular application.

It is essential for the operation of these couplers that the cross section be uniformly filled with air or some other low-loss dielectric material. Configurations (c) and (g) have enough rigidity so that they can be air-filled. With the other configurations it is usually necessary to use a rigid, low-loss dielectric material for mechanical support.

SEC. 13.02, DESIGN RELATIONS FOR TEM-MODE, COUPLED-TRANSMISSION-LINE DIRECTIONAL COUPLERS OF ONE SECTION

TEM-mode coupled-transmission-line directional couplers theoretically are perfectly matched and have infinite directivity at all frequencies. When a wave of voltage amplitude E is incident upon Port 1 of a quarter-wavelength coupler (Fig. 13.01-1), the voltage at Port 2 is given by

$$\frac{E_2}{E} = \frac{jc \sin\theta}{\sqrt{1-c^2}\cos\theta + j\sin\theta} \qquad (13.02\text{-}1)$$

while the voltage at Port 4 is given by

$$\frac{E_4}{E} = \frac{\sqrt{1-c^2}}{\sqrt{1-c^2}\cos\theta + j\sin\theta} \qquad (13.02\text{-}2)$$

where c is the *coupling factor*, which is the midband value of $|E_2/E|$.

Since the phases of E_2 and E_4 are usually of little interest, Eqs. (13.02-1) and (13.02-2) can be reduced to

$$\left|\frac{E_2}{E}\right|^2 = \frac{c^2\sin^2\theta}{1-c^2\cos^2\theta} \qquad (13.02\text{-}1a)$$

and

$$\left|\frac{E_4}{E}\right|^2 = \frac{1-c^2}{1-c^2\cos^2\theta} \qquad (13.02\text{-}2a)$$

For small values of c, Eq. (13.02-1) predicts that E_2/E varies as sin θ. For large values of c the coupling variation with frequency is as shown in Fig. 13.02-1. The voltage at Port 3 is zero. (These results are proved in Sec. 13.07.)

The electrical length, θ, of the coupled lines is related to the physical length l by means of the relation $\theta = 2\pi l/\lambda$, where λ is the wavelength in the medium surrounding the coupled lines. The midband voltage coupling factor c is

$$c = \frac{Z_{oe}/Z_{oo} - 1}{Z_{oe}/Z_{oo} + 1} \tag{13.02-3}$$

where Z_{oe} and Z_{oo} are the even- and odd-mode impedances discussed below. In order that the coupler be perfectly matched to its terminating transmission line of characteristic impedance Z_0, it is necessary that

$$Z_0 = \sqrt{Z_{oe} Z_{oo}} \quad . \tag{13.02-4}$$

The even-mode impedance Z_{oe} is the characteristic impedance of a single coupled line to ground when equal currents are flowing in the two lines, while the odd-mode impedance Z_{oo} is the characteristic impedance of a single line to ground when equal and opposite currents are flowing in the two lines. Values of Z_{oe} and Z_{oo} for a number of cross sections are presented in Sec. 5.05.

Equations (13.02-3) and (13.02-4) can be rearranged to give the convenient design relations

$$Z_{oe} = Z_0 \sqrt{\frac{1 + c}{1 - c}} \tag{13.02-5}$$

and

$$Z_{oo} = Z_0 \sqrt{\frac{1 - c}{1 + c}} \quad . \tag{13.02-6}$$

The physical dimensions necessary to give the required even- and odd-mode impedances Z_{oe} and Z_{oo} can then be obtained from Sec. 5.05.

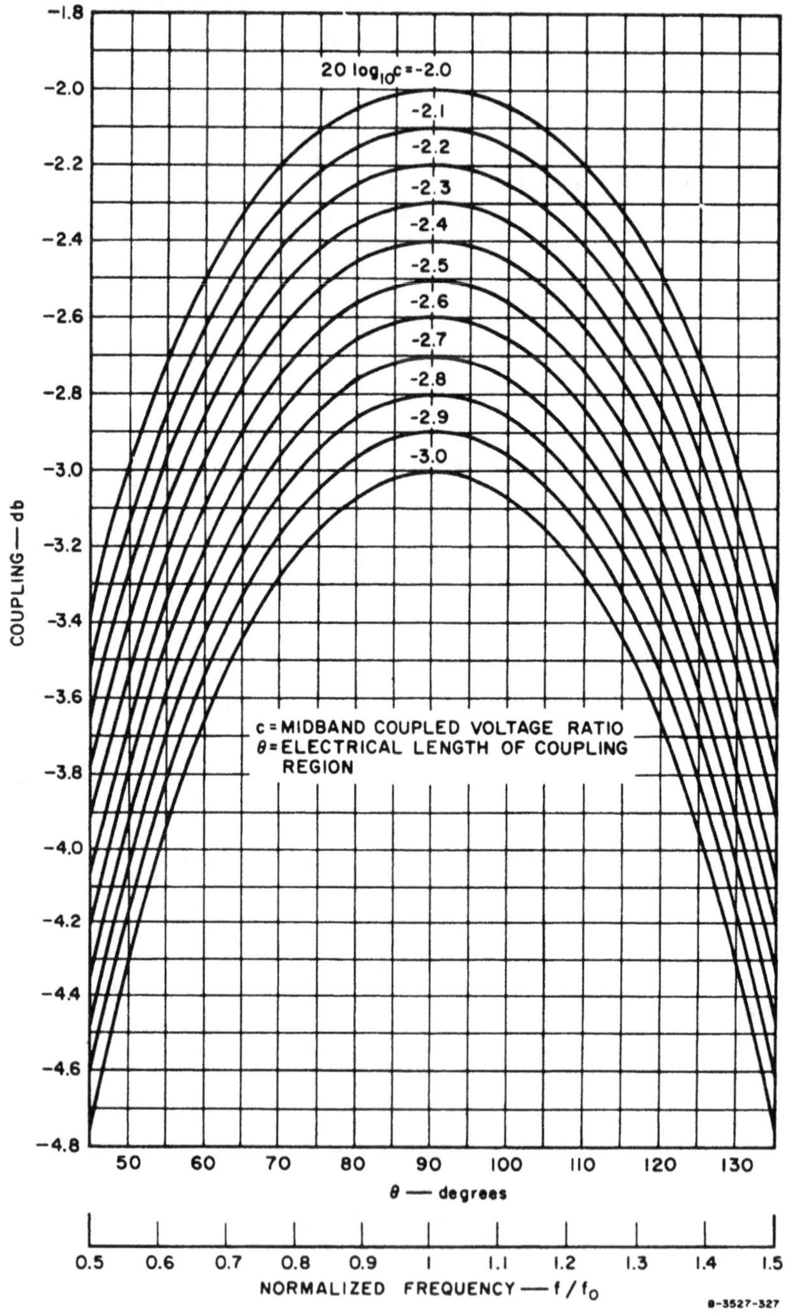

FIG. 13.02-1 COUPLING AS A FUNCTION OF FREQUENCY FOR TEM-MODE, COUPLED-TRANSMISSION-LINE DIRECTIONAL COUPLERS OF ONE SECTION, HAVING TIGHT COUPLING

SEC. 13.03, DESIGN RELATIONS FOR TEM-MODE, COUPLED-TRANSMISSION-LINE DIRECTIONAL COUPLERS OF THREE SECTIONS

The analysis of TEM-mode couplers of more than one section soon leads to long and complicated expressions. The problem simplifies somewhat when one recognizes that the TEM-mode coupler is analytically similar to the quarter-wave transformer. That is to say, the even and odd modes of operation are duals of each other (impedance in one corresponds to admittance in the other, and vice versa), and the mathematics of each mode then reduces to the analysis of cascaded transmission-line sections, each of which is one-quarter wavelength long at center frequency. There is, however, one major difference between the desired performance of the TEM-mode coupler and that of the quarter-wave transformer of Chapter 6: Whereas the quarter-wave transformer is required to have *low* reflection in the operating band, the "quarter-wave filter"* on which the TEM coupler is based is required to have a sizeable specified and nearly constant reflection coefficient across the band of operation. Thus the transducer loss functions are not the same, and the numerical data in Chapter 6 then do not apply to TEM-mode directional couplers. A new analysis has to be undertaken for the quarter-wave filter prototype. Figure 13.03-1(a) shows a three-section coupler and Fig. 13.03-1(b) shows a quarter-wave-filter prototype from which it can be designed.

The results of an analysis of a symmetrical three-section coupler (Fig. 13.03-1) will be given without proof in this section. It will be followed in Sec. 13.04 with a first-order formula for a symmetrical coupler of any (odd) number of sections, and some particular solutions for five-section couplers.

Since this type of coupler has matched resistive impedances at all of its ports, $|E/E_4|^2$ corresponds to the transducer loss ratio. For three-section couplers it is given by a polynomial in $\cos^2\theta$ (compare with Sec. 6.02),

$$\left|\frac{E}{E_4}\right|^2 = A_0 + A_1 \cos^2\theta + A_2 \cos^4\theta + A_3 \cos^6\theta \qquad (13.03\text{-}1)$$

---

*The term "quarter-wave filter" is introduced to distinguish the resulting physically symmetrical structures from quarter-wave impedance transformers with a monotone sequence of steps, as in Chapter 6.

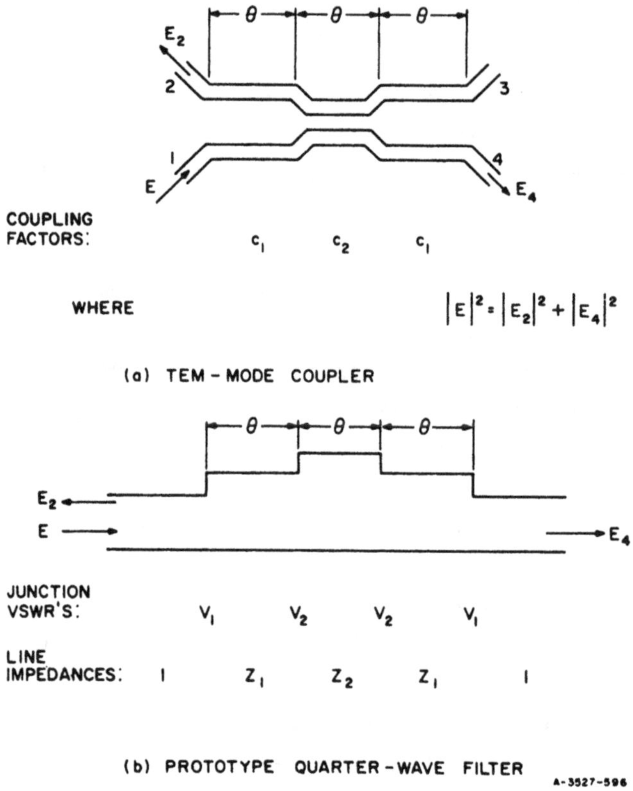

SOURCE: Proc. IEE (see Ref. 24 by L. Young)

FIG. 13.03-1 THREE-SECTION TEM-COUPLER AND QUARTER-WAVE FILTER PROTOTYPE

where A_0, A_1, A_2, and A_4 are functions of the end- and middle-section coupling factors c_1 and c_2. The A's are given by

$$A_0 = L_0^2 \tag{13.03-2}$$

$$A_1 = (L_2 - 1)^2 - L_0^2 - L_0 L_1 L_2 \tag{13.03-3}$$

$$A_2 = \left(\frac{L_1 L_2}{2}\right)^2 + L_0 L_1 L_2 - 2L_2(L_2 - 1) \tag{13.03-4}$$

$$A_3 = L_2^2 - \left(\frac{L_1 L_2}{2}\right)^2 \tag{13.03-5}$$

in which

$$\left.\begin{aligned} L_0 &= \frac{1 + c_1^2 - 2c_1c_2}{(1 - c_1^2)\sqrt{1 - c_2^2}} \\ L_1 &= \frac{2}{\sqrt{1 - c_1^2}} \\ L_2 &= 2 + \frac{2(1 - c_1c_2)}{\sqrt{(1 - c_1^2)(1 - c_2^2)}} \quad . \end{aligned}\right\} \qquad (13.03\text{-}6)$$

The coupling voltage E_2 is then found (from conservation of energy) to be related to E and E_4 by

$$\left|\frac{E_2}{E}\right|^2 = 1 - \left|\frac{E_4}{E}\right|^2 \quad . \qquad (13.03\text{-}7)$$

Design Formulas—Since the three-section coupler shown in Fig. 13.03-1 is symmetrical, one has to determine two even-mode and two odd-mode impedances, so that its physical dimensions can be determined from Sec. 5.05. The even- and odd-mode impedances $(Z_{oe})_1$, $(Z_{oo})_1$, $(Z_{oe})_2$, and $(Z_{oo})_2$ are related to the coupling factors c_1 and c_2 by

$$(Z_{oe})_i = Z_0\sqrt{\frac{1 + c_i}{1 - c_i}} \qquad (i = 1,2) \qquad (13.03\text{-}8)$$

and

$$(Z_{oo})_i = Z_0\sqrt{\frac{1 - c_i}{1 + c_i}} \qquad (i = 1,2) \quad . \qquad (13.03\text{-}9)$$

Design formulas for c_1 and c_2 will now be stated, and their derivation will be indicated afterwards. The midband over-all coupling is denoted by c_0, which is the value of $|E_2/E|$ at midband. The coupler performance is usually specified by the midband coupling c_0 and one other parameter which determines the shape of the coupling response against frequency, for instance whether it is to be maximally flat or to have a specified ripple. This parameter is here denoted by V_1 and is determined from Figs. 13.03-2 and 13.03-3 as will be explained below. The design procedure for a three-section coupler (Fig. 13.03-1) is then as follows:

1. Determine the midband coupling, c_0. For instance, for a three-section 3-db coupler with 0.3-db ripple, $10 \log_{10} (c_0^2) = 3.3$ db.

2. Then determine $V_{1,MF}$ from Fig. 13.03-2. For a maximally flat coupler, $V_1 = V_{1,MF}$. For an equal-ripple coupler, determine V_1 from Fig. 13.03-3. Use the 10-db curve for all couplers with coupling weaker than 10 db.

   If the ripple is specified (for instance, if the coupling is to lie between 2.7 and 3.3 db, the ripple is 0.3 db), then the lower curves in Fig. 13.03-3 are used. If the fractional bandwidth w is specified, the upper curves in Fig. 13.03-3 are used. (The fractional bandwidth w is here defined, as usual, by

$$w = \frac{f_2 - f_1}{f_0} = 2\left(\frac{f_2 - f_1}{f_2 + f_1}\right) , \tag{13.03-10}$$

   where f_1 and f_2 are the band-edge frequencies of the equal-ripple pass band. Either the ripple or the bandwidth may be specified, and they are related as shown in Fig. 13.03-4.*

3. Determine R from

$$R = \frac{1 + c_0}{1 - c_0} \tag{13.03-11}$$

   where c_0 is the coupling factor. [The midband coupling in decibels is $10 \log_{10}(c_0^2)$ db.]

4. Determine V_2 from

$$V_2 = V_1\sqrt{R} \quad . \tag{13.03-12}$$

5. Determine Z_1 and Z_2 from

$$Z_1 = V_1 \quad , \quad Z_2 = V_1 V_2 \tag{13.03-13}$$

6. Finally, determine c_1 and c_2 from

$$c_1 = \frac{Z_1^2 - 1}{Z_1^2 + 1} \quad \text{and} \quad c_2 = \frac{Z_2^2 - 1}{Z_2^2 + 1} \quad . \tag{13.03-14}$$

---

* There is a slight inconsistency in these curves owing to a lack of more complete numerical data. The 10-db curve refers to 10 db at center frequency; the 3-db curve refers to 3-db average over the band.

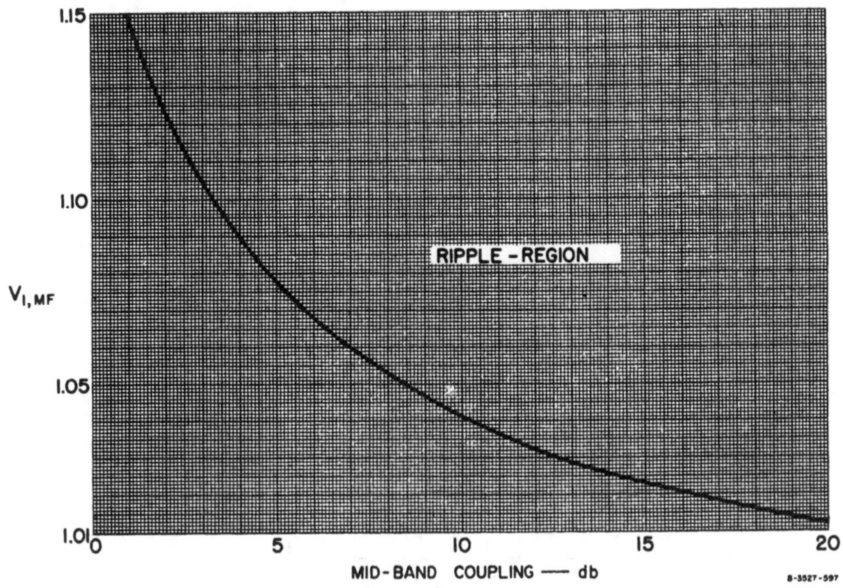

SOURCE: Proc. IEE (see Ref. 24 by L. Young)

FIG. 13.03-2 $V_1 = V_{1,MF}$ OF MAXIMALLY FLAT, THREE-SECTION, TEM-MODE COUPLER AS A FUNCTION OF MIDBAND COUPLING

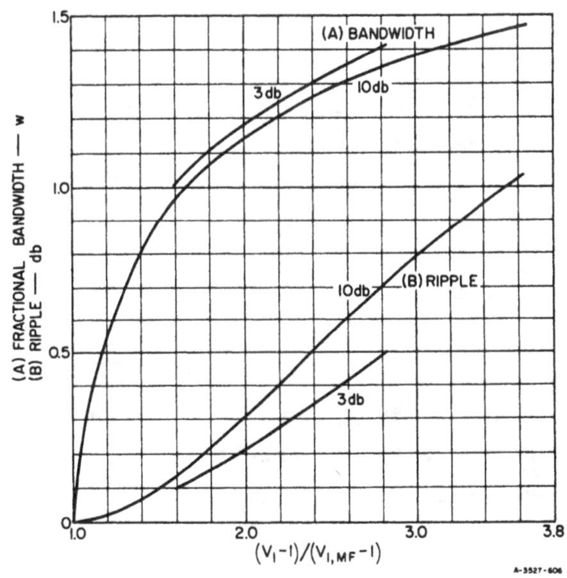

SOURCE: Proc. IEE (see Ref. 24 by L. Young)

FIG. 13.03-3 FRACTIONAL BANDWIDTH AND RIPPLE AS V_1 IS INCREASED, FOR 3-db AND 10-db, THREE-SECTION TEM-MODE COUPLERS

The $(Z_{oe})_i$ and $(Z_{oo})_i$ are then given by Eqs. (13.03-8) and (13.09-9) and the physical dimensions can be obtained from Sec. 5.05, as already pointed out.

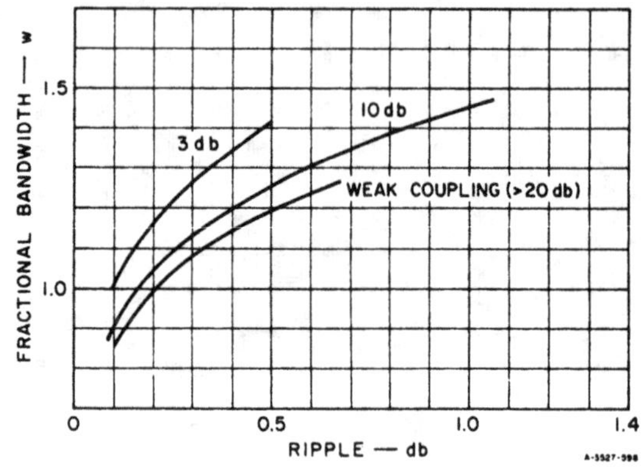

SOURCE: Proc. IEE (see Ref. 24 by L. Young)

FIG. 13.03-4 FRACTIONAL BANDWIDTH AS A FUNCTION OF RIPPLE FOR THREE-SECTION TEM-MODE COUPLERS WITH 3-db, 10-db, AND WEAK COUPLING FACTORS

Quarter-Wave Filter Parameters—The parameters, V_1, V_2, Z_1, Z_2, and R just introduced have meaning in terms of the quarter-wave filter prototype circuit, which will be explained more fully in Sec. 13.08. The essence of this design concept is that the backward-coupled wave of the TEM-mode coupler (E_2 in Fig. 13.03-2) corresponds analytically to the reflected wave of the quarter-wave filter, as indicated in Fig. 13.03-1(a) and (b). One therefore has to synthesize only a two-port (instead of a four-port), making its reflected wave behave as one would want the coupling to be. The V_i are the step-VSWR's, and the Z_i are the normalized impedances [Fig. 13.03-1(b)] which are turned into the $(Z_{oe})_i$ and $(Z_{oo})_i$ by means of Eqs. (13.03-8) through (13.03-14). Equation (13.03-6) also simplifies to

$$
\left.\begin{aligned}
L_0 &= \frac{1}{2}\left(\frac{V_1}{V_2} + \frac{V_2}{V_1}\right) \\
L_1 &= V_1 + \frac{1}{V_1} \\
L_2 &= V_2 + 2 + \frac{1}{V_2} \quad .
\end{aligned}\right\} \qquad (13.03\text{-}15)
$$

The parameter R corresponds to the midband VSWR of the quarter-wave filter, and c_0 is its midband reflection coefficient. Furthermore, at midband its section lengths are $\theta = \pi/2$, and the reflected and transmitted powers are then given by

$$
\left|\frac{E_2}{E}\right|^2 = c_0^2 = \left(\frac{R-1}{R+1}\right)^2 \qquad (13.03\text{-}16)
$$

and

$$
\left|\frac{E_4}{E}\right|^2 = \frac{1}{A_0} = \frac{1}{L_0^2} = \frac{4R}{(R+1)^2} \quad . \qquad (13.03\text{-}17)
$$

Maximally Flat Coupling—To obtain a maximally flat response at $\theta = \pi/2$, we may specify

$$
A_1 = 0 \qquad (13.03\text{-}18)
$$

in Eq. (13.03-1). Eliminating V_2 leaves a quartic equation in V_1:

$$
\left(V_1\sqrt{R} + 1 + \frac{1}{V_1\sqrt{R}}\right)^2 - \frac{R+1}{2\sqrt{R}}\left(2V_1 + \frac{2}{V_1} + V_1^2\sqrt{R} + \frac{1}{V_1^2\sqrt{R}}\right) = \frac{3(R+1)^2}{4R} \quad . \qquad (13.03\text{-}19)
$$

The solution of this equation is graphed in Fig. 13.03-2 where $V_1 = V_{1,MF}$ is plotted against the midband coupling in decibels. Selecting $V_1 = V_{1,MF}$ from Fig. 13.03-2 or Eq. (13.03-19) will give a maximally flat response. (It turns out that when A_1 is thus made to vanish, A_2 is not equal to zero. To obtain a flatter response with both A_1 and A_2 equal to zero, the coupler or filter cannot be symmetrical between ends.)

When the midband coupling is maintained constant while V_1 is increased above the curve in Fig. 13.03-2, the coupling-versus-frequency characteristic turns from maximally flat to equal-ripple. The ripple region for V_1 is indicated in Fig. 13.03-2.

10-db Couplers—A maximally flat, 10-db coupler has the quarter-wave filter parameters $V_1 = Z_1 = 1.041$, $V_2 = Z_2/Z_1 = 1.446$ (so $Z_2 = 1.505$), and hence $c_1 = 0.0403$, $c_2 = 0.388$. As V_1 is increased, the coupling response develops ripples, as can be seen from Fig. 13.03-5 which shows six curves of coupling against frequency for V_1 = 1.04, 1.06, 1.08, 1.10, 1.12, and 1.14. The coupling is symmetrical about the center frequency in Fig. 13.03-5, and is therefore plotted only for frequencies above midband. The midband coupling has been maintained at 10 db by satisfying Eq. (13.03-12) with $R = 1.926$. [This R is obtained from Eq. (13.03-11) with $c_0 = 0.3162$, corresponding to 10-db coupling.] For any given V_1, the parameters c_1 and c_2, and the even- and odd-mode impedances, $(Z_{oe})_i$ and $(Z_{oo})_i$, can be derived from Eqs. (13.03-8) through (13.03-14).

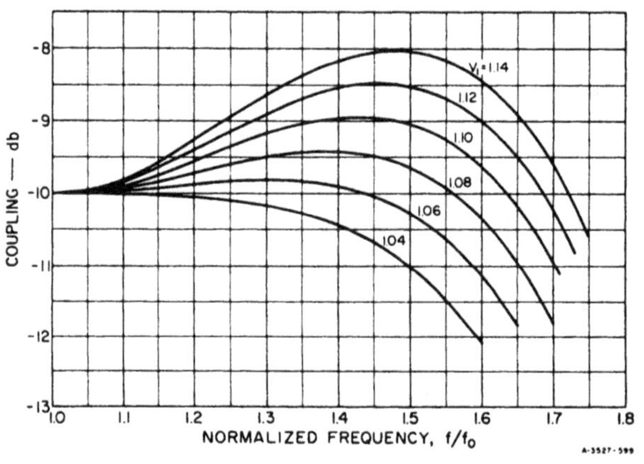

SOURCE: Proc. IEE (see Ref. 24 by L. Young)

FIG. 13.03-5 COUPLING AS A FUNCTION OF FREQUENCY FOR SIX THREE-SECTION TEM-MODE COUPLERS HAVING 10-db MIDBAND COUPLING

3-db Couplers[1,5]—Similar curves for 3-db couplers are reproduced in Fig. 13.03-6, except that here the *average* coupling is maintained constant at 3 db. (In Fig. 13.03-5 the *midband* coupling was maintained constant at 10 db.) The five curves shown in Fig. 13.03-6 have coupling ripples of 0.1, 0.2, 0.3, 0.4, and 0.5 db, respectively, and their coupling factors are given in Table 13.03-1.

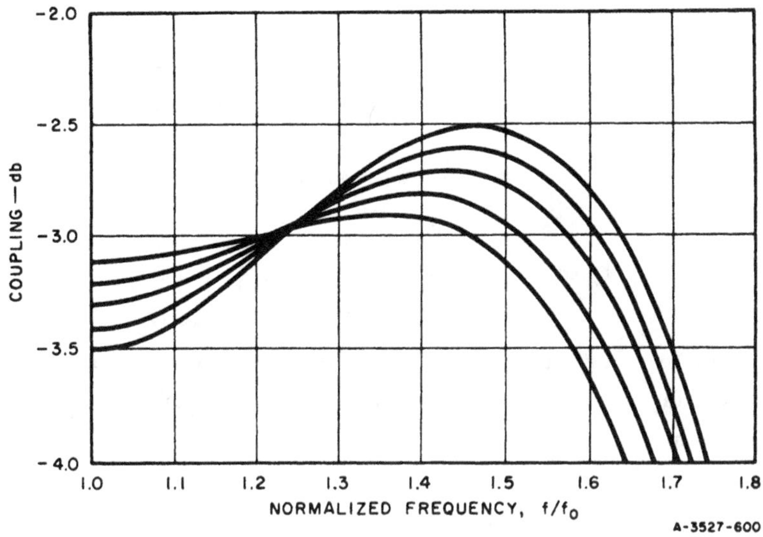

SOURCE: Final Report, Contract DA-36-039 SC-63232, SRI; reprinted in *IRE Trans. PGMTT* (see Ref. 5 by J. K. Shimizu and E. M. T. Jones)

FIG. 13.03-6 COUPLING AS A FUNCTION OF FREQUENCY FOR FIVE THREE-SECTION TEM-MODE COUPLERS HAVING 3-db AVERAGE COUPLING

0-db Coupling—It can be shown for the maximally flat, three-section coupler that in the limit as R tends to infinity, and as the coupling ratio therefore tends to unity (0-db), V_1 tends ultimately to $\sqrt{3/2}$. (V_2 tends to infinity.) Thus V_1 for the maximally flat three-section coupler never exceeds $\sqrt{3/2} = 1.224745$, or equivalently, c_1 never exceeds 0.2. (Of course it is not possible to ever attain 0-db coupling, just as it is impossible to ever attain complete reflection in a stepped-impedance filter.)

Table 13.03-1

DESIGN PARAMETERS FOR THREE-SECTION, 3-db, TEM-MODE COUPLERS

db RIPPLE	c_1	c_2	c_0
±0.1 db	0.15505	0.8273	0.6998
±0.2 db	0.18367	0.8405	0.6918
±0.3 db	0.21104	0.85241	0.6839
±0.4 db	0.23371	0.86119	0.6760
±0.5 db	0.25373	0.86838	0.6683

SOURCE: Final Report, Contract DA-36-039 SC-63232, SRI; reprinted in *IRE Trans. PGMTT* (see Ref. 5 by J. K. Shimizu and E. M. T. Jones)

SEC. 13.04, RELATIONS FOR TEM-MODE COUPLED-TRANSMISSION-LINE DIRECTIONAL COUPLERS OF FIVE AND MORE SECTIONS

First-Order Design Theory—It is not difficult to show that for weak couplings the amplitude coupling ratio, $c(\theta)$, of an n-section coupler as a function of θ, is

$$\left|\frac{E_2}{E}\right| = c_1 \sin n\,\theta + (c_2 - c_1) \sin (n - 2)\,\theta + \ldots$$

$$+ (c_i - c_{i-1}) \sin (n - 2i + 2)\,\theta + \ldots$$

$$+ (c_{[(n+1)/2]} - c_{[(n-1)/2]}) \sin \theta \qquad (13.04\text{-}1)$$

in the notation of Fig. 13.04-1. If, for example, it is required to obtain maximally flat coupling, then the c_i must satisfy a set of $(n - 1)/2$ linear equations

$$\left.\begin{aligned} \left[\frac{d^p c(\theta)}{d\theta^p}\right]_{\theta=\pi/2} &= 0 \\ \text{for } p = 2, 4, 6, \ldots, n - 1 & \end{aligned}\right\} \qquad (13.04\text{-}2)$$

in addition to one equation, determining the center-frequency coupling. The solutions of the system of equations (13.04-2) for $n = 3$ and $n = 5$ can be shown to be,

$$\text{for } n = 3, \quad c_2 = 10\,c_1\ ; \qquad c_0 = c_2 - 2c_1 = 8c_1 \qquad (13.04\text{-}3)$$

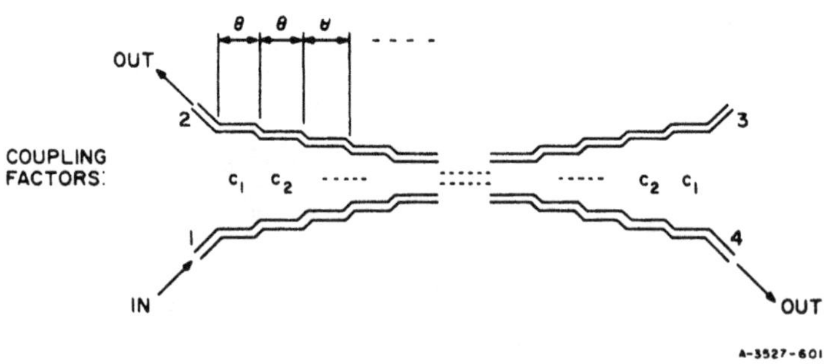

SOURCE: Proc. IEE (see Ref. 24 by L. Young)

FIG. 13.04-1 TEM-MODE COUPLER OF n SECTIONS

and for $n = 5$,

$$\left.\begin{aligned} \frac{c_3}{536} &= \frac{c_2}{84} = \frac{c_1}{9} \\ c_0 &= c_3 - 2c_2 + 2c_1 = 386c_1/9 \end{aligned}\right\} \quad . \qquad (13.04\text{-}4)$$

It can also be shown that Eq. (13.03-19) for the maximally flat, three-section coupler reduces in the first-order approximation to

$$V_1 = \frac{7 + \sqrt{R}}{8} \quad . \qquad (13.04\text{-}5)$$

Solutions for Five-Section Couplers—As was the case with the three-section coupler in Sec. 13.03, the solution of the five-section coupler is most easily expressed in terms of the normalized quarter-wave filter impedances Z_1, Z_2, and Z_3, which are again related to the coupling factors c_i (Fig. 13.04-1) by

$$Z_i^2 = \frac{1 + c_i}{1 - c_i} \quad , \quad \text{or} \quad c_i = \frac{Z_i^2 - 1}{Z_i^2 + 1}$$

$$(i = 1,\ 2,\ 3) \qquad (13.04\text{-}6)$$

corresponding to Eq. (13.03-14) for the three-section coupler. Again the $(Z_{oe})_i$ and $(Z_{oo})_i$ are obtained as in Eqs. (13.03-8) and (13.03-9), but with $i = 1, 2, 3$.

The first-order solution in Eq. (13.04-4) for maximally flat coupling must become less accurate as the c_i approach unity, since they can never exceed unity. Equation (13.04-4) becomes more accurate for appreciable couplings if we substitute $\log Z_i$ for c_i. [Compare the similarly expedient substitution used in Eq. (6.06-11) in the first-order theory of quarter-wave transformers.] Then for maximally flat couplers, to a good approximation,

$$\frac{\log Z_3}{536} = \frac{\log Z_2}{84} = \frac{\log Z_1}{9} \quad . \tag{13.04-7}$$

The Z_1, Z_2, and Z_3 are completely determined when the midband coupling factor c_0 is also specified, from the formula

$$\log Z_3 - 2 \log Z_2 + 2 \log Z_1 = \frac{1}{2} \log \left(\frac{1 + c_0}{1 - c_0} \right) \tag{13.04-8}$$

(which is a special case of the more general formulas for n-section couplers given later in Sec. 13.08). The first-order equations, Eqs. (13.04-7) and (13.04-8), for maximally flat coupling have been found to hold very well for coupling as tight as 10-db ($c_0^2 = 0.1$). The coupling becomes weaker as the frequency deviates from midband. The following fractional bandwidths are obtained for various coupling deviations from the midband c_0 (when c_0 corresponds to 10-db coupling or less): $w = 0.82$ for 0.1 db deviation; $w = 1.01$ for 0.25 db deviation; $w = 1.13$ for 0.5 db deviation; $w = 1.32$ for 1.0 db deviation; and $w = 1.51$ for 2.0 db deviation.

The first-order relations in Eqs. (13.04-7) and (13.04-8) for maximally flat coupling were even solved for a 3-db coupler ($c_0^2 = 0.5$) with the following results. The solution is easily found to be

$$\left. \begin{aligned} Z_1 &= 1.0207 \\ Z_2 &= 1.2114 \\ Z_3 &= 3.4004 \quad . \end{aligned} \right\} \tag{13.04-9}$$

As the frequency departs from midband, the coupling at first becomes stronger, going from 3.01 db to 2.97 db, and then becomes weaker. It is 3.11 db at the edges of a 103-percent band ($w = 1.03$), and 3.51 db at the edges of a 124-percent band ($w = 1.24$).

An exact solution for a particular 3-db coupler has been given by Cohn and Koontz.[12] They found by trial-and-error that with

$$Z_1 = 1.098$$

$$Z_2 = 1.417$$

$$Z_3 = 4.130$$

an equal-ripple characteristic (with three ripples) was obtained, the coupling being 3.01 ± 0.163 db over a fractional bandwidth of $w = 1.41$. (The midband coupling is 2.847 db.)

In all cases the coupler is determined by the Z_i, which give the c_i from Eq. (13.04-6), and the $(Z_{oe})_i$ and $(Z_{oo})_i$ from Eqs. (13.03-8) and (13.03-9), the physical dimensions then being obtained from Sec. 5.05.

SEC. 13.05, TYPICAL DESIGNS FOR TEM-MODE, COUPLED-TRANSMISSION-LINE DIRECTIONAL COUPLERS OF ONE SECTION WITH APPROXIMATELY 3 db AVERAGE COUPLING

Figure 13.05-1 shows what has proved to be a very popular method of construction for a completely shielded quarter-wavelength coupler having tight coupling.[6] In this structure the individual lines are not symmetrically located with respect to the upper and lower ground planes; hence there is a tendency for the connections at the ends to excite ground-plane modes. However, if the structure is completely closed in at the sides as shown in Fig. 13.05-1, such modes are cut off, and no difficulty should be encountered. The coupler illustrated was designed to have 2.9 db midband coupling at Port 2. The actual measured coupling, directivity and VSWR of the coupler is also shown in the figure. The excellent values shown were obtained without recourse to any special matching techniques since the center frequency was only 225 Mc. For operation at frequencies on the order of 1000 Mc and higher it is usually necessary to place capacitive tuning screws at either end of the coupled lines. These compensate the discontinuities there, and allow satisfactory values of directivity and VSWR to be achieved. The coupling response is usually little affected by these matching devices.

The procedure used to design this coupler which uses 0.002-inch-thick copper strips supported by polystyrene dielectric having a relative dielectric constant $\epsilon_r = 2.56$ is as follows. The midband coupling is 2.9 db, so $c = 0.716$. The terminating impedance Z_0 is equal to 50 ohms, and therefore Eqs. (13.02-5) and (13.02-6) yield $\sqrt{\epsilon_r}\, Z_{oe} = 197$ ohms and $\sqrt{\epsilon_r}\, Z_{oo} = 32.5$ ohms. The spacing between the strips was chosen to be

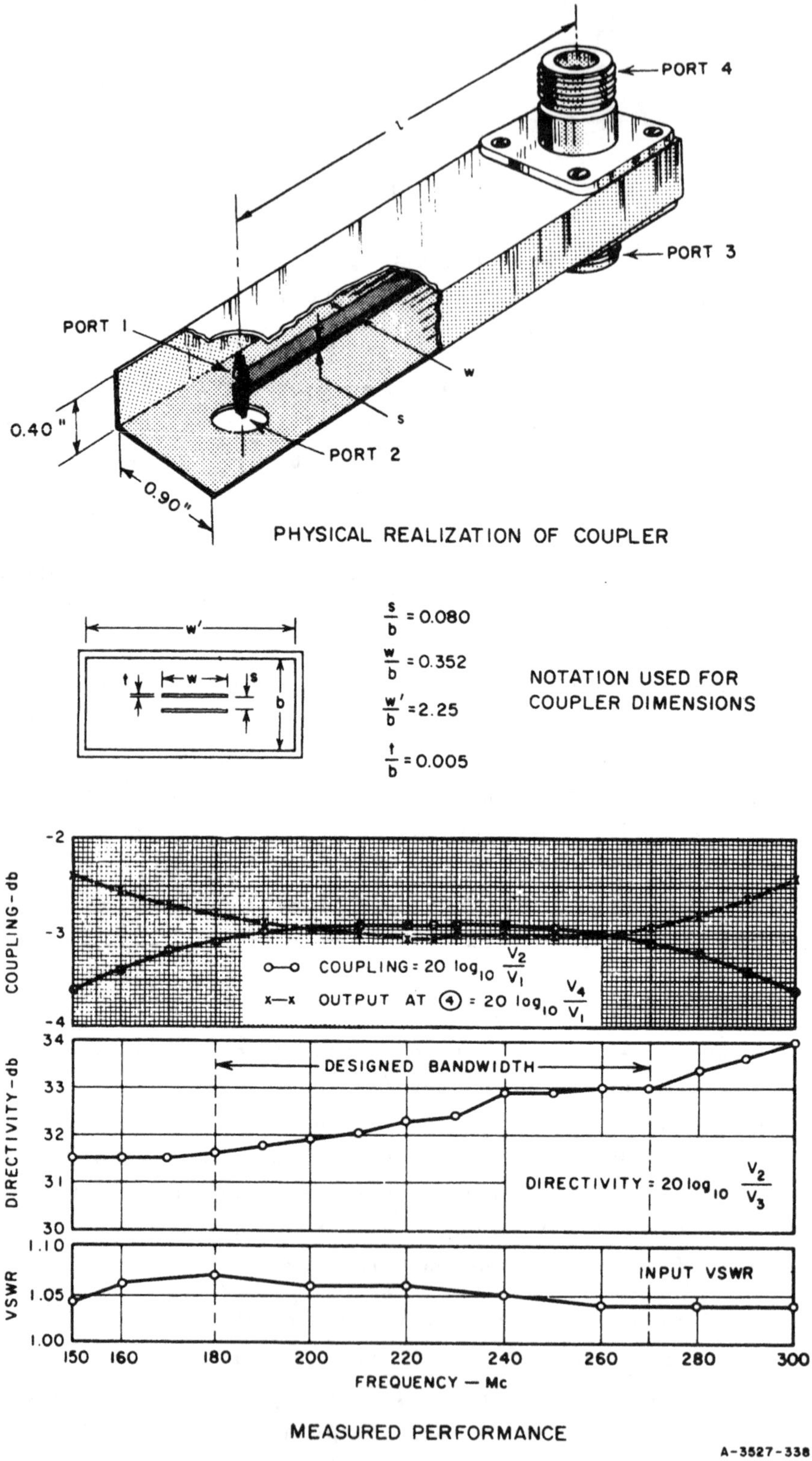

SOURCE: Scientific Report 1, Contract AF 19(604)-1571, SRI (see Ref. 6 by J. K. Shimizu)

FIG. 13.05-1 CONSTRUCTION DETAILS AND MEASURED PERFORMANCE OF A PRINTED-CIRCUIT, 3-db COUPLER

0.032 inch so that with $b = 0.400$ inch in Fig. 13.05-1, $s/b = 0.08$, and $t/b = 0.005$. The value of w/b is then determined from either Eq. (5.05-16) or Eq. (5.05-17), using the values of $C'_{fe}/\epsilon = 0.545$ or $C'_{fo}/\epsilon = 1.21$ determined from Fig. 5.05-4. In using Eq. (5.05-16) one should use the value $(s + 2t)/b = 0.09$ for s/b in that formula, which assumes $t/b = 0$, since it is the fields external to the strips that are significant in determining Z_{oe}. One finds that $w/b = 0.373$. Using Eq. (5.05-17) with $s/b = 0.08$, since the fields between the strips are of most significance in determining Z_{oo}, one finds $w/b = 0.337$. The fact that different values of w/b are obtained using these two formulas is a reflection of the fact that they are only approximately correct.

The actual value of w/b used in this coupler was 0.352, which lies between these two theoretical values. It was computed by an approximate technique before Eqs. (5.05-16) and (5.05-17) were available. The spacing $(w' - w)/2$ between the edges of the strips and the side walls of the metal case is great enough so that it does not have any appreciable effect on the even- and odd-mode impedances of the strips.

Another useful type of printed-circuit configuration[13] for tight coupling is illustrated in Fig. 13.05-2. The measured response of a coupler constructed with these dimensions is shown in Fig. 13.05-3. This coupler was designed to operate at a 50-ohm impedance level and to have 2.8-db coupling at midband. However, the absorption in the dielectric material having relative dielectric constant $\epsilon_r = 2.77$ was sufficient to reduce the transmission by 0.2 db.

The dimensions of the strips are determined as follows. Equations (13.02-5) and (13.02-6) multiplied by $\sqrt{\epsilon_r}$ yield $\sqrt{\epsilon_r}\, Z_{oe} = 205$ ohms and $\sqrt{\epsilon_r}\, Z_{oo} = 32.8$ ohms. A value of $\Delta C/\epsilon = 4.82$ is then determined from Eq. (5.05-20). Next, a value of $g/b = 0.096$ was chosen and $d/g = 0.445$ read from Fig. 5.05-5. Figure 5.05-6 and Fig. 5.05-7 then yield $C'_{c\infty}/\epsilon = 0.55$ and $C'_{ce}/\epsilon = 0.115$, respectively. Using these values and the value of $C_{oe}/\epsilon = 1.81$ determined from Eq. (5.05-21) one finds an initial value of $c/b = 0.108$. Figure 5.05-8 yields $C'_{ae}/\epsilon = -0.105$, and Eq. (5.05-22a) yields $a/b = 0.284$. Since the strip width c is narrow enough compared to b so that the fringing fields at either edge interact and decrease its total strip capacitance, it is necessary to increase the value of c as given in Eq. (5.05-22c), yielding the final values of $c/b = 1.42$.

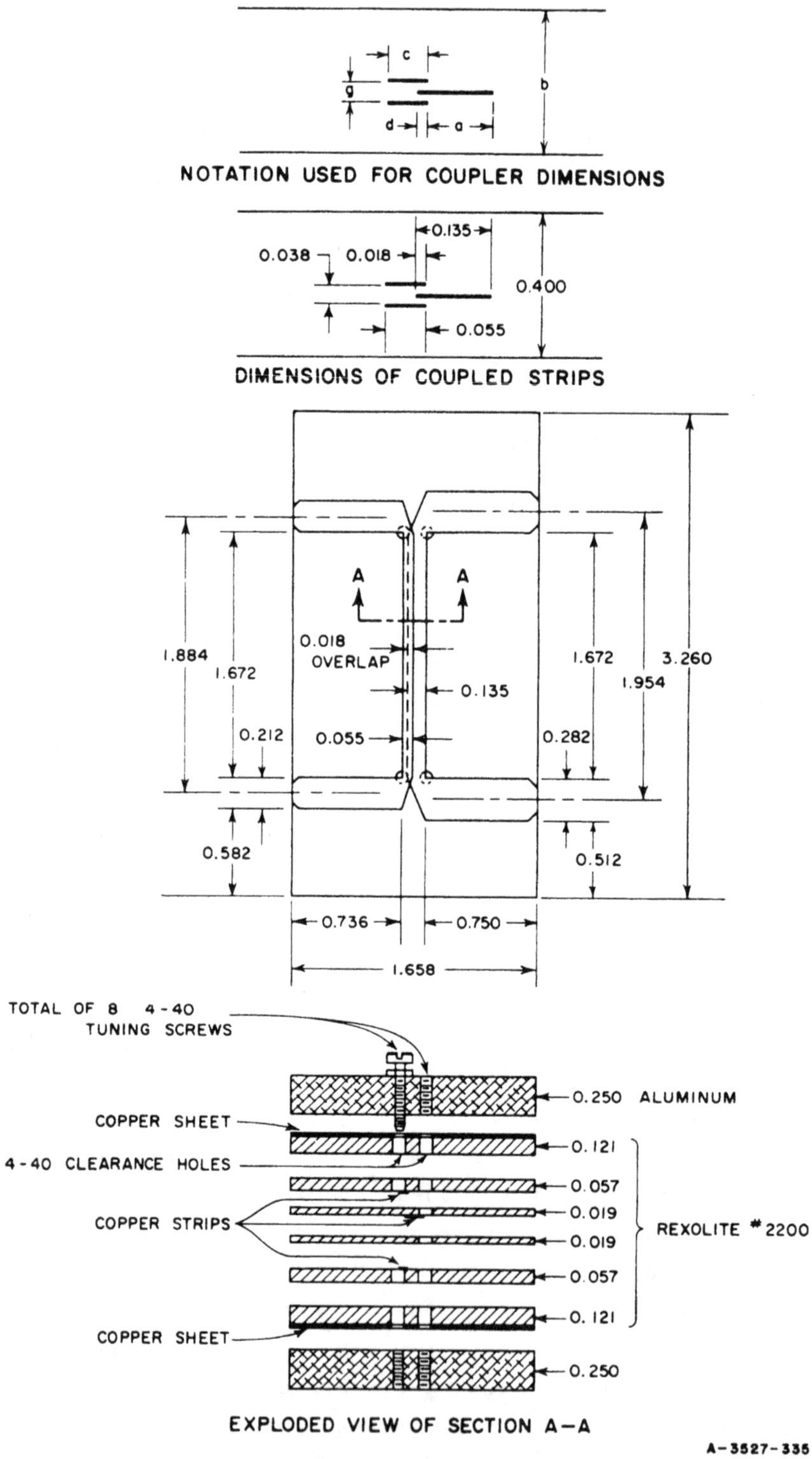

SOURCE: Final Report, Contract DA 36-039 SC-74862, SRI; reprinted in *IRE Trans. PGMTT* (see Ref. 13 by W. J. Getsinger)

FIG. 13.05-2 DETAILS OF CONSTRUCTION OF 1000-Mc, 3-db COUPLED-TRANSMISSION-LINE DIRECTIONAL COUPLIER USING INTERLEAVED STRIPS

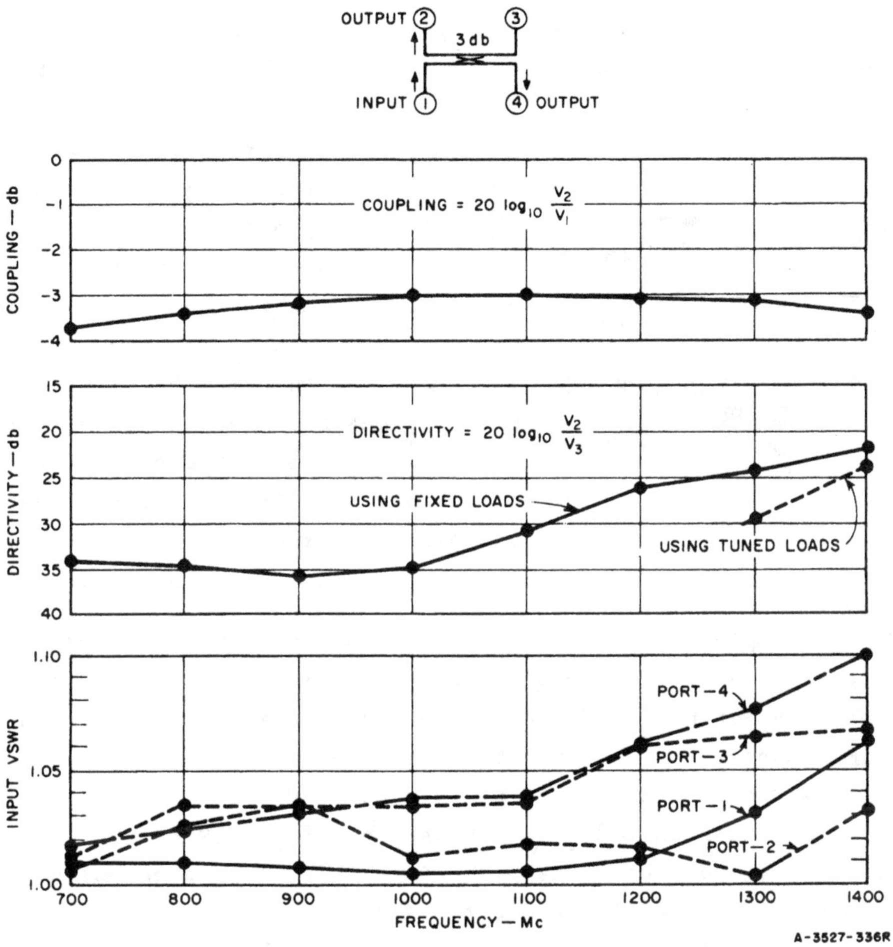

SOURCE: Final Report, Contract DA-36-039 SC-74862, SRI; reprinted in *IRE Trans. PGMTT* (see Ref. 13 by W. J. Getsinger)

FIG. 13.05-3 MEASURED PERFORMANCE OF 1000-Mc, 3-db BACKWARD COUPLER USING INTERLEAVED STRIPS

The method of construction used for this coupler is clearly indicated in Fig. 13.05-2. The series inductive discontinuities introduced by the mitered corners at each port of the coupler were compensated by means of the capacitive screws shown in the figure. These screws were introduced in a symmetric fashion about the midplane of the coupler to prevent unwanted excitation of the parallel-plate TEM mode. This method of compensating the coupler was found to be much more effective than adding capacitive tabs directly to the strips themselves.

A directional coupler having tight coupling can also be constructed using thick strips[14] of the type illustrated in Fig. 13.05-4. The strips in this method of construction are thick enough so that they have a substantial amount of mechanical rigidity. Hence, they may be supported at discrete points by means of low-reflection dielectric spacers, similar to those used to support the center conductors of coaxial lines. In some cases they may be simply held in place by means of the connectors at each of the four ports. Capacitive tuning screws are usually necessary at each port of this type of coupler to compensate the inevitable discontinuities.

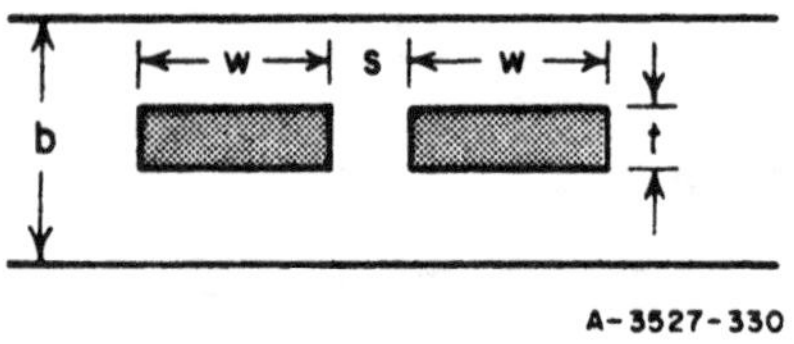

FIG. 13.05-4 NOTATION FOR 3-db COUPLER USING THICK COUPLED STRIPS

As an example of the method of choosing the dimensions of the coupled strips for such a coupler the theoretical design of a quarter-wavelength coupler having 2.7 db midband coupling and operating at a 50-ohm impedance level will now be considered. (The coupling over an octave bandwidth will then be 3.0 ± 0.3 db.) Reference to Eqs. (13.02-5) and (13.02-6) shows that Z_{oe} = 127.8 ohms and Z_{oo} = 19.6 ohms, and Eq. (5.05-20) yields the value of $\Delta C/\epsilon = 8.13$. The thickness of the strips is then arbitrarily set so that $t/b = 0.1$. Figure 5.05-9 then yields the values of $s/b = 0.015$ and $C'_{fe}/\epsilon = 0.578$. From Fig. 5.05-10(b) one finds $C'_f/\epsilon = 1.2$, and from Eq. (5.05-18), $C_{oe}/\epsilon = 2.95$. Then, using Eq. (5.05-23), one finds $w/b = 0.396$.

In order to minimize discontinuity effects it is usually best to choose $b \le 0.1\lambda$, where λ is the wavelength in the dielectric material filling the cross section. The length of the coupled strips is of course 0.25λ at midband.

An interesting structure due to Cohn[7]* which is suitable for strong coupling was indicated in Fig. 13.01-2(h); a completely shielded version is shown in Figs. 13.05-5 and 13.05-6. Two coaxial transmission lines are placed side-by-side. Their inner conductors *A* and *B* are entirely separate, but their outer conductors touch and become the conductor *C* in Fig. 13.05-5. The double-barreled conductor *C* is made one-quarter wavelength long and its potential "floats" between that of the two

* Patent applied for.

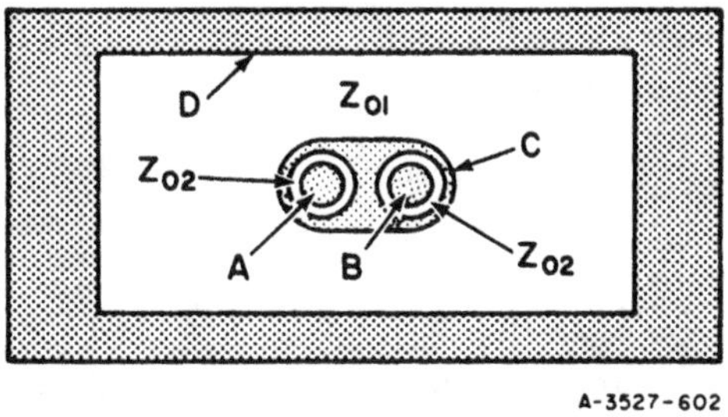

SOURCE: Rantec Report, Contract DA-36-239 SC-8735 (see Ref. 7 by S. B. Cohn and S. L. Wehn)

FIG. 13.05-5 RE-ENTRANT COUPLED CROSS SECTION

conductors *A* and *B*, and that of the metal case *D*. The conductor *C* is held in position by dielectric supports (not shown in the figures). The conductor *A* in Fig. 13.05-5 connects Ports 3 and 4 in Fig. 13.05-6, while conductor *B* connects Ports 1 and 2. The characteristic impedance of *C* within *D* is denoted by Z_{01} (Fig. 13.05-5), and the two concentric lines within *C* each have characteristic impedance Z_{02}. The analysis is again carried out in terms of an even and an odd mode. For the odd mode, *A* and *B* are at equal and opposite potentials, and *C* is at ground potential. Therefore

$$Z_{oo} = Z_{02} \quad . \tag{13.05-1}$$

For the even mode, each inner coaxial line (*A* or *B* to *C*) is in series with half the outer line (*C* to *D*), because the floating conductor *C* passes the currents flowing from its inner surface onto its outer surface which is therefore in series with it. This leads to[7]

$$Z_{oe} = Z_{02} + 2Z_{01} \quad . \tag{13.05-2}$$

The coupling is then determined as before from Eqs. (13.02-4) through (13.02-6).

For example, it was shown above that for a 2.7 db coupler with Z_0 = 50 ohms, one requires Z_{oe} = 127.8 ohms and Z_{oo} = 19.6 ohms.

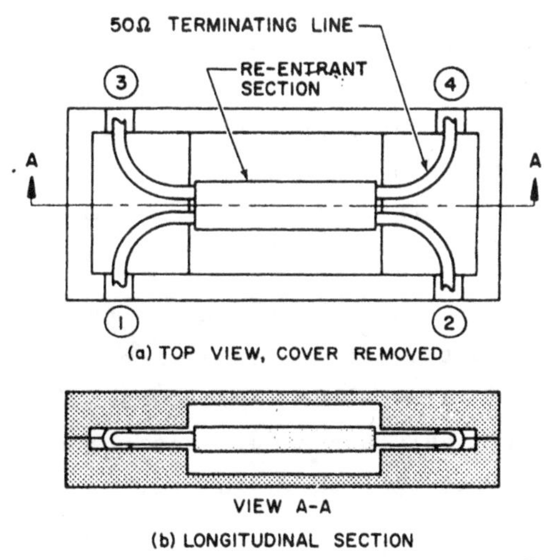

SOURCE: Rantec Report, Contract DA-36-239 SC-8735 (see Ref. 7 by S. B. Cohn and S. L. Wehn)

FIG. 13.05-6 VIEWS OF THE RE-ENTRANT DIRECTIONAL COUPLER

Solving for Z_{01} and Z_{02} from Eqs. (13.05-1) and (13.05-2),

$$Z_{01} = 54.1 \text{ ohms} , \quad Z_{02} = 19.6 \text{ ohms} \quad .$$

In an experimental model[7], two tubes of 0.250-inch outside diameter were soldered together to form C, and the space between the tubes was filled with solder. The tube wall thickness was 0.022 inch, leaving an inner diameter of 0.206 inch. The inner conductors A and B were of diameter 0.149 inch to yield an impedance of $Z_{02} = 19.6$ ohms. The impedance Z_{01} of the outer line was obtained from data such as that in Fig. 5.04-2 for a rectangular inner conductor, by taking $t = 0.250$ inch; the effective width w was determined by equating cross-sectional areas. Then $w = (1 + \pi/4)t = 1.786t$. From Fig. 5.04-2 the plate-to-plate spacing is then $b = 0.746$ inch.

The impedance of the four lines emerging from the coupler is to be 50 ohms. (This applied to the region in which the inner conductors are curved 90-degrees toward the four ports in Fig. 13.05-6.) The dimensions for this "slab-line" region (circular inner conductor between ground planes) were chosen to be: diameter of inner conductor = 0.125 inch, plate-to-plate spacing $b = 0.228$ inch. (This gives a 50-ohm line.[7]) Since this plate-to-plate spacing (0.228 inch),is less than the height of the double-barreled conductor C (0.250 inch), a small gap has to be allowed between the step in b from 0.746 inch to 0.228 inch and the re-entrant section. This is clearly shown in Fig. 13.05-6. A gap of 0.050 inch was allowed.

Directivities better than 29 db over an octave bandwidth were reported for two experimental couplers, one centered at 500 megacycles and one at 1500 megacycles.[7] A three-section coupler was also built, with only the middle section re-entrant, the two outer sections being of the type shown in Fig. 13.01-2(a).[7]

SEC. 13.06, TYPICAL DESIGNS FOR TEM-MODE, COUPLED-TRANSMISSION-LINE DIRECTIONAL COUPLERS OF ONE SECTION AND WITH WEAK COUPLING

This section describes the theoretical design of TEM-mode couplers operating at a 50-ohm impedance level and having 20-db coupling at midband. Each coupler has a length $\ell = 0.25\lambda$ at midband, where λ is the wavelength in the dielectric surrounding the coupled lines. In order to minimize discontinuity effects, the spacing b between ground planes in these couplers should be no more than a tenth of a wavelength at the highest operating frequency.

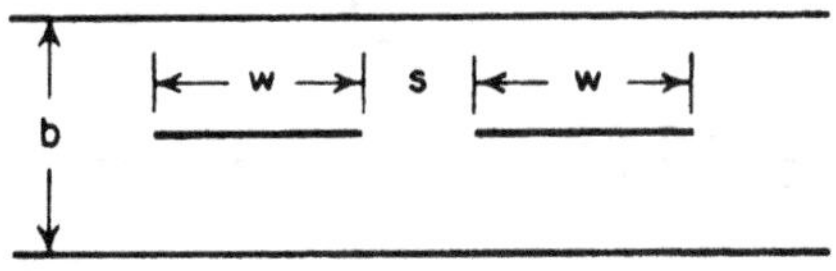

FIG. 13.06-1 NOTATION FOR 20-db COUPLER USING EDGE-COUPLED FLAT STRIPS

Figure 13.06-1 illustrates the cross section of the first coupler to be considered. The coupled strips usually have a negligible thickness, t, and are assumed to be supported by means of a dielectric of relative dielectric constant $\epsilon_r = 2.56$ completely filling the cross section. Values of $\sqrt{\epsilon_r}\, Z_{oe} = 88.5$ ohms and $\sqrt{\epsilon_r}\, Z_{oo} = 72.3$ ohms are obtained using Eqs. (13.02-5) and (13.02-6). Then using Fig. 5.05-3(a) one finds $s/b = 0.32$ and using Fig. 5.05-3(b) one finds $w/b = 0.715$.

The cross section of the second type of coupler to be considered is illustrated in Fig. 13.06-2. The thickness, t, of these coupled strips will be assumed to be negligible as well as the thickness of the slot of width d. The length of this slot is of course a quarter wavelength at midband. The dielectric material supporting these strips has $\epsilon_r = 2.56$. Equations (13.02-5) and (13.02-6) yield $\sqrt{\epsilon_r}\, Z_{oe} = 88.5$ ohms and $\sqrt{\epsilon_r}\, Z_{oo} = 72.3$ ohms. The value of $\sqrt{\epsilon_r}\, Z_{oo}$ is equal to $\sqrt{\epsilon_r}\, Z_0$, the characteristic impedance of an uncoupled line, and therefore the value of $w/b = 0.86$ can be obtained from Fig. 5.04-1. Then by the use of Eqs. (5.05-7) through (5.05-10) one finds $d/b = 1.05$.

The third type of coupler to be considered is shown in cross section in Fig. 13.06-3. In this type of coupler the conductors do not need to be surrounded with dielectric for support. In addition, the round conductors make this type of coupler ideal for high-power applications. Using Eqs. (13.02-5) and (13.02-6) we find $Z_{oe} = 55.3$ ohms and $Z_{oo} = 45.2$ ohms. Then from Eq. (5.05-11) one finds $s/b = 1.01$ and from Eq. (5.05-12) one finds $d/b = 0.59$.

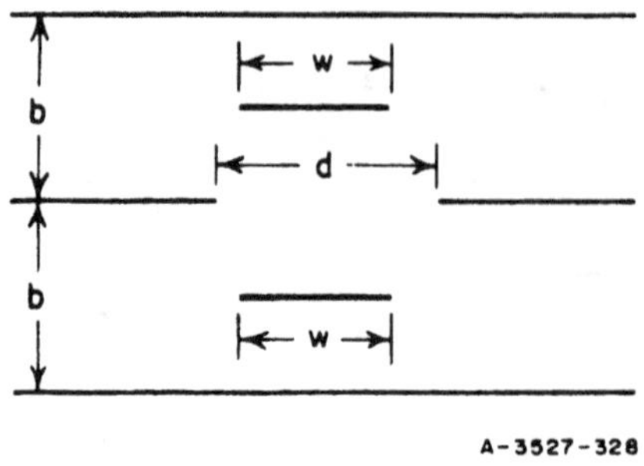

FIG. 13.06-2 NOTATION FOR 20-db COUPLER USING SLOT-COUPLED FLAT STRIPS

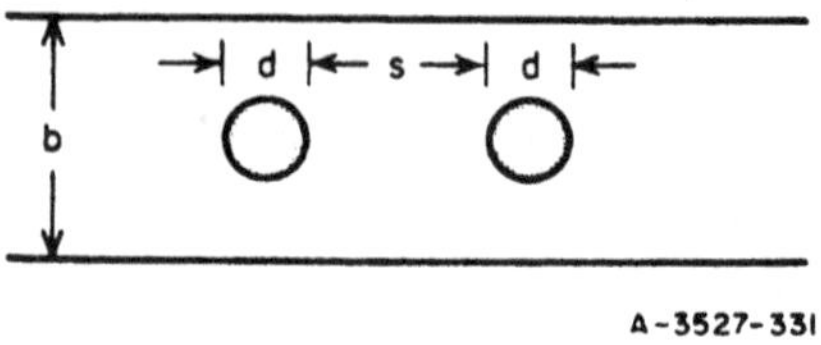

FIG. 13.06-3 NOTATION FOR 20-db COUPLER USING ROUND CONDUCTORS

SEC. 13.07, DERIVATION OF DESIGN FORMULAS FOR TEM-MODE, COUPLED-TRANSMISSION-LINE COUPLERS OF ONE SECTION

The basic coupled-transmission-line directional coupler is shown schematically at the top of Fig. 13.07-1 with resistance terminations Z_0 at each port. It is excited with a voltage 2E in series with Port 1. At the bottom of Fig. 13.07-1 the same directional coupler is shown in two different states of excitation. In the odd excitation, out-of-phase voltages are applied in series with Ports 1 and 2, while the even excitation applies in-phase voltages in series with these ports. Through the principle of superposition it may be seen that the behavior of the directional coupler with voltage 2E applied in series with Port 1 can be obtained from its behavior with even- and odd-voltage excitations.

Symmetry considerations show that a vertical electric wall may be placed between the strips of the coupler when it is excited with voltages having odd symmetry, and a vertical magnetic wall may be inserted between the strips when it is excited with voltages having even symmetry. The characteristic impedance of one of these strips in the presence of an electric wall is Z_{oo}, while the characteristic impedance of a strip in presence of a magnetic wall is Z_{oe}.

In order for the directional coupler to be perfectly matched at all frequencies it is necessary that the input impedance Z_{in} be equal to Z_0. Applying the principle of superposition, one sees that the input impedance of the directional coupler terminated in Z_0 can be written as

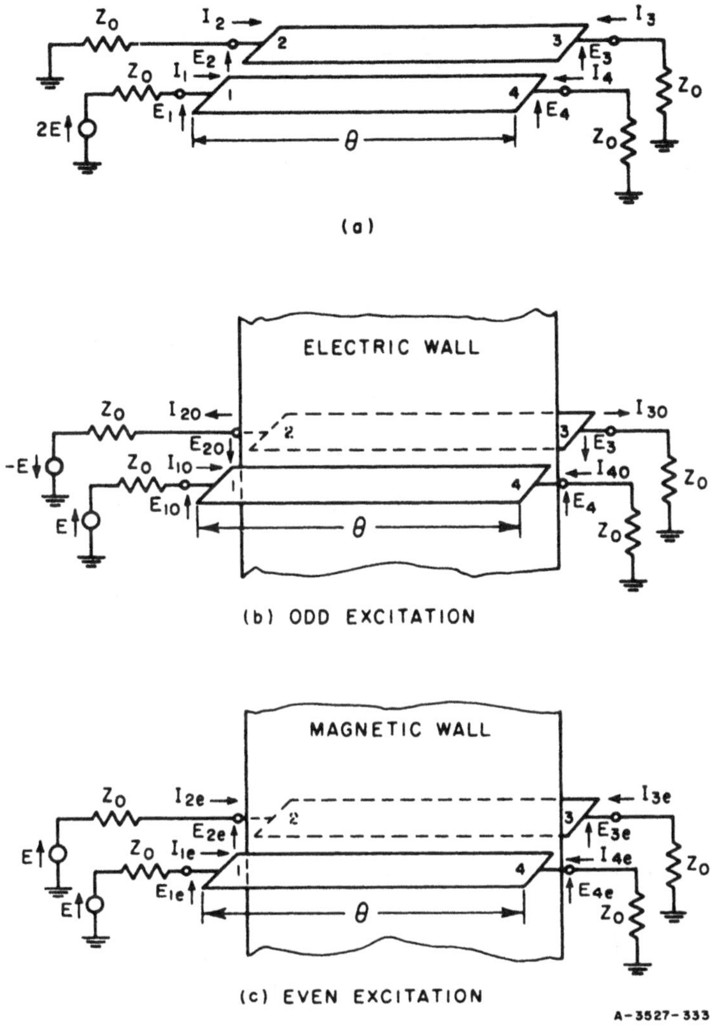

FIG. 13.07-1 COUPLED-STRIP DIRECTIONAL COUPLER WITH EVEN AND ODD EXCITATION

$$Z_{in} = \frac{E_{1o} + E_{1e}}{I_{1o} + I_{1e}} = \frac{\dfrac{Z_{1o}}{Z_0 + Z_{1o}} + \dfrac{Z_{1e}}{Z_0 + Z_{1e}}}{\dfrac{1}{Z_0 + Z_{1o}} + \dfrac{1}{Z_0 + Z_{1e}}} \qquad (13.07\text{-}1)$$

where

$$Z_{1o} = Z_{oo} \frac{Z_0 + jZ_{oo} \tan\theta}{Z_{oo} + jZ_0 \tan\theta} \qquad (13.07\text{-}2)$$

and

$$Z_{1e} = Z_{oe} \frac{Z_0 + jZ_{oe} \tan\theta}{Z_{oe} + jZ_0 \tan\theta} \quad . \qquad (13.07\text{-}3)$$

When Eq. (13.07-2) and Eq. (13.07-3) are substituted into Eq. (13.07-1) it is found that $Z_{in} = Z_0$ when

$$Z_0 = \sqrt{Z_{oo} Z_{oe}} \quad , \qquad (13.07\text{-}4)$$

for all values of θ.

Under the condition that $Z_0 = \sqrt{Z_{oe} Z_{oo}}$, the voltage appearing at Port 1 of the coupler is $E_1 = E$. The voltages $E_{2e} = E_{1e}$ and $E_{3e} = E_{4e}$ may be determined from the straightforward analysis of a transmission line of length θ, and characteristic impedance Z_{oe} terminated at either end by an impedance $\sqrt{Z_{oe} Z_{oo}}$ and fed by a voltage E. Likewise the normal-mode voltages $E_{2o} = -E_{1o}$ and $E_{3o} = -E_{4o}$ may be determined from a similar analysis of a transmission line of length θ and characteristic impedance Z_{oo} terminated by $\sqrt{Z_{oe} Z_{oo}}$. The results of this analysis show that

$$\frac{E_2}{E} = \frac{E_{2e} - E_{2o}}{E} = j \frac{\sin\theta \left(\sqrt{\dfrac{Z_{oe}}{Z_{oo}}} - \sqrt{\dfrac{Z_{oo}}{Z_{oe}}} \right)}{2\cos\theta + j\sin\theta \left(\sqrt{\dfrac{Z_{oe}}{Z_{oo}}} + \sqrt{\dfrac{Z_{oo}}{Z_{oe}}} \right)} \qquad (13.07\text{-}5)$$

$$= \frac{jc \sin \theta}{\sqrt{1 - c^2} \cos \theta + j \sin \theta} \qquad (13.07\text{-}6)$$

$$E_3 = E_{3e} - E_{3o} = 0 \qquad (13.07\text{-}7)$$

$$\frac{E_4}{E} = \frac{E_{4e} - E_{4o}}{E} = \frac{2}{2 \cos \theta + j \sin \theta \left(\sqrt{\frac{Z_{oe}}{Z_{oo}}} + \sqrt{\frac{Z_{oo}}{Z_{oe}}} \right)} \qquad (13.07\text{-}8)$$

$$= \frac{\sqrt{1 - c^2}}{\sqrt{1 - c^2} \cos \theta + j \sin \theta} \quad . \qquad (13.07\text{-}9)$$

The maximum coupled voltage E_2 occurs when the coupler is a quarter-wavelength long (*i.e.*, $\theta = 90°$). Therefore, from Eq. (13.07-5) a maximum coupling factor c may be defined as in Eq. (13.02-3). With this substitution, the voltage at Port 2 can be written as Eq. (13.02-1) while the voltage at Port 4 can be written as Eq. (13.02-2). The maximum coupling occurs at center frequency ($\theta = \pi/2$) and is given by (Eq. (13.02-3).

SEC. 13.08, QUARTER-WAVE FILTER PROTOTYPE CIRCUITS FOR TEM-MODE, COUPLED-TRANSMISSION-LINE DIRECTIONAL COUPLERS

A single-section, quarter-wave filter prototype circuit is shown in Fig. 13.08-1. It consists of two impedance steps (one up and one down), separated by an electrical distance θ. All characteristic impedances are normalized with respect to the characteristic impedances of the two terminating lines, which are taken as unity. The wave amplitudes a_1, b_1, and a_2 (Fig. 13.08-1) are defined in terms of the power flow,[16] *i.e.* $|a_1|^2$ and $|a_2|^2$ represent power carried forward, while $|b_1|^2$ represents reflected power. With the notation of Fig. 13.08-1 it can be shown[16] that

$$\frac{b_1}{a_1} = \frac{\Gamma(1 - e^{-2j\theta})}{1 - \Gamma^2 e^{-2j\theta}} \quad , \qquad (13.08\text{-}1)$$

or equivalently,

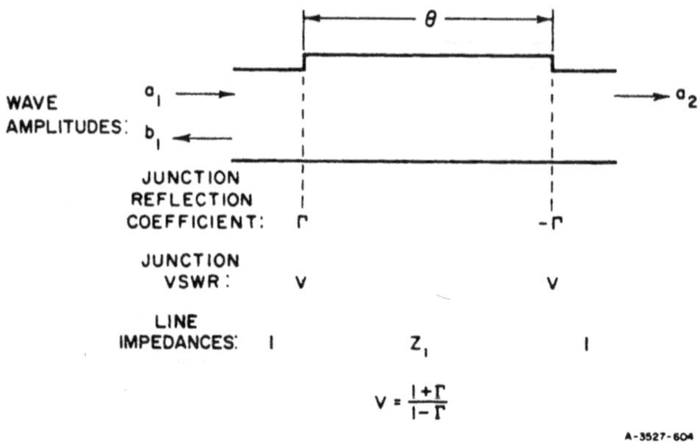

SOURCE: Proc. IEE (see Ref. 24 by L. Young)

FIG. 13.08-1 QUARTER-WAVE FILTER PROTOTYPE FOR SINGLE-SECTION TEM-MODE COUPLER

$$\frac{b_1}{a_1} = \frac{\left(\dfrac{2\Gamma}{1+\Gamma^2}\right) j \sin\theta}{\left(\dfrac{1-\Gamma^2}{1+\Gamma^2}\right)\cos\theta + j\sin\theta} , \qquad (13.08\text{-}2)$$

and

$$\frac{a_2}{a_1} = \frac{(1-\Gamma^2)e^{-j\theta}}{1-\Gamma^2 e^{-2j\theta}} \qquad (13.08\text{-}3)$$

$$= \frac{\left(\dfrac{1-\Gamma^2}{1+\Gamma^2}\right)}{\left(\dfrac{1-\Gamma^2}{1+\Gamma^2}\right)\cos\theta + j\sin\theta} \qquad (13.08\text{-}4)$$

Equations (13.07-6) and (13.07-9) with Eqs. (13.08-2) and (13.08-4) now show that b_1/a_1 corresponds to E_2/E, and a_2/a_1 corresponds to E_4/E, and that these equations are mathematically equivalent when we set

$$c = \frac{2\Gamma}{1+\Gamma^2} = \frac{Z_1^2 - 1}{Z_1^2 + 1} \qquad (13.08\text{-}5)$$

when Z_1 is chosen greater than unity. Conversely

$$\Gamma = \frac{1 - \sqrt{1 - c^2}}{c} \quad . \tag{13.08-6}$$

It follows that quarter-wave, coupled TEM-mode directional couplers can be solved as quarter-wave filters as discussed at the beginning of Sec. 13.03. It can be shown that for the single-section coupler of Fig. 13.01-1(a), [compare Eqs. (6.02-5) and (6.02-6)]

$$\left|\frac{E}{E_4}\right|^2 = 1 + \frac{(R - 1)^2}{4R} \sin^2 \theta \tag{13.08-7}$$

where

$$R = Z_1^2 = \frac{1 + c}{1 - c} \tag{13.08-8}$$

is the VSWR of the prototype filter (Fig. 13.08-1) at center frequency ($\theta = \pi/2$). The over-all power coupling ratio is then

$$\left.\begin{aligned} \left|\frac{E_2}{E}\right|^2 &= 1 - \left|\frac{E_4}{E}\right|^2 \\ &= \frac{\dfrac{(R - 1)^2}{4R} \sin^2 \theta}{1 + \dfrac{(R - 1)^2}{4R} \sin^2 \theta} \\ &= \frac{c^2 \sin^2 \theta}{1 - c^2 \cos^2 \theta} \quad . \end{aligned}\right\} \tag{13.08-9}$$

Table 13.08-1 summarizes the principal formulas for TEM-mode couplers (with and without end-to-end symmetry) in terms of the parameters of the prototype quarter-wave filter.

Table 13.08-1

GENERAL FORMULAS FOR THE DESIGN OF TEM-MODE COUPLED-TRANSMISSION-LINE DIRECTIONAL COUPLERS FROM THE QUARTER-WAVE FILTER PROTOTYPE

The midband VSWR R of the quarter-wave filter is related to the over-all coupling factor c_0 of the coupler by

$$R = \frac{1 + c_0}{1 - c_0} \quad \text{or} \quad c_0 = \frac{R - 1}{R + 1} \quad .$$

The midband VSWR R of the quarter-wave filter is

$$R = \left(\frac{Z_1 Z_3 Z_5 \cdots}{Z_2 Z_4 Z_6 \cdots}\right)^{\pm 2} > 1$$

where the Z_i are the normalized impedances of the quarter-wave filter prototype sections.

The coupling factors c_i of the several coupler sections are related to the normalized impedances Z_i of the quarter-wave filter prototype sections by

$$Z_i^2 = \frac{1 + c_i}{1 - c_i} \ , \ c_i = \frac{Z_i^2 - 1}{Z_i^2 + 1} \quad .$$

The even-mode and odd-mode impedances, $(Z_{oe})_i$ and $(Z_{oo})_i$, of the coupler sections are given by

$$(Z_{oe})_i = Z_0 \sqrt{\frac{1 + c_i}{1 - c_i}}$$

$$(Z_{oo})_i = Z_0 \sqrt{\frac{1 - c_i}{1 + c_i}} \quad .$$

Certain simplifications result for couplers with end-to-end symmetry. In that case n is odd, and

$$c_i = c_{n+1-i}$$

$$Z_i = Z_{n+1-i}$$

$$V_i = V_{n+2-i} = \text{VSWR of } i\text{th step in prototype}$$

and

$$R = (V_1 V_3 V_5 \cdots V_n)^2$$

$$= (V_2 V_4 V_6 \cdots V_{n+1})^2$$

Figure 13.03-1 showed a symmetrical TEM-mode coupler of three sections together with its quarter-wave filter prototype, and the associated notation. Notice that the filter has end-to-end symmetry, unlike the transformers in Chapter 6. This prototype filter is not matched at center frequency, as zero reflection in the prototype corresponds to zero coupling in the corresponding directional coupler. The TEM-mode coupled-transmission-line directional coupler performs exactly as a quarter-wave filter in which the forward and backward waves have been separated and have become two traveling waves (in opposite directions) on two parallel transmission lines, the over-all reflection coefficient of the prototype becoming the over-all coupling factor of the coupler. This is indicated in Fig. 13.03-1.

No general solutions have been obtained to date for the *symmetrical* n-section TEM-mode coupler, having equal-ripple characteristics as shown in Fig. 13.08-2. (Solutions for three-section and five-section couplers were presented in Secs. 13.03 and 13.04.) However, Levy[25] has presented the general solution for optimum *unsymmetrical* couplers having equal-ripple characteristics.

Phase Relation Between Outputs—For a symmetrical coupler the two outputs E_2 and E_4 (shown for a three-section coupler in Fig. 13.03-1) are in quadrature at all frequencies (*i.e.*, differ by 90 degrees when measured at the extremities of the coupler). This result follows at once from the quarter-wave filter analogy, since for any symmetrical dissipation-free 2-port, the reflected and transmitted wave amplitudes are orthogonal.[17]

SEC. 13.09, CONSIDERATIONS AND GENERAL FORMULAS FOR BRANCH-LINE COUPLERS

Branch-line couplers are directional couplers consisting of two parallel transmission lines coupled through a number of branch lines (Fig. 13.09-1). The lengths of the branch lines and their spacings are all one-quarter guide wavelength at center frequency, as shown in Fig. 13.09-1. The characteristic impedances of the two parallel main lines may be changed from section to section, and the branch impedances may be adjusted also to improve the electrical performance.

Figure 13.09-1 could represent the cross section of the inner conductor of a coaxial line, or the printed center conductor of a strip

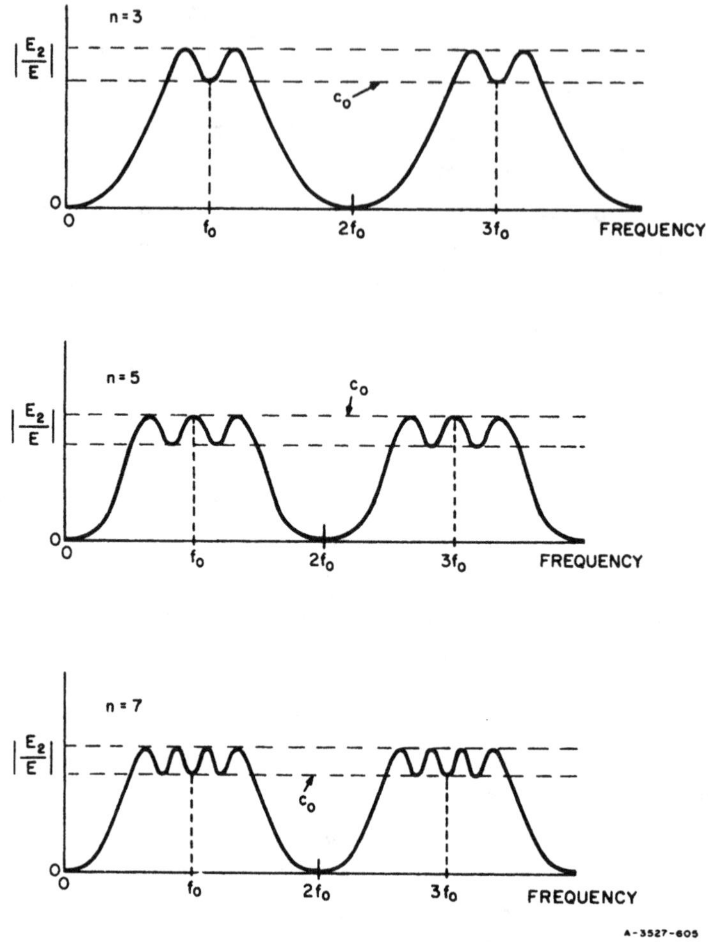

SOURCE: Proc. IEE (see Ref. 24 by L. Young)

FIG. 13.08-2 DESIRED COUPLING CHARACTERISTICS OF THREE-, FIVE-, AND SEVEN-SECTION TEM-MODE SYMMETRICAL COUPLERS

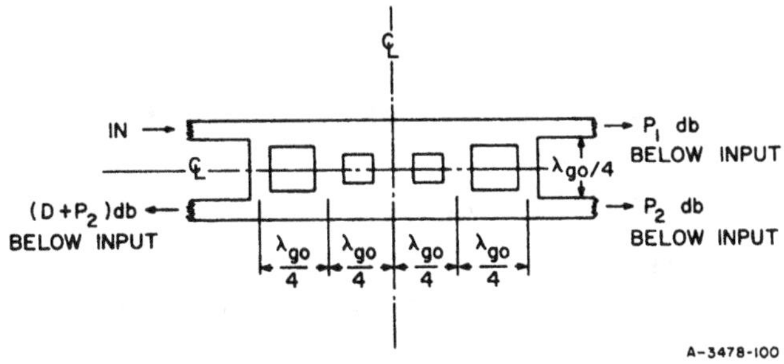

SOURCE: Tech. Note 3, Contract AF 30(602)-2392, SRI; (See Ref. 22 by Leo Young)

FIG. 13.09-1 BRANCH-LINE COUPLER SCHEMATIC

transmission line, or the E-plane cross section of a waveguide (Sec. 13.14). The output at the in-line termination is denoted by P_1 db (below the input), the coupling is P_2 db, and the directivity is D db (see Fig. 13.09-1).

The branches may be in shunt with the main line (as in coaxial and strip transmission line) or in series with it (as in waveguide). For shunt junctions it is more convenient to use admittances, and for series junctions it is more convenient to use impedances. The term "immittance" will be used to denote either (shunt) admittances or (series) impedances. The notation giving the characteristic immittances is depicted in Fig. 13.09-2. The coupler is supposed to have end-to-end symmetry, so that $H_{n+1} = H_1$, etc., and $K_1 = K_n$, etc. It will be convenient to normalize with respect to the terminating immittances $K_0 = K_{n+1} = 1$ throughout this chapter.

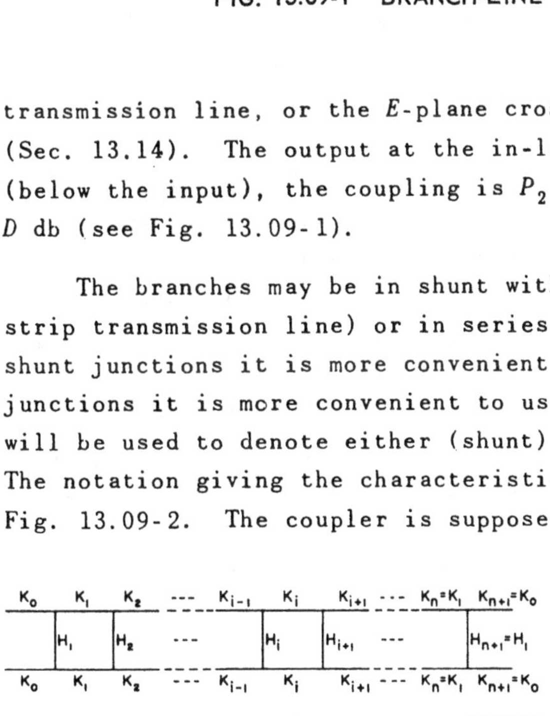

SOURCE: *IRE Trans. PGMTT* (see Ref. 11 by Leo Young)

FIG. 13.09-2 BRANCH-LINE COUPLER NOTATION

All the junctions will be assumed to be ideal shunt or series junctions in the design procedure, and each line length between junctions will be assumed to vary as a line one-quarter guide wavelength along at the center frequency. The effect of the junction reference planes (Sec. 13.14) is included only in calculating the final physical dimensions; the effect of junction discontinuity reactances is often not large, and in many cases can be compensated using the data in Sec. 5.07 and in Refs. 18, 19, and 20, as will be discussed in an example in Sec. 13.14. The appearance of branch-line couplers (Sec. 13.11) in rectangular wave-guide frequently works out as indicated in Fig. 13.09-1 (see Sec. 13.14), where the two outer walls (top and bottom of Fig. 13.09-1) have been shown as straight lines. However, this result should not necessarily be assumed to hold in all cases.

Even- and Odd-Mode Analysis—Branch-line couplers may be analyzed in terms of an even mode (two in-phase inputs) and an odd mode (two out-of-phase inputs) which are then superposed, adding at one port and cancelling at the other, thus yielding only one actual input. This is summarized in Fig. 13.09-3. For shunt-connected branches the two inputs of the odd mode produce zero voltage across the center of all the branches, and thus a short-circuit may be placed there; the two halves may therefore be "separated", each half consisting of a transmission line with 45-degree shorted stubs at 90-degree spacings. Meanwhile the even mode similarly yields 45-degree open-circuited stubs (Fig. 13.09-3). The 45-degree short-circuited stubs, and the 45-degree open-circuited stubs produce equal and opposite phase shifts $\pm\theta$ at center frequency in the even- and odd-mode inputs. For a matched coupler with perfect directivity the two outputs are then $\cos^2\theta$ and $\sin^2\theta$ in power, when the input power (sum of even and odd modes) is unity. For the case of series-connected stubs, the same reasoning applies, only the roles of the even and odd modes are reversed. For a fuller explanation, the references should be consulted.[8-11]

Cascaded Matched Directional Couplers—When matched couplers are cascaded, as shown in Fig. 13.09-4, they act as a single directional coupler, and the over-all couplings of the combination, $P_{1,com}$ and $P_{2,com}$ are readily calculated from the individual couplings $P_{1,i}$ db and $P_{2,i}$ db by the following formula:

$$P_{1,com} = 20 \log_{10} \left[\cos\left(\sum_{i=1}^{N} \theta_i\right)\right] \quad \text{db} \qquad (13.09\text{-}1)$$

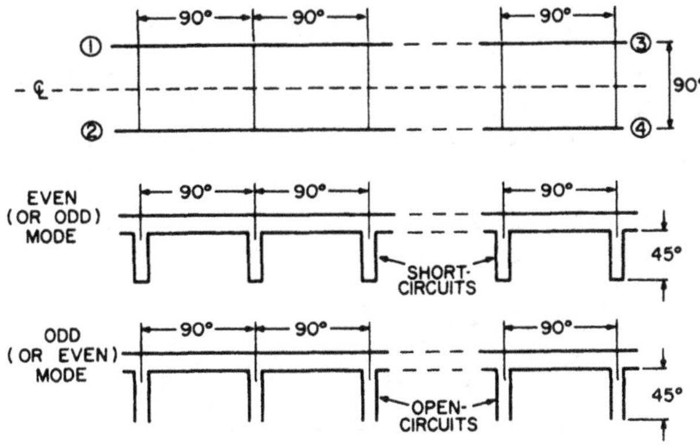

PORT	1	2	3	4
EVEN MODE:	↑	↑	θ	θ
ODD MODE:	↑	↓	θ	θ
SUM:	↑	ZERO	↓	→
AMPLITUDE:	1	0	COS θ	SIN θ
POWER:	1	0	$COS^2\ \theta$	$SIN^2\ \theta$

B-3478-102

SOURCE: Tech. Note 3, Contract AF 30(602)-2392, SRI; (see Ref. 22 by Leo Young)

FIG. 13.09-3 SUMMARY OF EVEN- AND ODD-MODE ANALYSIS

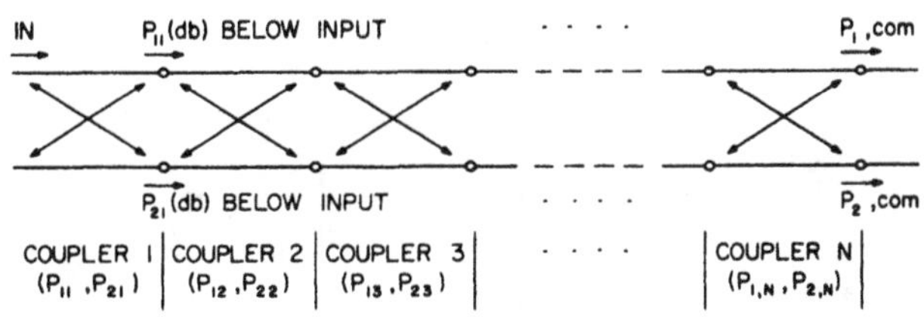

A-3478-103

SOURCE: Tech. Note 3, Contract AF 30(602)-2392, SRI; (see Ref. 22 by Leo Young)

FIG. 13.09-4 SEVERAL MATCHED DIRECTIONAL COUPLERS IN CASCADE

Table 13.09-1

MIDBAND FORMULAS FOR BRANCH-LINE COUPLERS UP TO SIX BRANCHES

	2-BRANCHES	3-BRANCHES	4-BRANCHES	5-BRANCHES	6-BRANCHES
Condition for Perfect Match and Directivity	$K_1^2 = 1 + H_1^2$	$H_2 = \dfrac{2H_1K_1^2}{1 + H_1^2}$	$K_2^2 = H_2^2 + \dfrac{K_1^2(K_1^2 - 2H_1H_2)}{1 + H_1^2}$	$\dfrac{H_3}{K_2^2} = \dfrac{2(H_2 + H_1^2H_2 - H_1K_1^2)}{K_1^4 + H_2^2(1 + H_1^2) - 2H_1H_2K_1^2}$	$K_2^4 - H_2^2K_3^2 + H_2^2H_3^2 + H_1^2K_2^4 - K_1^4K_3^2$ $+ H_3^2K_1^4 - H_1^2H_2^2K_3^2 + H_1^2H_2^2H_3^2$ $+ 2(-H_2H_3K_2^2 + H_1H_2K_1^2K_3^2 + H_1H_3K_1^2K_2^2$ $- H_1H_2H_3^2K_1^2 - H_1^2H_2H_3K_2^2) = 0$
$\dfrac{E_2}{E_1} = \text{antilog}\left(\dfrac{P_1 - P_2}{20}\right)$ (P_1 and P_2 positive)	$\dfrac{2H_1}{1 - H_1^2 + K_1^2}$	$\dfrac{2H_1K_1^2 + H_2(1 - H_1^2)}{2(K_1^2 - H_1H_2)}$	$\dfrac{2(H_1K_2^2 - H_1H_2^2 + H_2K_1^2)}{K_1^4 + (1 - H_1^2)(K_2^2 - H_2^2) - 2H_1H_2K_1^2}$	$\dfrac{\left\{\begin{array}{l}(1 - H_1^2)H_2(2K_2^2 - H_2H_3)\\ + 2H_1K_1^2(K_2^2 - H_2H_3) + H_3K_1^4\end{array}\right\}}{2[K_1^2K_2^2 - H_2H_3K_1^2 - H_1H_2(2K_2^2 - H_2H_3)]}$	$\left\{\begin{array}{l}2(H_1K_2^4 - 2H_1H_2H_3K_2^2 + H_3K_1^2K_2^2 - H_1H_2^2K_3^2\\ + H_2K_1^2K_3^2 + H_1H_2^2H_3^2 - H_2H_3^2K_1^2)\end{array}\right\}$ divided by $[K_2^4 - H_2^2K_3^2 + H_2^2H_3^2 - H_1^2K_2^4 - H_3^2K_1^4$ $+ K_1^4K_3^2 + H_1^2H_2^2K_3^2 - H_1^2H_2^2H_3^2$ $+ 2(H_1^2H_2H_3K_2^2 + H_1H_2H_3^2K_1^2$ $- H_1H_2K_1^2K_3^2 - H_1H_3K_1^2K_2^2 - H_2H_3K_2^2)]$
Conditions for 3-db Directional Coupler	$H_1 = 1$ $K_1 = \sqrt{2}$	$H_1 = \sqrt{2} - 1$ $K_1^2 = H_2\sqrt{2}$	$H_2 = \dfrac{K_1^2(1 - H_1)}{1 + 2H_1 - H_1^2}$ $K_2^2 = \dfrac{2H_2^2}{(1 - H_1)^2}$ $\Big\vert$ $K_2 = \dfrac{H_2\sqrt{2}}{1 - H_1}$ if $H_1 < 1$	$\dfrac{K_1^2}{1 + H_1} = \dfrac{H_2(2K_2^2 - H_2H_3)}{K_2^2 - H_2H_3}$ $\dfrac{K_1^4}{1 - 2H_1 - H_1^2} = \dfrac{H_2(2K_2^2 - H_2H_3)}{H_3}$	$\dfrac{K_1^2(1 - H_1)}{1 + 2H_1 - H_1^2} = H_2 + \dfrac{H_3K_2^2}{K_3^2 - H_3^2}$ $\dfrac{K_1^2}{1 - H_1} = \dfrac{(K_2^2 - H_2H_3 - H_2K_3)(K_2^2 - H_2H_3 + H_2K_3)}{H_2(K_3^2 - H_3^2) + H_3K_2^2}$
Conditions for 0-db Directional Coupler	not possible (H_1, $K_1 \to \infty$)	$H_1 = 1$ $H_2 = K_1^2$	$K_2 = H_2 = \dfrac{K_1^2}{2H_1}$	$\dfrac{K_1^2}{H_1} = \dfrac{H_2(2K_2^2 - H_2H_3)}{K_2^2 - H_2H_3}$ $\dfrac{K_1^4}{1 - H_1^2} = \dfrac{H_2(2K_2^2 - H_2H_3)}{H_3}$	$K_3 = \dfrac{H_3K_1^2}{K_1^2 - 2H_1H_2}$ $K_2^2 = H_2(K_3 + H_3)$ for $K_1^2 > 2H_1H_2$, $K_2 > H_2H_3$

SOURCE: *Proc. Nat. Electronics Conf.* (see Ref. 10 by Leo Young).

$$P_{2,\mathrm{com}} = 20 \log_{10} \left[\sin\left(\sum_{i=1}^{N} \theta_i\right)\right] \quad \text{db} \tag{13.09-2}$$

where

$$\theta_i = \cos^{-1}\left[\text{antilog}\left(\frac{P_{1,i}}{20}\right)\right] \tag{13.09-3}$$

$$= \sin^{-1}\left[\text{antilog}\left(\frac{P_{2,i}}{20}\right)\right] \quad . \tag{13.09-4}$$

For instance, a phase shift of θ = 30 degrees yields a 6-db coupler, and a phase shift of θ = 90 degrees yields a 0-db coupler; therefore, a cascade of three matched 6-db couplers results in a 0-db coupler.

Midband Formulas—There are certain relations which ensure perfect match and perfect directivity at center frequency. Furthermore, at center frequency the ratio of the two output voltages $E_1 \propto$ antilog $(-P_1/20)$ and $E_2 \propto$ antilog $(-P_2/20)$ can be written down fairly simply in terms of the H_i and K_i in Fig. 13.09-2. Table 13.09-1 gives these formulas for up to six branches. Also given are simpler formulas which determine the conditions for 3-db and 0-db coupling, together with perfect match and perfect directivity, at center frequency.

SEC. 13.10, PERIODIC BRANCH-LINE COUPLERS

Additional restrictions may be placed on the immittances of symmetrical couplers (Fig. 13.09-2). In particular when

$$\left.\begin{aligned} K_i &= K_0 \ , \quad i = 1, 2, \ldots, n+1 \ , \\ H_i &= H_2 \ , \quad 2 \le i \le n \ , \end{aligned}\right\} \tag{13.10-1}$$

and further, when $H_1 = H_{n+1}$ is chosen to give a perfect input match at center frequency, then such couplers will herein be called *periodic*. Thus the through-lines are uniform, and all the interior branches are the same. Such couplers have been analyzed by Reed and Wheeler.[8,9] Let the power coupling ratio at center frequency be denoted by

$$c_0^2 = \text{antilog}\left(-\frac{P_2}{10}\right) \qquad (13.10\text{-}2)$$

where P_2 is in decibels and was defined in Fig. 13.09-1. (For example, for a 6-db coupler, $c_0^2 = 1/4$.) The solution of periodic couplers involves the Tchebyscheff polynomials $S_n(x)$, which have been tabulated numerically,[21] and which are defined by

$$\left.\begin{aligned} S_0(x) &= 1 \\ S_1(x) &= x \\ S_2(x) &= x^2 - 1 \\ S_3(x) &= x^3 - 2x \\ S_4(x) &= x^4 - 3x^2 + 1 \\ S_5(x) &= x^5 - 4x^3 + 3 \\ S_6(x) &= x^6 - 5x^4 + 6x^2 - 1 \\ S_n(x) &= xS_{n-1}(x) - S_{n-2}(x) \quad . \end{aligned}\right\} \qquad (13.10\text{-}3)$$

The immittances H_1 and H_2 of a periodic coupler then have to satisfy[9]

$$S_{n-1}(-H_2) = \left\{\begin{array}{ll} \sqrt{1 - c_0^2} \;, & \text{when } n = \text{odd} \\ \text{or } c_0 \;, & \text{when } n = \text{even} \end{array}\right\} \qquad (13.10\text{-}4)$$

and

$$H_1 = \left| \frac{|\sqrt{1 - S_{n-1}^2(-H_2)}| - |S_{n-2}(-H_2)|}{S_{n-1}(-H_2)} \right| \quad . \qquad (13.10\text{-}5)$$

Table 13.10-1[9] gives solutions for some 0-db, 3-db, and 10-db couplers.

The performances of four different 3-db periodic couplers given by Table 13.10-1 are summarized in Tables 13.10-2 and 13.10-3. The maximum VSWR over three specified fractional bandwidths

$$w_b = 2\left(\frac{\lambda_{g1} - \lambda_{g2}}{\lambda_{g1} + \lambda_{g2}}\right) \qquad (13.10\text{-}6)$$

Table 13.10-1

EXAMPLES OF 0-. 3-, AND 10-db PERIODIC COUPLERS

Number of Sections (n)	0-db Coupler		3-db Coupler		10-db Coupler	
	H_1	H_2	H_1	H_2	H_1	H_2
2	1.000	1.000	0.4141	0.7071	0.162	0.3162
3	0.500	1.000	0.2346	0.5412	0.0945	0.2265
4	0.618	0.618	0.2088	0.3810	0.0811	0.1602
5	0.309	0.618	0.1464	0.3179	0.0592	0.1312
6	0.445	0.445	0.1374	0.2583		
7	0.2225	0.445	0.1064	0.2257		
8	0.3473	0.3473	0.1022	0.1948		
9	0.1736	0.3473	0.0837	0.1752		
10	0.2846	0.2846				
11	0.1423	0.2846				
12	0.2410	0.2410				
13	0.1205	0.2410	0.0587	0.1210		

SOURCE: *IRE Trans. PGMTT* (see Ref. 9 by J. Reed)

Table 13.10-2

MAXIMUM VSWR OF SEVERAL 3-db PERIODIC COUPLERS OVER STATED BANDWIDTHS

FRACTIONAL BANDWIDTH w_b	COUPLERS			
	Four-branch	Five-branch	Six-branch	Eight-branch
0.16	1.02	1.05	1.02	1.01
0.28	1.08	1.10	1.05	1.04
0.48	1.32	1.12	1.15	1.07

SOURCE: Tech. Note 3, Contract AF30(602)-2392, SRI (see Ref. 22 by Leo Young)

Table 13.10-3

COUPLING UNBALANCE (db) OF SEVERAL 3-db PERIODIC COUPLERS OVER STATED BANDWIDTHS

FRACTIONAL BANDWIDTH w_b	COUPLERS			
	Four-branch	Five-branch	Six branch	Eight-branch
0.16	0.20	0.18	0.16	0.14
0.28	0.67	0.51	0.49	0.43
0.48	2.3	1.6	1.5	1.4

SOURCE: Tech. Note 3, Contract AF30(602)-2392, SRI (see Ref. 22 by Leo Young)

is shown in Table 13.10-2, where λ_{g1} is the longest guide wavelength, and λ_{g2} is the shortest guide wavelength over the band under consideration. The guide wavelength at band-center is given by

$$\lambda_{g0} = \left(\frac{2\lambda_{g1}\lambda_{g2}}{\lambda_{g1} + \lambda_{g2}}\right) \quad . \tag{13.10-7}$$

The coupling P_2 becomes stronger, and the coupling P_1 becomes weaker, than 3 db at about the same rate, as the frequency changes from band-center. The total variation $|\Delta P_1| + |\Delta P_2|$, referred to as the "coupling unbalance," is shown in Table 13.10-3 for the same 3-db couplers.

The performances of five different 0-db periodic couplers given by Table 13.10-1 are summarized in Tables 13.10-4 and 13.10-5 over the same

Table 13.10-4

MAXIMUM VSWR OF SEVERAL 0-db PERIODIC COUPLERS OVER STATED BANDWIDTHS

FRACTIONAL BANDWIDTH w_b	COUPLERS				
	Six-branch	Eight-branch	Ten-branch	Twelve-branch	Fourteen-branch
0.16	1.10	1.06	1.04	1.03	1.02
0.28	1.19	1.09	1.05	1.03	1.02
0.48	1.21	1.10	1.12	1.09	1.05

SOURCE: Tech. Note 3, Contract AF30(602)-2392, SRI (see Ref. 22 by Leo Young)

Table 13.10-5

MAXIMUM INSERTION LOSS (db) OF SEVERAL 0-db PERIODIC COUPLERS OVER STATED BANDWIDTHS

FRACTIONAL BANDWIDTH w_b	COUPLERS						
	Three-branch	Four-branch	Six-branch	Eight-branch	Ten-branch	Twelve-branch	Fourteen-branch
0.16	0.7	0.1	0.02	0.01	0.006	0.004	0.003
0.28	2.0	0.7	0.11	0.04	0.02	0.013	0.011
0.48	3.1	1.9	0.33	0.22	0.19	0.15	0.113

SOURCE: Tech. Note 3, Contract AF30(602)-2392, SRI (see Ref. 22 by Leo Young)

three specified fractional bandwidths. More detailed characteristics of some of these and other couplers are given in Refs. 9 and 22. Only couplers with an even number of branches (n odd) are given in

Tables 13.10-4 and 13.10-5; the performance of couplers with an odd number of branches (n even) is not as good. This is probably due to the fact that all the branches have the same immittances, including the two end ones, when the number of branches is odd, and thus there is no end-matching effect. For instance, a twelve-branch, 0-db periodic coupler gives better electrical performance than a thirteen-branch 0-db periodic coupler.[22]

SEC. 13.11, THE CHARACTERISTICS AND PERFORMANCE OF SYNCHRONOUS BRANCH-LINE COUPLERS

The *synchronous tuning condition* as it applies to transmission-line filters was formally defined toward the end of Sec. 6.01. Now recall that the even- and odd-mode analysis reduces a branch-guide coupler to two transmission-line filters, as indicated in Fig. 13.09-3. When the through-line impedance is uniform, neither of these two filters is synchronously tuned, since the reflections from any two finite reactances spaced one-quarter wavelength apart are not phased for maximum cancellation. They can be so phased, however, by adjusting the through-line impedances, and it turns out that both the even- and the odd-mode "filters" (Fig. 13.09-3) then become synchronously-tuned with the same choice of through-line impedances.[11] A quarter-wave transformer (Chapter 6) can then be used as a prototype circuit, resulting in a clearly marked pass band, in which the VSWR and directivity are nearly optimized. To meet a specified electrical performance over a given bandwidth, fewer branches will then be required with such optimized couplers than with periodic couplers (Sec. 13.10); from numerical examples, it appears that generally about half as many branches[22] will do. On the other hand, some of these branches will be wider than any in the corresponding periodic design, and so may depart more from ideal junction behavior. Such factors should be taken into account before deciding on the type of coupler. The chief advantages of synchronous couplers over periodic couplers are that they are more compact (for a specified performance), since there are fewer branches, and that the performance can be predicted fairly accurately.

The first step in designing a synchronous coupler with a specified center frequency coupling $P_{2\theta}$ decibels and a specified fractional bandwidth w_b over which the performance is to be optimum, is to determine the output-to-input impedance ratio R and the fractional bandwidth w_q of the quarter-wave transformer prototype. They are generally obtained closely enough by the approximate relations

$$P_{2,0} \approx 20 \log_{10}\left(\frac{R + 1}{R - 1}\right) \quad \text{db} \tag{13.11-1}$$

and

$$w_b \approx 0.6 w_q \tag{13.11-2}$$

Equation (13.11-1) is plotted in Fig. 13.11-1. It is exact for couplers which are perfectly matched at center frequency (such as all couplers with even n, and any maximally flat couplers). It is accurate enough for nearly all practical cases, but overestimates the numerical value of $P_{2,0}$ in proportion to the reflection loss. (An exact formula, necessary only when the center-frequency VSWR is high, is given in Ref. 11). Notice that 0-db coupling would require $R \to \infty$, and can, therefore, not be realized in a single synchronous design. However, two synchronous 3-db couplers, or three synchronous 6-db couplers, etc., can be cascaded to give a 0-db coupler.

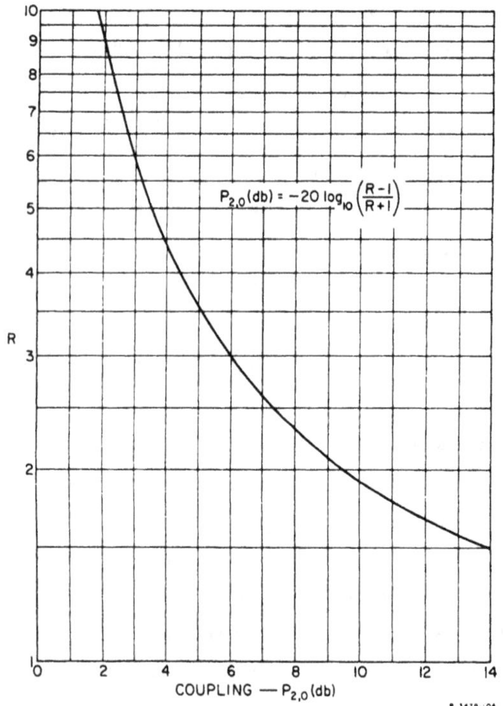

SOURCE: *IRE Trans. PGMTT* (see Ref. 11 by Leo Young)

FIG. 13.11-1 PLOT OF CENTER-FREQUENCY COUPLING $P_{2,0}$ vs. IMPEDANCE-RATIO PARAMETER R FOR A MATCHED BRANCH-LINE COUPLER

Equation (13.11-2) is less accurate, the numerical factor generally lying between 0.5 and 0.7. A more accurate determination can be made from Fig. 13.11-2 after the number of branches has been selected. In this figure are plotted the best estimates of the bandwidth contraction factor

$$\beta = w_b / w_q \tag{13.11-3}$$

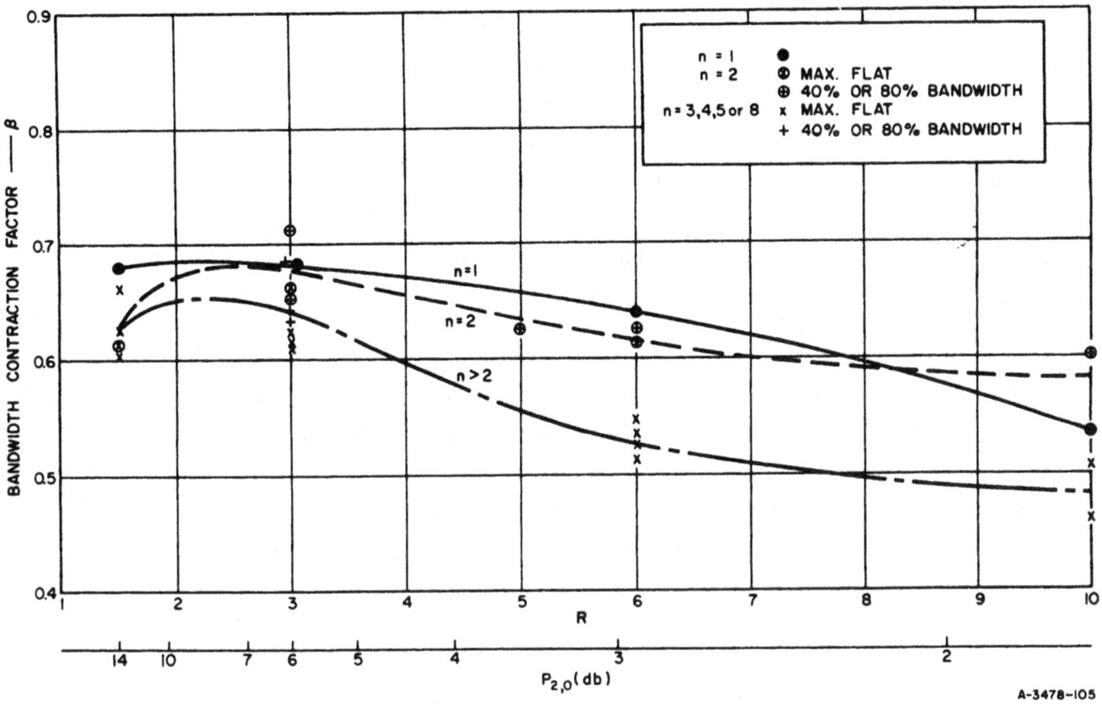

SOURCE: *IRE Trans. PGMTT* (see Ref. 11 by Leo Young)

FIG. 13.11-2 BEST ESTIMATES FOR BANDWIDTH CONTRACTION FACTOR β, BASED ON 27 INDIVIDUAL SOLUTIONS

against R, or $P_{2,0}$ and with n as a parameter, based on a large number of numerical solutions.[11]

Having determined R and w_q from the specified coupling $P_{2,0}$ and the coupler fractional bandwidth w_b, the only remaining independent parameter is the number of sections, n, or number of branches, $n + 1$ (Fig. 13.09-2). For a given R and w_q, the ripple VSWR V_r of the transformer decreases with increasing n, and is readily determined from Sec. 6.02. The maximum pass-band VSWR of the coupler, V_{max}, and its minimum directivity D_{min}, may then be obtained to a good approximation by

$$V_{max} \approx 1 + \frac{V_r - 1}{\text{antilog } (P_2/20)} \qquad (13.11\text{-}4)$$

$$D_{min} \approx -20 \log_{10} \left[\frac{V_r - 1}{2 \text{ antilog } (P_1/20)} \right] - P_2 \quad \text{db} \quad . \quad (13.11\text{-}5)$$

(The directivity D is defined as the ratio of power out of the "decoupled" arm to power out of the coupled arm. See Fig. 13.09-1.) Exact formulas (necessary only for large V_r) are given in Ref. 11. The number of sections n must be made large enough to meet the desired performance.

The frequency variation of the input VSWR V, the directivity D (db) and the two couplings (the in-line coupling P_1 and the cross-over coupling P_2, both in db), are plotted in Figs. 13.11-3, 4 and 5 for couplers based on *maximally flat* quarter-wave transformer prototypes for n = 1, 2, 4, and 8, and for R = 1.5, 3 and 6, corresponding to

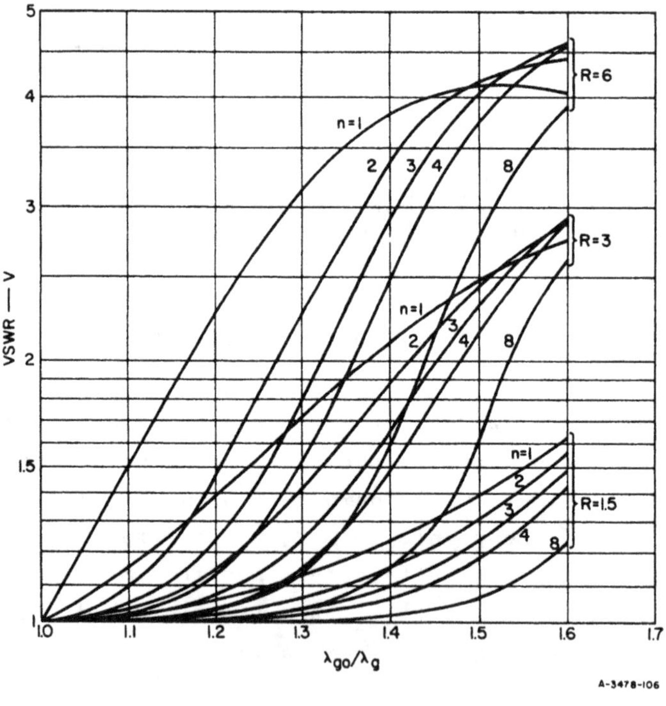

SOURCE: *IRE Trans. PGMTT* (see Ref. 11 by Leo Young)

FIG. 13.11-3 VSWR CHARACTERISTICS OF SOME MAXIMALLY FLAT, BRANCH-LINE COUPLERS
Immittances of these couplers are given in Tables 13.12-1 and 13.12-2

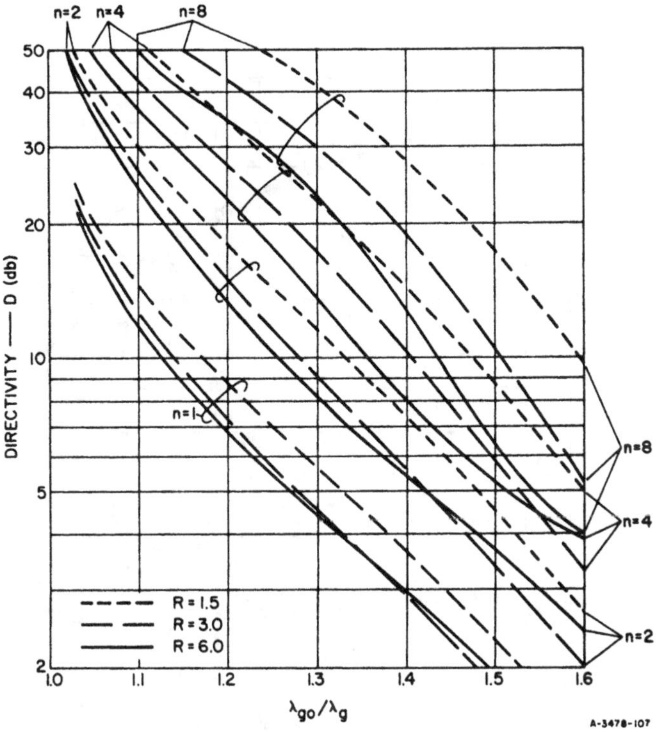

SOURCE: *IRE Trans. PGMTT* (see Ref. 11 by Leo Young)

FIG. 13.11-4 DIRECTIVITY CHARACTERISTICS OF SOME MAXIMALLY FLAT, BRANCH-LINE COUPLERS
Immittances of these couplers are given in Tables 13.12-1 and 13.12-2

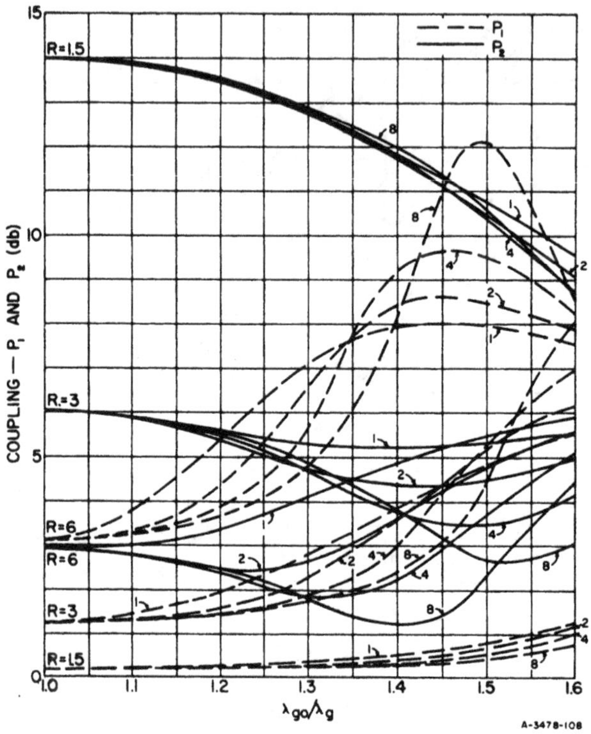

SOURCE: *IRE Trans. PGMTT* (see Ref. 11 by Leo Young)

FIG. 13.11-5 COUPLING CHARACTERISTICS OF SOME MAXIMALLY FLAT, BRANCH-LINE COUPLERS
Immittances of these couplers are given in Tables 13.12-1 and 13.12-2

couplings $P_{2,0}$ of approximately 14, 6, and 3 db. The graphs are plotted against the quantity (λ_{g0}/λ_g), where λ_g is the guide wavelength, and λ_{g0} is its value at band-center. The response is symmetrical about $(\lambda_{g0}/\lambda_g) = 1$, so that Figs. 13.11-3, 4, and 5 actually cover the range from 0.4 to 1.6, although only the portion from 1.0 to 1.6 is shown. For non-dispersive (TEM mode) lines, the guide wavelength λ_g reduces to the free-space wavelength λ, and then λ_{g0}/λ_g reduces to $\lambda_0/\lambda = f/f_0$, where f is the frequency, and f_0 is its value at band-center (also called the center frequency).

It can be seen from Fig. 13.11-5 that the coupling P_2 generally becomes stronger (P_2 measured in decibels decreases) on either side of center frequency (the curves are symmetrical about $\lambda_{g0}/\lambda_g = 1$), and correspondingly P_1 becomes weaker (P_1 measured in decibels increases).

The frequency variations of P_1 and P_2 for couplers based on Tchebyscheff transformer prototypes have been found to be very similar to the curves shown in Fig. 13.11-5 for couplers based on maximally flat transformer prototypes.

SEC. 13.12, TABLES OF IMMITTANCES OF SYNCHRONOUS BRANCH-LINE COUPLERS

It was shown in Sec. 13.11 how to determine the quarter-wave transformer prototype parameters R, w_q, and the number of sections n (corresponding to $n + 1$ branches). The computation of the coupler immittances H_i and K_i (Fig. 13.09-2) is quite tedious, and will not be treated here. The design process is simplified by the use of a special chart,[11] which can handle any transformer impedances. In practice, it is likely that maximally flat and Tchebyscheff transformers (Chapter 6) will usually be taken for prototypes; the immittances H_i and K_i derived from such transformers were computed on a digital computer[22] and are tabulated in Tables 13.12-1 to 13.12-8. Most cases of practical interest can be obtained from these tables by interpolation.

Table 13.12-1

BRANCH-LINE COUPLER IMMITTANCES FOR $n = 1$ SECTION (TWO BRANCHES)

$n = 1$		
R	K_1	H_0
1.25	1.006	0.1119
1.50	1.021	0.2040
2.00	1.061	0.3535
2.50	1.108	0.4732
3.00	1.155	0.5775
4.00	1.250	0.7500
5.00	1.341	0.894
6.00	1.429	1.021
8.00	1.592	1.238
10.00	1.739	1.423

SOURCE: *IRE Trans. PGMTT* (see Ref. 11 by Leo Young)

Table 13.12-2

IMMITTANCES OF MAXIMALLY FLAT BRANCH-LINE COUPLERS FOR n = 2 TO 8 SECTIONS (THREE TO NINE BRANCHES)

R	K_1	K_2	K_3	K_4	H_1	H_2	H_3	H_4	H_5
n = 2									
1.50	1.0153				0.1010	0.2062			
2.00	1.0449				0.1715	0.3639			
2.50	1.0783				0.2251	0.4983			
3.00	1.1124				0.2679	0.6188			
4.00	1.1785				0.3333	0.8333			
5.00	1.2399				0.3819	1.0249			
6.00	1.2965				0.4202	1.2008			
8.00	1.3978				0.4775	1.5196			
10.00	1.4861				0.5194	1.8070			
n = 3									
1.50	1.0089	1.0206			0.0501	0.1539			
2.00	1.0258	1.0606			0.0840	0.2694			
2.50	1.0446	1.1067			0.1086	0.3656			
3.00	1.0634	1.1546			0.1274	0.4498			
4.00	1.0988	1.2499			0.1542	0.5957			
5.00	1.1307	1.3416			0.1732	0.7220			
6.00	1.1594	1.4288			0.1854	0.8351			
8.00	1.2087	1.5909			0.2029	1.0344			
10.00	1.2501	1.7392			0.2138	1.2091			
n = 4									
1.50	1.0047	1.0176			0.0249	0.1017	0.1548		
2.00	1.0137	1.0517			0.0415	0.1750	0.2738		
2.50	1.0235	1.0907			0.0533	0.2331	0.3758		
3.00	1.0333	1.1307			0.0620	0.2814	0.4676		
4.00	1.0514	1.2093			0.0740	0.3596	0.6326		
5.00	1.0675	1.2835			0.0817	0.4220	0.7811		
6.00	1.0816	1.3530			0.0870	0.4743	0.9185		
8.00	1.1057	1.4795			0.0935	0.5589	1.1698		
10.00	1.1256	1.5924			0.0971	0.6265	1.3986		
n = 5									
1.50	1.0024	1.0124	1.0206		0.0124	0.0630	0.1285		
2.00	1.0070	1.0363	1.0606		0.0206	0.1069	0.2259		
2.50	1.0121	1.0630	1.1067		0.0264	0.1400	0.3079		
3.00	1.0170	1.0902	1.1546		0.0306	0.1662	0.3804		
4.00	1.0262	1.1421	1.2499		0.0363	0.2058	0.5078		
5.00	1.0342	1.1899	1.3416		0.0398	0.2348	0.6196		
6.00	1.0413	1.2335	1.4288		0.0422	0.2573	0.7210		
8.00	1.0532	1.3106	1.5909		0.0450	0.2904	0.9019		
10.00	1.0630	1.3769	1.7392		0.0464	0.3140	1.0625		
n = 6									
1.50	1.0012	1.0079	1.0185		0.0062	0.0376	0.0957	0.1290	
2.00	1.0035	1.0230	1.0544		0.0103	0.0631	0.1658	0.2284	
2.50	1.0061	1.0397	1.0956		0.0131	0.0817	0.2225	0.3138	
3.00	1.0086	1.0564	1.1380		0.0152	0.0959	0.2707	0.3908	
4.00	1.0132	1.0878	1.2216		0.0180	0.1162	0.3509	0.5295	
5.00	1.0172	1.1161	1.3010		0.0197	0.1301	0.4170	0.6549	
6.00	1.0208	1.1414	1.3758		0.0208	0.1402	0.4738	0.7712	
8.00	1.0268	1.1852	1.5128		0.0221	0.1539	0.5689	0.9846	
10.00	1.0317	1.2220	1.6361		0.0228	0.1626	0.6477	1.1796	
n = 7									
1.50	1.0006	1.0047	1.0143	1.0206	0.0031	0.0218	0.0665	0.1126	
2.00	1.0018	1.0137	1.0419	1.0606	0.0051	0.0364	0.1135	0.1983	
2.50	1.0030	1.0236	1.0731	1.1067	0.0065	0.0468	0.1499	0.2709	
3.00	1.0043	1.0334	1.1048	1.1546	0.0076	0.0546	0.1795	0.3355	
4.00	1.0066	1.0516	1.1662	1.2499	0.0089	0.0653	0.2258	0.4497	
5.00	1.0086	1.0678	1.2233	1.3416	0.0098	0.0724	0.2612	0.5509	
6.00	1.0104	1.0822	1.2759	1.4288	0.0103	0.0772	0.2898	0.6431	
8.00	1.0134	1.1067	1.3699	1.5909	0.0110	0.0834	0.3339	0.8090	
10.00	1.0159	1.1270	1.4520	1.7392	0.0113	0.0869	0.3672	0.9574	
n = 8									
1.50	1.0003	1.0027	1.0101	1.0190	0.0015	0.0124	0.0440	0.0895	0.1129
2.00	1.0009	1.0079	1.0293	1.0559	0.0025	0.0207	0.0743	0.1558	0.2000
2.50	1.0015	1.0135	1.0508	1.0982	0.0032	0.0265	0.0969	0.2100	0.2748
3.00	1.0021	1.0191	1.0724	1.1419	0.0037	0.0308	0.1146	0.2567	0.3425
4.00	1.0033	1.0295	1.1134	1.2282	0.0044	0.0366	0.1408	0.3357	0.4644
5.00	1.0043	1.0386	1.1508	1.3104	0.0049	0.0402	0.1595	0.4021	0.5749
6.00	1.0052	1.0467	1.1846	1.3881	0.0051	0.0427	0.1737	0.4601	0.6775
8.00	1.0067	1.0603	1.2436	1.5308	0.0055	0.0457	0.1939	0.5591	0.8661
10.00	1.0080	1.0715	1.2940	1.6598	0.0056	0.0474	0.2078	0.6426	1.0388

SOURCE: *IRE Trans. PGMTT* (see Ref. 11 by Leo Young)

Table 13.12-3

IMMITTANCES OF BRANCH-LINE COUPLERS FOR n = 2, 3, AND 4 SECTIONS (THREE, FOUR, AND FIVE BRANCHES) WHEN w_q = 0.20

R	K_1	K_2	H_1	H_2	H_3
			n = 2		
1.50	1.0155		0.1022	0.2036	
2.00	1.0453		0.1738	0.3594	
2.50	1.0790		0.2282	0.4922	
3.00	1.1135		0.2717	0.6111	
4.00	1.1803		0.3385	0.8229	
5.00	1.2425		0.3883	1.0121	
6.00	1.3000		0.4276	1.1859	
8.00	1.4029		0.4870	1.5007	
10.00	1.4929		0.5307	1.7845	
			n = 3		
1.50	1.0090	1.0206	0.0511	0.1530	
2.00	1.0262	1.0606	0.0856	0.2678	
2.50	1.0454	1.1067	0.1108	0.3634	
3.00	1.0645	1.1546	0.1300	0.4473	
4.00	1.1006	1.2499	0.1574	0.5925	
5.00	1.1331	1.3416	0.1761	0.7182	
6.00	1.1624	1.4288	0.1897	0.8309	
8.00	1.2129	1.5909	0.2078	1.0295	
10.00	1.2554	1.7392	0.2193	1.2036	
			n = 4		
1.50	1.0048	1.0177	0.0256	0.1017	0.1535
2.00	1.0140	1.0519	0.0426	0.1751	0.2715
2.50	1.0241	1.0909	0.0547	0.2332	0.3727
3.00	1.0341	1.1311	0.0636	0.2817	0.4638
4.00	1.0527	1.2099	0.0760	0.3602	0.6274
5.00	1.0692	1.2844	0.0839	0.4230	0.7747
6.00	1.0837	1.3543	0.0894	0.4756	0.9110
8.00	1.1085	1.4814	0.0962	0.5610	1.1603
10.00	1.1290	1.5949	0.1000	0.6293	1.3872

SOURCE: Tech. Note 3, Contract AF30(602)-2392, SRI (see Ref. 22 by Leo Young)

Table 13.12-4

IMMITTANCES OF BRANCH-LINE COUPLERS FOR n = 2, 3, AND 4 SECTIONS (THREE, FOUR, AND FIVE BRANCHES) WHEN w_q = 0.40

R	K_1	K_2	H_1	H_2	H_3
n = 2					
1.50	1.0159		0.1061	0.1958	
2.00	1.0464		0.1807	0.3456	
2.50	1.0811		0.2376	0.4733	
3.00	1.1166		0.2835	0.5876	
4.00	1.1857		0.3543	0.7913	
5.00	1.2503		0.4078	0.9731	
6.00	1.3102		0.4506	1.1400	
8.00	1.4181		0.5162	1.4424	
10.00	1.5132		0.5655	1.7148	
n = 3					
1.50	1.0095	1.0206	0.0540	0.1500	
2.00	1.0275	1.0606	0.0907	0.2627	
2.50	1.0477	1.1067	0.1176	0.3567	
3.00	1.0679	1.1546	0.1381	0.4391	
4.00	1.1061	1.2499	0.1678	0.5821	
5.00	1.1407	1.3416	0.1882	0.7062	
6.00	1.1719	1.4288	0.2031	0.8174	
8.00	1.2260	1.5909	0.2235	1.0139	
10.00	1.2718	1.7392	0.2367	1.1862	
n = 4					
1.50	1.0052	1.0178	0.0276	0.1016	0.1497
2.00	1.0150	1.0523	0.0459	0.1751	0.2647
2.50	1.0259	1.0917	0.0590	0.2335	0.3634
3.00	1.0367	1.1323	0.0688	0.2823	0.4522
4.00	1.0567	1.2119	0.0823	0.3617	0.6117
5.00	1.0745	1.2873	0.0911	0.4255	0.7554
6.00	1.0904	1.3580	0.0972	0.4791	0.8884
8.00	1.1173	1.4869	0.1048	0.5667	1.1315
10.00	1.1397	1.6023	0.1093	0.6373	1.3528

SOURCE: Tech. Note 3, Contract AF30(602)-2392, SRI (see Ref. 22 by Leo Young)

Table 13.12-5

IMMITTANCES OF BRANCH-LINE COUPLERS FOR n = 2, 3, AND 4 SECTIONS (THREE, FOUR, AND FIVE BRANCHES) WHEN w_q = 0.60

R	K_1	K_2	H_1	H_2	H_3
n = 2					
1.50	1.0165		0.1128	0.1824	
2.00	1.0483		0.1925	0.3220	
2.50	1.0846		0.2539	0.4408	
3.00	1.1218		0.3037	0.5471	
4.00	1.1945		0.3818	0.7363	
5.00	1.2630		0.4418	0.9051	
6.00	1.3269		0.4906	1.0598	
8.00	1.4430		0.5674	1.3398	
10.00	1.5464		0.6271	1.5917	
n = 3					
1.50	1.0103	1.0206	0.0594	0.1446	
2.00	1.0299	1.0606	0.1000	0.2535	
2.50	1.0518	1.1067	0.1299	0.3444	
3.00	1.0739	1.1546	0.1530	0.4243	
4.00	1.1159	1.2499	0.1867	0.5632	
5.00	1.1542	1.3416	0.2104	0.6839	
6.00	1.1889	1.4288	0.2281	0.7925	
8.00	1.2497	1.5909	0.2529	0.9845	
10.00	1.3018	1.7392	0.2696	1.1533	
n = 4					
1.50	1.0058	1.0181	0.0313	0.1012	0.1432
2.00	1.0169	1.0530	0.0522	0.1746	0.2533
2.50	1.0292	1.0930	0.0672	0.2332	0.3476
3.00	1.0413	1.1342	0.0785	0.2824	0.4327
4.00	1.0642	1.2152	0.0942	0.3630	0.5853
5.00	1.0845	1.2920	0.1046	0.4283	0.7228
6.00	1.1027	1.3641	0.1119	0.4836	0.8500
8.00	1.1339	1.4960	0.1214	0.5746	1.0826
10.00	1.1600	1.6143	0.1271	0.6486	1.2944

SOURCE: Tech. Note 3, Contract AF30(602)-2392, SRI (see Ref. 22 by Leo Young)

Table 13.12-6

IMMITTANCES OF BRANCH-LINE COUPLERS FOR n = 2, 3, AND 4 SECTIONS (THREE, FOUR, AND FIVE BRANCHES) WHEN w_q = 0.80

R	K_1	K_2	H_1	H_2	H_3
n = 2					
1.50	1.0173		0.1225	0.1630	
2.00	1.0508		0.2097	0.2875	
2.50	1.0891		0.2776	0.3933	
3.00	1.1286		0.3333	0.4879	
4.00	1.2062		0.4221	0.6557	
5.00	1.2798		0.4919	0.8049	
6.00	1.3491		0.5499	0.9413	
8.00	1.4760		0.6437	1.1872	
10.00	1.5905		0.7192	1.4075	
n = 3					
1.50	1.0114	1.0206	0.0679	0.1361	
2.00	1.0334	1.0606	0.1147	0.2388	
2.50	1.0580	1.1067	0.1495	0.3247	
3.00	1.0830	1.1546	0.1768	0.4004	
4.00	1.1309	1.2499	0.2175	0.5324	
5.00	1.1749	1.3416	0.2468	0.6476	
6.00	1.2153	1.4288	0.2692	0.7513	
8.00	1.2870	1.5909	0.3020	0.9353	
10.00	1.3491	1.7392	0.3254	1.0975	
n = 4					
1.50	1.0069	1.0184	0.0374	0.0997	0.1339
2.00	1.0199	1.0540	0.0626	0.1724	0.2368
2.50	1.0344	1.0947	0.0808	0.2309	0.3251
3.00	1.0489	1.1368	0.0947	0.2803	0.4045
4.00	1.0763	1.2196	0.1143	0.3620	0.5472
5.00	1.1009	1.2983	0.1276	0.4288	0.6757
6.00	1.1230	1.3725	0.1372	0.4860	0.7946
8.00	1.1614	1.5083	0.1501	0.5813	1.0119
10.00	1.1939	1.6306	0.1582	0.6598	1.2097

SOURCE: Tech. Note 3, Contract AF30(602)-2392, SRI (see Ref. 22 by Leo Young)

Table 13.12-7

IMMITTANCES OF BRANCH-LINE COUPLERS FOR n = 2, 3, AND 4 SECTIONS (THREE, FOUR, AND FIVE BRANCHES) WHEN w_q = 1.00

R	K_1	K_2	H_1	H_2	H_3
			n = 2		
1.50	1.0183		0.1354	0.1373	
2.00	1.0537		0.2326	0.2418	
2.50	1.0944		0.3091	0.3303	
3.00	1.1364		0.3728	0.4090	
4.00	1.2195		0.4759	0.5480	
5.00	1.2990		0.5590	0.6708	
6.00	1.3742		0.6294	0.7824	
8.00	1.5133		0.7463	0.9821	
10.00	1.6397		0.8432	1.1595	
			n = 3		
1.50	1.0131	1.0206	0.0807	0.1233	
2.00	1.0383	1.0606	0.1369	0.2165	
2.50	1.0668	1.1067	0.1795	0.2948	
3.00	1.0958	1.1546	0.2134	0.3639	
4.00	1.1521	1.2499	0.2652	0.4847	
5.00	1.2047	1.3416	0.3039	0.5904	
6.00	1.2534	1.4288	0.3347	0.6859	
8.00	1.3412	1.5909	0.3817	0.8556	
10.00	1.4189	1.7392	0.4173	1.0057	
			n = 4		
1.50	1.0084	1.0188	0.0472	0.0961	0.1214
2.00	1.0246	1.0552	0.0794	0.1667	0.2147
2.50	1.0426	1.0969	0.1030	0.2240	0.2946
3.00	1.0607	1.1400	0.1211	0.2728	0.3665
4.00	1.0952	1.2251	0.1476	0.3545	0.4956
5.00	1.1267	1.3062	0.1661	0.4223	0.6118
6.00	1.1554	1.3828	0.1799	0.5803	0.7191
8.00	1.2057	1.5237	0.1994	0.5802	0.9152
10.00	1.2490	1.6510	0.2127	0.6635	1.0935

SOURCE: Tech. Note 3, Contract AF30(602)-2392, SRI (see Ref. 22 by Leo Young)

Table 13.12-8

IMMITTANCES OF BRANCH-LINE COUPLERS FOR n = 2, 3, AND 4 SECTIONS (THREE, FOUR, AND FIVE BRANCHES) WHEN w_q = 1.20

R	K_1	K_2	H_1	H_2	H_3
			n = 2		
1.50	1.0192		0.1513	0.1056	
2.00	1.0565		0.2607	0.1855	
2.50	1.0995		0.3479	0.2527	
3.00	1.1441		0.4212	0.3121	
4.00	1.2325		0.5419	0.4160	
5.00	1.3174		0.6410	0.5067	
6.00	1.3982		0.7263	0.5884	
8.00	1.5481		0.8708	0.7331	
10.00	1.6852		0.9929	0.8601	
			n = 3		
1.50	1.0152	1.0206	0.0994	0.1046	
2.00	1.0445	1.0606	0.1696	0.1838	
2.50	1.0780	1.1067	0.2238	0.2504	
3.00	1.1125	1.1546	0.2680	0.3092	
4.00	1.1800	1.2499	0.3376	0.4123	
5.00	1.2439	1.3416	0.3918	0.5025	
6.00	1.3040	1.4288	0.4364	0.5841	
8.00	1.4140	1.5909	0.5082	0.7291	
10.00	1.5133	1.7392	0.5657	0.8572	
			n = 4		
1.50	1.0108	1.0192	0.0631	0.0886	0.1047
2.00	1.0315	1.0566	0.1066	0.1542	0.1851
2.50	1.0548	1.0994	0.1393	0.2080	0.2539
3.00	1.0785	1.1438	0.1650	0.2544	0.3157
4.00	1.1243	1.2316	0.2037	0.3331	0.4263
5.00	1.1667	1.3156	0.2320	0.3995	0.5256
6.00	1.2059	1.3951	0.2541	0.4578	0.6171
8.00	1.2761	1.5419	0.2873	0.5581	0.7837
10.00	1.3381	1.6753	0.3119	0.6437	0.9345

SOURCE: Tech. Note 3, Contract AF30(602)-2392, SRI (see Ref. 22 by Leo Young)

The notation for the immittances H_i and K_i is shown in Fig. 13.09-2. For a coupler with series T-junctions, H_i and K_i are both characteristic impedances (*e.g.* for E-plane junctions in waveguide); for a coupler with shunt T-junctions, H_i and K_i are both characteristic admittances (*e.g.*, for coaxial or strip transmission lines).

SEC. 13.13, EXAMPLES ILLUSTRATING THE DESIGN AND PERFORMANCE OF SYNCHRONOUS BRANCH-LINE COUPLERS

Example—A 3-db coupler is to have an input VSWR of less than 1.10 and a directivity in excess of 20 db over a 24-percent fractional bandwidth.

From Fig. 13.11-1, when $P_{2,0}$ = 3 db, then by Eq. (13.11-1), R = 5.84. Try a two-branch coupler first, corresponding to a single-section, quarter-wave transformer prototype (n = 1). From Fig. 13.11-2, β = 0.64 for n = 1, so that the prototype fractional bandwidth must be w_b/β = 24/0.64 percent, or nearly 40 percent by Eq. (13.11-3). The maximum VSWR of a single-section quarter-wave transformer of R = 6, and w_q = 0.40, is 1.86, from Table 6.02-2. It follows from Eq. (13.11-4) that V_{max} of the coupler would then be considerably greater than the 1.10 specified.

Try a three-branch coupler next, corresponding to a two-section, quarter-wave transformer (n = 2). From Fig. 13.11-2, β = 0.62 for n = 2, so that the prototype fractional bandwidth must be 24/0.62 percent, or almost 40 percent. Now the maximum VSWR of a two-section quarter-wave transformer of R = 6, w_q = 0.40, is 1.11 from Table 6.02-2. For a 3-db coupler, Eq. (13.11-4) yields $V_{max} \approx 1 + 0.11/1.414 = 1.08$, which is below the 1.10 specified. The directivity from Eq. (13.11-5) will be better than $-20 \log_{10} (0.04) - P_2$ = 25 db, which exceeds the 20 db specified.

Thus the prototype quarter-wave transformer will in this case have two sections (n = 2), R = 5.84, and bandwidth w_q = 0.40. The corresponding branch-guide coupler design can be obtained by interpolation from Table 13.12-4.

If we may use Fig. 13.11-5 as a guide, then over the 24-percent band specified, this three-branch, 3-db coupler would be expected to change each of its couplings, P_1 and P_2, by a little under 0.3 db. Thus if the coupler were designed to have 3-db coupling at center frequency (corresponding to $R = 5.84$ picked before), then P_2 would go to 2.7 db and P_1 to 3.3 db at the 24-percent band edges. If the specification required both P_1 and P_2 to be maintained to within ±0.15 db of 3 db over the 24-percent band, or generally to optimize the balance over the band as a whole, then the coupler would be designed with $P_{2,0} = 3.15$ db, corresponding to $R = 5.7$, by Fig. 13.11-1 or Eq. (13.11-1). With this choice, and from Table 13.12-4,

$$K_0 = 1 \quad , \quad K_1 = 1.2902 \quad , \quad H_1 = 0.4363 \quad , \quad H_2 = 1.0844 \quad .$$

The performance of this coupler was analyzed by computer and is reproduced in Fig. 13.13-1. It is found to conform very closely to the specifications. (Again, this is plotted only for one side of band-center, since the response is symmetrical as plotted.) From Fig. 13.13-1, the analyzed performance over the 24-percent bandwidth is: maximum VSWR, 1.07 (1.08 was predicted); minimum directivity, 26 db (25 db was predicted); couplings P_1 and P_2 both within ±0.2 db of 3 db (±0.15 db was predicted).

Comparison of Synchronous and Periodic Couplers—The maximum VSWR and the coupling unbalance of several 3-db periodic couplers was given in

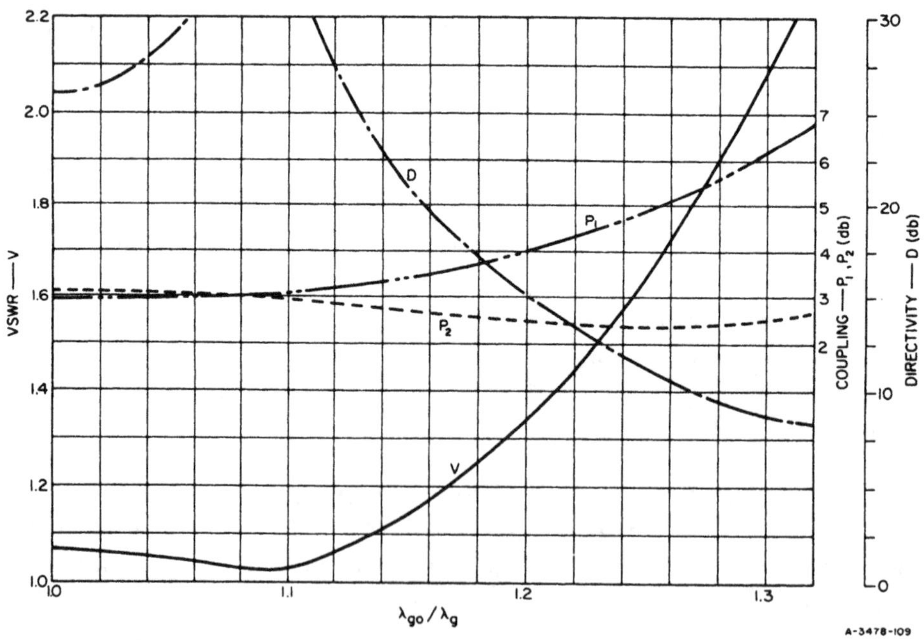

SOURCE: *IRE Trans. PGMTT* (see Ref. 11 by Leo Young)

FIG. 13.13-1 COMPUTED PERFORMANCE OF A THREE-BRANCH (n = 2) COUPLER

Tables 13.10-2 and 13.10-3 for three particular bandwidths w_b. None of the periodic couplers could be designed for any particular bandwidth, and the performance of any one coupler simply degenerates slowly as its operating bandwidth is increased. (For instance, the VSWR ripples keep getting bigger.) To improve the performance over any particular bandwidth one can increase the number of branches (but avoid an increase from an even to an odd number of branches).

Table 13.13-1

PERFORMANCES OF THREE SYNCHRONOUS 3-db COUPLERS OF FOUR BRANCHES

CASE	FRACTIONAL BANDWIDTH w_b	MAXIMUM VSWR	COUPLING UNBALANCE (db)
1	0.16	1.01	0.19
2	0.28	1.02	0.57
3	0.48	1.07	1.9

SOURCE: Tech. Note 3, Contract AF 30(602)-2392, SRI (see Ref. 22 by Leo Young)

By contrast, synchronous couplers have a clearly defined pass band over which at least their VSWR and directivity is optimized. Three different four-branch synchronous couplers were obtained from the tables in Sec. 13.12. Each has $n = 3$ and $R = 5.5$, and their several transformer prototype bandwidths are $w_q = 0.40$ (Case 1), $w_q = 0.60$ (Case 2), and $w_q = 0.80$ (Case 3). These three couplers give about optimum performance over bandwidths of $w_b = 0.16$, $w_b = 0.28$ and $w_b = 0.48$, respectively, corresponding to the bandwidths in Tables 13.10-2 and 13.10-3. The maximum VSWR and coupling unbalance over their respective pass bands is shown in Table 13.13-1. A comparison with Tables 13.10-2 and 13.10-3 shows that the VSWR of the four-branch synchronous couplers is as good as (or better than) that of the eight-branch periodic coupler. There is also a slight improvement in the coupling unbalance over the four-branch periodic coupler.

SEC. 13.14, DESIGN OF AN EXPERIMENTAL BRANCH-LINE COUPLER IN WAVEGUIDE

The design of a 6-db coupler in waveguide will now be discussed. A prototype transformer having $n = 4$ (five branches), $R = 3$ and $w_q = 0.40$ was selected. At that time[11] the tables in Sec. 13.12 were not available and the following immittances were calculated by charts;[11] the more exact values from Table 13.12-4 (not used in the construction) are shown in parentheses:

$$\left.\begin{array}{llll} K_0 = 1.0 & (1.0) & H_1 = 0.070 & (0.0688) \\ K_1 = 1.036 & (1.0367) & H_2 = 0.274 & (0.2823) \\ K_2 = 1.127 & (1.1323) & H_3 = 0.450 & (0.4522) \end{array}\right\} \quad (13.14\text{-}1)$$

It was verified by analysis on a digital computer that the immittances from the tables (in parentheses) would give a slightly improved coupler; the improvement was, however, not large enough to warrant the expense of building a new model. It was also demonstrated[11,22] that the performance of this

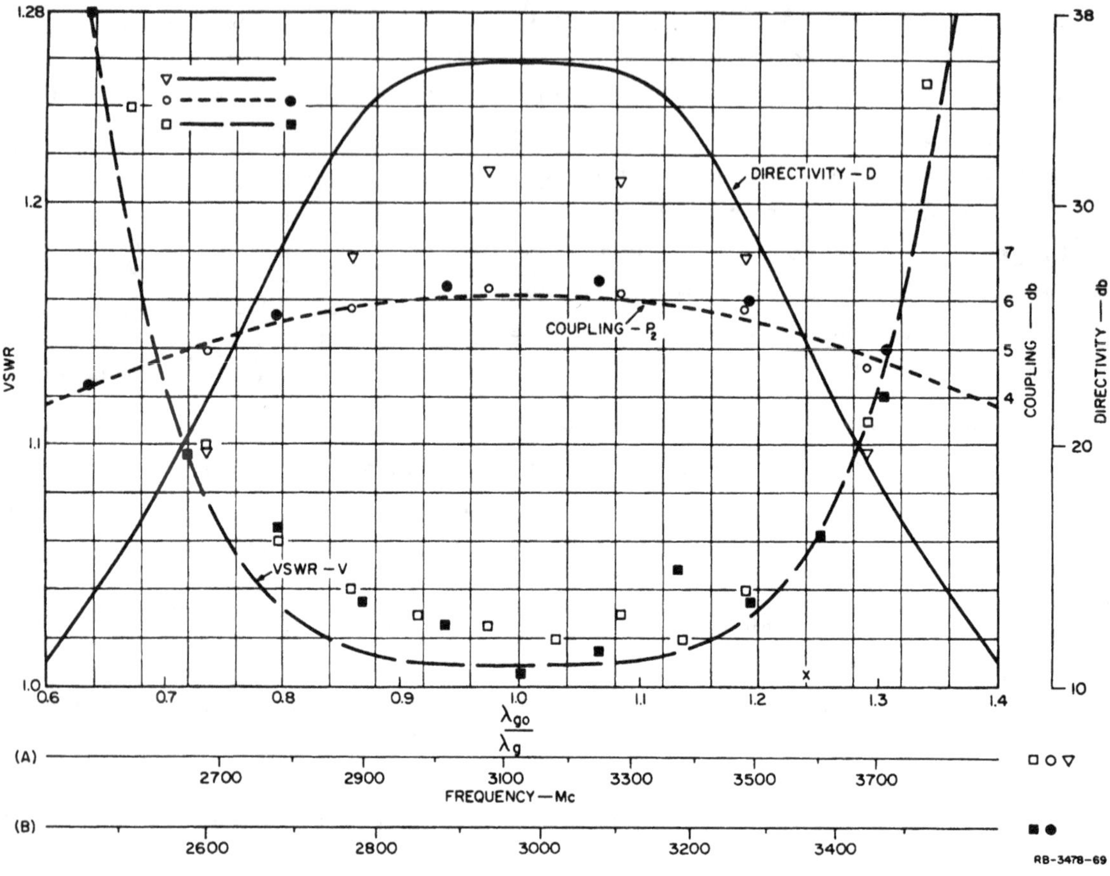

SOURCE: *IRE Trans. PGMTT* (see Ref. 11 by Leo Young)

FIG. 13.14-1 PERFORMANCE OF FIVE-BRANCH (n = 4) COUPLER
Lines are computed; points are measured on experimental models shown in Figs. 13.14-3, 13.14-4, and 13.14-5

coupler could be predicted accurately. The details of the calculation are similar to the example in Sec. 13.13, and are not reproduced here.

The computed performance of this coupler is shown in Fig. 13.14-1. The experimental points are plotted in the same figure, and will be explained below.

Construction and Experimental Performance—The coupler was to be constructed in S-band, with rectangular waveguide outputs of 2.840 inches by 1.420 inches. The center frequency was to be 2975 Mc. (Therefore, λ_{g0} = 5.54 inches.)

Since waveguide T-junctions are series junctions, the immittances K_i and H_i are impedances; however, since waveguide T-junctions are not perfect series junctions, they can be represented by an equivalent circuit such as is shown in Fig. 13.14-2. (This is the same circuit as Fig. 6.1-2 on p. 338 of Ref. 20.) At any T-junction of the coupler, K_{i-1} and K_i are generally not equal, although they differ only slightly, so that K_{av} in Fig. 13.14-2(a) was set equal to their arithmetic mean $(K_{i-1} + K_i)/2$. Similarly the wave-guide height substituted into the graphs of Ref. 20 to obtain the T-junction properties was $b_{av} = (b_{i-1} + b_i)/2$, and the junction was treated as if it were symmetrical [see Fig. 13.14-2(a)].

The waveguide heights (their b-dimensions) were first taken as proportional to the respective K or H values (Eq. 13.14-1); they were fixed by the b-dimensions of the four ports, which were each equal to b_0 = 1.420 inches. This determined the b-dimensions in the through-guides without further adjustments, but the b-dimensions of the branch guides had to be further increased to allow for the transformer factor denoted by n^2 in Ref. 20 (no connection with the number of sections, n). This factor was found from Table 6.1-10 on p. 346 in Ref. 20, and the branch b-dimensions, were increased by a factor $1/n^2$. Since this changed the junction dimensions, the quantity n^2 had to be worked out again, and more times if necessary (usually this is not necessary) until each product of n^2 and the branch guide b-dimension was proportional to the appropriate impedance H. Finally, the reference plane positions, d and d', of each junction were determined[20] (Fig. 13.14-2).

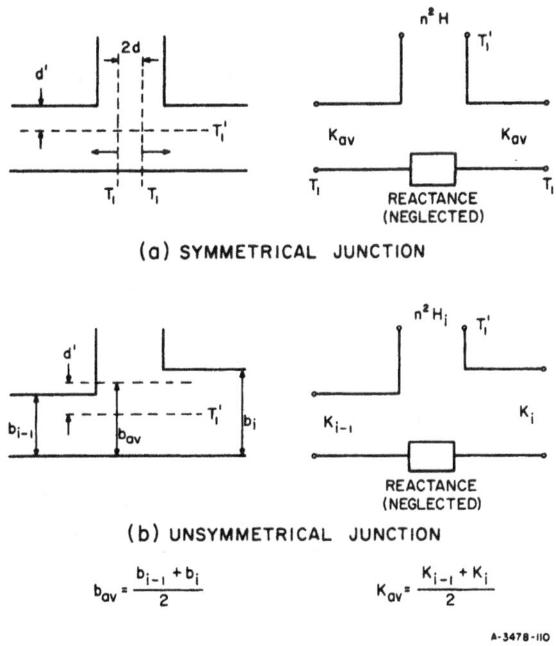

SOURCE: Tech. Note 3, Contract AF 30(602)-2392, SRI; (see Ref. 22 by Leo Young)

FIG. 13.14-2 EQUIVALENT CIRCUITS OF T-JUNCTIONS

The dimensions of this coupler were calculated, using the impedances of Eq. (13.14-1) and the *T*-junction equivalent circuits in Ref. 20. (The series reactances were included in several of the earlier computations analyzing the coupler performance. Their presence was found to affect the performance only slightly, and has been ignored in all later computations.)

The coupler *E*-plane cross section might be expected to require stepped top and bottom walls (Fig. 13.09-1) to obtain the changes in impedance K_i called for in Eq. (13.14-1). However, the reference planes (T_1' in Fig. 13.14-2) move in such a manner that the branches have to be shortened more where the impedances K are larger, with the result that the top and bottom walls (Fig. 13.09-1) come out almost straight. In order to obtain straight top and bottom walls as shown in Fig. 13.09-1, and yet maintain the correct impedances K_i, branch lengths some of which differ slightly from the lengths calculated[20] have to be accepted.

The dimensions so calculated are shown in Fig. 13.14-3. The coupler was constructed in two halves as shown in Fig. 13.14-4. Three jig-plates of aluminum were used to make a *U*-shaped channel, in which six aluminum blocks were placed and bolted down to form the waveguide channels. The end blocks contain transformers from the waveguide height of 1.420 inches

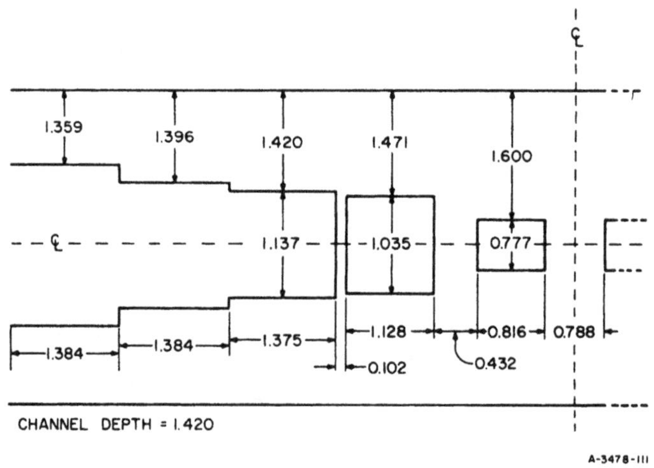

SOURCE: *IRE Trans. PGMTT* (see Ref. 11 by Leo Young)

FIG. 13.14-3 DIMENSIONS OF FIRST S-BAND, FIVE-BRANCH, 6-db COUPLER

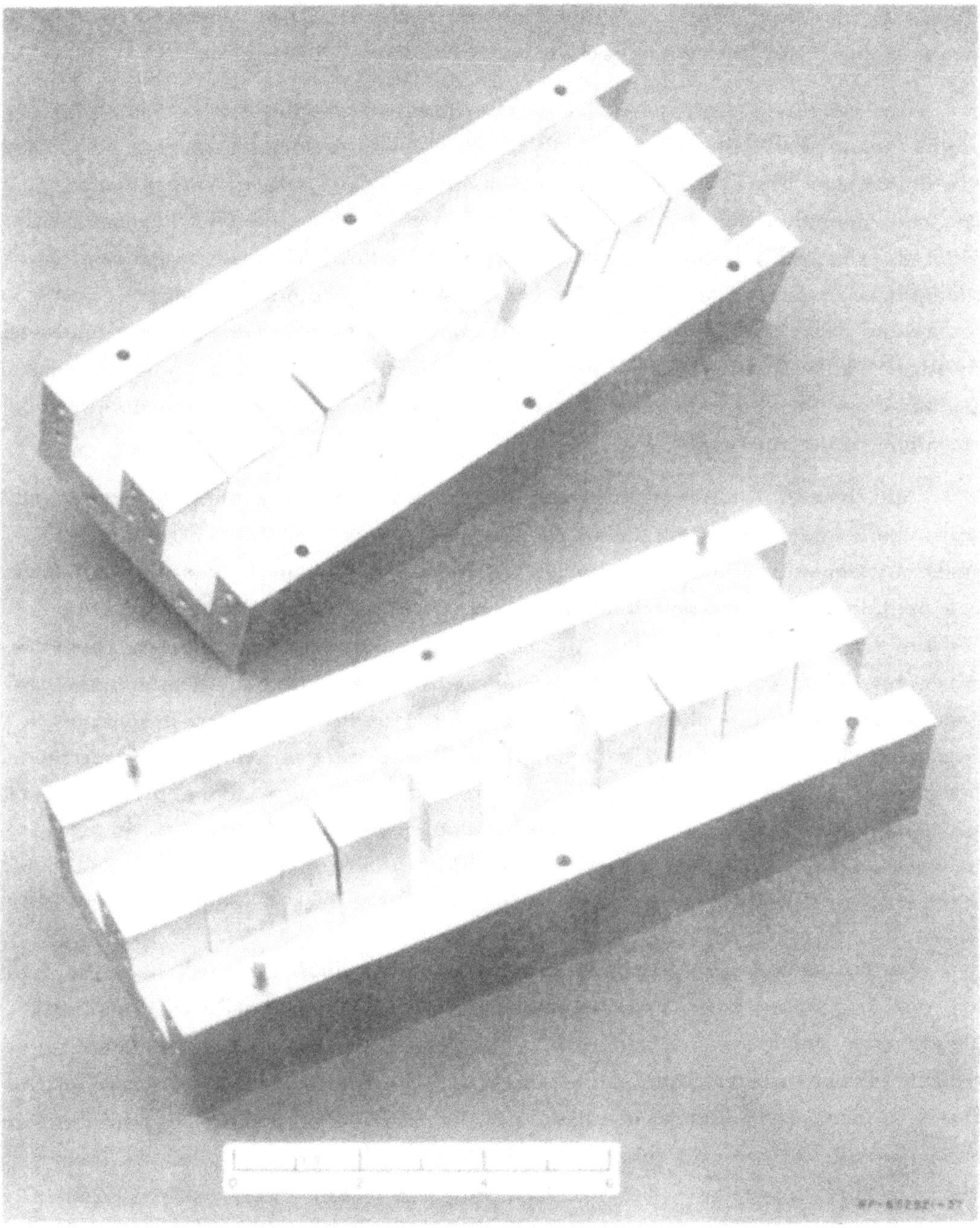

SOURCE: *IRE Trans. PGMTT* (see Ref. 11 byLeo Young)

FIG. 13.14-4 EXPLODED VIEW OF S-BAND, EXPERIMENTAL FIVE-BRANCH, 6-db COUPLER

to 1.340 inches, so that standard WR-284 waveguide test equipment could be connected. The depth of all the channels is half of 2.840 inches, or 1.420 inches. The two pieces shown in Fig. 13.14-4 were finally superimposed and bolted together to form the 6-db coupler.

The measured performance of the completed coupler is shown by the light points in Fig. 13.14-1, which go with the frequency scale (*A*) near the bottom of Fig. 13.14-1. Plotted on a λ_{g0}/λ_g scale, the points fit the computed curves very closely; however, the center frequency is 3125 Mc instead of the design value of 2975 Mc. This discrepancy is thought to result from the relatively large *b*-dimensions which, for instance, make the length of an outline edge on the two center squares in Fig. 13.14-3 only about one-seventh wavelength. Thus, non-propagating higher-order modes could be set up, giving rise to interaction effects at such close spacings.

All branch lengths and spacings, nominally one-quarter wavelength, were then scaled in the ratio of the guide wavelengths to reduce the center frequency from 3125 to 2975 Mc, and the coupler was tested again. Its center frequency moved down as expected, but the coupling there became stronger, going from 6.1 to 5.8 db. Since the coupling becomes stronger still at off-center frequencies, it was decided to reduce the branch heights to weaken the coupling by 0.5 db at center frequency, changing the 5.8 db to 6.3 db coupling. The new dimensions calculated are shown in Fig. 13.14-5. The measured results are shown by the black points in Fig. 13.14-1 which go with the frequency scale (*B*) at the bottom of Fig. 13.14-1. It is seen that this coupler gives the desired center frequency, and its performance closely follows the computed curves.

Power-Handling Capacity—All edges orthogonal to the electric field of the TE_{10} mode were then rounded off to increase the power-handling capacity. The edges of the eight rectangular blocks were rounded to a radius of one-eighth inch; the edges of the four outside blocks, on the faces defining the outside edges of the narrowest branches, were rounded to a radius of one-sixteenth inch. These radii were estimated from Fig. 15.02-5 to reduce the field strength at the edges to less than double the field strength in the waveguides.[23] Any further rounding would have helped very little. The performance of the coupler with rounded edges was measured, and found to reproduce the performance shown in Fig. 13.14-1 to within experimental accuracy.

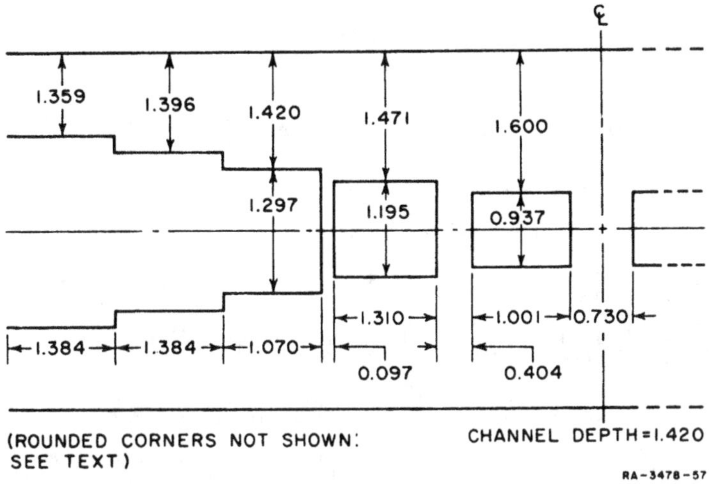

SOURCE: *IRE Trans. PGMTT* (see Ref. 11 by Leo Young)

FIG. 13.14-5 DIMENSIONS OF S-BAND, FIVE-BRANCH, 6-db COUPLER AFTER MODIFICATIONS

It was estimated[22] that this coupler should handle about 39-percent of the peak power handled by the waveguide. This is chiefly determined by the radii of the *T*-junction corners. In WR-284 waveguide, this corresponds roughly to 1.0 megawatt. The coupler was tested[11] at high power using a 3.8-microsecond pulse at 60 pps, at a frequency of 2857 Mc. With air at atmospheric pressure, no arcing was observed with 1 megawatt peak power.

REFERENCES

1. S. B. Cohn, J. K. Shimizu, E. M. T. Jones and P. M. Sherk, "Strip Transmission Lines and Components," Final Report, Contract DA 36-039 SC-63232, Stanford Research Institute, Menlo Park, California (February 1957).
2. B. M. Oliver, "Directional Electromagnetic Couplers," *Proc. IRE* 42, pp. 1686-1692 (November 1954).
3. W. L. Firestone, "Analysis of Transmission Line Directional Couplers," *Proc. IRE* 42, pp. 1529-1538 (October 1954).
4. E. M. T. Jones and J. T. Bolljahn, "Coupled-Strip-Transmission Line Filters and Directional Couplers," *IRE Trans. PGMTT*-4, pp. 75-81 (April 1956).
5. J. K. Shimizu and E. M. T. Jones, "Coupled-Transmission Line Directional Couplers," *IRE Trans. PGMTT*-6, pp. 403-410 (October 1958).
6. J. K. Shimizu, "A Strip-Line 3 db Directional Coupler," Scientific Report 1, Contract AF 19(604)-1571, Stanford Research Institute, Menlo Park, California (June 1957).
7. S. B. Cohn and S. L. Wehn, "Microwave Hybrid Coupler Study Program," Second Quarterly Progress Report, Contract DA 36-2395c-87435, Rantec Corporation, Calabasas, California (November 1961).
8. J. Reed and G. Wheeler, "A Method of Analysis of Symmetrical Four-Port Networks," *IRE Trans. PGMTT*-4, pp. 246-252 (October 1956).
9. J. Reed, "The Multiple Branch Waveguide Coupler," *IRE Trans. PGMTT*-6, pp. 398-403 (October 1958).
10. Leo Young, "Branch Guide Directional Couplers," *Proc. Nat. Electronics Conf.*, Vol. 12, pp. 723-732 (1956).
11. Leo Young, "Synchronous Branch Guide Directional Couplers for Low and High Power Applications," *IRE Trans. PGMTT*-10, pp. 459-475 (November 1962).
12. S. B. Cohn and R. H. Koontz, "Microwave Hybrid Coupler Study Program," Third Quarterly Progress Report, Contract DA 36-239-SC-87435, Rantec Corporation, Calabasas, California (February 1962).
13. W. J. Getsinger, "A Coupled Strip-Line Configuration Using Printed-Circuit Construction that Allows Very Close Coupling," *IRE Trans. PGMTT*-9, pp. 535-544 (November 1961).
14. W. J. Getsinger, "Coupled Rectangular Bars Between Parallel Plates," *IRE Trans. PGMTT*-10, pp. 65-72 (January 1962).
15. A. L. Fel'dshtein, "Synthesis of Stepped Directional Couplers," *Radiotekhnika i Elektronika*, Vol. 6, No. 2, pp. 234-240 (February 1961); English translation, Pergamon Press, pp. 75-85 (1961).
16. Leo Young, "Analysis of a Transmission Cavity Wavemeter," *IRE Trans. PGMTT*-8, pp. 436-439 (July 1960). [The exponent in the numerator of Eq. (20) should be $-J\theta$ and not $-2\ J\theta$].
17. Leo Young, "A Theorem on Lossless Symmetrical Networks," *IRE Trans. PGCT*-7, p. 75 (March 1960).
18. S. B. Cohn, W. J. Getsinger, E. M. T. Jones, B. M . Schiffman, J. K. Shimizu, "Design Criteria for Microwave Filters and Coupling Structures," Tech. Report 3, Contract DA 36-039 SC-74862, Stanford Research Institute, Menlo Park, California (August 1958).
19. A. G. Franco and A. A. Oliner, "Symmetric Strip Transmission Line Tee Junction," *IRE Trans. PGMTT*-10, pp. 118-124 (March 1962).
20. N. Marcuvitz, *Waveguide Handbook*, pp. 336-350, MIT Rad. Lab. Series, Vol. 10 (McGraw-Hill Book Co., Inc., New York, N. Y. 1951).
21. "Tables of Chebyshev Polynomials $S_n(x)$ and $C_n(x)$," National Bureau of Standards Applied Mathematics Series No. 9, United States Government Printing Office, Washington 25, D. C. (1952).
22. Leo Young, "The Design of Branch Guide Directional Couplers for Low and High Power Applications," Tech. Note 3, Contract AF 30(602)-2392, Stanford Research Institute, Menlo Park, California. RADC-TDR-62-130 (February 1962).
23. S. B. Cohn, "Rounded Corners in Microwave High-Power Filters and Other Components," *IRE Trans. PGMTT*-9, pp. 389-397 (September 1961).
24. Leo Young, "The Analytical Equivalence of TEM-Mode Directional Couplers and Transmission-Line Stepped-Impedance Filters," *Proc. IEE*, 110, pp. 275-281 (February 1963).
25. R. Levy, "General Synthesis of Asymmetric Multi-Element Coupled-Transmission-Line Directional Couplers," *IEEE Trans. PTGMTT*-11, pp. 226-237 (July 1963).

CHAPTER 14

DIRECTIONAL, CHANNEL-SEPARATION FILTERS AND TRAVELING-WAVE RING-RESONATORS*

SEC. 14.01, INTRODUCTION

This chapter discusses a particularly useful type of multiplexing filter, called a directional channel-separation filter,[1,2,3] which is used to combine or separate signals of different frequencies. In addition, information is presented on a closely related structure, the traveling-wave ring-resonator. This is a passive power-multiplying device which has found wide acceptance for the high-power testing of components.[4,5,6,7]

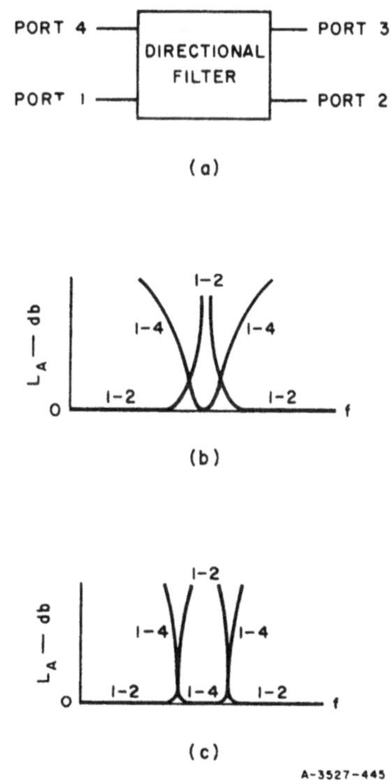

SOURCE: Final Report, Contract DA 36-039 SC-64625, SRI; reprinted in *Proc. IRE* (see Ref. 1 by S. B. Cohn and F. S. Coale)

FIG. 14.01-1 ATTENUATION PROPERTIES OF DIRECTIONAL FILTERS

The directional filter is a four-port device having the theoretical insertion loss characteristics defined in Fig. 14.01-1 when each port is terminated in its characteristic impedance. Power incident at Port 1 emerges from Port 4 with the frequency response of a band-pass filter, while the remaining power emerges from Port 2 with the complementary frequency response of a band-reject filter. No power emerges from Port 3 and none is reflected from Port 1. This type of performance is obtained in an analogous way no matter which of the four ports is used as the input port [*i.e.*, the ports can be renumbered analogously with the new Port 1 as the new input port, and Figs. 14.01-1(b),(c) would still apply]. The frequency response shown in Fig. 14.01-1(b) is typical of a single-resonator directional filter, while that

---

* Some further discussion of directional filters will be found in Sec. 16.02.

of Fig. 14.01-1(c) is typical of a multi-resonator directional filter.

The midband insertion loss, between the ports having the band-pass frequency response, of an actual directional filter employing resonators with a finite unloaded Q is the same as that of a two-port band-pass filter having the same frequency response and utilizing resonators with the same unloaded Q. The midband insertion loss can be computed by the methods given in Secs. 4.13 and 11.06.

Figure 14.01-2 gives an example of the use of directional filters. In the case of Filters *a, b,* and *d,* the isolated port is not used. However, in the case of Filter *c*, it is used to permit f_c and f_d to appear at a common port. Because the directional filter is a reciprocal device, Fig. 14.01-2 can also be used as an example of a frequency combining device if all the arrows are reversed. Countless other interconnections of directional filters are feasible to meet specific requirements.

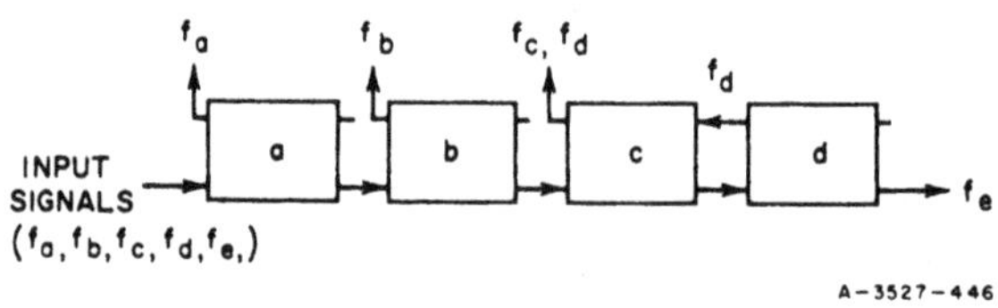

SOURCE: Final Report, Contract DA 36-039 SC-64625, SRI; reprinted in *Proc. IRE* (see Ref. 1 by S. B. Cohn and F. S. Coale)

FIG. 14.01-2 EXAMPLE OF DIRECTIONAL-FILTER APPLICATION TO CHANNEL SEPARATION

A variety of directional filters have been devised, although only a limited number have found wide application. The most useful types are discussed in detail in this chapter and are illustrated in Fig. 14.01-3.

The directional filter shown in Fig. 14.01-3(a) utilizes two rectangular waveguides operating in the dominant TE_{10} mode connected by means of cylindrical direct-coupled cavity resonators operating in the circularly polarized TE_{11} mode. Any number of circularly polarized

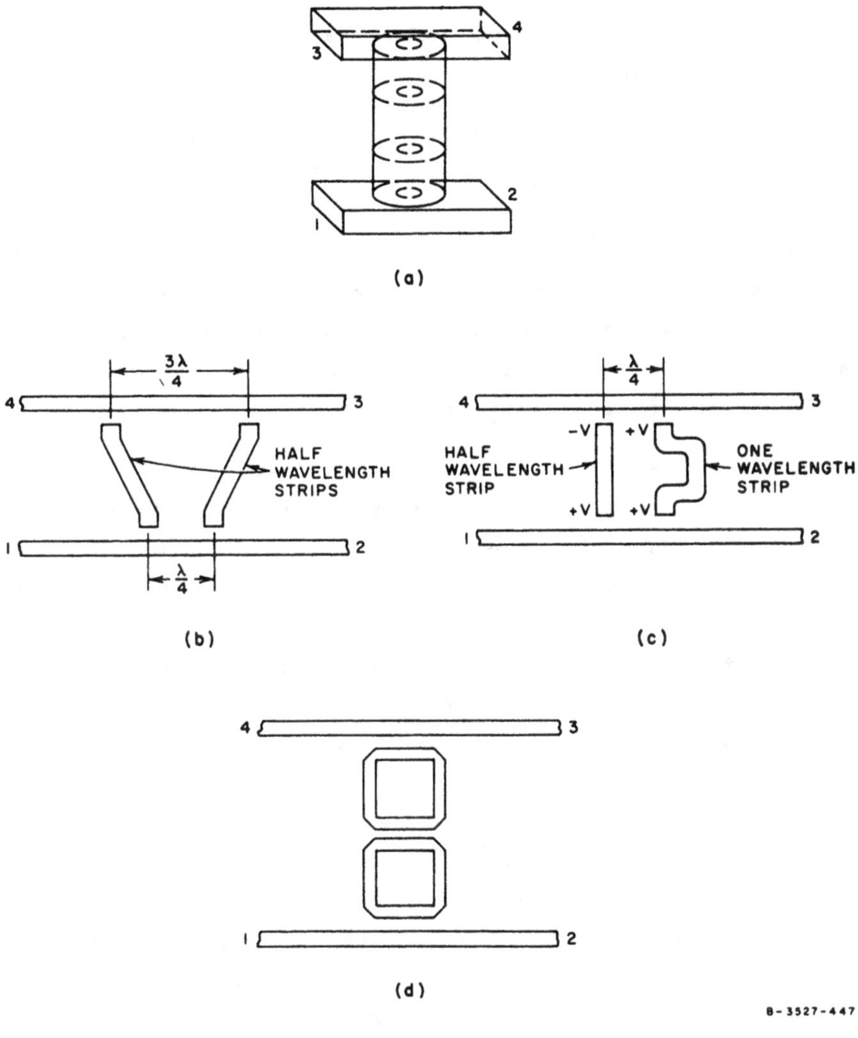

SOURCE: Final Report, Contract DA 36-039 SC-64625, SRI; reprinted in *Proc. IRE* (see Ref. 1 by S. B. Cohn and F. S. Coale)

FIG. 14.01-3 TYPES OF DIRECTIONAL FILTERS DISCUSSED IN THIS CHAPTER

TE_{11} resonators can be used in this directional filter to obtain increased off-channel rejection. The pass band of this type of filter is typically only a fraction of one percent wide. However, by using special coupling apertures between the rectangular waveguides, bandwidths on the order of 3 percent can be obtained for single cavity filters and bandwidths of about 2 percent can be obtained for multi-cavity filters. The high-*Q* cavities used in this type of filter permit relatively low-loss performance for a given pass-band width.

The two forms of strip-line directional filters shown in Fig. 14.01-3(b) and -3(c) have essentially the same frequency response. The resonators in these directional filters cannot be cascaded to obtain increased off-channel rejection as could those in the previously described waveguide directional filter.[3] Therefore, this type of directional filter is suitable for use when only a single-resonator band-pass frequency response is required between Ports 1 and 4. Pass-band widths of several percent are easily obtained with this type of directional filter.

The strip-line directional filter shown in Fig. 14.01-3(d) utilizes traveling-wave ring resonators which are typically one wavelength in mean circumference at midband. They are coupled to one another, and to the terminating strip lines, by means of quarter-wavelength directional couplers of the type discussed in Chapter 13. As in the case of the waveguide directional filter, any number of traveling-wave ring resonators can be utilized to obtain increased off-channel rejection between Ports 1 and 4. Pass-band widths on the order of 10 percent or more can in principle be obtained with this type of directional filter because it is possible to tightly couple the traveling-wave rings to the terminating lines and to each other. However, in practice, these devices are usually designed for more modest bandwidths up to only a few percent. Strip-line construction for directional filters, of course, has an advantage of ease of manufacture. However, the resonator Q's are lower so that the losses will be greater for a given pass-band width than if cavities are used.

This latter type of directional filter could also be realized in waveguides using any of a variety of waveguide couplers to connect a waveguide ring to the external waveguide, but it appears that this form of filter has not been used in practice. A closely related form of waveguide traveling-wave ring structure which has been used extensively as a power multiplier is shown in Fig. 14.01-4. This structure is formed from a single-resonator traveling-wave ring directional filter in which the output waveguide has been removed. When the ring, which typically has a length of several guide wavelengths, is resonant, most of the power entering Port 1 is delivered to the large amplitude traveling wave within the ring, while the remainder is delivered to the load. The average power level of the wave circulating within the ring can be made quite high. As an example of

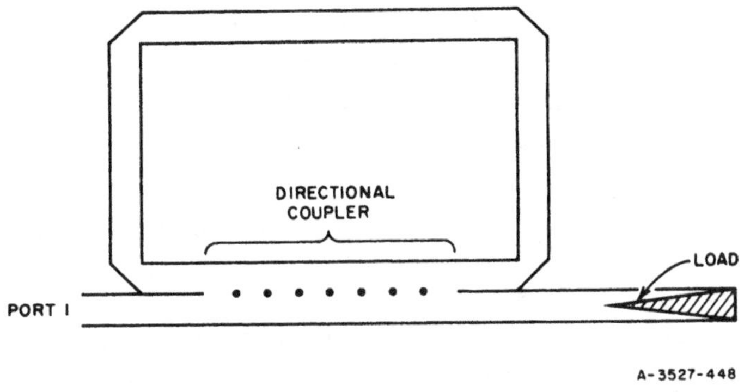

FIG. 14.01-4 WAVEGUIDE TRAVELING-WAVE-RING POWER MULTIPLIER

the power multiplication that can be obtained it is found that if the attenuation of a wave traveling once around the ring is 0.2 db and the directional coupler has a coupling of −16 db, the average power level of the wave circulating within the ring is 20 times the input power. Thus, microwave devices such as waveguide windows which absorb little power but must be able to operate with high power passing through them, can be tested by inserting them in the ring circuit—while the ring circuit itself is excited by a relatively low-power source.

Another circuit which operates as a directional filter is discussed in Sec. 16.02. This circuit uses two hybrid junctions and two conventional filters to obtain a directional filter type of operation.

SEC. 14.02, WAVEGUIDE DIRECTIONAL FILTERS

This section presents design information and measured performance for the type of waveguide directional filter illustrated in Fig. 14.02-1. This filter utilizes two rectangular waveguides operating in the TE_{10} mode connected by means of cylindrical, direct-coupled cavity resonators operating in the circularly polarized TE_{11} mode. Coupling between the cavities is obtained by means of circular apertures, while coupling between the cavities and the external waveguide is obtained either by means of the circular apertures shown or by other types of apertures described later.

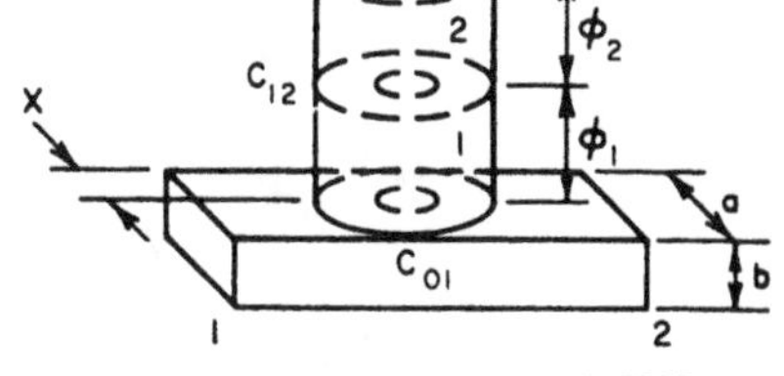

FIG. 14.02-1 A WAVEGUIDE DIRECTIONAL FILTER HAVING n RESONATORS

At the midband frequency a signal incident on Port 1 excites circularly polarized TE_{11} modes in the various cavities. The circularly polarized wave in the *n*th cavity then excites a TE_{10} mode in the external waveguide traveling toward Port 4. No power is transmitted to Port 3. The frequency response at Port 4 is equivalent to that of a band-pass filter, while the response at Port 2 is complementary and equivalent to that of a band-reject filter.

The shape of the attenuation response for transmission between Port 1 and Port 4 is governed by the number of resonant cavities in the directional filter. In fact, any shape of band-pass frequency response, such as those shown at the right in Fig. 14.02-2, that can be obtained with an n-cavity band-pass waveguide filter (Sec. 8.06) can also be obtained with this type of n-cavity directional filter. The width of the pass-band response at Port 4 is typically only a fraction of a percent. However, as explained later the width of the pass band can be increased to several percent by using specially shaped coupling apertures between the rectangular waveguides and the cylindrical cavities.

Table 14.02-1 is a design chart for the waveguide directional filter shown in Fig. 14.02-1. In this chart the design parameters for waveguide

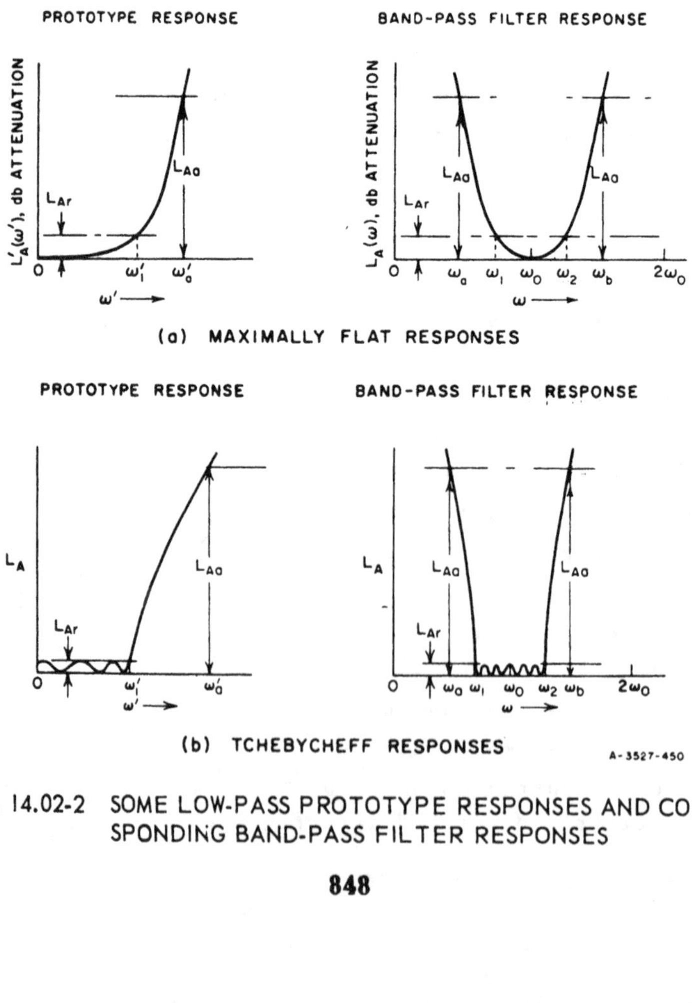

FIG. 14.02-2 SOME LOW-PASS PROTOTYPE RESPONSES AND CORRESPONDING BAND-PASS FILTER RESPONSES

Table 14.02-1

DESIGN EQUATIONS FOR CIRCULAR-WAVEGUIDE DIRECTIONAL FILTERS

The parameters $g_0, g_1, \ldots, g_{n+1}$ are as defined in Sec. 4.04, while ω_1', ω_0, ω_1, and ω_2 are as defined in Fig. 14.02-2.

$$(Q_e)_A = \frac{g_0 g_1 \omega_1'}{w} = \text{external } Q \tag{1}$$

$$k_{i,i+1}\Big|_{i=1 \text{ to } n} = \frac{w}{\omega_1' \sqrt{g_i g_{i+1}}} = \text{coupling coefficient} \tag{2}$$

$$(Q_e)_B = \frac{\omega_1' g_n g_{n+1}}{w} = \text{external } Q \tag{3}$$

where

$$w = \frac{\omega_1 - \omega_1}{\omega_0} \quad \text{and} \quad \omega_0 \approx \frac{\omega_1 + \omega_2}{2} \tag{4}$$

$$c_{01} = 1\Bigg/\sqrt{\frac{(Q_e)_A}{2\pi p}\left(\frac{\lambda_0}{\lambda'_{g0}}\right)^2 + \frac{1}{2}} \tag{5}$$

$$c_{i,i+1}\Big|_{i=1 \text{ to } n-1} = p\pi k_{i,i+1}\left(\frac{\lambda'_{g0}}{\lambda_0}\right)^2 \tag{6}$$

$$c_{n,n+1} = 1\Bigg/\sqrt{\frac{(Q_e)_B}{2\pi p}\left(\frac{\lambda_0}{\lambda'_{g0}}\right)^2 + \frac{1}{2}} \tag{7}$$

where $(c_{i,i+1})^2$ is the power coupling factor, and where p is the number of half-wavelengths in the resonators.

To determine aperture sizes first compute the polarizabilities (the equations given are most accurate for very small apertures):

$$M_{i,i+1}\Big|_{i=0 \text{ and } n} = \sqrt{\frac{3a^3 bD^2\lambda'_{g0}(c_{i,i+1})^2}{32\pi^2\lambda_{g0}\cos^2\left(\frac{\pi x}{a}\right)}} \tag{8}$$

$$M_{i,i+1}\Big|_{i=1 \text{ to } n-1} = \frac{3c_{i,i+1}D^2\lambda'_{g0}}{32\pi} \tag{9}$$

(continued on p. 850)

Table 14.02-1 *concluded*

where a, b, and D are defined in Fig. 14.02-1, and

$$x = \frac{a}{\pi} \cot^{-1} \frac{2a}{\lambda_{g0}} = \text{distance of a small aperture from side of waveguide} \tag{10}$$

$$\lambda'_{g0} = \frac{\lambda_0}{\sqrt{1 - \left(\frac{\lambda_0}{1.706D}\right)^2}} = \text{guide wavelength in cylindrical waveguide of diameter } D \text{ at frequency } \omega_0 \tag{11}$$

$$\lambda_{g0} = \frac{\lambda_0}{\sqrt{1 - \left(\frac{\lambda_0}{2a}\right)^2}} = \text{guide wavelength in rectangular waveguide of width } a \text{ at frequency } \omega_0 \tag{12}$$

λ_0 = wavelength of a plane wave at frequency ω_0 .

For small circular apertures, the required hole diameters are approximately

$$d_{i,i+1}\Big|_{i=0 \text{ to } n} = \sqrt[3]{6M_{i,i+1}} \tag{13}$$

Approximate corrections for aperture thickness t and size can be obtained by computing compensated polarizabilities

$$(M_{i,i+1})_c = M_{i,i+1}\left[1 - \left(\frac{1.7d_{i,i+1}}{\lambda_0}\right)^2\right] 10^{\frac{1.6t}{d_{i,i+1}}\sqrt{1-\left(\frac{1.7d_{i,i+1}}{\lambda_0}\right)^2}} \tag{14}$$

and then recomputing the $d_{i,i+1}$ using Eq. (13) and the $(M_{i,i+1})_c$.

$$\phi_i = \pi p - \frac{1}{2}\left[\tan^{-1} \frac{2X_{i-1,i}}{Z'_0} + \tan^{-1} \frac{2X_{i,i+1}}{Z'_0}\right] \quad \text{radians}$$

$$= \text{electrical distance between irises} \tag{15}$$

where

$$\frac{X_{i,i+1}}{Z'_0} = \frac{16M_{i,i+1}}{0.955D^2\lambda'_{g0}}$$

$$= \text{normalized shunt reactance of iris}$$

A suggested low-pass to band-pass mapping (Sec. 8.04) is

$$\frac{\omega'}{\omega'_1} = \frac{2}{w}\left(\frac{\omega - \omega_0}{\omega_0}\right) \tag{16}$$

directional filters are specified in terms of the band-pass frequency response at Port 4. This band-pass response is related in turn to the coupling coefficients $k_{i,i+1}$ and external Q's, $(Q_e)_A$ and $(Q_e)_B$, of a two-port band-pass filter having an identical frequency response. It is also related to the response of a low-pass prototype filter having element values g_0, g_1, g_2, . . . , g_{n+1}. As discussed in Secs. 4.13 and 11.06, the increase in midband attenuation $(\Delta L_A)_0$ at Port 4 due to dissipation in the circuit can be computed from the prototype element values and the Q's of the resonators.

The quantity c in this chart is the square root of a power coupling factor. It is related to the square of the magnitude of a scattering coefficient (Sec. 2.12) as defined in Figs. 5.10-7 and 5.10-10. For example, the power coupling factor $(c_{i,i+1})^2$ between the cavities i and $i + 1$ is defined as follows. Let cavity $i + 1$ be extended in length and a matched load* inserted inside. Let cavity i be cut in two and power fed to iris i, $i + 1$ from a linearly or circularly polarized matched generator.* Then the ratio of the power delivered to the load to that available from the generator is $(c_{i,i+1})^2$, which is the same as $|S_{12}|^2$ in Fig. 5.10-7.

The power coupling factors $(c_{01})^2$ and $(c_{n,n+1})^2$ are defined in a similar way. Again we assume that cavity number 1 is extended and a matched load* inserted inside. Likewise, a matched load* is placed at Port 2 and a matched generator* is placed at Port 1. The ratio of the total power in the circularly polarized wave delivered to the load in the cylindrical guide to that available from the generator is $(c_{01})^2$, which in Fig. 5.10-10 is $|S_{14}|^2 + |S_{13}|^2$, where for the case of circular polarization $|S_{14}| = |S_{13}|$.

In practically all cases, the length of the cavities is approximately one-half guide wavelength, $\lambda'_{g0}/2$, at midband. However, the cavities can in principle be designed to be approximately an integral number p of half guide-wavelengths long. The exact electrical length ϕ_i of the cavities, which is given in the chart, is reduced somewhat from the value of 180 p degrees because of the stored magnetic energy in the coupling apertures.

In order that a pure circularly polarized wave be induced in the cylindrical cavities, it is necessary, when using circular coupling apertures between the end cavities and the rectangular waveguides, to offset the axis of the cavities a distance x from the side of the rectangular guide as given in Table 14.02-1. In this case a perfectly circularly

* "Matched" as used here refers to matching the guide impedance.

polarized wave is induced only at the design center frequency, although the wave is nearly circularly polarized over the range of pass-band frequencies. A type of coupling arrangement using three slots that produces an essentially circularly polarized wave in the cavities[8] over a much wider frequency interval is shown in Fig. 14.02-3. In this arrangement the axis of the cylindrical cavities coincides with the axis of the rectangular guides.

Waveguide directional filters having larger bandwidths can be obtained by using large coupling apertures of the types shown in Fig. 14.02-4. The dimensions of these apertures are too large to be computed very accurately using the small-aperture coupling data presented in Chapter 5. However, in the case of the two-slot configuration in Fig. 14.02-4(b) Bethe's theory can be used to obtain a first approximation for the coupling slots, and then corrections can be made after experimental tests. If M_1 is the magnetic polarizability of the transverse slot in the transverse direction while M_2 is the magnetic polarizability of the longitudinal slot in the longitudinal direction, then according to Bethe's small aperture theory

$$M_2 = \frac{Rck_c r_2}{4\pi\sqrt{2}\left(\sin\frac{\pi x_2}{a}\right)J_1(k_c r_2)}\sqrt{\frac{3a^3 b\lambda'_g}{\lambda_g}} \qquad (14.02\text{-}1)$$

and

$$M_1 = \frac{M_2\lambda_g \sin\left(\frac{\pi x_2}{a}\right)\frac{J_1(k_c r_2)}{k_c r_2}}{2a\cos\left(\frac{\pi x_1}{a}\right)\left[J_0(k_c r_1) - \frac{J_1(k_c r_1)}{k_c r_1}\right]} \qquad (14.02\text{-}2)$$

where

λ_g = guide wavelength in rectangular guide

λ'_g = guide wavelength in circular guide

a = width of rectangular guide

b = height of rectangular guide

R, r_1, r_2, x_1, and x_2

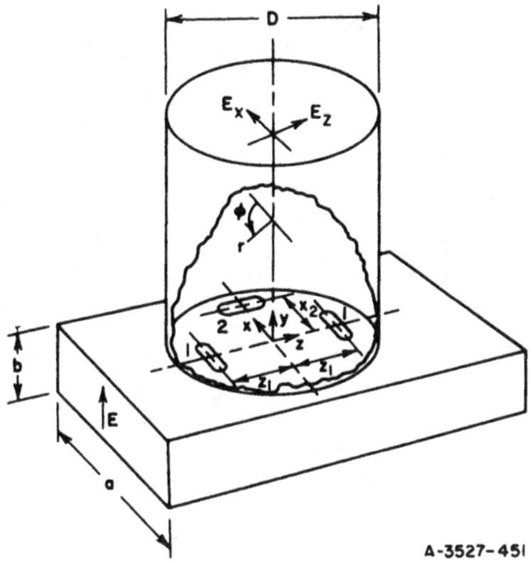

Assuming small apertures, the condition for circular polarization is

$$\frac{M_2}{M_1} = \frac{4x_2 a J_1(\alpha z_1)\cos(360 z_1/\lambda_{g0})}{\xi z_1 \lambda_{g0} J_1(\alpha x_2)\sin(180 x_2/a)}$$

M_1 = magnetic polarizability of each transverse slot [Fig. 5.10-4(a)]

M_2 = magnetic polarizability of the longitudinal slot [Fig. 5.10-4(a)]

λ_{g0} = guide wavelength in rectangular guide at midband frequency

α = $2\pi/1.705D$

ξ = tolerance factor: when $\xi = 1.02$ the axial ratio in the circular guide is less than 1.02 from $\lambda_g = 8.82z_1$ to $6.28z_1$. The design center value of λ_{g0} in this case is $7.30z_1$.

Under this condition

$$c^2 = \frac{128\pi^2 \lambda_{g0} M_2^2 \sin^2(180x_2/a)[J_1(\alpha x_2)]^2}{3a^3 b D^2 \lambda'_{g0} \alpha^2 x_2^2}$$

λ'_{g0} = guide wavelength in circularly polarized cavity at midband

SOURCE: *Proc. IRE* (see Ref. 8 by S. B. Cohn)

FIG. 14.02-3 APPROXIMATE DESIGN EQUATIONS FOR A BROADBAND, CIRCULAR-POLARIZATION COUPLER
The axial ratio is the ratio of the field strengths E_X and E_Z at right angles to each other

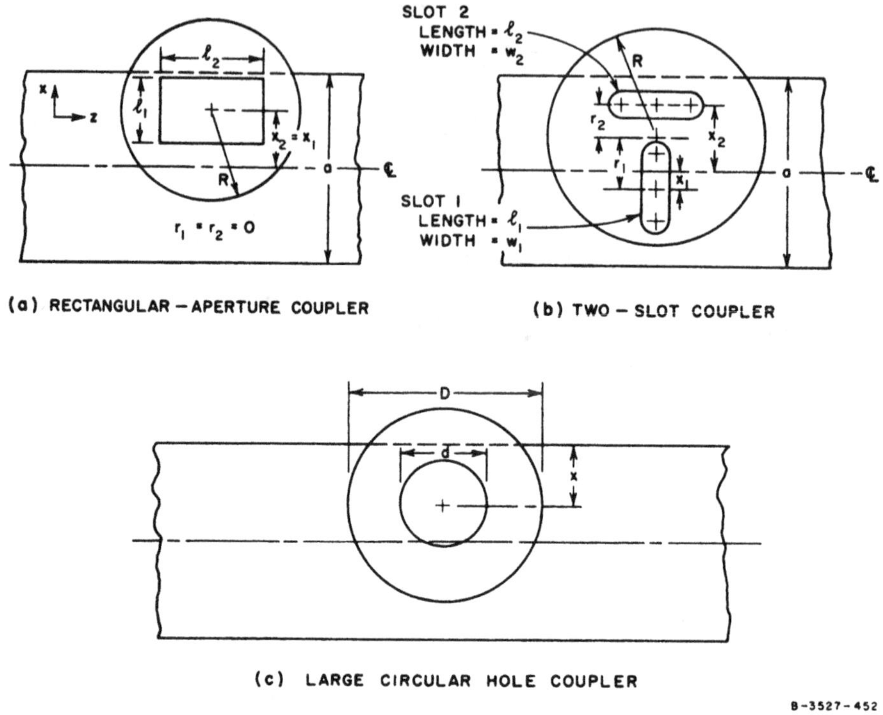

FIG. 14.02-4 APERTURE CONFIGURATIONS HAVING LARGE COUPLING FACTORS

are distances as indicated in Fig. 14.02-4(b), $k_c = 1.84/R$, and c is the square root of the power coupling factor and is equal to c_{01} or $c_{n,n+1}$ obtained from Eq. (5) or Eq. (7) of Table 14.02-1. The trial dimensions of the slots can be obtained from Fig. 5.10-4(a) along with the thickness and size correction given by Eq. (5.10-6). Cut and try refinement of the slot dimensions can then be achieved using experimental techniques described later in this section.

Table 14.02-2 shows the final dimensions and measured performance for several aperture designs of the types in Fig. 14.02-4. These designs all have relatively tight coupling, but that of the two-slot configurations is decidedly the tightest. The *axial ratio* referred to in these data is the ratio of field strengths at right angles to each other in a plane perpendicular to the axis of the circular guide. For perfect circular polarization the axial ratio is one. Figure 14.02-5 shows the measured power coupling and axial ratio as a function of frequency for the two-slot configuration whose dimensions are given in the column on the far right in Table 14.02-2.

Table 14.02-2

TYPICAL COUPLING PROPERTIES OF LARGE APERTURES FOR WAVEGUIDE DIRECTIONAL FILTERS AT f = 9780 Mc

	CIRCULAR HOLE [Fig. 14.02-4(c)]	RECTANGULAR APERTURE [Fig. 14.02-4(a)]	TWO-SLOT COUPLER [Fig. 14.02-4(b)]	TWO-SLOT COUPLER [Fig. 14.02-4(b)]
Dimensions (inches)	d = 0.440 x = 0.207 from side wall $r_1 = r_2 = 0$ D = 1.114	l_1 = 0.442 l_2 = 0.450 $x_1 = x_2$ = 0.229 from center line $r_1 = r_2 = 0$ R = 0.554	l_1 = 0.461 w_1 = 0.136 l_2 = 0.518 w_2 = 0.142 x_1 = 0 x_2 = 0.320 r_1 = 0.243 r_2 = 0.077 R = 0.554	l_1 = 0.510 w_1 = 0.136 l_2 = 0.565 w_2 = 0.142 x_1 = 0 x_2 = 0.320 r_1 = 0.243 r_2 = 0.077 R = 0.554
Aperture Thickness (t) (inches)	0.023	0.025	0.025	0.025
Axial Ratio	1.03	1.04	1.04	1.10
Total Coupling Factor (db) $10 \log_{10} c^2$	-17.3	-12.8	-8.8	-8.2
VSWR in Rectangular Waveguide	1.05	1.08	1.25	1.23
$(Q_e)_A$ =	556	194	73.5	63.5

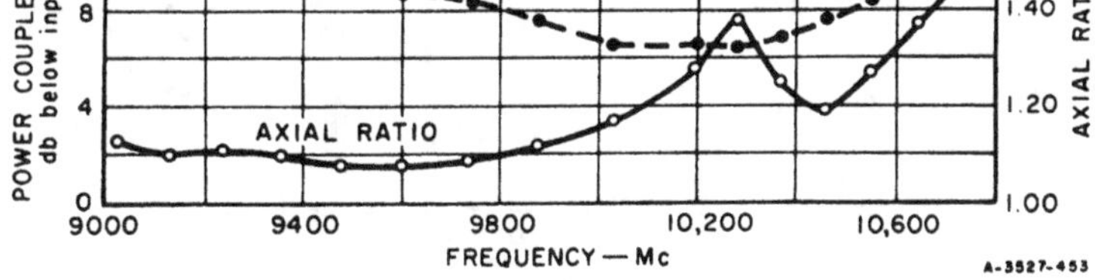

FIG. 14.02-5 POWER COUPLING AND AXIAL RATIO OF A PAIR OF SLOTS HAVING LARGE COUPLING FACTORS
The slot dimensions are shown in the column at the far right in Table 14.02-2

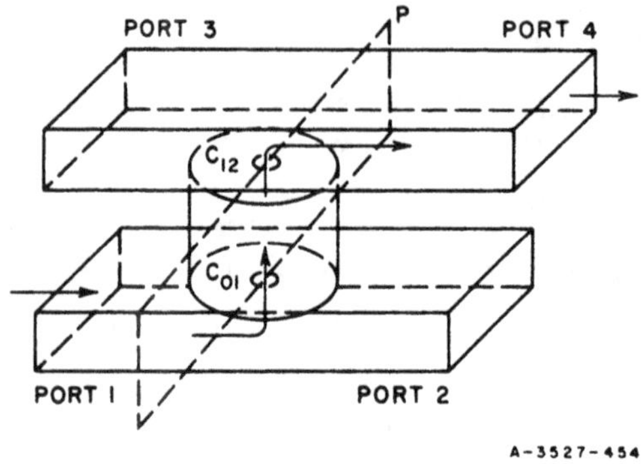

FIG. 14.02-6 SINGLE-CAVITY WAVEGUIDE DIRECTIONAL FILTER

A single-cavity *X*-band waveguide directional filter of the type in Fig. 14.02-6 is photographed in assembled and disassembled form in Fig. 14.02-7. In this filter the cylindrical cavity (which was made of brass) was split along its mid-plane—a position of minimum longitudinal wall current—so that the unloaded Q of the cavity was not appreciably degraded by the joint. The two halves of the experimental filter were held together by a C-clamp; however, they could also be easily soldered together. The capacitive tuning screws shown in the picture were adjusted so that the two spacially orthogonal TE_{11} modes that can exist in the cavity resonated at the same frequency. The necessity for this tuning procedure can be made plausible if one remembers that the circularly polarized mode induced in the cavity can be resolved into spacially orthogonal TE_{11} modes excited in time quadrature.

In this filter the terminating guides have inside dimensions $a = 0.900$ inch and $b = 0.400$ inch while the coupling apertures have a thickness $t = 0.020$ inch. The cavity diameter $D = 1.114$ inch, the coupling hole diameters $d = 0.414$ inch and the cavity height $h = 0.716$ inch. The axis of the cylindrical cavities was offset a distance $x = 0.207$ inch from the sidewall of the rectangular waveguides. The measured performance of this single cavity waveguide directional filter is shown in Fig. 14.02-8. The measured loaded Q, Q'_L, of this cavity *loaded at both ends* was 249 and the midband attenuation $(L_A)_0 = 0.72$ db or $(P_{avail}/P_{out}) = 1.19$. Using the relation

$$Q'_e = Q'_L \sqrt{\frac{P_{avail}}{P_{out}}} = \frac{Q_e}{2} \qquad (14.02\ 3)$$

which applies for single-cavity directional filters as well as single-cavity two-port filters, we find that the external Q, Q'_e, of the cavity

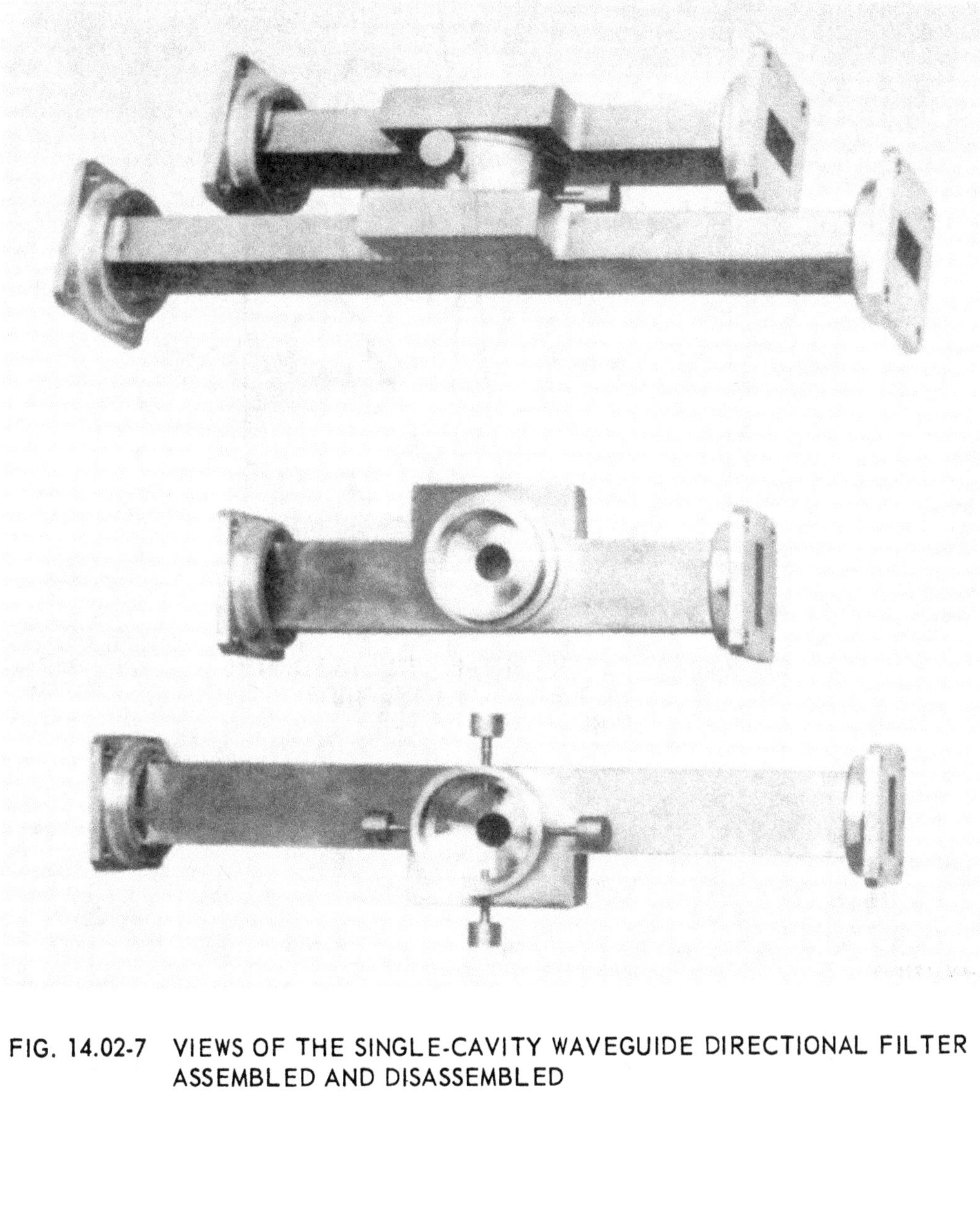

FIG. 14.02-7 VIEWS OF THE SINGLE-CAVITY WAVEGUIDE DIRECTIONAL FILTER ASSEMBLED AND DISASSEMBLED

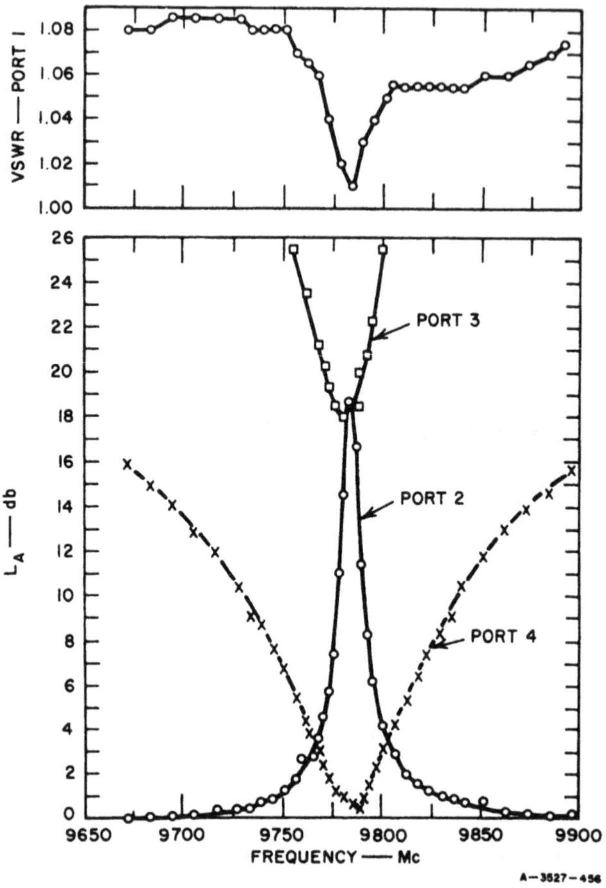

FIG. 14.02-8 EXPERIMENTAL RESULTS FOR THE SINGLE-CAVITY WAVEGUIDE DIRECTIONAL FILTER

loaded at both ends is 271. Therefore, the external Q, Q_e, of the cavity *loaded at one end* as given in Table 14.02-1 is 542. Again using the relation that

$$\frac{1}{Q'_L} = \frac{1}{Q_e} + \frac{1}{Q_u} \qquad (14.02\text{-}4)$$

the unloaded Q, Q_u, of the cavity is 3030. This value would be approximately doubled if the cavity had been made of copper or if it were silver-plated.

The value of the coupling factor $(c_{01})^2$ computed from the measured value $Q_e = (Q_e)_A = 542$, using Eq. (5) in Table 14.02-1, is $(c_{01})^2 = 0.0195$, or −17.1 db. Using $d_{01} = 0.414$ inch and solving for $(c_{01})^2$ using Eqs. (13) and (8) of Table 14.02-1 gives $(c_{01})^2 = -16.6$ db, which is in unexpectedly good agreement with the value obtained from (Q_e), considering that the hole is quite large and is close to the side wall. Technically, both a large-aperture correction and a thickness correction as in Eq. (14) of Table 14.02-1 should also be applied, but these corrections are very rough approximations and are in this case made to be quite uncertain due to the close proximity of the side wall. In such cases, experimental checking of the apertures is desirable. (This was done in the case of this design by experimentally determining Q_e.)

A two-cavity waveguide directional filter of the type illustrated in Fig. 14.02-9 was also constructed by modifying the single-cavity filter described above. The modification was accomplished by inserting a length of cylindrical guide containing a coupling iris at its midplane, between the two halves of the original filter. The diameter d_{12} of the central iris in this filter was progressively enlarged to obtain first an undercoupled response, then a critically coupled response and finally an overcoupled response. The measured response of these filters when critically coupled, and overcoupled is shown in Figs. 14.02-10(a) and 14.02-10(b), respectively. A tabular summary of the filter dimensions and performance is presented in Table 14.02-3.

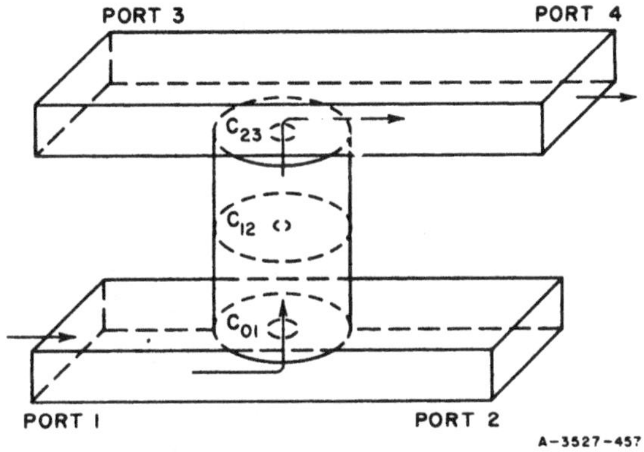

FIG. 14.02-9 A TWO-CAVITY WAVEGUIDE DIRECTIONAL FILTER

The experimentally determined central-iris diameter, $d_{12} = 0.190$ inch, for critical coupling agreed reasonably well with the $d_{12} = 0.178$ inch value computed using the formulas in Table 14.02-1 and the correction in Eq. (5.10-6). The measured midband attenuation $(\Delta L_A)_0$ due to dissipation in the two-cavity directional filter also agreed well with that calculated

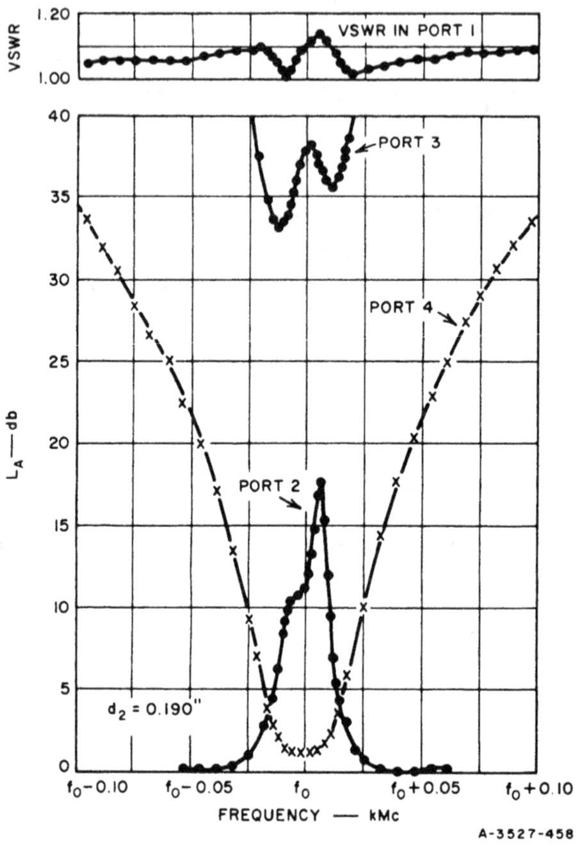

FIG. 14.02-10(a) MEASURED PERFORMANCE OF A CRITICALLY COUPLED TWO-CAVITY WAVEGUIDE DIRECTIONAL FILTER

from Eq. (4.13-11). For example, consider the critically coupled or maximally flat case having the low-pass prototype elements $g_0 = 1$, $g_1 = g_2 = 1.414$, $g_3 = 1$, and 3-db point $\omega_1' = 1$, $w = 0.00317$, and the measured unloaded Q, $Q_u = 3030$, as determined from measurements on the single-cavity filter, then the dissipation factor (Sec. 4.13) is $d = \omega_1'/wQ_u = 0.0104$. Equation (4.13-11) yields a value of $(\Delta L_A)_0 = 1.28$ db as compared with the measured midband loss of 1.25 db. Nearly all of this loss was due to dissipation since the midband VSWR was very low.

Four tuning screws were used in each of the cavities for tuning purposes. The technique used to adjust these tuning screws is described

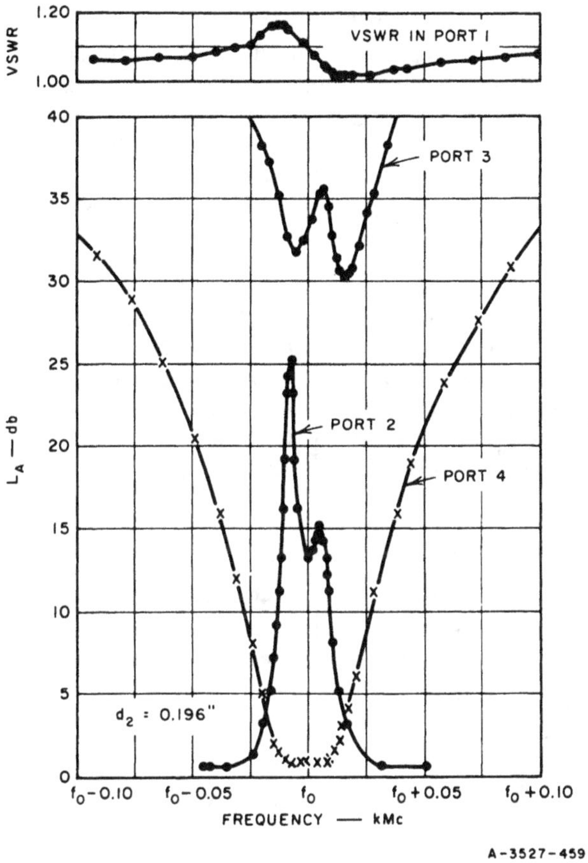

FIG. 14.02-10(b) MEASURED PERFORMANCE OF A SLIGHTLY OVER-COUPLED TWO-CAVITY WAVEGUIDE DIRECTIONAL FILTER

below. Tuning of waveguide single-cavity directional filters can be accomplished in a straightforward manner. The proper procedure is to place matched detectors at Port 4 and Port 2, a matched load at Port 3 and a generator at Port 1. Then one adjusts the four equally spaced tuning screws for maximum signal at Port 4 and minimum signal at Port 2.

In multiple-cavity filters this simple technique is difficult to apply and it is better to use another technique. One such technique consists of tuning each cavity separately using the apparatus illustrated in Fig. 14.02-11. In this apparatus a linearly polarized signal is fed into the cylindrical cavity through a rectangular- to cylindrical-waveguide

Table 14.02-3

DIMENSIONS AND SUMMARY OF THE MEASURED PERFORMANCE OF A TWO-CAVITY WAVEGUIDE DIRECTIONAL FILTER

	UNDER-COUPLED	CRITICALLY COUPLED*	OVER-COUPLED†
Resonant frequency, f_0	9775 Mc	9774 Mc	9774 Mc
Fractional 3 db bandwidth	0.00256	0.00317	0.00338
Insertion loss (at f_0)§	0.90 db	1.25 db	0.95 db
VSWR at f_0	--	1.10	1.04
Maximum VSWR (off resonance)	--	1.14	1.17
Diameter d_{12} of central iris	0.179 in.	0.190 in.	0.196 in.
Thickness t_{12} of central iris	0.025 in.	0.025 in.	0.025 in.
Diameter $d_{01} = d_{23}$ of end irises	0.414 in.	0.414 in.	0.414 in.
Thickness $t_{01} = t_{23}$ of end irises	0.023 in.	0.023 in.	0.023 in.
Inside diameter of each cavity, D	1.114 in.	1.114 in.	1.114 in.
Height of each cavity, h	0.770 in.	0.770 in.	0.770 in.
Displacement of cylindrical guide axis from side wall of the rectangular guide	0.207 in.	0.207 in.	0.207 in.
Height of rectangular guide	0.400 in.	0.400 in.	0.400 in.
Width of rectangular guide	0.900 in.	0.900 in.	0.900 in.

* Shown in Fig. 14.02-10(a)

† Shown in Fig. 14.02-10(b)

§ The inside surfaces of the cavities were machined brass. By polishing and plating the inside surfaces it is believed that midband insertion losses of 0.5 db could easily be obtained.

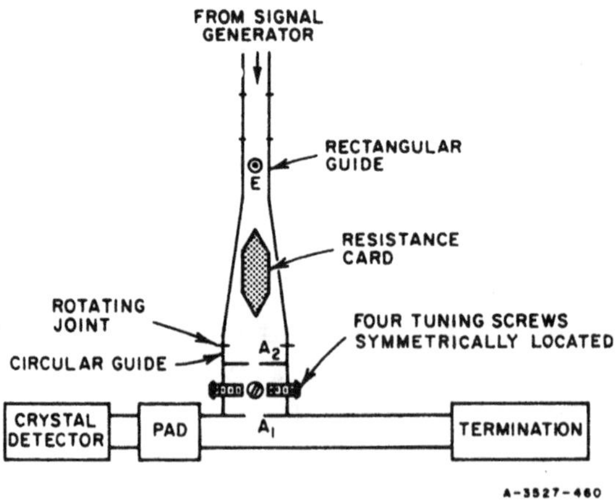

FIG. 14.02-11 SCHEMATIC DIAGRAM OF SETUP FOR PRETUNING CAVITIES OF THE TWO-CAVITY DIRECTIONAL FILTER

transition, past a rotating joint. A resistance card within the transition serves the purpose of damping out any cross-polarized wave that might be reflected from the cavity. When the electric field in the exciting guide is oriented as shown, the two tuning screws parallel to the electric field are adjusted for maximum signal in the detector. Next the exciting guide is rotated by 90 degrees and the other pair of screws is adjusted for maximum signal in the detector. Since the pairs of tuning screws do not furnish completely independent adjustment, this process is repeated until the signal at the detector is independent of the polarization orientation of the exciting wave. The measured performance shown in Fig. 14.02-10 was obtained by pretuning the cavities in this fashion, and without further adjustment of the assembled filter.

Another technique for tuning a multicavity filter of this sort is to use the setup in Fig. 14.02-11 to feed a linearly polarized signal through all the cavities simultaneously. In this case a slotted line is inserted in the rectangular guide at the top and the cavities may be tuned for each of the two orthogonal linear polarizations by observing the quarter-wavelength shift of the voltage minimum in the input waveguide as successive cavities are brought to resonance. This procedure is the same as that described in Sec. 11.05 for tuning direct-coupled cavity filters. Alternatively, if one has a sweep signal source available, it may be possible to tune the cavities by simply maximizing the signal at the detector in the rectangular waveguide for the two orthogonal linear polarizations.

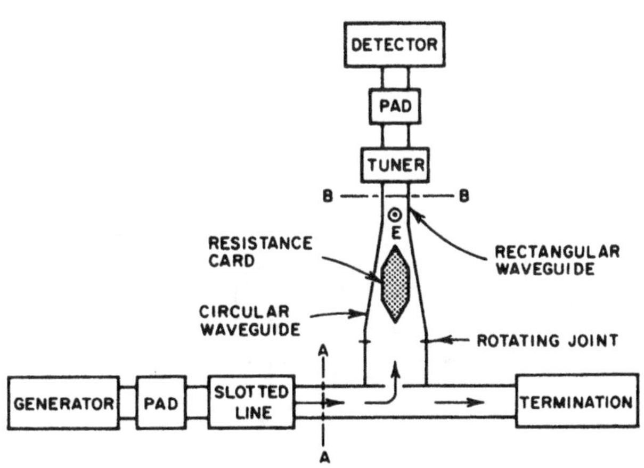

FIG. 14.02-12 SCHEMATIC DIAGRAM OF SETUP FOR MEASURING THE PERFORMANCE OF APERTURES HAVING LARGE COUPLING FACTORS

The technique for measuring the coupling and axial ratio of the wave excited by the large apertures of Fig. 14.02-4 is illustrated in Fig. 14.02-12. The power coupling factor c^2 is equal to the sum of the power received by the detector in the position shown and that received when it is rotated 90 degrees, divided by the available power from the generator. The axial ratio is the square root of the ratio of the maximum to the minimum power received by the detector as it is rotated.

SEC. 14.03, STRIP TRANSMISSION LINE DIRECTIONAL FILTERS USING HALF- OR FULL-WAVELENGTH STRIPS

A form of strip-transmission-line directional filter that has proved to be very useful is illustrated in Fig. 14.03-1 together with pertinent design information. It might seem that the resonators in this type of filter could be cascaded as shown in Fig. 14.03-2 to obtain a directional filter having a multiple resonator response, but this is not the case.[3]

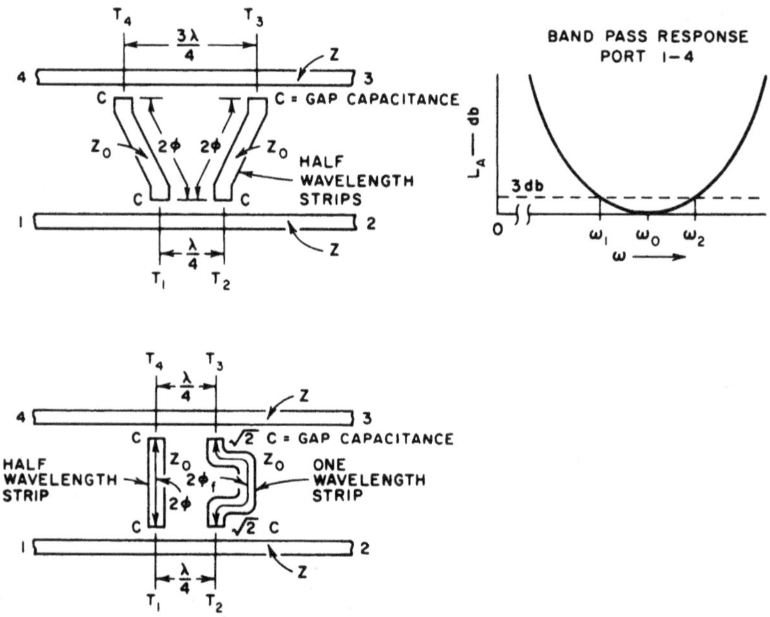

$$\frac{1}{Q} = \frac{\omega_2 - \omega_1}{\omega_0} = \frac{2\omega_0^2 C^2 Z Z_0}{\pi} = \frac{1}{\left(\text{Q OF DISSIPATIONLESS CIRCUIT LOADED AT PORTS 1 AND 4}\right)}$$

$$2\phi = \pi - 2\omega_0 C Z_0 \quad \text{RADIANS}$$

$$2\phi_f = 2\pi - 2\sqrt{2}\,\omega_0 C Z_0 \quad \text{RADIANS}$$

B-3527-462

FIG. 14.03-1 DESIGN INFORMATION FOR HALF-WAVELENGTH-STRIP AND ONE-WAVELENGTH-STRIP DIRECTIONAL FILTERS

Therefore, this type of directional filter is suitable to use only when a single resonator band-pass frequency response is desired between Port 1 and Port 4.

The action of the filters in Fig. 14.03-1 can be understood by invoking the principle of superposition. Excitation of Port 1 with a wave of amplitude V is equivalent to excitation of Port 1 and Port 4 at the reference planes T_1 and T_4 with waves both of amplitude $+V/2$ (even-mode case), and excitation of Ports 1 and 4 at the same reference planes with waves of amplitude $+V/2$ and $-V/2$, respectively (odd-mode case). Even-mode excitation of Ports 1 and 4 at the center frequency causes only the right-hand strip to resonate, which reflects waves having amplitude $+V/2$ at Port 1 and Port 4. On the other hand, odd-mode excitation causes only the left-hand strip to resonate, reflecting waves having amplitude $-V/2$ at Port 1 and $+V/2$ at Port 4. Therefore, at the resonant frequency of the strips the amplitude of the reflected wave at Port 4 is V while that at Port 1 is zero. Or, in other words, at the center frequency a signal incident on Port 1 is ideally completely transferred to Port 4. At frequencies off resonance a signal incident at Port 1 passes through unattenuated to Port 2.

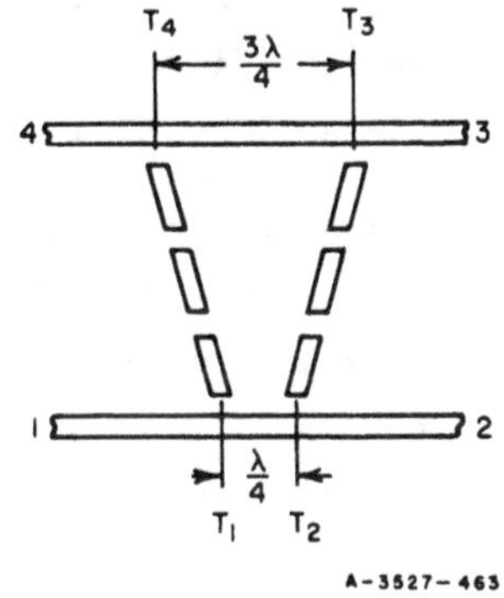

FIG. 14.03-2 EXAMPLE OF A MULTI-RESONATOR STRUCTURE THAT DOES NOT HAVE DIRECTIONAL FILTER PROPERTIES

The reason multiple-strip-resonator structures of the type shown in Fig. 14.03-2 cannot be used as directional filters can also be qualitatively understood by applying the principle of superposition. It can be seen that when either odd-mode waves or even-mode waves are incident on Ports 1 and 4, some of the resonators in both the left- and right-hand string of resonators will be excited. Therefore, the necessary destructive interference between the even- and odd-mode reflected waves at Port 1 and the necessary constructive interference between even- and odd-mode reflected waves at Port 4 is not obtained and directional filter action does not result. On the other hand, Wanselow and Tuttle[9] have reported that marginal directional-filter performance is obtained using multiple-resonator structures of the type shown in Fig. 14.03-2. The explanation of this apparent paradox appears to be that Wanselow and Tuttle placed a short-circuit at Port 3, presumably at reference plane T_3, thus reducing it to a three-port device. Under these conditions when a signal is incident at Port 1 the right-hand chain of resonators would reflect a relatively high shunt impedance at Port 1 while the short-circuit at reference plane T_3 would reflect a shunt open-circuit at reference plane T_4. Therefore, the transmission characteristic between Port 1 and Port 4 roughly approximates that of the multi-resonator chain at the left of the structure in Fig. 14.03-2.

The dimensions of the TEM lines having the characteristic impedances Z and Z_0 shown in Fig. 14.03-1 can be determined from Figs. 5.04-1 or 5.04-2. The gap spacing at the ends of the resonators necessary to realize the coupling capacitances C can be determined approximately from Fig. 5.05-9 by making use of the relation that C is approximately equal to ΔC times the width of the resonator strip.

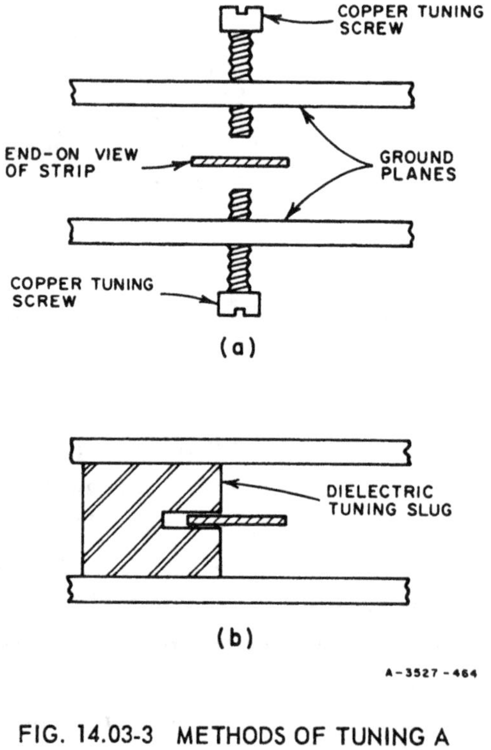

FIG. 14.03-3 METHODS OF TUNING A STRIP-LINE DIRECTIONAL FILTER

In order to tune the two resonators in the directional filter to the same frequency, copper tuning screws or dielectric tuning slugs of the types shown in cross section in Fig. 14.03-3 may be employed. For maximum effectiveness these tuners should be used near either end of a resonator where the electric field is high. In addition, if copper tuning screws are used they should be inserted symmetrically from either ground plane so that the parallel plate TEM mode is not excited.

The measured response of a typical strip-line directional filter of this type is shown in Fig. 14.03-4.

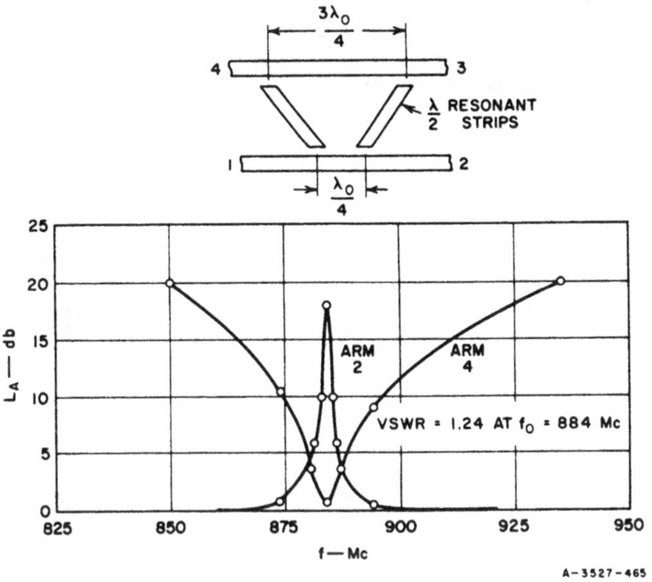

SOURCE: Final Report, Contract DA 36-039 SC-64625, SRI; reprinted in *IRE Trans. PGMTT* (see Ref. 1 by S. B. Cohn and F. S. Coale)

FIG. 14.03-4 RESPONSE OF STRIP-LINE HALF-WAVELENGTH-RESONATOR DIRECTIONAL FILTER

SEC. 14.04, TRAVELING-WAVE-LOOP DIRECTIONAL FILTER

A type of strip-transmission-line directional filter that can be made with multiple resonators to obtain increased off-channel rejection is illustrated in Fig. 14.04-1. The resonators in this directional filter are traveling-wave loops whose mean circumference is a multiple of 360 degrees at the midband frequency. Coupling between the loops is obtained by the use of quarter-wavelength directional couplers of the type described in Chapter 13. At the midband frequency a signal incident on Port 1 excites clockwise traveling waves in each of the loops. The traveling wave in the *n*th loop excites a signal at Port 4. At frequencies well removed from the resonant frequency of the loops a signal incident on Port 1 is transferred to Port 2.

The design chart in Table 14.04-1 summarizes the design of traveling-wave-loop filters from low-pass prototypes so as to give a desired band-pass response between Ports 1 and 4. The voltage coupling factors $c_{i,i+1}$ are the midband voltage couplings of the strip-line directional couplers

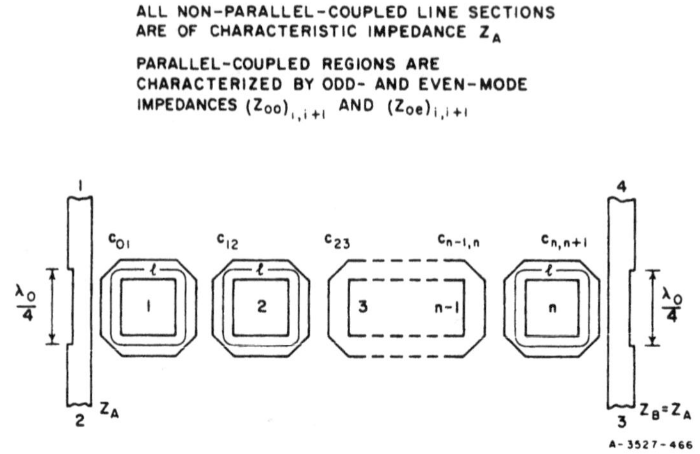

FIG. 14.04-1 A TRAVELING-WAVE-LOOP DIRECTIONAL FILTER HAVING n RESONATORS

as defined in Eq. (13.02-2), while the coupling coefficients $k_{i,i+1}$ are the resonator coupling coefficients in Fig. 8.02-3. It would also be possible to build this type of traveling-wave-loop directional filter in waveguide form. In this case the design parameters in Table 14.04-1 must be modified by inclusion of the factor $(\lambda'_{g0}/\lambda_0)^2$ in exactly the same fashion as is done in Table 14.02-1.

The layout of two, single-loop, directional filters, one of which is tuned to 1024 Mc and the other of which is tuned to 1083 Mc, is shown in Fig. 14.04-2. The measured performance of the device is shown in Fig. 14.04-3. The filter having a center frequency of 1024 Mc was designed to have a 3 db bandwidth of 8.2 Mc [Q_e = 250 and $(c_{01})^2 = (c_{12})^2$ = 0.0251] while the other filter was designed to have a 3 db bandwidth of 6.9 Mc [Q_e = 314 and $(c_{01})^2 = (c_{12})^2$ = 0.020].

The length of the ring resonators along the coupled transmission lines was made to be a quarter wavelength measured in the dielectric at the midband frequency in each case. The length of each of the other two sides of the resonator plus the equivalent length of the two matched, mitered corners, was also made approximately a quarter wavelength, measured in the dielectric at midband. Data on the equivalent length of matched mitered corners is presented in Fig. 5.07-4, although the design under discussion was made before these data were available.

Table 14.04-1

DESIGN EQUATIONS FOR TRAVELING WAVE-LOOP DIRECTIONAL FILTERS AS SHOWN IN FIG. 14.04-1

The parameters $g_0, g_1, \ldots, g_{n+1}$ are as defined in Sec. 4.04, and ω_1', ω_0, ω_1, and ω_2 are as defined in Fig. 14.02-2.

$$(Q_e)_A = \frac{g_0 g_1 \omega_1'}{w} = \text{external } Q \tag{1}$$

$$k_{i,i+1}\Big|_{i=1 \text{ to } n-1} = \frac{w}{\omega_1' \sqrt{g_i g_{i+1}}} = \text{coupling coefficient} \tag{2}$$

$$(Q_e)_B = \frac{\omega_1' g_n g_{n+1}}{w} = \text{external } Q \tag{3}$$

where

$$w = \frac{\omega_2 - \omega_1}{\omega_0} \quad \text{and} \quad \omega_0 = \frac{\omega_1 + \omega_2}{2} \tag{4}$$

$$c_{01} = 1\Big/\sqrt{\frac{(Q_e)_A}{2\pi m} + \frac{1}{2}} \tag{5}$$

$$c_{i,i+1}\Big|_{i=1 \text{ to } n-1} = m\pi k_{i,i+1} \tag{6}$$

$$c_{n,n+1} = 1\Big/\sqrt{\frac{(Q_e)_B}{2\pi m} + \frac{1}{2}} \tag{7}$$

where the $c_{i,i+1}$ are *voltage* coupling factors, and m is the number of full wavelengths in the resonator loops at resonance.

The odd- and even-mode impedances of the directional-coupler sections are

$$(Z_{oe})_{i,i+1}\Big|_{i=0 \text{ to } n} = Z_0\sqrt{\frac{1 + c_{i,i+1}}{1 - c_{i,i+1}}} \tag{8}$$

$$(Z_{oo})_{i,i+1}\Big|_{i=0 \text{ to } n} = Z_0\sqrt{\frac{1 - c_{i,i+1}}{1 + c_{i,i+1}}} \tag{9}$$

The directional-coupler sections are a quarter-wavelength long at ω_0.

A suggested low-pass to band-pass mapping (Sec. 8.04) is

$$\frac{\omega'}{\omega_1'} = \frac{2}{w}\left(\frac{\omega - \omega_0}{\omega_0}\right) \quad . \tag{10}$$

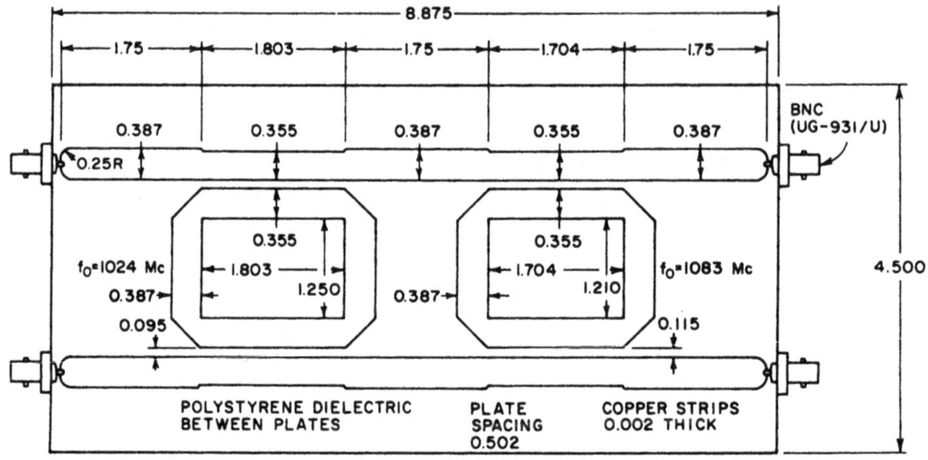

FIG. 14.04-2 LAYOUT OF A DUAL-LOOP DIRECTIONAL FILTER CIRCUIT

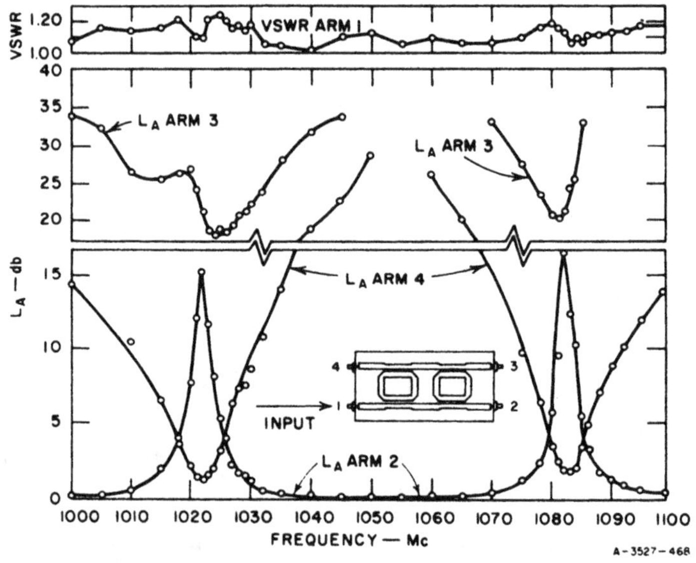

SOURCE: Final Report, Contract DA 36-039 SC-64625, SRI; reprinted in *IRE Trans. PGMTT* (see Ref. 2 by F. S. Coale)

FIG. 14.04-3 MEASURED RESPONSE OF DUAL-LOOP DIRECTIONAL FILTER CIRCUIT

The region of each loop that is coupled to the terminating lines is designed according to the formulas in Chapter 13 as a matched, 50-ohm directional coupler having the coupling specified above. The necessary even- and odd-mode impedances Z_{oe} and Z_{oo} are determined from Eqs. (13.02-5) and (13.02-6) or Eqs. (8) and (9) of Table 14.04-1. The widths of the strips necessary to realize these impedances are determined from Fig. 5.05-3 in the case of typical printed-circuit construction. The width of the strips forming the remainder of the loops is a value appropriate for a 50-ohm transmission line as determined from Fig. 5.04-1 or 5.04-2. Thus theoretically, a 50-ohm impedance level is preserved throughout the complete circumference of the loop.

In practice, it is found to be difficult to design the loops accurately enough so that there are no reflections on the loops and some sort of tuning is required if a single traveling-wave is to be established on each loop. Even a quite small discontinuity in any loop will generate a wave of substantial magnitude traveling in a direction opposite to that of the desired wave since it is repeatedly excited by the wave traveling in the desired direction. One method of tuning the loops that has been found to be appropriate when large quantities of directional filters are required consists of empirically adjusting the width of the loops and the width of the corners on a prototype model, and then mass-producing this design. Another effective technique for tuning each loop is to use four pairs of tuning screws, of the type shown in Fig. 14.03-3(a), placed at quarter-wavelength intervals around the perimeter as shown in Fig. 14.04-4.

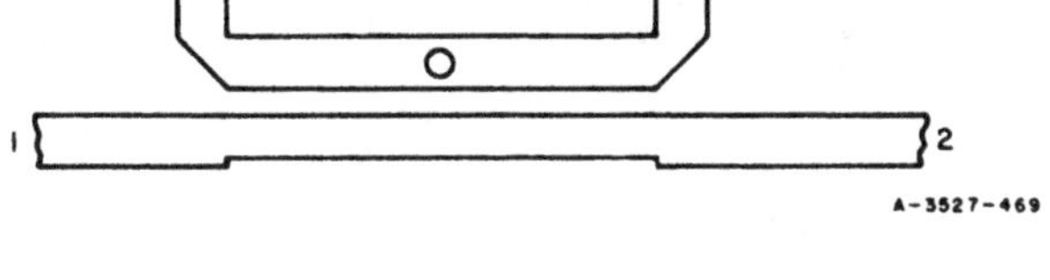

FIG. 14.04-4 SKETCH ILLUSTRATING THE PLACEMENT OF TUNING SCREWS IN A TRAVELING-WAVE-LOOP DIRECTIONAL FILTER

In practice it has been found that, for the filters shown in Fig. 14.04-2, size 4-40 screws can be used to tune the loops over a 6-percent frequency band.

Proper adjustment of the tuning screws for a single-resonator traveling-wave-loop directional filter can be accomplished by observing the response of matched detectors at Port 2 and Port 4 when a matched load is placed at Port 3 and a signal source is placed at Port 1. If a sweeping signal source is available to feed Port 1 this same procedure can be used to tune two- and three-resonator traveling-wave-loop directional filters.

SEC. 14.05, TRAVELING-WAVE RING RESONATOR

The traveling-wave ring resonator of the type illustrated in Fig. 14.05-1 which is closely related to the traveling-wave loop directional filters has found wide acceptance as a passive power multiplying device, for use in testing components that must carry high power. This device which is usually constructed in waveguide can be thought of as being formed by removing the output coupler from a single-loop waveguide directional filter. The traveling-wave ring is usually several wavelengths long to allow the insertion of test pieces within the ring. By adjusting the variable phase shifter and variable impedance matching transformer shown in Fig. 14.05-1 a pure traveling wave can be excited in the ring even though the component under test is slightly mismatched.

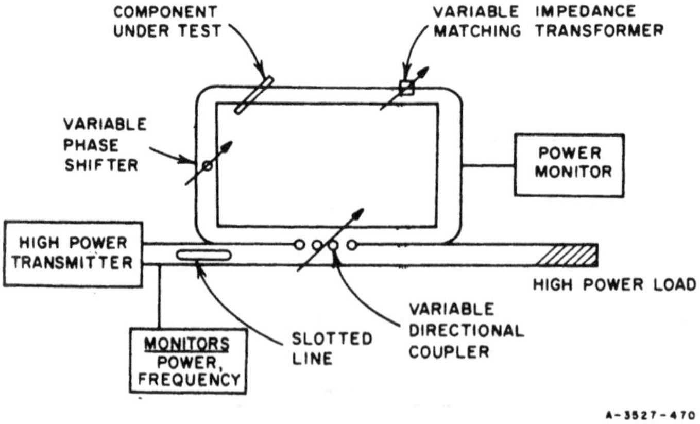

SOURCE: *IRE Trans. PGMTT* (see Ref. 4 by K. Tomiyasu)

FIG. 14.05-1 A RESONANT RING CIRCUIT

Most of the power delivered by the generator is used to set up a high-amplitude traveling wave within the ring and only a small amount of it is delivered to the load. The expression for the multiplication or power gain, of the wave within the loop is given by

$$\text{Power Gain} = \left[\frac{c}{1 - 10^{-\alpha/20}\sqrt{1 - c^2}}\right]^2 \qquad (14.05\text{-}1)$$

where

c = voltage coupling factor of the directional coupler

α = one-way attenuation around the ring measured in db.

Equation (14.05-1) has been plotted in two different ways in Figs. 14.05-2(a), (b) to facilitate the determination of the power gain in the traveling-wave

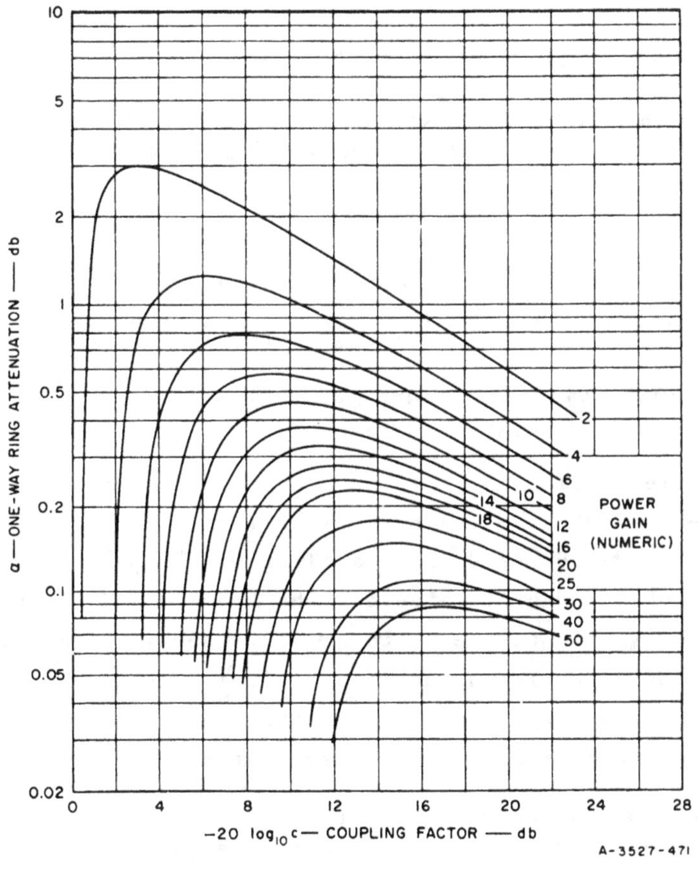

SOURCE: *IRE Trans. PGMTT* (see Ref. 4 by K. Tomiyasu)

FIG. 14.05-2(a) RESONANT RING CHARACTERISTICS WITH POWER GAIN AS A PARAMETER

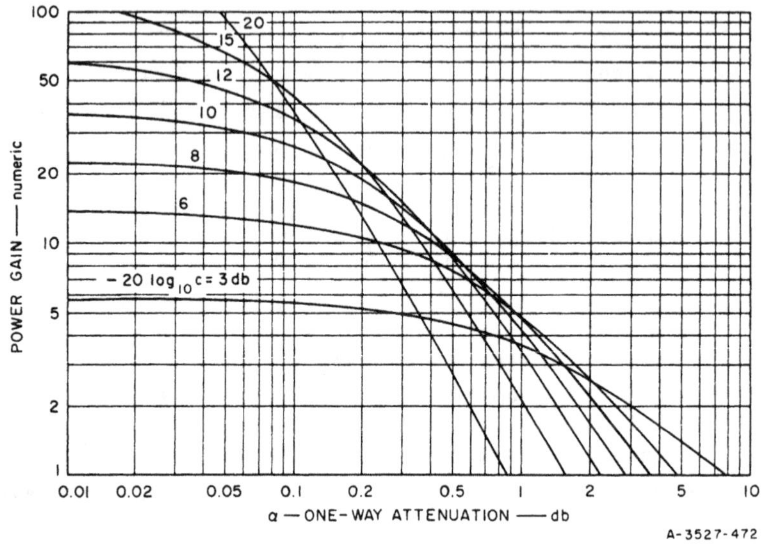

SOURCE: *IRE Trans. PGMTT* (see Ref. 4 by K. Tomiyasu)

FIG. 14.05-2(b) RESONANT RING POWER GAIN WITH COUPLING FACTOR IN db AS A PARAMETER

ring. It is seen that substantial power gains can be obtained even for relatively larger values of the one-way ring attenuation, α.

SEC. 14.06, DERIVATION OF FORMULAS FOR WAVEGUIDE DIRECTIONAL FILTERS

The detailed performance of the waveguide directional filters described in Sec. 14.02, the traveling-wave loop directional filters described in Sec. 14.04, and the traveling-wave ring resonator can be most easily understood by considering the behavior of the multiply-reflected waves within the resonators. This section presents such an analysis for the waveguide directional filter. Consider first the single-cavity waveguide directional filter of the type shown in Fig. 14.02-6 and assume that the following conditions are satisfied:

(1) The coupling apertures excite a pure circularly polarized TE_{11} wave in the circular waveguide and introduce no reflection in the rectangular waveguide.

(2) The cavity is perfectly symmetrical about its axis, so that the two orthogonal linearly polarized modes in the cavity resonate at the same frequency.

(3) Each port of the filter is terminated in its characteristic impedance.

The reference plane, *P*, is chosen to be perpendicular to the axes of the rectangular waveguides at the center points of the coupling apertures. With a wave of unit amplitude incident on Port 1, the amplitudes in the four ports of the filter at the instant just after the wave coupled into the cavity has been reflected for the first time by the upper end wall are as follows:

$$\begin{aligned} &\text{Port 1} \quad E_1 = 1 && \text{Port 3} \quad E_3 = 0 \\ &\text{Port 2} \quad E_2 = \sqrt{1 - c_1^2} && \text{Port 4} \quad E_4 = c_1 c_2 e^{-\alpha h - j\theta} \end{aligned} \tag{14.06-1}$$

where $(c_{01})^2$, which will be replaced here by c_1^2, is the power coupling factor (as defined in Table 14.02-1) of the lower aperture, and $c_2^2 = (c_{12})^2$ is that of the upper aperture, h is the axial height of the cavity, α is the attenuation constant in nepers per unit length in the circular waveguide, and θ is the electrical length in radians of the cavity plus the phase shifts introduced by the apertures.

In the neighborhood of resonance, and in the case of narrow bandwidth,

$$\theta = p\pi \frac{\lambda'_{g0}}{\lambda'_g} - \pi \tag{14.06-2}$$

where

λ'_{g0} is the value of λ'_g at resonance, and

p is the number of half-wave field variations along the axis of the cavity.

Because of multiple reflections between the end walls of the cavity, the steady-state wave amplitude in Port 4 is composed of an infinite summation of components, as follows:

$$E_4 = c_1 c_2 e^{-(\alpha h + j\theta)} + c_1 c_2 (1 - c_1^2)^{1/2} (1 - c_2^2)^{1/2} \rho_0^2 e^{-3(\alpha h + j\theta)} + \dots$$

$$+ c_1 c_2 (1 - c_1^2)^{\frac{p-1}{2}} (1 - c_2^2)^{\frac{p-1}{2}} \rho_0^{2(p-1)} e^{-(2p-1)(\alpha h + j\theta)} + \dots \tag{14.06-3}$$

where ρ_0 is the factor by which the reflected wave amplitude at an end wall is reduced because of dissipation loss on that wall. ρ_0 is approximately equal to the magnitude of the voltage reflection coefficient of the end wall in the absence of the aperture. Therefore,

$$\rho_0 \approx \left|\frac{Z_s - Z_0}{Z_s + Z_0}\right| = \left|\frac{R_s + jR_s - Z_0}{R_s + jR_s + Z_0}\right| \tag{14.06-4}$$

where

Z_s is the surface impedance of the end wall in ohms per square,

R_s is the surface resistivity of the end wall in ohms per square, and

Z_0 = $377\lambda_g'/\lambda$ is the wave impedance of the circular waveguide in ohms per square.

Since $R_s << Z_0$, ρ_0 may be approximated by

$$\rho_0 \approx 1 - \frac{R_s\lambda}{60\pi\lambda_g'} \quad . \tag{14.06-5}$$

The amplitude E_4 may be expressed in closed form, as follows:

$$E_4 = \frac{c_1 c_2 e^{-(ah+j\theta)}}{1 - (1 - c_1^2)^{1/2}(1 - c_2^2)^{1/2}\rho_0^2 e^{-2(ah+j\theta)}} \quad . \tag{14.06-6}$$

In a similar manner, a formula for E_2 may be obtained:

$$\begin{aligned}E_2 &= (1 - c_1^2)^{1/2} - c_1^2(1 - c_2^2)^{1/2}\rho_0 e^{-2(ah+j\theta)} \\ &\quad - c_1^2(1 - c_2^2)(1 - c_1^2)^{1/2}\rho_0^3 e^{-4(ah+j\theta)} \\ &\quad - c_1^2(1 - c_2^2)^{3/2}(1 - c_1^2)\rho_0^5 e^{-6(ah+j\theta)} - \ldots \\ &= (1 - c_1^2)^{1/2} - \frac{c_1^2(1 - c_2^2)^{1/2}\rho_0 e^{-2(ah+j\theta)}}{1 - (1 - c_1^2)^{1/2}(1 - c_2^2)^{1/2}\rho_0^2 e^{-2(ah+j\theta)}}\end{aligned} \tag{14.06-7}$$

or

$$E_2 = \frac{(1 - c_1^2)^{1/2} - (1 - c_2^2)^{1/2}[\rho_0^2 - c_1^2\rho_0(\rho_0 - 1)]e^{-2(ah+j\theta)}}{1 - (1 - c_1^2)^{1/2}(1 - c_2^2)^{1/2}\rho_0^2 e^{-2(ah+j\theta)}} . \quad (14.06\text{-}8)$$

At resonance $\theta = (p - 1)\pi$, so that

$$e^{-2(ah+j\theta)} = e^{-2ah} \quad (14.06\text{-}9)$$

Also, in the usual case of light coupling, $c_1^2\rho_0(\rho_0 - 1) <<< \rho_0^2$. Therefore, the condition for a perfect null on Port 2 at resonance is

$$\frac{e^{4ah}}{\rho_0^4} = \frac{(1 - c_2^2)}{(1 - c_1^2)} . \quad (14.06\text{-}10)$$

In the usual design problem cavity losses are sufficiently low that c_1^2 and c_2^2 determined from Eq. (14.06-9) will be almost equal. In that case if one makes the couplings equal, a deep, although imperfect, rejection on Port 2 will generally be obtained. A formula for the loaded Q, Q_L', of the cavity loaded at both ends, will now be derived for $c_1^2 = c_2^2 = c^2$.

Q_L' is defined as $f_0/2(f_0 - f_1)$ where f_1 is the frequency at which the transmitted power has fallen to one half of the value it has at the center frequency, f_0. It is necessary, therefore, to find the angle θ_1 corresponding to the frequency f_1. The first step is to form

$$|E_4E_4^*| = \frac{1}{2} = \frac{c^4e^{-2ah}e^{j\theta_1}e^{-j\theta_1}}{|1 - (1 - c^2)\rho_0^2e^{-2ah}(\cos 2\theta_1 - j \sin 2\theta_1)|^2} . \quad (14.06\text{-}11)$$

This yields

$$\cos 2\theta_1 = 1 - \frac{[1 - (1 - c^2)\rho_0^2e^{-2ah}]^2}{2(1 - c^2)\rho_0^2e^{-2ah}} . \quad (14.06\text{-}12)$$

At resonance, $\theta = \theta_0 = (p - 1)\pi$ and therefore, in the case of narrow bandwidth, $\cos 2\theta_1$ may be replaced by $\cos 2(\delta\theta)_1$, where $(\delta\theta)_1 = \theta_0 - \theta_1$, and hence

$$\cos 2\theta_1 = 1 - \frac{[2(\delta\theta)_1]^2}{2} + \ldots . \qquad (14.06\text{-}13)$$

Thus

$$(\delta\theta)_1 = \frac{1 - (1 - c^2)\rho_0^2 e^{-2\alpha h}}{2(1 - c^2)^{1/2}\rho_0 e^{-\alpha h}} . \qquad (14.06\text{-}14)$$

Next a relationship between $(\delta\theta)_1$ and $(\delta f)_1 = f_0 - f_1$ is needed. With the aid of the well-known formula for guide wavelength in terms of frequency, we obtain

$$\delta\theta = \frac{-p\pi\lambda'_{g0}}{(\lambda'_g)^2}\,\delta\lambda'_g \approx -p\pi\frac{\delta\lambda'_g}{\lambda'_g} = p\pi\frac{\delta f}{f}\left(\frac{\lambda'_g}{\lambda}\right)^2 . \qquad (14.06\text{-}15)$$

Now we make use of the definition of Q'_L given above to obtain

$$Q'_L = \frac{p\pi(1 - c^2)^{1/2}\rho_0 e^{-\alpha h}}{1 - (1 - c^2)\rho_0^2 e^{-2\alpha h}} \cdot \left(\frac{\lambda'_g}{\lambda}\right)^2 . \qquad (14.06\text{-}16)$$

This is the complete doubly loaded Q of the cavity including the effects of internal losses and external loading. If internal losses are negligible compared to external loading, Q'_L becomes equal to the doubly loaded external Q, Q'_e as follows

$$Q'_e = \frac{p\pi(1 - c^2)^{1/2}}{c^2} \cdot \left(\frac{\lambda'_g}{\lambda}\right)^2 \approx p\pi\left(\frac{\lambda'_g}{\lambda}\right)^2\left(\frac{1}{c^2} - \frac{1}{2}\right). \qquad (14.06\text{-}17)$$

However, in most cases one can replace the numerator factor $(1 - c^2)^{1/2}$ by unity. If c_1 and c_2 are slightly unequal, the formulas for Q will hold sufficiently closely with $c^2 = c_1 c_2$. The external Q, Q_e, of the end resonator of a multiple cavity directional filter which is externally loaded at only one end as defined in Table 14.02-1 is $2Q'_e$.

Since $E_1 = 1$, the relative power transmitted out of Port 4 at resonance is simply $|t|^2 = E_4^2$ with $\theta = 0$. Therefore

$$|t|^2 = \frac{c_1^2 c_2^2 e^{-2\alpha h}}{[1 - (1 - c_1^2)^{1/2}(1 - c_2^2)^{1/2}\rho_0^2 e^{-2\alpha h}]^2} \qquad (14.06\text{-}18)$$

or, if $c_1 = c_2 = c$,

$$|t|^2 = \frac{c^4 e^{-2\alpha h}}{[1 - (1 - c^2)\rho_0^2 e^{-2\alpha h}]^2} \quad . \qquad (14.06\text{-}19)$$

The attenuation in db is 10 $\log_{10}$ $(1/|t|^2)$, or

$$L_A = 20 \log_{10} \left[\frac{1 - (1 - c^2)\rho_0^2 e^{-2\alpha h}}{c^2 e^{-\alpha h}} \right] \text{ db} \quad . \qquad (14.06\text{-}20)$$

If Eq. (14.06-17) is combined with Eqs. (14.06-14) and (14.06-15) we obtain $Q_e' = Q_L'/(\rho_0 |t|)$. However, ρ_0 is of the order of 0.9999, and hence we may relate the external Q and complete loaded Q by the following simple expression:

$$Q_e' = \frac{Q_L'}{|t|} \quad . \qquad (14.06\text{-}21)$$

It may also be shown that, in the case of high Q and narrow bandwidth, the above equations reduce to the following relationship between Q_L', Q_e', and the unloaded Q, Q_u, of the cavity:

$$\frac{1}{Q_L'} = \frac{1}{Q_u} + \frac{1}{Q_e'} \quad . \qquad (14.06\text{-}22)$$

This is the same well-known expression that applies to a simple two-port filter consisting of a single resonator. The formula for unloaded Q in terms of the waveguide-cavity parameters is

$$Q_u = \frac{1}{\dfrac{\alpha\lambda^2}{\pi\lambda'_g} + \dfrac{R_s}{30\pi^2 p}\left(\dfrac{\lambda}{\lambda'_g}\right)^3} \quad . \tag{14.06-23}$$

The value of the power coupling factors $(c_{01})^2$ and $(c_{n,n+1})^2$ for the end apertures in Fig. 14.02-1 in terms of the low-pass prototype elements are obtained by combining Eq. (14.06-17) with either Eq. (6) or Eq. (7) in Fig. 8.02-3, using the approximation $\sqrt{1 - (c_{01})^2/c^2} \approx (1/c^2) - 1/2$.

For multiple-resonator directional filters the values $c_{i,i+1}$ for the internal apertures in Fig. 14.02-1 in terms of the low-pass prototype elements are obtained by using the relation

$$c_{i,i+1} \approx \left| \frac{2Y'_0}{B_{i,i+1}} \right| \tag{14.06-24}$$

in conjunction with Eqs. (2) and (4) in Fig. 8.06-1. The coefficient of coupling $k_{i,i+1}$ is then obtained by the use of Eq. (8) in Fig. 8.02-3.

SEC. 14.07, DERIVATION OF FORMULAS FOR TRAVELING-WAVE LOOP DIRECTIONAL FILTERS AND THE TRAVELING-WAVE RING RESONATOR

An analysis similar to that presented in Sec. 14.06 can be carried through for traveling-wave loop directional filters. By particularizing the analysis the power gain for the traveling-wave ring resonator can also be obtained. This analysis is restricted to the case of TEM-mode propagation on the loop, although the results can be modified to apply to the case where the loop is a dispersive transmission line as explained below and in Sec. 14.04.

It is further assumed in this analysis that: (1) all transmission lines have the same characteristic impedance, Z_A, (2) all directional couplers are designed to be perfectly matched if terminated in Z_A, and they have infinite directivity, and (3) no points of reflection exist in the loop and therefore a pure traveling wave exists. A schematic diagram of the filter is given in Fig. 14.07-1. The phases of the voltages shown at the terminals of the directional couplers are, for convenience, referred to the midplanes of the directional couplers. The amplitudes shown, however, apply at the terminals of the couplers. The total length of the loop at resonance, including the lengths of the coupling regions, is $l = m\lambda_0$, where m is an integer and λ_0 is the plane-wave wavelength at the resonant frequency. In Fig. 14.07-1 it is seen that the following voltages are excited during the first traversal of the loop by the traveling wave. For simplicity, we will again replace c_{01} by c_1 and c_{12} by c_2:

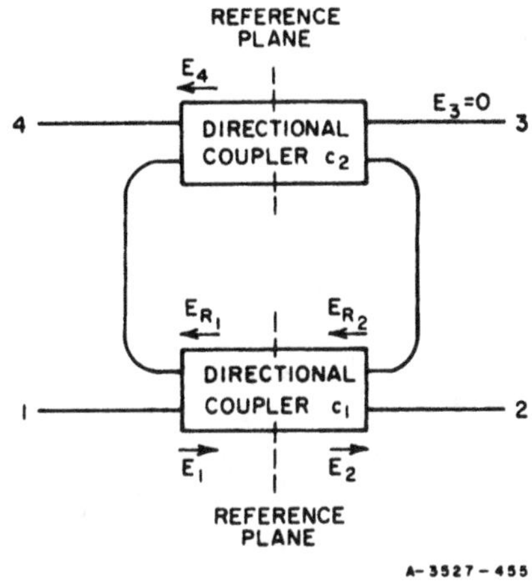

SOURCE: Final Report, Contract DA 36-039 SC-64625, SRI; reprinted in *IRE Trans. PGMTT* (see Ref. 2 by F. S. Coale)

FIG. 14.07-1 SCHEMATIC DIAGRAM OF TRAVELING-WAVE-LOOP DIRECTIONAL FILTER

$$E_1 = 1 \quad \angle 0 \tag{14.07-1}$$

$$E_2 = (1 - c_1^2)^{1/2} \quad \angle 0 \tag{14.07-2}$$

$$E_3 = 0 \tag{14.07-3}$$

$$E_4 = c_1 c_2 e^{-\frac{\alpha l}{2}} \quad \angle 0 \tag{14.07-4}$$

$$E_{R_1} = c_1 \quad \angle +90 \tag{14.07-5}$$

$$E_{R_2} = c_1 (1 - c_2^2)^{1/2} e^{-\alpha l} \quad \angle +90 \tag{14.07-6}$$

where c_1 and c_2 are the voltage coupling factors of the directional couplers, and α is the attenuation in nepers per unit length of the loop transmission line.

For the mth traversal of the loop, the voltages may be written in terms of E_{R_2} as follows:

$$E_2 = (1 - c_1^2)^{1/2} - c_1 E_{R_2} \quad \angle 0 \tag{14.07-7}$$

$$E_3 = 0 \tag{14.07-8}$$

$$E_4 = \frac{c_2 e^{\frac{\alpha l}{2}}}{(1 - c_2^2)^{½}} E_{R_2} \quad \angle 0 \tag{14.07-9}$$

$$E_{R_1} = \frac{e^{\alpha l}}{(1 - c_2^2)^{½}} E_{R_2} \quad \angle{+90} \tag{14.07-10}$$

where

$$E_{R_2} = c_1 (1 - c_2^2)^{½} e^{-\alpha l} [1 + \sum_{j=1}^{j=m} (1 - c_1^2)^{j/2} (1 - c_2^2)^{j/2} e^{-j\alpha l}] \quad \angle{+90} \quad . \tag{14.07-11}$$

Letting $m \to \infty$ and summing this expression in closed form, we obtain

$$E_{R_2} = \frac{c_1 (1 - c_2^2)^{½} e^{-\alpha l}}{1 - (1 - c_1^2)^{½} (1 - c_2^2)^{½} e^{-\alpha l}} \quad \angle{+90} \quad . \tag{14.07-12}$$

The condition for perfect rejection between Ports 1 and 2 may be obtained by setting $E_2 = 0$, which yields

$$e^{2\alpha l} = \frac{1 - c_2^2}{1 - c_1^2} \quad . \tag{14.07-13}$$

For perfect rejection, Eq. (14.07-13) requires $c_2 < c_1$ if $\alpha > 0$.

The doubly loaded Q of the filter may be calculated assuming $c_1 = c_2 = c$. At resonance the output voltage is

$$E_4 = \frac{c^2 e^{-\frac{\alpha l}{2}}}{1 - (1 - c^2) e^{-\alpha l}} \quad . \tag{14.07-14}$$

At a different frequency which corresponds to an electrical length around the loop equal to $2\pi m + \theta$, the output voltage is

$$E_4 = \frac{c^2 e^{-(\alpha l + j\theta)/2}}{1 - (1 - c^2)e^{-(\alpha l + j\theta)}} \quad . \qquad (14.07\text{-}15)$$

Defining $Q_L' = f_L/2(f_0 - f_1) = m\pi/\theta_1$, where f_1 and θ_1 are values at the half-power point, we have from Eqs. (14.07-14) and (14.07-15),

$$E_4 = \left| \frac{c^2 e^{-(\alpha l + j\theta_1)/2}}{1 - (1 - c^2)e^{-(\alpha l + j\theta_1)}} \right| = \frac{\sqrt{2}}{2} \frac{c^2 e^{\frac{\alpha l}{2}}}{[1 - (1 - c^2)e^{-\alpha l}]} \quad . \qquad (14.07\text{-}16)$$

Taking absolute magnitudes of each side and simplifying, we may solve for θ_1 in terms of c and αl:

$$\cos\theta_1 = 1 - \frac{[1 - (1 - c^2)e^{-\alpha l}]^2}{2(1 - c^2)e^{-\alpha l}} \quad . \qquad (14.07\text{-}17)$$

If θ_1 is small, we may replace $\cos\theta_1$ by $1 - \theta_1^2/2$ which yields

$$\theta_1 = \frac{1 - (1 - c^2)e^{-\alpha l}}{(1 - c^2)^{1/2} e^{-\frac{\alpha l}{2}}} \quad . \qquad (14.07\text{-}18)$$

The doubly loaded Q of this filter* is

$$Q_L' = \frac{m\pi(1 - c^2)^{1/2} e^{-\frac{\alpha l}{2}}}{1 - (1 - c^2)e^{-\alpha l}} \quad . \qquad (14.07\text{-}19)$$

It should be noted that this expression gives the true loaded Q of the resonator as it would be measured, and takes account of the external loading due to both couplers combined with the losses in the transmission line of the loop.

The attenuation at resonance is given by $20 \log_{10} E_1/E_4$, and since $E_1 = 1$ we have

---

* When a waveguide loop is used in the filter, Q_L' as given in Eq. (14.07-19) must be multiplied by $(\lambda_g/\lambda)^2$ where λ_g is the guide wavelength and λ is the free-space wavelength. Equation (14.07-20) for attenuation applies without change to the waveguide case.

$$L_A = 20 \log_{10} \frac{1 - (1 - c^2)e^{-\alpha l}}{c^2 e^{-\frac{\alpha l}{2}}} . \qquad (14.07\text{-}20)$$

Equations (14.07-19) and (14.07-20) were derived assuming equal characteristic impedances of the transmission lines and assuming c to be the voltage coupling, but they apply correctly to unequal impedances of the ring and input and output Ports if c^2 is defined as the power coupling factor of the directional couplers.

The design equations in Table 14.04-1 are derived from the above analysis in exactly the same manner as the design equations in Table 14.02-1 for the waveguide directional filter were derived from the analysis in Sec. 14.06.

The traveling-wave ring resonator power gain is numerically equal to $|E_{R_1}|^2$ when $c_2 = 0$. The value of power gain given in Eq. (14.05-1) is obtained when Eq. (14.07-10) is evaluated for this condition.

SEC. 14.08, DERIVATION OF FORMULAS FOR STRIP-TRANSMISSION-LINE DIRECTIONAL FILTERS USING HALF-WAVELENGTH AND ONE-WAVELENGTH STRIPS

The strip transmission line directional filters using half-wavelength and one-wavelength strips are most easily analyzed by considering the even- and odd-mode excitation of Ports 1 and 4 as described in Sec. 14.03. The present analysis will assume that the electrical spacing between resonators is constant over the pass band of the filter, which is essentially correct for the usual narrow-band directional filters.

Referring to Fig. 14.03-1, when an even mode wave is incident on Ports 1 and 4 at the reference planes T_1 and T_4 shown in the figure, the wave will pass by the left-hand resonator but will be reflected at reference planes T_3 and T_2, since it excites only the right-hand strip in a resonant mode. Similarly, when an odd mode wave is incident on Ports 1 and 4, it will be reflected at reference planes T_1 and T_4, since it excites only the left-hand strip in a resonant mode. The voltage reflection coefficient Γ in each case at the appropriate reference planes will be the same and can be written as

$$\Gamma = \frac{1}{1 + j\dfrac{2X}{Z}} . \qquad (14.08\text{-}1)$$

In this equation Z is the characteristic impedance of the main transmission lines and X is the shunt reactance introduced by a half-wavelength strip resonator with its midpoint grounded, or a full-wavelength resonator with its midpoint open-circuited. The shunt reactance presented by the half-wavelength strip is

$$X = -\frac{1}{\omega C} + Z_0 \tan \phi \qquad (14.08\text{-}2)$$

where C is the capacitance of the gap at either end of the half-wavelength strip and 2ϕ is its electrical length. The shunt reactance presented by the one-wavelength strip is

$$X_f = -\frac{1}{\omega C_f} - Z_0 \cot \phi_f \qquad (14.08\text{-}3)$$

where C_f is the capacitance at either end of the one-wavelength strip and $2\phi_f$ is its electrical length. In order that the half-wavelength and the one-wavelength resonators have the same frequency response it is necessary that $X_f = X$ over the pass band: or stated another way, the reactance slope parameter $\varkappa$ must be the same for the two resonators. Since $\phi_f \approx 2\phi$ it can be shown that this condition obtains when $C_f = \sqrt{2}\, C$.

Because the even- and odd-mode reflected waves add at Port 4, and cancel at Port 1, the attenuation L_A, between Port 1 and Port 4, can be written either as

$$L_A = 10 \log_{10} \frac{1}{|\Gamma|^2} = 10 \log_{10} \left[1 + 4\left(\frac{Z_0}{Z} \tan \phi - \frac{1}{\omega CZ}\right)^2\right] \quad \text{db} \qquad (14.08\text{-}4)$$

using the parameters for the half-wavelength strip or as

$$L_A = 10 \log_{10} \left[1 + 4\left(\frac{Z_0}{Z} \cot \phi_f + \frac{1}{\sqrt{2}\omega CZ}\right)^2\right] \quad \text{db} \qquad (14.08\text{-}5)$$

using the parameters for the one-wavelength strip. Equation (14.08-4) shows that the electrical length ϕ_0 of the half-wavelength strip at midband is given by

$$\tan \phi_0 = \frac{1}{\omega C} \quad . \tag{14.08-6}$$

Similarly, the electrical length ϕ_{0f} of the one-wavelength strip at mid-band is given by

$$\cot \phi_{0f} = -\frac{1}{\sqrt{2}\omega C Z_0} \quad . \tag{14.08-7}$$

The bandwidth of either directional filter will now be computed using the parameters of the half-wavelength strip. Inspection of Eq. (14.08-4) shows that the attenuation increases to 3 db when

$$\left| \frac{Z_0}{Z} \tan \phi - \frac{1}{\omega C Z} \right| = \frac{1}{2} \quad . \tag{14.08-8}$$

For narrow bandwidth and infinite unloaded Q, the doubly loaded external Q, Q_e' is determined from Eq. (14.08-5), using Taylor series expansion techniques similar to those used in deriving Eqs. (14.06-17) and (14.07-19), to be

$$Q_e' = \frac{f_0}{(\Delta f)_{3\,db}} = \frac{\pi}{2} \frac{Z_0}{Z} \frac{1}{(\omega_0 C Z_0)^2} \quad . \tag{14.08-9}$$

REFERENCES

1. S. B. Cohn and F. S. Coale, "Directional Channel Separation Filters," *Proc. IRE* 44, pp. 1018-1024 (August 1956).

2. F. S. Coale, "A Traveling-Wave Directional Filter," *IRE Trans. MTT-4,* pp. 256-260 (October 1956).

3. R. L. Sleven, P. E. Dorato, "Strip Transmission Line Directional Filter," paper presented at Institute of Radio Engineers PGMTT National Symposium 9-11 May 1960, San Diego, California.

4. K. Tomiyasu, "Attenuation in a Resonant Ring Circuit," *IRE Trans. MTT-8,* pp. 253-254 (March 1960).

5. L. Young, "A Hybrid-Ring Method of Simulating Higher Powers than Are Available in Waveguide," *Proc. IEE* (London) Vol. 101, Part III, pp. 189-190 (May 1954).

6. L. J. Milosevic and R. Vautey, "Traveling-Wave Resonator," *IRE Trans. MTT-6,* pp. 126-143 (April 1958).

7. S. J. Miller, "The Traveling Wave Resonator and High-Power Microwave Testing," *The Microwave Journal,* pp. 50-58 (September 1960).

8. S. B. Cohn, "Impedance Measurements by Means of a Broadband Circular Polarization Coupler," *Proc. IRE* 42, pp. 1554-1558 (October 1954).

9. R. D. Wanselow and L. P. Tuttle, Jr., "Practical Design of Strip-Transmission Line Half-Wavelength Resonator Directional Filters," *IRE Trans. MTT-7,* No. 1, pp. 168-173 (January 1959).

CHAPTER 15

HIGH-POWER FILTERS

SEC. 15.01, INTRODUCTION AND GENERAL CONSIDERATIONS

High-power microwave radar transmitters radiate an appreciable amount of spurious output power—that is, frequency components other than the carrier frequency and the normal sideband modulation components.[1-21] At the outputs of these transmitters it is necessary to use high-power filters, that will pass the carrier frequency and normal modulation components but will suppress the spurious emissions, and thus keep them from causing interference in neighboring equipments.

Another application of high-power filters is to systems in which two (or more) transmitters operating at different frequencies are required to be connected to a single antenna. This is analogous to the more common diplexing (or multiplexing) with several receivers connected to a single antenna, with the additional requirement of high power-handling capacity (either high pulse power, or high average power, or both).

This section presents the design considerations for high-power filters and illustrates the manner in which various types of filters can be utilized to achieve specified objectives. Later sections in the chapter present more detailed information on the most common types of high-power filters.

Transmitter Spurious Emissions—Common types of high-power microwave transmitters in use at the present time are the magnetron oscillator, the klystron amplifier and the traveling-wave-tube amplifier. Each of these tubes produces an appreciable amount of power at frequencies which are harmonics of the fundamental frequency. In addition, the magnetron often emits so-called moding frequencies near the carrier frequency as is illustrated in Fig. 15.01-1, which are produced when the magnetron oscillates weakly in modes other than the desired mode. Figure 15.01-1 shows that the spurious harmonic power from the magnetron decreases as the harmonic frequency increases but that the power at even harmonics tends to be lower than that from adjacent odd harmonics. The spurious harmonic power from

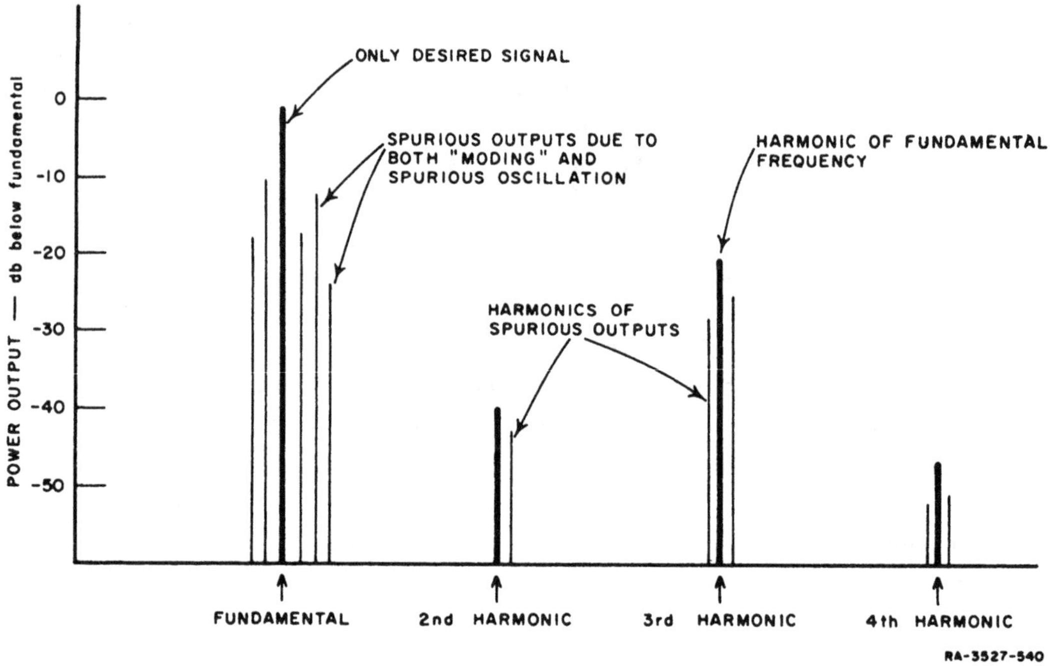

SOURCE: Final Report, Contract AF 30(602)-1670, General Electric Company (See Ref. 11 by G. Novick and V. G. Price)

FIG. 15.01-1 TYPICAL FREQUENCY SPECTRUM OF SIGNALS GENERATED BY A MAGNETRON OSCILLATOR
Note: Fine structure representing the cross modulation product for each of the above lines is not shown.

klystron and traveling-wave tube amplifiers usually decreases monotonically as the order of the harmonic increases. The harmonic power output from a klystron transmitter is shown in Fig. 15.01-2. These particular data[20] were taken on a CW klystron amplifier, but appear to be representative of pulsed tubes also. Here the klystron was first set up without regard to harmonic output (as it would be normally) and the harmonic output was measured (Lines A, Fig. 15.01-2). Then various electrode voltages were adjusted and the drive power reduced by 3 db to minimize the second harmonic, without changing the fundamental power output, and the harmonic output was measured again (Lines B, Fig. 15.01-2). There is a considerable improvement (at least 13 db, and up to 24 db), showing that much "filtering" can be performed by appropriate adjustments on the transmitter alone.

The relative harmonic power levels from a high-power traveling-wave tube may be as much as 20 db greater than those from a klystron. Several

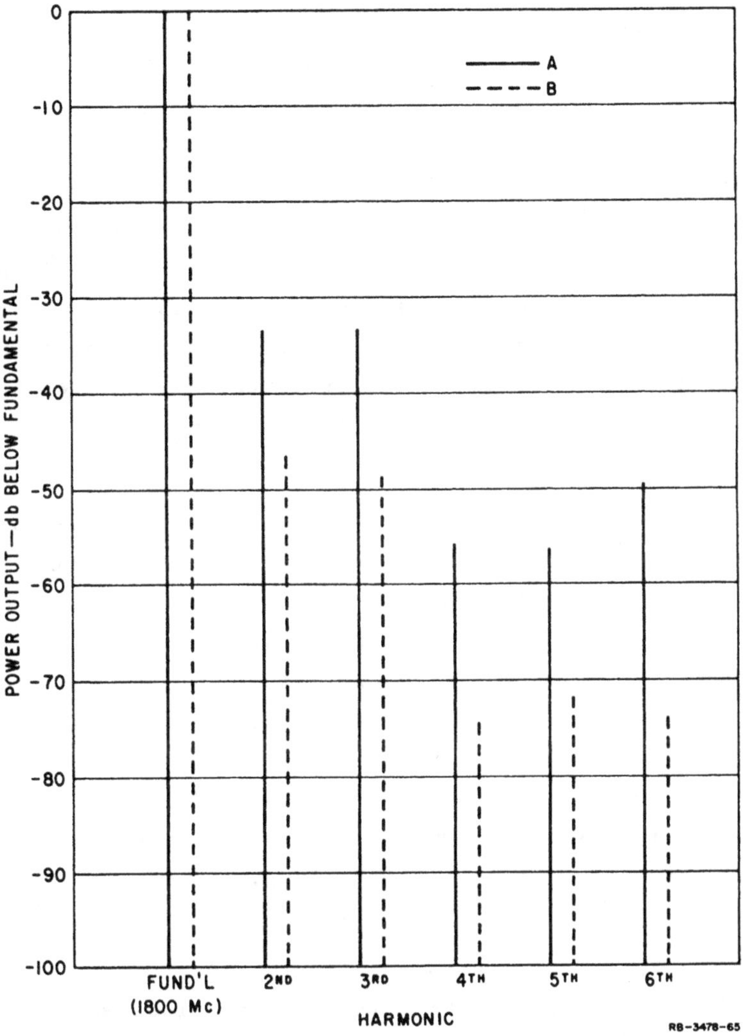

SOURCE: Final Report, Contract AF 30(602)-2392,SRI
(See Ref. 20 by L. Young, E. G. Cristal, E. Sharp
and J. F. Cline)

FIG. 15.01-2 FREQUENCY SPECTRUM OF A KLYSTRON AMPLIFIER OUTPUT; (A) AND (B) BEFORE AND AFTER TUBE ADJUSTMENT TO MINIMIZE HARMONIC OUTPUT WITH CONSTANT FUNDAMENTAL OUTPUT

other types of spurious emissions from high-power tubes can occur and they have been summarized by Tomiyasu.[21]

Filters for Suppressing the Spurious Emissions—Filters for this purpose should ideally have a low-attenuation pass band just wide enough to pass the carrier frequency and its modulation components, and they should have a stop band that extends upward over many octaves to suppress the harmonic frequencies. Furthermore, it is always desirable and often mandatory to preclude the possibility of having resonances at the spurious frequencies occur in the transmission line between the high-power source and the filter. The prevention of such resonances is important if breakdown problems are to be minimized and optimum operation of the tube is to be assured. It is usually impossible to obtain all these characteristics simultaneously with a single filter, and therefore various combinations of filters and auxiliary structures of the type described below are required.

Figure 15.01-3 is a schematic diagram of the most common filter combinations for high-power applications. Filter combination (a) utilizes two reflective band-pass filters and two 3-db hybrids. In the pass band, power is transferred unattenuated from input to output. However in the stop band, power is reflected from the filters. Over the operating frequency band of the hybrids, most of this reflected power is dissipated in the load on the left and a good match is presented to the transmitter. (The termination at the upper right would ideally receive no power; it is included to absorb any stray power due to mismatch of the output load, etc.) At frequencies outside the operating band of the hybrids, an appreciable mismatch may exist at the transmitter.

Filter combination (b) uses a reflective low-pass filter, a 3-db hybrid and resistively terminated tapered waveguides which function as high-pass filters.[22] The pass band of this combination occurs when the low-pass filter is operating in its pass band and the high-pass filters are operating in their stop band and reflect all the power incident on them. As frequency is increased, the attenuation of the combination increases because the high-pass filters begin to propagate and most of the power incident on them is dissipated in the matched loads. However, not all of the unwanted power is so absorbed, since some modes which propagate in the regular waveguide will be cutoff in the tapered guides. Any reflected power in the stop band is also attenuated by the low-pass filter. A relatively good match is obtained over the operating band of the 3-db

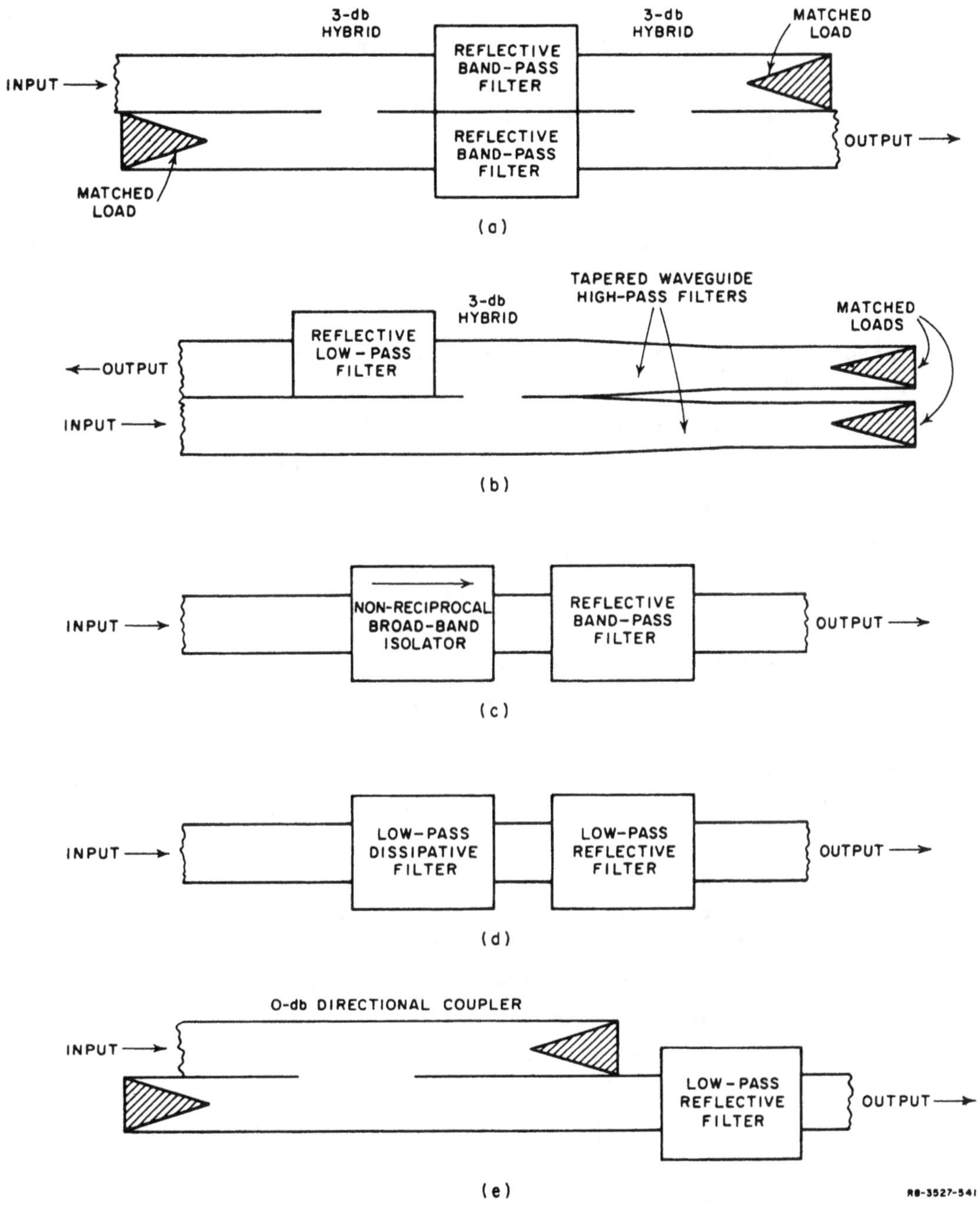

FIG. 15.01-3 COMMON FILTER AND COUPLER COMBINATIONS USEFUL IN SUPPRESSING SPURIOUS EMISSIONS FROM HIGH-POWER TRANSMITTERS

hybrids, even though outside their operating band an appreciable mismatch may exist. Branch-guide directional couplers of the type described in Chapter 13 are quite suitable for use in the circuits shown in Fig. 15.01.3(a) and 15.01-3(b), as are short-slot hybrids[23] in which the coupling aperture is in the side wall.

The filter combination in Fig. 15.01-3(c) employs a non-reciprocal broadband high-power isolator[24] in conjunction with a reflective band-pass filter. The bandwidth of the isolator would be ideally much greater than the pass-band width of the reflective filter. Thus, a good match would be presented to the transmitter over the operating bandwidth of the isolator. The operating bandwidth of presently available isolators is not great enough for applications where higher harmonics must be absorbed, although isolators can be used to absorb the spurious emissions near the fundamental frequency.

The filter combination in Fig. 15.01-3(d) cascades a low-pass reflective filter and a low-pass dissipative filter (*i.e.*, a filter that attenuates by absorbing power incident upon it and which is typically quite well matched both in its pass band and in its stop band). In the pass band, both of these filters have low attenuation. The dissipative filter will typically have relatively high attenuation at the second and third harmonic frequencies and less attenuation at the lower frequencies. If the reflective filter is of the waffle-iron type described in Sec. 15.05, its high-attenuation stop band may extend from the second harmonic up through the tenth harmonic of the fundamental frequency. In this case, the combination shown would be designed so that the waffle-iron filter provides most of the attenuation and the dissipative filter acts primarily as a pad, thus presenting a good match to the transmitter at the harmonic frequencies. For applications where large attenuation at harmonics above the third is not required, the reflective filter shown in Fig. 15.01-3(d) may be omitted and the length of the dissipative filter may be increased to provide the necessary attenuation at the second and third harmonic frequencies.

The filter combination in Fig. 15.01-3(e) replaces the low-pass dissipative filter in Fig. 15.01-3(d) by a coupler which, in the fundamental band, is directional with 0-db coupling, and whose geometry is such that power at the second harmonic and higher frequencies goes mostly straight through into the dummy load, instead of to the reflective filter at the output. Transvar couplers, branch-guide couplers[23,25] (Chapter 13), and side-wall and top-wall short-slot hybrids have been tried in the circuit of Fig. 15.01-3(e). Branch-guide couplers have the advantage that filters can be built into the branches to improve the separation between the fundamental and the harmonic frequencies. Experiments indicate that two 3-db side-wall short-slot couplers cascaded to form a 0-db coupler give relatively good performance in the circuit in Fig. 15.01-3(e), besides being compact and relatively inexpensive.[23]

A dissipative filter may also be placed between the transmitter and the filter combinations in Figs. 15.01-3(a) and 15.01-3(b) to improve the match presented to the transmitter at frequencies outside the operating band of the hybrids.

Placement of Filter—A filter to suppress harmonic frequencies should be placed near the output of the system past the last nonlinear circuit element. For instance there is some evidence that TR-switches generate substantial amounts of harmonics, so that it is preferable to put the spurious-frequency-suppression filter between the duplexer and the antenna, rather than immediately after the transmitter and before the duplexer.

SEC. 15.02, POWER-HANDLING CAPACITY OF VARIOUS TRANSMISSION-LINES

The power-handling capacity of transmission lines propagating pulsed signals having short duration (less than about 5 microseconds) and having a low average power level, but a high pulse-power level, is usually limited by breakdown due to ionization of the gas that fills the guide. Under these conditions there is no appreciable heating of the waveguide. The voltage gradient at which such a breakdown occurs in air is approximately 29 kv/cm per atmosphere when the distance between electrodes is much greater than the oscillation distance of free electrons, which in turn is much greater than the mean free path of the electrons. These conditions are satisfied for the common air-filled waveguides and coaxial lines operating over the frequency range from 1 to 100 Gc and at pressures from 0.1 to 10 atmospheres. The breakdown which occurs under these circumstances is an electrode-less discharge, since most of the free electrons execute many cycles of oscillation before reaching an electrode. Gould[26] and others[27] have made extensive studies of gaseous breakdown phenomena and the interested reader should consult their publications for more detailed information. Hart, *et al.*,[28, 29] have also measured the pulse-power capacity of a variety of waveguide components.

The other condition that is necessary for the initiation of discharge is the presence of a free electron. Such an electron will always be supplied eventually by cosmic ray bombardment or natural radioactivity. However, it is usually found expedient when making voltage breakdown measurements to supply electrons from a radioactive source such as cobalt-60. Such a source might typically have a strength of 100 millicuries (*i.e.*, emit 3.7×10^9 electrons per second).

A minimum value of the breakdown field strength occurs at a pressure where the radio frequency is equal to the gas collision frequency. For air at microwave frequencies, this pressure varies linearly from a pressure of approximately 1 mm of mercury at 1 Gc to approximately 10 mm of mercury at 10 Gc. At lower pressures, the breakdown field strength increases rapidly. At pressures on the order of 10^{-6} mm of mercury, corresponding to a good vacuum, breakdown is no longer due to ionization of the remaining gas molecules, but to other mechanisms. The exact mechanism of breakdown at high vacuum is not completely understood; however, it seems likely that the most important process that occurs is field emission whereby electrons are pulled loose from metal surfaces by electric fields having strengths of megavolts per cm.[30] Such field strengths can arise from minute irregularities on the surface even when the average field strength over the surface is much lower. Thus, it is very important in evacuated[31] high-power filters and transmission lines that the inside surfaces be quite smooth. For smooth surfaces a breakdown dc-field strength of 350 kv/cm is often used as a design value in high-power tubes and it seems that this is a reasonable value to use for microwave breakdown field strengths in a high vacuum until more data on this subject become available.

Another phenomenon, which under ordinary circumstances will not appreciably reduce the power transmitted through an evacuated high-power transmission line or filter, is called *multipactor*. This is a resonant secondary-emission phenomenon which occurs when an electron under the action of an electric field has a transit time between opposite electrodes equal to one-half the period of an RF cycle. It typically occurs at RF voltages of the order of 1000 volts. This phenomenon has recently been utilized to make TR-switches.[32]

Figure 15.02-1 gives the pulse-power capacity[33] of a variety of standard rectangular waveguides over their normal frequency bands when filled with air at atmospheric pressure, which can be computed from Eq. (5.06-14) on the assumption that the peak electric field for breakdown is 29 kv/cm. The various waveguides are given their RETMA or EIA designation (which is the width of the waveguide, to three significant

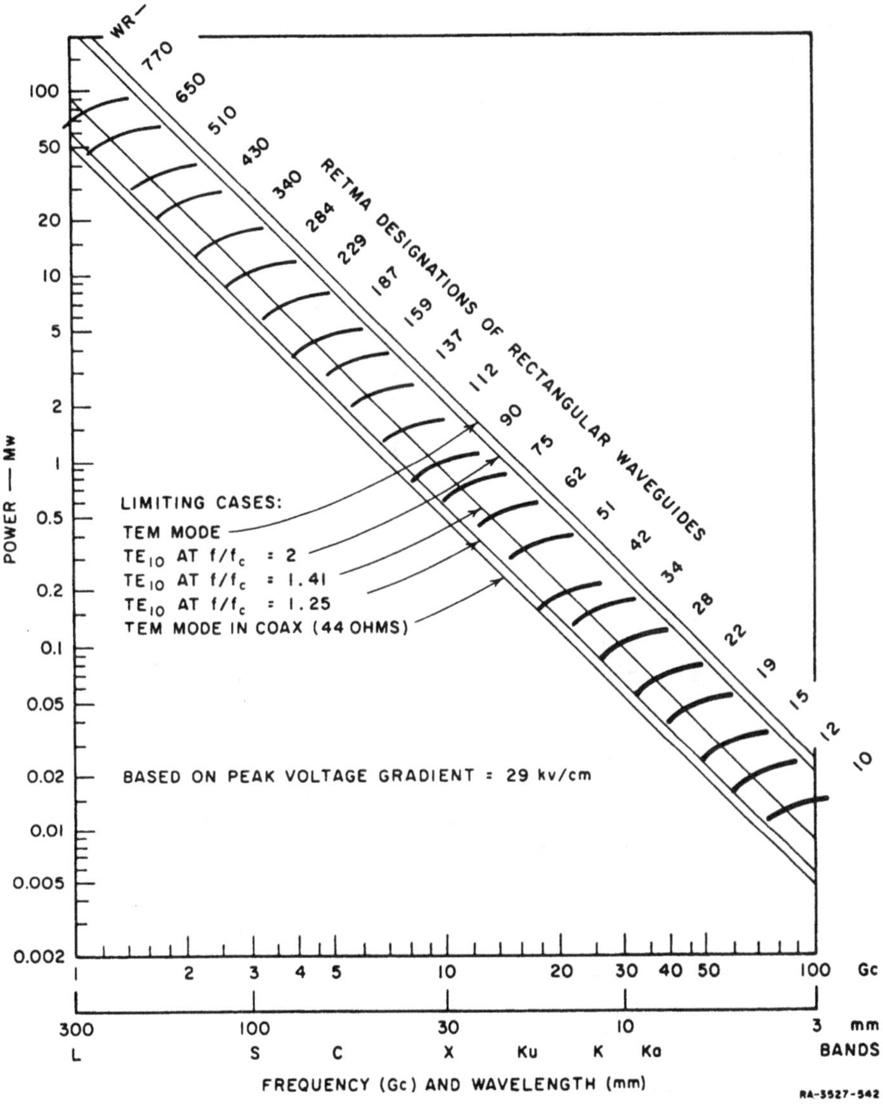

SOURCE: Wheeler Monographs (See Ref. 33 by H. A. Wheeler)

FIG. 15.02-1 CHART OF PULSE-POWER CAPACITY OF WAVEGUIDES

digits, measured in inches and multiplied by 100). Also shown is the pulse-power capacity of a 44-ohm coaxial line [which can be computed from Eq. (5.03-7)], for the case of a line having a mean circumference of one wavelength (outer diameter of 0.430 wavelength and inner diameter of 0.206 wavelength). This is the coaxial line with maximum pulse-power capacity[34] when only the dominant TEM mode is to be allowed to propagate. The highest line on the chart is computed for a hypothetical parallel-strip waveguide propagating a TEM mode having a width and height equal to one-half wavelength. This is the largest size guide that will support the TEM mode to the exclusion of higher modes.

The chart in Fig. 15.02-2 (which looks like a nomogram but is not) gives a series of power-adjusting factors[33] for waveguides and coaxial

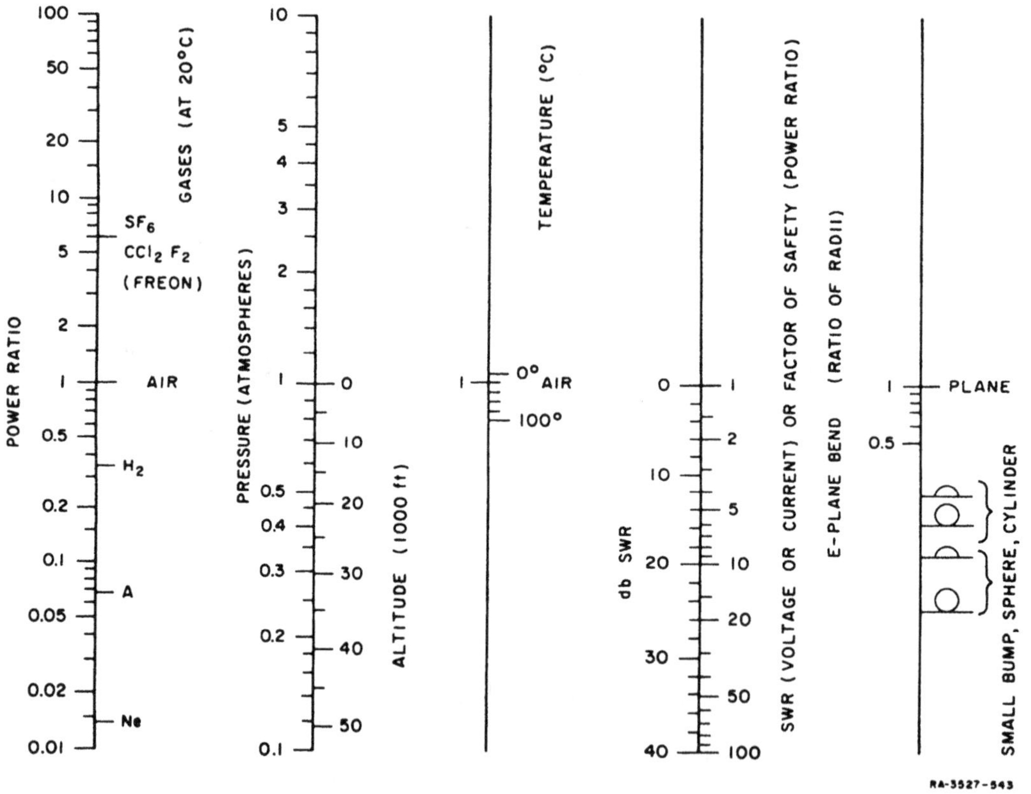

SOURCE: Wheeler Monographs (See Ref. 33 by H. A. Wheeler)

FIG. 15.02-2 CHART OF FACTORS FOR ADJUSTING POWER CAPACITY ESTIMATES
The scale of the ordinates is the same as that in Fig. 15.02-1

lines in the frequency range of 1-100 Gc. These factors are plotted on the same logarithmic scale as that used in Fig. 15.02-1. Therefore, the scale reading on the chart of Fig. 15.02-1 can be adjusted by the distance from the center on one or more of the scales in Fig. 15.02-2, in a manner similar to multiplication on a slide rule. The reference condition for each scale is noted at the center. The standard conditions are air, 20°C, one atmosphere (zero altitude), pure traveling wave (unity VSWR or 0-db VSWR), straight, smooth waveguide (no bend, no bump).

As an example of the use of these charts, we will determine the amount of power that can be delivered to a load through a WR90 waveguide at 10 Gc for a specified set of conditions. We assume that the waveguide is filled with air at 20°C, and is operating at 5,000 feet altitude. A VSWR of 3 db exists on the line, the factor of safety is two, and all other conditions are normal. From Fig. 15.02-1, we find that the pulse power capacity of the guide is 1 Mw, the power ratio for operation at 5,000 feet is 0.7, the power ratio for the 3 db VSWR is 0.7 and the power ratio for the specified factor of safety is 0.5. Therefore, the pulse-power capacity becomes 1 Mw × 0.7 × 0.7 × 0.5 = 0.25 Mw.

Cohn[35] has computed the static electric field for two-dimensional rounded-corner geometries. These results may be applied to actual filter structures with sufficient accuracy for practical purposes if the following two conditions are satisfied: (1) the rounded-corner geometry of the high-power filter may be considered to be composed of infinite cylindrical surfaces, and (2) the essential portions of the rounded corners are small in terms of wavelength so that the field distributions in these regions approximate the static field distributions.

Cohn's results[35] are presented in terms of E_{max}/E_0, where E_{max} is the maximum electric field at a rounded corner and E_0 is the reference electric field at some position well removed from the rounded corner. Since the pulse-power capacity of the filter is proportional to the square of the electric field, the adjusting factor for filters with rounded corners is $(E_0/E_{max})^2$. Two types of two-dimensional curved boundaries are considered: (1) the optimum boundary shape which yields constant electric field over the curved surface, and (2) an approximately circular cross-section shape which would be used in most practical filter structures. A curve for a three-dimensional spherical-corner geometry will be given in Sec. 15.05.

One type of curved boundary that arises in high-power filters is illustrated in Fig. 15.02-3; it consists of an array of 180-degree corners. The shapes of these corners are adjusted for constant electric field strength along their curved portions. Curves giving the shapes of these corners in more detail are presented in Fig. 15.02-4. The constant electric-field strength E_{max} along the curved boundaries turns out to be very simply related to the uniform field E_0 well below the array as follows:

$$\frac{E_{max}}{E_0} = \sqrt{\frac{l}{t}} \quad . \tag{15.02-1}$$

On the straight vertical boundaries the electric field drops off very rapidly.

Another type of rounded corner configuration often encountered is the rounded corner near an electric wall. The ratio E_{max}/E_0 for this type of corner is shown in Fig. 15.02-5 both for the optimum- or uniform-field-strength corner and the approximately circular corner. Also shown is the defining radius for the uniform-field-strength corner. It is seen that the E_{max}/E_0 for the approximately circular boundary is not very much greater than for the uniform-field-strength boundary. Since the uniform-field-strength boundary is much harder to machine, the circular boundary would probably be used in most applications.

An abrupt change in height of a waveguide (case of radius $r = 0$), or in diameter of a coaxial line (cases of $r = 0$), has an equivalent circuit consisting simply of a shunt capacitive susceptance $B|_{r=0}$ at reference planes corresponding with the step itself. Graphical data for such a waveguide discontinuity are given in Fig. 5.07-11, while the equivalent discontinuity capacitance in a coaxial line is given in Fig. 5.07-2. A rounded corner of the type shown in Fig. 15.02-5 will have the shunt susceptance of a sharp corner reduced by the amount ΔB. The value of ΔB for a rounded corner in waveguide is plotted in Fig. 15.02-6. Thus, the total susceptance B of the rounded corner is

$$B = B|_{r=0} + \Delta B \quad . \tag{15.02-2}$$

Note that ΔB is a negative number. Rounding the corner also results in an increase in stored magnetic energy. This may be taken into account by

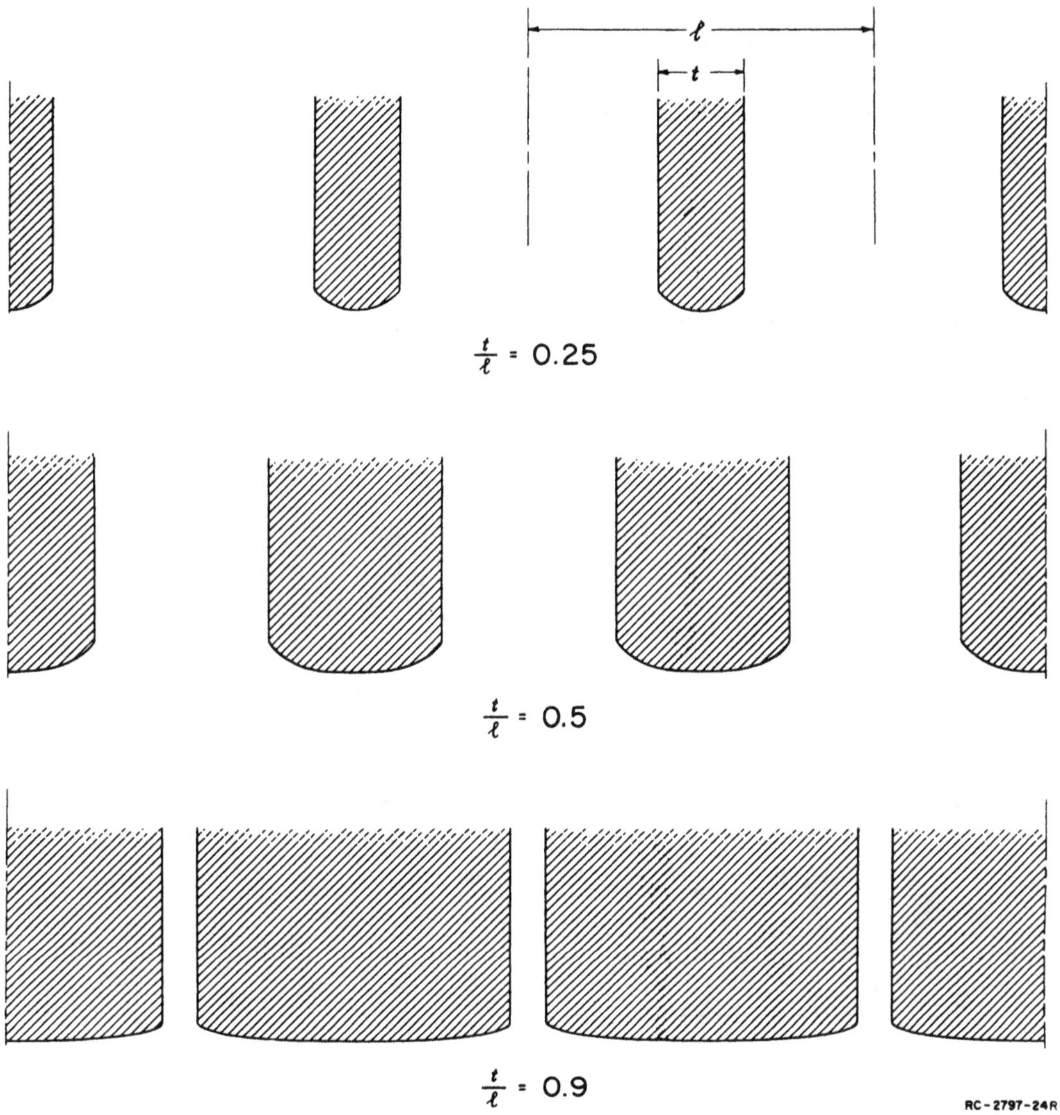

SOURCE: *IRE Trans. PGMTT* (See Ref. 35 by S. B. Cohn)

FIG. 15.02-3 ARRAYS OF ROUNDED 180-DEGREE CORNERS SHAPED FOR CONSTANT ELECTRIC FIELD
The shapes of the curved surfaces are given in Fig. 15.02-4

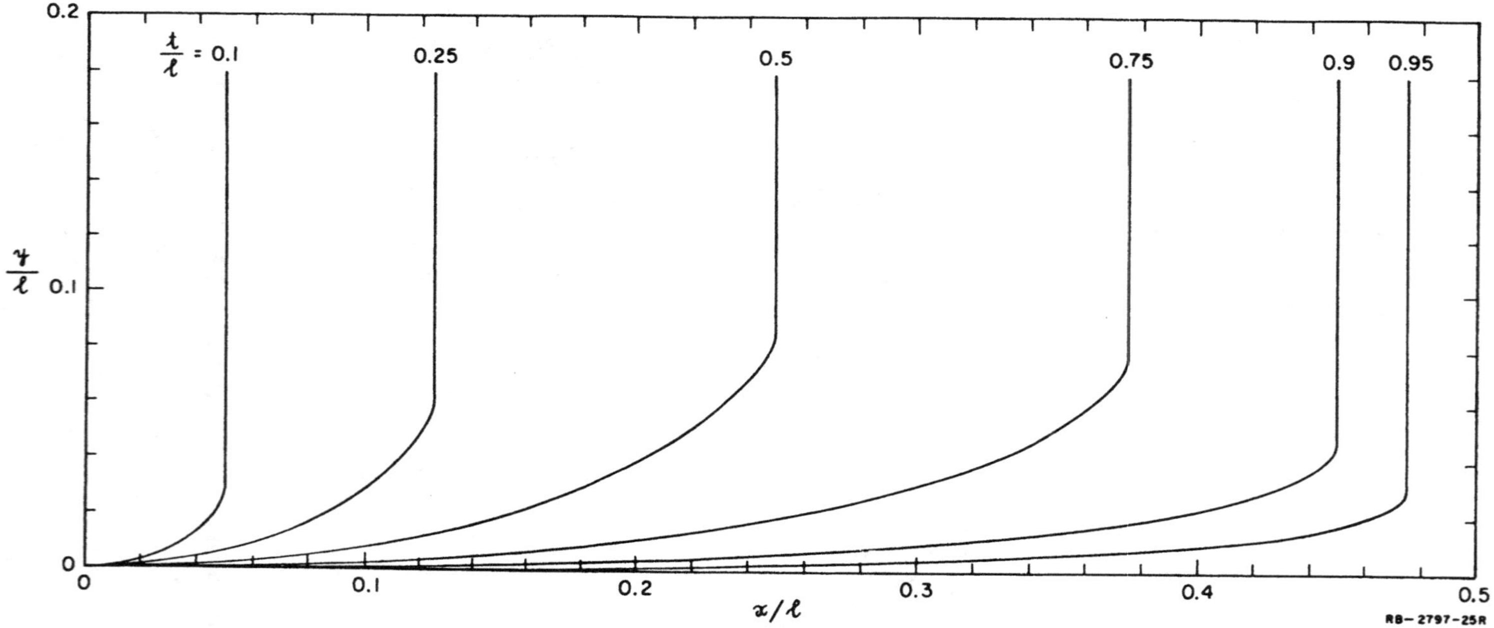

SOURCE: *IRE Trans. PGMTT* (See Ref. 35 by S. B. Cohn)

FIG. 15.02-4 SHAPES OF ROUNDED 180-DEGREE CORNERS FOR VARIOUS VALUES OF t/l

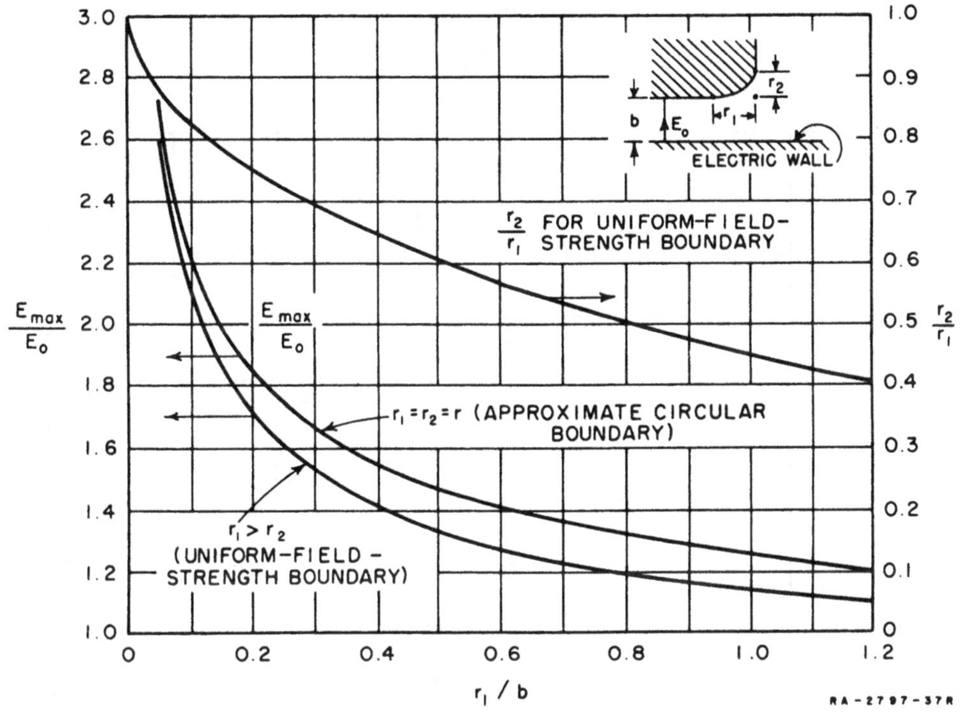

SOURCE: *IRE Trans. PGMTT* (See Ref. 35 by S. B. Cohn)

FIG. 15.02-5 PLOT OF E_{max}/E_0 FOR ROUNDED 90-DEGREE CORNER NEAR AN ELECTRIC WALL; ALSO, r_2/r_1 FOR A UNIFORM-FIELD-STRENGTH BOUNDARY

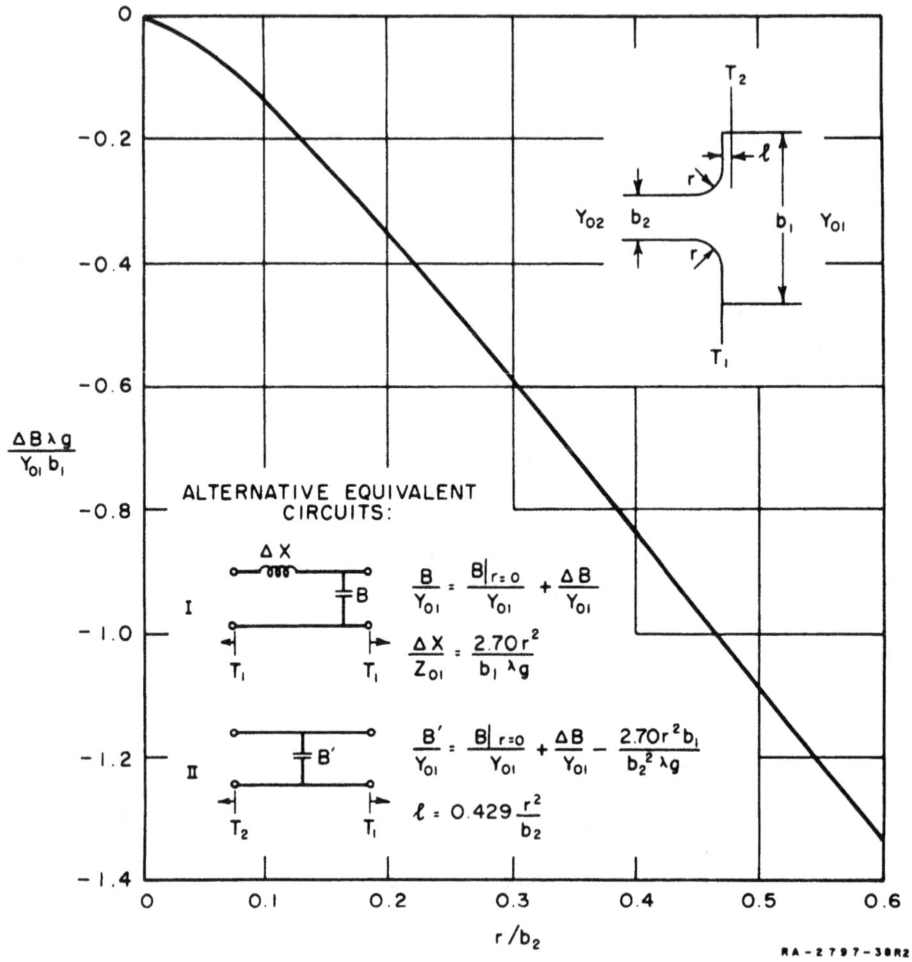

SOURCE: *IRE Trans. PGMTT* (See Ref. 35 by S. B. Cohn)

FIG. 15.02-6 CAPACITIVE SUSCEPTANCE CORRECTION AND EQUIVALENT CIRCUITS FOR ROUNDED STEP IN WAVEGUIDE

adding a series-inductive element in the equivalent circuit at the reference plane of the step, or by a shift of one of the reference planes and modification of B. These alternative equivalent circuits are shown in Fig. 15.02-6. Values of ΔB shown are accurate for $r \leq (b_1 - b_2)/4$, and for $X/Z_{02} < 0.3$, where $Z_{02} = 1/Y_{02}$. The final type of rounded-corner geometry considered is the rounded corner near a magnetic wall. A graph giving the field strengths and defining radii for the uniform field strength boundary is given in Fig. 15.02-7. Comparing this figure with Fig. 15.02-5, it is seen that the field strengths near the magnetic wall are slightly higher than those near the electric wall.

The average power rating of transmission lines is determined by the permissible temperature rise in their walls.[36, 37] Curves giving the average power rating for copper waveguide for various conductor temperature rises are given in Fig. 15.02-8. In deriving these curves, it was assumed that the heat is transferred from the waveguide only by thermal convection and thermal radiation, and that the emissivity of the waveguide walls is 0.3. An additional assumption is that the dissipation of power per unit area in the walls is uniform for all walls. When the ambient temperature is different from the assumed reference of 40°C the curves of Fig. 15.02-9 can be used to determine the correction factor F for the average power rating.

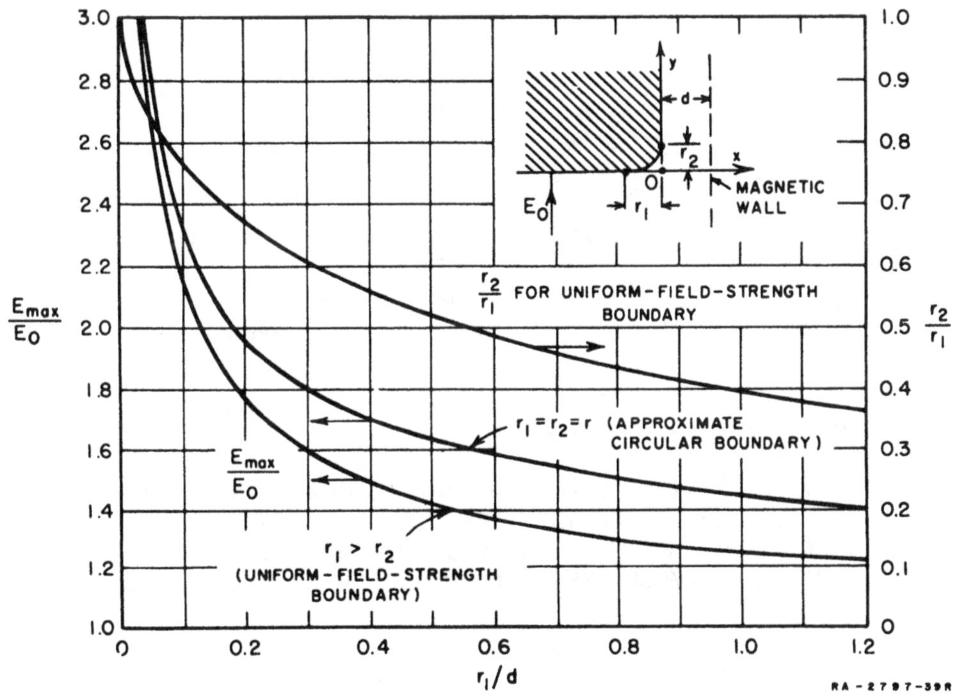

SOURCE: *IRE Trans. PGMTT* (See Ref. 35 by S. B. Cohn)

FIG. 15.02-7 PLOT OF E_{max}/E_0 FOR ROUNDED 90-DEGREE CORNER NEAR A MAGNETIC WALL; ALSO, r_2/r_1 FOR A UNIFORM-FIELD-STRENGTH BOUNDARY

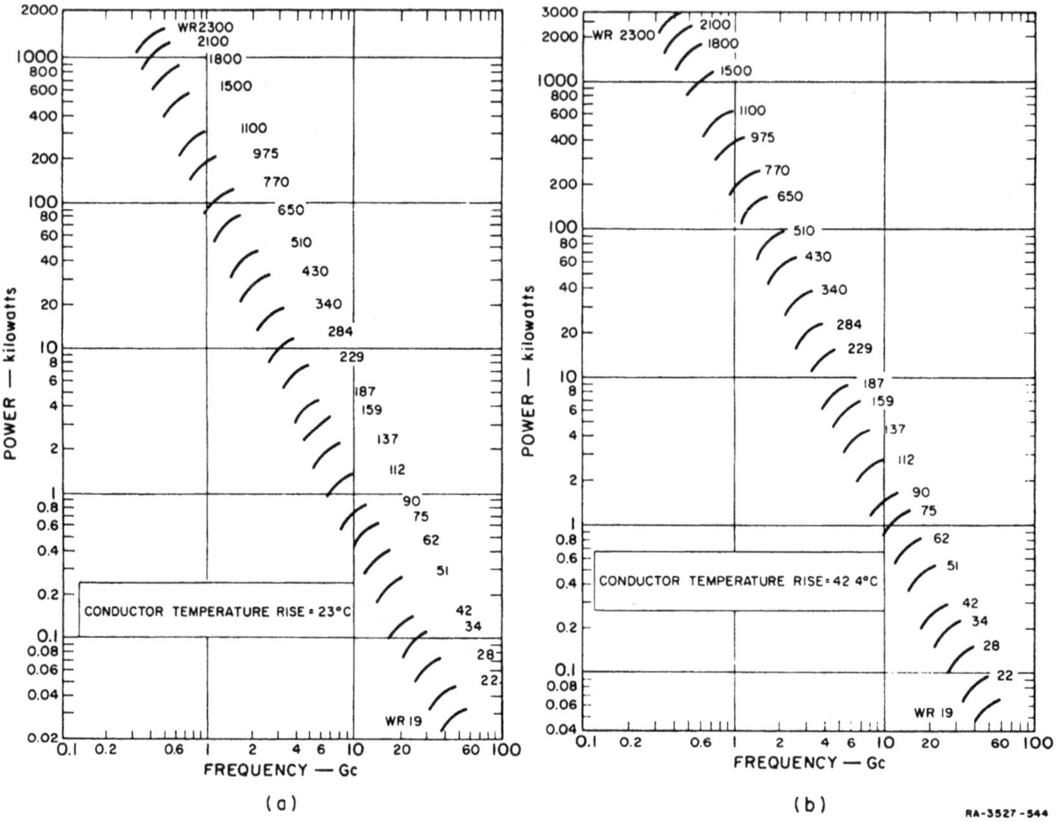

SOURCE: *IRE Trans. PGMTT* (See Ref. 36 by H. E. King)

FIG. 15.02-8 THEORETICAL CURVES OF THE AVERAGE POWER RATING FOR COPPER RECTANGULAR WAVEGUIDE OPERATING IN THE TE_{10} MODE WITH UNITY VSWR AT AN AMBIENT TEMPERATURE OF 40°C, FOR VARIOUS TEMPERATURE RISES (THE WAVEGUIDES ARE LISTED BY THEIR RETMA DESIGNATION)

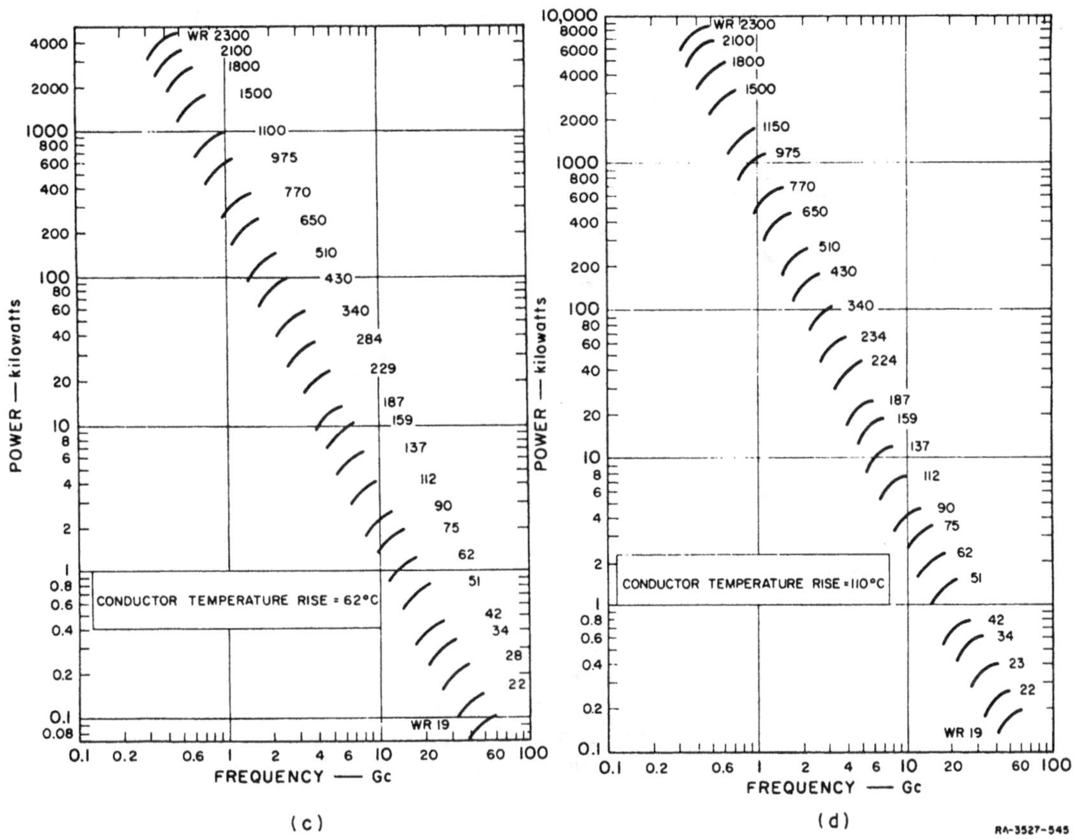

FIG. 15.02-8 Concluded

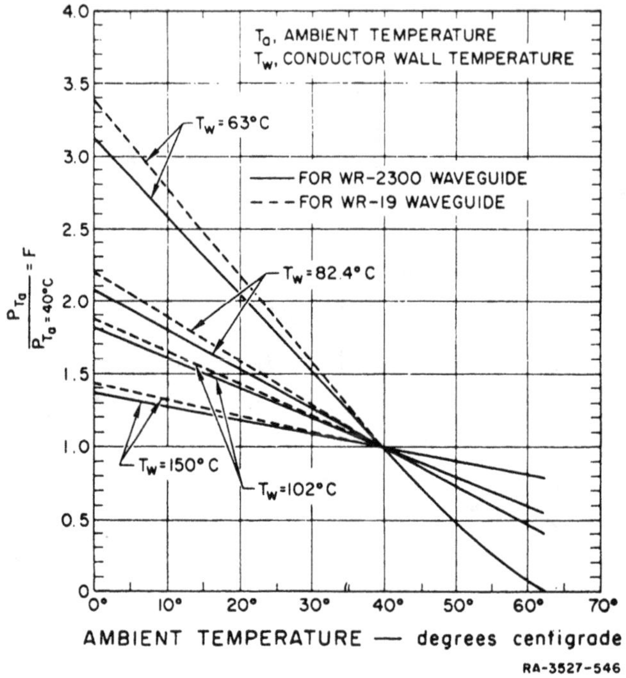

SOURCE: *IRE Trans. PGMTT* (See Ref. 36 by H. E. King)

FIG. 15.02-9 CORRECTION FACTOR CURVE FOR AVERAGE POWER RATING OF RECTANGULAR WAVEGUIDE FOR VARIOUS AMBIENT TEMPERATURES AND CONDUCTOR WALL TEMPERATURES

Standing waves in the waveguide reduce the average power rating of waveguides for a given permissible temperature rise because they produce local hot spots along the wall.

The heat conductivity of the copper waveguides is sufficiently great that an appreciable amount of heat flows axially from these hot spots, thus reducing their temperature. The derating factor that must be applied to waveguide for various values of VSWR is shown in Fig. 15.02-10 and Fig. 15.02-11 for both a copper waveguide and a hypothetical waveguide in which there is no axial heat transfer. The curves in Fig. 15.02-10 apply to the case where the amount of power delivered to the load is a constant while those in Fig. 15.02-11 apply when the amount of power incident on the waveguide is constant.

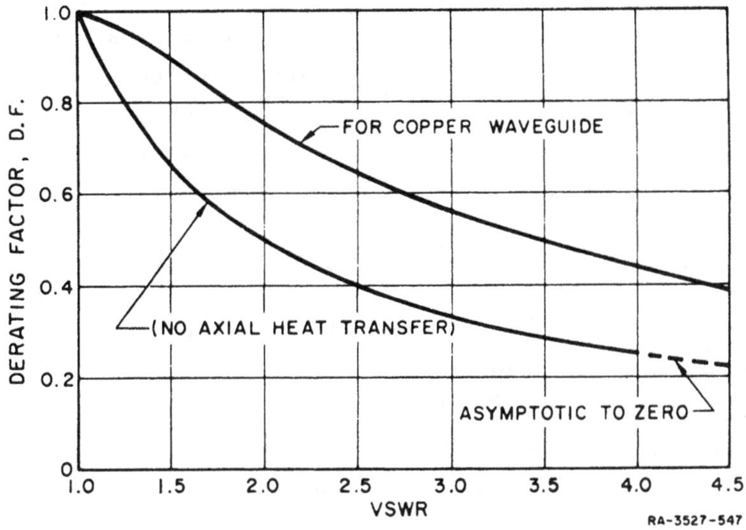

SOURCE: *IRE Trans. PGMTT* (See Ref. 36 by H. E. King)

FIG. 15.02-10 POWER DERATING FACTOR DUE TO VSWR FOR THE CONDITION THAT A CONSTANT POWER IS TO BE DELIVERED TO THE LOAD

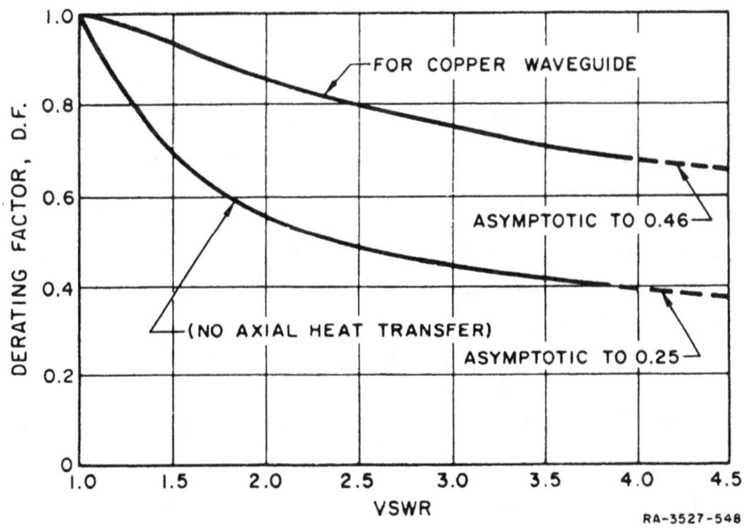

SOURCE: *IRE Trans. PGMTT* (See Ref. 36 by H. E. King)

FIG. 15.02-11 POWER DERATING FACTOR DUE TO VSWR FOR THE CONDITION THAT A CONSTANT POWER IS INCIDENT ON THE WAVEGUIDE

Directional Couplers—Directional couplers are shown in several of the circuits in Fig. 15.01-3. Not much is known about the power-handling capacity of short-slot couplers. Estimates* for the 3-db side-wall coupler range from full line power down to one-third line power. The 3-db top-wall coupler is less well-suited for handling high power and will break down at about one-quarter full line power.

The power-handling capacity of branch-line couplers is approximately independent of the coupling ratio and the branch-to-through-guide impedance ratios. In waveguide it is determined largely by the radii of the corners at the *T*-junctions, and can be estimated from Fig. 15.02-5. The power-handling capacity can readily be made equal to 40 percent of line power as in the example in Sec. 13.14, and could probably be increased to 60 or 70 percent by doubling the corner radius used in that example.

SEC. 15.03, THEORETICAL PULSE-POWER CAPACITY OF DIRECT-COUPLED-RESONATOR FILTERS

This section will discuss the pulse-power capacity of band-pass filters of the type shown in Fig. 15.03-1(a) having a frequency response such as the band-pass response in Fig. 15.03-1(b). The design information will be presented primarily for narrow-band filters in terms of the element values g_k of the low-pass prototype filter shown in Fig. 15.03-1(c). The cavities constituting the filter can be fabricated from transmission lines having any cross-sectional shape, and the terminating transmission lines can have the same, or a different cross section.

The pulse-power capacity is limited by voltage breakdown at the positions of high electric field within the filter. When the usual inductive coupling apertures are used between the cavities, the peak electric fields do not occur in the apertures but rather within the cavities themselves as indicated schematically in Fig. 15.03-1(a).

Pulse-Power Capacity at Midband of Narrow-Band Filters—Power handling may become a severe problem in narrow-band filters. Most narrow-band filter designs are based on a low-pass prototype circuit (Chapter 4), and they are usually well-matched at midband. We shall therefore first restrict ourselves to narrow-band filters which are reflectionless at midband. An exact general formula will be given later [Eq. (15.03-4)].

* Private communication from H. J. Riblet to L. Young (July 9, 1962). (Note added in proof: A recent measurement by SRI, in a direct comparison with uniform waveguide, gives about 70-percent of line power for the side-wall coupler, and about 40-percent for the top-wall coupler.)

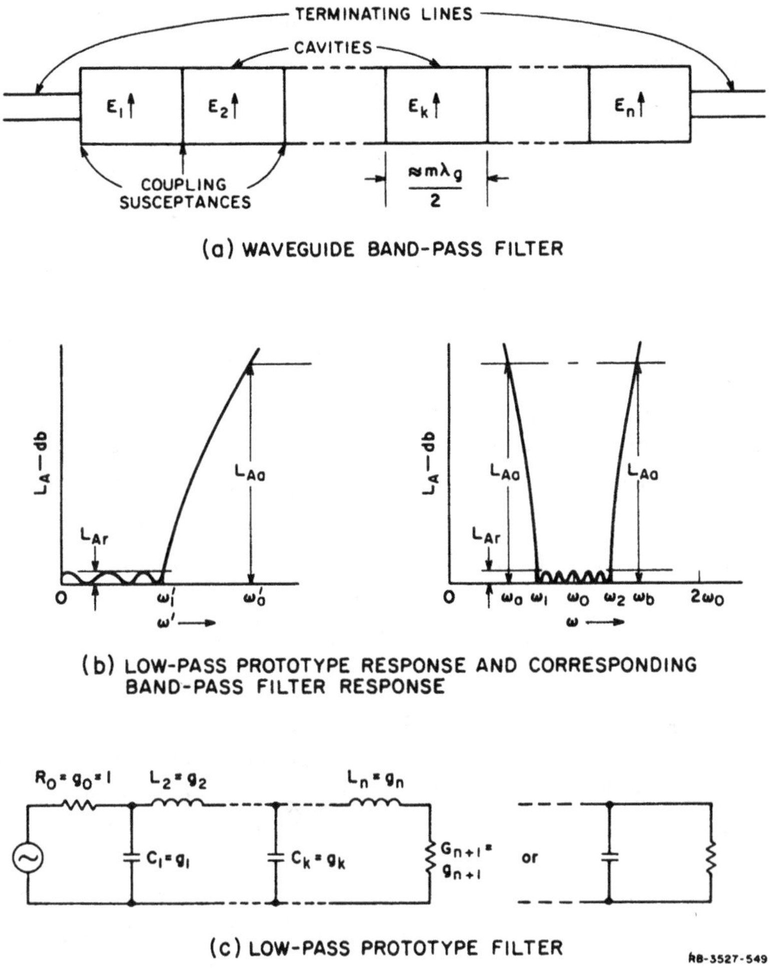

FIG. 15.03-1 WAVEGUIDE BAND-PASS FILTER AND EQUIVALENT LOW-PASS PROTOTYPE

We will define P_{qk} as the pulse power (*i.e.*, the power level during a short pulse) that could be transmitted through an n-cavity filter at the onset of voltage breakdown in Cavity k. Then it will be shown at the end of this section (an independent proof is also given in Ref. 38) that at the midband frequency f_0,

$$(P_{qk})_0 = \frac{P'_{qk}\pi m w}{2g_k\omega'_1}\left(\frac{\lambda_{g0}}{\lambda_0}\right)^2 \tag{15.03-1}$$

where P'_{qk} is the pulse-power capacity of a matched waveguide having the same cross section as the cavity, $(P_{qk})_0$ is P_{qk} for $f = f_0$, λ_{g0} is the guide wavelength in the cavity at f_0, λ_0 is the free-space wavelength at f_0, m is the length of the cavity measured in half guide wavelengths, the fractional bandwidth is $w = (\omega_2 - \omega_1)/\omega_0$, and g_k is the corresponding element in the equivalent low-pass prototype circuit, *having element values normalized so that* $g_0 = 1$. Because the values of g_k are not necessarily the same for each element in the low-pass prototype, the pulse-power capacity of the over-all filter is limited by voltage breakdown in the cavity associated with the largest value of g_k. When all the elements in the low-pass prototype are equal, the midband pulse-power capacity of the filter is maximized for a given off-channel selectivity. Use of the equal-element low-pass prototype also results in approximately the minimum midband dissipation loss $(\Delta L_A)_0$, for given resonator Q's and off-channel rejection, as discussed in Sec. 11.07 and Sec. 6.14.

Table 15.03-1 gives formulas for the pulse-power ratings P'_q of some common transmission lines. The midband pulse-power ratings $(P_{qk})_0$ of the cavity resonators constructed from them as computed from Eq. (15.03-1) are also presented. As is pointed out later in this section, the power capacity near band edge is less than that at midband. Table 15.03-1 gives all dimensions in centimeters, frequencies in gigacycles and powers in megawatts. The components are assumed to be filled with air at atmospheric pressure so that the peak voltage gradient is taken as 29 kv/cm. The factor $\lambda_g/\{\lambda[1 - (\lambda/\lambda_g)^2]\}$ which occurs in the right-hand column of Table 15.03-1 is a dimensionless function of the wavelength-to-cutoff wavelength ratio, and will generally lie between 3 and 4.

The midband pulse-power capacity $(P_{qk})_0$ of the air-filled cavities operating at atmospheric pressure is also plotted for convenience in

Table 15.03-1

PULSE-POWER RATING OF SOME COMMON TRANSMISSION LINES AND CAVITY RESONATORS CONSTRUCTED FROM THEM

(Based on a Peak Voltage Gradient of 29 kv/cm)

Note: Dimensions and wavelengths are in cm.
Frequencies f, and f_c are in gigacycles.
P'_q and $(P_{qk})_0$ are in megawatts.
f_c = cutoff frequency of guide.

w = fractional bandwidth = $(\omega_2 - \omega_1)/\omega_0$ = $(f_2 - f_1)/f_0$.
m = number of half-wavelengths in resonators.
g_k, and ω'_1 are defined in Fig. 15.03-1, where the g_k are assumed to be normalized so that $g_0 = 1$.

	P'_q, PULSE-POWER CAPACITY OF A MATCHED LINE	λ_c, CUTOFF WAVELENGTH	$(P_{qk})_0$, MIDBAND PULSE-POWER CAPACITY OF AN n-SECTION FILTER AT THE ONSET OF BREAKDOWN IN CAVITY k
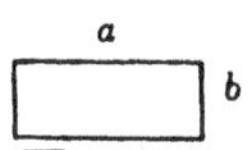 TE_{10}	$0.14 \times \frac{\lambda}{\lambda_g} \lambda_c^2 \frac{b}{a}$ TE_{10} mode	$2a$	$198 \frac{mbw}{a g_k \omega'_1 f^2} \cdot \frac{\lambda_g}{\lambda\left[1 - \left(\frac{\lambda}{\lambda_g}\right)^2\right]}$
 TE_{11}	$0.144 \times \frac{\lambda}{\lambda_g} \lambda_c^2$ TE_{11} mode	$3.415a$	$204 \frac{mw}{g_k \omega'_1 f^2} \cdot \frac{\lambda_g}{\lambda\left[1 - \left(\frac{\lambda}{\lambda_g}\right)^2\right]}$
 TE_{01}	$0.625 \times \frac{\lambda}{\lambda_g} \lambda_c^2$ TE_{01} mode	$1.64a$	$886 \frac{mw}{g_k \omega'_1 f^2} \cdot \frac{\lambda_g}{\lambda\left[1 - \left(\frac{\lambda}{\lambda_g}\right)^2\right]}$
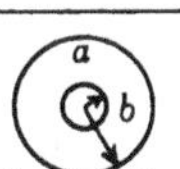 TEM Z_0 = 30 ohms	$1.29\ b^2$ TEM mode	None	$2.03 \frac{mwb^2}{g_k \omega'_1}$
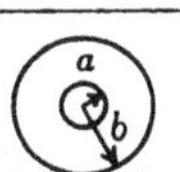 TEM Z_0 = 50 ohms	$1.105\ b^2$ TEM mode	None	$1.74 \frac{mwb^2}{g_k \omega'_1}$
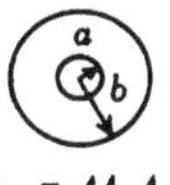 TEM Z_0 = 44.4 ohms	$0.055\ \lambda_c^2$ Maximum pulse-power rating when operating at cutoff wavelength λ_c of first higher order TE mode	$\lambda_c = \frac{30}{f_c} \approx \pi(a + b)$ (Cutoff wavelength of first higher order TE mode)	$77.8 \frac{mw}{g_k \omega'_1 f_c^2} = 1.87 \frac{mwb^2}{g_k \omega'_1}$

SOURCE: Technical Note 2, Contract AF 30(602)-1998, SRI (See Ref. 38 by E. M. T. Jones)

Fig. 15.03-2. In this figure frequencies are in gigacycles and powers in megawatts. If the cavities are filled with some other gas, or are operated at a different pressure the power adjusting factors in Fig. 15.02-2 may be employed in conjunction with Fig. 15.03-2 to determine the midband power capacity of the filter.

Pulse-Power Capacity of Narrow-Band Filters as a Function of Frequency—The peak electric fields in the various cavities of a band-pass filter vary with frequency. A particularly simple way to calculate this variation in narrow-band filters is to make use of the fact that the peak electric fields in the cavities are proportional to the voltages across the capacitances and the currents through the inductances in the low-pass prototype.[39]

Figure 15.03-3 is a plot of the square of the normalized peak electric fields in the cavities of a three-cavity filter, whose low-pass prototype has element values g_0, g_1, g_2, g_3, and g_4 all equal to unity. The curves are plotted against the angular frequency variable ω' of the low-pass prototype. Also included for reference is the insertion-loss characteristic of the low-pass prototype. The analytic expressions for the square of the normalized peak electric fields in the three cavities are

$$\left.\begin{aligned} E_1^2 &= \frac{1 - \omega'^2 + \omega'^4}{P_{avail}/P_{out}} \\ E_2^2 &= \frac{1 + \omega'^2}{P_{avail}/P_{out}} \\ E_3^2 &= \frac{1}{P_{avail}/P_{out}} \quad . \end{aligned}\right\} \tag{15.03-2}$$

The transducer loss ratio P_{avail}/P_{out} is given by

$$\frac{P_{avail}}{P_{out}} = 1 + \frac{\omega'^2(1 - \omega'^2)^2}{4} \tag{15.03-3}$$

Reference to Fig. 15.03-3 shows that the square of the normalized peak electric field in Cavity 2 for $\omega' = 1.24$ rises to 2.25 times the midband value. Therefore, the pulse-power capacity of this filter is only 0.445 times the midband value of P_{q2} determined from Fig. 15.03-2.

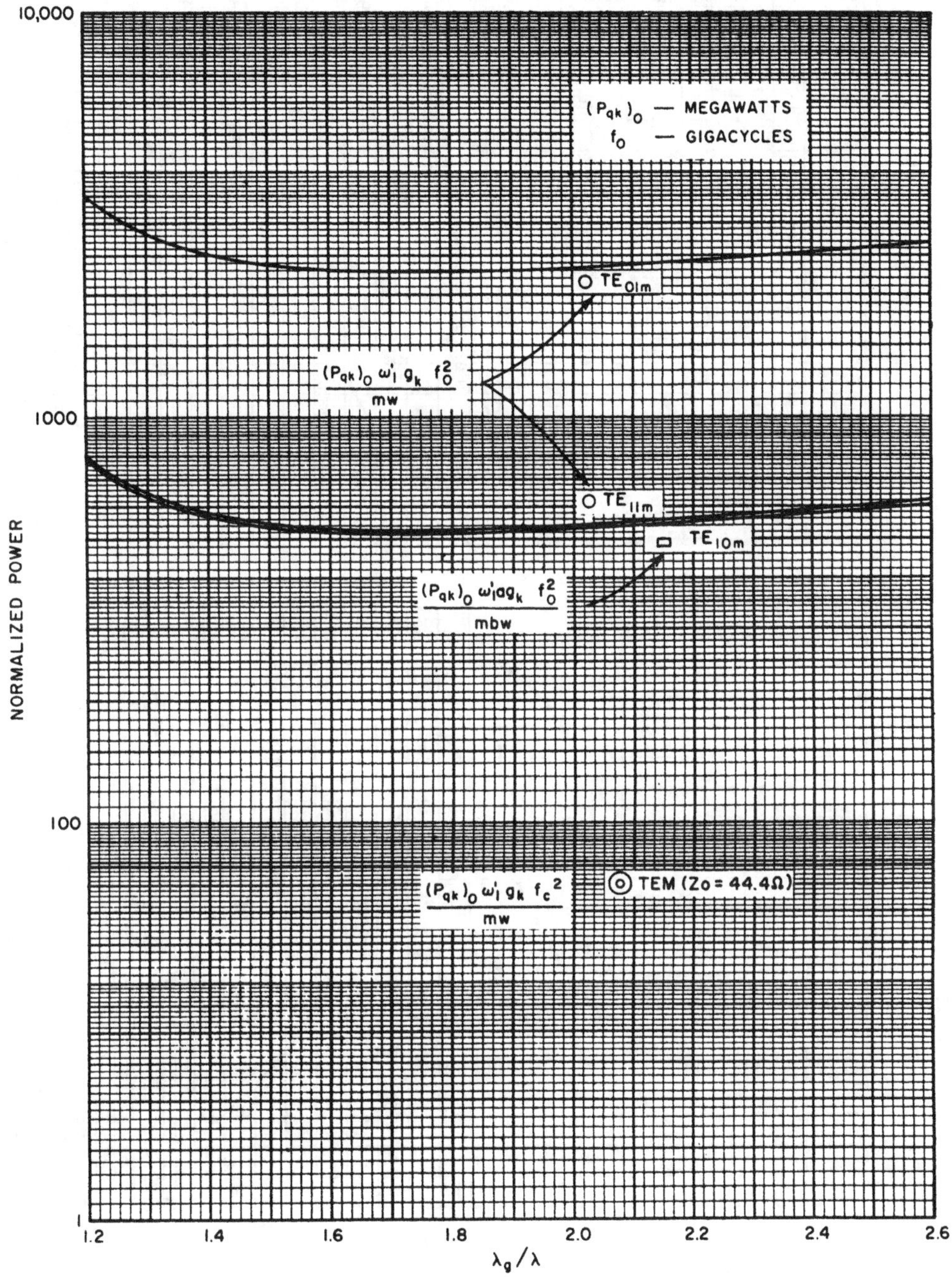

SOURCE: Technical Note 2, Contract AF 30(602)-1998, SRI
(See Ref. 38 by E. M. T. Jones)

FIG. 15.03-2 MIDBAND PULSE-POWER CAPACITY $(P_{qk})_0$ OF AN n-CAVITY FILTER AT THE ONSET OF BREAKDOWN IN CAVITY k (BASED ON PEAK VOLTAGE GRADIENT OF 29 kv/cm)
See Table 15.03-1 for definitions of parameters

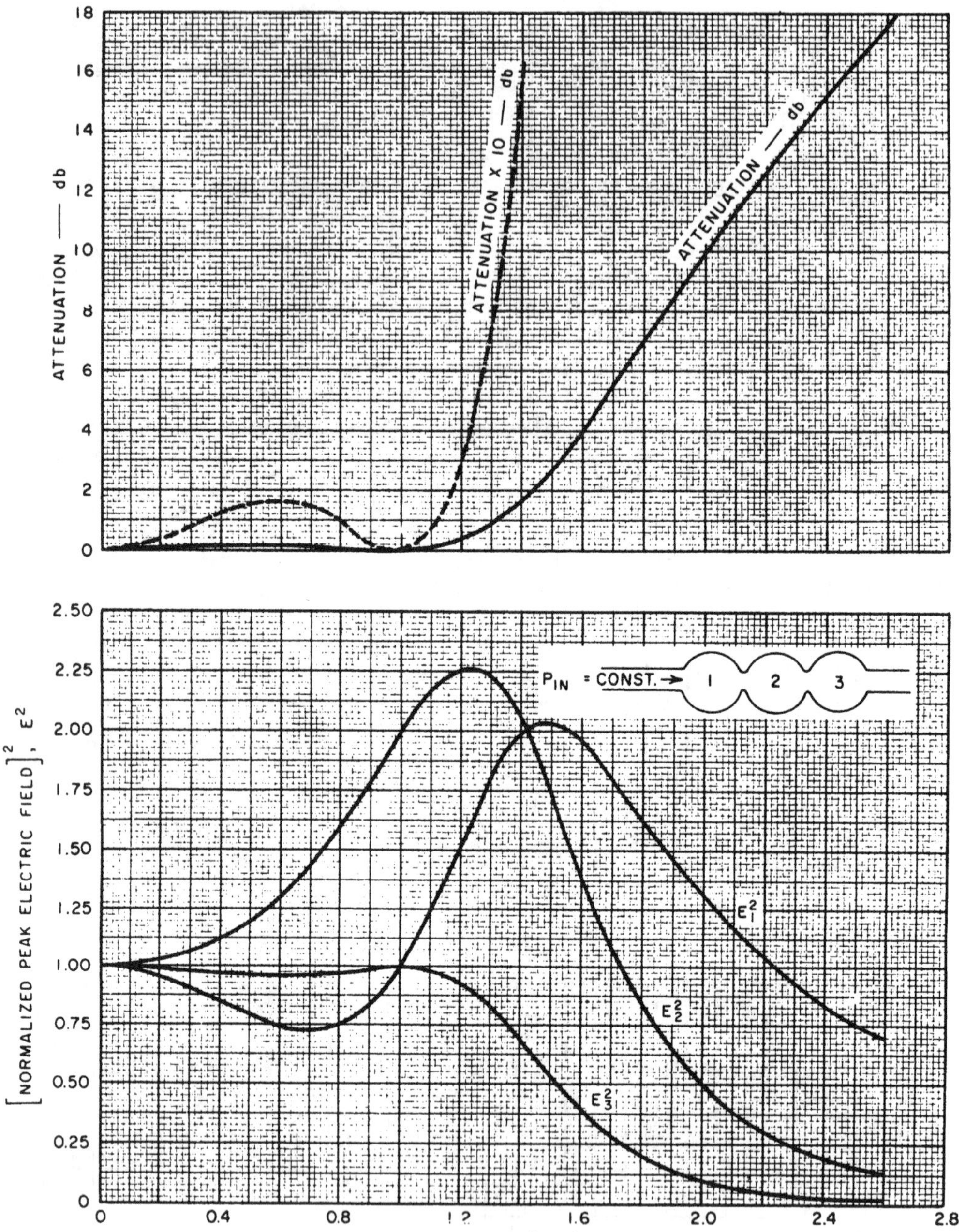

SOURCE: Technical Note 2, Contract AF 30(602)-1998 SRI
(See Ref. 38 by E. M. T. Jones)

FIG. 15.03-3 THE SQUARE OF THE PEAK ELECTRIC FIELD AS A FUNCTION OF FREQUENCY IN A THREE-CAVITY FILTER HAVING AN EQUAL-ELEMENT LOW-PASS PROTOTYPE

Filters of Arbitrary Bandwidth—It is possible to determine exactly the peak electric fields in transmission-line resonators of band-pass filters having arbitrary bandwidth, both at midband and at any other frequency.[40] One computes the internal voltage standing-wave ratios S_k (Sec. 6.14) seen looking toward the load in each cavity. Then one computes the ratio P'_{qk}/P_{qk} in each cavity from the relation

$$\frac{P'_{qk}}{P_{qk}} = S_k \tag{15.03-4}$$

where P'_{qk} is again the power rating of a matched waveguide with the same cross section as cavity resonator k. The equivalent power ratio (E.P.R.) is defined as the ratio of P'_{qk} to the maximum incident (or available) power which can be handled without breakdown,

$$\text{E.P.R.} = \frac{P'_{qk}}{P_{qk}}(1 - |\rho_0|^2) = S_k(1 - |\rho_0|^2) \tag{15.03-5}$$

since the power, P_{qk}, transmitted to the load, is the incident power times $(1 - |\rho_0|^2)$, ρ_0 being the input reflection coefficient.

The midband values of the internal VSWR's S_k are easily determined for synchronous filters as in Sec. 6.14, and the same S_k then give the pulse-power handling capacity directly from Eq. (15.03-4) or (15.03-5).

Figure 15.03-4 shows the equivalent power ratio (E.P.R.) calculated from Eq. (15.03-5) for a six-cavity shunt-inductively-coupled (or series-capacitively-coupled) filter (such as that in Fig. 8.06-1) in which the transmission lines terminating the filter have the same cross section as the cavities. This filter was designed in Sec. 9.04 to have a 10-percent fractional bandwidth, measured on a reciprocal guide wavelength basis, with a 0.01-db attenuation ripple in the pass band. In this filter, Cavity 1 is nearest the generator while Cavity 6 is nearest the load.

Connection with Group Delay and Dissipation Loss—The universal delay curves, Figs. 6.15-1 through 6.15-10 may be used to obtain an estimate of the frequency variation of the power-handling capacity of filters satisfying Eq. (6.09-1), which includes most filters up to about 20-percent bandwidth. It is supposed that the midband power-handling capacity has

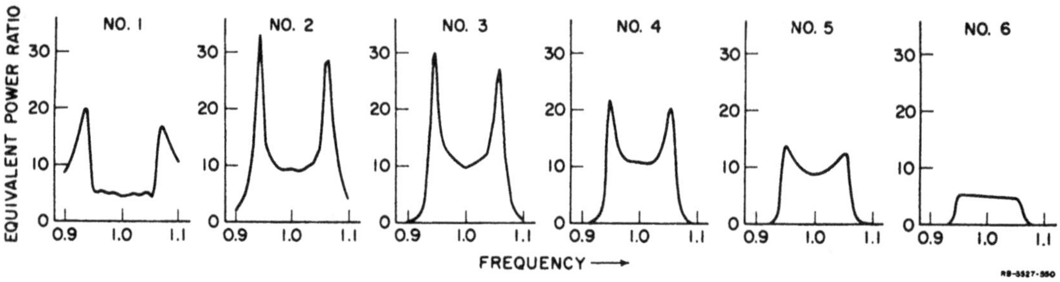

SOURCE: *IRE Trans. PGMTT* (See Ref. 40 by L. Young)

FIG. 15.03-4 EQUIVALENT POWER RATIO IN THE SIX CAVITIES OF THE FILTER DESIGNED IN SECTION 9.04 TO HAVE A 10 PERCENT BANDWIDTH WITH 0.01-db PASS-BAND RIPPLE
The cavities and terminating waveguides all have the same cross section and Cavity 1 is nearest the generator

first been calculated, as from Eq. (15.03-1) or (15.03-4). The equivalent power ratio is proportional to the stored energy, and so is proportional to the group delay, over the pass-band region (Sec. 6.15).

The dissipation loss and the group delay both increase with the *average* increase in stored energy, so that they are closely proportional throughout the pass band. Thus, only the *average* increase in stored energy over the pass-band region may be deduced from the universal curves, Figs. 6.15-1 through 6.15-10. On the other hand, the power-handling capacity is linked with the *greatest* increase in stored energy in any section or resonator. The universal curves indicate in a general way how the equivalent power rises toward the band edges, and the frequency at which the maximum occurs (which is almost the same frequency for each cavity). The cavity nearest the load does not have a maximum near the band edge (Figs. 15.03-3 and 15.03-4), while the greatest increase in equivalent power relative to midband occurs generally in the cavity nearest the generator. It is probably safe to assume that the ratio of the maximum-to-the-midband-equivalent-power-ratio is never more than twice the ratio of the maximum-to-the-midband-group-delay. For example in Fig. 15.03-3 the equivalent power peaks at 2.25 times its value at midband in the second cavity, whereas the group delay rises by a factor of 0.6/0.38 = 1.6 (Fig. 6.15-2) from its midband value to its maximum value. A similar comparison for the filter in Fig. 15.03-4 with the corresponding group delay curves (Fig. III-19, Ref. 43) gives factors of 4.1 and 2.25.

It seems reasonable to assume then that the group delay curves Figs. 6.15-1 to 6.15-10 can be used to predict the equivalent power rise toward the band-edges, provided that a safety factor is used to allow for the greater rise in the cavities nearer the generator; this safety factor is probably always less than 2. (Compare 2.25/1.6 = 1.4 and 4.1/2.25 = 1.8 in the above two examples.)

Connection with External Q of a Single Cavity, Q_e—A symmetrical single-cavity resonator is shown in Fig. 15.03-5. The quantity c^2 is the power coupling coefficient defined by

$$c^2 = \frac{P}{P'} \tag{15.03-6}$$

$$= \frac{4S}{(S+1)^2} \tag{15.03-7}$$

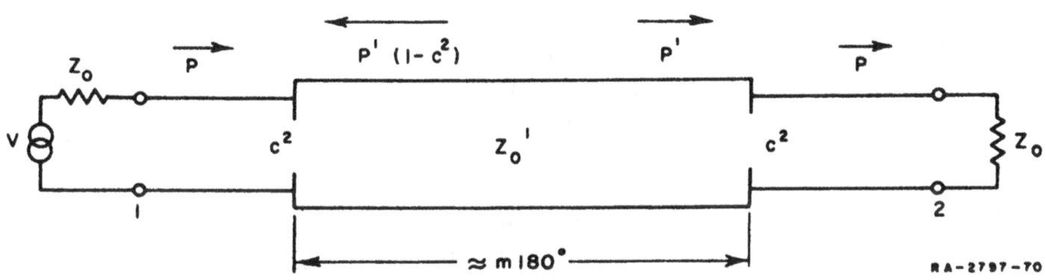

SOURCE: Technical Note 2, Contract AF 30(602)-1998, SRI (See Ref. 38 by E. M. T. Jones)

FIG. 15.03-5 SINGLE-CAVITY FILTER

where S is the internal VSWR, equal to the discontinuity-VSWR $V_1 = V_2$ in this case (Chapter 6). Defining Q as the ratio of 2π-times-the-energy-stored to the energy-dissipated-per-cycle,[41, 42] the external Q due to loading by only *one* coupling aperture can be shown to be

$$Q_e = \pi m \frac{(S+1)^2}{2S}\left(\frac{\lambda_{g0}}{\lambda_0}\right)^2 = \frac{2\pi m}{c^2}\left(\frac{\lambda_{g0}}{\lambda_0}\right)^2 \tag{15.03-8}$$

where λ_{g0} is the midband guide wavelength of the cavity mode and λ_0 is the free-space wavelength. The number m is again the length of the cavity measured in half guide wavelengths; m is an integer for stepped-impedance filters (Chapter 6), and Eq. (15.03-8) is then exact; m is close to an integer for narrow-band, reactance-coupled filters, and Eq. (15.03-8) is then a close approximation.

Note that Q_e is *defined* for a singly loaded cavity. The doubly loaded external Q, Q'_e, of the cavity shown in Fig. 15.03-5 is just half as much:

$$Q'_e = \frac{Q_e}{2} \quad . \tag{15.03-9}$$

Equation (15.03-8) can be solved for S in terms of Q_e, and the resulting expression substituted into Eq. (15.03-4) then yields P'_q/P_q. A particularly simple formula results for narrow-band filters ($S >> 1$, $c << 1$), and then

$$\frac{P'_q}{P_q} \approx \frac{2Q_e}{\pi m}\left(\frac{\lambda_0}{\lambda_{g0}}\right)^2 \quad . \tag{15.03-10}$$

Proof of Eq. (15.03-1)—It can be seen from Eqs. (6.09-2) and (6.14-10), which hold for narrow-band filters that when $g_0 = 1$

$$S_k = \frac{4}{\pi}\frac{\omega'_1}{w_q} g_k \quad . \tag{15.03-11}$$

For narrow-band filters of fractional bandwidth w, whose line-resonators are m half-wavelengths long,

$$w = \frac{w_q}{2m}\left(\frac{\lambda_0}{\lambda_{g0}}\right)^2 \quad . \tag{15.03-12}$$

Substituting Eqs. (15.03-11) and (15.03-12) into Eq. (15.03-4) yields Eq. (15.03-1).

SEC. 15.04, A HIGH-POWER TUNABLE NARROW-BAND TE_{011}-MODE FILTER

This section discusses the design and performance of a narrow-band, high-power filter that uses cylindrical TE_{011} mode resonators.[38] This type of filter can be designed to have a narrow pass band together with a very low midband insertion loss. Furthermore, as can be seen by reference to Fig. 15.03-2, it has a considerably higher midband pulse-power capacity than filters constructed from other types of resonators. Hence, it is ideal for use with high-power transmitters in suppressing spurious emissions that have frequencies close to the center frequency. Its main disadvantage is that it has spurious pass bands at frequencies relatively close to the main pass band; these spurious pass bands occur when the cavities resonate in other modes. However, by using the techniques described below, the insertion loss of these spurious pass bands can be kept quite high over an appreciable frequency band.

Figure 15.04-1 shows the measured insertion loss of the experimental three-cavity TE_{011} mode filter which was constructed from aluminum and is tunable from 1250 to 1350 Mc. The coupling apertures in the filter were adjusted so that the product of the external Q of the end cavities and the coupling coefficient between cavities was approximately equal to unity. Thus the frequency response of the filter is approximately equal to that of a filter designed from an equal-element low-pass prototype. In air at atmospheric pressure, the theoretical midband pulse-power capacity of the filter is limited by voltage breakdown within the cavities and can be determined from Fig. 15.03-2 to be about 2.25 Mw. The pulse-power capacity at the edges of the pass band is reduced to about 1 Mw for the reasons discussed in Sec. 15.03. A dimensioned drawing of the filter is shown in Fig. 15.04-2.

The measured midband attenuation of the filter as a function of tuning frequency is shown in Fig. 15.04-3. These values of attenuation were measured by the substitution method with the filter terminated at either end with pads whose VSWR was about 1.10. Hence it is believed that the mismatch loss would introduce a maximum error of 0.04 db in the measurements. The theoretical unloaded Q for the cavities, assuming a conductivity for the aluminum of 40 percent of that of annealed copper, ranges from 53,000 at 1250 Mc to 47,000 at 1350 Mc. The measured values of Q_u as determined from the midband attenuation and the 3-db bandwidth of the filter using Eqs. (4.13-11) and (4.13-3) are 27,300, 55,700 and 28,400 at

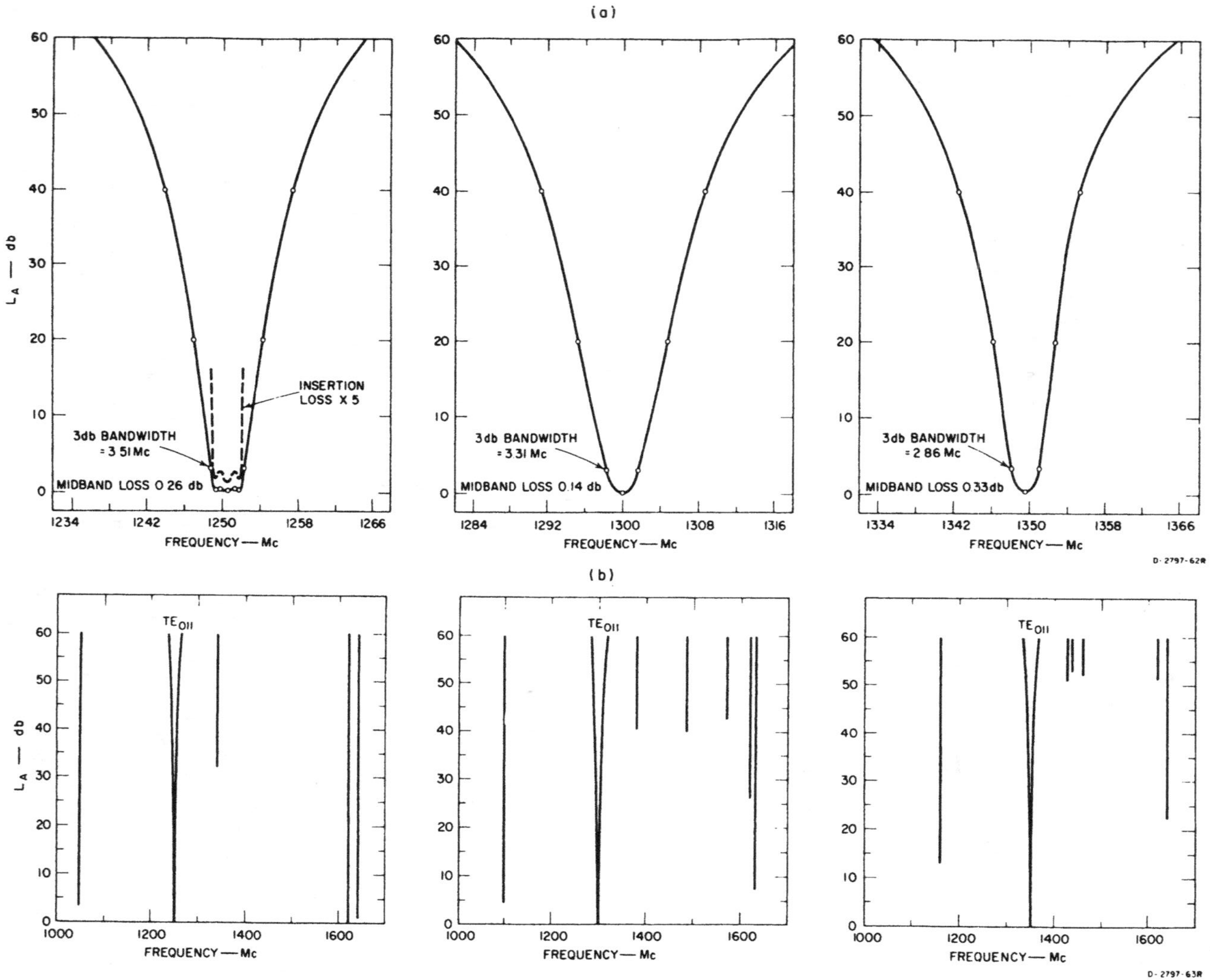

SOURCE: Technical Note 2, Contract AF 30(602)-1998, SRI
(See Ref. 38 by E. M. T. Jones)

FIG. 15.04-1 MEASURED ATTENUATION OF THE EXPERIMENTAL THREE-CAVITY TE_{011} MODE FILTER

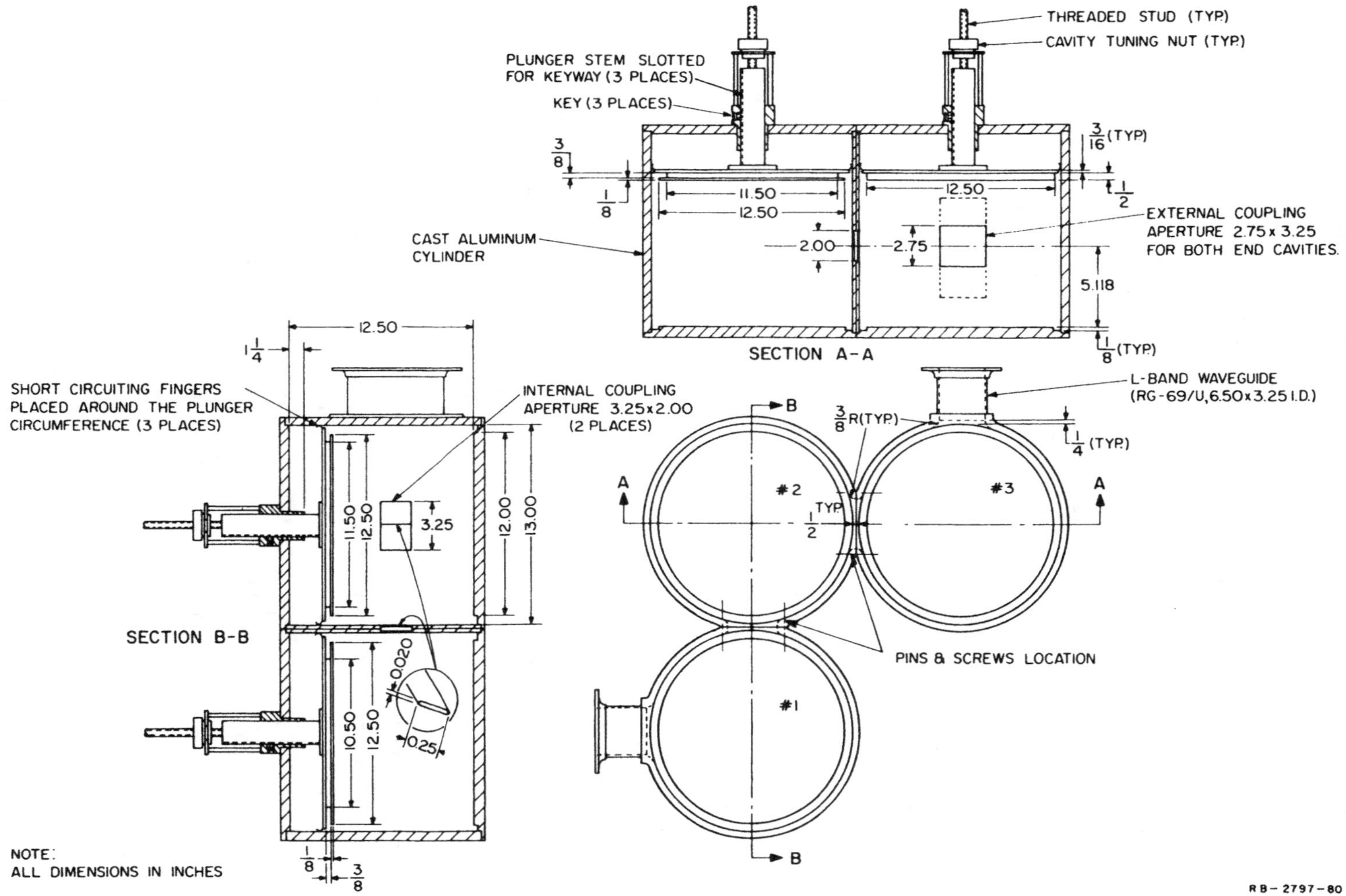

SOURCE: Technical Note 2, Contract AF 30(602)-1998, SRI (See Ref. 38 by E. M. T. Jones)

FIG. 15.04-2 DIMENSIONED DRAWINGS OF THE EXPERIMENTAL THREE-CAVITY TE_{011} MODE FILTER

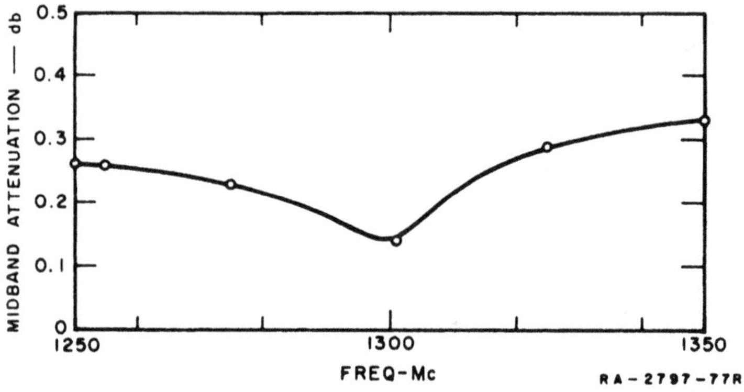

SOURCE: Technical Note 2, Contract AF 30(602)-1998, SRI
(See Ref. 38 by E. M. T. Jones)

FIG. 15.04-3 MEASURED MIDBAND ATTENUATION VERSUS TUNING FREQUENCY

1250, 1300, and 1350 Mc, respectively. The variation in the measured value of Q_u may be due to measurement inaccuracies or to unsuspected losses in the absorbing material mounted behind the tuning plungers as described later on in this section.

Design Technique for TE_{011} *Mode Filters*—The method used in designing this type of filter involves a combination of theoretical and experimental techniques. The low-pass prototype of the filter is shown in Fig. 15.03-1(c); a schematic diagram of the filter is shown in Fig. 15.04-4; and finally, the narrow-band design equations are presented in Table 15.04-1. These design relations were derived using Bethe's small-aperture coupling theory by a procedure similar to that outlined in Secs. 8.07 and 8.14 for the design of narrow-band rectangular-cavity filters. In the present filter, the apertures used are rectangular in shape and have a length $l_{k,k+1}$ measured in a direction parallel to the unperturbed magnetic field at the aperture, a height $h_{k,k+1}$, measured in a direction perpendicular to the unperturbed magnetic field at the aperture, and a thickness $t_{k,k+1}$.

The theoretical static magnetic polarizabilities $(M_1)_{k,k+1}$ of the various apertures, assuming zero thickness and no large-aperture effects, were determined by extrapolating the values for the rectangular aperture given in Fig. 5.10-4(a). The extrapolated value was taken as

$$\frac{(M_1)_{k,k+1}}{(l_{k,k+1})^3} = 0.061 + 0.197\frac{h_{k,k+1}}{l_{k,k+1}} \quad . \qquad (15.04\text{-}1)$$

The estimated actual magnetic polarizability $(M_1')_{k,k+1}$ taking into account both the thickness and frequency corrections were then determined from Eq. (5.10-6), which in the present instance takes the form

$$(M_1')_{k,k+1} = \frac{(M_1)_{k,k+1}}{1 - \left(\frac{2l_{k,k+1}}{\lambda}\right)^2} \times 10^{-\left[\frac{1.346t_{k,k+1}}{l_{k,k+1}} \sqrt{1 - \left(\frac{2l_{k,k+1}}{\lambda}\right)^2}\right]} \tag{15.04-2}$$

Because the apertures are cut in cylindrical walls, they are thicker at the edges than in the center. A thickness $t_{k,k+1}$, which is an average of these two values, is used in Eq. (15.04-2).

The accuracy of the values of magnetic polarizability computed from Eq. (15.04-2) were found to be quite high. This can be seen from an inspection of Fig. 15.04-5 which shows the measured external Q of the cavities compared with the theoretical values computed by solving Eq. (5) in Table 15.04-1 for $(Q_e)_A$. It is also seen from an inspection of Fig. 15.04-6 which shows the measured coupling coefficient k between the

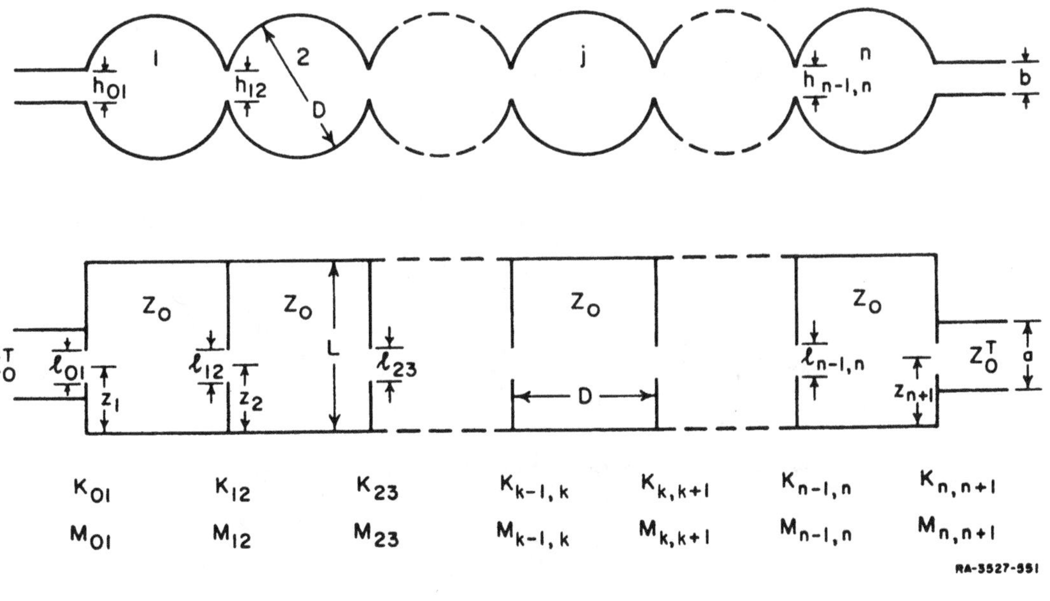

FIG. 15.04-4 A DIRECT-COUPLED TE_{011} MODE FILTER HAVING n RESONATORS

Table 15.04-1

TE$_{011}$-MODE DIRECT-COUPLED CAVITY FILTERS AND THEIR RELATION TO LOW-PASS PROTOTYPES

The prototype parameters $g_0, g_1, \ldots, g_{n+1}$ are as defined in Fig. 15.03-1(c) (or Sec. 4.04), while ω_1', ω_0, ω_1, and ω_2 are as defined in Fig. 15.03-1(b). Figure 15.04-4 defines the filter dimensions.

$$(Q_e)_A = \frac{g_0 g_1 \omega_1'}{w} = \text{external } Q \tag{1}$$

$$k_{k,k+1}\Big|_{k=1 \text{ to } n} = \frac{w}{\omega_1' \sqrt{g_k g_{k+1}}} = \text{coupling coefficient} \tag{2}$$

$$(Q_e)_B = \frac{g_n g_{n+1} \omega_1'}{w} = \text{external } Q \tag{3}$$

where

$$w = \frac{\omega_2 - \omega_1}{\omega_0} \quad \text{and} \quad \omega_0 = \frac{\omega_2 + \omega_1}{2} \tag{4}$$

$$(M_1')^2_{01} = \frac{L\lambda^T_{g0}\, abD^4}{(Q_e)_A\, 47.5\lambda_0^2 \sin^2\left(\frac{\pi z_1}{L}\right)} \tag{5}$$

$$(M_1')_{k,k+1} = \frac{LD^4 k_{k,k+1}}{3.78\lambda_0^2 \sin^2\left(\frac{\pi z_k}{L}\right)} \tag{6}$$

$$(M_1')^2_{n,n+1} = \frac{L\lambda^T_{g0}\, abD^4}{(Q_e)_B 47.5\lambda_0^2 \sin^2\left(\frac{\pi z_{n+1}}{L}\right)} \tag{7}$$

or alternatively, in terms of normalized impedance inverter parameters

$$\frac{K_{01}}{Z_0^T} = \sqrt{\frac{w\pi Z_0 \lambda^2_{g0}}{2\omega_1' Z_0^T \lambda_0^2 g_1 g_0}} \tag{8}$$

continued

Table 15.04-1 *concluded*

$$\frac{K_{k,k+1}}{Z_0} = \frac{w\pi\lambda_{g0}^2}{2\omega_1'\lambda_0^2\sqrt{g_k g_{k+1}}} \tag{9}$$

$$\frac{K_{n,n+1}}{Z_0^T} = \sqrt{\frac{w\pi Z_0\lambda_{g0}^2}{2\omega_1' Z_0^T\lambda_0^2 g_n g_{n+1}}} \tag{10}$$

where

$$\frac{Z_0}{Z_0^T} = \frac{0.528D^4}{L\lambda_{g0}^T ab \sin^2\left(\frac{\pi z_1}{L}\right)} \tag{11}$$

$$(M_1')_{01} = \left(\frac{K_{01}}{Z_0^T}\right)\frac{\lambda_{g0}^T ab}{4\pi} \tag{12}$$

$$(M_1')_{k,k+1} = \left(\frac{K_{k,k+1}}{Z_0}\right)\frac{D^4}{23.7L \sin^2\left(\frac{\pi z_k}{L}\right)} \tag{13}$$

$$(M_1')_{n,n+1} = \left(\frac{K_{n,n+1}}{Z_0^T}\right)\frac{\lambda_{g0}^T ab}{4\pi} \tag{14}$$

where a, b, D and L are defined in Fig. 15.04-4, and

$$\lambda_{g0} = \frac{\lambda_0}{\sqrt{1 - \left(\frac{\lambda_0}{0.821D}\right)^2}} = \left[\begin{array}{l}\text{guide wavelength in a}\\ \text{cylindrical waveguide of}\\ \text{diameter } D \text{ operating in the}\\ \text{TE}_{01} \text{ mode at frequency } \omega_0\end{array}\right] \tag{15}$$

$$\lambda_{g0}^T = \frac{\lambda_0}{\sqrt{1 - \left(\frac{\lambda_0}{2a}\right)^2}} = \left[\begin{array}{l}\text{guide wavelength in a rec-}\\ \text{tangular terminating waveguide}\\ \text{of width } a \text{ operating in the}\\ \text{TE}_{10} \text{ mode at frequency } \omega_0\end{array}\right] \tag{16}$$

λ_0 = free space wavelength at frequency ω_0

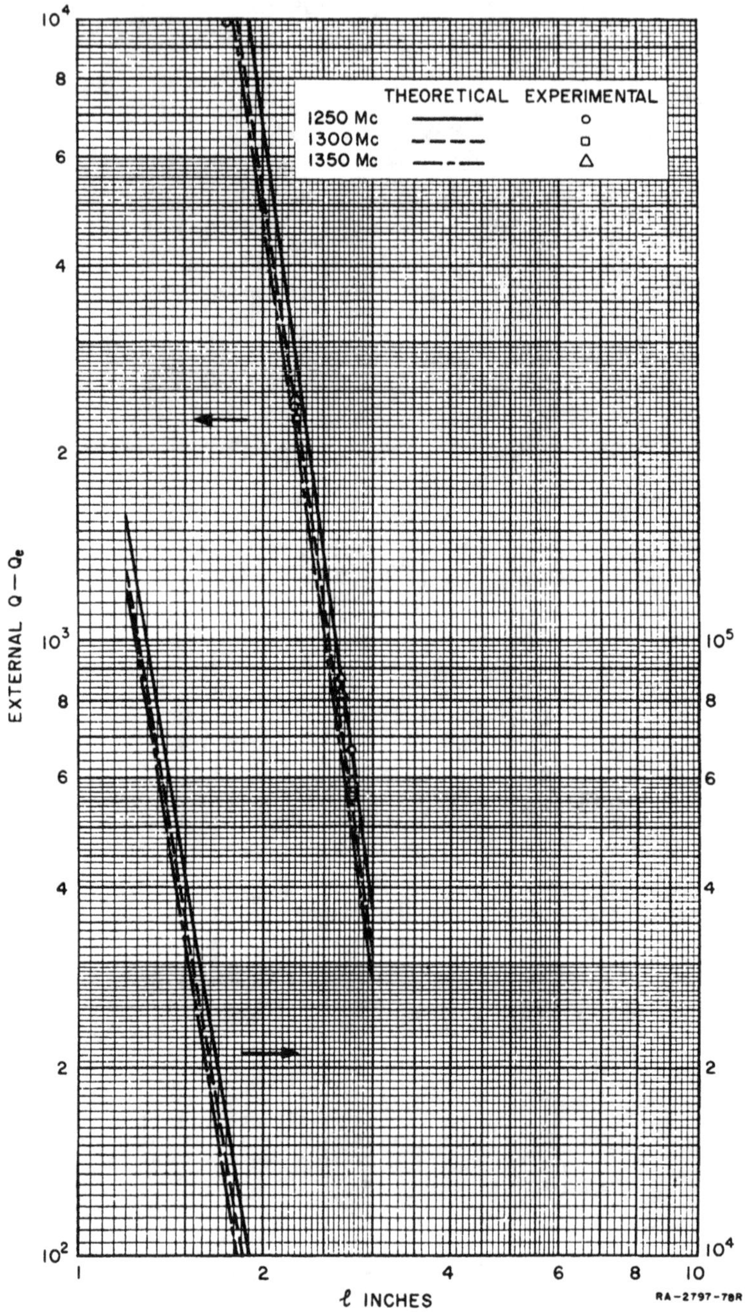

SOURCE: Technical Note 2, Contract AF 30(602)-1998 SRI
(See Ref. 38 by E. M. T. Jones)

FIG. 15.04-5 THEORETICAL AND EXPERIMENTAL VALUES OF EXTERNAL Q, Q_e

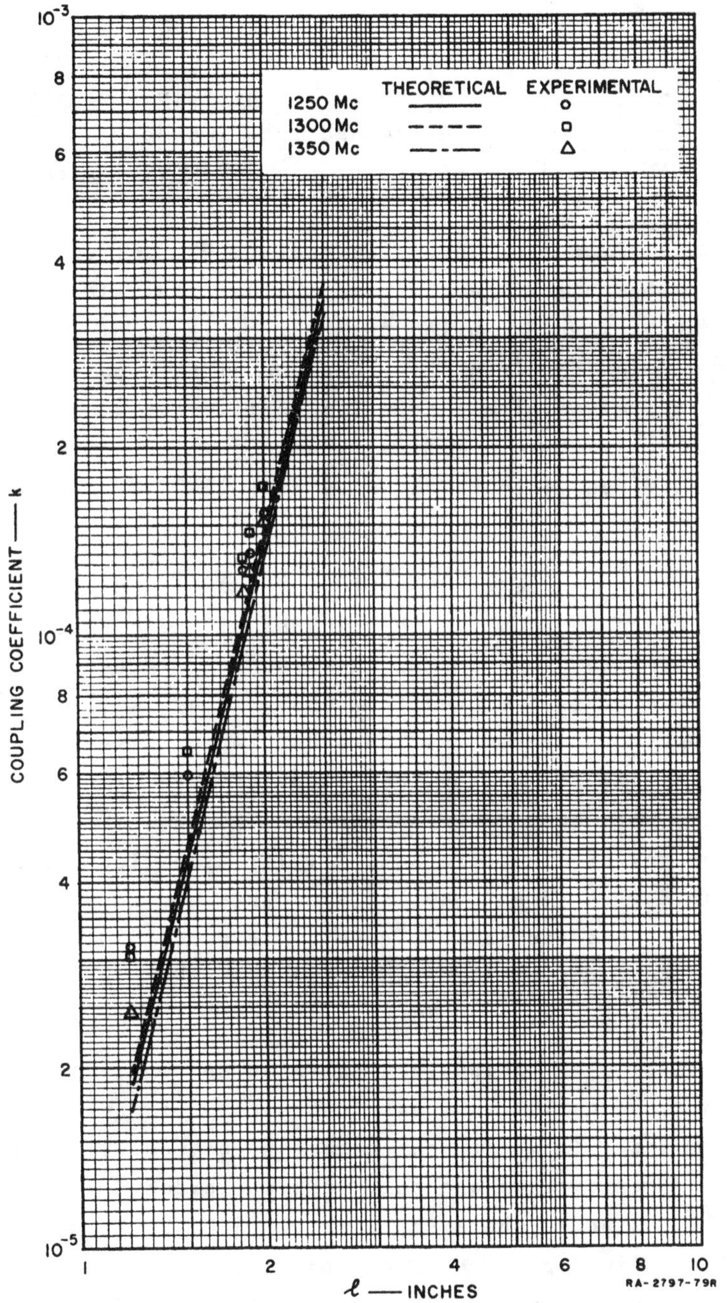

SOURCE: Technical Note 2, Contract AF 30(602)-1998 SRI (See Ref. 38 by E. M. T. Jones)

FIG. 15.04-6 THEORETICAL AND EXPERIMENTAL VALUES OF THE COUPLING COEFFICIENT k

Table 15.04-2

MEASURED VALUES OF k AND Q_e

f(Mc)	Q_e	k	kQ_e
1250	677	0.00137	0.93
1300	562	0.00176	0.99
1350	594	0.00180	1.07

cavities compared with the theoretical value computed by solving Eq. (6) in Table 15.04-1 for $k_{k,k+1}$. The values of external Q were determined from impedance measurements made on one of the end cavities operating as a singly-loaded resonator using the prescriptions in Sec. 11.02. After the apertures in the end cavities were adjusted to give the desired value of external Q the two end cavities were coupled together by means of a small iris. Measurements made on the frequency response of this pair of resonators then yielded the values of the coupling coefficient k using the prescriptions in Sec. 11.04. The final measured values of Q_e and k for the apertures whose dimensions are shown in Fig. 15.04-2 are listed in Table 15.04-2.

Techniques for Suppressing Spurious Modes—The diameter of the cavities was chosen so that, as the lengths of the cavities were varied to tune them from 1250 to 1350 Mc, the resonant frequencies of the other cavity modes were as far removed from that of the TE_{011} mode as possible. The inside diameter selected was 13 inches, which fixes the length of the unperturbed cavities at 10.23 inches when they are resonant at 1250 Mc, and 7.64 inches when they are resonant at 1350 Mc. The actual measured lengths of the cavities were found to differ significantly from these values, indicating that the coupling apertures had an appreciable perturbing effect on the resonant frequency. Figure 15.04-7 shows a mode chart for an unperturbed cylindrical resonator showing all the possible modes that can exist from 1050 Mc to 1725 Mc.

The terminating waveguides are oriented to couple strongly to those TE modes within the cavity that have components of magnetic field parallel to the cavity axes. TM modes, which have no components of magnetic field parallel to the cavity axes are only weakly excited from the terminating waveguides. It is noted, however, in Fig. 15.04-2 that the length l of the internal coupling apertures along the circumference of the cavity is greater than the aperture height parallel to the cavity axes. Therefore, TM modes inadvertently excited in the end cavities are very strongly coupled together through the internal coupling apertures. To reduce the internal coupling of the TM modes, and thereby reduce the spurious transmissions through the filter via these modes, a metal bar 0.25 inch wide and 0.020 inch thick with rounded edges is placed across each internal coupling aperture parallel to the cavity axes as shown. This bar has negligible effect on the coupling of the apertures to TE modes.

Several techniques are used to minimize coupling through the filter at frequencies other than the design frequency via the many TE modes which may be coupled by the apertures. One technique consists of positioning the coupling apertures so that they lie midway between the top

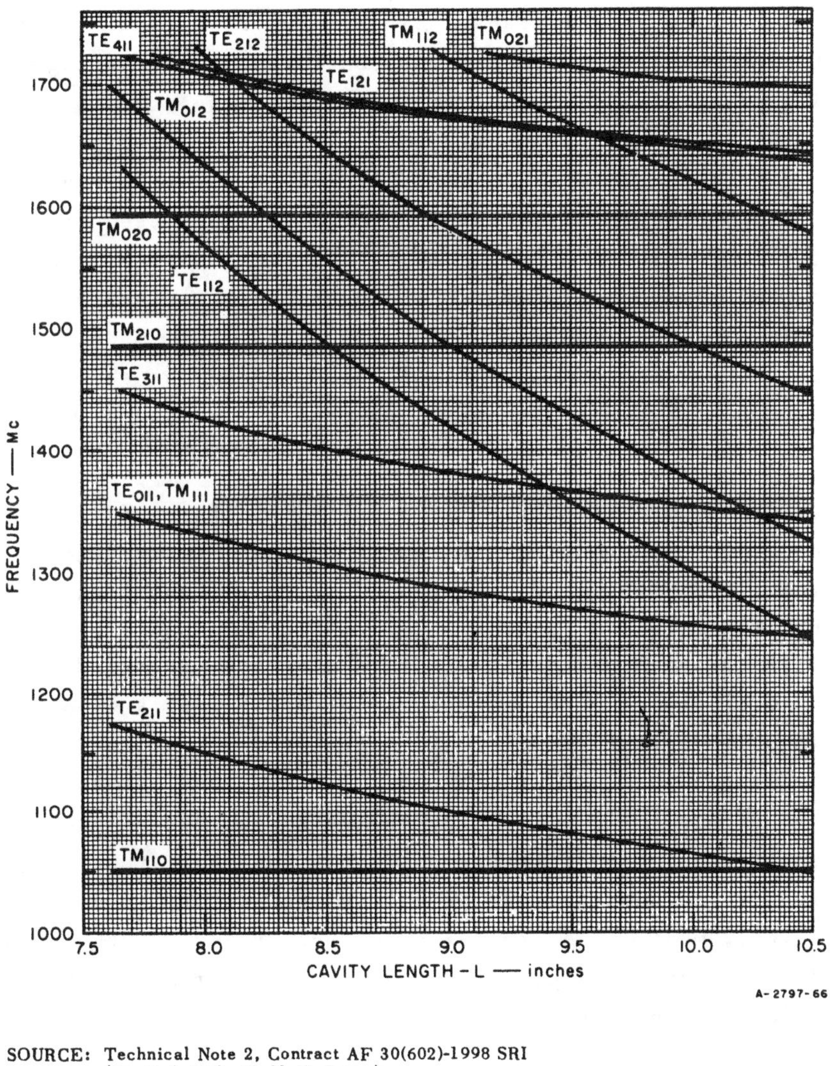

SOURCE: Technical Note 2, Contract AF 30(602)-1998 SRI
(See Ref. 38 by E. M. T. Jones)

FIG. 15.04-7 MODE CHART FOR UNPERTURBED CYLINDRICAL RESONATOR

and bottom of the cavities when they are tuned to 1250 Mc. This procedure minimizes the coupling to the TE_{211} mode from the external waveguides, and also minimizes the coupling of this mode between cavities. In order to reduce the coupling of the TE_{311} mode between cavities, the coupling slots in each cavity are oriented at right angles to one another, resulting in the positioning of the cavities shown in Fig. 15.04-2.

In addition, radial transmission lines of different lengths are formed on the tuning plungers of each of the cavities as shown in Fig. 15.04-2. The radial transmission lines are formed by attaching undercut aluminum disks to the brass,contacting tuning plungers. In the two tuning plungers having the deepest chokes, two disks of resistive paper 0.0035 inch thick and having a resistance of 2000 ohms per square are placed between the brass,contacting tuning plungers and the undercut aluminum disks. In these cavities nylon screws are used to hold the two parts of the tuning plungers together. The resistive sheets lowered the unloaded Q of the undesired modes in these two cavities, without appreciably affecting the unloaded Q of the desired TE_{011} mode. The resistive paper was not placed in the tuning plunger having the shallowest radial line because in that case the paper lowered the unloaded Q of the TE_{011} mode also.

These radial transmission lines shift by different amounts the resonant frequencies of the undesired modes in each cavity, and, as explained above, lower the unloaded Q of unwanted modes in two of the cavities. Hence, the transmission via these higher-order modes is greatly reduced. The shift in resonant frequency in an end cavity, produced by the two deepest radial line chokes on some of the modes that are near in frequency to the TE_{011} mode is illustrated in Fig. 15.04-8. It is seen that these chokes do not cause the tuning curves of any of the unwanted modes to cross that of the TE_{011} mode. The chokes in the tuning plungers also have the advantage that they slightly shift the resonant frequency of the TM_{111} mode which, in an unperturbed cylindrical resonator, is always degenerate with the TE_{011} mode. The grooves at the outer edge of the bottom plates of the cavities have a depth of 0.125 inch and a width of 0.500 inch and are also for the purpose of shifting the frequency of the TM_{111} mode.

The usual technique for suppressing the effects of spurious modes on echo boxes or frequency meters operating in the TE_{01m} mode is to place absorbing material in the cavities in such a position that it couples strongly to all modes except the TE_{01m} modes[44] and therefore greatly reduces their unloaded Q, Q_u. Hence, such a dissipatively loaded cavity when used as a transmission device has a very low transmission for all modes except the TE_{01m} modes, or when used as a reaction device produces only a negligible reaction in the feeding line except at the resonant frequency of the TE_{01m} mode. One place where the lossy material is often put is behind the tuning plunger which is made non-contacting. The walls of the cavity can also be made in the form of a helix with lossy material between the turns. This lossy material, particularly when placed behind

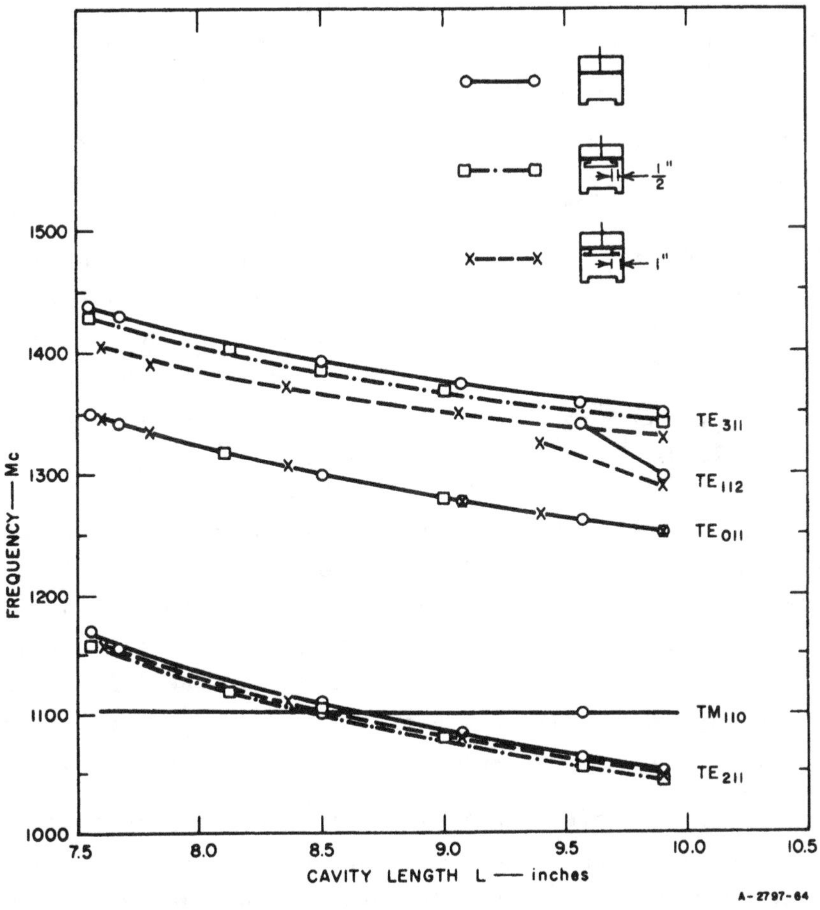

SOURCE: Technical Note 2, Contract AF 30(602)-1998 SRI (See Ref. 38 by E. M. T. Jones)

FIG. 15.04-8 MODE CHART FOR THE END CAVITY ILLUSTRATING THE SHIFT IN MODE RESONANT FREQUENCY CAUSED BY COUPLING APERTURES AND RADIAL-LINE CHOKES

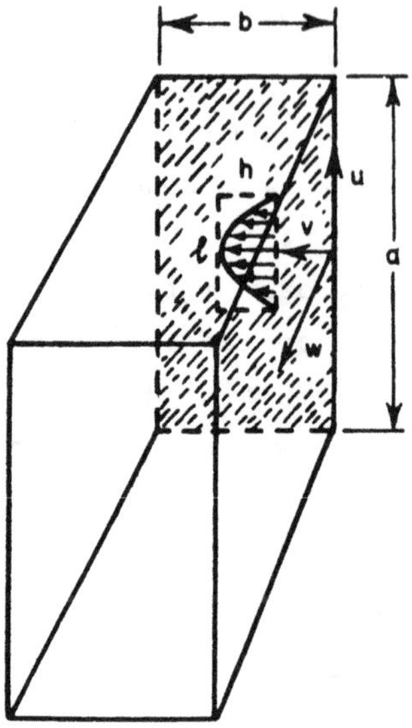

SOURCE: Technical Note 2, Contract AF 30(602)-1998 SRI (See Ref. 38 by E. M. T. Jones)

FIG. 15.04-9 RECTANGULAR WAVEGUIDE EXCITED BY A RECTANGULAR APERTURE

the tuning plunger, drastically shifts the resonant frequencies of all modes except the TE_{01m} modes so that their tuning curves are vastly different than they would be in an unperturbed cylindrical cavity as shown in Fig. 15.04-7, and probably cross the tuning curve of the TE_{01m} mode. The amount of power absorbed by these higher-order crossing modes depends on the ratio of Q_e to Q_u for the particular crossing mode. It is maximum when Q_e is equal to Q_u and minimum when Q_e is vastly different from Q_u.

In the case of weakly coupled TE_{01m} cavities, such as echo boxes and frequency meters, the ratio of Q_e and Q_u is always very large for the higher order crossing modes, and therefore they do not absorb an appreciable amount of power. The situation is entirely different in low-insertion-loss filters operating in the TE_{01m} mode because the ratio of Q_e to Q_u for this mode is quite small, and hence the ratio of Q_e to Q_u of higher-order crossing modes is lower and may be quite close to unity. In this latter unfortunate case a great deal of power is absorbed by the crossing modes and the effective unloaded Q of the desired TE_{01m} mode is lowered, resulting in a large insertion loss for the filter. This effect was observed experimentally in the present filter and is the reason that a minimum of dissipative elements were used to suppress spurious modes.

Approximate Peak Electric Field in the Coupling Apertures—The peak electric field in the apertures coupling the cavities to each other and to the external waveguides can be calculated approximately in terms of an assumed sinusoidal distribution of electric field in the apertures. In the discussion presented here, infinitesimally thin rectangular apertures will be considered first. Figure 15.04-9 illustrates a rectangular waveguide operating in the TE_{10} mode excited by a rectangular aperture having an assumed sinusoidal electric field distribution (in the plane of the aperture) along its length, given by

$$E_1 \cos \frac{\pi u}{l} \qquad (15.04\text{-}3)$$

where E_1 is the peak electric field in the plane of the aperture. The exact expression for the amplitude E, the peak electric field excited in the rectangular guide, is

$$E = \frac{4lh \cos \frac{\pi l}{2a}}{\pi ab\left(1 - \frac{l^2}{a^2}\right)} E_1 \quad . \qquad (15.04\text{-}4)$$

The peak electric field in the plane of such an aperture can also be calculated approximately from another formula which has more utility in the calculation of electric fields in the internal apertures. The expression in mks units is

$$E_1 = \frac{\pi\omega_0\mu_0 M_1' \Delta H}{2hl} \quad \text{volts/meter} \qquad (15.04\text{-}5)$$

where, ω_0 is the angular frequency, $\mu_0 = 1.257 \times 10^{-6}$ henries/meter, M_1' is the magnetic polarizability of the aperture and ΔH is the difference between the magnetic fields on either side of the aperture. The magnetic field H_0 on the cavity side of an external aperture is much greater than the magnitude of the H-field of the external waveguide. Therefore, for the external apertures, $H_0 \approx \Delta H$.

The peak electric field $E_{k,k+1}$ in the plane of an internal window of polarizability $M'_{k,k+1}$ connecting Cavities k and $k+1$, is given by

$$E_{k,k+1} = \frac{\pi\omega_0\mu_0 M'_{k,k+1} \Delta H_{k,k+1}}{2l_{k,k+1}h_{k,k+1}} \qquad (15.04\text{-}6)$$

where $l_{k,k+1}$ and $h_{k,k+1}$ are the length and height of the rectangular coupling aperture between Cavity k and Cavity $k+1$. It is tedious and difficult to evaluate $\Delta H_{k,k+1}$ exactly, since its value depends on both the magnitude and phase of the magnetic field on either side of the aperture. A conservative design estimate is to assume that $\Delta H_{k,k+1}$ is equal to the difference between the magnetic field in the two cavities when they are in phase opposition.

In order that the fringing fields perpendicular to the plane of the aperture and at its edges do not become excessive, it is necessary to use thick coupling apertures, and to round the edges of the apertures. As shown in Fig. 15.04-2, it is the aperture edges parallel to the u axis (in Fig. 15.04-9) that must be rounded. The aperture edges parallel to the v axis need not be rounded. The magnetic polarizability of the apertures is quite insensitive to the rounding of the corners.

For an infinitesimally thin aperture with dimensions of l = 2.75 inches and h = 3.25 inches, exciting a rectangular waveguide with dimensions of a = 6.5 inches and b = 3.25 inches, Eq. (15.04-4) predicts a peak electric field E_1 in the plane of the aperture equal to 1.94 E. The actual thick aperture in the experimental filter had the same l and h dimensions and, in addition, the inner edge of each aperture was rounded with a 0.325-inch radius along the l dimensions. Reference to Fig. 15.02-5 shows that the maximum electric field E_1' near the rounded corners is 1.85 E_1. (Note that E_1 is the peak field in the *plane* of the aperture, while E_1' is the peak field regardless of direction.) Therefore, the maximum electric field E_1' near the external coupling apertures when exciting a wave of amplitude E in the external guide is E_1' = 3.59 E. The pulse-power capacity of the external coupling aperture is thus only 0.0776 times that of the external guide. Using a relation taken from Table 15.03-1 (that the pulse-power capacity P_m' of the external guide is 76.4 λ/λ_g megawatts for air at standard temperature and pressure) it is found that the peak pulse-power capacity of the external coupling apertures at midband, where the filter is matched, is 4.06, 4.24, and 4.37 megawatts at frequencies of 1250, 1300, and 1350 Mc, respectively, which is greater than that of the resonators.

The ratio of the peak electric field in infinitesimally thin internal coupling apertures (*i.e.*, apertures between resonators) to that in infinitesimally thin external coupling apertures (*i.e.*, apertures between end resonators and the terminating guides) can be calculated from Eq. (15.04-6). Since the edges of the actual apertures in the filter were all rounded by the same amount, this ratio should also apply approximately for the actual apertures. The result is, using the final values of $l = 2.75$ and $h = 3.25$ for the external apertures and $l = 2.00$ and $h = 3.25$ for the internal apertures,

$$\frac{E'_{int}}{E'_{ext}} = \frac{M_{int}}{M_{ext}} \frac{(\Delta H)_{int}}{(\Delta H)_{ext}} 1.375 \quad . \qquad (15.04\text{-}7)$$

In this expression E'_{int} and E'_{ext} are the peak electric fields at the internal and external apertures, while $(\Delta H)_{int}$ and $(\Delta H)_{ext}$ are the differences between the H-fields on opposite sides of the internal apertures and on opposite sides of the external apertures. To an excellent approximation, $(\Delta H)_{ext} = H_{ext}$ where H_{ext} is the peak H-field in the center of the external apertures. In Eq. (15.04-7) M_{ext} is the polarizability of the external aperture which has calculated values of 4.60, 4.67, and 5.17 at frequencies of 1250, 1300, and 1350 Mc, respectively. The polarizability M_{int} of the internal apertures has calculated values of 1.275, 1.288, and 1.347 at frequencies of 1250, 1300, and 1350 Mc, respectively. Making the conservative assumption that $(\Delta H)_{int}$ is equal to $2H_{ext}$, one finds that the pulse-power capacity of the internal apertures is 7.0, 7.35, and 8.53 megawatts at 1250, 1300, and 1350 Mc, respectively. Reference to Fig. 15.03-3 shows that the fields in the second and third cavity rise to about 1.4 times their midband values, for ω' of 1.4. Therefore at this frequency $(\Delta H)_{int}$ increases to about 1.4 times its assumed midband value. Consequently the pulse-power capacities of the internal apertures would be reduced to about one half their midband value, which still is higher than the pulse-power capacity of the filter cavities themselves.

SEC. 15.05, HIGH-POWER WAFFLE-IRON FILTERS[43,45,58,59]

This section describes the design and measured performance of two L-band model waffle-iron filters (Sec. 7.05). In the one developed first,[45] the principal requirement was for a very wide stop band, to stop all harmonics from the second to the tenth, inclusive. This structure consisted of three waffle-iron filters in cascade with overlapping stop bands. In the second model[43] the emphasis was on power handling and pass-band width. This was achieved by changing from square to circular teeth, changing the end sections, and paralleling four identical filters. This increased the power handling capacity more than five times. There is no reason why the best features of both models could not be combined to produce a filter with a wide stop band (second to tenth harmonic), wide pass band (VSWR $<$ 1.2 over almost all of L-band), and a pulse-power capacity of approximately 8 megawatts in air at atmospheric pressure, but such a model has not yet been built. The two separate models will now be described.

Waffle-Iron Filter with Stop Band up to Tenth Harmonic[45,59]—In this structure three different waffle-iron filters with overlapping stop bands are connected in series to give a combined stop band which extends from 2.2 Gc to 13.7 Gc where the attenuation is 60 db or greater for all possible propagating modes. This stop band includes the second through tenth harmonic of frequencies in the pass band of 1.25-1.35 Gc. In the pass band, the waffle-iron filter is matched to a WR650 waveguide using quarter-wavelength stepped transformers (Chapter 6). The pulse-power capacity without breakdown was measured in the presence of a cobalt-60 radioactive source to be 1.4 megawatts in air at atmospheric pressure using pulses 2 microseconds long and a repetition rate of 60 pulses per second.

Figure 15.05-1 is a photograph of the filter in the disassembled condition showing the quarter-wavelength transformers used to connect the various waffle-iron sections together. The transformers used to connect the waffle-iron filter to the WR650 guide are not shown.

The theoretical staggered image stop bands of the filters are shown in Fig. 15.05-2. The design of Waffle-Iron A has been previously described as the first example in Sec. 7.05 under the heading: "*Design Using Cohn's Corrugated Filter Data.*" It was made ten sections long in order to achieve a theoretical stop-band attenuation of about 80 db.

Waffle-Iron B and Waffle-Iron C were designed using the technique described in Sec. 7.05 under the heading: "*Design Using the T-Junction Equivalent Circuit of Marcuvitz.*" Waffle-Iron B was made seven sections long in order to achieve a theoretical stop-band attenuation of more than 60 db above 5.75 Gc while Waffle-Iron C was made nine sections long in order to achieve a theoretical stop-band attenuation of more than 60 db above 9.25 Gc. Figure 15.05-3 shows a typical high-power waffle-iron filter section illustrating the notation used to specify the dimensions, while Table 15.05-1 shows the dimensions of each of the three waffle-iron filters.

SOURCE: Technical Note 2, Contract AF 30(602)-2392, SRI
(See Ref. 45 by E. D. Sharp)

FIG. 15.05-1 DISASSEMBLED WAFFLE-IRON FILTER AND CONNECTING QUARTER-WAVE TRANSFORMER SECTIONS

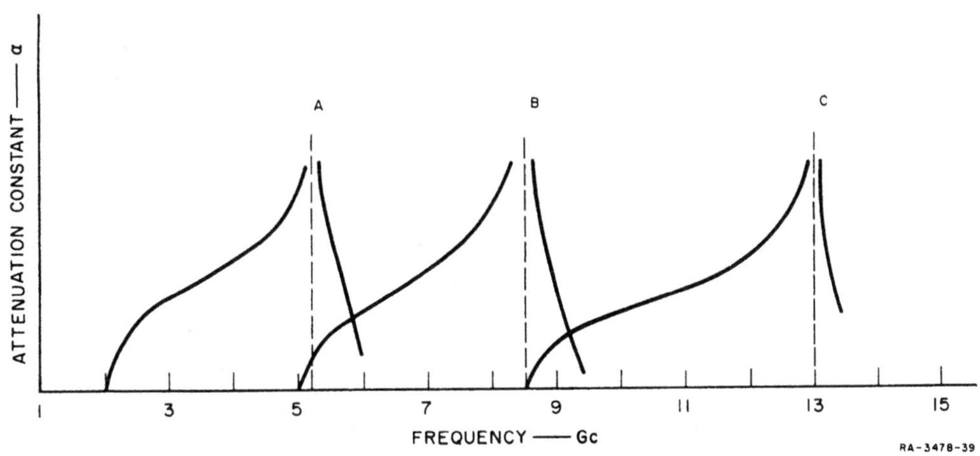

SOURCE: Technical Note 2, Contract AF 30(602)-2392, SRI
(See Ref. 45 by E. D. Sharp)

FIG. 15.05-2 ATTENTUATION CONSTANTS OF WAFFLE-IRON FILTERS

Table 15.05-1

DIMENSIONS OF WAFFLE-IRON FILTERS

WAFFLE IRON FILTER	DIMENSIONS (Inches)					NUMBER OF SECTIONS		LENGTH (Inches)	RADIUS OF ROUNDED CORNERS (Inches)
	b'	b''	b	l	l'	In Width	In Length		
A	0.280	0.210	1.61	0.510	0.790	5	10	13.0	0.063
B	0.280	0.210	0.986	0.231	0.360	11	7	4.137	0.063
C	0.280	0.210	0.700	0.150	0.256	16	9	3.654	0.045

SOURCE: Technical Note 2, Contract AF 30(602)-2392, SRI (See Ref. 45 by E. D. Sharp)

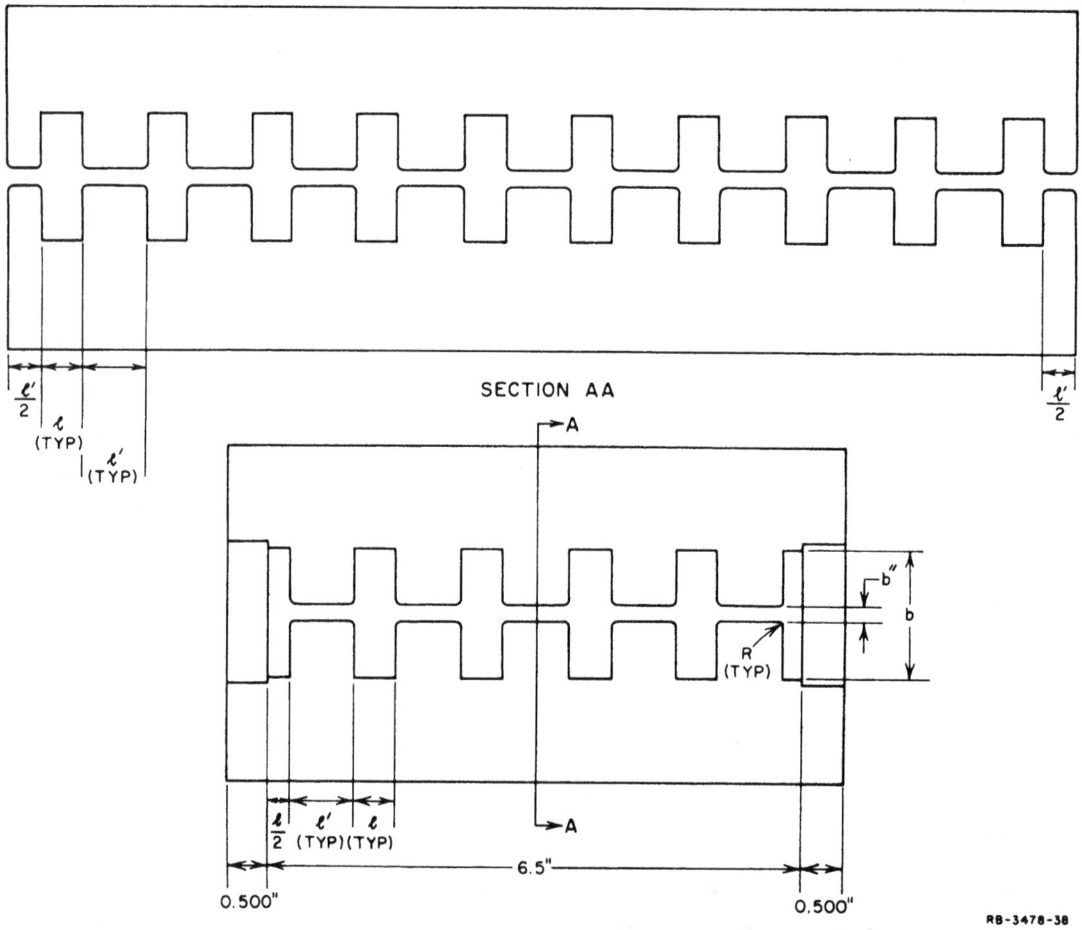

SOURCE: Technical Note 2, Contract AF 30(602)-2392, SRI (See Ref. 45 by E. D. Sharp)

FIG. 15.05-3 TYPICAL L-BAND WAFFLE-IRON FILTER CROSS-SECTIONS

The corners of the bosses in each of the filters were rounded to reduce the electric field at these positions. The radius of curvature used in each case is shown in the table. For Waffle-Iron A and Waffle-Iron B which have $b'' < l$, it is appropriate to use Fig. 15.02-5 to determine the field strength at the rounded corners. Reference to this figure shows that the maximum electric field at the rounded corners is only 1.4 times its value at the center of the tooth. For Waffle-Iron C, $l < b''$ and it is more appropriate to use Fig. 15.02-7 to determine the field strength. Reference to this figure shows that the maximum electric field at the corners is 1.58 times its value at the center of the boss.

The height of the waveguides necessary to present an image match to each waffle-iron section was computed using the procedures outlined in Sec. 7.05. In addition, measurements of the image impedance of each filter were made using Dawir's method, which is described in Sec. 3.09. The results of these measurements are presented in Fig. 15.05-4 in terms of the height of the terminating guide necessary to achieve an image match. The height of the quarter-wave transformers necessary to match the waffle-iron sections to each other and to the terminating waveguides over a 1250- to 1350-Mc band were determined from these data and the quarter-wave-transformer tables of Sec. 6.04. The height of the single-section quarter-wave transformer between Waffle-Iron B and Waffle-Iron C is 0.341 inch. The dimensions of the two-section quarter-wave transformer between Waffle-Iron A and Waffle-Iron B are shown in Fig. 15.05-5, while the dimensions of the transformers at either end of the filter are shown in Fig. 15.05-6.

The measured VSWR of the assembly is shown in Fig. 15.05-7, while the stop-band insertion loss is shown in Fig. 15.05-8.

The performance of a single section S-band waffle-iron operating in both the evacuated and pressurized condition has also been described by Guthart and Jones.[46]

Waffle-Iron Filter with Increased Pass-Band Width and Increased Power-Handling Capacity[43]—The principle of the waffle-iron filter with wide pass-band width (and wide stop-band width) was explained at the end of Sec. 7.05. A photograph of this filter was shown in Fig. 7.05-10, and its dimensions were given in Sec. 7.05. The principle behind the increased power-handling capacity of the round (as opposed to square) teeth will now be explained.

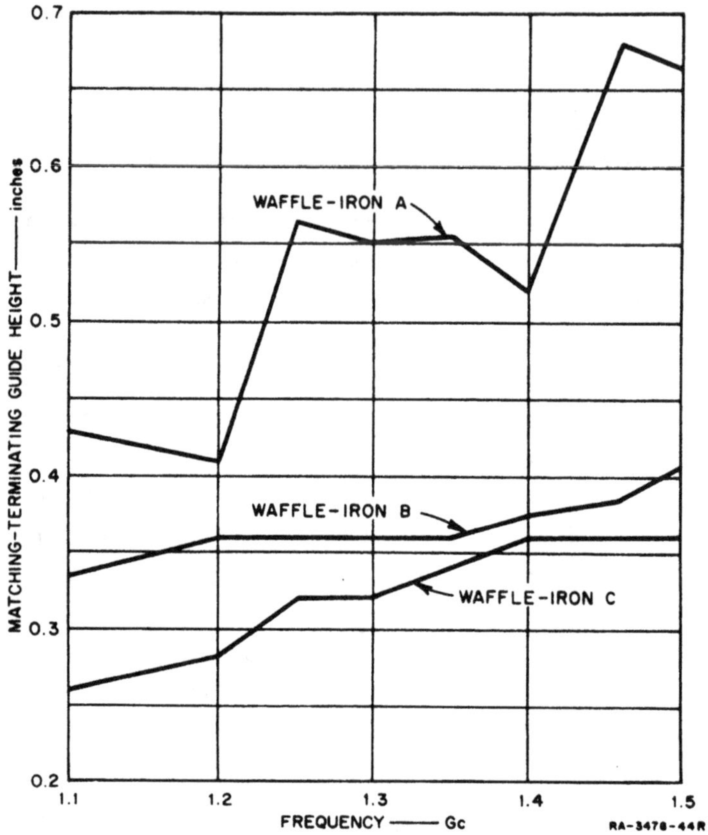

SOURCE: Technical Note 2, Contract AF 30(602)-2392, SRI (See Ref. 45 by E. D. Sharp)

FIG. 15.05-4 MEASURED WAFFLE-IRON IMAGE IMPEDANCE

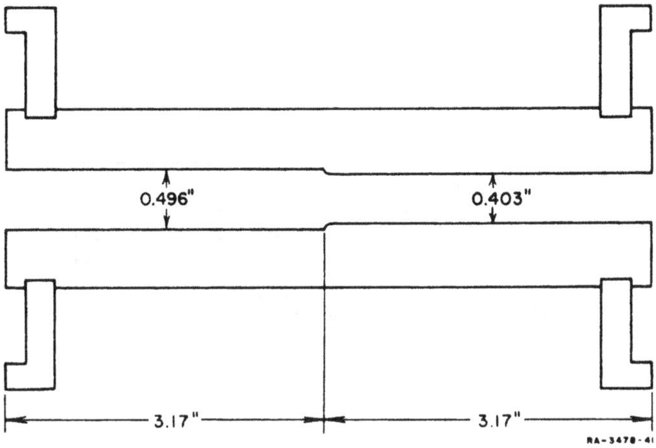

SOURCE: Technical Note 2, Contract AF 30(602)-2392, SRI
(See Ref. 45 by E. D. Sharp)

FIG. 15.05-5 CROSS-SECTION OF TRANSFORMER BETWEEN WAFFLE IRON FILTERS A AND B

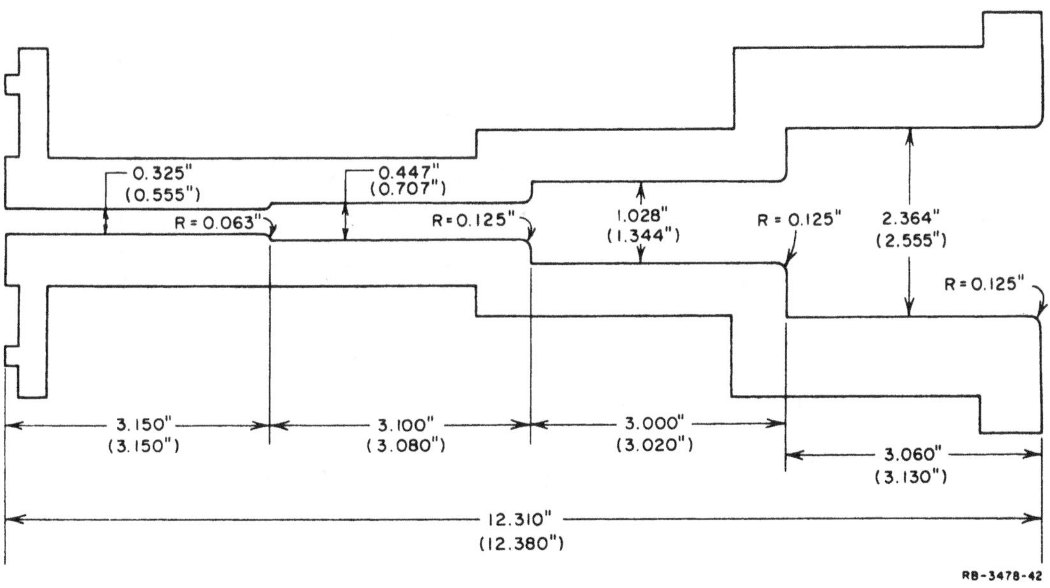

SOURCE: Technical Note 2, Contract AF 30(602)-2392, SRI
(See Ref. 45 by E. D. Sharp)

FIG. 15.05-6 CROSS-SECTIONS OF INPUT AND OUTPUT TRANSFORMERS (DIMENSIONS WITH AND WITHOUT PARENTHESES ARE FOR TRANSFORMERS THAT MATCH WAFFLE IRONS A AND C RESPECTIVELY, TO L-BAND WAVEGUIDE)

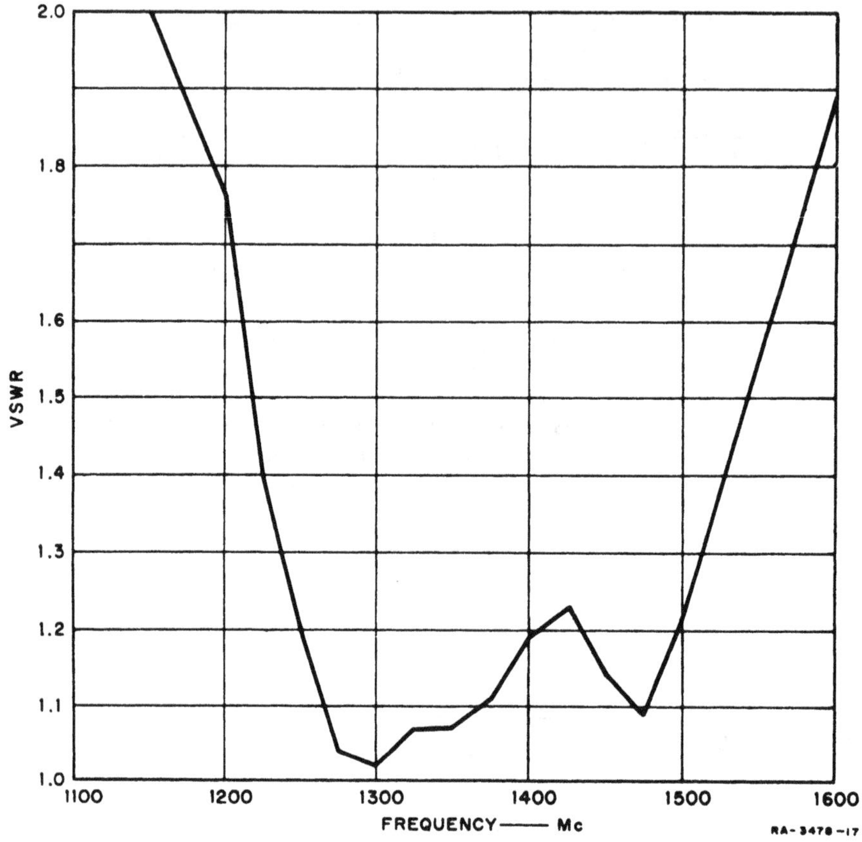

SOURCE: Technical Note 2, Contract AF 30(602)-2392, SRI
(See Ref. 45 by E. D. Sharp)

FIG. 15.05-7 VSWR OF WAFFLE-IRON FILTER ASSEMBLY

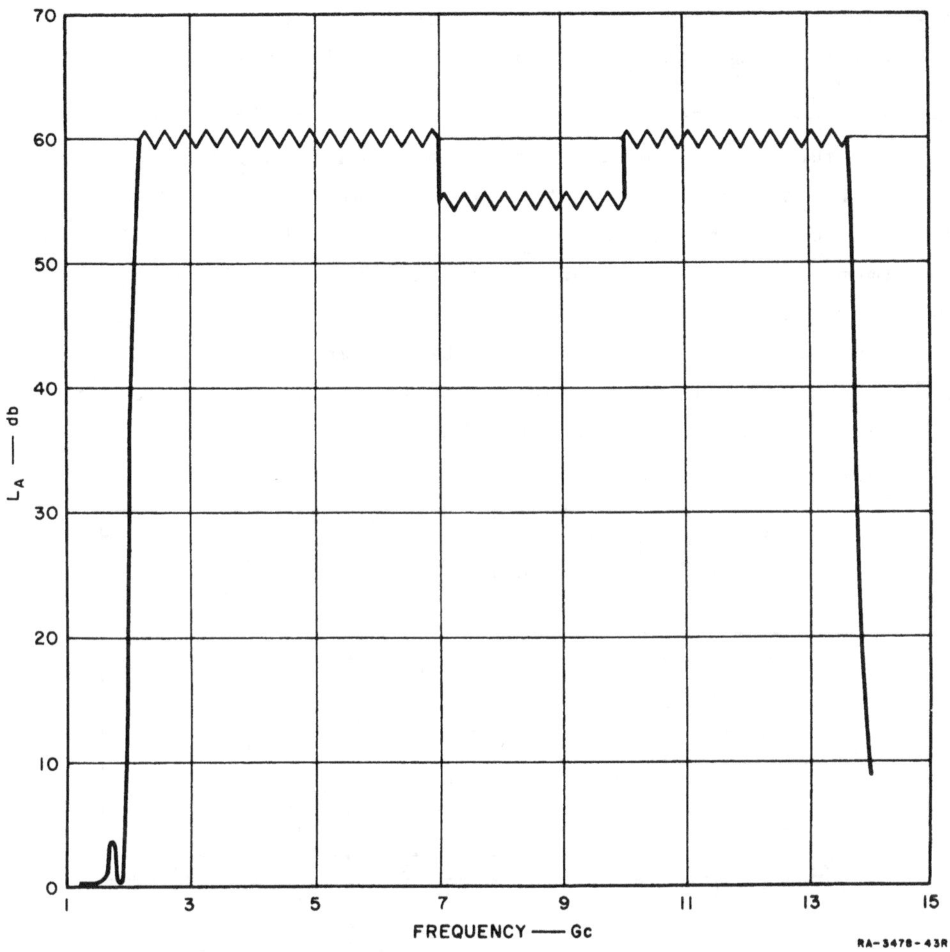

SOURCE: Technical Note 2, Contract AF 30(602)-2392, SRI
(See Ref. 45 by E. D. Sharp)

FIG. 15.05-8 ATTENUATION OF WAFFLE-IRON FILTER ASSEMBLY

The filter shown in Fig. 15.05-1 was tested at high power until arcing occurred. It was then opened and examined: strong burn marks were found on the four corners of the teeth in the center three longitudinal rows. (There were no burn marks on the two rows of teeth along the side wall.) Rounding the edges alone would leave a ridge at the four corners, which had therefore been rounded off into approximately spherical corners; but this still remained the weakest part of the filter from the standpoint of power-handling capacity. To spread the field more evenly, it was argued that a circular tooth should be better than a square tooth, since it has only one edge and no corners. To arrive at a quantitative comparison, the following reasoning was used.

Cohn[35] has analyzed the fields near two-dimensional rounded corners. With the dimensions of both the round and square teeth, his results should still hold closely for the fields near the line-edges of the teeth. One curve in Fig. 15.05-9 is taken from Fig. 15.02-5, and shows the maximum field near a rounded edge with circular boundary. It is shown as a function of the ratio $2R/b''$, where R is the radius of the rounded edge and b'' is the spacing between teeth (Fig. 15.05-3). To calculate the maximum field near the spherically rounded corners of the square teeth, we proceed as follows.[43]

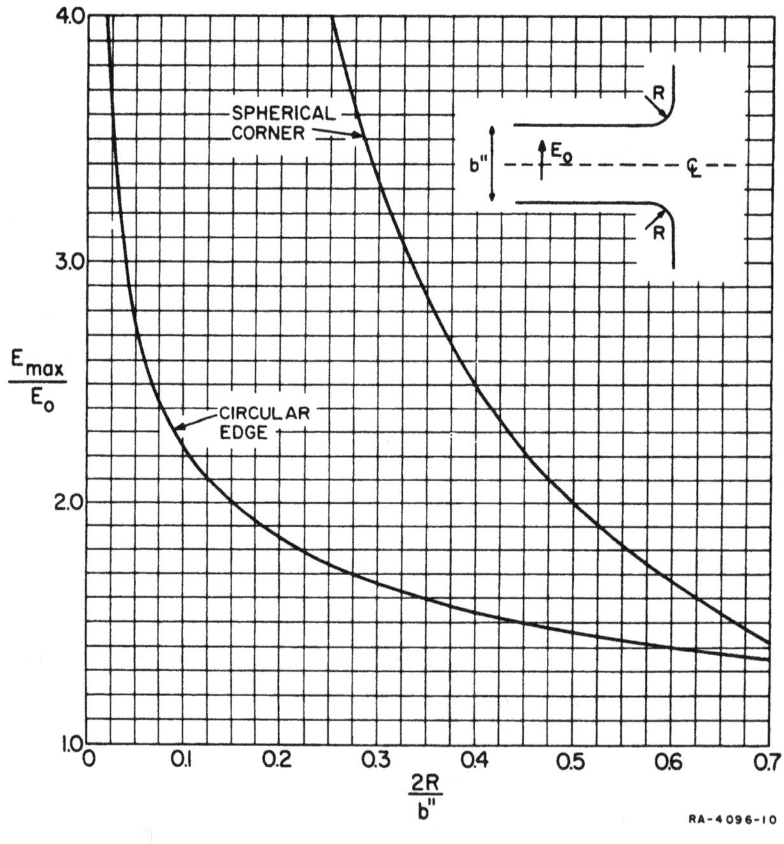

SOURCE: Quarterly Progress Report 1, Contract AF 30(602)-2734, SRI (See Ref. 43 by L. Young)

FIG. 15.05-9 MAXIMUM FIELD INTENSITIES NEAR CIRCULAR EDGE OF ROUND TEETH AND NEAR SPHERICAL CORNERS OF SQUARE TEETH

The plane of symmetry between teeth may be regarded as being at ground potential. The field intensity well inside the region between teeth is denoted by E_0. Then the potential V of a tooth is

$$V = \pm E_0 b''/2 \quad . \qquad (15.05\text{-}1)$$

The field E_{max} on the surface of an isolated sphere of radius R and at a potential V above ground is

$$E_{max} = \frac{V}{R} = \frac{E_0 b''}{2R} \quad . \qquad (15.05\text{-}2)$$

Therefore,

$$\frac{E_{max}}{E_0} \approx \frac{b''}{2R} \qquad (15.05\text{-}3)$$

approximately, for the three-dimensional corners on the square teeth, when the corners have spherical contours. Equation (15.05-3) is also plotted against $2R/b''$ in Fig. 15.05-9 and is a rectangular hyperbola. For the filters under consideration

$$\frac{2R}{b''} = \frac{0.126}{0.210} = 0.6 \quad , \qquad (15.05\text{-}4)$$

and Fig. 15.05-9 shows that E_{max}/E_0 is 1.415 for the circular edge, and 1.667 for the spherical corner. Therefore, the waffle-iron filter with circular teeth should handle about

$$\left(\frac{1.667}{1.415}\right)^2 = 1.39 \qquad (15.05\text{-}5)$$

times as much power as the waffle-iron filter with square teeth. With four waffle-iron filters in parallel and circular teeth, as in Fig. 15.05-10, the filter should handle about 4 × 1.39 = 5.56 times the power of the single filter in Fig. 7.05-10, corresponding to about 15 percent of the power handled by WR650 L-band rectangular waveguide.

SOURCE: Quarterly Progress Report 1, Contract AF 30(602)-2734, SRI (See Ref. 43 by L. Young)

FIG. 15.05-10 HIGH-POWER WAFFLE-IRON FILTER USING ROUND TEETH AND FOUR FILTERS IN PARALLEL

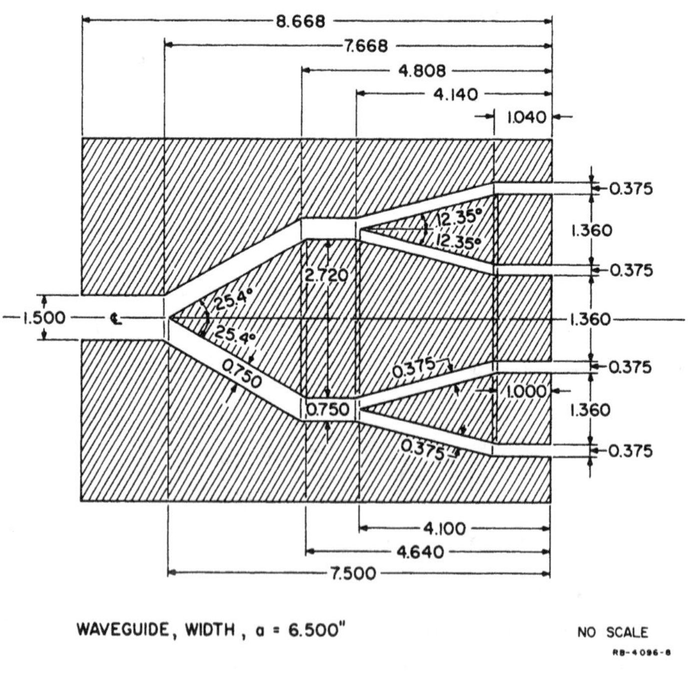

SOURCE: Quarterly Progress Report 1, Contract AF 30(602)-2734, SRI (See Ref. 43 by L. Young)

FIG. 15.05-11 DIMENSIONS OF POWER-DIVIDERS SHOWN IN FIG. 15.05-10

This four-waffle-iron filter was tested in the presence of a cobalt-60 radioactive source in air at atmospheric pressure, using pulses 2.5 microseconds long with a repetition rate of 200 pulses per second. At 6.3 megawatts pulse power (the maximum power available) there was still no sign of arcing. Based on the above calculation and 1.4 megawatts for the single-waffle-iron square-toothed filter of Fig. 15.05-1, the filter in Fig. 15.05-10 should handle $5.56 \times 1.4 = 7.8$ megawatts under the same conditions.

It was pointed out at the beginning of Sec. 15.02 that the pulse-power capacity referred to short pulses. The reader should use caution when applying these results to longer pulses. Both the square-toothed and the round-toothed waffle-iron filters were subjected to pulses 110 microseconds long, at a rate of 30 pulses per second. The pulse-power capacity with the 110-microsecond pulses was only about one-third of that with the 2-microsecond pulses. Furthermore, it appeared that arcing started about 50 microseconds after the beginning of the pulse.

The dimensions of the power dividers for the high-power waffle-iron filter in Fig. 15.05-10 are given in Fig. 15.05-11. Each of the four parallel filters has the same dimensions as were given in Sec. 7.05 for the single filter shown in Fig. 7.05-10.

The VSWR of the filter shown in Fig. 15.05-10 is shown in Fig. 15.05-12, including both power dividers in the measurements. The solid line is for the original filter in which the three one-eighth-inch plates supporting the round teeth (Fig. 15.05-10) were solid plates from end to end. Later these plates were modified by cutting large circular holes between teeth, as shown in Fig. 15.05-13, which also shows the shelf-like construction of the filter in more detail. The reason for the holes was to allow coupling between the four parallel filter units, so that propagation through the various waffle iron filters could not get out of phase, and thus to forestall any possible trapped resonance around a plate due to poor tolerances or other imperfections. (No such resonance was ever actually observed with or without coupling holes.) The VSWR of the filter with coupling holes is shown by the dotted line in Fig. 5.05-12. The VSWR is better than 1.2 over almost the whole of L-band, with or without coupling holes, and better than 1.1 over two considerable portions of it.

The stop-band performance is shown in Fig. 15.05-14 together with that of Waffle-Iron Filter A (which has square teeth, has ten sections,

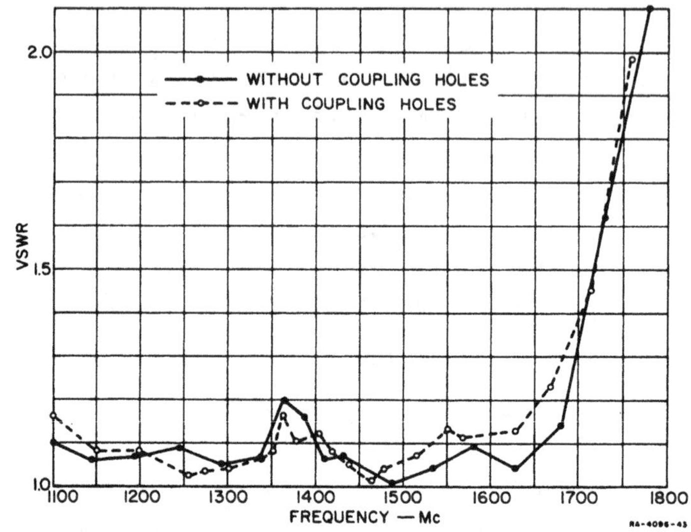

SOURCE: Quarterly Progress Report 2, Contract AF 30(602)-2734, SRI
(See Ref. 43 by L. Young)

FIG. 15.05-12 VSWR OF FILTERS IN FIGS. 15.05-10 AND 15.05-13

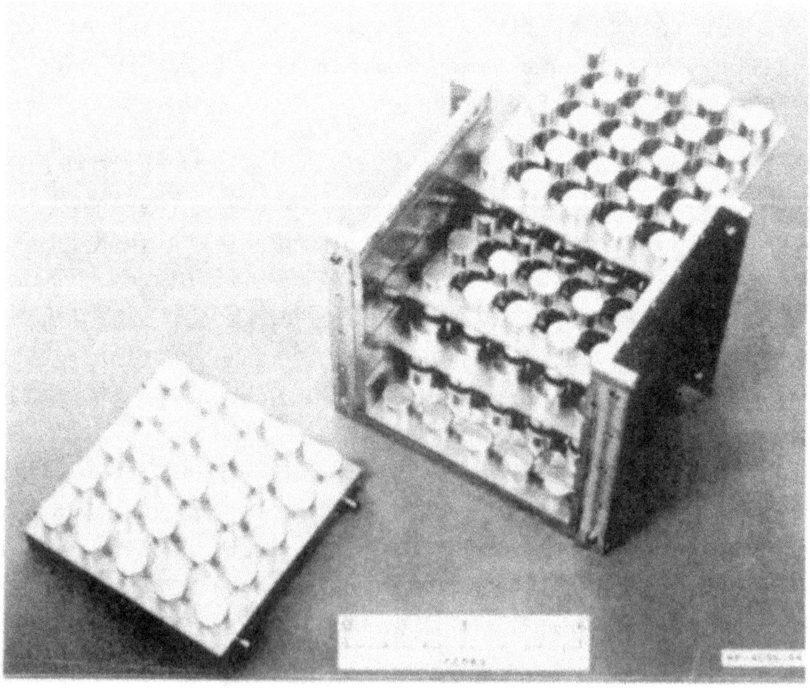

SOURCE: Quarterly Progress Report 2, Contract AF 30(602)-2734, SRI
(See Ref. 43 by L. Young)

FIG. 15.05-13 SHELF-LIKE CONSTRUCTION OF FILTER IN FIG. 15.05-10

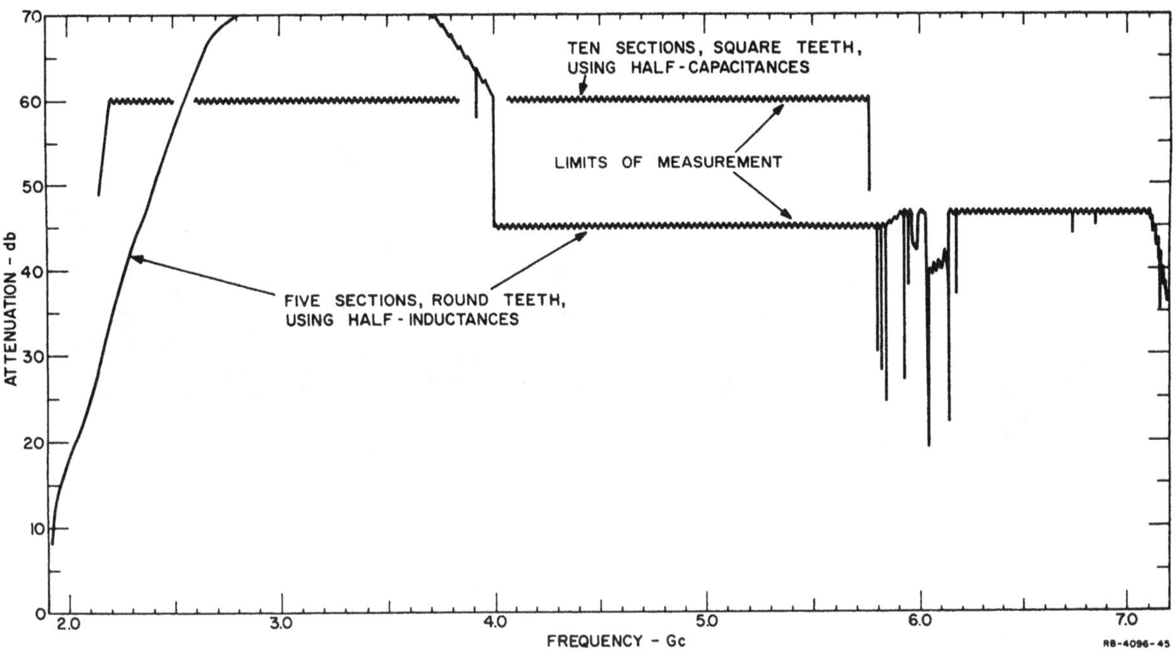

SOURCE: Quarterly Progress Report 2, Contract AF 30(602)-2734, SRI
(See Ref. 43 by L. Young)

FIG. 15.05-14 ATTENUATION OF TWO WAFFLE-IRON FILTERS

and begins with half-teeth). After allowance is made for the fact that one is a five-section and one is a ten-section filter, there is no noticeable difference in the stop-band performance. However, some difficulty was experienced as a result of imperfect contact at the coupling flange. Since TE_{01} and other TE_{0n} modes can propagate through the waffle-iron filter in the higher part of the stop band, one must make sure that TE_{01} and other TE_{0n} modes are not excited where the waffle-iron sections begin. Normally these modes are not excited because of the symmetry of the structure, and they are cut off in the low-height waveguide matching sections so that none can reach the filter proper. However if the waffle-iron filter and the low-height waveguide matching sections are made in separate pieces and then connected, imperfect contact at the flanges may set up the unwanted modes. Then any conversion from propagating TE_{m0} modes in the exterior waveguide to propagating TE_{0n} modes in the filter proper, and subsequent re-conversion at the output, will cause spurious pass bands. A few very sharp and narrow spurious responses were observed, and were found to improve with better flange fitting. To avoid the need for excessive care in mating flanges, six one-eighth-inch-diameter rods were passed across the filter, as shown in Fig. 15.05-15, which completely eliminated all spurious responses in the stop band.

SOURCE: Quarterly Progress Report 2, Contract AF 30(602)-2734, SRI
(See Ref. 43 by L. Young)

FIG. 15.05-15 WAFFLE-IRON FILTER WITH THIN RODS TO SUPPRESS TE_{01} MODE EXCITED BY IMPERFECT CONTACT OR MISALIGNMENT OF FLANGES

SEC. 15.06, DISSIPATIVE WAVEGUIDE FILTERS

Dissipative low-pass waveguide filters are extensively used to suppress the spurious harmonic radiation from high-power transmitters because they are well matched in both the pass and stop bands and hence preclude resonances between the transmitter and the filter that might be damaging to the transmitting tube.[47,48] They usually have maximum attenuation at the second and third harmonic frequencies with lesser attenuations at higher frequencies. Most designs can handle essentially the full waveguide pulse-power without recourse to pressurization. It is also relatively easy to pressurize them if necessary in order to make them compatible with other components in a system.

Figure 15.06-1 is a sketch of a portion of a typical dissipative waveguide filter. It is seen to consist of a central waveguide that is aperture coupled to an array of secondary waveguides, which stand on all four sides of the central waveguide, and each of which is terminated in a matched load. The width a_2 of the secondary waveguides is small enough so that they are below cutoff in the pass band of the filter and hence at these frequencies power can propagate down the central waveguide with an attenuation that is ordinarily less than 0.1 db. In order to achieve this low, pass-band attenuation the tips of the spear-shaped loads in the secondary waveguides are placed approximately a distance a_2 (equal to one-half of the TE_{10}-mode cutoff wavelength in the secondary waveguide) from the apertures so that the fringing fields at the apertures do not interact with the loads. In the stop band, which exists at frequencies above the pass band, the side waveguides can propagate and harmonic power

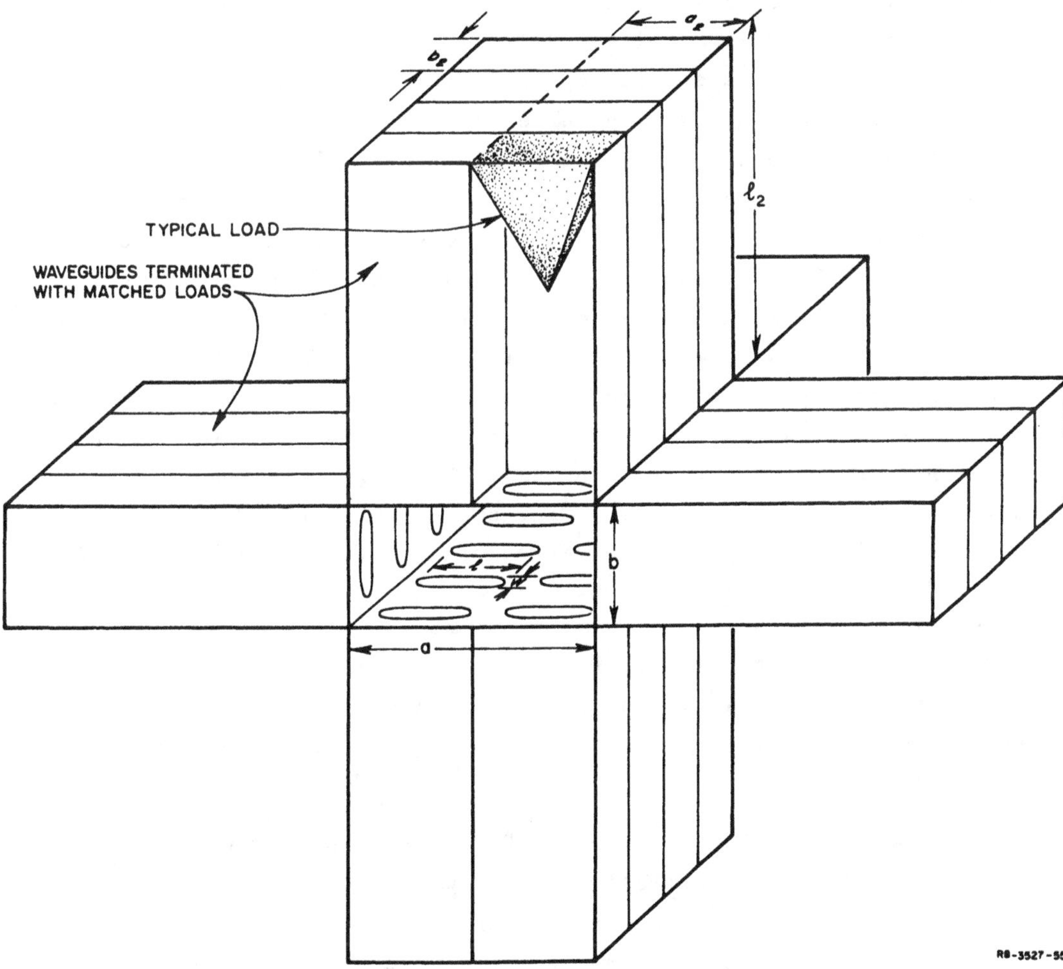

FIG. 15.06-1 TYPICAL CONSTRUCTION OF A DISSIPATIVE WAVEGUIDE FILTER

propagating down the central waveguide is coupled into the side waveguides where it is dissipated in the matched terminations.

The coupling apertures for the secondary waveguides are usually long and narrow slots as shown, and maximum attenuation in the central waveguide occurs at the frequency at which the aperture lengths l are about one-half a free-space wavelength. The center-to-center spacing of the apertures, along the length of the waveguide, is usually made less than half a free-space wavelength at the highest frequency of operation of the filter in order to avoid multiple reflections from the apertures which would raise the stop-band VSWR. The coupling apertures cannot couple equally well to all possible modes in the central waveguide, which can propagate a rapidly increasing number of modes as the frequency is increased. When slots are used, as shown in Fig. 15.06-1, the slots attenuate most strongly those modes with currents trying to cross the slots (*i.e.*, having transverse magnetic fields in the case of Fig. 15.06-1); the secondary waveguides are then effectively in series with the central waveguide. Furthermore, the attenuation is greatest when the centers of the coupling apertures are located at the positions of maximum transverse magnetic field (maximum longitudinal wall current). The attenuation of a mode measured in db at a particular frequency is also approximately inversely proportional to its impedance in the central waveguide (which for TE modes and a given a/b ratio is proportional to the ratio of guide wavelength to free-space wavelength). Therefore, a higher-order mode having the coupling apertures of the secondary waveguides arranged so that the mode will suffer maximum attenuation may still be less strongly attenuated than a lower-order mode having a lower guide impedance. Greater attenuation by the secondary waveguides along the broad wall of the central guide is obtained when the impedance of the guide is reduced by reducing its height, b.[47] However, this procedure also reduces the pulse-power capacity of the filter since the pulse-power capacity is directly proportional to b.

In the filter shown in Fig. 15.06-1 the secondary waveguides along the broad wall of the central waveguide would couple strongly to the second-harmonic power propagating in the TE_{20} mode. They also couple just about as strongly to the second-harmonic power propagating in the TE_{10} mode. The waveguides along the narrow wall of the central waveguide couple most strongly to second-harmonic power propagating in the TE_{01} mode.

If the double array of side waveguides along each broad wall of the central waveguide in Fig. 15.06-1 are replaced by triple arrays, the filter would be most effective in suppressing power propagating at the third harmonic frequency. For this reason some dissipative filters have a double array of waveguides along a portion of the broad wall and a triple array along the rest of the wall.

In order to provide a good match between the dissipative filter and the terminating waveguides it is customary to linearly taper the lengths of the coupling apertures at each end of the filter. When about a dozen apertures are tapered at each end of the filter the resulting pass-band VSWR is usually less than 1.20.

An S-Band Dissipative Filter—A filter which is typical of dissipative filters in general, has been developed by V. Price, *et al.*[49] The configuration of the side waveguides in this filter is as shown in Fig. 15.06-1; however, a cylindrical pressure vessel surrounds the filter so that it can be operated at a pressure of 30 psig (pounds per square inch, gauge). The filter, which is made of aluminum, is fabricated in the following fashion. First, the coupling apertures are milled in the central waveguide. Then a number of aluminum stampings are made which, when assembled in the fashion of an egg-crate, form the side waveguides. The side waveguides and the central waveguides are then clamped together and permanently joined to each other by immersion in a dip brazing tank. The specifications for such a filter are given in Table 15.06-1, and its attenuation characteristics are shown in Figs. 15.06-2, 3 and 4.

Table 15.06-1

SPECIFICATIONS OF AN *S*-BAND DISSIPATIVE-FILTER EXAMPLE

Rows of slots on each broad wall	2
Rows of slots on each narrow wall	1
Number of full-length slots on both broad walls	272
Number of full-length slots on both narrow walls	136
Number of linearly tapered matching slots at each end of each row of slots	16
b_2, height of side waveguides	0.26 inch
a_2, width of side waveguides	1.39 inches
l_2, length of narrow wall side waveguides	2.6 inches
l_2, length of broad wall side waveguides	2.4 inches
l, length of coupling apertures	1.30 inches
w, width of coupling apertures	0.19 inch
Over-all filter length including flanges	36 inches
Outside diameter of pressure vessel surrounding filter	9.25 inches

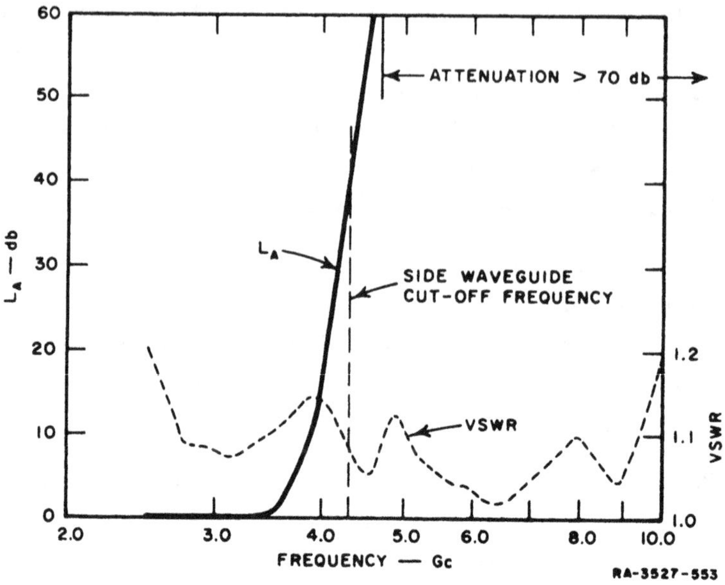

SOURCE: Final Report, Change A, Contract AF 30(602)-1670, General Electric Company (See Ref. 49 by V. G. Price, J. P. Rooney and R. H. Stone)

FIG. 15.06-2 TE_{10}-MODE ATTENUATION AND VSWR OF S-BAND DISSIPATIVE FILTER WHOSE DIMENSIONS ARE GIVEN IN TABLE 15.06-1

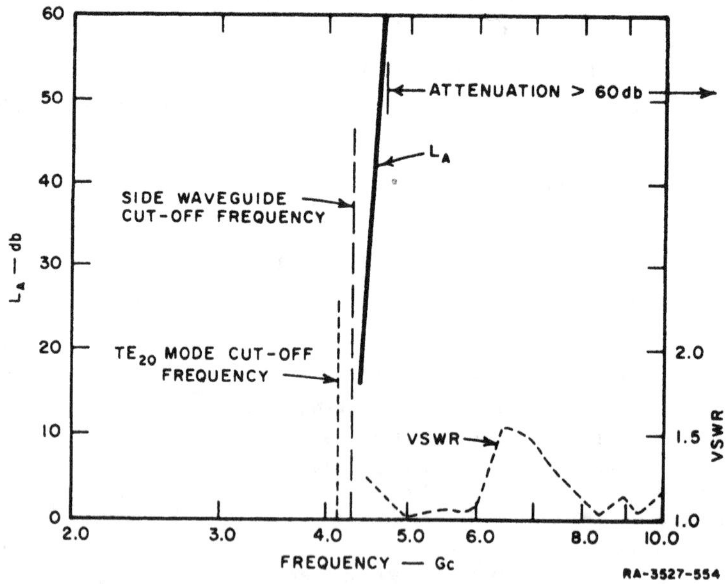

SOURCE: Final Report, Change A, Contract AF 30(602)-1670 General Electric Company (See Ref. 49 by V. G. Price, J. P. Rooney and R. H. Stone)

FIG. 15.06-3 TE_{20}-MODE ATTENUATION AND VSWR OF DISSIPATIVE FILTER WHOSE DIMENSIONS ARE GIVEN IN TABLE 15.06-1

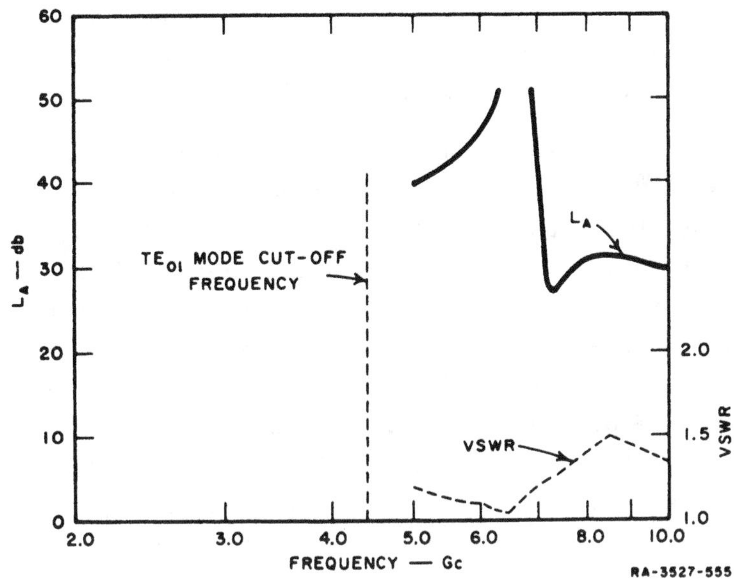

SOURCE: Final Report, Change A, Contract AF 30(602)-1670 General Electric Company (See Ref. 49 by V. G. Price, J. P. Rooney and R. H. Stone)

FIG. 15.06-4 TE_{01}-MODE ATTENUATION AND VSWR OF DISSIPATIVE FILTER WHOSE DIMENSIONS ARE GIVEN IN TABLE 15.06-1

Other types of harmonic-absorbing waveguide filters with secondary waveguides have also been built. One of the more promising variations on this theme has been the use of circular instead of rectangular secondary waveguides, open-ended into the rectangular central waveguide. The secondary guides may or may not be loaded with dielectric material, but in any case must be cutoff in the pass band, and have to start propagating at some frequency below the second harmonic.

SEC. 15.07, DISSIPATIVE COAXIAL-LINE FILTERS

Dissipative low-pass filters can also be constructed in coaxial line to suppress the spurious power from high-power transmitters. Their principle of operation is essentially the same as that of the waveguide dissipative filters discussed in Sec. 15.06. The stop-band attenuation per unit of length of the TEM mode in a typical coaxial dissipative filter is greater than that of the lowest mode in a typical waveguide dissipative filter because the characteristic impedance of the central transmission line is lower. However, the pulse-power capacity is less for a coaxial filter because the pulse-power capacity of the best coaxial line is less than that of standard waveguide in the same frequency range.

Experimental models of coaxial dissipative filters have been built by E. G. Cristal.[50] One model has a pass-band cutoff frequency of 1.7 Gc and a characteristic impedance of 50 ohms. The inside diameter of the outer conductor is 1.527 inches while the diameter of the inner conductor is 0.664 inch. Twenty-one rectangular-slot pairs were cut in the 0.049-inch-thick coaxial line wall and were spaced 0.400 inch apart center to center. The width of each slot was 0.25 inch, and each slot subtended an angle of 150 degrees at the axis. In addition, ten smaller slots were cut in the coaxial line in the regions ahead of and behind the twenty-one main slots; the added slots were arranged to form a tapered sequence so as to improve the impedance match in the pass band.

The secondary waveguides were 0.375 inch high and 3.5 inches wide, corresponding to a cutoff frequency of 1.7 Gc. The length of the secondary waveguides is about 12 inches, but this length could be reduced. Each of the secondary waveguides was terminated with resistive Teledeltos* paper. (Each load was found to be able to dissipate at least three watts average power without cooling.) Figure 15.07-1 shows a sketch of the filter (without the tapered slots). Its measured attenuation to the dominant TEM mode is shown in Fig. 15.07-2, for several values of the coaxial line

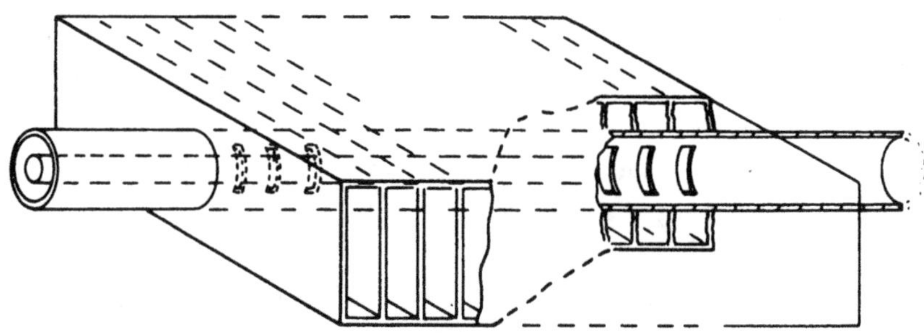

SOURCE: Technical Note 4, Contract AF 30(602)-2392, SRI (See Ref. 50 by E. G. Cristal)

FIG. 15.07-1 COAXIAL LEAKY-WAVE FILTER

* Obtainable from Micro Circuits Co., New Buffalo, Michigan.

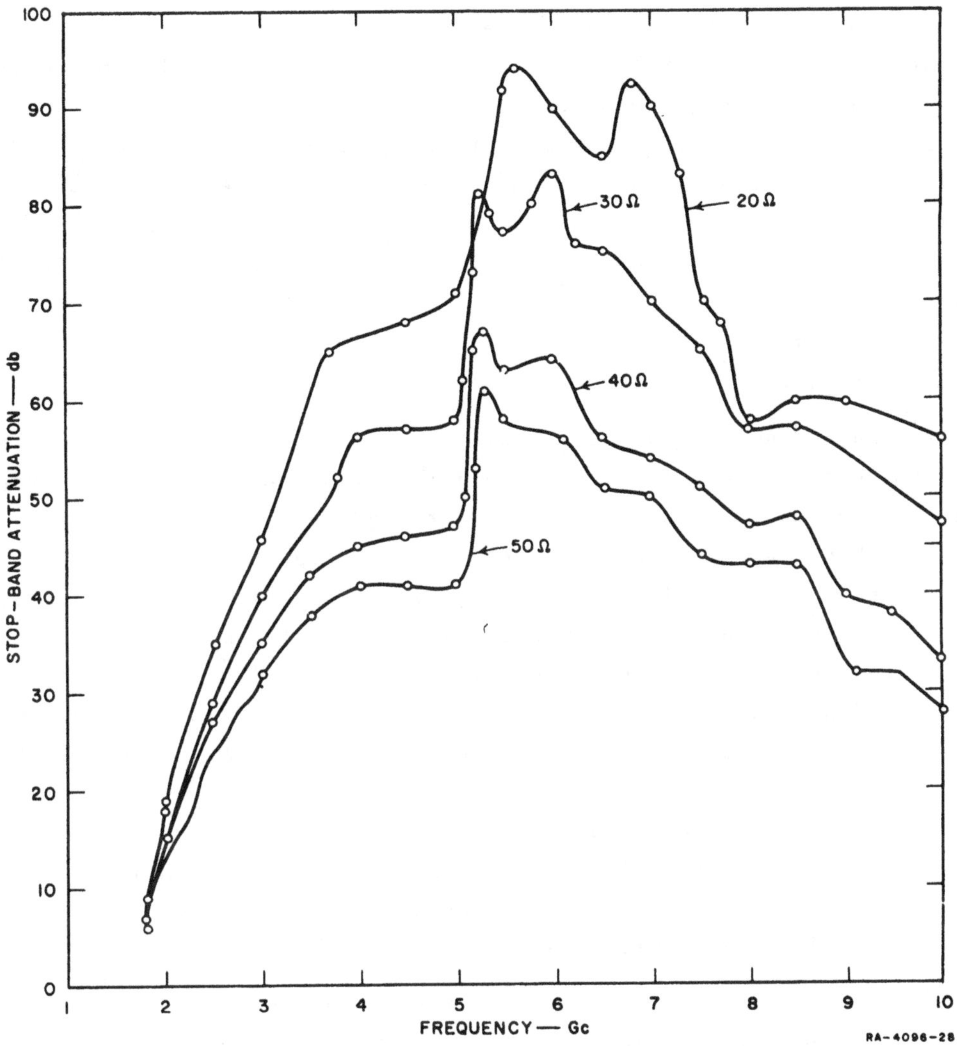

SOURCE: Quarterly Progress Report 3, Contract AF 30(602)-2734, SRI (See Ref. 23 by E. G. Cristal et al.)

FIG. 15.07-2 ATTENUATION OF EXPERIMENTAL COAXIAL LEAKY-WAVE FILTER TO A TEM WAVE FOR VARIOUS LINE IMPEDANCES

impedance, from 50 ohms to 20 ohms. The impedance was adjusted by changing the diameter of the inner conductor from 0.664 inch (for 50 ohms) on up. It can be seen from Fig. 15.07-2 that the attenuation to the TEM mode is substantial over a very wide frequency band, for all line impedances, and that the attenuation increases as the impedance decreases.

The TE_{11} mode propagates above 3.5 Gc. A TE_{11} mode launcher was constructed and the attenuation of the coaxial leaky-wave filter was measured to two polarizations of the TE_{11} mode: in one polarization the currents try to cross the slots and are attenuated strongly, the over-all attenuation being about equal to that for the TEM mode. In the other polarization, the slots do not perturb the current flow so much, and the coupling through the slots is much weaker; the measured attenuation in db was about one-fifth of that for the orthogonal TE_{11} polarization and the TEM mode. One way to ensure high attenuation for the TE_{11} mode in all polarizations is to use two filters in cascade, but arranged with their transverse geometry orthogonal. At the time of this writing, other possibilities are still under consideration.

The shape and size of single slots were varied experimentally in order to determine the configuration for maximum coupling in the stop band. For example, dumbbell slots can be made to resonate at lower frequencies than plain rectangular slots, and thus give slightly stronger coupling near the cutoff frequency, but it is doubtful whether the small improvement is worth the mechanical complication for most applications. A numerical analysis for some idealized cases was also undertaken to discover the dependence of the stop-band attenuation on the various parameters involved. The slots were closely spaced in every case. Doubling the number of slots and secondary waveguides per unit length did not appreciably increase the attenuation, when the slot widths and waveguide heights had to be halved to allow for twice as many slots. The greatest attenuation and the flattest frequency response resulted from the widest slots (slot width equal to waveguide height).

The VSWR of the 50-ohm filter in the nominal pass band of 1.2 to 1.4 Gc was found to be less than 1.09. In the stop band, the VSWR for the TEM mode was less than 1.5 from the second to the fifth harmonic, inclusive.

REFERENCES

1. R. D. Campbell, "Radar Interference to Microwave Communication Services," *Elec. Eng.*, Vol. 77, pp. 916-921 (October 1958).

2. A. H. Ryan, "Control of Microwave Interference," *IRE Trans.*, *PGRFI*-1, pp. 1-10 (May 1959).

3. M. P. Forrer and K. Tomiyasu, "Effects and Measurements of Harmonics in High-Power Waveguide Systems," 1957 *IRE National Convention Record*, Part 1, pp. 263-269.

4. V. G. Price, "Measurement of Harmonic Power Generated by Microwave Transmitters," *IRE Trans.*, *PGMTT*-7, pp. 116-120 (January 1959).

5. C. C. Cutler, "Spurious Modulation of Electron Beams," *Proc. IRE*, Vol. 44, pp. 61-64 (January 1956).

6. R. L. Jepsen, "Ion Oscillations in Electron Beam Tubes; Ion Motion and Energy Transfer," *Proc. IRE*, Vol. 45, pp. 1069-1080 (August 1957).

7. A. D. Sutherland, "Relaxation Instabilities in High-Perveance Electron Beams," *IRE Trans.*, *PGED*-7, pp. 268-273 (October 1960).

8. Y. Koike and Y. Kumagai, "An Experiment of Ion Relaxation Oscillation in Electron Beams," *Proc. IRE*, Vol. 49, pp. 525-526 (February 1961).

9. O. Doehler, "Space Charge Effects in Traveling-Wave Tubes Using Crossed Electric and Magnetic Fields," *Proceedings of the Symposium on Modern Advances in Microwave Techniques*, Brooklyn, N.Y., November *8-10*, 1954, Polytechnic Institute of Brooklyn, N.Y., Vol. 4, pp. 101-121 (1955). See especially pp. 115-119.

10. R. W. Gould, "Space Charge Effects in Beam-Type Magnetrons," *J. Appl. Phys.*, Vol. 28, pp. 599-605 (May 1957).

11. G. Novick and V. G. Price, "Measurement and Control of Harmonic and Spurious Microwave Energy," General Electric Microwave Lab., Palo Alto, California, Final Report, Phase III, Contract AF 30(602)-1670, ASTIA Document No. AD-214430 (May 15, 1959). RADC TR-59-172.

12. V. G. Price, J. P. Rooney, R. H. Stone, "Measurement and Control of Harmonic and Spurious Microwave Energy," General Electric Microwave Laboratory, Palo Alto, California, Final Report, Change A, Contract AF 30(602)-1670 (March 1960).

13. V. G. Price, R. H. Stone, J. P. Rooney, "Measurement and Control of Harmonic and Spurious Microwave Energy," General Electric Microwave Laboratory, Palo Alto, California, Final Report, Phase II, ASTIA Document No. AD-214430 (March 1959).

14. L. A. MacKenzie, "Klystron Cavities for Minimum Spurious Output Power," Research Report No. EE418, School of Electrical Engineering, Ithaca, N.Y., Cornell University, (January 31, 1959).

15. K. Tomiyasu and M. P. Forrer, "Diode Oscillation in High-Voltage Klystrons," *IRE Trans.*, *PGED*-8, pp. 381-386 (September 1961).

16. M. Chodorow, *et al.*, "Design and Performance of a High-Power Pulsed Klystron," *Proc. IRE*, Vol. 41, pp. 1584-1602 (November 1953). See especially p. 1601.

17. K. Tomiyasu, "On the Possibility of Drift-Tunnel Oscillations in High Power Klystrons," *Proc. IRE*, Vol. 49, pp. 1207-1208 (July 1961).

18. J. A. Ruetz and W. H. Yocom, "High-Power Traveling-Wave Tubes for Radar Systems," *IRE Trans.*, *PGMIL*-5, pp. 39-45 (April 1961).

19. D. G. Dow, "Behavior of Traveling-Wave Tubes Near Circuit Cutoff," *IRE Trans.*, *PGED*-7, pp. 123-131 (July 1960).

20. L. Young, E. G. Cristal, E. Sharp, and J. F. Cline, "Techniques for the Suppression of Spurious Energy," Final Report, SRI Project 3478, Contract AF 30(602)-2392, Stanford Research Institute, Menlo Park, California (March 1962). RADC-TDR-62-164.

21. K. Tomiyasu, "On Spurious Outputs from High-Power Pulsed Microwave Tubes and Their Control," *IRE Trans., PGMTT-9*, pp. 480-484 (November 1961).

22. E. N. Torgow, "Hybrid Junction-Cutoff Waveguide Filters," *IRE Trans., PGMTT-7*, pp. 163-167 (January 1959).

23. E. G. Cristal, L. Young, and B. M. Schiffman, "Suppression of Spurious Frequencies," Quarterly Progress Report 3, SRI Project 4096, Contract AF 30(602)-2734, Stanford Research Institute, Menlo Park, California (January 1963).

24. B. J. Duncan and B. Vafiades, "Design of Full Waveguide Bandwidth High-Power Isolator," *IRE Trans., PGMTT-6*, pp. 411-414 (October 1958).

25. L. Young, "The Application of Branch-Guide Couplers to the Suppression of Spurious Frequencies," *Proceedings of the IRE-PGRFI* Fourth National Symposium on Radio Frequency Interference, San Francisco, California (June 1962).

26. L. Gould, "Handbook on Breakdown of Air in Waveguide Systems," Microwave Associates Report, Contract NObsr-63295 (Navy Department, Bureau of Ships), (April 1956). See also L. Gould and L. W. Roberts, "Breakdown of Air at Microwave Frequencies," *J. Appl. Phys.*, Vol. 27, pp. 1162-1170 (October 1956).

27. D. Dettinger and R. D. Wengenroth, "Microwave Breakdown Study," Final Engineering Report, Wheeler Laboratories, Great Neck, New York, Contract NObsr-52601, ASTIA No. AD 11121 (January 1953).

28. G. K. Hart and M. S. Tannenbaum, "High Power Breakdown of Microwave Components," IRE Convention Record, Part 8, pp. 62-67 (March 1955).

29. G. K. Hart, F. R. Stevenson and M. S. Tannenbaum, "High Power Breakdown of Microwave Structures," IRE Convention Record, Part 5, pp. 199-205 (March 1956).

30. W. S. Boyle, P. Kisliuk and L. Germer, "Electrical Breakdown in High Vacuum," *J. Appl. Phys.*, Vol. 26, pp. 720-725 (June 1955).

31. H. A. Wheeler and H. L. Bachman, "Evacuated Waveguide Filter for Suppressing Spurious Transmission from High-Power *S*-Band Radar," *IRE Trans., PGMTT-7*, pp. 154-162 (January 1959). Correction to above article, *IRE Trans., PGMTT-7*, **3**, p. 369 (July 1959).

32. M. P. Forrer and C. Milazzo, "Duplexing and Switching with Multipactor Discharges," *Proc. IRE* 50, pp. 442-450 (April 1962).

33. H. A. Wheeler, "Pulse Power Chart for Waveguides and Coaxial Line," Wheeler Monograph No. 16 (April 1953).

34. G. L. Ragan, "Microwave Transmission Circuits," M.I.T. Rad. Lab. Series, Vol. 9 (McGraw-Hill Book Co., New York, N.Y., 1946).

35. S. B. Cohn, "Rounded Corners in Microwave High Power Filters and Other Components," Tech. Note 1, SRI Project 2797, Contract AF 30(602)-1998, Stanford Research Institute, Menlo Park, California (June 1960). Also published as "Rounded Corners in Microwave High-Power Filters," *IRE Trans., PGMTT-9*, pp. 389-397 (September 1961).

36. H. E. King, "Rectangular Waveguide Theoretical CW Average Power Rating," *IRE Trans., PGMTT-9*, pp. 349-357 (July 1961).

37. W. W. Macalpine, "Heating of Radio Frequency Cables," *Electrical Communication 25*, pp. 84-99 (March 1948).

38. E. M. T. Jones, "A Three-Cavity High-Power *L*-Band Filter Tunable from 1250-1350 Mc," Tech. Note 2, SRI Project 2797, Contract AF 30(602)-1998, Stanford Research Institute, Menlo Park, California (February 1961).

39. S. B. Cohn, "Design Considerations for High Power Microwave Filters," *IRE Trans., PGMTT-7*, pp. 149-153 (January 1959).

40. L. Young, "Peak Internal Fields in Direct-Coupled-Cavity Filters," *IRE Trans., PGMTT-8*, pp. 612-616 (November 1960).

41. L. Young, "Analysis of a Transmission Cavity Wavemeter," *IRE Trans., PGMTT*-8, pp. 436-439 (July 1960).

42. L. Young, "*Q*-Factors of a Transmission Line Cavity," *IRE Trans., PGCT*-4, pp. 3-5 (March 1957).

43. L. Young, *et al.*, "Suppression of Spurious Frequencies," Quarterly Progress Reports 1 and 2, SRI Project 4096, Contract AF 30(602)-2734, Stanford Research Institute, Menlo Park, California (July and October 1962). See also Refs. 58 and 59.

44. E. L. Ginzton, *Microwave Measurements*, p. 366 (McGraw Hill Book Co., Inc., New York City, New York, 1957).

45. E. D. Sharp, "A High-Power, Wide-Band Waffle-Iron Filter," Tech. Note 2, SRI Project 3478, Contract AF 30(602)-2392, Stanford Research Institute, Menlo Park, California (January 1962). RADC-TDR-62-183. Also, *IEEE Trans. PTGMTT*-11, pp. 111-116 (March 1963).

46. H. Guthart and E. M. T. Jones, "A High-Power *S*-Band Waffle-Iron Filter," Tech. Note 4, SRI Project 2797, Contract AF 30(602)-1998, Stanford Research Institute, Menlo Park, California (March 1961). Also published as "A High Power *S*-Band Filter," *IRE Trans., PGMTT*-10, pp. 148-149 (March 1962).

47. V. Met, "Absorptive Filters for Microwave Harmonic Power," *Proc. IRE*. Vol. 47, pp. 1762-1769 (October 1959).

48. V. G. Price, *et al.*, "Harmonic Suppression by Leaky-Wall Waveguide Filter," 1959 IRE WESCON Convention Record, Part 1, pp. 112-118.

49. V. G. Price, J. P. Rooney, and R. H. Stone, "Measurement and Control of Harmonic and Spurious Microwave Energy," Final Report, Change A, Contract AF 30(602)-1670, RADC TR-60-77, General Electric Microwave Laboratory, Power Tube Department, Palo Alto, California (March 1960).

50. E. G. Cristal, "Some Preliminary Experimental Results on Coaxial Absorption Leaky-Wave Filters," *Proceedings of the IRE-PGRFI Fourth National Symposium on Radio Frequency Interference*, San Francisco, California (June 1962).

A more detailed treatment is given in "Coaxial Absorption Leaky-Wave Filter," Tech. Note 4, SRI Project 3478, Contract AF 30(602)-2392, Stanford Research Institute, Menlo Park, California (February 1962), RADC-TDR-62-184; and in Quarterly Progress Report 3, SRI Project 4096, Contract AF 30(602)-2734 (January 1963), Ref. 23.

OTHER REFERENCES ON HIGH-POWER FILTERS

A high-power filter using radial lines and uniform line discontinuities to increase the power capacity, is described by:

51. J. H. Vogelman, "High-Power Microwave Filters," *IRE Trans., PGMTT*-6, pp. 429-439 (October 1958). See also *IRE Trans., PGMTT*-7, pp. 461-465 (October 1959).

A branching filter with the circuit of Fig. 15.01-3(a), but using waveguide high-pass filters, is described by:

52. L. Young and J. Q. Owen, "A High Power Diplexing Filter," *IRE Trans., PGMTT*-7, pp. 384-387 (July 1959).

The measurement of harmonic power should precede the choice and design of filter. One method is described by:

53. E. D. Sharp and E. M. T. Jones, "A Sampling Measurement of Multimode Waveguide Power," *IRE Trans., PGMTT*-10, pp. 73-82 (January 1962), and another method is given in Ref. 4.

The former gives the harmonic power levels; the latter gives the spectral as well as the mode distribution, but requires a digital computer to compile the data.

A way to change the mode distribution with time, and then by time-averaging measure the spectral distribution of the harmonic output from a transmitter, has been described by:

54. V. G. Price and W. A. Edson, "Mode Scattering Technique for Interference Measurements," *IRE-PGRFI Fourth National Symposium on Radio Frequency Interference,* San Francisco, California (June 1962).

Electrical Breakdown in Vacuum is treated by:

55. L. Cranberg, "The Initiation of Electrical Breakdown in Vacuum," *J. Appl. Phys.,* Vol. 23, pp. 518-522 (May 1952), and

56. W. S. Boyle, P. Kisliuk, and L. H. Germer, "Electrical Breakdown in High Vacuum," *J. Appl. Phys.,* pp. 720-725 (June 1958).

The following is a tutorial paper:

57. S. C. Brown, "High-Frequency Gas-Discharge Breakdown," *Proc. IRE,* Vol. 39, pp. 1493-1501 (December 1951).

Further data on high-power, wide-band waffle-iron filters will be found in:

58. L. Young and B. M. Schiffman, "New and Improved Types of Waffle-Iron Filters," *Proc. IEE* (London)-110, pp. 1191-1198 (July 1963), and

59. Leo Young, "Postscript to Two Papers on Waffle-Iron Filters," *IEEE Trans. PTGMTT*-11, pp. 555-557 (November 1963).

Recent papers on high-power *absorbing* filters include:

60. E. G. Cristal, "Analytical Solution to a Waveguide Leaky-Wave Structure," *IEEE Trans. PTGMTT*-11, pp. 182-190 (May 1963), and

61. E. G. Cristal, "A 1-5/8 Inch Coaxial Leaky-Wave Filter for the Suppression of Spurious Energy," *The Microwave Journal*, Vol. 6, pp. 72-76 (September 1963).

62. Leo Young, "Waveguide 0-db and 3-db Directional Couplers as Harmonic Pads," to be published in *The Microwave Journal*. (Compare Fig. 15.01-3).

The design of high-power, low-loss waveguide filters treated in Sec. 15.03 is further considered in:

63. Leo Young, "Some Considerations in the Design of Narrow-Band Waveguide Filters," to be published in the *IEEE Trans. PTGMTT*.

CHAPTER 16

MULTIPLEXER DESIGN

SEC. 16.01, INTRODUCTION

As was discussed in Sec. 1.02, filters connected in parallel or in series are often needed in order to split a single channel carrying many frequencies into a number of separate channels carrying narrower bands of frequencies. Similar filter groups are often required also for the inverse process of summing a number of channels carrying different bands of frequencies so that all of the frequencies can be put in a single broad-band channel without loss of energy (which would otherwise occur due to leakage of energy from any one of the input channels into the other input channels).

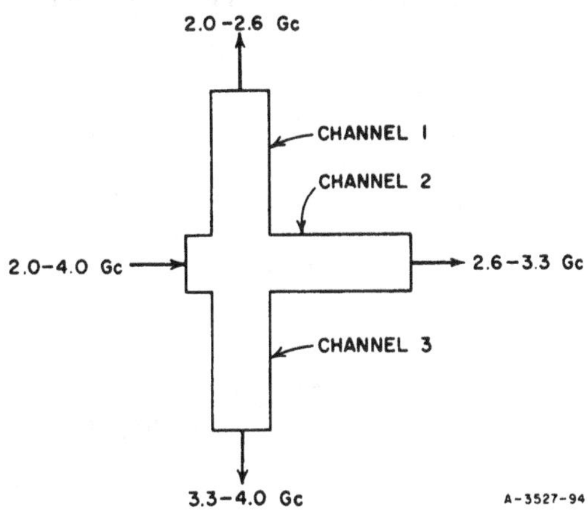

FIG. 16.01-1 A THREE-CHANNEL MULTIPLEXING FILTER GROUP

Figure 16.01-1 shows a three-channel multiplexer which would use three separate band-pass filters, one to provide the 2.0- to 2.6-Gc channel, a second to provide the 2.6- to 3.3-Gc channel, and a third to provide the 3.3- to 4.0-Gc channel. It might at first appear that the design of this multiplexer could easily be accomplished by designing the filters using any of the band-pass filter design procedures previously discussed in this book, and then connecting the filters in parallel. However, though the procedures previously discussed are useful for multiplexer design, they should be used along with special techniques in order to avoid undesirable interaction between the filters, which could result in very poor performance.

Most of the discussions in this chapter relate to the problem of design so as to eliminate such undesirable interaction effects. However, the constant-resistance-filter approach discussed in the next section avoids this problem entirely by using directional-filter units which have matched, constant-resistance input impedances at all frequencies so that in theory no interaction effects should occur.

SEC. 16.02, MULTIPLEXERS USING DIRECTIONAL FILTERS

The directional filters discussed in Chapter 14 have a constant-resistance input impedance provided that their output ports are terminated in their proper resistance terminations. Filters of this sort, when all designed for the same terminating resistance, can be cascaded as shown in Fig. 16.02-1 to form a multiplexer which in theory completely avoids the filter interaction problem mentioned in Sec. 16.01.

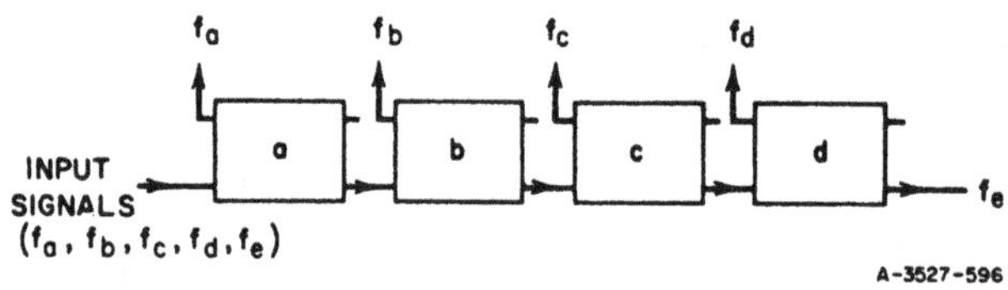

FIG. 16.02-1 EXAMPLE OF DIRECTIONAL FILTERS USED FOR MULTIPLEXING

Each filter provides the proper termination for its neighbor, so that to the extent that there are no residual VSWR's due to design and manufacturing imperfections, the system is reflectionless. In Fig. 16.02-1, Filter *a* removes all of the energy at frequency f_a, but passes on the energy at all other frequencies. Filter *b* removes the energy at frequency f_b and so on.

The use of directional filters for multiplexing is a conceptually simple and elegant way of solving the multiplexer problem. In many cases it is also a very practical way of dealing with multiplexing problems, though by no means always the most practical way. Each filter will generally have some parasitic VSWR, which will affect the system significantly if many filters are to be cascaded. However, probably the greatest practical drawback of directional filters is that each resonator of each filter has two different orthogonal modes, and if more than one or two resonators are required per filter, the tuning of the filters may be difficult. Although an effort is made to tune the two modes of each resonator independently with separate sets of tuning screws, the tuning screws for one mode generally have some effect on the other mode, so that some iteration in the tuning process is generally required. Thus, if there are more than, say, two resonators per filter, the fact that the two modes in each resonator must be tuned separately, along with unavoidable interaction effects in the tuning mechanism, can make the tuning process quite time consuming.

Figure 16.02-2 shows another form of directional filter which also has the constant-resistance input properties exhibited by the filters in Chapter 14.[1,2,3] In this circuit two hybrid junctions are used, such as the 3-db couplers in Sec. 13.03, the 3-db branch-guide couplers treated in Secs. 13.12 and 13.13, short-slot hybrids, or Magic-T's. In the case of the Magic-T, an extra quarter-wavelength of line is required on one of the side ports of the hybrid in order to give the desired 90-degree phase difference between the outputs of the side ports of the hybrid. In Fig. 16.02-2, the required phase relations for transmission between the various ports is indicated by arrows and numbers.

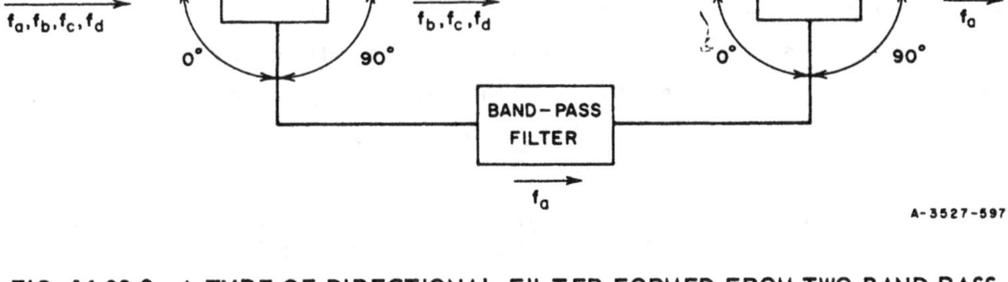

FIG. 16.02-2 A TYPE OF DIRECTIONAL FILTER FORMED FROM TWO BAND-PASS FILTERS AND TWO HYBRIDS

Besides the hybrid junctions, the circuit in Fig. 16.02-2 also uses two identical band-pass filters which are designed to pass some frequency f_a, which the circuit is to separate from other frequencies. Thus, if four frequencies f_a, f_b, f_c, and f_d are introduced at Port 1, the energy at frequency f_a is divided equally between the two side arms of the hybrid on the left, it passes through the two band-pass filters, and on into the side arms of the hybrid on the right. Because of the phase relations of the two components, the signal cancels to zero at Port 3, and all of the energy at frequency f_a emerges from Port 4. Meanwhile the other frequencies f_b, f_c, and f_d are reflected by the band-pass filters and they re-enter the side arms of the hybrid on the left. Because of the phase relation, the components cancel at Port 1 and all of the energy at these frequencies emerges from Port 2. The over-all performance of this circuit is just like that of the directional filters in Chapter 14, and filter circuits of this type may also be cascaded as indicated in Fig. 16.02-1.

The circuit in Fig. 16.02-2 will usually be much easier to adjust than the directional filters in Chapter 14, if there are very many resonators per filter. The two identical band-pass filters can be tuned separately, and the performance of the hybrid junctions is not particularly critical. For example, if the power split of the hybrid junction is not equal, and, say, is so bad that twice as much power comes out one side arm as comes out the other, the net result if both hybrids are identical will be to introduce only 0.5 db of attenuation in the transmission from Port 1 to Port 4.[3] The circuit in Fig. 16.02-2 has the drawback, however, of being physically quite complicated and in many cases quite bulky.

In summary, directional filters with their constant-resistance input properties provide attractive possibilities for use in multiplexers. However, they also have appreciable practical drawbacks. In any given design situation their merits and drawbacks should be weighed against those of other possible multiplexer techniques.

SEC. 16.03, MULTIPLEXERS USING REFLECTING NARROW-BAND FILTERS, WITH GUARD BANDS BETWEEN CHANNELS

If the channels of a multiplexer are quite narrow (say, of the order of 1 percent bandwidth or less) and if the channels are separated by guard bands which are several times the pass-band width of the individual filters (or more), then various relatively simple decoupling techniques should work quite well for preventing harmful interaction between filters.

One method is shown in Fig. 16.03-1.[1,4] In this figure a three-channel waveguide multiplexer is shown; the individual filters are of the iris-coupled type discussed in Secs. 8.06 and 8.07. At frequency f_a, the filter on the right has a pass band while the other two filters are in their stop bands. Since the filters are of relatively narrow bandwidth, the first resonator of both the upper and the lower filter is relatively loosely coupled to the main waveguide, hence the input coupling irises of these two filters introduce only small reactive discontinuities in the main waveguide. These discontinuities will disturb the response of the filter on the right a little, and if this is objectionable, it can be compensated for by small corrections in the tuning and input coupling of the first resonator of the filter on the right.

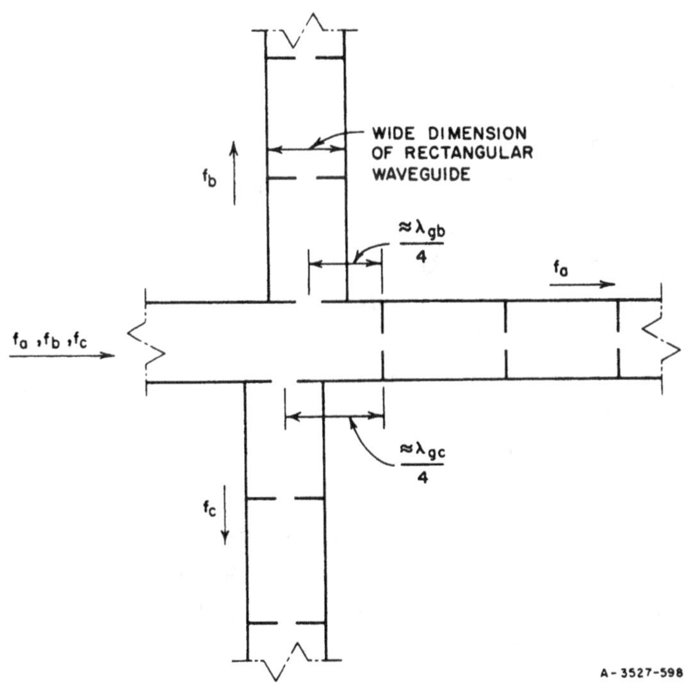

FIG. 16.03-1 A MULTIPLEXER WITH NARROW-BAND WAVEGUIDE FILTERS MOUNTED SO AS TO ELIMINATE FILTER INTERACTION

At frequency f_b the upper filter in Fig. 16.03-1 has a pass band hile the other filters have stop bands and reflect small reactances to the main quide. If the upper filter is placed approximately one-quarter guide-wavelength (at frequency f_b) from the input iris of the filter on the right, the filter on the right will reflect an open circuit to the plane of the coupling iris. (Actually, because of the residual reactances, somewhat less than a quarter wavelength should be best.) Thus, the filter on the right will be completely decoupled, and the energy at frequency f_b will flow relatively undisturbed out through the upper filter. There will usually be some residual junction effects to disturb the performance of the upper filter a little. Again, these effects can be compensated for by small corrections in the input coupling iris size and in the tuning of the first resonator of the upper filter.

The lower filter is coupled to the main waveguide in a similar manner. If it is desired to add more channels, they can be mounted on the main waveguide at additional points corresponding to various odd multiples of $\lambda_g/4$ from the filter on the right in Fig. 16.03-1 (where λ_g is in each case the guide wavelength at the midband frequency of the filter in question).

Figure 16.03-2 shows another decoupling technique, which has been proposed by J. F. Cline.[6] Although the circuits have been drawn in lumped-element form, they can be realized in a variety of more or less equivalent microwave forms. In this case the filters are again assumed to be narrow band, with guard bands between channels. However, in this case decoupling between filters is achieved by a decoupling resonator adjacent to each filter. Since the filters are narrow band, their coupling to the main transmission line is quite loose, and the coupling between the decoupling resonator and the main line is also quite loose. Each decoupling resonator is tuned to the pass-band center frequency of its adjacent filter. Because of their loose couplings the filters and the decoupling resonators have very little effect on transmission when they are off resonance. Suppose that a signal of frequency f_2 enters the circuit. It will pass the f_1 filter and the f_1 decoupling resonator with very little reflection and go on to the f_2 filter and decoupling resonator. The decoupling resonator is a band-stop resonator and in the upper circuit of Fig. 16.03-2 the f_2 decoupling resonator circuit will short-circuit the entire circuit to the right, while in the case of the lower circuit of Fig. 16.03-2 the decoupling resonator open-circuits the entire circuit

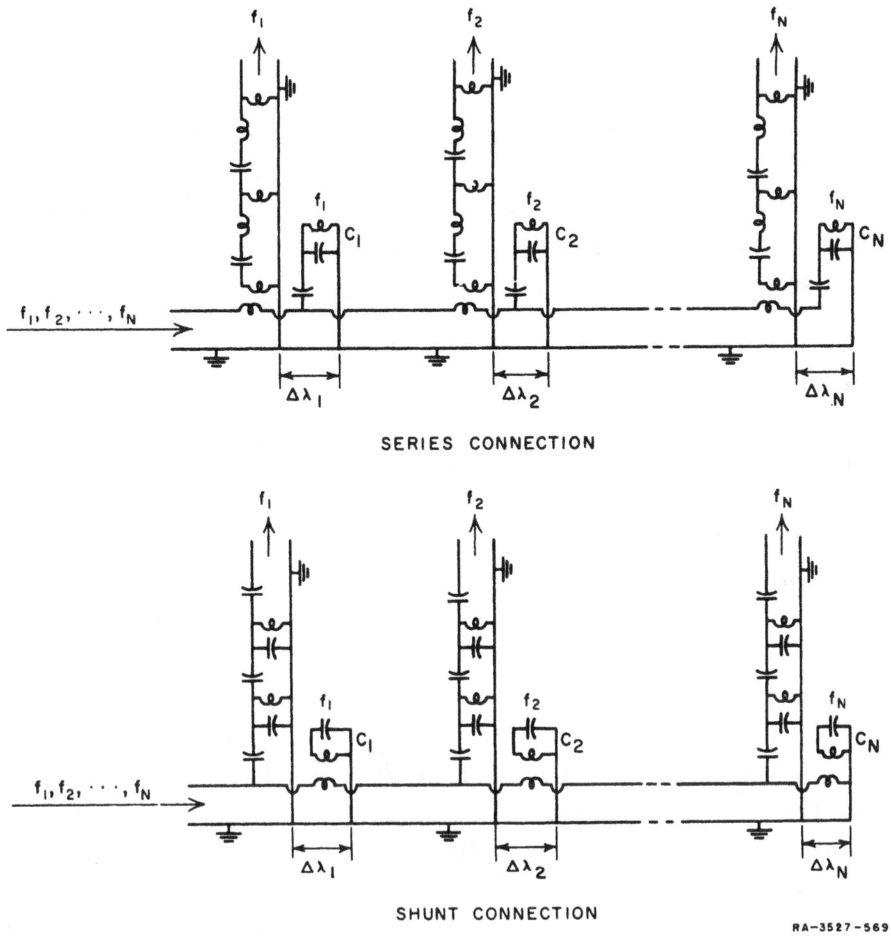

SOURCE: **Final Report, Contract AF 19(606)-2247, SRI (see Ref. 6 by J. F. Cline, *et al.*)**

FIG. 16.03-2 MULTIPLEXERS USING DECOUPLING RESONATORS
In the upper structure filters with inductively coupled resonators are used, while the decoupling resonators are capacitively coupled. In the lower circuit the reverse is true.

to the right. In this manner the energy is channelled to the desired filter with a minimum of interaction effects. Note that in this case the spacing between filters is not critical, so that this multiplexing technique is particularly useful where it must be possible to tune the various filters over a range of frequencies. In the figure the spacing $\Delta\lambda_k$ between each filter and its decoupling resonator is assumed to be made as small as possible in order to avoid undesirable transmission-line effects. Residual reflections in the system can be largely compensated for by small adjustments of the tuning of the first resonator of each filter, and of the coupling between the transmission line and the input of each filter.

A possible way for utilizing the principle in Fig. 16.03-2 for waveguide filters is suggested in Fig. 16.03-3. The band-pass filter shown is of the iris-coupled type discussed in Secs. 8.06 and 8.07. The decoupling resonator is of the waveguide band stop type discussed in Chapter 12. The band-stop resonator presents a very large series reactance when it is resonant, but this resonator is located a quarter-guide-wavelength from the band-pass filter so that when looking right from the band-pass filter a very large shunt susceptance will be seen when the filter and band-stop resonator are resonant. The filter, being series-connected will then receive all of the energy at frequency f_a, and any circuits to the right will be completely decoupled. Just outside of the pass band of the band-pass filter, the first resonator of the band-pass filter will present reactances to the main waveguide which are similar in nature to those presented by the band-stop resonator. Thus at frequencies off f_a, the structure functions almost exactly like a two-resonator band-stop filter of the form in Fig. 12.08-1, operating in its pass band. It should be possible to design the band-stop resonator using the band-stop filter techniques discussed in Chapter 12. (The band-pass filter is replaced mathematically by a band-stop resonator having a similar off-resonance reactance characteristic, and the decoupling resonator represents the second band-stop resonator.) Using this approach the band-pass filter and band-stop resonator combination could present a well-controlled low VSWR at frequencies off f_a, just as occurs in properly designed two-resonator band-stop filters.

The approach suggested in Fig. 16.03-2 and 16.03-3 looks especially attractive if very many channels are to be multiplexed, because using this approach the individual band-pass filter and decoupling resonator

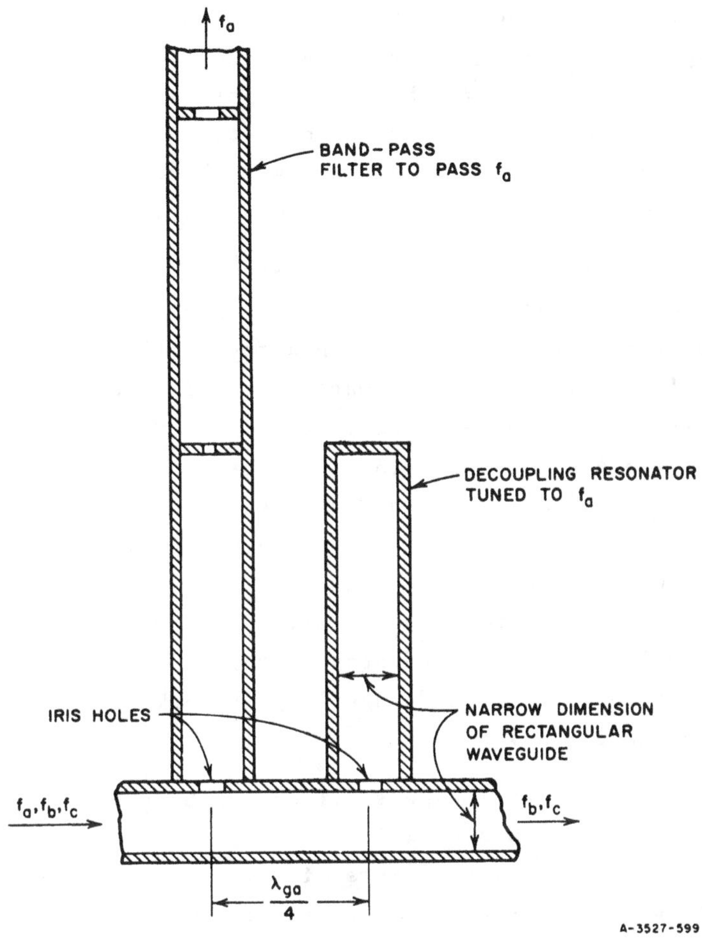

FIG. 16.03-3 A POSSIBLE METHOD FOR REALIZING DECOUPLING RESONATORS IN A WAVEGUIDE MULTIPLEXER

units can be developed and tested separately, and then connected together more or less as is convenient. Strip-line, coaxial-line, and semi-lumped-element forms of these structures are also possible.

SEC. 16.04, MULTIPLEXERS WITH CONTIGUOUS PASS BANDS*

The multiplexers discussed in the preceding section were assumed to have guard bands separating the various operating channel bands.

---

*The design equations and trial designs in this section were worked out by E. G. Cristal.

The multiplexers to be discussed in this section, however, are assumed to have contiguous channels so that channels which are adjacent have attenuation characteristics which typically cross over at their 3 db points. Most diplexers are simply two-channel versions of multiplexers, so the same principles apply to them. However, some additional comments with respect to diplexer design will be found in Sec. 16.05.

Figure 16.04-1 shows a schematic drawing which consists of N channels composed of specially designed band-pass filters. The channels are connected in parallel, and a susceptance-annulling network is added in shunt to help provide a nearly constant total input admittance Y_{TN} which approximates the generator conductance G_B, across the operating band of the multiplexer. Figure 16.04-2 shows the analogous case of a series-connected multiplexer. Since the series case is the exact dual of the shunt case, so that the same principles apply to both, our attention will be confined largely to the shunt case. However, it should be understood that the same remarks apply to series-connected multiplexers simply by replacing admittances by impedances, parallel connections by series connections, and the filters and annulling networks used for parallel connections by their duals.

The singly terminated low-pass filter Tchebyscheff prototypes discussed and tabulated in Sec. 4.06 are very useful for use in the design of band-pass filters for multiplexers of this type. Consider the singly terminated low-pass filter circuit in Fig. 16.04-3, which is driven by a zero-impedance generator at the right end and which has a resistor termination only at the left end. As was discussed in Sec. 4.06, the power delivered to the load on the left in this circuit is given by

$$P = |E_g|^2 \, \mathrm{Re}\, Y_k' \qquad (16.04\text{-}1)$$

where Y_k' is the admittance seen from the generator. Thus if the filter has a Tchebyscheff transmission characteristic, Re Y_k' must also have a Tchebyscheff characteristic. Figure 16.04-4 shows typical Re Y_k' characteristics for Tchebyscheff filters designed to be driven by zero-impedance generators. Notice that in the figure a low-pass prototype filter parameter g''_{n+1} is defined. This parameter will be referred to later with regard to the details of multiplexer design. However, it should be noted at this point that either g''_{n+1} or its reciprocal corresponds to the geometric mean between the value of Re Y_k' at the top of the

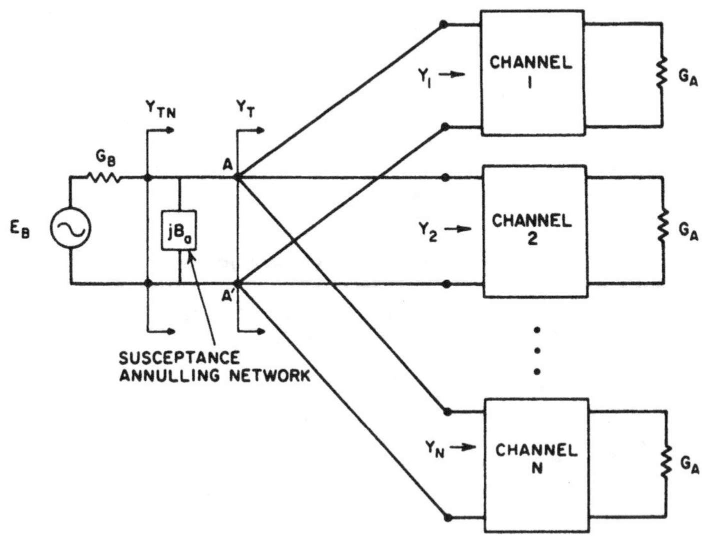

SOURCE: Quarterly Progress Report 6, Contract DA 36-039 SC-87398, SRI; reprinted in the *IEEE Trans. PTGMTT* (see Ref. 10 by E. G. Cristal and G. L. Matthaei)

FIG. 16.04-1 A PARALLEL-CONNECTED MULTIPLEXER

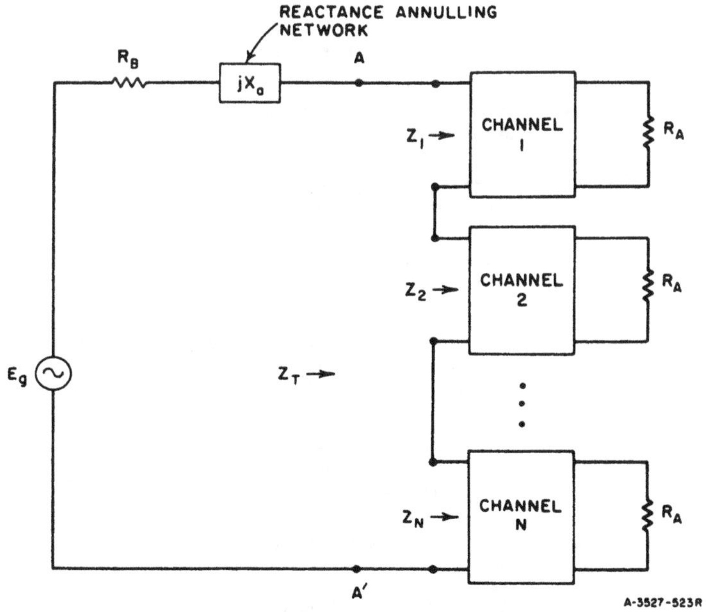

FIG. 16.04-2 A SERIES-CONNECTED MULTIPLEXER

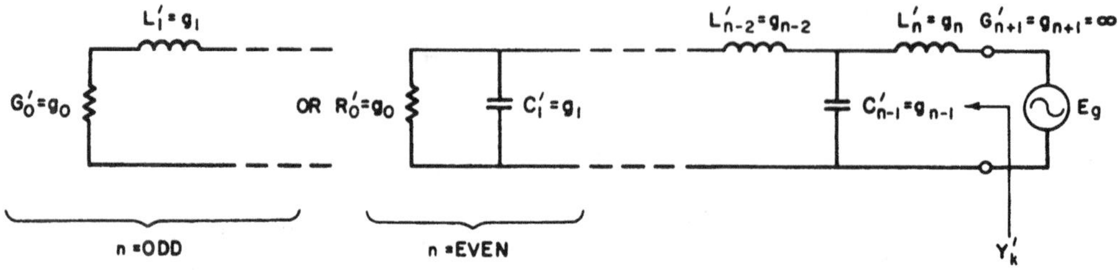

ω_1' = PASS-BAND EDGE FREQUENCY

A-3527-527R

SOURCE: Quarterly Progress Report 6, Contract DA 36-039 SC-87398, SRI; reprinted in the *IEEE Trans. PTGMTT* (see Ref. 10 by E. G. Cristal and G. L. Matthaei)

FIG. 16.04-3 A LOW-PASS PROTOTYPE FILTER DRIVEN BY A ZERO-IMPEDANCE GENERATOR

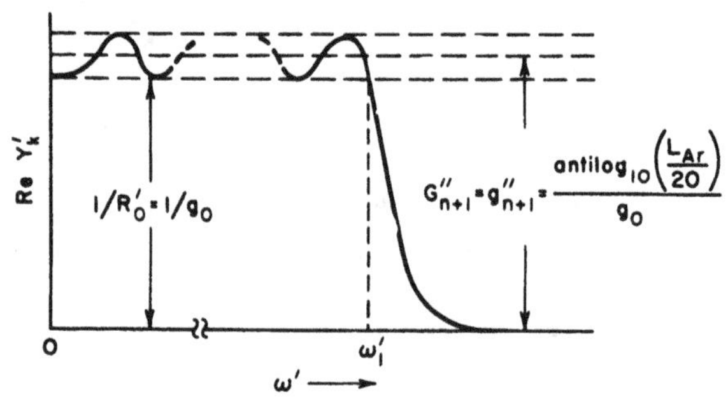

(a) CASE OF n EVEN

L_{Ar} = TCHEBYSCHEFF ATTENUATION RIPPLE OF SINGLY LOADED PROTOTYPE IN db

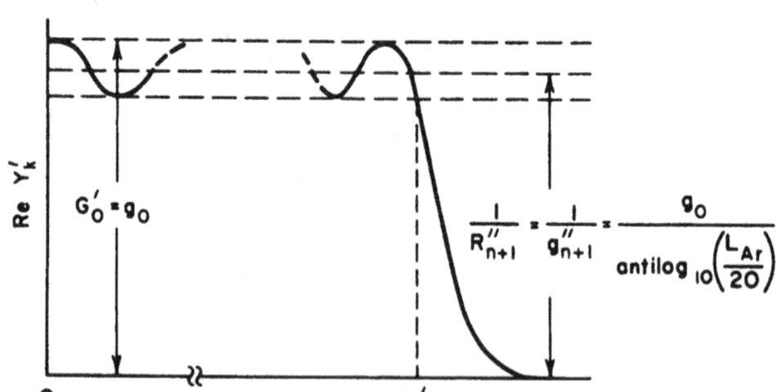

(b) CASE OF n ODD

A-3527-528R

SOURCE: Quarterly Progress Report 6, Contract DA 36-039 SC-87398, SRI; reprinted in the *IEEE Trans. PTGMTT* (see Ref. 10 by E. G. Cristal and G. L. Matthaei)

FIG. 16.04-4 REAL PART OF THE INPUT ADMITTANCE FOR TCHEBYSCHEFF FILTERS OF THE TYPE IN FIG. 16.04-3

ripples and Re Y'_k at the bottom of the ripples. This admittance level for the low-pass prototype is analogous to the terminating admittance G_B in Fig. 16.04-1, for the actual multiplexer with band-pass filters. That is, by the methods of this section the multiplexer is to be designed to have a Tchebyscheff Re Y_{TN} characteristic with G_B equal to the mean value of the ripples.

The principles of this design procedure are most easily understood in terms of an example. Consider the comb-line filter in Fig. 16.04-5 (which can be designed by a modified version of the comb-line filter design techniques discussed in Secs. 8.13 and 8.14). In this figure the filter has been designed to include a fine-wire high-impedance line at the right end, for coupling to the common junction of the multiplexer. This type of coupling to the main junction has the advantage of relieving possible crowding of the filters at the common junction. Other coupling methods, such as series, capacitive-gap coupling, could also be used. However, it should be noted that although numerous types of filter structures and structures for coupling to the common junction are possible, these structures must be of the sort which tends to give a small susceptance in the stop bands, if the filters are to be connected in parallel. If the filters were of the type to present large susceptances in their stop bands, they would tend to short-circuit the other filters.

A four-resonator, 10-percent-bandwidth filter design was worked out for a comb-line filter as shown in Fig. 16.04-5 using an n = 4 singly

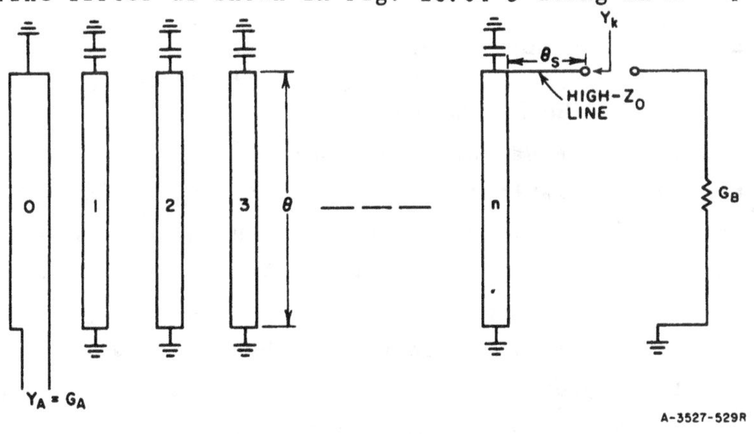

SOURCE: Quarterly Progress Report 6, Contract DA 36-039 SC-87398, SRI; reprinted in the *IEEE Trans. PTGMTT* (see Ref. 10 by E. G. Cristal and G. L. Matthaei)

FIG. 16.04-5 A POSSIBLE COMB-LINE FILTER CONFIGURATION FOR USE IN MULTIPLEXERS

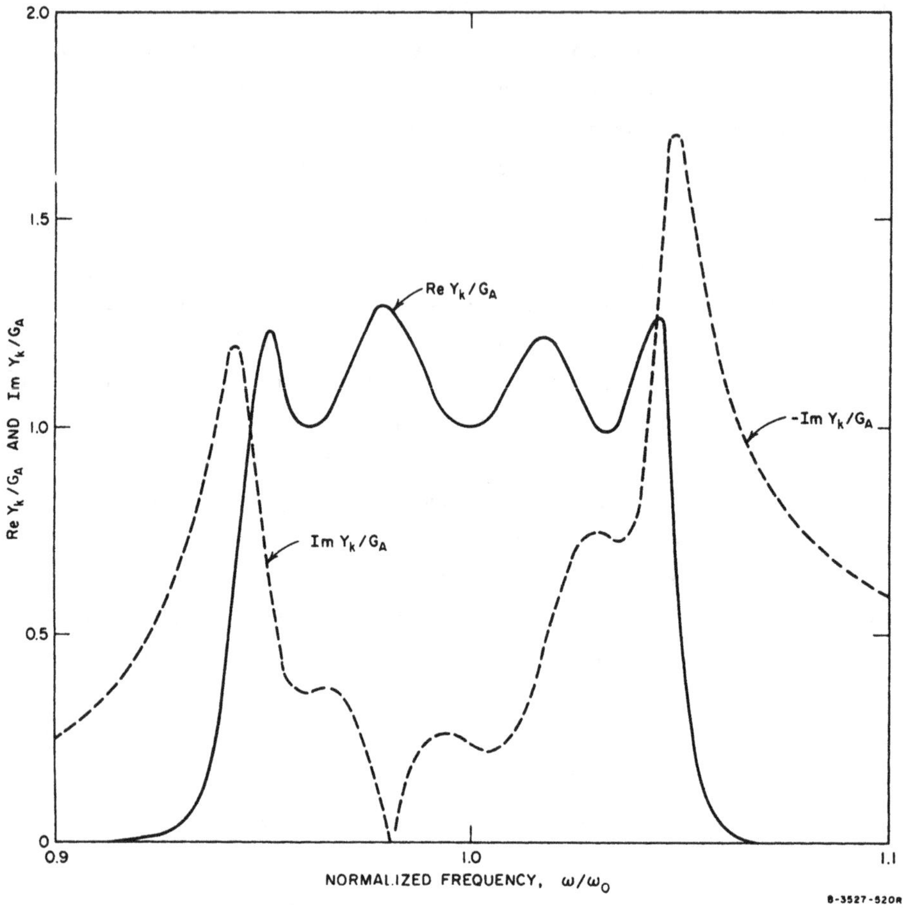

SOURCE: Quarterly Progress Report 6, Contract DA 36-039 SC-87398, SRI; reprinted in the *IEEE Trans. PTGMTT* (see Ref. 10 by E. G. Cristal and G. L. Matthaei)

FIG. 16.04-6 COMPUTED INPUT ADMITTANCE CHARACTERISTIC OF A COMB-LINE MULTIPLEXER FILTER AS SHOWN IN FIG. 16.04-5
This filter has four resonators

terminated prototype having 1-db ripple. (The prototype element values were obtained from Table 4.06-2). This filter design was expressed in an approximate form similar to that in Fig. 8.14-1, and its input admittance Y_k was computed using a digital computer. The results are shown in Fig. 16.04-6, normalized with respect to G_A. Note that Re Y_k/G_A is very nearly perfectly Tchebyscheff, while in the pass band the slope of $Im\ Y_k/G_A$ is negative on the average. (Note that the right half of the dashed curve is for negative values.)

If band-pass filters with input admittance characteristics such as that in Fig. 16.04-6 are designed to cover contiguous bands, and if they are designed so that the Re Y_k/G_A characteristics of adjacent filters overlap at approximately their Re $Y_k/G_A = 0.5$ points or slightly below, then their total input admittance $Y_T = Y_1 + Y_2 + Y_3 \ldots + Y_N$, when they are paralleled, will also have an approximately Tchebyscheff real-part characteristic. This was done in the example in Fig. 16.04-7(a). This figure shows the computed real-part characteristic for three paralleled, comb-line filters with contiguous pass bands, where the individual filters have input admittance characteristics as shown in Fig. 16.04-6 except for a shift in frequency. Figure 16.04-7(b) shows the corresponding *Im* Y_T/G_A characteristic for this three-channel design.

Note that the *Im* Y_T/G_A characteristic in Fig. 16.04-7(b) is negative on the right side of the figure, and that on the average the slope of the curve is negative throughout the operating band of the multiplexer. Since, by Foster's reactance theorem, the susceptance slope of a lossless network is always positive, the operating-band susceptance in Fig. 16.04-7(b)with its average negative slope can be largely cancelled by adding an appropriate lossless shunt branch (which will have a positive susceptance slope). In this case a susceptance-annulling branch could consist of a short-circuited stub of such a length as to give resonance at the normalized frequency $\omega/\omega_0 = 1.02$, where the *Im* Y_T/G_A curve in Fig. 16.04-7(b) is approximately zero. Estimates indicate that if the stub had a normalized characteristic admittance of $Y_s/G_A = 3.8502$, the susceptance slope of the annulling network should be about right. Figure 16.04-8 shows the normalized susceptance Y_{TN}/G_A (Fig. 16.04-1) after the susceptance-annulling network has been added. Note that although *Im* Y_{TN} has not been completely eliminated, it has been greatly reduced.

Since Re Y_{TN}/G_A = Re Y_T/G_A, the total input admittance Y_{TN} with the annulling network in place is given by Figs. 16.04-7(a) and 16.04-8. To the extent that Y_{TN} approximates a constant, pure conductance across the operating band, G_B and Y_{TN} in Fig. 16.04-1 will act as a resistive voltage divider, and the voltage developed across terminals $A - A'$ will be constant with frequency. Under these conditions the response of the individual filters would be exactly the same as if they were driven by zero-impedance generators (as they were designed to be driven). However, since Y_{TN} only approximates a pure constant conductance, the performance will be altered somewhat from this idealized performance.

The driving generator conductance G_B for the trial multiplexer design was given a normalized value of $G_B/G_A = 1.15$, which makes the

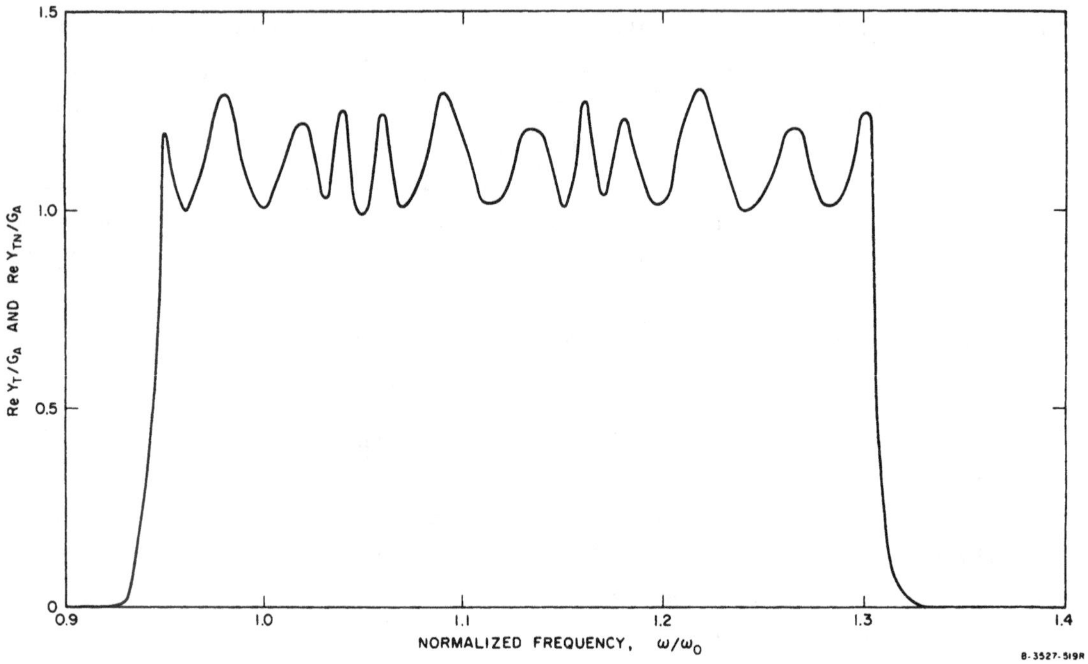

SOURCE: Quarterly Progress Report 6, Contract DA 36-039 SC-87398, SRI; reprinted in the *IEEE Trans. PTGMTT* (see Ref. 10 by E. G. Cristal and G. L. Matthaei)

FIG. 16.04-7(a) Re Y_T/G_A vs. FREQUENCY FOR THREE, PARALLELED, COMB-LINE MULTIPLEXER FILTERS

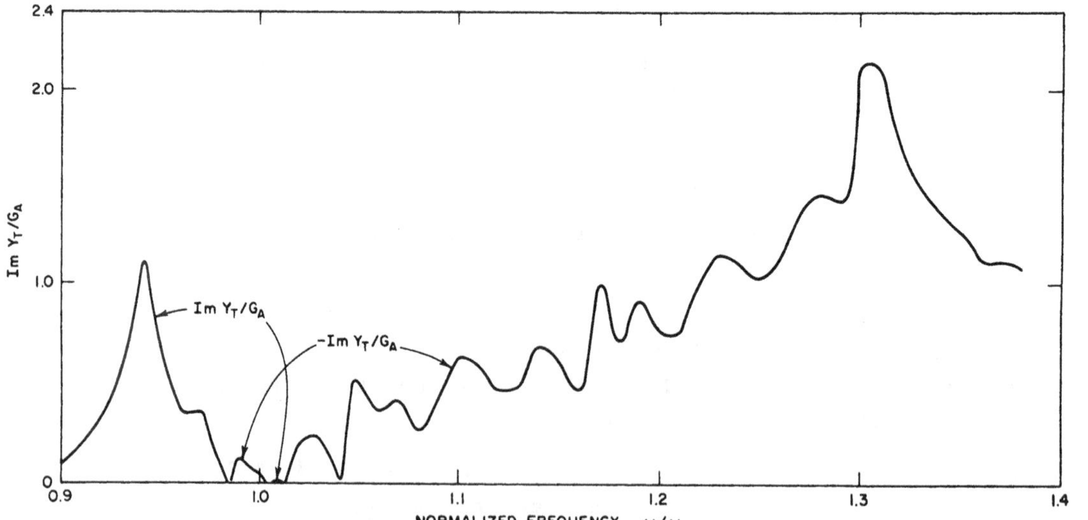

SOURCE: Quarterly Progress Report 6, Contract DA 36-039 SC-87398, SRI; reprinted in the *IEEE Trans. PTGMTT* (see Ref. 10 by E. G. Cristal and G. L. Matthaei)

FIG. 16.04-7(b) Im Y_T/G_A vs. FREQUENCY FOR THREE, PARALLELED, COMB-LINE FILTERS

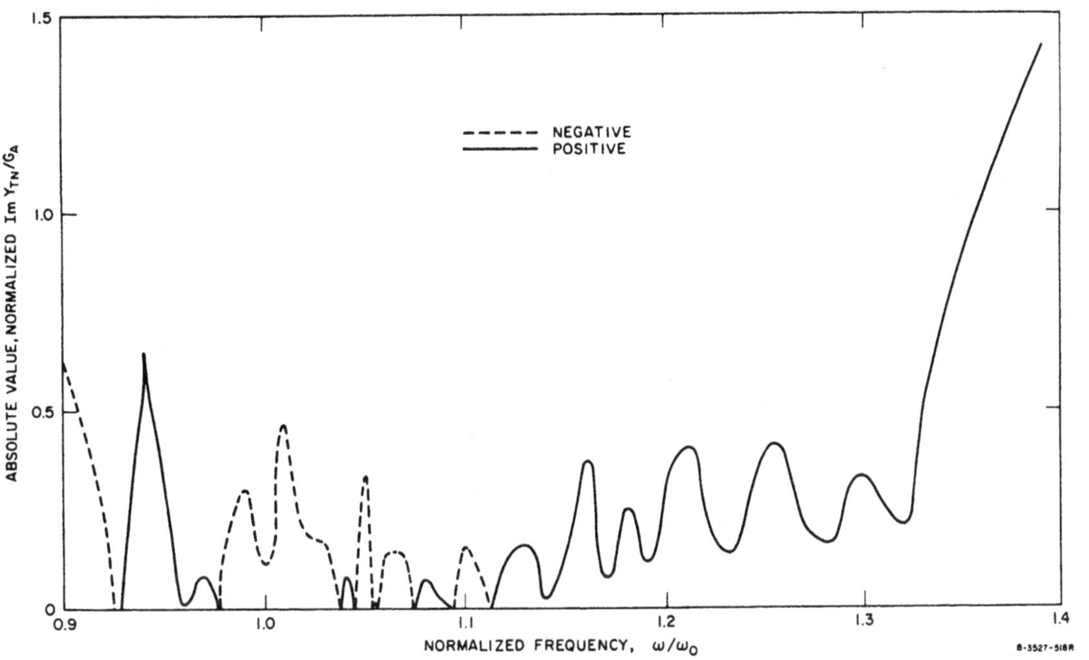

SOURCE: Quarterly Progress Report 6, Contract DA 36-039 SC-87398, SRI; reprinted in the *IEEE Trans. PTGMTT* (see Ref. 10 by E. G. Cristal and G. L. Matthaei)

FIG. 16.04-8 Im Y_{TN}/G_A vs. FREQUENCY FOR MULTIPLEXER WITH COMB-LINE FILTERS

generator conductance equal to the mean value of the ripples in Fig. 16.04-7(a). Figure 16.04-9(a) shows the computed response of the multiplexer, while Fig. 16.04-9(b) shows the details of the pass-band response in enlarged scale. Note that the attenuation characteristics cross over at about the 3-db points, and that though the filters were designed to have 1-db Tchebyscheff ripple when driven by a zero-impedance generator, the pass-band attenuation is much less than that in the completed multiplexer. This is due largely to the fact that adding the generator internal conductance G_B tends to mask out the variations in Re Y_{TN}. Also, choosing G_B to be equal to the mean value of Re Y_{TN} in the operating band tends to reduce the amount of mismatch that will occur.

Some Practical Details—The discussion of the example above has laid out the principles of the multiplexer design technique under consideration. Some steps will now be retraced in order to treat design details.

When designing the band-pass channel filters using singly terminated prototype filters such as that in Fig. 16.04-3, the prototype element values g_0, g_1, ..., g_n, g_{n+1} can be obtained from the tables

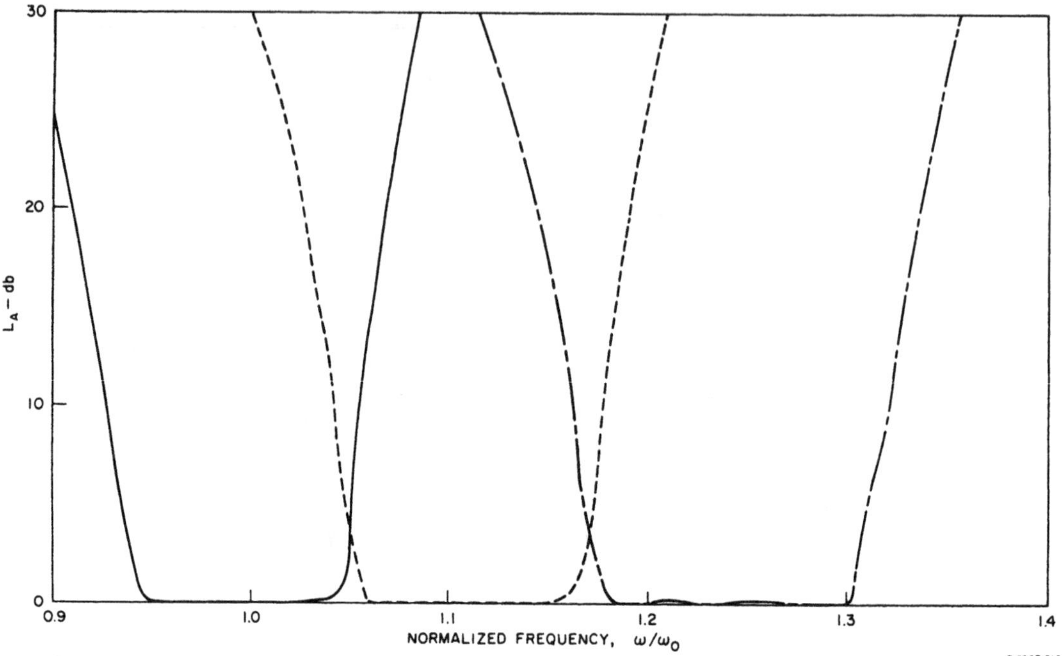

SOURCE: Quarterly Progress Report 6, Contract DA 36-039 SC-87398, SRI; reprinted in the *IEEE Trans. PTGMTT* (see Ref. 10 by E. G. Cristal and G. L. Matthaei)

FIG. 16.04-9(a) COMPUTED CHANNEL ATTENUATION OF MULTIPLEXER WITH COMB-LINE FILTERS

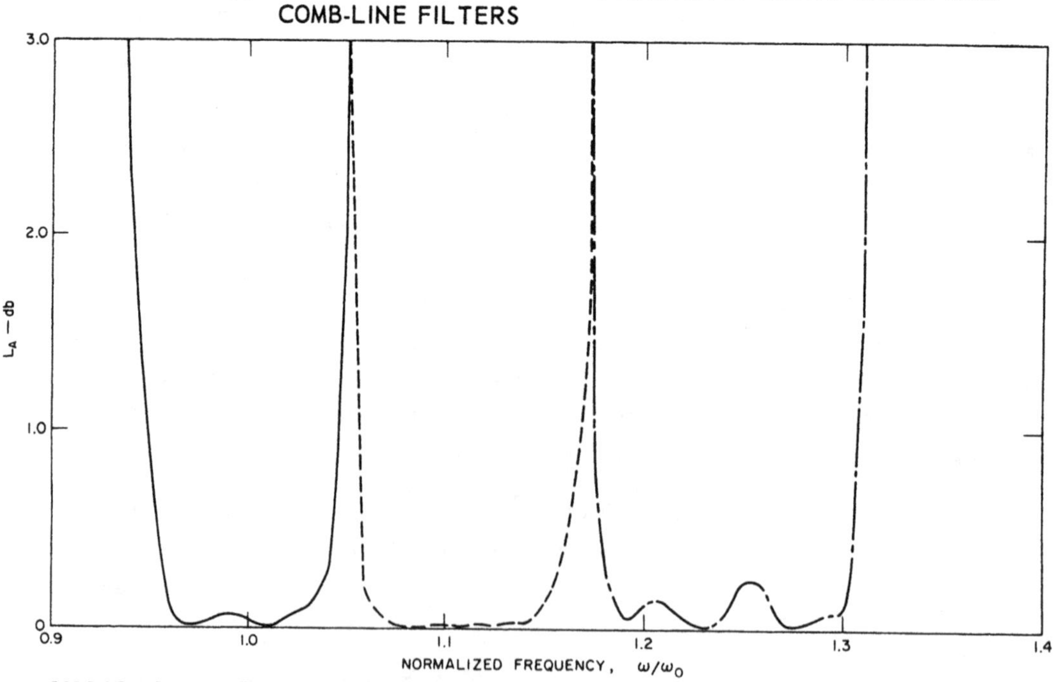

SOURCE: Quarterly Progress Report 6, Contract DA 36-039 SC-87398, SRI; reprinted in the *IEEE Trans. PTGMTT* (see Ref. 10 by E. G. Cristal and G. L. Matthaei)

FIG. 16.04-9(b) DETAILS OF PASS-BAND COMPUTED PERFORMANCE OF COMB-LINE MULTIPLEXER

in Sec. 4.06. Note then in this case $g_{n+1} = \infty$ which corresponds to the infinite-internal-conductance of a zero-impedance voltage generator (or for the dual case, the impedance of an infinite-internal-impedance current generator). In the actual multiplexer the filters will be driven by a finite-internal-impedance generator, so it is convenient to replace $g_{n+1} = \infty$ by g''_{n+1} defined in Fig. 16.04-4. As has been previously discussed, g''_{n+1} (or its reciprocal) is a prototype parameter corresponding to the conductance G_B (or the resistance R_B for the dual series-connected case) of the termination to be used at the common junction of the multiplexer.

The band-pass filters for a multiplexer can be designed by the methods of Chapter 8 or Chapter 10, using any of a variety of structures. For example, if a lumped-element structure were desired it could be designed directly from the prototype in Fig. 16.04-3 using the mapping procedure summarized in Fig. 8.02-2(b). The resulting filter would be as shown in Fig. 16.04-10. Note that this filter starts out with a *series* resonator, which will cause the input admittance Y_k to be small in the stop band. This is necessary for a filter to be used in a parallel-connected multiplexer. The dual form of filter which has a *shunt* resonator at its input is appropriate for a series-connected multiplexer. In the cases of filters such as those in Fig. 8.02-3 or 8.02-4 in which the resonators are coupled by J- or K-inverters (Secs. 4.12, 8.02 and 8.03), it is necessary that the inverters next to the common junction of the multiplexer be of the J-inverter type (Fig. 8.03-2) if the filters are to be connected in parallel, and that they be of the K-inverter type (Fig. 8.03-1) if the filters are to be connected in series.

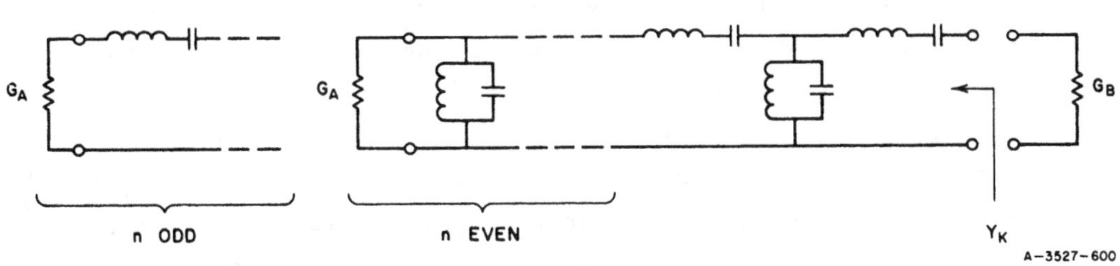

FIG. 16.04-10 A LUMPED-ELEMENT BAND-PASS CHANNEL FILTER

Most of the design data in Chapters 8 and 10 applies directly to the design of multiplexers without any changes, at least for the case where the desired terminations of the multiplexer are all equal. Some exceptions are the design data in Figs. 8.09-1 and 8.13-1(b), and in Tables 10.02-1 and 10.06-1. These design procedures incorporate special matching end sections which have an effect like referring the desired Y_k (or Z_k) input characteristic through a length of transmission line, which will cause the real part of the input admittance (or impedance) to not have the desired shape. This problem can be overcome by replacing the matching section at the end of the filter which is to be connected to the multiplexer junction by a semi-lumped-element inverter. This is what was done in the case of the comb-line filter in Fig. 16.04-5, where a fine-wire coupling is shown at the right end. This point will be discussed further, later in this section.

After the low-pass prototype filter parameters have been specified there is no choice in the ratio G_A/G_B of the terminations if lumped-element filter designs of the form in Fig. 16.04-10 are to be used. Also, the ratio G_A/G_B is not unity for that case—a situation that could be inconvenient. However, in the case of designs using impedance inverters (as in the generalized cases in Figs. 8.02-3 and 8.02-4) the terminations can be specified as desired, and the J- or K-inverters will provide the necessary impedance transformations. For example, using the data in Fig. 8.02-4, G_A, G_B and the resonator slope parameters b_j can be chosen arbitrarily, and the impedance-matching conditions will automatically come out as is called for by the prototype parameters, g_0, g_1, g_2, ... g_n, and g''_{n+1}.

Determination of Fractional Bandwidths So As To Give Desired Cross-Over Frequencies—Since it is usually desirable in the type of multiplexer under discussion for adjacent filters to cross over at their 3-db points, care must be taken in choosing the filter bandwidths to provide for this condition. By adapting Eqs. (4.03-4) and (4.03-5) an explicit expression for Re Y'_k for the singly loaded low-pass prototype filter can be obtained as a function of ω'. Then for n even

$$\left.\frac{\omega'}{\omega'_1}\right|_{\omega'>\omega'_1} = \cosh\left[\frac{1}{n}\cosh^{-1}\sqrt{\frac{1}{\epsilon}\left(\frac{1+\epsilon}{R'_0 \text{ Re } Y'_k} - 1\right)}\right] \qquad (16.04\text{-}2)$$

where

$$\epsilon \doteq \left[\operatorname{antilog}_{10}\left(\frac{L_{Ar}}{10}\right)\right] - 1 \quad , \tag{16.04-3}$$

L_{Ar} is the db Tchebyscheff ripple of the prototype filter, R_0' is the termination in Fig. 16.04-3, and Re Y_k' and ω_1' are as indicated in Fig. 16.04-4. Having specified R_0' Re Y_k' = 0.5, corresponding approximately to the 3-db point, then the designer can solve for the corresponding normalized frequency ω'/ω_1'. From this the required fractional bandwidth can be obtained by using the low-pass to band-pass transformation appropriate for the filter structure to be used. For example, suppose that the desired cross-over points are f_a and $f_b > f_a$, and an appropriate mapping function is

$$\frac{\omega'}{\omega_1'} = \frac{2}{w}\left(\frac{f - f_0}{f_0}\right) \tag{16.04-4}$$

where

$$f_0 = \frac{f_a + f_b}{2} \quad .$$

Then the required fractional bandwidth is

$$w = 2\left(\frac{f_b - f_0}{f_0}\right)\bigg/\left(\frac{\omega'}{\omega_1'}\right) \quad , \tag{16.04-5}$$

where ω'/ω_1' is specified for the desired 3-db point by use of Eq. (16.04-2), provided that n is even. If n is odd, Eq. (16.04-2) becomes

$$\left.\frac{\omega'}{\omega_1'}\right|_{\omega'>\omega_1'} = \cosh\left[\frac{1}{n}\cosh^{-1}\sqrt{\frac{1}{\epsilon}\left(\frac{G_0'}{\operatorname{Re} Y_k'} - 1\right)}\right] \quad . \tag{16.04-6}$$

Coupling to the Common Junction—Although other means of coupling would have also been possible in the comb-line multiplexer example of this section it was assumed that the filters were coupled to the center junction by series inductances formed from high-impedance wire.

Figure 16.04-11 shows a high-impedance wire and its equivalent circuit when used as a J-inverter. The comb-line filter design equations discussed in Secs. 8.13 and 8.14 were modified to use this type of inverter at one end of the filter. The susceptance B at one side of the inverter in Fig. 16.04-11(b) was compensated for by absorbing it into Resonator n, while the susceptance B on the other side of the inverter was effectively absorbed into the susceptance-annulling network. In this case the high-impedance-wire type of coupling was suggested because it would help prevent the common junction from being crowded. However, it is desirable to keep the high-impedance wire quite short in order to avoid unwanted resonances, while not having it so small in diameter that it would increase the losses. Thus, in some narrow-band cases, capacitive-gap couplings to approximate inverters as in Fig. 8.03-2(b) will be preferable. Table 16.04-1 summarizes the equations used in the design of the filters for the example of Figs. 16.04-5 to 16.04-9(b).

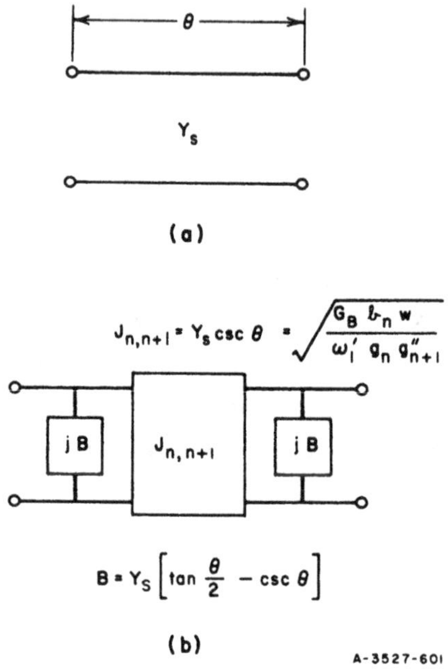

SOURCE: Quarterly Progress Report 6, Contract DA 36-039 SC-87398, SRI; reprinted in the *IEEE Trans. PTGMTT* (see Ref. 10 by E. G. Cristal and G. L. Matthaei)

FIG. 16.04-11 USE OF A HIGH-IMPEDANCE WIRE AS A J-INVERTER

Figure 16.04-12 shows the general form of another type of multiplexer structure once built by one of the authors. This multiplexer split an octave band into three parts, and used capacitively coupled, strip-line filters of the form in Fig. 8.05-4(a). Although the filters used series-capacitance couplings everywhere except at the common junction of the multiplexer, high-impedance wire coupling (*i.e.*, series-inductance coupling) was used at the common junction, as indicated in Fig. 16.04-12. (In this case the wire was bent into a rectangle as shown.) The wire plus a small section of the adjacent 50-ohm line were regarded as

Table 16.04-1

DESIGN EQUATIONS FOR COMB-LINE FILTERS WITH HIGH-Z_0 WIRE COUPLING AT ONE END AS SHOWN IN FIG. 16.04-5

Choose a low-pass prototype filter and values for $G_A = Y_A$, $G_B = Y_B$, $\mathscr{b}_1/Y_A$, ..., $\mathscr{b}_n/Y_A$, θ_0, and θ_s.

NORMALIZED MUTUAL CAPACITANCES PER UNIT LENGTH

$$\frac{C_{01}}{\epsilon} = Y_A \frac{376.7}{\sqrt{\epsilon_r}} \left[\frac{w}{\omega_1'} \frac{(\mathscr{b}_1/Y_A)}{g_0 g_1}\right]^{1/2} \quad . \tag{1}$$

$$\frac{C_{i,i+1}}{\epsilon} = Y_A \frac{376.7}{\sqrt{\epsilon_r}} \frac{w}{\omega_1'} \left[\frac{(\mathscr{b}_i/Y_A)(\mathscr{b}_{i+1}/Y_A)}{g_i g_{i+1}}\right]^{1/2} \tan\theta_0 , \quad i = 1,2, \ldots, n-1 \quad . \tag{2}$$

NORMALIZED ADMITTANCE OF HIGH-Z_0 COUPLING WIRE

$$\frac{Y_S}{Y_A} = \left[\frac{w}{\omega_1'} \frac{(\mathscr{b}_n/Y_A)(Y_B/Y_A)}{g_n g_{n+1}''}\right]^{1/2} \sin\theta_s \tag{3}$$

NORMALIZED SELF-CAPACITANCES PER UNIT LENGTH

$$\frac{C_0}{\epsilon} = \frac{376.7}{\sqrt{\epsilon_r}} \left(Y_A - \frac{C_{01}}{\epsilon} \frac{\sqrt{\epsilon_r}}{376.7}\right) \quad . \tag{4}$$

$$\frac{C_1}{\epsilon} = Y_A \left[\frac{376.7}{\sqrt{\epsilon_r}} \frac{2\mathscr{b}_1}{Y_A} [\cot\theta_0 + \theta_0 \csc^2\theta_0]^{-1} - \frac{(C_{12}/\epsilon)}{Y_A} - \frac{\sqrt{\epsilon_r}}{376.7} \frac{(C_0/\epsilon)(C_{01}/\epsilon)}{Y_A^2}\right] \quad . \tag{5}$$

$$\left(\frac{C_i}{\epsilon}\right) = \frac{Y_A\, 376.7}{\sqrt{\epsilon_r}} \left[2\frac{\mathscr{b}_i}{Y_A} [\cot\theta_0 + \theta_0 \csc^2\theta_0]^{-1}\right]$$

$$- \left[(C_{i,i+1}/\epsilon) + (C_{i-1,i}/\epsilon)\right] \qquad \text{for} \quad i = 2,3, \ldots, (n-1) \quad . \tag{6}$$

$$\frac{C_n}{\epsilon} = \frac{-C_{n-1,n}}{\epsilon} + \frac{376.7\, Y_A}{\sqrt{\epsilon_r}} \left[\frac{2(\mathscr{b}_n/Y_A) - (Y_s/Y_A)\,[\cot\theta_s + \theta_s \csc^2\theta_s]}{\cot\theta_0 + \theta_0 \csc^2\theta_0}\right] \quad . \tag{7}$$

NORMALIZED LUMPED LOADING SUSCEPTANCES

$$\frac{\omega_0 C_i^s}{Y_A} = 2\frac{\mathscr{b}_i}{Y_A} [\cot\theta_0 + \theta_0 \csc^2\theta_0]^{-1} \cot\theta_0 , \quad i = 1,2, \ldots, n-1 \quad . \tag{8}$$

$$\frac{\omega_0 C_n^s}{Y_A} = \frac{Y_s}{Y_A} \cot\theta_s + \frac{2(\mathscr{b}_n/Y_A) - (Y_s/Y_A)\,[\cot\theta_s + \theta_s \csc^2\theta_s]}{\cot\theta_0 + \theta_0 \csc^2\theta_0} \cot\theta_0 \quad . \tag{9}$$

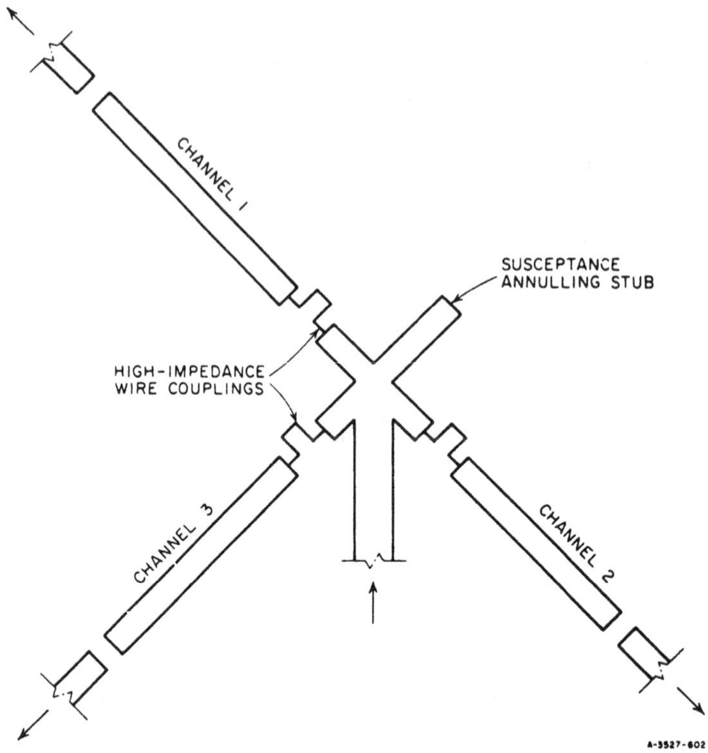

FIG. 16.04-12 A SCHEMATIC DRAWING OF A THREE-CHANNEL MULTIPLEXER USING FILTERS OF THE TYPE IN SEC. 8.05
Capacitive coupling is used except at the common junction

comprising an inverter of the form in Fig. 8.03-2(c). This type of coupling and point of view was used because it eliminated the undesirably small capacitive coupling gaps that would otherwise have been needed, and because in this particular type of filter it provided a convenient reference plane for paralleling the filters. In this case only a small capacitive stub was found to be required for susceptance annulling. This result was partly due to the large band covered by the multiplexer and probably partly due to the rather sizeable junction effects present in this particular design at the common junction. The multiplexer was designed for 1 db or less transducer loss in the pass bands, and achieved this performance quite well. The input VSWR at the common junction was 2:1 or less across the entire octave operating band of the multiplexer, which indicated that the desired contiguous pass-bands had been achieved.

Design of Annulling Networks—Solution of integral equations given by Bode[8] shows that if the real part of an admittance has the rectangular form defined by the equations

$$\begin{aligned} \mathrm{Re}\, Y_T &= 0 \text{ for } 0 \leqq \omega < \omega_a \\ &= G_A \text{ for } \omega_a < \omega < \omega_b \\ &= 0 \text{ for } \omega_b < \omega < \infty \end{aligned} \qquad (16.04\text{-}7)$$

then the minimum imaginary part is given by

$$Im\, Y_T = \frac{G_A}{\pi}\, ln \left| \frac{1 + \dfrac{\omega}{\omega_0}\, w - \left(\dfrac{\omega}{\omega_0}\right)^2}{1 - \dfrac{\omega}{\omega_0}\, w - \left(\dfrac{\omega}{\omega_0}\right)^2} \right| \qquad (16.04\text{-}8)$$

where

$$\omega_0 = \sqrt{\omega_a \omega_b}$$

and

$$w = \frac{\omega_b - \omega_a}{\omega_0} \quad .$$

The imaginary part above is minimum in the sense that no shunt susceptance could be removed from the network without making the network non-physical.[8] When the filters for a parallel-connected multiplexer are designed by the methods of this section, after the filters are connected, the real part of the total input admittance can be approximated by Eq. (16.04-7), and to the extent that there is no shunt susceptance in the couplings to the common junction, the imaginary part can be approximated by Eq. (16.04-8). For example, Fig. 16.04-13 shows computed curves of Re Y_T and $Im\ Y_T$ for a four-channel, lumped-element filter multiplexer design (the solid lines), along with corresponding approximations obtained using Eqs. (16.04-7) and (16.04-8). Note that the degree of agreement is quite good. In the case of Figs. 16.04-7(a),(b) the ripples in Re Y_T are larger so that the accuracy would be less good on that score. Also note from Fig. 16.04-7(b) that $Im\ Y_T$ does not pass through zero near the middle of the operating

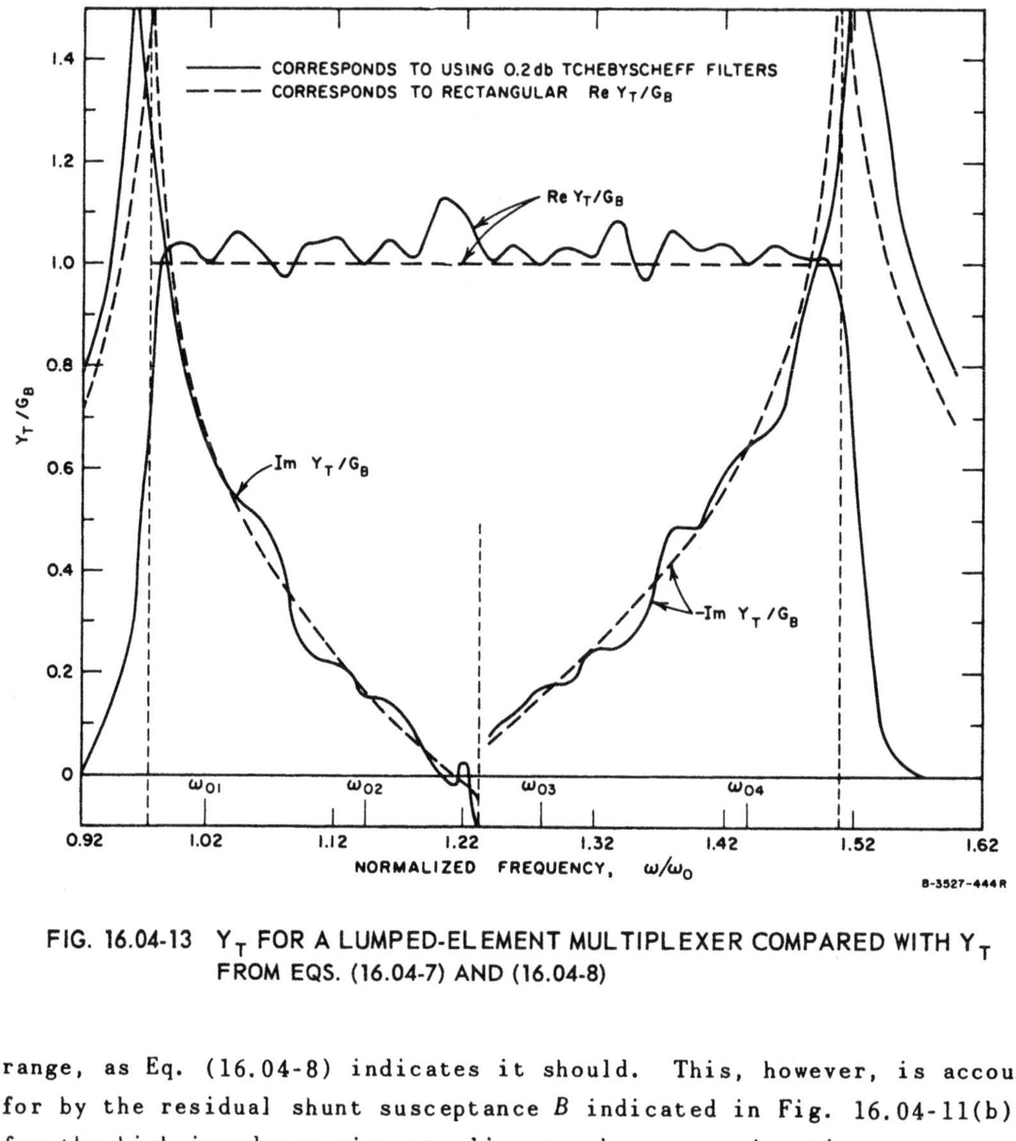

FIG. 16.04-13 Y_T FOR A LUMPED-ELEMENT MULTIPLEXER COMPARED WITH Y_T FROM EQS. (16.04-7) AND (16.04-8)

range, as Eq. (16.04-8) indicates it should. This, however, is accounted for by the residual shunt susceptance B indicated in Fig. 16.04-11(b), for the high-impedance wire coupling to the common junction.

In many practical cases it may be desirable to design the susceptance-annulling network after assembling the multiplexer and measuring the admittance characteristic at the common junction (or the impedance characteristic if the multiplexer is series-connected). In this manner all significant effects due to the physical size of the common junction can be accounted for. However, when such junction effects are relatively small, Eq. (16.04-8) (plus a possible correction for any shunt susceptance known to be present across the inputs of the filters) should give a very useful estimate of what the imaginary part of Y_T will be. Having this estimate, the susceptance-annulling network can then be designed. The required annulling network for a parallel-connected multiplexer is generally a parallel-resonant circuit having zero susceptance at the frequency where $Im\ Y_T$ is zero.

In much of the above discussion the comments have been in terms of admittances and parallel-connected multiplexers. It should be recalled that the same remarks apply in an analogous way to impedances in a dual series-connected multiplexer.

SEC. 16.05, DIPLEXERS WITH CONTIGUOUS PASS BANDS

Diplexers are basically two-channel multiplexers, hence they can be designed by the same techniques discussed in Sec. 16.04. However, since diplexers are frequently composed of low-pass and high-pass filters rather than two band-pass filters, this case will be given some special attention. Also, an alternate procedure for the design of diplexers will be suggested.

Figure 16.05-1 shows a diplexer consisting of a low-pass filter and a high-pass filter. Using the design viewpoint of Sec. 16.04, a singly loaded prototype from Sec. 4.06 is used for designing the low-pass filter, while the same prototype used with the mapping procedure in Fig. 7.07-2 can be used for design of the high-pass filter. The proper equal-ripple band-edge frequencies needed to cause the filters to cross over at their 3-db points can be determined with the aid of Eqs. (16.04-2), (16.04-3),

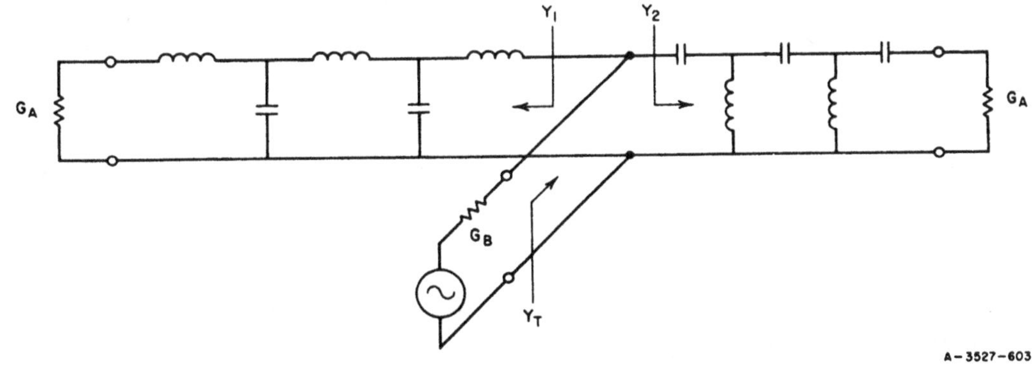

FIG. 16.05-1 A PARALLEL-CONNECTED DIPLEXER COMPOSED OF A LOW-PASS AND A HIGH-PASS FILTER

and (16.04-6), along with the low-pass to high-pass mapping in Sec. 7.07. It can be shown that in this case the two filters will annul each others susceptances about as well as can be done, without any additional susceptance-annulling network. In fact, if the filters were designed from maximally flat, singly loaded prototypes, the imaginary part of the input admittance Y_T in Fig. 16.05-1 could be made to be exactly zero at all frequencies (at least for the idealized lumped-element case). The actual filters for a microwave diplexer such as this could in many cases be designed using the methods of Secs. 7.03 to 7.07.

In some cases it will be convenient to use a wide-band band-pass filter for the high-frequency channel, rather than a high-pass structure of the form in Fig. 16.05-1. In this case the susceptances of the two filters will not annul each other as well. Using Bode's integral equations,[8] if the two filters are designed so that the real part of their input admittance after parallel connection is approximately

$$\begin{aligned} \mathrm{Re}\, Y_T &= G_B \text{ for } 0 \leq \omega \leq \omega_b \\ &= 0 \quad \text{for } \omega_b < \omega < \infty \end{aligned} \tag{16.05-1}$$

then the minimum imaginary part is given by

$$Im\ Y_T = \frac{G_B}{\pi}\ ln\left|\frac{\omega_b - \omega}{\omega_b + \omega}\right| \tag{16.05-2}$$

where the significance of the term "minimum" is as discussed with reference to Eq. (16.04-8). In this case the susceptance can be largely annulled by a series-resonant shunt branch having a susceptance of the form

$$B(\omega) = \frac{-1}{\omega_c L\left(\dfrac{\omega}{\omega_c} - \dfrac{\omega_c}{\omega}\right)} \tag{16.05-3}$$

where ω_c is slightly larger than ω_b and L is a constant (*i.e.*, an inductance for a lumped circuit).

An Alternative Point of View for Diplexer Design—We will now consider an alternative approach to design of diplexers. This approach requires more guess work than is required for the approach suggested above, however, it has a possible advantage for some situations in that it can be used to modify conventional doubly terminated filters for use in diplexers. The approach about to be described is in many respects equivalent to the "fractional termination" method used for paralleling filters designed on the image basis.[9] In the discussion to follow series connection of the filters will be assumed, though the same technique applies analogously to the dual case of parallel-connected diplexers.

Figure 16.05-2(a) shows a low-pass filter and a band-pass filter with dotted lines indicating wires for connecting these filters in series.

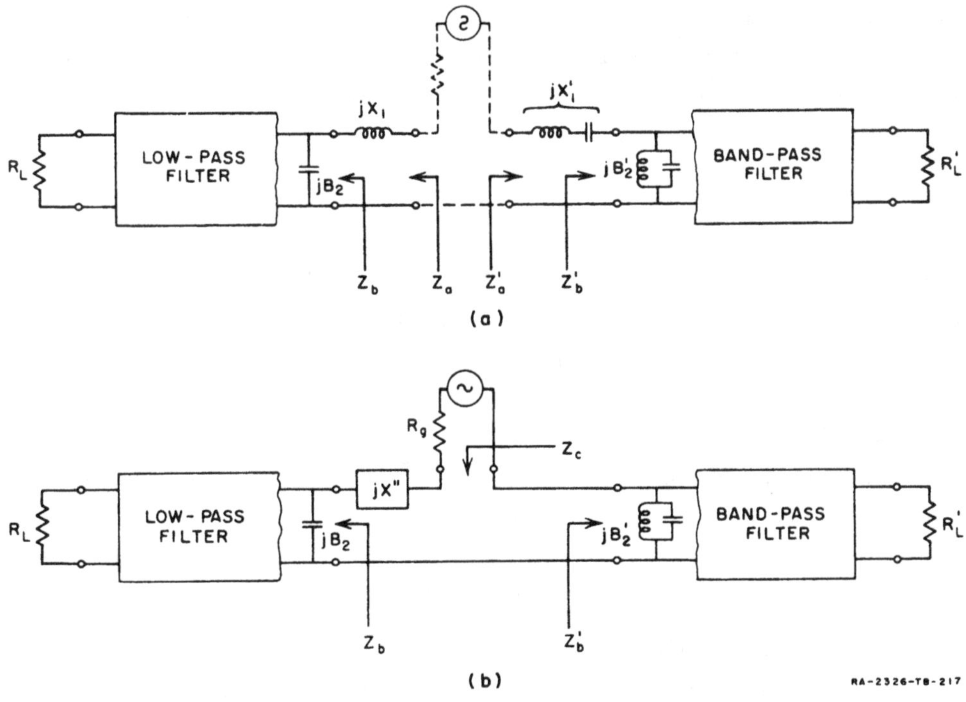

FIG. 16.05-2 FILTERS CONNECTED IN SERIES

If each of these filters were designed to operate normally with pure resistance terminations at both ends, the two filter's performance would be greatly disrupted by this connection. This disruption would be due to the fact that although each filter exhibits a nominally resistive input impedance in its individual pass band, each filter also exhibits large reactive impedances in its stop band. Thus if the filters shown in Fig. 16.05(a) have contiguous pass bands, the band-pass filter will introduce a large reactive component in the terminating impedance seen by the low-pass filter in the pass band of the low-pass filter, while the low-pass filter will introduce a large reactive component to the terminating impedance seen by the band-pass filter in the pass band of the band-pass filter.

The large stop-band reactive components of the input impedances Z_a and Z_a' of the filters in Fig. 16.05-2(a) are due largely to the series reactances X_1 and X_1' shown. If these are removed, the remaining input impedances Z_b and Z_b' have the same real part characteristics but much smaller stop-band reactances. Thus, to form a diplexer, jX_1 and jX_1' are removed and the remaining circuits are connected as shown in Fig. 16.05-2(b). Next, an additional reactance jX'' is introduced to further adjust the imaginary part of Z_c in order to cause

$$Z_c = Z_b + Z_b' + jX'' \tag{16.05-4}$$

to approximate a pure resistance equal to R_g as nearly as possible.* In the pass band of the low-pass filter $Z_b' + jX''$ represents, under the conditions described above, a reactance about equivalent to jX_1 in that band, so that the low-pass filter will operate very nearly in its normal fashion throughout its pass band. Likewise, in the pass band of the band-pass filter, $Z_b + jX''$ represents a reactance about equivalent to jX_1' in that band, so that the band-pass filter will operate very nearly in its normal fashion throughout its pass band. However, the removal of the series branches jX_1 and jX_1' may result in some reduction in stop-band attenuation of one or both filters.†

Figure 16.05-3 shows a diplexer that was designed by use of this approach. The low-pass channel was composed of the filter in Fig. 7.03-3(a) cut off just to the right of the last capacitive disk on the right of the filter. This corresponded to removing jX_1 in Fig. 16.05-2(a). The upper-channel used a band-pass filter of the type discussed in Secs. 10.03 and 10.05. This filter was designed as though it were to have a series stub at one end (making a series resonator). This was accomplished by designing that end of the filter using the equations in Table 10.03-1, while the other half of the filter was designed using the equations in

---

*This assumes that the normal terminating impedance next to Z_a and Z_a' for the individual filters is R_g. It also assumes that the frequency responses of the filters are scaled to intersect close to the 3-db points.

†The decrease in stop-band attenuation resulting from the removal of jX_1 and jX_1' is largely compensated for as a result of the input voltage reduction in the stop band of each filter due to the series loading of the other filter.

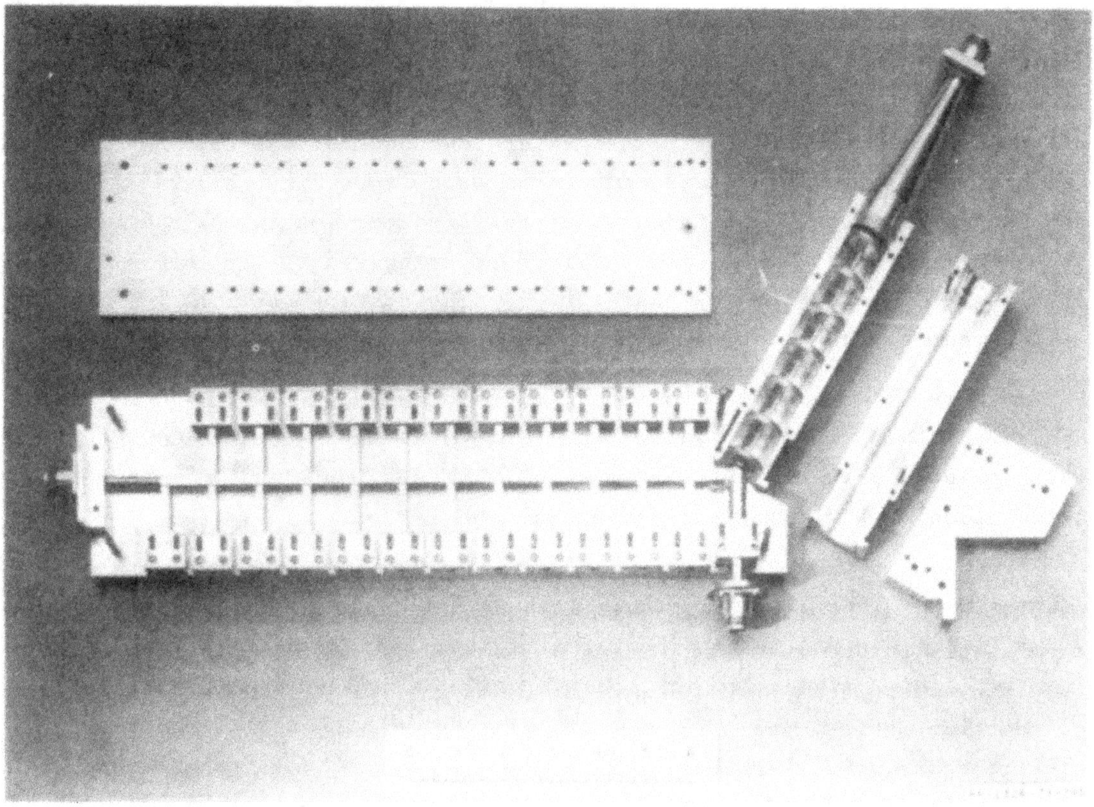

FIG. 16.05-3 PHOTOGRAPH OF A SERIES-CONNECTED DIPLEXER WITH THE COVER PLATES REMOVED
This diplexer uses a low-pass and a band-pass filter

Table 10.05-1. In the actual diplexer the series stub was not included, which was analogous to removing the series resonator jX_1' in Fig. 16.05-2(a). Then the impedances of the two filters connected in series (with reactances jX_1 and jX_1' removed) was computed, and the desired form of reactance-annulling network jX'' was determined. It was found that a desirable annulling network should have a reactance of the form

$$X'' = \frac{-1}{\omega_c C\left(\frac{\omega}{\omega_c} - \frac{\omega_c}{\omega}\right)} \quad . \tag{16.05-5}$$

Note that Eq. (16.05-5) is the dual of that in Eq. (16.05-3), which is to be expected since the former case is for a parallel connection while the later is for a series connection.

Figure 16.05-4 shows how the series connection was achieved in the diplexer in Fig. 16.05-3. The input line to the common junction passes within the end shunt stub of the band-pass filter. The center conductor of the input line connects to the center conductor of the coaxial low-pass filter, while the outer conductor of the input line connects to the band-pass filter. In this way a series connection is obtained.* The annulling reactance jX'' is obtained by using a disk to form a radial line in series with the low-pass filter. The common-junction for the series connection of the diplexer is the region marked A in the figure.

Figure 16.05-5 shows the measured performance of the low-pass channel of this diplexer, while Fig. 16.05-6 shows the performance of the high-pass channel. As can be seen from the figures, low pass-band attenuation with a sharp cross-over was obtained.

Either of the two diplexer design methods described above can be used with good results. The method where the filters are designed from singly loaded low-pass prototype filters has an advantage of involving less guess work for precise designs, and it should be capable of superior performance when very low pass-band attenuation and very precise cross-over characteristics are required. The alternative point of view for diplexer design described above may not be capable of quite as good results because the individual filters were originally designed to be doubly terminated and were later modified. As such, the filters do not have quite as favorable impedance properties for interconnection as do singly terminated filters. However, the alternative method does have the possible advantage that the individual filters can be constructed and tested first as conventional doubly terminated filters.

---

* This technique of feeding the multiplexer by a coaxial line within the shunt stub of the band-pass filter was suggested by S. B. Cohn.

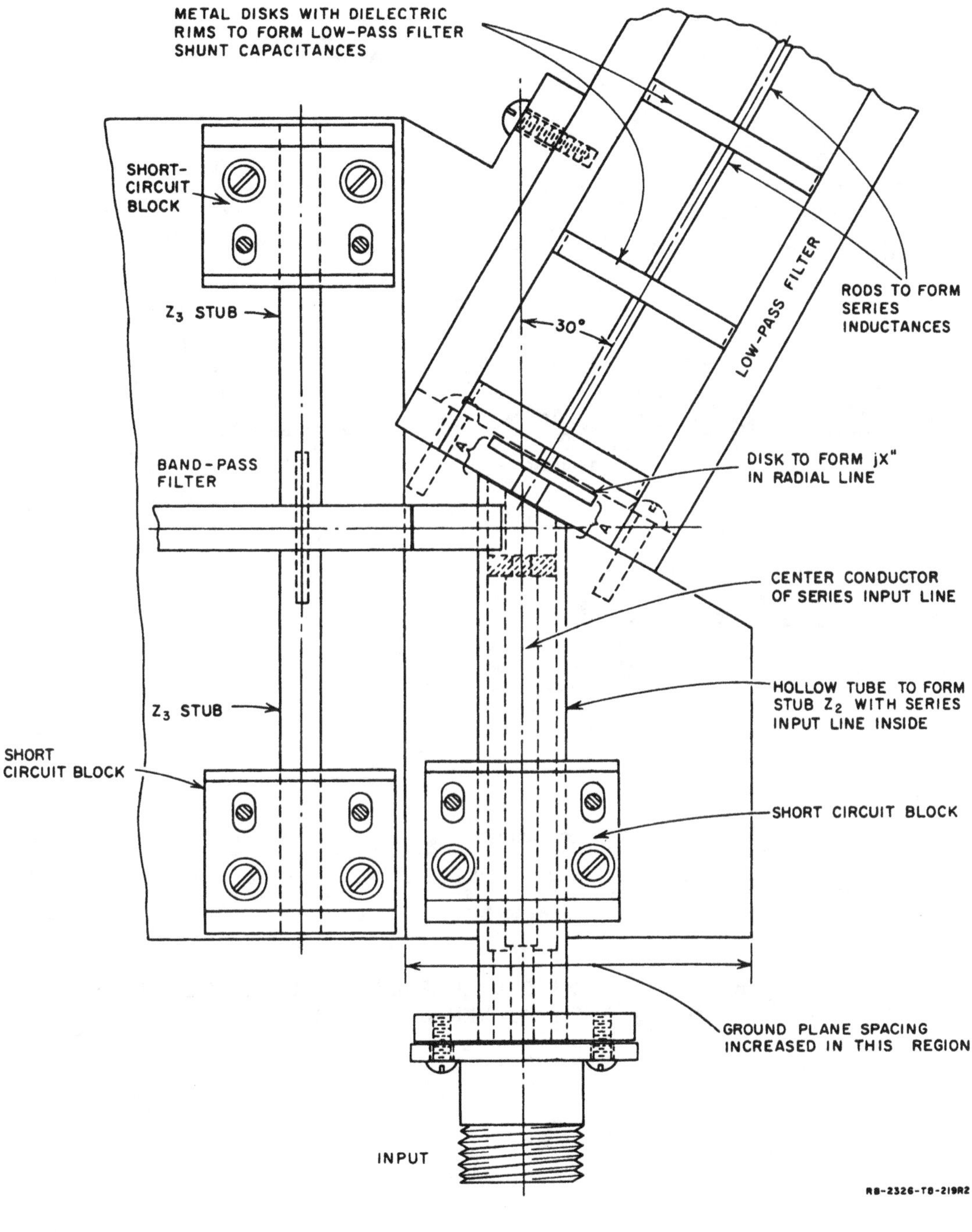

FIG. 16.05-4 A PARTIAL VIEW OF A SERIES-CONNECTED DIPLEXER CONSTRUCTION

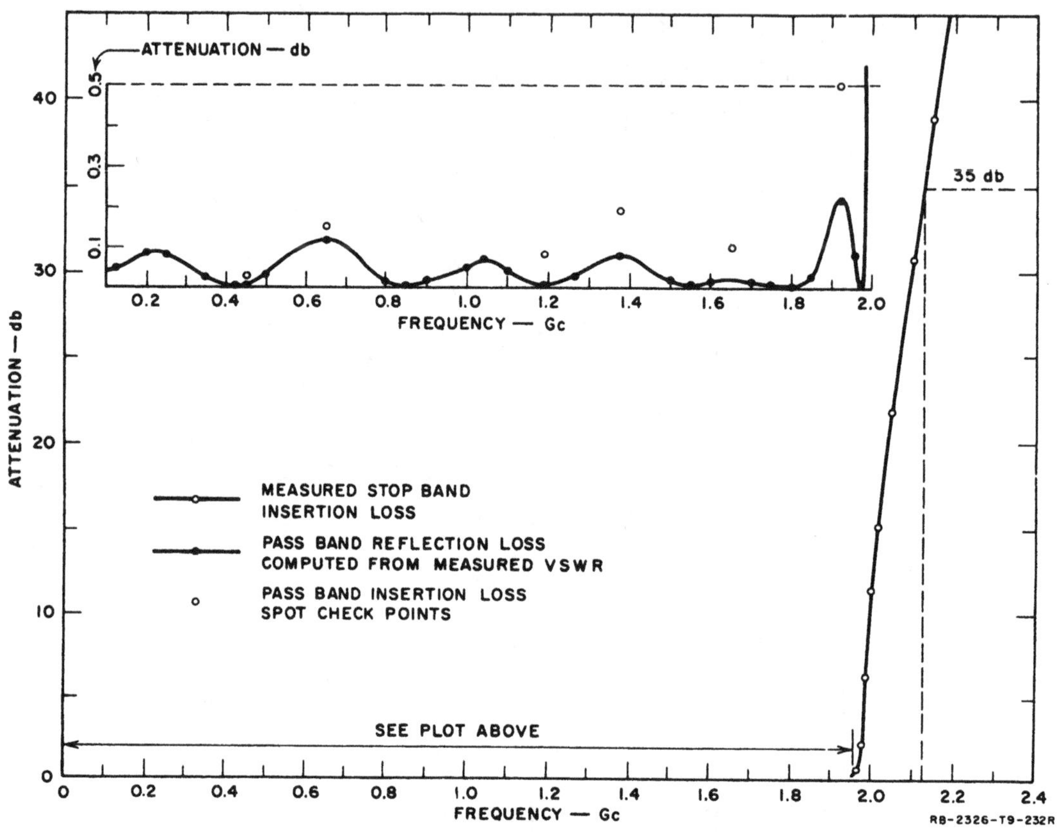

FIG. 16.05-5 MEASURED TRANSMISSION CHARACTERISTICS OF THE LOW-PASS CHANNEL OF THE DIPLEXER IN FIG. 16.05-3

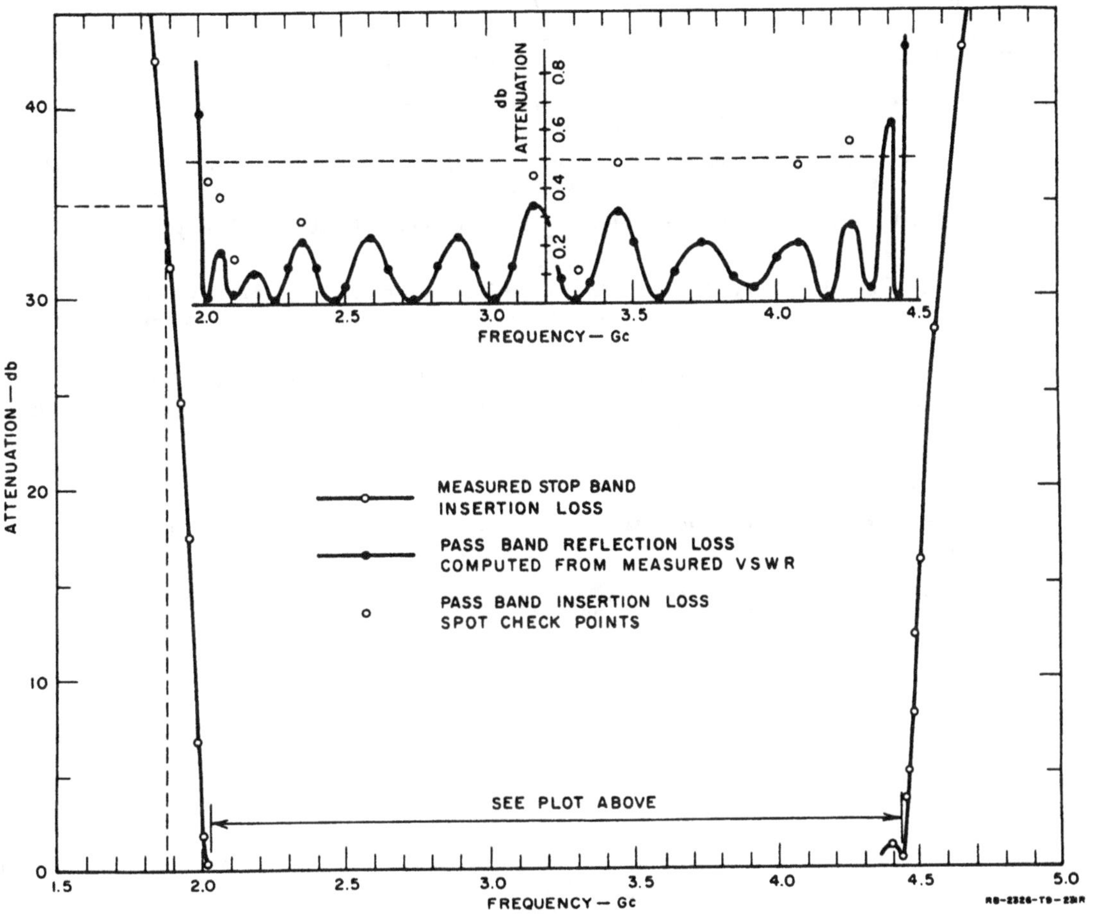

FIG. 16.05-6 MEASURED TRANSMISSION CHARACTERISTICS OF THE BAND-PASS CHANNEL OF THE DIPLEXER IN FIG. 16.05-3

REFERENCES

1. G. C. Southworth, *Principles and Applications of Waveguide Transmission*, pp. 306-317, D. Van Nostrand Co., Inc., New York City (1950).

2. D. Alstadter and E. O. Houseman, Jr., "Some Notes on Strip Transmission Line and Waveguide Multiplexers," *1958 IRE Wescon Convention Record,* Part 1, pp. 54-69.

3. E. A. Marcatili and D. L. Bisbee, "Band-Splitting Filter," *BSTJ*, Vol. XL, pp. 197-212 (January 1961).

4. G. L. Ragan, *Microwave Transmission Circuits*, MIT Rad. Lab. Series, Vol. 9, pp. 708-709, McGraw-Hill Book Co., Inc., New York City (1948).

5. N. Marcuvitz, *Waveguide Handbook*, MIT Rad. Lab. Series, Vol. 10, pp. 360-363, McGraw-Hill Book Co., Inc. (1951).

6. J. F. Cline et al, "Design Data for Antenna-Multicoupler Systems," Final Report, SRI Project 2183, Contract AF 19(604)-2247, Stanford Research Institute, Menlo Park, California (September 1959).

7. W. J. Getsinger, E. G. Cristal, and G. L. Matthaei, "Microwave Filters and Coupling Structures," Quarterly Report 6, Part III, SRI Project 3527, Contract DA 36-039 SC-87398, Stanford Research Institute, Menlo Park, California (August 1962).

8. H. W. Bode, *Network Analysis and Feedback Amplifier Design*, p. 335, D. Van Nostrand Co., Inc., New York City (1945).

9. E. A. Guillemin, *Communication Networks*, Vol. II, pp. 356-371, John Wiley and Sons, Inc., New York City (1935).

10. E. G. Cristal and G. L. Matthaei, "A Technique for the Design of Multiplixers Having Contiguous Channels," *IEEE Trans. PTGMTT-12*, pp. 88-93 (January 1964).

CHAPTER 17

MECHANICALLY AND MAGNETICALLY TUNABLE MICROWAVE FILTERS

SEC. 17.01, INTRODUCTION

This chapter will deal mainly with band-pass filters such as might be desired for use as receiver preselectors; however, some discussion of band-stop magnetically tunable filters will be included. Although there are a variety of simple means for shifting the resonant frequency of the resonators of a filter, if it is desired to maintain constant bandwidth and response shape as the filter is tuned, the problem becomes fairly complex (Sec. 17.02). The design data given in this chapter are slanted primarily toward designing filters with relatively constant response shape and bandwidth as the filter is tuned. Fortunately, in the case of magnetically tunable filters, the resonators inherently tend to give constant bandwidth, although there will be some change in response shape as the filter is tuned.

Figure 17.01-1 shows the two types of mechanically tunable filters that are specifically discussed in this chapter. The filter shown at (a) uses coaxial resonators that are tuned by sliding their inner conductors out or in, while the filter shown at (b) is tuned by a sliding wall on one side of each resonator cavity. The constancy of the bandwidth and response shape for filters of these types is improved by proper choice of the locations of the coupling apertures and loops, as is discussed in Secs. 17.03 and 17.04.

Certain materials that exhibit ferrimagnetic resonance have a high-Q resonance and can be easily coupled in to and out of using strip-line, coaxial-line, or waveguide circuitry. The resonant frequency of such ferrimagnetic resonators is controlled by a DC biasing magnetic field. Thus, it becomes possible to tune such resonators electronically by placing the filter in the field of an electromagnet, and then controlling the current in the electromagnet by electronic means. Section 17.05 outlines the general properties of ferrimagnetic resonators that are important to their use in microwave filters. Section 17.06 discusses means for determining the crystal axes of ferrimagnetic resonators—a problem of

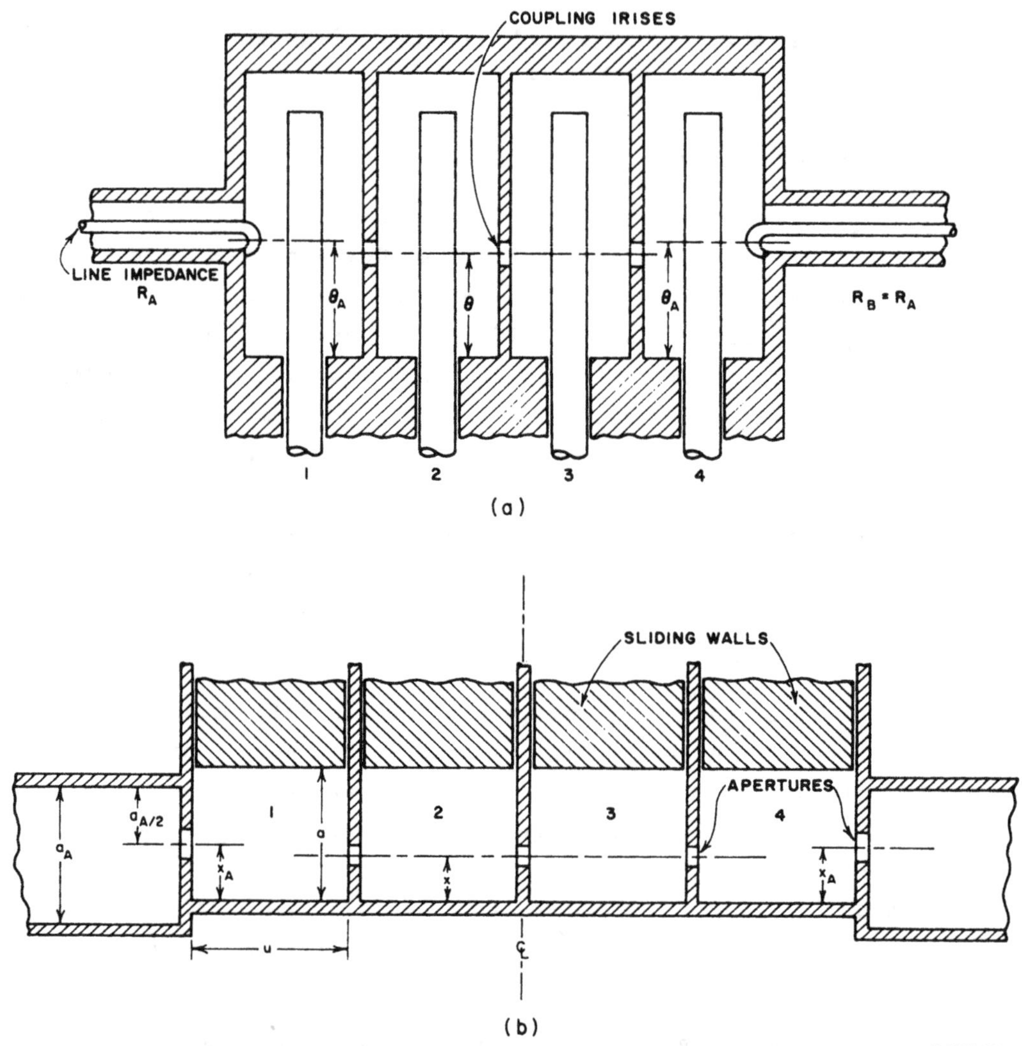

FIG. 17.01-1 COAXIAL AND WAVEGUIDE MECHANICALLY TUNABLE BAND-PASS FILTERS

considerable practical importance since the resonant frequency of ferrimagnetic resonators is influenced by the orientation of the crystal axes of the material, with respect to the biasing DC magnetic field.

Figure 17.01-2 shows several of the magnetically tunable filter structures discussed in Secs. 17.07 to 17.09. The structure shown at (a) utilizes input and output strip lines that are completely separated by a dividing wall except for a small coupling slot next to the short-circuited ends of the strip lines. There is negligible coupling between the two strip lines when they are without the ferrimagnetic resonators; however, when ferrimagnetic resonators such as spheres of single-crystal yttrium-iron-garnet (YIG) are added as shown in the figure, good transmission from one strip line to the other occurs, provided that the proper biasing magnetic field H_0 is applied. The filter shown at (a) will give a two-resonator response, while that shown at (b) will give a three-resonator response. A waveguide two-resonator filter is shown at (c).

A magnetically tunable directional filter is shown in Fig. 17.01-3. This filter is very similar in principle to the waveguide directional filters discussed in Sec. 14.02, except that the fixed-tuned electromagnetic resonators used in the filters in Sec. 14.02 are here replaced by magnetically tunable ferrimagnetic resonators consisting of spheres of YIG (or of some other suitable material). Besides being magnetically tunable, the filter in Fig. 17.01-3 differs from those in Sec. 14.02 in one other important way. As a result of the non-reciprocal properties of ferrimagnetic resonators, the filter in Fig. 17.01-3 has circulator action at resonance as well as directional filter action. For example, at resonance, energy entering Port 1 will go to Port 4, but energy entering at Port 4 will go to Port 3. (The filters in Fig. 17.01-2 do not have this property.) Directional filters of this type are discussed in Sec. 17.10.

Magnetically tunable band-stop filters are also of practical interest. One possible form for such filters is the strip-line filter structure shown in Fig. 17.01-4. This filter uses three YIG resonators placed between the strip line and one of the ground planes. This filter could be designed to produce a narrow stop band that can be moved in frequency by controlling the biasing magnetic field. Band-stop magnetically tunable filters are discussed in Sec. 17.11.

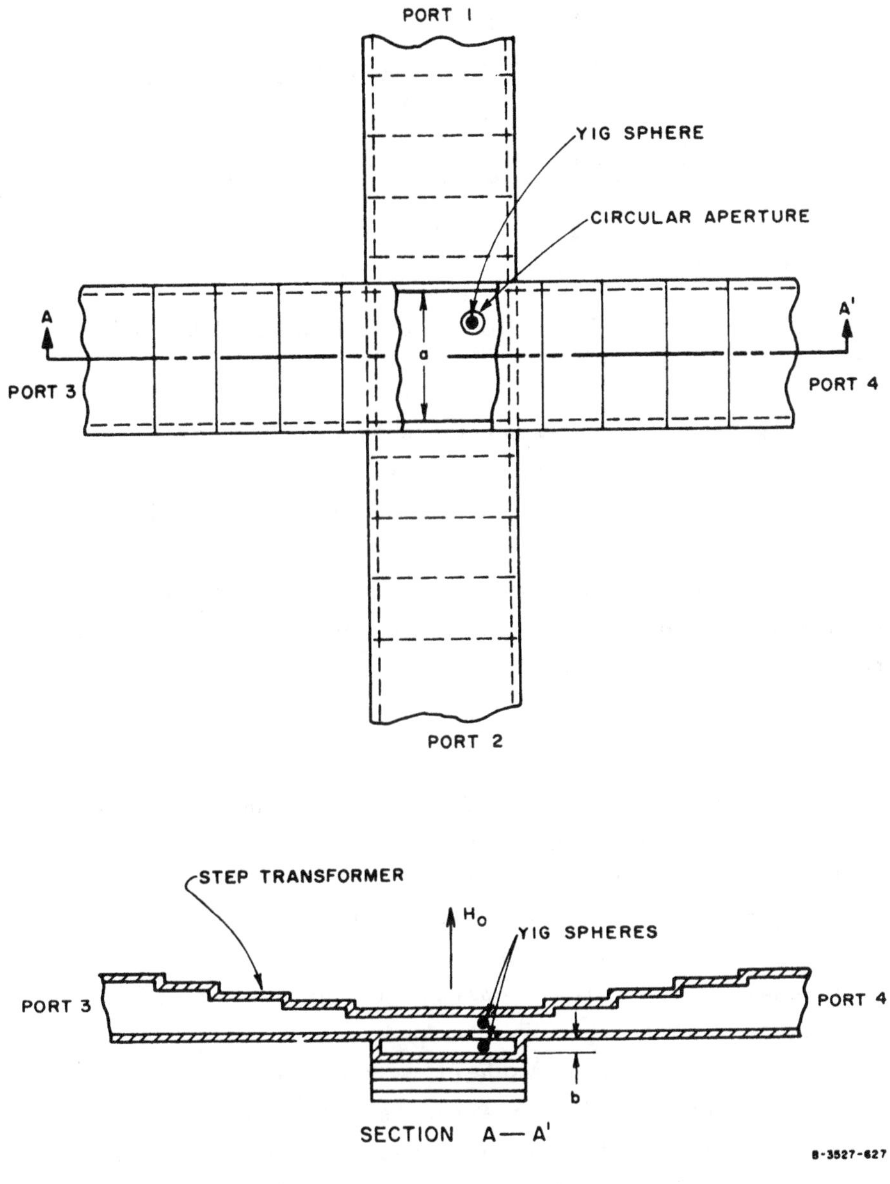

FIG. 17.01-3 A MAGNETICALLY TUNABLE DIRECTIONAL FILTER HAVING CIRCULATOR ACTION

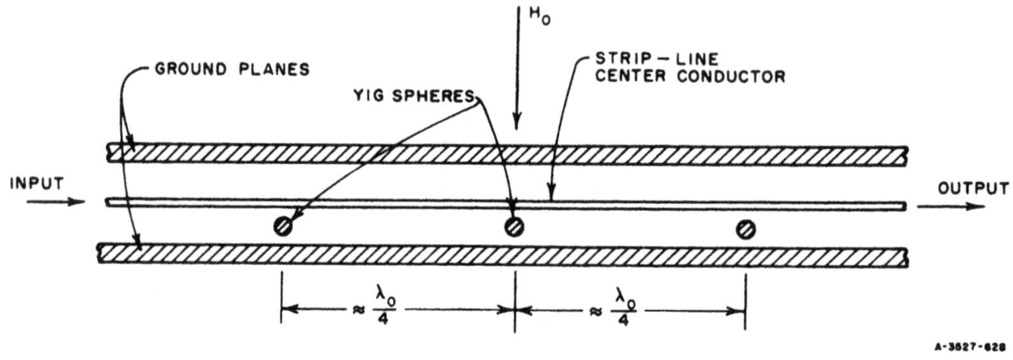

FIG. 17.01-4 A MAGNETICALLY TUNABLE STRIP-LINE BAND-STOP FILTER

The tuning of microwave resonators by use of variable-capacitance diodes or by use of ferroelectric materials might at first seem to be attractive. However, these possibilities are not considered in this chapter because at the time of this writing the Q's of available varactor diodes and of ferroelectric materials are not sufficiently high at microwave frequencies to provide very satisfactory means for electronic tuning of microwave-filter resonators.

SEC 17.02, THEORY OF IDEAL, TUNABLE BAND-PASS FILTERS

Although almost any kind of filter can be tuned by varying the lengths of the resonators, or by introducing variable capacitive or inductive loading of some form in the resonators, there are considerations besides resonator tuning that limit the types of structures desirable for use as tunable filters. Generally, it will be desired that the filter response shape will remain more or less constant as the filter is tuned. If steps are not taken to assure constant response shape, it is possible that a filter might have a good Tchebyscheff response at one end of the tuning range, but at the other end have a response of considerably different shape and bandwidth (and possibly with a sizeable reflection loss in the pass band). Thus, we shall now want to explore the factors that must be considered if constant response shape (and also constant bandwidth) are to be maintained as a filter is tuned. Since tunable filters are generally of narrow bandwidth, it will be convenient to use the external Q's of the end resonators and the coupling coefficients between adjacent resonators as the basic design parameters (Sec. 8.02).

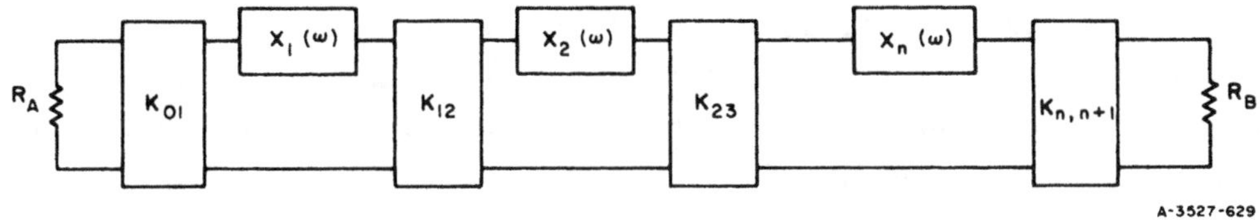

FIG. 17.02-1 A GENERALIZED FILTER CIRCUIT USING SERIES RESONATORS AND IMPEDANCE INVERTERS

Figure 17.02-1 shows a generalized filter circuit with series resonators coupled by impedance inverters (Sec. 8.02). By Fig. 8.02-3, the external Q's of the end resonators of this circuit are

$$(Q_e)_A = \frac{\mathscr{X}_1}{K_{01}^2/R_A} = \frac{g_0 g_1 \omega_1'}{w} \tag{17.02-1}$$

$$(Q_e)_B = \frac{\mathscr{X}_n}{K_{n,n+1}^2/R_B} = \frac{\omega_1' g_n g_{n+1}}{w} \tag{17.02-2}$$

and the coupling coefficients between resonators are

$$k_{j,j+1}\Big|_{j=1 \text{ to } n-1} = \frac{K_{j,j+1}}{\sqrt{\mathscr{X}_j \mathscr{X}_{j+1}}} = \frac{w}{\omega_1'\sqrt{g_j g_{j+1}}} \quad . \tag{17.02-3}$$

In Eqs. (17.02-1) to (17.02-3), the $K_{j,j+1}$ are impedance inverter parameters, the $\mathscr{X}_j$ are resonator slope parameters defined by

$$\mathscr{X}_j = \frac{\omega_0}{2} \frac{dX_j(\omega)}{d\omega}\Bigg|_{\omega=\omega_0} \tag{17.02-4}$$

where X_j is the reactance of resonator j, and ω_0 is the resonant frequency. The parameters g_0, g_1, ..., g_{n+1}, ω_1' are parameters of the low-pass prototype filter from which the band-pass filter is designed, and w is the fractional bandwidth of the band-pass filter as measured to the points ω_1 and ω_2 on its response, corresponding to ω_1' on the low-pass prototype response (Sec. 8.02).

Examining Eqs. (17.02-1) to (17.02-3) we see that for the type of response to be preserved (*i.e.*, for maximally flat, Tchebyscheff, or some other response properties to be preserved) the external Q's must vary inversely with the fractional bandwidth w, while the coupling coefficients must be directly proportional to w. Usually it is desired that the absolute bandwidth be kept constant. Thus, if Δf is the desired *fixed* bandwidth in cycles per second, f_0 is any given tuning frequency of the filter, and $(f_0)_m$ is the mean tuning frequency (*i.e.*, the center frequency of the tuning range), then it is usually desired that

$$w = \frac{\Delta f}{f_0} = \frac{w_m (f_0)_m}{f_0} \tag{17.02-5}$$

where

$$w_m = \frac{\Delta f}{(f_0)_m} \tag{17.02-6}$$

is the mean fractional bandwidth. Inserting Eq. (17.02-5) in Eqs. (17.02-1) to (17.02-3) gives

$$(Q_e)_A = \frac{g_0 g_1 \omega_1' f_0}{w_m (f_0)_m} \tag{17.02-7}$$

$$(Q_e)_B = \frac{\omega_1' g_n g_{n+1} f_0}{w_m (f_0)_m} \tag{17.02-8}$$

and

$$k_{j,j+1}\Big|_{j=1 \text{ to } n-1} = \frac{w_m (f_0)_m}{\omega_1' f_0 \sqrt{g_j g_{j+1}}} \tag{17.02-9}$$

where all quantities on the right are constants except for f_0. From these equations we see that in order to give a constant response shape and bandwidth as the filter is tuned, the external Q's must vary directly with the tuning frequency f_0, while the coupling coefficients must vary inversely with the tuning frequency.

The most commonly used tunable filters use inductive couplings, and the impedance inverters may be regarded as being of the form in Fig. 17.02-2. In the case of a narrow-band filter (tunable filters are usually narrow band), all of the reactances in the inverter circuit of Fig. 17.02-2 would be very small. The negative, series reactances are therefore small compared with the reactances that they are connected to, and hence have little effect in the filter except to cause a slight shift in the resonant frequency of the resonators. Thus, to a good approximation, we may represent the circuit in Fig. 17.02-1 by the circuit in Fig. 17.02-3 if the K-inverters are of the form in Fig. 17.02-2 and if the bandwidth is reasonably narrow (say, of the order of a few percent or less). Then by Fig. 17.02-2,

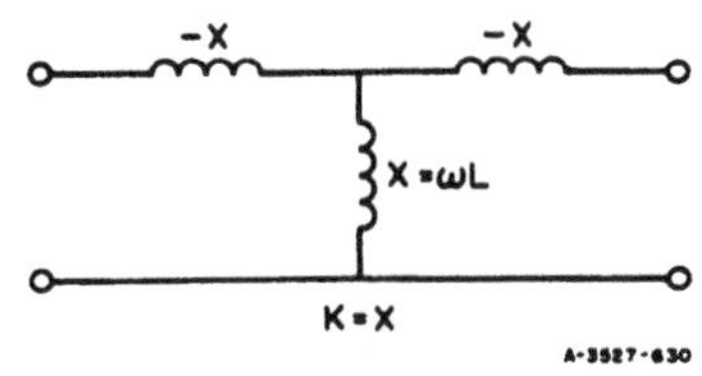

FIG. 17.02-2 A FORM OF K-INVERTER CORRESPONDING TO SHUNT-INDUCTIVE COUPLINGS

$$K_{j,j+1} = X_{j,j+1} = (X_{j,j+1})_m \frac{f_0}{(f_0)_m} \qquad (17.02\text{-}10)$$

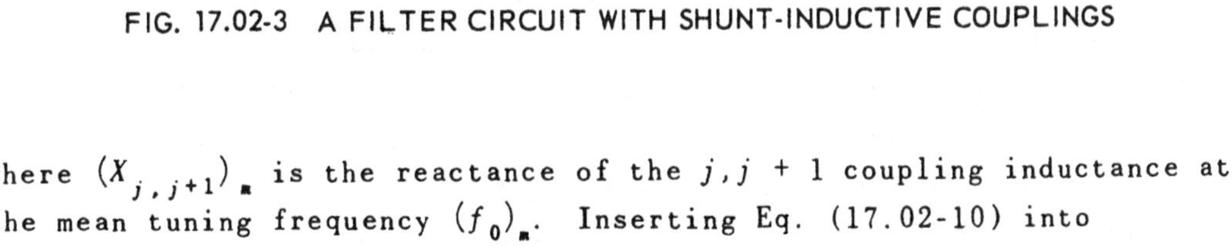

FIG. 17.02-3 A FILTER CIRCUIT WITH SHUNT-INDUCTIVE COUPLINGS

where $(X_{j,j+1})_m$ is the reactance of the $j, j+1$ coupling inductance at the mean tuning frequency $(f_0)_m$. Inserting Eq. (17.02-10) into Eqs. (17.02-1) to (17.02-3) gives

$$(Q_e)_A = \frac{\alpha_1 R_A (f_0)^2_m}{(X_{01})^2_m f_0^2} \qquad (17.02\text{-}11)$$

$$(Q_e)_B = \frac{\mathscr{x}_n R_B (f_0)^2_m}{(X_{n,n+1})^2_m f_0^2} \tag{17.02-12}$$

$$k_{j,j+1}\Big|_{j=1 \text{ to } n-1} = \frac{(X_{j,j+1})_m f_0}{\sqrt{\mathscr{x}_j \mathscr{x}_{j+1}}\,(f_0)_m} \tag{17.02-13}$$

Comparing Eqs. (17.02-11) and (17.02-12) with Eqs. (17.02-7) and (17.02-8) we see that, to maintain constant response shape and bandwidth, the resonator slope parameters $\mathscr{x}_1$ and $\mathscr{x}_n$ must vary as

$$\mathscr{x}_1 = (\mathscr{x}_1)_m \left[\frac{f_0}{(f_0)_m}\right]^3$$

$$\mathscr{x}_n = (\mathscr{x}_n)_m \left[\frac{f_0}{(f_0)_m}\right]^3 \tag{17.02-14}$$

where $(\mathscr{x}_1)_m$ and $(\mathscr{x}_n)_m$ are the slope parameter values at the mean tuning frequency $(f_0)_m$. Now, comparing Eqs. (17.02-9) and (17.02-13), we see that to maintain constant response shape and bandwidth

$$\mathscr{x}_j\Big|_{j=1 \text{ to } n} = (\mathscr{x}_j)_m \left[\frac{f_0}{(f_0)_m}\right]^2 \tag{17.02-15}$$

where the $(\mathscr{x}_j)_m$ are again the slope parameter values at $(f_0)_m$. Note that Eqs. (17.02-14) and (17.02-15) appear to be contradictory since Eq. (17.02-14) says that $\mathscr{x}_1$ and $\mathscr{x}_n$ must vary as $(f_0)^3$ while Eq. (17.02-15) says they must vary as $(f_0)^2$. These two requirements are not, however, irreconcilable, since Eq. (17.02-14) refers to the slope parameters of Resonators 1 and n as seen from the coupling reactances X_{01} and $X_{n,n+1}$, while Eq. (17.02-15) relates to the slope parameters of Resonators 1 and n as seen from the between-resonator reactances X_{12} and $X_{n-1,n}$. Thus, as will be evident from the discussions in Sec. 17.03 and 17.04, it may be possible to approximately satisfy both Eqs. (17.02-14) and (17.02-15) at the same time by making the X_{01} and $X_{n,n+1}$ coupling reactances at the ends of the filter couple into Resonators 1 and n at points on the resonator structures different from those of couplings X_{12} and $X_{n-1,n}$. In this manner the slope parameters seen by the end couplings will be different from those seen by the couplings between resonators.

Perhaps the most familiar type of series resonator is that consisting of an inductance and capacitance in series, which gives

$$X_j = \omega L_j - \frac{1}{\omega C_j} = \omega L_j - \frac{L_j \omega_0^2}{\omega} \quad . \tag{17.02-16}$$

By Eq. (17.02-4)

$$\mathscr{X}_j = \omega_0 L_j \quad . \tag{17.02-17}$$

If the resonators are tuned by varying the C_j while the L_j remains fixed, then

$$\mathscr{X}_j = (\mathscr{X}_j)_m \frac{f_0}{(f_0)_m} \tag{17.02-18}$$

which does not have the frequency variation of either Eq. (17.02-14) or Eq. (17.02-15). It is easily shown that if C_j is held fixed and the filter is tuned by varying the L_j, then

$$\mathscr{X}_j = (\mathscr{X}_j)_m \frac{(f_0)_m}{f_0} \tag{17.02-19}$$

which deviates even more from the desired type of frequency variation.

From the above discussion we see that a simple series L-C resonator cannot give the desired $(f_0)^2$ or $(f_0)^3$ variation of the resonator slope parameters unless the resonators are tuned by varying both the L_j and the C_j at the same time. However, Fig. 17.02-4 shows a resonator circuit that could approximate the desired frequency variations while being tuned by varying only a series capacitance. In this case a parallel resonant circuit has been added, which adds a pole of reactance at frequency ω_∞. Thus, as the resonator is tuned to higher frequencies, the resonator slope parameter increases more rapidly than it would if the pole of reactance were not there.

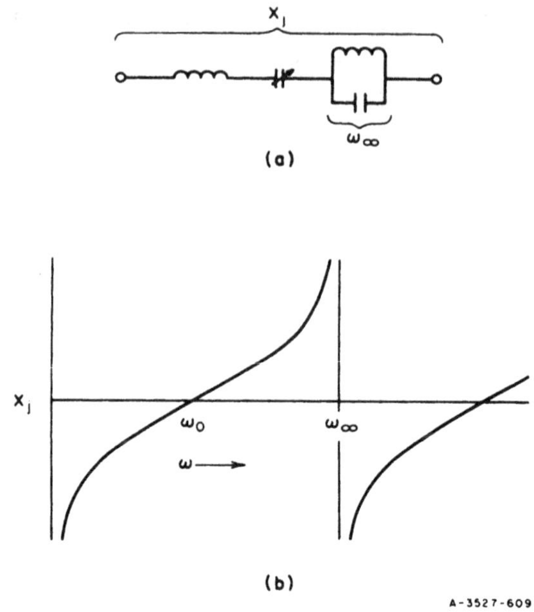

FIG. 17.02-4 A TUNABLE RESONATOR AND ITS REACTANCE CHARACTERISTIC

Although the form of the resonator circuit in Fig. 17.02-4 is not of much interest for microwave filters, the discussion illustrates the types of considerations and possible solutions involved in the design of tunable filters with constant response shape and bandwidth. The discussion above has been in terms of filters with series resonators and shunt inductive couplings; however, by use of the data in Figs. 8.02-3, 8.02-4, 8.03-1, 8.03-2 and 8.03-3, the same type of reasoning can be applied for the derivation of the required parameter properties for numerous other types of filters. Of course, Eqs. (17.02-7) to (17.02-9) apply to all filters as long as constant response shape and bandwidth are required.

Another factor to be considered in the design of tunable filters is the choice of low-pass prototype filters to be used. Since the filters used are usually of narrow bandwidth, the equal-element low-pass prototypes discussed in Sec. 11.07 are often a desirable choice. As is discussed in Sec. 11.07, filters designed from equal-element prototypes can give very nearly the absolute minimum midband dissipation loss for given resonator unloaded Q's, and for a given amount of attenuation required at some frequency a specified frequency interval from the mid-pass-band frequency.

For many narrow-band inductively coupled filters such as that in Fig. 17.02-3, a useful low-pass to band-pass mapping (Sec. 8.04) to use is

$$\frac{\omega'}{\omega'_1} = \frac{2}{w}\left(\frac{\omega - \omega_0}{\omega}\right) \tag{17.02-20}$$

where

$$w = 2\left(\frac{\omega_2 - \omega_1}{\omega_2 + \omega_1}\right) \quad , \tag{17.02-21}$$

$$\omega_0 = \frac{2\omega_2\omega_1}{\omega_2 + \omega_1} \quad , \tag{17.02-22}$$

ω' and ω are the frequency variables of the low-pass prototype and band-pass filters, respectively, and ω_1 and ω_2 are the band-pass filter band-edge frequencies corresponding to ω'_1 for the low-pass prototype.

SEC. 17.03, MECHANICALLY TUNABLE COAXIAL BAND-PASS FILTERS

Figure 17.03-1 shows a form of coaxial, mechanically tunable filter, which has been used a good deal.[1] The resonators operate in the TEM mode and are a quarter-wavelength long at resonance. One end of each resonator is open-circuited while the other is short-circuited, and tuning is accomplished by sliding the round center conductor back and forth through the short-circuiting region at the lower end of each resonator. The coaxial input and output lines are coupled to the first and last resonators, respectively, by magnetic coupling loops, while the resonators are coupled to their neighbors by inductive irises. The cross-sectional shape of the resonators is approximately coaxial; however, the region between resonators is flattened so that the coupling irises will not be so thick.

The filter in Fig. 17.03-1 is of the inductively coupled type shown in generalized form in Fig. 17.02-3 and analyzed in Sec. 17.02. As was discussed in connection with Eqs. (17.02-14) and (17.02-15), a somewhat different frequency variation of the resonator slope parameters is usually

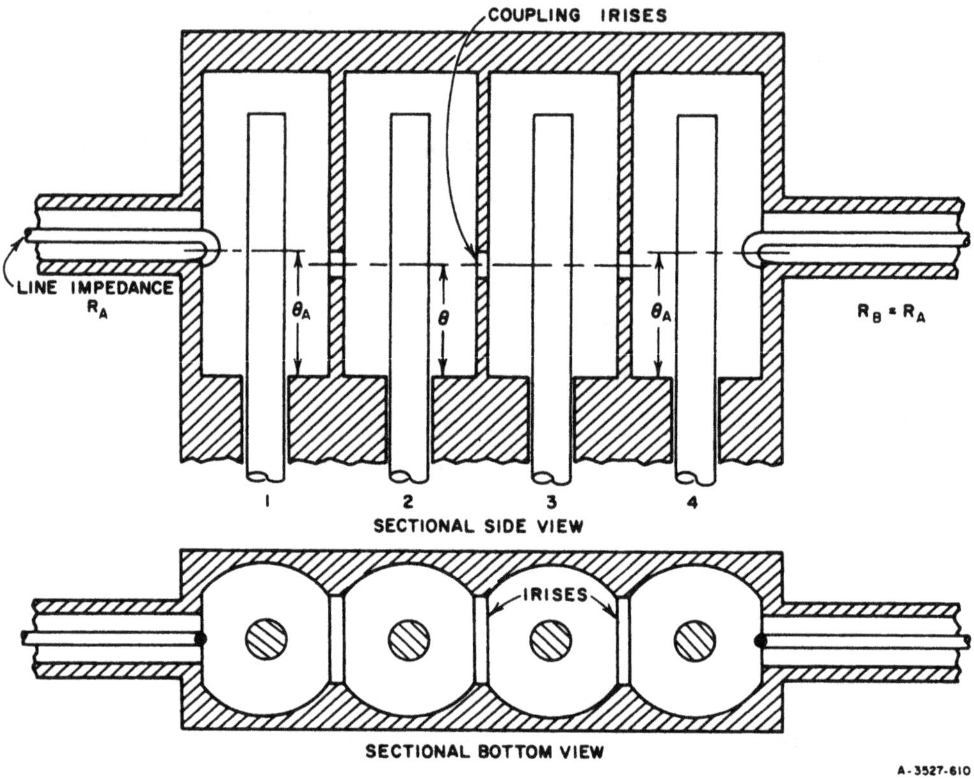

FIG. 17.03-1 A FOUR-RESONATOR MECHANICALLY TUNABLE TEM-MODE FILTER

desired in determining $(Q_e)_A$ and $(Q_e)_B$ as compared with the frequency variation of the resonator slope parameters desired for determining the coupling coefficients $k_{j,j+1}$. In the filter structure in Fig. 17.03-1 the desired difference is achieved in an approximate fashion by locating the input and output coupling loops an electrical distance θ_A from the short-circuited ends of the resonators which is different from the electrical distance θ which the inductive coupling irises are located from the short-circuited ends of the resonators.

Figure 17.03-2(a) shows an equivalent circuit used for calculating $(Q_e)_A$. In the figure

$$X_1 = Z_0 \left[\tan \theta_A - \cot \left(\frac{\pi}{2} - \theta_A\right)\right] \tag{17.03-1}$$

where Z_0 is the characteristic impedance of the resonator line. The resonator reactance slope parameter is then

$$\mathscr{x}_1 = \frac{\omega_0}{2} \frac{dX}{d\omega}\bigg|_{\omega_0} = \frac{\pi}{4} Z_0 \sec^2 \theta_A \tag{17.03-2}$$

and by Eq. (17.02-11) the external Q at end A is

$$(Q_e)_A = \frac{\pi}{4} \frac{Z_0 R_A}{(X_{01})^2 \dfrac{f_0^2}{(f_0)_m^2} \cos^2 \left[(\theta_A)_m \dfrac{f_0}{(f_0)_m}\right]} \tag{17.03-3}$$

where the subscript m indicates a quantity evaluated at the mid-tuning-range frequency $(f_0)_m$. Of course, an analogous equation applies for $(Q_e)_B$.

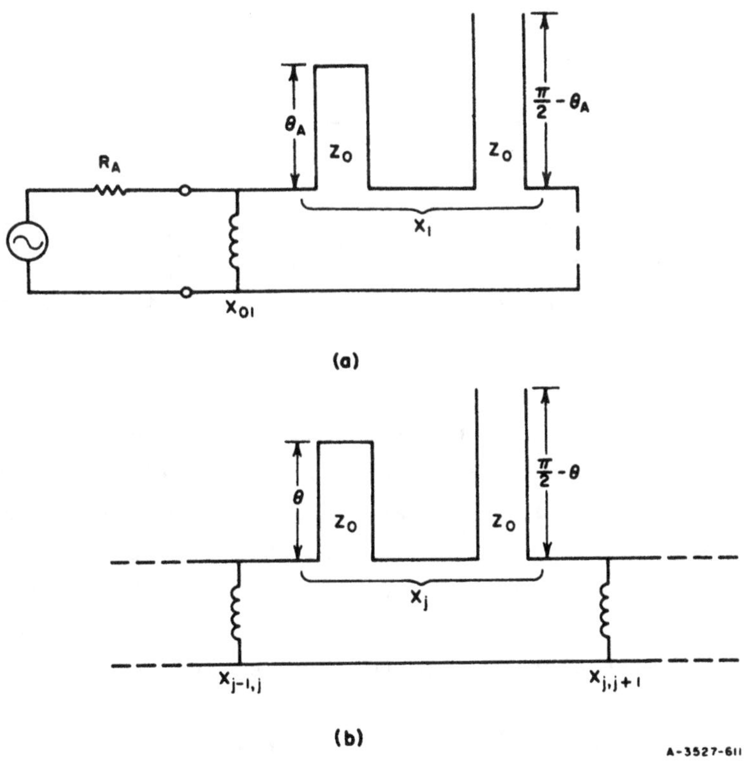

FIG. 17.03-2 RESONATOR EQUIVALENT CIRCUITS USED IN COMPUTING $(Q_e)_A$ AND $k_{j,j+1}$

Figure 17.03-2(b) shows the resonator equivalent circuit used in computing the coupling coefficients between resonators. Using Eq. (17.02-13), analogously to the derivation above we obtain

$$k_{j,j+1} = \frac{4(X_{j,j+1})_m f_0 \cos^2 [(\theta)_m f_0/(f_0)_m]}{\pi Z_0 (f_0)_m} \quad . \qquad (17.03\text{-}4)$$

Now as indicated by Eqs. (17.02-7) to (17.02-9), for constant response shape and bandwidth we desire that

$$(Q_e)_A = \left[(Q_e)_A\right]_m \frac{f_0}{(f_0)_m} \quad , \qquad (17.03\text{-}5)$$

$$(Q_e)_B = \left[(Q_e)_B\right]_m \frac{f_0}{(f_0)_m} \quad , \qquad (17.03\text{-}6)$$

and

$$k_{j,j+1} = (k_{j,j+1})_m \frac{(f_0)_m}{f_0} \quad . \qquad (17.03\text{-}7)$$

If we equate Eqs. (17.03-5) and (17.03-3) for $f_0/(f_0)_m = 0.7$ and for $f_0/(f_0)_m = 1.3$, and solve for $(\theta_A)_m$, we find that $(\theta_A)_m = 0.9675$ radian. If we equate Eqs. (17.03-4) and (17.03-7) at the same frequencies and solve for $(\theta)_m$ we then find that $(\theta)_m = 0.8534$ radian.

From Eqs. (17.03-5) and (17.03-7) we see that both $(Q_e)_A$ and $1/k_{j,j+1}$ should ideally be directly proportional to the frequency, if constant response shape and bandwidth are desired. Figure 17.03-3 presents normalized curves for these quantities *vs*. $f_0/(f_0)_m$ for the values of $(\theta_A)_m$ and $(\theta)_m$ given above. We see that the external Q curve deviates most from the desired linear variation, which means that the response shape will change somewhat as the filter is tuned. The quantity $N/k_{j,j+1}$ follows the desired linear variation more closely, but it also deviates some from the desired line. Since the coupling coefficients between resonators have the dominant effect on bandwidth, the deviation in the $N/k_{j,j+1}$ curve from the desired line will cause some variation of the bandwidth. Although the deviations from ideal performance indicated in Fig. 17.03-3 are sizeable, they are probably acceptable for most applications. Also, the curves in

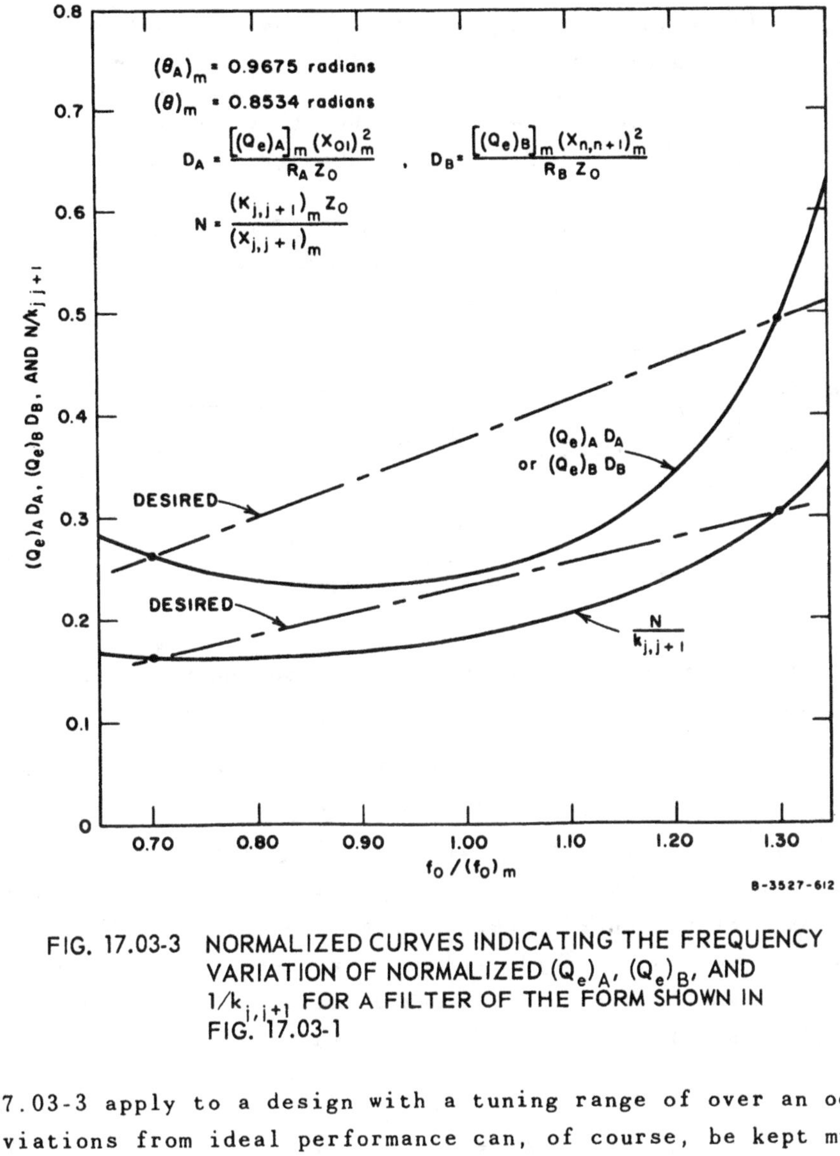

FIG. 17.03-3 NORMALIZED CURVES INDICATING THE FREQUENCY VARIATION OF NORMALIZED $(Q_e)_A$, $(Q_e)_B$, AND $1/k_{j,j+1}$ FOR A FILTER OF THE FORM SHOWN IN FIG. 17.03-1

Fig. 17.03-3 apply to a design with a tuning range of over an octave. The deviations from ideal performance can, of course, be kept much smaller if smaller tuning ranges are sufficient.

Summary of Design Procedure—Let us now summarize how filters of the sort in Fig. 17.03-1 can be designed.

The design process is carried out at the mid-tuning-range frequency $(f_0)_m$. In order to obtain nearly optimum resonator Q's, the resonators should have a line impedance of approximately Z_0 = 76 ohms (see Fig. 5.03-2). Use the values of $(\theta_A)_m$ and $(\theta)_m$ in Fig. 17.03-3, or if tighter control on the response is required and the required tuning range is smaller than an octave, solve for values of $(\theta_A)_m$ and $(\theta)_m$ as described above, in order to give a closer approximation over a smaller tuning range.

The coupling reactances can be computed by combining Eqs. (17.03-3) and (17.03-4) with Eqs. (17.02-1) to (17.02-3) to obtain

$$(X_{01})_m = \frac{1}{\cos (\theta_A)_m} \sqrt{\frac{\pi Z_0 w_m R_A}{4 g_0 g_1 \omega_1'}} \tag{17.03-8}$$

$$(X_{j,j+1})_m \bigg|_{j=1 \text{ to } n-1} = \frac{\pi Z_0 w_m}{4 \cos^2 (\theta_m)\, \omega_1' \sqrt{g_j g_{j+1}}} \tag{17.03-9}$$

$$(X_{n,n+1}) = \frac{1}{\cos (\theta_A)_m} \sqrt{\frac{\pi Z_0 w_m R_A}{4 g_n g_{n+1} \omega_1'}} \tag{17.03-10}$$

where w_m is the fractional bandwidth $\Delta f/(f_0)_m$ to the pass-band edge points corresponding to ω_1' for the low-pass prototype.

As a rough guide in the design of the loop couplings at each end of the filter, the following reasoning can be used.[1] If I is the current in Resonator 1 or n at an electrical distance $(\theta_A)_m$ from the short-circuit end of the resonator, then the magnetic field at a radial distance r from the center axis of the coaxial resonator is $H = I/(2\pi r)$. This field will excite a voltage of $jA\mu_0\omega H = jA\mu_0\omega I/2\pi r$ in a loop of area A located at a mean radius r from the axis of the coaxial line. Thus, the mutual reactance of such a loop in a coaxial line is

$$X = \frac{A\mu_0\omega}{2\pi r} \tag{17.03-11}$$

where, if r is in inches and A is in square inches, $\mu_0 = (0.0254)\, 4\pi 10^{-7}$ henries/inch. This formula can serve as a guide for the initial design of the coupling loop but in practice experimental adjustments as described in Sec. 11.02 will probably be desirable. Reference 1 points out that in order to reduce the self reactance of the coupling loops (excess reactance can cause the tuning of the end resonators to track differently than that of the interior resonators), it is desirable for the coupling loop to be faired into the side wall as much as possible, as is suggested in

Fig. 17.03-1; under such conditions Eq. (17.03-11) may be a poor approximation.

Using Bethe's small-aperture theory the apertures between the resonators should have magnetic polarizabilities of approximately

$$M_{j,j+1} = \frac{\lambda_m a^2 (X_{j,j+1})_m}{60} \tag{17.03-12}$$

where a is the nominal radius of the inside of the outer conductor of the resonators and λ_m is the wavelength at the mean tuning frequency $(f_0)_m$. The sizes of the apertures can then be determined from the M_1 polarizability data in Fig. 5.10-4(a), (the M_1 rather than the M_2 data is used since the lengths of the coupling irises are to be in the circumferential direction). Approximate corrections for the size and thickness t of the apertures can be made by obtaining trial aperture lengths $\ell_{j,j+1}$ using Eq. (17.03-12) and Fig. 5.10-4(a), and then computing compensated polarizabilities

$$(M_{j,j+1})_{\text{comp.}} = M_{j,j+1} \left[1 - \left(\frac{2\ell_{j,j+1}}{\lambda_m}\right)^2\right] 10^{\left(\frac{1.36t}{\ell_{j,j+1}}\right)\sqrt{1-\left(\frac{2\ell_{j,j+1}}{\lambda_m}\right)^2}} \tag{17.03-13}$$

from which improved aperture lengths are obtained using Fig. 5.10-4(a). For greatest accuracy the experimental procedures described in Sec. 11.02 to 11.04 should be used to check the aperture sizes.

One possible source of trouble in the practical operation of filters of the type under discussion lies in the fact that the resonator rods must slide freely while still maintaining a good short-circuit at one end. Sliding contact fingers can be used to help ensure a good short-circuit connection, but the non-contacting short-circuit connections in Fig. 17.03-4 are found to be preferable.[1] The design in Fig. 17.03-4(a) uses a very low impedance section, which is a quarter-wavelength long at frequency $(f_0)_m$. In this design the resonator rod is supported by a dielectric bearing surface at the left. The low-impedance line section is effectively open-circuited at its left end and, as a result, tends to reflect a very large susceptance at its right end. The design in Fig. 17.03-4(b) is similar, but it uses an additional high-impedance line

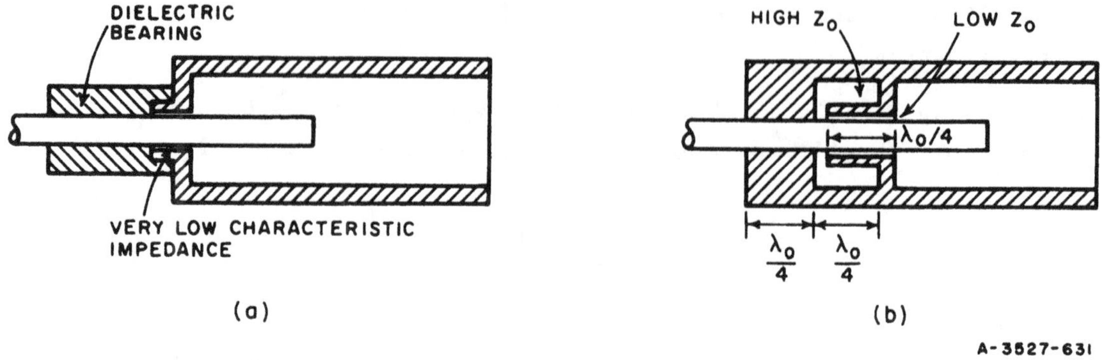

SOURCE: *Very High Frequency Techniques*, Vol. II (see Ref. 1 by Harvard Radio Research Laboratory Staff)

FIG. 17.03-4 TWO POSSIBLE DESIGNS FOR THE SHORT-CIRCUIT AND BEARING PORTION OF MECHANICALLY TUNABLE COAXIAL RESONATORS

section to reflect a very high impedance to the open-circuited end of the low-impedance line section. In this type of design the resonator rod can be supported by a metal bearing. Although this type of choke joint design is frequency sensitive to some extent, it can be made to work very well over as much as a five-to-one tuning range.[1]

SEC. 17.04, WAVEGUIDE MECHANICALLY TUNABLE BAND-PASS FILTERS

Figure 17.04-1 shows a waveguide mechanically tunable filter which is in many respects analogous to the coaxial type of filter discussed in Sec. 17.03. This filter consists of rectangular cavity resonators that are tuned by moving one of the side walls. The resonators are coupled by apertures which are located so as to come as close as possible to giving the desired frequency variations of the external Q's and the coupling coefficients. As in the filter of Sec. 17.03, the coupling apertures between resonators are positioned differently from the apertures coupling to the input and output guides.

The analysis of the filter in Fig. 17.04-1 proceeds similarly to that discussed in Sec. 17.03. In this case the resonator slope parameters are given by

$$\mathscr{X}_j = \frac{\pi}{2} Z_0 \left(\frac{\lambda_g}{\lambda}\right)^2 \tag{17.04-1}$$

when the analysis is carried out on a frequency basis. (See Sec. 8.14 for a discussion of the differences between waveguide filter design on a frequency basis *vs.* design on a reciprocal-guide-wavelength basis.) In Eq. (17.04-1) Z_0 is the guide impedance and λ_g and λ are the guide and free-space wavelengths at resonance. The coupling reactances and the effects of different guide dimensions for the terminating guides as compared to the dimensions of the cavity resonators can all be obtained from Fig. 5.10-6. Thus, using Eqs. (17.04-1) and (17.02-10) to (17.02-13), along with the data in Fig. 5.10-6 gives

$$(Q_e)_A = \frac{a b a_A b_A \lambda_{gA} u^3}{4\pi\lambda^2 M_{01}^2 \sin^2 \frac{\pi x_A}{a}} \tag{17.04-2}$$

$$k_{j,j+1}\Big|_{j=1 \text{ to } n-1} = \frac{M_{j,j+1}\lambda^2 \sin^2 \frac{\pi x}{a}}{a b u^3} \tag{17.04-3}$$

$$(Q_e)_B = \frac{a b a_A b_A \lambda_{gA} u^3}{4\pi\lambda^2 M_{n,n+1}^2 \sin^2 \frac{\pi x_A}{a}} \tag{17.04-4}$$

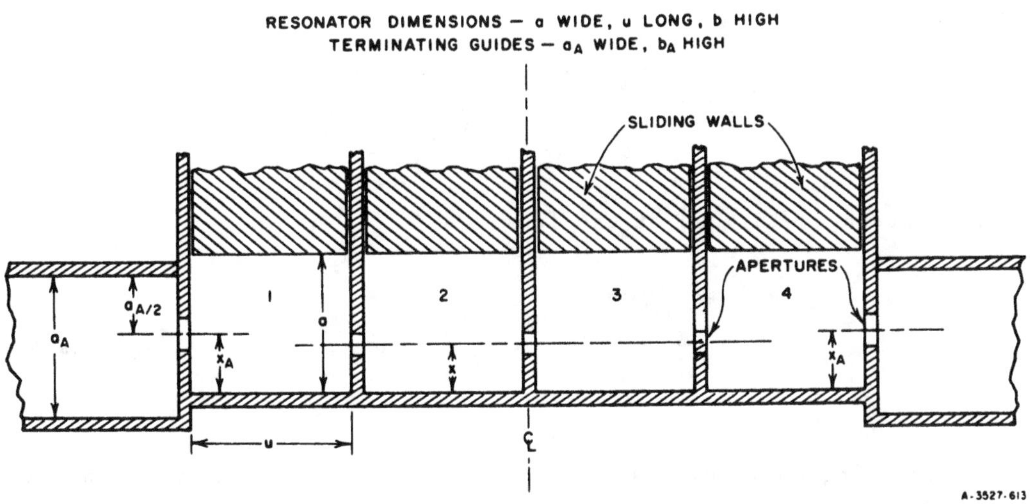

FIG. 17.04-1 A FOUR-RESONATOR MECHANICALLY TUNABLE WAVEGUIDE FILTER

where the $M_{j,j+1}$ are the horizontally directed magnetic polarizabilities of the coupling apertures, λ is the free-space wavelength at the resonant frequency f_0,

$$\lambda_{gA} = \frac{\lambda}{\sqrt{1 - \left(\frac{\lambda}{2a_A}\right)^2}} \tag{17.04-5}$$

is the wavelength in the terminating guides, a_A, b_A, a, b, u, x_A, and x are dimensions defined in Fig. 17.04-1, and where

$$a = \frac{u}{\sqrt{\left(\frac{2u}{\lambda}\right)^2 - 1}} \tag{17.04-6}$$

is required in order to give resonance at the tuning frequency f_0 corresponding to the free-space wavelength λ.

As for the filter in Sec. 17.03, for constant response shape and bandwidth we desire the external Q's and coupling coefficients to vary with frequency as in Eqs. (17.03-5) to (17.03-7). Forcing Eqs. (17.03-5) to (17.03-7) to agree with Eqs. (17.04-2) to (17.04-4) at two frequencies $(f_0)_1$ and $(f_0)_2$ leads to equations of the form

$$\frac{Q_e\big|_{f_0=(f_0)_1}}{Q_e\big|_{f_0=(f_0)_2}} = \frac{(f_0)_1}{(f_0)_2} \tag{17.04-7}$$

for the external Q's, and of the form

$$\frac{k_{j,j+1}\Big|_{f_0=(f_0)_1}}{k_{j,j+1}\Big|_{f_0=(f_0)_2}} = \frac{(f_0)_2}{(f_0)_1} \tag{17.04-8}$$

for the coupling coefficients. These can be solved to obtain x_A and x just as θ_A and θ were solved for in the case of the filter in Sec. 17.03. However, in the present case there are other degrees of freedom available, which result from the fact that there is a choice in the proportions of the cavities and of the terminating guides.[2] If, however, we specify $(f_0)_1/(f_0)_m$, $(f_0)_2/(f_0)_m$, a_A, u and the wavelength λ_m at the mean tuning frequency $(f_0)_m$, then the required x_A and x values can be computed.

Table 17.04-1

PARAMETERS WHICH CAUSE EQS. (17.04-2) TO (17.04-4) TO SATISFY EQS. (17.04-7) AND (17.04-8) FOR $(f_0)_1/(f_0)_m$ = 0.90, $(f_0)_2/(f_0)_m$ = 1.1, AND a_A = 0.76 λ_m

$\frac{\lambda_m}{u}$	$\frac{x_A}{u}$	$\frac{x}{u}$	a/u FOR $f_0 = (f_0)_m$
0.6	0.198	0.160	0.314
0.9	0.315	0.262	0.504
1.2	0.463	0.403	0.750
1.5	0.694	0.635	1.134

Table 17.04-1 shows normalized values for x_A and x which will cause Eqs. (17.04-7) and (17.04-8) to be satisfied for $(f_0)_1/(f_0)_m$ = 0.90 and $(f_0)_2/(f_0)_m$ = 1.10. The choice of cavity length u is controlled by the λ_m/u parameter at the left. At the right is shown the corresponding ratio a/u for the filter when tuned to the mean tuning range frequency $(f_0)_m$.

Figure 17.04-2 shows plots of normalized external Q *vs.* $f_0/(f_0)_m$ for the various cases listed in Table 17.04-1. Notice that the curves approach the desired linear variation most closely when λ_m/u is made small.

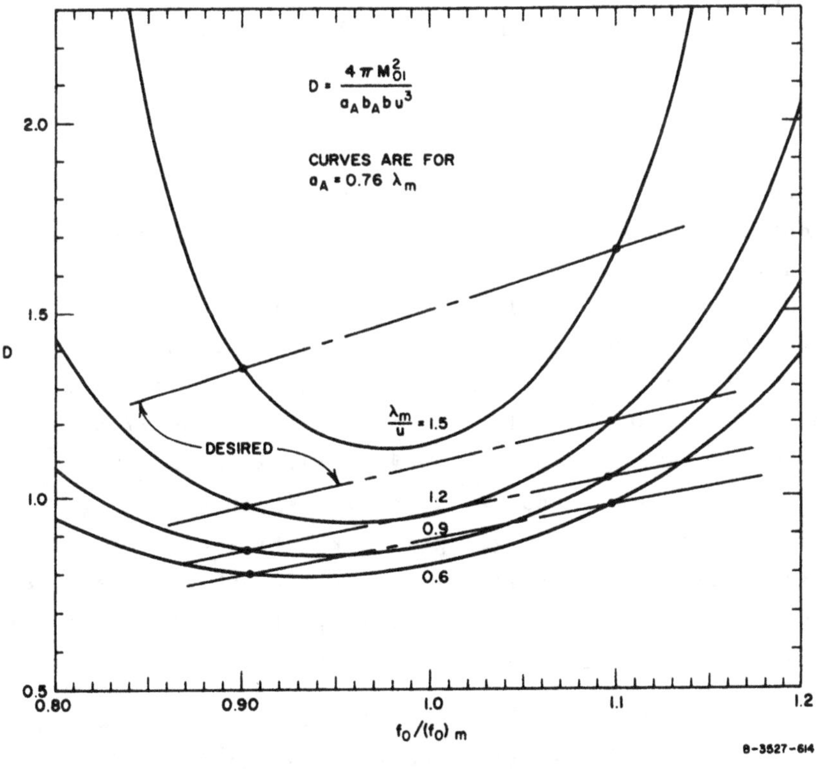

FIG. 17.04-2 CURVES OF NORMALIZED EXTERNAL Q vs. $f_0/(f_0)_m$ FOR PARAMETER VALUES IN TABLE 17.04-1

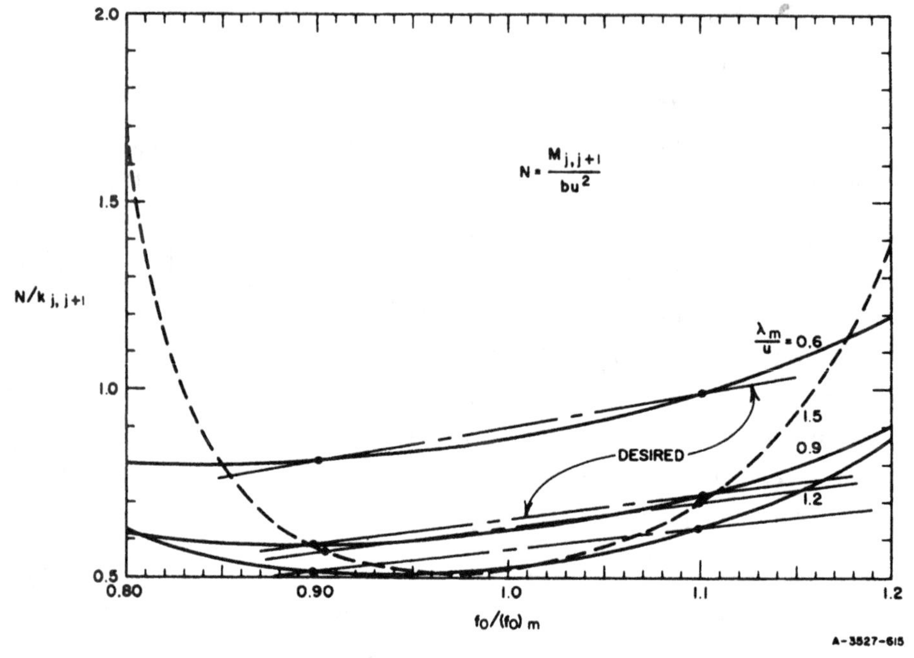

FIG. 17.04-3 CURVES OF NORMALIZED RECIPROCAL COUPLING COEFFICIENT vs. $f_0/(f_0)_m$ FOR RESONATOR PARAMETER VALUES IN TABLE 17.04-1

Figure 17.04-3 shows a corresponding plot of normalized reciprocal coupling coefficient *vs.* $f_0/(f_0)_m$. Here again the desired linear variation is approximated most closely if λ_m/u is small. However, Fig. 17.04-4 shows a plot of the corresponding values of a/u *vs.* $f_0/(f_0)_m$ where it should be recalled that u is constant while the cavity width a is varied to achieve the desired tuning frequency. Note that the small values of λ_m/u, which gave the most desirable results in Figs. 17.04-2 and 17.04-3, correspond to tuning characteristics that have very large changes in resonant frequency f_0 for very small changes in cavity width a. Thus, as has been discussed by Sleven,[2] in designing a tunable filter of this type one should not insist on any greater uniformity of bandwidth and response shape than is really necessary, since this uniformity is bought at the price of criticalness in the tuning adjustment of the resonators.

Summary of Design Procedure—The first step in the design of a filter of the type in Fig. 17.04-1 is to select a value for λ_m/u using Figs. 17.04-2 to 17.04-4 while keeping the above points in mind. Some idea of the variations in bandwidth that will result from the deviations of the resonator coupling coefficients can be gained from the fact that the $k_{j,j+1}$ are directly proportional to the *fractional* bandwidth of the filter. The choice of the external Q's of a multi-resonator filter affects the shape of the pass band of a filter much more than it affects the bandwidth. With a two-resonator filter, steadily increasing the size of the external Q's tends to result in a Tchebyscheff response with an increasingly large hump in the middle. Steadily decreasing the external Q's of such a filter decreases and finally eliminates the hump in the middle of the response and eventually leads to an "undercoupled" response (called *undercoupled* because the coupling between resonators is quite loose as compared to the coupling between the terminations and the resonators). Analogous effects occur in filters with more resonators. The above points should be of some help in assessing the effects of deviations from the ideal external Q and coupling characteristics *vs.* frequency. More exact pictures of these effects can be obtained by working back from the external Q and coupling coefficient values at various frequencies by solving for the corresponding sets of low-pass prototype element values g_0, g_1, ..., g_{n+1} using

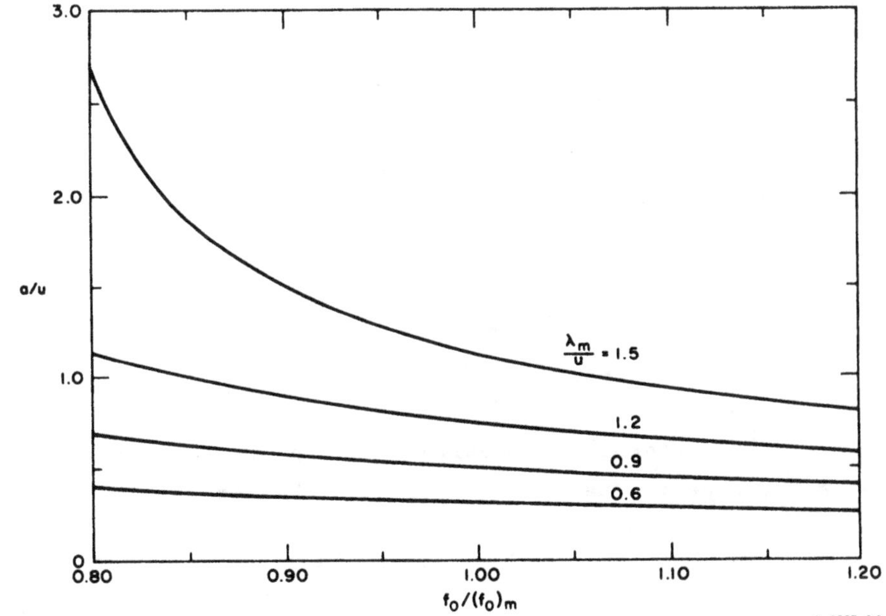

FIG. 17.04-4 a/u vs. $f_0/(f_0)_m$ FOR RESONATOR PARAMETER VALUES IN TABLE 17.04-1

Eqs. (17.02-1) to (17.02-3) and then comparing these values with the various tabulated designs in Sec. 4.05, or by actually computing the responses corresponding to these sets of element values.

Having arrived at a suitable compromise choice for λ_m/u, then x_A/u and x/u are obtained from Table 17.04-1 or by calculation as described above. Having specified a mean tuning frequency $(f_0)_m$, the corresponding free-space wavelength λ_m can be computed along with u, x_A, and x. The desired external Q's and coupling coefficients are computed from the low-pass prototype parameters by use of Eqs. (17.02-7) to (17.02-9). Then using Eqs. (17.04-2) to (17.04-6) with $\lambda = \lambda_m$, the magnetic polarizabilities M_{01}, M_{12}, ..., $M_{n,n+1}$ are obtained.

After the designer has obtained the required polarizabilities, the dimensions of the apertures can be obtained from the data in Sec. 5.10. If rectangular or elongated apertures are used, their length should be in a direction parallel to the a dimension of the cavities in Fig. 17.04-1. Corrections for aperture length $\ell_{j,j+1}$ and thickness t can be made as discussed at the end of Sec. 17.03 by use of Eq. (17.03-13). If round apertures are used, the same procedure applies, except that the polarizability is given by the approximate formula

$$M_{j,j+1} = \frac{d^3_{j,j+1}}{6} \tag{17.04-9}$$

(where $d_{j,j+1}$ is the diameter of the aperture) and Eq. (17.03-13) becomes

$$(M_{j,j+1})_{comp} = M_{j,j+1}\left[1 - \left(\frac{1.7d_{j,j+1}}{\lambda_m}\right)^2\right] 10^{\left(\frac{1.6t}{d_{j,j+1}}\right)\sqrt{1-\left(\frac{1.7d_{j,j+1}}{\lambda_m}\right)^2}} \tag{17.04-10}$$

If high accuracy in the design is desired, the aperture sizes can be checked using the experimental procedures described in Secs. 11.02 to 11.04.

SEC. 17.05, PROPERTIES OF FERRIMAGNETIC RESONATORS*

There are a number of single-crystal materials that have possible use as ferrimagnetic resonators in magnetically tunable microwave filters. Some materials of interest at the time of this writing are

(1) Yttrium-iron-garnet (YIG)

(2) Gallium-substituted yttrium-iron-garnet (GaYIG)

(3) Lithium ferrite

(4) Barium ferrite.

The YIG material listed above has proved the most useful to date and has been successfully used in constructing a variety of magnetically tunable microwave filters such as those described by Carter.[3,4,5,6]

In this book the detailed theory of ferrimagnetic resonance will not be treated. Such theory can be found in various references.[7,8,29] However, in this section a qualitative description of ferrimagnetic resonance will be presented along with some basic formulas and concepts useful in the design of magnetically tunable filters.

Description of the Resonance Phenomenon—Let us suppose that a dc H-field of strength H_0 is applied to a single-crystal YIG sphere in a horizontal direction, and then the direction of the H_0 field is rapidly switched to the vertical direction as shown in Fig. 17.05-1(a). The YIG material contains unpaired electrons which yield magnetic moments as a result of their spins. When the dc H-field is rapidly switched to the vertical position, these spin magnetic moments will precess about the vertical H-field H_0 at a rate of roughly (if H_0 is in oersteds)

$$(f_0)_{Mc} \approx 2.8\, H_0 \qquad \text{Mc} \tag{17.05-1}$$

As time elapses the trajectory of the electron-spin magnetic moments will spiral in toward the direction of the H_0 field until the spins will ultimately end up parallel to the H_0 field. If the fields around the sphere are sampled while this process is going on, a circularly polarized RF field will be observed about the sphere like that which would be created if the

* To this writer's knowledge the first microwave filter structure using a ferrimagnetic resonator was due to R. W. DeGrasse (see Ref. 20). His device was primarily a fixed-frequency narrow-band limiter.

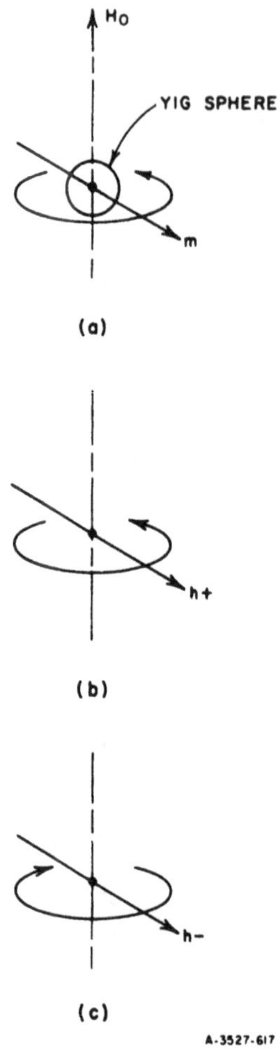

FIG. 17.05-1 A YIG SPHERE WITH A CIRCULARLY POLARIZED RF MAGNETIC MOMENT IS SHOWN AT (a). EXTERIOR CIRCULARLY POLARIZED RF MAGNETIC FIELDS ARE SHOWN AT (b) AND (c)

sphere contained a magnetic dipole rotating at the resonance rate given approximately by Eq. (17.05-1). This rotating dipole moment is pictured by Fig. 17.05-1(a). The circularly polarized magnetic field seen to be emanating from the sphere would die out exponentially with time in the same way that transient voltages and currents die out in a resonant circuit having dissipation loss.

Let us now consider another experiment. If the H_0-field is on and a circularly polarized RF H-field is applied as indicated by the rotating h^+ vector in Fig. 17.05-1(b), the RF field will have no effect on the sphere unless the frequency is at or very near the ferrimagnetic resonance frequency, which was given approximately by Eq. (17.05-1). When the circularly polarized applied field is at or very near the ferrimagnetic resonance frequency, circularly polarized RF H-fields will build up about the sphere in much the same way as the H-field will build up in the inductance of an L-C tuned circuit excited at its resonant frequency.

If a circularly polarized field with the circular polarization in the opposite direction, as indicated by the rotating vector in Fig. 17.05-1(c), is applied, the sphere will not respond even if the signal is at the ferrimagnetic resonance frequency. Thus it is seen that the resonance phenomenon is nonreciprocal.

If a linearly polarized RF H-field is applied to the sphere in a direction perpendicular to the biasing field H_0, it will be found that the sphere will respond by emanating a circularly polarized RF H-field, provided that the linearly polarized field is at or very near the ferrimagnetic resonance frequency. The reason for this is that any linearly polarized field may be regarded as being the sum of two circularly polarized fields of equal strength rotating in opposite directions. Thus one of the circularly polarized components of the linear field will excite the sphere. Because of this, depending on the manner in which ferrimagnetic resonators are used, the transmission characteristics of the filter may be either reciprocal or nonreciprocal.

Parameters of Ferrimagnetic Resonator Materials—Several parameters characterize the various types of materials that can be used to construct ferrimagnetic resonators:

(1) Saturation magnetization, M_s

(2) Line width ΔH, or unloaded Q, Q_u

(3) Anisotropy field constant, K_1/M_s

(4) Curie temperature, T_c.

The saturation magnetization M_s is a function of the number of electron spins in the material per unit volume. The larger the M_s, the easier it is to couple from an exterior strip-line or waveguide circuit to a ferrimagnetic resonator.

The line width ΔH is defined in various ways, but it is basically the width of the resonance in oersteds as the signal frequency is held constant and the applied dc field is varied. For filter applications the line width ΔH is usually defined as the difference between the two values of biasing field (at each side of resonance) for which the imaginary part of the intrinsic susceptibility of the material equals the real part, while frequency is held constant. Looking at the same resonance phenomenon from the viewpoint of holding the biasing field constant and varying the frequency, we measure the unloaded Q of the resonator. With ΔH defined as above, for a spherical resonator

$$Q_u = \frac{f_0(10^{-6})}{2.8\ \Delta H} \qquad (17.05\text{-}2)$$

where f_0 is the resonant frequency in cycles per second and ΔH is the line width in oersteds. The unloaded Q of YIG will increase with frequency up to around 5 or 10 Gc, but the curve of Q_u *vs.* frequency then levels out.[9] However, useful Q's appear to be possible up to at least 60 Gc.[10]

The line width (or unloaded Q) that will be measured for a ferrimagnetic resonator will depend upon both the material itself and the shape and surface finish of the material. In order for the line width to be narrow (and for the unloaded Q to be high) the internal field within the material must be uniform. Assuming that the applied H_0-field is uniform before the material is inserted, in order for the resonator material to have a uniform H-field within itself after being inserted within the biasing field, the resonator must be spheroidal or ellipsoidal.[7,8,29] Further, in order to achieve the narrowest possible line width (and highest possible unloaded Q), it is necessary that the surface of the resonator be very highly polished. Even when a ferrimagnetic resonator is spheroidal or ellipsoidal and is highly polished, its line width may still be degraded by the presence of metallic walls near the resonator. Some of this degradation, however, is unavoidable in most filter structures.

The first-order anisotropy field constant K_1/M_s is important because of the lattice structure of single-crystal material. The lattice structure makes the material easier to magnetize along some crystal axes than along others. Because of this, the ferrimagnetic resonance frequency of a ferrimagnetic resonator will be influenced to some extent by the orientation of the crystal axes with respect to the biasing field H_0. As will be discussed below, the first-order anisotropy field constant K_1/M_s is used in computing the resonant frequency for various orientations of the crystal lattice with respect to the biasing field. There is also a second-order constant, but it is so small as to be unimportant for the applications herein.

The Curie temperature T_c is the temperature at which the saturation magnetization drops to zero.* Resonator operation at temperatures close to or above T_c is not possible.

Table 17.05-1 presents values for the parameters discussed above for various materials. Note that the saturation magnetization is given in gaussian units and is listed as $4\pi M_s$ gauss. This is the most commonly used unit for saturation magnetization. These values could be converted to mks units by use of the conversion

$$(\mu_0 M_s \text{ in webers/meter}^2) = (4\pi M_s \text{ in gauss})10^{-4} \qquad (17.05\text{-}3a)$$

* For ferrimagnetic materials this temperature is also known as the Néel temperature.[29]

or

$$(M_s \text{ in ampere-turns/meter}) = 79.5(4\pi M_s \text{ in gauss}) \quad . \quad (17.05\text{-}3b)$$

Also recall that

$$(H_0 \text{ in ampere-turns/meter}) = 79.5(H_0 \text{ in oersteds}) \quad . \quad (17.05\text{-}4)$$

Note that in Table 17.05-1 the ΔH = 0.22 oersted line width listed (which is for a very high-quality YIG resonator) corresponds to an unloaded Q of 6500 at 4 Gc. This line width was measured in a waveguide with the spherical resonator some distance from any metallic walls. In practical filters the YIG resonators must be closer to metal walls, and the disturbing effects of the currents in the walls may reduce the unloaded Q to 2000 or less.

Table 17.05-1

PROPERTIES OF SINGLE-CRYSTAL FERRIMAGNETIC MATERIALS FOR MAGNETICALLY TUNABLE FILTERS[6]

MATERIAL	$4\pi M_s$ (AT ROOM TEMPERATURE) (gauss)	K_1/M_s (AT ROOM TEMPERATURE) (oersteds)	ΔH^* (AT ROOM TEMPERATURE) (oersteds)	T_c (°C)
Yttrium-Iron-Garnet† (YIG)	1750	-43	0.22 (4 Gc), (Ref. 3)	292
Gallium-Substituted† Yttrium-Iron-Garnet (GaYIG)	50 - 1750 600 950 ± 50	-- -55.8 -41.7	-- -- 0.7 - 2.0 (at 4.4 Gc)	-- 160 206
Lithium Ferrite	3550 ± 40 (Ref. 11)	--	3§(5 Gc)	--
"Planar" Ferrite "Zn_2Y" ($Ba_{12}Zn_2Fe_{12}O_{22}$)	2850 (Ref. 12)	4950 (Ref. 12)	16 (X-band), (Ref. 12)	--

* These values of ΔH were measured in cavities. The line width may vary considerably from sample to sample and will be larger when measured in a closed-in filter structure.

† These materials were supplied by Microwave Chemical Laboratory, New York, N.Y.

§ Private communication to P. S. Carter, Jr., from J. W. Nielsen, Airtron Division of Litton Industries, Morris Plains, New Jersey.

The value of ΔH may vary considerably from sample to sample of a given type of material, depending on how perfect the crystal structure is. However, M_s, K_1/M_s, and T_c are physical constants for a given type of material, and should vary only slightly from sample to sample. As will be discussed later, M_s and K_1/M_s do vary with temperature.

Determination of the Resonant Frequency—Three factors may cause the resonant frequency of a ferrimagnetic resonator to differ from the frequency indicated by Eq. (17.05-1):

(1) The shape of the resonator

(2) Crystalline anisotropy

(3) Interaction with currents in metallic walls close to the ferrimagnetic resonator.

The shape of the resonator can affect the resonant frequency because the shape affects the intensity of the demagnetizing field within the resonator.[7,8,29] Assuming that the biasing field H_0 is in the z direction, and x and y are rectangular coordinates perpendicular to H_0, then for any ellipsoidal resonator the resonant frequency (neglecting anisotropy effects) is given by

$$(f_0)_{Mc} = 2.8\sqrt{[H_0 - (N_z - N_x)(4\pi M_s)][H_0 - (N_z - N_y)(4\pi M_s)]} \quad \text{Mc} \tag{17.05-5}$$

where N_x, N_y, and N_z are demagnetizing factors[7,8] in the x, y, and z directions, H_0 is in oersteds and $4\pi M_s$ is in gauss.

For a sphere

$$N_x = N_y = N_z = \frac{1}{3} \quad , \tag{17.05-6}$$

and Eq. (17.05-5) becomes

$$(f_0)_{Mc} = 2.8\, H_0 \quad \text{Mc} \quad , \tag{17.05-7}$$

which is identical to Eq. (17.05-1).

We may consider a long thin rod and a very thin disk as limiting cases of ellipsoids. For an infinitely thin rod parallel to H_0,

$$N_x = N_y = \frac{1}{2} \quad \text{and} \quad N_z = 0 \quad . \tag{17.05-8}$$

For an infinitely thin disk in the plane of H_0 and the y axis,

$$N_x = 1, \quad \text{and} \quad N_y = N_z = 0 \quad . \tag{17.05-9}$$

If the disk lies in the plane of the x and y axis, then

$$N_x = N_y = 0 \quad \text{and} \quad N_z = 1 \quad . \tag{17.05-10}$$

Thus, it can be seen that the shape of the resonator can have considerable effect on the resonant frequency. However, of the possible ellipsoidal shapes that can be used, the sphere is generally the most practical since it is the easiest to prepare with precision.

Materials such as YIG and GaYIG, which have cubic crystal structure, have three types of principal crystal axes: the [100], the [110], and the [111] axes.[13] Even though there are only three kinds of principal axes, there are three [100] axes, six [110] axes, and four [111] axes in each single-crystal. For a given applied biasing field H_0, the resonant frequency of a ferrimagnetic resonator will be influenced by the direction of these axes relative to the direction of the applied field. The opposite extremes of this effect occur when the [111] or the [100] axis is parallel to the H_0 field, when the [110] axis is parallel to H_0 an intermediate effect results. For a *sphere* of cubic crystal material with a [111] axis parallel to the H_0 field, Eq. (17.05-7) becomes

$$(f_0)_{Mc} = 2.8 \left(H_0 - \frac{4}{3}\frac{K_1}{M_s}\right) \quad \text{Mc} \quad , \tag{17.05-11}$$

while if a [100] axis is parallel to H_0

$$(f_0)_{Mc} = 2.8 \left(H_0 + 2\frac{K_1}{M_s}\right) \quad \text{Mc} \tag{17.05-12}$$

where H_0 and K_1/M_s are in oersteds. For YIG or GaYIG, K_1/M_s is negative and Eq. (17.05-11) gives resonance for a lower field strength than does Eq. (17.05-12). As a result, for materials such as YIG and GaYIG which have negative K_1/M_s, the [111] axes are known as *easy* axes while the [100] axes are known as *hard* axes. If K_1/M_s is positive, the roles of these axes are reversed. For YIG, the difference in resonant frequencies given by Eqs. (17.05-11) and (17.05-12) is about 401 Mc, which is, of course, a significant difference. Figure 17.05-2 shows curves of the resonant frequency *vs*. biasing field strength for the [100], [110], and [111] axes parallel to the biasing field.

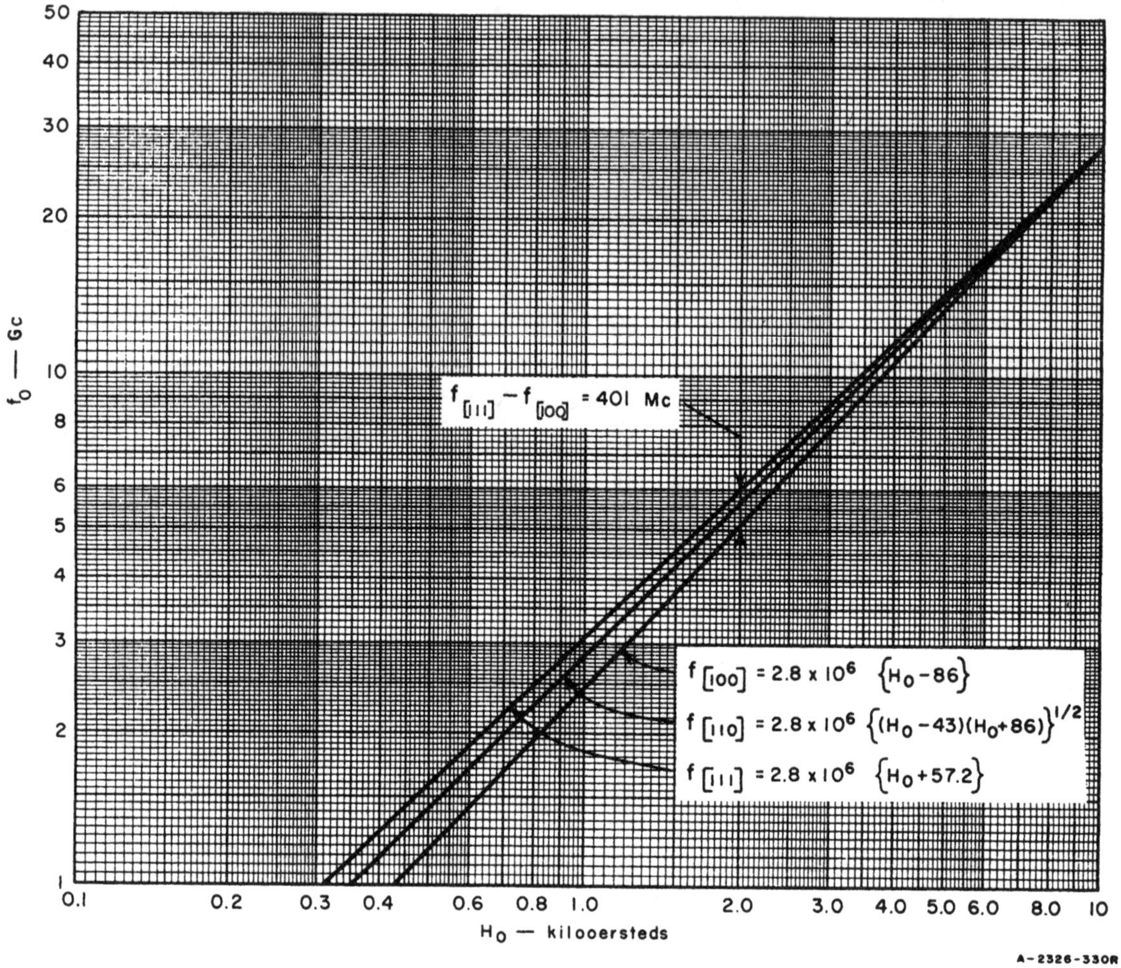

FIG. 17.05-2 RESONANT FREQUENCY OF YIG SPHERE vs. APPLIED dc FIELD WITH FIELD ALONG THE [100], [110], OR [111] PRINCIPAL AXES

If a sphere of ferrimagnetic material with cubic crystal structure is rotated about a [110] crystal axis that is perpendicular to H_0, the value of biasing field H_0 required to give resonance at $(f_0)_{Mc}$ megacycles is

$$H_0 = \frac{(f_0)_{Mc}}{2.8} - \left(2 - \frac{5}{2}\sin^2\theta - \frac{15}{8}\sin^2\theta\right)\frac{K_1}{M_s} \qquad \text{oersteds} \tag{17.05-13}$$

where θ is the angle between H_0 and that [100] axis which becomes parallel to H_0 as the sphere is rotated about the given [110] axis. This manner of rotation will cause the sphere at different times to have a [100] or a

[111] axis parallel to H_0, and the field strength for resonance will cover the largest possible range as the resonator is rotated. Figure 17.05-3 presents measured data, due to Y. Sato and P. S. Carter, Jr.,[6] which show how the measured field strength for resonance at 3000 Mc varies as a YIG sphere is rotated about an [110] axis that is perpendicular to H_0.

As was mentioned above, the metallic boundary conditions seen by ferrimagnetic resonators can also alter their resonant frequencies. If all of the resonators see the same boundary conditions, this effect should cause no trouble. However, if one or more of the resonators see different boundary conditions (such as generally occurs in a filter with three or more YIG resonators) then some means for compensating for the detuning effects of the metallic boundaries is desirable. A successful way of doing this has been found: spherical resonators are mounted in such a way that they can be rotated about [110] axes that are perpendicular to H_0, as was done in making the measurements in Fig. 17.05-3. Using this technique, the resonators can be tuned by rotating their mounting shafts until angular positions relative to each other are obtained that will give synchronous tuning.

Minimum Tuning Frequency—As a ferrimagnetic resonator is tuned to lower and lower frequencies, the applied field H_0 to give resonance gets

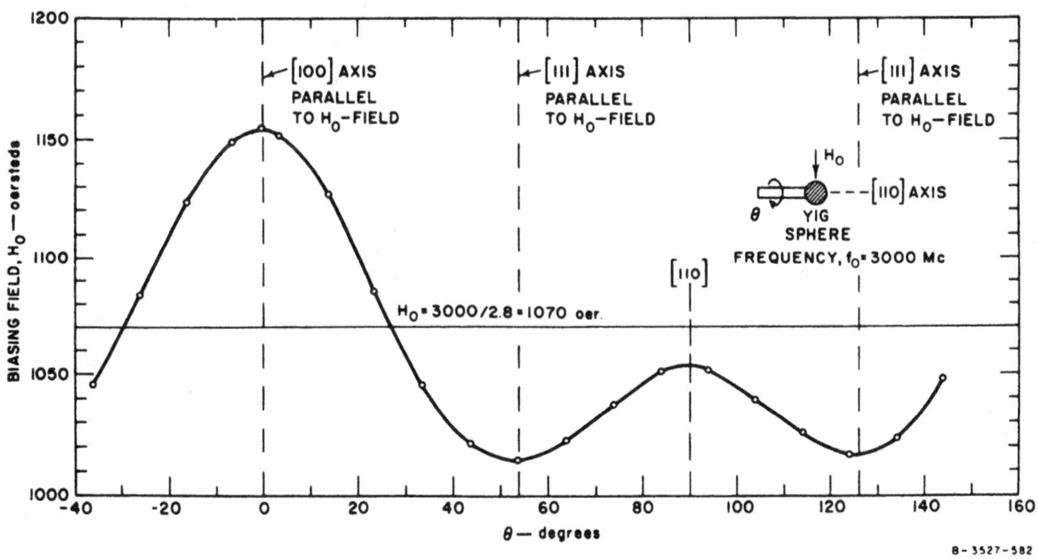

FIG. 17.05-3 FIELD STRENGTH TO GIVE RESONANCE AT 3000 Mc AS A YIG SPHERE IS ROTATED ABOUT A [110] AXIS WHICH IS PERPENDICULAR TO H_0

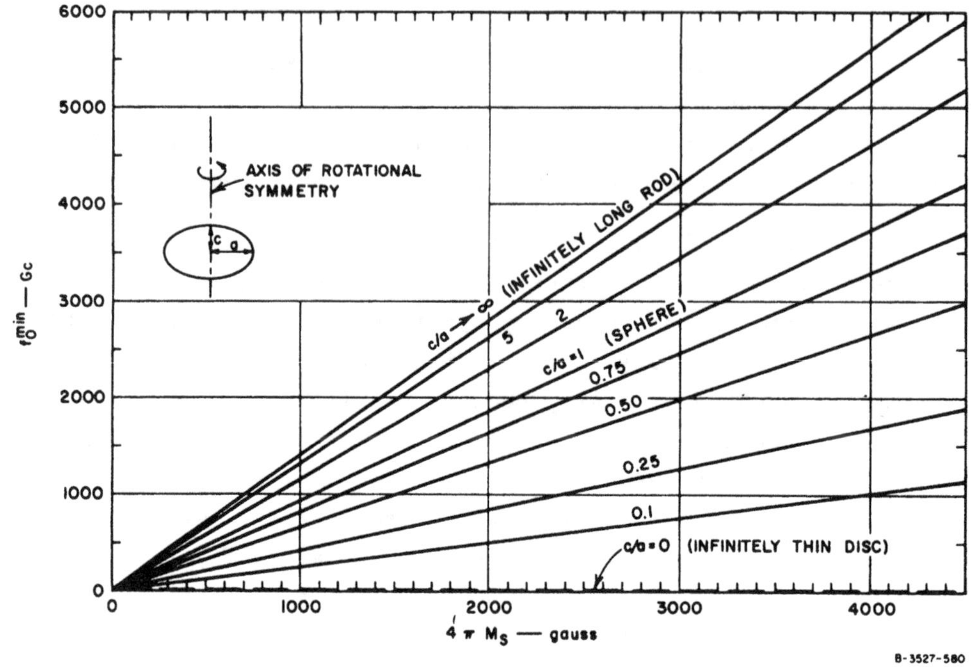

FIG. 17.05-4 APPROXIMATE MINIMUM RESONANT FREQUENCIES OF FERRIMAGNETIC ELLIPSOIDS HAVING VARIOUS AXIS RATIOS

smaller and smaller. When the applied field becomes so low as to be approximately equal to or less than the demagnetizing field, the resonator will cease to function. Since the demagnetizing field is determined by M_s and the demagnetizing factors (which are a function of the shape of the resonator), the minimum resonant frequency is determined by M_s and the resonator shape. Figure 17.05-4 shows plots of the minimum resonant frequency f_0^{min} in gigacycles *vs.* $4\pi M_s$ in gauss for various ellipsoidal shapes. Note that for YIG, which has $4\pi M_s$ = 1750 gauss, the minimum resonant frequency for a spherical sample is approximately 1630 Mc. (In practice, the minimum frequency is found to be a little higher than this.) In theory, by using flat, disk-like ellipsoidal shapes the minimum resonant frequency could be reduced greatly. However, in practice, disk-shaped resonators do not appear to work very well, possibly because of the difficulty in obtaining disk-like resonators that are sufficiently perfect ellipsoids.

A more practical way of obtaining lower minimum resonant frequencies appears to be to use spheroidal resonators of single-crystal material having a lower value of $4\pi M_s$. At this time the gallium-substituted YIG materials appear to be the most promising in this respect. For example, the GaYIG material with $4\pi M_s$ = 600 gauss listed in Table 17.05-1 should give a minimum resonant frequency of around 700 Mc for a spherical sample. The possible disadvantages of such materials are that the lower values of M_s will make coupling from the external circuit to the spheres more difficult, and the line widths of the material may not be as narrow as those obtainable with ordinary YIG.

Temperature Effects—Temperature will affect the values of K_1/M_s, M_s, and ΔH. Of these effects probably the changes in K_1/M_s are potentially the most troublesome. Figure 17.05-5 shows some measured data due to Dillon[14] for $-K_1/M_s$ *vs.* temperature in degrees Kelvin for YIG. Note that the anisotropy constant varies considerably with temperature. By Eq. (17.05-11), if the [111] axis is parallel to H_0 the change in resonant frequency Δf_0 due to a change $\Delta|K_1/M_s|$ in the anisotropy field constant is

$$(\Delta f_0)_{Mc} = 3.73\,\Delta\left|\frac{K_1}{M_s}\right| \quad \text{Mc} \qquad (17.05\text{-}14)$$

while if the [100] axis is parallel to H_0

$$(\Delta f_0)_{Mc} = 5.6\,\Delta\left|\frac{K_1}{M_s}\right| \quad \text{Mc} \qquad (17.05\text{-}15)$$

where $\Delta|K_1/M_s|$ is in oersteds. From Fig. 17.05-5, for a change in temperature from 100°F to 200°F (*i.e.*, 311°K to 466°K), $\Delta|K_1/M_s|$ = 31 oersteds. Then the changes in resonant frequency given by Eqs. (17.05-14) and (17.05-15), are 115 and 173 Mc, respectively. Thus, we see that the shift in resonant frequency can be quite significant if the change in temperature is large.

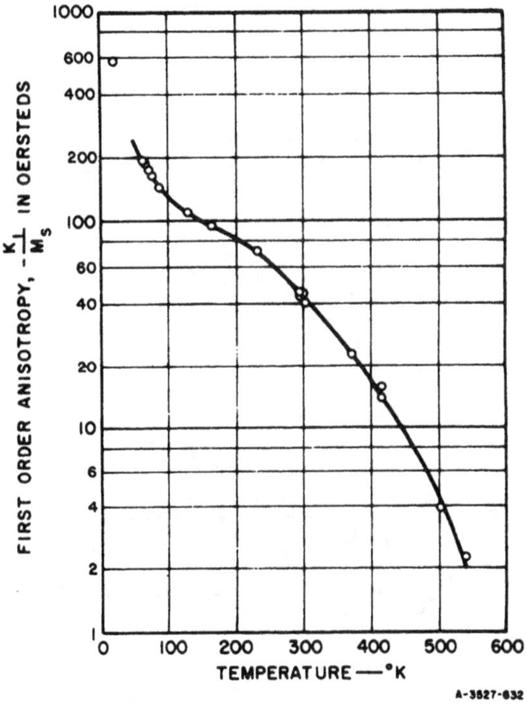

SOURCE: *Physical Review* (see Ref. 14 by J. F. Dillon, Jr.)

FIG. 17.05-5 MEASURED VARIATION OF $-K_1/M_s$ vs. TEMPERATURE FOR YIG

It should be possible to eliminate the effect of temperature on the resonant frequency by orienting the crystal axes in certain directions. Note that in Fig. 17.05-3 the resonant frequency in megacycles is given by 2.8 H_0 when the sphere is oriented at the plus or minus 27 degree points. With these orientations, the anisotropy effects are cancelled out, and the resonant frequency should be independent of temperature.

Figure 17.05-6 shows some measured data of $4\pi M_s$ *vs.* temperature, which was obtained from work of Kooi, Stinson, Moss, Bradley, and Freiberg.[15]

Note that the saturation magnetization decreases as the temperature increases. As will be seen from the discussion in Sec. 17.07, the external Q's of a filter will be proportional to $1/M_s$, while the coupling coefficients between resonators are proportional to M_s. As a result, the bandwidth of the filter will be proportional to M_s while the response shape (*i.e.*, its Tchebyscheff character, for example) should remain unchanged as M_s is varied. Of course, if M_s is decreased (and the bandwidth

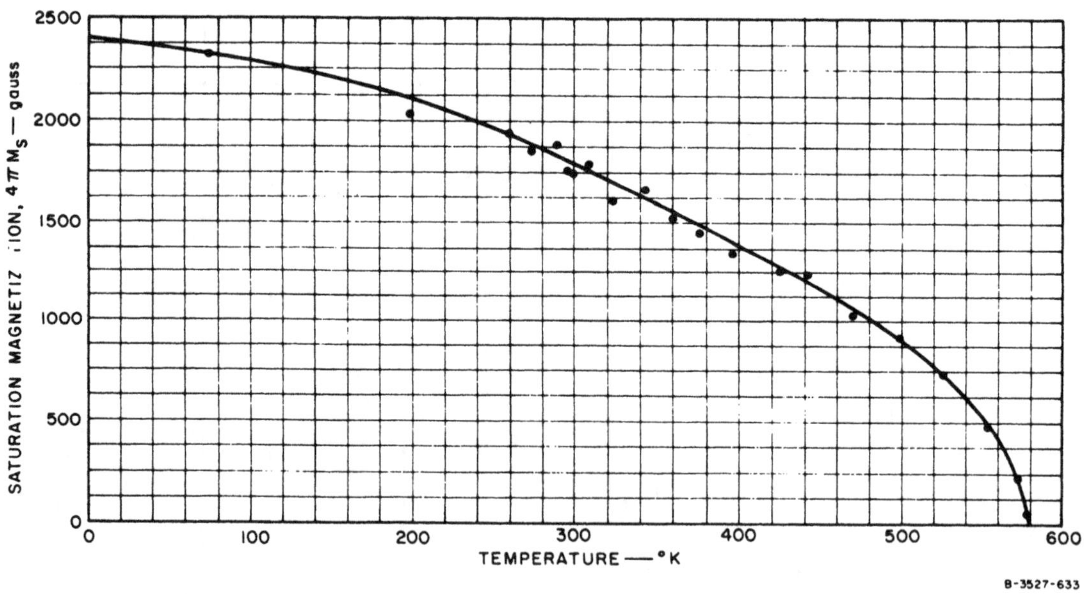

SOURCE: Lockheed LMSD Semi-Annual Report No. 8712 (see Ref. 15 by C. Kooi, D. Stinson, R. Moss, F. Bradley, and L. Freiberg)

FIG. 17.05-6 MEASURED VARIATION OF SATURATION MAGNETIZATION WITH TEMPERATURE FOR YIG

is decreased), the pass-band dissipation loss will increase, as is always the case when a filter's bandwidth is narrowed; if the resonator unloaded Q's are held constant.

Data from Spencer, LeCraw, and Linares[16] indicates that for very pure, well-polished YIG, the line width should vary approximately proportionally to the *absolute* temperature, at least in the room temperature range. Since the line width controls the unloaded Q of the resonators, this will have some influence on the dissipation loss of the filter. However, unless the temperature ranges covered are very large, this effect should not be of great importance in most practical situations.

Higher-Order Magnetostatic Modes—In the desired "free-precessional" resonant mode of ferrimagnetic resonators, all of the electron spins precess with the same phase. However, if either the applied biasing H-field or the RF H-field is not uniform, higher-order modes can occur in which the phases of the precessions in parts of the sphere will be different from the phases in other parts. This causes the ferrimagnetic resonator to have more than one resonant frequency for a given value of biasing H-field.[17] Because of the metallic material required near ferrimagnetic resonators in order to couple to them, there are almost always some disturbing effects which will tend to excite higher-order magnetostatic modes. However, if care is taken to keep the magnetic fields as uniform as possible, it should in most cases be possible to keep spurious responses due to higher-order modes at a level of 30 or more db below the level of the main response.

Besides designing the structure so as to keep the fields as uniform as possible, another important measure is to keep the ferrimagnetic resonator as small as possible. Fletcher and Solt[18] have found that if other factors are held equal, the coupling to higher-order magnetostatic modes depends on the ratio D_m/λ where D_m is the diameter of the sphere and λ is the wavelength at the frequency in question. The smaller D_m/λ, the less coupling there will be to higher-order modes. Thus, it is desirable to make ferrimagnetic resonators as small as possible—consistent with considerations of obtaining adequate coupling from the external circuit to the desired resonance of the resonators.

Power Limiting Effects—When the power passing through a ferrimagnetic resonator becomes sufficiently large, the insertion loss of the resonator will begin to increase greatly as a result of nonlinear effects.[19] Because of this property of ferrimagnetic resonators, ferrimagnetic resonance filters also find application as limiters. Depending on the circuit design, type of resonator material, and frequency range, the limiting level may be much less than a milliwatt or up to around 10 watts.[20,21,22,23,3,4] A treatment of the theory of ferrimagnetic limiting is, however, beyond the scope of this discussion.

SEC. 17.06, DETERMINATION OF THE CRYSTAL AXES OF SPHERICAL FERRIMAGNETIC RESONATORS

The crystal axes of ferrimagnetic resonators can be determined by X-ray techniques, but fortunately there are simpler and easier methods. When a spherical ferrimagnetic resonator is placed within a strong magnetic field, it will automatically try to align itself so that one of its "easy" axes will be parallel to the applied field. If all of the resonators of a ferrimagnetic-resonator filter see the same boundary conditions (such as is usually the case in two-resonator ferrimagnetic filters) then aligning the resonators on their easy axes by this method is usually sufficient. After the easy axis of a resonator is determined by use of a strong magnetic field, the resonator is cemented to a small dielectric mount in the desired position.

As was discussed in Sec. 17.05, the easy axis is the [111] axis for yttrium-iron-garnet (YIG) or gallium-substituted yttrium-iron-garnet (GaYIG), both of which have negative anisotropy constants. In some cases, it may be desirable to rotate a resonator sphere about a [110] axis in order to obtain a tuning effect of maximum size, or in order to orient the resonator to a direction where anisotropy effects will be cancelled out, to make the tuning of the resonator independent of temperature. In order to locate a [110] axis, special procedures are required.

A technique for determining crystal axes other than the easy axes has been described by M. Auer.[24] His procedure involves geometric construction to determine the location of other axes after two easy axes have been determined. P. S. Carter, Jr. and Y. Sato[5,28] have extended Auer's method by devising an aligning jig which accomplishes the desired results very quickly and simply.

Briefly, the device of Carter and Sato[5,28] works as follows. The magnet is placed on a rotating mount as shown in Fig. 17.06-1 so that the field can be oriented in any desired direction with respect to the jig that holds

SOURCE: Quarterly Progress Report 5, Contract DA 36-039 SC-87398, SRI; reprinted in *IRE Trans. PGMTT* (see Ref. 28 by Y. Sato and P. S. Carter, Jr.)

FIG. 17.06-1 DEVICE FOR ORIENTING FERRIMAGNETIC CRYSTALS USING A ROTATABLE ELECTROMAGNET AND ALIGNING JIG

the spherical crystal. The crystal is first mounted so that it is free to turn until an easy axis (of which there are four in a cubic crystal with negative anisotropy) comes into coincidence with the direction of the magnetic field. After this first orientation is completed, the sphere is attached (using some easily soluble glue, wax, etc.) to a wire which is in line with the determined easy axis and which is placed in a radial hole

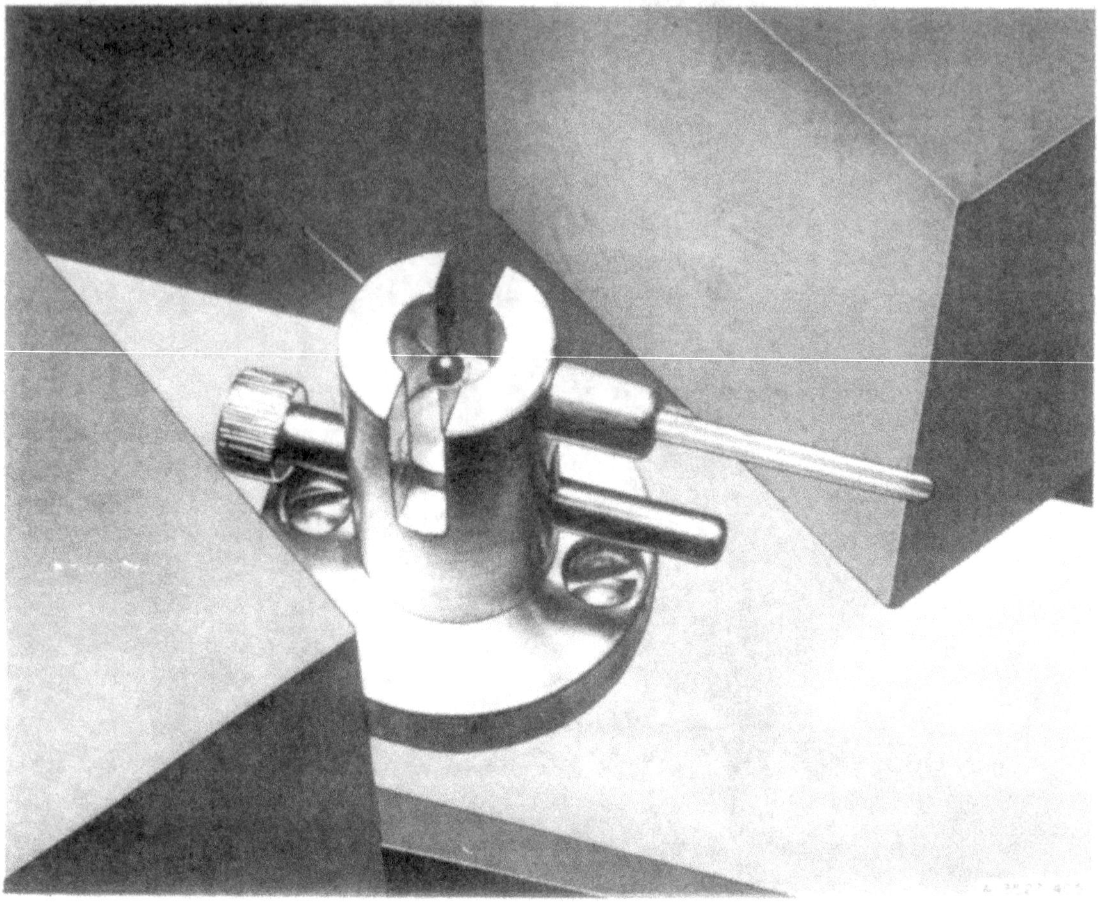

SOURCE: Quarterly Progress Report 5, Contract DA 36-039 SC-87398, SRI; reprinted in *IRE Trans. PGMTT* (see Ref. 28 by Y. Sato and P. S. Carter, Jr.)

FIG. 17.06-2 ALIGNING JIG SHOWING YIG SPHERE ATTACHED TO WIRE ALONG ONE EASY AXIS AND QUARTZ ROD ALONG [110] AXIS

in the side of the aligning jig as shown in Fig. 17.06-2. This wire is free to turn in the radial hole under the influence of the small torques exerted on the ferrimagnetic sample by an applied dc field. This step locates and retains one of the easy axes, and permits the sphere to rotate about this axis.

The next step is to rotate the magnet by an amount that depends on the crystal axis which one is trying to locate. To determine the [110] axis (the face diagonal) requires (as shown in Fig. 2 of Ref. 20) that we locate a second easy axis displaced from the first by 70½ degrees. This second angle is accurately located by means of the milling head protractor. The ferrimagnetic sphere rotates on the wire to which it was attached in the previous step, so that the second easy axis is now placed along the dc field. The final step is to attach the sphere to a dielectric rod along the [110] axis. A radial hole is located in the side of the alignment jig along the bisector of the angle between the two easy axes. The radial hole, and the dielectric rod that it holds in position, are shown in Fig. 17.06-2. The sample is attached to the dielectric rod (or other holder) with a drop of cement and the wire is taken off.

SEC. 17.07, DESIGN OF MAGNETICALLY TUNABLE BAND-PASS FILTERS WITH TEM-MODE INPUTS AND OUTPUTS, AND A DISCUSSION OF GENERAL DESIGN PRINCIPLES

Figure 17.07-1 shows a possible form of single-resonator magnetically tunable filter. This filter uses two coupling loops, oriented perpendicularly to each other to minimize the coupling between them. One loop is within the other, and a spherical ferrimagnetic resonator is so placed as to be at the center of both loops. (In the picture shown, the loops would have to be somewhat egg-shaped, with the long dimension of the y loop in the x direction and the long dimension of the x loop in the z direction, in order that both loops will have equal coupling to the sphere and still

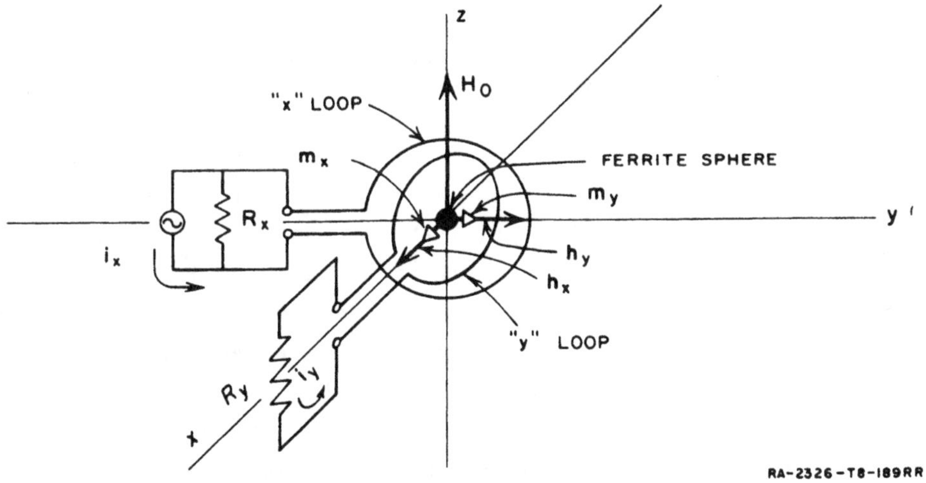

SOURCE: Final Report, Contract DA 36-039 SC-74862, SRI; reprinted in *IRE Trans. PGMTT* (see Ref. 3 by P. S. Carter, Jr.)

FIG. 17.07-1 A SINGLE-RESONATOR MAGNETICALLY TUNABLE FILTER USING LOOP COUPLING

not touch each other.) The biasing H_0-field is parallel to the plane of both loops. When a signal at the ferrimagnetic resonant frequency is applied to the x loop, resonance will be excited in the sphere causing resulting RF magnetic dipole moments in both the x and y directions. The dipole moment in the y direction will create a magnetic field, which will couple to the y loop and result in transmission to the R_y termination. When the signal applied to the x loop is not at the ferrimagnetic resonance frequency, there will be no coupling between the x and y loops except for a small unavoidable amount of stray coupling.

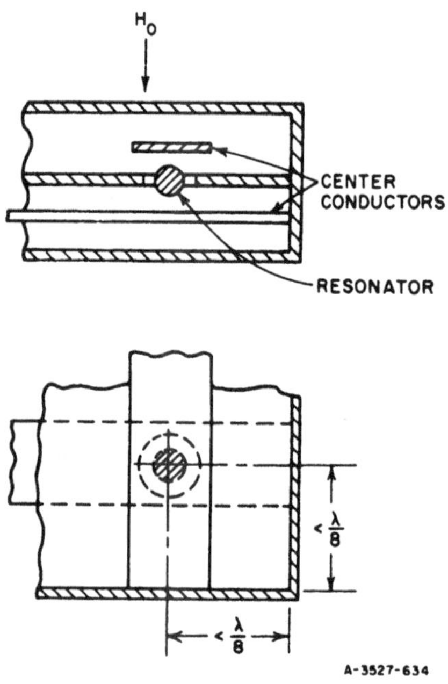

FIG. 17.07-2 A SINGLE-RESONATOR STRIP-LINE MAGNETICALLY TUNABLE FILTER

Successful single-resonator filters have been built in the form shown in Fig. 17.07-1, but this type of structure has certain disadvantages. One is that the stray coupling is larger than in some other possible structures, and the other is that the RF magnetic fields are not very uniform, which will tend to excite higher-order magnetostatic modes. Both of these disadvantages can be combated by making the loops and the sphere as small as is practically feasible.

Figure 17.07-2 shows a strip-line single-resonator magnetically tunable filter structure. In this structure a dividing wall separates the two strip-lines and decreases the coupling between them. Note that the lines are at right angles to each other, which will further tend to reduce any stray coupling. In this case the sphere is placed in a hole in the wall separating the two strip-lines, and at resonance the circularly polarized magnetic moment excited in the sphere will cause coupling from one strip-line to the other. This structure still has a non-uniform RF field in the sphere as a result of being half on one side of the dividing wall and half on the other. For this reason, in order to minimize difficulties with higher-order modes, the sphere should be quite small (*i.e.*, of the order of a fiftieth to a hundredth of a wavelength in diameter). Kotzebue[25] has constructed coaxial filters which are very similar to the strip-line filter shown in Fig. 17.07-2.

Figure 17.07-3 shows a type of two-resonator strip-line structure with which Carter has obtained very good results.[4] In this structure, each sphere is mounted close to the short-circuited end of the input or output strip-line structure. Thus, the large RF H-field in the vicinity of the short-circuits gives relatively good coupling between the strip lines and the spheres. Coupling between spheres is obtained by use of a long slot having its long dimension parallel to the axis of the strip lines. This orientation of the slot causes minimum disturbance of the currents and fields about the strip lines, while furnishing maximum isolation between the strip lines. Thus, the coupling between strip lines is extremely small when the spheres are not resonant. However, when the spheres become resonant, the circularly polarized RF dipole moment in

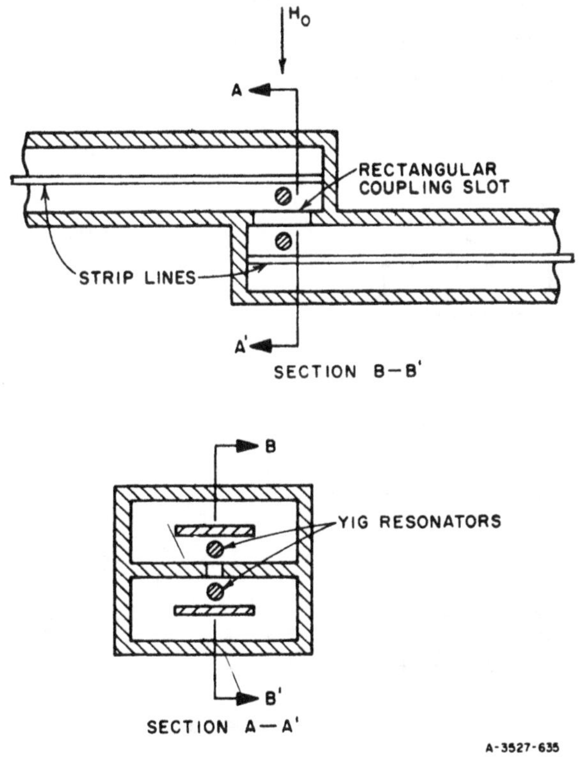

FIG. 17.07-3 A TWO-RESONATOR MAGNETICALLY TUNABLE FILTER WITH COUPLING THROUGH TOP AND BOTTOM WALLS

the spheres provides a component of RF H-field which will couple through the elongated slot very easily, and good coupling is obtained between spheres. This structure has the advantage of providing relatively uniform RF H-fields so that higher-order magnetostatic modes are not so easily excited, and high off-resonance isolation is not difficult to obtain. As compared with the structures in Fig. 17.07-4, which are about to be discussed, the structure in Fig. 17.07-3 has the disadvantages of requiring a larger magnet air gap and of not being quite as suitable for the design of filters with more than two resonators.

Figure 17.07-4 shows two side-wall coupled filter configurations which are similar to that in Fig. 17.07-3 except that the spheres are coupled through the side wall of the structure instead of through the top and bottom wall. One advantage of this structure is that the input and output strip-lines can lie in the same plane, and as a result the magnet air gap required can be minimized. Another advantage of this structure is that additional spheres can be added as shown in Fig. 17.07-4(b) to give a filter with more than two resonators, while all of the resonators see very nearly the same boundary conditions. As can be seen from the discussion

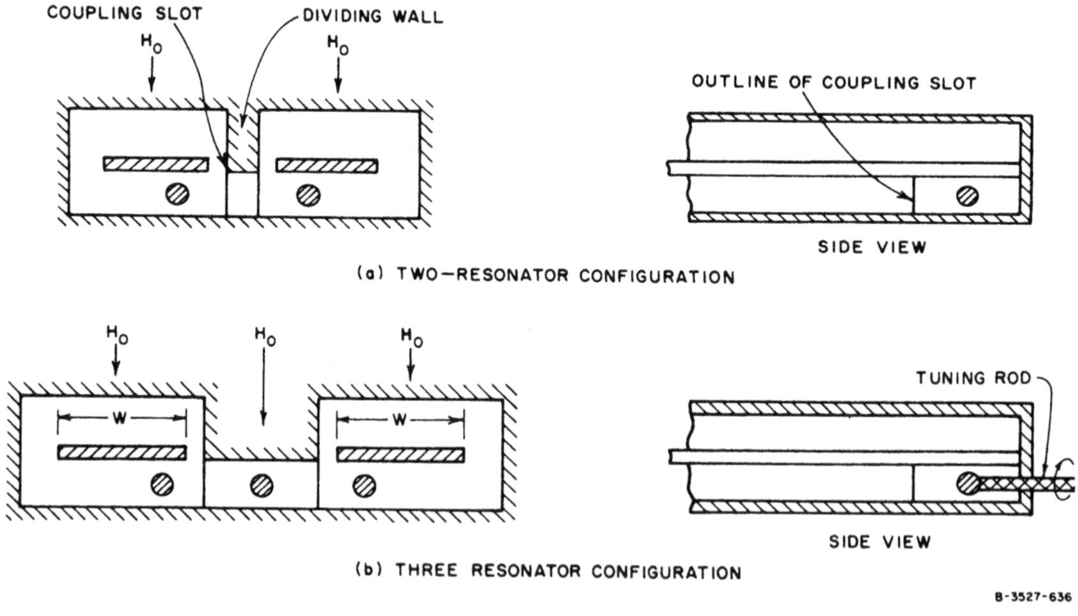

FIG. 17.07-4 TWO- AND THREE-RESONATOR MAGNETICALLY TUNABLE FILTERS WITH COUPLING THROUGH THE SIDE WALLS

in Sec. 17.05, it is necessary for all spheres to see the same boundary conditions, and for all spheres to have the same orientation of their crystal axes with respect to H_0 if the spheres are all to have the same resonant frequencies for any given H_0 and given temperature. Although the structure in Fig. 17.07-4(b) gives relatively uniform boundary conditions for all spheres, some means for tuning the spheres is desirable in order to correct the small tuning errors that may be present. In Fig. 17.07-4(b) provision for rotating the spheres about a [110] axis (as discussed in Sec. 17.05) is suggested. A disadvantage of the structure in Fig. 17.07-4(a) as compared to that in Fig. 17.07-3 is that in the structure in Fig. 17.07-4(a) it may be necessary to place the spheres toward the inner edges of the strip-lines in order to obtain adequate coupling between spheres. This puts the spheres in a region of less uniform RF H-field, which makes them more vulnerable to higher-order modes.

All of the filters in Figs. 17.07-1 to 17.07-4 are reciprocal, except for a gyrator action* which is present in the circuits in Fig. 17.07-1 and 17.07-2.[3,6] For purposes of analysis the circuits in Figs. 17.07-1 to 17.07-4 may be thought of as operating like the filter with magnetically coupled resonators shown in Fig. 17.07-5. When the coupled coils are replaced by their T-equivalents, the circuit in Fig. 17.07-5 becomes of the form in Fig. 17.02-3, except for a small amount of residual self-inductance in series with the terminations R_A and R_B. Fortunately, the properties of ferrimagnetic resonators are such that they give nearly constant bandwidth as the filter is tuned, without resorting to special measures such as were required in the design of the filters in Secs. 17.03 and 17.04. There will be some change in response shape, however.

As with most any kind of narrow-band, band-pass filter, the desired response shape and bandwidth can be obtained by starting with a low-pass prototype filter and a specified desired fractional bandwidth w, and then computing the external Q's and coupling coefficients required for the band-pass filter, by use of Eqs. (17.02-1) to (17.02-3). The tabulated maximally flat or Tchebyscheff low-pass prototypes in Sec. 4.05 can be used, but the equal-element prototypes discussed in Sec. 11.07 should be of special interest for this application. The rate of cutoff of the filter can be

* This causes 180 degrees more phase shift in one direction of transmission than in other directions of transmission. This gyrator action results from the fact that in the circuits in Figs. 17.01-1 and 17.01-2 the input and output lines couple to orthogonal components of the circularly polarized magnetic moment within the sphere.

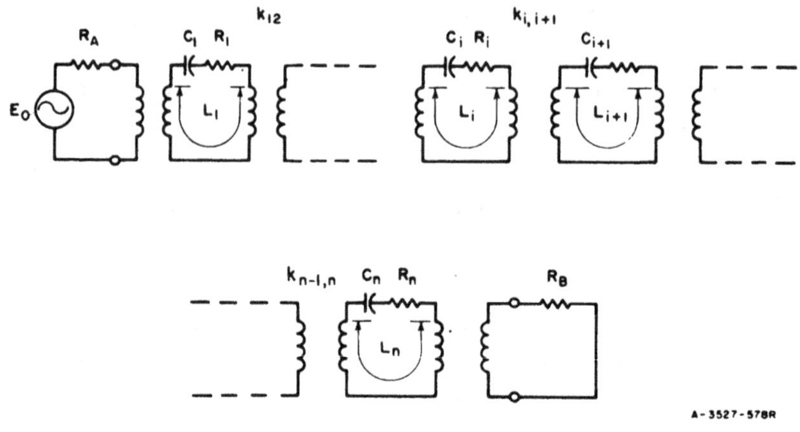

FIG. 17.07-5 EQUIVALENT CIRCUIT OF A FILTER HAVING n FERRIMAGNETIC RESONATORS

estimated by use of the mapping defined by Eqs. (17.02-20) to (17.02-22) along with the charts in Sec. 4.03, or by use of the data in Sec. 11.07. The midband dissipation loss to be expected can be estimated by use of the methods of Secs. 11.06 or 11.07.

Once the filter design has been fixed in terms of determining the required number of resonators, determining the required external Q's of the end resonators, and determining the coupling coefficients of the couplings between resonators, the designer can focus his attention on the physical parameters of the circuit required to realize these coupling properties. Let us now consider the design of the structure after the desired values for the external Q's and coupling coefficients have been determined.

Design for Prescribed External Q's—Carter[3] has derived approximate expressions for the external Q of ferrimagnetic resonators in various coupling structures. For a loop of radius r meters having a sphere of volume V_m cubic meters at its center, the external Q is

$$Q_e = \frac{4r^2 R_A}{\mu_0 V_m \omega_m}\left[1 + \left(\frac{\omega L_s}{R_0}\right)^2\right] \qquad (17.07\text{-}1)$$

where R_A is the terminating resistance connected to the loop, L_s is the self-inductance of the loop in henries, $\mu_0 = 1.256 \times 10^{-6}$ henries/meter is the permeability of air, and

$$\omega_m = \gamma \mu_0 M_s \tag{17.07-2}$$

where $\gamma = 1.759 \times 10^{11}$ (mks units) is the gyromagnetic ratio. The quantity $\mu_0 M_s$ in webers/meter^2 can be obtained from $4\pi M_s$ in gauss by use of Eq. (17.05-3a). For YIG, ω_m is about 3.08 (10^{10}).

A corresponding equation for a YIG sphere mounted near a short-circuiting wall in a strip-line structure of impedance Z_0' is given in Fig. 17.07-6 along with curves that apply when $Z_0' = 50$ ohms and when the resonator is made from YIG. Note that the curves give Q_e for a given center conductor to ground-plane spacing d and sphere diameter D_m, both given in inches. If these curves are to be used for a resonator of some material other than YIG, or with a uniform strip-line of some impedance other than 50 ohms, the values obtained from the chart should be scaled as indicated by the equation

$$Q_e\Big|_{\text{scaled}} = Q_e \frac{(M_s)_{\text{YIG}}}{(M_s)_{\text{used}}} \frac{50}{(Z_0')_{\text{used}}} \tag{17.07-3}$$

where $(Z_0')_{\text{used}}$ is the actual line impedance used and $(M_s)_{\text{YIG}}/(M_s)_{\text{used}}$ is the ratio of the saturation magnetization of YIG to that of the material actually used. The equation in Fig. 17.07-6 (upon which the curves are based) was computed assuming that the strip-line was bounded on each side by a magnetic wall (which implies that there are no fringing fields), and that the fields about the strip-line are perfectly uniform. Of course, if the strip-line is so narrow that the sphere projects into fringing fields, this approximation will not be good.

Figure 17.07-7 shows computed and experimental data obtained by Carter[3] in order to check out the equation in Fig. 17.07-6. These data were taken at 3000 Mc using the structure shown. Note that the sphere for each test was mounted about 0.125 inch from the short-circuit. It was found that moving the sphere too close to the short-circuiting wall would disrupt the sphere's performance, while putting the sphere too far from the wall would introduce excess reactance that would decouple the sphere. Note in Fig. 17.07-7 that the agreement between theory and measurement is reasonably good.

Other experimental work due to Carter[4] showed that the presence of an adjacent coupling slot such as that in the filter in Fig. 17.07-3 has a decoupling effect on the input circuit which tends to raise the external Q about 20 percent or somewhat more from its value when the slot is not there. Therefore, some allowance should be made for this effect when designing for the strip-line and sphere dimensions tc realize a required external Q.

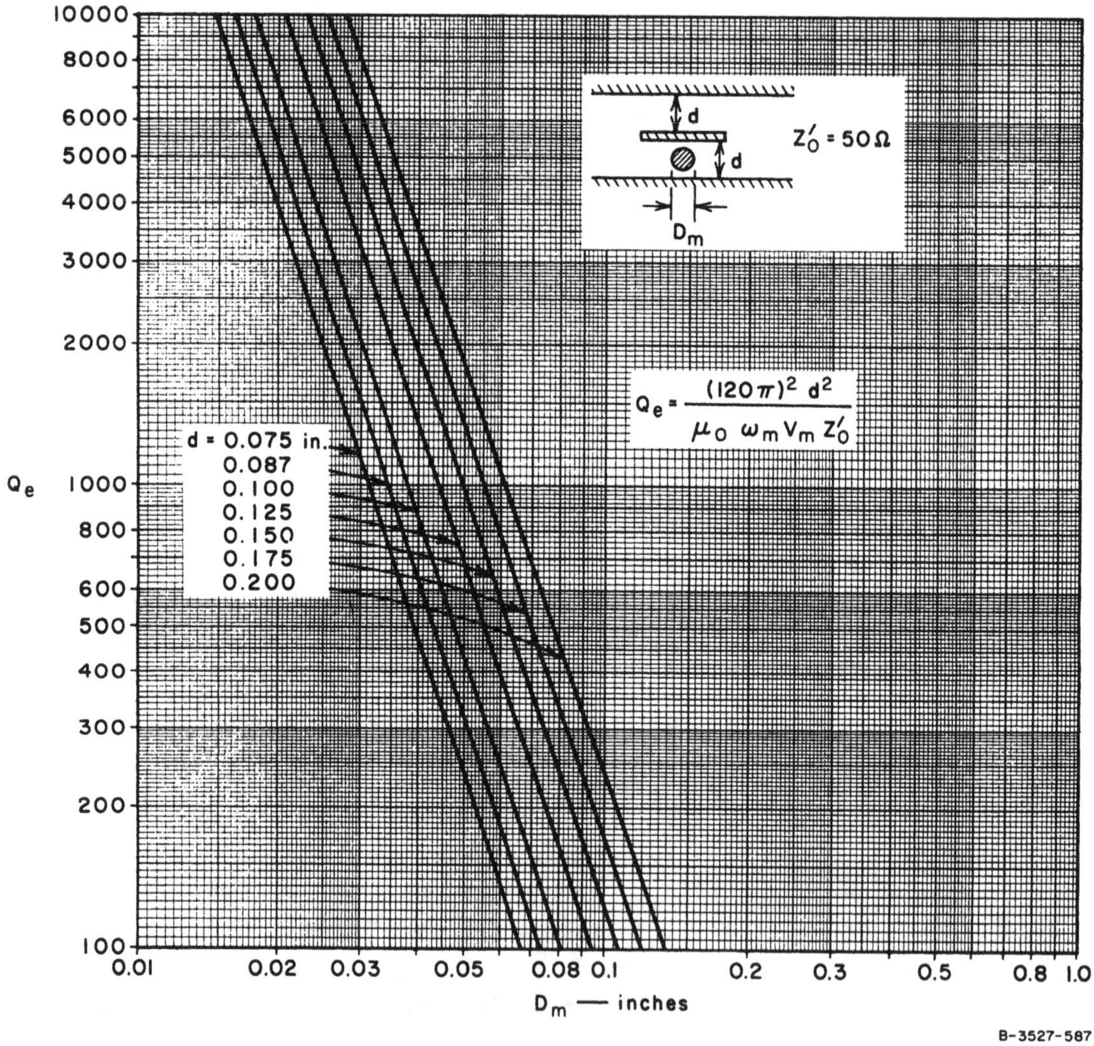

FIG. 17.07-6 Q_e vs. SPHERE DIAMETER FOR SPHERICAL YIG RESONATOR IN SYMMETRICAL STRIP TRANSMISSION LINE
In the equation shown μ_0 is the permeability of the region about the sphere, $\omega_m = \gamma \mu_0 M_s$, $\gamma = 1.759 \times 10^{11}$ in mks units, V_m is the volume of the sphere, and Z_0' is the impedance of the strip line ($Z_0' = 50$ ohms for the graph).

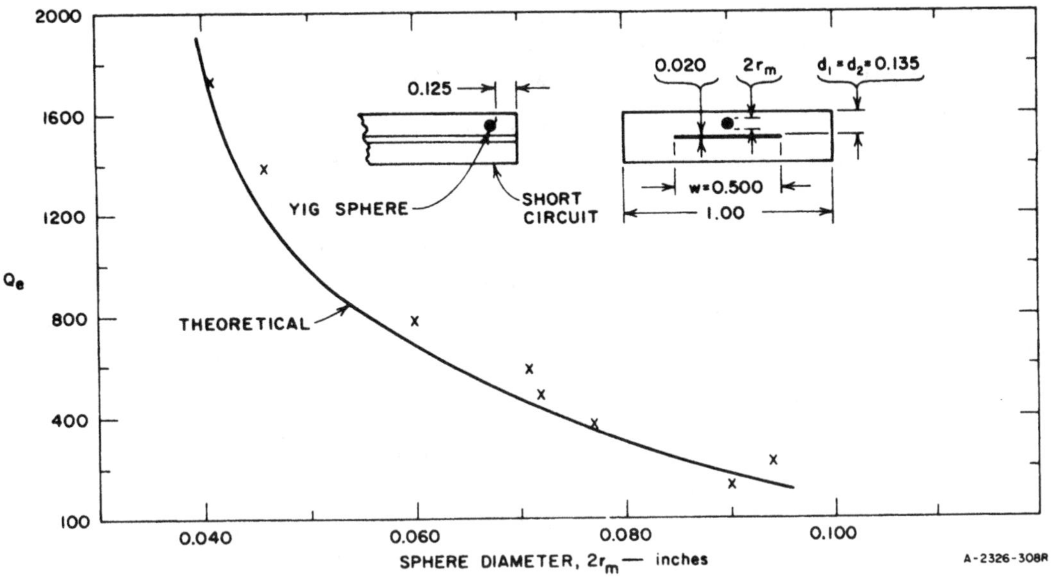

SOURCE: Final Report, Contract DA 36-039 SC-74862, SRI; reprinted in *IRE Trans. PGMTT* (see Ref. 3 by P. S. Carter, Jr.)

FIG. 17.07-7 THEORETICAL AND EXPERIMENTAL VALUES OF Q_e OF YIG RESONATOR USING STRIP-TRANSMISSION-LINE

Note in the equation in Fig. 17.07-6 that the external Q is inversely proportional to the line impedance Z_0'. This is because for a given sphere size and ground-plane-to-center-conductor spacing d, the narrower the strip-line, the more concentrated the RF H-field is assumed to be in the vicinity of the sphere. (The presence of fringing fields disrupts this picture when the strip becomes narrow, however.) The equation and curves in Fig. 17.07-6 assume that the strip line is uniform and that it is terminated in its characteristic impedance. However, if a short length of line of impedance Z_0' is used in the vicinity of the sphere, and this line is terminated in a much lower resistance R_A, it should be possible to obtain lower values of external Q than would be possible otherwise. Thus, if Q_e is the external Q when the line is terminated in a resistance $R = Z_0'$, then the external Q when termination is changed to R_A is approximately

$$Q_e\Big|_{\text{corrected}} = Q_e \frac{R_A}{Z_0'} \quad . \tag{17.07-4}$$

This, of course, assumes that the line section of impedance Z_0' (to which the sphere is coupled) is short compared to a quarter wavelength, so that there will not be excessive reactance effects due to the Z_0' line section not being matched. As seen by Eq. (17.07-4), the coupling can be made tighter (*i.e.*, the corrected value of Q_e can be made lower) if R_A is made small. This result will be seen to be in agreement with Eq. (17.07-1) Thus, in cases where difficulty is being experienced in obtaining a sufficiently low value of Q_e, decreasing the size of R_A may be a satisfactory way of obtaining the desired result. If the external load impedance is not suitable for this, a step transformer can be used to present the desired R_A value to the resonator.

The procedures described in Sec. 11.02 will be found to be convenient for experimentally checking external Q values, as well as unloaded Q values.

Design for Prescribed Coupling Coefficients Between Resonators—The calculation of the spacing of ferrimagnetic resonators and the dimensions of intervening apertures in order to obtain specified coupling coefficients is a relatively difficult problem and no such procedures are presently available. However, it is quite practical to experimentally determine the proper sphere spacings and aperture sizes by use of the two-resonator test procedures described in Sec. 11.04. Some of the results shown in Sec. 17.08 should serve as a useful guide in estimating the approximate sphere spacings and aperture sizes to be expected.

Besides the consideration of obtaining adequate coupling between spheres, other considerations are the obtaining of sufficiently high isolation off resonance, and the maintenance of sufficient separation between the spheres so that the spheres will not cause each other to see non-uniform biasing H-fields (which could help excite unwanted modes). The experimental results obtained from the examples in Sec. 17.08 should be helpful guides with regard to both of these points. With regard to the spacing of the spheres, spacing their centers by about three times their diameters appears to be satisfactory.

Effects of Scaling the Filter Parameters—It should be useful to know the effect on the external Q's of the end resonators and on the coupling coefficient between resonators of changing the dimensional scale of the circuit, the saturation magnetization M_s, and the operating frequency range. Such information should help the designer in relating information obtained from a given filter design to other filter designs having somewhat different parameters.

As previously discussed in this section, the equivalent circuits in Fig. 17.02-3 can be used in analyzing filters with ferrimagnetic resonators. Hence, Eqs. (17.02-11) to (17.02-13) also apply and we may restate them here in the form

$$(Q_e)_A = \frac{R_A \mathscr{x}_1}{(X_{01})^2} \tag{17.07-5}$$

$$k_{j,j+1} = \frac{X_{j,j+1}}{\sqrt{\mathscr{x}_j \mathscr{x}_{j+1}}} \tag{17.07-6}$$

$$(Q_e)_B = \frac{R_B \mathscr{x}_n}{(X_{n,n+1})^2} \tag{17.07-7}$$

where the $X_{j,j+1}$ are mutual reactances and the $\mathscr{x}_j$ are resonator slope parameters. The mutual reactances can be assumed to be of the form

$$X_{j,j+1} = \omega_0 \mu_0 W_{j,j+1} \tag{17.07-8}$$

where ω_0 is the radian frequency at resonance, μ_0 is the permeability of air in henries/unit length, and $W_{j,j+1}$ is a scale factor depending on the dimensions of the circuit. The parameter $W_{j,j+1}$ has the dimension of length. Then by Eq. (17.07-5) and (17.07-8)

$$(Q_e)_A = \frac{R_A \mathscr{x}_1}{\omega_0^2 \mu_0^2 (W_{01})^2} \quad . \tag{17.07-9}$$

Now Carter[26] has shown that in general, if a ferrimagnetic resonator is coupled by way of a linearly polarized RF magnetic field h, then

$$Q_e = \frac{R_A}{\mu_0 V_m (U_{01})^2 \omega_m} \tag{17.07-10}$$

where V_m is the volume of the sphere,

$$\omega_m = \mu_0 \gamma M_s \quad , \tag{17.07-11}$$

γ is the gyromagnetic ratio, and

$$U_{01} = \frac{h}{I_A} \quad , \tag{17.07-12}$$

where h is the RF field seen by the resonator due to current I_A flowing in the load R_A. The parameter U_{01} is seen to have dimensions of 1/(length). Equating Eqs. (17.07-10) and (17.07-9), and solving for the resonator slope parameter $\mathscr{x}_1$ gives

$$\mathscr{x}_1 = \frac{\omega_0^2 \mu_0 (W_{01})^2}{V_m (U_{01})^2 \omega_m} \quad . \tag{17.07-13}$$

Assuming that $\mathscr{x}_j = \mathscr{x}_1$, by Eqs. (17.07-6), (17.07-8), and (17.07-13) we obtain

$$k_{j,j+1} = \frac{\omega_m (U_{01})^2 W_{j,j+1} V_m}{\omega_0 (W_{01})^2} \quad . \tag{17.07-14}$$

Checking the dimensionality of Eq. (17.07-14) we find that

$$k_{j,j+1} = \frac{A\omega_m}{\omega_0} = \frac{A\mu_0 \gamma M_s}{\omega_0} \tag{17.07-15}$$

where A is a dimensionless factor which is determined only by the *relative* proportions of the various parts of the circuit, but which is independent of the dimensional scale of the circuit.

From a somewhat different point of view, it would appear that changing the M_s of the resonators ought to be like changing the number of turns in the coils in the circuit in Fig. 17.07-5. In that case the mutual reactance X_{01} for coupling to the load ought to be proportional to M_s, while the mutual reactances $X_{j,j+1}$ for coupling between resonators ought to be proportional to $(M_s)^2$. An analysis from this point of view was worked out in a manner similar to that described above. This analysis also led to Eq. (17.07-15), which helps to verify its correctness.

A number of useful conclusions can be drawn from the above analysis:

(1) Scaling the circuit up or down in size will have no effect on the coupling coefficients between resonators, although by Eqs. (17.07-10) to (17.07-12) it can be seen that the external Q's of the end resonators will be affected. The Q_e will vary as 1/(size).

(2) The coupling coefficients between resonators will vary as $1/\omega_0$ but the external Q's are independent of frequency. Since the coupling coefficients predominate in determining bandwidth, the bandwidth of a filter with ferrimagnetic resonators will tend to remain constant as it is tuned (see Sec. 17.02), but the response shape will change. A filter with moderate-sized Tchebyscheff ripples at the center of the tuning range will tend to have larger ripples when tuned at the lower edge of the tuning range, and may tend toward a nearly maximally flat response when tuned at the upper edge of the tuning range.

(3) The coupling coefficients are directly proportional to M_s while the external Q's are inversely proportional to M_s. Thus as can be seen with the aid of Sec. 17.02, any change in M_s will alter the filter bandwidth in proportion to the change in M_s while the response shape should remain unaffected. One qualification to this statement needs to be made with regard to dissipation loss. Since the dissipation loss effects are influenced by bandwidth (see Secs. 11.06 and 11.07), there may be some change in response shape for this reason.

(4) Because the external Q varies as 1/(size), as the circuit is scaled in size, and because the coupling coefficients vary as $1/\omega_0$, the obtaining of adequate coupling between the end resonators and the external circuits, and the obtaining of adequate coupling between adjacent resonators, will become increasingly difficult as the operating frequency is increased. Ferrimagnetic materials with especially large values of M_s should be helpful for use in the higher microwave ranges.

At the time of this writing, the above conclusions have not been systematically checked by experimental means; however, they do appear to be consistent with available experimental evidence. In particular, the measured responses of the filters worked on by Carter[3,4,5,6] bear out the second of the above conclusions very well.

SEC. 17.08, RESULTS OF SOME TRIAL MAGNETICALLY TUNABLE BAND-PASS FILTER DESIGNS HAVING STRIP-LINE INPUTS AND OUTPUTS*

Several trial magnetically tunable filter designs will now be described. Since part of the design process involves cut and try (namely the determination of the spacings of the resonators and the sizes of the coupling apertures), these designs should provide helpful guide lines for additional designs.

A Trial Two-Resonator YIG Filter Design Using Top and Bottom Wall Coupling—Figure 17.08-1 shows the cross section of the strip-lines which were used in the design of a magnetically tunable filter of the form in Fig. 17.07-3. In this case, the strip line was made unsymmetrical in an effort to obtain tighter coupling to the spheres for given line impedance. The smaller spacing d_1 between the strip line and the bottom wall should give a greater flux density and tighter coupling to the sphere for a given line impedance. However, using a larger upper wall spacing d_2 increases the magnet air gap required, and other means for obtaining a tight coupling to the sphere appear at this time to be preferable. (The use of a step transformer to lower the terminating impedance seen by the resonator is recommended, as discussed in Sec. 17.07.)

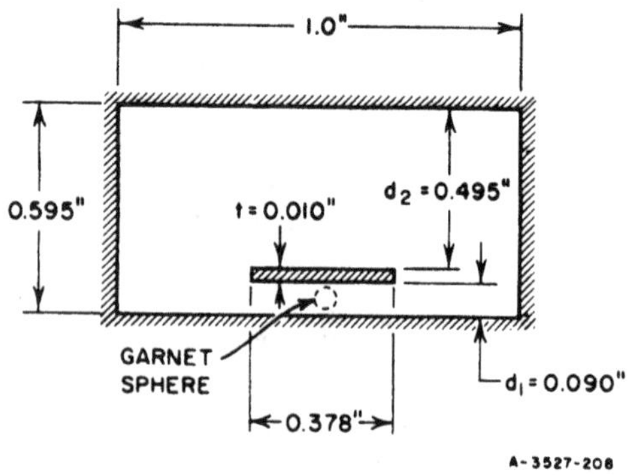

FIG. 17.08-1 STRIP-LINE DIMENSIONS FOR AN EXPERIMENTAL FILTER OF THE FORM SHOWN IN FIG. 17.07-3

In this case the external Q's were computed using the formula

$$Q_e = \frac{120\,\pi W'(d_1/d_2)(d_1 + d_2)}{\mu_0 \omega_m V_m} \qquad (17.08\text{-}1)$$

where

$$W' = \frac{377}{Z_0'} \frac{d_1 d_2}{(d_1 + d_2)} \qquad (17.08\text{-}2)$$

---

* These designs are due to P. S. Carter, Jr.[4,6]

is the *effective* width of the line if it had no fringing capacitance and Z_0' is the impedance of the line. All other quantities are as defined in Fig. 17.08-1, or as were defined in connection with Eq. (17.07-1), (17.07-2), and Fig. 17.07-6. The actual strip width W to give a Z_0' = 50-ohm line with fringing capacitance was estimated using the fringing capacitance data in Sec. 5.05.

This trial filter design used 0.074-inch-diameter YIG spheres mounted on 0.010-inch-thick dielectric plates, as indicated in Fig. 17.08-2. The coupling slot between the spheres (see Fig. 17.07-3) was 0.105 inch wide and 0.320 inch long, while the metal wall separating the input and output strip line structures was 0.125 inch thick, as indicated in Fig. 17.08-2. It was found desirable to use a relatively thick dividing wall in order to separate the spheres enough that they would not disturb the biasing H_0 field seen by each other. Each sphere was mounted with its center 0.160 inch

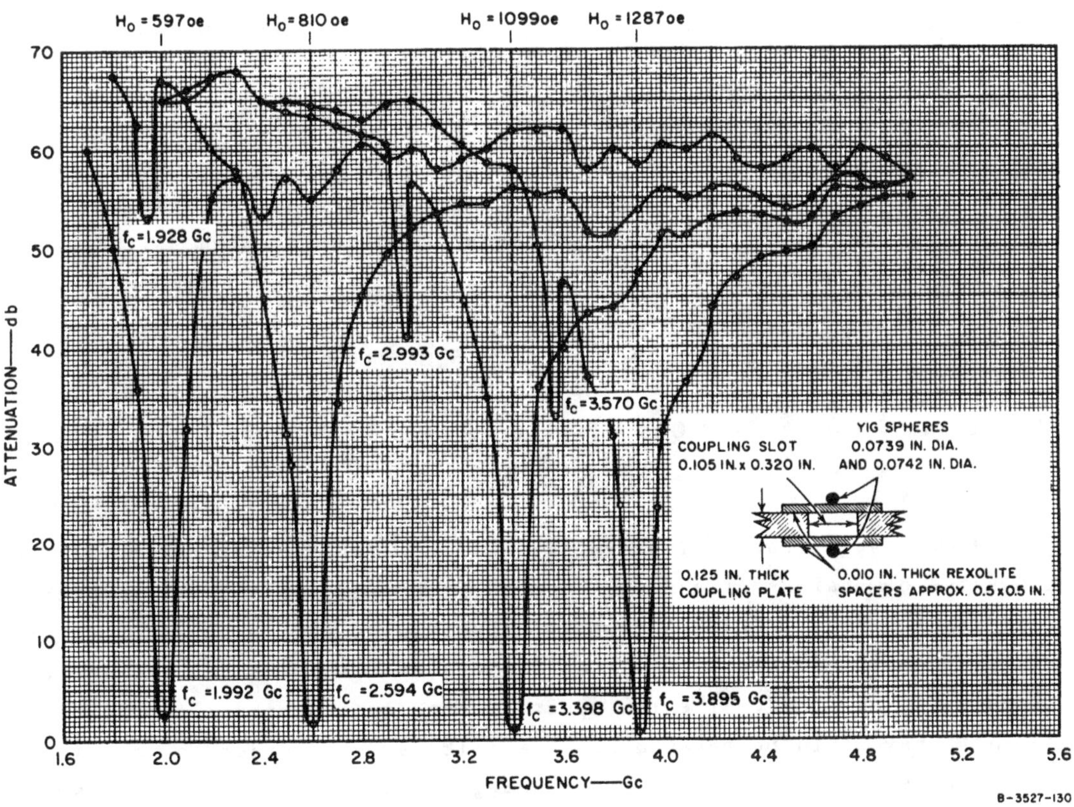

FIG. 17.08-2 MEASURED RESPONSE FOR A TRIAL TWO-RESONATOR FILTER OF THE FORM SHOWN IN FIG. 17.07-3
The responses for four different values of biasing H_0 field are superimposed.

from the vertical short-circuit wall terminating its strip line. The estimated external Q computed using Eqs. (17.08-1) and (17.08-2) was 105, while measured values obtained without the coupling slot averaged about 125. As might be expected, cutting the coupling slot in the wall between the two spheres tended to increase the external Q's, so that the measured values averaged around 170. The size of the coupling slot was determined by cut and try increases in the coupling-slot width until a Tchebyscheff response with small ripple was obtained.

Tests on one of the resonators gave a measured unloaded Q value which ranged from about 550 at 2 Gc to 1255 at 3.90 Gc, when tested in the strip-line structure before the coupling slot was cut. After the 0.105-inch by 0.302-inch slot was present, these values were, interestingly enough, raised to 710 and 1810, respectively. This same resonator when tested in waveguide had an unloaded Q of 3800 at 2.6 Gc, and 6,600 at 4.0 Gc.[4] The considerably lower Q's in the strip-line structure are due to the deleterious effects of the relatively confining metallic boundaries seen by the sphere.

The spheres were mounted so that they had a [111] axis (*i.e.*, an easy axis) parallel to H_0. Figure 17.08-2 shows the measured attenuation characteristics of the filter obtained for four values of biasing field H_0. Note that the off-resonance attenuation is quite high, although there is a spurious response below the pass band in each case. The higher the pass-band frequency, the lower the attenuation at the spurious response, until when the filter is tuned around 4.5 Gc, the spurious response merges with the pass-band response. The minimum attenuation of this filter was 2.8 db when tuned at 2.0 Gc, 1.8 db at 2.6 Gc, 1.3 db at 3.4 Gc, and 0.8 db at 3.9 Gc.

A Side-Wall Coupled Two-Resonator YIG Filter Design—Figure 17.08-3 shows the dimensions of an experimental filter of the sidewall-coupled type in Fig. 17.07-4(a). Note that the dividing wall with its aperture was made easily removable, so that various wall thicknesses and aperture sizes could be tried. The YIG spheres were mounted on dielectric rods so that they could be rotated about a [110] axis in order to achieve synchronous tuning. This type of tuning was provided mainly because it was planned that this filter would later be converted to a three-resonator filter, and it appeared desirable to include means for tuning the resonators with respect to each other.

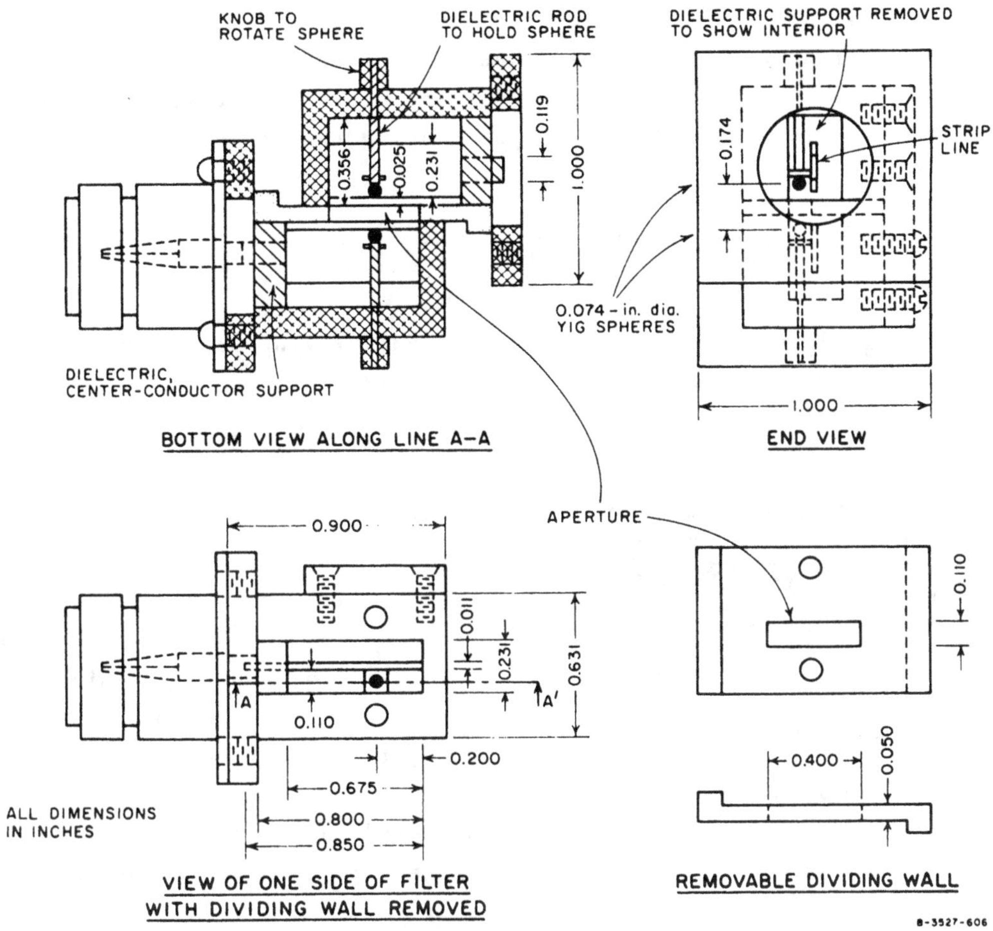

FIG. 17.08-3 DETAILS OF A TWO-RESONATOR YIG FILTER OF THE FORM SHOWN IN FIG. 17.07-4(a)

Note that the strip lines in Fig. 17.08-3 are closer to the dividing wall than to the outer walls. This was done so that adequate coupling could be maintained between spheres while still keeping the spheres under the strip lines. The strip lines were designed to give a desired impedance of $Z_0' = 50$ ohms by first computing C_0/ϵ, the required normalized capacitance per unit length, by use of

$$\frac{C_0}{\epsilon} = \frac{376.7}{\sqrt{\epsilon_r} Z_0'} \qquad (17.08\text{-}3)$$

where ϵ_r is the relative dielectric constant of the medium of propagation. Then, referring to Fig. 17.08-4, desired values for t, b, $S_1/2$ and $S_2/2$ were selected, and the corresponding line width W required to give the desired line impedance Z_0' was computed using the formula

$$\frac{W}{b} = \frac{1}{2}\left(1 - \frac{t}{b}\right)\left[\frac{C_0}{2\epsilon} - \left(\frac{C'_{f0}}{\epsilon}\right)_1 - \left(\frac{C_{f0}}{\epsilon}\right)_2\right] , \qquad (17.08\text{-}4)$$

where $(C_{f0}/\epsilon)_1$ and $(C_{f0}/\epsilon)_2$ are normalized fringing capacitances obtained from Fig. 5.05-10(a) using $s/b = S_1/b$ and $s/b = S_2/b$, respectively.

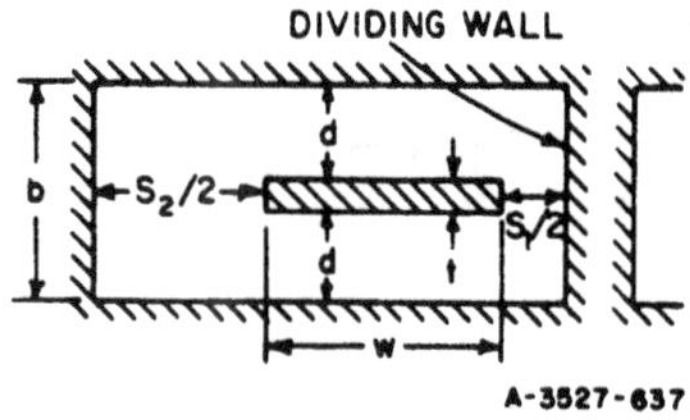

FIG. 17.08-4 DEFINITION OF THE STRIP-LINE PARAMETERS FOR THE FILTER SHOWN IN FIG. 17.08-3

This trial filter used 0.074-inch-diameter YIG spheres, while the spacing d between the strip-line and the adjacent ground planes was 0.110 inch. By Fig. 17.07-6, these dimensions are seen to yield a theoretical value of external Q of about 150, which is satisfactorily low. Measured values of Q_e in this structure ranged from 212 to 236. These relatively high measured values are no doubt due largely to the spheres being located near the edges of the strip lines instead of at their centers. The size of the coupling aperture was held constant while various thicknesses of the dividing wall in Fig. 17.08-3 were tried until a response with a small Tchebyscheff ripple was obtained.

Table 17.08-1 along with Figs. 17.08-5 and 17.08-6 show the measured performance of this filter. The data in Fig. 17.08-6 were taken by holding the frequency constant and varying the biasing magnetic field. As can be seen from Fig. 17.08-6, the spurious response below the pass band is more severe than is the

Table 17.08-1

MEASURED CHARACTERISTICS OF TWO-RESONATOR SIDE-WALL-COUPLED FILTER IN FIG. 17.08-3

f_c (Gc)	$(L_A)_{min}$ (db)	VSWR (minimum)	Q_u (CALCULATED FROM MEASURED DATA)	H_0 (oersteds)	$\Delta f_{3\ db}$* (Mc)	$\Delta f_{30\ db}$* (Mc)
2.000	4.0	1.9	--	714.7	26	117
3.000	1.2	1.01	1500	1070	20	113
4.000	1.0	1.48	2050	1427	25	121

* The bandwidths given are measured to the 3-db or 30-db points below $(L_A)_{min}$.

main spurious response in the case of the top- and bottom-wall coupled filter whose response is shown in Fig. 17.08-2. The larger spurious response in Fig. 17.08-6 is believed to be due to having the spheres so close to the edge of the strip-lines. By using a thinner dividing wall between strip lines, it is possible to move the spheres further back under the strip lines, but a thinner dividing wall was found to reduce the off-resonance isolation of the filter. Note from Fig. 17.08-5 and Table 17.08-1 that the 3-db and 30-db bandwidths of the filter are nearly constant as the filter is tuned, while the response changes from an over-coupled response at the lower end of the tuning range to an under-coupled response at the upper end of the tuning range. These results are in agreement with the conclusions listed at the end of Sec. 17.07.

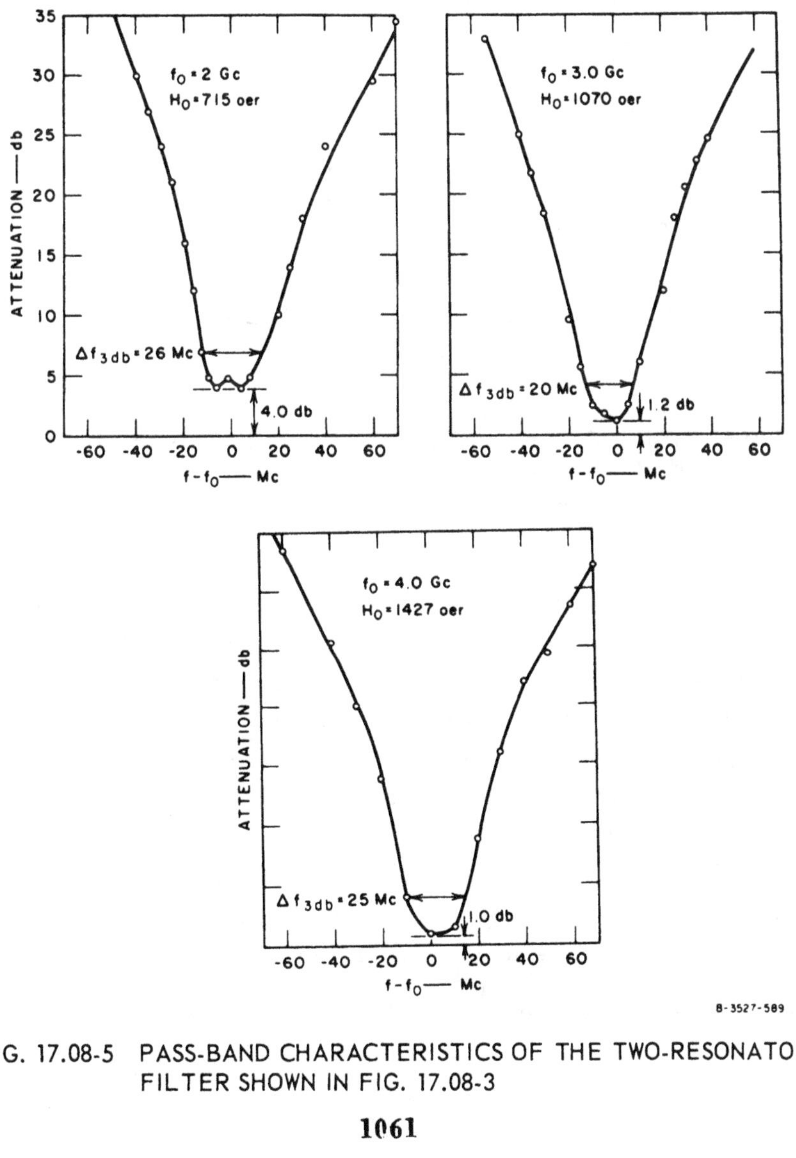

FIG. 17.08-5 PASS-BAND CHARACTERISTICS OF THE TWO-RESONATOR FILTER SHOWN IN FIG. 17.08-3

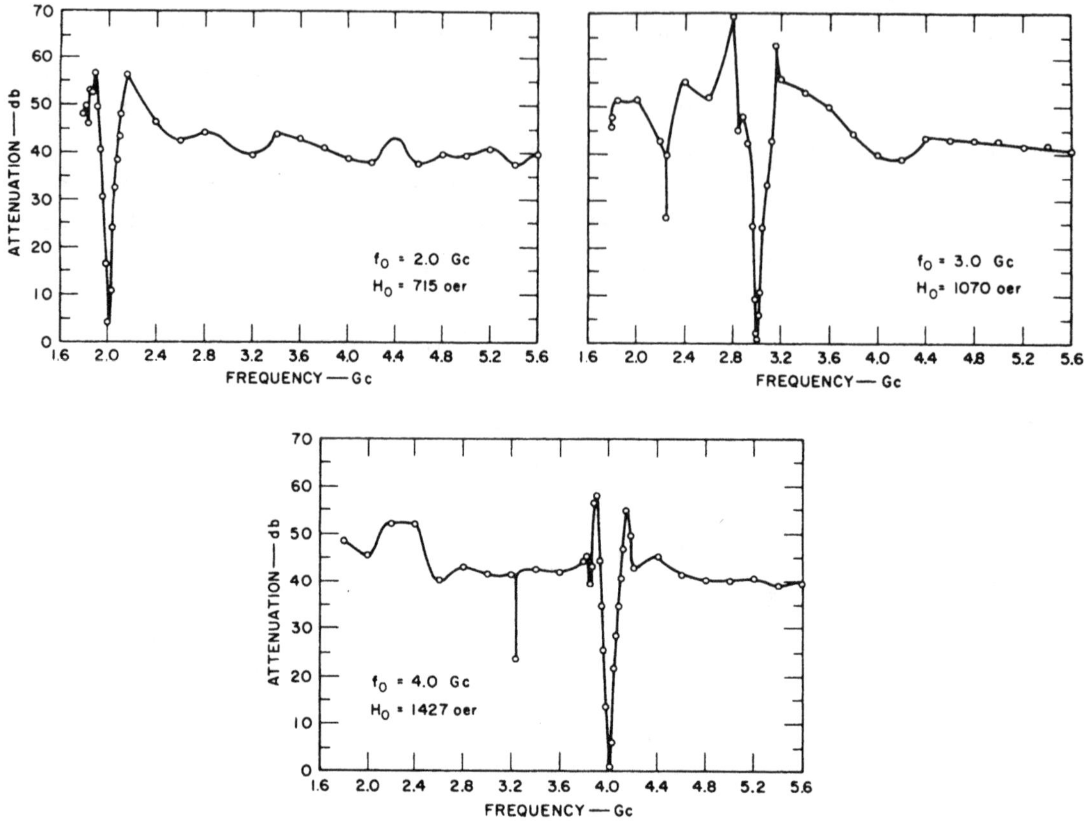

FIG. 17.08-6 STOP-BAND CHARACTERISTICS OF THE TWO-RESONATOR FILTER SHOWN IN FIG. 17.08-3

One decided advantage of side-wall coupled filters of this type is that they make possible the use of a magnet air gap of minimum size. A more refined version of the filter in Fig. 17.08-3 has been built which requires a magnet air gap of only 0.260 inch. This filter was constructed to split in half horizontally in the plane of the strip-lines. The upper and lower half were machined from separate blocks of brass, and the dividing wall was machined integral with the rest of the structure. Figure 17.08-7 shows this filter with half partially removed. The rectangular tabs protruding from both sides of the middle of the filter were included for possible use in mounting the filter in a magnet.

A Trial Three-Resonator YIG Filter—The filter in Fig. 17.08-3 was modified to make it into a three-resonator filter of the form in

Fig. 17.07-4(b). This was done by removing the original dividing wall and inserting the thicker wall shown in Fig. 17.08-8. Note that this wall contains an additional 0.074-inch-diameter YIG sphere, and there is also provision for rotating the added sphere about a [110] axis. The height of the aperture in this dividing wall is 0.110 inch, the same as the spacing *d* between the strip lines and the adjacent ground planes. Thus, this third sphere will see boundary conditions very similar to those seen by the spheres under the strip lines. The thickness of the dividing wall was chosen so that the center-to-center spacings of the three spheres would be the same as the center-to-center spacings of the spheres in the two-resonator version. In order to optimize the response shape, some adjustment of this spacing of the spheres would be desirable, but the spacing used was found to give reasonably good results. Figure 17.08-9 shows the assembled trial three-resonator filter without a magnet.

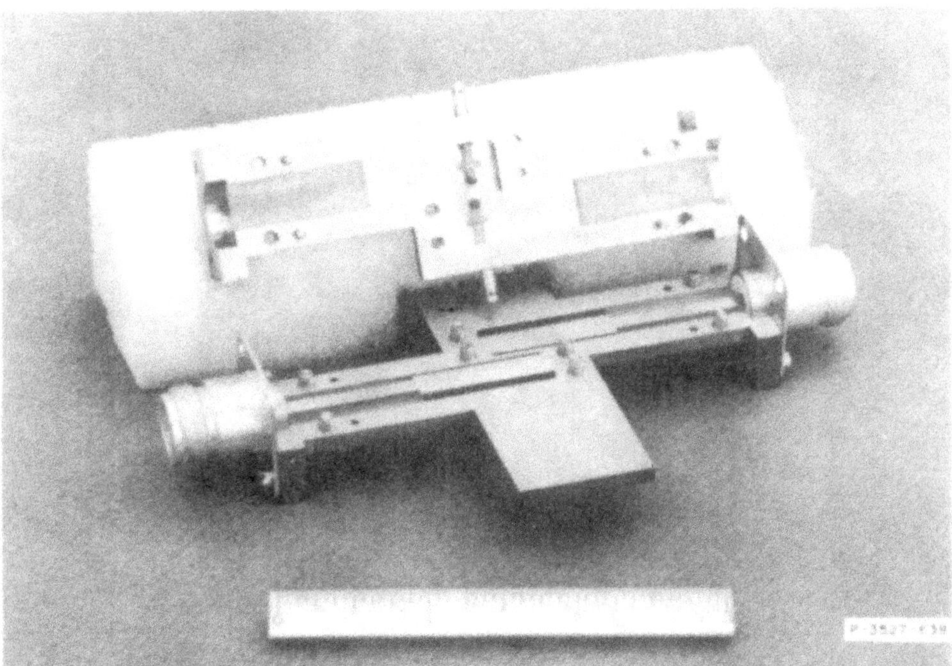

FIG. 17.08-7 A MORE REFINED VERSION OF THE FILTER SHOWN IN FIG. 17.08-3 WITH ITS UPPER HALF PARTIALLY REMOVED
This filter requires a magnet air gap of only 0.260 inch

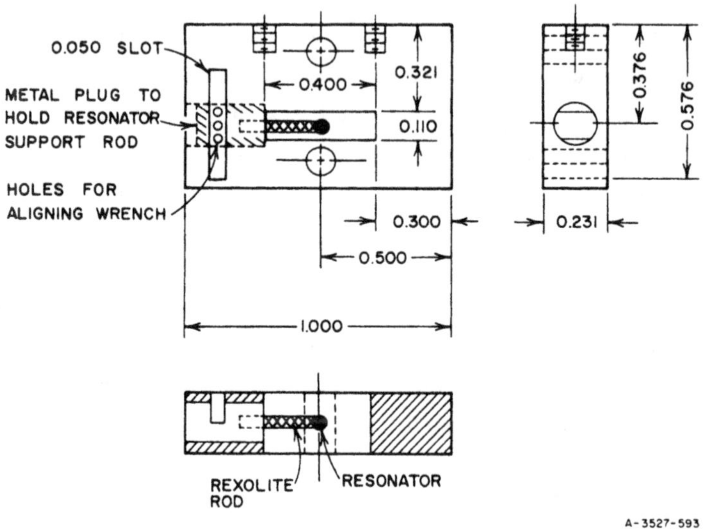

FIG. 17.08-8 DIVIDING WALL FOR THE THREE-RESONATOR VERSION OF THE FILTER SHOWN IN FIG. 17.08-3

FIG. 17.08-9 PHOTOGRAPH OF THE ASSEMBLED THREE-YIG-RESONATOR FILTER SHOWN WITHOUT A MAGNET
The required orientation of the biasing field H_0 is indicated by an arrow.

The two end resonators were first tuned to be in about the middle of their tuning range, when the filter was operated as a two-resonator filter. The tuning rods were then cemented, and a third resonator was added and tuned by rotating it so as to give a symmetrical response. In this type of filter it would be desirable to adjust the resonators in such a way as to be close to the orientation which would cancel out the temperature effects of anisotropy. Otherwise, if all of the spheres do not have the same orientations of the crystal axes, and if there are sizeable changes in temperature, the resonators may suffer appreciable mistuning with respect to each other.

Table 17.08-2 along with Figs. 17.08-10 and 17.08-11 show the measured performance of the filter. The data in these figures were again taken by holding the frequency constant and varying the magnetic field. This procedure is in most respects equivalent to holding the field constant and varying the frequency, and the use of this procedure permitted much more rapid testing of the filter. Note that the stop-band attenuation level has been greatly improved in this filter while the pass-band attenuation has been increased only a little.

Table 17.08-2

MEASURED CHARACTERISTICS OF THE THREE-RESONATOR YIG FILTER SHOWN IN FIG. 17.08-9

f_c (Gc)	$(L_A)_{min}$ (db)	$\Delta f_{3\ db}$* (Mc)	$\Delta f_{30\ db}$* (Mc)	VSWR (minimum)	H_0 (oersteds)
2.0	3.5	32.9*	91.4	2.50	610
2.5	1.7	34.2	90.1	--	867.1
3.0	1.7	33.6	90.6	1.23	990
4.0	1.4	30.9	87.2	1.10	1310
4.5	1.6	24.0	89.1	--	1586.8
7.0	1.6	29.6	--	--	--

* The bandwidths given are measured to the 3-db or 30-db point below $(L_A)_{min}$.

One of the important things demonstrated by this trial three-resonator filter is that the resonators can be kept in tune with each other as the H_0-field is varied. An earlier trial three-resonator filter[4] was not so successful in this regard. It is believed that the similar boundary conditions seen by all three spheres in filters of the type in Fig. 17.07-4(b) are important for maintaining synchronous tuning.

The 30-db bandwidth of the filter stays quite constant over the tuning range, just as for the two-resonator version. The ratio of the 30-db bandwidth to the 3-db bandwidth runs around 2.8 for the three-resonator filter design as compared to around 5 for the two-resonator filter design. Although the cutoff rate was considerably sharper for the three-resonator

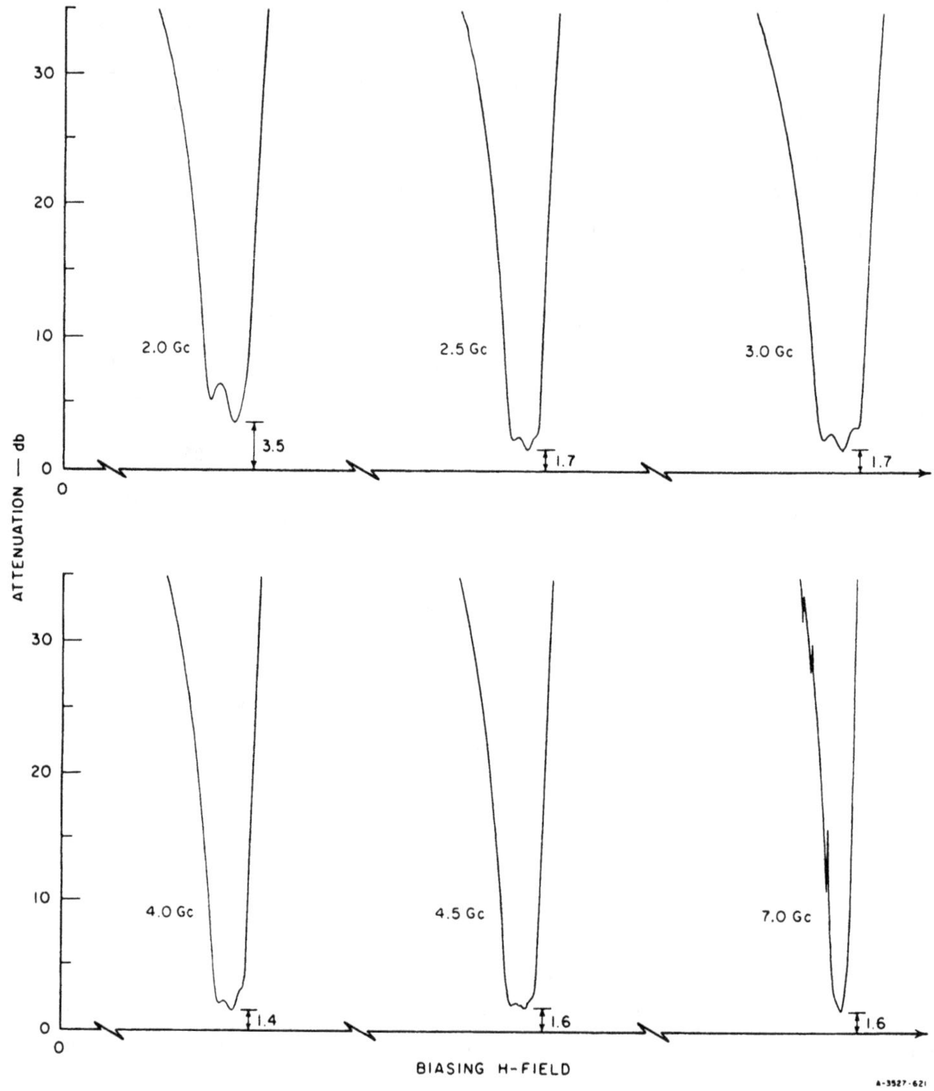

FIG. 17.08-10 ATTENUATION vs. BIASING H-FIELD FOR THREE-YIG-RESONATOR FILTER SHOWN IN FIG. 17.08-9
These curves were obtained using a recorder and a swept magnet power supply.

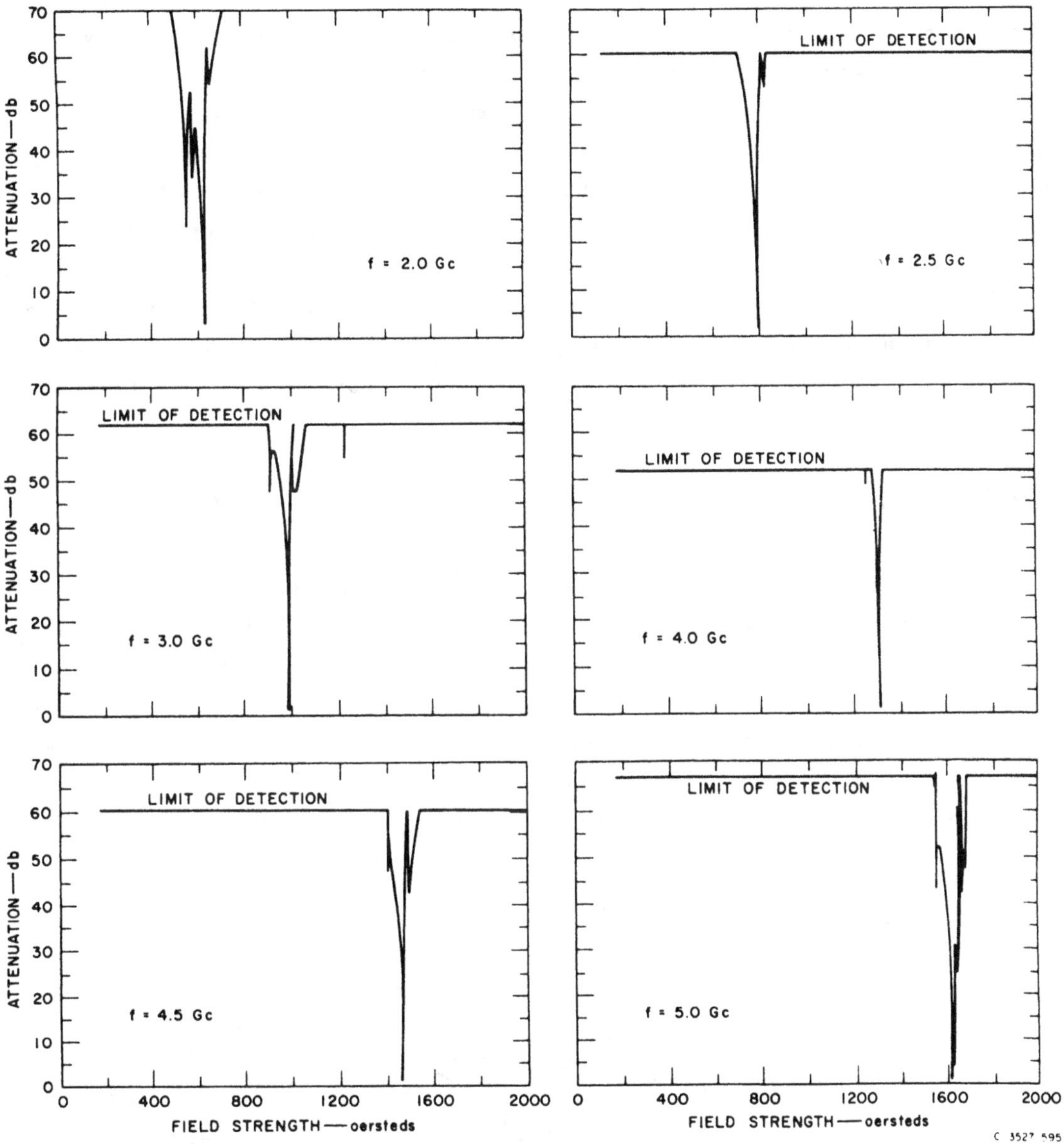

FIG. 17.08-11 THE STOP-BAND ATTENUATION CHARACTERISTICS vs. BIASING H-FIELD FOR THE FILTER SHOWN IN FIG. 17.08-9

design, the performance of this particular filter was probably not optimum in this respect because of a small amount of mistuning and because no effort was made to optimize the shape of the pass-band response for this condition. Filters designed to correspond to equal-element prototypes (Sec. 11.07) would probably be desirable for most tunable filter applications. Such filters would result in very nearly minimum midband dissipation loss for a given specified 30-db (or other specified level) stop-bandwidth. In the case of the two- and three-resonator filters discussed herein, the three-resonator design has a 30-db bandwidth which is about 23 percent less than that for the two-resonator design, but with somewhat larger midband attenuation. Further optimization of the design to correspond to an equal-element prototype at a given tuning frequency, should make it possible to reduce the midband loss somewhat at least at that tuning frequency.

SEC. 17.09, MAGNETICALLY TUNABLE BAND-PASS FILTERS WITH WAVEGUIDE INPUTS AND OUTPUTS

Figure 17.09-1 shows a two-resonator waveguide filter which is analogous to the strip-line filter in Fig. 17.07-3. Ferrimagnetic resonator spheres are placed near short-circuit walls in the waveguides so that the spheres see strong magnetic fields. The spheres are coupled through an elongated coupling slot with its length parallel to the waveguide axes. With this orientation of the slot, there is very little coupling between guides except when resonance in the spheres is excited.

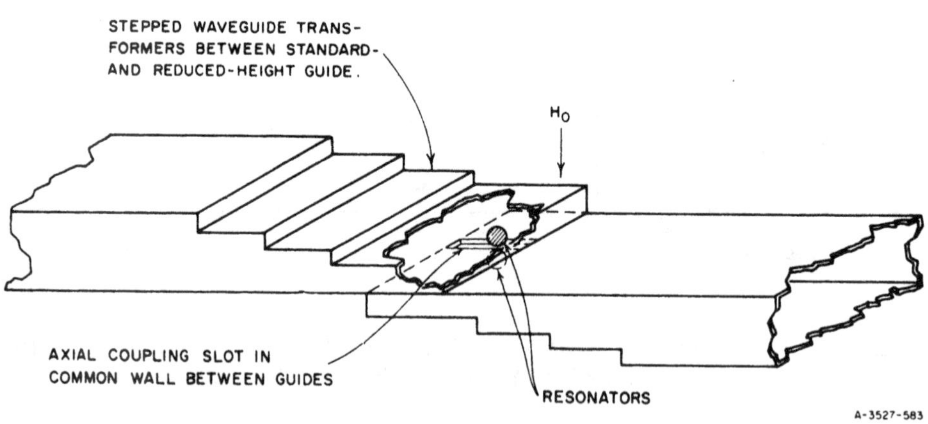

FIG. 17.09-1 TWO-RESONATOR FILTER USING OVERLAPPING WAVEGUIDES AND ELONGATED AXIAL COUPLING SLOT

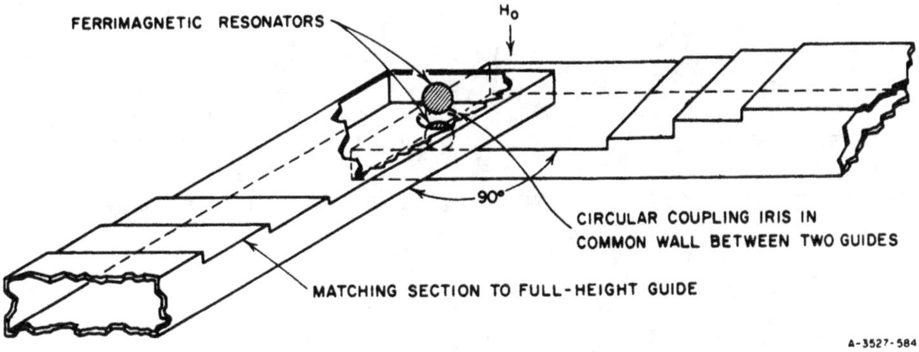

FIG. 17.09-2 TWO-RESONATOR FILTER USING WAVEGUIDES AT RIGHT ANGLES AND A CIRCULAR COUPLING IRIS

The spheres are placed in reduced-height waveguide in order to obtain adequate coupling between the spheres and the waveguides, and also to reduce the size of the magnet air gap required. Step transformers are used to transform the impedance of the standard terminating guides to that of the reduced-height guide sections.

Figure 17.09-2 shows another possible waveguide configuration, a single-resonator version of which has been discussed by Kotzebue.[25] This configuration differs from that in Fig. 17.09-1 mainly in that off-resonance isolation is obtained in this case by using a round coupling hole which will permit both linear components of the circularly polarized fields generated by the spheres at resonance (see Sec. 17.05) to be coupled from one sphere to the other. In this case the guides are placed at right angles to each other in order to increase the off-resonance isolation. With the guides so oriented, the filter will have the same attenuation characteristic for both directions of transmission, but it will have 180 degrees more phase shift for one direction of transmission than for the other.[6,3] In the case of the filter in Fig. 17.09-1, both the attenuation and phase characteristics are independent of the direction of transmission.

The design process for a waveguide filter with ferrimagnetic resonators closely parallels that for the strip-line cases discussed in Sec. 17.07. Carter[3] has derived an equation for the external Q of a ferrimagnetic resonator mounted in waveguide, and this equation is given in Fig. 17.09-3. In this equation λ_g is the guide wavelength, λ is the free-space wavelength, a and b are guide dimensions indicated in the figure,

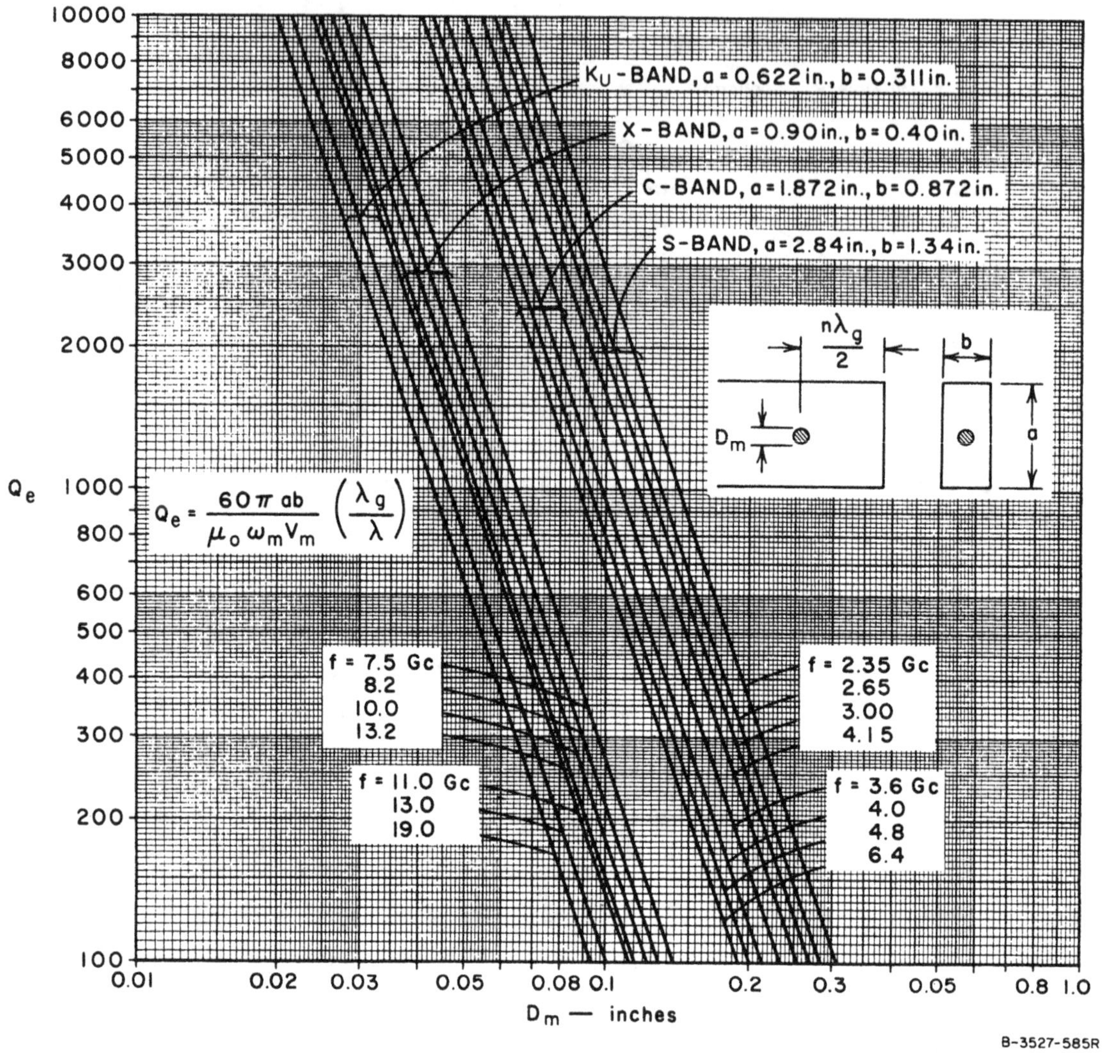

FIG. 17.09-3 Q_e vs. SPHERE DIAMETER OF SPHERICAL YIG RESONATOR LOCATED AT A HIGH-CURRENT POSITION IN SHORT-CIRCUITED TE_{10} RECTANGULAR WAVEGUIDE
In the equation shown, μ_0 is the permeability of the region about the sphere, $\omega_m = \gamma\mu_0 M_s$, $\gamma = 1.759 \times 10^{11}$ in mks units, and V_m is the volume of the sphere.

and the other parameters are the same as was discussed in connection with Eqs. (17.07-1) and (17.07-2). Figure 17.09-3 also shows curves of Q_e *vs*. sphere diameter for YIG spheres ($4\pi M_s$ = 1750 gauss) in various standard height waveguides. To scale the values of Q_e in Fig. 17.09-3 to correspond to other than standard height waveguides use the relation

$$Q_e\big|_{b'} = Q_e\big|_b \frac{b'}{b} \qquad (17.09\text{-}1)$$

where b is the height of the standard guide, b' is the actual height to be used, and $Q_e|_b$ and $Q_e|_{b'}$ are the external Q's for the guides of height b and b', respectively. To scale data in Fig. 17.09-3 to correspond to materials other than YIG use

$$Q_e' = Q_e \frac{M_s}{M_s'} \qquad (17.09\text{-}2)$$

where M_s/M_s' is the ratio of the saturation magnetization of YIG to that of the material to be used, Q_e is the external Q for YIG as obtained from Fig. 17.09-3, and Q_e' is the external Q for the ferrimagnetic material actually used.

Figure 17.09-4 shows computed and measured data for the external Q of YIG spheres in one-quarter height X-band waveguide. Notice that the agreement between theory and experiment is quite good. However, as for the strip-line cases discussed in Secs. 17.07 and 17.08, the presence of an adjacent coupling hole or slot can be expected to increase the external Q of a resonator, typically by a factor of 20 percent or somewhat more.

The determination of the proper spacing between spheres and the size of the intervening coupling aperture can be determined as discussed in Secs. 17.07 and 17.08. Approximate designs can be obtained by scaling the sphere spacings and aperture sizes in the examples in Sec. 17.08, using the principles discussed in Sec. 17.07. It is generally desirable to make the aperture somewhat undersized to start out with (*i.e.*, somewhat narrow if it is an elongated slot), and then its size can be increased until the desired pass-band shape (or coupling coefficient) is obtained.

As was discussed in Sec. 17.05, the coupling to magnetostatic modes in a ferrimagnetic resonator tends to increase as the frequency increases. The work of Fletcher and Solt[18] indicates that this coupling depends strongly on the sphere diameter as compared to a wavelength. Thus, for the same level of spurious-response activity, filters designed to operate at X-band or higher should use smaller resonator spheres than filters designed for S-band, if other factors are equal.

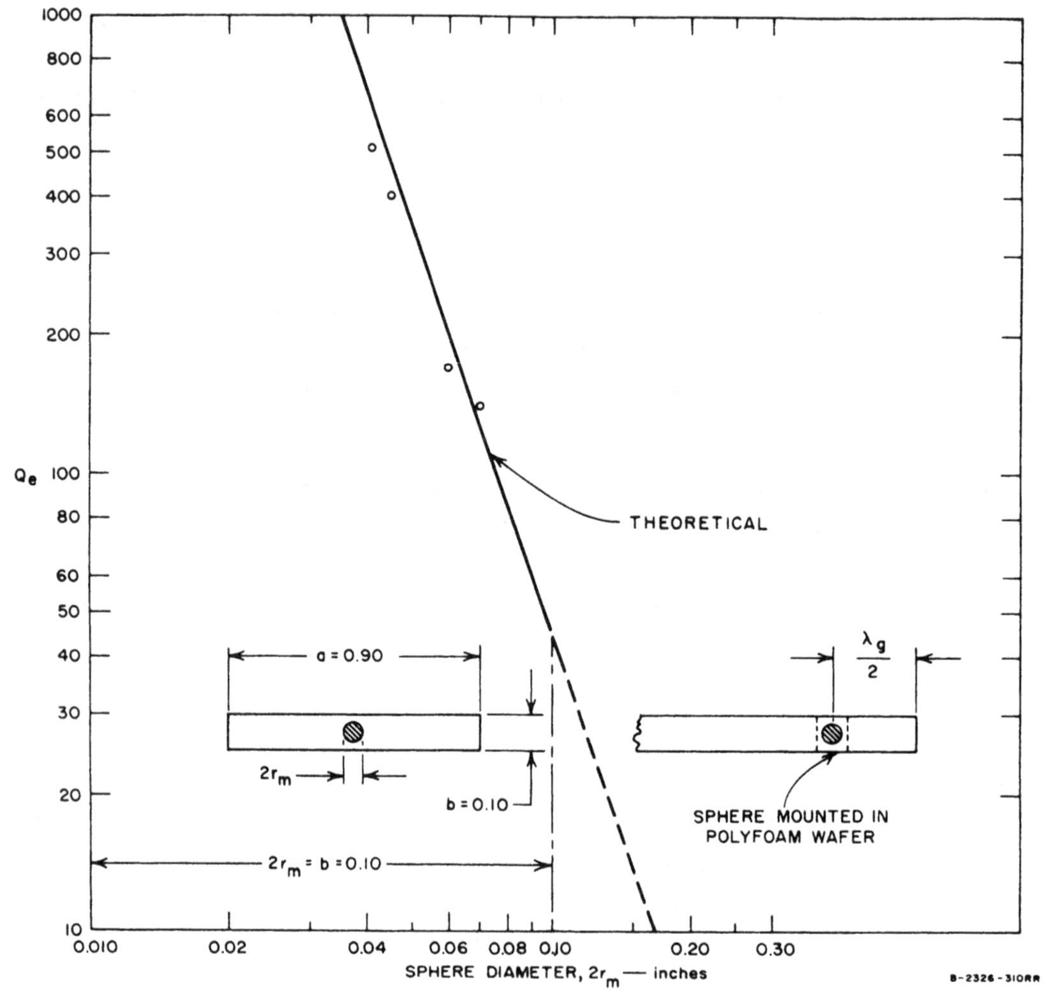

SOURCE: Final Report, Contract DA 36-039 SC-74862, SRI; reprinted in *IRE Trans. PGMTT* (see Ref. 3 by P. S. Carter, Jr.)

FIG. 17.09-4 THEORETICAL AND EXPERIMENTAL Q_e OF YIG RESONATORS IN REDUCED (One-Quarter) HEIGHT, STANDARD-WIDTH X-BAND WAVEGUIDE AT 10 Gc

SEC. 17.10, MAGNETICALLY TUNABLE DIRECTIONAL FILTERS HAVING CIRCULATOR ACTION

Figure 17.10-1 shows a waveguide directional filter of a type discussed in Chapter 14. This filter uses cylindrical resonators with circular apertures, which all together can propagate circularly polarized TE_{11} modes at resonance. The first and last apertures are cut in the adjacent TE_{10}-mode rectangular waveguide at points where the *H*-field is circularly polarized at the resonant frequency of the cavities. As a result of the circular polarization in the resonators and coupling apertures, when the cavities become resonant, power flows from Port 1 through the resonators and out Port 4, while in theory no power emerges from Port 2 or 3. Off of resonance all of the power flows straight through from Port 1 to Port 2. Figure 17.10-2 shows the measured attenuation characteristics for transmission from Port 1 to Ports 2, 3, and 4 for a two-resonator filter of the type in Fig. 17.10-1. The input VSWR at Port 1 is also shown. In theory this type of filter will always present a matched impedance looking into any one of its ports if the other three ports are terminated in waveguide loads of unity VSWR.

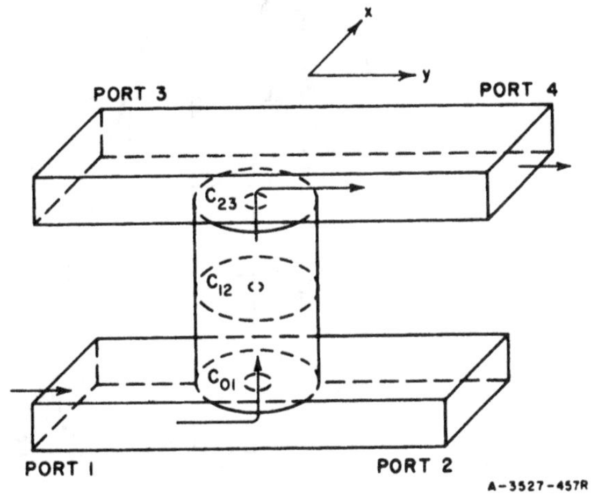

FIG. 17.10-1 A TWO-RESONATOR, FIXED-TUNED WAVEGUIDE DIRECTIONAL FILTER

Figure 17.10-3 shows a magnetically tunable two-resonator directional filter which is potentially capable of producing the same attenuation characteristics shown in Fig. 17.10-2 when the input is at Port 1. As was indicated in the case of the filters in Figs. 17.09-1 and 17.09-2, the spherical YIG ferrimagnetic resonators are placed in reduced-height waveguide in order to increase the coupling between the waveguides and the spheres, and step transformers are used to match between the reduced-height guide and the standard height terminating guides. The YIG spheres are placed in the waveguide at points in the guides where the RF *H*-field will be circularly polarized. Now at ferrimagnetic resonance the circularly polarized RF *H*-field in the first guide will excite a strong circularly polarized magnetic moment in the first sphere (see Sec. 17.05). This magnetic moment will in turn produce an intense circularly polarized *H*-field which will couple through the circular aperture to the second sphere. An intense circular polarized magnetic moment will then be excited in the second sphere which will cause energy to be radiated out Port 4. Thus we see that the filter in

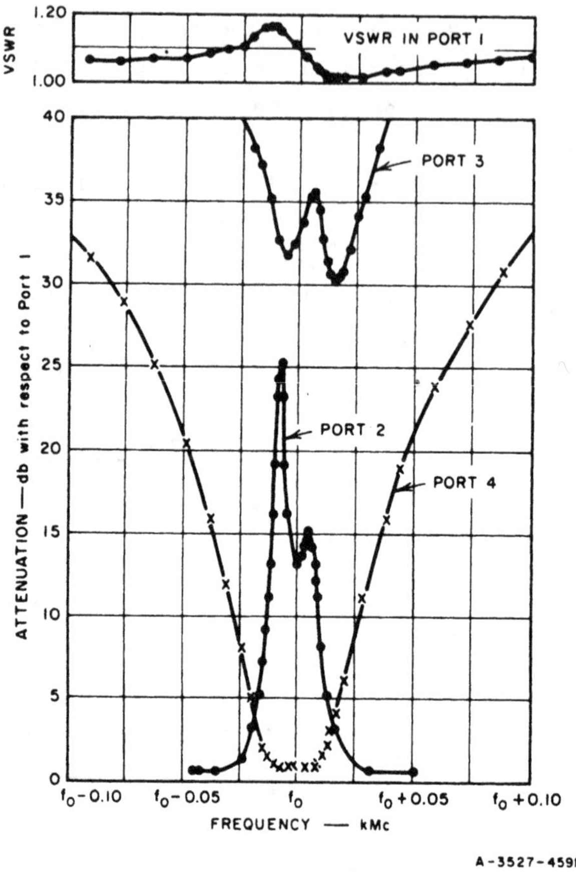

FIG. 17.10-2 MEASURED PERFORMANCE OF A TWO-RESONATOR WAVEGUIDE DIRECTIONAL FILTER OF THE FORM SHOWN IN FIG. 17.10-1

Fig. 17.10-3 will operate essentially the same as the filter in Fig. 17.10-1 when power is fed in Port 1. The main difference is that the filter in Fig. 17.10-1 uses circularly polarized resonances in *electromagnetic* resonators while the filter in Fig. 17.10-3 uses circularly polarized resonances in *ferrimagnetic* resonators.

If power is fed into Port 4 of the filter in Fig. 17.10-1 it will exhibit the same general transmission properties indicated in Fig. 17.10-2, except for the curve marked for Port 4 now applies to Port 1, etc. This is necessary since the filter in Fig. 17.10-1 is a reciprocal device. However, as was noted in Sec. 17.05, ferrimagnetic resonators are not reciprocal circuit elements. As a result, if power is fed into Port 4

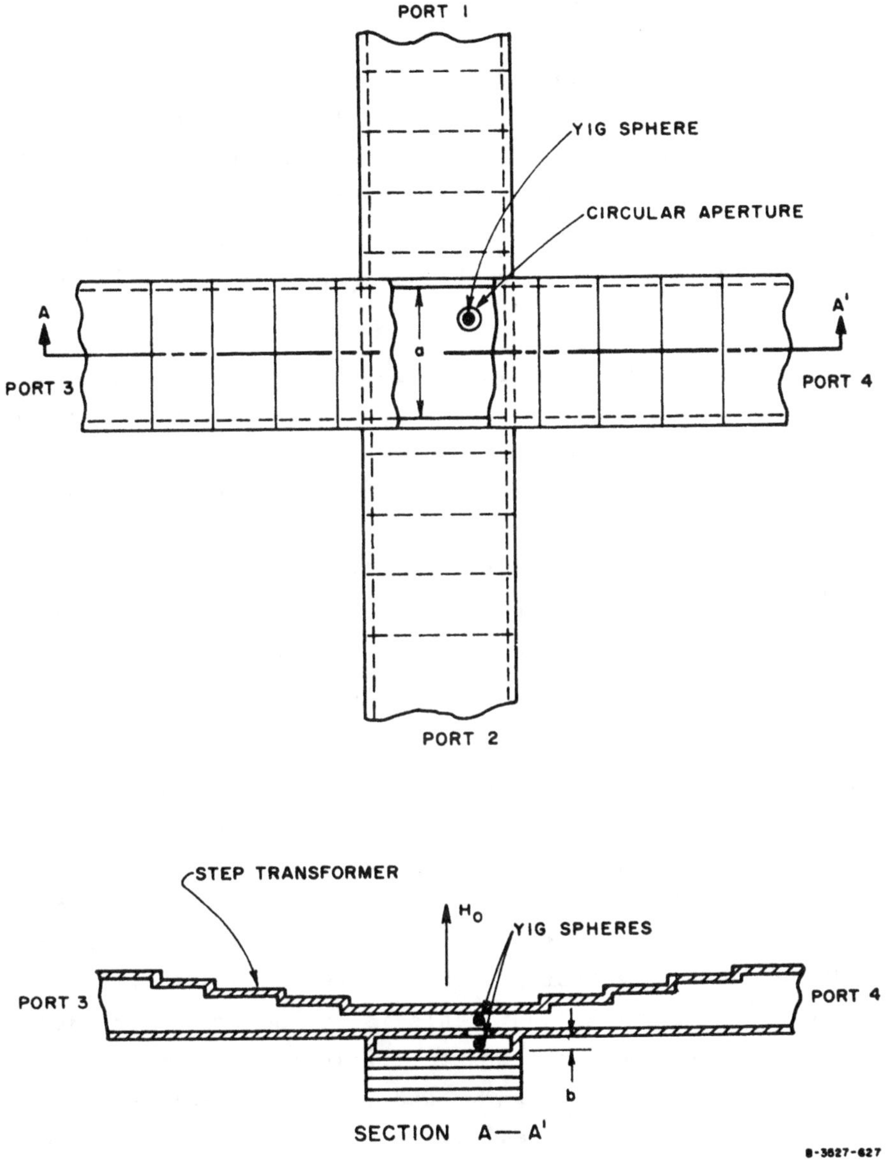

FIG. 17.10-3 A MAGNETICALLY TUNABLE DIRECTIONAL FILTER THAT HAS CIRCULATOR ACTION

of the filter in Fig. 17.10-3, the circularly polarized H-fields will have the wrong direction of polarization to excite resonance, and the power will emerge from Port 3 instead of going to Port 1 as it does in the case of the filter in Fig. 17.10-1. Thus, at resonance we can get transmission from Port 1 to Port 4, from Port 4 to Port 3, from Port 3 to Port 2, and from Port 2 to Port 1. Therefore at resonance this device can be used as a circulator. If the direction of the biasing H_0 field is reversed, then the directions for transmission at resonance would all be reversed, *i.e.*, power could flow from Port 4 to Port 1 at resonance, etc. Thus, this device can also be used as a magnetically controlled switch.

The amount of reverse isolation between, say, Ports 4 and 1 at resonance depends on how nearly perfectly polarized the H-fields seen by the resonators are. For example, if the resonators are located at points in the waveguides so that they see elliptically polarized fields, if the main transmission is from Port 4 to Port 3 there will also be some transmission through the resonators to Port 1. This is because any elliptically polarized field can be regarded as being composed of two circularly polarized fields with opposite directions of rotation, and the component with the proper direction of circular polarization to excite the spheres will pass through to the other guide.

Patel[27] has made an extensive analysis of single-resonator filters of the type in Fig. 17.10-3. In his filters a single YIG sphere was so placed in the center of the coupling aperture between guides as to be half in the upper guide and half in the lower guide. Among other things, he made a study to determine the optimum location of the sphere in the waveguide to obtain the best possible circular polarization of the RF H-fields over a waveguide band. This is an important consideration since for any given position of the sphere, perfect circular polarization can only be obtained at one frequency, and it is desirable that the ellipticity of the H-fields at the upper and lower edges of the desired frequency tuning range be minimized. Patel's study shows that the optimum compromise location for the ferrimagnetic resonator is approximately $a/4$ from the side wall of the waveguide, where a is the width of the waveguide. Figure 17.10-4 shows experimental points and theoretical curves obtained by Patel for a single-resonator filter of the type in Fig. 17.10-3, which illustrates how the forward and reverse insertion loss at resonance vary. The upper dashed curve shows how the theoretical reverse isolation at

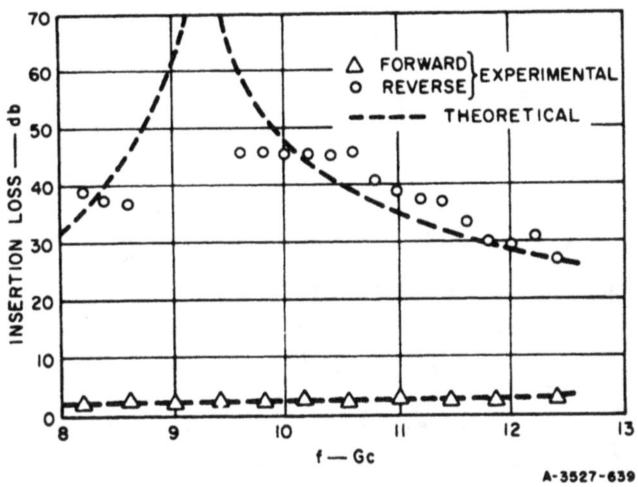

SOURCE: *IRE Trans. PGMTT* (see Ref. 27 by C. N. Patel)

FIG. 17.10-4 THEORETICAL AND EXPERIMENTAL INSERTION LOSS IN FORWARD DIRECTION (Port 1 to Port 4) AND REVERSE DIRECTION (Port 4 to Port 1) FOR A SINGLE-RESONATOR FILTER OF THE FORM SHOWN IN FIG. 17.10-3
The magnetic field was adjusted to give resonance at each test frequency.

resonance varies as a function of the tuning frequency, as a result of the ellipticity of the *H*-field polarization. Note that the theoretical isolation goes to infinity at around 9.3 Gc where the polarization is circular. The lower dashed curve shows the forward insertion loss at resonance as the filter is tuned across the band. Even though, when the *H*-field does not have perfect circular polarization, there will be some leakage of power out Port 2 at resonance (when power is fed in Port 1), this source of power loss is small, and the insertion loss indicated in this figure is mostly due to dissipation loss in the resonator.

The design of magnetically tunable directional filters can be carried out in much the same fashion as was described for the cases treated in Secs. 17.07 to 17.09. Patel[27] gives, for the external Q of a ferrimagnetic resonator in a directional filter, the formula

$$Q_e = \frac{Z_0}{\mu_0 \omega_m V_m (k_x^2 + k_y^2)} \qquad (17.10\text{-}1)$$

where

$$k_x = \frac{\pi}{2a} \sin \frac{\pi x}{a} \quad , \qquad k_y = -\frac{\pi}{2a}\left(\frac{\lambda_g}{2a}\right) \cos \frac{\pi x}{a} \quad , \tag{17.10-2}$$

and the other quantities are as defined for Fig. 17.09-3. The parameters k_x and k_y indicate the relative RF H-field strength seen by the resonator in the transverse and longitudinal directions, respectively, when the resonator is located a distance x from the side wall of the waveguide. For circular polarization $|k_x| = |k_y|$.

Patel does not give an explicit definition of the guide impedance Z_0 in Eq. (17.10-1). However, it can be shown that for $x = a/2$ (which gives $k_x = \pi/2a$ and $k_y = 0$), Eq. (17.10-1) should reduce to two times Carter's equation shown in Fig. 17.09-3.* Making use of this fact, we conclude that Eq. (17.10-1) can be restated as

$$Q_e = \frac{120\,\pi ab}{\mu_0 \omega_m V_m \left[\left(\sin \frac{\pi x}{a}\right)^2 + \left(\frac{\lambda_g}{2a} \cos \frac{\pi x}{a}\right)^2\right]} \left(\frac{\lambda_g}{\lambda}\right) \tag{17.10-3}$$

Using $x = a/4$, as appears to be desirable from Patel's work, and letting $\lambda_g/(2a) = 0.870$ as is typical for the center frequency of a waveguide operating band, then

$$Q_e = 2.28\ (Q_e \text{ obtained from Sec. 17.09}) \tag{17.10-4}$$

Thus, using Eq. (17.10-4), along with Fig. 17.09-3 used in the manner described in Sec. 17.09, the proper waveguide dimensions and sphere sizes required to give desired external Q values can be estimated. As was

---

* When $x = a/2$ and the guide is short circuited as in Fig. 17.09-3, the equivalent circuit is like that in Fig. 17.11-2(b) with the line to the right shorted out. Then $Q_e = \mathscr{b}/Y_0$. When the short circuit is removed, as for the case under consideration, then $Q_e = 2\mathscr{b}/Y_0$ since the parallel-resonant circuit is in series with the two lines.

discussed in Sec. 17.07, the presence of the coupling aperture between resonators will tend to raise the external Q somewhat (perhaps about 20 percent), so some allowance should be made for this fact when determining the sphere dimensions.

When the designer has determined the diameter of the ferrimagnetic resonators and the desired waveguide dimensions, if step transformers such as those shown in Fig. 17.10-3 are required, the desired design can be obtained using the data in Tables 6.02-2 through 6.02-5, and Tables 6.04-1 through 6.04-4. Correction for the fringing capacitances at the step discontinuities should be made using the procedure described in Sec. 6.08.

The required diameter of coupling aperture between resonators for a filter such as that in Fig. 17.10-3 can be determined experimentally by making the hole relatively small to start out with, and then increasing its size until desired shape of response is obtained. If the value of external Q for the resonators was chosen so as to correspond to a given low-pass prototype filter and a specified fractional bandwidth, then it should be possible to obtain the desired response shape and bandwidth by increasing the size of the coupling aperture between the resonators until the proper coupling coefficient is obtained between the resonators.

SEC. 17.11, MAGNETICALLY TUNABLE BAND-STOP FILTERS

Figure 17.11-1 shows a possible form of strip-line, magnetically tunable, band-stop filter having a narrow stop band with high attenuation.

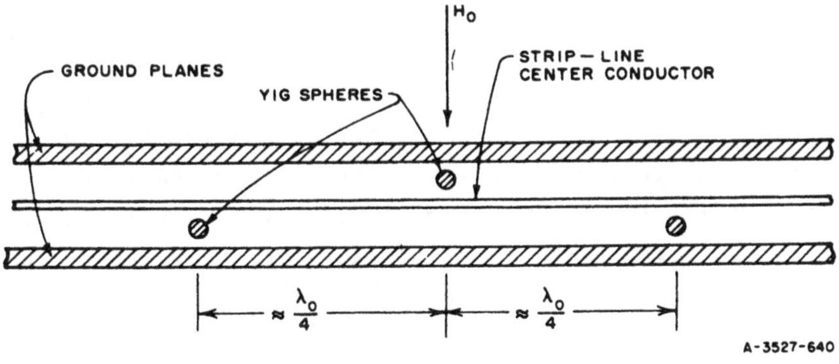

FIG. 17.11-1 A MAGNETICALLY TUNABLE STRIP-LINE BAND-STOP FILTER

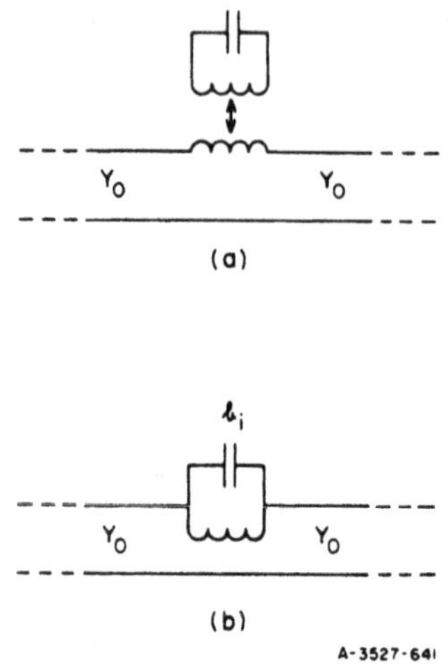

FIG. 17.11-2 EQUIVALENT CIRCUITS OF A YIG SPHERE COUPLED TO A STRIP LINE AS SHOWN IN FIG. 17.11-1
In this figure resonator losses are neglected

In this filter spheres of ferrimagnetic material such as YIG are placed between the strip-line center conductor and an adjacent ground plane, so that the magnetic field about the strip line will couple to the spheres. Each sphere then has mutual inductance coupling to the transmission line as suggested by the equivalent circuit in Fig. 17.11-2(a). (Note that this equivalent circuit neglects resonator dissipation loss.) When the signal frequency and the biasing magnetic field strength H_0 are proper to excite resonance in the spheres the equivalent circuit of the coupled resonator is as shown in Fig. 17.11-2(b), where within practical limits the slope parameter b_i of the resonator can be controlled by the choice of sphere diameter and by the magnitude of the saturation magnetization of the ferrimagnetic material. The ferrimagnetic resonators of the filter in Fig. 17.11-1 can be tuned by varying the biasing magnetic field strength H_0. The second resonator is shown above the strip-line while its neighbors are shown below the strip-line in order to minimize possible undesirable interaction between resonators. Whether this is necessary or not will depend on the relative size of the spheres and of the ground-plane spacing as compared to the wavelengths in the operating frequency range.

From the equivalent circuit in Fig. 17.11-2(b) it is seen that the magnetically tunable filter under consideration is of the type discussed in Sec. 12.04 and shown in Fig. 12.04-2, which is repeated in Fig. 17.11-3. Filters designed from the data in Fig. 17.11-3 will usually be most practical if they are designed to have a constant main-line characteristic admittance which equals the termination admittance Y_0. Under this condition, in Fig. 17.11-3(a)

$$Y_0 = Y_1 = \frac{1}{g_0 g_{n+1}} , \qquad (17.11\text{-}1)$$

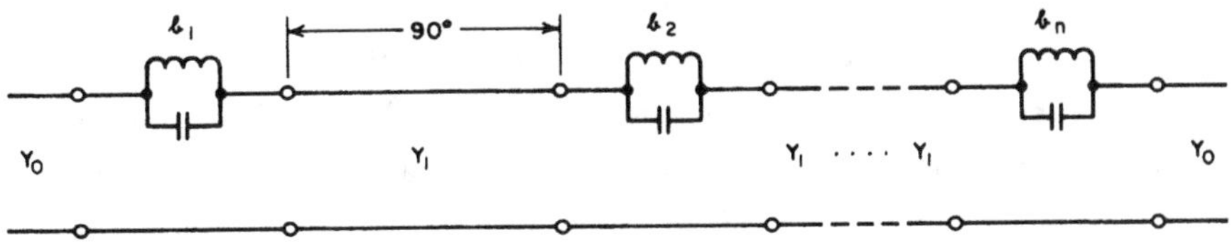

$$\frac{b_1}{Y_0} = \frac{1}{\omega_1' g_0 g_1 w}$$

$$\left.\frac{b_i}{Y_0}\right|_{i=\text{even}} = \left(\frac{Y_1}{Y_0}\right)^2 \frac{g_0}{\omega_1' g_i w}$$

$$\left.\frac{b_i}{Y_0}\right|_{i=\text{odd}} = \frac{1}{\omega_1' g_0 g_i w}$$

$$\text{If } n = \text{even}, \quad \left(\frac{Y_1}{Y_0}\right)^2 = \frac{1}{g_0 g_{n+1}}$$

$$\text{If } n = \text{odd}, \quad Y_1 = Y_0$$

(a)

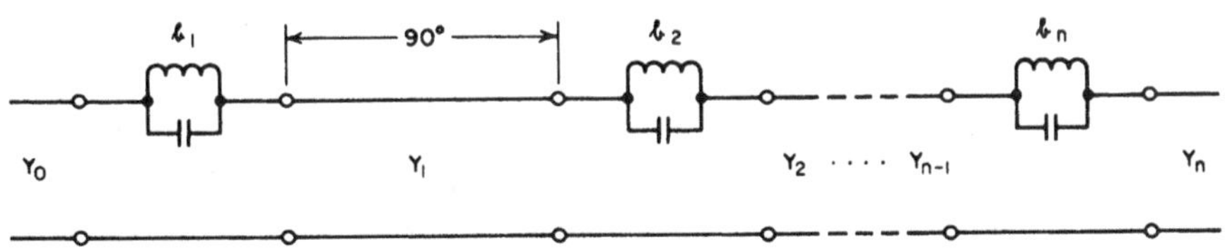

$$\frac{b_1}{Y_0} = \frac{1}{\omega_1' g_0 g_1 w}$$

$$\left.\frac{b_i}{Y_0}\right|_{i=\text{even}} = \left(\frac{Y_1 Y_3 \cdots\cdots Y_{i-1}}{Y_0 Y_2 \cdots\cdots Y_{i-2}}\right)^2 \frac{g_0}{\omega_1' g_i w}$$

$$\left.\frac{b_i}{Y_0}\right|_{\substack{i=\text{odd}\\ i\neq 1}} = \left(\frac{Y_2 Y_4 \cdots\cdots Y_{i-1}}{Y_1 Y_3 \cdots\cdots Y_{i-2}}\right)^2 \frac{1}{\omega_1' g_0 g_i w}$$

$$\text{If } n = \text{even}, \quad \frac{Y_0}{Y_n} = \left(\frac{Y_0 Y_2 \cdots\cdots Y_{n-2}}{Y_1 Y_3 \cdots\cdots Y_{n-1}}\right)^2 \frac{1}{g_0 g_{n+1}}, \text{ or}$$

$$\text{If } n = \text{odd}, \quad \frac{Y_0}{Y_n} = \left(\frac{Y_1 Y_3 \cdots\cdots Y_{n-2}}{Y_2 Y_4 \cdots\cdots Y_{n-1}}\right)^2 \frac{g_0}{g_{n+1}}$$

(b)

B-3527-229

SOURCE: Quarterly Report 3, Contract DA 36-039 SC-87398, SRI; reprinted in *IRE Trans. PGMTT* (see Ref. 1 of Chapter 12, by Young, Matthaei, and Jones)

FIG. 17.11-3 BAND-STOP FILTER WITH SERIES BRANCHES AND QUARTER-WAVE COUPLINGS: (a) EQUAL LINE ADMITTANCES, Y_1; (b) GENERAL CASE OF UNEQUAL CONNECTING-LINE ADMITTANCES

a condition which is obtained for any of the maximally flat low-pass prototypes in Sec. 4.05, or for any of the Tchebyscheff prototypes having n odd in that section. Using the data in Fig. 17.11-3(a) the normalized resonator slope parameters $\mathscr{b}_i/Y_0$ can be calculated, where in the figure w is the fractional bandwidth of the stop band to the band-edge points corresponding to ω_1' for the low-pass prototype (see Sec. 12.02).

After the normalized resonator slope parameters $\mathscr{b}_i/Y_0$ have been computed, the required ferrimagnetic resonator parameters can be computed by use of data given earlier in this chapter. Note in Fig. 17.11-2 that if a short-circuit were placed across the transmission line just to the right side of the resonator, assuming that the line to the left of the resonator is terminated in its characteristic admittance Y_0, the Q of the circuit would be

$$(Q_e)_i = \frac{\mathscr{b}_i}{Y_0} \quad . \tag{17.11-2}$$

Note that operating the resonators in the circuit in Fig. 17.11-1 individually with a short-circuit at one side of the sphere under test, and with a matched transmission line at the other side, is exactly the condition under which the external Q data in Fig. 17.07-6 and Eqs. (17.07-3) and (17.07-4) apply. Thus, using the reference external Q values given by Eq. (17.11-2) along with the data in Sec. 17.07, the required resonator diameters for given M_s values can be determined. If the unloaded Q's of the resonators are known, the peak stop-band attenuation of the filter can be computed by use of the data in Sec. 12.03. If the unloaded Q's are not known, they can be measured by placing the resonator spheres one at a time in a short-circuited strip-line structure close to or $\lambda/2$ from a short-circuit and making measurements as described in Sec. 11.02. This procedure can, of course, also provide an experimental verification of $(Q_e)_i$.

One obvious disadvantage of the type of filter in Fig. 17.11-1 is that its design is based on the assumption that its resonators are spaced a quarter-wavelength at resonance. If the filter is tuned over a very large range its response may deviate appreciably from the ideal since the resonator spacings are $\lambda_0/4$ at only one frequency. The most serious effect may be that the peak attenuation might fall below acceptable levels in some parts of the tuning range. However, if the performance requirements are not very severe, or if the tuning range is relatively small, filters of the type in Fig. 17.11-1 may be quite satisfactory.

Figure 17.11-4 shows a suggested magnetically tunable filter structure which may overcome the drawbacks of the structure in Fig. 17.11-1 by giving increased coupling to the spheres. Without the ferrimagnetic resonators the structure shown is basically a semi-lumped-element low-pass filter structure such as can be designed by the methods discussed in Sec. 7.03. The structure consists of low-impedance line sections which operate predominantly like shunt capacitors, alternating with relatively high-impedance line sections which function predominantly like series inductances. This structure must be designed so that its cutoff frequency will be above the highest transmission frequency of interest. When ferrimagnetic resonators (such as YIG spheres) are introduced in the circuit, they should typically have little effect except at or near ferrimagnetic resonance. At resonance they introduce very high series impedances in the circuit (as can be seen from Fig. 17.11-2). Since these very high series impedances will alternate with very low shunt impedances, very nearly the maximum possible attenuation will be achieved from the resonators. Another advantage of this type of structure is that the structure would be considerably shorter than that in Fig. 17.11-1. This could be particularly important at the lower microwave frequencies where, if the spheres were spaced a quarter-wavelength apart, a very large magnet face would be required. In Fig. 17.11-4 the dielectric pieces have been placed in the interior of the metal capacitor blocks so that the RF H-fields seen by the spheres will be as uniform as possible.*

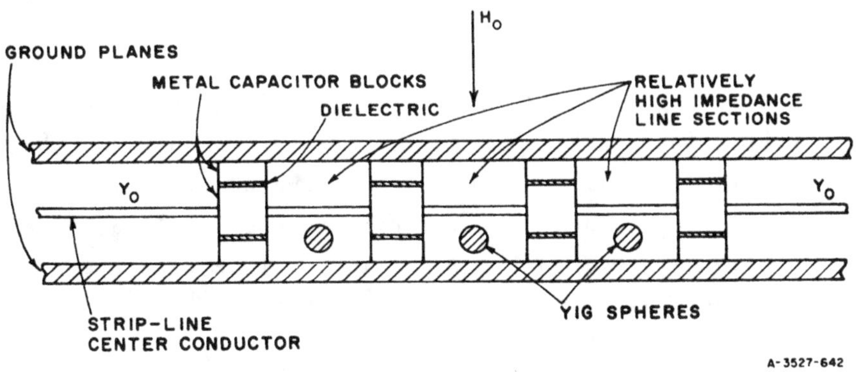

FIG. 17.11-4 A SUGGESTED MAGNETICALLY TUNABLE BAND-STOP FILTER CONFIGURATION FOR LARGE TUNING RANGES

*However, it may be preferable to put the dielectric pieces next to the ground planes in order to minimize possible direct magnetic coupling between spheres.

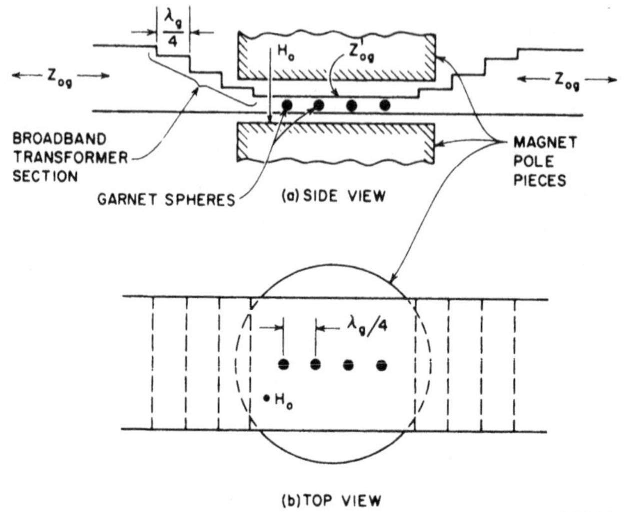

FIG. 17.11-5 A FOUR-RESONATOR MAGNETICALLY TUNABLE WAVEGUIDE BAND-STOP FILTER

Though a detailed design procedure for filters of the type in Fig. 17.11-4 has not been worked out at the time of this writing, the design equations in Fig. 17.11-3(b) should prove helpful. The quarter-wavelength connecting lines of admittance Y_i operate as impedance inverters (Sec. 8.03) having inverter parameter $K_i = 1/Y_i$. The equations in Fig. 17.11-3(b) allow for a range of choice for these inverter parameters. Now by Fig. 8.03-1(d) we see that a length of line with a shunt capacitor in the middle can operate as an impedance inverter. Thus, for purposes of determining the proper sphere sizes, etc., the circuits between spheres in Fig. 17.11-4 can be characterized in terms of impedance inverters of the form in Fig. 8.03-1(d). After the designer has worked out the basic low-pass filter structure design he can compute values for the K_i and $Y_i = 1/K_i$. Then the equations in Fig. 17.11-3(b) can be used to compute the required normalized resonator slope parameters. Finally, the required sphere sizes can be computed from the $\mathscr{b}_i/Y_0$ by procedures such as were discussed above.

The discussion in this section has so far been phrased in terms of strip-line structures. However, the same principles also will apply to the design of waveguide band-stop filters. Figure 17.11-5 shows a multi-resonator waveguide band-stop filter. Judging from experience with cavity-type waveguide band-stop filters (Sec. 12.08), some difficulty might be expected due to coupling directly from one resonator to the next. In order to avoid this problem, the waveguide height and sphere size should both be kept very small compared to one-quarter wavelength. It should be possible to design waveguide filters of the form in Fig. 17.11-5 using the same procedure described for the filter in Fig. 17.11-1 except that Fig. 17.09-3 and Eqs. (17.09-1) and (17.09-2) should be used in place of Fig. 17.07-6 and Eqs. (17.07-3) and (17.07-4).

Waveguide filters analogous to the strip-line filter in Fig. 17.11-4 are also possible. These filters could take the general form of the corrugated waveguide filters discussed in Sec. 7.04, with YIG spheres inserted in the high-impedance sections of the corrugations.

REFERENCES

1. Harvard Univ. Radio Research Laboratory Staff, *Very High-Frequency Techniques*, McGraw-Hill Book Co., Inc., New York (1947), pp. 769-795.

2. R. L. Sleven, "Design of a Tunable Multi-Cavity Waveguide Band-Pass Filter," *1959 IRE National Convention Record,* Part 3, pp. 91-112.

3. G. L. Matthaei, ..., P. S. Carter, Jr., *et al.*, "Design Criteria for Microwave Filters and Coupling Structures," Chap. 28, Final Report, SRI Project 2326, Contract DA 36-039 SC-74862, Stanford Research Institute, Menlo Park, California (January 1961). An abbreviated version of this material will be found in the paper: P. S. Carter, Jr., "Magnetically Tunable Microwave Filters Using Single-Crystal Yttrium-Iron-Garnet Resonators," *IRE Trans. PGMTT*-9, pp. 252-260 (May 1961).

4. P. S. Carter, Jr., L. Young, G. L. Matthaei, and E. M. T. Jones, "Microwave Filters and Coupling Structures," Third Quarterly Progress Report, Part II, SRI Project 3527, Contract DA 36-039 SC-87398, Stanford Research Institute, Menlo Park, California (October 1961).

5. G. L. Matthaei, Leo Young, and P. S. Carter, Jr., "Microwave Filters and Coupling Structures," Fifth Quarterly Progress Report, Part IV, SRI Project 3527, Contract DA 36-039 SC-87398, Stanford Research Institute, Menlo Park, California (May 1962).

6. B. M. Schiffman, P. S. Carter, Jr., and G. L. Matthaei, "Microwave Filters and Coupling Structures," Seventh Quarterly Progress Report, Part III, SRI Project 3527, Contract DA 36-039 SC-87398, Stanford Research Institute, Menlo Park, California (October 1962).

7. J. Smit and H. P. Wijn, *Ferrites*, John Wiley and Sons, Inc., New York (1959).

8. R. F. Soohoo, *Theory and Application of Ferrites,* Prentice-Hall, Inc., Englewood Cliffs, N. J. (1960).

9. P. S. Carter, Jr., and C. Flammer, "Unloaded Q of Single Crystal Yttrium-Iron-Garnet Resonators as a Function of Frequency," *IRE Trans. PGMTT*-8, pp. 570-571 (September 1960).

10. D. Douthett and I. Kaufman, "The Unloaded Q of a YIG Resonator from X-Band to 4 Millimeters," *IRE Trans. PGMTT*-9, pp. 261-262 (May 1961).

11. J. W. Nielsen, D. A. Lepore, J. Zneimer, and G. B. Townsend, "Effect of Mechanical, Thermal, and Chemical Treatment of the Ferrimagnetic Resonance Linewidth on Lithium Ferrite Crystals," *Jour. Appl. Phys.*, Vol. 33S, pp. 1379S-1380S (March 1962).

12. C. R. Buffler, "Resonance Properties of Single-Crystal Hexagonal Ferrites," *Jour. Appl. Phys.*, Vol. 33S, pp. 1360S-1362S (March 1962).

13. J. E. Artman, "Microwave Resonance Relations in Anisotropic Single Crystal Ferrites," *Proc. IRE,* Vol. 44, pp. 1284-1293 (October 1956).

14. J. F. Dillon, Jr., "Ferrimagnetic Resonance in Yttrium Iron Garnet," *Phys. Rev.*, Vol. 105, Second Series, pp. 759-760 (January 15, 1957).

15. C. Kooi, D. Stinson, R. Moss, F. Bradley, L. Freiberg, "Microwave Ferrites," Report LMSD-5036, Lockheed Missile Systems Division, Sunnyvale, California (1 July 1958).

16. E. G. Spencer, R. C. LeCraw, and R. C. Linares, Jr., "Low-Temperature Ferromagnetic Relaxation in Yttrium Iron Garnet," *Phys. Rev.*, Vol. 123, Second Series, pp. 1937-1938 (September 15, 1961).

17. P. C. Fletcher and R. O. Bell, "Ferrimagnetic Resonance Modes in Spheres," *Jour. Appl. Phys.*, Vol. 30, pp. 687-698 (May 1959).

18. P. C. Fletcher and I. H. Solt, Jr., "Coupling of the Magnetostatic Modes," *Jour. Appl. Phys.*, Vol. 30, pp. 181S-182S (April 1959).

19. H. Suhl, "The Nonlinear Behavior of Ferrites at High Microwave Signal Levels," *Proc. IRE,* Vol. 44, pp. 1270-1284 (October 1946).

20. R. W. DeGrasse, "Low-Loss Gyromagnetic Coupling Through Single Crystal Garnets," *J. Appl. Phys.* Vol. 30, pp. 155S-156S (April 1959).

21. F. J. Sansalone and E. G. Spencer, "Low-Temperature Microwave Power Limiter," *IRE Trans. MTT*-9, pp. 272-273 (May 1961).

22. F. C. Rossol, "Power Limiting in the 4 kMc to 7 kMc Frequency Range Using Lithium Ferrite," *Proc. IRE,* Vol. 49, p. 1574 (October 1961).

23. F. R. Arams, M. Grace, and S. Okwit, "Low-Level Garnet Limiters," *Proc. IRE,* Vol. 49, pp. 1308-1313 (August 1961).

24. Martin Auer, "Novel Method to Orient Ferrimagnetic Single-Crystal Spheres," *IRE Trans. PGMTT*-10, p. 88 (January 1962).

25. K. L. Kotzebue, "Broadband Electronically-Tunable Microwave Filters," *1960 IRE WESCON Convention Record,* Part 1, pp. 21-27.

26. P. S. Carter, Jr., G. L. Matthaei, and W. J. Getsinger, "Design Criteria for Microwave Filters and Coupling Structures," Technical Report 8, Part II, SRI Project 2326, Contract DA 36-039 SC-74862, Stanford Research Institute, Menlo Park, California (October 1959).

27. C. N. Patel, "Magnetically Tunable Nonreciprocal Band-Pass Filters Using Ferrimagnetic Resonators," *IRE Trans. PGMTT*-10, pp. 152-161 (May 1962).

28. Y. Sato and P. S. Carter, Jr. "A Device for Rapidly Aligning and Mounting Ferromagnetic Single Crystals Along any Desired Axis," *IRE Trans. PGMTT*-10, pp. 611-612 (November 1962).

29. B. Lax and K. J. Button, *Microwave Ferrites and Ferrimagnetics,* McGraw-Hill Book Co., Inc., New York (1961).

INDEX

Recent Titles in the Artech House Microwave Library

Advanced Automated Smith Chart Software and User's Manual, Version 3.0, Leonard M. Schwab

Behavioral Modeling of Nonlinear RF and Microwave Devices, Thomas R. Turlington

Computer-Aided Analysis of Nonlinear Microwave Circuits, Paulo J. C. Rodrigues

Design of FET Frequency Multipliers and Harmonic Oscillators, Edmar Camargo

Design of RF and Microwave Amplifiers and Oscillators, Pieter L. D. Abrie

EPFIL: Waveguide E-plane Filter Design, Software and User's Manual, Djuradj Budimir

Feedforward Linear Power Amplifiers, Nick Pothecary

Generalized Filter Design by Computer Optimization, Djuradj Budimir

GSPICE for Windows, Sigcad Ltd.

Introduction to Microelectromechanical (MEM) Microwave Systems, Hector J. De Los Santos

Microwave Engineers' Handbook, Two Volumes, Theodore Saad, editor

Microwave Filters, Impedance-Matching Networks, and Coupling Structures, George L. Matthaei, Leo Young, and E.M.T. Jones

Microwave Mixers, Second Edition, Stephen Maas

Microwaves and Wireless Simplified, Thomas S. Laverghetta

Microwave Radio Transmission Design Guide, Trevor Manning

PACAD: RF Linear Power Amplifier Design, Software and User's Manual, Ramin Fardi, Keyvan Haghighat, and Afshin Fardi

RF Design Guide: Systems, Circuits, and Equations, Peter Vizmuller

The RF and Microwave Circuit Design Handbook, Stephen A. Maas

RF and Microwave Coupled-Line Circuits, Rajesh Mongia, Inder Bahl, and Prakash Bhartia

RF Power Amplifiers for Wireless Communications, Steve C. Cripps

RF Systems, Components, and Circuits Handbook, Ferril Losee

TRAVIS Pro: Transmission Line Visualization Software and User's Manual, Professional Version, Robert G. Kaires and Barton T. Hickman

www.ingramcontent.com/pod-product-compliance
Lightning Source LLC
LaVergne TN
LVHW080056220926
848484LV00006BB/1585